PHYTOPLANKTON PIGMENTS

Characterization, Chemotaxonomy and Applications in Oceanography

Pigments act as tracers to elucidate the composition and fate of phytoplankton in the world's oceans and are often associated with important biogeochemical cycles related to, for example, carbon dynamics in the oceans. They are increasingly used in *in situ* and remote-sensing applications, detecting algal biomass and major taxa through changes in water colour (associated with changes in algal pigments).

This book is a follow-up to the 1997 volume *Phytoplankton Pigments in Oceanography*, edited by Jeffrey, Mantoura and Wright (UNESCO Press). Since then, there have been many advances and discoveries concerning phytoplankton pigments and it is widely recognized – as concluded by a recent meeting supported by the Scientific Committee on Oceanic Research (SCOR) – that these should be brought together in a new book to update the user community. This book includes recent discoveries on several new algal classes, particularly for the picoplankton, and on new pigments. It also includes many advances in methodologies, including liquid chromatography-mass spectrometry (LC-MS) and developments and updates on the mathematical methods used to exploit pigment information and extract the composition of phytoplankton communities. The book includes seven sections: (1) Algal chlorophylls and carotenoids, (2) Methodology guidance, (3) Water-soluble 'pigments', (4) Selected pigment applications in oceanography, (5) Future perspectives, (6) Aids for practical laboratory work, and (7) Phytoplankton pigments data sheets. Electronic versions of the data sheets, plus extra and extended Appendices, are also available online at www.cambridge.org/phytoplankton.

The book is invaluable primarily as a reference for students, researchers and professionals in aquatic science, biogeochemistry and remote sensing.

SUZANNE ROY is a Professor of biological oceanography at the Institut des Sciences de la Mer of the Université du Québec à Rimouski (Canada) and a member of Québec-Océan. Over the last 20 years, Professor Roy has developed an expertise in the ecology and physiology of marine and estuarine phytoplankton, focusing on various aspects such as population dynamics of harmful algae, environmental impacts of aquaculture and ozone-related ultraviolet radiation effects. She also runs an analytical laboratory for the HPLC determination of algal pigments and UV-screening compounds. Her current research interests include the combined influence of climate warming and enhanced UV on phytoplankton communities, photoprotection and cell mortality in Arctic phytoplankton, and the transport of non-indigenous dinoflagellates in ships' ballast tanks. Several of these projects are part of Canada's major NSERC Research Networks such as CAISN and CFL. Professor Roy is a member of the Scientific Committee for the international Global Ecology and Oceanography of Harmful Algal Blooms (GEOHAB) programme.

CAROLE A. LLEWELLYN is a microbial biogeochemist at the Plymouth Marine Laboratory, UK. She has over 20 years' experience in phytoplankton pigments and UV absorbing compounds. Her research interests are focused on understanding the role of phytoplankton in the ocean and more specifically on microbial and food web dynamics, microbial biodiversity, community composition and photophysiology. At an applied level, her research contributes to eutrophication and pollution studies and links with satellite remote-sensing and bio-optics. More recently she has used her knowledge on algae and pigments to contribute to the rapidly growing area of algal biotechnology.

EINAR SKARSTAD EGELAND is an Associate Professor in the Faculty of Biosciences and Aquaculture at University of Nordland (formerly Bodø University College), Norway. He has broad experience in organic chemical analysis (chromatography and spectroscopy). He is an internationally recognised scientist on carotenoid analysis from natural sources (mostly prasinophyte algae, but also other algal classes). Currently, he is involved in several cross-disciplinary research projects related to marine ecology, aquaculture and seafood quality. He is an active member of the Marine Ecology Group at Bodø University College.

GEIR JOHNSEN is a Professor of marine biology at the Norwegian University of Science and Technology, Trondheim (NTNU), and an Adjunct Professor in marine bio-optics at the University Centre in Svalbard (UNIS), Longyearbyen, Norway. His major interests are the use of bio-optical methods in taxonomy, ecology and physiology of micro- and macroalgae. His main focus in the last 20 years has been on photosynthesis, light harvesting and utilization in algae and marine invertebrates with photosynthetic endosymbionts. Current interests include new approaches in *in situ* and remote-sensing techniques for monitoring and mapping of planktonic and benthic organisms in the water surface, water column and sea floor.

International Council for Science

SCOR

Scientific Committee on Oceanic Research

UQAR ISMER

Université du Québec à Rimouski
Institut des sciences de la mer de Rimouski

NTNU – Trondheim
Norwegian University of
Science and Technology

PML | Plymouth Marine Laboratory

UNIVERSITY OF NORDLAND

DHI

IAEA
International Atomic Energy Agency
Atoms For Peace

CAMBRIDGE ENVIRONMENTAL CHEMISTRY SERIES

Series editors

All books available in the series

P. Brimblecombe *Air Composition and Chemistry, Second Edition*

A. C. Chamberlain *Radioactive Aerosols*

M. Cresser, K. Killham, and A. Edwards *Soil Chemistry and its Applications*

A. Edwards and M. Cresser *Acidification of Freshwaters*

R. M. Harrison and S. J. de Mora *Introductory Chemistry for the Environmental Sciences, Second Edition*

T. D. Jickells and J. E. Rae *Biogeochemistry of Intertidal Sediments*

S. J. de Mora *Tributylin: Case Study of an Environmental Contaminant*

S. J. de Mora, S. Demers, and M. Vernet *The Effect of UV Radiation in the Marine Environment*

S. Roy, E. S. Egeland, G. Johnsen, and C. A. Llewellyn *Phytoplankton Pigments: Characterization, Chemotaxonomy and Applications in Oceanography*

E. Tipping *Cation Binding by Humic Substances*

D. A. Wright and P. Welbourn *Environmental Toxicology*

PHYTOPLANKTON PIGMENTS

Characterization, Chemotaxonomy and Applications in Oceanography

Edited by

SUZANNE ROY
Université du Québec à Rimouski, Canada

CAROLE A. LLEWELLYN
Plymouth Marine Laboratory, UK

EINAR SKARSTAD EGELAND
University of Nordland, Norway

GEIR JOHNSEN
Norwegian University of Science and Technology, Trondheim, Norway
University Centre in Svalbard, Norway

CAMBRIDGE
UNIVERSITY PRESS

CAMBRIDGE UNIVERSITY PRESS
Cambridge, New York, Melbourne, Madrid, Cape Town,
Singapore, São Paulo, Delhi, Tokyo, Mexico City

Cambridge University Press
The Edinburgh Building, Cambridge CB2 8RU, UK

Published in the United States of America by Cambridge University Press, New York

www.cambridge.org

Information on this title: www.cambridge.org/9781107000667

First published 2011

Printed in the United Kingdom at the University Press, Cambridge

A catalogue record for this publication is available from the British Library

Library of Congress Cataloging-in-Publication Data

Phytoplankton pigments : characterization, chemotaxonomy, and applications in oceanography /
edited by Suzanne Roy [et al.].
p. cm. – (Cambridge environmental chemistry series) ISBN 978-1-107-00066-7 (Hardback)
1. Phytoplankton–Composition. 2. Phytoplankton–Chemotaxonomy. 3. Photosynthetic pigments.
4. Algae–Classification. 5. Oceanography–Methodology. I. Roy, Suzanne, 1955–
II. Title. III. Series.

QK933.P496 2011
579.8′1776–dc22
2011009229
ISBN 978-1-107-00066-7 Hardback

Additional resources for this publication at www.cambridge.org/phytoplankton

This volume is dedicated to Dr S. W. Jeffrey, a pioneer in the development of tools and knowledge on pigments in ocean environments, and an inspiration and great help in the production of the present volume.

Contents

Part IV Selected pigment applications in oceanography

Part V Future perspectives

Part VI Aids for practical laboratory work

The colour plates are to be found between pages 230 and 231.

Contributors

R. Airs
Plymouth Marine Laboratory, Prospect Place, Plymouth PL1 3DH, UK
e-mail: ruai@pml.ac.uk

K. Andresen
Trondhjem Biological Station, Dept. Biology, Norwegian University of Science and Technology, N-7491 Trondheim, Norway
e-mail: kjersti.andresen@bio.ntnu.no

R. R. Bidigare
Department of Oceanography, University of Hawai'i at Manoa, 1680 East-West Road POST 105, Honolulu, HI 96822, USA
e-mail: bidigare@hawaii.edu

A. Bricaud
Laboratoire d'Océanographie de Villefranche, B.P. 8, Quai de la Darse, Villefranche-sur-Mer, CEDEX 06238 France
e-mail: annick@obs-ulfr.fr

C. Brunet
Stazione Zoologica A. Dohrn, Villa Comunale, 80121 Napoli, Italy
e-mail: christophe.brunet@szn.it

M. Carignan
Instituto Nacional de Investigación y Desarrollo Pesquero (INIDEP), Paseo Victoria Ocampo No. 1, B7602HSA, 7600 Mar del Plata, Argentina
e-mail: marioc@inidep.edu.ar

J. Carreto
Instituto Nacional de Investigación y Desarrollo Pesquero (INIDEP),
Paseo Victoria Ocampo No. 1, B7602HSA, 7600 Mar del Plata, Argentina
e-mail: jcarreto@inidep.edu.ar

M. Chauton
Department of Biotechnology, Norwegian University of Science and Technology,
N-7491, Norway
e-mail: matilde.chauton@biotech.ntnu.no

L. Clementson
CSIRO Marine Research, PO Box 1538, Hobart, Tasmania, 7007, Australia
e-mail: Lesley.Clementson@marine.csiro.au

Y. Dandonneau
LOCEAN (Laboratoire d'océanographie et du climat), Université Pierre et Marie
Curie, Case 100, Tour 45–55, 4ème étage 75252 PARIS CEDEX 05, France
e-mail: Yves.Dandonneau@locean-ipsl.upmc.fr

E. S. Egeland
Faculty of Biosciences and Aquaculture, University of Nordland, N-8049,
Bodø, Norway
e-mail: einar.skarstad.egeland@uin.no

J. Garrido
Instituto de Investigaciones Marinas, CSIC, Eduardo Cabello 6,
E-36208 Vigo, Spain
e-mail: garrido@iim.csic.es

A. Gutiérrez-Rodríguez
Institut de Ciències del Mar (CSIC), Passeig Marítim de la Barceloneta 37–49,
E-08003 Barcelona, Spain
e-mail: ang008@mail.ucsd.edu

G. Hallegraeff
University of Tasmania, School of Plant Science, Private Bag 55, Hobart Tasmania
7001, Australia
e-mail: Hallegraeff@utas.edu.au

H. Higgins
CSIRO Marine Research, PO Box 1538, Hobart, TAS 7001, Australia
e-mail: Harry.Higgins@csiro.au

S. B. Hooker
NASA Ocean Biology and Biogeochemistry, GSFC Calibration and Validation Office, 1450 S. Rolling Road, Halethorpe, MD 21227, USA
e-mail: stanford.b.hooker@nasa.gov

S. W. Jeffrey
CSIRO, GPO Box 1538, Hobart, TAS 7001, Australia
e-mail: Shirley.Jeffrey@csiro.au

G. Johnsen
Trondhjem Biological Station, Dept. Biology, Norwegian University of Science and Technology, N-7491 Trondheim, Norway
University Centre in Svalbard, N-9171 Longyearbyen, Norway
e-mail: geir.johnsen@bio.ntnu.no

M. Latasa
Centro Oceanografíco de Gijón, Instituto Español de Oceanografía (IEO), Auda. Príncipe de Asturias 70 bis, E-33212 Gijón, Asturias, Spain
e-mail: latasa@gi.ieo.es

J. Lavaud
UMR CNRS 6250 'LIENSA', Institute for Coastal and Environmental Research, University of La Rochelle, 2 rue Olympe deGouges, 17000 La Rochelle, France
e-mail: johann.lavaud@univ-lr.fr

C. A. Llewellyn
Plymouth Marine Laboratory, Prospect Place, Plymouth PL1 3DH, UK
e-mail: call@pml.ac.uk

M. Lohr
Johannes Gutenberg-Universität, Institut für Allgemeine Botanik, D-55099 Mainz, Germany
e-mail: lohr@uni-mainz.de

D.F. Millie
Florida Institute of Oceanography, University of South Florida, 100 8th Ave. SE, St. Petersburg, FL 33701, USA
e-mail: David.Millie@MyFWC.com

M. Moline
Biological Sciences Dept. & Center for Marine and Coastal Sciences, California Polytechnic State University, San Luis Obispo, CA 93407, USA
e-mail: mmoline@calpoly.edu

A. Neeley
Science Systems and Applications, Inc., NASA Goddard Space Flight Center, Ocean Ecology Branch, Mail Code 614.2, 8800 Greenbelt Rd., Greenbelt, MD 20771, USA
e-mail: aimee.neeley@nasa.gov

N. Nelson
Institute for Computational Earth System Science, Mail Code 3060, University of California, Santa Barbara, CA 93106, USA
e-mail: norm@icess.ucsb.edu

J. Neveux
Observatoire Océanologique de Banyuls, 66651 Banyuls-sur-mer, France
e-mail: jneveux@obs-banyuls.fr

G.L. Newman
Department of Chemistry, Durham University, South Road, Durham, DH1 3LE UK
e-mail: newman.gl@gmail.com

U. Oster
Dept. Biologie I: Botanik, Universität München, Großhadernerstr. 2, D-882152 Planegg-Martinsried, Germany
e-mail: oster@lrz.uni-muenchen.de

L.H. Pettersson
Nansen Environmental and Remote Sensing Centre, Thormøhlensgt. 47, N-5006 Bergen, Norway
e-mail: lasse.pettersson@nersc.no

J.L. Pinckney
Dept. Marine Science and Biological Sciences, University of South Carolina, EWS 603, 712 Main Street, Columbia, SC 29208, USA
e-mail: jpinckney@bid.sc.edu

R.J. Porra
CSIRO-Plant Industry, Clunies Ross Street, Black Mountain, Canberra, ACT 2601, Australia
e-mail: robert.porra@gmail.com

D.V. Pozdnyakov
Nansen International Environmental and Remote Sensing Centre, 14th line 7, St. Petersburg, Russia
e-mail: dmitry.pozdnyakov@miersc.spb.ru

B.B. Prézelin
Marine Science Institute, Department of Ecology, Evolution and Marine Biology, University of California, Santa Barbara, CA 93106, USA
e-mail: prezelin@lifesci.ucsb.edu

F. Rodríguez
Instituto Español de Oceanografía, Apdo. 1373, E-38120, Santa Cruz de Tenerife, Spain
e-mail: francisco.rodriguez@vi.ieo.es

S. Roy
ISMER, Université du Québec à Rimouski, 310 Allée des Ursulines, Rimouski, Québec, G5L 3A1, Canada
e-mail: suzanne_roy@ugar.gc.ca

H. Scheer
Department Biologie 1 – Botanik, Universität München, Menzinger Str. 67, D-80638 München, Germany
e-mail: H.Scheer@biologie.uni-muenchen.de

L. Schlüter
DHI Water & Environment, Agern Allé 11, DK-2970 Hørsholm, Denmark
e-mail: esc@dhigroup.com

O.M. Schofield
Coastal Ocean Observation Lab., 1 Dudley Road, Rutgers University, New Brunswick, NJ 08091, USA
e-mail: oscar@imcs.marine.rutgers.edu

J. Seppälä
Finnish Environment Institute SYKE, Marine Centre, Erik Palménin aukio 1, P O Box 140, FI-00251, Helsinki, Finland
e-mail: jukka.seppala@symparisto.fi

C. Thomas
Science Systems and Applications, Inc., NASA Goddard Space Flight Center, Ocean Ecology Branch, Mail Code 614.2, 8800 Greenbelt Rd., Greenbelt, MD 20771, USA
e-mail: cthomas@umces.edu

L. Van Heukelem
University of Maryland Center for Environmental Sciences, Horn Point Laboratory, 2020 Horns Point Rd, P.O. Box 775, Cambridge, MD 21613, USA
e-mail: laurievh@peoplepc.com

K. Whitehead
Institute for Systems Biology, 1441 North 34th Street, Seattle, WA 98103–8904, USA
e-mail: kwitehe@systemsbiology.org

S. Wright
Australian Antarctic Division, Channel Highway, Kingston, TAS 7050, Australia
e-mail: Simon.Wright@aad.gov.au

M. Zapata
Instituto de Investigaciones Marinas, CSIC, Eduardo Cabello 6, E-36208 Vigo, Spain
e-mail: zapata@iim.csic.es

K.-H. Zhao
College of Life Science and Technology, Huazhong University of Science and Technology, Wuhan 430074, Hubei, P.R. China
e-mail: khzhao@163.com

Preface

In 1997, the Scientific Committee on Oceanic Research (SCOR) (with support from the United Nations Educational, Scientific and Cultural Organization (UNESCO) and the editors' institutions) sponsored a volume on phytoplankton pigments entitled *Phytoplankton Pigments in Oceanography: Guidelines to Modern Methods*. This volume was edited by Drs S. W. Jeffrey, R. F. C. Mantoura and S. W. Wright and resulted from the activities of SCOR Working Group 78. The 1997 volume went out of print a few years after publication (about 2000 copies were sold), which prompted UNESCO Publishing to print another 500 copies in 2005.

In April 2006, SCOR sponsored a workshop of pigment specialists from around the world to examine updates in this field. This workshop was hosted by Dr R. Fauzi C. Mantoura and the International Atomic Energy Agency's Marine Environmental Laboratory in Monaco. The updates that were identified include new advances in the taxonomy of marine phytoplankton (several new algal groups have been described since 1997), improved analytical techniques (notably HPLC-linked mass spectrometry, not generally used for pigment analysis before 1997), and new applications for pigments. The outcome of this meeting was a consensus that an update of the original 1997 volume was urgently needed, and a new editorial team was nominated. The present volume is the result of this update. Two of the three former editors of the 1997 volume contributed to the present volume (S. W. Jeffrey and S. W. Wright). Their collaboration ensures a smoother transition between the two volumes and prevents repetition, focusing instead on developments since the 1997 volume.

Recent discoveries on several new algal classes particularly for the picoplankton category (smallest sized algae) and on new pigments are outlined in Chapter 1 of the present volume. These discoveries have benefited from improvements in culturing, microscopic and molecular methods. In particular, molecular methods have contributed to the recent advances in our understanding on the biosynthetic pathways for both chlorophylls and carotenoids (see Chapters 2 and 3). The present volume also includes overviews on water soluble 'pigments' used more extensively in oceanography, namely phycobiliproteins (Chapter 9) and mycosporine-like amino acids (Chapter 10).

The many recent advances in methodologies examined in the present volume include wider application of HPLC (Chapter 4), liquid chromatography-mass spectrometry (Chapter 7) and developments and updates on the mathematical methods used to exploit pigment information and extract the composition of phytoplankton communities (Chapter 6). The importance of high-quality chromatographic data for pigment determinations is highlighted in Chapter 5, particularly when pigments are used for remote-sensing applications and algorithm development. Mathematical tools have also been developed to extract information from absorption or fluorescence spectra without prior separation of the various pigments by a chromatographic technique – some applications are reviewed in Chapter 8.

A few selected applications in oceanography are included, notably on the use of pigments to provide information on the status of photoacclimation, through changes in photoprotective pigments (Chapter 11), as well as a review on the use of pigment labelling to infer rates of algal growth or the rate of grazing on algae, with highlights on the importance of microzooplankton in oceans (Chapter 12).

There is an increasing recognition of the impact that environmental change has on biological productivity, biodiversity and microbial cycling in the ocean. Knowledge on pigments in the aquatic environment is critical to understanding these fundamental aspects and is also a key complement to the rapidly advancing fields of remote sensing of pigments from space and environmental monitoring, particularly for coastal regions. Monitoring is particularly important for the study of phytoplankton bloom dynamics in general and harmful algal blooms (HABs). These often toxic blooms are a growing problem in many coastal regions of the world, for reasons that are not entirely clear, but which may be related to eutrophication, ballast transport, aquaculture, climate change, etc. Chapter 13 provides the background information on bio-optical properties of pigments, necessary for understanding the recently developed tools that make use of these properties, and Chapter 14 provides an outlook on the use of pigments for *in situ* and remote-sensing detection of phytoplankton blooms (including HABs) in coastal regions, with an Appendix containing information on pigments found in harmful algae (Appendix 14A). The final chapter, Chapter 15 presents a collection of perspectives on future directions for pigment research. The book also has further materials available online at www.cambridge.org/phytoplankton. Electronic versions of the data sheets in the book are supplied for easy reference, plus an extra Appendix on specific absorption coefficients, and an extended version of Appendix 14A.

The 1997 volume was considered an extremely useful handbook by most users. The book was not developed as a textbook for university students; it was addressed rather to aquatic scientists interested in analysing and using pigments to trace algae in their study systems, for example, in relation to environmental monitoring, climate change, remote sensing, biogeochemical, ecological and biodiversity studies. Our aim is the same with this present volume, making it an indispensable tool for professionals and students who wish to analyse and research all areas in relation to aquatic pigments. We hope you will find it useful.

The scientific opinions expressed in this volume are those of the authors and should not be interpreted as the views of SCOR or any other organizations. The publication of this volume has been supported financially in large part by SCOR, with additional support from the following institutions: Université du Québec à Rimouski (Canada), Plymouth Marine Laboratory (UK), Bodø University College (Norway), Norwegian University of Science and Technology (Norway), DHI Water and Environment (Denmark) and the International Atomic Energy Agency (Monaco).

Suzanne Roy
Carole Anne Llewellyn
Einar Skarstad Egeland
Geir Johnsen

Acknowledgements

We wish to express our gratitude to several people and organizations who contributed significantly to this volume. This includes first and foremost SCOR and its Executive Director, Dr Ed Urban, who facilitated discussion with the editors of the 1997 volume and supported the organization of the workshop on pigment updates in 2006 which led to the present volume. SCOR also supported this initiative financially, with the help of the co-editors' institutions: Université du Québec à Rimouski (Canada), Bodø University College (Norway), Plymouth Marine Laboratory (UK), Norwegian University of Science and Technology (Norway), as well as Dr Fauzi Mantoura of the International Atomic Energy Agency in Monaco (now director of the Villefranche Observatory in France and one of the editors of the 1997 volume).

We also warmly thank the external reviewers who contributed to the quality of all chapters in this volume, including Drs J. Aiken, R. A. Andersen, R. Barlow, A. Battersby, J. Dolan, N. Frankenberg-Dinkel, J. Gower, B. Karlberg, B. Keely, M. Latasa, L. Lazzara, R. F. C. Mantoura, D. Millie, T. Moore, K. Oubelkheir, J. Pinckney, E. Pfündel, M. Ragni, J. Raven, W. Ruediger, W. M. Schluchter, R. Sommaruga, A. Squier, D. Suggett, S. Takaichi, G. Tilstone, M. Vernet, N. Welschmeyer and S. W. Wright.

Ms Kelly-Marie Davidson (Plymouth Marine Laboratory, UK) and Dr Urs Neumeier (ISMER, Université du Québec à Rimouski) provided great help in preparing a design for the book cover. We are grateful for the support from the Plymouth Marine Laboratory (PML) for hosting an editorial meeting in November 2008 and for providing a meeting room, internet access and refreshments. Dr Stephen de Mora, director of PML, was very supportive and helped with the preparation of the book proposal to Cambridge University Press.

Finally, we thank our families and graduate students, who showed patience and understanding while we were striving to complete this volume.

Suzanne Roy
Carole A. Llewellyn
Einar Skarstad Egeland
Geir Johnsen

August, 2010

Abbreviations and symbols

The abbreviations shown below are common across several chapters. More specific abbreviations can be found in their respective chapters.

Pigment names

Allo	alloxanthin
Anth	antheraxanthin
APC	allophycocyanin
Aph	aphanizophyll
Asta	astaxanthin
ββ-Car	β,β-carotene (trivial name = β-carotene)
βε-Car	β,ε-carotene (trivial name = α-carotene)
βψ-Car	β,ψ-carotene (trivial name = γ-carotene)
BChl(s)	bacteriochlorophyll(s)
But-fuco	19′-butanoyloxyfucoxanthin
c_2-MGDG [14/14]	Chl c_2-monogalactosyldiacylglycerol [14:0/14:0] ester
c_2-MGDG [14/18]	Chl c_2-monogalactosyldiacylglycerol [14:0/18:4] ester
Calo	caloxanthin
Cantha	canthaxanthin
Car	carotene(s)
Chl	chlorophyll
Chl c_1+c_2	unresolved Chl $c_1 + c_2$
Chl c_{2Pg}	chlorophyll c_2-like *Pavlova gyrans*-type
Chlide	chlorophyllide
Cryp	cryptoxanthin
Diadino	diadinoxanthin
Diato	diatoxanthin
Dino	dinoxanthin
DV	divinyl
Echin	echinenone

Fuco	fucoxanthin
Gyr-de	gyroxanthin diester
Hex-fuco	19′-hexanoyloxyfucoxanthin
Hex-kfuco	19′-hexanoyloxy-4-ketofucoxanthin (also known as 4-keto-19′-hexanoyloxyfucoxanthin)
Kfuco	4-ketofucoxanthin
Kmyxo	4-ketomyxoxanthophyll
Kmyxoe	4-ketomyxoxanthophyll ester
Lut	lutein
MgDVP	Mg-2,4-divinyl pheoporphyrin a_5 monomethyl ester
MV	monovinyl
Myxo	myxoxanthophyll
Neo	neoxanthin
Oscil	oscillaxanthin
PC	phycocyanin
PCB	phycocyanobilin
PChlide	protochlorophyllide
PE	phycoerythrin
PEB	phycoerythrobilin
PEC	phycoerythrocyanin
Peri	peridinin
Pheide	pheophorbide
Phe	pheophytin
PPC	photoprotective carotenoids
Pras	prasinoxanthin
PSC	photosynthetic (light harvesting) carotenoids
PUB	phycourobilin
PVB	phycoviolobilin (also known as phycobiliviolin)
Siph	siphonaxanthin
Siph-e	siphonaxanthin ester (siphonein)
TChl	total chlorophyll (sum of Chl, allomers, epimers and Chlide)
Uri	uriolide
Vauch	vaucheriaxanthin
Viola	violaxanthin
XC	xanthophyll cycle
Zea	zeaxanthin

Other common abbreviations and symbols

CDOM	chromophoric dissolved organic matter
DAD	diode array detection
DIC	dissolved inorganic carbon
DNA	deoxyribonucleic acid

DOM	dissolved organic matter
ε	molar absorption coefficient
E	irradiance
HPLC	high performance liquid chromatography
MAAs	mycosporine-like amino acids
Me	methyl
MeOH	methanol
MS	mass spectrometry
NASA	National Aeronautics and Space Administration
NMR	nuclear magnetic resonance
PAR	photosynthetically active radiation, 400–700 nm
PS	photosystem
RC	photosynthetic reaction centre
SCOR	Scientific Committee on Oceanic Research
UV	ultraviolet radiation
Vis	visible range of wavelengths (400–700 nm)

Part I

Chlorophylls and carotenoids

1

Microalgal classes and their signature pigments

S. W. JEFFREY[1], SIMON W. WRIGHT AND MANUEL ZAPATA

1.1 Introduction

The microalgae that make up the extensive phytoplankton pastures of the world's oceans originated in ancient evolutionary times. They obtained their primitive 'plastids' from an unknown ancestral cyanobacterium with photosynthetic oxygen-evolving capabilities (Bhattacharya, 1997; Delwiche, 1999; McFadden, 2001; Palmer, 2003; Keeling, 2004a, b). Serial symbioses within heterotrophic hosts gave rise to the present wide diversity of photosynthetic microalgae, which evolved a range of photosynthetic pigments capable of collectively harvesting most of the wavelengths of light available to them in underwater marine habitats (Jeffrey and Wright, 2006).

At the present time, the marine phytoplankton contribute at least a quarter of the biomass of the world's vegetation, and constitute the base of the food web that supports either directly or indirectly all the animal populations of the open sea. Some microalgae also contribute significantly to climatic processes, providing nuclei for atmospheric water condensation (Aiken *et al.*, 1992). All microalgae, by their photosynthetic activities, contribute to atmospheric carbon dioxide 'draw-down' (Jeffrey and Mantoura, 1997), thus helping to ameliorate green-house gases, by removing nearly a third of the anthropogenic carbon released to the atmosphere (Sabine and Feely, 2007).

Because of the important global role of phytoplankton, monitoring their biomass by measuring ocean colour from space (Sathyendranath, 1986; Sathyendranath *et al.*, 2004; Nair *et al.*, 2008) and increasing the accuracy of *in situ* pigment measurements to determine algal types in the water column (Jeffrey *et al.*, 1997b), have become high priority areas for oceanographic research. Indeed, UNESCO and the Scientific Committee on Oceanic Research (SCOR) were pivotal in funding our first (1997) volume on this topic: *Phytoplankton Pigments in Oceanography: Guidelines to Modern Methods*.

[1] To whom correspondence should be addressed. (Shirley.Jeffrey@csiro.au)

Phytoplankton Pigments: Characterization, Chemotaxonomy and Applications in Oceanography, eds. Suzanne Roy, Carole A. Llewellyn, Einar Skarstad Egeland and Geir Johnsen. Published by Cambridge University Press.

Since 1997 a number of new algal classes have been erected (reviewed by Huisman and Saunders, 2007), and the availability of new algal cultures and improved HPLC techniques have allowed identification of many new pigments (e.g. Britton *et al.*, 2004; Garrido and Zapata, 2006). In the following sections, we indicate the range of these pigments across the known microalgal classes, and address their value in the developing science of phytoplankton pigment oceanography.

1.2 Algal classification

1.2.1 The 'protistan perspective'

An excellent guide to the classification of the tens of thousands of present-day and fossil phytoplankton species is given by Huisman and Saunders (2007) and Garcia and Playford (2007). These authors summarize a brief history of algal classification since the times of Kützing (1843), when the first microscopes revealed the astonishing world of minute plants and animals. The Kingdom Protista was subsequently proposed by Haeckel (1866) to accommodate all microscopic organisms (both pro- and eukaryotes), placing the protozoa, eukaryotic algae, slime moulds and some lower fungi in a coherent scheme. Acceptance of protistan concepts have fluctuated over time, but recently received additional support from evidence that eukaryotic cells gained mitochondria and plastids by endosymbiosis (Bhattacharya, 1997). A few 'lower' animals, such as protozoa 'enslaved' organelles such as plastids from other taxa and so became 'plants', but earlier classification schemes, which were devised in ignorance of this process, resulted in phylogenetically incoherent groups. The resurgence of 'protistology' (Corliss, 1986; 1994) as a field of study has provided a new means of putting these microscopic organisms and their inter-relationships into a phylogenetic perspective.

Adl *et al.* (2005) have published a new scheme of restructuring a higher-level classification of eukaryotes from a protistan perspective. They define 'protist' as a eukaryote with a unicellular level of organization, but accept that forms of vegetative cell differentiation may exist. We include here a short discussion of this topic since the new protistan classification schemes and terminology are starting to pervade the oceanographic literature (e.g. Gast *et al.*, 2006), and may cause confusion in relation to traditional pigment classification schemes. The 28 authors responsible for Adl *et al.* (2005) (phycologists, mycologists, parasitologists and protistologists) have adopted a different hierarchical system. Traditional 'Kingdoms', such as Metazoa, Fungi and Plantae are recognized as deriving from monophyletic protist lineages. The authors grouped molecular protistan phylogenies into six clusters (as shown below), which included both photosynthetic and non-photosynthetic (heterotrophic) organisms:

(1) ***Opisthokonta:*** animals, fungi, choanoflagellates and Mesomycetozoa;
(2) ***Amoebozoa:*** traditional amoebae, slime moulds etc.;
(3) ***Rhizaria:*** foraminifera, radiolaria, heterotrophic flagellates etc.;
(4) ***Archaeplastida:*** red algae, green algae, Glaucophyta and Plantae;

Table 1.1. *The protist perspective: highest ranks of photosynthetic eukaryotes are included, excluding all heterotrophic taxa. Adapted from the 'protist perspective' classification scheme of Adl* et al. *(2005).*

Super groups	First rank	Second rank (examples of photosynthetic eukaryotes)
Rhizaria	Cercozoa	Chlorarachniophyta, *Paulinella*
Archaeplastida	Glaucophyta	Glaucophyceae
	Rhodophyceae	Subdivisions uncertain according to Adl *et al.* (2005)
	Chloroplastida	Charophyta*, Chlorophyta, *Mesostigma*, Prasinophyta
Chromalveolata	Cryptophyceae	Cryptomonadales
	Haptophyta	Pavlovophyceae, Prymnesiophyceae
	Stramenopiles	Bacillariophyta, *Bolidomonas*, Chrysophyceae, Dictyochophyceae, Eustigmatales, Pelagophyceae, Phaeophyceae*, Phaeothamniophyceae, Pinguiochrysidales, Raphidophyceae, Synurales, Xanthophyceae
	Alveolata	Apicomplexa, Dinozoa
Excavata	Euglenozoa	Euglenida

* Clades with multicellular groups.

(5) ***Chromalveolata:*** alveolata, apicomplexa, stramenopiles, Cryptophyta, Haptophyta;
(6) ***Excavata:*** includes Euglenozoa.

For the purposes of pigment oceanography, we have modified this scheme of Adl *et al.* (2005) (see Table 1.1), to include only those photosynthetic microalgal groups relevant to the pigment oceanographer. This scheme necessarily omits the non-photosynthetic (heterotrophic) eukaryotes.

1.2.2 The classical 'pigment perspective'

Early studies using thin-layer chromatography (TLC) clearly showed pigments associated with several microalgal classes in phytoplankton field populations (e.g. Jeffrey, 1974), building on the previous recognition of the importance of pigments in algal taxonomy. Macro-chromatographic methods (large columns) had been used to separate pigments from seaweed extracts (Strain *et al.*, 1944; Smith and Benitez, 1955; Strain, 1958) which showed clearly defined pigment suites across the red, green and brown algae (see Jeffrey, 1997). A number of algal divisions/classes were erected on the basis of macrophyte pigment phylogeny. The colloquial term 'chromophyte' (Chadefaud, 1950; Bourrelly, 1957) at first referred to those algae that were 'coloured differently from green' (!) but was subsequently limited to those golden-brown algae containing chlorophylls *a* and *c* (Christensen, 1989). Patterson (1989) and others thought that 'the concept of chromophytes as pigment-based was restrictive and phylogenetically unsatisfactory'; he recommended that 'the concept of Chromophyta be relinquished in favour of an extended protistan assemblage'. The presence of tripartite flagellar hairs was chosen instead of pigments, as the defining characteristic of these groups (see the glossary at the end of this chapter).

Because of growing confusion about the classification of chromophyte algae, Dr J. C. Green of the Plymouth Marine Laboratory (UK) hosted a conference on Chromophyte Algae in 1987 (Green *et al.*, 1989). Key issues discussed included: an historical perspective (Christensen, 1989); carotenoids (Bjørnland and Liaaen-Jensen, 1989); chlorophyll *c* pigments (Jeffrey, 1989); the Kingdom Chromista (Cavalier-Smith, 1989); the protistan perspective (Patterson, 1989); flagellar-based ultrastructure (Preisig, 1989); flagellar hairs (Leadbeater, 1989) and many other topics. Patterson urged the adoption of the term 'stramenopile' based on flagellar structure (*stramen* (Latin) = straw; *pilus* = hair) for those golden-brown 'chromophyte' algae previously characterized by the plastid-based pigments, chlorophylls *a* and *c*. However, the character of tripartite flagellar hairs is not shared by all stramenopiles (e.g. bipartite hairs, *Pelagomonas*; absence of flagellar hairs, some Pinguiophyceae; Dr R. A. Andersen, pers. comm.). Before long, another conference followed, this time on the Haptophyte algae (Green and Leadbeater, 1994), which added to the resolution of this difficult polyphyletic group.

In recent years, the use of pigment data to map microalgal populations in the water column has become an established and convenient way of studying field phytoplankton populations (Jeffrey and Wright, 2006). The increasing success of microalgal culture has also allowed many more species to be isolated and studied, and the range of pigments available for characterizing algal groups, even down to genera (i.e. the Haptophyta) has escalated dramatically (Zapata *et al.*, 2004). The increased precision of current HPLC pigment methods (Garrido and Zapata, 2006 and Chapter 4, this volume) has also allowed many new members of the chlorophyll *c* and fucoxanthin families to be identified (Zapata *et al.*, 2006; Airs and Llewellyn, 2006).

Table 1.2 identifies 27 pigmented photosynthetic microalgal classes from 11 divisions, listed independently from their heterotrophic protistan counterparts. This scheme is based on that of Huisman and Saunders (2007). Several macrophyte classes could have been included, whose pigmented zoospores released seasonally into the water column include a brief spell in the planktonic mode. The green algal class Ulvophyceae is one such example (Van den Hoek *et al.*, 1995a). Several freshwater lineages are also included for completeness, thus recognizing the application of pigment techniques to freshwater and estuarine environments.

The above represents the authors' understanding of the rapidly changing field of algal taxonomy, in which further advances, largely driven by algal genomics, are already occurring (Brodie and Lewis, 2007).

Table 1.2. *Classification of photosynthetic microalgal classes found in marine phytoplankton. Size classes are indicated. + indicates presence and – indicates absence from the various size classes. (Adapted from Huisman and Saunders, 2007.)*

Algal division		Class	Common name	Microplankton >20 μm	Nanoplankton 2–20 μm	Picoplankton 0.2–2 μm
PROKARYOTES						
Division:	**Cyanophyta**					
Class:		Cyanophyceae, includes species previously known as Prochlorophyceae	Cyanophyte, cyanobacteria, blue-green algae, prochlorophytes	+	+	+
EUKARYOTES						
Division:	**Glaucocystophyta**					
Class:		Glaucocystophyceae	Glaucocystophyte	+	+	–
Division:	**Rhodophyta**					
Class:		Rhodophyceae*	Red algae	Macrophyte	+	–
Division:	**Heterokontophyta (≡ Chromista, Chromophyta, Stramenopiles)**					
Class:		Bacillariophyceae	Diatom	+	+	–
Class:		Bolidophyceae	Bolidophyte	–	–	+
Class:		Chrysophyceae	Golden-brown algae	+	+	+
Class:		Dictyochophyceae	Silicoflagellate	+	+	+**
Class:		Eustigmatophyceae	Eustigmatophyte	–	+	–
Class:		Pelagophyceae	Pelagophyte	–	+	+
Class:		Phaeothamniophyceae	Phaeothamniophyte	–	+	–
Class:		Pinguiophyceae	Pinguiophyte	+	+	–
Class:		Raphidophyceae	Raphidophyte	+	+	–
Class:		Synurophyceae	Synurophyte	+	–	–
Class:		Xanthophyceae	Xanthophyte	+	+	–
Division:	**Haptophyta**					
		Pavlovophyceae	Golden-brown algae	–	+	+
Class:		Prymnesiophyceae (≡ Coccolithophyceae[1])				

Table 1.2. (*cont.*)

Algal division		Class	Common name	Microplankton >20 μm	Nanoplankton 2–20 μm	Picoplankton 0.2–2 μm
			Golden-brown algae	–	+	+
			(coccolithophorid)	+	+	–
Division:	**Cryptophyta**					
Class:		Cryptophyceae	Cryptomonad	–	+	–
Division:	**Dinophyta**					
Class:		Dinophyceae	Dinoflagellate	+	+	–
Division:	**Euglenophyta**					
Class:		Euglenophyceae	Euglenophyte	+	+	–
Division:	**Chlorarachniophyta**					
Class:		Chlorarachniophyceae	Chlorarachniophyte	+	–	–
Division:	**Chlorophyta**					
Class:		Chlorophyceae	Chlorophyte/green algae	–	+	–
Class:		Prasinophyceae	Prasinophyte	–	+	+
Class:		Trebouxiophyceae	Green algae/lichen algae	+	+	–
Division:	**Streptophyta**					
Class:		Mesostigmatophyceae	Mesostigmatophyte	–	+	–
Division:	**Unknown**					
Class:		Unknown (1)	'Picobiliphytes'[2]	–	+	+
		Unknown (2)	Chlorophyll *d*-containing[3]	Unknown taxa		

* Includes macrophytes

** Includes one picoplanktonic species (Eikrem *et al.*, 2004)

[1] Silva *et al.* (2007)

[2] Not yet formally described (Not *et al.*, 2007);

[3] Kashiyama *et al.* (2008)

1.3 Origins of microalgal plastids

It is now generally accepted that plastids (chloroplasts) of eukaryotic algae are endosymbiotic organelles, originally derived from a previously free-living ancestral cyanobacterium that developed the capacity for oxygenic photosynthesis (Bhattacharya, 1997; Delwiche, 1999; McFadden, 2001; Palmer, 2003). The host cell was a non-photosynthetic (heterotrophic) protist of unknown origin. Development of this ancient symbiosis over eons of time eventually resulted in reduction of the size of the plastid genome, by gene transfer, loss and substitution, until the majority of the plastid proteins were encoded in the nuclear genome of the host. Further evolutionary development of this early endosymbiosis resulted in three major primary lineages, each clearly monophyletic: the glaucocystophytes, and the green and red radiations (Moreira *et al.*, 2000). The modern cyanobacterial radiation was derived directly from the ancestral cyanobacterium and its relatives without undergoing any further symbioses (Figure 1.1).

Many other photosynthetic algae arose from secondary (or even tertiary) endosymbioses of cells from these lineages (Delwiche and Palmer, 1997). In these cases an alga already equipped with a primary (or secondary) plastid was engulfed by a non-photosynthetic host cell, entering into a permanent or semi-permanent association with it. The history of these events can be seen in present day microalgae by the presence of vestigial nuclei (e.g. the nucleomorph), loss of cell compartments and organelles, variations in the number of residual membranes surrounding the plastid (two, three or four – Figure 1.1), and analysis of nuclear and plastid genomes. Such attributes supported the hypothesis (Gibbs, 1981; 1993) that plastids of heterokonts (diatoms, brown algae, chrysophytes etc.) and haptophytes, cryptophytes and dinoflagellates all arose from ancestral red algae by various secondary and tertiary endosymbioses (Figure 1.1). By a similar process euglenophytes, chlorarachniophytes and green dinoflagellates acquired their plastids from ancestral prasinophytes/chlorophytes. New evidence is also suggesting that in some groups (e.g. dinoflagellates), plastid losses and replacements have occurred during algal evolution (Saldarriaga *et al.*, 2001). Living dinoflagellates in modern oceans may be found with either no pigments (heterotrophs), their own unique plastid (containing the dinoflagellate carotenoid, peridinin), or with plastids derived from cryptomonads, pennate or centric diatoms, prymnesiophytes or prasinophytes/chlorophytes. Some dinoflagellates may even harbour intact symbiotic cells (e.g. cyanobacteria) rather than enslaved plastids (Hallegraeff and Jeffrey, 1984).

Within present-day phytoplankton populations, members of most microalgal lineages and their taxonomic branches may be found. In the ancient oceans, the fossil record clearly shows the initial dominance of the green algal superfamily, but in later evolutionary times, the balance has switched to ecological dominance of chromophyte algae from the red algal radiation. Establishing which forces have promoted these changes in the modern oceans is an active area of current research (Quigg *et al.*, 2003, Grzebyk *et al.*, 2003; Falkowski *et al.*, 2004a, b). Grzebyk *et al.* (2003) suggest that 'whereas all algal plastids possess a core set of genes, red plastids

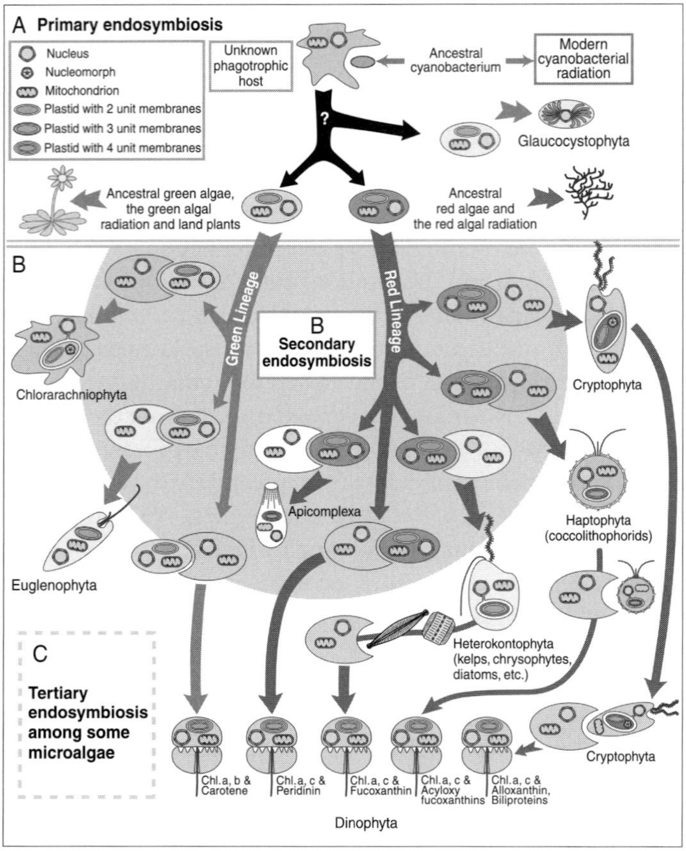

Figure 1.1. Diagram showing the hypothetical evolution of algal plastid diversity via serial endosymbiosis, based on Delwiche (1999), modified to show a common red algal origin of the plastids of apicomplexans, peridinin – containing dinoflagellates and heterokonts (Sanchez-Puerta and Delwiche, 2008; Janouškovec *et al.*, 2010). The evolutionary relationships among cryptophytes, haptophytes, heterokonts and alveolates are still controversial (Sanchez-Puerta and Delwiche, 2008), as are the number of tertiary endosymbioses (Bodył and Moszcyńksi, 2006). See colour plate section.

retain a complementary set of genes that potentially confer more capacity to express proteins regulating oxygenic photosynthetic and energy transduction pathways'.

1.4 Biological characteristics of currently recognized photosynthetic microalgal classes

Since publication of the SCOR-UNESCO volume: *Phytoplankton Pigments in Oceanography: Guidelines to Modern Methods* (Jeffrey *et al.*, 1997b), additional photosynthetic microalgal classes have been recognized that are relevant to pigment oceanography. Jeffrey and Vesk (1997) originally summarized the biological characteristics of twelve microalgal classes commonly encountered in phytoplankton field populations: diatoms, dinoflagellates, haptophytes, chrysophytes, raphidophytes, cryptomonads, chlorophytes, euglenophytes, eustigmatophytes, rhodophytes, cyanobacteria and prochlorophytes. In this chapter updated summaries of these groups are given, together with information on those classes not included previously: glaucocystophytes, bolidophytes, pelagophytes, phaeothamniophytes, pinguiophytes, synurophytes, xanthophytes, chlorarachniophytes, trebouxiophytes and mesostigmatophytes (Tables 1.3–1.27). These summaries will hopefully assist the oceanographer to recognize the less familiar microalgal classes. Three freshwater eukaryote groups – the glaucocystophytes, the phaeothamniophytes and mesostigmatophytes – are included, because of their relevance to freshwater and estuarine ecology and the endosymbiotic theory.

Figures 1.2 to 1.8 give a simple illustrative guide to some of these microalgae, but more complete information may be found in Margulis *et al.* (1989), Van den Hoek *et al.* (1995b), Jeffrey and Vesk (1997), Scott and Marchant (2005), Jeffrey and Wright (2006), Hallegraeff (2006), Brodie and Lewis (2007), McCarthy and Orchard (2007), Graham *et al.* (2009) and Hallegraeff *et al.* (2010).

Summaries of the microalgal classes are listed according to Table 1.2, following the nomenclature of Huisman and Saunders (2007) and the endosymbiotic theory of plastid evolution as shown in Figure 1.1 (adapted from Delwiche, 1999). This scheme has four major branches: the ancestral prokaryotes and three eukaryote branches: the glaucocystophytes, the red radiation and the green radiation.

The prokaryotes, which lack internal organelles (see Figure 1.2B), are described in Tables 1.3 and 1.4. The unique monophyletic eukaryote division, the Glaucocystophyta, is described in Table 1.5. Tables 1.6–1.21 summarize the major biological characteristics of eukaryotic photosynthetic microalgal divisions and classes within the red algal radiation, while Tables 1.22–1.27 describe microalgal divisions and classes of the green algal radiation. Two groups with unknown affinities were also identified from genomes of mixed phytoplankton samples (see Table 1.2). This suggests that many unknown algal types still remain to be identified from the global ocean.

The characteristics of a hypothetical eukaryotic microalgal cell are shown diagrammatically in Figure 1.2C, which illustrates the types of membrane-bound organelles present in most eukaryotes. Not every ultrastructural feature is found in all eukaryotic classes.

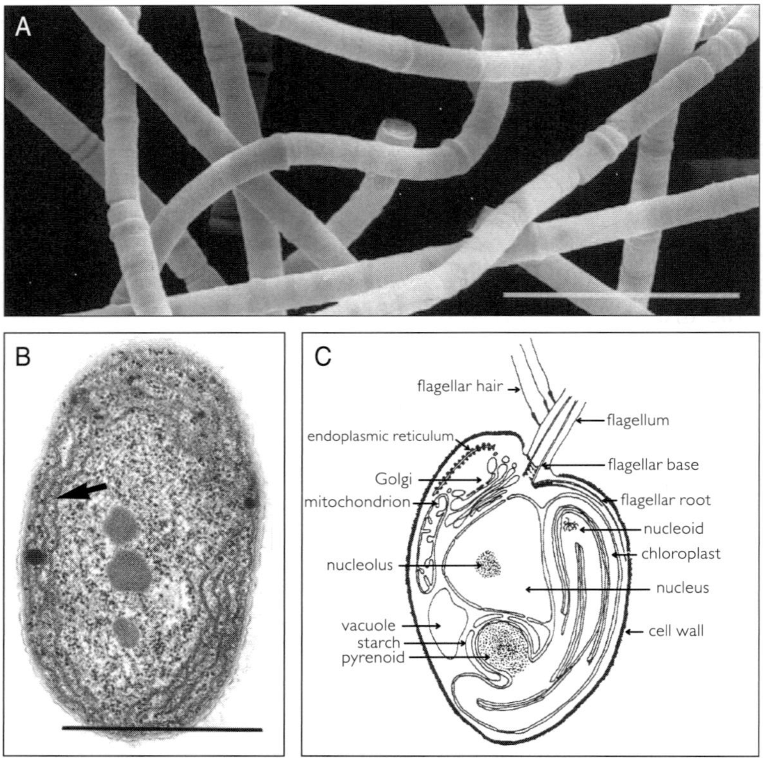

Figure 1.2. (A) Scanning electron micrograph (SEM) of the toxic cyanobacterium *Nodularia spumigena* from Australia. Scale bar = 50 μm; (B) Transmission electron micrograph (TEM) of the minute cyanobacterium *Synechococcus* sp. from the East Australian Current. Arrow points to single thylakoids lying in the peripheral cytoplasm. Note the simplicity of the cell ultrastructure of this prokaryotic microalga compared to that of a eukaryotic cell (Figure 1.2C). Scale bar = 1 μm; (C) Longitudinal section (LS) through a hypothetical eukaryotic algal cell showing the types of organelles often present.

1.4.1 Prokaryotes: the division Cyanophyta

1.4.1.1 Cyanophytes (Cyanophyceae); including species previously known as Prochlorophytes (Prochlorophyceae)

Cyanophyta are prokaryotic cells most commonly found in tropical and subtropical phytoplankton (Figure 1.2A, B). Their mixed nature as plants and bacteria justifies the other now commonly used name for these organisms, cyanobacteria (Stanier *et al.*, 1978). The most abundant filamentous species, *Trichodesmium* spp., form massive blooms both at the sea surface and at depth (Carpenter and Price, 1977). Coccoid, picoplanktonic cyanophytes (Figure 1.2B) are also widespread in the world's oceans (Waterbury *et al.*, 1979), excepting polar marine waters (Marchant *et al.*, 1987), and are often abundant in the dimly lit regions at the base of the euphotic zone. Intact symbiotic cyanobacteria are frequently found within large tropical centric diatoms and certain tropical dinoflagellates (Hallegraeff and Jeffrey, 1984; Vesk *et al.*, 1990; Gordon *et al.*, 1994).

A unique group of prokaryotic algal cells was discovered in the mid 1970s (Lewin, 1976) with a cell structure resembling that of the Cyanophyta (i.e. no cellular membrane-bound organelles, see Figure 1.2B) but with a different pigment composition. A new algal division, the Prochlorophyta, was erected to accommodate these forms (Lewin, 1976). This group is now recognized from genomic data to be within the polyphyletic Cyanophyta (Urbach *et al.*, 1992). For the present discussion we are retaining the distinction of two groups within the Cyanophyta (cyanophytes and prochlorophytes) because of their different pigmentation and oceanographic functions (Jeffrey and Wright, 2006) (Tables 1.3 and 1.4).

Table 1.3. *Characteristics of cyanobacteria (chloroxybacteria) (Class:* ***Cyanophyceae****; Division:* ***Cyanophyta****).*

Distribution	Found in most temperate and tropical aquatic (marine and freshwater) and terrestrial habitats; common in picoplankton, particularly in tropical and sub-tropical environments; some species (particularly freshwater forms) contain toxins (e.g. saxitoxins, microcystins and nodularin)
Anatomy, morphology	Cyanobacteria can be in the form of filaments (Figure 1.2A), coccoid unicells (Figure 1.2B) or colonies. Picoplanktonic forms are 1–2 μm in diameter; filamentous types can form colonies up to 2 mm in size
Cellular organization	No membrane-bound organelles (as found in eukaryotes) are present (Figure 1.2C); nuclear material exists as fibrils of DNA in the cytoplasm (e.g. as bacterial chromosomes); thylakoids lie free in the cytoplasm (no chloroplasts); phycobilisomes containing the biliproteins are attached to the thylakoids; the cell wall is a rigid murein peptidoglycan layer surrounded by an outer double-layered membrane and a mucilaginous sheath; rubisco-containing carboxysomes are present in the cytoplasm; gas vacuoles occur in some filamentous forms (e.g. *Trichodesmium* spp.) to aid flotation; specialized cells (heterocysts) fix atmospheric nitrogen in many genera of filamentous cyanobacteria (but are absent in the nitrogen-fixing *Trichodesmium* spp.); myxoxanthophyll stabilizes thylakoid membranes (Mohamed *et al.*, 2005)
Pigments	Two pigment types are recognized (Jeffrey and Wright, 2006): **CYANO-1** occurs in marine and freshwater filamentous forms such as *Trichodesmium* spp. and *Oscillatoria* sp. (Hertzberg *et al.*, 1971; Aakermann *et al.*, 1992; Carpenter *et al.*, 1993); **Chlorophylls:** Chl *a* and traces of Mg-DVP; **Carotenoids:** zeaxanthin, β,β-carotene and myxoxanthophyll (also known as myxol glycoside, Takaichi *et al.*, 2001); other carotenoids such as echinenone, canthaxanthin, oscillaxanthin, nostoxanthin, aphanizophyll, and 4-keto-myxoxanthophyll may be present;

Table 1.3. (*cont.*)

	CYANO-2 occurs in picoplanktonic forms such as *Synechococcus* spp.; **Chlorophylls:** Chl *a* and traces of Mg-DVP; **Carotenoids:** zeaxanthin and β,β-carotene (Jeffrey and Wright, 1997); **Phycobilins** occur in all Cyanophyceae: phycoerythrin, phycocyanin and allophycocyanin (Rowan, 1989); see Zhao *et al.* (this volume)
Colour	Blue-green, grey-green or red depending on the phycobilins present
Flagella	Absent
Phylogeny	Classification is continually changing as different phylogenetic techniques are applied to various taxonomic data sets; at least three orders are recognized (Reynolds, 2006): Chroococcales include unicellular and coenobial forms, mucilaginous colonies (freshwater) and picoplankton (marine); Oscillatoriales include uniseriate-filamentous forms (both freshwater and marine), and Nostocales include unbranched filamentous forms, some with heterocysts, from both fresh and brackish waters
References	Carpenter *et al.* (1993); Graham and Wilcox (2000a); Schlüter *et al.* (2004); Six *et al.* (2004); Kerfeld (2004); Mohamed *et al.* (2005); Reynolds (2006); Hayes *et al.* (2007)

Table 1.4. *Characteristics of prochlorophytes (now included in the polyphyletic division* ***Cyanophyta****).*

Distribution	Tropical and temperate oceans; Chl *d*-containing taxa (e.g. *Acaryochloris* sp.) are present in aquatic surface environments that receive near-infrared light, which is readily harvested by this pigment (see absorption spectrum, data sheets, this volume)
Anatomy, morphology and cellular organization	Prochlorophytes have a wide range of forms, habits and pigmentation, exemplified by four key genera: ***Prochloron***, ***Prochlorothrix***, ***Prochlorococcus*** and ***Acaryochloris***. Their characteristics are listed below: ***Prochloron*****:** symbiotic cells 10–25 μm diameter, found in association with marine invertebrates; cell wall resembles that of cyanobacteria, but lacks a mucilaginous sheath; no chloroplasts are found but parallel thylakoids around the cell periphery surround a vacuole formed from a dilated thylakoid; DNA is scattered as filaments in the cytoplasm (no nucleus is present); carboxysomes and crystalline or para-crystalline inclusions occur in the cytoplasm ***Prochlorothrix*****:** Filaments (trichomes) of this genus are only found in freshwater (dimensions 0.5–1.5 μm wide, 3–10 μm long); the cell wall resembles that of cyanobacteria; no chloroplasts are found, but

Table 1.4. (*cont.*)

	thylakoids occur in an irregular pattern in stacks of 2–8; DNA occurs in a central nucleoid; small carboxysomes (0.2 μm) and gas vacuoles are present in the cytoplasm ***Prochlorococcus*:** Coccoid or ellipsoid cells are 0.6–0.8 μm in diameter, and up to 1.6 μm long; they are a common component of tropical and subtropical marine picoplankton; the cell wall is surrounded by a sheath; no chloroplasts are found but thylakoids occur in a parallel band at the cell periphery with no dilation into a vacuole; DNA is found in a central nucleoid ***Acaryochloris*:** Coccoid or ellipsoid cells (1–1.5 μm in diameter, 1.5–3.0 μm long); free living in endolithic or epiphytic habitats; symbiotic in tropical ascidians; cell walls are surrounded by mucilage; thylakoids are stacked peripherally; gas vacuoles are absent, carboxysomes are present
Pigments	Three pigment types are recognized in the prochlorophytes (Jeffrey and Wright, 2006): **CYANO-3:** typical of *Prochloron* and *Prochlorothrix;* **Chlorophylls:** Chl *a*, Chl *b*, Mg-DVP; **Carotenoids:** β,β-carotene, zeaxanthin, cryptoxanthin (α or β form, see Table 1.28, Note a.), traces of β,β-carotene monoepoxide, echinenone; **Biliproteins** absent (Burger-Wiersma *et al.*, 1986; Foss *et al.*, 1987; Goericke *et al.*, 2000); **CYANO-4**: typical of *Prochlorococcus*; **Chlorophylls:** DV-Chl *a*, DV-Chl *b*, Mg-DVP (Goericke and Repeta, 1992) (Chl *a*, *b* not detected under normal conditions but Chl *b* detected under high light (see Partensky *et al.*, 1993; 1999); no **Biliproteins** except small amounts of a novel phycoerythrin; **Carotenoids:** β,ε-carotene, zeaxanthin; **Irradiance effects:** DV-Chl *b* and zeaxanthin ratios change markedly with irradiance (Partensky *et al.*, 1997); **CYANO-5:** typical of *Acaryochloris*; **Chlorophylls:** Chl *d*, with trace concentrations of Chl *a* and a Chl *c*-like pigment (Mg-DVP); **Carotenoids:** β,ε-carotene and zeaxanthin; traces of **Biliproteins** (phycocyanin and allophycocyanin, Miyashita *et al.*, 1996, 1997, 2003)
Flagella	Absent
Culture colour	Pale green
Phylogeny	Monophyletic genera within the polyphyletic Cyanophyta
References	Goericke and Repeta (1992); Partensky *et al.* (1993); Miyashita *et al.* (1996, 1997); Marquardt *et al.* (1997); Partensky *et al.* (1997, 1999); Moore and Chisholm (1999); Goericke *et al.* (2000); Rippka *et al.* (2000); Miyashita *et al.* (2003); Kashiyama *et al.* (2008)

1.4.2 Eukaryotes: the division Glaucocystophyta

The Glaucocystophyta (Figure 1.1 and Table 1.5) comprise a unique monophyletic group, whose cyanelle is clearly related to the Cyanophyta, but whose host cell is not. There are additional affinities to the Rhodophyta. Although this group is restricted to freshwater it is included here because of its unique ultrastructure, uncertain phylogeny, relevance to the understanding of symbiogenesis and possible importance in freshwater ecology (Graham and Wilcox, 2000a).

Table 1.5. *Characteristics of glaucocystophytes (Class:* ***Glaucocystophyceae****; Division:* ***Glaucocystophyta****).*

Distribution	Small group, poorly known; rarely encountered in nature; restricted to freshwater habitats such as bogs, ditches and swamps; unknown from marine plankton
Anatomy and morphology	Oblong bi-flagellated unicells (e.g. *Cyanophora*), 10–30 μm diameter; also coccoid and palmelloid cells
Cellular organization	Chloroplasts are 'cyanelles' within a special vacuole; starch grains lie outside the chloroplast; plastids retain an ancestral peptidoglycan cell wall between two plastid envelope membranes; carboxysomes are present
Pigments	**Chlorophylls:** Chl *a* (no Chl *b* or *c*); **Carotenoids**: β,β-carotene, zeaxanthin, β-cryptoxanthin; **Biliproteins**: phycocyanin and allophycocyanin are found in phycobilisomes attached to the thylakoids (pigment data from reviews, Kies and Kremer, 1990; Entwisle, 2007a)
Colour	Blue-green
Flagella	Unequal, paired, each having two rows of non-tubular hairs (e.g. *Cyanophora*)
Phylogeny	True diversity unknown, the most important genera being *Cyanophora, Glaucocystis* and *Gloeochaete*; 'plastids' (cyanelles) appear monophyletic and no longer considered an endosymbiotic cyanophyte; affinities of 'host cell' uncertain; cryptophytes and red algae are possible sister groups (Bhattacharya and Schmidt, 1997)
References	Kies and Kremer (1990); Bhattacharya *et al.* (1995); Bhattacharya and Schmidt (1997); Delwiche and Palmer (1997); Graham and Wilcox (2000a); Entwisle (2007a)

1.4.3 Eukaryotes: the red radiation

The algal division **Rhodophyta** (red algae) comprises primitive eukaryotes of very ancient lineage (Yoon *et al.*, 2004). These gave rise to a wide variety of pigmented unicells and macrophytes, known as the red algal radiation (see Figure 1.1 and Table 1.2). While only a few genera of truly planktonic red algal unicells are known in the present day ocean (Broadwater and Scott, 1994), which may represent reduced forms of reproductive cells (Garbary and Gabrielson, 1990), most are macrophytes (red seaweeds) with a high species diversity, particularly in southern Australia (over 4000 species known, West, 2007).

The red algae may be distinguished from other eukaryotic lineages by biochemical and ultrastructural features (Broadwater and Scott, 1994; Yoon *et al.*, 2006; Maggs *et al.*, 2007). These include pigments, polysaccharides and structural features of the cell (Table 1.6). Yoon *et al.* (2006) have published a comprehensive analysis of red algal genomics which clearly identifies seven major lineages of red algae recognizable at class level. The complete analysis of the plastid genome of the red algal macrophyte *Gracillaria* sp. is also providing insights into the evolution of rhodoplasts and their relationships to other plastids (Hagiopian *et al.*, 2004). There are no really definitive methods yet available for identifying the biomass of unicellular red algae in the phytoplankton of the global ocean, though new techniques, perhaps involving algal culturing procedures linked with other biochemical analyses (e.g. floridean starch), may hopefully soon be available.

1.4.3.1 The division Rhodophyta

Table 1.6. *Characteristics of red microalgae (Class:* ***Rhodophyceae****; Division:* ***Rhodophyta****).*

Distribution	Marine, also estuaries, freshwater and soil
Form	Coccoid unicells or colonies in a polysaccharide matrix
Size	5–15 μm; some reach 40 μm
Major pigments	**Chlorophylls:** Chl *a*; **Carotenoids:** β,β-carotene, zeaxanthin; **Biliproteins:** phycoerythrin, and lesser amounts of phycocyanin and allophycocyanin (e.g. *Porphyridium purpureum*; Grabowski *et al.*, 2001; Kopecký *et al.*, 2002). An earlier report of β,ε-carotene in this species was incorrect (Jeffrey and Wright, 1997)
Energy reserves	Floridean starch deposited in the cytoplasm outside the chloroplast
Chloroplasts	The plastid, usually one, is bound by two membranes: it may be stellate in shape, containing a pyrenoid; the single unstacked thylakoids have phycobilisomes on the surface; no chloroplast endoplasmic reticulum is present
Cell covering	Usually lack a cellulose cell wall, but may have a mucilaginous sheath of xylan with sulphated polysaccharides such as agar and carrageenan, which originate in the Golgi apparatus; unicellular forms are frequently cultured for their polysaccharides, which are useful commercially in food and pharmaceutical products
Flagella	None, but some cells exhibit gliding motility
Culture colour	Deep red or purple (phycoerythrin-containing), blue-green (phycocyanin-containing)

Table 1.6. (*cont.*)

Phylogeny	The division Rhodophyta has over 4,000 macrophyte species; unicellular genera also show genetic diversity: *Flintiella, Glaucosphaera, Porphyridium, Rhodella, Rhodospora, Cyanidium, Dixoniella* etc. (Broadwater and Scott, 1994). These are distributed across three of the seven red algal classes: Porphyridiophyceae, Rhodellophyceae and Stylomatophyceae (Yoon *et al.*, 2006)
References	Cole and Sheath (1990); Garbary and Gabrielson (1990); Broadwater and Scott (1994); Bhattacharya *et al.* (1995); Bernard *et al.* (1996); Burger *et al.* (1999); Yoon *et al.* (2004, 2006); Maggs *et al.* (2007); West (2007)

1.4.3.2 The division Heterokontophyta

The term 'heterokont' originally described cells having two different types of flagella with different beating patterns. The golden-brown Heterokontophyta (alternatively known as chromophytes, autotrophic stramenopiles etc; see the glossary, this chapter) were later identified by their flagellar morphology rather than their photosynthetic pigments (Patterson, 1989, see Section 1.3). Heterokont algae also have a similar chloroplast ultrastructure: lamellae with three appressed thylakoids, a girdle lamella (except for the Eustigmatophyceae), and two additional membranes surrounding the chloroplast. They lack Chl *b* but possess the light-harvesting Chls *a* and *c*. The carbohydrate storage product is not starch, but a β-1,3-linked glucan that is stored as a solute in the vacuole (Daugbjerg and Andersen, 1997).

As collated by Huisman and Saunders (2007), the Heterokontophyta include representatives of eleven major unicellular classes. These now dominate the phytoplankton of the global ocean, displacing the initially more successful green algal line, as seen from genetic and fossil evidence (Falkowski *et al.*, 2004a, b; Garcia and Playford, 2007). They are described briefly in the following summaries (Tables 1.7 to 1.17 and Figures 1.2 to 1.7).

1.4.3.2.1 Diatoms (class Bacillariophyceae; division Heterokontophyta) Diatoms are probably the best known of all the heterokont unicellular planktonic algae. The exquisite silica morphology of the frustule upon which diatom taxonomy is based has resulted in more than 10,000 living taxa being defined (Hostetter and Stoermer, 1971; Round *et al.*, 1990). Modern genome analyses (Williams, 2007) are now revising many classic taxonomic schemes. The biological characteristics of diatoms are listed in Table 1.7.

1.4.3.2.2 Bolidophytes (class Bolidophyceae; division Heterokontophyta) In the 1980s it was recognized that the phytoplankton of the world oceans contains vast numbers of minute prokaryotes – *Synechococcus* and *Prochlorococcus* spp. (Waterbury *et al.*, 1979; Chisholm *et al.*, 1988, 1992). Since then, many tiny eukaryotes (less than 2–3 μm) have also been detected in this size fraction (e.g. pelagophytes, Andersen *et al.*,

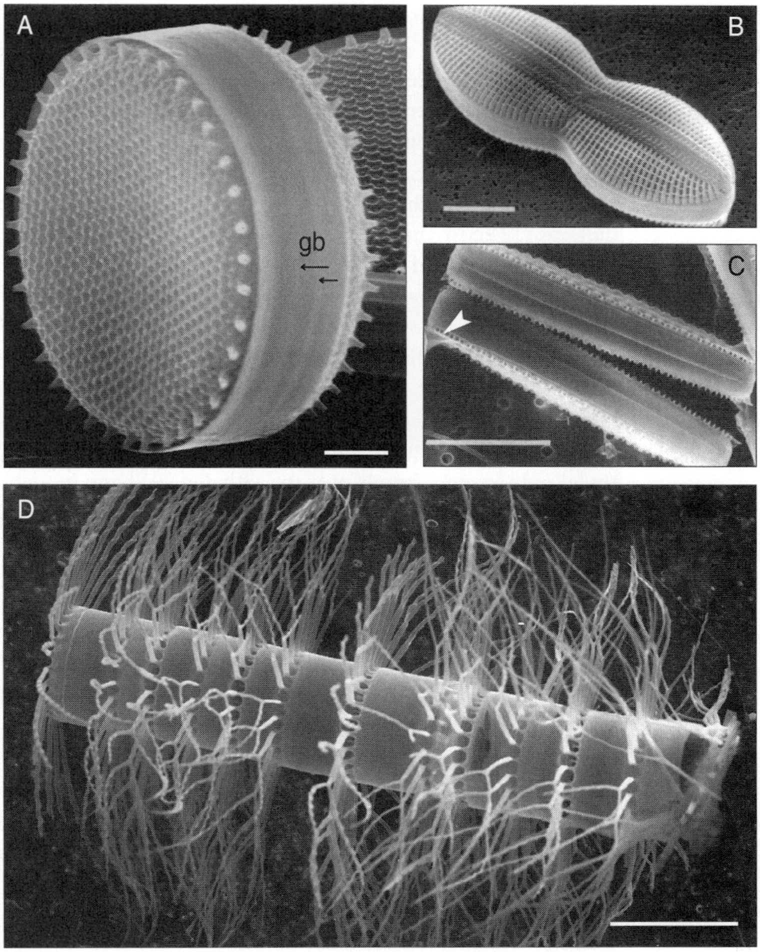

Figure 1.3. SEMs of centric and pennate diatoms showing valve morphology. (A) Centric diatom *Thalassiosira pacifica* with girdle bands (gb) connecting the epitheca and hypotheca. Scale bar = 10 μm; (B) Valve view of the benthic pennate diatom *Diploneis* sp. Scale bar = 10 μm; (C) Cells of the pennate diatom *Thalassionema nitzschiodes* connected by mucous pads (arrow head) at poles of each cell. Scale bar = 10 μm; (D) Chain-forming tropical diatom *Bacteriastrum furcatum* seen in girdle view; hair-like silica filaments (setae) allow rotation of the cell chain in nutrient-deficient tropical waters. Scale bar = 30 μm.

1993; Potter *et al.*, 1997a). Their isolation and cultivation has recently shown the presence of a new picoplanktonic algal class – the Bolidophyceae (Guillou *et al.*, 1999a, b). These are thought to represent primitive diatoms, being very similar to those in pigments, ultrastructure and SSU rDNA genomics. Members of the Parmales (Booth and Marchant, 1987) may also be life stages of the recently cultured bolidophytes (Dr R. A. Andersen, pers. comm.). Details are given in Table 1.8.

Table 1.7. *Characteristics of diatoms (Class:* ***Bacillariophyceae****; Division:* ***Heterokontophyta****).*

Distribution	Marine and freshwater; ubiquitous in the world's oceans, including sea ice; major phytoplankton group responsible for spring blooms in temperate waters; some tropical diatoms contain symbiotic cyanobacteria, which harvest nitrogen in oligotrophic waters; other diatoms are endosymbiotic within foraminifera and dinoflagellates
Anatomy and morphology	Unicellular or colonial forms (2–200 μm) with a highly intricate silica cell wall characterized by two overlapping halves (thecae) (Figures 1.3A–D); morphology of diatoms is based on radial or bilateral symmetry, e.g. centric or pennate diatoms
Cellular organization	The intricate siliceous cell wall is synthesized in special silica deposition vesicles beneath the cell membrane which often surrounds a large vacuole; chloroplasts (one to many) are bound by four membranes; thylakoids occur in bands of three with a girdle band lying beneath the chloroplast membrane; the pyrenoid is internal; chloroplast DNA is in a ring nucleoid; energy reserves are chrysolaminarin and oil
Pigments	**Chlorophylls:** three pigment types are distinguished on the basis of Chl *c* derivatives: **DIATOM-1**: Chl *a*, Chl c_1, Chl c_2, Mg-DVP (trace); **DIATOM-2**: Chl *a*, Chl c_2, Chl c_3, Mg-DVP (trace); **DIATOM-3**: Chl *a*, Chl c_1, Chl c_2, Chl c_3, Mg-DVP (trace); additional minor Chl *c* derivatives may also be found: e.g. MV-Chl c_3, and the well-established HPLC fractions: Chl $c_{2\ P.gyrans}$ and Chl $c_{1\ K.\ foliaceum}$, the structures of which are currently unknown **Carotenoids**: standard carotenoids are fucoxanthin, diadinoxanthin, diatoxanthin, β,β-carotene with occasionally 19′-butanoyloxyfucoxanthin; minor amounts of violaxanthin, antheraxanthin and zeaxanthin have been observed in some diatoms under high light conditions (Lohr and Wilhelm, 1999), interpreted by some as evidence of relict prasinophyte genes from an early endosymbiotic event (Frommolt *et al.*, 2008)
Colour	Orange to brown
Flagella	Absent, except in certain centric male gametes
Phylogeny	Pennate diatoms are monophyletic and evolved from centric forms (see Medlin *et al.*, 1993; Medlin and Kaczmarska, 2004); major lineages within centric diatoms are poorly resolved; the entire genomes of the diatoms, *Thalassiosira pseudonana* and *Phaeodactylum tricornutum*, have recently been sequenced (Armbrust *et al.*, 2004; Bowler *et al.*, 2008); molecular studies suggest diatoms may be a sister group to the bolidophytes, indeed *Bolidomonas* is now generally considered to be a primitive diatom (Dr R. A. Andersen, pers. comm.); for modern diatom classifications, please see Williams (2007)
References	Stauber and Jeffrey (1988); Medlin *et al.* (1993); Lohr and Wilhelm (1999); Graham and Wilcox (2000b); Armbrust *et al.* (2004); Medlin and Kaczmarska (2004); Zapata (2005); Williams (2007); Bowler *et al.* (2008); Frommolt *et al.* (2008)

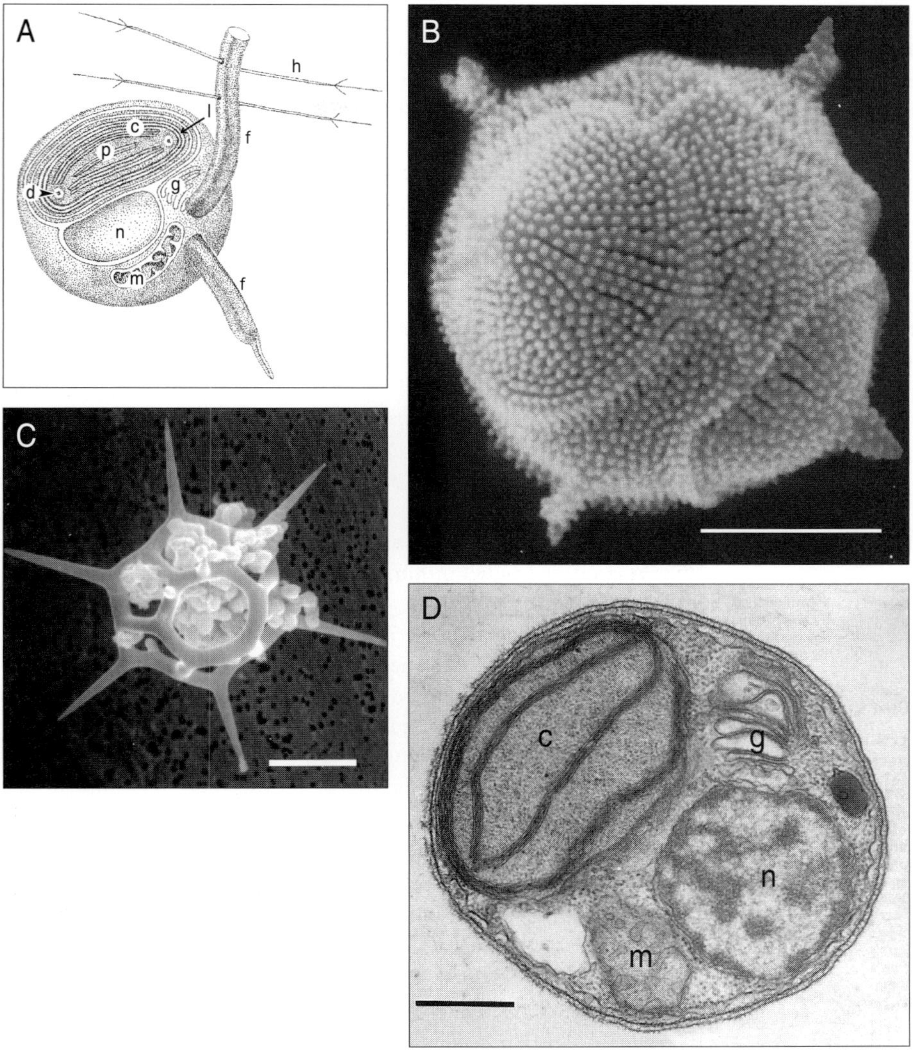

Figure 1.4. (A) Diagram of the bolidophyte *Bolidomonas* sp. showing two laterally inserted flagella (f), the longer one with tubular hairs (h). The cell contains one chloroplast (c), nucleus (n), Golgi body (g), and mitochondrion (m). The chloroplast has a ring DNA genophore (d) and girdle lamella (l) (from Guillou *et al.*, 1999a); (B) SEM of the Parmales *Tetraparma pelagicus* from the Southern Ocean (now thought to be a bolidophyte, see Table 1.8) showing silica plates. Scale bar = 1 μm; (C) SEM of the silicoflagellate *Dictyocha speculum*, showing the cell cytoplasm enclosed in a silica 'basket'. Scale bar = 10 μm; (D) TEM of a longitudinal section of the pelagophyte *Pelagococcus subviridis* showing a single chloroplast (c), nucleus (n), mitochondrion (m) and Golgi body (g). Scale bar = 1 μm.

Table 1.8. *Characteristics of the bolidophytes (Class:* ***Bolidophyceae****; Division:* ***Heterokontophyta****). Class erected by Guillou* et al. *(1999a).*

Distribution	First isolated from the picoplankton fraction of the tropical Pacific Ocean and Mediterranean Sea; bolidophyte sequences have also been recovered from mixed Arctic Ocean phytoplankton (Lovejoy *et al.*, 2006); bolidophytes are not abundant, comprising about 4% of the total picoeukaryote biomass
Anatomy and morphology	Round or heart-shaped picoplanktonic cells (1–1.7 μm in diameter; see Figure 1.4A)
Cellular organization	Single plastid with a ring genophore and girdle lamella; one mitochondrion with tubular cristae; one Golgi apparatus close to basal bodies; no pyrenoid, cell wall or theca; silicious structures present in Parmales (Fig. 1.4B)
Pigments	**Chlorophylls:** Chl *a* and *c* (Chl c_1 and c_2 were not separated by the HPLC method used; see Guillou *et al.*, 1999a); Chl c_3 and a Chl c_3-like pigment were also identified; **Carotenoids:** fucoxanthin and an unidentified fucoxanthin-like pigment, diadinoxanthin, diatoxanthin, and β,β-carotene
Colour	Golden-brown
Flagella	Heterokont, long flagellum (4–7 μm) with tubular hairs; short flagellum smooth (0.9–2.2 μm); two species are known (*Bolidomonas pacifica* and *B. mediterranea*), which differ in flagella insertion angles and swimming patterns; the long flagellum is engulfed and digested inside the cell prior to cell division
Phylogeny	Analyses of SSU rDNA indicate that *B. pacifica* and *B. mediterranea* are sister groups to diatoms; the ancestral heterokont that gave rise to the diatom lineage was probably a biflagellate unicell similar to a bolidophyte; one genus, two species known; species grouped within the Parmales (Figure 1.4B; Booth and Marchant, 1987; Ichinomiya *et al.*, 2011) may be a coccoid stage of the bolidophytes; strains of Parmales are now in culture (work in progress, Dr R. A. Andersen, pers. comm.)
References	Booth and Marchant (1987); Guillou *et al.* (1999a, b); Lovejoy *et al.* (2006); Ichinomiya *et al.* (2011)

1.4.3.2.3 Chrysophytes (class Chrysophyceae; division Heterokontophyta) The chrysophytes is a group of predominantly freshwater microalgae first recognized by Stein (1878) and Klebs (1893). In recent years the polyphyletic chrysophytes have been divided into several separate classes by Andersen *et al.* (1999) and Andersen (2007). Traditional classifications based on microscopic morphology are found to be incongruent with modern phylogenies. Studies based on molecular, ultrastructural and pigment data have supported the removal of the Haptophyta, Dictyochophyceae, Pelagophyceae and Phaeothamniophyceae from the Chrysophyceae (various authors, see Andersen *et al.*, 1999, who also list taxa considered to be Chrysophyceae *sensu stricto*). The Chrysophyceae are obviously a work-in-progress, but the availability of more chrysophyte species in culture should assist. Tables 1.9–1.17, which follow, give summaries of a number of related classes: Chrysophyceae, Dictyochophyceae, Eustigmatophyceae, Pelagophyceae and Phaeothamniophyceae, etc. Two very recently erected classes – Synchromophyceae (Horn *et al.*, 2007) and Aurearenophyceae (Kai *et al.*, 2008) – deserve further study.

Table 1.9. *Characteristics of chrysophytes (Class: **Chrysophyceae**; Division: **Heterokontophyta**.*

Distribution	Marine and freshwater
Form	Coccoid, ovoid, flagellate, colonial, loricate or amoeboid
Size	Most marine species 8–15 μm; some freshwater colonial species exceed 100 μm
Major pigments	**Chlorophylls:** Chl a, c_1, Chl $c_{2P.gyrans}$, Chl c_2 (variable); **Carotenoids**: fucoxanthin, violaxanthin, antheraxanthin, zeaxanthin and β,β-carotene present in all four species examined (see Notes, Table 1.29)
Energy reserves	Chrysolaminarin and lipid
Chloroplasts	One or two; thylakoids in bands of three with girdle band; chloroplast DNA usually in a ring nucleoid
Cell covering	Absent or with silica scales, or within a loose envelope (lorica) of cellulose
Flagella	Absent, or one or two; where two are present, they are unequal in length, one with tripartite hairs, the other smooth
Culture colour	Gold
Phylogeny	Chrysophyta previously encompassed many heterogeneous golden brown flagellates, but modern work has now defined several classes from within this group (see Andersen *et al.*, 1999; Horn *et al.*, 2007; Kai *et al.*, 2008)
References	Andersen *et al.* (1999); Jeffrey and Wright (2006); Zapata and Jeffrey (unpublished)

1.4.3.2.4 Silicoflagellates (class Dictyochophyceae; division Heterokontophyta)

Table 1.10. *Characteristics of silicoflagellates (Class: **Dictyochophyceae**; Division: **Heterokontophyta**).*

Distribution	Marine and freshwater; mostly temperate and polar regions (these cells prefer temperatures below 15 °C, Van Valkenburg 1980)

Table 1.10. (*cont.*)

Anatomy and morphology	Unicells (some 3–5 μm in diameter), may be naked or contained within a siliceous skeleton (20–100 μm in diameter) (see Figure 1.4C); one picoplanktonic species known (Eikrem *et al.*, 2004)
Cellular organization	Siliceous skeleton composed of a network of tubular elements supporting a cytoplasm of spongy appearance; mitochondria present; chloroplasts vary from two to many, with an internal pyrenoid, thylakoids in bands of three (Moestrup and Thomsen, 1990)
Pigments	Very few cultured species examined. The normal pigment complement comprises **Chlorophylls:** Chl *a*, Chl c_2; Chl c_1, c_3 variable; **Carotenoids:** fucoxanthin, β,β-carotene, diadinoxanthin, diatoxanthin, 19′-butanoyloxyfucoxanthin; violaxanthin and zeaxanthin variable; additional diatoxanthin-like, 19′-butanoyloxyfucoxanthin-like and Chl *a*-like pigments were detected in the silicoflagellate *Verrucophora farcimen* (Edvardsen *et al.*, 2007), now known as *Pseudochattonella farcimen* (Eikrem *et al.*, 2009), and the picoplanktonic *Florenciella parvula* (Eikrem *et al.*, 2004)
Colour	Gold
Flagella	Mostly contain a single flagellum, although some have two
Phylogeny	At least 13 genera (e.g. *Dictyocha*, *Apedinella*, *Pseudopedinella*, etc.); closely related to the Pelagophyceae and Chrysophyceae
References	van Valkenburg (1980); Daugbjerg (1996a, b); Daugbjerg and Henriksen (2001); Moestrup and O'Kelly (2002); Andersen (2004); Eikrem *et al.* (2004); Jeffrey and Wright (2006); Edvardsen *et al.* (2007); Hosoi-Tanabe *et al.* (2007)

1.4.3.2.5 Eustigmatophytes (class Eustigmatophyceae; division Heterokontophyta)

Table 1.11. *Characteristics of eustigmatophytes (Class:* ***Eustigmatophyceae****; Division:* ***Heterokontophyta****)*

Distribution	Mostly freshwater (soil, running water), but also found in coastal waters, tide pools and surface ocean waters; adapted to cool temperatures
Anatomy and morphology	Coccoid unicells or oval flagellates (zoospores); size 2–18 μm in diameter (marine species less than 10 μm)
Cellular organization	Many have a large chloroplast with bands of three thylakoids; the chloroplasts lack a girdle lamella; a prominent red eyespot (stigma), from which the class is named, is located outside the chloroplast
Pigments	**Chlorophylls:** Chl *a*, Mg-DVP (no Chl *b* or *c*); **Carotenoids:** antheraxanthin, vaucheriaxanthin and esters, violaxanthin, zeaxanthin, and β,β-carotene; in some species canthaxanthin is present

Table 1.11. (*cont.*)

Colour	Yellow-green
Flagella	Usually only one emergent flagellum inserted apically, with tripartite tubular hairs. If a second flagellum is present, it is short and smooth
Phylogeny	At least 12 genera are known (e.g. *Eustigmatos, Vischeria, Nannochloropsis etc.*); *N. oculata* is widely used in the mariculture industry as a highly nutritious live feed for animal larvae (Volkman *et al.*, 1993)
References	Hibberd and Leedale (1970); Norgård *et al.* (1974); Volkman *et al.* (1993); Fawley and Fawley (2007)

1.4.3.2.6 Pelagophytes (class Pelagophyceae; division Heterokontophyta)

Table 1.12. *Characteristics of the pelagophytes (Class:* ***Pelagophyceae****; Division:* ***Heterokontophyta****). Class erected by Andersen* et al. *(1993).*

Distribution	Abundant in the pico- and nano-plankton of the world's oceans (e.g. Potter *et al.*, 1997a); can cause toxic brown tides in littoral embayments
Anatomy and morphology	Coccoid unicells; flagellates oval; some filamentous sarcinoid and parmelloid colonies; coccoid forms 1.5–5.0 μm (Figure 1.4D); filamentous forms < 20 μm
Cellular organization	Plastids have thylakoids in stacks of three; girdle lamella present; pyrenoids stalked or embedded; eyespot absent; storage material (chrysolaminarin-like) in cytoplasmic vesicles
Pigments	**Chlorophylls:** Chl *a*, c_2 always present; Chl c_1, c_3 variable; **Carotenoids**: diadinoxanthin, diatoxanthin, fucoxanthin, 19′-butanoyloxyfucoxanthin, β,β-carotene; ε,ε-carotene, gyroxanthin diester, variable (Bjørnland *et al.*, 2003)
Colour	Pale-green (*Pelagococcus subviridis*); golden brown (*Pelagomonas, Aureococcus*)
Flagella	Generally two unequal flagella; forward flagellum with hairs, trailing flagellum, smooth (*Pelagomonas* has only a single flagellum)
Cell covering	Organic theca variable; three-layered wall (*Pelagococcus*); theca (*Pelagomonas*); wall (*Sarcinochryis, Aureoumbra*); wall absent (*Aureococcus, Chrysocystis*)
Phylogeny	Eight genera established (e.g. *Pelagococcus, Pelagomonas, Aureococcus, Sarcinochrysis, Aureoumbra*); work in progress suggests several species may be alternate stages of the same organism, with different phenotypic characters expressed (Dr R. A. Andersen, pers. comm.)
References	Lewin *et al.* (1977); Vesk and Jeffrey (1987); Andersen *et al.* (1993); Jeffrey and Wright (1997); Potter *et al.* (1997a); Andersen *et al.* (1999); Bjørnland *et al.* (2003)

1.4.3.2.7 Phaeothamniophytes (class Phaeothamniophyceae; division Heterokontophyta)

Table 1.13. *Characteristics of the phaeothamniophytes (Class:* ***Phaeothamniophyceae****; Division:* ***Heterokontophyta****). Class erected by Bailey* et al. *(1998).*

Distribution	Freshwater filamentous forms, widely distributed; included here because of their possible importance in freshwater ecology
Anatomy and morphology	Vegetative filaments, coccoid or capsoid unicells, motile zoospores; vegetative cells approximately 4–8 μm wide and 6–11 μm long (Figure 1.5D)
Cellular organization	One to three plastids in both vegetative cells and zoospores; lamellae with three appressed thylakoids, girdle lamella present; pyrenoids; chloroplast DNA as ring nucleoids; central nucleus with nucleolus; mitochondria, Golgi apparatus present; electron opaque vesicles beneath laminated cell wall
Colour	Golden
Pigments	**Chlorophylls:** Chl *a* and *c* (HPLC method used did not separate c_1 and c_2 but TLC showed both components); **Carotenoids:** fucoxanthin, diadinoxanthin, diatoxanthin, β,β-carotene; heteroxanthin assumed from retention time and absorption spectrum (Bailey *et al.*, 1998)
Flagella	Zoospores have laterally inserted flagella, with a multi-gyred flagellar transitional helix; tripartite flagellar hairs lack lateral filaments
Phylogeny	Possibly polyphyletic class established from filamentous genera formerly classified in the Chrysophyceae. Common genera include *Phaeothamnion*, *Stichogloea*, *Pleurochloridella*, *Phaeogloea*; but phylogeny is currently confused (Dr R. A. Andersen, pers. comm.)
References	Bailey *et al.* (1998); Andersen *et al.* (1998)

1.4.3.2.8 Pinguiophytes (class Pinguiophyceae; division Heterokontophyta) The class Pinguiophyceae was recently erected by Kawachi and co-workers (2002b) to denote a group of microalgae with a very high content of polyunsaturated fatty acids (namely eicosapentaenoic acid). This group is considered to be sister to a clade containing other closely related classes (e.g. chrysophytes, eustigmatophytes etc., see Phylogeny section, Table 1.14).

1.4.3.2.9 Raphidophytes (chloromonads) (class Raphidophyceae; division Heterokontophyta) This microalgal class appears related to chrysophytes and eustigmatophytes (Cavalier-Smith and Chao, 1996), but recent unpublished data suggest that Raphidophyceae may be sister to the Phaeophyceae/Xanthophyceae clade (Dr R. A. Andersen, pers. comm.). Raphidophytes are becoming a significant hazard in forming harmful algal blooms in highly polluted in-shore waters (see Table 1.15).

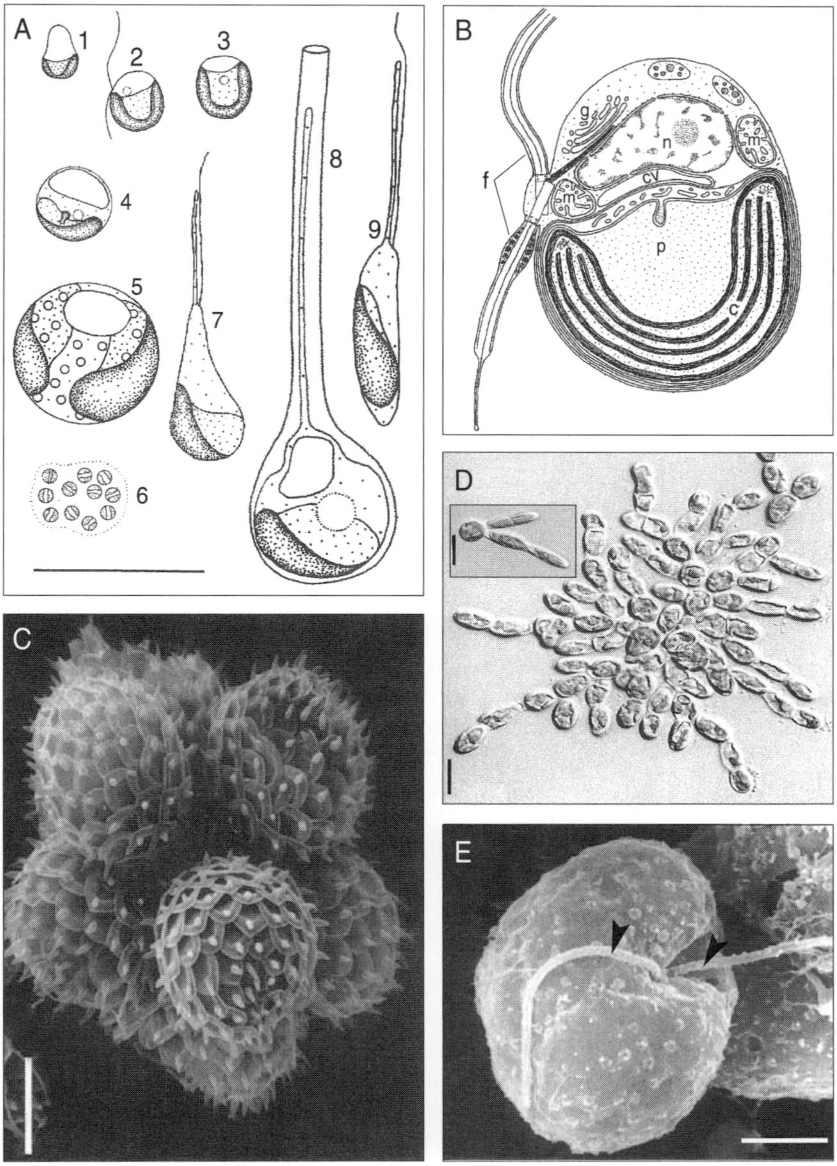

Figure 1.5. (A) Diagram of Pinguiophyceae taxa: 1. *Pinguiochrysis pyriformis*; 2, 3. *Phaeomonas parva*: 2, swimming cell; 3, non-motile cell; 4. *Pinguiococcus pyrenoidosus*. 5–7, *Glossomastix chrysoplasta*: 5, vegetative cell, 6, colony, 7, zoospore; 8, 9. *Polypodochrysis teissieri*; 8, loricate vegetative cell, 9, zoospore. Scale bar = 10 μm (1–5, 7–9) or = 1 μm (6) (from Kawachi *et al.*, 2002a); (B) Diagram of the pinguiophyte *Phaeomonas parva* showing cytoplasmic organelles: cv, capping vesicle; p, pyrenoid (from Honda and Inouye, 2002; other abbreviations as in Figure 1.4.; (C) The synurophyte *Synura mammillosa* showing external scales. Scale bar = 10 μm (from Andersen, 2007); (D) Filaments of the phaeothamniophyte, *Phaeothamnion confervicola* (Nomarski interference microscopy). Insert, young filament showing one basal spherical cell, and three elongate cells (from Andersen *et al.*, 1998); (E) SEM of the raphidophyte *Fibrocapsa japonica* with two flagella (arrowheads). Scale bar = 10 μm.

Table 1.14. *Characteristics of pinguiophytes (Class: **Pinguiophyceae**; Division: **Heterokontophyta**). Class erected by Kawachi* et al. *(2002*b*).*

Distribution	Marine, original culture was established from a tropical western Pacific Ocean sample
Anatomy and morphology	Single-celled microalga, size and form highly variable (2–40 μm, Figure 1.5A) in *Pinguiochrysis* (1), *Phaeomonas* (2, 3) and other genera
Cellular organization	One or two plastids, each with a girdle lamella, surrounded by the chloroplast endoplasmic reticulum; thylakoids consisting of three appressed lamellae; pyrenoids, stalked or embedded, with penetrating membranes; ultrastructure variable among the five genera that comprise the class (Figure 1.5B); capping vesicle function unknown (Figure 1.5B)
Pigments	**Chlorophylls:** Chls *a* and *c* (HPLC method used did not distinguish between Chl *c* types and needs re-examination); **Carotenoids:** fucoxanthin, violaxanthin, zeaxanthin, β,β-carotene
Colour	Golden-brown
Flagella	Biflagellate; forward-directed flagellum has tripartite flagellar hairs (mastigonemes)
Fatty acids	Eicosapentaenoic acid (20:5) comprises 23–56% of the total fatty acids, which is a very much higher content than other heterokonts; the fatty acids are considered a defining character of this new class
Phylogeny	Type genus *Pinguiochrysis* (Kawachi *et al.*, 2002a); genotype distinct; monophyletic; sister taxa to a clade containing Chrysophyceae, Eustigmatophyceae, Phaeophyta, Phaeothamniophyceae, Synurophyceae and Xanthophyceae. Closest relative could not be determined
References	Honda and Inouye (2002); Kawachi *et al.* (2002a, b, c)

Table 1.15. *Characteristics of raphidophytes (chloromonads) (Class: **Raphidophyceae**; Division: **Heterokontophyta**).*

Distribution	Predominantly freshwater (especially in acidic waters), but also found in coastal marine regions where they can form harmful blooms resulting in fish kills (Hara and Chihara, 1982)
Anatomy and morphology	Naked unicellular, coccoid or ovoid, often flattened dorsiventrally, size: 30–100 μm in diameter (Figure 1.5E, *Fibrocapsa japonica*); some freshwater species can be free-living or in a palmelloid condition where cells are surrounded by a gelatinous matrix
Cellular organization	No cell wall; numerous chloroplasts, thylakoids in bands of three, with a girdle band; pyrenoid protrudes from the chloroplast in marine species; chloroplast DNA in a ring nucleoid; mucocysts present

Table 1.15. (*cont.*)

	which readily discharge, destroying the cell; some species form cysts that can remain dormant in bottom sediments for several months before germinating (e.g. *Chattonella antiqua*)
Pigments	**Chlorophylls:** typical pigments include Chl *a*, c_1 and c_2, but c_2 may be absent (2 spp.); **Carotenoids:** fucoxanthin, violaxanthin, zeaxanthin, β,β-carotene; a sand-dwelling species (*Haramonas dimorpha)* contained 19′-butanoyloxyfucoxanthin (Mostaert *et al.*, 1998)
Colour	Gold to golden-brown in marine species; bright green in freshwater species
Flagella	Two, unequal in length, the anterior flagellum with tripartite hairs, the posterior smooth
Phylogeny	One order (Raphidomonales) with two families; related to eustigmatophytes and chrysophytes (Cavalier-Smith and Chao, 1996)
References	Hara and Chihara (1982); Fiksdahl *et al.* (1984b); Vesk *et al.* (1990); Hara *et al.* (1994); Cavalier-Smith and Chao (1996); Potter *et al.* (1997b); Mostaert *et al.* (1998)

1.4.3.2.10 Synurophytes (Class Synurophyceae; Division Heterokontophyta)

Table 1.16. *Characteristics of synurophytes (Class:* ***Synurophyceae****; Division:* ***Heterokontophyta****). Class erected by Andersen (1987) from the Chrysophyceae* sensu stricto.

Distribution	Flagellates found from polar to tropical regions, mostly in acidic or neutral freshwaters; two genera (e.g. *Synura and Mallomonas*) are sometimes found in seawater samples
Anatomy and morphology	Single cells (3–30 μm × 6–100 μm) or colonial (15–200 μm); palmelloid (e.g. *Mallomonas*), amoeboid or colonies (e.g. *Synura*) may be found
Cellular organization	External silica scales, often bilaterally symmetrical, are characteristic of the class (Figure 1.5C, *Synura mammillosa*); mostly two chloroplasts, each surrounded by four membranes; chloroplast lamellae formed from three appressed thylakoids, girdle lamella present; pyrenoids rare; no eyespot, but have phototactic responses; contractile vacuoles posterior to the nucleus and chloroplasts; silica deposition vesicles in which silica scales are formed
Storage products	Chrysolaminarin, oil droplets
Pigments	**Chlorophylls**: Chl *a* and c_1 (lack Chl c_2); **Carotenoids**: β,β-carotene, fucoxanthin and violaxanthin
Colour	Gold
Flagella	Two (sometimes one); subequal or very unequal; the longer flagellum beats in a sine wave, and bears tripartite tubular hairs; small organic scales may be present on the flagellum; numerous cytoskeletal microtubules provide structural elements for cell shape

Table 1.16. (*cont.*)

Phylogeny	Closely related to the Chrysophyceae; six genera and about 200 species in the Synurophyceae have been described; two families recognized but these are not strongly supported by molecular phylogenetics; important marine genera include *Mallomonas* (large genus of single-celled organisms) and *Synura* (colonial flagellates); a close relationship exists between Synurophyceae, Chrysophyceae, Phaeophyta and Bacillariophyta, but their phylogeny needs further attention
References	Andersen and Mulkey (1983); Andersen (1987, 1989, 2007)

1.4.3.2.11 Xanthophytes (Class Xanthophyceae; Division Heterokontophyta)

Table 1.17. *Characteristics of xanthophytes (Class:* ***Xanthophyceae****; Division:* ***Heterokontophyta****). Class emended by Hibberd (1990).*

Distribution	Widespread in freshwater and terrestrial habitats (damp soil, streams and salt marshes), with a few species estuarine and marine
Anatomy and morphology	Highly variable – coccoid unicells, as well as colonial, flagellated, amoeboid, siphonous or filamentous forms; size range 5–50 μm; filaments up to 150 μm
Cellular organization	Several discoid plastids per cell, red 'eyespots' embedded in the chloroplast; pyrenoids rare; ring-shaped nucleoid containing chloroplast DNA; cell walls consisting of cellulose, pectic material and silica
Pigments	**Chlorophylls:** Chl *a* with small amounts of Chl c_1 and c_2; **Carotenoids:** vaucheriaxanthin esters, diadinoxanthin, diatoxanthin, heteroxanthin and β,β-carotene
Colour	Grass-green or yellow-green
Flagella	Unequal in length (heterokont), inserted close to the cell apex; short, smooth flagellum sometimes absent
Phylogeny	Xanthophytes are monophyletic; probably a sister clade to the Phaeophyceae and Raphidophyceae; about 100 genera and 600 species known; previously included with the Chrysophyceae; important species include *Vaucheria*, *Botrydium*, *Tribonema*
References	Hibberd (1990); Van den Hoek *et al.* (1995b); Andreoli *et al.* (1999); Entwisle (2007b)

1.4.3.3 The division Haptophyta

The Haptophyta consist of unicellular, mostly photosynthetic flagellates (Hibberd, 1980), that comprise a major lineage of Chl *a* + *c* microalgae. Most occur in the nanoplankton of marine environments, although a few thrive in freshwater. The

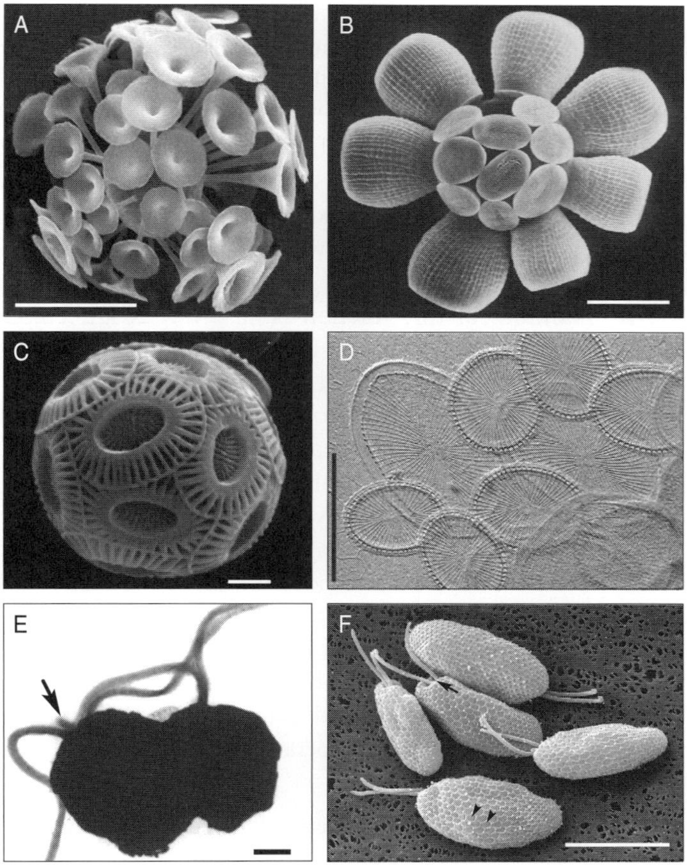

Figure 1.6. SEMs of coccolithophorids (Figs. A–C) showing the variety of coccolith sizes and shapes. (A) *Discosphaera tubifera*. Scale bar = 5 μm; (B) *Scyphosphaera aspteinii*. Scale bar = 10 μm; (C) *Emiliania huxleyi*. Scale bar = 1 μm; (D) Organic body scales of the prymnesiophyte *Chrysochromulina chiton*. Scale bar = 1 μm; (E) TEM of shadowed preparation of the prymnesiophyte *Pavlova* sp. with two flagella and a short haptonema (arrow). Scale bar = 1 μm; (F) SEM of the cryptophyte *Chroomonas* sp. showing hexagonal pattern of the pellicle (arrowheads) and two flagella emerging from the gullet (arrow). Scale bar = 10 μm.

variety of forms includes coccoid, colonial, amoeboid and filamentous stages. A distinguishing feature of the haptophytes is the filamentous haptonema, situated between the two flagella, either as a very long coiling appendage or a short flexible collection of microtubules (Figure 1.6E). Occasionally the haptonema may be absent. While the taxonomy of the Haptophyta has undergone many changes, two classes are now accepted: the Pavlovophyceae and the Prymnesiophyceae (see Medlin *et al*., 1997; Edvardsen *et al*., 2000). A phylogenetic reconstruction based on nucleotide sequences of 18S ribosomal DNA has also validated the erection of these classes

(Edvardsen *et al.*, 2000). Recently, Silva *et al.* (2007) recommended the class Prymnesiophyceae should be re-named Coccolithophyceae.

A defining characteristic of some members of the Prymnesiophyceae are the beautiful organic and calcified body scales (coccoliths, Figure 1.6A–C). Visible from space by reflectance spectroscopy (satellite instrumentation), widespread surface blooms can readily be mapped (Aiken *et al.*, 1992). A recent advance detected by *in situ* sampling has shown the unexpected dominance of tiny eukaryotic haptophytes in the picoplankton of the world oceans. This was inferred from the unexpected widespread occurrence of the so-called 'haptophyte' carotenoid, 19′-hexanoyloxyfucoxanthin, in this fraction (Liu *et al.*, 2009). However, for the presently known distribution of 19′-hexanoyloxyfucoxanthin across other microalgal classes please refer to Tables 1.18, 1.19, 1.21 (**DINO-2**) and 1.29.

Clearly, knowledge of this important group of microalgae is expanding rapidly, and the present summary can only be a stage-in-progress. In particular, the eight haptophyte pigment types identified in Table 1.29 do not have a biosynthetic basis (Zapata *et al.*, 2004) as do the five pigment types recommended for the prasinophytes (Table 1.30, Egeland *et al.*, 1997). However, the haptophyte types were observed in cultures growing in identical conditions and have proved useful for the oceanographer (e.g. Rodriguez *et al.*, 2006). For a clearer picture of the distribution of Chl *c* pigments and fucoxanthin derivatives in the eight haptophyte pigment types, please see Table 2.14 in Jeffrey and Wright (2006). We expect that greater understanding of the biosynthetic pathways and genomic compatibilities will assist in clarifying pigment relationships in this important division.

1.4.3.3.1 Pavlovophytes (class Pavlovophyceae; division Haptophyta)

Table 1.18. *Characteristics of golden-brown algae (Class: **Pavlovophyceae**; Division: **Haptophyta**). Class erected by Edvardsen* et al. *(2000).*

Distribution	Marine; found in oceanic, coastal, brackish and freshwater environments; common in coastal phytoplankton
Anatomy and morphology	Unicellular, 5–20 μm in diameter; almost exclusively flagellate unicells; cells elongate, often irregular and metabolic
Cellular organization	Body scales in the Pavlovophyceae are often knob-like, rather than plate-like; single chloroplast, with often a bulging basal pyrenoid; eyespots sometimes present; unusual dihydroxysterols (pavlovols, Véron *et al.*, 1996; Volkman *et al.*, 1997); unique mitosis that differs from that of the Prymnesiophyceae (Green and Hori, 1988; Hori and Green, 1994)
Pigments	Two pigment types **HAPTO-1** and **HAPTO-2** (see Tables 4, 5; Zapata *et al.*, 2004):
	HAPTO-1: Chlorophylls: Chl *a*, c_1, c_2, Mg-DVP; **Carotenoids**: fucoxanthin, diadinoxanthin, diatoxanthin, β,β-carotene;

Table 1.18. (*cont.*)

	HAPTO-2: Chlorophylls: Chl *a*, c_1, c_2, Mg-DVP; Chl $c_{2\ P.gyrans}$ (structurally unidentified but spectrum well known, Zapata *et al.*, 2004, 2006); **Carotenoids:** fucoxanthin, diadinoxanthin, diatoxanthin, β,β-carotene
Colour	Golden brown
Flagella	Two subequal flagella; longer flagellum inserted subapically; covered with cup-shaped hollow scales; long flagellum beats with an S-shaped wave; short flagellum beats with a stiff inflexible action; the haptonema (a flagellum-like appendage of variable size) is found between the two flagella (Figure 1.6E)
Phylogeny	Pavlovophyceae are considered more primitive than Prymnesiophyceae (Van Lenning *et al.*, 2003). One order exists (Pavlovales) with four genera (*Diacronema, Exanthemachrysis, Pavlova, Rebecca*) and 12 species (Jordan *et al.*, 2004)
References	Edvardsen *et al.* (2000); Van Lenning *et al.* (2003); Jordan *et al.* (2004); Zapata *et al.* (2004); Jeffrey and Wright (2006); Edvardsen and Medlin (2007)

1.4.3.3.2 Characteristics of golden-brown algae (class Prymnesiophyceae ≡ Coccolithophyceae)

Table 1.19. *Characteristics of golden-brown algae (Class:* ***Prymnesiophyceae*** *≡* ***Coccolithophyceae*** *(Silva* et al., *2007); Division:* ***Haptophyta****).*

Distribution	A major component of the nanoplankton (Marchant and Thomsen, 1994); abundant in tropical and subtropical oceans, with a few species abundant in polar waters (e.g. *Phaeocystis* spp.); some species may form extensive blooms (e.g. *Emiliania huxleyi* (Figure 1.6C); other species produce dimethyl sulphide, which can escape to the atmosphere and act as cloud-seeding nuclei, thus affecting climate (Malin *et al.*, 1992)
Anatomy and morphology	Normally coccoid, photosynthetic planktonic cells (mostly 5–20 μm in diameter), but palmelloid, amoeboid, filamentous, colonial and benthic forms also occur
Cellular organization	Most have one or two discoid peripheral chloroplasts (bounded by four membranes), with internal pyrenoid; thylakoids usually in bands of three, girdle band absent; chloroplast DNA occurs in scattered nucleoids; cells typically covered by one to several layers of unmineralized organic scales (Figure 1.6D); coccolithophorids also have exquisite calcified scales called coccoliths (Figure 1.6A–C); energy reserve: chrysolaminarin
Pigments	Six pigment types: **HAPTO 3–8** (see Tables 4, 5; Zapata *et al.*, 2004); **Chlorophylls:** Chl *a*, Mg-DVP; variable distributions of chlorophyll

Table 1.19. (*cont.*)

	c pigments: Chl c_3, Chl c_2, Chl c_1, Chl c_2-MGDG (18:4/14:0), Chl c_2-MGDG (14:0/14:0), unidentified non-polar Chl c_1 pigment and MV-Chl c_3;
	Carotenoids: HAPTO 3–8, all contain fucoxanthin, diadinoxanthin, diatoxanthin and β,β-carotene; **HAPTO 5**, 4-ketofucoxanthin; **HAPTO 6, 7** and **8**: also contain 19′- butanoyloxyfucoxanthin, 19′-hexanoyloxyfucoxanthin and 4-keto-19′-hexanoyloxyfucoxanthin (Zapata *et al.*, 2004; Jeffrey and Wright, 2006; Airs and Llewellyn, 2006) (see also Data sheets, this volume)
Colour	Gold to golden brown
Flagella	Two equal or subequal smooth flagella, separated by a third appendage, the haptonema, which differs structurally from the flagella and may serve for substrate or food attachment (Figure 1.6E); mature flagellum contains an autofluorescent substance (Kawai and Inouye, 1989)
Phylogeny	Prymnesiophyceae are currently divided into six orders (Jordan *et al.*, 2004; Edvardsen and Medlin, 2007), based on molecular information (SSU rDNA)
References	Fawley (1989); Kawai and Inouye (1989); Malin *et al.* (1992); Jeffrey and Wright (1994); Marchant and Thomsen (1994); Garrido *et al.* (2000); Jordan *et al.* (2004); Zapata *et al.* (2004, 2006); Jeffrey and Wright (2006); Airs and Llewellyn (2006); Edvardsen and Medlin (2007); Silva *et al.* (2007)

1.4.3.4 The division Cryptophyta

The Cryptophyta are a well-defined group of mainly photosynthetic nanoplanktonic flagellates. They can be blue-green, red or gold, depending on their complement of photosynthetic pigments (see below). Some colourless heterotrophic forms are also known (Throndsen, 1997). Cryptomonads are common in marine, estuarine and freshwater habitats, but they can be missed in field collections for microscopy unless non-destructive fixation methods are used. They are readily detected by their characteristic carotenoid, alloxanthin (Gieskes and Kraay, 1983).

1.4.3.4.1 Class Cryptophyceae; division Cryptophyta

Table 1.20. *Characteristics of cryptomonads (Class:* ***Cryptophyceae****; Division:* ***Cryptophyta****).*

Distribution	Ubiquitous in freshwater, brackish and marine environments, also soil, groundwater and snow
Anatomy and morphology	Ovoid asymmetrical unicells (6–20 μm), often flattened (Figure 1.6F). A unique cell covering or pellicle is made up of a ridged periplast

Table 1.20. (*cont.*)

	superimposed on an inner layer of thin proteinaceous plates (Figure 1.6F); no microtubular cytoskeleton
Cellular organization	One or (rarely) two large chloroplasts surrounding the posterior nucleus; thylakoids stacked in pairs with a pyrenoid protruding on the inner side, photosynthetic pigments, where present, are uniquely contained within the thylakoid lumen; the periplastidal space has ribosomes, the nucleomorph (relict of an ancestral host nucleus) and starch grains
Pigments	**Chlorophylls:** Chls *a* and c_2, Mg-DVP; **Carotenoids:** alloxanthin, crocoxanthin, monadoxanthin (characteristic of cryptomonads and organisms with cryptophyte endosymbionts); β,ε-carotene present; red or blue **Phycobiliproteins** present in thylakoid lumen (not contained in phycobilisomes, like other phycobiliprotein-containing algae); two unusual green algae also contain alloxanthin, with no evidence of cryptophyte endosymbionts (Lewin *et al.*, 2000; see note c, Table 1.30); many species are heterotrophic
Colour	Red or blue-green, depending on the dominance of cryptomonad-type phycoerythrins or phycocyanins
Flagella	Two, equal or nearly equal in length (4 to 10 μm; Figure 1.6F) with bipartite tubular hairs (different from heterokonts in which flagella are tripartite and hairs are only found on one of the flagella)
Symbioses	May be endosymbiotic in ciliates (e.g. *Mesodinium rubrum* Hibberd, 1977, now = *Myrionecta*) and dinoflagellates (Hackett *et al.*, 2003)
Phylogeny	About 200 photosynthetic species; others are heterotrophic; affinities exist with both the Rhodophyta (biliprotein accessory pigments) and the Heterokontophyta (Chl *c* accessory pigments); genomic data links cryptophytes and haptophytes as a joint group, recently termed Hacrobia *taxon nov.* (Okamoto *et al.*, 2009)
References	Hibberd (1977); Throndsen (1997); Clay *et al.* (1999); Lewin *et al.* (2000); Cavalier-Smith (2002); Hackett *et al.* (2003); Cerino and Zingone (2007)

1.4.3.5 The division Dinophyta

Dinoflagellates are a remarkably diverse and complex group of unicellular flagellates with at least 130 genera and 1200 living species (many more if fossil cyst species are included). About half the species are photosynthetic while the remainder have heterotrophic (animal-like) nutrition. Dinoflagellates can readily form symbiotic associations with other eukaryotic algae (see Figure 1.1). In addition to the basic peridinin-containing group (**DINO-1**), four endosymbiotic dinoflagellate pigment types also exist with pigments of haptophyte, diatom, cryptomonad and prasinophyte origin (e.g. **DINOs** types **2**, **3**, **4** and **5**, see Tables 1.21 and 1.29).

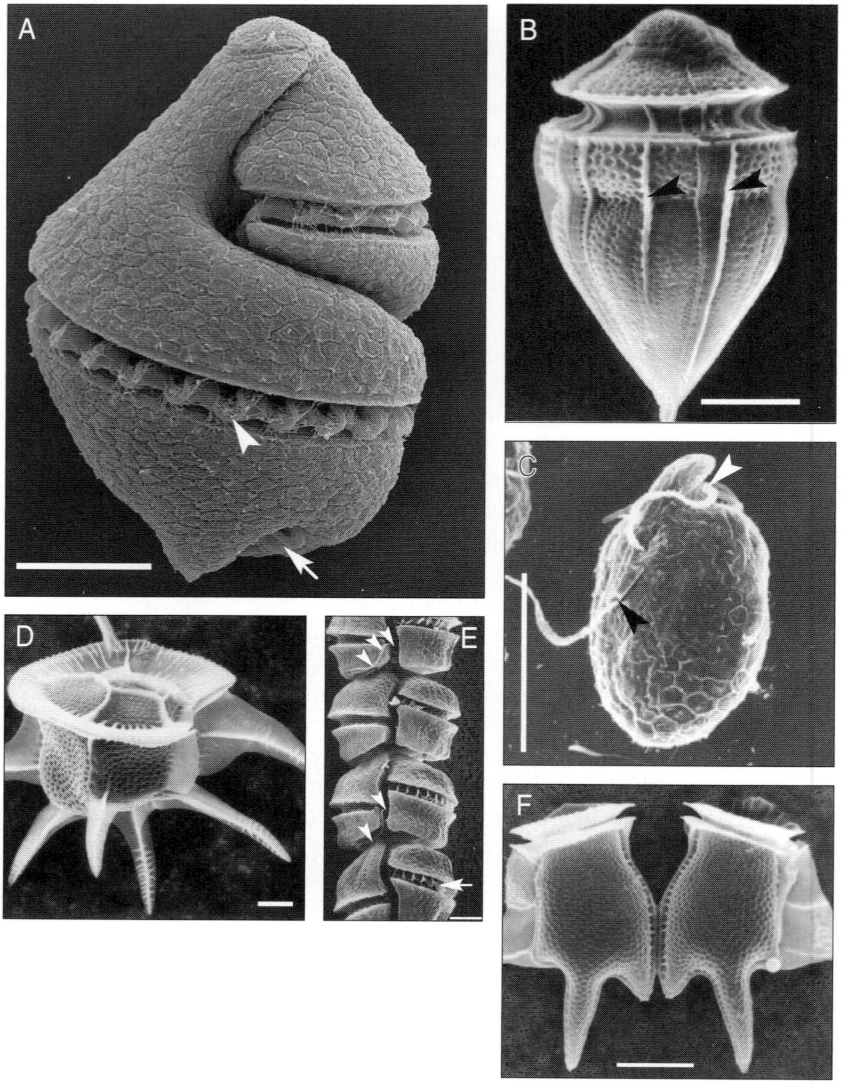

Figure 1.7. SEMs of dinoflagellates. (A) *Cochlodinium* sp., an unarmoured dinoflagellate, showing the coiled transverse flagellum within the girdle groove (arrow head); the longitudinal flagellum is just visible at the base of the sulcus (arrow). Scale bar = 10 μm; (B) *Oxytoxum constrictum* with longitudinal ridges on the hypotheca (arrow heads). Scale bar = 10 μm; (C) Naked (non-thecate) dinoflagellate *Amphidinium carterae* showing the two flagella (arrow heads) emerging from the flagellar pores. Scale bar = 10 μm; (D) Tropical *Ceratocorys horrida* with horn-like extensions of the hypotheca. Scale bar = 10 μm; (E) Chain-forming toxic *Gymnodinium catenatum* from Tasmanian estuaries showing longitudinal flagellum (arrow heads) and transverse flagellum (arrow). Scale bar = 10 μm; (F) Divided pair of *Dinophysis tripos*. Scale bar = 10 μm.

1.4.3.5.1 Class Dinophyceae; division Dinophyta

Table 1.21. *Characteristics of dinoflagellates (Class: **Dinophyceae**; Division: **Dinophyta**).*

Distribution	Mostly free-living, widely distributed in tropical, subtropical, temperate and polar oceans; also widely known from terrestrial freshwaters; dinoflagellate blooms occur mostly in coastal regions; non-motile cysts are formed as part of the life cycle, favouring survival and dispersal
Anatomy and morphology	Mostly unicellular (5–2000 μm), with each cell divided by a transverse girdle groove into an upper epitheca and a lower hypotheca (Figure 1.7A–F); unarmoured (e.g. *Amphidinium carterae* (Figure 1.7C) or armoured with cellulose plates (Figure 1.7B, D, E, F). Most species have a motile phase with two dissimilar flagella, one in a transverse girdle groove, the other longitudinal in the sulcus (Figure 1.7A, E)
Cellular organization	Large nucleus with abundant DNA from multiple permanently condensed chromosomes; trichocysts (ejectile structures) common at the cell periphery; several species contain toxins (e.g. saxitoxins, brevetoxins)
Pigments	Five pigment types are recognized (Jeffrey and Wright, 2006) with either the major dinoflagellate carotenoid peridinin (**DINO-1**) or pigments characteristic of their endosymbionts e.g. haptophytes (**DINO-2**), diatoms (**DINO-3**,), cryptophytes (**DINO-4**) or prasinophytes (**DINO-5**) **DINO-1: Photosynthetic peridinin-containing dinoflagellates**: **Chlorophylls:** Chl *a*, c_2, Mg-DVP; **Carotenoids:** peridinin, diadinoxanthin, diatoxanthin, dinoxanthin, peridininol, P-457 (7′,8′-dihydroneoxanthin-20′-al-3′-β-lactoside), pyrrhoxanthin and β,β-carotene **DINO-2: Haptophyte-containing**: **Chlorophylls:** Chl *a*, c_2, c_3, Mg-DVP; traces of c_1 and Chl c_2-MGDG [14:0/14:0] present in the dinoflagellate, *Gymnodinium breve* (now known as *Karenia brevis*, Daugbjerg *et al.*, 2000); **Carotenoids:** 19′-hexanoyloxyfucoxanthin, fucoxanthin, diadinoxanthin, 19′-butanoyloxyfucoxanthin, diatoxanthin, gyroxanthin diester, β,β-carotene, β,ε-carotene. No peridinin, C_{37}-xanthophylls or the glycoside P457 detected (Carreto *et al.*, 2001) **DINO-3**: **Diatom-containing**: **Chlorophylls:** Chl *a*, c_1, c_2; **Carotenoids**: fucoxanthin, diadinoxanthin, diatoxanthin, β,β-carotene; zeaxanthin, canthaxanthin and β,ψ-carotene (γ-carotene) variable; no acyloxyfucoxanthins detected; serial replacement of centric and pennate diatom endosymbionts has been observed in freshwater dinoflagellates (Kempton *et al.*, 2002; Takano *et al.*, 2008) **DINO-4**: **Cryptomonad-containing**: **Chlorophylls:** Chl *a*; c_2 present in *Dinophysis norvegica;* **Carotenoids:** alloxanthin (*D. norvegica;* Meyer-Harms and Pollehne, 1998); cryptomonad-type phycoerythrin (red), cryptomonad-type phycocyanin (blue-green) (Vesk *et al.*, 1996; Hewes *et al.*, 1998) **DINO-5**: **Prasinophyte-containing**: **Chlorophylls:** Chl *a, b;* no Chl *c* detected; **Carotenoids:** β,β-carotene,

Table 1.21. (*cont.*)

	neoxanthin, violaxanthin and zeaxanthin; prasinoxanthin was found in *Gymnodinium chlorophorum* (W. W. C. Gieskes, pers. comm. to Elbrächter and Schnepf, 1996) but not in a *G. chlorophorum* field sample which had an unknown major carotenoid (P. Henriksen, pers. comm. to present authors); no peridinin or fucoxanthin detected; this species needs to be re-examined with modern HPLC techniques
Colour	Reddish-brown (peridinin-containing); usually assumes the colour of the endosymbiont (e.g. gold [haptophyte/diatom] or green [prasinophyte] or red [cryptomonad])
Symbioses	Dinoflagellates occur as golden-brown endosymbionts within tropical reef animals e.g. corals, clams, radiolarians, acantharians or foraminiferans; cyanobacteria and diatoms may be symbiotic within some tropical dinoflagellates (Hallegraeff and Jeffrey, 1984)
Flagella	One ribbon-like flattened flagellum in the transverse girdle groove used for rotation and one smooth longitudinal flagellum in the sulcus used for propulsion (Figure 1.7A, E)
Phylogeny	Approximately 130 genera and 2000 living species (Taylor, 1980)
References	Jeffrey *et al.* (1975); Taylor (1980); Hallegraeff and Jeffrey (1984); Watanabe *et al.* (1990); Vesk *et al.* (1990, 1996); Elbrächter and Schnepf (1996); Daugbjerg *et al.* (2000); Carreto *et al.* (2001); Kempton *et al.* (2002); Moestrup and Daugbjerg (2007); Takano *et al.* (2008)

1.4.4 Eukaryotes: the green radiation

Green algae are photosynthetic eukaryotes which have chloroplasts bounded by two envelope membranes, stacked thylakoids, chlorophylls *a* and *b* and some unique carotenoids (see Table 1.30 for an overview of the pigment types of the green algal radiation). The green algal plastid 'arrived' prehistorically in a single endosymbiotic event (see Figure 1.1) where an ancestral cyanobacterium was 'enslaved' by a colourless eukaryotic host (Delwiche and Palmer, 1997) and subsequently, over time, the green radiation developed. Green algae are one of the most morphologically diverse groups of eukaryotes (see Tables 1.22–1.27). While they are abundantly distributed worldwide in almost every aquatic habitat, they have been superseded in biomass in the global ocean by the red algal radiation (Grzebyk *et al.*, 2003). In addition, they are found in soils of both arid and temperate areas (Pröschold and Leliaert, 2007). Green microalgae also form successful symbioses with other organisms, e.g. lichens (trebouxiophytes) and protozoa (e.g. euglenophytes, chlorarachniophytes etc.).

The green algal lineage comprises two major divisions and several minor divisions that include microalgae (Lewis and McCourt, 2004; Pröschold and Leliaert, 2007). The Chlorophyta contains the classes Chlorophyceae, Trebouxiophyceae, Prasinophyceae, Ulvophyceae, etc., while the Streptophyta contains the microalgal class Mesostigmatophyceae. The wide variety of green algal types may have developed from an ancestral green flagellate resembling a mesostigmatophyte (Lewis and McCourt, 2004). Other endosymbiotic events led to the chlorophyll *b*-containing divisions Chlorarachniophyta and Euglenophyta as well as the chlorophyll *b*-containing dinoflagellates (Figure 1.1; Table 1.21). Genomic studies are beginning to evaluate the order of branching events in the green lineage, and continued molecular phylogenetic investigations should provide further insights (Hall and Delwiche, 2007). Already a new class, the Mamiellophyceae classis nova, has been erected, which represents one of the most successful eukaryotic groups within the photosynthetic picoplankton (Marin and Melkonian, 2010).

Perhaps the most well-known unicellular marine green algae are species of *Chlorella, Dunaliella, Chlamydomonas, Micromonas* and *Carteria*. Some of these species (e.g. *Dunaliella, Haematococcus*) are mass cultured where conditions are appropriate for successful commercial production of their valuable carotenoids (Ben-Amotz *et al.*, 2008; Jeffrey and Egeland, 2008).

Six green microalgal classes that may be encountered in aquatic ecology are presented below (see Tables 1.22–1.27).

1.4.4.1 The division Euglenophyta

Table 1.22. *Characteristics of euglenophytes (Class:* ***Euglenophyceae****; Division:* ***Euglenophyta****).*

Distribution	Occur in most freshwater habitats (particularly those with decaying organic matter); also in brackish and marine environments (open sea, tidal zone or on beaches)
Anatomy and morphology	Unicellular, ovoid or fusiform; most species have a flexible pellicle, which allows movement by deformation (metaboly); other species have a rigid lorica; common marine forms are 40–60 μm long (sometimes up to 500 μm) and about 10 μm wide
Cellular organization	Mucilage-producing bodies lie beneath the proteinaceous pellicle; a large mesokaryotic nucleus is present with many condensed chromosomes; chloroplasts of green algal origin (one to many) vary in size, shape and pyrenoid structure; chloroplast bound by three membranes; thylakoids in extended and interconnected stacks of two or three; some species have an eyespot, located outside the chloroplast; storage product: a crystalline β-1,3 glucose polymer, paramylon; some species are phagotrophic
Pigments	**Chlorophylls:** Chls *a* and *b*, Mg-DVP; **Carotenoids:** eutreptiellanone, diadinoxanthin, diatoxanthin, 9′-*cis*-neoxanthin, β,β-carotene, β,ε-carotene (trace) (Bjørnland, 1982; Fiksdahl *et al.*, 1984a; Bjørnland *et al.*,1986; Jeffrey and Wright, 1997)
Colour	Grass green
Flagella	One or two, equal or unequal in length, with a single row of fine hairs

Table 1.22. (*cont.*)

Phylogeny	Nearly 20 years of molecular systematics have resulted in many taxonomic revisions (reviewed by Triemer and Farmer, 2007); four major clades of photosynthetic euglenoids are now recognized (Triemer and Farmer, 2007)
References	Bjørnland (1982); Fiksdahl *et al.* (1984a); Bjørnland *et al.* (1986); Jeffrey and Wright (1997); Marin *et al.* (2003); Triemer *et al.* (2006); Triemer and Farmer (2007)

1.4.4.2 The division Chlorarachniophyta

Table 1.23. *Characteristics of chlorarachniophytes ('green spiders') (Class: **Chlorarachniophyceae**; Division: **Chlorarachniophyta**). Erected by Geitler (1930); emended by Hibberd and Norris (1984).*

Distribution	Worldwide; marine habitats only, never freshwater; originally isolated from the Canary Islands, Penasco, Mexico and Dorset, UK. Amoeboid forms are found in sand, sediments and the ocean surface; pico-forms are probably planktonic (Figure 1.8A)
Anatomy and morphology	Amoeboid cells may be solitary, or connected by pseudopodia forming a mat-like continuum (15–30 μm Figure 1.8B); also exist as picoplanktonic flagellates (Figure 1.8A) and walled cysts (2 μm)
Cellular organization	Grass-green plastids (one to five per cell); internal bulbous pyrenoids; chloroplast matrix of lamellae composed of 1–3 thylakoids; girdle band absent; each chloroplast is bounded by four separate membranes (outer and inner membrane pairs); a cytoplasmic compartment separates the chloroplast envelope and the chloroplast endoplasmic reticulum which harbours a relict nucleus (the nucleomorph, similar to that of the Cryptophyta); nucleomorphs are structurally diverse across the class
Pigments	**Chlorophylls:** Chls *a* and *b* (no *c*); **Carotenoids:** neoxanthin, violaxanthin, lutein, β,β-carotene, antheraxanthin, zeaxanthin and loroxanthin ester (dodecanoate, Sasa *et al.*, 1992b)
Colour	Grass-green
Flagella	Basically bi-flagellate, but the second flagellum may be barely visible with only a short basal body; long flagellum ends in a hair point (Figure 1.8A)
Phylogeny	At least six genera and eleven species are known; the chlorarachniophyte endosymbiont, now greatly reduced, was once a green alga (on pigment evidence probably related to prasinophytes); the host cells resemble diverse amoeboid and flagellate heterotrophs (phylum Cercozoa) – see Table 1.1
References	Geitler (1930); Hibberd and Norris (1984); Hattakeyama *et al.* (1991); Sasa *et al.* (1992b); McFadden *et al.* (1994); Ishida *et al.* (1999, 2007); Ota *et al.* (2007, 2009a, b)

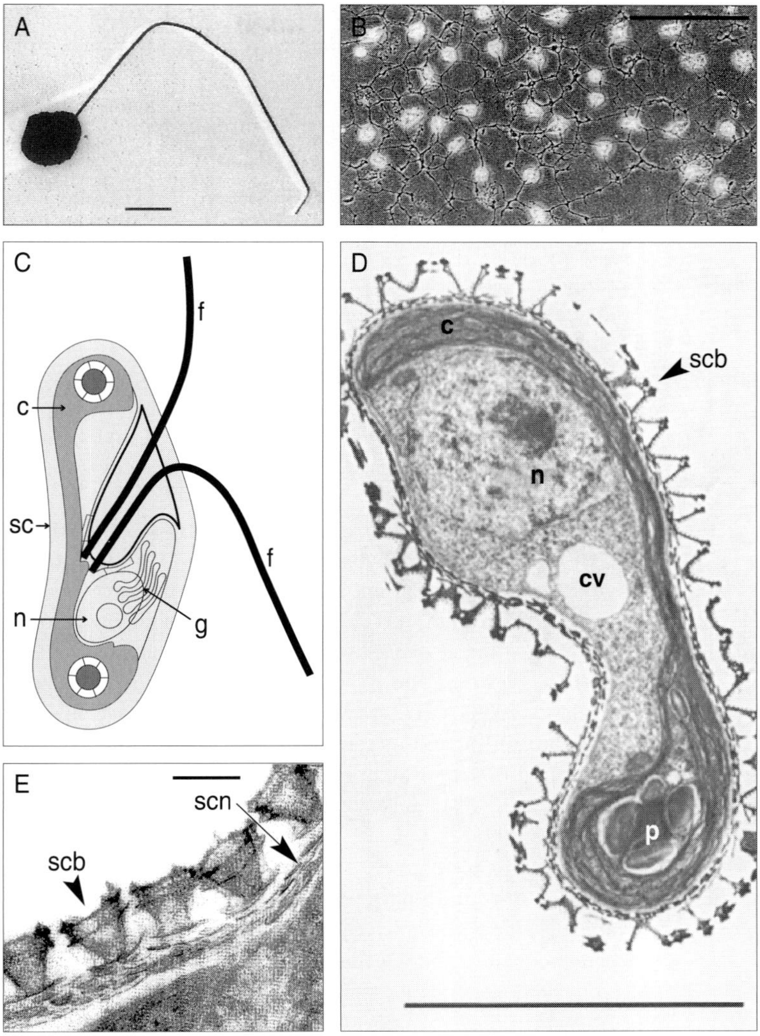

Figure 1.8. (A) Flagellate cell of the chlorarachniophyte *Bigelowiella natans* illustrating the emergent flagellum, with its terminal hairpoint. Scale bar = 2 μm (from Moestrup and Sengco, 2001); (B) Amoeboid stage of *Chlorarachnion reptans* ('green spider') showing the plasmodial reticulum. Scale bar = 20 μm (from Hibberd and Norris, 1984); (C) Diagram of the mesostigmatophyte *Mesostigma viride* showing a laterally compressed motile cell with chloroplast (c), nucleus (n), flagella (f), Golgi body (g) and the position of external scales (sc) (from Marin and Melkonian, 1999); (D) Electron micrograph of *Mesostigma viride* showing the chloroplast (c), nucleus (n), contractile vacuole (cv), pyrenoid (p), and a layer of basket scales (scb) on the cell surface. Scale bar = 5 μm (from Melkonian, 1989); (E) Higher magnification of spectacular conical surface body scales (scb) of *M. viride* with a square orifice directed outwards from the surface. The sides of these basket scales are ornamented with a network of fibrils. Beneath the basket scales, smaller naviculoid scales (scn) can be seen. Scale bar = 1 μm (from Graham and Wilcox, 2000c, personal permission from Professor D. Chappell).

1.4.4.3 The division Chlorophyta (Classes Chlorophyceae, Prasinophyceae and Trebouxiophyceae)

Table 1.24. *Characteristics of chlorophytes (Class:* ***Chlorophyceae****; Division:* ***Chlorophyta****).*

Distribution	Primarily freshwater; mostly occur in marine environments as planktonic unicells
Anatomy and morphology	Small green flagellates, naked, coccoid or ovoid in form; 10–40 μm in diameter (Ben-Amotz *et al*., 2008)
Cellular organization	Cellulose wall can be present or absent (with only a plasma membrane, as in *Dunaliella*); one large, cup-shaped chloroplast fills the posterior part of the cell; it is bound by two membranes and contains several interconnected and appressed thylakoids in bands of three to five, a large central pyrenoid and an eyespot; storage product is starch
Pigments	Two pigment types are recognized: **CHLORO-1** and **CHLORO-2: CHLORO-1: Chlorophylls:** Chls *a*, *b*, Mg-DVP; **Carotenoids:** lutein, violaxanthin, neoxanthin, antheraxanthin, β,β-carotene, β,ε-carotene (trace), zeaxanthin, β,ψ-carotene (trace); secondary carotenoids are formed under extreme environmental conditions (astaxanthin in *Haematococcus* and β,β-carotene in *Dunaliella* (Jeffrey and Egeland, 2008); **CHLORO-2** contains **CHLORO-1** pigments plus siphonaxanthin and its esters (Table 1.31); siphonaxanthin and esters are found in deep-water green macrophytes (Yokohama *et al*., 1977; Yokohama, 1981); loroxanthin has a disjunct distribution (Fawley, 1991)
Colour	Grass green or olive green
Flagella	2, 4, 8 smooth (or absent); when present equal flagella are inserted apically; the Chlorophyceae exhibit great variability in the arrangement of the flagellar apparatus
Phylogeny	Recent phylogenetic analyses (18S and 26S rDNA, Turmel *et al*., 2008) identify five major monophyletic groups among Chlorophyceae
References	Yokohama (1981); Fawley (1991); Lewis and McCourt (2004); Pröschold and Leliaert (2007); Turmel *et al*. (2008) Jeffrey and Egeland (2008); Ben-Amotz *et al*. (2008)

Table 1.25. *Characteristics of prasinophytes (Class:* ***Prasinophyceae****; Division:* ***Chlorophyta****).*

Distribution	Worldwide distribution in marine, estuarine and fresh waters; tide pools
Anatomy and morphology	Unicellular with a variety of cell shapes; both the flagella and cell body may be covered with minute organic scales; cells 1–40 μm in diameter
Cellular organization	One large cup-shaped chloroplast surrounding the central nucleus and containing an eyespot; thylakoids in bands of three
Pigments	Five prasinophyte pigment types are based on tentative biosynthetic pathways (Egeland *et al.*, 1997): **PRASINO-1** contains the common **CHLORO-1** pigments; **PRASINO-2A** contains **PRASINO-1** pigments plus loroxanthin and its esters (Garrido *et al.*, 2009); **PRASINO-2B** contains **PRASINO-1** pigments plus siphonaxanthin and its esters (see Table 1.30); **PRASINO-3A** contains **PRASINO-1** pigments and prasinoxanthin; **PRASINO-3B** contains **PRASINO-3A** pigments plus the uriolide/micromonal group (Egeland *et al.*, 1997; Garrido *et al.*, 2009). Some prasinophytes (e.g. types **PRASINO-3A, 3B** contain significant quantities of Mg-DVP as a light-harvesting pigment (see Ricketts, 1970)
Colour	Grass green or olive green
Flagella	One to eight flagella inserted in an apical depression of the cell
Phylogeny	Pigment composition shows good agreement with molecular phylogeny for the prasinoxanthin-containing groups only; the large groups have not changed substantially since the ultrastructure-based classification of Mattox and Stewart (1984); however, numerous revisions of taxa are occurring following the application of molecular methods e.g. see the erection of the Mamiellophyceae *class. nov.* (Marin and Melkonian, 2010)
References	Ricketts (1970); Mattox and Stewart (1984); Sasa *et al.* (1992a); Egeland *et al.* (1997); Latasa *et al.* (2004); Jeffrey and Wright (2006); Yoshii (2006); Pröschold and Leliaert (2007); Garrido *et al.* (2009); Marin and Melkonian (2010)

Table 1.26. *Characteristics of green algae/lichen algae (Class:* ***Trebouxiophyceae****; Division:* ***Chlorophyta****). Erected by Friedl (1995).*

Distribution	Freshwater, marine and soil habitats (*Chlorella*); algal photobiont of lichens (*Trebouxia*)
Type species	*Chlorella* (polyphyletic); *Chlorella vulgaris* (monophyletic); *Trebouxia* (green algal lichen photobiont)

Table 1.26. (*cont.*)

Anatomy and morphology	Coccoid, mostly non-motile cells; zoospores usually biflagellate, often elongated, flattened and asymmetrical; cell covering of polysaccharide material; motile cells lack both walls and scales
Cellular organization	Chloroplast cup-shaped, no clearly defined pyrenoid, two envelope membranes, stacked thylakoids; starch deposited inside the plastid
Pigments	**Chlorophylls**: Chls *a* and *b*; **carotenoids**: lutein, violaxanthin, neoxanthin, β,β-carotene (traces of astaxanthin and vaucheriaxanthin esters)
Colour	Grass green
Phylogeny	Polyphyletic, but most form independent lineages within the Trebouxiophyceae; Chlorophyta are the most diverse group of eukaryotes with at least 600 genera and 10,000 species
Symbioses	Found in many different symbioses with protozoa, as well as exosymbionts on lichens, foraminifera and parasites of tropical plants
References	Friedl (1995; 1997); Henley *et al.* (2004)

1.4.4.4 The division Streptophyta

The Mesostigmatophyceae is a unique freshwater microalgal class of green flagellates, which is included here because it is considered an important ancestral green alga (division Streptophyta), leading to land plants (Turmel *et al.*, 2002).

Table 1.27. *Characteristics of mesostigmatophytes (Class:* ***Mesostigmatophyceae****; Division:* ***Streptophyta****). Erected by Cavalier-Smith (1998); emended by Marin and Melkonian (1999) and Hall and Delwiche (2007).*

Distribution	Freshwater; no marine species
Anatomy and morphology	Flagellated unicells (15–20 μm in diameter, Figure 1.8C) covered in minute scales – basket scales (scb, 0.5 μm) and naviculoid scales (scn, 0.2 μm); cell shape is compressed with a concave inner surface (Melkonian, 1989)
Cellular organization	Mesostigmatophytes are unique from other green algae in having a multilayered structure at the base of the two subapically inserted flagellae (Figure 1.8C); an eyespot is present; the cell forms a phragmoplast during cell division
Pigments	**Chlorophylls:** Chls *a* and *b*; **Carotenoids:** antheraxanthin, violaxanthin, lutein, lycopene, *trans*-neoxanthin, siphonaxanthin and two siphonaxanthin fatty acid ester derivatives (C12:0 and C14:0), β,β-carotene and β,ψ-carotene; the presence of *trans*-neoxanthin is the first report for green algae, but it is also present in land plants (Yoshii *et al.*, 2003)
Colour	Green
Flagella	Two flagella inserted subapically; unilateral root of flagellar apparatus

Table 1.27. (*cont.*)

Phylogeny	The mesostigmatophytes have only recently been recognized as a group (Hall and Delwiche, 2007); previously thought to belong to the Prasinophyceae (Melkonian, 1989); they constitute one of the two primary lineages of green algae and are closest, phylogenetically, to land plants; Mesostigmatophyceae are the most primitive of the Streptophyta; two species are known
References	Manton and Ettl (1965); Marin and Melkonian (1999); Lemieux *et al.* (2000); Turmel *et al.* (2002); Yoshii *et al.* (2003); Hall and Delwiche (2007)

1.5 Pigment characteristics of currently recognized photosynthetic microalgal classes

1.5.1 Introduction

The pigment characteristics of 25 photosynthetic microalgal classes from the cyanobacterial, red algal and green algal radiations based on the taxonomic scheme of Huisman and Saunders (2007) are summarised in Tables 1.28–1.31. Previous pigment reviews of microalgae include the carotenoid distribution in chromophyte algae (Bjørnland and Liaaen-Jensen, 1989), a summary of marine phytoplankton pigments (Jeffrey and Vesk, 1997) published in the SCOR (1997) pigment volume, microalgal carotenoids (Liaaen-Jensen and Egeland, 1999), the pigments of picoplankton (Zapata, 2005), and a comparison of pigments in marine microalgal cultures and field samples (Jeffrey and Wright, 2006). Distributions of chlorophyll *a* and *b* are summarised in Garrido and Zapata (2006), and chlorophylls *c* in Zapata *et al.* (2006).

Advances in the field have resulted from:

- recent improvements in HPLC separation techniques (reviewed by Garrido and Zapata, 2006, and Chapter 4, this volume),
- improved HPLC-MS methods (Airs and Llewellyn, 2006 and Chapter 7, this volume),
- many more microalgae available in culture for experimentation (Andersen, 2003), and
- advances in understanding taxonomic relationships stimulated by the genomic literature (see Brodie and Lewis, 2007).

1.5.2 Signature pigments in the cyanobacterial lineage, the Glaucocystophyta and the red and green algal lineages

The pigment characteristics of the microalgal classes listed below are grouped according to 'pigment types' defined as pigment compositions typical of particular

Table 1.28. ***The cyanobacterial lineage and the Glaucocystophyta:*** *occurrence of chlorophylls, phycobiliproteins, carotenes and xanthophylls of the microalgal classes of the cyanobacterial lineage and the Glaucocystophyta with cross-references to their occurrence in red and green algal lineages (Tables 1.29 and 1.30).*

Key: ● = *dominant pigment;* • = *minor or variable pigment;* ○ = *significant but not always present; t = trace; Other lineages:* + = *present;* – = *absent; superscript letters: refer to Notes below; *prochlorophytes are now included in the polyphyletic Cyanophyta.*

	Cyanobacterial radiation						Present in other lineages	
	Cyanobacteria		Prochlorophytes*					
Pigment	CYANO-1	CYANO-2	CYANO-3	CYANO-4	CYANO-5	GLAUCO-1	RED	GREEN
Chlorophylls								
Chl *a*	●	●	●		t	●	+	+
DV-Chl *a* (Chl a_2)				●			–	–
Chl *b*			●	•			–	+
DV-Chl *b* (Chl b_2)				●			–	–
Chl *d*					●		–	–
MgDVP	t	t	t	t	•		+	+
Phycobiliproteins								
Allophycocyanin	●	●			t	●	+	–
Phycocyanin	●	●			t	●	+	–
Phycoerythrin	●	●		•			+	–
Carotenes								
β,β-carotene (β)	●	●	●			●	+	+
β,ε-carotene (α)				•	●		+	+

Xanthophylls								
Aphanizophyll	○						–	–
β,β-carotene-epoxide			t				–	–
Cryptoxanthin[a]	•		•			●	–	–
Iso-cryptoxanthin	•		•				–	–
Canthaxanthin	•						+	–
Echinenone	•		•				–	–
Myxoxanthophyll[b,c]	●						–	–
4-ketomyxoxanthophyll	○						–	–
Nostoxanthin	○						–	–
Oscillaxanthin[c]	○						–	–
Zeaxanthin	●	●	●	●	●	●	+	+
Source data (see below)	1	2	3	4	5	6		

Notes

[a] Cryptoxanthin is in the β form in glaucocystophytes, but the form in prochlorophytes is unclear: e.g. specified as β-cryptoxanthin by Foss *et al*. (1987), and α-cryptoxanthin by Goericke *et al*. (2000)

[b] 4-ketomyxol-2′-fucoside and its 1′-*O*-methyl derivative were diagnostic for the toxic colonial cyanobacterium, *Nodularia spumigens*, in the Baltic Sea (Schlüter *et al*., 2004, 2008)

[c] For discussion of cyanobacterial carotenoid glycosides, see Aakermann *et al*. (1992)

Source data: example taxa and key references

1. CYANO-1: Freshwater and marine colonial forms, e.g. *Aphanizomenon flos-aquae, Oscillatoria* sp., *Trichodesmium* sp. (Hertzberg *et al*., 1971; Carpenter *et al*., 1993, Schlüter *et al*., 2004)
2. CYANO-2: Marine coccoid planktonic species e.g. *Synechococcus* sp., *Synechocystis* sp. (Jeffrey and Wright, 1997; Six *et al*., 2004)
3. CYANO-3: *Prochloron didemni* (marine symbiont), *Prochlorothrix* strains (freshwater; free living) (Burger-Wiersma *et al*., 1986; Foss *et al*., 1987)
4. CYANO-4: *Prochlorococcus marinus* (marine, free living) (Goericke and Repeta, 1992)
5. CYANO-5: *Acaryochloris marina* (symbiotic; free living) (Miyashita *et al*., 1997, 2003)
6. GLAUCO-1: *Cyanophora* sp.: quantitative pigment assessment unknown to us (Kies and Kremer, 1990; Entwisle, 2007a)

Table 1.29. ***The red algal lineage***: *occurrence of chlorophylls, phycobiliproteins, carotenes and xanthophylls in microalgal classes of the red algal lineage, with cross references to their occurrence in the cyanobacterial and green algal lineages (Tables 1.28, 1.30).*
Key: ● = *dominant pigment;* • = *minor or variable pigment;* ○ = *significant but not always present; t = trace; superscript letters: refer to Notes below.*

		Chlorophyll *c*-containing classes of the red algal lineage																											Present in other lineages	
Pigments	RHODO-1	DIATOM-1	DIATOM-2	DIATOM-3	BOLIDO-1	CHRYSO-1	DICTYO-1	EUSTIG-1	PELAGO-1	PHAEOTHAM-1	PINGUIO-1	RAPHIDO-1	SYNURO-1	XANTHO-1	HAPTO-1	HAPTO-2	HAPTO-3	HAPTO-4	HAPTO-5	HAPTO-6	HAPTO-7	HAPTO-8	CRYPTO-1	DINO-1	DINO-2	DINO-3	DINO-4	DINO-5	CYANO	GREEN
Chlorophylls																														
Chl *a*	●	●	●	●	●	●[a]	●	●	●	●	●	●	●	●	●	●	●	●	●	●	●	●	●	●	●	●	●	See green algal lineage, Table 1.30	+	+
Mg-DVP	t	t	t	t		t	t	t	t		t		t		t	t	t	t	t	t	t	t	t	t	t	t	t		+	+
Chl *c* (unresolved)					●						●			•															−	−
Chl c_1		●		●		●	○		○	●		●[b]	●		●	●	●	●	●							●			−	−
Chl c_2		●	●	●		○	●		●	●		●[b]			●	●	●	●	●	●	●	●	●	●	●	●	●		−	−
Chl c_{2}-$_{P.gyrans}$						●										●													−	−
Chl c_3			●	●	●		○		○									●	●	●	●	●			●				−	−
MV-Chl c_3																				•									−	−
MGDG-Chl c_2-[18/14]																	•	•	•	●	●	●							−	−
MGDG-Chl c_2-[14/14]																					○				t				−	−
Phycobiliproteins[c]																														
Allophycocyanin	•																												+	−
Phycocyanin	•																						○				○		+	−
Phycoerythrin	●																						○				○		+	−

Carotenes																													
β,β-carotene (β)	●	•	•	•	•	•	•	•	•	•	•	•	•	•	•	•	•	•	•	•	•	•		•	•	•		+	+
β,ε-carotene (α)																							•		•		•	+	+
β,ψ-carotene (γ)																										○		−	+
ε,ε-carotene (ε)									○																			−	−
Xanthophylls																													
Alloxanthin																							●				●	−	+
Antheraxanthin						•		•																				−	+
Canthaxanthin								○																		○		+	−
Crocoxanthin																							•					−	−
Diadinoxanthin		●	●	●	●		●		●	●				•	●	●	●	●	●	●	●	●		●	●	●		−	+
Diatoxanthin		•	•	•	•		•		•	•				•	•	•	•	•	•	•	•	•		•	•	•		−	+
Dinoxanthin																								•				−	−
Fucoxanthin		●	●	●	●	●	●		●	●	●	●	●		●	●	●	●	●	●	●	●			●	●		−	−
4-ketofucoxanthin																			•									−	−
But-fuco			*d*				●		●			*e*								○	○	○			•			−	−
Pent-fuco																				•								−	−
Hex-fuco[f]																				●	○	●			●			−	−
Hex-kfuco																				●	●	●						−	−
Oct-fuco																				•								−	−
Gyroxanthin diester									•																●			−	−
Heteroxanthin										?				•														−	−
Monadoxanthin																							•					−	+
P-457[g]																								•				−	−
Peridinin																								●				−	−
Peridininol																								•				−	−
Pyrrhoxanthin																								•				−	−
Vaucheriaxanthin								●																				−	−

Table 1.29. (*cont.*)

Pigments	Chlorophyll *c*-containing classes of the red algal lineage																												Present in other lineages	
	RHODO-1	DIATOM-1	DIATOM-2	DIATOM-3	BOLIDO-1	CHRYSO-1	DICTYO-1	EUSTIG-1	PELAGO-1	PHAEOTHAM-1	PINGUIO-1	RAPHIDO-1	SYNURO-1	XANTHO-1	HAPTO-1	HAPTO-2	HAPTO-3	HAPTO-4	HAPTO-5	HAPTO-6	HAPTO-7	HAPTO-8	CRYPTO-1	DINO-1	DINO-2	DINO-3	DINO-4	DINO-5	CYANO	GREEN
Vauch. ester								●						•															−	+
Violaxanthin		*h*	*h*	*h*		●	○	●			•	•	●																−	+
Zeaxanthin	●	*h*	*h*	*h*		●	○	●	●		•	•														○			+	+
Source data (see below)	1	2	3	4	5	6	7	8	9	10	11	12	13	14	15	16	17	18	19	20	21	22	23	24	25	26	27	28		

Notes

[a] Since no recently published pigment data were available for the redefined class Chrysophyceae *sensu stricto* (Andersen *et al.*, 1999), unpublished HPLC data for four chrysophyte cultures were used (Jeffrey and Zapata, in prep.), see item 6 below

[b] Either Chl c_1 or Chl c_2 may occasionally be absent. (Mostaert *et al.*, 1998)

[c] Phycobiliproteins are also present in unknown 'picobiliphytes' (Not *et al.*, 2007)

[d] Detected in *Thalassiothrix heteromorpha* CS-132 (unconfirmed) (Stauber and Jeffrey, 1988)

[e] Detected in one species (Moestaert *et al.*, 1998)

[f] Hex-fuco is also present in unknown picoplanktonic haptophytes (Liu *et al.*, 2009)

[g] See Data Sheets for chemistry of this peridinin derivative (Jeffrey *et al.*, 1997b)

[h] Minor amounts of both violaxanthin and zeaxanthin occur under high light in some diatoms (Lohr and Wilhelm, 1999)

Source data: example taxa and key references

1. RHODO-1: *Porphyridium purpureum* CS-25 (Jeffrey and Wright, 1997; Kopecky *et al.*, 2002)
2. DIATOM-1: *Chaetoceros didymus* CS-2 (Stauber and Jeffrey, 1988)
3. DIATOM-2: *Rhizoselenia setigera* CS-62 (Stauber and Jeffrey, 1988)
4. DIATOM-3: *Nitzschia bilobata* CS-47 (Stauber and Jeffrey, 1988)
5. BOLIDO-1: *Bolidomonas pacifica* (Guillou *et al.*, 1999a)
6. CHRYSO-1: *Hibberdia* sp. CCMP456, *Chrysochaete brittanica* CCMP280, *Phaeoplaca thallosa* CCMP 634, *Ochromonas moestrupii* CCMP1278 (Zapata and Jeffrey, unpublished)
7. DICTYO-1: *Pseudochattonella (=Verrucophora) farcimen* (Edvardsen *et al.*, 2007; Eikrem *et al.*, 2009)
8. EUSTIG-1: *Nannochloropsis* spp. (Norgård *et al.*, 1974; Volkman *et al.*, 1993)
9. PELAGO-1: *Pelagococcus subviridis* CS-99 (Jeffrey and Wright, 1997)
10. PHAEOTHAM-1: *Stichogloea* sp. CCMP823 (Bailey *et al.*, 1998; Andersen *et al.*, 1998)
11. PINGUIO-1: *Pinguiochrysis pyriformis* (Kawachi *et al.*, 2002a)
12. RAPHIDO-1: *Chattonella marina* (Mostaert *et al.*, 1998)
13. SYNURO-1: *Mallomonas* sp. (Andersen and Mulkey, 1983)
14. XANTHO-1: *Vaucheria* sp. (Hibberd, 1990)
15. HAPTO-1: *Pavlova lutheri* CS-23 (Van Lenning *et al.*, 2003; Zapata *et al.*, 2004)
16. HAPTO-2: *Pavlova gyrans* CCMP608 (Van Lenning *et al.*, 2003; Zapata *et al.*, 2004)
17. HAPTO-3: *Dicrateria inornata* CS-254 (Zapata *et al.*, 2004)
18. HAPTO-4: *Prymnesium parvum* CS-345 (Zapata *et al.*, 2004)
19. HAPTO-5: *Ochrosphaera verrucosa* CCMP594 (Zapata *et al.*, 2004)
20. HAPTO-6: *Emiliania huxleyi* CS-57 (Zapata *et al.*, 2004; Airs and Llewellyn, 2006)
21. HAPTO-7: *Chrysochromulina hirta* CS-228 (Zapata *et al.*, 2004)
22. HAPTO-8: *Phaeocystis pouchetii* CS-165 (Zapata *et al.*, 2004)
23. CRYPTO-1: *Chroomonas salina* CS-174 (Jeffrey and Wright, 1997)
24. DINO-1: *Amphidinium carterae* CS-212 (Jeffrey and Wright, 1997)
25. DINO-2: *Karenia brevis* (formerly known as *Gymnodinium breve*; Daugbjerg *et al.*, 2000) (Carreto *et al.*, 2001)
26. DINO-3: *Kryptoperidinium foliaceum* (Jeffrey and Vesk, 1976; Kempton *et al.*, 2002)
27. DINO-4: *Dinophysis norvegica* (Vesk *et al.*, 1996; Meyer-Harms and Pollehne, 1998)
28. DINO-5: *Gymnodinium chlorophorum* (=*Lepidodinium viride*) (Watanabe *et al.*, 1990; Elbrächter and Schnepf, 1996) – see also the green algal lineage, Table 1.30

Table 1.30. ***The green algal lineage:*** *occurrence of chlorophylls, carotenes and xanthophylls in microalgal classes of the green algal lineage, with cross references to their occurrence in cyanobacterial and red algal lineages (Tables 1.28 and 1.29). DINO-5 has a prasinophyte endosymbiont and is listed here.*

Key: ● = *major pigment;* • = *minor or variable pigment;* ○ = *significant but not always present; t = trace;* † = *unusual rare occurrence, Unident. = unidentified.*

	Chlorophyll *b*-containing classes of the green algal lineage												Present in other lineages	
Pigment	EUGLENO-1	CHLORARACH-1	CHLORO-1	CHLORO-2	PRASINO-1	PRASINO-2A	PRASINO-2B	PRASINO-3A	PRASINO-3B	TREBOUXIO-1	MESOSTIG-1	DINO-5	CYANO	RED
Chlorophylls														
Chl *a*	●	●	●	●	●	●	●	●	●	●	●	●	+	+
Chl *b*	●	●	●	●	●	●	●	●	●	●	●	●	+	−
Chl *c* series														
Mg-DVP [a]	t	t	t		t	t	•	●	●	t		t	+	+
Chl c_{CS-170} [b]								○	○				−	−
Carotenes														
β,β-carotene (β) [c]	•	•	●	●	●	●	●	●	●	•	●	●	+	+
β,ε-carotene (α)	t		•	•	•	•	•	•	•				+	+
β,ψ-carotene (γ)			○		○						•		−	−
Lycopene											•		−	−

Carotenoids														
Alloxanthin [d]			†										–	+
Antheraxanthin		•	•	•	•	•	•	•	•	•	•		–	+
Astaxanthin [c]			○							t			–	–
Dihydrolutein									•				–	–
Diadinoxanthin	●												–	+
Diatoxanthin	•												–	+
Eutreptiellanone [e]	○												–	–
Loroxanthin			○	○		○							–	–
Loroxanthin ester		○				○							–	–
Lutein		●	●	●	●	•	●	•	•	●	•		–	–
Micromonal									●				–	–
Micromonol									•				–	–
Monadoxanthin [d]			†										–	+
cis-Neoxanthin	•	•	●	●	●	●	●	●	●	•		●	–	–
trans-Neoxanthin											●		–	–
Prasinoxanthin								●	●			○	–	–
Siphonaxanthin (Siph)				•			●				●		–	–
Siph ester(s) [f]				•			○				●		–	–
Unident. Car.-*M. pusilla* [g]									•				–	–
Uriolide									●				–	–
Violaxanthin		•	●	●	●	●	●	●	●	•	●	●	–	+
Zeaxanthin		•	•	•	•	•	•	•	•	•	*h*	●	+	+
Unident. carotenoid [i]												●	–	–
Source data (see below)	1	2	3	4	5	6	7	8	9	10	11	12		

Table 1.30. (*cont.*)

Notes

[a] Some prasinophytes contain significant amounts of Mg-DVP as a light-harvesting pigment (Ricketts, 1970)

[b] Chl $c_{CS\text{-}170}$ is a propionate derivative of Chl c_3 in *Micromonas pusilla* CS-170 (see Fookes and Jeffrey, cited in Data sheets, Jeffrey *et al.*, 1997a)

[c] Massive quantities of astaxanthin and β,β-carotene form under extreme environmental conditions in *Haematococcus pluvialis* and *Dunaliella salina* (Jeffrey and Egeland, 2008)

[d] Cryptomonad pigments have been found in two chlorophytes: *Picocystis salinarum* (Lewin *et al.*, 2000) and *Coccomyxa* sp. (Crespo *et al.*, 2009), with no evidence of a cryptomonad endosymbiont

[e] Where present, eutreptiellanone may be a major pigment

[f] For details of siphonaxanthin esters, see Table 1.31

[g] Prominent unidentified carotenoid in *Micromonas pusilla* CS-86 (pigment #44, Wright and Jeffrey, 1997)

[h] *Mesostigma viride* was kept in the dark before analysis to minimize zeaxanthin formation and its presence was not reported (Yoshii *et al.*, 2003)

[i] A major unidentified carotenoid is present in the green dinoflagellate *Gymnodinium (hepidodinium) chlorophorum* (Dr. P. Henriksen, pers. comm., see Table 1.21). DINO-5 species lack peridinin and fucoxanthin.

Source data: example taxa and key references

1. EUGLENO-1: *Euglena gracilis, Eutreptiella gymnastica* (Bjørnland, 1982; Fiksdahl *et al.*, 1984a; Jeffrey and Wright, 1997)
2. CHLORARACH-1: Unidentified chlorarachniophyte (Sasa *et al.*, 1992b)
3. CHLORO-1: *Dunaliella tertiolecta* (Fawley, 1991; Lewin *et al.*, 2000; Jeffrey and Egeland, 2008)
4. CHLORO-2: *Chlamydomonas parkeae* (Sasa *et al.*, 1992a)
5. PRASINO-1: *Nephroselmis olivaceae, Tetraselmis* sp. (Egeland *et al.*, 1997; Latasa *et al.*, 2004)
6. PRASINO-2A: *Pyramimonas parkeae, Tetraselmis* spp. (Sasa *et al.*, 1992b; Garrido *et al.*, 2009)
7. PRASINO-2B: *Nephroselmis pyriformis, Pyramimonas amylifera* (Egeland *et al.*, 1997; Latasa *et al.*, 2004; Yoshii *et al.*, 2005)
8. PRASINO-3A: *Pycnococcus provasolii* (Egeland *et al.*, 1997)
9. PRASINO-3B: *Micromonas pusilla* (Egeland *et al.*, 1997)
10. TREBOUXIO-1: *Picochlorum oklahomensis* (Henley *et al.*, 2004)
11. MESOSTIG-1: *Mesostigma viride* (Yoshii *et al.*, 2003)
12. DINO-5: *Gymnodinium (hepidodinium) chlorophorum, Lepidodinium viride* (Watanabe *et al.*, 1990; Elbrächter and Schnepf, 1996)

Table 1.31. *Distribution of siphonaxanthin (Siph) and its derivatives in microalgal classes* of the green algal lineage (% of total carotenoids, see Yoshii, 2006). Notation (e.g. C8:1) refers to the carbon chain length and number of unsaturated bonds in the esterified fatty acid.*

			Siph esters						6′-hydroxy Siph esters			
Taxa	Siph	19-Methoxy Siph	C8:1	C10:1	C12:0	C12:1	C14:0	C14:1	C10:1	C12:1	C14:1	References
Chlorophyceae												
Chlamydomonas parkeae			+	+		+						1
Prasinophyceae												
Nephroselmis sp.	<1	5–12	0–46	<1–7		<1–7		<1–23		1–7	10–26	2
Pyramimonas amylifera				~3		26			<1	5		3
Pterosperma cristatum								35			20	4
Mesostigmatophyceae												
Mesostigma viride	11				11		6					5

References: 1. Sasa *et al.* (1992b); 2. Yoshii *et al.* (2005); 3. Egeland *et al.* (1997); 4. Yoshii *et al.* (2002); 5. Yoshii *et al.* (2003)

* Siphonaxanthin is also characteristic of the deep water green algal macrophyte Siphonales (Yokohama *et al.*, 1977; Yokohama, 1981)

taxonomic groups (e.g. class, sub-class, genus or ecotype). These may be based on a unique pigment specific to that class or species (e.g. peridinin, **DINO-1**), or a suite of pigments such as combinations of various chlorophyll *c* and fucoxanthin derivatives within the Haptophyta (Van Lenning *et al*., 2003; Zapata *et al*., 2004, 2006; and Table 2.14 in Jeffrey and Wright, 2006) or similarities in biosynthetic pathways (e.g. Prasinophyceae, Egeland *et al*., 1997). At the request of the editors, the authors used only well-characterized pigments from published data sets for Tables 1.28–1.30. These employed the latest high-resolution chromatographic techniques and included chemical data, where available, confirming pigment identities. However, some pigments are included here because, although unidentified, they often occur in field samples and their taxonomic distribution is well known (e.g. the unidentified carotenoid from *Micromonas pusilla*, see note g, Table 1.30). In the general case of unknown pigments, the suffix '-like' was recommended to describe similar unknown pigments until chemically characterized (Professor Synnøve Liaaen-Jensen, SCOR Working Group 78, Plymouth, UK, 1986).

Because the pigment composition of microalgae can vary according to environmental factors such as light intensity, quality and day length, nutrient status particularly iron concentration and growth phase at harvest (Llewellyn and Gibb, 2000; Jeffrey and Wright, 2006), the summaries presented in Tables 1.28–1.31 should be seen only as 'typical' compositions. Occasionally these summaries are based on very sparse and inadequate data, particularly resolving chlorophylls c_1 and c_2 and other members of the chlorophyll *c* family. Many minor peaks seen on HPLC chromatograms were not identified in the original publications but may have been taxonomically significant. Thus, 'presence' data in the accompanying tables is more reliable than 'absence' data.

The red algal lineage (Table 1.29) shows the increasing number and importance of members of the chlorophyll *c* and fucoxanthin families (Airs and Llewellyn, 2006; Zapata *et al*., 2006). Carotenoids form a variety of glycosides (e.g. myxoxanthophyll, oscillaxanthin), many of which have different HPLC retention times based on their different chemistry. Esters also introduce further variety to carotenoids through incorporation of various acyl groups that often produce multiple HPLC peaks (e.g. gyroxanthin, vaucheriaxanthin, fucoxanthin, loroxanthin and siphonaxanthin, Table 1.31). Gyroxanthin esters are potential indicators of toxic dinoflagellates (Millie *et al*., 1997), although they are also reported from some pelagophytes and haptophytes (Bjørnland *et al*., 2003; Zapata, 2005).

The 74 pigments listed across the 25 microalgal classes (Tables 1.28–1.31) provide an array of chemotaxonomic markers for oceanographers. This possibly daunting challenge for field researchers may be reduced by focusing on a smaller number of selected pigments, for example, as used in the CHEMTAX methodology, guidelines for which are presented in Wright and Jeffrey (2006) and Chapter 6, this volume.

1.5.3 Future developments

The application of pigment chemotaxonomy in oceanography will be more firmly established by advances in taxonomy and phylogenetics, coupled with improved pigment analysis (e.g. greater resolution with advanced HPLC and ultra high performance liquid chromatography – UPLC) and more rapid and secure chemical identifications (e.g. online mass spectrometry). With these improved techniques, which should also include reporting culture codes and Genebank references, more pigment and taxonomic diversity in oceanic phytoplankton populations may well be discovered and their role in the biogeochemical cycles of the ocean further understood. This is an urgent need in the current challenging environment of climate change (Nair *et al.*, 2008; Hallegraeff, 2010).

Acknowledgements

We thank our colleagues for kindly supplying the following figures:

Professor G. M. Hallegraeff: Figures 1.2A, 1.3A–D, 1.6A–D, 1.7B, E;
Dr Maret Vesk: Figures 1.2C, 1.4D, 1.6E;
Dr A. Rees: Figures 1.2B, 1.4C, 1.6F;
Professor H. Marchant: Figure 1.4B;
Mr Chris Puttock: Figure 1.5E;
Dr M. de Salas: Figure 1.7A;
Dr Peter Vesk: Figure 1.7C.

Copyright permissions to use published illustrations were received as listed below:

Australian Biological Resources Study, Figures 1.5C, 1.7A;
CSIRO, Figures 1.3B, 1.3C, 1.6A, 1.6B, 1.6C, 1.7B, 1.7D, 1.7E, 1.7F;
Elsevier, Figure 1.8C;
John Wiley and Sons, Figures 1.4A, 1.4D, 1.5A, 1.5B, 1.5D, 1.8A, 1.8B,
Springer Science and Business, 1.4B;
Springer-Verlag Wien, 1.8D;
UNESCO Publishing, Figures 1.2A, 1.2B, 1.2C, 1.4C, 1.5E, 1.6D, 1.6E, 1.6F, 1.7C
University of Tasmania Publishing, Figures 1.3A, 1.3D;
Professor Dorothy Chappell, Wheaton College, Figure 1.8E.

The authors also thank Ms Denise Schilling, Ms Anne Pirrone and Ms Jillian le Patourel for typing assistance; Ms Louise Bell for the design of the figures; the CSIRO librarians, Ms Meredith Hepburn and Ms Joel MacKeen, for special help, and Dr Greg Ayers, former Chief of the CSIRO Division of Marine and Atmospheric Research and Dr John Volkman (CSIRO Marine and Atmospheric Research) for their interest in this project. This work was supported by the Australian Government's Cooperative Research Centres Programme through the Antarctic Climate and Ecosystems Cooperative Research Centre (ACE CRC). Dr Ed Urban of the Scientific Committee on Oceanic Research (SCOR) gave special assistance to S. W. Jeffrey towards the end of the project.

Finally, we are indebted to our colleagues Dr R. A. Andersen and Dr E. S. Egeland, and our senior editor, Professor S. Roy, whose insightful comments greatly improved the manuscript.

Glossary of important terms used in protistan taxonomy

Alveolae	Amphiesmal vesicles beneath the cell membrane, which may or may not contain cellulosic thecal plates
Alveolata	Major lineage of protists (see Table 1.1)
Alveolates	Organisms with such amphiesmal vesicles (see alveolae)
Apicomplexa	Major lineage of monophyletic obligate parasitic protists; the name is derived from 'anterior apical complex', an organ used to penetrate host cells. Most 'apicomplexa' have a relic non-photosynthetic plastid (apicoplast). Gene phylogenies show a common red algal origin of the apicomplexan, dinoflagellate and heterokont plastids (Janouškovec *et al.*, 2010)
Bikont	Cell with two flagella
Chlorarachniophytes	Green unicells with spider-like amoeboid and flagellate forms
Chromalveolate hypothesis	The mechanism by which Chl *c*-containing algae (chromophytes) gained their pigments from a red algal secondary endosymbiont
Chromalveolates	Major lineage of photosynthetic alveolates, closely related to the Dinophyta. The group includes chlorophyll c_2-containing chromophyte algae, plus numerous non-photosynthetic descendants
Chromapicomplexa	Photosynthetic Apicomplexa (one species only is known, *Chromera velia*, Moore *et al.*, 2008)
Chromista **Chromophyta** **Chromistophyte** **Chrysophyta** **Heterokontophyta** **Stramenopiles**	Synonyms for Chromophyte algae
Chromophyta	Coloured plants (Linnaeus, 1751); ≡ coloured algae excluding green ones
Chromophyte algae	Microalgae containing chlorophylls *a* and *c*
Cryptic species	Morphologically indistinguishable, but genetically distinct
Cryptophytes	This group probably diverged early in the red algal line from the main evolutionary pathway of the heterokont classes
Dinophyta	Dinoflagellate (Greek: *dinos*, whirling)
Flagellar hairs	Tubular appendages responsible for the reverse direction of movement in heterokont algae
Hacrobia	New taxon relating haptophytes and cryptophytes (Okamoto *et al.*, 2009)
Haptonema	Appendage found in haptophyte microalgae (Greek: *hapto*, to touch, attach to; *nema*, a thread)

Heterokont	Algae with usually two unequal flagella: one smooth and one covered with mastigonemes (Greek: *heterokontos*, with different oars)
Isokont	Algae with two equal flagella
Mastigoneme	Hair-like projection on the flagellar shaft
Metaboly	Cell has shape-changing contractions
Mixotroph	Photosynthetic organism capable of uptake of soluble or particulate nutrients for supplementary nutrition or a heterotroph with functional ingested chloroplasts
Monad	Flagellate stage in the life cycle of microalgae or protozoa
Nanoplankton	Phytoplankton in the size range 2–20 μm
Phytoplankton	Greek: *phytos*, plant; *planktos*, wandering.
Picoeukaryote, picoprokaryote	Eukaryotes/prokaryotes in the size range 0.2–2.0 μm
Picoplankton	Phytoplankton in the size range 0.2–2.0 μm
Protists	Heterogeneous group of mostly unicellular organisms including protozoa, photosynthetic and non-photosynthetic eukaryotic algae and fungi; belonging to the Protista
Stramenopile	Protists containing flagella with straw-like hairs (Latin: *stramen* = straw; *pilus* = hair); can include both heterotrophic and non-heterotrophic organisms; mitochondria have tubular cristae
Tripartite flagellar hair	Flagellar hairs with three segments: a short base, a long narrow tubular shaft and one or more terminal filaments
Tripartite heterokont flagellum	One flagellum carries tripartite tubular hairs
Unikont	Eukaryote with one flagellum
Xanthophytes	Yellow or yellow-brown algae (Greek: *xanthos* = yellow)

Abbreviations

Chloro	Chlorophyte
Dino	Dinoflagellate
Hapto	Haptophyte
Oct-fuco	19′-octanoyloxyfucoxanthin
Pent-fuco	19′-pentanoyloxyfucoxanthin
Prasino	Prasinophyte
rDNA	Ribosomal DNA
SSU	Small subunit
TLC	Thin-layer chromatography
UPLC	Ultra high performance liquid chromatography

References

Aakermann, T., Skulberg, O. M. and Liaaen-Jensen, S. (1992). A comparison of the carotenoids of strains of *Oscillatoria* and *Spirulina* (Cyanobacteria) *Biochem. Syst. Ecol.* **20**, 761–69.

Adl, S. M., Simpson, A. G. B., Farmer, M. A., Andersen, R. A., Anderson, O. R., Barta, J. R., Bowser, S. S., Brugerolle, G., Fensome, R. A., Fredericq, S., James, T. Y., Karpov, S., Kugrens, P., Krug, J., Lane, C. E., Lewis, L. A., Lodge, J., Lynn, D. H., Mann, D. G., McCourt, R. M., Mendoza, L., Moestrup, Ø., Mozley-Standridge, S. E., Nerad, T. A., Shearer, C. A., Smirnov, A. V., Spiegel, F. W. and Taylor, M. F. J. R. (2005). The new higher level classification of eukaryotes with emphasis on the taxonomy of protists. *J. Eukaryot. Microbiol.* **52**, 399–451.

Aiken, J., Moore, G. F. and Holligan, P. M. (1992). Remote sensing of ocean biology in relation to global climate change. *J. Phycol.* **28**, 579–90.

Airs, R. L. and Llewellyn, C. A. (2006). Improved detection and characterization of fucoxanthin-type carotenoids: Novel pigments in *Emiliana huxleyi* (Prymnesiophyceae). *J. Phycol.* **42**, 391–99.

Andersen, R. A. (1987). *Synurophyceae classis nov.*, a new class of algae. *Am. J. Bot.* **74**, 337–53.

Andersen, R. A. (1989). The Synurophyceae and their relationship to other golden algae. *Beih. Nova Hedwigia* **95**, 1–26.

Andersen, R. A. (2003). A world list of algal culture collections. In *Manual on Harmful Marine Microalgae*, ed. G. M. Hallegraeff, D. M. Anderson and A. D. Cembella. Paris: UNESCO Publishing, pp. 753–66.

Andersen, R. A. (2004). Biology and systematics of heterokont and haptophyte algae. *Am. J. Bot.* **91**, 1508–22.

Andersen, R. A. (2007). Heterokontophyta: Synurophyceae. In *Algae of Australia: Introduction*, ed. P. M. McCarthy and A. E. Orchard. Canberra, Australia: CSIRO Publishing/Australian Biological Resources Study, pp. 235–54.

Andersen, R. A. and Mulkey, T. J. (1983). The occurrence of chlorophylls c_1 and c_2 in the Chrysophyceae. *J. Phycol.* **19**, 289–94.

Andersen, R. A., Saunders, G. W., Paskind, M. P. and Sexton, J. P. (1993). Ultrastructure and 18S rRNA gene sequence for *Pelagomonas calceolata* gen. and sp. nov., and the description of a new algal class, the Pelagophyceae classis nov. *J. Phycol.* **29**, 701–15.

Andersen, R. A., Potter, D., Bidigare, R. R., Latasa, M., Rowan, K. and O'Kelly, C. J. (1998). Characterization and phylogenetic position of the enigmatic golden alga *Phaeothamnion confervicola*: ultrastructure, pigment composition and partial SSU rDNA sequence. *J. Phycol.* **34**, 286–98.

Andersen, R. A., Van de Peer, Y., Potter, D., Sexton, J. P., Kawachi, M. and LaJeunesse, T. (1999). Phylogenetic analysis of the SSU rRNA from members of the Chrysophyceae. *Protist* **150**, 71–84.

Andreoli, C., Moro, I., La Rocca, N., Rigoni, F., Dalla Valle, L. and Bargelloni, L. (1999). *Pseudopleurochloris antarctica* gen. et sp. nov., a new coccoid xanthophycean from pack-ice of Wood Bay (Ross Sea, Antarctica): ultrastructure, pigments and 18S rRNA gene sequence. *Eur. J. Phycol.* **34**, 149–59.

Armbrust, E. V., Berges, J. A., Bowler, C., Green, B. R., Martinez, D., Putnam, N. H., Zhou, S., Allen, A. E., Apt, K. E., Bechner, M., Brzezinski, M. A., Chaal, B. K., Chiovitti, A., Davis, A. K., Demarest, M. S., Detter, J. C., Glavina, T.,

Goodstein, D., Hadi, M. Z., Hellsten, U., Hildebrand, M., Jenkins, B. D., Jurka, J., Kapitonov, V. V., Kröger, N., Lau, W. W. Y., Lane, T. W., Larimer, F. W., Lippmeier, J. C., Lucas, S., Medina, M., Montsant, A., Obornik, M., Parker, M. S., Palenik, B., Pazour, G. J., Richardson, P. M., Rynearson, T. A., Saito, M. A., Schwartz, D. C., Thamatrakoln, K., Valentin, K., Vardi, A., Wilkerson, F. P. and Rokhsar, D. S. (2004). The genome of the diatom *Thalassiosira pseudonana*: ecology, evolution and metabolism. *Science* **306**, 79–86.

Bailey, J. C., Bidigare, R. R., Christensen, S. J. and Andersen, R. A. (1998). Phaeothamniophyceae *classis nova*: a new lineage of chromophytes based on photosynthetic pigments, *rbc*L sequence analysis and ultrastructure. *Protist* **149**, 245–63.

Ben-Amotz, A., Polle, J. E. W. and Subba Rao, D. V. (eds.) (2008). *The Alga Dunaliella: Biodiversity, Physiology, Genomics and Biotechonology*. Enfield: Science Publishers.

Bernard, C., Etienne, A.-L. and Thomas, J. C. (1996). Synthesis and binding of phycoerythrin and its associated linkers to the phycobilisome in *Rhodella violacea* (Rhodophyta): compared effects of high light and translation inhibitors. *J. Phycol.* **32**, 265–71.

Bhattacharya, D. (1997). *Origins of Algae and their Plastids*. New York: Springer.

Bhattacharya, D. and Schmidt, H. A. (1997). Division Glaucocystophyta. In *Origins of Algae and their Plastids*, ed. D. Bhattacharya. New York: Springer, pp. 139–48.

Bhattacharya, D., Helmchen, T., Bibeau, C. and Melkonian, M. (1995). Comparison of nuclear-encoded small sub-unit ribosomal RNAs reveal the evolutionary position of Glaucocystophyta. *Mol. Biol. Evol.* **12**, 415–20.

Bjørnland, T. (1982). Chlorophylls and carotenoids of the marine alga *Eutreptiella gymnastica*. *Phytochemistry* **21**, 1715–19.

Bjørnland, T. and Liaaen-Jensen, S. (1989). Distribution patterns of carotenoids in relation to chromophyte phylogeny and systematics. In *The Chromophyte Algae: Problems and Perspectives*, ed. J. C. Green, B. S. C. Leadbeater and W. L. Diver. Oxford, UK: Clarendon Press, pp. 37–61.

Bjørnland, T., Borch, G. and Liaaen-Jensen, S. (1986). Additional oxa-bicyclo [2.2.1]heptane carotenoids from *Eutreptiella gymnastica*. *Phytochemistry* **25**, 201–05.

Bjørnland, T., Haxo, F. T. and Liaaen-Jensen, S. (2003). Carotenoids of the Florida red tide dinoflagellate *Karenia brevis*. *Biochem. Syst. Ecol.* **31**, 1147–62.

Bodył, A. and Moszcyńksi, K. (2006). Did the peridinin plastid evolve through tertiary endosymbiosis? A hypothesis. *Eur. J. Phycol.* **41**, 435–48.

Booth, B. C. and Marchant, H. (1987). Parmales, a new order of marine chrysophytes, with descriptions of three new genera and seven new species. *J. Phycol.* **23**, 245–60.

Bourrelly, P. (1957). Recherches sur les Chrysophycées: morphologie, phylogénie, systématique. *Rev. Algol. Mém. Hors-Séries* **1**, 1–412.

Bowler, C. and 76 colleagues (2008). The *Phaeodactylum* genome reveals the evolutionary history of diatom genomes. *Nature* **456**, 239–44.

Britton, G., Liaaen-Jensen, S. and Pfander, H. (eds.), Mercadante, A.Z. and Egeland, E. S. (compilers) (2004). *Carotenoids. Handbook*. Basel: Birkhäuser.

Broadwater, S. T. and Scott, J. L. (1994). Ultrastructure of unicellular red algae. In *Evolutionary Pathways and Enigmatic Algae*, ed. J. Sechbach. Boston: Kluwer Academic Publishers, pp. 215–30.

Brodie, J. and Lewis, J. (eds.) (2007). *Unravelling the Algae: the Past, Present and Future of Algal Systematics*. The Systematics Association Special Volume Series 75, Boca Raton: CRC Press.

Burger, G., Saint-Louis, D., Gray, M. W. and Lang, B. F. (1999). Complete sequence of the mitochondrial DNA of the red alga *Porphyra purpurea*: cyanobacterial introns and shared ancestry of red and green algae. *Plant Cell* **11**, 1675–94.

Burger-Wiersma, T., Veenhuis, M., Korthals, H. J., Van de Wiel, C. C. M. and Muir, L. R. (1986). A new prokaryote containing chlorophylls *a* and *b*. *Nature* **320**, 262–63.

Carpenter, E. J. and Price, C. C. (1977). Nitrogen fixation, distribution and production of *Oscillatoria (Trichodesmium)* spp. in the western Sargasso and Caribbean Seas. *Limnol. Oceanogr*. **22**, 60–72.

Carpenter, E. J., O'Neil, J. M., Dawson, R., Capone, D. G., Siddiqui, P. J. A., Roenneberg, T. and Bergman, B. (1993). The tropical diazotrophic phytoplankter *Trichodesmium*: biological characteristics of two common species. *Mar. Ecol. Progr. Ser*. **95**, 295–304.

Carreto, J. L., Seguel, M., Montoya N. G., Clement A. and Carignan, M. J. (2001). Pigment profile of the ichthyotoxic dinoflagellate *Gymnodinium* sp. from a massive bloom in southern Chile. *J. Plankton Res*. **23**, 1171–75.

Cavalier-Smith, T. (1989). The kingdom Chromista. In *The Chromophyte Algae: Problems and Perspectives*, ed. J. C. Green, B. S. C. Leadbeater and W. L. Diver. The Systematics Association Special Volume no. 38. Oxford: Clarendon Press, pp. 381–407.

Cavalier-Smith, T. (1998). A revised six-kingdom system of life. *Biol. Rev*. **73**, 203–66.

Cavalier-Smith, T. (2002). Chloroplast evolution: secondary symbiogenesis and multiple losses. *Curr. Biol*. **12**, R62–64.

Cavalier-Smith, T. and Chao, E. E. (1996). 18S rRNA sequence of *Heterosigma carterae* (Raphidophyceae), and the phylogeny of heterokont algae (Ochrophyta). *Phycologia* **35**, 500–10.

Cerino, F. and Zingone, A. (2007). Decrypting cryptomonads: a challenge for molecular taxonomy. In *Unravelling the Algae, the Past, Present, and Future of Algal Systematics*. ed. J. Brodie and J. Lewis. Boca Raton: CRC Press, pp. 197–214.

Chadefaud, M. (1950). Les cellules nageuses des algues dans l'embrachment des Chromophycées. *C. R. Acad. Sci*. **231**, 788–90.

Chisholm, S. W., Olson, R. J., Zettler, E. R., Goericke, R., Waterbury, J. B. and Welschmeyer, N. A. (1988). A novel free-living prochlorophyte abundant in the oceanic euphotic zone. *Nature* **334**, 340–43.

Chisholm, S. W., Frankel, S. L., Goericke, R. Olson, R. J., Palenik, B., Waterbury, J. B., West-Johnsrud, L. and Zettler, E. R. (1992). *Prochlorococcus marinus* nov. gen. nov. sp.: an oxyphototrophic marine prokaryote containing divinyl chlorophyll *a* and *b*. *Arch. Microbiol*. **157**, 297–300.

Christensen, T. (1989). The Chromophyta, past and present. In *The Chromophyte Algae: Problems and Perspectives*, ed. J. C. Green, B. S. C. Leadbeater and W. L. Diver. The Systematics Association Special Volume no. 38. Oxford: Clarendon Press, pp. 1–12.

Clay, B. I., Kugrens, P., and Lee, R. E. (1999). A revised classification of Cryptophyta. *Bot. J. Linn. Soc*. **131**, 131–51.

Cole, K. M. and Sheath, R. G. (eds.) (1990). *Biology of the Red Algae*. Cambridge: Cambridge University Press.

Corliss, J. O. (1986). Progress in protistology during the first decade following re-emergence of the field as a respectable interdisciplinary area in modern biological research. In *Progress in Protistology*, vol. 1, ed. J. O. Corliss and D. J. Patterson. Bristol: Biopress Ltd., pp. 11–63.

Corliss, J. O. (1994). An interim utilitarian ('user-friendly') hierarchical classification and characterization of the protists. *Acta Protozool.* **33**, 1–51.

Crespo, C., Rodríguez, M., Segade, P., Iglesias, R. and García-Estévez, J. N. (2009). *Coccomyxa* sp. (Chlorophyta: Chlorococcales), a new pathogen in mussels (*Mytilus galloprovincialis)* of Vigo estuary (Galicia N.W. Spain). *J. Invertebr. Pathol.* **102**, 214–19.

Daugbjerg, N. (1996a). *Mesopedinella arctica* gen. et sp. nov. (Pedinellales, Dictyochophyceae) I: fine structure of a new marine phytoflagellate from Arctic Canada. *Phycologia* **35**, 435–45.

Daugbjerg, N. (1996b). *Mesopedinella arctica* (Pedinellales) II: Phylogeny of *Mesopedinella*, including a cladistic analysis of Dictyochophyceae. *Phycologia* **35**, 563–68.

Daugbjerg, N. and Andersen, R. A. (1997). A molecular phylogeny of the heterokont algae based on analyses of chloroplast-encoded *rbc*L sequence data. *J. Phycol.* **33**, 1031–41.

Daugbjerg, N. and Henriksen, P. (2001). Pigment composition and *rbc*L sequence data from the silicoflagellate *Dictyocha speculum*: a heterokont alga with pigments similar to some haptophytes. *J. Phycol.* **37**, 1110–20.

Daugbjerg, N., Hansen, G., Larsen, J. and Moestrup, Ø. (2000). Phylogeny of some of the major genera of dinoflagellates based on ultrastructure and partial LSU and DNA sequence data, including the erection of three new genera of unarmoured dinoflagellates. *Phycologia.* **39**, 302–17.

Delwiche, C. F. (1999). Tracing the thread of plastid diversity through the tapestry of life. *Am. Nat.* **154**, S164–S177.

Delwiche, C. F. and Palmer, J. D. (1997). The origin of plastids and their spread via secondary symbiosis. In *Origins of Algae and their Plastids*, ed. D. Bhattacharya. New York: Springer, pp. 53–86.

Edvardsen, B. and Medlin, L. K. (2007). Molecular systematics of Haptophyta. In *Unravelling the Algae: the Past, Present and Future of Algal Systematics*, ed. J. Brodie and J. M. Lewis. Boca Raton: CRC Press, pp. 183–96.

Edvardsen, B., Eikrem, W., Green, J. C., Andersen, R. A., Moon-van der Staay, S. Y. and Medlin, L. K. (2000). Phylogenetic reconstructions of the Haptophyta inferred from 18S ribosomal DNA sequence and available morphological data. *Phycologia* **39**, 19–35.

Edvardsen, B., Eikrem, W., Shalchian-Tabrizi, K., Riisberg, I., Johnsen, G., Naustvoll, L. and Throndsen, J. (2007). *Verrucophora farmicen* gen. et sp. nov. (Dictyophyceae, Heterokonta) – a bloom-forming ichthyotoxic flagellate from the Skagerrak, Norway. *J. Phycol.* **43**, 1054–70.

Egeland, E. S., Guillard, R. R. L. and Liaaen-Jensen, S. (1997). Additional carotenoid prototype representatives and a general chemosystematic evaluation of carotenoids in Prasinophyceae (Chlorophyta). *Phytochemistry* **44**, 1087–97.

Eikrem, W., Romari, K., Latasa, M., Le Gall, F., Throndsen, J. and Vaulot, D. (2004). *Florenciella parvula* gen. et sp. nov. (Dictyochophyceae, Heterokontophyta), a small flagellate isolated from the English Channel. *Phycologia* **43**, 658–68.

Eikrem, W., Edvardsen, B. and Throndsen, J. (2009). Renaming *Verrucophora farcimen* Eikrem, Edvardsen et Throndsen. *Phycol. Res.* **57**, 170.

Elbrächter, M. and Schnepf, E. (1996). *Gymnodinium chlorophorum*, a new, green, bloom-forming dinoflagellate (Gymnodiniales, Dinophyceae) with a vestigial prasinophyte endosymbiont. *Phycologia* **35**, 381–93.

Entwisle, T. J. (2007a). Glaucocystophyta. In *Algae of Australia: Introduction*, ed. P. M. McCarthy and A. E. Orchard. Melbourne: CSIRO Publishing/ Australian Biological Resources Study, pp. 207–08.

Entwisle, T. J. (2007b). Xanthophyceae. In *Algae of Australia: Introduction*, ed. P. M. McCarthy and A. E. Orchard. Melbourne: CSIRO Publishing/Australian Biological Resources Study, pp. 255–56.

Falkowski, P. G., Katz, M. E., Knoll, A. H., Quigg, A., Raven, J. A., Schofield, O. and Taylor, F. J. R. (2004a). The evolution of modern eukaryotic phytoplankton. *Science* **305**, 354–60.

Falkowski, P. G., Schofield, O., Katz, M. E., van de Schootbrugge, B. and Knoll, A. H. (2004b). Why is the land green and the ocean red? In *Coccolithophores: From Molecular Processes to Global Impacts*, ed. H. R. Thierstein and J. R. Young. Berlin: Springer, pp. 429–53.

Fawley, M. W. (1989). A new form of Chlorophyll *c* involved in light harvesting. *Plant Physiol.* **91**, 727–32.

Fawley, M. W. (1991). Disjunct distribution of the xanthophyll loroxanthin in the green algae (Chlorophyta) *J. Phycol.* **27**, 544–48.

Fawley, K. P. and Fawley, M. W. (2007). Observations on the diversity and ecology of freshwater *Nannochloropsis* (Eustigmatophyceae), with descriptions of new taxa. *Protist* **158**, 325–36.

Fiksdahl, A., Bjørnland, T. and Liaaen-Jensen, S. (1984a). Algal carotenoids with novel end groups. *Phytochemistry*. **23**, 649–55.

Fiksdahl, A., Withers, N., Guillard, R. R. L. and Liaaen-Jensen, S. (1984b). Carotenoids of the Raphidophyceae – a chemosystematic contribution. *Comp. Biochem. Physiol.* **78**B, 265–71.

Foss, P., Lewin, R. A. and Liaaen-Jensen, S. (1987). The carotenoids of *Prochloron* sp. (Prochlorophyta). *Phycologia* **26**, 142–44.

Friedl, T. (1995). Inferring taxonomic positions and testing genus level assignments in coccoid green lichen algae: a phylogenetic analysis of 18S ribosomal RNA sequences from *Dictyochloropsis reticulata* and from members of the genus *Myrmecia* (Chlorophyta, Trebouxiophyceae *cl. nov.*). *J. Phycol.* **31**, 632–39.

Friedl, T. (1997). The evolution of the green algae. In *Origins of Algae and their Plastids*, ed. D. Bhattacharya. New York: Springer, pp. 87–101.

Frommolt, R., Werner, S., Paulsen, H., Goss, R., Wilhelm, C., Zauner, C., Maier, U. G., Grossman, A. R., Bhattacharya, D. and Lohr, M. (2008). Ancient recruitments by chromists of green algal genes encoding enzymes for carotenoid biosynthesis. *Mol. Biol. Evol.* **25**, 2653–67.

Garbary, D. J. and Gabrielson, P. W. (1990). Taxonomy and evolution. In *Biology of the Red Algae*, ed. K. M. Cole and R. G. Sheath. Cambridge: Cambridge University Press, pp. 477–98.

Garcia, A. and Playford, G. (2007). The fossil record of algae. In *Algae of Australia: Introduction*, ed. P. M. McCarthy and A. E. Orchard. Melbourne: CSIRO Publishing/Canberra: Australian Biological Resources Study, pp. 104–28.

Garrido, J. L. and Zapata, M. (2006). Chlorophyll analysis by new high performance liquid chromatography methods. In *Chlorophylls and Bacteriochlorophylls: Biochemistry, Biophysics, Functions and Applications*, ed. B. Grimm, R. J. Porra, W. Rüdiger and H. Scheer. Dordrecht: Kluwer Academic Publishers, pp. 109–21.

Garrido, J. L., Otero, J., Maestro, M. A. and Zapata, M. (2000). The main non-polar chlorophyll *c* from *Emiliania huxleyi* (Prymnesiophyceae) is a chlorophyll c_2-monogalactosyl-diacylglyceride ester: a mass spectrometry study. *J. Phycol.* **36**, 497–505.

Garrido, J. L., Rodríguez, F. and Zapata, M. (2009). Occurrence of loroxanthin, loroxanthin decenoate, and loroxanthin dodecenoate in *Tetraselmis* species (Prasinophyceae, Chlorophyta). *J. Phycol.* **45**, 366–74.

Gast, R. J., Moran, D. M., Beaudoin, D. J., Blythe, J. N. and Dennett, M. R. (2006). Abundance of a novel dinoflagellate phylotype in the Ross Sea, Antarctica. *J. Phycol.* **42**, 233–42.

Geitler, L. (1930). Ein Grünes Filarplasmodium und andere neue Protisten. *Arch. Protistenk.* **69**, 615–36.

Gibbs, S. P. (1981). The chloroplasts of some algal groups may have evolved from endosymbiotic eukaryotic algae. *Ann. N.Y. Acad. Sci.* **361**, 193–208.

Gibbs, S. P. (1993). The evolution of algal chloroplasts. In *Origins of Plastids*, ed. R. A. Lewin. New York: Chapman and Hall, pp. 101–21.

Gieskes, W. W. C. and Kraay, G. W. (1983). Dominance of Cryptophyceae during the phytoplankton spring bloom in the central North Sea detected by HPLC analysis of pigments. *Mar. Biol.* **75**, 179–85.

Goericke, R. and Repeta, D. J. (1992). The pigments of *Prochlorococcus marinus*: the presence of divinyl chlorophyll *a* and *b* in a marine prokaryote. *Limnol. Oceanogr.* **37**, 425–33.

Goericke, R., Olson, R. J. and Shalapyonok, A. (2000). A novel niche for *Prochlorococcus* sp. in low light sub-oxic environments in the Arabian Sea and the Eastern Tropical North Pacific. *Deep-Sea Res.* **47**, 1183–205.

Gordon, N., Angel, D. L., Neori, A., Kress, N. and Kimor, B. (1994). Heterotrophic dinoflagellates with symbiotic cyanobacteria and nitrogen limitation in the Gulf of Aqaba. *Mar. Ecol. Progr. Ser.* **107**, 83–88.

Grabowski, B., Cunningham, F. X. and Gantt, E. (2001). Chlorophyll and carotenoid binding in a simple red algal light-harvesting complex crosses phylogenetic lines. *Proc. Natl. Acad. Sci. USA* **98**, 2911–16.

Graham, L. E. and Wilcox, L. W. (2000a). Cyanobacteria (chloroxybacteria). In *Algae*, ed. L. E. Graham and L. W. Wilcox. Upper Saddle River: Prentice-Hall, pp. 97–131.

Graham, L. E. and Wilcox, L. W. (2000b). Ochrophytes I: Introduction to the Ochrophytes and a focus on diatoms. In *Algae*, ed. L. E. Graham and L. W. Wilcox. Upper Saddle River: Prentice-Hall, pp. 232–68.

Graham, L.E. and Wilcox, L.W. (2000c). Green Algae V. Charophyceans. In *Algae*, ed. L. E. Graham and L. W. Wilcox, Upper Saddle River: Prentice-Hall, pp. 494–543.

Graham, J. E., Wilcox, L. W. and Graham, L. E. (2009). *Algae*, 2nd edn. San Francisco: Benjamin Cummings.

Green, J. C. and Hori, T. (1988). The fine structure of mitosis in *Pavlova* (Prymnesiophycese). *Can. J. Bot.* **66**, 1497–509.

Green, J. C. and Leadbeater, B. S. C. (eds.) (1994). *The Haptophyte Algae*. The Systematics Association Special Volume no. 51. Oxford: Clarendon Press.

Green, J. C., Leadbeater, B. S. C. and Diver, W. L. (eds.) (1989). *The Chromophyte Algae: Problems and Perspectives*. The Systematics Association Special Volume no. 38. Oxford: Clarendon Press.

Grzebyk, D., Schofield, O., Vetriani, V. and Falkowski P. G. (2003). The Mesozoic radiation of eukaryotic algae: the portable plastid hypothesis. *J. Phycol.* **39**, 259–67.

Guillou, L., Chrétiennot-Dinet, M.-J., Medlin, L. K., Claustre, H., Loiseaux-de Goer, S. and Vaulot, D. (1999a). *Bolidomonas*: a new genus with two species belonging to a new algal class, the Bolidophyceae (Heterokonta). *J. Phycol.* **35**, 368–81.

Guillou, L., Moon-van der Staay, S.-Y., Claustre, H., Partensky, F. and Vaulot, D. (1999b). Diversity and abundance of Bolidophyceae (Heterokonta) in two oceanic regions. *Appl. Environ. Microbiol.* **65**, 4528–36.

Hackett, J. D., Maranda, L., Su Yoon, H. and Bhattacharya, D. (2003). Phylogenetic evidence for the cryptophyte origin of the plastid of *Dinophysis* (Dinophysiales, Dinophyceae). *J. Phycol.* **39**, 440–48.

Haeckel, E. (1866). *Generelle Morphologie der Organismem*, 2 vols. Berlin: G. Reimer.

Hagopian, J. C., Reis, M., Kitajima, J. P., Bhattacharya, D. and de Oliveira, M. C. (2004). Comparative analysis of the complete plastid genome sequence of the red alga *Gracilaria tenuistipitata* var. *liui* provides insights into the evolution of rhodoplasts and their relationship to other plastids. *J. Mol. Evol.* **59**, 464–77.

Hall, J. D. and Delwiche, C. F. (2007). In the shadow of giants: systematics of the charophyte green algae. In *Unravelling the Algae, the Past, Present and Future of Algal Systematics*, ed. J. Brodie and J. Lewis. Boca Raton: CRC Press, pp. 155–69.

Hallegraeff, G. M. (2006). *Plankton – a Critical Creation*. Hobart: University of Tasmania Press.

Hallegraeff, G. M. (2010). Ocean climate change, phytoplankton community responses, and harmful algal blooms: a formidable predictive challenge. *J. Phycol.* **46**, 220–35.

Hallegraeff, G. M. and Jeffrey, S.W. (1984). Tropical phytoplankton species and pigments of continental shelf waters of north and north-west Australia. *Mar. Ecol. Progr. Ser.* **20**, 59–74.

Hallegraeff, G. M., Bolch, C. J. S., Hill, D. R. A., Jameson, I., Le Roi, J.-M., McMinn, A., Murray, S., De Salas, M. F., Saunders, K. (2010). *Algae of Australia: Phytoplankton of Temperate Coastal Waters*. ABRS, Canberra; CSIRO Publishing, Melbourne.

Hara, Y. and Chihara, M. (1982). Ultrastructure and taxonomy of *Chattonella* (Class Raphidophyceae) in Japan. *Jpn. J. Phycol.* **30**, 47–56.

Hara, Y., Doi, K. and Chihara, M. (1994). Four new species of *Chattonella* (Raphidophyceae, Chromophyta) from Japan. *Jpn. J. Phycol.* **42**, 407–20.

Hattakeyama, N., Sasa, T., Watanabe, M. M. and Takaichi, S. (1991). Structure and pigment composition of *Chlorarachnion* sp. *J. Phycol.* **27** (Suppl. No. 3), 29.

Hayes, P. K., El Semary, N. A. and Sanchez-Baracaldo, P. (2007). The taxonomy of cyanobacteria: molecular insights into a difficult problem. In *Unravelling the Algae: the Past, Present and Future of Algal Systematics*, ed. J. Brodie and J. Lewis. The Systematics Association Special Volume Series 75, Boca Raton: CRC Press, pp. 93–101.

Henley, W. J., Hironake, J. L., Guillou, L., Buchheim, M. A., Buchheim, J. A., Fawley, M. W. and Fawley, K. P. (2004). Phylogenetic analysis of *Nannochloris*-like algae and diagnoses of *Picochlorum oklahomensis* gen. et sp. nov. (Trebouxiophyceae, Chlorophyta). *Phycologia* **43**, 641–52.

Hertzberg, S., Liaaen-Jensen, S. and Siegelman, H. W. (1971). The carotenoids of blue-green algae. *Phytochemistry* **10**, 3121–27.

Hewes, C. D., Mitchell, B. G., Moisan, T. A., Vernet, M. and Reid, F. M. (1998). The phycobilin signatures of chloroplasts from three dinoflagellate species: a microanalytical study of *Dinophysis cordata*, *D. fortii*, and *D. acuminata* (Dinophysiales, Dinophyceae). *J. Phycol.* **34**, 945–51.

Hibberd, D. J. (1977). Observations on the ultrastructure of the cryptomonad endosymbiont of the red-water ciliate *Mesodinium rubrum*. *J. Mar. Biol. Assoc. UK* **57**, 45–61.

Hibberd, D. J. (1980). Prymnesiophytes (Haptophytes). In *Phytoflagellates*, ed. E. R. Cox. New York: Elsevier, North-Holland, pp. 273–317.

Hibberd, D. J. (1990). Phylum Xanthophyta. In *Handbook of Protoctista*, ed. M. Margulis, J. O. Corliss, M. Melkonian and D. J. Chapman. Boston: Jones and Bartlett Publishers, pp. 686–97.

Hibberd, D. J. and Leedale, G. F. (1970). Eustigmatophyceae – a new algal class with unique organization of the motile cell. *Nature* **225**, 758–60.

Hibberd, D. J. and Norris, R. E. (1984). Cytology and ultrastructure of *Chlorarachnion reptans* (Chlorarachniophyta *divisio nova*, Chlorarachniophyceae *classis nova*). *J. Phycol.* **20**, 310–30.

Honda D. and Inouye, I. (2002). Ultrastructure and taxonomy of a marine photosynthetic stramenopile *Phaeomonas parva* gen. et sp. nov. (Pinguiophyceae) with emphasis on the flagellar apparatus architecture. *Phycol. Res.* **50**, 75–89.

Hori, T. and Green, J. C. (1994). Mitosis and cell division. In *The Haptophyte Algae*, ed. J. C. Green and B. S. C. Leadbeater. Oxford: Clarendon Press, pp. 91–109.

Horn, S., Ehlers, K., Fritzsch, G., Gil-Rodríguez, M. C., Wilhelm, C. and Schnetter, R. (2007). *Synchroma grande* spec. nov. (Synchromophyceae class. nov., Heterokontophyta): An amoeboid marine alga with unique plastid complexes. *Protist* **158**, 277–93.

Hosoi-Tanabe, S., Honda, D., Fukaya, S., Otake, I., Inagaki Y. and Sako, Y. (2007). Proposal of *Pseudochattonella verruculosa* gen. nov., comb. nov. (Dictyochophyceae) for a former raphidophycean alga *Chattonella verruculosa*, based on 18S rDNA phylogeny and ultrastructural characteristics. *Phycol. Res.* **55**, 185–92.

Hostetter, H. P. and Stoermer, E. F. (1971). Bibliography of the Bacillariophyceae. In *Selected Papers in Phycology*, ed. J. R. Roskowski and B. C. Parker. Lincoln: Dept of Botany, Univ. of Nebraska, pp. 784–90.

Huisman, J. M. and Saunders, G. W. (2007). Phylogeny and classification of the algae. In *Algae of Australia: Introduction*, ed. P. M. McCarthy and A. E. Orchard. Melbourne: CSIRO Publishing/ Canberra: Australian Biological Resources Study, pp. 66–103.

Ichinomiya, M., Yoshikawa, S., Kamiya, M., Ohki, K., Takaichi, S. and Kuwata, A. (2011). Isolation and characterization of Parmales (Heterokonta/ Heterokontophyta/Stramenopiles) from the Oyashio region, Western North Pacific. *J. Phycol.* **47**, 114–51.

Ishida, K., Green, B. R. and Cavalier-Smith, T. (1999). Diversification of a chimaeric algal group, the chlorarachniophytes: phylogeny of nuclear and nucleomorph small sub-unit rRNA genes. *Mol. Biol. Evol.* **16**, 321–31.

Ishida, K., Yabuki, A. and Ota, S. (2007). The Chlorarachniophytes: evolution and classification. In *Unravelling the Algae: The Past, Present and Future of Algal Systematics*, ed. J. Brodie and J. Lewis. The Systematics Association Special Volume Series 75. Boca Raton: CRC Press, pp. 171–82.

Janouškovec, J., Horák, A., Obornik, M., Lukeš, J. and Keeling, P. J. (2010). A common red algal origin of the apicomplexan, dinoflagellate, and heterokont plastids. *Proc. Natl. Acad. Sci. USA* **107**, 10949–54.

Jeffrey, S. W. (1974). Profiles of photosynthetic pigments in the ocean using thin-layer chromatography. *Mar. Biol.* **26**, 101–10.

Jeffrey, S. W. (1989). Chlorophyll *c* pigments and their distribution in the chromophyte algae. In *The Chromophyte Algae: Problems and Perspectives*, ed. J. C. Green, B. S. C. Leadbeater and W. L. Diver. The Systematics Association Special Volume no. 38, Oxford: Clarendon Press, pp. 13–36.

Jeffrey, S. W. (1997). Preparation of chlorophyll standards. In *Phytoplankton Pigments in Oceanography: Guidelines to Modern Methods*, ed. S. W. Jeffrey, R. F. C. Mantoura and S. W. Wright. Paris: UNESCO Publishing, pp. 207–38.

Jeffrey, S. W. and Egeland, E. S. (2008). Pigments of green and red forms of *Dunaliella*, and related chlorophytes. In *The Alga Dunaliella: Biodiversity, Physiology, Genomics and Biotechonology*, ed. A. Ben-Amotz, J. E. W. Polle and D. V. Subba Rao. Enfield: Science Publishers, pp. 111–45.

Jeffrey, S. W. and Mantoura, R. F. C. (1997). Development of pigment methods for oceanography: SCOR supported Working Groups and Objectives. In *Phytoplankton Pigments in Oceanography: Guidelines to Modern Methods*, ed. S. W. Jeffrey, R. F. C. Mantoura and S. W. Wright. Paris: UNESCO Publishing, pp 19–36.

Jeffrey, S. W. and Vesk, M. (1976). Further evidence for a membrane-bound endosymbiont within the dinoflagellate *Peridinium foliaceum*. *J. Phycol.* **12**, 450–55.

Jeffrey, S. W. and Vesk, M. (1997). Introduction to marine phytoplankton and their pigment signatures. In *Phytoplankton Pigments in Oceanography: Guidelines to Modern Methods*, ed. S. W. Jeffrey, R. F. C. Mantoura and S. W. Wright. Paris: UNESCO Publishing, pp. 37–84.

Jeffrey, S. W. and Wright, S. W. (1994). Photosynthetic pigments in the Prymnesiophyceae. In *The Haptophyte Algae*, ed. J. C. Green and B. S. C. Leadbeater. Oxford: Clarendon Press, pp. 111–32.

Jeffrey, S. W. and Wright, S. W. (1997). Qualitative and quantitative HPLC analysis of SCOR reference algal cultures. In *Phytoplankton Pigments in Oceanography: Guidelines to Modern Methods*, ed. S. W. Jeffrey, R. F. C. Mantoura and S. W. Wright. Paris: UNESCO Publishing, pp. 343–60.

Jeffrey, S. W. and Wright, S. W. (2006). Photosynthetic pigments in marine microalgae: insights from cultures and the sea. In *Algal Cultures, Analogues of Blooms and Applications*, ed. D. V. Subba Rao. Enfield: Science Publishers, pp. 33–90.

Jeffrey, S. W., Sielicki, M. and Haxo, F. T. (1975). Chloroplast pigment patterns in dinoflagellates. *J. Phycol.* **11**, 374–84.

Jeffrey, S. W., Mantoura, R. F. C. and Bjørnland, T. (1997a). Data for the identification of 47 key phytoplankton pigments. In *Phytoplankton Pigments in*

Oceanography: Guidelines to Modern Methods, ed. S.W. Jeffrey, R.F.C. Mantoura and S.W. Wright. Paris: UNESCO Publishing, pp. 449–559.

Jeffrey, S.W., Mantoura, R.F.C. and Wright, S.W. (eds.) (1997b). *Phytoplankton Pigments in Oceanography: Guidelines to Modern Methods*. Paris: UNESCO Publishing.

Jordan, R.W., Cros, L. and Young, J.R. (2004). A revised classification scheme for living haptophytes. *Micropaleontology* **50**, 55–79.

Kai, A., Yoshii, Y., Nakayama, T. and Inouye, I. (2008). Aurearenophyceae *classis nova*, a new class of Heterokontophyta based on a new marine unicellular alga *Aurearena cruciata* gen. et sp. nov. inhabiting sandy beaches. *Protist* **159**, 435–57.

Kashiyama, Y., Miyashita, H., Okhubo, S., Ogawa, N.O., Chikaraishi, Y., Takano, Y., Suga, H., Toyofuku, T., Nomaki, H., Kitazako, H., Nagata, T. and Ohkouchi, N. (2008). Evidence of global chlorophyll *d*. *Science* **321**, 658.

Kawachi, M.N., Atsumi, H., Ikemoto, H. and Miyachi, S. (2002a). *Pinguiochrysis pyriformis* gen. et sp. nov. (Pinguiophyceae), a new picoplanktonic alga isolated from the Pacific Ocean. *Phycol. Res.* **50**, 49–56.

Kawachi, M.N., Inouye, I., Honda, D., O'Kelly, C.J., Bailey, J.C., Bidigare, R.R. and Andersen, R.A. (2002b). The Pinguiophyceae *classis nova*, a new class of photosynthetic stramenopiles whose members produce large amounts of omega-3 fatty acids. *Phycol. Res.* **50**, 31–47.

Kawachi, M., Noe, L. and Andersen, R.A. (2002c). Re-examination of the marine 'chrysophyte' *Polypodochrysis teissieri* (Pinguiophyceae). *Phycol. Res.* **50**, 91–100.

Kawai, H. and Inouye, I. (1989). Flagella autofluorescence in forty-four chlorophyll *c* containing algae. *Phycologia*. **28**, 222–27.

Keeling, P.J. (2004a). A brief history of plastids and their hosts. *Protist* **155**, 3–7.

Keeling, P.J. (2004b). Diversity and evolutionary history of plastids and their hosts. *Am. J. Bot.* **91**, 1481–93.

Kempton, J.W., Wolny, J., Tengs, T., Rizzo, P., Morris, R., Tunnell, J., Scott, P., Steidinger, K., Hymel, S.N. and Lewitus, A.J. (2002). *Kryptoperidinium foliaceum* blooms in Southern Carolina: a multi-analytical approach to identification. *Harmful Algae* **1**, 383–92.

Kerfeld, C.A. (2004). Water-soluble carotenoid proteins of cyanobacteria. *Arch. Biochem. Biophys.* **430**, 2–9.

Kies, L. and Kremer, B.P. (1990). Phylum Glaucocystophyta. In *Handbook of Protoctista*, ed. L. Margulis, J.O. Corliss, M. Melkonian and D.J. Chapman. Boston: Jones and Bartlett Publishers, pp. 152–66.

Klebs, G. (1893). Flagellaten studien II. *Z. Wiss. Zool.* **55**, 353–445.

Kopecký, J., Riederer, M. and Pfündel, E. (2002). *Porphyridium purpureum* (formerly *P. cruentum*) contains β-carotene but no α-carotene. *Algol. Studies* **104**, 189–95.

Kützing, F.T. (1843). *Phycologia Generalis oder Anatomie, Physiologie und System Kunda der Tange*. Leipzig: F.A. Brockhaus.

Latasa, M., Scharek, R., Le Gall, F. and Guillou, L. (2004). Pigment suites and taxonomic groups in the Prasinophyceae. *J. Phycol.* **40**, 1149–55.

Leadbeater, B.S.C. (1989). The phylogenetic significance of flagellar hairs in the Chromophyta. In *The Chromophyte Algae: Problems and Perspectives*, ed. J.C. Green, B.S.C. Leadbeater and W.L. Diver. The Systematics Association Special Volume no. 38, Oxford: Clarendon Press, pp. 145–65.

Lemieux, C., Otis, C. and Turmel, M. (2000). Ancestral chloroplast genome in *Mesostigma viride* reveals an early branch of green plant evolution. *Nature* **403**, 649–52.

Lewin, J., Norris, R. E., Jeffrey, S. W. and Pearson, B. (1977). An aberrant chrysophycean alga, *Pelagococcus subviridis* Norris, gen. nov., sp. nov., from the North Pacific. *J. Phycol.* **13**, 259–66.

Lewin, R. A. (1976). Prochlorophyta as a proposed new division of algae. *Nature* **261**, 697–98.

Lewin, R. A., Krienitz, L., Goericke, R., Takeda, H. and Hepperle, D. (2000). *Picocystis salinarum* gen. et sp. nov. (Chlorophyta) – a new picoplanktonic green alga. *Phycologia* **39**, 560–65.

Lewis, L. A. and McCourt, R. M. (2004). Green algae and the origin of land plants. *Am. J. Bot.* **91**, 1535–56.

Liaaen-Jensen, S. and Egeland, E. S. (1999). Microalgal carotenoids. In *Chemicals from Microalgae*, ed. Z. Cohen. London: Taylor and Francis, pp. 145–72.

Linnaeus, C. (1751). *Philosophia Botanica*. Stockholm: Kiesewetter.

Liu, H., Probert, I., Uitz, J., Claustre, H. Aris-Brosou, S., Frada, M., Not, F. and de Vargas, C. (2009). Extreme diversity in noncalcifying haptophytes explains a major pigment paradox in open oceans. *Proc. Natl. Acad. Sci. USA* **106**, 12803–08.

Llewellyn, C. A. and Gibb, S. W. (2000). Intra-class variability in the carbon, pigment and biomineral content of prymnesiophytes and diatoms. *Mar. Ecol. Progr. Ser.* **193**, 33–44.

Lohr, M. and Wilhelm, C. (1999). Algae displaying the diadinoxanthin cycle also possess a violaxanthin cycle. *Proc. Natl. Acad. Sci. USA* **96**, 8784–89.

Lovejoy, C., Massana, R. and Pedrós-Alió, C. (2006). Diversity and distribution of marine microbial eukaryotes in the Arctic Ocean and adjacent seas. *Appl. Environ. Microbiol.* **72**, 3085–95.

Maggs, C. A., Verbruggen, H. and de Clerck, O. (2007). Molecular systematics of red algae: building future structures on firm foundations. In *Unravelling the Algae: The Past, Present and Future of Algal Systematics*, ed. J. Brodie and J. Lewis. The Systematics Association Special Volume 75. Boca Raton: CRC Press, pp.103–21.

Malin, G., Turner, S. M. and Liss, P. S. (1992). Sulfur: The plankton/climate connection. *J. Phycol.* **28**, 590–97.

Manton, I. and Ettl, H. (1965). Observations on the fine structure of *Mesostigma viride* Lauterborn. *J. Linn. Soc. (Bot.)* **59**, 175–84.

Marchant, H. J. and Thomsen, H. A. (1994). Haptophytes in polar waters. In *The Haptophyte Algae*, ed. J. C. Green and B. S. C. Leadbeater. Oxford: Clarendon Press, pp. 209–28.

Marchant, H. J., Davidson, A. and Wright, S. W. (1987). The distribution and abundance of chroococcoid cyanobacteria in the Southern Ocean. Proc. NIPR Symp. *Polar Biol.* **1**, 1–9.

Margulis, L., Corliss, J. O., Melkonian, M. and Chapman, D. J. (eds.) (1989). *Handbook of Protoctista*. Boston: Jones and Bartlett Publishers.

Marin, B. and Melkonian, M. (1999). Mesostigmatophyceae, a new class of streptophyte green algae revealed by SSU rRNA sequence comparisons. *Protist* **150**, 399–417.

Marin, B. and Melkonian, M. (2010). Molecular phylogeny and classification of the Mamiellophyceae class. nov. (Chlorophyta), based on sequence comparisons of the nuclear- and plastid- encoded rRNA operons. *Protist* **161**, 304–36.

Marin, B., Palm, A., Klingberg, M. and Melkonian, M. (2003). Phylogeny and taxonomic revision of plastid-containing euglenophytes based on SSU rDNA sequence comparisons and synapomorphic signatures in the SSU rRNA secondary structure. *Protist* **154**, 99–145.

Marquardt, J., Senger, H., Miyashita, H., Miyachi, S. and Mörschel, E. (1997). Isolation and characterization of biliprotein aggregates from *Acaryochloris marina*, a *Prochloron*-like prokaryote containing mainly chlorophyll *d*. *FEBS Lett.* **410**, 428–32.

Mattox, K. R. and Stewart, K. D. (1984). Classification of the green algae: a concept based on comparative cytology. In *Systematics of the Green Algae*, ed. D. E. G. Irvine and D. M. John. London: Academic Press, pp. 29–72.

McCarthy, P. M. and Orchard A. E. (2007). *Algae of Australia: Introduction.* Melbourne: CSIRO Publishing/Canberra: Australian Biological Resources Study.

McFadden, G. I. (2001). Primary and secondary endosymbiosis and the origin of plastids. *J. Phycol.* **37**, 951–59.

McFadden, G. I., Gilson, P. R., Hofman, C. J. B., Adcock, G. J. and Maier, U. G. (1994). Evidence that an amoeba acquired a chloroplast by retaining part of an engulfed eukaryotic alga. *Proc. Natl. Acad. Sci. USA* **91**, 3690–94.

Medlin, L. K. and Kaczmarska, I. (2004). Evolution of the diatoms. V. Morphological and cytological support for the major clades and a taxonomic revision. *Phycologia* **43**, 245–70.

Medlin, L. K., Williams, D. M. and Sims, P. A. (1993). The evolution of diatoms (Bacillariophyta). 1. Origin of the group and assessment of the monophyly of its major divisions. *Eur. J. Phycol.* **28**, 261–75.

Medlin, L. K., Kooistra, W. H. C. F., Potter, D., Saunders, G. W. and Andersen, R. A. (1997). Phylogenetic relationships of the 'golden algae' (haptophytes, heterokont chromophytes) and their plastids. *Pl. Syst. Evol.* [Suppl.] **11**, 187–219.

Melkonian, M. (1989). Flagellar apparatus ultrastructure in *Mesostigma viride* (Prasinophyceae). *Pl. Syst. Evol.* **164**, 93–122.

Meyer-Harms, B. and Pollehne, F. (1998). Alloxanthin in *Dinophysis norvegica* (Dinophysiales, Dinophyceae) from the Baltic Sea. *J. Phycol.* **34**, 280–85.

Millie, D. F., Schofield, O. M., Kirkpatrick, G. J., Johnsen, G., Tester, P. A. and Vinyard, B. T. (1997). Detection of harmful algal blooms using photopigments and absorption signatures: A case study of the Florida red tide dinoflagellate, *Gymnodinium breve*. *Limnol. Oceanogr.* **42**, 1240–51.

Miyashita, H., Ikemoto, H., Kurano, N., Adachi, K., Chihara, M. and Miyachi, S. (1996). Chlorophyll *d* as a major pigment. *Nature* **383**, 402–03.

Miyashita, H., Adachi, K., Kurano, N., Ikemoto, H., Chihara, M. and Miyachi, S. (1997). Pigment composition of a novel oxygenic photosynthetic prokaryote containing chlorophyll *d* as the major chlorophyll. *Plant Cell Physiol.* **38**, 274–81.

Miyashita, H., Ikemoto, H. Kurano, N, Miyachi, S. and Chihara M. (2003). *Acaryochloris marina* gen. et. sp. nov. (Cyanobacteria) an oxygenic photosynthetic prokaryote containing Chl *d* as a major pigment. *J. Phycol.* **39**, 1247–53.

Moestrup, Ø. and Daugbjerg, N. (2007). On dinoflagellate phylogeny and classification. In *Unravelling the Algae: The Past, Present and Future of Algal Systematics*, ed. J. Brodie and J. Lewis. The Systematics Association Special Volume 75. Boca Raton: CRC Press, pp. 215–30.

Moestrup, Ø. and O'Kelly, C. J. (2002). Class Silicoflagellata. In *An Illustrated Guide to the Protozoa*, ed. J. J. Lee, G. F. Leedale and P. Bradbury. 2nd edn. vol.2 Lawrence: Society of Protozoologists, pp. 775–82.

Moestrup, Ø. and Sengco, M. (2001). Ultrastructural studies on *Bigelowiella natans* gen. et sp. nov., a chlorarachniophyte flagellate. *J. Phycol.* **37**, 624–46.

Moestrup, Ø. and Thomsen, H. A. (1990). *Dictyocha speculum* (Silicoflagellata, Dictyochophyceae), studies on armoured and unarmoured stages. *Biologiske Skrifter* **37**, 5–57.

Mohamed, H. E., van der Meene, A. M. L., Roberson, R. W. and Vermaas, W. F. J. (2005). Myxoxanthophyll is required for normal cell wall structure and thylakoid organization in the cyanobacterium *Synechocystis* sp. strain PCC 6803. *J. Bacteriol.* **187**, 6883–92.

Moore, L. R. and Chisholm, S. W. (1999). Photophysiology of the marine cyanobacterium *Prochlorococcus*: ecotypic differences among cultured isolates. *Limnol. Oceanogr.* **44**, 628–38.

Moore, R. B., Obornik, M., Janouŝkovec, J., Chrudimský, T., Vancová, M., Green, D. H., Wright, S. W., Davies, N. W., Bolch, C. J. S., Heimann, K., Ŝlapeta, S., Hoegh-Guldberg, O., Logsdon, J. M. Jr. and Carter, D. A. (2008). A photosynthetic alveolate closely related to apicomplexan parasites. *Nature* **451**, 959–63.

Moreira, D., LeGuyader, H. and Phillipe, H. (2000). The origin of red algae and the evolution of chloroplasts. *Nature* **405**, 69–72.

Mostaert, A. S., Karsten, U., Hara, Y. and Watanabe, M. M. (1998). Pigments and fatty acids of marine raphidophytes: a chemotaxonomic re-evaluation. *Phycol. Res.* **46**, 213–20.

Nair, A., Sathyendranath, S., Platt, T., Morales, J., Stuart, V., Forget, M.-H., Devred, E. and Bouman, H. (2008). Remote sensing of phytoplankton functional types. *Remote Sens. Environ.* **112**, 3366–75.

Norgård, S., Svec, W. A., Liaaen-Jensen, S., Jensen, A. and Guillard, R. R. L. (1974). Chloroplast pigments and algal systematics. *Biochem. Syst. Ecol.* **2**, 3–6.

Not, F., Valentin, K., Romari, K., Lovejoy, C., Massana, R., Tőbe, K., Vaulot, D. and Medlin, L. K. (2007). Picobiliphytes: a marine picoplanktonic algal group with unknown affinities to other eukaryotes. *Science* **315**, 253–55.

Okamoto, N., Chantangsi, C., Horák, A., Leander, B. S. and Keeling, P. J. (2009). Molecular phylogeny and description of the novel Katablepharid *Roombia truncata* gen. et sp. nov., and establishment of the Hacrobia taxon nov. *PLoS ONE* 4(9), e7080.

Ota, S., Ueda, K. and Ishida, K. (2007). Taxonomic study of *Bigelowiella longifila* sp. nov. (Chlorarachniophyta) and a time-lapse video observation of the unique migration of amoeboid cells. *J. Phycol.* **43**, 333–43.

Ota, S., Silver, T. D., Archibald, J. M. and Ishida, K. (2009a). *Lotharella oceanic* sp. nov. – a new planktonic chlorarachniophyte studied by light and electron microscopy. *Phycologia* **48**, 315–23.

Ota, S., Vaulot, D., Le Gall, F., Yabuki, A. and Ishida, K. (2009b). *Partenskyella glossopodia* gen. et sp. nov., the first report of a chlorarachniophyte that lacks a pyrenoid. *Protist* **160**, 137–50.

Palmer, J. D. (2003). The symbiotic birth and spread of plastids: How many times and whodunit? *J. Phycol.* **39**, 4–11.

Partensky, F., Hoepffner, N., Li, W. K. W., Ulloa, O. and Vaulot, D. (1993). Photoacclimation of *Prochlorococcus* sp. (Prochlorophyta) strains isolated from the north Atlantic and the Mediterranean Sea. *Plant Physiol.* **101**, 285–96.

Partensky, F., La Roche, J., Wyman, K. and Falkowski, P. G. (1997). The divinyl-chlorophyll *a*/*b*-protein complexes of two strains of the oxyphototrophic marine prokaryote *Prochlorococcus* – characterization and response to changes in growth irradiance. *Photosynth. Res.* **51**, 209–22.

Partensky, F., Hess, W. R. and Vaulot, D. (1999). *Prochlorococcus*, a marine photosynthetic prokaryote of global significance. *Microbiol. Mol. Biol. Rev.* **63**, 106–27.

Patterson, D. J. (1989). Stramenopiles: chromophytes from a protistan perspective. In *The Chromophyte Algae: Problems and Perspectives*, ed. J. C. Green, B. S. C. Leadbeater and W. L. Diver. The Systematics Association Special Volume no. 38, Oxford: Clarendon Press, pp. 357–79.

Potter, D., Lajeunesse, T. C., Saunders, G. W. and Andersen, R. A. (1997a). Convergent evolution masks extensive biodiversity among marine coccoid picoplankton. *Biodiversity Conserv.* **6**, 99–107.

Potter, D., Saunders, G. W. and Andersen, R. A. (1997b). Phylogenetic relationships of the Raphidophyceae and Xanthophyceae as inferred from nucleotide sequences of the 18S ribosomal RNA gene. *Am. J. Bot.* **84**, 966–72.

Preisig, H. R. (1989). The flagellar base ultrastructure and phylogeny of chromophytes. In *The Chromophyte Algae: Problems and Perspectives*, ed. J. C. Green, B. S. C. Leadbeater and W. L. Diver. The Systematics Association Special Volume no. 38, Oxford: Clarendon Press, pp. 167–87.

Pröschold, T. and Leliaert, F. (2007). Systematics of the green algae: conflict of classic and modern approaches. In *Unravelling the Algae: the Past, Present and Future of Algal Systematics*, ed. J. Brodie and J. Lewis. The Systematics Association Special Volume 75. Boca Raton: CRC Press, pp. 123–53.

Quigg, A., Finkel, Z. V., Irwin, A. J., Rosenthal, T., Ho, T. Y., Reinfelder, J. R., Schofield, O., Morel, F. M. M. and Falkowski, P. G. (2003). The evolutionary inheritance of elemental stoichiometry in marine phytoplankton. *Nature* **425**, 291–94.

Reynolds, C. (2006). *Ecology of Phytoplankton*. Cambridge: Cambridge University Press.

Ricketts, T. R. (1970). The pigments of the Prasinophyceae and related organisms. *Phytochemistry* **9**, 1835–42.

Rippka, R., Coursin, T., Hess, W., Lichtlé, C., Scanlan, D. J., Palinska, K. A., Iteman, I., Partensky, F., Houmard, J. and Herdman, M. (2000). *Prochlorococcus marinus* Chisholm *et al.* 1992 subsp. *pastoris* subsp. nov. strain PCC 9511, the first axenic chlorophyll a_2/b_2-containing cyanobacterium (Oxyphotobacteria). *Int. J. Syst. Evol. Microbiol.* **50**, 1833–47.

Rodríguez, F., Garrido, J. L., Crespo, B. G., Arbones, B. and Figueiras, F. G. (2006). Size-fractionated phytoplankton pigment groups in the NW Iberian upwelling system: impact of the Iberian Poleward Current. *Mar. Ecol. Progr. Ser.* **323**, 59–73.

Round, F. E., Crawford, R. M. and Mann D. G. (1990). *The Diatoms, Biology and Morphology of the Genera*. Cambridge: Cambridge University Press.

Rowan, K. S. (1989). *Photosynthetic Pigments of Algae*. Cambridge: Cambridge University Press.

Sabine, C. L. and Feely, R. A. (2007). The oceanic sink for carbon dioxide. In *Greenhouse Gas Sinks*, ed. D. Reay, N. Hewitt, J. Grace and K. Smith. Oxfordshire: CABI Publishing, pp. 31–49.

Saldarriaga, J. F., Taylor, F. J. R., Keeling, P. J. and Cavalier-Smith, T. (2001). Dinoflagellate nuclear SSU rRNA phylogeny suggests multiple plastid losses and replacements. *J. Mol. Evol.* **53**, 204–13.

Sanchez-Puerta, M. V. and Delwiche, C. F. (2008). A hypothesis for plastid evolution in chromalveolates. *J. Phycol.* **44**, 1097–107.

Sasa, T., Suda, S., Watanabe, M. M. and Takaichi, S. (1992a). A yellow marine *Chlamydomonas*: morphology and pigment composition. *Plant Cell Physiol.* **33**, 527–34.

Sasa, T., Takaichi, S., Hatakeyama, N. and Watanabe, M. M. (1992b). A novel carotenoid ester, loroxanthin dodecanoate, from *Pyramimonas parkeae* (Prasinophyceae) and a Chlorarachniophycean alga. *Plant Cell. Physiol.* **33**, 921–25.

Sathyendranath, S. (1986). Remote sensing of phytoplankton: a review, with special reference to picoplankton. In *Photosynthetic Picoplankton*, ed. T. Platt and W. K. W. Li. *Can. Bull. Fish. Aquat. Sci.* **214**, 561–83.

Sathyendranath, S., Watts, L., Devred, E., Platt, T., Caverhill, C. and Maass, H. (2004). Discrimination of diatoms from other phytoplankton using ocean-colour data. *Mar. Ecol. Prog. Ser.* **272**, 59–68.

Schlüter, L., Garde, K. and Kaas, H. (2004). Detection of the toxic cyanobacteria *Nodularia spumigena* by means of a 4-keto-myxoxanthophyll-like pigment in the Baltic Sea. *Mar. Ecol. Progr. Ser.* **275**, 69–78.

Schlüter, L., Lutnæs, B. F., Liaaen-Jensen, S., Garde, K., Kaas, H., Jameson, I. and Blackburn, S. I. (2008). Correlation of the content of hepatotoxin nodularin and glycosidic carotenoids, 4-ketomyxol-2′-fucoside and novel 1′-*O*-methyl-4-ketomyxol-2′-fucoside, in 20 strains of the cyanobacterium *Nodularia spumigena*. *Biochem. Syst. Ecol.* **36**, 749–57.

Scott, F. J. and Marchant, H. J. (2005). *Antarctic Marine Protists*. Canberra: Australian Biological Resources Study and Australian Antarctic Division.

Silva, P. C., Throndsen, J. and Eikrem, W. (2007). Revisiting the nomenclature of haptophytes. *Phycologia* **46**, 471–75.

Six, C., Thomas, J. C., Brahamsha, B., Lemoine, Y. and Partensky, F. (2004). Photophysiology of the marine cyanobacterium *Synechococcus* sp. WH8102, a new model organism. *Aquat. Microb. Ecol.* **35**, 17–29.

Smith, J. H. C. and Benitez, A. (1955). Chlorophylls: Analysis in plant materials. In *Modern Methods of Plant Analysis*, vol. 4. ed. K. Paech and M. Tracey, Heidelberg: Springer, pp. 142–96.

Stanier, R. Y., Sistrom, W. R., Hansen, T. A., Whitton, B. A., Castenholz, R. W., Pfennig, N., Gorlenko, V. N., Kondratieva, E. N., Eimhjellen, K. E., Whittenbury, R., Gherna, R. J. and Trüper, H. G. (1978). Proposal to place the nomenclature of the cyanobacteria (blue-green algae) under the rules of the international code of nomenclature of bacteria. *Int. J. System. Bacteriol.* **28**, 335–36.

Stauber, J. L. and Jeffrey, S. W. (1988). Photosynthetic pigments in fifty-one species of marine diatoms. *J. Phycol.* **24**, 158–72.

Stein, F. R. von (1878). *Der Organismus der Infusionsthiere. III Der Organismus der Flagellaten*. Leipzig: Wilhelm Engelmann.

Strain, H. H. (1958). *Chloroplast Pigments and Chromatographic Analysis. Thirty-Second Annual Priestly Lecture*. University Park: Pennsylvania State University Press.

Strain, H. H., Manning, W. M. and Hardin, G. (1944). Xanthophylls and carotenes of diatoms, brown algae, dinoflagellates and sea anemones. *Biol. Bull.* **86**, 169–91.

Takaichi, S., Maoka, T. and Masamoto, K. (2001). Myxoxanthophyll in *Synechocystis* sp. PCC 6803 is myxol 2′-dimethyl-fucoside, (3*R*, 2′*S*)-myxol 2′-(2,4-di-*O*-methyl-α-L-fucoside), not rhamnoside. *Plant Cell Physiol.* **42**, 756–62.

Takano, Y., Hansen, G., Fujita, D. and Horiguchi, T. (2008). Serial replacement of diatom endosymbionts in two freshwater dinoflagellates, *Peridiniopsis* spp. (Peridiniales, Dinophyceae). *Phycologia* **47**, 41–53.

Taylor, F. J. R. (1980). On dinoflagellate evolution. *Biosystems* **13**, 65–108.

Throndsen, J. (1997). The planktonic marine flagellates. In *Identifying Marine Phytoplankton: a Guide to Naked Flagellates and Coccolithophorids*, ed. C. R. Tomas. San Diego: Academic Press, pp. 591–729.

Triemer, R. E. and Farmer, M. A. (2007). A decade of euglenoid molecular phylogenetics. In *Unravelling the Algae: Past, Present, and the Future of Algal Systematics*, ed. J. Brodie and J. Lewis. The Systematics Association Special Volume 75. Boca Raton: CRC Press, pp. 315–30.

Triemer, R. E., Linton, E., Shin, W., Nudelman, A., Monfils, A., Bennett, M. and Brosnan, S. (2006). Phylogeny of the Euglenales based upon combined SSU and LSU rDNA sequence comparisons and description of *Discoplastis* gen. nov. (Euglenophyta). *J. Phycol.* **42**, 731–40.

Turmel, M., Otis, C. and Lemieux, C. (2002). The complete mitochondrial DNA sequence of *Mesostigma viride* identifies this green alga as the earliest green plant divergence and predicts a highly compact mitochondrial genome in the ancestor of all green plants. *Mol. Biol. Evol.* **19**, 24–38.

Turmel, M., Brouard, J-S., Gagnon, C., Otis, C. and Lemieux, C. (2008). Deep division in the Chlorophyceae (Chlorophyta) revealed by chloroplast phylogenomic analyses. *J. Phycol.* **44**, 739–50.

Urbach, E., Robertson, D. L. and Chisholm, S. W. (1992). Multiple evolutionary origins of prochlorophytes within the cyanobacterial radiation. *Nature* **355**, 267–70.

Van den Hoek, C., Mann, D. G. and Jahns, H. M. (1995a). Chlorophyta: Class 3. Ulvophyceae. In *Algae – an Introduction to Phycology*, ed. C. Van den Hoek, D. G. Mann and H. M. Jahns. Cambridge: Cambridge University Press, pp. 391–408.

Van den Hoek, C., Mann, D. G. and Jahns, H. M. (1995b). *Algae – an Introduction to Phycology*. Cambridge: Cambridge University Press.

Van Lenning, K., Latasa, M., Estrada, M., Sáez, A. G., Medlin, L., Probert, I., Véron, B. and Young, B. (2003). Pigment signatures and phylogenetic relationships of the Pavlovophyceae (Haptophyta). *J. Phycol.* **39**, 379–89.

Van Valkenburg, S. G. (1980). Silicoflagellates. In *Phytoflagellates*, ed. E. R. Cox. New York: Elsevier, North-Holland, pp. 335–50.

Véron, B., Dauguet, J-C. and Billard, C. (1996). Sterolic biomarkers in marine phytoplankton. 1. Free and conjugated sterols of *Pavlova lutheri* (Haptophyta). *Eur. J. Phycol.* **31**, 211–15.

Vesk, M. and Jeffrey, S. W. (1987). Ultrastructure and pigments of two strains of the picoplanktonic alga *Pelagococcus subviridis* (Chrysophyceae). *J. Phycol.* **23**, 322–36.

Vesk, M., Jeffrey, S. W. and Hallegraeff, G. M. (1990). Golden-brown algae: Prymesiophyta and Chrysophyta (includes Raphidophyceae). In *Biology of Marine Plants*, ed. M. N. Clayton and R. J. King. Melbourne: Longman Cheshire, pp. 96–114.

Vesk, M., Dibbayawan, T. P. and Vesk, P. A. (1996). Immunogold localization of phycoerythrin in chloroplasts of *Dinophysis acuminata* and *D. fortii* (Dinophysiales, Dinophyta). *Phycologia* **35**, 234–38.

Volkman, J. K., Brown, M. R., Dunstan, G. A. and Jeffrey, S. W. (1993). The biochemical composition of marine microalgae from the class Eustigmatophyceae. *J. Phycol.* **29**, 69–78.

Volkman, J. K., Farmer, C. L., Barrett, S. M. and Sikes, E. L. (1997). Unusual dihydroxysterols as chemotaxonomic markers for microalgae from the order Pavlovales (Haptophyceae). *J. Phycol.* **33**, 1016–23.

Watanabe, M. M., Suda, S., Inouye, I., Sawaguchi, T. and Chihara, M. (1990). *Lepidodinium viride* gen. et. sp. nov. (Gymnodiniales, Dinophyta), a green dinoflagellate with a chlorophyll *a* and *b*-containing endosymbiont. *J. Phycol.* **26**, 741–51.

Waterbury, J. B., Watson, S. W., Guillard, R. R. L. and Brand, L. E. (1979). Widespread occurrence of a unicellular marine planktonic cyanobacterium. *Nature* **277**, 293–94.

West, J. A. (2007). Rhodophyta: Cyanidiophyceae and Bangiophyceae. In *Algae of Australia: Introduction*, ed. P. M. McCarthy and A. E. Orchard. Melbourne: CSIRO Publishing/Canberra: Australian Biological Resources Study, pp. 209–23.

Williams, D. M. (2007). Classification and diatom systematics: the past, the present and the future. In *Unravelling the Algae: Past, Present, and the Future of Algal Systematics*, ed. J. Brodie and J. Lewis. Boca Raton: CRC Press, pp. 57–91.

Wright, S. W. and Jeffrey, S. W. (1997). High resolution HPLC system for chlorophylls and carotenoids of marine phytoplankton. In *Phytoplankton Pigments in Oceanography: Guidelines to Modern Methods*, ed. S. W. Jeffrey, R. F. C. Mantoura and S. W. Wright. Paris: UNESCO Publishing, pp. 327–41.

Wright, S. W. and Jeffrey, S. W. (2006). Pigment markers for phytoplankton production. In *Marine Organic Matter: Biomarkers, Isotopes and DNA*, ed. J. K. Volkman. The Handbook of Environmental Chemistry vol. 2, Part N. Berlin: Springer, pp. 71–104.

Yokohama, Y. (1981). Distribution of the green light-absorbing pigments siphonaxanthin and siphonein in marine green algae. *Bot. Mar.* **24**, 637–40.

Yokohama, Y., Kageyama, A., Ikawa, T. and Shimura, S. (1977). A carotenoid characteristic of chlorophycean seaweeds living in deep coastal waters. *Bot. Mar.* **20**, 433–36.

Yoon, H. S., Hackett, J. D., Ciniglia, C., Pinto, G. and Bhattacharya, D. (2004). A molecular time-line for the origin of photosynthetic eukaryotes. *Mol. Biol. Evol.* **21**, 809–18.

Yoon, H. S., Müller, K. M., Sheath, R. G., Ott, F. D. and Bhattacharya, D. (2006). Defining the major lineages of red algae (Rhodophyta). *J. Phycol.* **42**, 482–92.

Yoshii, Y. (2006). Diversity and evolution of photosynthetic antenna systems in green plants. *Phycol. Res.* **54**, 220–29.

Yoshii, Y., Takaichi, S., Maoka, T. and Inouye, I. (2002). Characterization of two unique carotenoid fatty acid esters from *Pterosperma cristatum* (Prasinophyceae, Chlorophyta). *J. Phycol.* **38**, 297–303.

Yoshii, Y., Takaichi, S., Maoka, T. and Inouye, I. (2003). Photosynthetic pigment composition in the primitive green alga *Mesostigma viride* (Prasinophyceae): phylogenetic and evolutionary implications. *J. Phycol.* **39**, 570–76.

Yoshii, Y., Takaichi, S., Maoka, T., Suda, S., Sekiguchi, H., Nakayama, T. and Inouye, I. (2005). Variation of siphonaxanthin series among the genus *Nephroselmis* (Prasinophyceae, Chlorophyta), including a novel primary methoxy carotenoid. *J. Phycol.* **41**, 827–34.

Zapata, M. (2005). Recent advances in pigment analysis as applied to picoplankton. *Vie et Milieu* **55**, 233–48.

Zapata, M., Jeffrey, S. W., Wright, S. W., Rodríguez, F., Garrido, J. L. and Clementson, L. (2004). Photosynthetic pigments in 37 species (65 strains) of Haptophyta: implications for oceanography and chemotaxonomy. *Mar. Ecol. Prog. Ser.* **270**, 83–102.

Zapata, M., Garrido, J. L. and Jeffrey, S. W. (2006). Chlorophyll *c* pigments: current status. In *Chlorophylls and Bacteriochlorophylls: Biochemistry, Biophysics and Biological Function*, ed. B. Grimm, R. J. Porra, W. Rüdiger and H. Scheer. Dordrecht: Springer, pp. 39–53.

2

Recent advances in chlorophyll and bacteriochlorophyll biosynthesis

ROBERT J. PORRA, ULRIKE OSTER AND HUGO SCHEER

2.1 Introduction

In 1997, the first edition of *Phytoplankton Pigments in Oceanography* was published (Jeffrey *et al.*, 1997) in which a chapter was presented by Porra *et al.* (1997) on photosynthetic pigments, namely, chlorophylls (Chls), phycobilins and carotenoids. In this current volume, the phycobilins and carotenoids are discussed elsewhere (Chapters 9 and 3, this volume). The earlier presentation (Porra *et al.*, 1997) only briefly described the Chl biosynthetic pathway while addressing the functions and locations of Chls in protein complexes of both the light-harvesting antenna complexes and reaction centres of the two photosystems present in the chloroplasts of higher plants and green algae. A comprehensive survey of Chl biodegradation was also described in the earlier presentation (Porra *et al.*, 1997) but more recent information is now available (Kräutler and Hörtensteiner, 2006). In this current chapter, a more comprehensive account of the Chl biosynthetic pathway is presented together with the structures of many of the naturally occurring Chls, with a special focus on recently discovered Chls and their possible syntheses.

2.2 Structures of chlorophylls

The structures and properties of naturally occurring Chls have been extensively reviewed (Scheer, 1991, 2003, 2006). The Chls are mostly magnesium coordination complexes, but also rarely Zn-coordination complexes (see Sections 2.2.3 and 2.4.10.3), of cyclic tetrapyrroles which contain a fifth isocyclic (cyclopentanone) ring E constructed enzymically from the 13-propionate side chain of Mg-protoporphyrin IX (see Section 2.4.3): for the IUPAC-IUB tetrapyrrole atom numbering and ring labelling systems, see Figure 2.1B. All Chls possess a 13^1-oxo group and mostly, but not always, a 13^2 methylcarboxylate substituent while the 17-propionate side chain is usually, but not always, esterified with a long-chain isoprenoid alcohol such as phytol, farnesol or geranyl-geraniol.

Phytoplankton Pigments: Characterization, Chemotaxonomy and Applications in Oceanography, eds. Suzanne Roy, Carole A. Llewellyn, Einar Skarstad Egeland and Geir Johnsen. Published by Cambridge University Press.

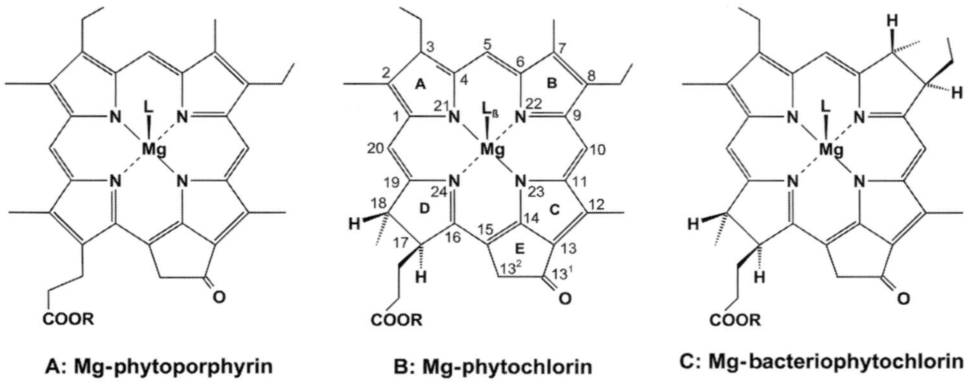

Figure 2.1. General structures of chlorophylls: oxidation states of the three types of chlorophyll macrocycles. Atom numbering and ring labelling systems according to IUPAC-IUB are indicated in the Mg-phytochlorin structure B (Moss, 1988), the omnipresent ligand (L) to the central Mg is shown above the macrocycle plane (i.e. the β-position).

The IUPAC-IUB nomenclature (Moss, 1988) distinguishes the 5-ring macrocycle of the three Chl types (Figure 2.1) from the 4-ring porphyrin or chlorin macrocycles by adding the prefix 'phyto' to form 'phytoporphyrin' and 'phytochlorin', where 'chlorin' denotes a 4-ring [17,18-dihydro]-porphyrin. Bacteriochlorin is a still further reduced compound, defined by IUPAC-IUB as a 4-ring [7,8-dihydro]-chlorin: the 5-ring macrocycle of bacteriochlorophyll (BChl) *a*, which is a [7,8-dihydro]-phytochlorin derivative, therefore, logically becomes a 'bacteriophytochlorin' macrocycle. As defined by IUPAC-IUB, the phytoporphyrin, phytochlorin and bacteriophytochlorin-type macrocycles possess a 13^1-oxo group but no 13^2 methylcarboxylate side chain (Moss, 1988). Thus, most members of the three naturally occurring Chl groups are more correctly described as the 13^2-methylcarboxylates of Mg-phytoporphyrin, Mg-phytochlorin or Mg-bacteriophytochlorin derivatives (see Figure 2.1); for convenience, they will be referred to in the text by their basic macrocycle type: phytoporphyrin, phytochlorin and bacteriophytochlorin. The BChls *c, d* and *e* are a special case, as their names relate to their origin from green photosynthetic bacteria; nonetheless, BChls *c, d* and *e* are phytochlorins not bacteriophytochlorins and, further, they do not possess a 13^2-methylcarboxylate group (see Section 2.2.2).

2.2.1 Phytoporphyrin-type chlorophylls

The phytoporphyrin Chls (see Figure 2.1A) comprise the Chl *c* family which are Mg-phytoporphyrin-13^2-methylcarboxylate derivatives possessing a fully unsaturated macrocycle and they have absorption spectra absorbing moderately at ~620 nm and approximately 10-fold more strongly in the Soret region (400–450 nm). The Chls *c* function biologically as antenna Chls in marine chromophyte algae and some prokaryotes (certain cyanobacteria). In the Chl *c* family, the 17-propionate side chain

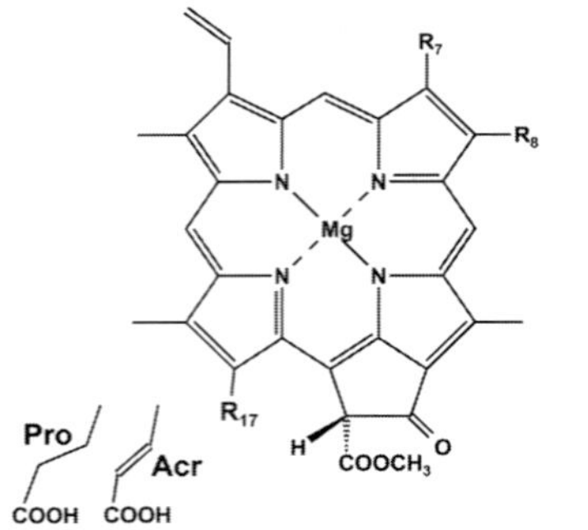

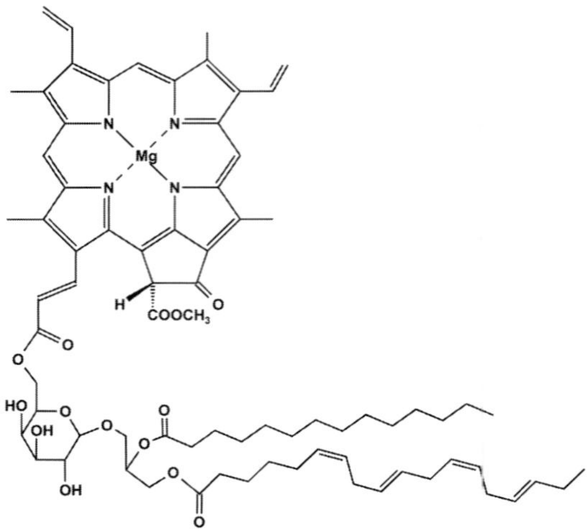

(14:0,18:4)MGDG-Chl c_2 [3)]

Pigment	R_7	R_8	R_{17}	Remarks
Chl c_1	CH_3	C_2H_5	Acr	1
Chl c_2	CH_3	C_2H_3	Acr	1
Chl c_3	$COOCH_3$	C_2H_3 or C_2H_5	Acr	1
[8-Vinyl]-PChlide a	CH_3	C_2H_3	Pro	1,2
Chl $c_{3(CS\text{-}170)}$	CH_3	C_2H_3	Pro	1
PChlide a	CH_3	C_2H_5	Pro	-

Figure 2.2. Structures of chlorophylls of the phytoporphyrin type. In the *c*-type chlorophylls, as in the chlorophyll precursors, divinyl protochlorophyllide *a* and proto chlorophyllide *a*, ring D is not reduced. The substituent at C-17 is an acrylic (Acr) or (rarely) a propionic acid group (Pro); they are generally unesterified; therefore, the pigments mostly resemble protochlorophyllides. An exception is the Chl c_2 monogalatosyl-diacylglyceride ester, a protochlorophyll, shown at right. Chl c_3 exists in mono- and divinyl forms. *Remarks*: 1, Unknown stereochemistry at C-13^2; 2. Also known as divinyl-PChlide *a* or Mg-DVP; 3. Also with two myristic (14:0) fatty acid chains.

is mostly, but not always, dehydrogenated to an acrylate which is mostly, but not always, unesterified thus partially qualifying the chemical status of 'chlorophyllide' (Chlide) to most of the Chl *c* family. Moreover, as all Chls *c*, without exception, are not yet hydrogenated at C-17/C-18 (compare Section 2.2.2), they are more accurately described chemically as 'protochlorophyllides' (PChlide, if unesterified) or 'protochlorophylls' (PChl, when esterified). Examples of the latter are the non-polar forms of Chl c_2 which have been isolated by new HPLC techniques and identified as 17^3-monogalactosyldiacylglyceride (MGDG) esters of the 17-acrylic acid side chain: the diacyl moieties of the galactolipid are octadecatetraenoic (18:4) and/or myristic (14:0) fatty acids (see Figure 2.2). The Chl *c* family is an ever-growing family but some of the known structures are presented in Figure 2.2. For a contemporary review of the Chl *c* family see Zapata *et al.* (2006).

2.2.2 Phytochlorin-type chlorophylls

The phytochlorin-type Chls (see Figure 2.1B) are Mg-phytochlorin-13^2-methyl-carboxylate derivatives in which ring D is *trans*-reduced at C-17 and -18. This Chl group includes the well-known Chls *a* and *b* of higher plants, cyanobacteria and green algae, Chl *d* of some free-living or epiphytic (living on red algae) oxygenic prokaryotes (see Figure 2.3A) as well as the recenlty discovered Chl *f* (see Addendum). As already discussed, the green BChls *c*, *d* and *e* from the green photosynthetic bacteria (see Figure 2.3B) are phytochlorins but have no 13^2-methylcarboxylate group (see Section 2.2): this suggests that the green photosynthetic bacteria possess a decarboxymethylase (Frigaard *et al.*, 2006). The phytochlorin-type Chls, in solution, have two strong absorption bands at 400–470 nm and 640–700 nm. Figures 2.3A and B show the structures of some of the Chls and BChls of this group, respectively.

2.2.3 Bacteriophytochlorin-type chlorophylls

The bacteriophytochlorin-type chlorophylls (see Figures 2.1C and 2.4) are Mg-bacteriophytochlorin-13^2-methylcarboxylate derivatives in which rings B and D are *trans*-reduced at C-7, -8, and C-17, -18, respectively. The BChls *a*, *b* and *g* (see Figure 2.4; M = Mg) belong to this group and are found mainly in the purple photosynthetic bacteria and heliobacteria. The bacteriophytochlorin-type Chls possess two strong absorption bands in the near ultraviolet (~360–400 nm) and the near infrared (~770–795 nm). The BChl *a* in which the central Mg has been replaced by Zn (M = Zn in Figure 2.4) has been isolated from *Acidiphilium (Acp.) rubrum* in acidic habitats (see below) and, to date, is the only transmetallated Chl which is functional in photosynthesis (see Kobayashi *et al.*, 2006a, b; Küpper *et al.*, 2006).

Some spectrophotometric data for the many chlorophylls of the three macrocycle types are presented in Table 2.1, and for more data see Jeffrey *et al.* (1997), Porra (2006) and Zapata *et al.* (2006).

2.3 Biosynthesis of protoporphyrin IX

Many reviews of the biosynthesis of various naturally occurring tetrapyrrole pigments are available (Jordan, 1991a; Battersby, 1994; Chadwick and Ackrill, 1994; Porra, 1997; Bollivar, 2006; see Chapters 10–15 in Grimm *et al.*, 2006; Willows and Kriegel, 2008). The formation of these pigments, including the iron-containing haems, the linear-tetrapyrrolic bile pigments (such as phycobilins formed by oxidative cleavage of the haem macrocycle) and the magnesium-containing Chls and BChls, all share common enzymic steps converting 5-aminolevulinic acid (ALA), the first committed intermediate in tetrapyrrole biosynthesis, to uroporphyrinogen (urogen) III which is a branch point for formation of Ni-containing coenzyme F430, of sirohaem and of cobalt containing vitamin B_{12} (Figure 2.5). The largely common pathway continues to protoporphyrin (proto) IX, where a short iron branch leads through haem, and on to bilins, and a longer Mg branch leads to the Chls (Figure 2.5). Together, the ALA to proto IX reaction sequence followed by the Mg-branch of the pathway forms a long series of reactions (Figure 2.5).

(A)

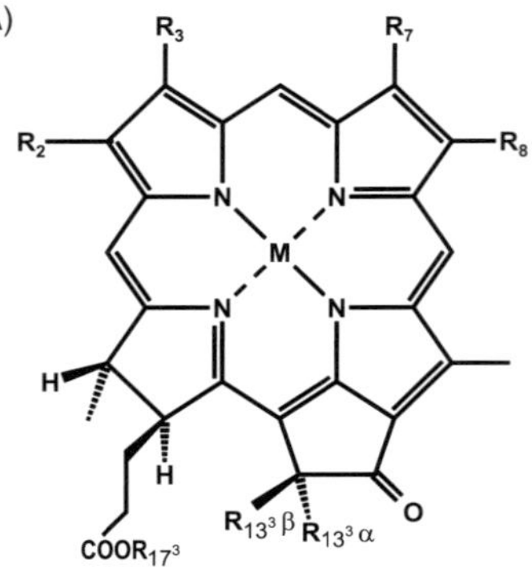

Phy =

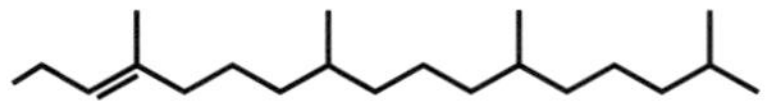

Pigment	R_3	R_3	R_7	R_8	$R_{13^3\beta}$	$R_{13^3\alpha}$	R_{17^3}	M
Chl a	CH_3	C_2H_3	CH_3	C_2H_5	H	$COOCH_3$	Phy	Mg
Chl a'	CH_3	C_2H_3	CH_3	C_2H_5	$COOCH_3$	H	Phy	Mg
Chlide a	CH_3	C_2H_3	CH_3	C_2H_5	H	$COOCH_3$	H	Mg
Chl b	CH_3	C_2H_3	CHO	C_2H_5	H	$COOCH_3$	Phy	Mg
Chl d	CH_3	CHO	CH_3	C_2H_5	H	$COOCH_3$	Phy	Mg
Chl f	CHO	C_2H_3	CH_3	C_2H_5	H	$COOCH_3$	Phy	Mg
[8-Vinyl]-Chl a	CH_3	C_2H_3	CH_3	C_2H_3	H	$COOCH_3$	Phy	Mg
[8-Vinyl]-Chl b	CH_3	C_2H_3	CHO	C_2H_3	H	$COOCH_3$	Phy	Mg
8^2-OH-Chl a	CH_3	C_2H_3	CH_3	C_2H_4OH	H	$COOCH_3$	Phy	Mg
Phe a	CH_3	C_2H_3	CH_3	C_2H_5	H	$COOCH_3$	Phy	H_2

(B)

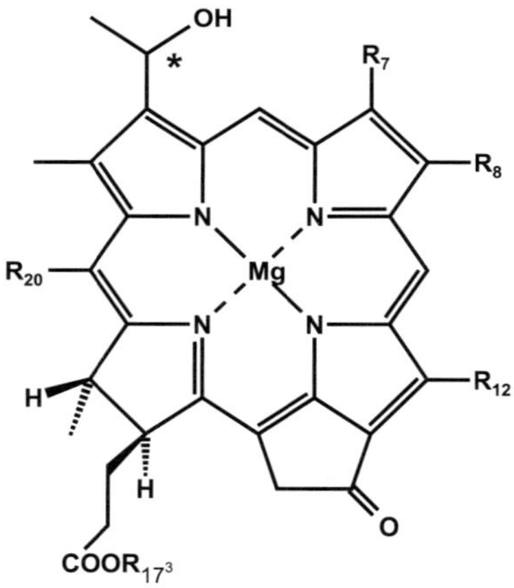

Pigment	R_7	R_8	R_{12}	R_{17^3}	R_{20}
BChl c	CH_3	et,pr,ib,np	me,et	Farn,18:0,many others	CH_3
BChl d	CH_3	et,pr,ib,np	me,et	Farn,18:0,many others	H
BChl e	CHO	et,pr,ib,np	me,et	Farn,18:0,many others	CH_3

Figure 2.3. Structures of chlorophylls of the phytochlorin type (with reduced ring D). (A) Plant and green algal chlorophylls belong to this group, they are mostly esterified with phytol (Phy); (B) BChls *c*, *d* and *e* from the chlorosomes of green photosynthetic bacteria (sometimes designated 'Chlorobium Chls'), they lack the 13^2-methylcarboxylate group and are mainly esterified to farnesol. *Abbreviations*: Farn, farnesyl; 18:0, stearyl; et, ethyl; ib, isobutyl; np, neopentyl; pr, propyl. $C3^1$ (see asterisk) can occur as R and S epimers.

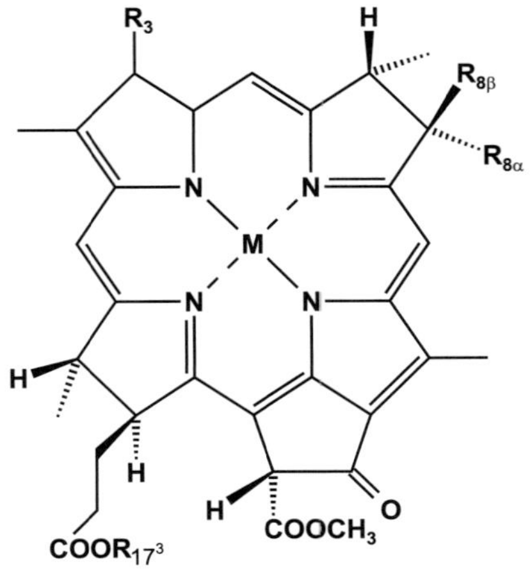

Pigment	R_3	$R_{8\alpha}$	$R_{8\beta}$	R_{17^3}	M
BChl a	$COCH_3$	H	CH_3	Phy or other	Mg (or Zn)
BPhe a	$COCH_3$	H	CH_3	Phy or other	H_2
BChl b	$COCH_3$	=CH-CH_3		Phy or other	Mg
BPhe b	$COCH_3$	=CH-CH_3		Phy or other	H_2
BChl g	C_2H_3	=CH-CH_3		Phy or other	Mg

Figure 2.4. Structures of chlorophylls of the bacteriophytochlorin type (with reduced rings B and D). The 17-propionic acid side chain may be esterified to phytol Phy), farnesol (Farn) or other alcohols.

Many of the porphyrin- and chlorin-type intermediates are phototoxic, but this long pathway is tightly regulated and intermediates occur only at very low concentrations in healthy cells.

2.3.1 Formation of 5-aminolevulinic acid and its regulation

Figures 2.5 and 2.6 show that ALA can be formed *either* by condensation of succinyl-CoA and glycine (Gibson *et al.*, 1958) in the C_{4+1} pathway (i.e. the Shemin or ALA synthase pathway), *or* from the intact carbon skeleton of glutamate by the C_5-pathway (Beale *et al.*, 1975).

Figure 2.6A shows that in the C_{4+1} pathway, the methylamino group of glycine is condensed with succinyl-CoA from the tricarboxylic acid (TCA) cycle in a pyridoxal-phosphate-dependent reaction catalysed by ALA-synthase thus forming ALA and releasing CO_2 and CoASH (Jordan, 1991b). In this pathway, the C-5 of ALA arises from the C-2 of glycine (Oh-hama *et al.*, 1985b; see Figure 2.6A). The C_{4+1} pathway was discovered long before the C_5 pathway, but it is far less common in nature (see Sections 2.3.2 and 2.3.3).

Table 2.1. *Spectroscopic properties of some major chlorophylls of all three macrocycle types.*

Pigment	Solvent	Major absorption bands: λ_{max} [nm]; (ε_{mM} [l·mM^{-1}·cm^{-1}])						A_{Soret}:A_{Qy}	Reference
A	**Phytoporphyrin-type pigments**								
Chl c_1	Diethyl ether	444 [9.93][a]			578 [0.67]	628 [1.00]		9.93	Jeffrey (1969)
	Acetone/pyridine[c]	446.1 (213)[b]				629.1 (23.9)		8.91	Jeffrey (1972)
Chl c_2	Diethyl ether	448 [14.1][a]			582 [1.15]	629 [1.00]		14.1	Jeffrey (1969)
	Acetone/pyridine[c]	444.6 (195)[b]				629.6 (22.7)		8.59	Jeffrey (1972)
Chl c_3	Diethyl ether	451.3 [32.1][a]			584.5 [3.79]	625.9 [1.00]		32.1	Jeffrey and Wright (1987)
B	**Phytochlorin-type pigments**								
Chl *a*	Diethyl ether	410 (76.1)	430 (117.5)	533.5 (3.8)	578 (8.3)	615 (14.6)	662 (90.2)	1.30	Smith and Benitez (1955)
	Acetone	410.6 (74.3)	429.6 (100.4)	534.2 (3.9)	579 (8.9)	616 (15.4)	661.6 (82.6)	1.22	Lichtenthaler (1987)
DVChl *a*[g]	Diethyl ether		436 [1.4][a]				661 [1.0]	1.4	Bazzaz (1981)
	Diethyl ether		435			615	660		Goericke and Repeta (1992)
Chl *b*	Diethyl ether	430 (56.9)	455 (158.62)	549 (6.42)	595 (11.52)		644 (56.3)	2.82	Smith and Benitez (1955)
	Acetone		455.8 (131.9)		595.8 (11.0)		644.8 (46.9)	2.81	Lichtenthaler (1987)
DV-Chl *b*[g]	Diethyl ether		462 [3.4][a]				648 [1.0]	3.4	Bazzaz (1981)
	Diethyl ether		460		595		644		Goericke and Repeta (1992)
Chl *d*	Diethyl ether	447 (87.6)					688 (98.9)	0.86	Smith and Benitez (1955)
		445.6 [0.85][a]					686.2 [1.0]	0.85	Kobayashi *et al.* (2006b)
BChl *c*	Diethyl ether	412 (72.6)	432 (14)			622 (15.9)	660 (91.0)	1.56	Stanier and Smith (1960)
(809)[d]	Acetone	413 (71.8)	433 (115.5)			625 (13.7)	662.5 (74.9)	1.54	Stanier and Smith (1960)
BChl *d*	Diethyl ether	406 (69)	425 (115.6)	530 (2.7)	575 (6.8)	612 (12.5)	650 (89.9)	1.28	Stanier and Smith (1960)
(792)[d]	Acetone	406 (69.6)	427 (99.7)	530 (3.1)	575 (7.3)	612.5 (13.0)	654 (77.6)	1.28	Stanier and Smith (1960)
BChl *e* (835.1)[e]	Acetone		462 (185.0)				649 (48.9)	3.78	Borrego *et al.* (1999)

C	Bacteriophytochlorin type pigments							
BChl *a*	Diethyl ether	357 (73.4)	392 (47.1)	573 (22.0)		770 (96.0)	0.76	Sauer *et al.* (1966)
	Acetone	358 (65.7)		576.5 (19.4)		770 (69.2)	0.94	Sauer *et al.* (1966)
[Zn]-BChl *a*	Ether	353 (58.9)	389 (39.7)	558 (18)		762 (67.7)		Hartwich *et al.* (1998)
		353.5 [0.79][a]				763.2 [1.0]		Kobayashi *et al.* (2006b)
BChl *b*	Diethyl ether	368 (81)	408 (78)	578 (25)	676 (18)	794 (100)	0.81	Steiner (1981)
						794 (106)		Scheer and Steiner (1985)[f]
	Acetone	368 (94)	407 (82)	580 (27)		794 (100)	0.94	Baumgarten (1970)
BChl *g*	Diethyl ether[a]	364.4 [0.94][a]				767.4 [1.0]	0.94	Kobayashi *et al.* (2006b)
		364 (90)				767.2 (96)	0.94	Kobayashi *et al.* (1991)
	Acetone	364.8	404.8	566.4		761.6 (76)		van de Meent *et al.* (1991)

[a] Band ratios relative to Q_y band in square brackets []

[b] ε_{mM} values in round brackets ()

[c] Acetone (99%)/pyridine (1%)

[d] These Mr values were used to calculate ε_{mM} coefficients from the a coefficients of Stanier and Smith (1960)

[e] To calculate ε_{mM} values, Borrego *et al.* (1999) used Mr = 835.1 for the [8-propyl-12-ethyl-17^3-farnesyl]-homologue

[f] This ε_{mM} coefficient, determined by Scheer and Steiner in 1985, was cited as a private communication in Oelze (1985).

[g] DV-Chls *a* and *b* also known as [8-vinyl]-Chls *a* and *b*

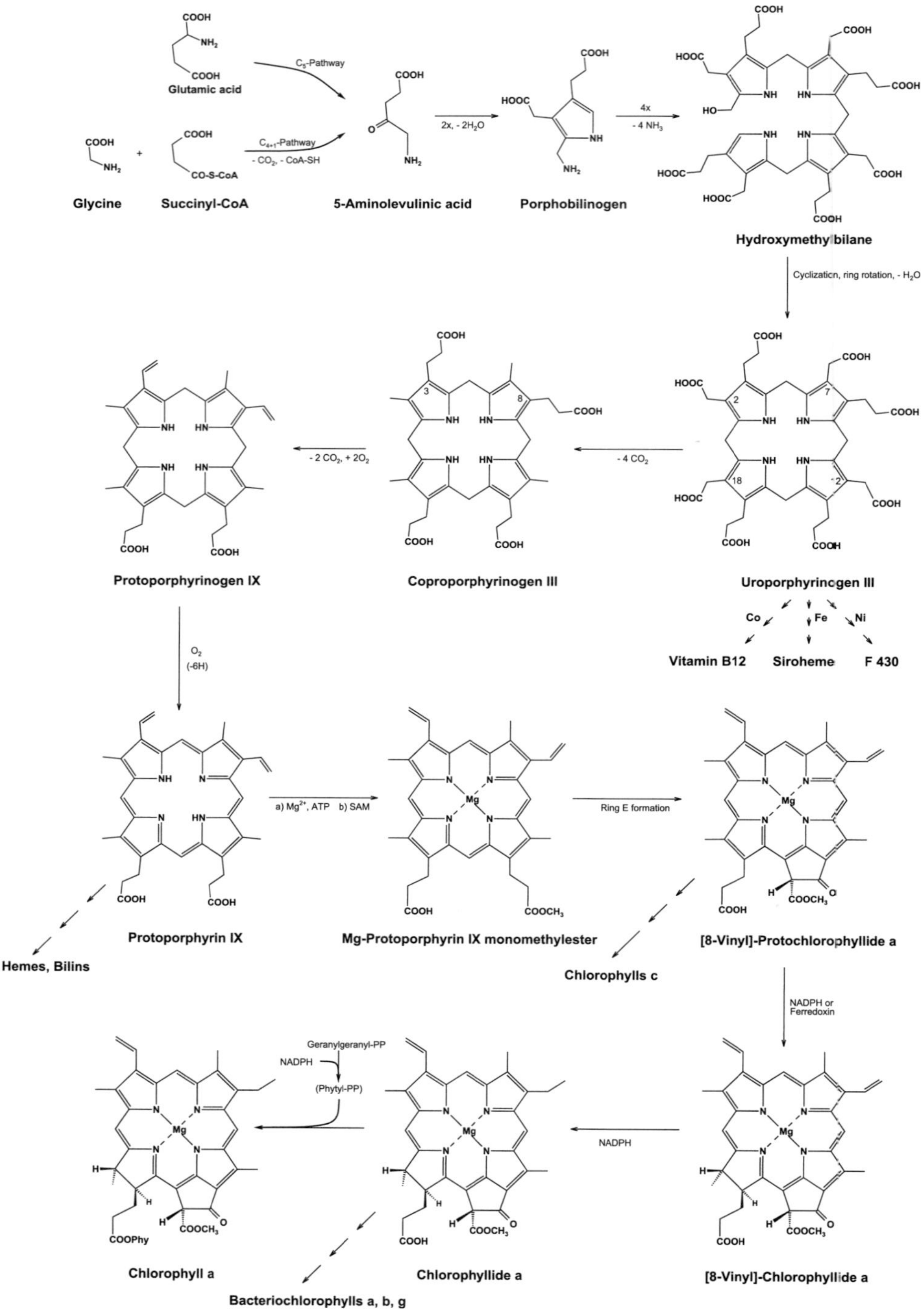

Figure 2.5. The pathway of tetrapyrrole biosynthesis. Overview of chlorophyll biosynthesis with bifurcation points for the formation of other tetrapyrrolic pigments.

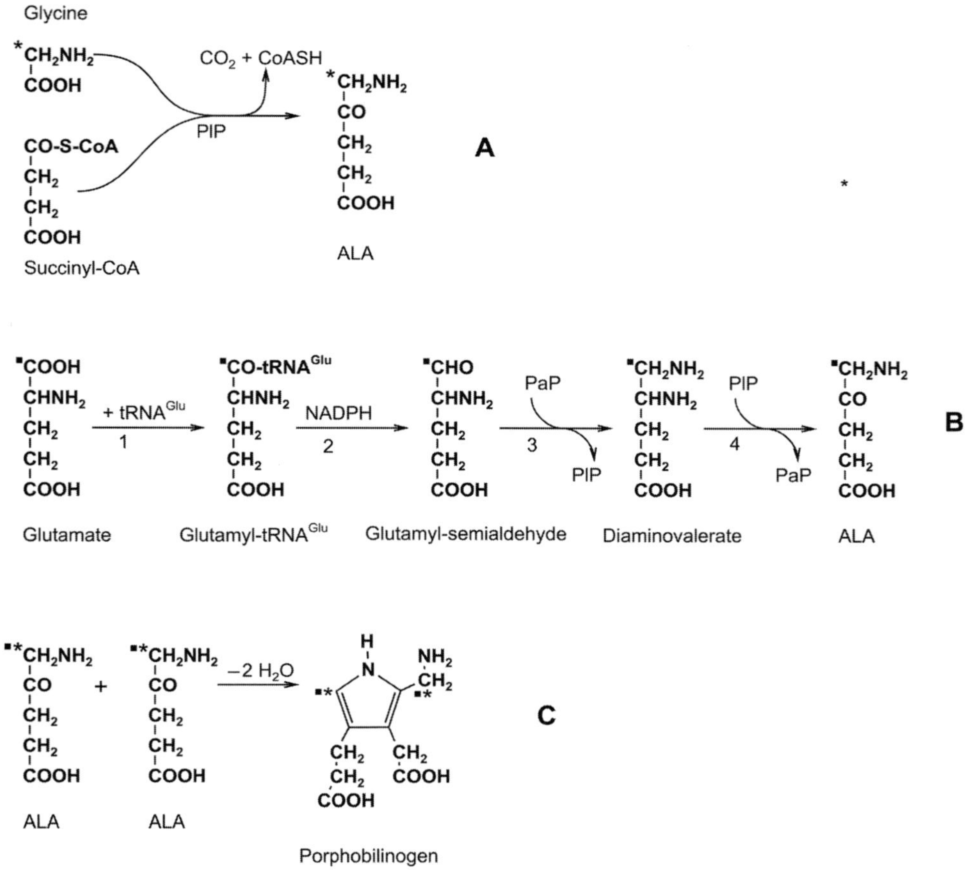

Figure 2.6. Formation of 5-aminolevulinate (ALA) and monopyrrolic porphobilinogen (PBG). (A) C_{4+1} (Shemin) pathway from glycine and succinyl-CoA. (B) C_5 pathway from glutamate; for both pathways, the origin of the methylamino and α-pyrrole carbons of PBG is indicated. (C) Conversion of ALA to PBG. *Abbreviations*: PAP, pyridoxal phosphate; PIP, pyridoxamine phosphate. From Porra and Scheer (2000), with kind permission from Springer Science+Business Media.

In the C_5 pathway (Figure 2.6B), all five carbons of glutamate are incorporated intact into ALA. The C-5 of ALA in this pathway arises from the C-1 of glutamate (Oh-hama *et al*., 1982, 1985a; Porra *et al*., 1982, 1983; see Figure 2.6B). The asterisks (C-2 of glycine) and solid squares (C-1 of glutamate) show the different origins of the methylamino- and α-pyrrole-carbons of porphobilinogen (PBG) depending on the operation of the C_{4+1} or C_5 pathways (see Figure 2.6C). The C_5 pathway requires ATP, pyridoxal phosphate, $tRNA^{Glu}$ and NADPH, and, unlike the C_{4+1} route, it is inhibited by gabaculine, an inhibitor of glutamyl-semialdehyde aminotransferase (see Figure 2.6B; reactions 3 and 4). Thus gabaculine provides a simpler method than isotope labelling for distinguishing the two pathways (Kannangara *et al.*, 1994).

Importantly, both pathways of ALA formation are rate-limiting steps in the formation of haems and Chls thus contributing to the economy and safety of pigment formation referred to above. In the C_{4+1} and C_5 pathways, the rate-limiting enzymes are ALA-synthase (Figure 2.6A) and glutamyl-tRNA^{Glu}-reductase (see Figure 2.6B; reaction 2), respectively. In both pathways, these enzymes experience feedback inhibition by biosynthetic end-products (haem) and to a lesser extent by Mg-protoporphyrin IX (Mg-proto), protochlorophyllide (PChlide) and chlorophyllide (Chlide), and they are transcriptionally regulated by environmental conditions (light/dark and ambient O_2 tension) and by other metabolites (Beale, 2006; Yaronskaya and Grimm, 2006). Induction of ALA-synthase formation was observed in the presence of some drugs, hormones and foreign chemicals (Granick, 1966) including industrial pollutants like polyhalogenated aromatic compounds (Strik, 1973) leading to excessive phototoxic porphyrin formation. Frequently, non-limiting ALA formation leads to formation of uro-and coproporphyrins of the isomer I series because uroporphyrinogen co-synthase then becomes a rate-limiting step (see Section 2.3.6).

2.3.2 Distribution of the C_{4+1} pathway

The distribution of the C_{4+1} and C_5 pathways has been discussed in detail by Kannangara *et al.* (1988), Avissar *et al.* (1989), Avissar and Moberg (1995) and Beale (2006): a brief summary follows.

The C_{4+1} pathway of ALA formation operates in many *eukaryotic* cells including man and all mammalian cells, in yeast cells and in at least two algae, *Euglena gracilis* (Beale *et al.*, 1981) and *Scenedesmus obliquus* (mutant C-2A′) (Drechsler-Thielmann *et al.*, 1993) where ALA-synthase is involved in mitochondrial haem biosynthesis but not in Chl formation (compare Section 2.3.3). The pre-enzyme form of ALA-synthase is cytosolic, but the mature enzyme is located in the mitochondria. The C_{4+1} pathway is also present in many *prokaryotic* organisms. It is found in the α- but not the β-, γ- and δ-subdivisions of the purple photosynthetic bacteria which use the C_5 pathway (Avissar *et al.*, 1989): the α-group includes *Rhodobacter (Rba.) sphaeroides* and *Rhodospirillum (Rsp.) rubrum* where it is involved in both haem and BChl formation. This pathway is also found in non-photosynthetic bacteria including species of *Bradyrhizobium, Rhizobium, Spirillum, Pseudomonas, Propionobacter* and *Streptomyces* as well as in Archeae.

2.3.3 Distribution of the C_5 pathway

The more ubiquitous C_5 pathway of ALA formation (Kannangara *et al.*, 1988; Avissar *et al.*, 1989; Avissar and Moberg, 1995; Beale, 2006) is thought evolution-wise to be the older pathway and to have developed in primitive prokaryotic (cyanobacterial) cells from which chloroplasts are thought to have been derived: these cells existed in a primordial O_2-sparse atmosphere and with an incomplete TCA cycle, unable to supply succinyl-CoA from oxoglutarate oxidation, for the ALA-synthase pathway. This C_5 pathway operates in both *eukaryotic* and *prokaryotic* cells. In higher plants, the C_5 pathway participates in the formation of both Chl formation and all cellular haem. In all orders of eukaryotic photosynthetic algal cells so far examined, the chloroplast-located C_5 route operates for Chl and haem formation. There are, however, two exceptions known: *Euglena gracilis* and *Scenedesmus obliquus* (mutant C-2A′) use the C_5 pathway for Chl and chloroplast haem biosynthesis but the C_{4+1} pathway provides ALA for mitochondrial haem (compare Section 2.3.2). Indirect support for the C_5 pathway in phytoplankton comes from a study by Sachs *et al.* (1999). They observed a Chl *a* nitrogen isotope (^{15}N) depletion of 5% relative to total cellular nitrogen in eight phytoplankton species and proposed that this occurred during transamination of glutamate in ALA formation by the C_5 pathway. The C_5 route has been demonstrated in *Cyanidium caldarium* (Rhodophyta), *Camptothecium* sp. (Bryophyta), *Prochlorothrix hollandica* (Prochlorotrichaceae), *Synechococcus* sp. and *Synechocystis* sp. (Cyanobacteria) and most photosynthetic bacteria including *Chlorobium, Chloroflexus, Heliospirillum* and *Chromatium.* The C_5 pathway is also found in many non-photosynthetic Eubacteria including *Escherichia coli, Salmonella typhimurium, Bacillus subtilis, Clostridium thermoaceticum, Clostridium tetanomorphum* actinobacteria and in Archeae (*Methanobacterium* sp.).

2.3.4 Conversion of 5-aminolevulinate to monopyrrolic porphobilinogen

Condensation of two molecules of ALA, by a Knorr-type condensation, is catalysed by ALA-dehydratase (Jordan, 1991b) to form monopyrrolic porphobilinogen (PBG) with acetate and propionate side chains (see Figure 2.6C): PBG is common to the formation of all tetrapyrrolic pigments. ALA dehydratase is octameric in mammalian cells, in yeast and in *E. coli* and binds two Zn^{2+} ions per unit while many bacterial enzymes including *Rba. sphaeroides* require Mg^{2+} ions. The chloroplast enzyme is a hexamer and also binds Mg^{2+} ions.

2.3.5 Formation of uroporphyrinogen III from porphobilinogen

This conversion has been extensively reviewed (Battersby and Leeper, 1990; Jordan, 1991b). The reaction is catalysed by two interesting enzymes, PBG-deaminase and uroporphyrinogen (urogen) III-cosynthetase. The PBG-deaminase is unusual because it synthesizes its own enzyme-bound di-pyrrolic (C1 and C2) co-enzyme (see Figure 2.7) with the release of two molecules of NH_3. A further four molecules of PBG (A, B, C and D) are then sequentially added by deamination (see Figure 2.7), releasing four molecules of NH_3, forming an enzyme-bound hexapyrrole which splits between rings A and C2 to regenerate the enzyme-bound dipyrrole complex and

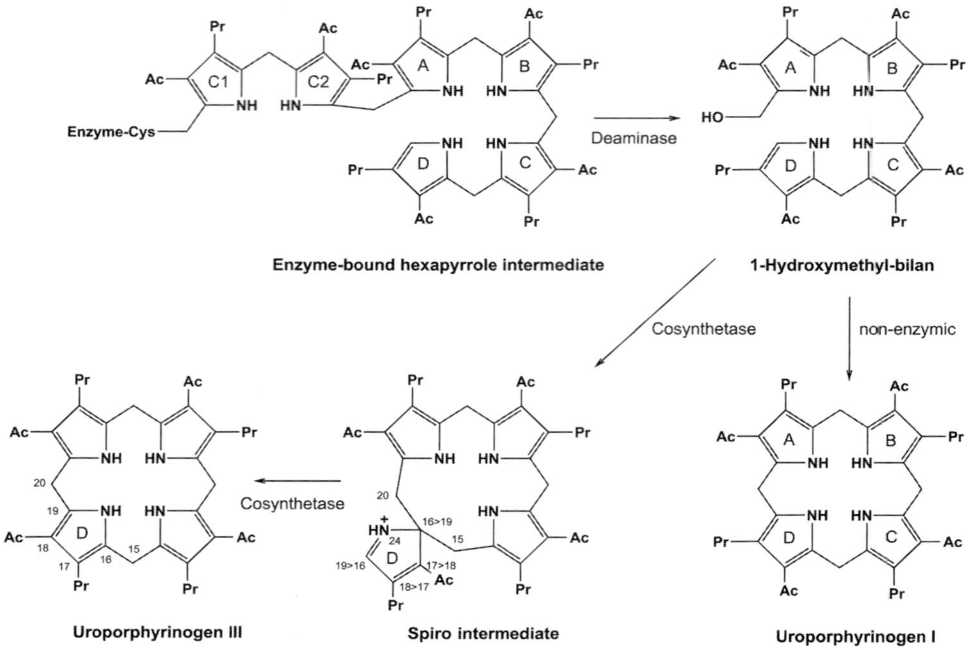

Figure 2.7. Formation of uroporphyrinogen III by PBG-deaminase and uroporphyrinogen III cosynthase. By stepwise deamination, six pyrroles are covalently attached to PBG-deaminase, of which two remain bound (C1, C2). The linear tetrapyrrole, hydroxymethylbilane I (rings A–D) is released to uroporphyrinogen (urogen) III cosynthase which converts it, via a spiro intermediate, to urogen III. *Abbreviations*: Ac, acetate; Pr, propionate

release the linear octacarboxylic tetrapyrrole, hydroxymethylbilane (HMB) I. This series I isomer, HMB, possesses a regular, alternating sequence of acetate and propionate side chains from rings A to D (see Figure 2.7). Urogen III-cosynthetase is closely associated with PBG-deaminase; it binds the released HMB I and converts it to cyclic urogen III by rotating ring D about *spiro* C-16 so reversing C-16 and C-19 of HMB I to become C-19 and C-16 of urogen III after they bond to C-20 and C-15 (see Figure 2.7). Urogen III, because the acetate and propionate side chains on ring D of HMB I are reversed during cyclization, is an asymmetric isomer. In the absence of the cosynthetase, HMB I spontaneously cyclizes non-enzymically to urogen I. When ALA formation becomes non-limiting, urogen III cosynthetase becomes the limiting step and urogens I and III are formed which can be enzymically converted to coproporphyrinogens (coprogens) I and III. Coprogen I accumulates since it can be metabolized no further; however, like all porphyrinogens, it is readily oxidized under aerobic conditions to form phototoxic coproporphyrin (copro) I.

Urogen III is the bifurcation point for the formation of the cobalt, nickel and certain iron tetrapyrroles, namely, the cobalamines, coenzyme-F_{430} and sirohaem (see Figure 2.5)

2.3.6 Decarboxylation of uroporphyrinogen III to coproporphyrinogen III

This conversion has been reviewed by Akhtar (1991, 1994). The structures of hepta-, hexa-, penta-carboxylic porphyrins from the urine of porphyric patients suggests that the acetate substituents of urogen III at C-18, C-2, C-7 and C-12 on rings D, A, B and C, respectively, were enzymically decarboxylated to methyl groups in that clockwise order to form tetracarboxylic coprogen III: the porphyrins in the urine were aerobic oxidation products of the corresponding porphyrinogens (i.e. hexahydroporphyrins). When erythrocyte lysates were offered urogen III as substrate a random pattern of decarboxylations was observed (Lash, 1991) but, with PBG as substrate, the clockwise pattern re-emerged (Luo and Lim, 1994). This suggests that the PBG-deaminase and urogen III cosynthase are closely associated with the urogen III decarboxylase and, in the presence of PBG, offer urogen III to the decarboxylase at or below the limiting rate so that the enzyme-substrate complexes are formed consistent with their affinity constants and thereby providing an ordered decarboxylation pattern; however, with high exogenous concentrations of urogen III, the decarboxylase could form complexes of lesser affinity so generating a random series of decarboxylations. Urogen III decarboxylase has a very low substrate specificity because it not only decarboxylates all isomers of hepta-, hexa- and penta-carboxylic porphyrinogen intermediates, but can also decarboxylate urogen I (formed from HMB I in the absence of urogen III cosynthetase) to coprogen I which then accumulates because it is not a substrate for coprogen III oxidase but is, instead, rapidly and non-enzymically oxidized under aerobic conditions to copro I.

2.3.7 Oxidative decarboxylation of coproporhyrinogen III to protoporphyrinogen IX

Coprogen III oxidase oxidatively decarboxylates the propionate side chain first at C-3 (ring A) and then at C-8 (ring B) to form vinyl groups (Akhtar, 1991; 1994). This reaction occurs aerobically in mammalian, avian and higher-plant cells and in some bacteria. In anaerobic organisms containing tetrapyrrolic pigments, the reaction still occurs, indicating alternative electron-acceptors to O_2. Sano and Granick (1961) proposed the removal of a hydride ion from the ß-carbon of the propionate group to an acceptor resulting in simultaneous decarboxylation and double bond formation. The enzyme has no known cofactors.

2.3.8 Oxidation of protoporphyrinogen IX to protoporphyrin IX

The non-fluorescing porphyrinogen intermediates, namely, urogen III, coprogen III and protogen IX have little or no characteristic absorption in the near UV/visible spectrum but, by shaking aqueous solutions aerobically, their orange-red fluorescing porphyrin oxidation products, uro III, copro III and proto IX, respectively, are formed. They have sharp characteristic absorptions in the visible spectral range and the near-UV, especially in monodisperse solution in organic solvents (see Table 2.1). For protogen IX, nonetheless, there is a flavoprotein enzyme, protogen IX oxidase (PPOX), which significantly accelerates this six-electron oxidation process. Protogen IX oxidase activity is found in both plant mitochondria and plastids (Jacobs and Jacobs, 1987; Smith *et al.*, 1993). The PPOX is inhibited by diphenyl ether and phthalimide herbicides. It is encoded by two genes forming two enzymes, PPOX I and PPOX II. While PPOX I, a plastid enzyme, is located on the stromal side of thylakoid membranes, PPOX II exists in two forms of 57 kDa (large) and 55 kDa (small). The PPOX IIL is located in the inner plastid envelope on the stromal side and PPOX IIS in the inner mitochondrial membrane. Protogen IX is converted in the chloroplast to proto IX by PPOX I and by PPOX IIL for Mg-proto IX and haem formation, respectively, within the chloroplast; however, protogen IX, formed in the chloroplast and transported to the mitochondrion is oxidized by mitochondrial PPOX IIS for mitochondrial haem (Rüdiger and Grimm, 2006). The herbicide inhibitors cause accumulation of protogen IX which leaks out of the plastid to be oxidized by non-specific, herbicide-resistant peroxidases forming excessive amounts of proto IX which are phototoxic in the light, generating reactive oxygen species (ROS) which cause cell destruction and death.

Proto IX is the bifurcation point for the iron and Mg branches of tetrapyrrole pigment formation. Proto IX with Fe^{2+} ions is converted by ferrochelatase to protohaem either for cytochrome prosthetic groups or for oxidative cleavage between C-4 and C-5 to form linear tetrapyrrolic phycobilins with liberation of C-5 as CO (see Figure 2.5). Phycobilins are distinguished from the bile pigments of Chl catabolism where cleavage also occurs between C-4 and C-5 but the C-5 is retained as an aldehyde group on ring B (compare Porra *et al.*, 1997). Proto IX is also the substrate of Mg-chelatase for Chl formation.

2.4 Biosynthesis of chlorophylls

This magnesium branch of tetrapyrrole biosynthesis (see Figure 2.5) culminates in the formation of all three groups of Chls possessing the phytoporphyrin, phytochlorin and bacteriophytochlorin macrocycles (see Section 2.2). The steps beginning with proto IX have recently been reviewed by Willows (2003), Frigaard *et al.* (2006), Rüdiger (2006) and Yaronskaja and Grimm (2006).

2.4.1 Formation of Mg-protoporphyrin IX by Mg-chelatase

While ferrochelatase is a single mitochondrial membrane-associated protein, encoded by the *hem15* gene, and requires no further cofactors, the enzymic insertion of Mg^{2+} into proto IX is far more complex. It is an energy-requiring process using Mg^{2+} and

ATP as cofactors. The complexity for Mg^{2+} insertion is reflected by the ease of insertion of biologically relevant metals into porphyrins, based on the order of stability, as follows: $Ni^{2+} > Co^{2+} > Cu^{2+} > Fe^{2+} > Zn^{2+} > Mg^{2+}$ (compare Phillips, 1963).

Insertion of Mg^{2+} into a porphyrin requires four steps: deprotonation of the pyrrole nitrogen atoms, removal of water molecules coordinated to the Mg^{2+} ion, coordination of the Mg^{2+} ion with the four pyrrole nitrogens and finally completion of the magnesium penta (or hexa)-coordination sphere with axial ligands (see Walker and Willows, 1997).

Magnesium-chelatase, unlike ferrochelatase, is a soluble enzyme system comprising three protein subunits encoded by three genes designated *bchI, bchD* and *bchH* in purple bacteria like *Rba. sphaeroides*, or *chlI, chlD* and *chlH* in higher plants and green algae. The reaction takes place in two steps: an activation step followed by a chelation step. It is suggested (Walker and Willows, 1997; Reid and Hunter, 2004) that the activation step involves a cofactor Mg^{2+} ion and an ATP-dependent dissociation of a D-subunit aggregate (possibly followed by phosphorylation of the single D subunit) which then associates with possibly two I subunits. In the chelation step, the H subunit, containing the bound proto IX and Mg^{2+} substrates, associates with the I_2–D complex. The newly formed H–I_2–D complex catalyses insertion of the Mg^{2+} ion which is accompanied by ATP hydrolysis. Walker and Willows (1997) propose that the required H_2O molecule comes from the coordination sphere of the Mg^{2+} ion, thus facilitating chelation.

The Mg-chelatase reaction is the first step in the Mg branch of Chl biosynthesis and is, therefore, in a prime position for regulation of the Mg branch. The Mg^{2+} ion and high ATP dependency as well as the complexity of the organization of the multi-subunit enzyme, together with the possible phosphorylation of the D subunit (see above) all offer many possibilities for regulatory processes. Substrate channelling of the phototoxic Mg-porphyrins may be another important aspect in the process (see Walker and Willows, 1997; Reid and Hunter, 2004; Yaronskaya and Grimm, 2006).

2.4.2 Formation of Mg-protoporphyrin IX-monomethylester by S-adenosylmethionine: Mg-protoporphyrin IX methyltransferase

Enzymically active preparations from wheat (Ellsworth *et al.*, 1974; Yee *et al.*, 1989), *Euglena gracilis* (Ebbon and Tait, 1969) and *Rba. sphaeroides* (Gibson *et al.*, 1963) catalysed methylation of the 13-propionate side chain of Mg-protoporphyrin IX (Mg-proto) in the presence of ATP and methionine forming Mg-protoporphyrin IX-monomethylester (Mg-protoMME). Examination by ^{13}C-NMR spectroscopy of Chls or BChls synthesized in the presence of [2–^{13}C]-glycine in green algae (Oh-hama *et al.*, 1982, 1985a) and during greening of etiolated higher plants (Porra *et al.*, 1982; 1983) or from *Rba. sphaeroides* (Oh-hama *et al.*, 1985b) adapting from respiratory to photosynthetic conditions, showed that the 13^4-methyl ester carbon was derived from the C-2 of glycine via S-adenosylmethionine derived by C_1 pathway activity. The enzyme is encoded by the bacterial gene *bchM*.

2.4.3 Formation of the isocyclic ring E by Mg-protoporphyrin IX-monomethylester cyclase

The isocyclic (cyclopentanone) ring E is formed from the 13-propionate side chain of Mg-protoMME.

2.4.3.1 The hydratase pathway

A mass spectrometric study of BChl *a*, formed during adaptation of *Rba. sphaeroides* from aerobic (respiratory) to anaerobic (photosynthetic) conditions in the presence of $^{18}O_2$ or $H_2{}^{18}O$, showed that the 13^1-oxo group was labelled only by $H_2{}^{18}O$ and so the four-step hydratase pathway described below was proposed (Porra *et al.*, 1995; 1996):

(1) the 13-propionate was desaturated to form the 13-acrylate derivative;
(2) the acrylate derivative was hydrated to form the 13^1-hydroxy-propionate derivative;
(3) the 13^1-hydroxy derivative was oxidized to the 13^1-oxo derivative; and,
(4) bond formation between C-13^2 and C-15 by dehydrogenase activity completed ring E.

In this experiment, the 3^1-acetyl oxygen was also labelled by $H_2{}^{18}O$ indicating enzymic hydration of the 3-vinyl group of Mg-protoMME. A requirement for vitamin B_{12} has been established for the anaerobic cyclase in *Rhodobacter capsulatus* (Gough *et al.*, 2000).

Ellsworth and Aronoff (1968, 1969), incorrectly (see Section 2.4.3.2) described the same four-step hydration mechanism for green algae after detecting accumulations of the 13-acrylate-, 13^1-hydroxy- and 13^1-oxo-propionate derivatives of Mg-protoMME in *Chlorella* mutants. The 13-acrylate derivative of Mg-protoMME, detected by Ellsworth and Aronoff (1968, 1969) in *Chlorella* mutants, probably arose from non-specific dehydratase activity on the 13^1-hydroxy derivative, arising from an oxygenase (see Section 2.4.3.2) not a hydratase activity. Supporting this view, the 13^1-hydroxy and 13^1-oxo derivatives of Mg-protoMME, but not the acrylate, were substrates for cyclase activity in a higher-plant preparation (Walker *et al.*, 1988).

The *bchE* gene has been described in the purple bacterium, *Rubrivivax gelatinosus*, to code for the anaerobic cyclase, which is probably a non-haem iron-sulfur protein (Ouchane *et al.*, 2004).

2.4.3.2 The oxygenase pathway

In contrast to the results with *Rba. sphaeroides*, ^{18}O-labelling experiments with growing *Chlorella vulgaris* cells (Schneegurt and Beale, 1992) and in greening of etiolated maize leaves (Porra *et al.*, 1993, 1994) showed that the 13^1-oxo groups of Chls *a* and *b* were formed from molecular oxygen, not from water. The following three-step mono-oxygenase mechanism was proposed (Porra *et al.*, 1996) for the conversion of Mg-protoMME to Mg-3,8-divinylprotochlorophyllide (Mg-DVP):

(1) a mono-oxygenase converted the 13-propionate side chain to a 13^1-hydroxy derivative;
(2) the 13^1-hydroxy derivative was dehydrogenated to the 13^1-oxo derivative; and,
(3) a bond between C-13^2 and C-15, formed by dehydrogenase activity, completed ring E formation.

Labelling experiments with $^{18}O_2$ or $H_2{}^{18}O$ showed that in *Roseobacter (Rsb.) denitrificans*, an obligate aerobic, BChl *a*-containing bacteria, unlike *Rba. sphaeroides*, but like higher plants and green algae, incorporated only $^{18}O_2$ into the 13^1-oxo group of BChl *a* (Porra *et al.*, 1996); again the 3^1-acetyl-group oxygen was labelled by $H_2{}^{18}O$. Similar ^{18}O-labelling experiments were performed on *Rhodovulum (Rhv.) sulfidophilum* which forms BChl *a* anaerobically in the light like *Rba. sphaeroides*, but, unlike *Rba. sphaeroides*, also forms BChl *a* aerobically in the dark. In *Rhv. sulfidophilum* the 13^1-oxo group was formed by a hydratase activity under anaerobic conditions but by both hydratase and oxygenase mechanisms in aerobic conditions (Porra *et al.*, 1998). How the two activities are regulated in *Rhv. sulfidophilum* under aerobic conditions is not yet known. The 3^1-acetyl oxygen, however, was still derived from water both anaerobically and aerobically.

The oxidative cyclase is probably a multi-subunit enzyme, of which two subunits are membrane bound, and one is water soluble The gene coding for one of the membrane-bound subunits (*acsF*, *Xantha 1*, *Chl27*; depending on the organism) is conserved from purple bacteria to higher plants (Rzeznicka *et al.*, 2005). The cyanobacterium *Synechocystis* PCC6803, contains two *acsF*-like and three *bchE*-like genes (anaerobic synthase); of which one *acsF*-like gene is required for aerobic growth, the other for microaerophilic growth (Minamizaki *et al.*, 2008). In *Rubrivivax gelatinosus*, the anaerobic cyclase, BchE, is required for anaerobic and microaerophilic growth, and AcsF for aerobic growth, indicating a different regulation. Like BchE, AsfE shows a characteristic sequence motif for a non-haem iron-sulfur cluster, but the genes are otherwise unrelated (Ouchane *et al.*, 2004). A regulatory element, RegA, has been identified in another purple bacterium, *Rhodobacter capsulatus*, to control BchE (Willett *et al.*, 2007).

The MgDVP is a possible bifurcation point for the formation of all Chls *c*, but especially those which are 3,8-divinyl derivatives such as Chl c_2, Chl c_2-MGDG [18:4/14:0], Chl c_2-MGDG [14:0/14:0], Chl c_3 and Chl $c_{3\ (CS\text{-}170)}$ (see Figure 2.2): Chl $c_{3\ (CS\text{-}170)}$ possesses a 17-propionate side chain as does MgDVP and PChlide *a*. The latter pigment may be the precursor of monovinyl Chls *c* such as Chl c_1 and MV-Chl c_3. The MgDVP is not only an intermediate in Chl biosynthesis (see Figure 2.5) but also a light-harvesting Chl *c*-type pigment in phototrophic prokaryotes and algae (Helfrich *et al.*, 1999) and, principally, PChlide *a* might also serve both functions.

2.4.4 *Protochlorophyllide* a *formation by 8-vinyl reductase*

Rebeiz *et al.* (1983) reported that some Chls are 3,8-divinyl (DV, or [8-vinyl]) derivatives but most are 3-vinyl (MV) compounds and so they proposed that 8-vinyl reductase has a wide substrate specificity so that 8-vinyl group reduction might occur anywhere between the conversion of protogen IX to Chlide *a*; thus, two parallel linear MV- and DV-Chl biosynthetic sequences were proposed rather than the usual single linear MV-Chl pathway. Furthermore, it was proposed that the proportion of DV- and MV-Chls depended on the organism examined and the ambient light regime (Rebeiz *et al.*, 1983). A broad substrate specificity for 8-vinyl reductase was also supported by the detection of MV and DV forms of the 13-acrylate-, 13^1-hydroxy- and 13^1-oxo-derivatives of Mg-protoMME in *Chlorella* mutants (Ellsworth and Aronoff, 1968; 1969). This reductase appears to be an NADPH-dependent enzyme. The reductase gene has been identified in *Arabidopsis thaliana* and functionally over-expressed in *E. coli*; as expected from the exclusive presence of [8-vinyl]-Chls ('DV-Chls'), it is missing in *Prochlorococcus marinus* (Nagata *et al.*, 2005).

While Chl c_1 and MV-Chl c_3 could both arise from PChlide *a* after reduction of the 8-vinyl group of Mg-DVP, both these Chls would then require oxidation of the 17-propionate to a 17-acrylate. As the 8-vinyl reductase has a broad specificity, the 8-ethyl group of Chl c_1 could also arise from reduction of Chl c_2, which is a 3,8-divinyl-17-acrylate compound.

2.4.5 *Reduction of ring D by protochlorophyllide* a *oxidoreductases*

PChlide *a* is converted to Chlide *a* by PChlide *a* oxidoreductases (POR) which exist in light-dependent and light-independent (dark) forms, namely LPOR and DPOR (Fujita, 1996; Reinbothe *et al.*, 1996; Rüdiger, 2006).

2.4.5.1 Light-dependent protochlorophyllide oxidoreductases

In angiosperms, light-dependent LPOR combines with its substrates PChlide *a* and NADPH to form a photoactive ternary complex, LPOR:NADPH:PChlide *a* which when illuminated undergoes a stereoselective *trans*-reduction of the C-17/C-18 double bond of ring D forming an LPOR:$NADP^+$:Chlide *a* complex. The $NADP^+$ is quickly re-reduced to NADPH with release of Chlide *a* and the remaining LPOR:NADPH complex then readily accepts more PChlide *a*. The LPOR from mesophilic and thermophilic purple and cyanobacteria have recently been investigated in some detail (Heyes and Hunter, 2005). PChlide and [8-vinyl]-PChlide were equally good substrates for LPOR from the purple bacterium, *Rhodobacter capsulatus* (Heyes *et al.*, 2006b). The first step is a light-induced charge (probably hydride) transfer that is followed by several dark reactions. Radical species are probably by-products of the reaction and not formed in stoichiometric amounts

(Heyes *et al.*, 2006a). A light-activation of the enzyme before the actual photoreduction by a second photon has been described by Hunter (see Sytina *et al.*, 2008).

In *Arabidopsis (A). thaliana* three differently regulated LPORs are described by Su *et al.* (2001). The first, PORA is labile in light and therefore occurs in large amounts in etioplasts of dark-grown plant tissue where large amounts of phytotoxic PChlide *a* accumulate. The phototoxicity is reduced by both the ATP-dependent combination of PChlide *a* with PORA and the subsequent PORA: NADPH:PChlide ternary complex formation which, on illumination, releases Chlide *a* as described above.

Secondly, PORB, a long-lived LPOR, exists in both etiolated (etioplasts) and green tissue (chloroplasts), and its concentration remains constant in light and dark. There is about 75% sequence identity between PORA and PORB and both form ternary complexes with NADPH and PChlide *a*.

A third form of NADPH-dependent LPOR, PORC, is not present in etiolated *A. thaliana* and is detected only after 6 h illumination and although surviving high illumination it soon disappears on return to darkness. These three LPOR forms may allow plants more selective adaptation to different light regimes (Su *et al.*, 2001).

2.4.5.2 Light-independent (dark) protochlorophyllide oxidoreductases

Unlike angiosperms that etiolate in darkness, gymnosperms, mosses, green algae and cyanobacteria remain green and form Chls in the dark. The dark oxidoreductases (DPOR) are encoded in photosynthetic bacteria by three genes *bchB, bchL* and *bchB*, and by *chlB, chlL and chlN* in cyanobacteria and green algae. They show considerable homology with the three nitrogenase genes of *Azotobacter vinelandii*. The first electron transfer reactions also seem to be similar but they differ in the second part where the substrate, PChlide or N_2, respectively, is reduced. The *chlN* (*bchN*) gene encodes a protein with an ATP binding site and four cysteines which, unlike the nitrogenase homologue, binds a Fe-S cluster (Fujita *et al.*, 1989, 1991; Bröcker *et al.*, 2008a): it is suggested that this protein is involved in the DPOR stepwise transport of two electrons to PChlide *a* (Fujita, 1996; Bröcker *et al.*, 2008b). The substrate specificity of the enzyme is considerably different from that of LPOR (Bröcker *et al.*, 2008b).

Chlide *a* and 3,8-divinyl-Chlide *a* are possible bifurcation points for the formation of bacteriophytochlorin type BChls *a*, *b* and *g*.

2.4.6 Conversion of chlorophyllide* a *to chlorophyllide* b *by chlorophyllide* a *oxygenase

Labelling experiments with $^{18}O_2$ and $H_2{}^{18}O$ proved that the 7-formyl oxygen of Chl *b* in *Chlorella vulgaris* (Schneegurt and Beale, 1992) and in greening maize leaves (Porra *et al.*, 1993; 1994) was derived only from $^{18}O_2$, indicating a mono- or di-oxygenase mechanism. Later, Tanaka *et al.* (1998) discovered a Chlide *a*

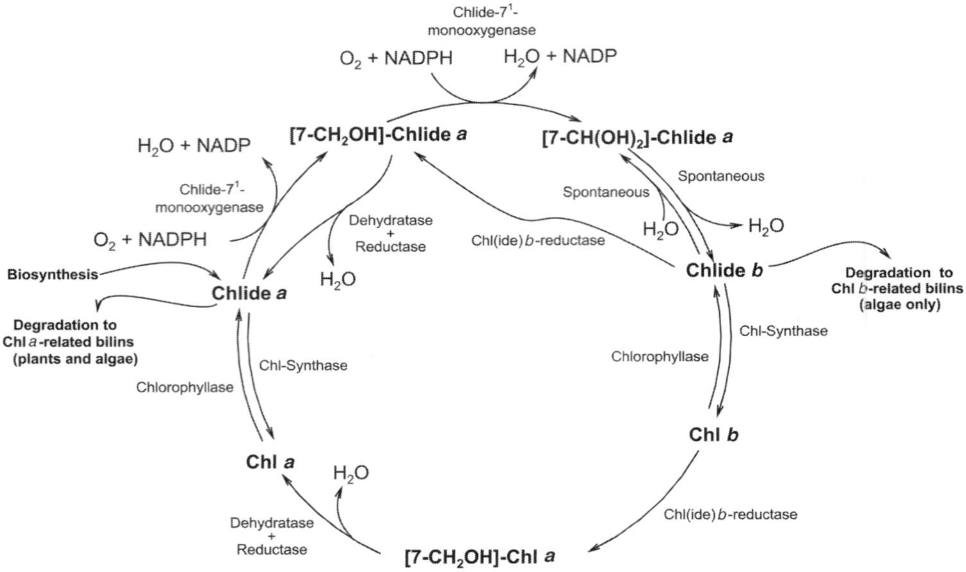

Figure 2.8. The chlorophyll cycle linking chlorophyll biosynthesis and degradation. This scheme, modified from Oster *et al.* (2000), shows the conversion of Chlide *a* to Chlide *b* by two successive and irreversible monooxygenase reactions, and back reactions to Chlide *a* by a reductase/dehydratase/reductase sequence. This cycle is not a metabolic cycle for continuous biosynthesis of Chlide *b* from Chlide *a*; rather, it is a switch linking Chl biosynthesis to Chl degradation, possibly, to fine-tune Chl *a*/*b* ratios during adaptation to new light regimes. From Porra and Scheer (2000), with kind permission from Springer Science+Business Media.

oxygenase (*CAO*) gene which was essential to reinstate Chl *b* formation in Chl *b*-less mutants of *Chlamydomonas reinhardtii*. The enzyme protein encoded by *CAO* was found to be an oxygenase containing a Rieske-type [2Fe-2S] cluster and another non-haem Fe binding site: this enzyme converted Chlide *a* to Chlide *b* and Chl *a* was not a substrate. Oster *et al.* (2000) showed that Chlide *a* oxygenase converted Chlide *a*, in two sequential steps, firstly to [7-CH_2OH]-Chlide *a* and then, via the *gem*-diol, [7-$CH(OH)_2$]-Chlide *a* and subsequent spontaneous dehydration, to Chlide *b* (see Figure 2.8).

2.4.7 Reduction of chlorophyll(ide) **b** *to chlorophyll(ide)* **a***: the chlorophyll cycle*

All oxygenase reactions are strongly exothermic and irreversible (Hayaishi, 1974) but reduction of Chl *b* to Chl *a* by two reductases was demonstrated in etioplast membranes (Ito *et al.*, 1996). Successful substrates for these membranes included Chl *a*, Chlide *a* and their 7^1-hydroxy derivatives. Reduction to the 7^1-hydroxyl derivative required NADPH and further reduction (and dehydration) to form the 7-methyl group required reduced ferredoxin (Scheumann *et al.*, 1998).

Following these discoveries, Oster *et al.* (2000) proposed a 'chlorophyll cycle' (see Figure 2.8) as a mechanism for fine-tuning the changes in Chl *a*/*b* ratios observed in photosynthetic cells adapting to new light regimes, as an alternative to the relatively simplistic notion of using either Chl degradation and/or *de novo* Chl biosynthesis. Through the 'Chl cycle' Chl *b* can be re-converted to Chl *a* via Chlide *b* reductase and a [7-CH_2OH]-Chlide *a* dehydratase and reductase (see Figure 2.8). Both Chls are degraded via dephytylation to Chlides (see Figure 2.8) followed by Mg-dechelatase activity (not shown) to form pheophorbides (Pheides). Pheide oxygenases then cleave the Pheide macrocycle at C5 to form linear-tetrapyrrolic bilins. In higher-plants this oxygenase attacks only Pheide *a* (Hörtensteiner *et al.*, 1995) so that Chl *b* in plants is degraded via Chlide *a* formed by Chl cycle activity (see Figure 2.8); Pheide oxygenase from the green alga, *Chlorella protothecoides*, however, attacks both Pheide *a* and Pheide *b* (Iturraspe *et al.*, 1994).

2.4.8 Esterification of chlorophyllides a *and* b *to chlorophylls* a *and* b *by chlorophyll synthase*

The 17-propionate of most Chls (and BChls) are esterified to the long-chain (C_{20}) branched polyisoprenoid alcohol (phytol) catalysed by Chl-synthase which is encoded by the *chlG* gene and found in both etioplast and chloroplast membranes: geranylgeraniol, farnesol or other esters are less frequently encountered. Chl-synthase preferentially converts Chlide *a* or Chlide *b*. Chls a_{GG} and b_{GG} are formed in the presence of geranylgeranyl-diphosphate (GG-PP): GG and ATP, and phytyldiphosphate are less favoured substrates, and PChlide *a* was inert indicating that esterification of Chlide *a* occurs after reduction of ring D (Rüdiger, 2006).

A reductase, encoded by the *chlP* gene then reduces the double bonds of the geranylgeranyl moiety at C-6, C-10 and C-14 leaving a single double bond at C-2. The reductase reduces Chl_{GG} stepwise to the phytylated form (Chl_P) but can also reduce GG-diphosphate to phytol-diphosphate which is also essential for tocopherol formation in tobacco (Rüdiger, 2006).

The biosyntheses of the unusual side chains of the BChls *c, d* and *e*, which are also phytochlorins are discussed in Section 2.4.10.2.

2.4.9 Formation of bacteriochlorophyll a*: reduction of ring B by chlorin-reductase*

The C-7/C-8 double bond of ring B of Chlide *a* is reduced by chlorin reductase producing BChlide *a*. This reductase is encoded by three genes *bchX, bchY and bchZ.* The 3-vinyl group is then subjected to hydratase (encoded by *bchF*) and dehydrogenase (encoded by *bchC*) activities to form the 3-acetyl group of BChlide *a* which is then phytylated by BChl-synthase (*bchG*).

A putative pathway to form BChlides *b* and *g* involves an isomerase catalysing an outward migration of the 7, 8 double bond forming an 8-ethylidene (=CH-CH_3) group (see Scheer and Katz, 1974): BChlides *b* and *g* contain 3-acetyl- and 3-vinyl groups, respectively (see Figure 2.4) and are then phytylated to their respective BChls by a BChl-synthase encoded by *bchG*. The biosynthetic relationships between the BChls of the phytochlorin and bacteriophytochlorin types are discussed by Porra (1997) and Frigaard *et al*. (2006).

2.4.10 Unusual chlorophylls and unusual side chains

Many new Chls are being isolated and identified (see Kobayashi *et al*., 2006b) using new high performance liquid chromatographic methods (Garrido and Zapata, 2006) and sensitive spectrometric techniques such as NMR, MS, etc. (Kobayashi *et al*., 2006a). Some of these unusual Chl pigments are discussed below.

2.4.10.1 In the phytoporphyrin family

Figure 2.2 shows that Chl c_3 and Chl $c_{3\ (\mathrm{CS\text{-}170})}$ possess a 7-methylcarboxylate derived from the 7-CH_3 of Chlide *a*. The 7-formyl group of Chl *b* arises from the spontaneous dehydration of a *gem*-diol derivative of Chlide *a* formed by two successive oxygenation steps catalysed by CAO (Oster *et al*., 2000; see Section 2.4.6). Hypothetically, a 7-*gem*-diol of MgDVP similarly formed, could be oxidized to a 7-carboxylate which, following methylation by a SAM-methyltransferase-type enzyme, could form a 7-methylcarboxylate group (compare Porra, 1997; Porra *et al*., 1997).

Since CAO from *A. thaliana* could not use PChlide *a* in vitro as a substrate (see Oster *et al*., 2000), the existence of PChlide *b* remains controversial. Nonetheless, it has been reported in barley etioplasts (Reinbothe *et al*., 2003), while Kolossov and Rebeiz (2003) reported it in green but not in etiolated plants. Xu *et al*. (2002) reported that the *Synechocystis* PCC6803 mutant, expressing the genes for plant LHC II protein and CAO, formed small amounts of PChlide *b* but long after PChlide *a* accumulated, suggesting that CAO can convert PChlide *a*, but very ineffectively. This mutant produced large amounts of functional Chl *b*.

Noteworthy in Figure 2.2 is the large MGDG-type ester of the 17^3-acrylate side chain producing a Chl with a very high mass molecular ion ($m/z = 1313$). MGDG, the main galactolipid of thylakoid membranes, is produced by acylation of the *sn*-1 and *sn*-2 positions of glycerol-3-phosphate by two fatty-acyl-CoAs, followed by phosphatase activity to form diacylglycerol (DAG): the nature of the fatty acids is species specific. The DAG then reacts with UDP-galactose, formed from galactose-1-phosphate and UDP-glucose, producing MGDG. The Mg-branch enzymes of Chl formation are located on the prolamellar or thylakoid membranes of etioplasts and chloroplasts, respectively (Rüdiger and Grimm, 2006), and Jeffrey and Anderson (2000) have proposed that esterification of Chl c_2 with MGDG may facilitate its transport to its correct position in the fucoxanthin-Chl *a/c* protein of the light-harvesting antenna. Alternatively, formation of the MGDG ester group of Chl c_2 may lock the chromophore into its final position with correct orientation of the Chl, also providing a space-filling function between antenna monomers or within the antenna monomers where the long acyl chains of the MGDG could line hydrophobic clefts to accept light-harvesting carotenoids conveniently close to the Chl *c* chromophore (Liu *et al.*, 2004).

2.4.10.2 *In the phytochlorin family*

The most abundant unusual Chls of this family are probably the [8-vinyl]-Chls *a* and *b* (i.e. 3,8-divinyl-Chls or DV-Chls) shown in Figure 2.3A. As discussed earlier they derive from a divinyl-type biosynthetic pathway running parallel to the monovinyl route (see Section 2.4.4). Chisholm *et al.* (1988) and Goericke and Repeta (1992) found that marine *Prochlorococcus* species contained only DV and no MV Chls. Because of the abundance of these organisms in the oceans, they are responsible for a very significant proportion of global photosynthesis.

Chlorophyll *d* is 3-*des*-vinyl-3-formyl-Chl *a* (i.e. 3-formyl-Chl *a*; see Figure 2.3A): the mechanism of conversion of the 3-vinyl- to a 3-formyl side chain is still unknown. It is found as the major pigment in the cyanobacterium *Acaryochloris (Acc.) marina* (Akiyama *et al.*, 2001) and as a minor pigment in an epiphytic prokaryote on red algae (Murakami *et al.*, 2004). The novel Chl *f* has recently been found in stromatolites (see *Addendum*).

Reaction centres (RCs) often contain unusual Chls with specific functions. One molecule of Chl *a′* ('*a*-prime'), the C-13^2 epimer of Chl *a* (see Figure 2.3A) is found in photosystem (PS) I RCs of plants and cyanobacteria, where it forms, together with one molecule of Chl *a*, the primary electron donor in photosynthetic electron transport (Kobayashi *et al.*, 1988, 2006b; Jordan *et al.*, 2001). Other 13^2-epimers include Chl *d′*, found in *Acc. marina* (Akiyama *et al.*, 2001), and BBChl *g′*, present in *Heliobacter* (Kobayashi *et al.*, 1991): both are likely to have similar functions as both organisms contain PSI-type RCs. Another unusual pigment, 8^1-hydroxy-Chl *a* (see Figure 2.3A), is found in PS I RCs of green bacteria, where it supposedly acts as primary electron acceptor (Van de Meent *et al.*, 1991). All type II RCs contain pheophytins (see Figure 2.3A) as electron acceptors (Akiyama *et al.*, 2001; Scheer, 2006; Kobayashi *et al.*, 2006b): pheophytins lack the central Mg. The demetalation step in the RC is poorly understood but may occur during the binding reaction (compare Scheer, 2006); however, a non-proteinaceous demetalating factor has been proposed (Suzuki and Shioi, 2002).

The BChls *c*, *d* and *e*, of the green photosynthetic bacteria (see Figure 2.3B), all lack a 13^2-methylcarboxylate group (as mentioned earlier) but all possess a 3-hydroxyethyl group indicating hydration of the 3-vinyl side chain of Chlide *a*. The 17-propionate is mainly esterified by farnesol but also by geranylgeraniol and other alcohols (Frigaard *et al.*, 2006; Tamiaki *et al.*, 2007). The stepwise elongation of the 8-ethyl group to propyl, butyl, neopentyl etc., and of the 12-methyl group to an ethyl (see Figure 2.3B) occurs by enzymic methylation reactions involving methyl-cobalamine as donor: cobalamine is re-methylated in a SAM-dependent enzymic reaction after each C_1 elongation step (see Frigaard *et al.*, 2006). The C-20 methylation of BChls *c* occurs by a different methyltransferase but the mechanism is unknown (Frigaard *et al.*, 2006).

Another unusual phytochlorin group containing Ni as the central metal atom and designated 'tunichlorins', has been isolated from tunicates and is possibly derived from microsymbiont algal Chls (Bible *et al.*, 1988): they may be involved in electron transfer or other metabolic processes (Pettit *et al.*, 1993).

2.4.10.3 In the bacteriophytochlorin family

The most unusual member of this group is the acid stable form of BChl *a*, namely Zn-BChl *a*, in which the central Mg is replaced by Zn (see Figure 2.4; M = Zn). This BChl occurs in *Acp. rubrum* which lives in a low pH habitat where the Mg form of BChl *a* might be quickly demetallated (Kobayashi *et al.*, 1998). Most of the physical and photochemical properties of Zn-BChl *a* are similar to those of BChl *a* which permit it to function both in antennae and reaction centres (Kobayashi *et al.*, 2006b). No Zn-chelatase or Zn-protoMME has been detected so Zn may replace Mg at a later step; however, no transmetallase has been discovered so metal replacement may be non-enzymic (see Küpper *et al.*, 2006). Type II-RCs of anoxygenic phototrophic bacteria contain bacteriopheophytins *a* (or *b* in the case of BChl *b*-containing species) as electron acceptors (Akiyama *et al.*, 2001; Scheer, 2006; Kobayashi *et al.*, 2006b): bacteriopheophytins lack the central Mg.

2.5 Concluding remarks

Much progress has been made in the last decade in our understanding of Chl biosynthesis and its regulation and of the structure of new Chls. This has been achieved by using newly developed spectrometric and chromatographic methods, by the application of genomic methods and by X-ray crystallographic studies; nevertheless, as described in this review, many problems and questions remain. The continued use and development of advanced techniques, however, may help solve many of these questions.

Addendum A new chlorophyll, Chl *f*, was recently discovered in a filamentous cyanobacterial component in the lower layers of stromatolites found at Shark Bay, Western Australia, and identified as [2-formyl]-Chl *a* (Chen, M., Schliep, M., Willows, R.D., Cai, Z.-L., Neilan, B.A. and Scheer, H. (2010). *Science* **329**, 1318–19). In methanol, Chl *f* has red-shifted fluorescence emission and Q_y absorption maxima at 722 and 706 nm, respectively, and a blue-shifted Soret maximum at 406 nm when compared with Chl *a*; the Q_y red-shift is even larger than in Chl *d* (= [3-formyl]-Chl *a*). This suggests that oxygenic photosynthesis can operate further into the near-infrared region than previously thought, which is particularly advantageous under a dense cover of overlaying algae or cyanobacteria using Chls *a* or *d*.

The missing Chl *e* was isolated in trace amounts from feral algal populations of two members of the Xanthophyceae, *Tribonema bombycinum* and *Vaucheria hamata*, but has not been detected in these species when cultured: in methanol, Chl *e* absorbs at 415 and 654 nm. The chemical structure has not been determined and its status as a natural Chl or artefact is still uncertain. Research on this Chl was done in the 1940s and 50s by the pigment chemist, H. H. Strain, and reviewed by Allen (1966) (Allen, M.B. (1966). Distribution of the chlorophylls. In *The Chlorophylls*, ed. L. P. Vernon and G. B. Seely. New York: Academic Press, pp. 511–19).

Very recently, Jaschke *et al.* (2011) (Jaschke, P. R., Hardjasa, A., Digby, E. L., Hunter, C. N. and Beatty, J. T. (2011). A *bchD* (Magnesium chelatase) mutant of *Rhodobacter sphaeroides* synthesizes zinc bacteriochlorophyll through novel zinc-containing intermediates. *J. Biol. Chem.* **286**, 20313–22) have demonstrated that the insertion of Zn^{2+} into proto is catalysed by ferrochelatase, and the resulting Zn-proto is converted to [Zn]-BChl *a*: they also isolated two intermediates in which the commonly present Mg^{2+} was replaced by Zn^{2+}. While these data have been obtained with a mutant of *Rhodobacter sphaeroides*, it is very possible that this metalation reaction is also the first dedicated step in the biosynthesis of [Zn]-BChl *a* in *Acidiphilium rubrum*, where it is the major pigment (see Kobayashi *et al.*, 1998).

Acknowledgements

R. J. Porra (RJP) thanks CSIRO-Plant Industry for the use of the Black Mountain Library and communications facilities. Research by RJP and H. Scheer (HS) at the University of Munich was funded by the Deutsche Forschungsgemeinschaft, Bonn (SFB 533), the Volkswagen Stiftung, Hannover and the Hans-Fischer-Gesellschaft, München.

Abbreviations

ALA	5-aminolevulinic acid
CAO	Chlorophyllide *a* oxygenase
CoASH	Coenzyme A
Copro	Coproporphyrin
Coprogen	Coproporphyrinogen
DPOR	Light-independent POR
HMB	Hydroxymethyl-bilin
IUPAC-IUB	International Union of Pure and Applied Chemistry-International Union of Biochemistry
LHC(s)	Light-harvesting complex(es)
LPOR	Light-dependent POR

Mg-proto	Mg-protoporphyrin IX
Mg-protoMME	Mg-protoporphyrin IX monomethylester
MGDG	Monogalactosyldiacylglycerol
PBG	Porphobilinogen
PChl	Protochlorophyll
POR	Protochlorophyllide oxidoreductase
PPOX	Protoporphyrinogen IX oxidase
Proto	Protoporphyrin IX
Protogen	Protoporphyrinogen IX
Uro	Uroporphyrin
Urogen	Uroporphyrinogen

References

Akhtar, M. (1991). Mechanism and stereochemistry of the enzymes involved in the conversion of uroporphyrinogen III to heme. In *Biosynthesis of Tetrapyrroles*, ed. P.M. Jordan. Amsterdam: Elsevier, pp. 67–99.

Akhtar, M. (1994). The modification of acetate and propionate side chains during the biosynthesis of haem and chlorophylls: mechanistic and chemical studies. In *The Biosynthesis of the Tetrapyrrole Pigments*, ed. D.J. Chadwick and K. Ackrill. CIBA Symposium 180. Chichester: Wiley, pp. 131–55.

Akiyama, M., Miyashita, H., Kise, H., Watanabe, T., Miyachi, S. and Kobayashi, M. (2001). Detection of chlorophyll *d′* and pheophytin *a* in a chlorophyll *d*-dominating oxygenic photosynthetic prokaryote *Acaryochloris marina*. *Anal. Sci.* **17**, 205–08.

Avissar, Y.J. and Moberg, P.A. (1995). The common origins of the pigments of life: early steps of chlorophyll biosynthesis. *Photosynth. Res.* **44**, 221–24.

Avissar, Y.J., Ormerod, J.G. and Beale, S.I. (1989). Distribution of δ-aminolevulinic acid biosynthetic pathways among phototropic bacterial groups. *Arch. Microbiol.* **151**, 513–19.

Battersby, A.R. (1994). How nature builds the pigments of life: the conquest of vitamin B_{12}. *Science* **264**, 1551–57.

Battersby, A.R. and Leeper, F.J. (1990). Biosynthesis of the pigments of life: mechanistic studies on the conversion of porphobilinogen to uroporphyrinogen III. *Chem Rev.* **90**, 1261–74.

Baumgarten, D. (1970). *The Structure of Bacteriochlorophyll b*. M.Sc. thesis, University of California, Berkeley.

Bazzaz, M.B. (1981). New chlorophyll chromophores isolated from a chlorophyll-deficient mutant of maize. *Photobiochem. Photobiophys.* **2**, 199–207.

Beale, S.I. (2006). Biosynthesis of 5-aminolevulinic acid. In *Chlorophylls and Bacteriochlorophylls: Biochemistry, Biophysics, Functions and Applications*, ed. B. Grimm, R.J. Porra, W. Rüdiger and H. Scheer. Dordrecht: Springer, pp. 147–58.

Beale, S.I., Castelfranco, P.A. and Granick, S. (1975). The biosynthesis of δ-aminolevulinic acid from the intact carbon skeleton of glutamate in greening barley. *Proc. Natl. Acad. Sci. USA* **72**, 2719–23.

Beale, S.I., Foley, T. and Dzelzkalns, V. (1981). δ-aminolevulinic acid synthase from *Euglena gracilis*. *Proc. Natl. Acad. Sci. USA* **78**, 1666–69.

Bible, K.C., Buytendorp, M., Zierath, P.D. and Rinehart, K.L. (1988). Tunichlorin: A nickel chlorine isolated from the Caribbean tunicate *Trididemnum solidum*. *Proc. Natl. Acad. Sci. USA* **85**, 4582–86.

Bollivar, D.W. (2006). Recent advances in chlorophyll biosynthesis. *Photosynth. Res.* **90**, 173–94.

Borrego, C.M., Arellano, J.B., Abella, C.A., Gilbro, T. and Garcia-Gil, J. (1999). The molar extinction coefficient of bacteriochlorophyll *e* and the pigment stoichiometry in *Chlorobium phaeobacteroides*. *Photosynth. Res.* **60**, 257–64.

Bröcker, M.J., Virus, S., Ganskow, S., Heathcote, P, Heinz, D.W., Schubert, W., Jahn, D. and Moser, J. (2008a). ATP-driven reduction by dark operative protochlorophyllide oxidoreductase from *Chlorobium tepidum* mechanistically resembles nitrogenase catalysis. *J. Biol. Chem.* **283**, 10559–67.

Bröcker, M.J., Wätzlich, D., Uliczka, F., Virus, S., Saggu, M., Lendzian, F., Scheer, H., Rüdiger, W., Moser, J. and Jahn, D. (2008b). Substrate recognition of nitrogenase-like dark operative protochlorophyllide oxidoreductase from *Prochlorococcus marinus*. *J. Biol. Chem.* **283**, 29873–81.

Chadwick, D.H. and Ackrill, K. (eds.) (1994). *The Biosynthesis of the Tetrapyrrole Pigments*. CIBA Symposium 180. Chichester: John Wiley and Sons.

Chisholm, S.W., Olson, R.J., Zettler, E.R., Goericke, R. Waterbury, J.B and Welschmeyer, N.A. (1988). A novel free-living prochlorophyte abundant in the oceanic euphotic zone. *Nature* **334**, 340–43.

Drechsler-Thielmann, B., Dörnemann, D. and Senger, H. (1993). Synthesis of protoheme via both the C5 and the Shemin pathways in the pigment mutant C-2A' of *Scenedesmus obliquus*. *Z. Naturforsch. C.* **48**, 584–89.

Ebbon, J.G. and Tait, G.H. (1969). Studies on S-adenosyl-methionine-magnesium-protoporphyrin methyl-transferase in *Euglena gracilis* strain Z. *Biochem. J.* **111**, 573–82.

Ellsworth, R.K. and Aronoff, S. (1968). Investigations on the biogenesis of chlorophyll *a*. III. Biosynthesis of Mg-vinylporphine a_5 methylester from Mg-protoporphine IX monomethylester as observed in *Chlorella* mutants. *Arch. Biochem. Biophys.* **125**, 269–77.

Ellsworth, R.K. and Aronoff, S. (1969). Investigations on the biogenesis of chlorophyll *a*. IV. Isolation and partial characterization of some biosynthetic intermediates between Mg-protoporphine IX monomethylester and Mg-vinylporphine a_5 methylester obtained from *Chlorella* mutants. *Arch. Biochem. Biophys.* **130**, 374–83.

Ellsworth, R.K., Dullaghan, J.P. and St. Pierre, M.E. (1974). The reaction mechanism of S-adenosyl-L-methionine: Mg-protoporphyrin IX methyltransferase of wheat. *Photosynthetica* **8**, 375–83.

Frigaard, N.-U., Gomez Maqueo Chew, A., Maresca, J.A. and Bryant, D.A. (2006). Bacteriochlorophyll biosynthesis in green bacteria. In *Chlorophylls and Bacteriochlorophylls: Biochemistry, Biophysics, Functions and Applications*, ed. B. Grimm, R.J. Porra, W. Rüdiger and H. Scheer. Dordrecht: Springer, pp. 201–21.

Fujita, Y. (1996). Protochlorophyllide reduction: a key step in the greening of plants. *Plant Cell Physiol.* **34**, 411–12.

Fujita, Y., Takahashi, Y., Kohchi, T., Ozeki, H., Ohyama, K. and Matsubara, H. (1989). Identification of a novel *nifH*-like (*frxC*) protein in chloroplasts of the liverwort *Marchantia polymorpha*. *Plant Mol. Biol.* **13**, 551–61.

Fujita, Y., Takahashi, Y., Shonai, F., Ogura, Y. and Matsubara, H. (1991). Cloning nucleotide sequences and differential expression of *nifH* and *nifH*- (*frxC*) genes from the filamentous nitrogen-fixing cyanobacteria *Plectonema boryanum*. *Plant Cell Physiol.* **32**, 1093–106.

Garrido, J.L. and Zapata, M. (2006). Chlorophyll analysis by new high performance liquid chromatography methods. In *Chlorophylls and Bacteriochlorophylls: Biochemistry, Biophysics, Functions and Applications*, ed. B. Grimm, R.J. Porra, W. Rüdiger and H. Scheer. Dordrecht: Springer, pp. 109–21.

Gibson, K.L., Laver, W.G. and Neuberger, A. (1958). Initial stages in the biosynthesis of porphyrins. 2. The formation of δ-aminolaevulic acid from glycine and succinyl-coenzyme A by particles from chicken erythrocytes. *Biochem. J.* **70**, 71–81.

Gibson, K.L., Neuberger, A. and Tait, G.H. (1963). Studies on the biosynthesis of porphyrin and bacteriochlorophyll by *Rhodopseudomonas sphaeroides*. 4. S-adenosylmethionine-magnesium-protoporphyrin methyltransferase. *Biochem. J.* **88**, 325–34.

Goericke, R. and Repeta, D.J. (1992). The pigments of *Prochlorococcus marinus*: the presence of divinyl chlorophylls *a* and *b* in a marine prokaryote. *Limnol. Oceanogr.* **37**, 425–33.

Gough, S.P., Petersen, B.O. and Duus, J.O. (2000). Anaerobic chlorophyll isocyclic ring formation in *Rhodobacter capsulatus* requires a cobalamin cofactor. *Proc. Natl. Acad. Sci. USA* **97**, 6908–13.

Granick, S. (1966). The induction *in vitro* of the synthesis of δ-aminolevulinic acid synthetase in chemical porphyria: a response to certain drugs, sex hormones and foreign chemicals. *J. Biol. Chem.* **241**, 1359–75.

Grimm, B., Porra, R.J., Rüdiger, W. and Scheer, H. (eds.) (2006). *Chlorophylls and Bacteriochlorophylls: Biochemistry, Biophysics, Functions and Applications*. Dordrecht: Springer.

Hartwich, G., Feodor, L., Simonin, I., Cmiel, E., Schäfer, W., Noy, D., Scherz, A. and Scheer, H. (1998). Metal-substituted bacteriochlorophylls. 1. Preparation and influence of metal and coordination on spectra. *J. Am. Chem. Soc.* **120**, 3675–83.

Hayaishi, O. (1974). General properties and biological functions of oxygenases. In *Molecular Mechanisms of Oxygen Activation*, ed. O. Hayaishi. London: Academic Press, pp. 1–28.

Helfrich, M., King, G.C., Turner, A.G. and Larkum, A.W.G. (1999). Identification of [8-vinyl]-protochlorophyllide *a* in phototrophic prokaryotes and algae: chemical and spectroscopic properties. *Biochim. Biophys. Acta* **1410**, 262–72.

Heyes, D.J. and Hunter, C.N. (2005). Making light work of enzyme catalysis: protochlorophyllide oxidoreductase. *Trends Biochem. Sci.* **30**, 642–49.

Heyes, D.J., Heathcote, P., Rigby, S.E.J., Palacios, M.A., van Grondelle, R. and Hunter, C.N. (2006a). The first catalytic step of the light-driven enzyme protochlorophyllide oxidoreductase proceeds via a charge transfer complex. *J. Biol. Chem.* **281**, 26847–53.

Heyes, D.J., Kruk, J. and Hunter, C.N. (2006b). Spectroscopic and kinetic characterization of the light-dependent enzyme protochlorophyllide oxidoreductase (POR) using monovinyl and divinyl substrates. *Biochem. J.* **394**, 243–48.

Hörtensteiner, S., Vincentini, F. and Matile, P. (1995). Chlorophyll breakdown in senescent cotyledons of rape, *Brassica napus* L.: Enzymatic cleavage of phaeophorbide *a in vitro*. *New Phytol.* **129**, 237–46.

Ito, H., Ohtsuka, T. and Tanaka, A. (1996). Conversion of chlorophyll *b* to chlorophyll *a* via 7-hydroxymethyl chlorophyll. *J. Biol. Chem.* **271**, 1475–79.

Iturraspe, J., Engel, N. and Gossauer, A. (1994). Chlorophyll catabolism: Isolation and structure elucidation of chlorophyll *b* catabolites in *Chlorella protothecoides*. *Phytochemistry* **35**, 1387–90.

Jacobs, J.M. and Jacobs, N.J. (1987). Oxidation of protoporphyrinogen to protoporphyrin: a step in chlorophyll and haem biosynthesis. *Biochem. J.* **244**, 219–24.

Jeffrey, S.W. (1969). Properties of two spectrally different components in chlorophyll *c* preparations. *Biochim. Biophys. Acta* **177**, 456–67.

Jeffrey, S.W. (1972). Preparation and some properties of crystalline chlorophyll c_1 and c_2 from marine algae. *Biochim. Biophys. Acta* **279**, 15–33.

Jeffrey, S.W. and Wright, S.W. (1987). A new spectrally distinct component in preparations of chlorophyll *c* from the micro-alga *Emiliania huxleyi* (Prymesiophyceae). *Biochim. Biophys. Acta* **894**, 180–88.

Jeffrey, S.W. and Anderson, J.M. (2000). *Emiliania huxleyi* (Haptophyta) holds promising insights for photosynthesis. *J. Phycol.* **36**, 449–52.

Jeffrey, S.W., Mantoura, R.F.C. and Wright, S.W. (eds.) (1997). *Phytoplankton Pigments in Oceanography: Guidelines to Modern Methods*. Paris: UNESCO Publishing.

Jordan, P.M. (ed.) (1991a). *Biosynthesis of Tetrapyrroles*. Amsterdam: Elsevier.

Jordan, P.M. (1991b). The biosynthesis of 5-aminolaevulinic acid and its transformation into uroporphyrinogen III. In *Biosynthesis of Tetrapyrroles*, ed. P.M. Jordan. Amsterdam: Elsevier, pp. 1–66.

Jordan, P., Fromme, P., Witt, H.T., Klukas, O., Saenger, W. and Krauss, N. (2001). Three dimensionsal structure of cyanobacterial photosystem I at 2.5Å resolution. *Nature* **411**, 909–17.

Kannangara, C.G., Gough, S.P., Bruyant, P., Hoober, J.K., Kahn, A. and von Wettstein, D. (1988). $tRNA^{Glu}$ as a cofactor in δ-aminolevulinate biosynthesis: steps that regulate chlorophyll synthesis. *Trends in Biochem. Sci.* **13**, 139–43.

Kannangara, C.G., Andersen, R.V., Pontoppidan, B., Willows, R. and von Wettstein, D. (1994). Enzymic and mechanistic studies on the conversion of glutamate to 5-aminolaevulinate. In *The Biosynthesis of the Tetrapyrrole Pigments*, ed. D.J. Chadwick and K. Ackrill. CIBA Symposium **180**. Chichester: Wiley, pp. 3–25.

Kobayashi, M., Watanabe, T., Nakazato, M., Ikegame, I., Hiyama, T. and Matsunaga, T. (1988). Chlorophyll *a'*/P680 stoichiometries in higher plants and cyanobacteria determined by HPLC analysis. *Biochim. Biophys. Acta* **936**, 81–89.

Kobayashi, M., Van de Meent, E.J., Amesz J., Ikigami, I. and Watanabe, T. (1991). Bacteriochlorophyll *g* epimer as a possible reaction center component of Heliobacteria. *Biochim. Biophys. Acta* **1057**, 89–96.

Kobayashi, M., Yamamura, M., Akiyama, M., Kise, H., Inoue, K., Hara, M., Wakao, N., Yahara, K. and Watanabe, T. (1998). Acid resistance of

Zn-Bacteriochlorophyll *a* from an acidophilic bacterium *Acidiphilium rubrum*. *Anal. Sci.* **14**, 1149–52.

Kobayashi, M., Akiyama, M., Kano, H. and Kise, H. (2006a). Spectroscopy and structure determination. In *Chlorophylls and Bacteriochlorophylls: Biochemistry, Biophysics, Functions and Applications*, ed. B. Grimm, R. J. Porra, W. Rüdiger and H. Scheer. Dordrecht: Springer, pp. 79–94.

Kobayashi, M., Akiyama, M., Kise, H. and Watanabe, T. (2006b). Unusual tetrapyrrole pigments of photosynthetic antennae and reaction centers: specially tailored chlorophylls. In *Chlorophylls and Bacteriochlorophylls: Biochemistry, Biophysics, Functions and Applications*, ed. B. Grimm, R.J. Porra, W. Rüdiger and H. Scheer. Dordrecht: Springer, pp. 55–66.

Kolossov, V.L. and Rebeiz, C.A. (2003). Protochlorophyllide *b* occurs in green but not in etiolated plants. *J. Biol. Chem.* **278**, 49675–78.

Kräutler, B. and Hörtensteiner, S. (2006). Chlorophyll catabolites and the biochemistry of chlorophyll breakdown. In *Chlorophylls and Bacteriochlorophylls: Biochemistry, Biophysics, Functions and Applications*, ed. B. Grimm, R.J. Porra, W. Rüdiger and H. Scheer. Dordrecht: Springer, pp. 223–35.

Küpper, H., Küpper, F.C. and Spiller, M. (2006). [Heavy metal]-chlorophylls formed *in vivo* during heavy metal stress and degradation products formed during digestion, extraction and storage of plant material. In *Chlorophylls and Bacteriochlorophylls: Biochemistry, Biophysics, Functions and Applications*, ed. B. Grimm, R.J. Porra, W. Rüdiger and H. Scheer. Dordrecht: Springer, pp. 67–77.

Lash, T.D. (1991). Action of uroporphyrinogen decarboxylase on uroporphyrinogen III: a reassessment of the clockwise decarboxylation hypothesis. *Biochem. J.* **278**, 901–02.

Lichtenthaler, H.K. (1987). Chlorophylls and carotenoids: pigments of photosynthetic membranes. *Methods Enzymol.* **148**, 350–82.

Liu, Z., Yan, H., Wang, K., Kuang, T., Zang, J., Gul, L., An, X. and Chang, W. (2004). Crystal structure of spinach major light-harvesting complex at 2.72 Å resolution. *Nature* **428**, 287–92.

Luo, J. and Lim, C.K. (1994). Order of uroporphyrinogen III decarboxylation on incubation of porphobilinogen and uroporphyrinogen III with erythrocyte uroporphyrinogen decarboxylase. *Biochem. J.* **289**, 529–32.

Minamizaki, K., Mizoguchi, T., Goto, T., Tamiaki, H. and Fujita, Y. (2008). Identification of two homologous genes, $chlA_I$ and $chlA_{II}$, that are differently involved in isocyclic ring formation of chlorophyll *a* in the cyanobacterium *Synechocystis* sp. PCC6803. *J. Biol. Chem.* **283**, 2684–92.

Moss, G.P. (1988). IUPAC-IUB Joint Commission on Biochemical Nomenclature (JCBN): Nomenclature of tetrapyrroles (Recommendations 1986). *Eur. J. Biochem.* **178**, 277–328.

Murakami, A., Miyashita, H., Iseki, M., Adachi, K. and Mimuro, M. (2004). Chlorophyll *d* in an epiphytic cyanobacterium of red algae. *Science* **303**, 1633.

Nagata, N., Tanaka, R., Satoh, S. and Tanaka, A. (2005). Identification of a vinyl reductase gene for chlorophyll synthesis in *Arabidopsis thaliana* and implications for the evolution of *Prochlorococcus* species. *Plant Cell* **17**, 233–40.

Oelze, J. (1985). Analysis of bacteriochlorophylls. *Methods Microbiol.* **18**, 257–84.

Oh-hama, T.H., Seto, H., Otake, N. and Miyachi, S. (1982). ^{13}C-NMR evidence for the pathway of chlorophyll biosynthesis in green algae. *Biochem. Biophys. Res. Commun.* **105**, 647–52.

Oh-hama, T.H., Seto, H. and Miyachi, S. (1985a). ^{13}C-nuclear magnetic resonance studies of the biosynthesis of 5-aminolevulinic acid destined for chlorophyll biosynthesis in dark-grown *Scenedesmus obliquus. Plant Sci.* **42**, 153–58.

Oh-hama, T.H., Seto, H. and Miyachi, S. (1985b). ^{13}C-nuclear magnetic resonance studies on bacteriochlorophyll *a* biosynthesis in *Rhodopseudomonas spheroides S. Arch. Biochem. Biophys.* **237**, 72–79.

Oster, U., Tanaka, R., Tanaka, A. and Rüdiger, W. (2000). Cloning and functional expression of the gene encoding the key enzyme for chlorophyll *b* biosynthesis (CAO) from *Arabidopsis thaliana. Plant J.* **21**, 306–10.

Ouchane, S., Steunou, A.-S., Picaud, M. and Astier, C. (2004). Aerobic and anaerobic Mg-protoporphyrin monomethyl ester cyclases in purple bacteria: A strategy adopted to bypass the repressive oxygen control system. *J. Biol. Chem.* **279**, 6385–94.

Pettit, G.R., Kantoci, D., Doubec, D.L., Tucker, B.E., Petit, W.E. and Schroll, E. M. (1993). Isolation of the nickel-chlorin chelate, tunichlorin, from the South Pacific Ocean sea hare, *Dolabella auricularia. J. Nat. Prod.* **56**, 1981–84.

Phillips, J.N. (1963). Physico-chemical properties of porphyrins. In *Comprehensive Biochemistry*, vol. 9, ed. M. Florkin and E.H. Stotz. Amsterdam: Elsevier, pp. 34–72.

Porra, R.J. (1997). Recent progress in porphyrin and chlorophyll biosynthesis. *Photochem. Photobiol.* **65**, 492–516.

Porra, R.J. (2006). Spectrometric assays for plant, algal and bacterial chlorophylls. In *Chlorophylls and Bacteriochlorophylls: Biochemistry, Biophysics, Functions and Applications*, ed. B. Grimm, R.J. Porra, W. Rüdiger and H. Scheer. Dordrecht: Springer, pp. 95–107.

Porra, R.J. and Scheer, H. (2000). ^{18}O and mass spectrometry in chlorophyll research: derivation and loss of oxygen atoms at the periphery of the chlorophyll macrocycle during biosynthesis, degradation and adaptation. *Photosynth. Res.* **66**, 159–75.

Porra, R.J., Klein, O. and Wright, P.E. (1982). ^{13}C-NMR studies of chlorophyll biosynthesis in higher plants: an unequivocal proof of the participation of the C_5 pathway and evidence of a new route for the incorporation of glycine. *Biochem. Int.* **5**, 345–50.

Porra, R.J., Klein, O. and Wright, P.E. (1983). The proof by ^{13}C-NMR spectroscopy of the predominance of the C_5 pathway over the Shemin pathway in chlorophyll biosynthesis in higher plants and of the formation of the methyl ester group of chlorophyll from glycine. *Eur. J. Biochem.* **130**, 509–16.

Porra, R.J., Schäfer, W., Cmiel, E., Katheder, I. and Scheer, H. (1993). Derivation of the formyl-group oxygen of chlorophyll *b* from molecular oxygen in greening leaves of a higher plant *(Zea mays). FEBS Lett.* **323**, 31–34.

Porra, R.J., Schäfer, W., Cmiel, E., Katheder, I. and Scheer, H. (1994). The derivation of the formyl-group oxygen of chlorophyll *b* in higher plants from molecular oxygen: Achievement of high enrichment of the 7-formyl group oxygen from $^{18}O_2$ in greening maize leaves. *Eur. J. Biochem.* **219**, 671–79.

Porra, R.J., Schäfer, W., Katheder, I. and Scheer, H. (1995). The derivation of the oxygen atoms of the 13^1-oxo and 3-acetyl groups of bacteriochlorophyll *a* from water in *Rhodobacter sphaeroides* cells adapting from respiratory to photosynthetic conditions: evidence for an anaerobic pathway for the formation of *iso*cyclic ring E. *FEBS Lett.* **371**, 21–24.

Porra, R.J, Schäfer, W., Gad'on, N., Katheder, I., Drews, G. and Scheer, H. (1996). Origin of the two carbonyl oxygens of bacteriochlorophyll *a*: Demonstration of two different pathways for the formation of ring E in *Rhodobacter sphaeroides* and *Roseobacter denitrificans*, and of a common hydratase mechanism for 3-acetyl group formation. *Eur. J. Biochem.* **239**, 83–92.

Porra, R.J., Pfündel, E.P. and Engel, N. (1997). Metabolism and function of photosynthetic pigments. In *Phytoplankton Pigments in Oceanography: Guidelines to Modern Methods*, ed. S.W. Jeffrey, R.F.C. Mantoura and S.W. Wright. Paris: UNESCO Publishing, pp. 85–126.

Porra, R.J., Urzinger, M., Winkler, J., Bubenzer, C. and Scheer, H. (1998). Biosyntheis of the 3-acetyl and 13^1-oxo groups of bacteriochlorophyll *a* in the facultative aerobic bacterium, *Rhodovulum sulfidophilum*: The presence of both oxygenase and hydratase pathways for isocyclic ring formation. *Eur. J. Biochem.* **257**, 185–91.

Rebeiz, C.A., Wu, S.M., Kuhadja, M., Daniell, H. and Perkins, E.J. (1983). Chlorophyll *a* biosynthetic routes and chlorophyll *a* chemical heterogeneity in plants. *Mol. Cell. Biochem.* **57**, 97–125.

Reid, J.D. and Hunter, C.N. (2004). Magnesium-dependent ATPase activity and cooperativity of magnesium chelatase from *Synechocystis* sp. PCC6803. *J. Biol. Chem.* **279**, 26893–99.

Reinbothe, S., Reinbothe, C., Lebedev, N. and Apel, K. (1996). PORA and PORB, two light-dependent protochlorophyllide reducing enzymes of angiosperm chlorophyll biosynthesis. *Plant Cell* **8**, 763–69.

Reinbothe, S., Pollmann, S. and Reinbothe, C. (2003). *In situ* conversion of protochlorophyllide *b* to protochlorophyllide *a* in barley: Evidence for a novel role of 7-formyl reductase in the prolamellar bodies of etioplasts. *J. Biol. Chem.* **278**, 800–06.

Rüdiger, W. (2006). Biosynthesis of chlorophylls *a* and *b*: the last steps. In *Chlorophylls and Bacteriochlorophylls: Biochemistry, Biophysics, Functions and Applications*, ed. B. Grimm, R.J. Porra, W. Rüdiger and H. Scheer. Dordrecht: Springer, pp. 189–200.

Rüdiger, W. and Grimm, B. (2006). Chlorophyll metabolism, an overview. In *Chlorophylls and Bacteriochlorophylls: Biochemistry, Biophysics, Functions and Applications*, ed. B. Grimm, R.J. Porra, W. Rüdiger and H. Scheer. Dordrecht: Springer, pp. 133–46.

Rzeznicka, K., Walker, C.J., Westergren, T., Kannangara, C.G., von Wettstein, D., Merchant, S., Gough, S.P. and Hansson, M. (2005). *Xantha-l* encodes a membrane subunit of the aerobic Mg-protoporphyrin IX monomethyl ester cyclase involved in chlorophyll biosynthesis. *Proc. Natl. Acad. Sci. USA* **102**, 5886–91.

Sachs, J.P., Repeta, D.J. and Goericke, R.F. (1999). Nitrogen and carbon isotopic ratios of chlorophyll from marine phytoplankton. *Geochim. Cosmochim. Acta* **63**, 1431–41.

Sano, S. and Granick, S. (1961). Mitochondrial coproporphyrinogen oxidase and protoporphyrin formation. *J. Biol. Chem.* **236**, 1173–80.

Sauer, K., Smith, J.R.L. and Schultz, A.J. (1966). The dimerization of chlorophyll *a*, chlorophyll *b* and bacteriochlorophyll in solution. *J. Am. Chem. Soc.* **66**, 2681–88.

Scheer, H. (1991). Chemistry of chlorophylls. In *Chlorophylls*, ed. H. Scheer. Boca Raton: CRC Press, pp. 3–30.

Scheer, H, (2003). The pigments. In *Light-harvesting Antennas in Photosynthesis*, ed. B.R. Green and W.W. Parsons. Dordrecht: Springer, pp. 29–81.

Scheer, H. (2006). An overview of chlorophylls and bacteriochlorophylls: biochemistry, biophysics, functions and applications. In *Chlorophylls and Bacteriochlorophylls: Biochemistry, Biophysics, Functions and Applications*, ed. B. Grimm, R.J. Porra, W. Rüdiger and H. Scheer. Dordrecht: Springer, pp. 1–26.

Scheer, H. and Katz, J.J. (1974). Structure of bacteriochlorophyll *b*. *J. Am. Chem. Soc.* **96**, 3714–16.

Scheumann, V., Schoch, S. and Rüdiger, W. (1998). Chlorophyll *a* formation in the chlorophyll *b* reductase requires reduced ferredoxin. *J. Biol. Chem.* **273**, 35102–08.

Schneegurt, M.A. and Beale, S.I. (1992). Origin of the chlorophyll *b* formyl oxygen in *Chlorella vulgaris*. *Biochemistry* **31**, 11677–83.

Smith, J.H.C. and Benitez, A. (1955). Chlorophylls: analysis in plant materials. In *Modern Methods of Plant Analysis*, vol. IV, ed. K. Paech and M.V. Tracey. Berlin: Springer, pp. 142–96.

Smith, A.G., Marsh, O. and Elder, G.H. (1993). Investigation of the subcellular location of the tetrapyrrole-biosynthesis enzyme coproporphyrinogen oxidase in higher plants. *Biochem. J.* **292**, 503–08.

Stanier, R.Y. and Smith, J.H.C. (1960). The chlorophylls of green bacteria. *Biochim. Biophys. Acta* **41**, 478–84.

Steiner, R. (1981). *Bacteriochlorophyll b aus Ectothiorhodospira halochloris*. Zulassungsarbeit, University of Munich.

Strik, J.J. T. W.A. (1973). Chemical porphyria in Japanese quail *(Coturnix c, Japonica)*. *Enzyme* **16**, 211–23.

Su, Q., Frick, G., Armstrong, G. and Apel, K. (2001). POR C of *Arabidopsis thaliana*: a third light- and NADPH-dependent protochlorophyllide oxidoreductase that is differentially regulated by light. *Plant Mol. Biol.* **47**, 805–13.

Suzuki, T. and Shioi, Y. (2002). Re-examination of Mg-dechelation reaction in the degradation of chlorophylls using chlorophyllin *a* as a substrate. *Photosynth. Res.* **73**, 217–23.

Sytina, O.A., Heyes, D.J., Hunter, C.N., Alexandre, M.T., van Stokkum, I.H.M., van Grondelle, R. and Groot, M.L. (2008). Conformational changes in an ultrafast light-driven enzyme determine catalytic activity. *Nature* **456**, 1001–04.

Tamiaki, H., Shibata, R., and Mizoguchi, T. (2007). The 17-propionate function of (bacterio)chlorophylls: biological implication of their long esterifying chains in photosynthetic systems. *Photochem. Photobiol.* **83**, 152–62.

Tanaka, A., Ito, H., Tanaka, R., Tanaka, N.K. and Okada, K. (1998). Chlorophyll *a* oxygenase (CAO) is involved in chlorophyll *b* formation from chlorophyll *a*. *Proc. Natl. Acad. Sci. USA* **95**, 12719–23.

Van de Meent, E.J., Kobayashi, M., Erkelens, C., van Veelen, P.A., Amesz, J. and Watanabe, T. (1991). Identification of 8^1-hydroxychlorophyll *a* as a functional reaction center pigment in *Heliobacteria*. *Biochim. Biophys. Acta* **1058**, 356–62.

Walker, C.J. and Willows, R.D. (1997). Mechanism and regulation of Mg-chelatase. *Biochem. J.* **327**, 321–33.

Walker, C.J., Mansfield, K.E., Rezzano, I.N., Hanamoto, C.M. Smith, K.M. and Castelfranco, P.A. (1988). The magnesium-protoporphyrin IX (oxidative) cyclase system: studies on the mechanism and specificity of the reaction. *Biochem. J.* **255**, 685–92.

Willett, J., Smart, J.L. and Bauer, C.E. (2007). RegA control of bacteriochlorophyll and carotenoid synthesis in *Rhodobacter capsulatus*. *J. Bacteriol.* **189**, 7765–73.

Willows, R.D. (2003). Biosynthesis of chlorophylls from protoporphyrin IX. *Nat. Prod. Rep.* **20**, 327–41.

Willows, R.D. and Kriegel, A.M. (2008). Molecular structure and biosynthesis of pigments and cofactors. In *The Purple Photosynthetic Bacteria*, ed. C.N. Hunter, F. Daldal, M.C. Thurnauer and J.T. Beatty. Dordrecht: Springer, pp. 57–79.

Xu, H., Vavilin, D. and Vermaas, W. (2002). The presence of chlorophyll *b* in *Synechocystis* sp. PCC 6803 disturbs tetrapyrrole biosynthesis and enhances chlorophyll degradation. *J. Biol. Chem.* **277**, 42726–32.

Yaronskaya, E. and Grimm, B. (2006). The pathway of 5-aminolevulinic acid to protochlorophyllide and protoheme. In *Chlorophylls and Bacteriochlorophylls: Biochemistry, Biophysics, Functions and Applications*, ed. B. Grimm, R.J. Porra, W. Rüdiger and H. Scheer. Dordrecht: Springer, pp. 173–88.

Yee, W.C., Eglsaer, S.J. and Richards, W.S. (1989). Confirmation of a ping-pong mechanism for S-adenosyl-L-methionine:Mg-protoporphyrin methyltransferase of etiolated wheat by an exchange mechanism. *Biochem. Biophys. Res. Commun.* **162**, 483–90.

Zapata, M., Garrido, G.L. and Jeffrey, S.W. (2006). Chlorophyll *c* pigments: current status. In *Chlorophylls and Bacteriochlorophylls: Biochemistry, Biophysics, Functions and Applications*, ed. B. Grimm, R.J. Porra, W. Rüdiger and H. Scheer. Dordrecht: Springer, pp. 39–53.

3

Carotenoid metabolism in phytoplankton

MARTIN LOHR

3.1 Introduction

Carotenoids are among the natural products with the highest diversity. To date more than 700 different naturally occurring carotenoids have been described (Britton *et al.*, 2004), and they are virtually ubiquitous in living organisms. Carotenoids belong to the compound class of isoprenoids, and the majority of carotenoids are tetraterpenoids with a C_{40} skeleton as the basic molecular structure. Formally, they can be divided into the carotenes, which are pure hydrocarbons, and the xanthophylls, which are derived from carotenes by introduction of oxygen functions. The ability of *de novo* biosynthesis of carotenoids is not limited to land plants and algae, but is frequently encountered among prokaryotes (both eubacteria and archaebacteria) and fungi (Britton, 1998). To meet their metabolic demands, animals rely on food-borne uptake of carotenoids which then, however, can be further metabolized. An important example is β-carotene (provitamin A; trivial names for carotenes are used in this chapter – see Data sheets for alternative names such as β,β-carotene in this case) as a precursor of the visual pigment retinal in mammals and animals in general (Goodwin, 1984; von Lintig *et al.*, 2005) or – together with its hydroxylated derivatives – as precursors of ketocarotenoids used as colourants by many crustaceans, various carp species and birds like flamingos or finches (Goodwin, 1984; McGraw *et al.*, 2006). This chapter summarizes our current view on the biosynthesis of carotenoids, termed carotenogenesis, in land plants, cyanobacteria and algae with a focus on recent advances in the genetics of carotenoid biosynthesis. The scope of this review will be limited to oxygenic phototrophs; for details on the carotenogenesis of anoxygenic phototrophic bacteria, readers are referred to an excellent recent review by Maresca, Graham and Bryant (Maresca *et al.*, 2008).

When the chapter on carotenoids in the volume by Jeffrey *et al.* (1997) was written, only the enzymes of the cytosolic mevalonic acid pathway of isoprene

Phytoplankton Pigments: Characterization, Chemotaxonomy and Applications in Oceanography, eds. Suzanne Roy, Carole A. Llewellyn, Einar Skarstad Egeland and Geir Johnsen. Published by Cambridge University Press.

formation and those catalysing the early steps of carotenoid biosynthesis up to the carotenes were known. During the last decade, most of the genes involved in carotenogenesis in cyanobacteria and in land plants have been identified, and a new pathway of isoprene formation that is operative in many bacteria and in plastids has been discovered. Most of our current knowledge about the biosynthesis of carotenoids in oxygenic phototrophs stems from work on seed plants and cyanobacteria, and to some extent on green algae. Almost no experimental data are available from other algal groups, but increasingly more whole-genome data are generated enabling homology-based identification of potential carotenogenic genes in these algae. As will be discussed, the emerging picture is rather complex with a significant number of reactions being catalysed by multiple enzymes which are often phylogenetically unrelated. Other enzymes share the same ancestor, but yet catalyse different reactions. In the following, the known enzymatic steps of carotenogenesis in oxygenic phototrophs will be summarized, with special reference to carotenoid biosynthesis in algae and cyanobacteria.

3.2 Biosynthesis of carotenes

A general outline of the biosynthetic pathway of carotenoids common to land plants, algae and cyanobacteria is depicted in Figure 3.1. The formation of carotenoids can be formally divided into five stages: (1) generation of active isoprene which is the building block of all isoprenoids (steps 1–8; Section 3.2.1), (2) formation of the first carotenoid, phytoene, by stepwise condensation of eight isoprene units (steps 9–10; Section 3.2.2), (3) consecutive desaturation and isomerization reactions which extend the mesomeric π-electron system of conjugated double bonds and thus the absorbance spectrum of the molecule into the visible light range, leading to lycopene (steps 11–14; Section 3.2.3), (4) cyclization of the linear ends of lycopene yielding cyclic carotenes like α-carotene or β-carotene (steps 15–16; Section 3.2.4) and the aromatic carotenes in cyanobacteria (step 26; Section 3.2.5), and finally (5) the stepwise introduction of molecular oxygen creating a vast array of xanthophylls (Figure 3.2 and Section 3.3).

The presence of β-carotene is shared by all photosynthetic eukaryotes and cyanobacteria, implying that the general biosynthetic pathway of carotenoids was inherited from the cyanobacterial ancestor of all extant plastids. This idea is further supported by the high degree of homology between carotenogenic genes from cyanobacteria, algae and land plants (Hirschberg *et al.*, 1997). In accordance with the postulated endosymbiotic origin of the pathway in plants and algae, carotenogenesis is mostly confined to the plastids in photosynthetic eukaryotes (but see Section 3.3.4). In the land plants and algae that have been examined, however, all known enzymes of the pathway are nucleus-encoded (Lange and Ghassemian, 2003; Lohr *et al.*, 2005; Coesel *et al.*, 2008). Consequently, they are synthesized on cytoplasmic ribosomes as preproteins with N-terminal targeting signals that govern their post-translational import into the plastids (Inaba and Schnell, 2008; Bolte *et al.*, 2009).

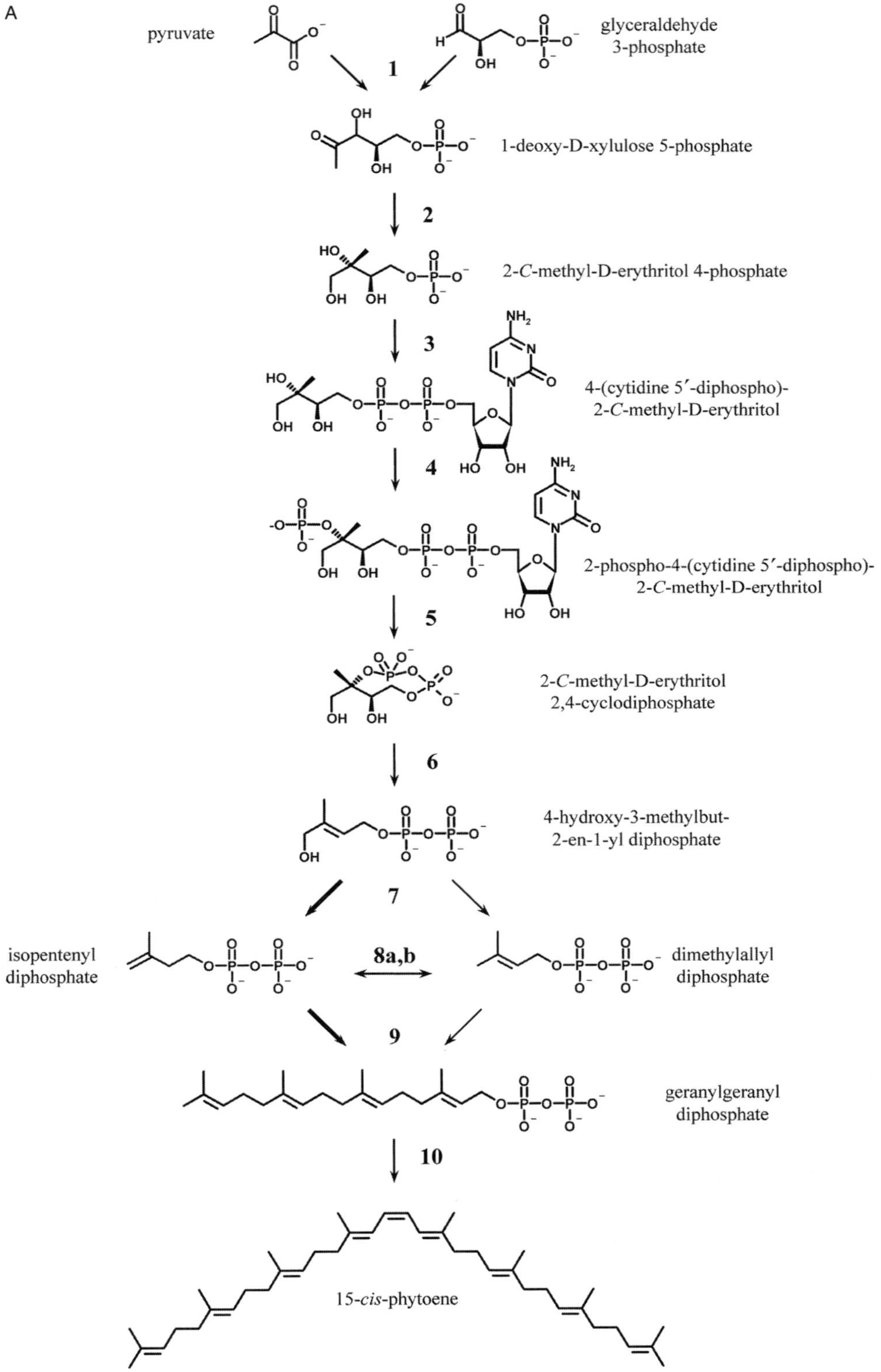

Figure 3.1. (*cont.*)

B

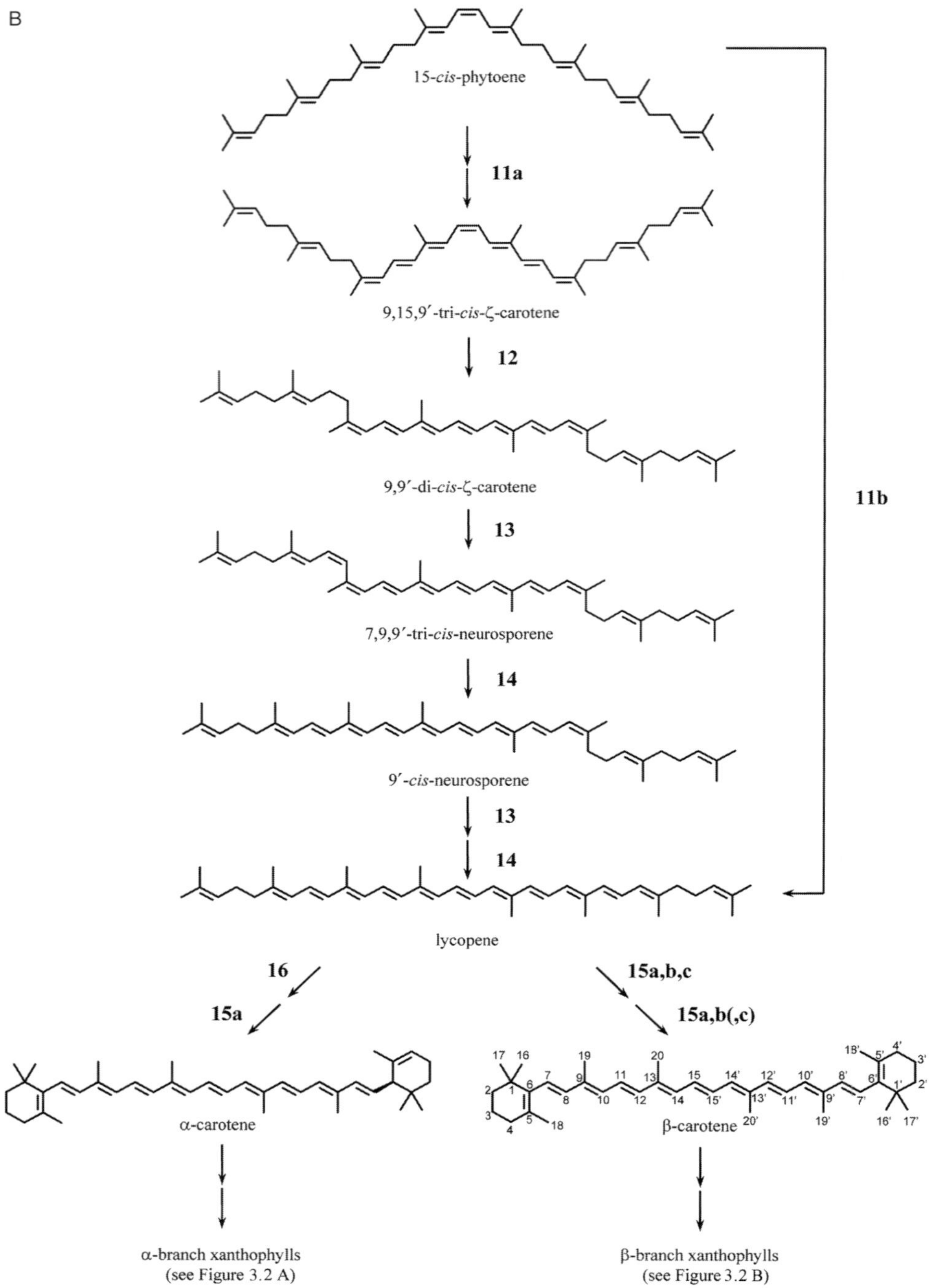

Figure 3.1. Biosynthetic pathway of carotenes in cyanobacteria and algae. Depicted are (A) the formation of the first carotenoid 15-*cis*-phyotene via the methylerythritol phosphate (MEP) pathway and (B) the subsequent generation of the cyclic carotenes. Numbers denote the reaction steps that are catalysed by the equivalently numbered enzymes in Table 3.1. Chemical structures of intermediates are according to Hirschberg (2001), Rodríguez-Concepción and Boronat (2002), and Isaacson *et al.* (2004). In (B) the numbering scheme for the carotenoid skeleton according to the rules of the International Union of Pure and Applied Chemistry (IUPAC) is exemplified using β-carotene.

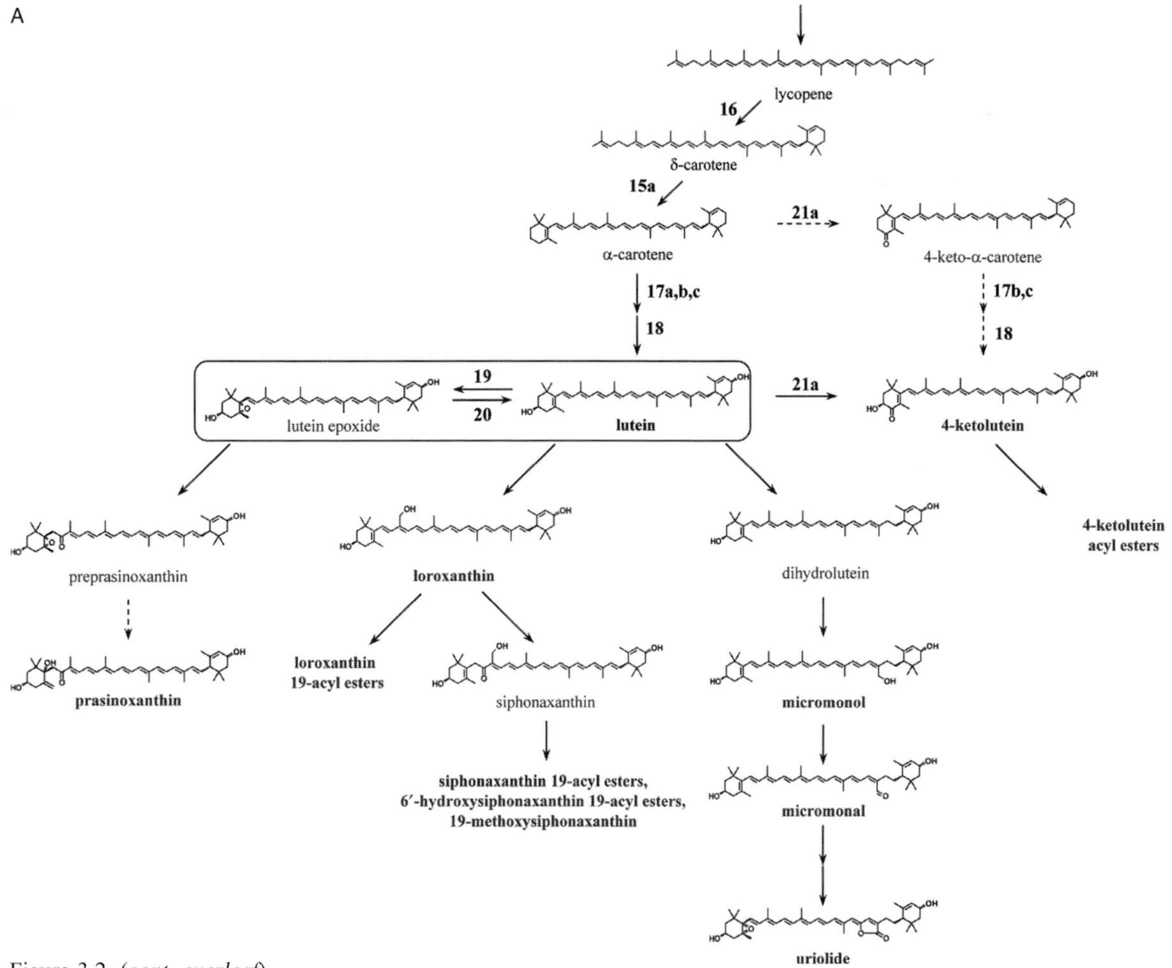

Figure 3.2. (*cont. overleaf*)

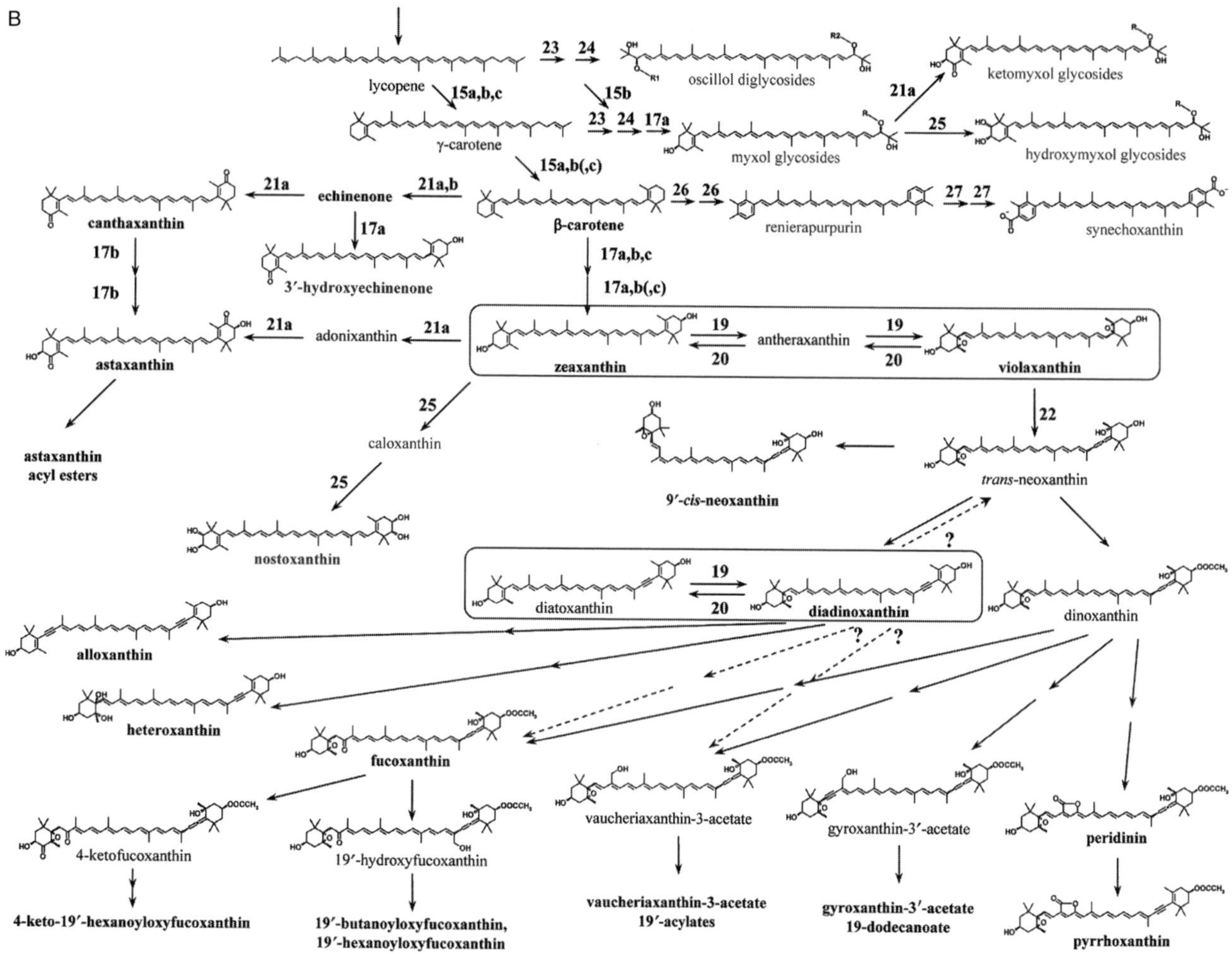

Figure 3.2. Biosynthetic pathways of the major xanthophylls in cyanobacteria and algae. Shown are pathways that are postulated to give rise to xanthophylls which are derived either (A) from α-carotene or (B) from lycopene, γ-carotene and β-carotene. The postulated biosynthetic relations are based on chemosystematic and experimental data as discussed in the text. Names of major products in the pathways are in bold, with carotenoids restricted to green algae in green, to cyanobacteria in blue and to chromalveolates in brown. Xanthophyll cycles are boxed. Dashed arrows indicate potential alternate paths. Numbers denote the reaction steps that are catalysed by the equivalently numbered enzymes in Table 3.1. Chemical structures of pigments are according to the *Carotenoids Handbook* (Britton *et al*., 2004). See colour plate section.

3.2.1 Formation of the active isoprene

The 'active isoprene' unit isopentenyl diphosphate and its isomer dimethylallyl diphosphate are the basic building blocks not only of carotenoids, but also of other isoprenoids like sterols and the phytol or geranylgeraniol moieties of (bacterio-) chlorophylls and have long been thought to be generated solely by the cytosolic mevalonic acid (MVA) pathway. During the last decade, however, it was recognized that cyanobacteria and many other bacteria as well as the plastids of eukaryotic phototrophs synthesize isoprene from pyruvate and glyceraldehyde-3-phosphate by an alternative route, which is now known as the deoxy-xylulose phosphate pathway or methylerythritol phosphate (MEP) pathway. The method of analysing the ^{13}C-labelling pattern of isoprenoids by NMR after feeding of ^{13}C-glucose proved to be crucial to the discovery of this non-mevalonate pathway (Lichtenthaler, 1999; Eisenreich *et al.*, 2004). The results indicated that in land plants and algae, carotenoids and related isoprenoids like phytol are synthesized from pyruvate and glyceraldehyde phosphate, while sterols or the polyprenyl chain of the mitochondrial ubiquinone are derived from acetyl-coenzyme A via mevalonic acid. Cloning and sequencing of the genes encoding the first enzyme of the non-mevalonate pathway, 1-deoxy-D-xylulose 5-phosphate synthase (DXS, Figure 3.1A, step 1), from *Mentha piperita*, *Arabidopsis thaliana* and *Chlamydomonas reinhardtii*, and the discovery that the proteins contained an N-terminal plastid targeting sequence proved that the MEP pathway occurs in the plastids of plants and algae (Lichtenthaler, 1999; Rodríguez-Concepción and Boronat, 2002).

Thus, in plants and algae two different pathways of isoprene formation coexist, the cytosolic MVA pathway and the plastidic MEP pathway. In *A. thaliana*, however, perturbation of the MEP pathway by inhibition of the enzyme 1-deoxy-D-xylulose 5-phosphate reductoisomerase (step 2) by fosmidomycin cannot be compensated by the cytosolic MVA pathway (Laule *et al.*, 2003). Mutants of *A. thaliana* defective in the reductoisomerase, the 2-*C*-methyl-D-erythritol 4-phosphate cytidylyltransferase (step 3) or the 4-hydroxy-3-methylbut-2-enyl diphosphate reductase (step 7) are either seedling-lethal (Budziszewski *et al.*, 2001) or display an albino phenotype (Guevara-García *et al.*, 2005) indicating that at least in seed plants the MVA pathway does not complement the MEP pathway. Furthermore, a downregulated expression of 1-deoxy-D-xylulose 5-phosphate reductoisomerase in *A. thaliana* resulted in a variegated phenotype (Carretero-Paulet *et al.*, 2006), again suggesting that there is only limited exchange of active isoprene between plastids and cytosol.

Exceptions found so far are green algae and euglenids. The chlorophytes *C. reinhardtii*, *Chlorella fusca* and *Scenedesmus obliquus* have been shown by analysis of the isotope-labelling patterns of carotenoids and sterols (Disch *et al.*, 1998) and by comparative genomics (Lohr *et al.*, 2005), to use only the MEP pathway for biosynthesis of all cellular isoprenoids. The available genomes from chlorophytes of the genus *Chlorella* and from prasinophyte algae of the genera *Ostreococcus* and *Micromonas* also lack the genes of the MVA pathway (Lohr, unpublished), suggesting that they also synthesize isoprenoids exclusively via the MEP pathway. In the case of the euglenophyte *Euglena gracilis*, isotope-labelling studies indicated that both the MVA pathway and the MEP pathway contribute to the synthesis of carotenoids (Kim *et al.*, 2004).

Meanwhile, the MEP pathway has been elucidated, and all genes/enzymes involved (Table 3.1) have been identified (for detailed reviews see Rodríguez-Concepción and Boronat, 2002; Eisenreich *et al.*, 2004; Hunter, 2007). 1-Deoxy-D-xylulose 5-phosphate synthase (DXS, Figure 3.1A, step 1) catalyses a transketolase reaction in which pyruvate is decarboxylated and the C_2-residue is transferred to glyceraldehyde 3-phosphate yielding 1-deoxy-D-xylulose 5-phosphate. This molecule serves as a precursor not only of isoprenoids but also of thiamine (Julliard and Douce, 1991) and pyridoxol (Hill *et al.*, 1989). As the committed step of isoprene biosynthesis, 1-deoxy-D-xylulose 5-phosphate is converted to 2-*C*-methyl-D-erythritol 4-phosphate (MEP) by an intramolecular rearrangement and subsequent saturation, which are catalysed by 1-deoxy-D-xylulose 5-phosphate reductoisomerase (DXR or IspC, step 2).

The next three steps lead to cyclization of MEP resulting in an activated cyclodiphosphate. First, the phosphocytidyl moiety of CTP is transferred to the phosphate at C-4 of MEP by 2-*C*-methyl-D-erythritol 4-phosphate cytidylyltransferase (MCT/IspD, step 3), then the product is phosphorylated at C-2 by 4-(cytidine 5′-diphospho)-2-*C*-methyl-D-erythritol kinase (CMK/IspE, step 4) using ATP. Finally, elimination of CMP by 2-*C*-methyl-D-erythritol 2,4-cyclodiphosphate synthase (MDS/IspF, step 5) leads to formation of an intramolecular phosphoanhydride bond.

Reductive cleavage between C-2 and the phosphoanhydride and subsequent dehydration yield 4-hydroxy-3-methylbut-2-en-1-yl disphosphate, the reaction being catalysed by hydroxymethylbutenyl diphosphate synthase (HDS/IspG, step 6). In the last step, hydroxymethylbutenyl diphosphate is reduced by 4-hydroxy-3-methylbut-2-en-1-yl diphosphate reductase (HDR/IspH, step 7) to a stoichiometric ratio of five molecules isopentenyl diphosphate per one molecule dimethylallyl diphosphate (Adam *et al.*, 2002; Rohdich *et al.*, 2002). Hydroxymethylbutenyl diphosphate synthase and hydroxymethylbutenyl diphosphate reductase have been shown to contain iron-sulfur clusters as redox-active compounds (Wolff *et al.*, 2003; Seemann *et al.*, 2005) and in the case of hydroxymethylbutenyl diphosphate synthase it was demonstrated that ferredoxin is capable of donating electrons directly to the enzyme (Seemann *et al.*, 2006).

The stoichiometries of isopentenyl diphosphate and dimethylallyl diphosphate can readily be adjusted by the action of isopentenyl diphosphate:dimethylallyl diphosphate isomerase (IDI, step 8) which catalyses the rapid equilibration between the two pools. Two phylogenetically unrelated types of isopentenyl diphosphate isomerases are known; the type-I enzyme (8a) is common to part of the eubacteria and to phototrophic and heterotrophic eukaryotes while type-II (8b) is found in archaea and the other eubacteria including cyanobacteria (Table 3.1) (Steinbacher *et al.*, 2003).

Table 3.1. *Distribution of known carotenogenic genes/enzymes in the genomes of various algae and cyanobacteria.*

Data were compiled from the references in the text and were complemented by BLAST searches for homologues in GenBank (http://www.ncbi.nlm.nih.gov/GenBank) and the Genome Portal of the Joint Genome Institute (http://genome.jgi-psf.org). The presence of a unique homologue in a genome is indicated by '+', multiple highly similar homologues are denoted by the respective copy number; 'n.i.' = not identified; n.h. = no orthologue. See Figure 3.1, Figure 3.2 and text for function of the enzymes. Abbreviations of organisms are: *C.m.* = *Cyanidioschyzon merolae; C.r.* = *Chlamydomonas reinhardtii; E.h.* = *Emiliania huxleyi; G.s.* = *Galdieria sulphuraria; O.l.* = *Ostreococcus lucimarinus; O.t.* = *Ostreococcus tauri; P.t.* = *Phaeodactylum tricornutum; T.p.* = *Thalassiosira pseudonana; V.c.* = *Volvox carteri;* 6803 = *Synechocystis* sp. PCC 6803; 7002 = *Synechococcus* sp. PCC 7002; 7421 = *Gloeobacter violaceus* PCC 7421; 7942 = *Synechococcus elongatus* PCC 7942; 9313 = *Prochlorococcus marinus* MIT 9313.

Step	Enzyme (*gene*)[1]	Full name	EC	*C.r.* *V.c.*	*O.l.* *O.t.*	*C.m.* *G.s.*	*P.t.* *T.p.*	*E.h.*	7421	6803	7002	7942	9313
1	DXS	1-deoxy-D-xylulose 5-phosphate synthase	2.2.1.7	+	+	+	+	+	+	+	+	+	+
2	DXR/IspC	1-deoxy-D-xylulose 5-phosphate reductoisomerase	1.1.1.267	+	+	+	+	+	+	+	+	+	+
3	MCT/IspD	2-*C*-methyl-D-erythritol 4-phosphate cytidyltransferase	2.7.7.60	+	+	+	+	+	+	+	+	+	+
4	CMK/IspE	4-(cytidine 5′-diphospho)-2-*C*-methyl-D-erythritol kinase	2.7.1.148	+	+	+	+	+	+	+	+	+	+
5	MDS/IspF	2-*C*-methyl-D-erythritol 2,4-cyclodiphosphate synthase	4.6.1.12	+	+	+	+	+	+	+	+	+	+
6	HDS/IspG	4-hydroxy-3-methylbut-2-en-1-yl diphosphate synthase	1.17.4.3	+	+	+	+	+	+	+	+	+	+
7	HDR/IspH	4-hydroxy-3-methylbut-2-en-1-yl diphosphate reductase	1.17.1.2	+	+	+	+	+	+	+	+	+	+

Table 3.1. (*cont.*)

Step	Enzyme (*gene*)[1]	Full name	EC	*C.r.* *V.c.*	*O.l.* *O.t.*	*C.m.* *G.s.*	*P.t.* *T.p.*	*E.h.*	7421	6803	7002	7942	9313
8a	IDI type I	isopentenyl diphosphate:dimethylallyl diphosphate isomerase type I	5.3.3.2	+	+	+	2	+	–	–	–	–	–
8b	IDI type II	isopentenyl diphosphate:dimethylallyl diphosphate isomerase type II	5.3.3.2	–	–	–	–	–	–	+	+	+	–
9	GGPS/CrtE	geranylgeranyl diphosphate synthase	2.5.1.29	+	+	+	+	+	+	+	+	+	+
10	PSY/CrtB	phytoene synthase	2.5.1.32	+	2	+	+/2	+	+	+	+	+	+
11a	PDS/CrtP	phytoene desaturase		+	2	+	2	+	–	+	+	+	+
11b	CrtI	phytoene desaturase		–	–	–	–	–	+	–	–	–	–
12	Z-ISO	ζ-carotene isomerase		n.i.	n.i.	n.i.	n.i.	n.i.	–	n.i.	n.i.	n.i.	n.i.
13	ZDS/CrtQb	ζ-carotene desaturase	1.14.99.30	+	n.o.	+	+	+	–	+	+	+	+
14	CRTISO/CrtH	carotenoid isomerase		+	+	n.o.	n.o.	n.o.	n.o.	+	+	+	+
15a	LCYB/CrtL-b	lycopene β-cyclase		+	+	+	+	+	–	–	–	+	+
15b	CruA	lycopene β-cyclase		–	–	–	–	–	+	+	+	–	–
15c	CruP	lycopene β-cyclase		+[2]	+[2]	+[2]	+[2]	+[2]	+	+	+	+	–
16	LCYE/CrtL-e	lycopene ε-cyclase		+	+	–	–	–	–	–	–	–	+
17a	CrtR	carotene β-hydoxylase (non-haem iron)		–	–	+	–	–	–	+	+	+	+
17b	CHYB/CrtZ	carotene β-hydoxylase (non-haem iron)		+	+	–	–/+[3]	–	–	–	–	–	–
17c	CYP97A	carotene β-hydoxylase (cytochrome P450)		2/+	+	–	–	–	–	–	–	–	–
18	CYP97C	carotene ε-hydoxylase (cytochrome P450)		+	+	–	–	–	–	–	–	–	–
19	ZEP	zeaxanthin epoxidase	1.14.13.90	+	2	–	3/2	2	–	–	–	–	–

20	VDE	violaxanthin de-epoxidase	1.10.99.3	n.o.	+	−	+	+	−	−	−	−	−
21a	BKT/CrtW	carotene β-ketolase		+	−	−	−	−	+	−	+	−	−
21b	CrtO	carotene β-ketolase		−	−	−	−	−	+	+	−	−	−
22	NSY	neoxanthin synthase	5.3.99.9	+	+	−	+	+	+[4]	+[4]	+[4]	+[4]	−
23	CruF	carotenoid ψ-1-hydroxylase		−	−	−	−	−	+	+	+	−	−
24	CruG	carotenoid ψ-2′-*O*-glycosyltransferase		−	−	−	−	−	+	+	+	−	+
25	CrtG	2,2′-β-hydroxylase		−	−	−	−	−	−	+	−	+	−
26	CruE	carotenoid β-ring desaturase/methyltransferase		−	−	−	−	−	+	+	+	−	−
27	CruH	carotenoid χ-ring C18-hydroxylase		−	−	−	−	−	+	+	+	−	−

[1] Abbreviations of enzymes from eukaryotic/prokaryotic (if available) phototrophs. Abbreviations of eukaryotic enzymes are according to Hirschberg (2001), Cunningham (2002) and Phillips *et al.* (2008). In cases in which there are multiple synonyms, abbreviations indicative of catalytic functions of the proteins were preferred. For carotenogenic enzymes from bacteria, a different nomenclature is commonly used (Armstrong, 1997). Here, affiliation of a gene locus/enzyme to the pathway of carotenoid biosynthesis is indicated by the abbreviation 'Crt' for 'carotenoid' (or 'Isp' for 'isoprene', in the case of enzymes of the isoprene-forming MEP pathway), followed by a capital letter which is assigned in alphabetical order to newly discovered genes. Meanwhile, the number of known carotenogenic genes has become larger than the size of the alphabet. Consequently, by advancing the third letter in 'Crt' by one, new genes have been labelled with 'Cru' plus a capital letter.

[2] The function of the CruP gene in eukaryotic phototrophs is not known.

[3] CHYB in *T. pseudonana* is probably a pseudogene.

[4] As cyanobacteria do not synthesize neoxanthin, the function of the NSY orthologue in these organisms is unknown.

In general, the genes of the MEP pathway are highly conserved in oxygenic phototrophs. Likely orthologues of all genes from land plants are present in the available algal genomes of the green algae *Chlamydomonas reinhardtii*, *Volvox carteri*, *Ostreococcus tauri* and *O. lucimarinus*, the two red algae *Cyanidioschyzon merolae* and *Galdieria sulphuraria*, the two diatoms *Thalassiosira pseudonana* and *Phaeodactylum tricornutum*, the brown alga *Ectocarpus siliculosus* and the haptophyte *Emiliania huxleyi* (Table 3.1) (Lohr *et al*., 2005; Wilhelm *et al*., 2006; Frommolt *et al*., 2008). For several MEP enzymes, there is an apparently lower similarity between the plant/algal enzymes and the respective cyanobacterial homologues. It has been suggested that this stems from a replacement of the corresponding genes in the ancestor of extant cyanobacteria with homologues from other eubacteria sometime after the emergence of photosynthetic eukaryotes by primary endosymbiosis (Lange *et al*., 2000). Recent evidence, however, indicates that early in the establishment of the primary endosymbiosis some of the plant/algal genes were replaced by homologues from Chlamydia-like bacteria (Huang and Gogarten, 2007; Becker *et al*., 2008; Moustafa *et al*., 2008).

3.2.2 Formation of the first carotenoid phytoene

The formation of the diterpene geranylgeranyl diphosphate is catalysed by geranylgeranyl diphosphate synthase (GGPS or CrtE, step 9) through stepwise condensation of one dimethylallyl diphosphate with three molecules of isopentenyl diphosphate (Kuntz *et al*., 1992), resulting in gradual elongation of the isoprenoid backbone. Geranylgeranyl diphosphate synthase belongs to an extended family of prenyltransferases, with the other members catalysing the formation of polyprenyls of different chain lengths (Sandmann, 1994).

Phytoene, the first carotenoid, is synthesized from two molecules of geranylgeranyl diphosphate by the enzyme phytoene synthase (PSY/CrtB, step 10). Head-to-head condensation of the two geranylgeranyl diphosphates generates a central double bond in *cis*-configuration yielding 15-*cis*-phytoene as product (Misawa *et al*., 1994).

3.2.3 Formation of lycopene

All-*trans*-lycopene is synthesized from 15-*cis*-phytoene by a complex sequence of reactions involving multiple desaturations and isomerizations. In land plants, algae and most extant cyanobacteria, probably four different enzymes are involved: two desaturases and two *cis-trans*-isomerases (Li *et al*., 2007; Sandmann, 2009). The initial reaction (Figure 3.1B, step 11a) is the two-step-desaturation of 15-*cis*-phytoene to 9,15,9′-tri-*cis*-ζ-carotene that is catalysed by phytoene desaturase (PDS/CrtP) (Bartley *et al*., 1991; Linden *et al*., 1991; Pecker *et al*., 1992). The electrons liberated in this reaction are transferred to the cosubstrate plastoquinone (Norris *et al.*, 1995) and then funneled into the photosynthetic electron chain. In the dark or during the greening process of land plants, the electrons are consumed by a plastid terminal oxidase that reduces molecular oxygen to water (Carol *et al.*, 1999; Kuntz, 2004; Shahbazi *et al.*, 2007). Genes encoding CrtP-type phytoene desaturases have been identified in numerous plants and algae (Hirschberg, 1998).

Experimental data gained with recombinant phytoene desaturase (step 11a) and ζ-carotene desaturase (ZDS/CrtQ, step 13) (Bartley *et al.*, 1999; Matthews *et al.*, 2003; Breitenbach and Sandmann, 2005) and with mutants of *C. reinhardtii* (Nikulina *et al.*, 1999), *Euglena gracilis* (Cunningham and Schiff, 1985), maize (Li *et al.*, 2007) and tomato (J. Hirschberg, personal communication) which accumulate 9,15,9′-tri-*cis*-ζ-carotene in the dark, demonstrated that the 15-*cis* double bond of ζ-carotene needs to be isomerized to *trans* before further desaturations towards lycopene. The postulated ζ-carotene isomerase (step 12) has not yet been identified, but as noted above, mutant plants and algae with a likely defective isomerase have been described and the locus termed Z-ISO (Li *et al.*, 2007).

Next, a *cis*-double bond is introduced at C-7 of 9,9′-di-*cis*-ζ-carotene by ζ-carotene desaturase (step 13). As is the case for phytoene desaturase, ζ-carotene desaturase probably uses plastoquinone as electron acceptor in the oxidation of ζ-carotene (Norris *et al.*, 1995; Shahbazi *et al.*, 2007). The resulting 7,9,9′-tri-*cis*-neurosporene needs to be isomerized to 9′-*cis*-neurosporene by another carotenoid isomerase (step 14) termed CrtH in cyanobacteria (Breitenbach *et al.*, 2001; Masamoto *et al.*, 2001) and CRTISO in land plants (Isaacson *et al.*, 2002; Park *et al.*, 2002). This is followed by another sequence of desaturation at the C-7′ of 9′-*cis*-neurosporene by ζ-carotene desaturase and subsequent isomerization by CrtH/CRTISO, finally yielding all-*trans* lycopene (Isaacson *et al.*, 2004).

Phytoene desaturase and ζ-carotene desaturase are homologous proteins that most likely result from duplication of an ancient gene (Sandmann, 2002). Orthologues of both enzymes are present in the available genomes of cyanobacteria, algae and land plants with the exceptions of the basal cyanobacterium *Gloeobacter violaceus* and the prasinophyte algae *O. tauri* and *O. lucimarinus* (Table 3.1). The prasinophyte genomes lack genes for ζ-carotene desaturase, but instead contain two gene copies of CrtP-type phytoene desaturase suggesting that one of them might have acquired ζ-carotene desaturase activity. The genomes of the diatoms *T. pseudonana* and *P. tricornutum*, however, also contain two phytoene desaturase genes which are orthologous to the genes from the prasinophytes, but in addition have a gene for ζ-carotene desaturase (Coesel *et al.*, 2008; Frommolt *et al.*, 2008). The significance of the two putative phytoene desaturase genes in diatoms is not clear.

The cyanobacterium *G. violaceus* contains neither enzymes homologous to the CrtP-type phytoene desaturase nor to ζ-carotene desaturase (CrtQ). Instead, it utilizes a different type of phytoene desaturase termed CrtI that is able to catalyse all four desaturation steps as well as the isomerization reactions, yielding *trans*-lycopene (Figure 3.1, step 11b) (Steiger *et al.*, 2005; Tsuchiya *et al.*, 2005). CrtI is not related to the plant/algal carotenoid desaturases but instead shows significant similarity to CrtH/CRTISO. Likely orthologues of CrtI are found in all carotenogenic bacteria except modern cyanobacteria and thus are supposed to represent the most ancient type of phytoene desaturases (Sandmann, 2002, 2009). For CRTISO, orthologues are present in the green algal genomes (Lohr *et al.*, 2005; Lohr, unpublished). The available genomes of red algae and diatoms contain no obvious orthologues, but several potential open reading frames with similarity to CRTISO and CrtI were detected (Wilhelm *et al.*, 2006; Lohr, unpublished).

3.2.4 Formation of cyclic carotenes

At the stage of lycopene, the pathway of carotenoid biosynthesis in oxygenic phototrophs splits into two branches (Figures 3.1B and 3.2), the α-branch leading to the formation of cyclic carotenoids with one β-ionone ring (double bond between C-5 and C-6) and one ε-ionone ring (double bond between C-4 and C-5), and the β-branch with carotenoids containing two β-ionone rings. While β-carotene and its derived xanthophylls are ubiquitous among oxygenic phototrophs, α-carotene and related xanthophylls are mostly confined to land plants, green algae and some species of red algae and cyanobacteria (Egeland *et al.*, 1997; Hess *et al.*, 2001; Schubert *et al.*, 2006). In the case of chromalveolate algae, which originated by secondary endosymbiosis involving a red alga, the occurrence of xanthophylls with an ε-ionone ring has rarely been reported (see below).

Meanwhile, four different types of phylogenetically unrelated protein families are known to convert the two linear ψ ends of a lycopene molecule into ionone rings: (i) the CrtY cyclases of non-photosynthetic proteobacteria, actinobacteria and the photosynthetic Chloroflexaceae, (ii) the heterodimeric cyclases CrtYc/CrtYd restricted to gram-positive and a few gram-negative bacteria, (iii) the plant-type lycopene cyclases (LCY or CrtL), and (iv) the CruA-type cyclases from green sulfur bacteria (Sandmann, 2002; Maresca *et al.*, 2007, 2008).

Among cyanobacteria, some species have been found to possess LCY/CrtL-type cyclases (Cunningham *et al.*, 1994; Hess *et al.*, 2001) while all others contain genes with homology to the CruA-type cyclases. The cyanobacterium *Synechococcus* sp. PCC 7002 was shown to possess two paralogues named CruA (step 15b) and CruP (15c). Whereas CruP-deficient *Synechococcus* mutants had no pigment phenotype, deletion of the *cruA* gene resulted in a strong accumulation of lycopene and a concommitant decrease of β-carotene indicating that *cruP* can only partially compensate the loss of *cruA*. Expression of the *cruP* gene in lycopene-producing *E. coli* cells led to formation of γ-carotene, but very little β-carotene, suggesting that *cruP* may be involved preferentially in the formation of monocyclic carotenoids like myxoxanthophyll (Maresca *et al.*, 2007, see also Section 3.3.5).

In general, cyanobacteria synthesize only carotenoids containing β-ionone rings, like β-carotene, zeaxanthin, echinenone or myxol glycosides (Figure 3.2B) (Takaichi and Mochimaru, 2007). Exceptions are species of the genus *Prochlorococcus*, which contain α-carotene and derived xanthophylls with one β-ionone and one ε-ionone ring, indicating that they possess more than one lycopene cyclase. In accordance with this assumption, the genome of the cyanobacterium *Prochlorococcus marinus* strain MED4 (CCMP1986) was found to contain two highly similar genes encoding proteins with homology to plant-type LCY cyclases (Hess *et al.*, 2001). When expressed in *Escherichia coli* cells engineered to accumulate lycopene, one of the genes (*crtL-b*) induced the expected formation of β-ionone rings, whereas the protein encoded by the other paralogue (*crtL-e*) was able to synthesize both β- and ε-ionone rings (Stickforth *et al.*, 2003).

Green algae and land plants in general contain carotenoids that are derived from both α- and β-carotene and, accordingly, they were shown to contain two *LCY* genes as well (Cunningham *et al.*, 2007; Frommolt *et al.*, 2008). One (*LCYB* or *CRTL-B*, step 15a) encodes a protein displaying β-cyclase activity towards both ψ ends of lycopene, while the other (*LCYE*/*CRTL-E*; step 16) catalyses the formation of a single ε-ionone ring per lycopene molecule (Cunningham *et al.*, 1996).

Among rhodophytes, there are species with either zeaxanthin or lutein as the major dihydroxy-xanthophyll (Bjørnland and Aguilar-Martinez, 1976; Marquardt and Hanelt, 2004; Schubert *et al.*, 2006), but so far only the genomes of two zeaxanthin-containing red algae have been sequenced and found to encode a single LCY-type cyclase; the respective gene product from the red alga *C. merolae* catalysed the formation of β-carotene when expressed in lycopene-accumulating *E. coli* cells (Cunningham *et al.*, 2007). For chromalveolate algae, the occurrence of xanthophylls with an ε-ionone ring has not been reported, with the exception of some haptoflagellates (reviewed in Bjørnland and Liaaen-Jensen, 1989). In accordance with this observation, the genomes of the diatoms *T. pseudonana* and *P. tricornutum* contain only a single copy of the *LCY* gene.

Interestingly, the available genomes of land plants and algae (Table 3.1) contain a gene encoding a protein with high similarity to the *cruA*-type cyclases from green sulfur bacteria that is probably orthologous to *cruP* from *Synechococcus* sp. PCC 7002. This observation in combination with data on the EST-frequencies of LCYB and *cruP* in cDNA preparations from different tissues of rice, led Maresca and coworkers to speculate that the *cruP* gene in land plants might encode the principal β-cyclase (Maresca *et al.*, 2007). Recently, however, a mutant of *A. thaliana* with a strongly reduced content of carotenoids in the β-branch and a concomittant increase in the α-branch carotenoids was identified. This pigment phenotype was shown to result from a point mutation in the *LCYB* gene that led to an amino acid exchange in the protein and consequent reduction of its enzymatic activity, thus arguing for LCYB to be the major lycopene β-cyclase in land plants (Li *et al.*, 2009).

3.2.5 Formation of aromatic carotenes in cyanobacteria

In some carotenoid producers, β-ionone rings are further desaturated yielding aromatic rings. The introduction of two additional double bonds to the ionone ring, however, necessitates the dislocation of at least one of the two methyl groups at C-1. Displacement of one methyl group from C-1 to C-2 results in a ϕ-ring (methyl groups at C-1, C-2 and C-5), while additional displacement of the C-5-methyl group to C-3 gives rise to a χ-ring (methyl groups at C-1, C-2, C-3). Aromatic carotenoids are known from anoxygenic phototrophs, more specifically from some purple bacteria of the Chromatiaceae synthesizing the carotenoid okenone with one χ-ring, and from green sulfur bacteria (Chlorobiaceae) containing carotenoids with one or two ϕ-rings (Goodwin, 1980; Maresca *et al.*, 2008). For cyanobacteria and other oxygenic phototrophs, however, aromatic carotenoids had not been reported (Goodwin, 1980; Mohamed and Vermaas, 2006; Takaichi and Mochimaru, 2007). Recently, several cyanobacteria have been shown to synthesize derivatives of the χ-ring-containing carotenoid renierapurpurin, with the newly discovered synechoxanthin (Graham *et al.*, 2008) being the major aromatic carotenoid produced (Figure 3.2B and Section 3.3.5), and a gene encoding the β-ring desaturase/methyltransferase activity necessary for χ-ring formation (Figure 3.2B, step 26) has been identified by insertional mutagenesis in *Synechococcus* sp. PCC 7002 and *Synechocystis* sp. PCC 6803 (Graham and Bryant, 2008). The predicted gene product is homologous to the ϕ-ring-generating desaturase/methyltransferase CrtU from anoxygenic phototrophs and has been named CruE (Graham and Bryant, 2008). This has been found in the genomes of various cyanobacteria including the basal species *Gloeobacter violaceus* (Table 3.1) (Frigaard *et al.*, 2004; Graham and Bryant, 2008), indicating that the gene was already present in the ancestor of extant cyanobacteria and that some cyanobacteria may have secondarily lost it, rendering them unable to generate aryl carotenoids.

3.3 Biosynthesis of xanthophylls

Xanthophylls are oxygen-containing derivatives of the carotenes. They have important functions in photoprotection, and they are major light-harvesting pigments in most chlorophyll *a*/*c*-containing algae (chromalveolates) (Wilhelm, 1990). Land plants and many green algae display a rather uniform pattern of photosynthetic xanthophylls with lutein, violaxanthin and 9′-*cis*-neoxanthin as the major components, and most red algae contain either lutein or zeaxanthin as the sole xanthophyll. On the other hand, cyanobacteria, some prasinophytes, and in particular the chromalveolate algae synthesize a plethora of other xanthophylls, many of which are specific to defined taxonomic groups, allowing their use as chemosystematic markers (Bjørnland and Liaaen-Jensen, 1989). This bewildering diversity of pigments corresponds to an equally perplexing variety of pathways (Figure 3.2A and 3.2B), and while considerable progress has been made in understanding the genetic basis of xanthophyll diversity in land plants and in cyanobacteria, almost nothing is known about the enzymes that are involved in the synthesis of the more derived xanthophylls in chromalveolates and prasinophytes.

Xanthophylls with a hydroxyl group at C-3 of the ionone rings are ubiquitous among oxygenic phototrophs, and hydroxylation is the first step in the formation of the majority of xanthophylls. Other important oxygenation reactions that have been explored in some detail are ketolation at C-4 and epoxidation at C-5 and C-6 of β-ionone rings. Hydroxylation (Section 3.3.1) and epoxidation (Section 3.3.2) generate xanthophylls that are localized in the photosynthetic pigment-protein complexes, whereas ketolation (Section 3.3.4) is typical for secondary carotenoids which accumulate in algal resting stages under unfavourable environmental conditions. The reactions leading to the formation of the major light-harvesting xanthophylls in prasinophytes and chromalveolates are largely unexplored (Section 3.3.3), whereas considerable progress has recently been made in understanding the genetic basis of xanthophyll diversity in cyanobacteria (Section 3.3.5).

3.3.1 Hydroxylation

In most xanthophylls from cyanobacteria, algae and land plants the ionone rings are hydroxylated at C-3. This reaction can be accomplished by three unrelated types of enzymes as a result of convergent evolution. Two are non-haem iron monoxygenases, namely the CrtR-proteins and the CHYB- or CrtZ-type enzymes, while the third type of enzymes are members of the cytochrome P450 superfamily (Tian and DellaPenna, 2004; Kim and DellaPenna, 2006).

The CrtR-enzymes are found only in cyanobacteria. They share no homology with the hydroxylases from other bacteria and land plants, but are related to green algal and cyanobacterial carotenoid ketolases (Masamoto *et al.*, 1998; see also Section 3.3.4). As shown by pigment analyses of corresponding mutant strains of *Synechocystis* sp. PCC 6803 and *Nostoc* (*Anabaena*) sp. PCC 7120, the cyanobacterial CrtR (Figure 3.2, step 17a) appears to catalyse the hydroxylation of deoxymyxol or its glycosides to myxol or myxol glycosides, respectively (Lagarde and Vermaas, 1999; Mochimaru *et al.*, 2008). When heterologously expressed in β-carotene-accumulating *E. coli* cells, CrtR from zeaxanthin-containing cyanobacteria like *Synechocystis* sp. PCC 6803 also converted β-carotene to zeaxanthin, but CrtR from zeaxanthin-less species as *Nostoc* sp. PCC 7120 or *Anabaena variabilis* did not show significant hydroxylase activity (Makino *et al.*, 2008; Mochimaru *et al.*, 2008; Scaife *et al.*, 2009). 4-keto-β-ionone rings are probably not substrates of CrtR, because heterologous expression of CrtR from *Nostoc*, *Anabaena* and *Synechocystis* in *E. coli* cells engineered to accumulate canthaxanthin (4,4′-keto-β-carotene) did not lead to formation of astaxanthin (Makino *et al.*, 2008).

The second type of non-haem iron monoxygenase which is present in land plants, green algae (here abbreviated as CHYB; Sun *et al.*, 1996) and in non-photosynthetic carotenogenic bacteria (here termed CrtZ; Misawa *et al.*, 1995) not only catalyses the hydroxylation of β-carotene to zeaxanthin, but is also capable of hydroxylating the 4-keto-derivatives of β-carotene, i.e. echinenone and canthaxanthin (Fraser *et al.*, 1997; Fraser *et al.*, 1998; Linden, 1999) and therefore is likely to be involved in the formation of astaxanthin and other hydroxylated ketocarotenoids as well (Figure 3.2, step 17b; see also Section 3.3.4).

A third type of carotenoid hydroxylase, which was first identified in *A. thaliana* belongs to the large superfamily of cytochrome P450 enzymes (Tian *et al.*, 2004). Both *A. thaliana* (Tian *et al.*, 2004; Kim and DellaPenna, 2006) and rice (*Oryza sativa*; Quinlan *et al.*, 2007) have been shown to contain two closely related P450 enzymes of the CYP97 family, which catalyse the hydroxylation of ionone rings. One of them, CYP97A, strongly favours β-ionone rings as substrate (Figure 3.2, step 17c) (Fiore *et al.*, 2006; Kim and DellaPenna, 2006) whereas the other, CYP97C, is the only enzyme identified so far with hydroxylase activity specific towards ε-ionones (Figure 3.2A, step 18) (Tian *et al.*, 2004; Quinlan *et al.*, 2007; Kim *et al.*, 2009). Orthologues of the two P450 enzymes are present in the genomes of other land plants and green algae (Kim *et al.*, 2009). Inhibitor studies with intact cells of the green alga *Haematococcus pluvialis* suggested that a P450 enzyme may also be involved in the formation of astaxanthin (Schoefs *et al.*, 2001). It has not yet been examined, however, whether the CYP97-hydroxylases accept ketocarotenoids like echinenone, canthaxanthin or 4-keto-α-carotene as substrates.

The C-3-hydroxylated carotenoids zeaxanthin and lutein are ubiquitous among green algae, and in support of this observation, genes encoding putative carotenoid hydroxylases of both the CHYB and the P450 type are present in the genomes of the four green algae *C. reinhardtii*, *V. carteri*, *O. tauri* and *O. lucimarinus* (Table 3.1). The genomes of the two zeaxanthin-producing unicellular red algae *C. merolae* and *G. sulphuraria* encode neither homologues of CHYB nor cytochrome P450 enzymes belonging to the CYP97 family. Instead, the plastid genomes of both red algae and of a third basal rhodophyte, *Cyanidium caldarium*, contain genes encoding proteins with high similarity to the cyanobacterial CrtR-hydroxylases, but the plastid genomes from multicellular rhodophytes lack a respective homologue (Cunningham *et al.*, 2007). A recent study of the putative *crtR* gene from *C. merolae* failed to detect hydroxylase activity after heterologous expression in β-carotene producing *E. coli* cells (Cunningham *et al.*, 2007). It is possible that the red algal genes encode proteins that need a particular cofactor which is absent from *E. coli*, but without experimental proof, the nature of the carotenoid hydroxylases in red algae remains an open question.

In the genome of the centric diatom *T. pseudonana*, a gene model with homology to CHYB hydroxylases was predicted (Wilhelm *et al.*, 2006; Cunningham *et al.*, 2007). Yet the predicted ORF would encode a rather derived CHYB protein that might have acquired a different function or it may represent a pseudogene (Coesel *et al.*, 2008). The recently completed genome sequence of the pennate diatom *P. tricornutum* lacks genes with apparent homology to *CHYB*. Instead, two genes encoding P450 enzymes of the CYP97 family have tentatively been assigned as being involved in the hydroxylation of carotenoids and were designated as *LTL1* and *LTL2* (Coesel *et al.*, 2008). Orthologues of these genes are present in *T. pseudonana* and another diatom, *Skeletonema costatum*. In the latter, the single yet identified orthologue has been annotated as *CYP97E1* (Yang *et al.*, 2003). The respective proteins are most closely related to enzymes of the CYP97B family (Nelson, 2006) for which experimental evidence of function is still lacking.

3.3.2 Epoxidation, de-epoxidation and the xanthophyll cycles

The C-3-hydroxylated β-ionone rings of xanthophylls like zeaxanthin can be epoxidized at C-5/C-6 by zeaxanthin epoxidase (ZEP; Figure 3.2, step 19) and the reaction can be reverted by violaxanthin de-epoxidase (VDE; step 20). This so-called xanthophyll cycle is an important photoprotective mechanism in land plants and most algae, enabling them to respond to fluctuating light intensities on the time scale of minutes (Müller *et al.*, 2001; Matsubara *et al.*, 2003; García-Plazaola *et al.*, 2007; Lavaud, 2007; see also Chapter 11, this volume). Three different types of xanthophyll cycles have been described which are named after their major substrates, i.e. the violaxanthin/antheraxanthin/zeaxanthin cycle, the diadinoxanthin/diatoxanthin cycle and the lutein-epoxide/lutein cycle. They have in common the high-light-induced de-epoxidation of a light-harvesting xanthophyll (violaxanthin, diadinoxanthin or lutein-epoxide) to a photoprotective pigment species (zeaxanthin, diatoxanthin or lutein, respectively) that promotes the conversion of excess excitation energy as heat; the reaction is reversed in low light or darkness. While the lutein-epoxide cycle has only been found in a limited number of land plants (García-Plazaola *et al.*, 2007), the violaxanthin cycle is present in the majority of land plants, in green algae (Chlorophyta) and several groups of stramenopiles, i.e. the brown algae (Phaeophyceae), eustigmatophytes, chrysophytes, synurophytes and some raphidophytes (Hager and Stransky, 1970a, 1970b; Bjørnland and Liaaen-Jensen, 1989; Guillou *et al.*, 1999). Other stramenopiles, e.g. the diatoms (Bacillariophyceae), xanthophytes, bolidophytes and pelagophytes, contain the diadinoxanthin cycle, as do haptoflagellates and dinoflagellates (Hager and Stransky, 1970b; Stransky and Hager, 1970a; Bjørnland and Liaaen-Jensen, 1989; Guillou *et al.*, 1999; Dimier *et al.*, 2009). Cyanobacteria, cryptophytes and the majority of red algae, all using phycobiliproteins as light-harvesting proteins, lack epoxy-xanthophylls and consequently do not employ xanthophyll cycling for photoprotection (Hager and Stransky, 1970b; Stransky and Hager, 1970b; Andersson *et al.*, 2006). Some red algae have been described as containing antheraxanthin (Marquardt and Hanelt,

2004; Schubert *et al.*, 2006), and in one case the rapid cycling of xanthophylls in a red algal species has been reported (Rmiki *et al.*, 1996), but for several other antheraxanthin-accumulating rhodophytes no xanthophyll cycling was observed (Marquardt and Hanelt, 2004; Andersson *et al.*, 2006; Schubert and Garcia-Mendoza, 2008; Esteban *et al.*, 2009). Moreover, cyanobacterial photosystems have been shown to dissipate excess light energy by alternative mechanisms that are unrelated to xanthophyll cycle-induced photoprotection (Delphin *et al.*, 1998; Kirilovsky, 2007; Wilson *et al.*, 2008). In the light of current evidence, it is questionable whether the xanthophyll cycle mechanism would be compatible with photosynthetic antenna systems that are mainly comprised of phycobiliproteins.

Genes encoding zeaxanthin epoxidase have been cloned and their protein products functionally characterized from several land plants as, e.g. *Nicotiana plumbaginifolia* (Marin *et al.*, 1996) and *Gentiana lutea* (Zhua *et al.*, 2003), and a functional homologue has been identified in *C. reinhardtii* (Baroli *et al.*, 2003). By catalysing the formation of violaxanthin, zeaxanthin epoxidase not only participates in the xanthophyll cycle but also provides the substrate for formation of 9′-*cis*-neoxanthin and abscisic acid (see Sections 3.3.3 and 3.4.1).

Consistent with the absence of epoxy-xanthophylls in most red algae, no zeaxanthin epoxidase homologues are present in the genomes of *C. merolae* and *G. sulphuraria*. The two sequenced diatom genomes, however, contain multiple genes with homology to zeaxanthin epoxidase (Table 3.1) (Wilhelm *et al.*, 2006; Coesel *et al.*, 2008; Frommolt *et al.*, 2008). Zeaxanthin epoxidase in diatoms is likely to be involved in the epoxidation of diatoxanthin to diadinoxanthin as part of the photoprotective xanthophyll cycle and is probably also responsible for the formation of violaxanthin, which has been shown to be present in diatoms and other chromalveolate algae (Lohr and Wilhelm, 1999). Similarly to the situation in diatoms, available genomic and EST data from other chromalveolates indicate the presence of multiple copies of zeaxanthin epoxidase in haptoflagellates and in the brown alga *Ectocarpus siliculosus* (Frommolt *et al.*, 2008; and Table 3.1), but as yet no experimental proof of the catalytic function of the potential isozymes from these algae has been obtained.

A gene encoding violaxanthin de-epoxidase has first been isolated from lettuce (*Lactuca sativa*) (Bugos and Yamamoto, 1996) and homologues have been identified in several other land plants (Bugos *et al.*, 1998), but not in the green algae *C. reinhardtii* and *V. carteri* (Anwaruzzaman *et al.*, 2004; Lohr *et al.*, 2005). The genomes of two basal green algae of the Prasinophyceae, however, contain genes encoding putative violaxanthin de-epoxidases as do the available genomes of chromalveolates (Coesel *et al.*, 2008; Frommolt *et al.*, 2008); the two rhodophytes lack respective homologues as would be expected for algae that do not display xanthophyll cycling (Frommolt *et al.*, 2008; and Table 3.1).

Interestingly, both the zeaxanthin epoxidase and violaxanthin de-epoxidase sequences from stramenopiles and other chromalveolates are most closely related to the respective homologues from basal green algae of the Prasinophyceae

(Frommolt *et al.*, 2008), suggesting that these two groups share orthologous genes. Based on these observations, it has been proposed that chromalveolate algae inherited their photoprotective xanthophyll cycle from ancient green algae by lateral or even endosymbiotic gene transfer (Frommolt *et al.*, 2008). Chemosystematic considerations indicate that beside the genes encoding these two enzymes, chromalveolates and green algae may share other homologues involved in the biosynthesis of xanthophylls (Frommolt *et al.*, 2008). There are several striking examples of unusual oxygen functions in carotenoids both from groups that are generated by yet unknown enzymes, as for example, the keto group at position C-8 in fucoxanthin from diatoms which is also found at C-8 in preprasinoxanthin from prasinophytes, and the lactone ring that is present in peridinin and pyrrhoxanthin from dinoflagellates, as well as in uriolide from prasinophytes (Egeland *et al.*, 1995, 1996; Frommolt *et al.*, 2008).

3.3.3 Formation of light-harvesting xanthophylls

The hydroxylated xanthophylls lutein and zeaxanthin and the epoxy-xanthophyll violaxanthin are also precursors of a variety of further modified xanthophylls which serve as accessory light-harvesting pigment in the photosynthetic apparatus (Figure 3.2). In land plants and green algae, the allenic xanthophyll 9′-*cis*-neoxanthin is synthesized from all-*trans*-violaxanthin (step 22), since mutants with a defective zeaxanthin epoxidase lack both violaxanthin and neoxanthin (Marin *et al.*, 1996; Baroli *et al.*, 2003). The 9′-*cis*-neoxanthin has also been found in euglenophyte and chlorarachniophyte algae (Goodwin, 1980; Takaichi and Mimuro, 1998; Schagerl *et al.*, 2003). The enzymatic processes involved in the transformation of violaxanthin into 9′-*cis*-neoxanthin are not well understood. Genes that encode enzymes with neoxanthin synthase activity have been found in potato (Al-Babili *et al.*, 2000) and tomato (Bouvier *et al.*, 2000) and were identified as paralogues of the plant-type LCY lycopene cyclases. Other studies suggested, however, that the tomato paralogue serves mainly as a fruit-specific lycopene β-cyclase (Ronen *et al.*, 2000), and the genomes of land plants like rice or *A. thaliana* and of the green alga *C. reinhardtii* contain no orthologues of these particular cyclases (Hirschberg, 2001; Lohr *et al.*, 2005). Recently, a mutant of *A. thaliana* deficient in abscisic acid synthesis was identified which accumulates violaxanthin but lacks any isomers of neoxanthin, and the phenotype was shown to result from the knockout of a single gene encoding a protein with no significant similarity to any enzyme with known function in the databases (North *et al.*, 2007). While an effort to demonstrate by in vitro assay that the heterologously expressed protein has neoxanthin synthase activity failed, it probably is directly involved in the formation of neoxanthin in *A. thaliana*. Interestingly, putative orthologues are present in the genomes not only of green algae and diatoms but also of cyanobacteria which do not contain neoxanthin or other related allenic xanthophylls

(Table 3.1), suggesting that the protein has another as yet unknown function in cyanobacteria (North *et al.*, 2007).

Among green algae, ulvophytes and prasinophytes display a particularly complex pattern of xanthophylls which are all derived from α-carotene (Figure 3.2A). Chemosystematic analyses suggest that siphonaxanthin and its esters are derived from lutein via loroxanthin and that prasinoxanthin may be synthesized from lutein via lutein-epoxide (Egeland *et al.*, 1997; Yoshii *et al.*, 2005; Yoshii, 2006). In the case of micromonol, micromonal and uriolide, dihydrolutein is the likely precursor. However, while the most parsimonious chemosystematic scenario would predict that 7,8-dihydrolutein is synthesized from neurosporene (7,8-dihydrolycopene) (Egeland *et al.*, 1997), changes in the pool size of individual carotenoids in the prasinophyte *Mantoniella squamata* during treatment with different light regimes indicated that dihydrolutein may be derived from lutein (Böhme *et al.*, 2002).

Chromalveolate algae show a similar complexity of xanthophylls which, however, are all derived from β-carotene (Figure 3.2B). Early chemosystematic considerations based on the structure elucidation of carotenoids led to the proposal of violaxanthin as the common precursor to all carotenoids with either a 6,8-allenic or 7,8-acetylenic group (Bonnett *et al.*, 1969; Milborrow, 1982; Bjørnland and Liaaen-Jensen, 1989). Experimental support came from the analyses of pigment labelling kinetics in pulse-chase experiments. For the dinoflagellate *Amphidinium carterae*, a cell homogenate was shown to convert radioactive zeaxanthin into *trans*-neoxanthin and further into diadinoxanthin or peridinin which accumulated as the labelled end-products (Swift *et al.*, 1980). Similarly, the labelling kinetics of carotenoids in the diatom *Thalassiosira weissflogii* after pulse-chase with radioactive bicarbonate suggested the presence of two pools of diadinoxanthin, with one supplying precursors for the biosynthesis of fucoxanthin (Goericke and Welschmeyer, 1992). In the haptoflagellate *E. huxleyi*, fucoxanthin was proposed to be a precursor of 19′-hexanoyloxy fucoxanthin as inferred from biosynthetic rate calculations for steady-state growth under various culture conditions (Stolte *et al.*, 2000).

Studies of pigment conversion kinetics in the diatoms *P. tricornutum* and *C. meneghiniana* during treatment with inhibitors of *de-novo* carotenoid biosynthesis implied that both violaxanthin and antheraxanthin are intermediates in the formation of diadinoxanthin and fucoxanthin (Lohr and Wilhelm, 2001). Similar studies showed that in various groups of chromalveolate algae, e.g. the chrysophytes, raphidophytes, xanthophytes, eustigmatophytes, haptophytes and dinophytes, the xanthophyll cycle pigments can be converted into the major light-harvesting carotenoids (Lohr, 2001). This observation suggests that all these algae are able to efficiently 'recycle' the photoprotective carotenoids that have been accumulated in excess light to meet an increased demand in light-harvesting xanthophylls under subsequent light-limited growth conditions.

Based on the available experimental and chemosystematic data (see also Bjørnland and Liaaen-Jensen, 1989; Zapata, 2005; Yoshii, 2006; Garrido *et al.*,

2009), Figure 3.2B summarizes the putative biosynthetic relations of xanthophylls on the β-branch. It should be noted that the role of neoxanthin as an intermediate in the synthesis of the major allenic light-harvesting xanthophylls has been confirmed only in the case of peridinin from the dinoflagellate *A. carterae* (Swift *et al.*, 1980) and that the proposed involvement of dinoxanthin lacks experimental support. Furthermore, it is not known whether the transformation of diadinoxanthin into fucoxanthin proceeds via reconversion of its acetylenic group into an allenic group, i.e. the back reaction to neoxanthin, or instead via formation of a new allenic group at the other end of the diadinoxanthin molecule.

3.3.4 Formation of ketocarotenoids

The ability to synthesize ketocarotenoids is encountered frequently in green algae, but a significant number of cyanobacteria, non-photosynthetic bacteria and fungi are capable of ketocarotenoid formation as well (Goodwin, 1980; Takano *et al.*, 2006). In green algae, ketocarotenoids are located in cytosolic lipid globules where they accumulate only during unfavourable environmental conditions (Boussiba, 2000; Jin *et al.*, 2006). Consequently, they are not involved in photosynthesis, but are thought to protect cellular components like lipids and proteins from oxidative damage. In addition, they may shield the plastids from excess light (Jin *et al.*, 2006). In cyanobacteria, ketocarotenoids are found in both cytoplasmic and thylakoid membranes (Domonkos *et al.*, 2009) and a small amount is present even in the cytosol (Scherzinger and Al-Babili, 2008). In the cyanobacteria *Arthrospira maxima* and *Synechocystis* sp. PCC 6803, the ketocarotenoid 3′-hydroxyechinenone has been identified as a specific component of the orange carotenoid protein (OCP) which is involved in a photoprotective mechanism promoting thermal dissipation of excess energy in the photosynthetic antennae analogous to the xanthophyll cycle in most plants and algae (Kirilovsky, 2007; Wilson *et al.*, 2008; see also below).

The most popular ketocarotenoid is astaxanthin, which has raised significant commercial interest due to its use as food additive in fish farming, as a food colouring agent and as a potential cancer-preventing antioxidant (Boussiba, 2000; Makino *et al.*, 2008). Astaxanthin can be synthesized from β-carotene or zeaxanthin through the sequential addition of keto groups at C-4 of both ionone rings by β-carotene ketolase, *BKT* (Figure 3.2B, step 21) (Misawa *et al.*, 1995).

The algal ketolase gene, *BKT*, which was first identified in *H. pluvialis* (Kajiwara *et al.*, 1995; Lotan and Hirschberg, 1995), is homologous to the (cyano-) bacterial genes *crtW* (encoding carotenoid ketolase) and *crtR* (encoding carotenoid hydroxylase) and to the ER-localized omega-6 fatty acid desaturases from land plants and algae (Bouvier *et al.*, 2005b). *H. pluvialis* was found to contain three highly similar *BKT* genes that are differentially expressed in response to nitrogen limitation, salt stress or high light, indicating that the multiple copies of *BKT* allow for the rapid accumulation of large amounts of astaxanthin in a gene-dosage dependent manner (Huang *et al.*, 2006a). Interestingly, immunocytological studies of *H. pluvialis* detected *bkt* in both the chloroplast and cytosolic lipid droplets. After fractionation of the cells, however, *bkt* activity was observed only in the cytosolic lipid fraction, indicating that ketolation of carotenoids in *H. pluvialis* takes place outside the plastid (Grünewald *et al.*, 2001). Inhibitor studies implied that β-carotene is the metabolite being exported from the chloroplast in order to supply the cytosolic ketocarotenoid formation (Grünewald and Hagen, 2001). Recently, a putative *BKT* gene was also discovered in the genome of *C. reinhardtii* (Grossman *et al.*, 2004; Lohr *et al.*, 2005). Whereas *H. pluvialis* and other green algae accumulate ketocarotenoids as vegetative cells, the accumulation of ketocarotenoids in *C. reinhardtii* was found to be limited to the diploid zygospores, with 4-ketolutein and its fatty acyl esters being the major ketocarotenoids (Lohr, 2009; Lohr, unpublished). This suggests that the ability to synthesize ketocarotenoids may be more widespread among green algae than currently known, but has been overlooked in those cases where it is restricted to zygospores.

The *bkt* enzyme from *H. pluvialis* and CrtW from various non-photosynthetic bacteria and cyanobacteria have been shown to preferentially catalyse the formation of canthaxanthin from β-carotene (Breitenbach *et al.*, 1996; Fraser *et al.*, 1997; Fraser *et al.*, 1998), whereas other ketolases like the recently identified *bkt* from *Muriella (Chlorella) zofingiensis* and several cyanobacterial CrtW proteins displayed significant activity also with zeaxanthin as substrate (Huang *et al.*, 2006b; Makino *et al.*, 2008; Scaife *et al.*, 2009). Some cyanobacteria contain two CrtW proteins which differ in their substrate specificities, one of them being able to act on both β-carotene and zeaxanthin (Steiger and Sandmann, 2004; Makino *et al.*, 2008) as well as on the single hydroxy-ionone ring in myxol glycosides (Mochimaru *et al.*, 2005). Many cyanobacteria contain a second type of ketolase, CrtO, which is unrelated to CrtW and instead has significant sequence similarity to bacterial phytoene dehydrogenases of the CrtI-type (Fernández-González *et al.*, 1997; Tao and Cheng, 2004). Cyanobacterial CrtO-proteins have been found to act asymmetrically on β-carotene by introducing only a single keto group to one of the ionone rings, thus giving rise to echinenone (Fernández-González *et al.*, 1997; Mochimaru *et al.*, 2005; Scaife *et al.*, 2009). Hydroxylation of the unsubstituted β-ionone ring of echinenone by CrtR yields 3′-hydroxyechinenone (Figure 3.2B, step 17a) which has recently been identified in some cyanobacteria as an essential component of the photoactive protein OCP involved in photoprotection (Wilson *et al.*, 2008; Punginelli *et al.*, 2009).

Interestingly, there are several reports on the occurrence of ketocarotenoids in euglenophyte algae (Hager and Stransky, 1970b; Heelis *et al.*, 1979; Grung and Liaaen-Jensen, 1993) and in chromalveolates. Among the latter, e.g. eustigmatophyte algae of the genus *Nannochloropsis* have been shown to contain canthaxanthin and astaxanthin (Antia and Cheng, 1982; Lubián *et al.*, 2000), and small amounts of echinenone have been detected in two haptophytes of the genus *Isochrysis* (Hager and

Stransky, 1970b; Berger *et al.*, 1977). In addition, various haptophytes were shown to contain the novel ketocarotenoids 4-ketofucoxanthin and 4-keto-19′-hexanoyloxy-fucoxanthin (Egeland *et al.*, 2000; Zapata *et al.*, 2004). Finally, the occurrence of echinenone, canthaxanthin or astaxanthin and its esters was reported for several dinoflagellate species (Johansen *et al.*, 1974; Withers and Haxo, 1978; Bjørnland, 1990; Frassanito *et al.*, 2006). It will be interesting to identify the ketolase in these algae and to determine whether it is related to the enzyme of chlorophytes. The presence of CrtW-ketolases in extant cyanobacteria would suggest that algal ketolases trace back to the cyanobacterial endosymbiont and that the respective gene has been lost multiple times during evolution of the algae. However, several cyanobacteria have been found to contain both CrtW and CrtO, so it is possible that some of the euglenophyte and chromalveolate algae may contain a CrtO-type ketolase or even a third type of ketolase which is neither related to CrtW nor to CrtO.

3.3.5 Formation of xanthophylls specific to cyanobacteria

In recent years, increasing research on carotenoids in cyanobacteria has revealed that these organisms possess a larger biosynthetic capacity for carotenoids than was thought before and several new enzymes have been discovered. Among the major carotenoids unique to cyanobacteria (Figure 3.2B) are carotenoid glycosides like oscillol glycosides (also known as oscillaxanthin) and myxol glycosides (myxoxanthophyll), carotenoids with one or two 2,3-dihydroxy-β-ionone rings, namely caloxanthin and nostoxanthin, and the aromatic xanthophyll synechoxanthin (Graham *et al.*, 2008) that has been identified as a significant component in pigment extracts from several cyanobacteria (Graham and Bryant, 2008).

Formation of the oscillol or myxol moiety in the carotenoid glycosides necessitates several modifications at the linear ψ end(s) of lycopene. Firstly, a hydroxylase that has been termed CruF adds a hydroxy group at C-1 (step 23), which is the putative committed step in the synthesis of myxol and oscillol (Graham and Bryant, 2009). For the synthesis of oscillol, a second hydroxyl is added at C-1′ by CruF, whereas the formation of myxol proceeds with cyclization of the unmodified ψ end by CruA (step 15b), yielding the intermediate 1′-hydroxy-γ-carotene. This carotenoid may alternatively result from hydroxylation of γ-carotene by CruF. Two more modifications at the linear end(s) are necessary, the addition of another hydroxy group at C-2(′) and the introduction of a double bond between C-3(′) and C-4(′). The corresponding hydroxylase and desaturase have not yet been identified and a recent study failed to detect any intermediates between 1′-hydroxy-γ-carotene and myxol, leading the authors to propose that the reactions are catalysed by either a single protein or two tightly coupled enzymes (Graham and Bryant, 2009). In *Synechocystis* sp. PCC 7002, the final glycosylation of myxol has been shown to be accomplished by a protein named CruG displaying C-2′-O-glycosyltransferase activity (step 24). The CruG appears to accept various methyl pentose sugars like rhamnose, fucose or chinovose as substrates as was indicated by the identification of various sugars in cyanobacterial myxol glycosides (Foss *et al.*, 1986) and by a GDP-fucose-deficient mutant of *Anabaena* sp. PCC 7120 accumulating myxol-rhamnoside instead of myxol-fucoside (Mochimaru *et al.*, 2008).

Some cyanobacteria of the genera *Synechococcus* and *Thermosynechococcus* are able to introduce additional hydroxyl groups to zeaxanthin at C-2 and C-2′ yielding the trihydroxy-carotene caloxanthin and the tetrahydroxy-carotene nostoxanthin (Figure 3.2B, step 25). Similar reactions have been described to occur in proteobacteria of the genus *Brevundimonas* in which the ketocarotenoid astaxanthin was shown to be hydroxylated at C-2 and C-2′ by an enzyme named CrtG (Nishida *et al.*, 2005; Tao *et al.*, 2006). The genomes of nostoxanthin-producing cyanobacteria were found to contain putative open reading frames with homology to CrtG, suggesting that the respective gene products may be responsible for the formation of caloxanthin and nostoxanthin from zeaxanthin in these organisms (Takaichi and Mochimaru, 2007). This has recently been confirmed experimentally for the cyanobacterium *Thermosynechococcus elongatus* BP-1 by means of a CrtG-deficient mutant (Iwai *et al.*, 2008). Moreover, this mutant was no longer able to synthesize 2-hydroxymyxol glycosides, indicating that CrtG also catalyses the hydroxylation of myxol glycosides at C-2 (Figure 3.2B).

Finally, oxidation of the methyl groups (C-18/C-18′) at C-3 of the two χ-rings in the aryl carotene renierapurpurin to carboxyl groups yields the aromatic xanthophyll synechoxanthin. In *Synechococcus* sp. PCC 7002, the knockout of a gene termed *cruH* (step 27) resulted in the accumulation of renierapurpurin, suggesting that its translation product catalyses the oxidation of C-18/C-18′ of renierapurpurin as the committed step in the synthesis of synechoxanthin (Graham and Bryant, 2008). It is not known whether CruH is able to catalyse all reaction steps necessary for oxidation of the methyl groups to carboxyl groups, or whether additional oxidases are involved.

3.4 Carotenoid catabolism and carotenoids as precursors of other physiologically important metabolites

While most of the enzymes involved in the formation of carotenoids in cyanobacteria, green algae and land plants have now been identified, much less is known about carotenoid degradation. Most of our knowledge on carotenoid catabolism derives from studies on carotenoids as precursors of other molecules with important physiological functions in plants, algae or cyanobacteria, namely abscisic acid (ABA), retinal and strigolactones. The first step in the generation of these metabolites is an enzymatic cleavage of carotenoids by oxygenases, with the cleavage products being termed apocarotenoids. During the last decade, various closely related carotenoid cleavage oxygenases (CCO) with different substrate specificities have been characterized from land plants and more recently also from cyanobacteria (Auldridge *et al.*, 2006; Kloer and Schulz, 2006). The detection of ABA, retinal and other apocarotenoid-derived compounds in various algae (see below) suggests a corresponding cleavage activity in these organisms. In accordance with this assumption, the available genomes of green algae and diatoms have been found to encode multiple CCO homologues (Lohr, unpublished). However, none of the putative algal CCO enzymes has yet been examined.

3.4.1. Abscisic acid

Abscisic acid (ABA) is best known as a principal phytohormone in land plants that is involved in various processes like reserve accumulation, desiccation tolerance and dormancy in seeds or the regulation of stomatal closure and of drought- or cold-stress related gene expression during vegetative growth (Leung and Giraudat, 1998). Hence its metabolism in land plants has been the subject of numerous studies (for a recent review, see Nambara and Marion-Poll, 2005). However, detection of ABA has also been reported for cyanobacteria (Hirsch *et al.*, 1989; Zahradníčková *et al.*, 1991; Maršálek *et al.*, 1992; Esch *et al.*, 1994), green algae (Hirsch *et al.*, 1989; Tietz *et al.*, 1989; Cowan and Rose, 1991; Saradhi *et al.*, 2000; Jirásková *et al.*, 2009; Stirk *et al.*, 2009), brown algae (Boyer and Dougherty, 1988; Schaffelke, 1995; Nimura and Mizuta, 2002; Stirk *et al.*, 2009), and a diatom (Kentzer and Mazur, 1991), indicating that ABA biosynthesis may represent an ancient pathway. Although its physiological significance in unicellular organisms has been studied in less detail, ABA in green algae appears to be involved in the response to abiotic stress (Hirsch *et al.*, 1989). In the case of *H. pluvialis*, ABA has been shown to induce cyst formation during drought-stress (Kobayashi *et al.*, 1997), and in *C. reinhardtii* increased levels of ABA promoted the activity of antioxidant enzymes (Yoshida *et al.*, 2003; 2004) and the repair of photo-damaged photosystem II during light stress (Saradhi *et al.*, 2000).

In land plants, cleavage of 9′-*cis*-neoxanthin has been identified as the committed step in the formation of ABA. One type of CCO, the nine-*cis*-epoxy carotenoid dioxygenase (NCED), catalyses the asymmetric cleavage of the *cis*-epoxycarotenoids 9′-*cis*-neoxanthin and 9-*cis*-violaxanthin into a C_{25}-apoaldehyde and the C_{15} molecule xanthoxin (Figure 3.3); the respective enzyme from maize (also known as VP14) was the first CCO to be identified (Schwartz *et al.*, 1997). The subsequent reactions of ABA formation occur in the cytosol. In *A. thaliana*, a major part of xanthoxin was found to be oxidized by a short-chain dehydrogenase/reductase termed ABA2, yielding the intermediate abscisic aldehyde, which is further oxidized to ABA by the ABA-specific aldehyde oxidase AAO3 (Figure 3.3) (Taylor *et al.*, 2005; Wasilewska *et al.*, 2008).

The genome of *A. thaliana* has been found to contain a small family of *NCED* paralogues which are differentially expressed in a tissue-specific manner (Tan *et al.*, 2003), and multiple genes encoding functional NCED isozymes have been reported from other land plants, such as avocado (Chernys and Zeevaart, 2000) or potato (Destefano-Beltran *et al.*, 2006). The genome of the green alga *C. reinhardtii* contains at least four putative *CCO* genes, but none is a likely orthologue of the *NCED* family from land plants (Lohr, unpublished). Similarly, ABA2 belongs to a large family of short-chain dehydrogenases and has no obvious orthologue in *C. reinhardtii*. The involvement of enzymes which are members of large families of closely related proteins impedes elucidation of ABA synthesis in cyanobacteria and algae based solely on a comparative genomics approach. In the case of cyanobacteria, an additional complication is the lack of 9-*cis*-epoxy-xanthophylls, which are the precursors of ABA in plants. In contrast to green and brown algae, however, the chemical identity of ABA from cyanobacteria has not yet been rigorously established by mass spectrometry.

3.4.2 Retinal

Other enzymes of the CCO family are involved in the generation of retinal (Figure 3.3) which is the chromophore of rhodopsin-type photoreceptors and ion-transporters. Two unrelated types of rhodopsins are known: type I or archaeal rhodopsins have been found in archaea, eubacteria and various algae (see below), while type II rhodopsins are only known from animals and a few green algae (Spudich *et al.*, 2000; Andreeva and Kutuzov, 2001; Jung *et al.*, 2003; Hegemann, 2008). Type I rhodopsins include light-dependent ion pumps and sensory rhodopsins, whereas type II rhodopsins belong to the large family of G protein-coupled receptors (Andreeva and Kutuzov, 2001).

Based on their substrate preferences, retinal-forming carotenoid oxygenases can be classified into two categories. The first, represented by the animal β-carotene-15,15′-cleavage oxygenase BCO (von Lintig *et al.*, 2005) and the fungal enzymes CarX (Prado-Cabrero *et al.*, 2007) and UmCco1 (Estrada *et al.*, 2009), catalyses the symmetric cleavage of β-carotene at the C-15-C-15′ double bond into two molecules of retinal (Giuliano *et al.*, 2003). The second category, represented by the apocarotenoid cleavage oxygenases (ACO) from the cyanobacteria *Synechocystis* sp. PCC 6803 (Ruch *et al.*, 2005) and *Nostoc* sp. PCC 7120 (Marasco *et al.*, 2006; Scherzinger *et al.*, 2006), does not utilize C_{40}-carotenoids, but mediates the cleavage of the C-15-C-15′ site in apocarotenals or apocarotenols with variable chain length. Cleavage of these apocarotenoids leads to one molecule of retinal and an apocarotene dialdehyde with a chain length depending on that of the substrate (Figure 3.3). Genes encoding putative ACO enzymes have been found in the genomes of various cyanobacteria, and ACO from *Synechocystis* sp. PCC 6803 and *Nostoc* sp. PCC 7120 has

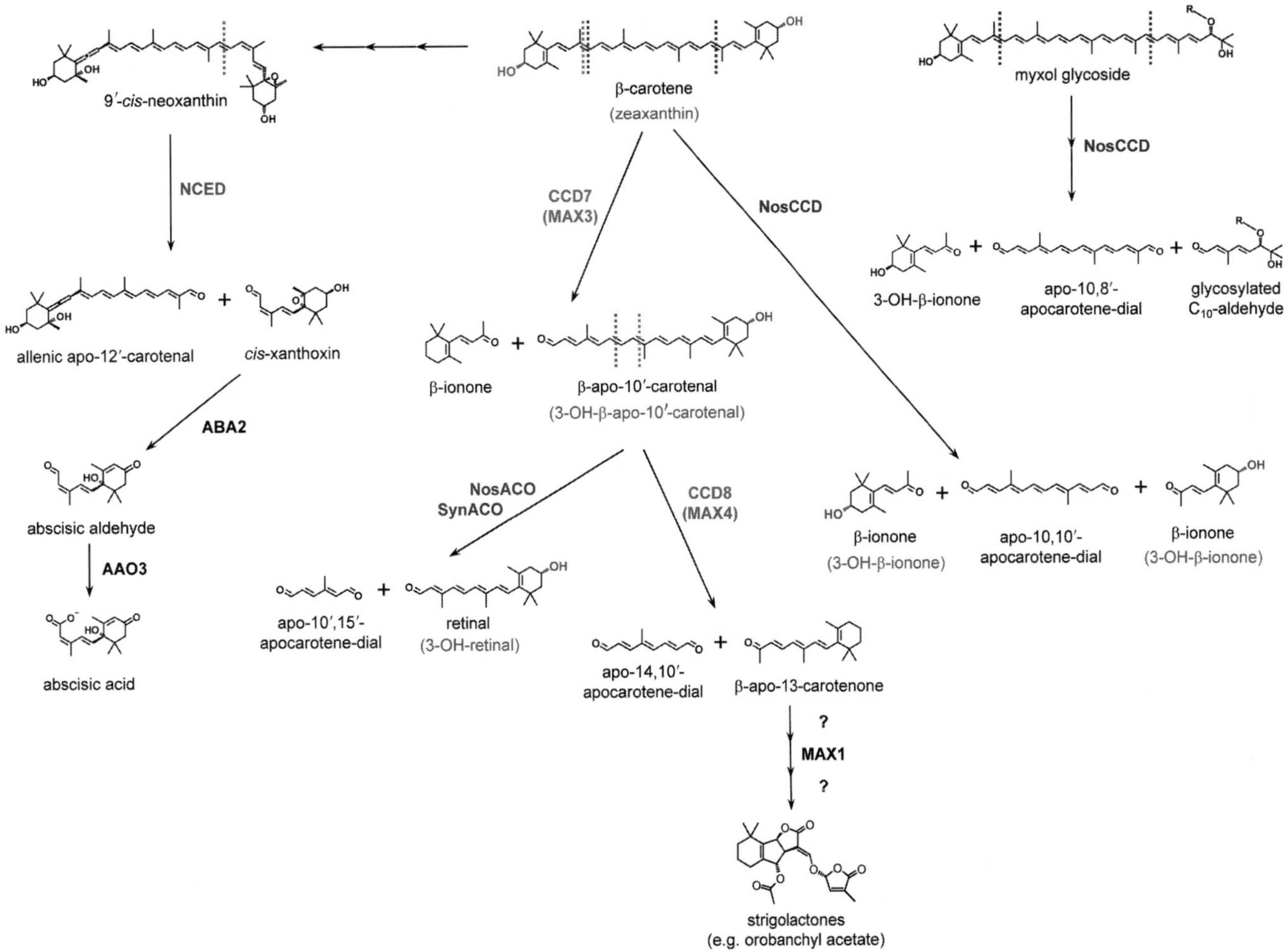

Figure 3.3. Catabolism of carotenoids in land plants, cyanobacteria and algae: formation of apocarotenoids by carotenoid cleavage oxygenases (CCO) and of metabolites derived there from with important physiological functions. Names of CCO enzymes and their corresponding cleavage site(s) have the same colour. For cyanobacterial CCO enzymes, examples of further substrates and the resulting products are in red. See text for further explanations. See colour plate section.

been characterized in detail demonstrating that the enzyme is membrane-bound and can also cleave 3-hydroxy-apocarotenoids to yield 3-hydroxy-retinal (Ruch *et al.*, 2005; Marasco *et al.*, 2006; Scherzinger *et al.*, 2006). Yet, neither the identity of the native substrate of ACO nor the reactions leading to its formation are known.

In accordance with the discovery of retinal-forming enzymes in cyanobacteria, *Nostoc* sp. PCC 7120 was shown to contain a rhodopsin that probably acts as a photoreceptor (Jung *et al.*, 2003). It has not been determined, however, whether both retinal and 3-hydroxy-retinal are synthesized in *Nostoc* and which of the two retinal species is present in its native rhodopsin. In contrast to *Nostoc* sp. PCC 7120, the genomes of *Synechocystis* sp. PCC 6803 and other cyanobacteria contain no obvious homologues of the known retinal-binding opsins. Therefore, it was supposed that (hydroxy-) retinal may have another as yet unknown function in cyanobacteria (Ruch *et al.*, 2005). Supporting the biological relevance of retinal and its derivatives in *Synechocystis*, a recent characterization of the sole cytochrome P450 (CYP120A1) occurring in this cyanobacterium revealed the enzyme as the first identified non-animal all-*trans*-retinoic acid hydroxylase (Alder *et al.*, 2009).

In the green alga *C. reinhardtii*, both type I and type II rhodopsins have been shown to be part of the eyespot apparatus that is governing phototaxis (Foster *et al.*, 1984; Derguini *et al.*, 1991; Kateriya *et al.*, 2004). In a carotenoid-deficient mutant of *C. reinhardtii* devoid of retinal, phototaxis could be restored with various exogenous substrates like β-carotene, α-carotene and apocarotenoids indicating that the alga contains enzymes that cleave these carotenoids (Saranak and Foster, 1994). It remains to be established, which of the putative *CCO* genes in *C. reinhardtii* encodes the enzyme responsible for retinal formation. As a promising candidate, peptide fragments of a particular CCO were recently identified in eyespot preparations of *C. reinhardtii* (Schmidt *et al.*, 2006). Various other algae have been found to contain type I rhodopsins, namely an euglenoid alga (Saranak and Foster, 2005), two cryptophytes (Sineshchekov *et al.*, 2005), and a dinoflagellate (Ruiz-González and Marín, 2004). However, nothing is known about the biosynthesis of retinal in these algae.

3.4.3 Strigolactones

Another group of important carotenoid-derived compounds are the strigolactones, a group of tricyclic lactones conjugated with a C_5-butyrolactone (Butler, 1995; Humphrey and Beale, 2006). Strigolactones had already been known to be released by plant roots in order to stimulate their colonization by arbuscular mycorrhizal fungi and to induce seed germination of parasitic weeds of the genera *Orobanche*, *Alectra* and the eponymous genus *Striga* (Butler, 1995; Humphrey and Beale, 2006; Parniske, 2008; Yoneyama *et al.*, 2009). Recently, it was shown that strigolactones or derivatives thereof are identical with the postulated shoot multiplication signal (SMS) which regulates apical dominance in vascular plants (Gomez-Roldan *et al.*, 2008;

Umehara *et al.*, 2008; Dun *et al.*, 2009). The discovery of this novel hormonal function was paved through the identification of mutants impaired in apical dominance from several plant species, like the Arabidopsis *more axillary branching 3* and *4* (*max3*, *max4*), the rice *high-tillering dwarf1* (*htd1*) and *dwarf10* (*d10*) and the pea *ramosus 5* and *1* (*rms5*, *rms1*), which were shown to be caused by lesions in two different *CCO* genes, namely *CCD7* and *CCD8* (Dun *et al.*, 2009; Leyser, 2009). Characterization of *A. thaliana* CCD7 (also termed MAX3) activity suggested asymmetric cleavage of β-carotene at the C-9,10 or C-9′/C-10′ double bond yielding β-ionone and β-apo-10′-carotenal (Booker *et al.*, 2004; Schwartz *et al.*, 2004) (Figure 3.3). In addition, coexpression of AtCCD7 and AtCCD8/MAX4 in β-carotene accumulating *E. coli* cells resulted in the formation of the C_{18}-compound β-apo-13-carotenone, indicating that AtCCD8 cleaves the β-apo-10′-carotenal product of AtCCD7 (Schwartz *et al.*, 2004). Recently, an in vitro study on CCD8 enzymes from *Arabidopsis*, rice and pea, confirmed β-apo-13-carotenone as the product synthesized from the C_{27}-intermediates β-apo-10′-carotenal or β-apo-10′-carotenol (Figure 3.3) through a highly conserved cleaving reaction, excluding the conversion of C_{40}-carotenoids (Alder *et al.*, 2008). The pathway from β-apo-13-carotenone to strigolactones is not yet elucidated, but it involves a cytochrome P450 enzyme (MAX1 in *Arabidopsis*) (Dun *et al.*, 2009).

It is also not known whether the ability to synthesize strigolatones is restricted to land plants or may have developed earlier in the evolution of eukaryotic phototrophs. In lichens, strigolactones have been suggested as potential candidates for the photobiont-derived factor which induces hyphal branching of the mycobiont after initial contact of alga and fungus (Harris, 2008); this hypothesis has yet to be experimentally substantiated. The genomes of green algae harbour genes encoding putative CCO enzymes related to CCD7 and CCD8 (Lohr, unpublished). However, it remains to be investigated whether algae are able to generate strigolactones and what their prospective functions would be.

3.4.4 Other products of carotenoid cleavage

In land plants, additional members of the CCO family have been characterized. The CCD1 subfamily comprises cytosolic enzymes which accept a rather broad range of substrates, generating C_{10} and C_{13} volatile isoprenoids like β-ionone, β-cyclocitral and safranal which are involved in flavours and fragrances (Simkin *et al.*, 2004; Bouvier *et al.*, 2005a; Rubio *et al.*, 2008; Huang *et al.*, 2009). In addition, the plant volatiles 6-methyl-5-hepten-2-one (C_8) (Vogel *et al.*, 2008) and geranial (C_{10}) (Ilg *et al.*, 2009) were shown to derive from carotenoids cleaved by CCD1. Several of these compounds have also been detected in natural lakes, and both cyanobacteria and eukaryotic algae have been implicated in their formation (Jüttner, 1984; Jüttner *et al.*, 1986; Jüttner, 1995).

In the cyanobacterium *Nostoc* sp. PCC 7120, a CCO termed NSC1 or NosCCD was identified that may be involved in the formation of volatile isoprenoids (Figure 3.3) (Marasco *et al.*, 2006; Scherzinger and Al-Babili, 2008). A detailed study of NosCCD revealed that it is a soluble enzyme with the ability to cleave a broad spectrum of mono- and bicyclic carotenoids. Bicyclic carotenoids like zeaxanthin, echinenone or canthaxanthin were cleaved symmetrically at both C-9–C-10 and C-9′–C-10′, whereas in the case of monocyclic carotenoids like γ-carotene and myxoxanthophyll the cyclic half was cleaved at C-9–C-10, but the linear half between C-8′–C-9′ (Scherzinger and Al-Babili, 2008). Because of its wide substrate specificity and marked induction under high light, NosCCD has been implicated in the scavenging of photodegraded carotenoids (Scherzinger and Al-Babili, 2008). *Nostoc* sp. PCC 7120 contains a third CCO named NSC3 which cleaves only apocarotenoids like β-8-apocarotenal (Marasco *et al.*, 2006) and may complement NosCCD to some extent. It will be interesting to learn to what extent the phytoplankton contributes to carotenoid-derived metabolites in lakes and oceans and whether these compounds have an as yet unknown biological function or are the result of an unavoidable metabolite leakage during carotenoid degradation (Watson, 2003).

3.5 Outlook

During the last two decades, the biosynthetic pathways leading to the formation of the major carotenoids in land plants and green algae like *C. reinhardtii* have largely been deciphered on the molecular level. More recently, considerable progress has been made in unravelling carotenogenesis in cyanobacteria as well, based on comparative genomics of an increasing number of bacterial genomes and powerful genetic tools. However, our picture of carotenoid metabolism in phytoplankton and more generally in oxygenic phototrophs is still far from complete. Major challenges will be the elucidation of the enzymes that are involved in the formation of the major xanthophylls in chromalveolate algae (e.g. diadinoxanthin, alloxanthin, fucoxanthin, vaucheriaxanthin or peridinin) and of the broad spectrum of unusual xanthophylls found in prasinophytes of the order Mamiellales (e.g. prasinoxanthin, micromonol or uriolide). Moreover, we have as yet only caught a glimpse of the catabolism of carotenoids and the distribution and potential roles of the resulting metabolites among the different groups of phytoplankton. Ongoing efforts in sequencing further genomes from eukaryotic algae together with an ever increasing molecular toolbox will lay a strong foundation for discovery of the respective genes and their regulation.

Acknowledgements

I wish to thank Salim Al-Babili, Matthias Bauch, Joseph Hirschberg, Sonja Werner and two reviewers for careful reading of the manuscript, critical comments and helpful suggestions, and the University of Mainz and the German Science Foundation for financial support.

Abbreviations

AAO	ABA-specific aldehyde oxidase
ABA	Abscisic acid
ACO	Apocarotenoid cleavage oxygenase
BCO	β-carotene-15,15′-cleavage oxygenase
BKT / *BKT*	Carotene β-ketolase (or β-carotene ketolase) enzyme / gene
CCD / *CCD*	Carotenoid cleavage dioxygenase enzyme / gene
CCO / *CCO*	Carotenoid cleavage oxygenase enzyme / gene
CHYB / *CHYB*	Carotene β-hydroxylase (or β-carotene hydroxylase) enzyme / gene
CMK	4-(cytidine 5′-diphospho)-2-*C*-methyl-D-erythritol kinase
Crt / *crt*	Carotenoid biosynthesis enzyme / gene
CRTISO	Carotenoid isomerase
Cru / *cru*	More recently discovered carotenoid biosynthesis enzyme / gene
CTP	Cytidine 5′-triphosphate
CYP / *CYP*	Cytochrome P450 enzyme / gene
DXR	1-deoxy-D-xylulose 5-phosphate reductoisomerase
DXS	1-deoxy-D-xylulose 5-phosphate synthase
ER	Endoplasmic reticulum
EST	Expressed sequence tag
GDP	Guanosine 5′-diphosphate
GGPS	Geranylgeranyl diphosphate synthase
HDR	4-hydroxy-3-methylbut-2-en-1-yl diphosphate reductase
HDS	4-hydroxy-3-methylbut-2-en-1-yl diphosphate synthase
htd	*High-tillering dwarf* mutant
IDI	Isopentenyl diphosphate:dimethylallyl diphosphate isomerase
Isp / *isp*	Isoprenoid biosynthesis enzyme / gene
LCY / *LCY*	Lycopene cyclase enzyme / gene
LTL	Lutein deficient-like
MAX / *MAX* / *max*	*More axillary branching* enzyme / gene / mutant
MCT	2-*C*-methyl-d-erythritol 4-phosphate cytidylyltransferase
MDS	2-*C*-methyl-d-erythritol 2,4-cyclodiphosphate synthase
MEP	Methylerythritol phosphate
MVA	Mevalonic acid
NCED / *NCED*	Nine-*cis*-epoxy carotenoid dioxygenase enzyme / gene
NSC (= NosCCD)	Nostoc carotenoid cleavage dioxygenase
OCP	Orange carotenoid protein
PDS	Phytoene desaturase
PSY	Phytoene synthase
SMS	Shoot multiplication signal
VDE	Violaxanthin de-epoxidase
ZDS	ζ-carotene desaturase
ZEP	Zeaxanthin epoxidase
Z-Iso	ζ-carotene isomerase

References

Adam, P., Hecht, S., Eisenreich, W. G., Kaiser, J., Grawert, T., Arigoni, D., Bacher, A. and Rohdich, F. (2002). Biosynthesis of terpenes: Studies on 1-hydroxy-2-methyl-2-(*E*)-butenyl 4-diphosphate reductase. *Proc. Natl. Acad. Sci. USA* **99**, 12108–13.

Al-Babili, S., Hugueney, P., Schledz, M., Welsch, R., Frohnmeyer, H., Laule, O. and Beyer, P. (2000). Identification of a novel gene coding for neoxanthin synthase from *Solanum tuberosum*. *FEBS Lett*. **485**, 168–72.

Alder, A., Holdermann, I., Beyer, P. and Al-Babili, S. (2008). Carotenoid oxygenases involved in plant branching catalyse a highly specific conserved apocarotenoid cleavage reaction. *Biochem. J.* **416**, 289–96.

Alder, A., Bigler, P., Werck-Reichhart, D. and Al-Babili, S. (2009). *In vitro* characterization of *Synechocystis* CYP120A1 revealed the first nonanimal retinoic acid hydroxylase. *FEBS J.* **276**, 5416–31.

Andersson, M., Schubert, H., Pedersen M., and Snoeijs, P. (2006). Different patterns of carotenoid composition and photosynthesis acclimation in two tropical red algae. *Mar. Biol.* **149**, 653–65.

Andreeva, A. V. and Kutuzov, M. A. (2001). Do plants have rhodopsin after all? A mystery of plant G protein-coupled signalling. *Plant Physiol. Biochem.* **39**, 1027–35.

Antia, N. J. and Cheng, J. Y. (1982). The keto-carotenoids of two marine coccoid members of the Eustigmatophyceae. *Br. Phycol. J.* **17**, 39–50.

Anwaruzzaman, M., Chin, B. L., Li, X. P., Lohr, M., Martinez, D. A. and Niyogi, K. K. (2004). Genomic analysis of mutants affecting xanthophyll biosynthesis and regulation of photosynthetic light harvesting in *Chlamydomonas reinhardtii*. *Photosynth. Res.* **82**, 265–76.

Armstrong, G. A. (1997). Genetics of eubacterial carotenoid biosynthesis: A colorful tale. *Annu. Rev. Microbiol.* **51**, 629–59.

Auldridge, M. E., McCarty, D. R. and Klee, H. J. (2006). Plant carotenoid cleavage oxygenases and their apocarotenoid products. *Curr. Opin. Plant Biol.* **9**, 315–21.

Baroli, I., Do, A. D., Yamane, T. and Niyogi, K. K. (2003). Zeaxanthin accumulation in the absence of a functional xanthophyll cycle protects *Chlamydomonas reinhardtii* from photooxidative stress. *Plant Cell* **15**, 992–1008.

Bartley, G. E., Viitanen, P. V., Pecker, I., Chamovitz, D., Hirschberg, J. and Scolnik, P. A. (1991). Molecular cloning and expression in photosynthetic bacteria of a soybean cDNA coding for phytoene desaturase, an enzyme of the carotenoid biosynthesis pathway. *Proc. Natl. Acad. Sci. USA* **88**, 6532–36.

Bartley, G. E., Scolnik, P. A. and Beyer, P. (1999). Two *Arabidopsis thaliana* carotene desaturases, phytoene desaturase and ζ-carotene desaturase, expressed in *Escherichia coli*, catalyze a poly-*cis* pathway to yield pro-lycopene. *Eur. J. Biochem.* **259**, 396–403.

Becker, B., Hoef-Emden, K. and Melkonian, M. (2008). Chlamydial genes shed light on the evolution of photoautotrophic eukaryotes. *BMC Evol. Biol.* **8**, 203.

Berger, R., Liaaen-Jensen, S., McAlister, V. and Guillard, R. R. L. (1977). Carotenoids of Prymnesiophyceae (Haptophyceae). *Biochem. Syst. Ecol.* **5**, 71–75.

Bjørnland, T. (1990). Chromatographic separation and spectrometric characterization of native carotenoids from the marine dinoflagellate *Thoracosphaera heimii*. *Biochem. Syst. Ecol.* **18**, 307–16.

Bjørnland, T. and Aguilar-Martinez, M. (1976). Carotenoids in red algae. *Phytochemistry* **15**, 291–96.

Bjørnland, T. and Liaaen-Jensen, S. (1989). Distribution patterns of carotenoids in relation to chromophyte phylogeny and systematics. In *The Chromophyte Algae: Problems and Perspectives*, ed. J. C. Green, B. S. C. Leadbeater and W. L. Diver. Oxford: Clarendon Press, pp. 37–60.

Böhme, K., Wilhelm, C. and Goss, R. (2002). Light regulation of carotenoid biosynthesis in the prasinophycean alga *Mantoniella squamata*. *Photochem. Photobiol. Sci.* **1**, 619–28.

Bolte, K., Bullmann, L., Hempel, F., Bozarth, A., Zauner, S. and Maier, U. -G. (2009). Protein targeting into secondary plastids. *J. Eukaryot. Microbiol.* **56**, 9–15.

Bonnett, R., Mallams, A. K., Spark, A. A., Tee, J. L., Weedon, B. C. L. and McCormick, A. (1969). Carotenoids and related compounds. Part XX. Structure and reactions of fucoxanthin. *J. Chem. Soc. C*: 429–54.

Booker, J., Auldridge, M., Wills, S., McCarty, D., Klee, H. and Leyser, C. (2004). MAX3/CCD7 is a carotenoid cleavage dioxygenase required for the synthesis of a novel plant signaling molecule. *Curr. Biol.* **14**, 1232–38.

Boussiba, S. (2000). Carotenogenesis in the green alga *Haematococcus pluvialis*: Cellular physiology and stress response. *Physiol. Plant.* **108**, 111–17.

Bouvier, F., D'Harlingue, A., Backhaus, R. A., Kumagai, M. H. and Camara, B. (2000). Identification of neoxanthin synthase as a carotenoid cyclase paralog. *Eur. J. Biochem.* **267**, 6346–52.

Bouvier, F., Isner, J. C., Dogbo, O. and Camara, B. (2005a). Oxidative tailoring of carotenoids: a prospect towards novel functions in plants. *Trends Plant Sci.* **10**, 187–94.

Bouvier, F., Rahier, A. and Camara, B. (2005b). Biogenesis, molecular regulation and function of plant isoprenoids. *Prog. Lipid Res.* **44**, 357–429.

Boyer, G. L. and Dougherty, S. S. (1988). Identification of abscisic acid in the seaweed *Ascophyllum nodosum*. *Phytochemistry* **27**, 1521–22.

Breitenbach, J. and Sandmann, G. (2005). ζ-carotene *cis* isomers as products and substrates in the plant poly-*cis* carotenoid biosynthetic pathway to lycopene. *Planta* **220**, 785–93.

Breitenbach, J., Misawa, N., Kajiwara, S. and Sandmann, G. (1996). Expression in *Escherichia coli* and properties of the carotene ketolase from *Haematococcus pluvialis*. *FEMS Microbiol. Lett.* **140**, 241–46.

Breitenbach, J., Vioque, A. and Sandmann, G. (2001). Gene *sll0033* from *Synechocystis* 6803 encodes a carotene isomerase involved in the biosynthesis of all-*E* lycopene. *Z. Naturforsch. C* **56**, 915–17.

Britton, G. (1998). Overview of Carotenoid Biosynthesis. In *Carotenoids, vol. 3, Biosynthesis and Metabolism*, ed. G. Britton, S. Liaaen-Jensen and H. Pfander, Basel: Birkhäuser, pp. 13–147.

Britton, G., Liaaen-Jensen, S. and Pfander, H. (eds.), Mercadante, A. Z. and Egeland, E. S. (compilers) (2004). *Carotenoids. Handbook.* Basel: Birkhäuser Verlag.

Budziszewski, G. J., Lewis, S. P., Glover, L. W., Reineke, J., Jones, G., Ziemnik, L. S., Lonowski, J., Nyfeler, B., Aux, G., Zhou, Q., McElver, J., Patton, D. A., Martienssen, R., Grossniklaus, U., Ma, H., Law, M. and Levin, J. Z. (2001). *Arabidopsis* genes essential for seedling viability: Isolation of insertional mutants and molecular cloning. *Genetics* **159**, 1765–78.

Bugos, R. C. and Yamamoto, H. Y. (1996). Molecular cloning of violaxanthin de-epoxidase from romaine lettuce and expression in *Escherichia coli*. *Proc. Natl. Acad. Sci. USA* **93**, 6320–25.

Bugos, R. C., Hieber, A. D. and Yamamoto, H. Y. (1998). Xanthophyll cycle enzymes are members of the lipocalin family, the first identified from plants. *J. Biol. Chem.* **273**, 15321–24.

Butler, L. G. (1995). Chemical communication between the parasitic weed *Striga* and its crop host – a new dimension in allelochemistry. In *Allelopathy – Organisms, Processes, and Applications*, ed. A. Inderjit, K. M. M. Dakshini and F. A. Einhellig. Washington: American Chemical Society, pp. 158–68.

Carol, P., Stevenson, D., Bisanz, C., Breitenbach, J., Sandmann, G., Mache, R., Coupland, G. and Kuntz, M. (1999). Mutations in the *Arabidopsis* gene *IMMUTANS* cause a variegated phenotype by inactivating a chloroplast terminal oxidase associated with phytoene desaturation. *Plant Cell* **11**, 57–68.

Carretero-Paulet, L., Cairó, A., Botella-Pavía, P., Besumbes, O., Campos, N., Boronat, A. and Rodríguez-Concepción, M. (2006). Enhanced flux through the methylerythritol 4-phosphate pathway in *Arabidopsis* plants overexpressing deoxyxylulose 5-phosphate reductoisomerase. *Plant Mol. Biol.* **62**, 683–95.

Chernys, J. T. and Zeevaart, J. A. D. (2000). Characterization of the 9-*cis*-epoxycarotenoid dioxygenase gene family and the regulation of abscisic acid biosynthesis in avocado. *Plant Physiol.* **124**, 343–53.

Coesel, S., Oborník, M., Varela, J., Falciatore, A. and Bowler, C. (2008). Evolutionary origins and functions of the carotenoid biosynthetic pathway in marine diatoms. *PLoS ONE* **3**, e2896.

Cowan, A. K. and Rose, P. D. (1991). Abscisic acid metabolism in salt-stressed cells of *Dunaliella salina*: possible interrelationship with β-carotene accumulation. *Plant Physiol.* **97**, 798–803.

Cunningham, F. X. (2002). Regulation of carotenoid synthesis and accumulation in plants. *Pure Appl. Chem.* **74**, 1409–17.

Cunningham, F. X. and Schiff, J. A. (1985). Photoisomerization of ζ-carotene stereoisomers in cells of *Euglena gracilis* mutant W3BUL and in solution. *Photochem. Photobiol.* **42**, 295–308.

Cunningham, F. X., Sun, Z. R., Chamovitz, D. Hirschberg, J. and Gantt, E. (1994). Molecular structure and enzymatic function of lycopene cyclase from the cyanobacterium *Synechococcus* sp. strain PCC7942. *Plant Cell* **6**, 1107–21.

Cunningham, F. X., Pogson, B., Sun, Z. R., McDonald, K. A., DellaPenna, D. and Gantt, E. (1996). Functional analysis of the β and ε lycopene cyclase enzymes of *Arabidopsis* reveals a mechanism for control of cyclic carotenoid formation. *Plant Cell* **8**, 1613–26.

Cunningham, F. X., Lee, H. and Gantt, E. (2007). Carotenoid biosynthesis in the primitive red alga *Cyanidioschyzon merolae*. *Eukaryot. Cell* **6**, 533–45.

Delphin, E., Duval, J. C., Etienne, A. L. and Kirilovsky, D. (1998). Delta pH-dependent photosystem II fluorescence quenching induced by saturating, multiturnover pulses in red algae. *Plant Physiol.* **118**, 103–13.

Derguini, F., Mazur, P., Nakanishi, K., Starace, D. M., Saranak, J. and Foster, K. W. (1991). All-*trans* retinal is the chromophore bound to the photoreceptor of the alga *Chlamydomonas reinhardtii*. *Photochem. Photobiol.* **54**, 1017–22.

Destefano-Beltran, L., Knauber, D., Huckle, L. and Suttle, J. C. (2006). Effects of postharvest storage and dormancy status on ABA content, metabolism, and expression of genes involved in ABA biosynthesis and metabolism in potato tuber tissues. *Plant Mol. Biol.* **61**, 687–97.

Dimier, C., Giovanni, S., Ferdinando, T. and Brunet, C. (2009). Comparative ecophysiology of the xanthophyll cycle in six marine phytoplanktonic species. *Protist* **160**, 397–411.

Disch, A., Schwender, J., Müller, C., Lichtenthaler, H. K. and Rohmer, M. (1998). Distribution of the mevalonate and glyceraldehyde phosphate/pyruvate pathways for isoprenoid biosynthesis in unicellular algae and the cyanobacterium *Synechocystis* PCC 6714. *Biochem. J.* **333**, 381–88.

Domonkos, I., Malec, P., Laczko-Dobos, H., Sozer, O., Klodawska, K., Wada, H., Strzalka, K. and Gombos, Z. (2009). Phosphatidylglycerol depletion induces an increase in myxoxanthophyll biosynthetic activity in *Synechocystis* PCC6803 cells. *Plant Cell Physiol.* **50**, 374–82.

Dun, E. A., Brewer, P. B. and Beveridge, C. A. (2009). Strigolactones: discovery of the elusive shoot branching hormone. *Trends Plant Sci.* **14**, 364–72.

Egeland, E. S., Eikrem, W., Throndsen, J., Wilhelm, C., Zapata, M. and Liaaen-Jensen, S. (1995). Carotenoids from further prasinophytes. *Biochem. Syst. Ecol.* **23**, 747–55.

Egeland, E. S., Eikrem, W., Throndsen, J., Wilhelm, C., Zapata, M. and Liaaen-Jensen, S. (1996). Erratum to 'Carotenoids from further prasinophytes'. *Biochem. Syst. Ecol.* **24**, 179–80.

Egeland, E. S., Guillard, R. R. L. and Liaaen-Jensen, S. (1997). Additional carotenoid prototype representatives and a general chemosystematic evaluation of carotenoids in Prasinophyceae (Chlorophyta). *Phytochemistry* **44**, 1087–97.

Egeland, E. S., Garrido, J. L., Zapata, M., Maestro, M. A. and Liaaen-Jensen, S. (2000). Algal carotenoids. Part 64. Structure and chemistry of 4-keto-19'-hexanoyloxyfucoxanthin with a novel carotenoid end group. *J. Chem. Soc. Perkin Trans. 1*, 1223–30.

Eisenreich, W., Bacher, A., Arigoni, D. and Rohdich, F. (2004). Biosynthesis of isoprenoids via the non-mevalonate pathway. *Cell. Mol. Life Sci.* **61**, 1401–26.

Esch, H., Hundeshagen, B., Schneiderpoetsch, H. and Bothe, H. (1994). Demonstration of abscisic acid in spores and hyphae of the arbuscular mycorrhizal fungus *Glomus* and in the N_2-fixing cyanobacterium *Anabaena variabilis*. *Plant Sci.* **99**, 9–16.

Esteban, R., Martínez, B., Fernández-Marín, B., Becerril, J. M. and García-Plazaola, J. I. (2009). Carotenoid composition in Rhodophyta: insights into xanthophyll regulation in *Corallina elongata*. *Eur. J. Phycol.* **44**, 221–30.

Estrada, A. F., Brefort, T., Mengel, C., Díaz-Sánchez, V., Alder, A., Al-Babili, S. and Avalos, J. (2009). *Ustilago maydis* accumulates β-carotene at levels determined by a retinal-forming carotenoid oxygenase. *Fungal Genet. Biol.* **46**, 803–13.

Fernández-González, B., Sandmann, G. and Vioque, A. (1997). A new type of asymmetrically acting β-carotene ketolase is required for the synthesis of echinenone in the cyanobacterium *Synechocystis* sp. PCC 6803. *J. Biol. Chem.* **272**, 9728–33.

Fiore, A., Dall'Osto, L., Fraser, P. D., Bassi, R. and Giuliano, G. (2006). Elucidation of the β-carotene hydroxylation pathway in *Arabidopsis thaliana*. *FEBS Lett.* **580**, 4718–22.

Foss, P., Skulberg, O. M., Kilaas, L. and Liaaen-Jensen, S. (1986). The carbohydrate moieties bound to the carotenoids myxol and oscillol and their chemosystematic applications. *Phytochemistry* **25**, 1127–32.

Foster, K. W., Saranak, J., Patel, N., Zarilli, G., Okabe, M., Kline, T. and Nakanishi, K. (1984). A rhodopsin is the functional photoreceptor for phototaxis in the unicellular eukaryote *Chlamydomonas*. *Nature* **311**, 756–59.

Fraser, P. D., Miura, Y. and Misawa, N. (1997). *In vitro* characterization of astaxanthin biosynthetic enzymes. *J. Biol. Chem.* **272**, 6128–35.

Fraser, P. D., Shimada, H. and Misawa, N. (1998). Enzymic confirmation of reactions involved in routes to astaxanthin formation, elucidated using a direct substrate *in vitro* assay. *Eur. J. Biochem.* **252**, 229–36.

Frassanito, R., Flaim, G., Mancini, I. and Guella, G. (2006). High production of unexpected carotenoids in Dinophyceae. Astaxanthin esters from the freshwater dinoflagellate *Tovellia sanguinea*. *Biochem. Syst. Ecol.* **34**, 843–53.

Frigaard, N. U., Maresca, J. A., Yunker, C. E., Jones, A. D. and Bryant, D. A. (2004). Genetic manipulation of carotenoid biosynthesis in the green sulfur bacterium *Chlorobium tepidum*. *J. Bacteriol.* **186**, 5210–20.

Frommolt, R., Werner, S., Paulsen, H., Goss, R., Wilhelm, C., Zauner, S., Maier, U. G., Grossman, A. R., Bhattacharya, D. and Lohr, M. (2008). Ancient recruitment by chromists of green algal genes encoding enzymes for carotenoid biosynthesis. *Mol. Biol. Evol.* **25**, 2653–67.

García-Plazaola, J. I., Matsubara, S. and Osmond, C. B. (2007). The lutein epoxide cycle in higher plants: its relationships to other xanthophyll cycles and possible functions. *Funct. Plant Biol.* **34**, 759–73.

Garrido, J. L., Rodríguez, F. and Zapata, M. (2009). Occurrence of loroxanthin, loroxanthin decenoate, and loroxanthin dodecenoate in *Tetraselmis* species (Prasinophyceae, Chlorophyta). *J. Phycol.* **45**, 366–74.

Giuliano, G., Al-Babili, S. and von Lintig, J. (2003). Carotenoid oxygenases: cleave it or leave it. *Trends Plant Sci.* **8**, 145–49.

Goericke, R. and Welschmeyer, N. A. (1992). Pigment turnover in the marine diatom *Thalassiosira weissflogii*. II. The $^{14}CO_2$-labeling kinetics of carotenoids. *J. Phycol.* **28**, 507–17.

Gomez-Roldan, V., Fermas, S., Brewer, P. B., Puech-Pagès, V., Dun, E. A., Pillot, J. P., Letisse, F., Matusova, R., Danoun, S., Portais, J. C., Bouwmeester, H., Bécard, G., Beveridge, C. A., Rameau, C. and Rochange, S. F. (2008). Strigolactone inhibition of shoot branching. *Nature* **455**, 189–94.

Goodwin, T. W. (1980). *The Biochemistry of the Carotenoids, vol. I: Plants*, 2nd edn. London: Chapman & Hall.

Goodwin, T. W. (1984). *The Biochemistry of the Carotenoids, vol. II: Animals*, 2nd edn. London: Chapman & Hall.

Graham, J. E. and Bryant, D. A. (2008). The biosynthetic pathway for synechoxanthin, an aromatic carotenoid synthesized by the euryhaline, unicellular cyanobacterium *Synechococcus* sp. strain PCC 7002. *J. Bacteriol.* **190**, 7966–74.

Graham, J. E. and Bryant, D. A. (2009). The biosynthetic pathway for myxol 2'-fucoside (myxoxanthophyll) in the cyanobacterium *Synechococcus* sp. strain PCC 7002. *J. Bacteriol.* **191**, 3292–300.

Graham, J. E., Lecomte, J. T. J. and Bryant, D. A. (2008). Synechoxanthin, an aromatic C_{40} xanthophyll that is a major carotenoid in the cyanobacterium *Synechococcus* sp. PCC 7002. *J. Nat. Prod.* **71**, 1647–50.

Grossman, A. R., Lohr, M. and Im, C. S. (2004). *Chlamydomonas reinhardtii* in the landscape of pigments. *Annu. Rev. Genet.* **38**, 119–73.

Grünewald, K. and Hagen, C. (2001). β-carotene is the intermediate exported from the chloroplast during accumulation of secondary carotenoids in *Haematococcus pluvialis*. *J. Appl. Phycol.* **13**, 89–93.

Grünewald, K., Hirschberg, J. and Hagen, C. (2001). Ketocarotenoid biosynthesis outside of plastids in the unicellular green alga *Haematococcus pluvialis*. *J. Biol. Chem.* **276**, 6023–29.

Grung, M. and Liaaen-Jensen, S. (1993). Algal carotenoids 52; Secondary carotenoids of algae 3; Carotenoids in a natural bloom of *Euglena sanguinea*. *Biochem. Syst. Ecol.* **21**, 757–63.

Guevara-García, A., San Román, C., Arroyo, A., Cortés, M. E., Gutíerrez-Nava, M. D. and León, P. (2005). Characterization of the *Arabidopsis clb6* mutant illustrates the importance of posttranscriptional regulation of the methyl-D-erythritol 4-phosphate pathway. *Plant Cell* **17**, 628–43.

Guillou, L., Chrétiennot-Dinet, M. J., Medlin, L. K., Claustre, H., Loiseaux-de Goër, S. and Vaulot, D. (1999). *Bolidomonas*: A new genus with two species belonging to a new algal class, the Bolidophyceae (Heterokonta). *J. Phycol.* **35**, 368–81.

Hager, A. and Stransky, H. (1970a). The carotenoid pattern and the occurrence of the light-induced xanthophyll cycle in various classes of algae. III. Green algae. *Arch. Mikrobiol.* **72**, 68–83.

Hager, A. and Stransky, H. (1970b). The carotenoid pattern and the occurrence of the light-induced xanthophyll cycle in various classes of algae. V. A few members of Cryptophyceae, Euglenophyceae, Bacillariophyceae, Chrysophyceae and Phaeophyceae. *Arch. Mikrobiol.* **73**, 77–89.

Harris, S. D. (2008). Branching of fungal hyphae: regulation, mechanisms and comparison with other branching systems. *Mycologia* **100**, 823–32.

Heelis, D. V., Kernick, W., Phillips, G. O. and Davies, K. (1979). Separation and identification of the carotenoid pigments of stigmata isolated from light-grown cells of *Euglena gracilis* strain Z. *Arch. Microbiol.* **121**, 207–12.

Hegemann, P. (2008). Algal sensory photoreceptors. *Annu. Rev. Plant Biol.* **59**, 167–89.

Hess, W. R., Rocap, G., Ting, C. S., Larimer, F., Stilwagen, S., Lamerdin, J. and Chisholm, S. W. (2001). The photosynthetic apparatus of *Prochlorococcus*: Insights through comparative genomics. *Photosynth. Res.* **70**, 53–71.

Hill, R. E., Sayer, B. G. and Spenser, I. D. (1989). Biosynthesis of vitamin B_6: incorporation of D-1-deoxyxylulose. *J. Am. Chem. Soc.* **111**, 1916–17.

Hirsch, R., Hartung, W. and Gimmler, H. (1989). Abscisic acid content of algae under stress. *Bot. Acta* **102**, 326–34.

Hirschberg, J. (1998). Molecular biology of carotenoid biosynthesis. In *Carotenoids, vol. 3, Biosynthesis and Metabolism*, ed. G. Britton, S. Liaaen-Jensen and H. Pfander. Basel: Birkhäuser, pp. 149–94.

Hirschberg, J. (2001). Carotenoid biosynthesis in flowering plants. *Curr. Opin. Plant Biol.* **4**, 210–18.

Hirschberg, J., Cohen, M., Harker, M., Lotan, T., Mann, V., Pecker, I. (1997). Molecular genetics of the carotenoid biosynthesis pathway in plants and algae. *Pure Appl. Chem.* **69**, 2151–58.

Huang, F. C., Horvath, G., Molnar, P., Turcsi, E., Deli, J., Schrader, J., Sandmann, G., Schmidt, H. and Schwab, W. (2009). Substrate promiscuity of

RdCCD1, a carotenoid cleavage oxygenase from *Rosa damascena*. *Phytochemistry* **70**, 457–64.

Huang, J. L. and Gogarten, J. P. (2007). Did an ancient chlamydial endosymbiosis facilitate the establishment of primary plastids? *Genome Biol.* **8**, R99.

Huang, J. C., Chen, F. and Sandmann, G. (2006a). Stress-related differential expression of multiple β-carotene ketolase genes in the unicellular green alga *Haematococcus pluvialis*. *J. Biotechnol.* **122**, 176–85.

Huang, J. C., Wang, Y., Sandmann, G. and Chen, F. (2006b). Isolation and characterization of a carotenoid oxygenase gene from *Chlorella zofingiensis* (Chlorophyta). *Appl. Microbiol. Biotechnol.* **71**, 473–79.

Humphrey, A. J. and Beale, M. H. (2006). Strigol: Biogenesis and physiological activity. *Phytochemistry* **67**, 636–40.

Hunter, W. N. (2007). The non-mevalonate pathway of isoprenoid precursor biosynthesis. *J. Biol. Chem.* **282**, 21573–77.

Ilg, A., Beyer, P. and Al-Babili, S. (2009). Characterization of the rice carotenoid cleavage dioxygenase 1 reveals a novel route for geranial biosynthesis. *FEBS J.* **276**, 736–47.

Inaba, T. and Schnell, D. J. (2008). Protein trafficking to plastids: one theme, many variations. *Biochem. J.* **413**, 15–28.

Isaacson, T., Ronen, G., Zamir, D. and Hirschberg, J. (2002). Cloning of *tangerine* from tomato reveals a carotenoid isomerase essential for the production of β-carotene and xanthophylls in plants. *Plant Cell* **14**, 333–42.

Isaacson, T., Ohad, I., Beyer, P. and Hirschberg, J. (2004). Analysis *in vitro* of the enzyme CRTISO establishes a poly-*cis*-carotenoid biosynthesis pathway in plants. *Plant Physiol.* **136**, 4246–55.

Iwai, M., Maoka, T., Ikeuchi, M. and Takaichi, S. (2008). 2,2′-β-Hydroxylase (CrtG) is involved in carotenogenesis of both nostoxanthin and 2-hydroxymyxol 2′-fucoside in *Thermosynechococcus elongatus* strain BP-1. *Plant Cell Physiol.* **49**, 1678–87.

Jeffrey, S. W., Mantoura, R. F. C. and Wright, S. W. (1997). *Phytoplankton Pigments in Oceanography: Guidelines to Modern Methods*. Paris: UNESCO Publishing.

Jin, E., Lee, C. G. and Polle, J. E. W. (2006). Secondary carotenoid accumulation in *Haematococcus* (Chlorophyceae): Biosynthesis, regulation, and biotechnology. *J. Microbiol. Biotechnol.* **16**, 821–31.

Jirásková, D., Poulíčková, A., Novák, O., Sedláková, K., Hradecká, V. and Strnad, M. (2009). High-throughput screening technology for monitoring phytohormone production in microalgae. *J. Phycol.* **45**, 108–18.

Johansen, J. E., Svec, W. A., Liaaen-Jensen, S. and Haxo, F. T. (1974). Algal carotenoids. X. Carotenoids of the Dinophyceae. *Phytochemistry* **13**, 2261–71.

Julliard, J. and Douce, R. (1991). Biosynthesis of the thiazole moiety of thiamin (vitamin B_1) in higher plant chloroplasts. *Proc. Natl. Acad. Sci. USA* **88**, 2042–45.

Jung, K. H., Trivedi, V. D. and Spudich, J. L. (2003). Demonstration of a sensory rhodopsin in eubacteria. *Mol. Microbiol.* **47**, 1513–22.

Jüttner, F. (1984). Dynamics of the volatile organic substances associated with cyanobacteria and algae in a eutrophic shallow lake. *Appl. Environ. Microbiol.* **47**, 814–20.

Jüttner, F. (1995). Physiology and biochemistry of odorous compounds from freshwater cyanobacteria and algae. *Water Sci. Technol.* **31**, 69–78.

Jüttner, F., Hoflacher, B. and Wurster, K. (1986). Seasonal analysis of volatile organic biogenic substances (VOBS) in fresh-water phytoplankton populations dominated by *Dinobryon*, *Microcystis* and *Aphanizomenon*. *J. Phycol.* **22**, 169–75.

Kajiwara, S., Kakizono, T., Saito, T., Kondo, K., Ohtani, T., Nishio, N., Nagai, S. and Misawa, N. (1995). Isolation and functional identification of a novel cDNA for astaxanthin biosynthesis from *Haematococcus pluvialis*, and astaxanthin synthesis in *Escherichia coli*. *Plant Mol. Biol.* **29**, 343–52.

Kateriya, S., Nagel, G., Barnberg, E. and Hegemann, P. (2004). 'Vision' in single-celled algae. *News Physiol. Sci.* **19**, 133–37.

Kentzer, T. and Mazur, H. (1991). Abscisic acid as endogenous inhibitor of the marine diatom *Coscinodiscus granii*. *Acta Physiol. Plant.* **13**, 153–57.

Kim, D., Filtz, M. R. and Proteau, P. J. (2004). The methylerythritol phosphate pathway contributes to carotenoid but not phytol biosynthesis in *Euglena gracilis*. *J. Nat. Prod.* **67**, 1067–69.

Kim, J. and DellaPenna, D. (2006). Defining the primary route for lutein synthesis in plants: The role of *Arabidopsis* carotenoid β-ring hydroxylase CYP97A3. *Proc. Natl. Acad. Sci. USA* **103**, 3474–79.

Kim, J., Smith, J. J., Tian, L. and DellaPenna, D. (2009). The evolution and function of carotenoid hydroxylases in *Arabidopsis*. *Plant Cell Physiol.* **50**, 463–79.

Kirilovsky, D. (2007). Photoprotection in cyanobacteria: the orange carotenoid protein (OCP)-related non-photochemical-quenching mechanism. *Photosynth. Res.* **93**, 7–16.

Kloer, D. P. and Schulz, G. E. (2006). Structural and biological aspects of carotenoid cleavage. *Cell. Mol. Life Sci.* **63**, 2291–303.

Kobayashi, M., Hirai, N., Kurimura, Y., Ohigashi, H. and Tsuji, Y. (1997). Abscisic acid-dependent algal morphogenesis in the unicellular green alga *Haematococcus pluvialis*. *Plant Growth Regul.* **22**, 79–85.

Kuntz, M. (2004). Plastid terminal oxidase and its biological significance. *Planta* **218**, 896–99.

Kuntz, M., Römer, S., Suire, C., Hugueney, P., Weil, J. H., Schantz, R. and Camara, B. (1992). Identification of a cDNA for the plastid-located geranylgeranyl pyrophosphate synthase from *Capsicum annuum*: correlative increase in enzyme activity and transcript level during fruit ripening. *Plant J.* **2**, 25–34.

Lagarde, D. and Vermaas, W. (1999). The zeaxanthin biosynthesis enzyme β-carotene hydroxylase is involved in myxoxanthophyll synthesis in *Synechocystis* sp. PCC 6803. *FEBS Lett.* **454**, 247–51.

Lange, B. M. and Ghassemian, M. (2003). Genome organization in *Arabidopsis thaliana*: a survey for genes involved in isoprenoid and chlorophyll metabolism. *Plant Mol. Biol.* **51**, 925–48.

Lange, B. M., Rujan, T., Martin, W. and Croteau, R. (2000). Isoprenoid biosynthesis: The evolution of two ancient and distinct pathways across genomes. *Proc. Natl. Acad. Sci. USA* **97**, 13172–77.

Laule, O., Fürholz, A., Chang, H. S., Zhu, T., Wang, X., Heifetz, P. B., Gruissem, W. and Lange, B. M. (2003). Crosstalk between cytosolic and plastidial pathways of isoprenoid biosynthesis in *Arabidopsis thaliana*. *Proc. Natl. Acad. Sci. USA* **100**, 6866–71.

Lavaud, J. (2007). Fast regulation of photosynthesis in diatoms: mechanisms, evolution and ecophysiology. *Funct. Plant Sci. Biotechnol.* **1**, 267–87.

Leung, J. and Giraudat, J. (1998). Abscisic acid signal transduction. *Annu. Rev. Plant Physiol. Plant Mol. Biol.* **49**, 199–222.

Leyser, O. (2009). The control of shoot branching: an example of plant information processing. *Plant Cell Environ.* **32**, 694–703.

Li, F., Murillo, C. and Wurtzel, E. T. (2007). Maize *Y9* encodes a product essential for 15-*cis*-ζ-carotene isomerization. *Plant Physiol.* **144**, 1181–89.

Li, Z. R., Ahn, T. K., Avenson, T. J., Ballottari, M., Cruz, J. A., Kramer, D. M., Bassi, R., Fleming, G. R., Keasling, J. D. and Niyogi, K. K. (2009). Lutein accumulation in the absence of zeaxanthin restores nonphotochemical quenching in the *Arabidopsis thaliana npq1* mutant. *Plant Cell* **21**, 1798–812.

Lichtenthaler, H. K. (1999). The 1-deoxy-D-xylulose 5-phosphate pathway of isoprenoid biosynthesis in plants. *Annu. Rev. Plant Physiol. Plant Mol. Biol.* **50**, 47–65.

Linden, H. (1999). Carotenoid hydroxylase from *Haematococcus pluvialis*: cDNA sequence, regulation and functional complementation. *Biochim. Biophys. Acta* **1446**, 203–12.

Linden, H., Misawa, N., Chamovitz, D., Pecker, I., Hirschberg, J. and Sandmann, G. (1991). Functional complementation in *Escherichia coli* of different phytoene desaturase genes and analysis of accumulated carotenes. *Z. Naturforsch. C* **46**, 1045–51.

Lohr, M. (2001). *Beziehungen zwischen den Xanthophyllzyklen und der Biosynthese von Lichtsammelxanthophyllen in Chlorophyll a/c-haltigen Algen*. Ph.D. thesis, Johannes Gutenberg-Universität, Mainz, Germany.

Lohr, M. (2009). Carotenoids. In *The Chlamydomonas Sourcebook, vol. 2: Organellar and Metabolic Processes*, ed. D. Stern, 2nd edn. Amsterdam: Academic Press, pp. 799–817.

Lohr, M. and Wilhelm, C. (1999). Algae displaying the diadinoxanthin cycle also possess the violaxanthin cycle. *Proc. Natl. Acad. Sci. USA* **96**, 8784–89.

Lohr, M. and Wilhelm, C. (2001). Xanthophyll synthesis in diatoms: quantification of putative intermediates and comparison of pigment conversion kinetics with rate constants derived from a model. *Planta* **212**, 382–91.

Lohr, M., Im, C. S. and Grossman, A. R. (2005). Genome-based examination of chlorophyll and carotenoid biosynthesis in *Chlamydomonas reinhardtii*. *Plant Physiol.* **138**, 490–515.

Lotan, T. and Hirschberg, J. (1995). Cloning and expression in *Escherichia coli* of the gene encoding β-C-4-oxygenase, that converts β-carotene to the ketocarotenoid canthaxanthin in *Haematococcus pluvialis*. *FEBS Lett.* **364**, 125–28.

Lubián, L. M., Montero, O., Moreno-Garrido, I., Huertas, I. E., Sobrino, C., González-del Valle, M. and Parés, G. (2000). *Nannochloropsis* (Eustigmatophyceae) as source of commercially valuable pigments. *J. Appl. Phycol.* **12**, 249–55.

Makino, T., Harada, H., Ikenaga, H., Matsuda, S., Takaichi, S., Shindo, K., Sandmann, G., Ogata, T. and Misawa, N. (2008). Characterization of cyanobacterial carotenoid ketolase CrtW and hydroxylase CrtR by complementation analysis in *Escherichia coli*. *Plant Cell Physiol.* **49**, 1867–78.

Marasco, E. K., Vay, K. and Schmidt-Dannert, C. (2006). Identification of carotenoid cleavage dioxygenases from *Nostoc* sp. PCC 7120 with different cleavage activities. *J. Biol. Chem.* **281**, 31583–93.

Maresca, J. A., Graham, J. E., Wu, M., Eisen, J. A. and Bryant, D. A. (2007). Identification of a fourth family of lycopene cyclases in photosynthetic bacteria. *Proc. Natl. Acad. Sci. USA* **104**, 11784–89.

Maresca, J. A., Graham, J. E. and Bryant, D. A. (2008). The biochemical basis for structural diversity in the carotenoids of chlorophototrophic bacteria. *Photosynth. Res.* **97**, 121–40.

Marin, E., Nussaume, L., Quesada, A., Gonneau, M., Sotta, B., Hugueney, P., Frey, A. and Marion-Poll, A. (1996). Molecular identification of zeaxanthin epoxidase of *Nicotiana plumbaginifolia*, a gene involved in abscisic acid biosynthesis and corresponding to the ABA locus of *Arabidopsis thaliana*. *Embo J.* **15**, 2331–42.

Marquardt, J. and Hanelt, D. (2004). Carotenoid composition of *Delesseria lancifolia* and other marine red algae from polar and temperate habitats. *Eur. J. Phycol.* **39**, 285–92.

Maršálek, B., Zahradníčková, H. and Hronková, M. (1992). Extracellular abscisic acid produced by cyanobacteria under salt stress. *J. Plant Physiol.* **139**, 506–08.

Masamoto, K., Misawa, N., Kaneko, T. Kikuno, R. and Toh, H. (1998). β-carotene hydroxylase gene from the cyanobacterium *Synechocystis* sp. PCC6803. *Plant Cell Physiol.* **39**, 560–64.

Masamoto, K., Wada, H., Kaneko, T. and Takaichi, S. (2001). Identification of a gene required for *cis*-to-*trans* carotene isomerization in carotenogenesis of the cyanobacterium *Synechocystis* sp. PCC 6803. *Plant Cell Physiol.* **42**, 1398–402.

Matsubara, S., Morosinotto, T., Bassi, R., Christian, A. L., Fischer-Schliebs, E., Lüttge, U., Orthen, B., Franco, A. C., Scarano, F. R., Förster, B., Pogson, B. J. and Osmond, C. B. (2003). Occurrence of the lutein-epoxide cycle in mistletoes of the Loranthaceae and Viscaceae. *Planta* **217**, 868–79.

Matthews, P. D., Luo, R. B. and Wurtzel, E. T. (2003). Maize phytoene desaturase and ζ-carotene desaturase catalyse a poly-Z desaturation pathway: implications for genetic engineering of carotenoid content among cereal crops. *J. Exp. Bot.* **54**, 2215–30.

McGraw, K. J., Nolan, P. M. and Crino, O. L. (2006). Carotenoid accumulation strategies for becoming a colourful House Finch: analyses of plasma and liver pigments in wild moulting birds. *Funct. Ecol.* **20**, 678–88.

Milborrow, B. V. (1982). Stereochemical aspects of carotenoid biosynthesis. In *Carotenoid Chemistry and Biochemistry*, ed. G. Britton, G. and T-W. Goodwin. Oxford: Pergamon Press, pp. 279–95.

Misawa, N., Truesdale, M. R., Sandmann, G., Fraser, P. D., Bird, C., Schuch, W. and Bramley, P. M. (1994). Expression of a tomato cDNA coding for phytoene synthase in *Escherichia coli*, phytoene formation *in vivo* and *in vitro*, and functional analysis of the various truncated gene products. *J. Biochem.* **116**, 980–85.

Misawa, N., Satomi, Y., Kondo, K., Yokoyama, A., Kajiwara, S., Saito, T., Ohtani, T. and Miki, W. (1995). Structure and functional analysis of a marine bacterial carotenoid biosynthesis gene cluster and astaxanthin biosynthetic pathway proposed at the gene level. *J. Bacteriol.* **177**, 6575–84.

Mochimaru, M., Masukawa, H. and Takaichi, S. (2005). The cyanobacterium *Anabaena* sp. PCC 7120 has two distinct β-carotene ketolases: CrtO for echinenone and CrtW for ketomyxol synthesis. *FEBS Lett.* **579**, 6111–14.

Mochimaru, M., Masukawa, H., Maoka, T., Mohamed, H. E., Vermaas, W. F. J. and Takaichi, S. (2008). Substrate specificities and availability of fucosyltransferase and β-carotene hydroxylase for myxol 2′-fucoside synthesis in *Anabaena* sp. strain PCC 7120 compared with *Synechocystis* sp. strain PCC 6803. *J. Bacteriol.* **190**, 6726–33.

Mohamed, H. E. and Vermaas, W. F. J. (2006). Sll0254 ($CrtL^{diox}$) is a bifunctional lycopene cyclase/dioxygenase in cyanobacteria producing myxoxanthophyll. *J. Bacteriol.* **188**, 3337–44.

Moustafa, A., Reyes-Prieto, A. and Bhattacharya, D. (2008). Chlamydiae has contributed at least 55 genes to Plantae with predominantly plastid functions. *PLoS ONE* **3**, e2205.

Müller, P., Li, X. P. and Niyogi, K. K. (2001). Non-photochemical quenching. A response to excess light energy. *Plant Physiol.* **125**, 1558–66.

Nambara, E. and Marion-Poll, A. (2005). Abscisic acid biosynthesis and catabolism. *Annu. Rev. Plant Biol.* **56**, 165–85.

Nelson, D. R. (2006). Plant cytochrome P450s from moss to poplar. *Phytochem. Rev.* **5**, 193–204.

Nikulina, K., Chunaev, A. S. and Boschetti, A. (1999). Accumulation of ζ-carotene in *Chlamydomonas reinhardtii* under control of the *ac5* nuclear gene. *Plant Cell Rep.* **19**, 37–42.

Nimura, K. and Mizuta, H. (2002). Inducible effects of abscisic acid on sporophyte discs from *Laminaria japonica* Areschoug (Laminariales, Phaeophyceae). *J. Appl. Phycol.* **14**, 159–63.

Nishida, Y., Adachi, K., Kasai, H., Shizuri, Y., Shindo, K., Sawabe, A., Komemushi, S., Miki, W. and Misawa, N. (2005). Elucidation of a carotenoid biosynthesis gene cluster encoding a novel enzyme, 2,2′-β-hydroxylase, from *Brevundimonas* sp. strain SD212 and combinatorial biosynthesis of new or rare xanthophylls. *Appl. Environ. Microbiol.* **71**, 4286–96.

Norris, S. R., Barrette, T. R. and DellaPenna, D. (1995). Genetic dissection of carotenoid synthesis in *Arabidopsis* defines plastoquinone as an essential component of phytoene desaturation. *Plant Cell* **7**, 2139–49.

North, H. M., De Almeida, A., Boutin, J. P., Frey, A., To, A., Botran, L., Sotta, B. and Marion-Poll, A. (2007). The Arabidopsis ABA-deficient mutant *aba4* demonstrates that the major route for stress-induced ABA accumulation is via neoxanthin isomers. *Plant J.* **50**, 810–24.

Park, H., Kreunen, S. S., Cuttriss, A. J., DellaPenna, D. and Pogson, B. J. (2002). Identification of the carotenoid isomerase provides insight into carotenoid biosynthesis, prolamellar body formation, and photomorphogenesis. *Plant Cell* **14**, 321–32.

Parniske, M. (2008). Arbuscular mycorrhiza: the mother of plant root endosymbioses. *Nat. Rev. Microbiol.* **6**, 763–75.

Pecker, I., Chamovitz, D., Linden, H., Sandmann, G. and Hirschberg, J. (1992). A single polypeptide catalyzing the conversion of phytoene to ζ-carotene is transcriptionally regulated during tomato fruit ripening. *Proc. Natl. Acad. Sci. USA* **89**, 4962–66.

Phillips, M. A., León, P., Boronat, A. and Rodríguez-Concepción, M. (2008). The plastidial MEP pathway: unified nomenclature and resources. *Trends Plant Sci.* **13**, 619–23.

Prado-Cabrero, A., Scherzinger, D., Avalos, J. and Al-Babili, S. (2007). Retinal biosynthesis in fungi: characterization of the carotenoid oxygenase CarX from *Fusarium fujikuroi*. *Eukaryot. Cell* **6**, 650–57.

Punginelli, C., Wilson, A., Routaboul, J. M. and Kirilovsky, D. (2009). Influence of zeaxanthin and echinenone binding on the activity of the Orange Carotenoid Protein. *Biochim. Biophys. Acta* **1787**, 280–88.

Quinlan, R. F., Jaradat, T. T. and Wurtzel, E. T. (2007). *Escherichia coli* as a platform for functional expression of plant P450 carotene hydroxylases. *Arch. Biochem. Biophys.* **458**, 146–57.

Rmiki, N. E., Brunet, C., Cabioch, J. and Lemoine, Y. (1996). Xanthophyll cycle and photosynthetic adaptation to environment in macro- and microalgae. *Hydrobiologia* **327**, 407–13.

Rodríguez-Concepción, M. and Boronat, A. (2002). Elucidation of the methylerythritol phosphate pathway for isoprenoid biosynthesis in bacteria and plastids. A metabolic milestone achieved through genomics. *Plant Physiol.* **130**, 1079–89.

Rohdich, F., Hecht, S., Gartner, K., Adam, P., Krieger, C., Amslinger, S., Arigoni, D., Bacher, A. and Eisenreich, W. (2002). Studies on the nonmevalonate terpene biosynthetic pathway: Metabolic role of IspH (LytB) protein. *Proc. Natl. Acad. Sci. USA* **99**, 1158–63.

Ronen, G., Carmel-Goren, L., Zamir, D. and Hirschberg, J. (2000). An alternative pathway to β-carotene formation in plant chromoplasts discovered by map-based cloning of *Beta* and *old-gold* color mutations in tomato. *Proc. Natl. Acad. Sci. USA* **97**, 11102–07.

Rubio, A., Rambla, J. L., Santaella, M., Gómez, M. D., Orzaez, D., Granell, A. and Gómez-Gómez, L. (2008). Cytosolic and plastoglobule-targeted carotenoid dioxygenases from *Crocus sativus* are both involved in β-ionone release. *J. Biol. Chem.* **283**, 24816–25.

Ruch, S., Beyer, P., Ernst, H. and Al-Babili, S. (2005). Retinal biosynthesis in Eubacteria: *in vitro* characterization of a novel carotenoid oxygenase from *Synechocystis* sp. PCC 6803. *Mol. Microbiol.* **55**, 1015–24.

Ruiz-González, M. X. and Marín, I. (2004). New insights into the evolutionary history of type 1 rhodopsins. *J. Mol. Evol.* **58**, 348–58.

Sandmann, G. (1994). Carotenoid biosynthesis in microorganisms and plants. *Eur. J. Biochem.* **223**, 7–24.

Sandmann, G. (2002). Molecular evolution of carotenoid biosynthesis from bacteria to plants. *Physiol. Plant.* **116**, 431–40.

Sandmann, G. (2009). Evolution of carotene desaturation: The complication of a simple pathway. *Arch. Biochem. Biophys.* **483**, 169–74.

Saradhi, P. P., Suzuki, I., Katoh, A., Sakamoto, A., Sharmila, P., Shi, D. J. and Murata, N. (2000). Protection against the photo-induced inactivation of the photosystem II complex by abscisic acid. *Plant Cell Environ.* **23**, 711–18.

Saranak, J. and Foster, K. W. (1994). The *in vivo* cleavage of carotenoids into retinoids in *Chlamydomonas reinhardtii*. *J. Exp. Bot.* **45**, 505–11.

Saranak, J. and Foster, K. W. (2005). Photoreceptor for curling behavior in *Peranema trichophorum* and evolution of eukaryotic rhodopsins. *Eukaryot. Cell* **4**, 1605–12.

Scaife, M. A., Burja, A. M. and Wright, P. C. (2009). Characterization of cyanobacterial β-carotene ketolase and hydroxylase genes in *Escherichia coli*, and their application for astaxanthin biosynthesis. *Biotechnol. Bioeng.* **103**, 944–55.

Schaffelke, B. (1995). Abscisic acid in sporophytes of three *Laminaria* species (Phaeophyta). *J. Plant Physiol.* **146**, 453–58.

Schagerl, M., Pichler, C. and Donabaum, K. (2003). Patterns of major photosynthetic pigments in freshwater algae. 2. Dinophyta, Euglenophyta, Chlorophyceae and Charales. *Ann. Limnol. – Int. J. Lim.* **29**, 49–62.

Scherzinger, D. and Al-Babili, S. (2008). *In vitro* characterization of a carotenoid cleavage dioxygenase from *Nostoc* sp. PCC 7120 reveals a novel cleavage pattern, cytosolic localization and induction by highlight. *Mol. Microbiol.* **69**, 231–44.

Scherzinger, D., Ruch, S., Kloer, D. P., Wilde, A. and Al-Babili, S. (2006). Retinal is formed from apo-carotenoids in *Nostoc* sp. PCC7120: *in vitro* characterization of an apo-carotenoid oxygenase. *Biochem. J.* **398**, 361–69.

Schmidt, M., Gessner, G., Matthias, L., Heiland, I., Wagner, V., Kaminski, M., Geimer, S., Eitzinger, N., Reissenweber, T., Voytsekh, O., Fiedler, M., Mittag, M. and Kreimer, G. (2006). Proteomic analysis of the eyespot of *Chlamydomonas reinhardtii* provides novel insights into its components and tactic movements. *Plant Cell* **18**, 1908–30.

Schoefs, B., Rmiki, N. E., Rachadi, J. and Lemoine, Y. (2001). Astaxanthin accumulation in *Haematococcus* requires a cytochrome P450 hydroxylase and an active synthesis of fatty acids. *FEBS Lett.* **500**, 125–28.

Schubert, N. and García-Mendoza, E. (2008). Photoinhibition in red algal species with different carotenoid profiles. *J. Phycol.* **44**, 1437–46.

Schubert, N., García-Mendoza, E. and Pacheco-Ruiz, I. (2006). Carotenoid composition of marine red algae. *J. Phycol.* **42**, 1208–16.

Schwartz, S. H., Qin, X. Q. and Loewen, M. C. (2004). The biochemical characterization of two carotenoid cleavage enzymes from *Arabidopsis* indicates that a carotenoid-derived compound inhibits lateral branching. *J. Biol. Chem.* **279**, 46940–45.

Schwartz, S. H., Tan, B. C., Gage, D. A., Zeevaart, J. A. D. and McCarty, D. R. (1997). Specific oxidative cleavage of carotenoids by VP14 of maize. *Science* **276**, 1872–74.

Seemann, M., Wegner, P., Schünemann, V., Bui, B. T. S., Wolff, M., Marquet, A., Trautwein, A. X. and Rohmer, M. (2005). Isoprenoid biosynthesis in chloroplasts via the methylerythritol phosphate pathway: the (*E*)-4-hydroxy-3-methylbut-2-enyl diphosphate synthase (GcpE) from *Arabidopsis thaliana* is a [4Fe-4S] protein. *J. Biol. Inorg. Chem.* **10**, 131–37.

Seemann, M., Bui, B. T. S., Wolff, M., Miginlac-Maslow, M. and Rohmer, M. (2006). Isoprenoid biosynthesis in plant chloroplasts via the MEP pathway: Direct thylakoid/ferredoxin-dependent photoreduction of GcpE/IspG. *FEBS Lett.* **580**, 1547–52.

Shahbazi, M., Gilbert, M., Laboure, A. M. and Kuntz, M. (2007). Dual role of the plastid terminal oxidase in tomato. *Plant Physiol.* **145**, 691–702.

Simkin, A. J., Schwartz, S. H., Auldridge, M., Taylor, M. G. and Klee, H. J. (2004). The tomato *carotenoid cleavage dioxygenase 1* genes contribute to the formation of the flavor volatiles β-ionone, pseudoionone, and geranylacetone. *Plant J.* **40**, 882–92.

Sineshchekov, O. A., Govorunova, E. G., Jung, K. H., Zauner, S., Maier, U. G. and Spudich, J. L. (2005). Rhodopsin-mediated photoreception in cryptophyte flagellates. *Biophys. J.* **89**, 4310–19.

Spudich, J. L., Yang, C. S., Jung, K. H. and Spudich, E. N. (2000). Retinylidene proteins: Structures and functions from archaea to humans. *Annu. Rev. Cell Dev. Biol.* **16**, 365–92.

Steiger, S. and Sandmann, G. (2004). Cloning of two carotenoid ketolase genes from *Nostoc punctiforme* for the heterologous production of canthaxanthin and astaxanthin. *Biotechnol. Lett.* **26**, 813–17.

Steiger, S., Jackisch, Y. and Sandmann, G. (2005). Carotenoid biosynthesis in *Gloeobacter violaceus* PCC4721 involves a single crtI-type phytoene desaturase instead of typical cyanobacterial enzymes. *Arch. Microbiol.* **184**, 207–14.

Steinbacher, S., Kaiser, J., Gerhardt, S., Eisenreich, W., Huber, R., Bacher, A. and Rohdich, F. (2003). Crystal structure of the type II isopentenyl diphosphate: dimethylallyl diphosphate isomerase from *Bacillus subtilis*. *J. Mol. Biol.* **329**, 973–82.

Stickforth, P., Steiger, S., Hess, W. R. and Sandmann, G. (2003). A novel type of lycopene ε-cyclase in the marine cyanobacterium *Prochlorococcus marinus* MED4. *Arch. Microbiol.* **179**, 409–15.

Stirk, W. A., Novák, O., Hradecká, V., Pěnčík, A., Rolčík, J., Strnad, M. and Van Staden, J. (2009). Endogenous cytokinins, auxins and abscisic acid in *Ulva fasciata* (Chlorophyta) and *Dictyota humifusa* (Phaeophyta): towards understanding their biosynthesis and homoeostasis. *Eur. J. Phycol.* **44**, 231–40.

Stolte, W., Kraay, G. W., Noordeloos, A. A. M. and Riegman, R. (2000). Genetic and physiological variation in pigment composition of *Emiliania huxleyi* (Prymnesiophyceae) and the potential use of its pigment ratios as a quantitative physiological marker. *J. Phycol.* **36**, 529–39.

Stransky, H. and Hager, A. (1970a). The carotenoid pattern and the occurrence of the light-induced xanthophyll cycle in various classes of algae. II. Xanthophyceae. *Arch. Mikrobiol.* **71**, 164–90.

Stransky, H. and Hager, A. (1970b). The carotenoid pattern and the occurrence of the light-induced xanthophyll cycle in various classes of algae. IV. Cyanophyceae and Rhodophyceae. *Arch. Mikrobiol.* **72**, 84–96.

Sun, Z. R., Gantt, E. and Cunningham, F. X. (1996). Cloning and functional analysis of the β-carotene hydroxylase of *Arabidopsis thaliana*. *J. Biol. Chem.* **271**, 24349–52.

Swift, I. E., Milborrow, B. V. and Jeffrey, S. W. (1980). Formation of neoxanthin, diadinoxanthin and peridinin from [^{14}C]zeaxanthin by a cell-free system from *Amphidinium carterae*. *Phytochemistry* **21**, 2859–64.

Takaichi, S. and Mimuro M. (1998). Distribution and geometric isomerism of neoxanthin in oxygenic phototrophs: 9′-*cis*, a sole molecular form. *Plant Cell Physiol.* **39**, 968–77.

Takaichi, S. and Mochimaru, M. (2007). Carotenoids and carotenogenesis in cyanobacteria: unique ketocarotenoids and carotenoid glycosides. *Cell. Mol. Life Sci.* **64**, 2607–19.

Takano, H., Asker, D., Beppu, T. and Ueda, K. (2006). Genetic control for light-induced carotenoid production in non-phototrophic bacteria. *J. Ind. Microbiol. Biotechnol.* **33**, 88–93.

Tan, B. C., Joseph, L. M., Deng, W. T., Liu, L. J., Li, Q. B., Cline, K. and McCarty, D. R. (2003). Molecular characterization of the *Arabidopsis* 9-*cis* epoxycarotenoid dioxygenase gene family. *Plant J.* **35**, 44–56.

Tao, L. and Cheng, Q. (2004). Novel β-carotene ketolases from non-photosynthetic bacteria for canthaxanthin synthesis. *Mol. Genet. Genomics* **272**, 530–37.

Tao, L., Rouvière, P. E. and Cheng, Q. O. (2006). A carotenoid synthesis gene cluster from a non-marine *Brevundimonas* that synthesizes hydroxylated astaxanthin. *Gene* **379**, 101–08.

Taylor, I. B., Sonneveld, T., Bugg, T. D. H. and Thompson, A. J. (2005). Regulation and manipulation of the biosynthesis of abscisic acid, including the supply of xanthophyll precursors. *J. Plant Growth Regul.* **24**, 253–73.

Tian, L. and DellaPenna, D. (2004). Progress in understanding the origin and functions of carotenoid hydroxylases in plants. *Arch. Biochem. Biophys.* **430**, 22–29.

Tian, L., Musetti, V., Kim, J., Magallanes-Lundback, M. and DellaPenna, D. (2004). The *Arabidopsis LUT1* locus encodes a member of the cytochrome P450 family that is required for carotenoid ε-ring hydroxylation activity. *Proc. Natl. Acad. Sci. USA* **101**, 402–07.

Tietz, A., Ruttkowski, U., Koehler, R. and Kasprik, W. (1989). Further investigations on the occurrence and the effects of abscisic acid in algae. *Biochem. Physiol. Pflanzen* **184**, 259–66.

Tsuchiya, T., Takaichi, S., Misawa, N., Maoka, T., Miyashita, H. and Mimuro, M. (2005). The cyanobacterium *Gloeobacter violaceus* PCC 7421 uses bacterial-type phytoene desaturase in carotenoid biosynthesis. *FEBS Lett.* **579**, 2125–29.

Umehara, M., Hanada, A., Yoshida, S., Akiyama, K., Arite, T., Takeda-Kamiya, N., Magome, H., Kamiya, Y., Shirasu, K., Yoneyama, K., Kyozuka, J. and Yamaguchi, S. (2008). Inhibition of shoot branching by new terpenoid plant hormones. *Nature* **455**, 195–200.

Vogel, J. T., Tan, B. C., McCarty, D. R. and Klee, H. J. (2008). The carotenoid cleavage dioxygenase 1 enzyme has broad substrate specificity, cleaving multiple carotenoids at two different bond positions. *J. Biol. Chem.* **283**, 11364–73.

von Lintig, J., Hessel, S., Isken, A., Kiefer, C., Lampert, J. M., Voolstra, O. and Vogt, K. (2005). Towards a better understanding of carotenoid metabolism in animals. *Biochim. Biophys. Acta* **1740**, 122–31.

Wasilewska, A., Vlad, F., Sirichandra, C., Redko, Y., Jammes, F., Valon, C., Frey, N. F. D. and Leung, J. (2008). An update on abscisic acid signaling in plants and more. . . . *Mol. Plant* **1**, 198–217.

Watson, S. B. (2003). Cyanobacterial and eukaryotic algal odour compounds: signals or by-products? A review of their biological activity. *Phycologia* **42**, 332–50.

Wilhelm, C. (1990). The biochemistry and physiology of light-harvesting processes in chlorophyll *b*-containing and chlorophyll *c*-containing algae. *Plant Physiol. Biochem.* **28**, 293–306.

Wilhelm, C., Büchel, C., Fisahn, J., Goss, R., Jakob, T., LaRoche, J., Lavaud, J., Lohr, M., Riebesell, U., Stehfest, K., Valentin, K. and Kroth, P. G. (2006). The regulation of carbon and nutrient assimilation in diatoms is significantly different from green algae. *Protist* **157**, 91–124.

Wilson, A., Punginelli, C., Gall, A., Bonetti, C., Alexandre, M., Routaboul, J.-M., Kerfeld, C. A., van Grondelle, R., Robert, B., Kennis, J. T. M. and Kirilovsky, D. (2008). A photoactive carotenoid protein acting as light intensity sensor. *Proc. Natl. Acad. Sci. USA* **105**, 12075–80.

Withers, N. W. and Haxo, F. T. (1978). Isolation and characterization of carotenoid-rich lipid globules from *Peridinium foliaceum*. *Plant Physiol.* **62**, 36–39.

Wolff, M., Seemann, M., Bui, B. T. S., Frapart, Y., Tritsch, D., Estrabot, A. G., Rodríguez-Concepción, M., Boronat, A., Marquet, A. and Rohmer, M. (2003). Isoprenoid biosynthesis via the methylerythritol phosphate pathway: the (*E*)-4-hydroxy-3-methylbut-2-enyl diphosphate reductase (LytB/IspH) from *Escherichia coli* is a [4Fe-4S] protein. *FEBS Lett.* **541**, 115–20.

Yang, S., Wu, R. S. S., Mok, H. O. L., Zhang, Z. P. and Kong, R. Y. C. (2003). Identification of a novel cytochrome P450 cDNA, CYP97E1, from the marine diatom *Skeletonema costatum* Bacillariophyceae. *J. Phycol.* **39**, 555–60.

Yoneyama, K., Xie, X., Yoneyama, K. and Takeuchi, Y. (2009). Strigolactones: structures and biological activities. *Pest Manag. Sci.* **65**, 467–70.

Yoshida, K., Igarashi, E., Mukai, M., Hirata, K. and Miyamoto, K. (2003). Induction of tolerance to oxidative stress in the green alga, *Chlamydomonas reinhardtii*, by abscisic acid. *Plant Cell Environ.* **26**, 451–57.

Yoshida, K., Igarashi, E., Wakatsuki, E., Miyamoto, K. and Hirata, K. (2004). Mitigation of osmotic and salt stresses by abscisic acid through reduction of stress-derived oxidative damage in *Chlamydomonas reinhardtii*. *Plant Sci.* **167**, 1335–41.

Yoshii, Y. (2006). Diversity and evolution of photosynthetic antenna systems in green plants. *Phycol. Res.* **54**, 220–29.

Yoshii, Y., Takaichi, S., Maoka, T., Suda, S., Sekiguchi, H., Nakayama, T. and Inouye, I. (2005). Variation of siphonaxanthin series among the genus *Nephroselmis* (Prasinophyceae, Chlorophyta), including a novel primary methoxy carotenoid. *J. Phycol.* **41**, 827–34.

Zahradníčková, H., Maršálek, B. and Polišenská, M. (1991). High-performance thin-layer chromatographic and high-performance liquid chromatographic determination of abscisic acid produced by cyanobacteria. *J. Chromatogr.* **555**, 239–45.

Zapata, M. (2005). Recent advances in pigment analysis as applied to picophytoplankton. *Vie et Milieu* **55**, 233–48.

Zapata, M., Jeffrey, S. W., Wright, S. W., Rodríguez, F., Garrido, J. L. and Clementson, L. (2004). Photosynthetic pigments in 37 species (65 strains) of Haptophyta: implications for oceanography and chemotaxonomy. *Mar. Ecol. Prog. Ser.* **270**, 83–102.

Zhua, C. F., Yamamura, S., Nishihara, M., Koiwa, H. and Sandmann, G. (2003). cDNAs for the synthesis of cyclic carotenoids in petals of *Gentiana lutea* and their regulation during flower development. *Biochim. Biophys. Acta* **1625**, 305–08.

Part II

Methodology guidance

4

New HPLC separation techniques

JOSÉ L. GARRIDO, RUTH L. AIRS, FRANCISCO RODRÍGUEZ, LAURIE VAN HEUKELEM AND MANUEL ZAPATA

4.1 Introduction

The 1997 UNESCO monograph *Phytoplankton Pigments in Oceanography* covered the state of the art in most of the aspects of HPLC analysis of planktonic pigments, reviewing existing methods and their application to oceanography (Jeffrey, 1997a), giving guidelines for setting up an HPLC laboratory (Wright and Mantoura, 1997) and proposing new isocratic (Mantoura *et al.*, 1997) and gradient (Wright and Jeffrey, 1997) methods. Most of the information in that monograph is still of importance and its reading is strongly recommended.

Since then, three important reviews have accounted for developments in the field: Jeffrey *et al.* (1999) summarized advances in methods, with special attention to developments in stationary and mobile phases, and examined the taxonomic value of pigment suites and the methods for interpreting pigment data in natural samples in terms of percentages of algal types. A detailed review by Bidigare *et al.* (2005) addressed phytoplankton pigment analysis processes, accounting for laboratory methods (including sampling, extraction, HPLC methods and procedures) and quality assurance. Recently, Wright and Jeffrey (2006) exhaustively reviewed new methods for the analysis of pigments by HPLC and compared several of them with the original method proposed in the UNESCO monograph.

4.2 HPLC algal pigment methods published since the 1997 UNESCO monograph

In spite of continuous developments in instruments and columns, analysis of phytoplankton pigments is still a challenge for HPLC techniques. This is because many pigments vary greatly in structure and span a wide range of polarities, while others have similar chemical structures, with differences as small as the position of a double bond (Bjørnland, 1997; Jeffrey, 1997b). Most HPLC methods employ reversed-phase stationary phases, with C_8 to C_{30} chains chemically bonded to silica supports and gradient elution from partially aqueous mobile phases to non-aqueous organic mixtures (Table 4.1). Under such conditions, pigments are primarily resolved on the basis of their polarity.

Phytoplankton Pigments: Characterization, Chemotaxonomy and Applications in Oceanography, eds. Suzanne Roy, Carole A. Llewellyn, Einar Skarstad Egeland and Geir Johnsen. Published by Cambridge University Press.

Table 4.1. *Recent methods proposed for the analysis of phytoplankton pigments. F = flow rate, mL min^{-1}; ODS = octadecyl silane; RP = reversed phase; t = run time, min; T = temperature, °C.*

	HPLC method characteristics			
	Stationary phase	Mobile phase		Detection
Reference	Type Brand	Elution type t	Solvent composition (by volume) F, T	Equipment (wavelengths in nm)
Wright *et al.* (1991)	RP-C_{18} monomeric 5 μm 4.6 × 250 mm Spherisorb ODS-2	Ternary gradient t = 20	A: methanol: 0.5 M ammonium acetate (8:2) B: acetonitrile: water (9:1) C: ethyl acetate F = 1.0, T = nr	Diode array detector UV-Vis Fluorescence
Kraay *et al.* (1992)	RP-C_{18} Polymeric 5 μm 4.6 × 150 mm Rsil Biorad	Ternary gradient t = 30	A: methanol: 0.5 M ammonium acetate (8.5:1.5) B: acetonitrile: water 9:1 C: ethyl acetate F = 0.8, T = nr	Diode array detector
Van Heukelem *et al.* (1992)	RP-C_{18} polymeric 5 μm 4.6 × 250 mm Vydac 201TP 54	Binary gradient t = 15	A: methanol: 0.5 M ammonium acetate (8:2) B: methanol: acetone (7:3) F = 1.5, T = 40	Diode array detector
Garrido and Zapata (1993a)	RP-C_{18} polymeric 5 μm 4.6 × 150 mm Spherisorb PAH	Binary gradient t = 28	A: methanol: acetonitrile: 1 M ammonium acetate (5:3:2) B: acetonitrile: ethyl acetate (8:2) F = 1.0, T = 30	UV-Vis (430 and 450)
Garrido and Zapata (1993b)	RP-C_{18} polymeric 5 μm 4.6 × 250 mm Vydac 201TP 54	Binary gradient t = 35	A: methanol: 1 M ammonium acetate (8:2) B: acetonitrile: acetone (7:3) F = 1.0, T = 27	UV-Vis (430 and 450) Fluorescence (ex: 440; em: 660)
Goericke and Repeta (1993)	RP-C_8 monomeric 3 μm 4.6 × 100 mm Rainin, Dinamax	Binary gradient t = 40	A: methanol: 0.5 M ammonium acetate (7.5:2.5) B: methanol F = 1.0, T = nr	Diode array detector

Table 4.1. (*cont.*)

	HPLC method characteristics			
	Stationary phase	Mobile phase		Detection
Reference	Type Brand	Elution type t	Solvent composition (by volume) F, T	Equipment (wavelengths in nm)
Saitoh *et al.* (1993)	ODP 5 μm 6.0 × 150 mm Asahipak ODP 50	Isocratic t = 25	methanol: phosphate buffer pH = 3.0 (9:1) F=1.7, T=25	Diode array detector
Van Heukelem *et al.* (1994)	RP-C_{18} polymeric 5 μm 4.6 × 250 mm Vydac 201TP 54	Binary gradient t = 24	A: methanol: 0.5 M ammonium acetate (8:2) B: methanol: acetone (8:2) F = 1.0 (min 0–23), 1.5 (min 24), T = 60	Diode array detector Fluorescence (ex: 440; em: 650) Fluorescence (ex: 425; em: 670)
Van Heukelem *et al.* (1994)	RP-C_{18} polymeric 5 μm. 4.6 × 250 mm Vydac 201TP 54	Binary gradient t = 31	A: methanol: 0.5 M ammonium acetate (8:2) B: methanol: acetone (8:2) F = 1.0 (min 0–17), 1.5 (min 22), 2.0 (min 27), T=10	Diode array detector Fluorescence (ex: 440; em: 650) Fluorescence (ex: 425; em: 670)
Garrido *et al.* (1995)	RP-C_{18} polymeric 5 μm 4.6 × 250 mm Lichrospher PAH	Binary gradient t = 40	A: methanol: 1 M ammonium acetate 8:2 B: acetone F = 0.8, T = 27	UV-Vis (430 and 450) Fluorescence (ex: 440; em: 660)
Saitoh *et al.* (1995)	Column switching: i) ODP 5 μm 4.6 × 150 mm Asahipak ODP 50 ii) RP-C_{18} monomeric 5 μm 4.6 × 10 mm Inertsil ODS-2	Isocratic t = 60	methanol: phosphate buffer pH = 3.0 (9.2:0.8) F = 1.0, T = nr	Diode array detector
Van Lenning *et al.* (1995)	RP-C_{18} polymeric 5 μm 4.6 × 250 mm Lichrospher PAH	Binary gradient t = 48	A: methanol: 1 M ammonium acetate (8:2) B: acetone F = 0.8, T = Temperature gradient, t_0–t_{15} = 31 °C t_{17} = 8 °C	Diode array detector
Garrido and Zapata (1996)	RP-C_{18} monomeric 5 μm 4.6 × 250 mm Ultrasphere ODS	Binary gradient t = 40	A: methanol: 0.025 M aqueous pyridine, pH = 5.0 (8:2) B: acetonitrile: acetone (7:3) F = 1.2, T = 27	Fluorescence (ex: 440; em: 660)

Table 4.1. (*cont.*)

	HPLC method characteristics			
	Stationary phase	Mobile phase		Detection
Reference	Type Brand	Elution type t	Solvent composition (by volume) F, T	Equipment (wavelengths in nm)
Garrido and Zapata (1996)	RP-C_{18} polymeric 5 μm 4.6 × 250 mm Vydac 201TP 54	Binary gradient t = 35	A: methanol: 0.025 M aqueous pyridine, pH = 5.0 (8:2) B: acetonitrile: acetone (7:3) F = 1.2, T = 27	Fluorescence (ex: 440; em: 660)
Pinckney *et al.* (1996)	Column coupling: i) RP-C_{18} monomeric 3 μm 4.6 × 100 mm Rainin-Microsorb MV ii) Two RP-C_{18} polymeric columns 5μm 4.6 × 250 mm Vydac 201TP 54	Binary gradient t = 55	A: methanol: 0.5 M ammonium acetate (8:2) B: methanol: acetone (8:2) F = 0.8 (min 0–5), 1.25 (min 35), 1.50 (min 40), 0.8 (min 41–55), T = 40	Diode array detector Radioactivity detector
Vidussi *et al.* (1996)	RP-C_8 monomeric 3 μm 4.6 × 100 mm Hypersil MOS2	Binary gradient t = 20	A: methanol: 0.5 M ammonium acetate (7:3) B: methanol F = 1.0, T = nr	Diode array detector
Barlow *et al.* (1997)	RP-C_8 monomeric 3 μm 4.6 × 100 mm Hypersil MOS2	Binary gradient t = 30	A: methanol: 1.0 M ammonium acetate (7:3) B: methanol F = 1.0, T = nr	UV-Vis (440)
Garrido and Zapata (1997)	RP-C_{18} polymeric 5 μm 4.6 × 250 mm Vydac 201TP 54	Binary gradient t = 25	A: methanol: acetonitrile: 0.25 M aqueous pyridine pH = 5.0 (4.5:3.5:2) B: acetone F = 1.2, T = 15	Diode array detector Fluorescence (ex: 440; em: 660)
Miyashita *et al.* (1997)	RP-C_{18} monomeric 5 μm 4.6 × 150 mm TSKgel ODS 80Ts	Binary gradient t = 30	A: methanol: water (9:1) B: methanol F = 1.0 (min 0–9), 2.0 (min 9–30), T = nr	Diode array detector UV-Vis (440)
Garrido and Zapata (1998)	RP-C_{18} monomeric 5 μm 4.6 × 250 mm Beckman Ultrasphere	Binary gradient t = 40	A: methanol: 1 M ammonium acetate (8:2) B: acetonitrile: acetone (6:4) F = 1.0, T = 27	Diode array detector Fluorescence (ex: 440; em: 660)

Table 4.1. (*cont.*)

	HPLC method characteristics			
	Stationary phase	Mobile phase		Detection
Reference	Type Brand	Elution type t	Solvent composition (by volume) F, T	Equipment (wavelengths in nm)
Rodríguez *et al.* (1998)	RP-C_8 monomeric 3.5 μm 4.6 × 150 mm Waters Symmetry	Binary gradient t = 30	A: methanol: 1 M ammonium acetate (7.5:2.5) B: methanol F = 0.8, T = 25	Diode array detector Fluorescence (ex: 440; em: 660)
Goericke *et al.* (2000a)	RP-C_{30} polymeric 3 μm 4.6 × 150 mm YMC C30 *Carotenoid*	Binary gradient t = 25	A: methanol: 0.5 M ammonium acetate (8:2) B: acetone: methanol (8:2) F = 1.5, T = nr	Diode array detector UV-Vis (410 and 676)
Goericke *et al.* (2000b)	RP-C_8 monomeric 3 μm 4.6 × 100 mm Alltech Adsorbosphere	Binary gradient t = 36	A: methanol: 0.5 M ammonium acetate (7.5:2.5) B: methanol F = 1.5, T = nr	Diode array detector UV-Vis (410 and 676)
Zapata *et al.* (2000)	RP-C_8 monomeric 3.5 μm 4.6 × 150 mm Waters Symmetry	Binary gradient t = 40	A: methanol: acetonitrile: 0.25 M aqueous pyridine pH = 5.0 (5:2.5:2.5) B: methanol: acetonitrile: acetone (2:6:2) F = 1.0, T = 25	Diode array detector Fluorescence (ex: 440; em: 660)
Van Heukelem and Thomas (2001)	RP-C_8 monomeric 3.5 μm. 4.6 × 150 mm Eclipse XDB	Binary gradient t = 30	A: methanol: 28 mM tetrabutylammonium acetate (7:3) B: methanol F = 1.1, T = 60	Diode array detector
Inbaraj *et al.* (2006)	RP-C_{30} polymeric 5 μm 4.6 × 150 mm YMC C30 *Carotenoid*	Binary gradient t = 50	A: methanol: acetonitrile: water (84:14:2) B: methylene chloride F = 1.0, T = nr	Diode array detector

nr = not recorded

During the last ten years, recognition of the taxonomic and physiological importance of certain pigments not previously separated has fuelled attempts to develop new HPLC methods based on original approaches to the problem (Jeffrey *et al.*, 1999; Zapata, 2005; Wright and Jeffrey, 2006). Thus, prevalence of divinyl-chlorophyll *a* (DVChl *a*) and divinyl-chlorophyll *b* (DVChl *b*) in the euphotic zone of tropical oceans (Chisholm *et al.*, 1992; Goericke and Repeta, 1993) initiated the development of a series of methods focused on the separation of monovinylic pigments such as Chl *a*, Chl *b*, Chl c_1 and MVChl c_3 from their corresponding DV-analogues (DVChl *a*, DVChl *b*, Chl c_2 and Chl c_3) in algal samples (recently reviewed by Garrido and Zapata, 2006). These methods were subsequently extended to attempt the simultaneous resolution of these chlorophylls and of carotenoids (reviewed by Wright and Jeffrey, 2006).

On many occasions, however, these methodological efforts have been paralleled with the discovery of several new pigments (reviewed by Wright and Jeffrey, 2006 and Zapata, 2005), thus further complicating the problem and compelling the analysts to introduce new improvements to proposed methods. Examples of new pigments resolved include: MVChl c_3 (Garrido and Zapata, 1998), Chl *c*-MGDG esters (Garrido *et al.*, 2000; Zapata *et al.*, 2001), two new polar Chl *c*-pigments of unknown structure: a Chl c_2-like pigment detected in *Pavlova gyrans* by Fawley (1989) (Garrido and Zapata, 1997) and a Chl c_1-like pigment isolated from *Exanthemachrysis gayraliae* (Van Lenning *et al.*, 2003). A new fucoxanthin-type pigment was also detected in *Emiliania huxleyi* by new HPLC methods (Garrido and Zapata, 1998), then characterized as 4-keto-19′-hexanoyloxyfucoxanthin (Egeland *et al.*, 2000), and since detected in a number of other haptophytes (Zapata *et al.*, 2004; Airs and Llewellyn, 2006). Other fucoxanthin derivatives (19′-pentanoyloxy- and 19′-heptanoyloxy-fucoxanthin) have also been detected when improved methods were applied (Airs and Llewellyn, 2006).

In the last decade, strategies to manipulate separation selectivity have focused on the choice of stationary phase (i.e. aliphatic chain length and bonding phase chemistries), unique mobile phase modifiers and column temperature. For example, polymerically synthesized C_{18} and C_{30} bonded phases have been shown to discriminate similar compounds with different molecular shapes, while monomeric C_8 columns have shown special selectivity towards compounds with subtle differences in polarity. Methods employed for algal pigment separations since the early 1990s are summarized in Table 4.1.

4.3 Separation principles and applications of new HPLC pigment techniques

4.3.1 Polymeric bonded phases

Most bonded phases employed in reversed-phase HPLC are obtained by silane modification of silica surfaces and can be divided into two groups: monomeric phases and polymeric phases. Monomeric phases are prepared by reaction of monofunctional silanes with silica, and the resulting bond linkages are monomeric in nature. Polymeric phases are prepared using trifunctional silanes in the presence of water.

Hydrolysis results in the formation of silane silanols, which may subsequently react with other silane molecules to form silane polymers (Sander and Wise, 1990; Wright and Mantoura, 1997).

Silica-based polymeric stationary phases exhibit 'shape selectivity' (Sander *et al.*, 1999), i.e. very high selectivity towards very similar, even isomeric, compounds with rigid molecular structures. Study of these phases by NMR spectroscopy has demonstrated that organic chains are bonded as short unbranched oligomers that leave well-defined interstitial regions (Wirth, 1994) and recent studies by molecular dynamics simulation showed that alkyl chains comprising polymeric stationary phases contain a series of defined and rigid voids in which shape-constrained solutes can penetrate and hence be selectively retained (Lippa *et al.*, 2005a). These results support an intuitive model that represents polymeric bonded phases as a surface with slots into which the solute molecules penetrate during retention. The frequency and depth of this penetration (and, in consequence, the retention) depend on the overall solute polarity, shape, size, length and planarity (Sander and Wise, 1990), i.e. longer and more planar compounds will be better retained.

For example, MV and DV chlorophyll pairs are separable on polymerically synthesized bonded phases based on the differences in their molecular shape: in DV forms, the substituent and the tetrapyrrolic macrocycle are located in the same plane (as conjugated systems are greatly stabilized when double bonds are coplanar), whereas MV forms adopt a non-planar conformation to reduce steric hindrance (Figure 4.1). In consequence, the more planar DV forms are retained longer (or, in terms of the proposed model, would penetrate more easily and to a greater depth in the 'slots' of the stationary phase (Figure 4.1) than the corresponding MV forms (Garrido and Zapata, 1997). When a voluminous group is present at C7 of the macrocycle (i.e. the formyl (–CHO) in Chl *b*, or a methoxycarbonyl (–$COOCH_3$) in Chl c_3), the difference in molecular shape induced by an ethyl or a vinyl substituent at C8 is reduced. In these cases, the vinyl group appears slightly twisted from the macrocycle plane to relieve steric tension, which could explain the difficulties in separating Chl *b* from DVChl *b* in certain methods that however successfully resolve Chl *a* and DVChl *a* (Van Heukelem *et al.*, 1994; Van Lenning *et al.*, 1995; Garrido and Zapata, 1997). Carotenoids with two β rings (in which both terminal cycles are coplanar to the conjugated central chain, e.g. zeaxanthin) show higher retention (penetrate more frequently and/or to a greater depth in the slots) relative to their β,ε isomers (with the ε ring deviated from the plane defined by the central chain, e.g. lutein) (Figure 4.2).

Several variables (pore size, alkyl chain length, bonding density, temperature and mobile-phase composition) can influence shape selectivity (Sander and Wise, 1990; Sander *et al.*, 1999; Lippa *et al.*, 2005a, b). Several works have been published that explore the role of particle pore size (Garrido and Zapata, 1993b), and mobile phase

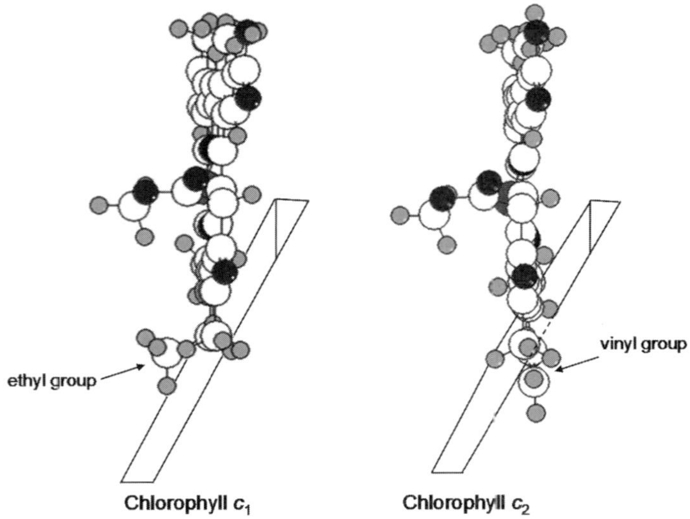

Figure 4.1. Molecular models of chlorophyll c_1 and chlorophyll c_2, showing their different capabilities in penetrating the interstitial voids in the polymeric bonded phase ('slots').

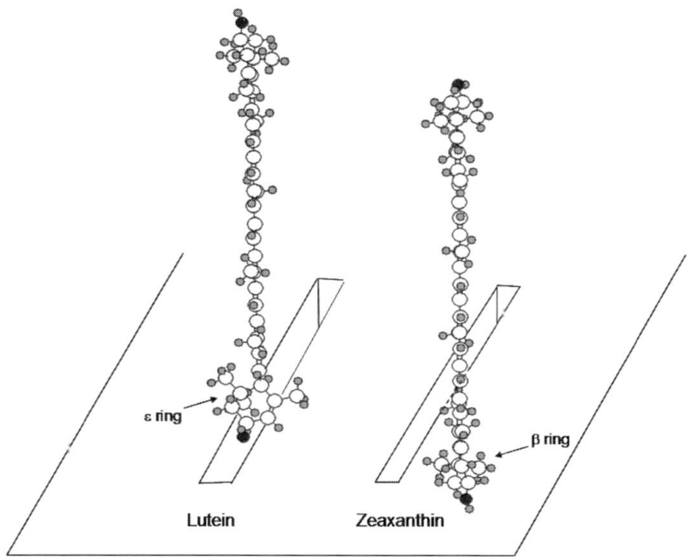

Figure 4.2. Molecular models of the carotenoids lutein and zeaxanthin, showing their different capabilities in penetrating the interstitial voids in the polymeric bonded phase ('slots').

composition (Jinno and Lin, 1995; Garrido and Zapata, 1996; Van Heukelem and Thomas, 2001) when polymeric C_{18} phases have been applied to pigment analysis. In general, shape selectivity of polymeric bonded phases increases with increasing alkyl chain length (Sander and Wise, 1990) as the size of the 'slots' increases (e.g. from approximate depths of 13 Å in C_{18} phases to cavities with depths up to ~24 Å in C_{30} phases, Lippa *et al.*, 2005b). Although C_{30} phases were engineered for optimizing the separation of carotenoids (Sander *et al.*, 1994), and have been employed in the analysis of pigments in food and tissues (Sander *et al.*, 2000), their application to algal photosynthetic pigments has been limited (Schmid and Stich, 1995; Goericke *et al.*, 2000a; Van Heukelem and Thomas, 2001; Inbaraj *et al.*, 2006).

4.3.2 Monomeric C_8 bonded phases

The variations in retention and selectivity observed among different reversed-phase columns can be related to two main parameters: column strength (J) and column polarity (P) (Snyder *et al.*, 1988). Column strength (a measure of the retention of non-polar solutes) affects band spacing because more hydrophobic (stronger) columns require stronger mobile phases. On the other hand, as column polarity increases, more polar compounds are retained preferentially compared with fewer polar compounds. Column strength can be correlated with the effective phase ratio of the column and affects the retention of all solutes (Antle *et al.*, 1985), whereas column polarity relates to hydrophobicity, dipolarity and hydrogen-bonding ability of the stationary phase (Izquierdo *et al.*, 2006). The increased value in column polarity of C_8 monomeric columns (P = 0.00, Antle *et al.*, 1985) relative to monomeric C_{18} phases (P = −0.55, Antle *et al.*, 1985) would explain the ability of C_8 phases to separate isomeric pairs of pigments with subtle differences in polarity, such as mono- and divinyl chlorophylls or lutein and zeaxanthin. The elution order for mono- and divinyl chlorophyll forms on C_8 columns, with the slightly more polar divinyl forms eluting before their corresponding monovinyl counterparts (and zeaxanthin eluting before lutein) indicates that, as could be expected for monomeric phases, the retention and selectivity are governed by a partitioning process (Snyder *et al.*, 1997).

Monomeric C_8 phases were first used by Goericke and Repeta (1993) to resolve Chl *a* and DVChl *a* and partially separate Chl *b* and DVChl *b*. Then, this method was modified to separate mono- and divinyl forms of Chls *a* (complete separation) and *b* (partial), and some taxon-specific carotenoids in less than 20 minutes (Vidussi *et al.*, 1996; Barlow *et al.*, 1997), but only two fractions of chlorophyll *c* pigments were separated. Rodríguez *et al.* (1998) recognized the cause of this lack of resolution: C_8 columns are weaker (J = 0.00) than C_{18} ones (J = 0.26) (Antle *et al.*, 1985), so they require weaker mobile phases to achieve adequate retentions of the most polar compounds. Rodríguez *et al.* (1998) showed that using i) adequate initial eluants (slightly

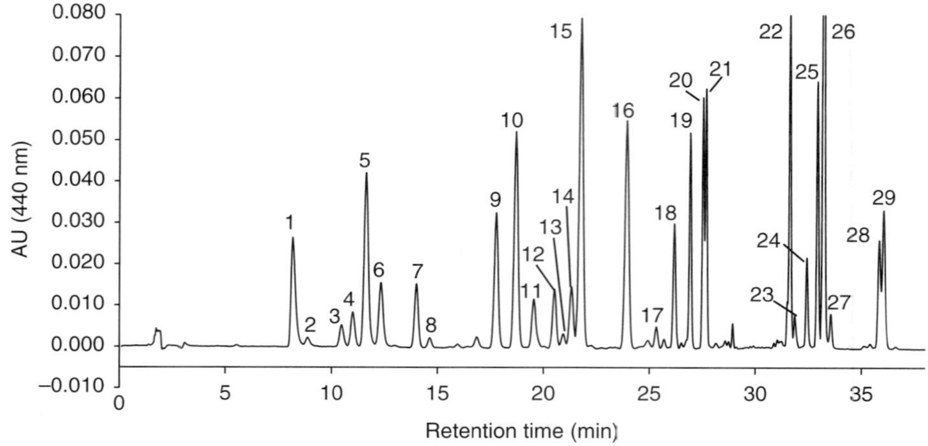

Figure 4.3. High-performance liquid chromatogram of pigments from marine phytoplankton (mixed pigment standard from DHI), obtained with the method of Zapata *et al.* (2000). Peak identification: 1, Chl c_3; 2, MV-Chl c_3; 3, Chlorophyllide *a*; 4, MgDVP; 5, Chl c_2; 6, Chl c_1; 7, Peridinin; 8, Peridinin-like unknown carotenoid; 9, 19′-butanoyloxyfucoxanthin; 10, Fucoxanthin; 11, Neoxanthin; 12, Prasinoxanthin; 13, 4-keto-19′-hexanoyloxyfucoxanthin; 14, Violaxanthin; 15, 19′-hexanoyloxyfucoxanthin; 16, Diadinoxanthin; 17, Antheraxanthin; 18, Diatoxanthin; 19, Alloxanthin; 20, Zeaxanthin; 21, Lutein; 22, DV-Chl *b* + Chl *b*; 23, DV-Chl *b* epimer; 24, MGDG-Chl c_2; 25, DV-Chl *a*; 26, Chl *a*; 27, Chl *a* epimer; 28, β,ε-Carotene; 29, β,β-Carotene.

more aqueous than those generally required for C_{18} phases), ii) gradient profiles in which the aqueous part is slowly removed from the eluant, and iii) correct injection conditions, monomeric C_8 columns can separate DV/MV pairs of acidic Chls (e.g. Chl c_3 from MVChl c_3, Chl c_2 from Chl c_1) simultaneously with other Chls and carotenoids.

A more recent method using a monomeric C_8 column thermostatted at 25 °C, a binary gradient and a pyridine-containing mobile phase separated 54 carotenoids (including taxon-specific ones), seven acidic chlorophyll *c* fractions, several non-polar chlorophyll fractions and Chl *a* from DVChl *a* (Zapata *et al.*, 2000; Zapata *et al.*, 2001) (Figure 4.3).

4.3.3 Mobile phase additives

The HPLC analysis of acidic pigments, such as some Chl *c* forms, chlorophyllides and pheophorbides, requires either ion-suppression or ion-pairing techniques to achieve sufficient retention and optimal resolution. Many reversed-phase pigment separation methods have used a gradient system where the first eluent contains ammonium acetate as the buffering and ion-pairing reagent (Table 4.1). Quaternary ammonium derivatives have also been used, as the selection of the chain length of the substituents (usually tetrabutyl ammonium salts) and changes in concentration influence the retention of acidic compounds (Mantoura and Llewellyn, 1983; Van Heukelem and Thomas, 2001). Regardless of what additive is used, the method should be thoroughly validated for appropriateness for the intended application. As an example, the use of ammonium acetate in the absence of other buffering or ion-pairing agents resulted in unpredictable allomerization of chlorophyll pigments in highly concentrated algal extracts and Chl *a* standards, with resulting non-linear calibration curves and high *y* intercepts during evaluation of a variety of stationary phases at differing column temperatures (40–60 °C) (Van Heukelem and Thomas, 2001). This problem was overcome by replacing 1.0 M ammonium acetate with 28 mM tetrabutylammonium acetate in solvent A and the injection buffer.

In a survey to assess the role of the molecular shape of the counter-ion in ion-pairing chromatography of acidic chlorophylls, Garrido and Zapata (1996) introduced pyridine (as pyridinium acetate). Several properties of pyridine make it a useful additive in the chromatographic separation of acidic Chls: (i) it is miscible with water and most organic solvents and shows adequate viscosity and boiling point values, (ii) although it absorbs strongly in the ultraviolet, it does not interfere with the visible pigment spectrum, and (iii) unlike the ammonium ion, it does not react with acetone (frequently employed as an eluent in RP-HPLC separation of Chls). The planar and rigid structure of the pyridinium ion makes it suitable as a counter-ion in paired-ion chromatography on polymeric bonded phases that exhibit special selectivity towards compounds with constrained molecular structure (Sander *et al.*, 1999).

The use of pyridine-containing mobile phases improves the chromatographic behaviour of acidic and esterified Chls both on monomeric and polymeric alkyl silica columns (Garrido and Zapata, 1996, 1997; Zapata *et al.*, 2000) because the pyridinium ion acts both as a more hydrophobic ion-pairing reagent (increasing the retention of acidic Chls), and as a real mobile phase modifier (co-solvent), affecting the selectivity towards both neutral and charged Chls. A possible explanation for this effect could rely on π,π interactions established between the aromatic ring of pyridine and the aromatic Chl macrocycle.

The combined use of polymeric C_{18} columns with pyridine-containing eluents, operating at low temperature (15 °C), has been applied to the simultaneous separation of polar and non-polar MV- and DV- Chl pairs of algal Chls (Garrido and Zapata, 1997)

4.3.4 Column temperature

Temperature can have a significant effect on phase selectivity and, in general, shape recognition increases dramatically at low temperatures, so that improved separations of isomers are usually possible at reduced column temperatures. This effect has been explained in terms of stationary phase order, which increases as the alkyl chain motion becomes reduced ('slots' become more rigid) with decreasing temperature

(Sander and Wise, 2001; Lippa *et al.*, 2005b). Temperature-mediated changes in shape selectivity of polymeric ODS columns have been exploited for the separation of isomeric carotenoids derived from β,β- and β,ε-carotene (Jinno and Lin, 1995), the discrimination of mono- and divinyl Chl forms (Garrido and Zapata, 1997) or the general separation of phytoplankton pigments (Van Heukelem *et al.*, 1994; Van Lenning *et al.*, 1995; Van Heukelem and Thomas, 2001). Van Heukelem *et al.* (1994) demonstrated that temperature changes can induce changes in the elution order of pigments, so the resolution of specific pigment pairs can be compromised at temperatures above or below the optimal.

Several studies have evaluated the combined effects of variations in column temperature and gradient steepness on changes in separation selectivity for a variety of analytes, some of which include algal pigments (Garrido and Zapata, 1993a; Van Lenning *et al.*, 1995; Zhu *et al.*, 1996a, b; Dolan *et al.*, 1998a, b; Dolan *et al.*, 2000; Van Heukelem and Thomas, 2001). More recently, the historical role of temperature selectivity in reversed-phase HPLC analysis was reviewed, with emphasis on the importance of mechanisms for achieving and maintaining accurate column temperatures (Dolan, 2002). The latter is especially important, as described in detail by Wolcott *et al.* (2000). The use of elevated column temperature and its successful application to separations was addressed by Zhu *et al.* (2005), and while it seems impractical to subject temperature-sensitive analytes to elevated column temperatures, the fact is that many successful separations have been achieved in this manner. A rigorous approach to evaluating analyte stability in high temperature HPLC analyses is given by Thompson and Carr (2002).

Van Heukelem and Thomas (2001) applied chromatography modelling software to enhance pigment separation through the combined use of gradient time and column temperature. Nine stationary phases were evaluated, including six monomeric and three polymeric phases of varying aliphatic chain length. Separation goals included methods for isolation of pigment standards and identification of an efficient method for field sample analyses. For the latter, a method was identified which uses a simple aqueous methanol to methanol gradient at 60 °C with tetrabutyl ammonium acetate as ion-paring agent and a highly efficient C_8 column. This method separates most of the taxonomically important carotenoids and chlorophylls with a separation time of approximately 27 minutes (Figure 4.4). The high efficiency of the column employed is enhanced by operation at an elevated temperature which reduces the resistance to mass transfer between phases, thus producing very sharp peaks. This fact is important in terms of good limits of detection, especially when natural samples are to be analysed.

4.4 Choice of HPLC method

Although some of the methods mentioned above have been recommended as general purpose methods for routine analysis, none of them are perfect. The best method is

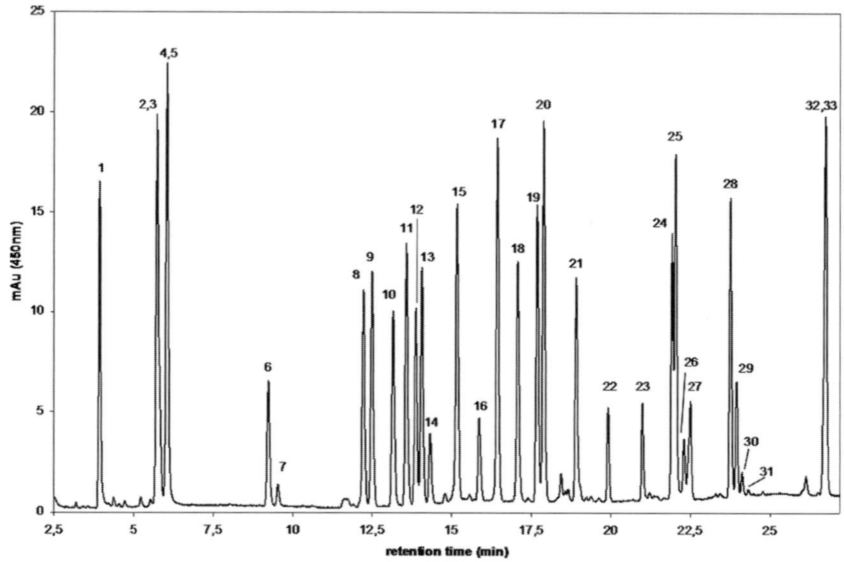

Figure 4.4. Chromatogram from the method of Van Heukelem and Thomas (2001) for the analysis of several combined algal cultures. Peak identification: 1, Chl c_3; 2, Chl c_2; 3, MgDVP; 4, Chl c_1; 5, Chlorophyllide *a*; 6, Peridinin; 7, Peridinin isomer; 8, 19′-butanoyoxyfucoxanthin; 9, Fucoxanthin; 10, Neoxanthin; 11, Prasinoxanthin; 12, Violaxanthin; 13, 19′-hexanoyloxyfucoxanthin; 14, Astaxanthin; 15, Diadinoxanthin; 16, Antheraxanthin; 17, Alloxanthin; 18, Diatoxanthin; 19, Zeaxanthin; 20, Lutein; 21, Canthaxanthin; 22, Gyroxanthin diester; 23, Gyroxanthin diester isomer; 24, DV-Chl *b*; 25, Chl *b*; 26, DV-Chl *b* epimer; 27, Chl *b* epimer; 28, DV-Chl *a*; 29, Chl *a*; 30, DV-Chl *a* epimer; 31, Chl *a* epimer; 32, β,ε-Carotene; 33, β,β-Carotene.

the one that approximates the best solution of an analytical problem with the lowest values of parameters important to the user such as analysis time, cost per analysis, risk to health or environment, etc. The analyst must be aware of the plasticity of the chromatographic techniques that allows us, in most cases, to adapt the analytical conditions to the nature of the sample to be studied. This not only refers to the actual separation of the compounds of interest (and hence, the consideration that those compounds that have no meaning for a certain purpose may not need to be separated), but also to the selection of the most adequate detection conditions (for example, employing selective UV/Visible (UV/Vis) wavelengths or monitoring only significant ions in mass spectrometric detection).

For example, general purpose methods, separating the maximum number of significant chlorophylls and carotenoids, should be used whenever samples of unknown qualitative composition are to be analysed. Thus, the methods of Zapata *et al.* (2000) and Van Heukelem and Thomas (2001) are good choices for unknown natural seawater samples. The reader is referred to the review by Wright and Jeffrey

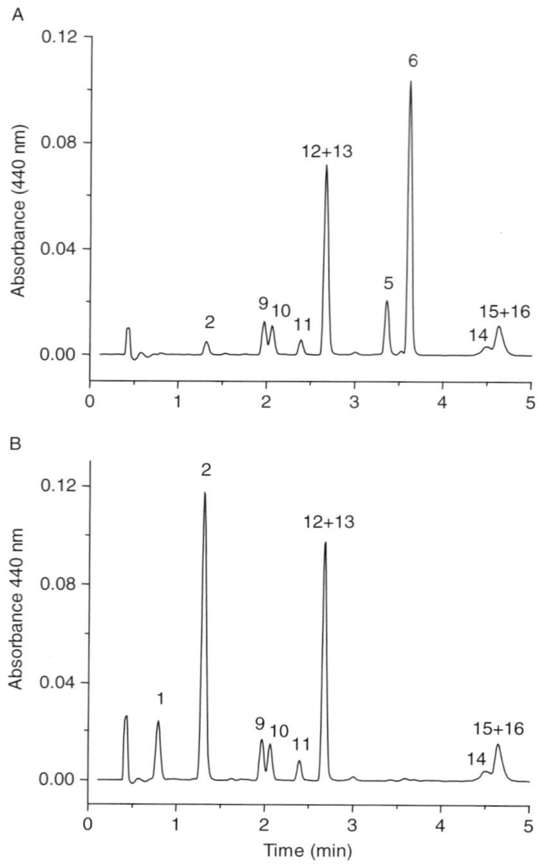

Figure 4.5. Absorbance (440 nm) chromatograms of pigment extracts from the marine microalga *Dunaliella salina* before (A), and after (B) incubation in acetone:water (1:1) to promote chlorophyllase activity. Peaks: 1 = Chlorophyllide *b*, 2 = Chlorophyllide *a*, 5 = Chl *b*, 6 = Chl *a*, 9 = Neoxanthin, 10 = Violaxanthin, 11 = Antheraxanthin, 12 = Lutein, 13 = Zeaxanthin, 14 = β,ψ-Carotene, 15 = β,ε-Carotene, 16 = β,β-Carotene (reprinted from Garrido, J.L., Rodríguez, F., Campaña, E. and Zapata, M. (2003). Rapid separation of chlorophyll *a* and *b* and their demetallated and dephytylated derivatives using a monolithic silica C18 column and a pyridine-containing mobile phase. *J. Chromatogr. A* **994**, 85–92, with permission from Elsevier).

(2006) for an exhaustive comparison of these methods, together with the C_{18} method of Wright *et al.* (1991) and the method based on polymeric columns by Garrido and Zapata (1997). If photosynthetic bacteria (e.g. Chlorobiaceae) are present, the method of Airs *et al.* (2001a) is recommended.

Short, specific methods, are indicated for particular purposes. Good examples are the detection of bacteriochlorophyll *a* in seawater with the method proposed by Goericke (2002), which employs very short gradient times (less than 10 minutes) and specific fluorescence (excitation at 360 nm, emission at 780 nm) and absorption (770 nm) detection; or the monitoring of chlorophyllase activity in chlorophytes in less than five minutes using a monolithic C_{18} column (Garrido *et al.*, 2003) (Figure 4.5).

Once a method is selected, analysts should be encouraged to modify it whenever necessary to achieve improved resolution under their particular circumstances. Slight changes in mobile phase composition, gradient profile or system temperature can be useful in achieving adequate separation of certain critical pairs or in saving analysis time, for example. It is important to remember, however, that any method implemented by an analyst should be adequately validated for the intended application, regardless of whether such validation steps have been implemented elsewhere to evaluate the method chosen.

It is always advisable to implement two alternative methods by having, for example, several different columns available. Thus, a two-step procedure for the isolation of pigments for use as standards has been proposed by Van Heukelem and Thomas (2001): extracts are chromatographed on a C_8 column and partially purified fractions are subsequently re-injected and the constituent pigments isolated from columns with different, primarily polymeric, selectivities. The use of columns with different selectivity principles (e.g. monomeric C_8 and polymeric C_{18}) can also be applied to fulfil one of the minimum criteria for identifying phytoplankton pigments: the co-elution of the unknown and the standard in different chromatographic systems (Jeffrey and Mantoura, 1997).

Irrespective of the separation principle, the chromatographic analysis of phytoplankton pigments still needs to be improved to achieve shorter analysis times, better sensitivity, less solvent consumption and good coupling with MS. In this regard, new chromatographic techniques hold promising results (Marston, 2007), especially those based on miniaturization (like capillary liquid chromatography, (Hernández-Borges *et al.*, 2007) or chromatography on chips (Reigner *et al.*, 1999, Ehlert *et al.*, 2008)) or in the use of high pressures and small particle size columns (Nguyen *et al.*, 2006). Interesting results for both charged (Chl *c*) and neutral pigments are being obtained with capillary electrophoresis techniques like micellar electrokinetic chromatography and non-aqueous capillary electrophoresis (Belin and Gülaçar, 2007).

4.5 Applications

4.5.1 Analysis of pigments from pelagic phytoplankton communities

The analysis of photosynthetic pigments from seawater samples is now frequently used in oceanographic studies as a tool to reconstruct the distribution and composition of algal groups based on their pigment diversity and characteristic chemotaxonomic markers (Wright and Jeffrey, 2006). The choice of a particular method depends on the scope of the study, the dominant algal groups and other practical considerations, particularly for large-scale surveys, including analysis time and ease of use.

Both C_{18} and C_8 methods are most commonly used for the analysis of photosynthetic pigments in phytoplankton, as they represent an optimal compromise in terms of efficiency, selectivity, analysis time and cost. Previous intercalibration studies using multiple HPLC methods (Hooker *et al.*, 2000; Claustre *et al.*, 2004) showed close agreement among their results for total chlorophyll *a* concentration (< 10% variability). Better correspondence was achieved if only values over the limit of quantitation were used, calibration procedures were standardized and Chl *a*/DVChl *a* were properly separated (Claustre *et al.*, 2004). For chemotaxonomic carotenoids the agreement was not as good, and inversely depended on pigment concentration (Claustre *et al.*, 2004).

The Wright *et al.* (1991) protocol was designed as the reference method in international programmes including JGOFS (Joint Global Ocean Flux Study protocols, 1994) and recommended by SCOR UNESCO Working Group 78 (Wright and Jeffrey, 1997) due to its overall resolution for the main marker pigments considered at that time. This method has been thoroughly evaluated and is still widely used (Bianchi *et al.*, 1996; Latasa and Bidigare, 1998; Wright and van den Enden, 2000; Schlüter and Møhlenberg, 2003; Estrada *et al.*, 2004; Havskum *et al.*, 2004; Muylaert *et al.*, 2006). The method proposed by Kraay *et al.* (1992) also continues to be employed in coastal and open ocean surveys (van Leeuwe *et al.*, 1998; Duineveld *et al.*, 2000; Riegman and Kraay, 2001; Croot *et al.*, 2002; Veldhuis and Kraay, 2004; Gameiro *et al.*, 2007). In spite of the good overall performance of these methods and their adequacy for chemotaxonomic analysis in many aquatic systems, it soon became obvious that it is crucial to resolve the pairs Chl *a*/DVChl *a* and Chl *b*/DVChl *b* from *Procholorococcus marinus* to obtain reliable estimates of its relative contribution in temperate and subtropical areas.

The polymeric C_{18} methods (Van Heukelem *et al.*, 1994; Van Lenning *et al.*, 1995; Garrido and Zapata, 1997) which combine the simultaneous separation of MV/DV Chl pairs and marker carotenoids, have found limited application in field studies until recently (Carreto *et al.*, 2003).

Because of their resolution of Chl *a*/DVChl *a*, the pioneering C_8 protocols (Goericke and Repeta, 1993; Vidussi *et al.*, 1996; Barlow *et al.*, 1997) have been adopted by many authors during the last decade (Gibb *et al.*, 2001; DiTullio *et al.*, 2003; Barlow *et al.*, 2004; Llewellyn *et al.*, 2005; Dandonneau *et al.*, 2006).

The improved C_8 protocol by Zapata *et al.* (2000) was applied to characterize the pigment composition of microalgal cultures (Van Lenning *et al.*, 2003; Latasa *et al.*, 2004; Zapata *et al.*, 2001, 2004) and of new algal isolates (De Salas *et al.*, 2003, 2005; Eikrem *et al.*, 2004; Lovejoy *et al.*, 2007). The method has been successfully applied to study phytoplankton assemblages in different marine provinces (Rodríguez *et al.*, 2002, 2003, 2006; DiTullio *et al.*, 2005; Not *et al.*, 2004, 2005; Seoane *et al.*, 2005; Roy *et al.*, 2006; Yu *et al.*, 2007).

Different oceanographic studies (Trice *et al.*, 2004; Adolf *et al.*, 2006; Montes-Hugo and Álvarez-Borrego, 2007) have employed the improved C_8 method of Van Heukelem and Thomas (2001). To date, this has been the method of choice for the simultaneous detection of chlorophyll *a* and bacteriochlorophyll *a* in studies on aerobic anoxygenic phototrophs in the ocean (Cottrell *et al.*, 2006; Koblizek *et al.*, 2007).

4.5.2 Analysis of pigments from microphytobenthos communities

The term microphytobenthos designates a complex community of unicellular and colonial organisms living on (or close to) the surface of intertidal and shallow subtidal sediments. Benthic primary producers constitute the basis of the marine foodweb in shallow environments and are involved in many biogeochemical processes including the production of labile matter, recycling of nutrients and production of photosynthetic oxygen. Analogous to phytoplankton studies, HPLC pigment analysis provides a fast and precise tool for reconstructing the composition of microphytobenthos. The chromatographic analysis of these samples handles a complex pool of pigments and alteration products from various autochthonous (microphytobenthos) and allochthonous sources (plants, grazing and diagenetic processes) (Brotas and Plante-Cuny, 2003).

Three C_{18} standard methods for oceanic samples have been applied to benthic samples with some modifications to separate the main marker carotenoids, chlorophyll degradation products and bacteriopigments: the method of Mantoura and Llewellyn (1983) (Lucas *et al.*, 2000), the protocol of Wright *et al.* (1991) (Chen *et al.*, 2001; Wysocki *et al.*, 2006) and in particular the method of Kraay *et al.* (1992) (Duineveld *et al.*, 2000; Brotas and Plante-Cuny, 2003; Serodio *et al.*, 2005; Jesus *et al.*, 2006; Cartaxana *et al.*, 2006; Brotas *et al.*, 2007; Buchaca and Catalán, 2007).

The new methodologies which employ different chromatographic columns have only recently been assayed using microphytobenthic samples. A few of these applications are, for example, the system employing a combination of monomeric and polymeric C_{18} columns (Pinckney *et al.*, 1996) applied to the analysis of benthic microalgal producers in trophic studies in sediments (Pinckney *et al.*, 2003), the protocol based on a C_{30} column (Wiltshire and Schroeder, 1994) applied to estuaries and benthic samples (Wiltshire *et al.*, 1996; Paterson *et al.*, 1998), and the C_8 method by Zapata *et al.* (2000) applied to the study of benthic microalgae (Grinham *et al.*, 2007) and to cyanobacterial mats in the top layers of sediments (Bonilla *et al.*, 2005).

4.5.3 Analysis of degradation products

The analysis of degradation products from photosynthetic pigments, together with other organic compounds, is essential in paleoceanographic studies to track the potential sources of organic carbon or as indicators of specific environmental conditions and processes (Keely, 2006). The reader is referred to the comprehensive review by Keely (2006) for a detailed description of the timing, processes and nature of chemical transformations that pigments suffer from the primary producers to the sediments.

The analytical methods to be employed in pigment geochemical analysis should provide good resolution over a wide polarity range, encompassing highly polar derivatives such as Chl *c* to highly apolar derivatives including chlorin steryl esters (Keely, 2006). Louda *et al.* (1998), aware of the need for additional 'space' or 'time' in the middle section of chromatograms for housing many of the derivatives that can be generated during degradation processes (pheophorbides, cyclic pheophorbides and their pyro derivatives, see Louda *et al.*, 2000), developed a modification of the method of Kraay *et al.* (1992) in which the times of all segments in the gradient were increased. This method was then employed in a series of studies on alteration products related to cellular senescence and death and early diagenetic processes (Louda *et al.*, 1998, 2000, 2002).

Goericke *et al.* (2000b) developed two HPLC methods (C_{18} and C_{30}) devised for the analysis of chlorophyll derivatives in a survey for the distribution and sources of cyclic pheophorbides in the ocean. These authors emphasize that some of these derivatives do not fluoresce, so fluorescence-based detection can underestimate total chlorins. In addition, this study detected several breakdown products originating from DVChl *a*, DVChl *b* and from Chl *c* compounds in marine sediments (Goericke *et al.*, 2000b) which can not be separated using common C_{18} procedures.

Although the C_{18} methods of Mantoura and Llewellyn (1983), Wright *et al.* (1991) and Kraay *et al.* (1992) are still currently used in many studies (Furlong and Carpenter, 1988; Louda *et al.*, 1998; Bianchi *et al.*, 2000; Reuss *et al.*, 2005), the procedure proposed by Airs *et al.* (2001a), using a small particle size C_{18} column in conjunction with ternary elution and long gradient times seems to be the method of choice for the analysis of very complex mixtures of pigment derivatives. This method achieves excellent resolution, sensitivity and compatibility with mass spectrometry.

Indeed, many studies of pigment breakdown products in marine environments are accomplished by coupling LC separation methods to mass spectrometry. In the chromatographic part of these methods, C_{18} columns are generally employed. The application of LC-MS to the analysis of photosynthetic pigments is covered in Chapter 7 in this volume.

4.5.4 Analysis of bacteriochlorophylls

Bacteriochlorophylls are antenna pigments used by photosynthetic bacteria that carry out anoxygenic photosynthesis. Purple bacteria contain bacteriochlorophyll *a* or *b* and various carotenoids, and are subdivided into the families Chromatoaceae (purple sulfur bacteria) and Rhodospirillaceae (purple non-sulfur bacteria). Green bacteria contain bacteriochlorophylls *c*, *d* or *e* as their main photosynthetic pigments (Pfennig, 1978). Recently, aerobic anoxygenic photoheterotrophs have been discovered in the open ocean, photosynthesizing using bacteriochlorophyll *a* (Kolber *et al.*, 2000). Green and purple bacteria, however, are confined to dimly lit anaerobic waters where a source of sulfide exists, for example the chemocline of the Black Sea, microbial mats, or meromictic or holomictic lakes. Recently, green sulfur bacteria were isolated from a hydrothermal vent (Beatty *et al.*, 2005), where the only source of light was geothermal radiation, illustrating spectacularly the low light conditions under which these organisms are adapted to grow.

The bacteriochlorophylls *c*, *d* and *e* present a unique analytical problem. They exist as a series of pseudo-homologues that differ in the degree of alkylation at C8 and/or C12, and in the alcohol esterified at position C17. Thus, even monocultures containing a single bacteriochlorophyll type can contain a complex array of structures (Airs *et al.*, 2001b; Glaeser *et al.*, 2002; Gich *et al.*, 2003). The separation of individual bacteriochlorophyll structures is important, as the bacteria change the proportion of bacteriochlorophylls according to growth conditions, including light, with a trend towards increased alkylation with low light intensity (Borrego and Garcia-Gil, 1995). Additionally, the unique bacteriochlorophyll signatures tend to be well preserved in the sediment record, providing sensitive environmental markers of past depositional environments (Mawson *et al.*, 2004; Wilson *et al.*, 2004; Squier *et al.*, 2005).

The complex distributions of structures of bacteriochlorophylls require high resolution methods for their chromatographic separation. One of the first methods specifically designed for HPLC analysis of bacteriochlorophylls was developed by Borrego and Garcia Gil (1994). They demonstrated the separation of bacteriochlorophyll homologues from species of green sulfur bacteria as two well-defined groups, corresponding to farnesyl esterified homologues, and homologues esterified by other alcohols, respectively. The reversed-phase method combined a Spherisorb C_{18} column with a mobile phase composition comprising methanol, ammonium acetate, ethyl acetate and acetonitrile. Improvements in resolution were obtained by coupling two RP columns together, and using a smaller particle size (Airs *et al.*, 2001a). This method permitted the resolution of additional components in the second group of homologues in *Chlorobium phaeobacteroides* than had been recognized previously, with indications that further peaks were yet unresolved. Coelutions could not be detected by mixed spectra within a peak, as the bacteriochlorophyll *e* pseudo-homologues exhibited indistinguishable UV/Vis spectra. The application of LC-MS^n confirmed the presence of additional unresolved bacteriochlorophylls and revealed that this strain of *Chl. phaeobacteroides* contained over 25 different bacteriochlorophyll *e* structures (Airs *et al.*, 2001b), including several sets of novel pseudo-homologues. A similar range of structures were later reported in a second strain of *Chl. phaeobacteroides* (Glaeser *et al.*, 2002). Thus, the complexity of bacteriochlorophyll structures found in species of green sulfur bacteria, combined with their indistinguishable UV/vis spectra require specialized high resolution methods for their analysis, often with structural confirmation by LC-MS^n (see Chapter 7).

References

Adolf, J. E., Yeager, C. L., Miller, W. D., Mallonee, M. E. and Harding Jr., L. W. (2006). Environmental forcing of phytoplankton floral composition, biomass and primary productivity in Chesapeake Bay, USA. *Estuar. Coast. Shelf Sci.* **67**, 108–22.

Antle, P. E., Goldberg, A. P. and Snyder, L. R. (1985). Characterization of silica-based reversed-phase columns with respect to retention selectivity. Solvophobic effects. *J. Chromatogr.* **321**, 1–32.

Airs, R. L. and Llewellyn, C. A. (2006). Improved detection and characterization of fucoxanthin-type carotenoids: novel pigments in *Emiliania huxleyi* (Prymnesiophyceae). *J. Phycol.* **42**, 391–99.

Airs, R. L., Atkinson, J. E. and Keely, B. J. (2001a). Development and application of a high resolution liquid chromatography method for the analysis of complex pigment distributions. *J. Chromatogr. A* **917**, 167–77.

Airs, R. L., Borrego, C. M., Garcia-Gil, J. and Keely, B. J. (2001b). Identification of the bacteriochlorophyll homologues of *Chlorobium phaeobacteroides* strain UdG6053 grown at low light intensity. *Photosynth. Res.* **70**, 221–30.

Barlow, R. G., Cummings, D. G. and Gibb, S. W. (1997). Improved resolution of mono-and divinyl chlorophylls *a* and *b* and zeaxanthin and lutein in phytoplankton extracts using reverse phase C-8 HPLC. *Mar. Ecol. Prog. Ser.* **161**, 303–07.

Barlow, R. G., Aiken, J., Moore, G. F., Holligan, P. M. and Lavender, S. (2004). Pigment adaptations in surface phytoplankton along the eastern boundary of the Atlantic Ocean. *Mar. Ecol. Prog. Ser.* **281**, 13–26.

Beatty, J. T., Overmann, J., Lince, M. T., Manske, A. K., Lang, A. S., Blankenship, R. E., Van Dover, C. L., Martinson, T. A. and Plumley, F. G. (2005). An obligately photosynthetic bacterial anaerobe from a deep-sea hydrothermal vent. *Proc. Natl. Acad. Sci. USA* **102**, 9306–10.

Belin, G. K. and Gülaçar, F. O. (2007). Separation of chlorines and carotenoids in capillary electrophoresis. *J. Chromatogr. Sci.* **45**, 593–99.

Bianchi, T. S., Demetropoulos, A., Hadjichristophorou, M., Argyrou, M., Baskaran, M. and Lamberta, C. (1996). Plant Pigments as Biomarkers of Organic Matter Sources in Sediments and Coastal Waters of Cyprus (eastern Mediterranean) *Estuar. Coast. Shelf Sci.* **42**, 103–15.

Bianchi, T. S., Johansson, B. and Elmgren, R. (2000). Breakdown of phytoplankton pigments in Baltic sediments: effects of anoxia and loss of deposit-feeding macrofauna. *J. Exp. Mar. Biol. Ecol.* **251**, 161–83.

Bidigare, R. R., Van Heukelem, L. and Trees, C. H. (2005). Analysis of algal pigments by high performance liquid chromatography. In *Algal Culturing Techniques*, ed. R. A. Andersen. Burlington: Elsevier Academic Press, pp. 327–45.

Bjørnland, T. (1997). Structural relationships between algal carotenoids. In *Phytoplankton Pigments in Oceanography: Guidelines to Modern Methods*, ed. S. W. Jeffrey, R. F. C. Mantoura and S. W. Wright. Paris: UNESCO Publishing, pp. 572–77.

Bonilla, S., Villeneuve, V. and Vincent, W. F. (2005). Benthic and planktonic algal communities in a high artic lake: pigment structure and contrasting responses to nutrient enrichment. *J. Phycol.* **41**, 1120–30.

Borrego, C. M. and Garcia-Gil, L. J. (1994). Separation of bacteriochlorophyll homologues from green photosynthetic sulphur bacteria by reversed-phase HPLC. *Photosynth. Res.* **41**, 157–63.

Borrego, C. M. and Garcia-Gil, L. J. (1995). Rearrangement of light harvesting bacteriochlorophyll homologues as a response of green sulphur bacteria to low light intensities. *Photosynth. Res.* **45**, 21–30.

Brotas, V., Mendes, C. R., Cartaxana, P. (2007). Microphytobenthic biomass assessment by pigment analysis: comparison of spectrophotometry and high performance liquid chromatography methods. *Hydrobiologia*, **587**, 19–24.

Brotas, V. and Plante-Cuny, M. R. (2003). The use of HPLC pigment analysis to study microphytobenthos communities. *Acta Oecol.* **24**, S109–15.

Buchaca, T. and Catalán, J. (2007). Factors influencing the variability of pigments in the surface sediments of mountain lakes. *Freshwat. Biol.* **52**, 1365–79.

Carreto, J. I., Montoya, N. G., Benavides, H. R., Guerrero, R. and Carignan, M. O. (2003). Characterization of spring phytoplankton communities in the Rio de la Plata maritime front using pigment signatures and cell microscopy. *Mar. Biol.* **143**, 1013–27.

Cartaxana, P., Mendes, C. R., van Leeuwe, M. A. and Brotas, V. (2006). Comparative study on microphytobenthic pigments of muddy and sandy intertidal sediments of the Tagus estuary. *Estuar. Coast. Shelf Sci.* **66**, 225–30.

Chen, N. H., Bianchi, T. S., McKee, B. A. and Bland, J. M. (2001). Historical trends of hypoxia on the Louisiana shelf: application of pigments as biomarkers. *Org. Geochem.* **32**, 543–61.

Chisholm, S. W., Frankel, S. L., Goericke, R., Olson, R. J., Palenik, B., Waterbury, J. B., West-Johnsrud, L. and Zettler, E. K. (1992). *Prochlorococcus marinus* nov. gen. nov. sp.: an oxyphototrophic marine prokaryote containing divinyl chlorophyll *a* and *b*. *Arch. Microbiol.* **157**, 297–300.

Claustre, H., Hooker, S. B., Van Heukelem, L., Berthon, J. F., Barlow, R., Ras, J., Sessions, H., Targa, C., Thomas, C. S., van der Linde, D. and Marty, J. C. (2004). An intercomparison of HPLC phytoplankton pigment methods using in situ samples: application to remote sensing and database activities. *Mar. Chem.* **85**, 41–61.

Cottrell, M. T., Mannino, A. and Kirchman, D. L. (2006). Aerobic anoxygenic phototrophic bacteria in the Mid-Atlantic Bight and the North Pacific Gyre. *Appl. Environ. Microbiol.* **72**, 557–64.

Croot, P. L., Karlson, B., Wulff, A., Linares, F. and Andersson, K. (2002). Trace metal/phytoplankton interactions in the Skagerrak. *J. Marine Syst.* **35**, 39–60.

Dandonneau, Y., Montel, Y., Blanchot, J., Giraudeau, J. and Neveux, J. (2006). Temporal variability in phytoplankton pigments, picoplankton and coccolithophores along a transect through the North Atlantic and tropical southwestern Pacific. *Deep-Sea Res. I* **53**, 689–712.

De Salas, M. F., Bolch, C. J. S., Botes, L., Nash, G., Wright, S. W. and Hallegraeff, G. M. (2003). *Takayama* gen. nov. (Gymnodiniales, Dinophyceae), a new genus of unarmoured dinoflagellates with sigmoid apical grooves, including the description of two new species. *J. Phycol.* **39**, 1233–46.

De Salas, M. F., Bolch, C. J. S. and Hallegraeff, G. M. (2005). *Karlodinium australe* sp. nov. (Gymnodiniales, Dinophyceae), a new potentially ichthyotoxic unarmoured dinoflagellate from lagoonal habitats of south-east Australia. *Phycologia* **44**, 640–50.

DiTullio, G. R., Geesey, M. E., Jones, D. R., Daly, K. L., Campbell, L. O. and Smith Jr., W. O. (2003). Phytoplankton assemblage structure and primary productivity along 170°W in the South Pacific Ocean. *Mar. Ecol. Prog. Ser.* **255**, 55–80.

DiTullio, G. R., Geesey, M., Maucher, J. M., Alm, M. B., Riseman, S. F. and Bruland, K. W. (2005). Influence of iron on algal community composition and physiological status in the Peru upwelling system. *Limnol. Oceanogr.* **50**, 1887–907.

Dolan, J. W. (2002). Temperature selectivity in reversed-phase high-performance liquid chromatography. *J. Chromatogr. A* **965**, 195–205.

Dolan, J. W., Snyder, L. R., Djordjevic, N. M., Hill, D. W., Saunders, D. L., Van Heukelem, L. and Waeghe, T. J. (1998a). Simultaneous variation of temperature and gradient steepness for reversed-phase high-performance liquid chromatography method development, I. Application to 14 different samples using computer simulation. *J. Chromatogr. A* **803**, 1–31.

Dolan, J. W., Snyder, L. R., Saunders, D. L. and Van Heukelem, L. (1998b). Simultaneous variation of temperature and gradient steepness for reversed-phase high-performance liquid chromatography method development, II. The use of further changes in conditions. *J. Chromatogr. A* **803**, 33–50.

Dolan, J. W., Snyder, L. R., Blanc, T. and Van Heukelem, L. (2000). Selectivity differences for C_{18} and C_8 reversed-phase columns as a function of temperature and gradient steepness: I. Optimizing selectivity and resolution. *J. Chromatogr. A* **897**, 37–50.

Duineveld, G. C. A., Tselepides, A., Witbaard, R., Bak, R. P. M., Berghuis, E. M., Nieuwland, G., van der Weele, J. and Kok, A. (2000). Benthic-pelagic coupling in the oligotrophic Cretan sea. *Progr. Oceanogr.* **46**, 457–80.

Eikrem, W., Romari, K., Gall, F. L, Latasa, M. and Vaulot, D. (2004). *Florenciella parvula* gen. and sp. nov. (Dictyochophyceae, Heterokontophyta) a small flagellate isolated from the English Channel. *Phycologia* **43**, 658–68.

Egeland, E. S., Garrido, J. L., Zapata, M., Maestro, M. A. and Liaaen-Jensen, S. (2000). Algal carotenoids. Part 64. Structure and chemistry of 4-keto-19'-hexanoyloxyfucoxanthin with a novel carotenoid end group. *J. Chem. Soc., Perkin Trans. 1*, **2000(8)**, 1223–30.

Ehlert, S., Kraiczek, K., Mora, J. A., Dittmann, M., Rozing, G. P. and Tallarek, U. (2008). Separation efficiency of particle-packed HPLC microchips. *Anal. Chem.* **80**, 5945–50.

Estrada, M., Henriksen, P., Gasol, J. M., Casamayor, E. O. and Pedrós-Alió, C. (2004). Diversity of planktonic photoautotrophic microorganisms along a salinity gradient as depicted by microscopy, flow-cytometry, pigment analysis and DNA-based methods. *FEMS Microbiol. Ecol.* **49**, 281–93.

Fawley, M. W. (1989). A new form of chlorophyll *c* involved in light-harvesting. *Plant Physiol.* **91**, 727–32.

Furlong, E. T. and Carpenter R. (1988). Pigment preservation and remineralization in oxic coastal marine sediments. *Geochim. Cosmochim. Acta* **52**, 87–99.

Gameiro, C., Cartaxana, P. and Brotas, V. (2007). Environmental drivers of phytoplankton distribution and composition in Tagus Estuary, Portugal. *Estuar. Coast. Shelf Sci.* **75**, 21–34.

Garrido, J. L. and Zapata, M. (1993a). High performance liquid chromatography of chlorophylls c_3, c_1, c_2 and *a* and of carotenoids of chromophyte algae on a polymeric octadecyl silica column. *Chromatographia* **35**, 543–47.

Garrido, J. L. and Zapata, M. (1993b). High performance liquid chromatographic separation of polar and non-polar chlorophyll pigments from algae using a wide pore polymeric octadecylsilica column. *J. High Resol. Chromatogr.* **16**, 229–33.

Garrido, J. L. and Zapata, M. (1996). Ion-pair reversed phase high-performance liquid chromatography of algal chlorophylls. *J. Chromatogr. A* **738**, 285–89.

Garrido, J. L. and Zapata, M. (1997). Reversed phase high performance liquid chromatographic separation of mono- and divinyl chlorophyll forms using pyridine-containing mobile phases and a polymeric octadecyl silica column. *Chromatographia* **44**, 43–49.

Garrido, J. L. and Zapata, M. (1998). Detection of new pigments from *Emiliania huxleyi* (Prymnesiophyceae) by high-performance liquid chromatography, liquid chromatography-mass spectrometry, visible spectroscopy and fast atom bombardment mass spectrometry. *J. Phycol.* **34**, 70–78.

Garrido, J. L. and Zapata, M. (2006). Chlorophyll analysis by new HPLC methods. In *Chlorophylls and Bacteriochlorophylls: Biochemistry, Biophysics, Functions and Applications*, ed. B. Grimm, R. J. Porra, W. Rüdiger and H. Scheer. Dordrecht: Springer, pp. 109–21.

Garrido, J. L., Zapata, M. and Muñiz, S. (1995). Spectral characterization of new chlorophyll *c* pigments isolated from *Emiliania huxleyi* (Prymnesiophyceae) by high-performance liquid chromatography. *J. Phycol.* **31**, 761–68.

Garrido, J. L., Otero, J., Maestro, M. A. and Zapata, M. (2000). The main non-polar chlorophyll *c* from *Emiliania huxleyi* (Prymnesiophyceae) is a chlorophyll c_2-monogalactosyldiacylglyceride ester: a mass spectrometry study. *J. Phycol.* **36**, 497–505.

Garrido, J. L., Rodríguez, F., Campaña, E. and Zapata, M. (2003). Rapid separation of chlorophyll *a* and *b* and their demetallated and dephytylated derivatives using a monolithic silica C18 column and a pyridine-containing mobile phase. *J. Chromatogr. A* **994**, 85–92.

Gibb, S. W., Cummings, D. G., Irigoien, X., Barlow, R. G. and Mantoura, F. R. C. (2001). Phytoplankton pigment chemotaxonomy of the northeastern Atlantic. *Deep-Sea Res. II* **48**, 795–823.

Gich, F., Airs, R. L., Danielsen, M., Keely, B. J., Abella, C. A., Garcia-Gil, J., Miller, M. and Borrego, C. M. (2003). Characterization of the chlorosome antenna of the filamentous anoxygenic phototrophic bacterium *Chloronema* sp. strain UdG9001. *Arch. Microbiol.* **180**, 417–26.

Glaeser, J., Baneras, L., Rutters, H. and Overmann, J. (2002). Novel bacteriochlorophyll *e* structures and species-specific variability of pigment composition in green sulfur bacteria. *Arch. Microbiol.* **177**, 475–85.

Goericke, R. (2002). Bacteriochlorophyll *a* in the ocean: Is anoxygenic photosynthesis important? *Limnol. Oceanogr.* **47**, 290–95.

Goericke, R. and Repeta, D. G. (1993). Chlorophylls *a* and *b* and divinyl chlorophylls *a* and *b* in the open Subtropical North Atlantic Ocean. *Mar. Ecol. Prog. Ser.* **101**, 307–13.

Goericke, R., Olson, R. J. and Shalapyonok, A. (2000a). A novel niche for *Prochlorococcus* sp. in low-light suboxic environments in the Arabian Sea and the Eastern Tropical North Pacific. *Deep-Sea Res.* **47**, 1183–205.

Goericke, R., Strom, S. L. and Bell, M. A. (2000b). Distribution and sources of cyclic pheophorbides in the marine environment. *Limnol. Oceanogr.* **45**, 200–11.

Grinham, A. R., Carruthers, T. J. B., Fisher, P. L., Udy, J. W. and Dennison, W. C. (2007). Accurately measuring the abundance of benthic microalgae in spatially variable habitats. *Limnol. Oceanogr. Methods* **5**, 119–25.

Havskum, H., Schlüter, L., Scharek, R., Berdalet, E. and Jacquet, S. (2004). Routine quantification of phytoplankton groups – microscopy or pigment analyses? *Mar. Ecol. Prog. Ser.* **273**, 31–42.

Hernández-Borges, J., Aturki, Z., Rocco, A. and Fanali, S. (2007). Recent applications in nanoliquid chromatography. *J. Sep. Sci.* **30**, 1589–610.

Hooker, S. B., Claustre, H., Ras, J., Van Heukelem, L., Berthon, J. F., Targa, C., van der Linde, D., Barlow, R. and Sessions, H. (2000). *The First SeaWiFS HPLC analysis round-robin experiment (SeaHARRE-1).* Sea WiFS Postlaunch Technical Report Series. S. B. Hooker and E. R. Firestone (eds). NASA Technical Memorandum 2000–206892, Vol. 14. Greenbelt: NASA Goddard Space Flight Center.

Inbaraj, S. B., Chien, J. T. and Chen, B. H. (2006). Improved high performance liquid chromatographic method for determination of carotenoids in the microalga *Chlorella pyrenoidosa*. *J. Chromatogr. A* **1102**, 193–99.

Izquierdo, P., Rosés, M. and Bosch, E. (2006). Polarity parameters of the Symmetry C18 and Chromolith Performance RP-18 monolithic chromatographic columns. *J. Chromatogr. A.* **1107**, 96–103.

Jeffrey, S. W. (1997a). Application of pigment methods in Oceanography. In *Phytoplankton Pigments in Oceanography: Guidelines to Modern Methods*, ed. S. W. Jeffrey, R. F. C. Mantoura and S. W. Wright. Paris: UNESCO Publishing, pp. 127–66.

Jeffrey, S. W. (1997b). Structural relationships between algal chlorophylls. In *Phytoplankton Pigments in Oceanography: Guidelines to Modern Methods*, ed. S. W. Jeffrey, R. F. C. Mantoura and S. W. Wright. Paris: UNESCO Publishing, pp. 566–71.

Jeffrey, S. W. and Mantoura, R. F. C. (1997). Minimum criteria for identifying phytoplankton pigments. In *Phytoplankton Pigments in Oceanography: Guidelines to Modern Methods*, ed. S. W. Jeffrey, R. F. C. Mantoura and S. W. Wright. Paris: UNESCO Publishing, pp. 631–32.

Jeffrey, S. W., Wright, S. W. and Zapata, M. (1999). Recent advances in HPLC pigment analysis of phytoplankton. *Mar. Freshwat. Res.* **50**, 879–96.

Jesus, B., Mendes, C. R., Brotas, V. and Paterson, D. M. (2006). Effect of sediment type on microphytobenthos vertical distribution: Modelling the productive biomass and improving ground truth measurements. *J. Exp. Mar. Biol. Ecol.* **332**, 60–74.

Jinno, K. and Lin, Y. (1995). Separation of carotenoids by high-performance liquid chromatography with polymeric and monomeric octadecylsilica stationary phases. *Chromatographia* **41**, 311–17.

Joint Global Ocean Flux Study (1994). *Protocols for the Joint Global Ocean Flux Study Core Measurements*. Intergovernmental Oceanographic Commission. *SCOR, Manual and Guides, UNESCO* **29**, 91–96.

Keely, B. J. (2006). Geochemistry of chlorophylls. In *Chlorophylls and Bacteriochlorophylls. Biochemistry, Biophysics, Functions and Applications*, ed. B. Grimm, R. J. Porra, W. Rüdiger and H. Scheer. Dordrecht: Springer, pp. 535–61.

Koblizek, M., Masin, M., Ras, J., Poulton, A. J. and Prasil, O. (2007). Rapid growth rates of aerobic anoxygenic phototrophs in the ocean. *Environ. Microbiol.* **9**, 2401–06.

Kolber, Z. S., Van Dover, C. L., Niederman, R. A. and Falkowski, P. G. (2000). Bacterial photosynthesis in surface waters of the open ocean. *Nature* **407**, 177–79.

Kraay, G. W., Zapata, M. and Veldhuis, M. J. W. (1992). Separation of chlorophylls c_1, c_2 and c_3 of marine phytoplankton by reversed-phase C_{18} high–performance liquid chromatography. *J. Phycol.* **28**, 708–12.

Latasa, M. and Bidigare, R. R. (1998). A comparison of phytoplankton populations of the Arabian Sea during the spring intermonsoon and southwest monsoon of 1995 as described by HPLC-analyzed pigments. *Deep-Sea Res. II* **45**, 2133–70.

Latasa, M., Scharek, R., Le Gall, F. and Guillou, L. (2004). Pigment suites and taxonomic groups in Prasinophyceae. *J. Phycol.* **40**, 1149–55.

Lippa, K. A., Sander, L. C. and Mountain, R. D. (2005a). Molecular dynamics simulations of alkylsilane stationary-phase order and disorder. 1. Effects of surface coverage and bonding chemistry. *Anal. Chem.* **77**, 7852–61.

Lippa, K. A., Sander, L. C. and Mountain, R. D. (2005b). Molecular dynamics simulations of alkylsilane stationary-phase order and disorder. 2. Effects of temperature and chain length. *Anal. Chem.* **77**, 7862–71.

Llewellyn, C. A., Fishwick, J. R. and Blackford, J. C. (2005). Phytoplankton community assemblage in the English Channel: a comparison using chl *a* derived from HPLC-CHEMTAX and carbon derived from microscopy cell counts. *J. Plankton Res.* **27**, 103–19.

Louda, J. W., Li, J., Liu, L., Winfree, M. N. and Baker, E. W. (1998). Chlorophyll *a* degradation during cellular senescence and death. *Org. Geochem.* **29**, 1233–51.

Louda, J. W., Loitz, J. W., Rudwick, D. T. and Baker, E. W. (2000). Early diagenetic alteration of chlorophyll *a* and bacteriochlorophyll *a* in a contemporaneous marine ecosystem: Florida Bay. *Org. Geochem.* **31**, 1561–80.

Louda, J. W., Liu, L. and Baker, E. W. (2002). Senescence- and death-related alteration of chlorophylls and carotenoids in marine phytoplankton. *Org. Geochem.* **33**, 1635–53.

Lovejoy, C., Vincent, W. F., Bonilla, S. E., Roy, S., Martineau, M. -J., Terrado, R., Potvin, M., Massana, R. and Pedrós-Alió, C. (2007). Distribution, phylogeny, and growth of cold-adapted picoprasinophytes in Arctic seas. *J. Phycol.* **43**, 78–89.

Lucas, C. H., Widdows, J., Brinsley, M. D., Salkeld, P. N. and Herman, P. M. J. (2000). Benthic-pelagic exchange of microalgae at a tidal flat. 1. Pigment analysis. *Mar. Ecol. Prog. Ser.* **196**, 59–73.

Mantoura, R. F. C. and Llewellyn, C. A. (1983). The rapid determination of algal chlorophyll and carotenoid pigments and their breakdown products in natural waters by reverse-phase high-performance liquid chromatography. *Anal. Chim. Acta* **151**, 297–314.

Mantoura, R. F. C., Barlow, R. G. and Head, E. J. H. (1997). Simple isocratic HPLC methods for chlorophyll and their degradation products. In *Phytoplankton Pigments in Oceanography: Guidelines to Modern Methods*, ed. S. W. Jeffrey, R. F. C. Mantoura and S. W. Wright. Paris: UNESCO Publishing, pp. 307–26.

Marston, A. (2007). Role of advances in chromatographic techniques in phytochemistry. *Phytochem.* **68**, 2785–97.

Mawson, D. H., Walker, J. S. and Keely, B. J. (2004). Variations in the distributions of sedimentary alkyl porphyrins in the Mulhouse basin in response to changing environmental conditions. *Org. Geochem.* **35**, 1229–41.

Miyashita, H., Adachi, K., Kurano, N., Ikemoto, H., Chihara, M. and Miyachi, S. (1997). Pigment composition of a novel photosynthetic prokaryote containing chlorophyll *d* as the major chlorophyll. *Plant Cell Physiol.* **38**, 272–81.

Montes-Hugo, M. A. and Álvarez-Borrego, S. (2007). Differences in photosynthetic pigment signatures between phytoneuston and phytoplankton communities in a coastal lagoon of Baja California. *Mar. Biol.* **151**, 1225–36.

Muylaert, K., Gonzales, R., Franck, M., Lionard, M., Van der Zee, C., Cattrijsse, A., Sabbe, K., Chou, L. and Vyverman, W. (2006). Spatial variation in phytoplankton dynamics in the Belgian coastal zone of the North Sea studied by microscopy, HPLC-CHEMTAX and underway fluorescence recordings. *J. Sea Res.* **55**, 253–65.

Nguyen, D. T. T., Guillarme, D., Rudaz, S. and Veuthey, J. L. (2006). Fast analysis in liquid chromatography using small particle size and high pressure. *J. Sep. Sci.* **29**, 1836–48.

Not, F., Latasa, M., Marie, D., Cariou, T., Vaulot, D. and Simon, N. (2004). A single species, *Micromonas pusilla* (Prasinophyceae), dominates the eukaryotic picoplankton in the western English Channel. *Appl. Environ. Microbiol.* **70**, 4064–72.

Not, F., Massana, R., Latasa, M., Marie, D., Colson, C., Eikrem, W., Pedrós-Alió, C., Vaulot, D. and Simon, N. (2005). Late summer community composition and abundance of photosynthetic picoeukaryotes in Norwegian and Barents seas. *Limnol. Oceanogr.* **50**, 1677–86.

Paterson, D. M., Wiltshire, K. H., Miles, A., Blackburn, J., Davidson, I., Yates, M. G., McGrorty, S. and Eastwood, J. A. (1998). Microbiological mediation of spectral reflectance from intertidal cohesive sediments. *Limnol. Oceanogr.* **43**, 1207–21.

Pfennig, N. (1978). General physiology and ecology of photosynthetic bacteria. In *The Photosynthetic Bacteria*, ed. R. K. Clayton and R. W. Sistrom. New York: Plenum Press, pp. 3–18.

Pinckney, J., Millie, D., Howe, K., Paerl, H. and Hurley, J. (1996). Flow scintillation counting of ^{14}C labeled microalgal photosynthetic pigments. *J. Plankton Res.* **18**, 1867–80.

Pinckney, J., Carman, K. R., Lumsden, S. E. and Hymel, S. N. (2003). Microalgal-meiofaunal trophic relationships in muddy intertidal sediments. *Aquat. Microb. Ecol.* **31**, 99–108.

Reigner, F. E., He, B., Lin, S. and Busse, J. (1999). Chromatography and electrophoresis on chips: critical elements of future integrated, microfluidic analytical systems for life science. *Trends Biotechnol.* **17**, 101–06.

Reuss, N., Conley, D. J. and Bianchi, T. S. (2005). Preservation conditions and the use of sediment pigments as a tool for recent ecological reconstruction in four Northern European estuaries. *Mar. Chem.* **95**, 283–302.

Riegman, R. and Kraay, G. W. (2001). Phytoplankton community structure derived from HPLC analysis of pigments in the Faroe-Shetland Channel during summer 1999: the distribution of taxonomic groups in relation to physical/chemical conditions in the photic zone. *J. Plankton Res.* **23**, 191–205.

Rodríguez, F., Garrido, J. L., Crespo, B. G., Arbones, B. and Figueiras, F. G. (2006). Size fractionated phytoplankton pigment groups in the NW Iberian upwelling system: impact of the Iberian poleward current. *Mar. Ecol. Prog. Ser.* **323**, 59–73.

Rodríguez, F., Pazos, Y., Maneiro, J. and Zapata, M. (2003). Temporal variation in phytoplankton assemblages and pigment composition at a fixed station of the Ría of Pontevedra (NW Spain). *Estuar. Coast. Shelf Sci.* **58**, 499–515.

Rodríguez, F., Varela, M. and Zapata, M. (2002). Phytoplankton assemblages in Gerlache and Bransfield Straits (Antarctic Peninsula) determined by light microscopy and CHEMTAX analysis of HPLC pigment data. *Deep-Sea Research II* **49**, 723–48.

Rodríguez, F., Zapata, M. and Garrido, J. L. (1998). High performance liquid chromatographic separation of chlorophyll *c* forms from marine phytoplankton on octylsilica bonded phases. *Chromatographia* **48**, 677–80.

Roy, S., Mohovic, B., Gianesella, S. M. F., Schloss, I., Ferrario, M. and Demers, S. (2006). Symposium-in-Print: UV effects on aquatic and coastal ecosystems. Effects of enhanced UV-B on pigment-based phytoplankton biomass and composition of mesocosm-enclosed natural marine communities from three latitudes. *Photochem. Photobiol.* **82**, 909–22.

Sander, L. C. and Wise, S. A. (1990). Evaluation of shape selectivity in liquid chromatography. *LC GC* **8**, 378–90.

Sander, L. C. and Wise, S. A. (2001). The influence of column temperature on selectivity in reversed-phase liquid chromatography for shape constrained solutes. *J. Sep. Sci.* **24**, 910–20.

Sander, L. C., Sharpless, K. E., Craft, N. E., Wise, S. (1994). Development of engineered stationary phases for the separation of carotenoid isomers. *Anal. Chem.* **66**, 1667–74.

Sander, L. C., Pursch, M. and Wise, S. A. (1999). Shape selectivity for constrained solutes in reversed-phase liquid chromatography. *Anal. Chem.* **71**, 4821–30.

Sander, L. C., Sharpless, K. E. and Pursch, M. (2000). C30 Stationary phases for the analysis of food by liquid chromatography. *J. Chromatogr. A* **880**, 189–202.

Saitoh, K., Awaka, I. and Suzuki, N. (1993). Separation of chlorophyll c_1 and c_2 by reversed-phase high-performance liquid chromatography. *J. Chromatogr. A* **653**, 247–51.

Saitoh, K., Awaka, I. and Suzuki, N. (1995). Determination of chlorophyll by reversed-phase high-performance liquid chromatography with isocratic elution and the column switching technique. *J. Chromatogr. A* **693**, 176–80.

Schlüter, L. and Møhlenberg, F. (2003). Detecting presence of phytoplankton groups with non-specific pigment signatures. *J. Appl. Phycol.* **15**, 465–76.

Schmid, H. and Stich, H. B. (1995). HPLC-analysis of algal pigments: comparison of columns, column properties and eluents. *J. Appl. Phycol.* **7**, 487–94.

Seoane, S., Laza, A., Urrutxurtu, I. and Orive, E. (2005). Phytoplankton assemblages and their dominant pigments in the Nervion River estuary. *Hydrobiologia* **549**, 1–13.

Serodio, J., Cruz, S., Vieira, S. and Brotas, V. (2005). Non-photochemical quenching of chlorophyll fluorescence and operation of the xanthophyll cycle in estuarine microphytobenthos. *J. Exp. Mar. Biol. Ecol.* **326**, 157–69.

Snyder, L. R., Glajch, J. L. and Kirkland, J. J. (1988). *Practical HPLC Method Development*. New York: John Wiley & Sons.

Snyder, L. R., Kirkland, J. J. and Glajch, J. L. (1997). *Practical HPLC Method Development*. 2nd edn. New York: John Wiley & Sons.

Squier, A. H., Hodgson, D. A. and Keely, B. J. (2005). Evidence of late quaternary environmental change in a continental east Antarctic lake from lacustrine sedimentary pigment distributions. *Antarct. Sci.* **17**, 361–76.

Thompson, J. D. and Carr, P. W. (2002). A study of the critical criteria for analyte stability in high-temperature liquid chromatography. *Anal. Chem.* **74**, 1017–23.

Trice, T. M., Glibert, P. M., Lea, C. and Van Heukelem, L. (2004). HPLC pigment records provide evidence of past blooms of *Aureococcus anophagefferens* in the coastal bays of Maryland and Virginia, USA. *Harmful Algae*, **3**, 295–304.

Van Heukelem, L. and Thomas, C. S. (2001). Computer-assisted high-performance liquid chromatography method development with applications to the isolation and analysis of phytoplankton pigments. *J. Chromatogr. A* **910**, 31–49.

Van Heukelem, L, Lewitus, A. J., Kana, T. M. and Craft, N. E. (1992). High-performance liquid chromatography of phytoplankton pigments using a polymeric reversed phase C_{18} column. *J. Phycol.* **29**, 867–72.

Van Heukelem, L., Lewitus, A. J., Kana, T. M. and Craft, N. E. (1994). Improved separations of phytoplankton pigments using temperature-controlled high-performance liquid chromatography. *Mar. Ecol. Prog. Ser.* **114**, 303–13.

Van Leeuwe, M. A., de Baar, H. J. W. and Veldhuis, M. J. W. (1998). Pigment distribution in the Pacific region of the Southern Ocean (Autumn 1995). *Polar Biol.* **19**, 348–53.

Van Lenning, K., Garrido, J. L., Arístegui, J. and Zapata, M. (1995). Temperature-programmed high-performance liquid chromatography separation of mono- and divinyl chlorophyll forms from marine phytoplankton. *Chromatographia* **41**, 539–43.

Van Lenning, K., Latasa, M., Estrada, M., Sáez, A. G., Medlin, L., Probert, I., Véron, B. and Young, J. (2003). Pigment signatures and phylogenetic relationships of the Pavlovophyceae (Haptophyta). *J. Phycol.* **39**, 379–89.

Veldhuis, M. J. W. and Kraay, G. W. (2004). Phytoplankton in the subtropical Atlantic Ocean: towards a better assessment of biomass and composition. *Deep-Sea Res. I* **51**, 507–30.

Vidussi, F., Claustre, H., Bustillos-Guzmán, J., Cailliau, C. and Marty, J. C. (1996). Determination of chlorophylls and carotenoids of marine phytoplankton: separation of chlorophyll *a* from divinyl chlorophyll *a* and zeaxanthin from lutein. *J. Plankton Res.* **18**, 2377–82.

Wilson, M. A., Hodgson, D. A. and Keely, B. J. (2004). Structural variations in derivatives of the bacteriochlorophylls of Chlorobiaceae: impact of stratigraphic resolution on depth profiles as revealed by methanolysis. *Org. Geochem.* **35**, 1299–307.

Wiltshire, K. and Schroeder, F. (1994). Pigment patterns in suspended matter from the Elbe estuary, Northern Germany. *Neth. J. Aquat. Ecol.* **28**, 255–65.

Wiltshire, K., Geissler, C. D., Schroeder, F. and Knauth, H. D. (1996). Pigments in suspended matter from the Elbe estuary and the German Bight. Their use as marker compounds for the characterisation of suspended matter and in the interpretation of heavy metal loadings. *Arch. Hydrobiol. Spec. Issues Adv. Limnol.* **47**, 53–63.

Wirth, M. J. (1994). Spectroscopic probing of reversed-phase chromatographic retention. *LC GC Int.* **7**, 626–30.

Wolcott, R. G., Dolan, J. W., Snyder, L. R., Bakalyar, S. R., Arnold, M. A. and Nichols, J. A. (2000). Control of column temperature in reversed-phase liquid chromatography *J. Chromatogr. A* **869**, 211–30.

Wright, S. W. and Jeffrey, S. W. (1997). High-resolution HPLC system for chlorophylls and carotenoids of marine phytoplankton. In *Phytoplankton Pigments in Oceanography: Guidelines to Modern Methods*, ed. S. W. Jeffrey, R. F. C. Mantoura and S. W. Wright. Paris: UNESCO Publishing, pp. 327–41.

Wright, S. W. and Jeffrey, S. W. (2006). Pigment markers for phytoplankton production. In *Marine Organic Matter: Biomarkers, Isotopes and DNA*, ed. J. K. Volkman. Berlin: Springer, pp. 71–104.

Wright, S. W. and Mantoura R. F. C. (1997). Guidelines for selecting and setting up an HPLC system and laboratory. In *Phytoplankton Pigments in Oceanography: Guidelines to Modern Methods*, ed. S. W. Jeffrey, R. F. C. Mantoura and S. W. Wright. Paris: UNESCO Publishing, pp. 383–406.

Wright, S. W. and van den Enden, R. L. (2000). Phytoplankton community structure and stocks in the East Antarctic marginal ice zone (BROKE survey, January–March 1996) pigment signatures. *Deep-Sea Res. II* **47**, 2363–2400.

Wright, S. W., Jeffrey, S. W., Mantoura, R. F. C., Llewellyn, C. A., Bjørnland, T., Repeta, D. and Welschmeyer, N. (1991). Improved HPLC method for the analysis of chlorophylls and carotenoids from marine phytoplankton. *Mar. Ecol. Prog. Ser.* **77**: 183–96.

Wysocki, L. A., Bianchi, T. S., Powell, R. T. and Reuss, N. (2006). Spatial variability in the coupling of organic carbon, nutrients, and phytoplankton pigments in surface waters and sediments of the Mississippi River plume. *Estuar. Coast. Shelf Sci.* **69**, 47–63.

Yu, Z.-G., Deng, C.-M., Yao, P., Zhen, Y. and Qian, S.-B. (2007). Prasinoxanthin-containing Prasinophyceae discovered in Jiaozhou Bay, China. *J. Integr. Plant Biol.* **49**, 497–506.

Zapata, M. (2005). Recent advances in pigment analysis as applied to picophytoplankton. *Vie et Milieu*, **55**, 233–48.

Zapata, M., Rodríguez, F., Garrido, J. L. (2000). Separation of chlorophylls and carotenoids from marine phytoplankton: A new HPLC method using a reversed-phase C_8 column and pyridine-containing mobile phases. *Mar. Ecol. Prog. Ser.* **195**, 29–45.

Zapata, M., Edvardsen, B., Rodríguez, F., Maestro, M. A. and Garrido J. L. (2001). Chlorophyll c_2 monogalactosyldiacylglyceride ester (chl c_2-MGDG). A novel marker pigment for *Chrysochromulina* species (Haptophyta). *Mar. Ecol. Prog. Ser.* **219**, 85–98.

Zapata, M., Jeffrey, S. W., Wright, S. W., Rodríguez, F., Garrido, J. L. and Clementson, L. (2004). Photosynthetic pigments in 37 species (65 strains) of Haptophyta: implications for oceanography and chemotaxonomy. *Mar. Ecol. Prog. Ser.* **270**, 83–102.

Zhu, P. L., Dolan, J. W. and Snyder, L. R. (1996b). Combined use of temperature and solvent strength in reversed phase gradient elution, II. Comparing selectivity for different samples and systems. *J. Chromatogr. A* **756**, 41–50.

Zhu, C., Goodall, D. M. and Wren, S. A. C. (2005). Elevated temperature HPLC: principles and applications to small molecules and biomolecules. *LCGC North America* **23**, 54–72.

Zhu, P. L., Snyder, L. R., Dolan, J. W., Djordjevic, N. M., Hill, D. W., Sander, L. C. and Waeghe, T. J. (1996a). Combined use of temperature and solvent strength in reversed-phase gradient elution, I. Predicting separation as a function of temperature and gradient conditions. *J. Chromatogr. A* **756**, 21–39.

5

The importance of a quality assurance plan for method validation and minimizing uncertainties in the HPLC analysis of phytoplankton pigments

LAURIE VAN HEUKELEM AND STANFORD B. HOOKER

5.1 Introduction

A quality assurance plan (QAP) describes a process to ensure an analytical method fulfils the agreed upon accuracy objectives at all points during the analysis of samples. A QAP includes such things as standardized procedures, method validation, quality control (QC) measurements, and quality assessment (QA); the latter quantitatively describes how results of QC measurements are used to determine whether a method is performing within expectations. By analogy, a QAP describes what can be considered, in a more general sense, a 'holistic' approach to sample analysis, whereby the 'whole is considered to be a result of the interdependence of all parts'. Here, the 'whole' represents the overall combined uncertainty of a final data product, and 'the interdependence of all parts' represents the uncertainties contributed by the many individual procedures required to produce that final data product.

An in depth discussion of uncertainty analysis in the chemical laboratory, given in EURACHEM (2000), is beyond the scope of this chapter, but the importance of a so-called holistic approach to sample analysis, whether this includes a formalized QAP or not, is necessary to provide knowledge of the uncertainties associated with measured values and, thus, facilitate confidence in the data products.[1] Such knowledge is important, because pigment data are often compiled in increasingly large databases, and end users of the data are remote (in space and time) from the data providers. Knowing the accuracy of the data facilitates their wider utility, for both current and unanticipated future applications. This chapter describes components of a QAP in the context of knowledge gained during intercomparisons sponsored by the National Aeronautics and Space Administration (NASA). A primary objective of these activities was to determine if the myriad providers of HPLC analyses to NASA researchers were satisfying the accuracy requirements for field observations.

[1] Appendix 5A contains equations and definitions of terms cited within this chapter. It is therefore necessary reading for a complete understanding of material presented here.

Phytoplankton Pigments: Characterization, Chemotaxonomy and Applications in Oceanography, eds. Suzanne Roy, Carole A. Llewellyn, Einar Skarstad Egeland and Geir Johnsen. Published by Cambridge University Press.

To ensure the required field measurements were in keeping with the remote sensing requirements, the Sea-viewing Wide Field-of-view Sensor (SeaWiFS) Project convened a workshop to draft the SeaWiFS Ocean Optics Protocols (hereafter referred to as the Protocols). The Protocols initially adhered to the Joint Global Ocean Flux Study (JGOFS) sampling procedures (JGOFS, 1994) and defined the standards for field observations to be used in SeaWiFS calibration and validation activities (Mueller and Austin, 1992).

Although the initial emphasis was on optical measurements, the first version of the Protocols established HPLC as the required methodology for pigment analyses. Over time, the Protocols were revised (Mueller and Austin, 1995), and then recurringly updated (essentially on an annual basis) to include a full suite of biogeochemical parameters (Mueller, 2000, 2002, 2003).

After the Protocols were established, intercalibration activities were conducted to assess uncertainties in field measurements of total Chlorophyll *a* concentrations, denoted [TChl *a*]. (Total Chlorophyll *a* is the sum of Chlorophyll *a* and divinyl Chlorophyll *a* and their allomers and epimers, and Chlorophyllide *a*.) The NASA Sensor Intercomparison and Merger for Biological and Interdisciplinary Oceanic Studies (SIMBIOS) and the SeaWiFS projects sponsored one and four intercomparisons, respectively, between 1999 and 2007. The SeaWiFS HPLC Analysis Round-Robin Experiment (SeaHARRE) has been the most extensive activity with three completed (Hooker *et al.*, 2000, 2005, and 2009), and the fourth one currently underway (Hooker *et al.*, 2010). The SIMBIOS round robin investigated causes for discrepancy between Chl *a* quantified fluorometrically and [TChl *a*] quantified by HPLC (Van Heukelem *et al.*, 2002). The SeaHARRE round robins were initiated to investigate the uncertainties over the [TChl *a*] dynamic range 0.05–50.0 mg m^{-3}, with a secondary emphasis on accuracy attained with accessory pigments. The results from these intercomparisons, which cover a wide range in trophic levels, show data quality is unequal across individual pigments, pigment concentrations or data providers.

Dispersion in data quality observed in NASA round robins was largely influenced by the lack of a) a common lexicon by which uncertainties unique to pigment analyses could be unambiguously described, b) a structured QAP emphasizing temporal monitoring of data quality, and c) a common statistical approach for quantitation of accuracy and precision. The latter is especially important, because, prior to these round robins, pigment analysts – who work largely in isolation – had no *benchmarks of achievement*, stated in a commonly used vernacular. The latter was found to be a significant disadvantage, because current procedures could not be adequately compared or improvements to methods properly evaluated. The NASA SeaHARRE technical reports attempted to fill these voids. Through interviews with round-robin participants, a common lexicon was synthesized which clarified procedures, defined pigments and higher-order data products, established a symbology, standardized terms used in calculation equations, and facilitated a common approach to accuracy and precision assessment (Appendix 5A).

Some information pertinent to a QAP for HPLC pigment analyses is available (JGOFS, 1994; Jeffrey *et al.*, 1997; Bidigare *et al.*, 2003; Bidigare *et al.*, 2005), but such information is largely procedural, with little emphasis on method validation and quality assessment. Other guidance documents describing quality assurance in the context of pharmaceutical and commercial applications, such as those from EURACHEM working groups (1998 and 2000), United States Department of Health and Human Services Food and Drug Administration (U.S. Department of Health and Human Services, Food and Drug Administration 1994, 1995, 1996 and 2001) and the International Conference on Harmonization (ICH) working groups (2005), are available. In addition, a quality assurance plan for monitoring in the Baltic Sea, as outlined by the Helsinki Commission (2006) for use with nutrients, trace metals and chlorinated organic compounds in marine environments is pertinent. Such validation procedures could have been adapted to HPLC pigment methods, but NASA round-robin participants had not done so. This deficiency compromised the interpretation of the results, because the very premise of a round robin is that all participants use a validated method. As such, each method is assumed to be equally capable of estimating a true result for each sample, and each sample is analysed no differently than any other normally analysed by the method. The result from each method is expected to be close to the truth, and the dispersion of the results will be equally expressed above and below the true value. A validated method has no inherent biases, because if one existed it would have been removed by the validation process. Computation of the accuracy (or uncertainty) for each method is based on computing the difference of each individual result from the truth.

To maintain the integrity of the round-robin approach, in spite of the fact that HPLC methods were not validated, data from measured performance parameters (similar to those obtained during normal validation processes) were collected *after* round-robin results had been submitted for SeaHARRE-1 and during SeaHARRE-2 (Claustre *et al.*, 2004; Hooker *et al.*, 2005). Furthermore, the quality of performance parameter results (hereafter referred to as *performance metrics*) allowed development of a classification scheme by which a laboratory's likelihood of producing results within specified accuracy objectives could be assessed. Consequently, field sample concentrations from laboratories with performance metrics of sufficient quality were used to compute *quality-assured* field sample reference values, to which the results of *all* participating laboratories were compared, and from which accuracy estimates were determined. In SeaHARRE-3 and -4, laboratories reported performance metrics results concurrent with field sample results, so the correlation between performance metrics and field sample accuracy could be further evaluated (Hooker *et al.*, 2009; Hooker *et al.* 2010). The performance metrics and associated performance categories provide *benchmarks of achievement* to analysts working in isolation and a means of surmising the accuracy capabilities of their methods.

This chapter emphasizes method validation, as summarized primarily from EURACHEM working groups, with significant overlap between the performance parameters described therein and *performance metrics* that emerged from SeaHARRE round robins. This chapter also suggests various QC measurements and how they can be used to verify, on a variety of time scales, that HPLC data products are consistently of sufficient quality to meet accuracy objectives. The latter effort is part of the generalized quality assessment category within a QAP; other important features of a QAP include detailed standardized procedures pertaining to accurate use of spectrophotometers, calibration of dilution devices, field sampling and transport, etc., which, due to space constraints, cannot be detailed here.

5.2 Method validation

Method validation for HPLC pigment analysis is performed before a method is put into service for its ultimate objective and includes a) defining the analytical requirement, and b) evaluating various performance parameters. The expression 'Fitness for Purpose' is used with method validation, as illustrated in Figure 5.1, and refers to the

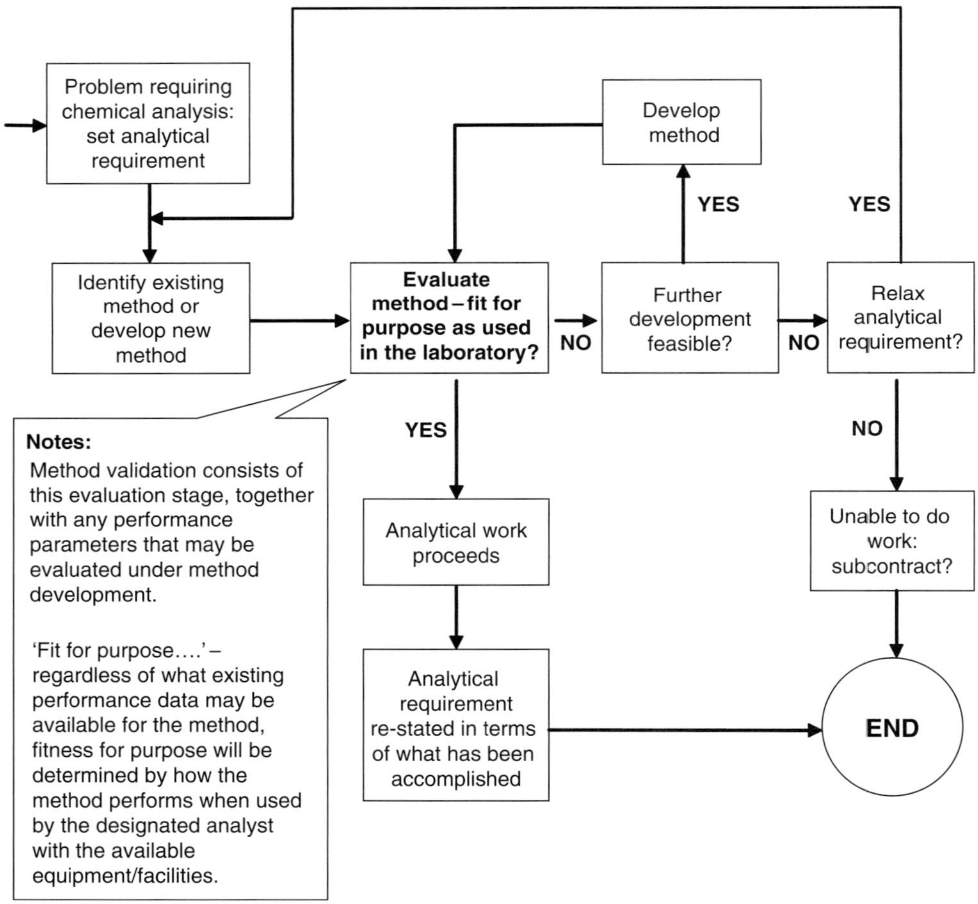

Figure 5.1. Choosing, developing and evaluating analytical methods. Modified from *The Fitness for Purpose of Analytical Methods: a Laboratory Guide to Method Validation and Related Topics* (EURACHEM, 1998).

process by which a method is deemed suitable (or not) for the analysis goals. In practice, some limitations of a method may not be known until it is used with all possible combinations of samples, a process that is not possible during method validation. As such, unforeseen co-elution problems may be encountered that require accuracy objectives to be relaxed (if possible) or, alternatively, some pigments originally intended for quantitation to be disregarded.

None of the laboratories participating in NASA round robins used validated HPLC methods and there was a lack of understanding regarding accuracy capabilities prior to participating in a round robin. While unvalidated methods may 'accidentally' meet analysis objectives, the very process of method validation facilitates a statistical defence of data products. Method validation steps, as detailed in EURACHEM guidelines (1998), are discussed in the context of their applicability to HPLC pigment analyses. Performance attained by laboratories for various performance parameters are given so that readers validating a method for the first time will have expectations of performance and can avoid problems encountered by others.

5.2.1 The analytical requirement

Defining the analytical requirement for pigment analyses includes identifying pigments to be quantified, determining the accuracy necessary to meet the experimental objectives, and selecting an appropriate method. In NASA round robins, the HPLC method and pigments quantified were at the discretion of each laboratory and no one HPLC method was considered inherently better than another. Prior to SeaHARRE-1, it was not evident that laboratories had identified accuracy requirements, so a key element in the validation process was missing.

Accuracy is defined as how well a measured value agrees with the *true* or expected value (Taylor, 1987). Assessing the accuracy of an HPLC method used to quantify pigments in field samples is complicated by the fact that *true* concentrations in field samples are not known, so proxies for *true* concentrations are represented by the average consensus concentration for a given pigment (from a specified sample) computed from individual results of the participating laboratories. Some laboratories will more closely approximate *true* concentrations than others, and the inclusion in the consensus average of results from laboratories using methods and procedures with known deficiencies, diminishes the accuracy with which true (or reference) concentrations are determined. This undermines the ability to quantify the accuracy of each laboratory, because the inaccuracy of the unvalidated methods is distributed across the results of all the participants. This problem was addressed in SeaHARRE round robins by computing consensus average concentrations using the results from laboratories meeting required performance criteria (Section 5.4). (The statistical approach used to compute laboratory accuracy and precision is given in Appendix 5A.)

In spite of the difficult nature of assessing accuracy with field samples, the accuracy requirement for [TChl *a*] was well established, because it had been set by the data quality requirements of field observations collected for satellite vicarious calibration and algorithm validation activities (Hooker and McClain, 2000). These so-called calibration and validation activities are instrumental in evaluating the attainment of accuracy thresholds set by the SeaWiFS project and most of the subsequent ocean colour satellite missions, like the moderate resolution imaging spectroradiometer sensors launched aboard the Terra and Aqua spacecraft as part of the Earth Observing System (EOS). For such missions, the satellite [TChl *a*] accuracy requirement is to within 35% over a range 0.05–50.0 mg m^{-3} (Hooker and Esaias, 1993). The *in situ* pigment observations will always be one of two variables to derive or validate the pigment relationships, so it is usually appropriate to reserve approximately half of the uncertainty budget for the *in situ* pigment observations (Hooker and McClain, 2000). The sources of uncertainty are assumed to combine independently (i.e. in quadrature), so an upper accuracy range of 25% is deemed acceptable, although 15% would presumably permit significant improvements in algorithm refinement, and this was ultimately adopted by SeaHARRE laboratories as the maximum, average accuracy threshold for quantitative analysis of [TChl *a*] in field samples (Hooker *et al.*, 2005).

Prior to SeaHARRE-1, there were no agreed accuracy thresholds for accessory pigments. In a round robin, the accuracy of a pigment can only be estimated if multiple laboratories report, so that a consensus average can be computed, and most methods report a wide diversity of accessory pigments. In NASA round robins there was no discrimination as to whether certain pigments were more important than others. This latter point is important because no one method can separate all pigments equally well. Based on the information provided by participating laboratories, analytical requirements evolved that categorized pigments according to two primary criteria: a) their utility with respect to biogeochemical investigations, and b) whether most laboratories quantified them. Pigments were referred to as *primary*, *secondary*, *tertiary* and *ancillary* pigments, as defined originally in Hooker *et al.* (2005) and summarized below, with full details in Appendix 5A:

- The *primary* pigments (PPig) are the total chlorophylls and the carotenoids most commonly used in chemotaxonomic or photophysiological studies in the open ocean or in coastal waters (Gieskes *et al.*, 1988; Barlow *et al.*, 1993; Claustre, 1994; Bidigare and Ondrusek, 1996);
- The *secondary* pigments are the individual pigments used to create a primary pigment composed of separate contributions (e.g. TChl *a*);
- The *tertiary* pigments are those pigments not included in the composition of the primary and secondary pigments for which three or more laboratories provided quantitations; and
- The *ancillary* pigments are those remaining pigments only analysed by one or two laboratories.

While this nomenclature implies some precedence or ranking of the pigments, this is only true from the current perspective of marine phytoplankton pigments for which certain pigments are routinely used more often than others (e.g. Chl *a*). The primary reason for establishing a unique vocabulary is to provide an appropriate categorization scheme for grouping the analytical results. The primary pigments were also used to define higher-order pigment products composed of sums, ratios and indices (Table 5A.1 in Appendix 5A). During SeaHARRE-2, participants agreed to apply the 15% accuracy objective originally adopted for [TChl *a*] to all twelve primary pigments. No accuracy requirements were set for the secondary, tertiary and ancillary pigments, for which 15% accuracy thresholds proved to be routinely difficult to achieve.

5.2.2 Performance parameters

Performance parameters describe the suitability of a method for the intended application by evaluating the following: a) the ability to quantify analytes in the presence of interferences (specificity); b) the minimum concentration that can be detected or quantified with a specified confidence, which are referred to hereafter as the limit of detection (LOD) and the limit of quantitation (LOQ), respectively; c) the linear ranges; d) the accuracy and precision of the method; e) the reproducibility and linearity of calibrations; and f) whether the method can be implemented successfully even with imperfect conditions (ruggedness). Performance parameters presented here are from EURACHEM (1998) and are a subset of parameters that can be used to evaluate a method. The measured values for many performance parameters are also useful as QC measurements during temporal quality assessment.

5.2.2.1 Specificity

An analytical method consists of a measurement stage and an isolation stage. The latter establishes that the signal produced at the measurement stage is due to the analyte of interest and does not arise from interference – arising perhaps from something chemically or physically similar or from an unfortunate coincidence (EURACHEM, 1998). Specificity describes the ability of the method to reliably quantify an analyte in the presence of interferences. With HPLC pigment analyses using photodiode array detectors (the most common detector in NASA round robins), the isolation phase consists of documenting retention time, resolution between pigments, and absorbance spectra of pigments to be quantified and pigments likely to be encountered in natural samples, but considered interferences. Jeffrey and Wright (1997) detail these procedures and recommend using the Scientific Committee on Oceanographic Research (SCOR) algal reference culture extracts for such inquiries. A mixed algal culture extract, which is commercially available from DHI (Hørsholm, Denmark), facilitates this process and has been a part of SeaHARRE activities since its introduction (Hooker *et al.*, 2005).

Metrics used in support of specificity include retention time (t_R) repeatability, resolution between pigments (R_S), and the percentage (or amount) of carryover between injections. Customizing a library of pigment absorbance spectra (including those of interferences) significantly enhances the interpretation of chromatograms from natural samples. Because many pigments have similar absorbance spectra, highly reproducible retention times and absorbance spectra collected with detector settings maximizing spectral selectivity can reduce instances of incorrect pigment identification. Low values of average t_R variability, observed as coefficient of variation (CV) for pigments in replicate injections of the same sample, were correlated in SeaHARRE-2 with high data quality and are, therefore, used as a performance metric (Section 5.4).

Adequate resolution between peaks is important to accurate quantitation, especially when based on peak area (Snyder and Kirkland, 1979), but few pigment analysts in NASA round robins documented R_S and, in fact, many quantified pigments with $R_S < 1.0$ ($R_S = 1.5$ for baseline resolution of symmetrical peaks, Eq. 5A.10, Appendix 5A). The negative effects of this practice were evaluated by having analysts quantify pigment concentrations and document R_S in a test mixture containing 20 pigments, including all 12 primary pigments (Hooker *et al.*, 2005). The most common co-elution problems affecting primary pigments were between a) Pheide *a* plus Peri, Fuco, or Hex-fuco; b) Hex-fuco plus Neo, Viola, or Pras; 3) Zea plus Lut or Diato; and 4) DVChl *a* plus MVChl *a*. The average accuracy (across all laboratories) for primary pigments in this mixture improved from 25% to 16% when pigments with $R_S < 1.0$ were excluded from the average. The correlation between data quality and R_S makes the latter an important performance metric (Section 5.4).

A useful *acceptance* criterion for performance of a new column (and during method validation for determining which pigments are quantifiable by peak area) is $R_S \geq 1.0$ between critical peak pairs. Resolution less than 1.0, for pigment pairs typically yielding greater resolution, is a useful *rejection* criterion for determining when to remove a used column from service. A critical pair refers to adjacent pigments, one of which is a primary or secondary pigment, and for which R_S is at least 1.0 but poorer than the R_S values observed for other pigment pairs involving either a primary or secondary pigment. For example, critical pairs of the HPL method are Zea plus Lut and Hex-fuco plus Viola (see Figure 4.4, Chapter 4, this volume). Quantitation by peak area is recommended for peaks with $R_S \geq 1.0$ and quantitation by peak height for pigments with $0.8 \leq R_S < 1.0$. For pigments with $R_S < 0.8$, quantitation based on a simultaneous equation may be appropriate (Latasa *et al.*, 1996; Hooker *et al.*, 2000; Claustre *et al.*, 2004).

It is not practical during method validation to anticipate all interferences possible in natural samples, but as more samples are processed from diverse environments, specificity limitations should be more completely documented. Pigments in low concentrations relative to [TChl *a*] are at great risk of interference from pigment degradation products, pigment isomers or adjacent peaks with large areas (Dolan,

2002). For example, 11–37% of selected pigments, when at concentrations less than 5% of [TChl *a*], deviated from reference concentrations by 100% or more (data derived from SeaHARRE-2 for But-fuco, Allo, Diato, Lut, Neo, Pras and Viola). Inaccuracies of this magnitude (and greater) can result from false reporting – pigments are reported present when they are not (a false positive) or reported absent when they are actually present (a false negative). False positives can also occur with pigments of high concentration if identification is based solely on retention time. For example, an interfering pigment is frequently present in high concentrations at the Pras retention time with the method of Van Heukelem and Thomas (2001), as described in Hooker *et al.* (2009). While specificity problems are not likely to be fully eliminated, even with ideal chromatographic conditions, a consensus of agreement regarding procedures for reporting small peaks with questionable identities will minimize their damaging effects on HPLC accuracy. Optimizing reporting practices is of paramount importance to SeaHARRE investigators and recommendations regarding such are anticipated.

Carryover is described by the cross contamination of a pigment from one analysis to a subsequent analysis. Carryover has many causes (Dolan, 1999, 2001a, 2001b, and 2006), all of which result in overestimation of the pigment in the subsequent analysis or, if it is really not present in the subsequent sample, a false positive result. SeaHARRE laboratories reported carryover was not present, or at amounts approximating 0.3 to 0.9% of the previous analysis (for Chl *a*), the latter of which is sufficient to cause up to 100% overestimation of Chl *a* in an oligotrophic extract. Carryover should be evaluated during method validation (and monitored during sample analysis) by injecting reagent blanks immediately following analysis of a sample that mimics the most concentrated expected.

An injection can deliver to the column more than the planned-for amounts if the sample loop or needle is contaminated – the latter can be contaminated either internally or externally. A type of internal contamination is possible with injector programs that require mixing sample with the buffering agent in a separate vial prior to injection, because the sample loop can retain a residual amount of material after the contents of the loop are expelled into the mixing vial. Contamination of this type can be evaluated during method development by interrupting the injector program and causing the injector to simply *draw and inject* a reagent blank (premixed on the laboratory bench with buffer or water in the necessary ratio) after the sample needle has delivered (and mixed) sample and buffer in the sample preparation vial. If possible, this test should be done after any wash functions are performed. To reiterate this test, the final volume specified for injection in the injector program is taken from the vial containing the premixed reagent blank and not from the sample preparation vial in which sample and buffer were combined (and mixed) by the injector program. If the needle and loop are contaminated by the sample extract when the reagent blank is withdrawn for injection, pigments will be present in the resulting chromatogram. Retention of pigments by the HPLC column after an analysis is complete can also produce carryover in a subsequent analysis. This type of carryover is evaluated by activating the gradient program with no injection and inspecting the resulting chromatogram for pigments and by also performing a blank injection with the strongest injection solvent typically used. Tests for column carry-over should be evaluated after typical sample and standard analyses.

5.2.2.2 Limits of detection and quantitation

Synoptic descriptions of the LOD and LOQ are presented, respectively, as the lowest concentration of the analyte value that can be a) confidently detected by the method, and b) determined with an acceptable level of repeatability precision and trueness (EURACHEM, 1998). Many approaches to determining LOD and LOQ are possible, for example, as detailed in the International Union of Pure and Applied Chemists (IUPAC) *Compendium of Analytical Nomenclature 1997* (IUPAC, 1998) and the International Conference on Harmonisation (ICH) *Guidelines for Validation of Analytical Procedures: Methodology* (2005). Procedures for determining LOD and LOQ can be characterized as 'instrumental or 'non-instrumental' (ICH, 2005), for which the latter reflects the combined effects of uncertainties resulting from instrumentation and method related procedures (e.g. calibration). According to ICH guidelines (2005), in a non-instrumental determination of LOD and LOQ, the standard deviation of the response is multiplied by a factor (e.g. 3 for LOD and 10 for LOQ), which is then divided by the slope, where slope is the linear regression of a detector response (y-axis) as a function of amount (x-axis). The standard deviation of the response is either determined from y-residuals of the regression line or the standard deviation of the y-intercept – clearly defined formulistically by Miller and Miller (2000). An instrumental approach does not take into account uncertainties associated with calibration, for example, but is based solely on the signal to noise ratio (SNR), which describes the relationship between instrument noise in the absence of sample and instrumental signal (produced by introduction of sample). Ultimately, the most rigorous evaluation of LOD and LOQ would include both an instrumental and a non-instrumental approach. The latter, while not having been recommended in SeaHARRE documents, requires inspection of the y-intercept, and as such, overcomes a common procedural deficiency – the y-intercept is forced through zero with little consideration for the effects of such on accuracy.

With the first SeaHARRE activity, it became evident that little attention was given to detectability by NASA round-robin participants. An attempt to understand the uncertainties of pigments in low concentrations led to recommendations that analysts calculate LOD and LOQ using an instrumental approach, wherein the former exhibits a SNR of 3 and the latter a SNR of 10 (Hooker *et al.*, 2005). Noise is defined as *very short-term detector noise* and is described by a statistical parameterization of the baseline fluctuations at times approximating the elution position of the

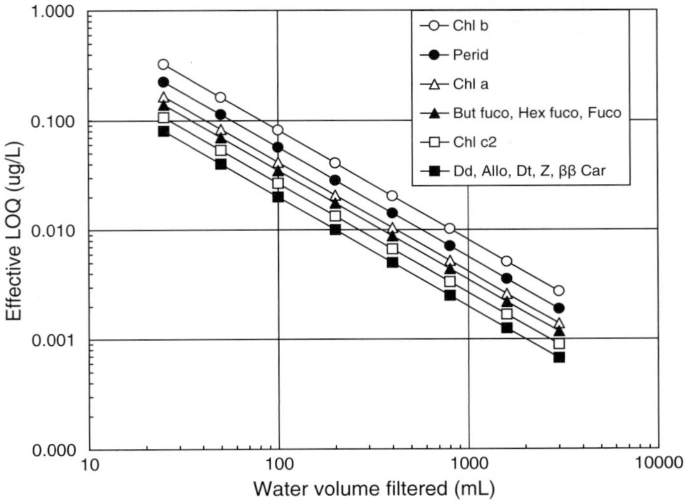

Figure 5.2. The *effective LOQ* (y-axis) as a function of filtration volume (x-axis) determined using constant extraction and injection volumes of 2.5 and 0.150 mL, respectively. The *ELOQ* represents pigment concentrations in seawater that, when using specified volumetric information, are detection-limited when analysed (e.g. the SNR is approximately 10).

pigments of interest. Values are expressed as nanograms per injection for each pigment, for which 0.06 to 0.23 ng are characteristic (Hooker *et al.*, 2005). The information from a detectability analysis, however, is most accessible if it is available in units of concentration (micrograms of pigment per litre of seawater). These so-called *effective* values are computed from the LOD and LOQ amounts and the volumetric information associated with the analysis of the samples. For example, with an LOD of 0.25 ng, and injected, extracted, and filtered volumes of 150 μL, 2.50 mL, 2.8 l, respectively, the *effective* LOD is 0.002 $\mu g\ l^{-1}$; assuming the same volumes except a filtered volume of 100 mL, the effective LOD is 0.042 $\mu g\ l^{-1}$. The end user of the data has no way of knowing that *both* of these two concentrations were analysed with detection-limiting conditions unless companion effective LOD and LOQ data are provided for each filtration, extraction and injection volume used, as graphically exemplified in Figure 5.2.

Currently, when describing uncertainties of pigments in field samples, there is no discrimination between pigments in high concentrations and pigments in low concentrations. This lack of discrimination can yield extraordinarily poor average estimates of accuracy for individual pigments predominantly in low concentrations (such as Diato) when accuracy is computed on a percentage basis (as in NASA round robins). Thus, LOD and LOQ values, with detailed procedures and associated formulae for their determination, and pertinent effective LOD and LOQ concentrations should be reported with pigment data as part of the permanent record.

5.2.2.3 Working and linear ranges

The linear range in a chromatographic system represents the range of concentrations (or amounts) over which the sensitivity of the detector is constant within a specified variation (IUPAC, 1998), keeping in mind that 'sensitivity' is defined as the slope of the calibration plot (Miller and Miller, 2000), and for which slope is determined from the regression of y (the detector response) on x (the injected amount). For clarity, such ranges are best described by pigment amount injected (e.g. in nanograms) and not concentration. The upper limit of linearity (ULOL) describes the point above which some limitations may occur as a function of the instrument's response system, and the lower limit of linearity (LLOL) describes the point below which uncertainties are exacerbated, by a low SNR, for example (adapted from EURACHEM, 1998).

The term 'working range' describes the range in concentrations over which samples – in this case, pigments in natural sample extracts – can be expected to provide acceptable results (EURACHEM, 1998, adapted from IUPAC, 1998). Thus, sample filters must be considered in the context of how dilute or concentrated pigments in the resulting extracts might be. Volumes used for extraction, filtration and injection should be specifically selected to ensure that resulting peak areas fall within a pigment's working range. To illustrate, anticipated concentrations of Fuco and Chl *a* in sample extracts are estimated using volumetric information from two different laboratories, as detailed in Table 5.1, for field samples representing oligotrophic and eutrophic extremes (0.05 to 50 $\mu g\ l^{-1}$ Chl *a*) and with an assumption that Fuco will be from 2% to 50% of Chl *a* concentrations. Ideally, endpoints of the

Table 5.1. *Volumetric information (from two example laboratories, A and HPL) needed to compute the anticipated amount of pigments in sample extracts for two example pigments in oligotrophic (less than 0.1 $\mu g\ l^{-1}$ Chl* a*) and eutrophic (greater than 1.0 $\mu g\ l^{-1}$ Chl* a*) samples. The lowest expected concentration for Fuco is set here at 0.001 $\mu g\ l^{-1}$.*

Laboratory	Pigment	Concentration in seawater	Filtration volume	Extraction volume	Volume injected	Amount on column
		($\mu g\ l^{-1}$)	(mL)	(mL)	(mL)	(ng)
A	Chl *a*	0.05	2000	4.00	0.073	1.8
A	Chl *a*	50	200	4.00	0.073	181
A	Fuco	0.001	2000	4.00	0.073	0.02
A	Fuco	25	200	4.00	0.073	91
HPL	Chl *a*	0.05	2000	2.50	0.150	6
HPL	Chl *a*	50	200	2.50	0.150	600
HPL	Fuco	0.001	2000	2.50	0.150	0.06
HPL	Fuco	25	200	2.50	0.150	300

working range bracket the minimum and maximum amounts loaded onto the column from anticipated sample extracts, which, as exemplified in Table 5.1 would be Chl *a* amounts between 2 and 181 ng for Laboratory A and amounts between 6 and 600 ng for Horn Point Laboratory (HPL). Clearly, the working range and anticipated amounts of pigments in sample extracts should be described before samples are analysed routinely.

A cost-effective means of evaluating a working range is to inject a dilution series prepared from a concentrated algal extract and to determine the linear regression of peak areas for a particular pigment (the y-variable) as a function of the corresponding dilution factor (the x-variable). Next, it is necessary to evaluate the quality of the linear regression, regardless of whether one uses the approach described here with algal culture dilutions or whether one uses discrete standards. Evaluating the quality of a linear regression is best done by first computing the y-residuals. For this, the fitted detector response ($\hat{y}$) is subtracted from the observed detector response (y), where the former is that predicted by the linear regression *with* a computed y-intercept, as exemplified in Miller and Miller (2000). The fitted detector response is associated with lower uncertainty relative to the observed response and is considered to be a better estimate of the true value and is therefore used as the reference when determining the y-residual, reiterated here as $(y-\hat{y})$. The residual, expressed as a percentage according to $100 * ((y-\hat{y})/\hat{y})$, yields a unitless *per cent residual*, which has utility across applications. Per cent residuals determined from linear regressions encompassing a working range, when converted to their absolute values and averaged – a statistic denoted as $|\bar{\psi}|_{\text{res}}$ and described in Table 5.2 – is a useful performance metric for comparing the quality of linear regressions within a single laboratory or among several laboratories (Hooker *et al.*, 2005).

Fucoxanthin, analysed by Laboratory A (see Table 5.1) in nine algal dilutions, is used here to exemplify evaluation of linearity. Initial inspection of the linear regression revealed a linear response was unattainable over the entire range of peak areas, with per cent residuals as high as 61%. However, by excluding the two most concentrated algal dilutions, a linear regression is produced with per cent residuals that range, from −2.6 to 1.2% and the average of the absolute per cent residuals, $|\bar{\psi}|_{\text{res}}$, is 0.9% (Figure 5.3a). A value of this magnitude for the latter is consistent with an ability to facilitate state-of-the-art accuracy, as detailed in Table 5.2.

Algal dilutions can be used to describe the range of peak areas over which linearity is attainable, but to describe the working range, the x-variables must be converted from dilution factors to amounts injected. For example, with the linearity of Fuco in the algal dilutions well described (Figure 5.3a), a solitary Fuco calibration standard (of known concentration) is injected in replicate and the observed average inverse response factor (Eq. 5A.5) is used to quantify Fuco amounts in each dilution – a process allowable because the peak area of the Fuco standard is within the linear range. The endpoints of the working range, as defined here, are now quantified as 0.2 and 19 ng of Fuco – the latter of which is less than the highest

Table 5.2. *The performance metrics for the four categories established during SeaHARRE-2 for validating the determination of marine pigments using an HPLC method (left to right): concentration (average precision, $\bar{\xi}$, and accuracy, $|\bar{\psi}|$, for TChl* a *and PPig); separation (minimum resolution, R_S, and average retention time precision, $\bar{\xi}\, t_R$); injection precision, $\bar{\xi}_{inj}$ (the average of an early- and late-eluting pigment standard, e.g. Peri and Chl* a*); and calibration (average residual, $|\bar{\psi}|_{res}$, for Chl* a *and the precision of the dilution devices, $\bar{\xi}_{cal}$). The PPig and TChl* a *performance metrics are based on using the analysis of a laboratory mixture of pigments and replicate field samples with approximately equal weights applied to each (remembering that uncertainties are assumed to combine in quadrature and that the latter presupposes the inclusion of replicate filter collection during field sampling). The corresponding values for method H are given as an example. The overall performance of H is considered 'state-of-the-art', because the average score of the weights is 3.7, (4+4+4+3+3+4+4+4+3+4)/10. From Hooker* et al. *(2009).*

Performance weight, category, and score		TChl *a*		PPig		Separation[a]		Injection[b] ($\bar{\xi}_{inj}$)		Calibration[c]	
		$\bar{\xi}$	$\lvert\bar{\psi}\rvert$	$\bar{\xi}$	$\lvert\bar{\psi}\rvert$	R_S	$\bar{\xi}\, t_R$	Peri	Chl *a*	$\lvert\bar{\psi}\rvert_{res}$	$\bar{\xi}_{cal}$
1. Routine	0.5	8%	25%	13%	40%	0.8	0.18%	10%	6%	5%	2.5%
2. Semiquantitative	1.5	5	15	8	25	1.0	0.11	6	4	3	1.5
3. Quantitative	2.5	3	10	5	15	1.2	0.07	4	2	2	0.9
4. State-of-the-art	3.5	≤2	≤5	≤3	≤10	≥1.5	≤0.04	≤2	≤1	≤1	≤0.5
Method H		*1*	*5*	*2*	*12*	*1.2*	*0.02*	*<1*	*<1*	*1.1*	*0.4*

[a] The R_S parameter is the minimum resolution determined from a critical pair for which one of the pigments is a primary pigment. The retention time precision, $\bar{\xi}\, t_R$, values are based on sequential replicate injections of pigments identified in a laboratory mix. In the absence of a diverse set of early- to late-eluting pigments, a practical alternative is to compute $\bar{\xi}\, t_R$ based on three sequential injections of Peri, Fuco, Diadino, Chl *a*, and ββ–Car.

[b] The $\bar{\xi}_{inj}$ terms are calculated from the average of replicate injections of an early- and late-eluting pigment in the same run. (Peri is chosen here to incorporate the possible effects of peak asymmetry which are not presented as a separate parameter.)

[c] The $|\bar{\psi}|_{res}$ values presented here are based on calibration points within the range of concentrations typical of the SeaHARRE-2 field samples. To determine this metric for an arbitrary sample set, $|\bar{\psi}|_{res}$ is computed using those calibration points within the range of concentrations expected in the field samples to be analysed.

amount (91 ng) expected in field sample extracts for volumetric information of Laboratory A (Table 5.1) and the former is 10 times higher than the 0.02 ng needed for the most weakly concentrated samples, 0.001 μg l^{-1}. Thus, some concentrated sample extracts may require dilution and subsequent reanalysis to ensure Fuco peak areas do not exceed the upper limit of the working range. Also, because these analyses do not include the lowest amount of Fuco expected in sample extracts, 0.02 ng, more dilute algal solutions should be injected to evaluate linearity at lower

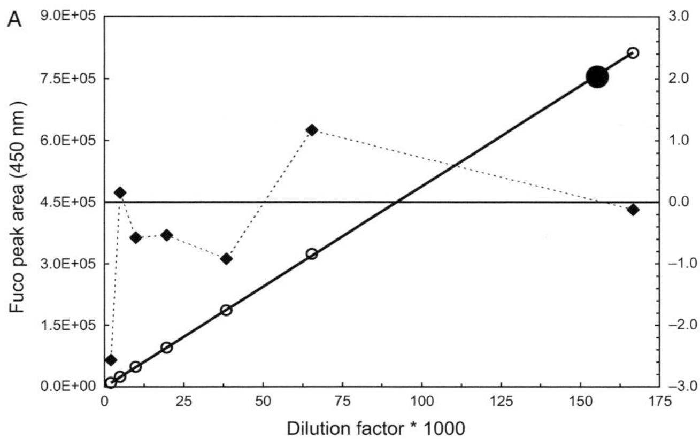

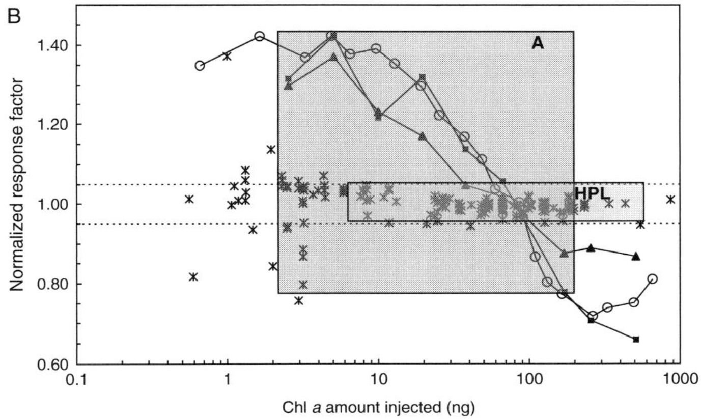

Figure 5.3. (A) Fucoxanthin peak areas observed from HPLC analysis (at 450 nm) of various dilutions of an acetone algal culture extract. Peak areas (open circles) yield a linear response over the range shown, as indicated by per cent residuals (solid diamonds) constrained within ± 3% – determined from a linear regression *not* forced through zero. A fucoxanthin calibration standard (large solid circle) yields a peak area within the linear range, from which an inverse response factor is computed and used to quantify the amounts of fucoxanthin in the algal solutions, which are determined to range over 0.2 to 19 ng. (B) A 'linearity plot', after King (1999), with normalized response factor as a function of Chl *a* amount injected acquired during analysis of multi-point Chl *a* calibration curves by Laboratory A (n = 3 calibration curves discriminated by symbol) and HPL (multiple curves not discriminated). Normalized response factors within ±5% (dashed lines) are expected for linear chromatographic systems. Shaded areas reflect amounts anticipated in natural sample extracts for Laboratory A and HPL.

amounts. This example exposes a commonly encountered problem with pigments routinely of low concentration, (e.g. Neo, Pras, Viola, Lut, Diato) – the amounts injected from sample extracts are frequently at low SNRs, are at risk of interference by adjacent, larger peaks or degradation products, and accurate peak integration is compromised. In fact, defining a working range and the quality requirements needed therewith, may not be possible at the lowest SNRs expected for these pigments in natural sample extracts. For example, the lowest amount of Fuco anticipated in natural sample extracts by HPL, 0.06 ng (Table 5.1), exhibits an approximate SNR of 3.

A more conventional determination of the linear range is through the use of the *linearity plot* (IUPAC, 1998), as expanded by King (1999) to describe the 'linear-through-zero range'. Here detector sensitivity is plotted as a function of amount and, for linear chromatographic systems, is expected to be constant within a specified variation, usually $\pm$ 5% (IUPAC, 1998). A linearity plot is illustrated in Figure 5.3b, where sensitivity is expressed as the normalized response factor and is plotted as a function of Chl *a* amount. In this illustration, normalized response factors were determined by first calculating the response factor (amount of Chl *a* injected divided by the resulting peak area) for each point within a multitude of different, multi-point Chl *a* calibration curves analysed by Laboratories A and HPL, including three analysed by Laboratory A and multiple calibration curves (not discriminated) analysed by HPL, all with a single method on the same HPLC over nine years. With HPL data, the observed response factors are normalized to the average response factor calculated from all data points from all calibration curves, spanning 0.55 to 546 ng. With Laboratory A data, the individual response factors within each calibration curve are normalized to a unique average response factor for each calibration curve. Also, average response factors of Laboratory A were limited to amounts injected between 19 to 263 ng. Normalization allows a realization of relative change in detector response as a function of amount injected across methods and instruments.

These data illustrate a linear response is attained by the HPL system and a nonlinear response is attained on the system of Laboratory A. With the results of Laboratory A, sensitivity changes with amount injected and the normalized response factor varies from 0.7 to 1.4 over the range of Chl *a* amounts anticipated in field sample extracts (2–181 ng). The data of HPL yield a linear response, all within a variation of $\pm$ 5% for amounts injected between 5 and 870 ng – a range that brackets the amounts expected in natural samples (6–600 ng). The data of HPL also reveal how amounts less than 5 ng exhibit large variations in normalized response factors ($\pm$ 25%) and an emerging negative bias – the normalized response factor is most often less than 1.0. Furthermore, the non-instrumental LOQ, computed by methods referred to in Section 5.2.2.2, is approximately 3 ng. These data should make evident the utility of a linearity plot and non-instrumental LOQ when choosing volumes for extraction, filtration and injection, so that amounts injected in sample extracts yield peak areas within regions of optimal linearity and detectability.

5.2.2.4 Calibration

Once the working range(s) and anticipated amounts in samples have been identified, the accuracy and precision of calibration over those ranges is evaluated. Calibration *repeatability* and *reproducibility* precision should both be evaluated during validation, for which the former describes replicate injections of calibration standards all performed on the same day, and the latter describes uncertainties associated with all steps in the calibration process, such as spectrophotometric analysis of stock standard solutions, dilutions, etc. as performed over multiple days. Calibration precision is assessed during validation with at least three independently formulated calibration curves. Calibration accuracy describes how exactly the calibration factors can measure the concentration of a pigment in an unknown solution. Many choices are made when selecting calibration procedures – all of which can affect accuracy. Such choices include (but are not limited to) the mode of calibration, e.g. whether single-point calibration is adequate or whether multi-point linear regression is needed, whether the *y*-intercept should be computed or forced through zero, which absorption coefficients are used for determining concentrations of stock solutions, and which statistical tools should be adopted for characterizing the quality of the calibration, keeping in mind that such criteria should be easily implemented by pigment analysts in a collective effort to establish standards of performance (as with performance metrics, Table 5.2).

Calibration should be considered in the context of previous achievements. For example, calibration repeatability precision averaging 1% and calibration reproducibility precision less than 5% are both attainable for many pigments (Van Heukelem *et al.*, 2002; Hooker *et al.*, 2005). Reproducibility precision for Chl *a* calibration is frequently better than for other pigments, and averages of 1.5% with 95% confidence levels of 3.2% over multiple columns and years have been cited (Hooker *et al.*, 2005). Calibration accuracy has been assessed in round robins with distribution of calibration standards (prepared by reference laboratories) that were quantified as unknowns by recipient laboratories using their pre-existing calibration factors and unique procedures. When reference and recipient laboratories used the same absorption coefficients to determine concentrations of pigment standards, calibration accuracy among quality-assured laboratories averaged 4.7% across six different pigments (Hooker *et al.*, 2005). Calibration accuracy for Chl *a* averaged 2.6% among all laboratories in SeaHARRE-2 and 1.9% for half of the laboratories in SIMBIOS (those with calibration repeatability to within 1%).

Poorly implemented calibration procedures degraded calibration accuracy and precision of some laboratories during SeaHARRE and SIMBIOS round robins and, because these problems are considered avoidable, the affected results were not included in the calibration achievements just cited. These problems included, but were not limited to: poor peak shape; analysing standards spectrophotometrically while still frozen or cold; using vial and cap combinations that allowed evaporation or prevented gas exchange during the withdrawal of viscous solutions (thereby producing inaccurate autoinjector draw volumes); using spectrophotometers with bandwidths greater than 2 nm or solutions yielding absorbance values outside the limits suggested for best accuracy; using uncalibrated pipettes; using single-point response factors when large y-intercepts were present; quantifying standard unknowns that produced peak areas beyond the upper limit of what had been described as yielding a linear response; poor analysis precision; and not accounting for chromatographic purity of standards when calculating response factors. Many of these problems could have been avoided if laboratories had used recommended procedures, such as those found in Mantoura and Repeta (1997), Bidigare *et al.* (2003 and 2005), and Hooker *et al.* (2005).

Estimates of calibration accuracy described here assume absorption coefficients are correct – a rather tenuous assumption given that most absorption coefficients were determined before the availability of analysis methods critical to determination of chemical purity. Other problems with absorption coefficients exist – not all laboratories use the same absorption coefficients, which introduces unwanted (and substantial) divergence in results (Hooker *et al.*, 2005). Furthermore, many commonly used absorption coefficients utilize solvents for standard dissolution that are different from those most commonly used for extracting natural samples – acetone and methanol. The often stronger solvents used with standards can cause peak-shape distortion of early eluting pigments if the injection conditions and gradient are not sufficiently rugged to accommodate the wide range in injection solvent polarities and viscosities (Zapata and Garrido, 1991; Castells and Castells, 1998). Evaporating the solvent in which pigment standards are dissolved (with re-dissolution in the solvent used for samples) may seem like a reasonable solution to this problem, but such actions may cause pigment degradation and erroneous conclusions regarding differences in observed response factors. Until well-validated absorption coefficients are available in the necessary solvents *and* they are adopted by all laboratories, a truly definitive assessment of calibration accuracy is not possible.

In most instances, HPLC pigment methods and HPLCs yield linear responses over a wide range of concentrations. If linearity is acceptable *and* the y-intercept is negligible and represents an amount well below the lowest anticipated in sample extracts, single-point calibration is appropriate. Data from the Chl *a* calibration curves of HPL (Figure 5.3b) support the use of single-point calibration because the response factor variation is all within a narrow range ($\pm$ 5% and 1% on average) *over the amounts anticipated in natural samples* (6 to 600 ng). Also supporting single-point calibration as suitable, amounts represented by the y-intercepts are always < 0.65 ng (and 0.03 ng on average) *and* a non-instrumental LOQ is 3 ng – *both* are below the lowest point of the working range, 5 ng. Furthermore, the quality of individual linear regressions over amounts encompassing the working range is consistent with state-of-the-art quantitation – the average of the absolute per cent residuals of a calibration curve ($|\bar{\psi}|_{\text{res}}$, Table 5.2), is 1.0% on average across 32 calibration curves.

Statistical tools documenting appropriateness of single-point calibration by HPL are used with data from Laboratory A to elucidate an appropriate mode of calibration for this atypical, nonlinear system. A calibration curve with 16 calibration levels and amounts that range from 1.6 to 263 ng is selected for discussion because the endpoints bracket the anticipated Chl *a* amounts in samples (1.8 to 181 ng) and calibration points are very close together, a necessary requirement for interpolation between data points when calibration is nonlinear (Snyder and Kirkland, 1979). Initial inspection of the linear regression equation using all 16 calibration levels reveals a large *y*-intercept, which represents 8.9 ng, the relative per cent residuals range from −242 to 554%, and the average of the absolute per cent residuals is 86%. Large per cent residuals, coupled with observations that sensitivity varies with amount injected (Figure 5.3b), indicate that a *y*-intercept forced through zero and single-point calibration are inappropriate choices. Note that the coefficient of determination (0.989 in this instance) is not an indicator of calibration quality and should not be used as such.

An alternative mode of calibration, using a so-called *continuous piecewise linear fit*, described by Hastie *et al.* (2001), is applied to the nonlinear calibration data (presented using open circles in Figure 5.3b). In this example, with this type of calibration, the calibrant concentrations are divided into three separate segments: 1.64 to 12.91 ng, 19.17 to 59.84 ng and 85.86 to 263.3 ng. Linear regression equations are computed for the data points within each segment. The intersections of the segments, at 19.07 and 76.67 ng, characterize the points at which adjacent linear regressions yield the same amounts injected for a given peak area and are used to connect the separate fits into a continuous function. Thus, linear regressions for the first, second and third segments are used for peak areas corresponding to amounts injected from 1.64 up to 19.07 ng, greater than 19.07 up to 76.67, and greater than 76.67 up to 263.31 ng, respectively (Figure 5.4). The performance statistics for the three linear regimes produce individual per cent residuals that are always to within 5%. The corresponding averages of the absolute per cent residuals are 1.8, 2.7 and 1.2% (from the lowest to highest concentration segments, respectively). Unlike performance statistics described in the preceding paragraph for these calibration data, the performance metrics based on a continuous piecewise linear fit indicate Chl *a* quantitation will be of sufficient quality for remote-sensing applications (see Table 5.2) and a satisfactory solution has been attained for a nonlinear system.

With a holistic approach to uncertainty analysis, the magnitudes of uncertainties from multiple sources are known and random variations in calibration factors are not likely to be misinterpreted as *real changes* in calibration response factors. Nonlinear calibration should be rechecked frequently over the entire range of amounts, while linear modes of calibration require only frequent checks with single-point measurements (Snyder and Kirkland, 1979). As an external check on calibration uncertainty (outside the context of a round robin), analysts are encouraged to purchase pigment standards in solid form (e.g. from companies such as Sigma or

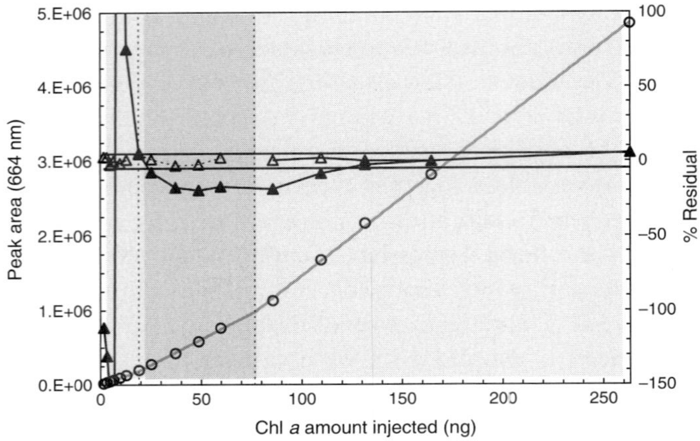

Figure 5.4. Chl *a* peak areas (left *y*-axis, open circles) and per cent residuals (right *y*-axis, triangles) as a function of Chl *a* amount injected. Per cent residuals are determined from a *linear regression not forced through zero* (shown with solid triangles) or from a *continuous piecewise linear fit* (open triangles). Note: for the latter, calibrants are grouped into segments demarcated by grey shading and vertical dashed lines, and all residuals are within $\pm$ 5%.

Carotenature), to prepare them with the same absorption coefficients used by vendors selling them already in solution, and to compare calibration factors determined from standards for various vendors. This effort may seem unnecessary, but knowledge pertaining to calibration variance supports discriminating expectations of method performance, can reduce the frequency of recalibrations and improve the likelihood of satisfying accuracy objectives.

5.2.2.5 Accuracy and precision

The purpose of accuracy assessment during method validation is to prove that the method is suitable for the intended application. Overall accuracy in the analysis of field samples derives from multiple uncertainty sources, as described by variables in the calculation equations (Appendix 5A) and includes volumetric terms (filtration, extraction and injection), calibration and peak areas. Uncertainties from sample collection (e.g. accuracy of filtration volumes, filter homogeneity, etc.) are not addressed here, but during method validation, should be minimized so that uncertainties from other calculation variables can be properly discerned.

These are no standard reference materials available to simulate field sample filters, so accuracy assessments with field samples requires participation in a round robin, which can, in fact, be initiated by individual laboratories and does not require a formalized activity, such as those sponsored by NASA. Participating laboratories should, however, qualify method performance through submission of performance metrics (Hooker *et al.*, 2005), otherwise the accuracy of reference concentrations may be compromised by results from laboratories using methods with questionable capabilities.

Spiked recoveries (Clesceri *et al.*, 1998) evaluate the accuracy of the analysis method and should be performed when validating a new HPLC pigment method, or when changes have occurred. Spiked recoveries are performed by preparing a solution that contains known amounts of pigments and the internal standard (if used) in the solvent typically added to filters for extraction, adding this solution to extraction tubes containing either field sample filters or blank filters and processing them through all the steps used for extraction and analysis of samples. The observed amounts of pigment per tube, as quantified by HPLC, are compared to the known amounts of pigments added to each tube, and the per cent recovery computed (Clesceri *et al.*, 1998; Hooker *et al.*, 2009).

Typically, spiked blank filters evaluate the accuracy associated with extraction plus analysis, and spiked sample filters do the same, but in the presence of the sample matrix (spiked recoveries do not evaluate extraction efficiency). Spiked recoveries are also used to evaluate utilization of an internal standard to accurately assess extraction volume (Hooker *et al.*, 2009). Spiked recoveries are not commonly conducted by NASA round-robin participants, although some data are available (Bidigare *et al.*, 2005; Hooker *et al.*, 2009). For example, average recovery observed by one laboratory for Chl c_1, Fuco, Allo, and Chl *a* was 97% from spiked blank filters and 99% from spiked sample filters, with a range of 93 to 102% for individual observations (Hooker *et al.*, 2009).

Accuracy with field samples, which relies on a consensus of agreement among laboratories, is sensitive to differences in extraction procedures, but in NASA round robins such variations contributed less to differences in results between laboratories than uncertainties due to improperly executed procedures (e.g. calibration). During the SIMBIOS round robin, HPLC extraction procedures differing primarily with respect to extraction volumes, disruption mode and soak times were evaluated with replicate filters from eutrophic, mesotrophic and oligotrophic samples. Results among the procedures differed on average by 2.8–8.1%, depending on procedure, when absolute per cent differences between Chl *a* values were computed (Van Heukelem *et al.*, 2002). The only SeaHARRE laboratory to use methanol instead of acetone as the extraction solvent produced results primarily within the 15% accuracy objectives in SeaHARRE round robins (Hooker *et al.*, 2000; Hooker *et al.*, 2009, 2010), except during SeaHARRE-2 (Hooker *et al.*, 2005) when they did not achieve optimal method performance metrics (unrelated to extraction solvent). Other extraction solvents have not been evaluated in NASA round robins simply because they were not used. An all too common source of uncertainty was inaccurate computation of extraction volumes (Van Heukelem *et al.*, 2002; Hooker *et al.*, 2005) and the use of an internal standard by some laboratories even though it degraded rather than improved analysis precision.

Inter-method comparisons can facilitate understanding of TChl *a* accuracy if the uncertainties of the methods being compared are known. In the SIMBIOS round robin fluorometric and HPLC analyses were performed on the same sample extract, and, during SeaHARRE-2, HPLC extracts were also evaluated spectrophotometrically using the trichromatic equation of Jeffrey and Humphrey (1975). Spectrophotometric and fluorometric analyses are not as selective for Chl *a* as HPLC analyses, so inter-method comparisons of results are not expected to be equivalent (Humphrey and Jeffrey, 1997). On average, HPLC TChl *a* concentrations were within −6% (with a 95% confidence limit of ± 9%) of fluorometric Chl *a* concentrations (Van Heukelem *et al.*, 2002). In SeaHARRE-2, the spectrophotometric Chl *a* to HPLC TChl *a* ratios averaged 1.2 (Hooker *et al.*, 2005). The procedure of standard addition, which, in this instance, is based on adding Chl *a* standard to HPLC sample extracts, with subsequent analysis of the recovery using the methods being compared, enhances the utility of inter-method comparisons. While underutilized, standard addition was evaluated with the trichromatic equation of Jeffrey and Humphrey (1975) during SeaHARRE-2 (data not reported) and recovery was approximately 100%.

In the context of method validation, repeatability precision is evaluated over a short time interval (e.g. multiple injections or multiple readings on a spectrophotometer, all executed on the same day). Reproducibility precision is evaluated over longer duration and includes precision associated with repetitive executions of a set of procedures, such as recalibrations. Repeatability precision might describe, for example, results from replicate filters analysed on the same day, and reproducibility precision would describe the results from replicate filters analysed over the course of several months. Both are important during method validation.

Overall method precision is described by the percentage coefficient of variation (CV%) of pigment results in replicate filters, which during SeaHARRE-2 averaged 5.4 and 3.6%, for PPig and TChl *a*, respectively, for laboratories producing quality-assured results (i.e. those achieving TChl *a* and PPig accuracy objectives). However, overall method precision is affected by many sources beyond the control of the HPLC analyst (e.g. collection, transport and storage of filters). It is possible, however, as achieved in SeaHARRE-2, to constrain these non-HPLC-related sources of uncertainty to low levels approximating 2%. Also, the contribution of HPLC imprecision to overall method precision requires documentation of such, which is most easily evaluated from replicate injections of the same sample extract at intervals that describe the minimum and maximum times a sample extract resides in the autosampler compartment.

5.2.2.6 Ruggedness

Ruggedness, as adapted from EURACHEM (1998), refers to how consistently a method performs when small changes in the environment or operating conditions occur. It is well known that the differences in strength (described by the polarity index) and viscosity of injection solvents, relative to initial mobile phases, can affect peak retention and shape, primarily of early-eluting pigments (Zapata and Garrido, 1991; Castells and Castells, 1998; Latasa *et al.*, 2001). A method should be developed to yield good peak shapes with all injection solvents to be used, including those in which the various available standards are suspended (e.g. ethanol and 90% acetone). During SeaHARRE-2, laboratories that had not adequately developed injection conditions for stronger solvents produced results that were inaccurate up to 30% and imprecise, with CV% as high as 12%, when they analysed standards in ethanol or 90% acetone.

In some instances, analysts have encountered difficulties when purchasing solvents and buffering agents not of HPLC grade quality or purchased from vendors that differ from those specified in the original method citations. Also, some vial and cap combinations will allow evaporation or, conversely, will not allow accurate volumes to be drawn from vials containing viscous solutions. Vial and cap combinations should, therefore, be tested for their ability to prevent evaporation and vials containing viscous solutions, such as buffers, should utilize pre-scored septa which permit gas exchange during solution withdrawal. This simple change improved injection precision of one SeaHARRE laboratory from 4 to 1.4% CV (Hooker *et al.*, 2009).

A multitude of other problems associated with ruggedness can be cited, many of which are related to adaptation of injection conditions from one method to another. Consequently, it is necessary for any analyst implementing a new method to investigate the sensitivity of all variables in the calculation equation to small changes in laboratory practices or hardware capabilities. The assumption cannot be made that because a method works well in one laboratory, it will easily be implemented in another. Differing gradient dwell volumes and adaption of a method developed with injectors capable of combining sample and buffering agent in the sample loop to injectors that only have a capability of mixing in a separate vial can require much effort to reproduce, and in some instances, such adaptations may not be possible.

5.3 Results from inter-laboratory comparisons

The initial objective of the SeaHARRE activity was simply to understand and estimate the uncertainty in HPLC pigment concentrations over the dynamic range of the remote-sensing problem, which spans three decades in TChl *a* concentration (0.05–50.0 mg m^{-3}). Because of this perspective, the emphasis was placed on analysing replicate natural samples spanning approximately two orders of magnitude, so that the necessarily small number of investigative opportunities would have maximum applicability to the world ocean as new round robins were executed (understanding that, at some level, this can never be adequately accomplished). Over time, being able to measure the performance of a method emerged as an equally important objective, because successive round robins revealed that analysts did not always have the tools to recognize when a method's performance was degrading. On more than one occasion, a method with an established proper QA and QC capability (hereafter referred to as being part of the QA subset), experienced poorer QA and QC capabilities in a subsequent round robin (sometimes without the analyst knowing it), with their results being demoted from the QA subset. As part of investigating how QA and QC criteria influence short- and long-term method performance, sample sets that were not part of the original distribution to the participants were analysed separately (in some cases as a reanalysis by the demoted laboratory).

From a program management perspective, performance metrics are a powerful product of investigation into uncertainties, because they have the potential for removing the burden of maintaining an overly diverse set of protocols (that have to be continually updated) or agreeing on a single protocol that satisfies the current suite of community problems. Community priorities will necessarily evolve, and at times rather rapidly, so it is appealing to be able to set performance metrics for each individual problem rather than revising one or more approved methods for each. The metrics can be applied to any candidate methodology, and provide all the evaluation criteria needed to determine whether or not it is suitable for the application. The evaluation of performance metrics – in particular the corresponding precision and accuracy parameters – resulted in some participants abandoning the methods they were using for a single (more modern) method with superior performance parameters: the Van Heukelem and Thomas (2001) method. Maximum diversity in methods was achieved during SeaHARRE-2 (four C_8 and four C_{18} methods), and the least during SeaHARRE-3 (six C_8 and two C_{18} methods).

A reduction in method diversity over time was not expected. Indeed, early discussions within the group of SeaHARRE participants suggested that performance metrics might be used as a motivation for investigating how best to improve a method with lesser performance capabilities. There was a strong feeling that selecting one method and recommending it as the most acceptable – as was done, for example, in the JGOFS program (JGOFS, 1994) – should not be repeated, because it stifles creativity and hinders advancement of the state of the art. The practical benefit of adopting a proven performer over investing an unknown amount of time and resource into trying to improve a method, however, was simply too alluring. The potential pitfall of this approach, which is quantified here in many aspects, is underestimating the difficulty of implementing a new method with all its attendant detail.

5.3.1 Precision

The average precision for the different methods across all three SeaHARRE activities is rather similar (Figure 5.5). Within each SeaHARRE activity, differences are partitioned between pigment types (or categories) and whether or not a method was part of the QA subset. The best average precision was associated with TChl *a*, and in all cases average precision for the QA subset was superior. The importance of this was well demonstrated during SeaHARRE-2 when a laboratory that was part of the QA subset also analysed an unequivocally damaged set of samples – the samples had been defrosted during shipping and arrived at room temperature. For the PPig

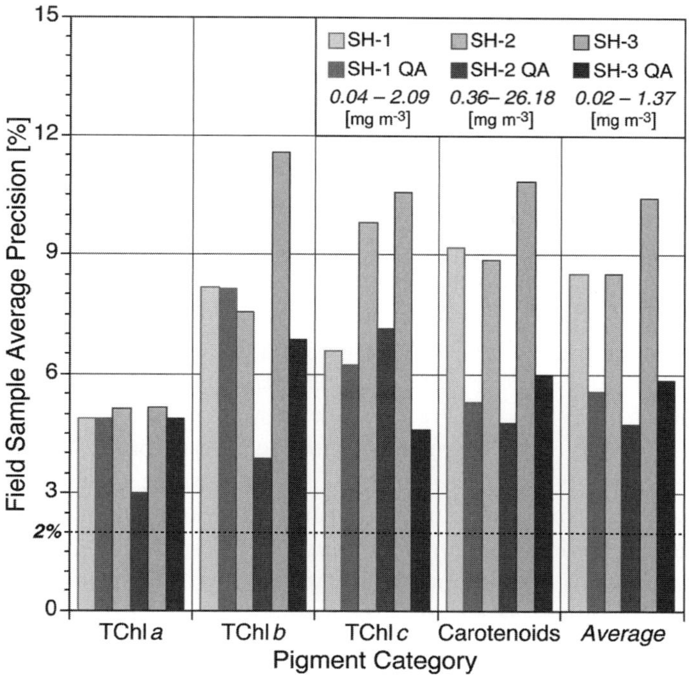

Figure 5.5. The average precision based on field samples for the laboratories participating in the first three SeaHARRE activities – which span a range in [TChl a] of 0.02–26.185 mg m^{-3} – and whether the methods were part of the QA subset (dark bars) or not (light bars). The range in [TChl a] for each individual SeaHARRE (SH) round robin is also shown. The pigment categories cover all the primary pigments and are organized as the separate total chlorophylls and the remaining nine carotenoids. The QA subset for each round robin is shown in the darker bar, and all the other methods in the corresponding lighter bars. The dotted line denotes the average 2% variability in filter homogeneity, which has been a recurring feature of the field sampling. See colour plate section.

results, the precision of the damaged samples was always better than the precision from the methods that were not part of the QA subset, except for TChl c. In fact, the average PPig precision (which includes TChl c) for the better laboratory analysing worse samples was 6.6%, whereas the corresponding average for the laboratories that were not part of the QA subset was 8.5%.

Some of the variability in precision based on natural samples arises from the data collection protocol used in the field, which was estimated to be approximately 2% for each SeaHARRE activity. This is an important distinction of round-robin sampling, because a) very significant effort is made during the field work to ensure that replicate samples (usually triplicates) are of the highest possible quality (one or more test filtrations are done to optimize the filtration volume while permitting all replicates to be filtered in a reasonable amount of time, all water is continuously mixed in a single carboy which is kept covered in black plastic and the selection of replicates for each investigator is randomized), and b) many investigators do not collect enough replicates to routinely estimate the influence of the sampling protocol on precision.

5.3.2 Accuracy

The SeaHARRE-1 results showed the determination of [TChl *a*] averaged to within 8.0% (Claustre *et al.*, 2004), which is well within the remote-sensing requirements. The average SeaHARRE-2 [TChl *a*] uncertainty was 11.4%, but only 7.8% for the QA subset. When the QA subset was used as the proxy (or reference) for truth in the uncertainty calculations, the average SeaHARRE-2 [TChl *a*] uncertainty for the QA methods was 5.9%, and 17.2% for the other methods. Correction of the SeaHARRE-1 data to ensure compliance with the criteria for the QA subset resulted in a similar uncertainty in [TChl *a*] of 5.5%. The importance of using only properly validated results as proxies for truth in the uncertainty calculations was also demonstrated during SeaHARRE-3. For this round robin, using only the QA methods in the proxies for truth yielded an uncertainty in [TChl *a*] of 6.3% for the QA methods and 33.2% for the others. When all of the methods were treated equally – as though they were all properly validated – the uncertainty in [TChl *a*] averaged 14.9%; so the QA subset was significantly degraded by the performance of the other laboratories.

Although the inclusion of methods without a proper QA and QC capability significantly degrades the TChl *a* estimate, it does not render the data useless for calibration and validation activities, on average. The same is not true for the quantitation of other (individual) primary pigments. The PPig accuracy for the QA subsets averaged 16.8, 16.0 and 12.3% for the first three SeaHARRE activities, respectively, and 16.8, 44.5 and 53.6%, respectively, for the other methods (all based on using the QA subset as the reference in the uncertainty computations). When the QA subset was not used for the referencing system, the average PPig accuracy results were 22.2, 22.9 and 24.9%, respectively. Because all these averages are close to the 25% threshold, a large portion of the individual accuracy estimates must have exceeded the threshold – in fact, almost half (45.1%) did. What this means is that the extra variance from the methods without a proper QA and QC capability is spread across the more capable methods which needlessly – and significantly – degrades their results.

Figure 5.6 summarizes the average accuracies estimated from the SeaHARRE activity and distinguishes three important aspects of pigment uncertainties: a) the robustness of TChl *a*, which has the lowest uncertainty and is the only pigment with an uncertainty that is invariant to the sampling regime and, thus, concentration; b) the QA subset of methods is distinguished by a significant decrease in uncertainties, which are almost always to within the 15% refinement objective (on average); and c) there is a functional decrease in uncertainties for the progression from

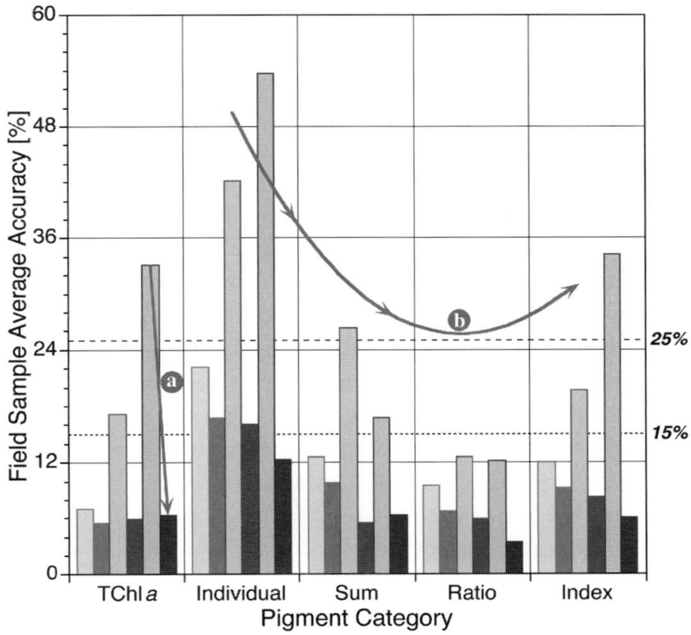

Figure 5.6. The average accuracy based on field samples for the laboratories participating in the first three SeaHARRE activities. The colour and coding scheme is patterned after Figure 5.5, but with the 25% remote-sensing requirement shown as a dashed line and the 15% algorithm refinement requirement shown as a dotted line. The pigment categories span TChl *a*, the individual pigments within PPig, and the higher-order variables: pigment sums, ratios and indices. The two grey curves denote a) a significant decrease in uncertainties associated with the QA subset, and b) a functional decrease in uncertainties for the progression from individual pigments to sums and ratios, followed by a small increase with the indices. See colour plate section.

individual pigments to sums and ratios, followed by a small increase in indices (which are a combination of sums and ratios). The functional form is seen for all round robins and for both data types (QA subset or not), and in almost all cases the uncertainty in pigment ratios is approximately equal to or less than the uncertainty in [TChl *a*]. Consequently, an accessible and effective way to mitigate the degradation in data quality from improper QA and QC procedures is to use the individual data to form higher-order variables (Claustre *et al.*, 2004).

The functional form of the uncertainties in the higher-order variables is reproduced in both types of data, the QA *and* non-QA subsets, but the amplitude of the function – that is the amount of change in accuracy within the progression from individual pigments to sums, ratios and indices – is larger with data from methods lacking a properly validated QA capability. Absence of the functional form establishes the possibility of an important failing in execution of a method. During SeaHARRE-3, one of the laboratories changed from a C_{18} to a C_8 method to chromatographically separate monovinyl and divinyl Chl *a*. In spite of a substantial effort, the new method could not be properly validated and a persistent and abnormal variability in the calibrations could not be explained. Because of the problems associated with getting the C_8 method operational, a second (duplicate) HPLC system was used to permit simultaneous analysis of the SeaHARRE-3 samples with both the new C_8 and the old C_{18} methods.

Implementation of the C_{18} method was carried out over a compressed time period, and this method was ultimately determined to be improperly validated. For the three total chlorophylls (*a*, *b* and *c*) and the nine carotenoids within PPig, the average C_8 method accuracies were 52.4 and 18.0%, respectively; the corresponding values for the C_{18} method were 15.0 and 44.6%, respectively. The anomalously bad chlorophyll results with the C_8 method resulted in poor average accuracies for the higher-order variables, which always exceeded 15% (whereas for the C_{18} method they did not), but more importantly, the C_8 uncertainties did not follow the functional form (and the C_{18} uncertainties did). Ultimately, it was the absence of the functional form that led to productive investigation as to why this laboratory could not produce results that satisfied accuracy objectives.

5.3.3 Validity of the round-robin approach

The validity of the referencing system used to compute accuracy (and, thus, uncertainties) can be investigated using standards, because there is an independent source of truth from the distributor of the reference material. Solitary pigments do not replicate the chromatographic complexity of natural samples, however, but mixed standards – a single solution containing a variety of standards all mixed together in known concentrations – are more similar. Validation occurs by sending the participants an unknown mixture and then comparing the uncertainties in the pigment concentrations from the various methods by using two different referencing systems: a) the known concentrations within the mix, and b) the average pigment concentrations derived from all of the methods.

This kind of experiment was conducted during SeaHARRE-2 and the results are shown in Figure 5.7. The two types of uncertainties are similar, distribute well along the one-to-one line, and the uncertainties from these two approaches agree for individual pigments to within 6.5% (except for one pigment) and to within 1.5% on average. In addition, ranking of the pigments from lowest to highest average uncertainty is very similar for the two referencing systems with almost equal numbers of points above and below the line, so there is no evidence of an overall bias. This indicates that the two independent approaches for estimating method uncertainties produce comparable results, which is a substantial validation of the round-robin approach: using a proxy for truth – referencing the accuracy calculations with respect to the overall pigment averages – provides the same basic result as using a laboratory standard, for which referencing is based on known concentrations.

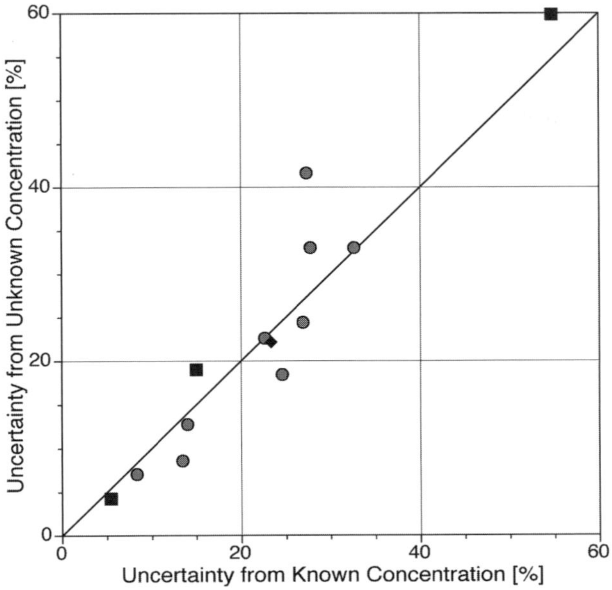

Figure 5.7. A comparison of the PPig individual uncertainties associated with using mixed standards as known versus unknown samples. The three grey squares denote the chlorophylls and the nine pale grey circles the carotenoids; the black diamond is the average for the two data sets.

A problem with the mixed standard approach is the difficulty of synthesizing a natural sample in the laboratory; with few exceptions, an individual laboratory will achieve lower uncertainties with a laboratory sample than with a natural sample. Another validation opportunity can be constructed by taking the best of the field sample results, that is, the best accuracy for each pigment from each method in the QA subset, to form a hypothetical best method. The accuracy of the hypothetical method can then be compared to the results achieved by the superior method in the QA subset when analysing laboratory standards. If this is done with the SeaHARRE-2 data set, the hypothetical method has an overall accuracy of 11.4% with field samples, which is very close to the 9.3% accuracy of the superior method (with the mixed standard) in the QA subset. This observation is consistent with other findings in SeaHARRE activities: the complexities of procedures associated with HPLC analysis (e.g. calibration, peak identification and reporting practices) frequently yield uncertainties far in excess of those associated with properly executed procedures not related to HPLC analysis (e.g. sample collection).

The convergence between a theoretical method and an established method demonstrates closure in the statistical description and is particularly important in defining realistic HPLC performance metrics. Another form of convergence that is a significant validation of the data analysis approach is the similarity in accuracies for the QA subsets across all three SeaHARRE activities. The average accuracies for each round robin are to within approximately 4.5% of one another across the entire range of data products (and more than three decades in trophic status). Additional evidence is provided by the increasing convergence of accuracy with method precision as the QA process is made more robust. For example, the accuracy for TChl *a* is very nearly the same as its precision.

5.4 Performance metrics

For marine pigment concentrations, the ocean colour community met part of the performance-based burden by agreeing on an accuracy metric for Chl *a* concentration, but there was no consensus for any other pigment or criteria other than accuracy. Consequently, the SeaHARRE participants arbitrarily adopted the Chl *a* metric for all data products, and developed a set of performance criteria for all the pigments, which are presented in Table 5.2 as an example of what form an approach based on performance metrics might take. The four different category labels were selected for convenience, and simply provide a scale of capabilities. In some cases the performance score may coincide with one of the chosen categories, like 'semiquantitative', but in other instances there may be reasons to develop a separate category (e.g. a 'validation' category). The term 'validation' was purposely not used to describe performance categories in the work presented here, because the validation, use and application of HPLC methods is more extensive than the narrower ocean colour (marine pigment) perspective adopted for the SeaHARRE activities.

Each category in a performance metric is assigned a weight and score, so the overall capability of a method is based on summing the applicable weights for each performance parameter, dividing by the number of parameters, and comparing the result to the category scores. This process permits any method to be evaluated against a) another method that is already properly validated, and b) the stated requirement for the type of work being pursued. For example, if calibration and validation activities require 'quantitative' data, then a method with an overall score of 2.5 or more, would be suitable for the task. The classification could also be recorded when data are submitted to a database, so future users could use only those data in keeping with their research objectives. In other words, if only 'state-of-the-art' data are applicable, then the database can be searched for only this quality of data.

As long as there is some range in performance thresholds and these are set so most methods qualify for the middle portion, the use of metrics allows analysts to understand which criteria associated with their individual methods need to be enhanced to advance the overall capability of their method. In some cases, this will be a rather straightforward exercise in discovering which procedures can be improved by using more accurate components or techniques; in other cases, the development of new approaches might be needed to overcome long-standing limitations. The latter represents new research that might not occur in the absence of a performance requirement.

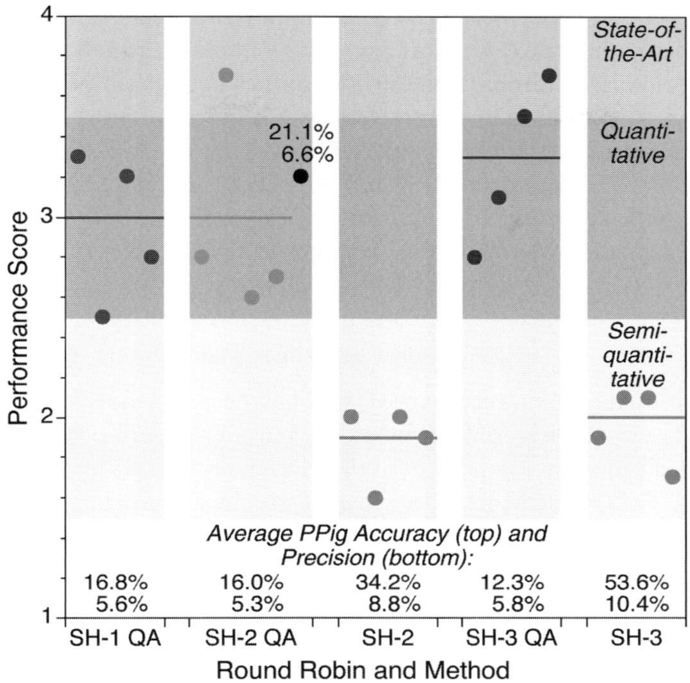

Figure 5.8. The average performance scores for laboratories participating in the first three SeaHARRE activities. The yellow, green and blue regions denote the semiquantitative, quantitative and state-of-the-art categories, respectively. The round robins are distinguished by whether or not the methods are grouped in the QA subset. Averages for the groups are shown as a solid line in the same colour as the group results (solid circles), except for the analysis of a defrosted set of samples during SeaHARRE-2 by a member of the QA subset, which is shown in black. The paired sets of numbers are the average accuracy (top number) and precision (bottom number) of the group results. See colour plate section.

Because performance metrics provide a quantitative assessment of quality, they can be used to establish what constitutes a properly validated capability within each subcategory (e.g. calibration) or across an entire method (e.g. the HPL method). Establishing individual parameters and scores is by necessity quantitative, but the detail of the underlying work remains hidden at the scoring level. Consequently, the step-by-step best practice, which is frequently presented in a protocol, is not available in a strictly performance-based approach. This might not be considered a significant loss of information for experienced analysts, but for new practitioners, it represents an important reason for maintaining detailed protocols. Protocols, uncertainty budgets and performance metrics represent a natural progression, the utility of which – should it be questioned – is shown in Figure 5.8. This figure summarizes the net performance score for all the methods participating in SeaHARRE activities with a further distinction in the QA subset. The corresponding average PPig accuracies and precisions make an unequivocal case for the importance of a carefully implemented QAP: on average, they are about a factor of 2.9 and 1.7 times better, respectively, than the corresponding results from the laboratories not in the QA subset. The most dramatic evidence for this is the aforementioned analysis of an unequivocally damaged set of samples by a SeaHARRE-2 laboratory that was part of the QA subset (the black circle). The data for these results are nonetheless at a 'quantitative' level of performance with an average PPig accuracy and precision that is better than the corresponding results from the laboratories not in the QA subset.

5.5 Quality assurance plan

Thus far, the quality assurance plan (QAP) has been discussed in the context of method validation and uncertainty reduction, but a QAP also collectively describes method-related procedures important to these. Such procedures include both QC and QA activities. The QC elements are associated with measurements performed during the analysis of samples and allow documentation of method performance. The QA elements describe how values attained for QC measurements are used to develop quantified limits of expected performance, and the processes undertaken to ascertain that a method is operating within those limits. Examples of QC and QA activities given here were implemented by HPL during NASA round robins and during analysis of thousands of samples for a multitude of investigators.

These particular examples are cited simply because they are accessible, have supported attainment of accuracy objectives in NASA round robins and represent a so-called holistic approach to uncertainty awareness.

5.5.1 QC measurements and QA applications

Taylor (1987) divides QC measurements into categories considered *external* and *internal*. Inter-laboratory comparisons with field samples (e.g. the SeaHARRE activities) are external and necessary to expose vulnerabilities not recognized when working in isolation. Internal QC measurements are more frequently implemented and evaluate whether the method can continually produce data that satisfies predefined accuracy objectives. Temporal monitoring of many QC results can yield *warning limits* (WL) and *control limits* (CL), the 95 and 99% confidence limits, respectively, of expected performance and thus provide a foundation for quality assessment. Recommendations vary regarding the use of such limits, but in general, if a number of consecutive points (e.g. 2 or 3) exceed the WL the method is considered potentially 'out of control' and investigations into the aberrant QC measurements should be investigated. If one point exceeds the CL, the method is out of control, analyses should be stopped and investigations into the abnormal performance initiated (adapted from Taylor, 1987; Clesceri *et al.*, 1998). Temporal

QC measurements should monitor a method's performance regarding repeatability *and* reproducibility precision, as described in Section 5.2.2.5. In the pharmaceutical industry, such quality assessment procedures are collectively described by the term *system suitability testing*, which establishes test parameters for the particular procedure being implemented and facilitates attainment of accuracy objectives by demonstrating that the system – in this case, a chromatographic system – is operating within expected tolerances (USDHHS-FDA, 1994; ICH, 2005).

Quality control measurements need to address all variables in the calculation equations (Appendix 5A), although many QC measurements simultaneously pertain to multiple variables. The ensuing subsections each describe the implementation of a QC measurement, how often it is made and the resulting QA application(s). This information is summarized in Table 5.3. Some QC measurements are performed physically, and others are processes applied to data already collected (e.g. inspection of pigment ratios). Readers should review these procedures as useful examples, and then develop QC and QA procedures pertinent to their unique situations (remembering to capture the uncertainties of all variables in their calculation equations).

5.5.1.1 Pigment resolution and retention time precision (daily)

An algal mixture containing all (or nearly all) pigments to be quantified, which must include critical pairs, is injected once per day during the analysis of samples, or more frequently if a new batch of solvent is added during the day. In this instance, a batch of solvent refers to a mobile phase formulated by the analyst on the lab bench from more than one solvent (e.g. methanol and water), and for which slight variations between formulations can cause retention time shifts greater than those caused by other environmental sources for an HPLC equipped with column temperature control.

This mix is analysed at the beginning of a sequence of analyses (all conducted on the same day) to document column performance ($R_S > 1.0$ between critical pairs) and to update retention times in the calibration table, which are expected to shift slightly if the solvent batch is changed, and less than 0.07% on average per day with no solvent change (Hooker *et al.*, 2005). As the column ages, late-eluting pigments may exhibit later retention times, so the elution position of ββ-Car must be documented as being within the pump and detector 'stop-times'.

5.5.1.2 Analysis precision and carryover (daily)

Batches of sample filters are extracted such that their analyses on the HPLC can be completed within approximately 24 h. Calculation of the results according to Eq. 5A.7 (Appendix 5A) utilizes the term $\hat{A}_{c1}$, which is the average peak area of the internal standard (ISTD) in the solution added to the sample filters for extraction. Replicate (at least duplicate) analyses of the ISTD solution, which are performed on the same day the ISTD is used in the analysis of the filter samples, are used to estimate $\hat{A}_{c1}$.

Table 5.3. *System suitability testing at HPL. Sections describe measurements specified; measurand is the material measured (*Any standard *is any primary or secondary pigment); Freq. is frequency of measurement (*D *is daily,* S *is per sequence,* Y *is yearly,* Ad *is analyst's discretion); parameters are* RPD *(relative per cent difference),* APD *(absolute per cent difference),* CV *(per cent coefficient of variation),* R_S *(resolution), with average value of measured parameter,* WL *and* CL *(95 and 99% confidence limits, respectively) in current use at HPL. Some values in* CL *column equal a threshold used in quality assessment – denoted by* t. *QC measurements describe uncertainties with all variables within an analyst's control in governing HPL equations (Appendix 5A, Eqs. 5A.1, 5A.6 and 5A.7).*

Section	Measurement	Typical Measurand	Freq	Parameter[a]				Variable
				Parameter	Average	WL	CL	
5.5.1.1	Resolution	Viola/Hex, Zea/Lut	D	R_S	–	–	<1.0 (t)	$\hat{A}_{Pi}$
5.5.1.2	Carryover	Chl c_2, Chl *a*, Caro	D	APD	–	–	≤0.1(t) [b]	
5.5.1.3	Calibration	Chl *a* standard	D	RPD	–	± 4.1	± 5.3	R_{Pi}
5.5.1.7	accuracy	Any standard	S			± 5		
5.5.1.7	Inter-laboratory calibration accuracy	Any standard	Y	APD	2	0.1 to 6 (t)		
5.5.1.8	Calibration linearity	Chl *a* standard	Ad	APD [c]	0.9	2.1	2.7	
5.5.1.6	Re-pipette calibration	Acetone	D	CV	0.3	0.5	0.8	V_m
5.5.1.2	ISTD precision	Vitamin E	D	CV	0.4	0.8	1.3	$\hat{A}_{c1}$, V_c
5.5.1.10	$Vx1$ vs. Vx'	All samples	D	APD	2	< 5		
5.2.2.5	Spiked recoveries	Any standard	Ad	% recovery	97	92 to 105		$\tilde{C}_{Pi}$
5.5.1.4	Pigment extract	TChl *a*	D	CV	0.4	1.6	2.2	
	analysis precision	PPig [d]		CV	1.8	4.7	6.2	
5.5.1.5	Method precision [e]	TChl *a*	S	CV	4	1–10		C_{Pi}
	(field samples)	PPig		CV	7	2–11		

[a] Values variable; [b] specific to carryover from sample extracts, not standards; [c] APD equals average of the absolute per cent residuals for a Chl *a* calibration curve representing amounts within the working range; [d] pigments detectable in one injection, but not the other, are excluded; [e] precision of duplicate filters varies among investigators.

Both injection precision and carryover are assessed with internal standard data described above. Horn Point Laboratory average injection precision approximates 0.4% CV, with 95 and 99% confidence limits of 0.8 and 1.3%, respectively. So if the CV in the replicate injections exceeds 1.3% or three sequential injections exceed 0.8%, remedial action is taken to find the cause of abnormal performance. Because the ISTD solution contains only vitamin E acetate (Fluka product no. 95250), the presence of any pigments in the chromatogram is evidence of carryover. An ISTD injection programmed to immediately follow analysis of a concentrated standard or sample is most effective for evaluating carryover, the threshold of which is 0.1% for samples. Carryover of the ISTD is periodically evaluated by inspecting chromatograms of solitary standards that immediately follow analysis of samples or the ISTD alone. Carryover can be caused with the HPL method by an aged column (e.g. an aged column produced 2.8% carryover for Phe *a*, which dropped to 0.03% with column replacement). The high concentrations of the HPL ISTD (about 3700 ng inj^{-1} versus typically less than 500 ng inj^{-1} for pigments) make it a sensitive indicator of injection valve and column performance, because the CV of $\hat{A}_{c1}$ often worsens before problems with pigments in samples are discernible.

5.5.1.3 Chl a *calibration accuracy (daily)*

A Chl *a* standard solution in 90% acetone is formulated quantitatively with calibrated dilution devices in volumes of sufficient quantity (e.g. 20 mL) to provide several injections and is stored at approximately −15 °C in amber glassware proven to prevent evaporation. The solution is used until expended, or until Chl *a* degradation occurs. The QC solution is injected once per day, and often immediately following the ISTD to evaluate ISTD carryover. The [Chl *a*] approximates midpoints within the Chl *a* working range.

To evaluate the accuracy of the Chl *a* calibration, the amount of Chl *a* in the QC injection is quantified using the current Chl *a* response factor and the relative per cent difference (RPD) between the observed and formulated amounts are expected to agree to within 4.1% (95% confidence limits). Any RPD values greater than 5.3% indicate that the method is *out of control*, with the most common cause being degradation of the QC standard. Newly prepared QC standards that are still out of control require investigation into causes, the most common of which are injection system malfunction or column age. Abnormal Chl *a* QC injections often coincide with other abnormal QC measurements.

5.5.1.4 Sample extract analysis precision (daily)

The extract of the first filter extracted in a daily batch of samples is placed into two HPLC vials, which are both placed in the autosampler compartment at the same time. One vial is the first to be analysed in the batch of sample extracts analysed on that day and the other vial is the last sample extract analysed on that day. These duplicate analyses bracket the minimum and maximum times the vials reside in the autosampler compartment before analysis.

The CV of the temporally separated duplicate injections sets an upper limit on the expected precision associated with the analysis of field sample extracts. The average CV, WL and CL for TChl *a* are 0.4, 1.6 and 2.2%, respectively. If all PPig data are included, the average CV and associated WL and CL are 3.2, 12.0 and 17.0%, respectively and if four of the 52 replicate injections used to compute these results are excluded because some pigments (notably Allo, Diato, Peri and But-fuco) were in low concentrations and had been reported as present in one injection, but not both, average CV, WL and CL improved to 1.8, 4.7 and 6.2%, respectively. These PPig data expose how damaging pigments with low SNRs are to estimates of analytical precision.

Precision with duplicate analyses conducted as such do not affect decisions regarding hardware maintenance or column changes, because other QC measurements are more sensitive (presumably because minor inaccuracies with an injection volume or draw volume, for example, are compensated for by the ISTD within each sample extract). These measurements are used primarily to defend analytical precision in the event that the precision of duplicate filters is poor in the particular batch being analysed, and as criteria affecting QC decisions (i.e. whether samples analysed during conditions where other QC measurements have failed need to be reanalysed).

5.5.1.5 Method precision (each sample batch or at least 5% of samples)

Duplicate filters should be provided with all samples submitted for analysis, so that they can be processed with each batch of samples (the minimum currently required of NASA investigators is 5% of the total). The analysis of duplicate filters describes overall method precision and is affected by such things as the field sampling protocol, handling, transport, storage, extraction and HPLC analysis. During routine analysis of samples for nine investigators (and 183 duplicate filter sets), average CV for each contributor ranged from 1–10% and 2–11%, for TChl *a* and PPig, respectively, with cumulative averages of 4 and 7%, respectively. As a quality assessment tool, performance expectations pertaining to precision of results with duplicate filters needs to be qualified (e.g. documented for each contributor), because many of the uncertainty components are outside the control of the HPLC analyst. In SeaHARRE activities, imprecision from the field sampling protocol was about 2% (remembering that many *identical* replicates need to be created for a round robin rather than only two for duplicate submissions). This, combined with a PPig field sample precision averaging 4.8% for the SeaHARRE-2 QA subset (Hooker *et al.*, 2005), suggests duplicate filters with a CV on the order of 5% is realistic. Consequently, contributors providing filters for which CV is poorer than 7% for PPig, for example, should investigate how to improve the field sampling protocol as well as the handling and storage conditions being used.

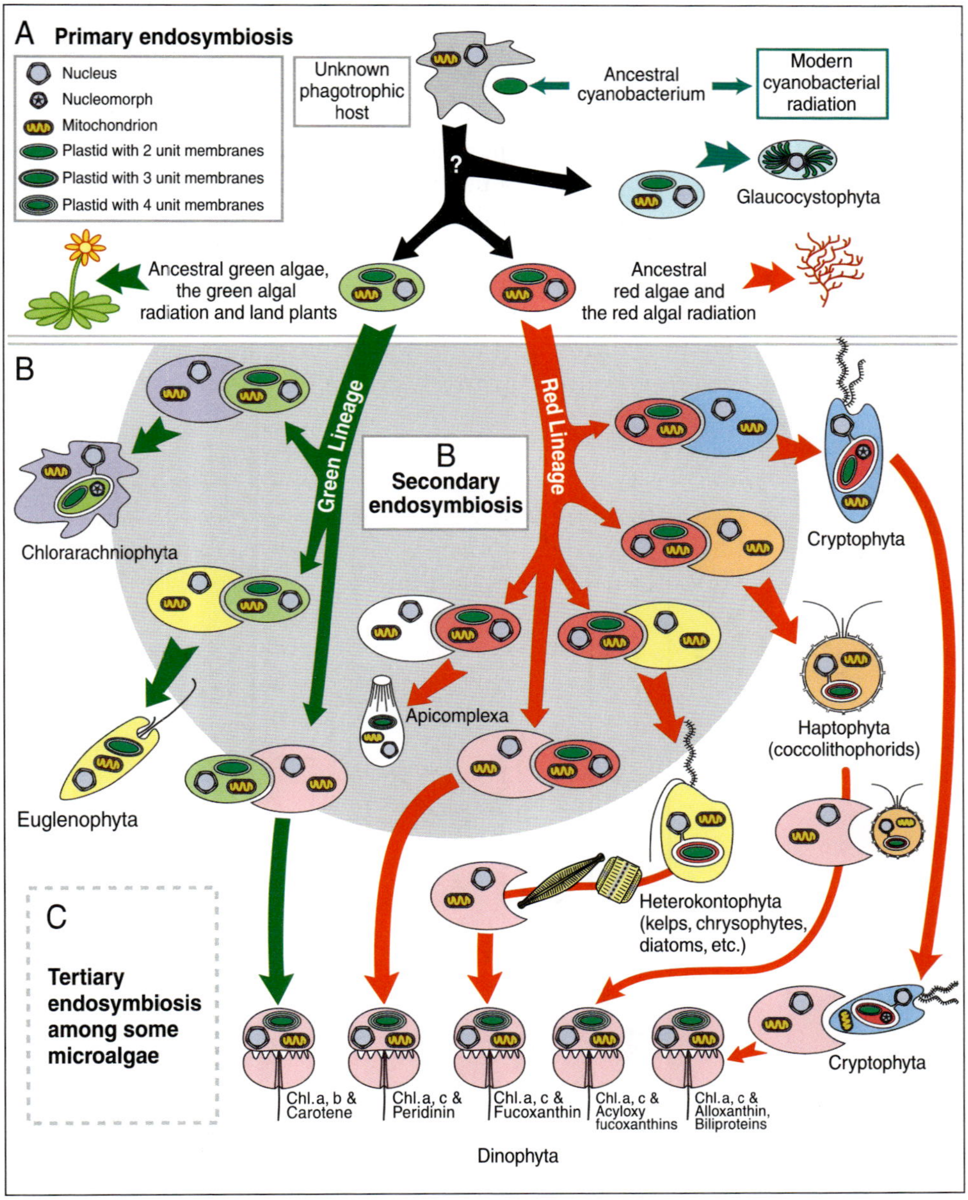

Figure 1.1. Diagram showing the hypothetical evolution of algal plastid diversity via serial endosymbiosis, based on Delwiche (1999), modified to show a common red algal origin of the plastids of apicomplexans, peridinin – containing dinoflagellates and heterokonts (Sanchez-Puerta and Delwiche, 2008; Janouškovec *et al.*, 2010). The evolutionary relationships among cryptophytes, haptophytes, heterokonts and alveolates are still controversial (Sanchez-Puerta and Delwiche, 2008), as are the number of tertiary endosymbioses (Bodył and Moszcyńksi, 2006).

A

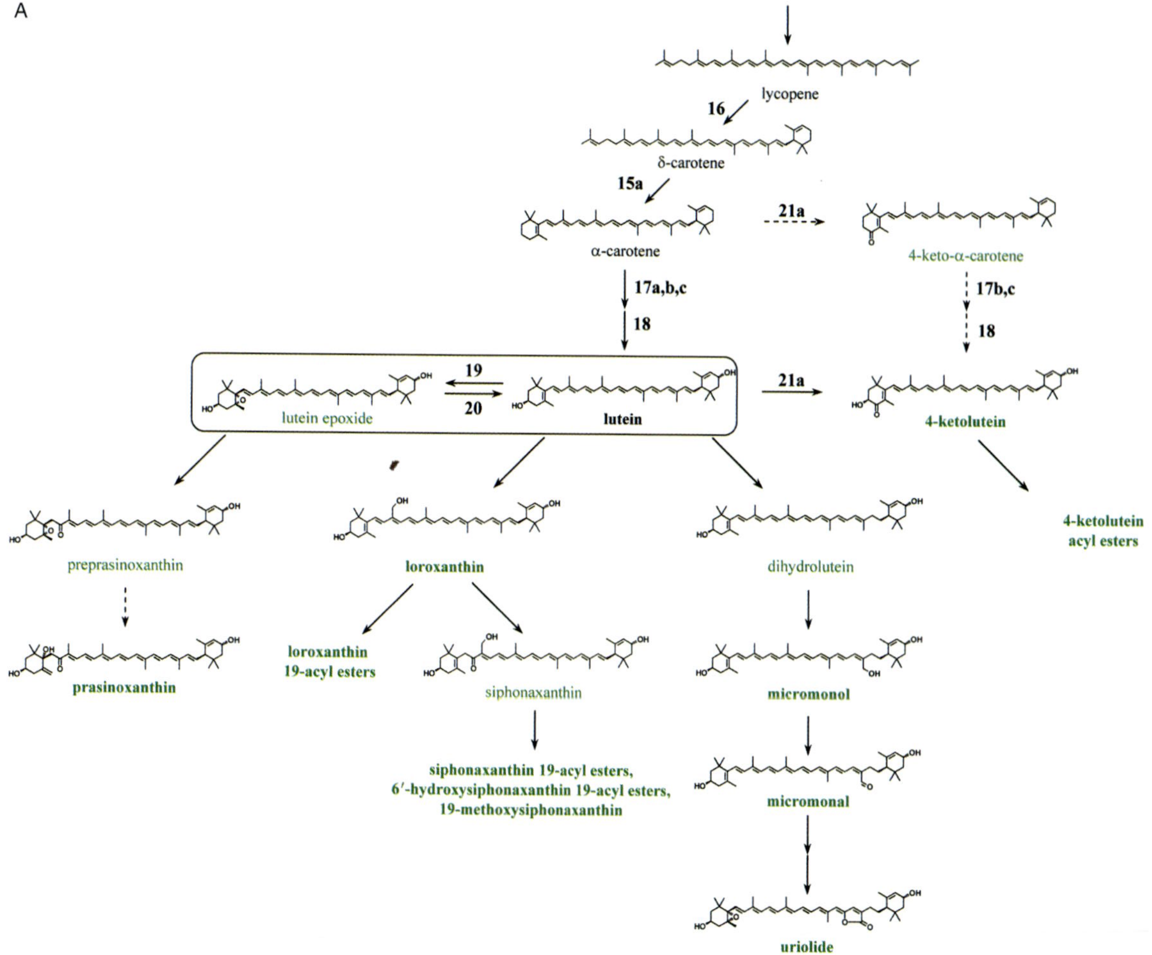

Figure 3.2. (*cont. overleaf*)

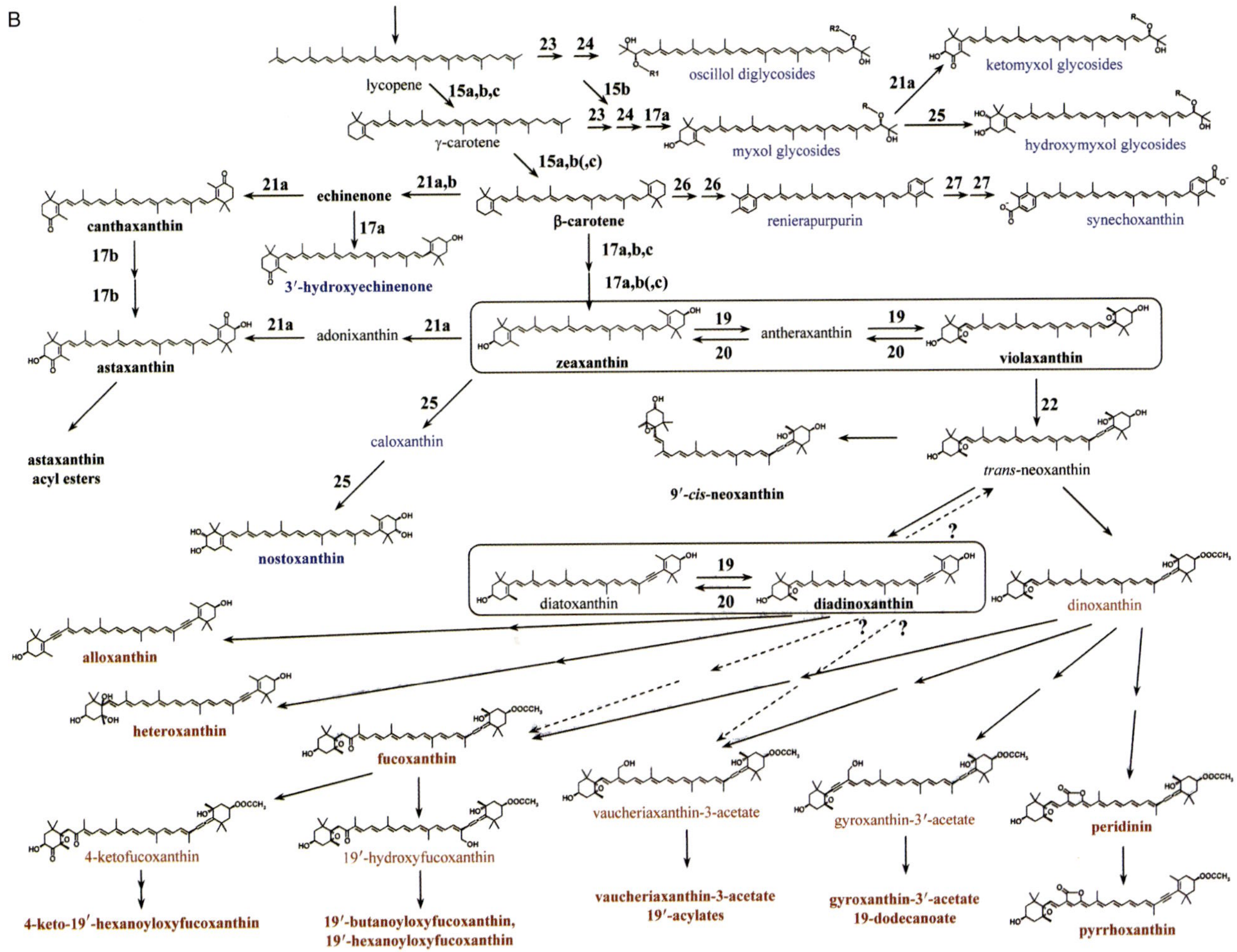

Figure 3.2. Biosynthetic pathways of the major xanthophylls in cyanobacteria and algae. Shown are pathways that are postulated to give rise to xanthophylls which are derived either (A) from α-carotene or (B) from lycopene, γ-carotene and β-carotene. The postulated biosynthetic relations are based on chemosystematic and experimental data as discussed in the text. Names of major products in the pathways are in bold, with carotenoids restricted to green algae in green, to cyanobacteria in blue and to chromalveolates in brown. Xanthophyll cycles are boxed. Dashed arrows indicate potential alternate paths. Numbers denote the reaction steps that are catalysed by the equivalently numbered enzymes in Table 3.1. Chemical structures of pigments are according to the *Carotenoids Handbook* (Britton *et al.*, 2004).

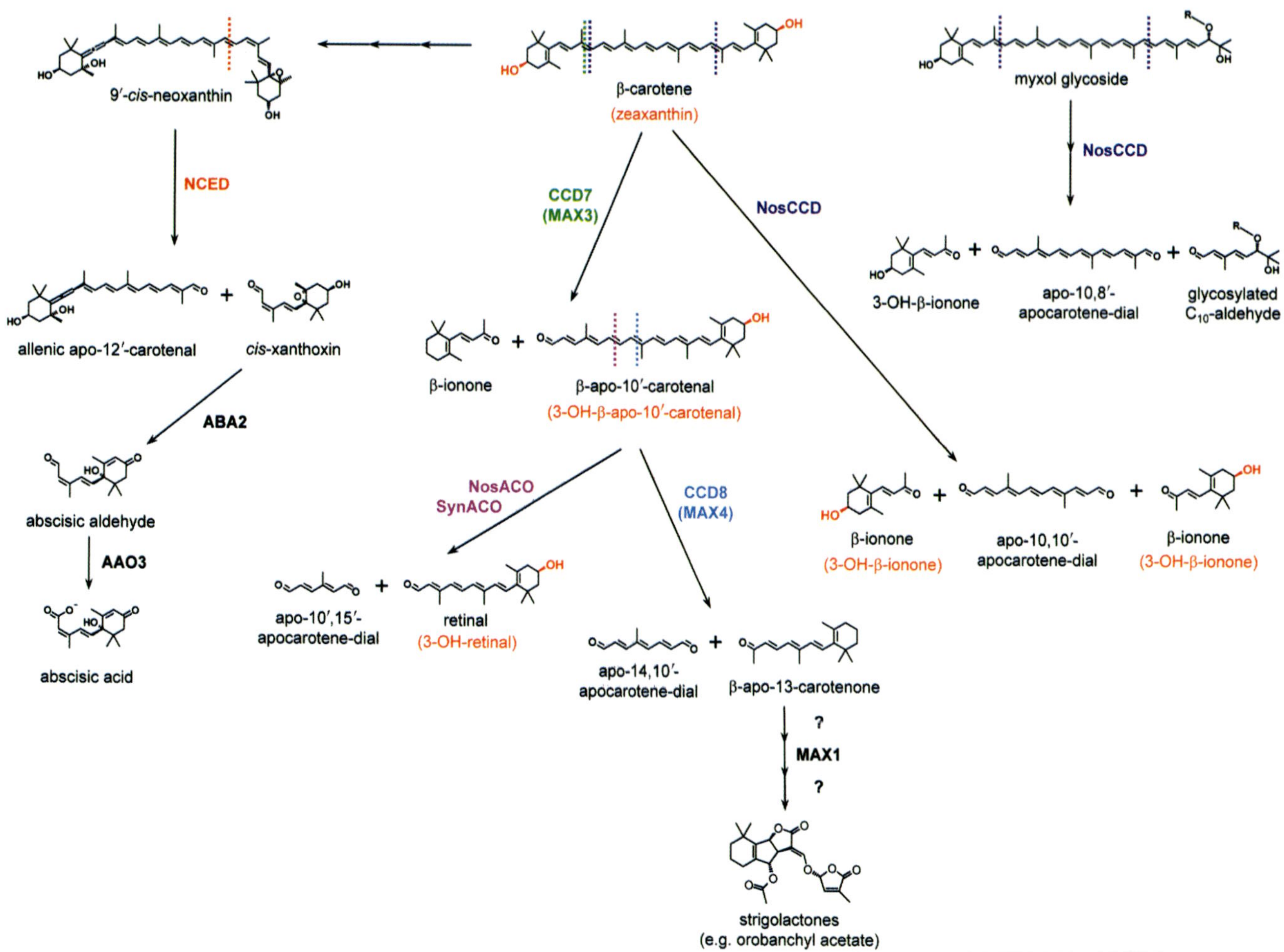

Figure 3.3. Catabolism of carotenoids in land plants, cyanobacteria and algae: formation of apocarotenoids by carotenoid cleavage oxygenases (CCO) and of metabolites derived there from with important physiological functions. Names of CCO enzymes and their corresponding cleavage site(s) have the same colour. For cyanobacterial CCO enzymes, examples of further substrates and the resulting products are in red. See text for further explanations.

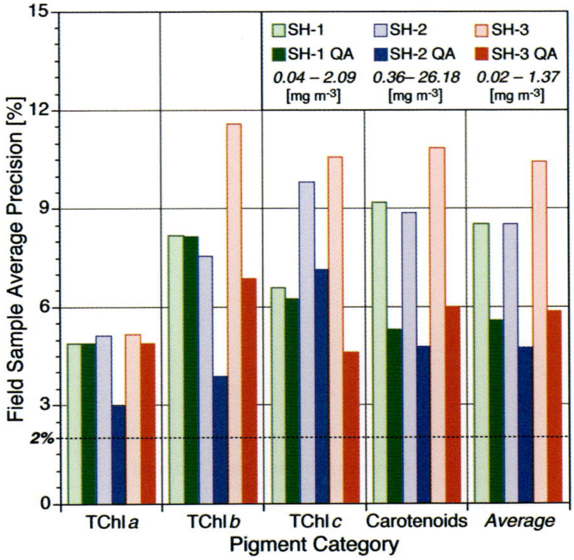

Figure 5.5. The average precision based on field samples for the laboratories participating in the first three SeaHARRE activities – which span a range in [TChl *a*] of 0.02–26.185 mg m^{-3} – and whether the methods were part of the QA subset (dark bars) or not (light bars). The range in [TChl *a*] for each individual SeaHARRE (SH) round robin is also shown. The pigment categories cover all the primary pigments and are organized as the separate total chlorophylls and the remaining nine carotenoids. The QA subset for each round robin is shown in the darker bar, and all the other methods in the corresponding lighter bars. The dotted line denotes the average 2% variability in filter homogeneity, which has been a recurring feature of the field sampling.

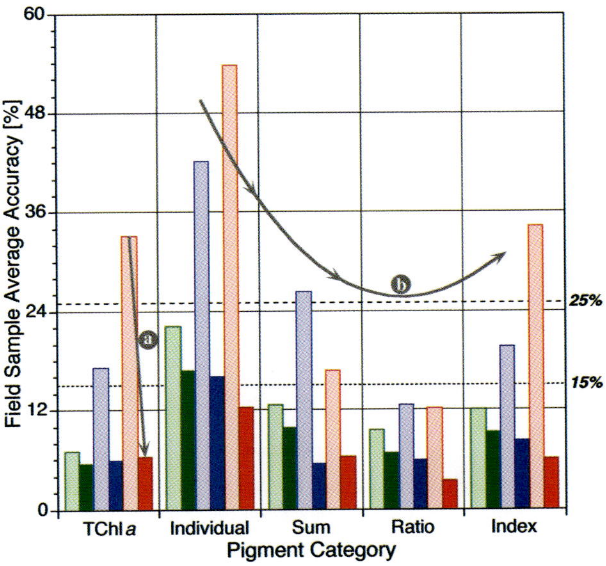

Figure 5.6. The average accuracy based on field samples for the laboratories participating in the first three SeaHARRE activities. The colour and coding scheme is patterned after Figure 5.5, but with the 25% remote-sensing requirement shown as a dashed line and the 15% algorithm refinement requirement shown as a dotted line. The pigment categories span TChl *a*, the individual pigments within PPig, and the higher-order variables: pigment sums, ratios and indices. The two grey curves denote a) a significant decrease in uncertainties associated with the QA subset, and b) a functional decrease in uncertainties for the progression from individual pigments to sums and ratios, followed by a small increase with the indices.

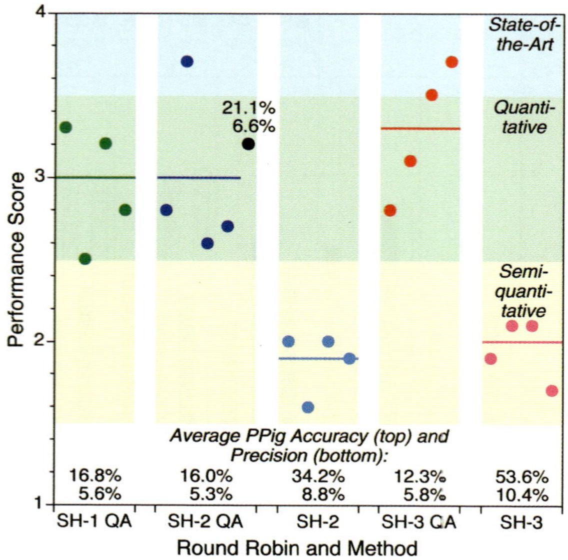

Figure 5.8. The average performance scores for laboratories participating in the first three SeaHARRE activities. The yellow, green and blue regions denote the semiquantitative, quantitative and state-of-the-art categories, respectively. The round robins are distinguished by whether or not the methods are grouped in the QA subset. Averages for the groups are shown as a solid line in the same colour as the group results (solid circles), except for the analysis of a defrosted set of samples during SeaHARRE-2 by a member of the QA subset, which is shown in black. The paired sets of numbers are the average accuracy (top number) and precision (bottom number) of the group results.

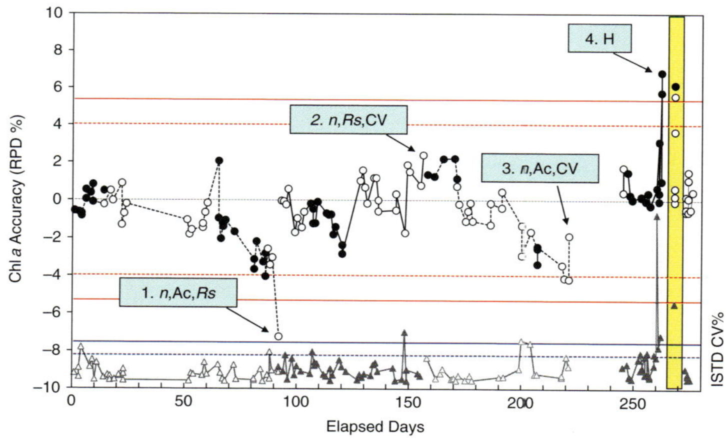

Figure 5.9. Control charts tracking accuracy of Chl *a* QC injections (left *y*-axis) and precision of daily ISTD replicate injections (right *y*-axis) over 275 days. Acceptable performance of Chl *a* QC injections, with alternating open and closed circles depicting different QC formulations, and all symbols connected by lines representing analyses on the same column, is constrained within red dashed lines (the warning limits WL) and solid red lines (the control limits CL). Acceptable ISTD CV%, with alternating open and solid grey triangles to discriminate analyses all performed on the same column, is constrained within the blue dashed line (the WL) and the blue solid line (the CL, which depicts 1.25 CV%). Columns 1, 2 and 3 were removed from service on days indicated by arrow positions for reasons based on multiple criteria, for which n = aged column, Ac = poor Chl *a* QC accuracy, $R_S = R_S$ between critical pairs was ≤ 1.0, CV = ISTD precision had exceeded CL. A hardware failure (H) occurred during analyses on column 4 on day indicated by arrow. Sample processing ceased during days highlighted in yellow while repairs and testing were conducted, after which sample processing on column 4 continued.

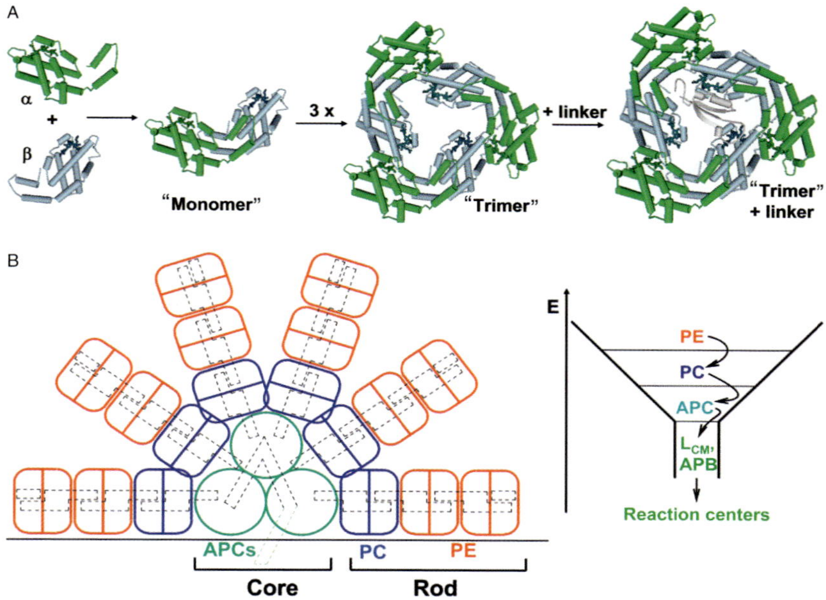

Figure 9.1. (A) Assembly of phycobiliprotein trimer. The model shown is from the only structure showing a phycobiliprotein-linker complex, i.e. APC L_C from *Mastigocladus laminosus* (Reuter *et al.*, 1999). Proteins are displayed schematically (α subunit light green, β-subunit light blue, linker L_C grey), the chromophores as stick models in darker colours. Note the proximity of L_C to two of the three β-chromophores. (B) Structural (left) and energetic funnel-models (right) of a phycobilisome. The structural model is of a hemidiscoidal PBS containing a tricylindrical core composed of APCs (greenish blue), and six rods, each containing a PC hexamer (blue) and two PE hexamers (red). The linkers in the center of the rods and in the core are indicated by the dashed shapes (black). The horizontal solid black line indicates the membrane surface. The same color coding is used in the energetic funnel model, where the vertical error indicates the ^1S excitation energy of the chromophores. Energy is transferred from the higher energy chromophores to the lower energetic ones in the PBS, and finally, via the terminal emitters, L_{CM} and allophycocyanin B (APB), to the reaction centers in the photosynthetic membrane.

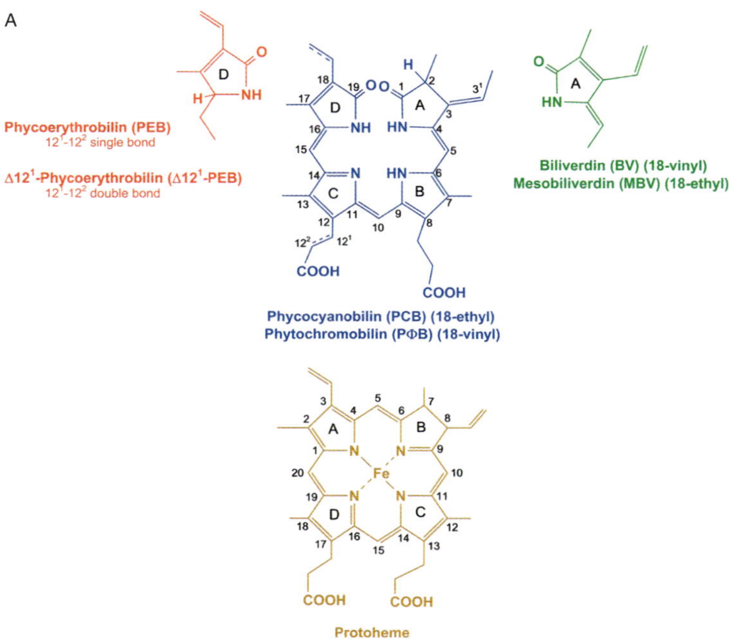

Figure 9.2. Structures of phycobiliprotein chromophores. (A) Free bilins in their thermodynamically most stable cyclic conformation; not shown is the helical character caused by steric hindrance of the terminal imine-oxygens, and protoheme. Note the different ring notation and carbon numbering for cyclic (protoheme) and open-chain tetrapyrroles (BV) as approved by the International Union of Pure and Applied Chemistry (IUPAC, (Moss, 1988) and http://www.chem.qmul.ac.uk/iupac/tetrapyrrole/). PCB and PΦB are shown in full, and only deviating partial structures for the other chromophores. $\Delta 12^1$-PEB and $\Delta 12^1$-DBV are also termed 'bilin 584' and 'bilin 618', respectively, according to the absorption maxima of the protein-bound chromophores in acidic urea (see Wedemayer *et al.*, 1996). Colours approximate those of the respective chromophores.

B

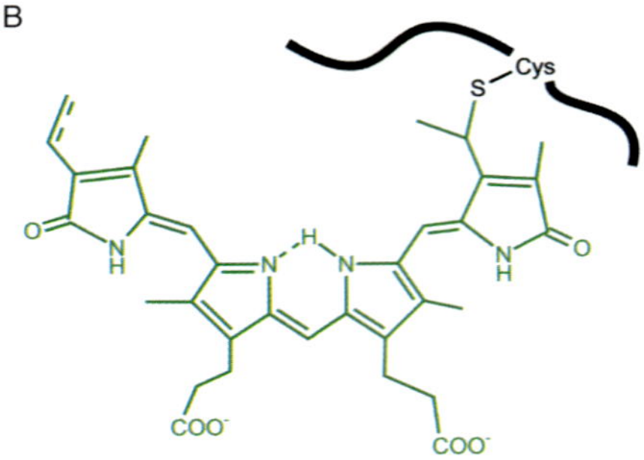

3^1-Cys - biliverdin (BV, vinyl at ring D)
3^1-Cys - mesobiliverdin (MBV, ethyl at ring D)

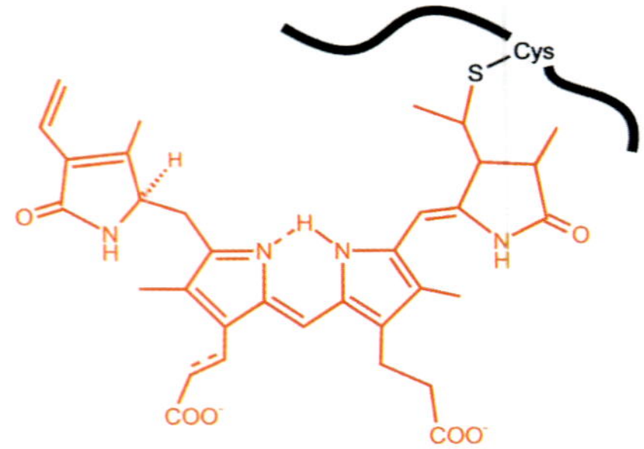

3^1-Cys - 15,16-dihydrobiliverdin (DBV)[a]
3^1-Cys - $\Delta 12^1$-15,16-dihydrobiliverdin ($\Delta 12^1$-DBV)

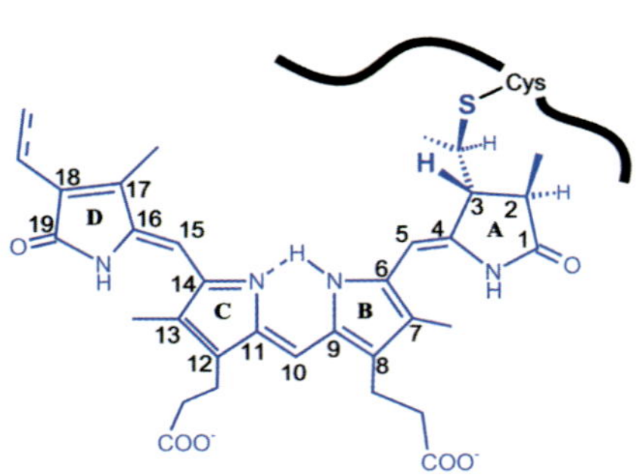

3^1-Cys-phycocyanobilin (PCB; ethyl at ring D)
3^1-Cys-phytochromobilin (PΦB; vinyl at ring D)

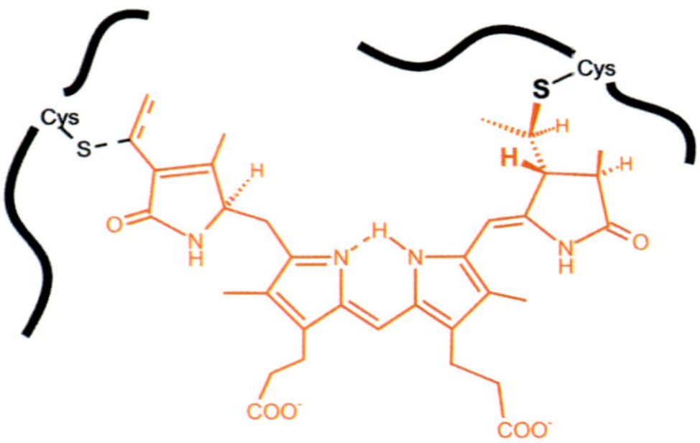

3^1(,18^1-Di)-Cys - phycoerythrobilin (PEB)
3^1(,18^1-Di)-Cys - $\Delta 12^1$-phycoerythrobilin ($\Delta 12^1$-PEB)

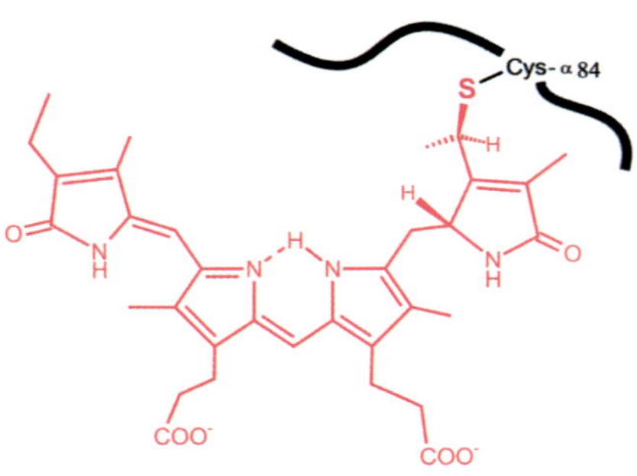

3^1-Cys - phycoviolobilin (PVB) (15-Z)

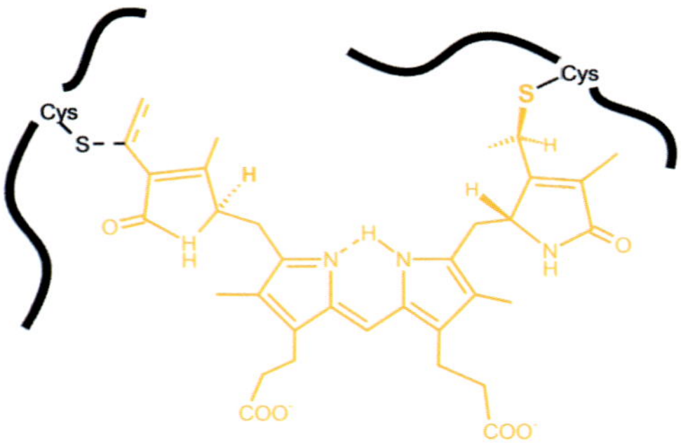

3^1(,18^1-Di)-Cys - phycourobilin (PUB)

Figure 9.2. (*cont.*) (B) Protein-bound chromophores with their thioether bond(s) to the apoprotein, in extended conformations that are typical for native phycobiliproteins. Colours approximate those of the respective chromophores. BV, the common biosynthetic precursor, is the chromophore of bacterial (class II) phytochromes. Phycobilisome-containing organisms (cyanobacteria, glaucocystophytes (cyanelles), red algae, *Galdieria sulphuraria*) have PCB, PEB, PVB and PUB as chromophores; cryptophytes contain also the others. $\Delta 12^1$-PEB and $\Delta 12^1$-DBV are also termed 'bilin 584' and 'bilin 618', respectively, according to the absorption maxima of the protein-bound chromophores in acidic urea. A doubly linked DBV is bound to cysteines at C-3^2 and C-18^1 (see Wedemayer *et al.*, 1996).

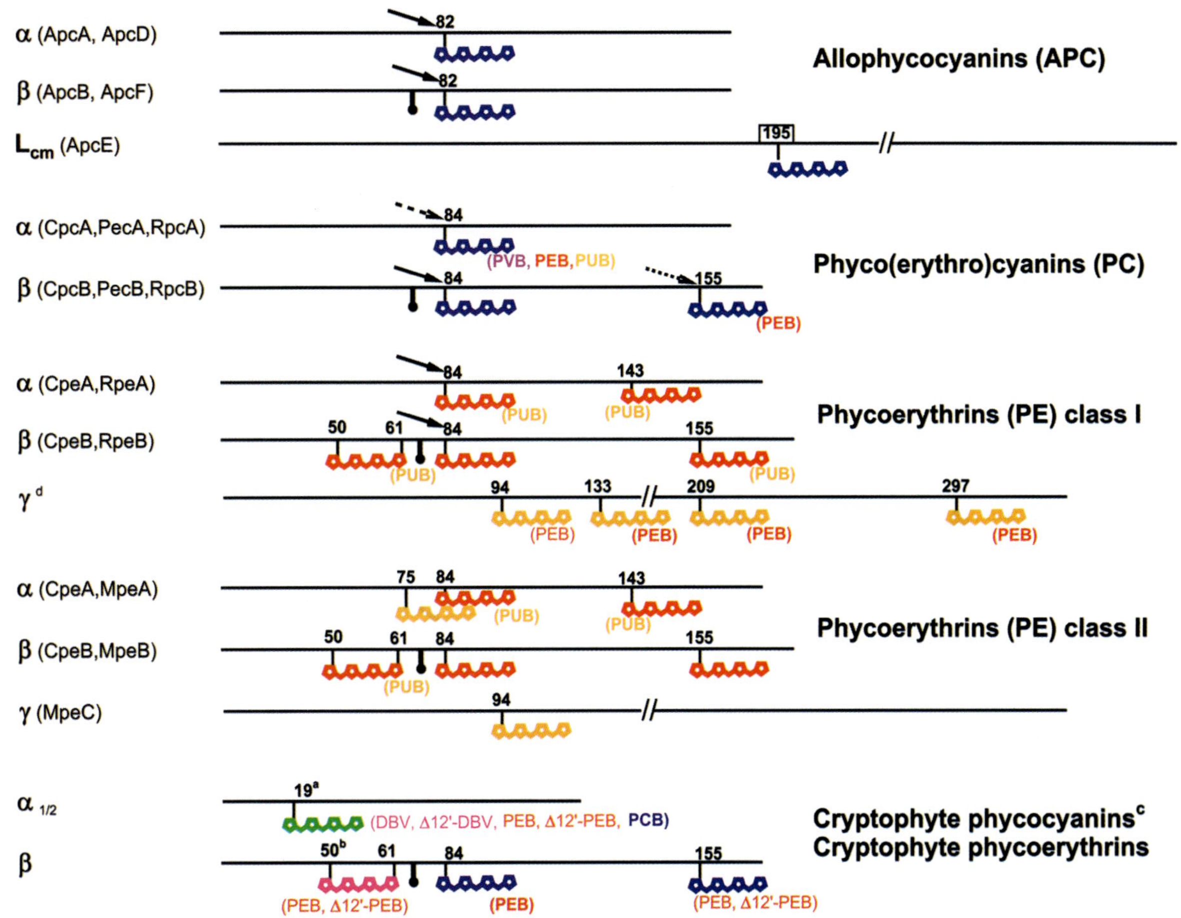

Figure 9.3. Post-translational modifications of biliproteins. Colours are indicative of the prevailing chromophores (blue = PCB, red = PEB, purple = PVB, orange = PUB), alternative chromophores are indicated by their abbreviations (see Figure 9.2). Numbers give the (consensus) position of binding cysteines. Solid arrows denote chromophore attachment by CpcS-type lyases, dotted arrows CpcT-type lyases and dashed arrows CpcE/F-type lyases (see text for details). The boxed number at ApcE indicates autocatalytic attachment. The vertical bars with knobs indicate γ-methylated asparagine-β72. Not shown is a third PE, termed PE III, that has been identified in a high-light *Prochlorococcus marinus* strain; it carries only a single chromophore on the α-subunit, and none on the β-subunit (Hess *et al*., 1996). a) MBV, b) DBV, c) main structure phycocyanin 645, d) only in red-algal b- (and possibly B-) phycoerythrin.

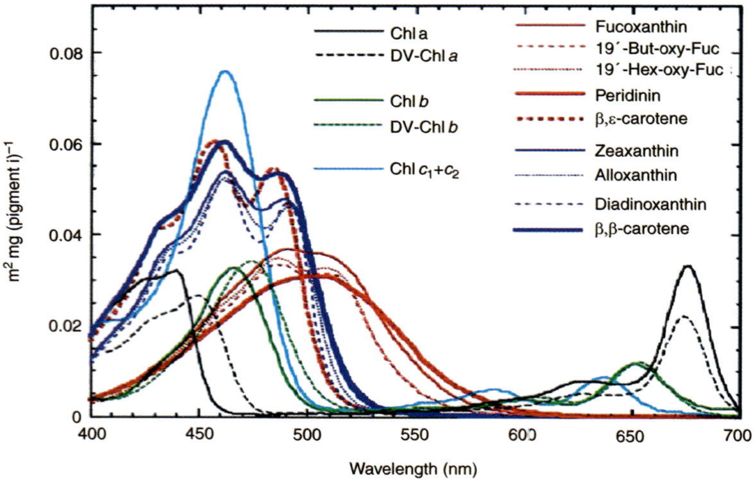

Figure 13.2. Assumed *in vivo* weight-specific absorption spectra of the main pigments, as derived from absorption spectra of individual pigments in solvent (Bricaud *et al.*, 2004).

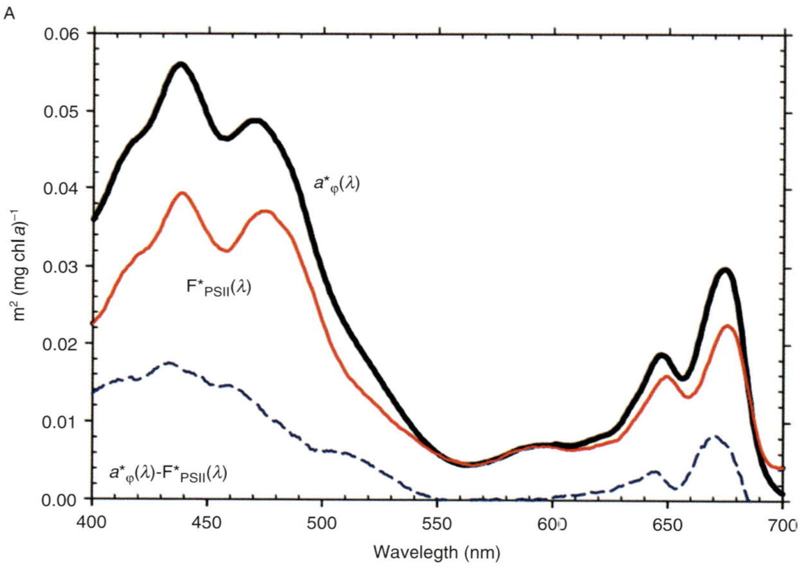

Figure 13.3. (A) Example of PSII-scaling of fluorescence excitation to absorption with average values for LL-acclimated prasinoxanthin-containing prasinophytes (*Bathycoccus prasinos, Micromonas pusilla* and *Pseudoscourfieldia marina*). *In vivo* Chl *a*-specific absorption spectra ($a^*_\varphi(\lambda)$), Chl *a*-specific PSII-scaled fl-ex spectra ($F^*_{PSII}(\lambda)$), and the corresponding difference spectra ($a^*_\varphi(\lambda)$-$F^*_{PSII}(\lambda)$) denoting the non-fluorescent fraction of photoprotective carotenoids and PSI. (B) Bio-optical characteristics of LL-acclimated chromophytes (7 PGs, upper panels), Chl *b*-containing phytoplankton (4 PGs, mid panels), and biliprotein-containing phytoplankton (2 PGs, lower panels). Left panels: $a^*_\varphi(\lambda)$; right panels: $F^*_{PSII}(\lambda)$. Pigment-groups (PG) are: PG 1 Bacillariophyceae (fucoxanthin, Chl c_{1+2}), PG 2 Dinophyceae I (peridinin, Chl c_2), PG 3 Dinophyceae II (acyl-oxy-fucoxanthins, gyroxanthin diester, Chl c_3), PG 4 Coccolithophyceae (acyl-oxy-fucoxanthins, Chl c_3), PG 5 Pavlovophyceae (fucoxanthin, Chl c_{1+2}), PG 6 Prasinophyceae I (prasinoxanthin, [3,8]-proto-chlorophyllide, Chl *b*), PG 7 Prasinophyceae II (lutein, Chl *b*), PG 8 Euglenophyceae (neoxanthin, Chl *b*), PG 9 Chlorophyceae (lutein, Chl *b*), PG 10 Chrysophyceae (fucoxanthin, Chl c_{1+2}), PG 11 Raphidophyceae (violaxanthin, Chl c_{1+2}), PG 12 Cryptophyceae (phycobiliprotein, alloxanthin, Chl c_2), and PG 13 Cyanophyceae (phycobiliproteins, zeaxanthin). (A) and (B) from Johnsen and Sakshaug (2007) with permission from Phycological Society of America.

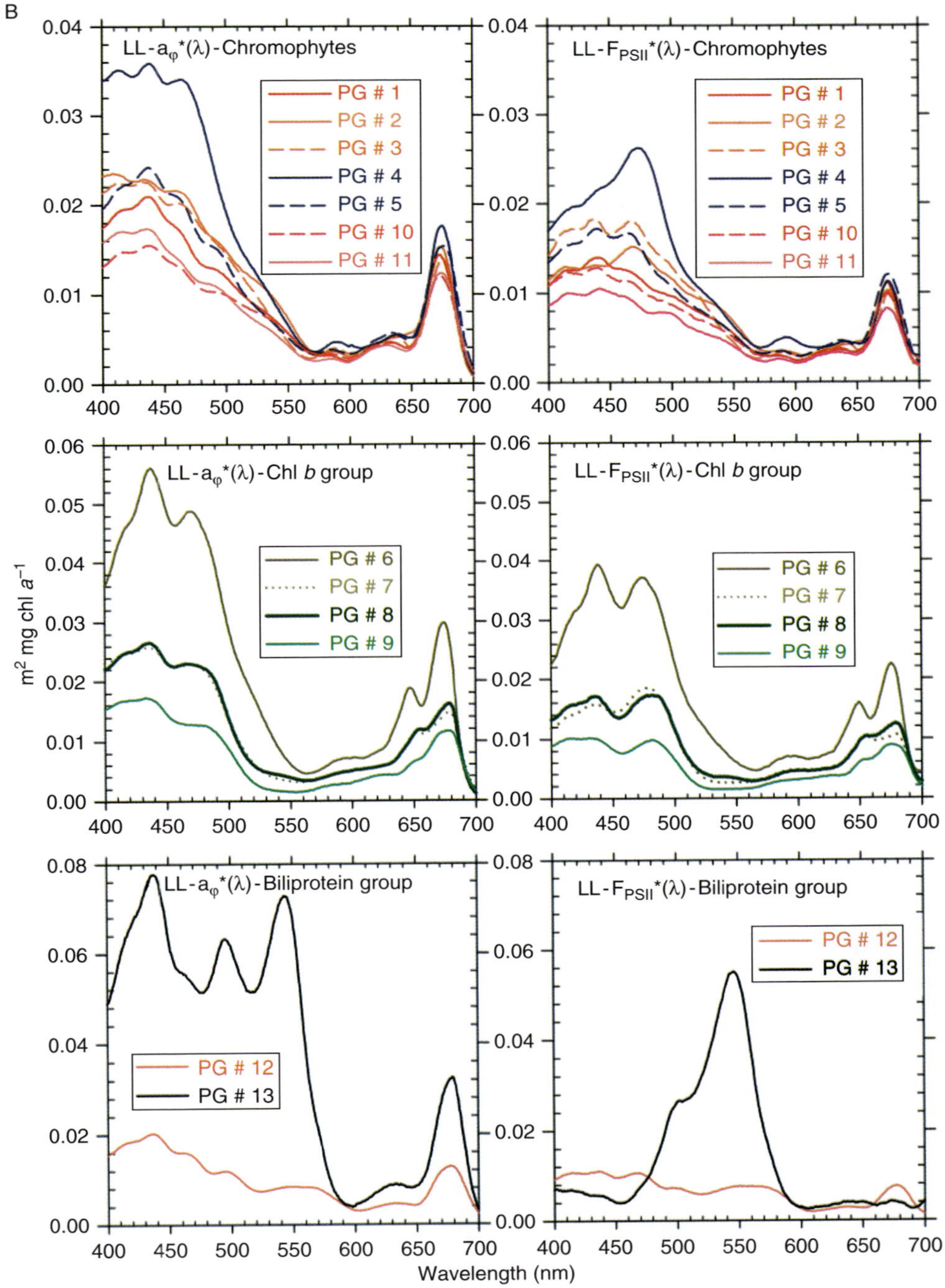

Figure 13.3. (*cont.*)

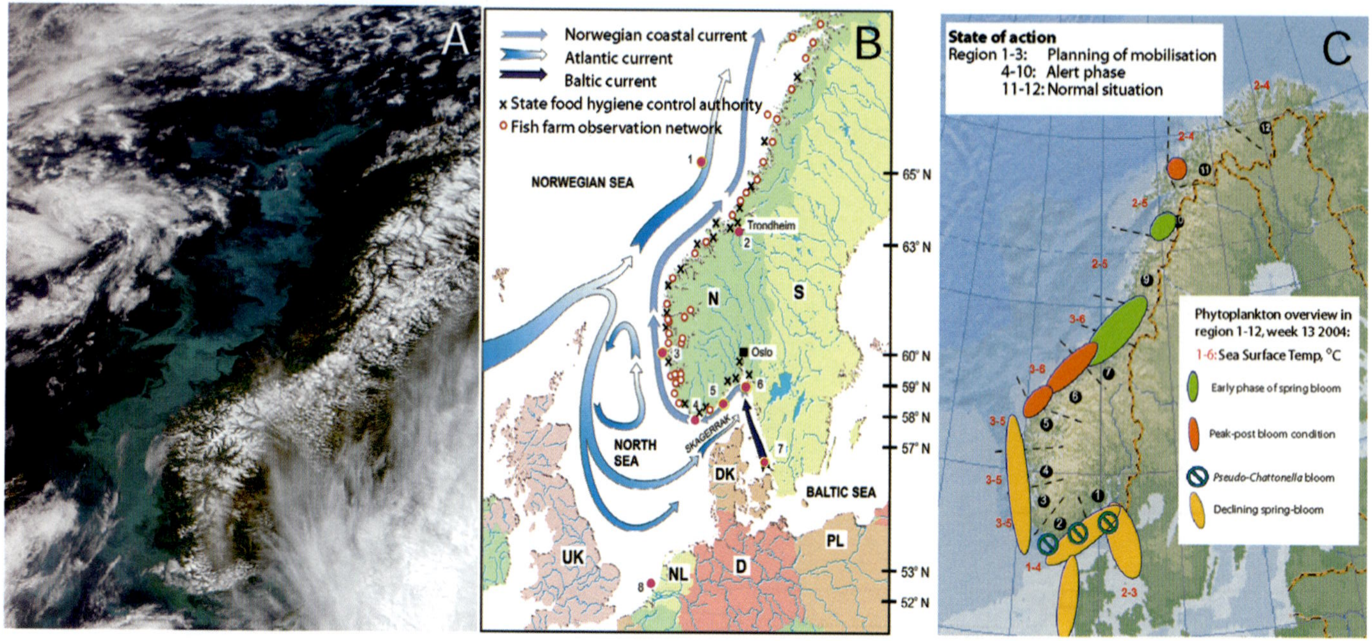

Figure 14.1. A: Phytoplankton bloom along the Norwegian coast 15 June 1998, apparently dominated by the coccolithophorid *Emiliania huxleyi*. Note that distribution of bloom follows the bottom topography and current patterns (B). SeaWiFS image (OrbView-2 satellite). NASA/Goddard Space Flight Centre. B: The Norwegian buoy (1–8) network system and sites for weekly collection of water samples for identification and enumeration of phytoplankton and mussel samples in 1993 (Johnsen *et al.*, 1997). The network is still operative, but the buoy system has been replaced by satellite images (D–G). C: Operative phytoplankton bloom situation in last week of March 2004 indicating spring-bloom in mid-Norway, see http://algeinfo.imr.no, providing the current phytoplankton (incl. HAB) situation in Norwegian waters. The coastline is divided into 12 different regions from south to north. Red numbers along the coast denote sea-surface temperature. The warning-sign denotes an ichthyotoxic bloom of *Pseudochattonella verruculosa* (Tanabe *et al.*, 2007, also known as *Verrucophora farcimen* (Edvardsen *et al.*, 2007), see D–G). D–G (time-series from 20–26 March 2001): Remotely sensed time-series of Sea WiFS images of Chl *a* biomass during bloom (ichthyotoxic) dominated by the dictyochophyte *Pseudochattonella verruculosa* (previously known as the raphidophyte *Chattonella* aff. *verruculosa* as indicated by C). The massive bloom occurred while seawater temperature was at its annual low, i.e. 2–5 °C. This species bloomed regularly in spring time in Scandinavia from 1998–2007. *Pseudochattonella verruculosa* (new species and genus) has a pigment signature similar to Chl c_3-containing coccolithophorids (Edvardsen *et al.*, 2007) and resembles the pigment-signature of the coccolithophorid genera *Chrysochromulina, Phaeocystis, Prymnesium* and *Emiliania*) and dinoflagellates with tertiary endosymbiont chloroplasts (*Karlodinium veneficum* and *Karenia mikimotoi*, Johnsen and Sakshaug, 1993; Rodríguez *et al.*, 2006; Tables 14.2–14.4), all characteristic HAB species of this region. These blooms have caused the death of approximately 2000 metric tons of farmed Atlantic salmon in the region. Images provided by Nansen Environmental and Remote Sensing Center (NERSC) using NASA OC4.4 Chl *a* algorithm to retrieve Chl *a* concentration (available at http://HAB.nersc.no).

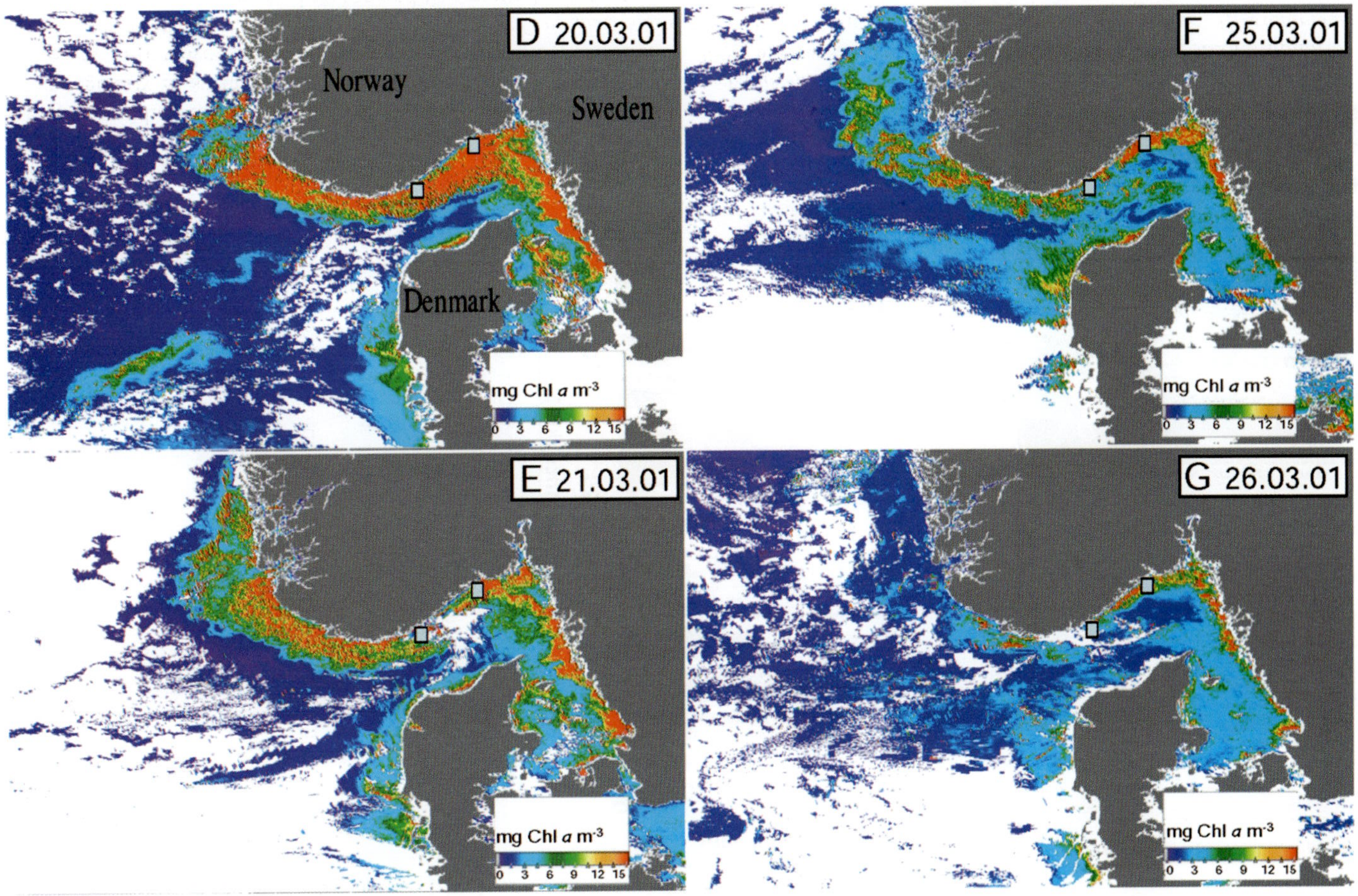

Figure 14.1. (*cont.*)

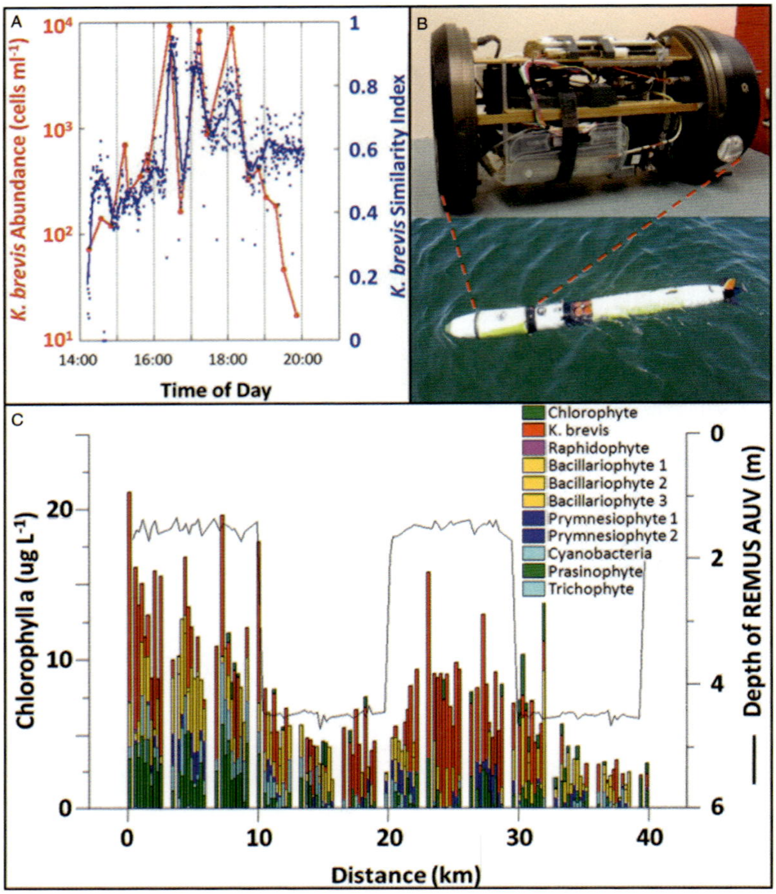

Figure 14.2. A: Surface measurements of *Karenia brevis* abundance on the West Florida Shelf taken over the day on 10 January 2003. Concurrent optically based measurements of the *K. brevis* similarity index (SI) were taken approximately every minute and corresponded well to the cell counts. The high temporal resolution of the SI measurement was able to capture rapid transitions in cell abundance. B: The 'Brevebuster' sensor for measuring the *K. brevis* SI. The sensor includes a capillary waveguide spectrometer and a series of pumps and standards for separating out the particulate fraction and CDOM. Here, the system is shown in a custom module which is integrated into a REMUS Autonomous Underwater Vehicle, AUV (below). C: Data collected on-board the REMUS AUV on 21 January, 2005. The vehicle conducted a near surface survey, first at ~2 m and then at 6 m depth and repeating this sequence again. The on-board Brevebuster collected the SI for *K. brevis* (red) and showed an enhanced concentration at 2 m (see Robbins *et al.*, 2006). In addition to *K. brevis*, a spectral library of multiple taxonomic groups grown at various light levels was applied to the full spectral data for each sample collected on the REMUS AUV to provide a breakdown of the major taxonomic groups of phytoplankton. This approach, based on optical differences in pigments, pigment concentrations and pigment packaging between taxonomic groups, provides information on phytoplankton community structure. The data illustrates the large differences in speciation that occurs over small spatial scales and demonstrates the benefit of optical approaches in delineating between phytoplankton taxa on relevant scales (see Kirkpatrick *et al.*, 2008).

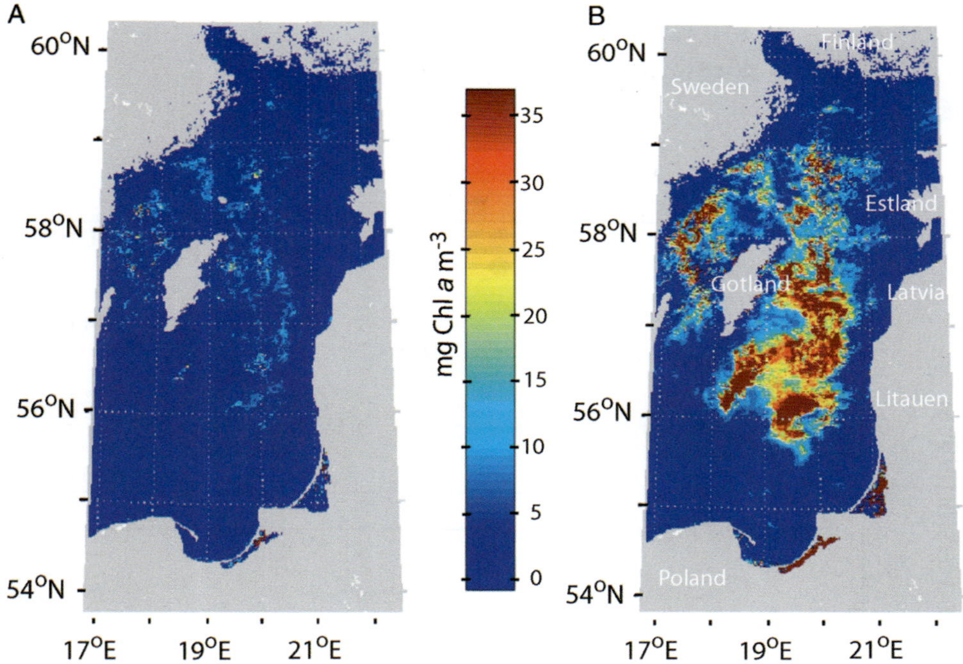

Figure 14.3. Phytoplankton biomass, seen as $C_{\mathrm{Chl}\,a}$ (mg Chl *a* m^{-3}), distributions across the central Baltic Sea as obtained from (A) MODIS standard product, (B) MODIS data processed with a regionally tuned bio-optical algorithm. The island in the centre of the image is Gotland. Images provided by Nansen Environmental and Remote Sensing Center (NERSC).

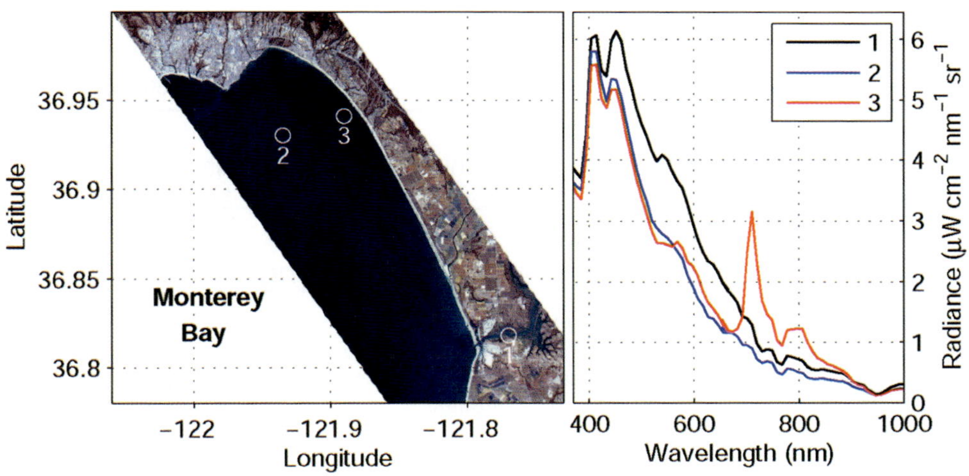

Figure 14.4. Left image: Pseudo true-colour image of an airborne remote-sensing swath across Monterey Bay, California. Data are from the Airborne Visible/Infrared Imaging Spectrophotometer (AVIRIS) on August 26, 2004. RGB bands were 711, 559 and 443 nm, respectively. Right image: The 711 nm band is effective at measuring the near-infrared signal of extreme 'red tide' blooms. Low radiance at 675 nm implies high phytoplankton biomass. Site 1 is from the river Elkhorn Slough, site 2 outside HAB and site 3 is in the middle of HAB dominated by peridinin-containing dinoflagellates such as *Akashiwo sanguiena*. Image and spectra from John Ryan, Monterey Bay Research Institute (MBARI).

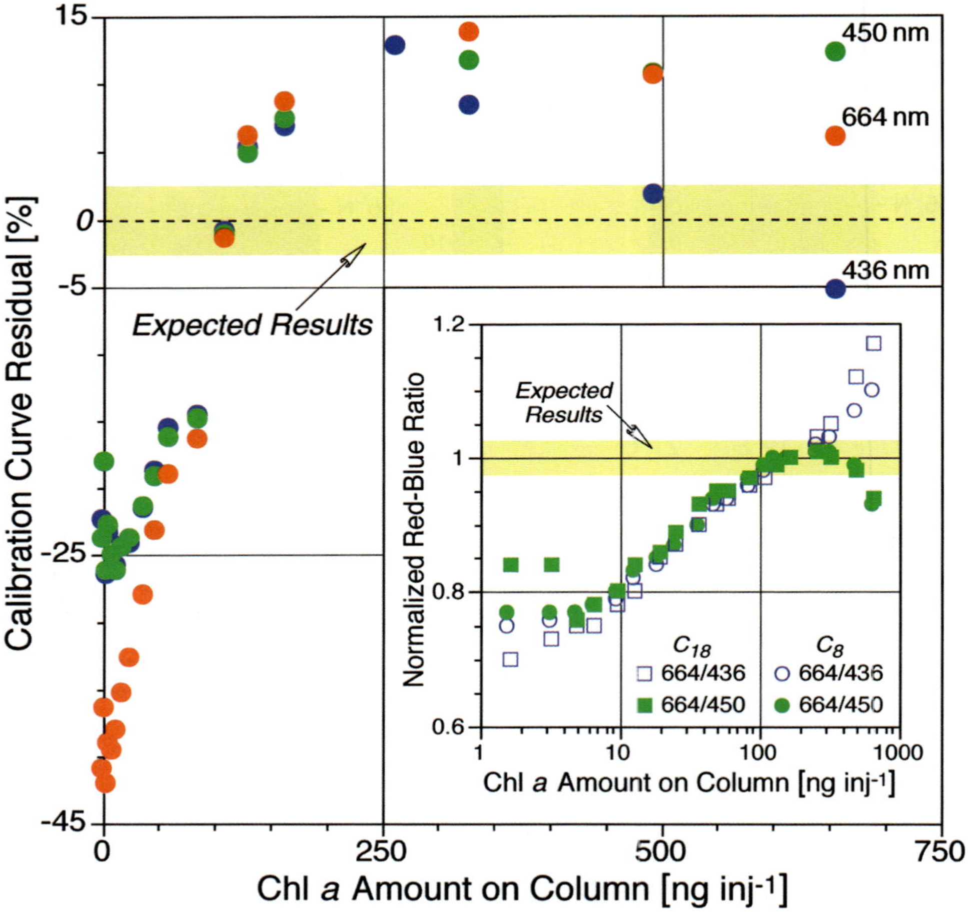

Figure B.1. Example of a detector with a nonlinear response. The main panel shows the calibration curve residuals as a function of the amount of Chl *a* on the column for three wavelengths used to quantitate pigments: 436, 450 and 664 nm. The dotted line shows expected results. The insert panel shows the degree of nonlinearity in a plot of normalized blue to red ratios (664/436 nm and 664/450 nm) for a C_8 (circles) and a C_{18} (squares) column.

5.5.1.6 Repipette accuracy and precision (daily)

At HPL, the solvent added to filters for extraction is typically 100% acetone containing the ISTD, as described in the 'one-step' extraction procedure (Eq. 5A.7 Appendix 5A). The repipette used to deliver this mixed volume of solvent and ISTD (described by V_m in Eq. 5A.7) is gravimetrically tested daily for accuracy and precision using said mixture and an analytical balance capable of measuring to within 10 μg – the gravimetrically calibrated volume is used for V_m. Precision is monitored by obtaining three replicate weights, unless the resulting CV exceeds control limits (0.83% CV), in which case the calibration is repeated that day with seven replicate weighings. Repipette imprecision exceeding a control limit requires repair and recertification by the manufacturer. The ISTD contributes a small amount to the solvent weight and is deducted before converting the weight of the acetone to volume, using the specific gravity of 100% acetone (at 25 °C). Balance performance is certified yearly with internal self-calibration performed at least daily.

As with all laboratory procedures, a series of choices must be made regarding how measurements will be conducted, and in a holistic approach to uncertainty reduction, more discriminating procedures could probably yield more accurate determinations of the variable, V_m, of note being the use of external, NIST-traceable calibration weights to complement the daily, internal calibration of the analytical balance. It would also be possible to use a specific gravity of acetone that is determined for the actual, not average, room temperature, which at HPL is 23 °C with seasonal fluctuation typically not less than 20 °C or greater than 30 °C. Riddick *et al.* (1986) cite the specific gravity of acetone as 0.7900, 0.7844 and 0.7803 at 20, 25 and 30 °C, respectively, while a commercial vendor of acetone cites a specific gravity of 0.7849 at 25 °C, the latter of which is used at HPL. So, there is some uncertainty in the reference value selected (at least 0.3%) and additional uncertainty contributed by the constant use of a specific gravity unique to 25 °C. Importantly, this latter uncertainty can be approximated as –0.6% at 20 °C and 0.4% at 30 °C. These examples, while seeming to be tedious, are important to the process of understanding all potential uncertainty sources, and from which informed decisions are made regarding laboratory procedures.

5.5.1.7 Calibration accuracy (spectrophotometric and HPLC analysis)

Calibration accuracy can be assessed *externally* by inter-laboratory comparisons with standards from various sources on annual or semi-annual time scales. Standards used may either be purchased through such vendors as DHI, Fluka and Carotenature or isolated from algal cultures. Intercomparisons can include both HPLC and spectrophotometric comparisons; the latter have proven to be very good for Chl *a*, with average uncertainties of 1.4% when correct spectrophotometric procedures are applied (Hooker *et al.*, 2005). Internal HPLC calibration checks should be implemented on a more regular time scale. It is important to keep uncertainties associated with processes of re-calibration low, for if this can be achieved, calibration factors should not vary greatly over time, and a cumulative average calibration factor (unique to each pigment) can be used in computations (Eq. 5A.1–5A.5, Appendix 5A). Subsequent analyses of standards are then used to document that uncertainty of recalibration is within expected limits (WL and CL) for the cumulative average response factor of that pigment.

Calibration accuracy is currently evaluated in annual inter-laboratory comparisons with pigments standards, in which HPL analyses DHI standards, and for which average absolute per cent differences (APDs) are 2% for primary and secondary pigments (with a range in APDs of 0.1 to 6%). With QC checks implemented exclusively at HPL, where average response factors are used to quantify pigments in natural samples and QC checks are injected (at least one per sequence) to test their validity, it is expected that the observed response factor will be within 5% of the average response factor at the 95% confidence limit for primary and secondary pigments. Other pigment standards may be less reproducible; in fact, the coefficient of variation for Phe *a* and Pheide *a* has been as poor as 17%.

5.5.1.8 Chl a *linearity (with new column installation)*

Five or more Chl *a* standards encompassing the working range are quantitatively prepared (in 90% acetone) using calibrated gas-tight glass syringes and Class A volumetric glassware. One injection per standard is performed after a retention time mixture has been injected on the new column to evaluate R_S and retention performance. The linear regression is expected to yield a *y*-intercept near zero (and well below the lowest point of the working range). The average of the absolute per cent residuals is 0.9% on average and not greater than 2.1% at the 95% confidence limits. (Data were acquired at HPL during analysis of 29 different Chl *a* calibration curves over nine years.) Residuals included in the average were limited to those from standards whose concentrations were within the working range. The reader is referred to Section 5.2.5 and Table 5.2 for further details on determination of per cent residuals.

Dismissal of a single calibration point is acceptable to bring calibration into expectations, or alternatively, dismissal of a single point can be subjected to a test for outliers, such as the Dixon's Q test (Miller and Miller, 2000). Dismissal of more than one point within the working range would require that the calibration curve be repeated. The slope is expected to be within 3.2% (95% confidence limits) of the average (previously observed slopes). These limits were observed across columns of varying lot numbers, but all from the same manufacturer for the method employed at HPL (Hooker *et al.*, 2005). Slopes have been observed to exceed these limits in instances where the concentration of the stock Chl *a* solution, determined spectrophotometrically, needed to be re-measured and when an aged column exhibited atypical peak tailing.

To evaluate critically the range of concentrations over which the best accuracy can be obtained, the per cent residuals are plotted as a function of amount injected (for all calibration curves for a particular pigment). With per cent residuals from multiple curves plotted on the same graph, a pattern should emerge where per cent residuals vary widely and are randomly distributed in sign at some lower concentration and at concentrations above this point, the per cent residuals will be constrained to a much narrower range, from which warning and control limits can be computed. This graphical representation is described by King (1999) as the *linear-through-zero* range and is extremely useful for identifying the reproducibility of calibration curve results and the lower concentrations at which prediction accuracy will be compromised. A similar graphical presentation is given in Figure 5.3b, showing normalized response factor as a function of Chl *a* amount injected (data from HPL).

5.5.1.9 Detector noise (variable frequency)

SeaHARRE participants collaborated to define very short-term detector noise (Hooker *et al.*, 2005) for use with determining LOD and LOQ values, which require computations of signal to noise ratios. When assessing noise, abnormal disturbances in the baseline should not be expected and should be investigated because they can exacerbate uncertainties, especially during integration of small peaks. Horn Point Laboratory detector noise is measured (in duplicate or triplicate) at wavelengths used for pigment quantitation with functions of the HPLC software and for which repeatability precision approximates 3.0 and 4.0% at 450 and 665 nm, respectively. The average absolute per cent difference in noise across days and between lamp changes is 7.7% at 450 nm and 8.5% at 665 nm. The frequency of noise determinations is unspecified, other than to perform such with installation of new detector lamp(s).

5.5.1.10 QC and QA elements from data manipulations

Several QC measurements with useful QA applications can be easily (and automatically) calculated when pigment data are converted to final data products. These types of inquiries include a) formulation of pigment ratios, b) assessment of water content in sample extracts, c) assessment of whether extraction volumes computed by the internal standard are realistic when compared to those expected had an internal standard not been used, d) automatic flagging of peak areas observed to be above a working range and e) automatic calculations of SNR for each peak quantified, which requires periodic updates of noise measurements for wavelengths used with quantitation.

Important pigment ratios include (Pheide *a* + Phe *a*)/TChl *a*, (Chl *a* allomers + Chl *a* epimers)/TChl *a*, and TAcc/TChl *a*. Other pigment ratios and sums are possible, and have applicability to the emerging importance of the functional form in uncertainties (Figure 5.6) as a QA tool. The two pheopigment and Chl *a* ratios are useful when discerning whether an aberrant HPLC result may have been caused, either by a damaged sample, or the sample represented unusually senescent cells high in detrital matter, in which case these two ratios are likely to be elevated. The utility of TAcc/TChl *a* ratios was originally introduced by Trees *et al.* (2000) and with its repeated application to waters from different locations has been useful for identifying abnormally analysed or aberrant samples. In fact, this ratio is often distorted when the pheopigment ratio is high. For the current HPL QAP, ratios are not used quantitatively to describe performance, but do yield information regarding why an individual result is extraordinary.

Extraction volume (V_{x1}) used in pigment quantitation is determined according to (Eq. 5A.7, Appendix 5A) at HPL, but during SeaHARRE-2, extraction volume precision for some laboratories was degraded through the use of an internal standard (Hooker *et al.*, 2005). A QC check can be implemented by calculating V_{x1} and $V_{x'}$ according to Eqs. 5A.7 and 5A.9 respectively, and computing the RPD (and APD) between the two results. Applying this comparison to sets of filters provided to HPL by six different contributors resulted in average differences (APDs) to within 2% with maximum RPDs less than 5%. This comparison is validated by previous tests that demonstrated solvent evaporation is minimal at HPL during extraction.

Another variable that emerges from these calculations is an estimate of the water content (V_w) in each filter (fully described in Hooker *et al.*, 2009). By setting Eq. 5A.7 equal to Eq. 5A.9, cancelling like terms and knowing V_m (volume of solvent added to the filter) and V_e, V_w can be computed. By comparing the computed V_w value to its nominal value, 0.2 mL for a 25 mm filter (Bidigare *et al.*, 2003), it is possible to identify sample extract solutions that have inadvertently reached water contents greater than 10%, and are therefore at risk of inaccurate quantitation (Latasa *et al.*, 2001).

5.5.2 Control charts

Control charts illustrate graphically the results of QC analyses and are extremely useful for identifying performance trends. They are developed after a sufficient number of QC measurements are acquired, so performance expectations can be quantified by 95% and 99% confidence limits – the warning limits (WL) and control limits (CL), respectively.

Trends in performance are best described by multiple QC measurements considered together. For example, indicators of column performance can include R_S, carryover, Chl *a* QC accuracy, and ISTD precision. These indicators may (or may not) simultaneously yield results that are out-of-control. To illustrate the complexity of monitoring column performance, control charts for accuracy of Chl *a* QC solutions and precision of ISTD daily injections (covering the same 275-day time period) are overlain in Figure 5.9. Accuracy of Chl *a* QC analysis is described by the RPD

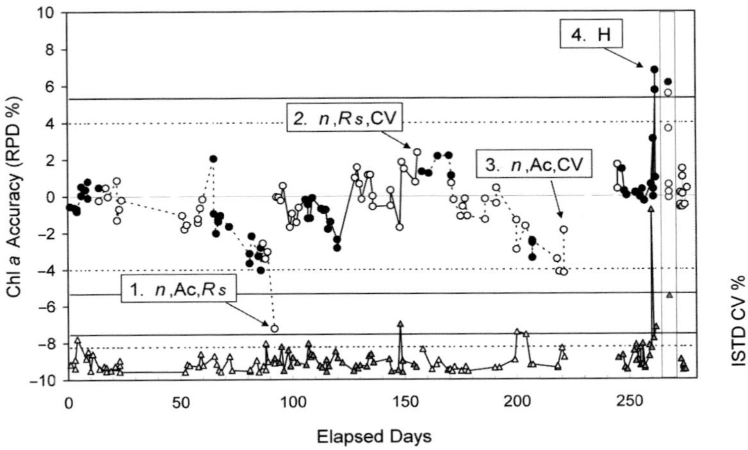

Figure 5.9. Control charts tracking accuracy of Chl *a* QC injections (left *y*-axis) and precision of daily ISTD replicate injections (right *y*-axis) over 275 days. Acceptable performance of Chl *a* QC injections, with alternating open and closed circles depicting different QC formulations, and all symbols connected by lines representing analyses on the same column, is constrained within red dashed lines (the warning limits WL) and solid red lines (the control limits CL). Acceptable ISTD CV%, with alternating open and solid grey triangles to discriminate analyses all performed on the same column, is constrained within the blue dashed line (the WL) and the blue solid line (the CL, which depicts 1.25 CV%). Columns 1, 2 and 3 were removed from service on days indicated by arrow positions for reasons based on multiple criteria, for which n = aged column, Ac = poor Chl *a* QC accuracy, $R_S = R_S$ between critical pairs was ≤ 1.0, CV = ISTD precision had exceeded CL. A hardware failure (H) occurred during analyses on column 4 on day indicated by arrow. Sample processing ceased during days highlighted in yellow while repairs and testing were conducted, after which sample processing on column 4 continued. See colour plate section.

between observed and formulated concentrations and instrument precision is described by the CV% of daily replicate injections of the ISTD.

Data in control charts exemplify how decisions are made regarding when to change columns or halt all analyses. A downward trend in the RPD of a Chl *a* QC solution is often remedied by simply formulating a new standard solution. However, new formulations of the QC standard did not eliminate the downward trend in RPDs on column 1 (Figure 5.9) – in fact, Chl *a* RPD exceeds the CL on day 92. These observations, the high number of injections (n), and a drop in R_S to 1.04 provided a rationale for removing column 1 from service. In contrast, the performance of column 2 became questionable with an ISTD CV result outside the CL on day 148, and although Chl *a* accuracy was within warning limits, R_S had dropped to 1.01 *and* the column had been used extensively, so this column was removed from service on day 156. Diminished Chl *a* QC accuracy and ISTD precision, and a large number of injections called for removal of column 3 on day 221, even though R_S was greater than 1.0 and still adequate. In these instances, with a simple need for column replacement, analyses were stopped only to replace the old column and evaluate the new column with typical QC injections.

The break in service between days 221 and 245 was caused partially by the need for component repairs, during which time samples were not analysed. On day 245, sample processing was reinitiated on column 4, but on day 262, injection valve failure occurred, which was detected because Chl *a* inaccuracy *and* ISTD imprecision both exceeded control limits. Analyses were halted and between days 262 and 273, repairs were implemented and QC measurements conducted to evaluate the effectiveness of such. Duplicate injections of a sample extract had bracketed the failed QC injections on day 262 *and* because TChl *a* precision (1.1%) was within the expected control limit (Section 5.5.1.4), an informed decision was made that it was not necessary to reinject other samples that had been analysed near the time of the QC failures. Sample processing was resumed after demonstrating QC measurements were again within expected limits.

5.6 Future directions

NASA round robins and the open dialogue they have fostered among analysts have exposed some practices that are detrimental to quantitative accuracy and precision of HPLC pigment results. While this process has (sometimes) been painful, knowledge of how to constrain uncertainties and where to best place future emphasis to improve quantitative accuracy is valuable beyond description. The larger community of HPLC analysts is indebted to the round-robin participants, because without their willingness to expose vulnerabilities, the current level of understanding or a clear direction of how best to proceed would not have been possible. Performance metrics and their utility in quantifying benchmarks of achievement are a beginning, but laboratories now need to develop and adopt commonly accepted procedures for method validation and ongoing QA and QC activities.

Data produced without uncertainty estimates are monetarily wasteful and use increasingly limited resources unwisely. While it is possible to observe trends from massive amounts of data procured with unknown and exacerbated uncertainties, this needlessly incomplete approach is inefficient and ultimately uses more resources than data acquired under quality-assured conditions. In an era where increasingly diverse observations are merged into ever expanding databases and exploited for a widening range of investigations affecting policy change pertaining to such important issues as global climate change, a naïve and incomplete understanding of data quality is not acceptable. Pigment analysts can make a difference by improving (and describing) their data quality, and indeed, they have the responsibility to advance the state of the art for new and heretofore unimagined applications.

While accuracy requirements for TChl *a* are prescribed for remote-sensing applications, accuracy objectives for accessory pigments (more specifically, the primary pigments) were rather arbitrarily adopted by the SeaHARRE community based on what had been achieved by quality-assured laboratories. For natural samples, many pigments are not quantifiable with equivalent accuracy, because they are usually in low concentrations for much of the world ocean and are quantified at levels below what would be likely to be a true limit of quantitation (if one was adequately described). Also, the reporting practices for pigments in low concentrations vary so much among laboratories, the practices themselves generate needlessly large uncertainties that overwhelm the uncertainty budgets in many cases. A consensus of agreement regarding such procedures will probably only occur if database managers or funding agencies impose mandatory guidelines.

An emerging application for quantitation of pigments in seawater is the discrimination of accessory pigments via hyperspectral and multispectral remote sensors. For success, the ground-truth, more properly sea-truth, observations needed for this new application will require consistently accurate results for pigments important to the discrimination of phytoplankton diversity. Presently, the accuracy of only one accessory pigment – Fuco – has mostly been equivalent to the accuracy observed for TChl *a*. Current performance metrics have proved satisfactory at the 'quantitative' level for achieving the Chl *a* remote-sensing accuracy objectives. It is likely, however, that the more difficult challenges of remotely sensed accessory pigments will require data produced with 'state-of-the-art' performance capabilities (or possible improvements to these).

As science advances and uncertainty requirements are pushed inexorably downward, absorption coefficients and the solvents used to determine the concentrations of pigment standards spectrophotometrically must be standardized. For six carotenoids within the SeaHARRE-2 QA subset, differences averaging approximately 8% – but as much as 15% – were observed when the absorption coefficients in ethanol were compared to those in acetone (Hooker *et al.*, 2005). While it seems sensible to suggest the use of acetone absorption coefficients (because most laboratories extract samples in acetone), it is not currently practical to do so, because many standards currently available in ethanol are not commercially available in acetone.

Acknowledgements

Round-robin intercomparisons make significant demands on analysts and their home institutions. The SeaHARRE activity was fortunate in having a large number of investigators who were willing to do all that was required to ensure the most comprehensive intercomparisons possible. The scientists who gave selflessly of themselves and prevailed upon their management to do the same are recognized here in alphabetical order: James Aiken (PML), Merete Allerup (DHI), Ray Barlow (MCM), Jean-François Berthon (JRC), Elisabetta Canuti (JRC), Hervé Claustre (LOV), Lesley Clementson (CSIRO), James Fishwick (PML), Erica Head (BIO), Dave Millie (FIO), Claire Normandeau (DU), Jason Perl (CHORS), Jay Pinckney

(USC), Joséphine Ras (LOV), Mary Elizabeth Russ (UMBC), Louise Schlüter (DHI), Heather Sessions (MCM), Venetia Stuart (BIO), Cristina Targa (JRC), Crystal Thomas (HPL), Charles Trees (CHORS), and Dirk van der Linde (JRC). The same gratitude and recognition is extended to the many anonymous investigators who participated in the SIMBIOS round robin.

Abbreviations and symbols

$\hat{A}_{c1}$	Average peak area of the internal standard
APD	Absolute per cent difference
CL	Control limits
CV	Coefficient of variation
EOS	Earth Observing System
HPL	Horn Point Laboratory
ICH	International Conference on Harmonization
ISTD	Internal standard
IUPAC	International Union of Pure and Applied Chemists
JGOFS	Joint Global Ocean Flux Study
LLOL	Lower limit of linearity
LOD	Limit of detection
LOQ	Limit of quantitation
n	Number of injections
$\lvert\bar{\psi}\rvert_{\mathrm{res}}$	Average of the absolute per cent residuals
PPig	Primary pigments (total chlorophylls and carotenoids)
QAP	Quality assurance plan
QA	Quality assessment
QC	Quality control
RPD	Relative per cent difference
R_S	Resolution between pigments
SeaHARRE	SeaWiFS HPLC Analysis Round-Robin Experiment
SeaWiFS	Sea-viewing Wide Field-of-view Sensor
SIMBIOS	Sensor Intercomparison and Merger for Biological and Interdisciplinary Oceanic Studies
SNR	Signal to noise ratio
t_R	Peak retention time
ULOL	Upper limit of linearity
V_e	Solvent volume added to the sample
V_m	Mixed volume of solvent and ISTD added to the filter
V_{x1}	Extraction volume
$V_{x'}$	$V_e + V_w$
V_w	Water content in the filter
WL	Warning limits

References

Barlow, R.G., Mantoura, R.F.C., Gough, M.A. and Fileman, T.W. (1993). Pigment signatures of the phytoplankton composition in the northeastern Atlantic during the 1990 spring bloom. *Deep-Sea Res. II* **40**, 459–77.

Bidigare, R.R. and Ondrusek, M.E. (1996). Spatial and temporal variability of phytoplankton pigment distributions in the central equatorial Pacific Ocean. *Deep-Sea Res. II* **43**, 809–33.

Bidigare, R.R., Van Heukelem, L. and Trees, C.C. (2003). HPLC phytoplankton pigments: sampling, laboratory methods and quality assurance procedures. In *Ocean Optics Protocols for Satellite Ocean Color Sensor Validation, Revision 4, Volume V: Biogeochemical and Bio-optical Measurements and Data Analysis Protocols*. NASA/TM–2003–211621/Rev4-Vol.V, ed. J.L. Mueller, G.S. Fargion and C.L. McClain. Greenbelt: NASA Goddard Space Flight Center, pp. 5–14.

Bidigare, R.R., Van Heukelem, L. and Trees, C.C. (2005). Analysis of algal pigments by high-performance liquid chromatography. In *Algal Culturing Techniques*, ed. R. Andersen. Burlington: Elsevier Academic Press, pp. 327–45.

Castells, C.B. and Castells, R.C. (1998). Peak distortion in reversed-phase liquid chromatography as a consequence of viscosity difference between sample solvent and mobile phase. *J. Chromatogr. A*. **805**, 55–61.

Claustre, H. (1994). The trophic status of various oceanic provinces as revealed by phytoplankton pigment signatures. *Limnol. Oceanogr.* **39**, 1206–10.

Claustre, H., Hooker, S.B., Van Heukelem, L., Berthon, J.F., Barlow, R., Ras, J., Sessions, H., Targa, C., Thomas, C.S., van der Linde, D., Marty, J-C. (2004). An intercomparison of HPLC phytoplankton pigment methods using *in situ* samples: application to remote sensing and database activities. *Mar. Chem.* **85**, 41–61.

Clesceri, L.S., Greenberg, A.E., and Eaton, A.D. (eds.) (1998). *Part 1020, Quality Assurance, Section 1020C, Quality Assessment in Standard Methods for the Examination of Water and Wastewater*. 20th edn. Baltimore: American Public Health Association, American Water Works Association, Water Environment Federation, Baltimore, Maryland, pp. 1–12.

Dolan, J.W. (1999). Readers' questions, carryover, mobile-phase temperature, and column care. *LC GC* **20**, 1006–08.

Dolan, J.W. (2001a). Autosampler carryover. *LC GC* **19**, 164–68.

Dolan, J.W. (2001b). Attacking carryover problems. *LC GC* **19**, 1050–54.

Dolan, J.W. (2002). Resolving minor peaks. *LC GC* **20**, 594–98.

Dolan, J.W. (2006). Autosampler carryover. *LC GC* **24**, 754–60.

EURACHEM Guide (1998). *The Fitness for Purpose of Analytical Methods, A Laboratory Guide to Method Validation and Related Topics*, Mr. David Holcombe, Drafting Secretary for EURACHEM working Group, LGC, Queens Rd. Teddington, Middlesex, TW11 0lY, United Kingdom. http://www.eurachem.org/

EURACHEM/CITAC Guide (2000). *Quantifying Uncertainty in Analytical Measurement*, 2nd edn, Ed. S.L.R. Ellison (LGC, UK), M. Rosslein (EMPA, Switzerland), A. Williams (UK). http://www.eurachem.org/

Gieskes, W.W.C., Kraay, G.W., Nontji, A., Setipermana, D. and Sutomo, D. (1988). Monsoonal alternation of a mixed and a layered structure in the phytoplankton of the euphotic zone of the Banda Sea (Indonesia):

A mathematical analysis of algal pigment fingerprints. *Neth. J. Sea Res.* **22**, 123–37.

Hastie, T., Tibshirani, R. and Friedman, J. (2001). *The Elements of Statistical Learning: Data Mining, Inference, and Prediction.* New York: Springer Science+Business Media.

Helsinki Commission (2006), Manual for Marine Monitoring in the COMBINE Programme of HELCOM, Part B. *General Guidelines on Quality Assurance for Monitoring in the Baltic Sea.* http://www.helcom.fi/groups/monas/CombineManual/PartB/en_GB/main/

Hooker, S.B. and Esaias, W.E. (1993). An overview of the SeaWiFS project. *Eos, Trans. AGU* **74**, 241–46.

Hooker, S.B. and McClain, C.R. (2000). The calibration and validation of SeaWiFS data. *Prog. Oceanogr.* **45**, 427–65.

Hooker, S.B., Claustre, H., Ras, J., Van Heukelem, L., Berthon, J.-F., Targa, C., van der Linde, D., Barlow, R. and Sessions, H. (2000). *The First SeaWiFS HPLC Analysis Round-Robin Experiment (SeaHARRE-1).* SeaWiFS Postlaunch Technical Report Series, S.B. Hooker and E.R. Firestone (eds.). NASA Technical Memorandum 2000–206892, vol. 14. Greenbelt: NASA Goddard Space Flight Center.

Hooker, S.B., Van Heukelem, L., Thomas, C.S., Claustre, H., Ras, J., Barlow, R., Sessions, H., Schlüter, L., Perl, J., Trees, C., Stuart, V., Head, E., Clementson, L., Fishwick, J., Llewellyn, C. and Aiken, J. (2005). *The Second SeaWiFS HPLC Analysis Round-Robin Experiment (SeaHARRE-2).* NASA Technical Memorandum 2005–212785. Greenbelt: NASA Goddard Space Flight Center.

Hooker, S.B., Van Heukelem, L., Thomas, C.S., Claustre, H., Ras, J., Schlüter, L., Clementson, L., van der Linde, D., Eker-Develi, E., Berthon, J.-F., Barlow, R., Sessions, H., Ismail, H. and Perl, J. (2009). *The Third SeaWiFS HPLC Analysis Round-Robin Experiment (SeaHARRE-3).* NASA Technical Memorandum 2009–215849. Greenbelt: NASA Goddard Space Flight Center.

Hooker, S.B., Van Heukelem, L., Thomas, C.S., Schlüter L., Russ, M.E., Ras, J., Claustre, H., Clementson, L., Canuti, E., Berthon, J-F., Perl J., Normandeau, C., Cullen J., Kienast, M., Pinckney, J.L., and Millie, D. (2010). *The Fourth SeaWiFS HPLC Analysis Round-Robin Experiment (SeaHARRE-4).* NASA Technical Memorandum 2010–215857. Greenbelt: NASA Goddard Space Flight Center.

Humphrey, G.G. and Jeffrey, S.W. (1997). Appendix G, Tests of accuracy of spectrophotometric equations for the simultaneous determination of chlorophylls a, b, c_1 and c_2. In *Phytoplankton Pigments in Oceanography: Guidelines to Modern Methods*, ed. S.W. Jeffrey, R.F.C. Mantoura and S.W. Wright. Paris: UNESCO Publishing, pp. 616–30.

International Conference on Harmonization of Technical Requirements for Registration of Pharmaceuticals for Human Use. *ICH Tripartite Guideline, Validation of Analytical Procedures: Text and Methodology.* Q2(R1) (2005). (Note: this document combines, with text unchanged, the two FDA documents Q2A (March 1995) and Q2B (November 1996)).

International Union of Pure and Applied Chemistry (IUPAC). (1998). Chapter 18, Quality Assurance of Analytical Processes, Section 18.4.3.7. In *Compendium of*

Analytical Nomenclature: Definitive Rules 1997. Oxford: Blackwell Science, and web edition: http://old.iupac.org/publications/analytical_compendium/

Jeffrey, S.W. and Humphrey, G.F. (1975). New spectrophotometric equations for determining chlorophylls a, b, c_1 and c_2 in higher plants, algae and natural phytoplankton. *Biochem. Physiol. Pflanzen*. **167**, 191–94.

Jeffrey, S.W. and Wright, S.W. (1997). Qualitative and quantitative HPLC analysis of SCOR reference algal cultures. In *Phytoplankton Pigments in Oceanography: Guidelines to Modern Methods*, ed. S.W. Jeffrey, R.F.C. Mantoura and S.W. Wright. Paris: UNESCO Publishing, pp. 343–60.

Jeffrey, S.W., Mantoura, R.F.C. and Wright, S.W. (eds.) (1997). *Phytoplankton Pigments in Oceanography: Guidelines to Modern Methods*. Paris: UNESCO Publishing.

JGOFS (1994). *Protocols for the Joint Global Ocean Flux Study Core Measurements*. Intergovernmental Oceanographic Commission, Scientific Committee on Oceanic Research. Manual and Guides, vol. **29**. Paris, France: UNESCO Publishing, pp. 91–96.

King, P.G. (1999). HPLC method development and validation: A direct procedure for determining the linear-through-zero range. *LC GC* **6**, 46.

Latasa, M., Bidigare, R.R., Ondrusek, M.E. and Kennicutt II, M.C. (1996). HPLC analysis of algal pigments: a comparison exercise among laboratories and recommendations for improved analytical performance. *Mar. Chem*. **51**, 315–24.

Latasa, M., van Lenning, K., Garrido, J.L., Scharek, R., Estrada, M., Rodríguez, F. and Zapata, M. (2001). Losses of chlorophylls and carotenoids in aqueous acetone and methanol extracts prepared for RPHPLC analysis of pigments. *Chromatographia* **53**, 385–91.

Mantoura, R.F.C. and Repeta, D.J. (1997). Calibration methods for HPLC. In *Phytoplankton Pigments in Oceanography: Guidelines to Modern Methods*, ed. S. W. Jeffrey, R.F.C. Mantoura and S.W. Wright. Paris: UNESCO Publishing, pp. 343–60.

Miller J.N. and Miller, J.C. (2000). *Statistics and Chemometrics for Analytical Chemistry*. 4th edn. Pearson Education Limited, Prentice Hall Publisher.

Mueller, J.L. (2000). Overview of measurement and data analysis protocols. In *Ocean Optics Protocols for Satellite Ocean Color Sensor Validation, Revision 2*, ed. G.S. Fargion and J.L. Mueller. NASA Technical Memorandum 2000–209966. Greenbelt, Maryland: NASA Goddard Space Flight Center, pp. 87–97.

Mueller, J.L. (2002). Overview of measurement and data analysis protocols. In *Ocean Optics Protocols for Satellite Ocean Color Sensor Validation, Revision 3*, vol. 1. ed. J.L. Mueller and G.S. Fargion. NASA Technical Memorandum 2002–210004/Rev. 3 – Vol. 1. Greenbelt: NASA Goddard Space Flight Center, pp. 123–37.

Mueller, J.L. (2003). Overview of measurement and data analysis methods. In *Ocean Optics Protocols for Satellite Ocean Color Sensor Validation, Revision 4, Volume III: Radiometric Measurements and Data Analysis Protocols*, ed. J.L. Mueller and 17 co-authors. NASA Technical Memorandum 2003–211621/Rev. 4 – Vol. III. Greenbelt: NASA Goddard Space Flight Center, pp. 1–20.

Mueller, J.L. and Austin, R.W. (1992). *Ocean Optics Protocols for SeaWiFS Validation*. NASA Technical Memorandum 104566, vol. 5, S.B. Hooker and E.R. Firestone. (eds.) Greenbelt: NASA Goddard Space Flight Center.

Mueller, J.L. and Austin, R.W. (1995). *Ocean Optics Protocols for SeaWiFS Validation, Revision 1*. NASA Technical Memorandum 104566, vol. 25, S.B. Hooker, E.R. Firestone, and J.G. Acker (eds). Greenbelt: NASA Goddard Space Flight Center.

Riddick, J.A., Bunger, W.B. and Sakano, T.K. (1986). *Organic Solvents: Physical Properties and Methods of Purification*, 4th edn. New York: John Wiley & Sons.

Snyder, L.R. and Kirkland, J.J. (1979). *Introduction to Modern Liquid Chromatography*, 2nd edn. New York: Wiley.

Taylor, J.K. (1987). *Quality Assurance of Chemical Measurements*. Chelsea: Lewis Publishers.

Trees, C.C., Clark, D.K., Bidigare, R.R., Ondrusek, M.E. and Mueller, J.L. (2000). Accessory pigments versus chlorophyll *a* concentrations within the euphotic zone: a ubiquitous relationship. *Limnol. Oceanogr.* **45**, 1130–43.

U.S. Department of Health and Human Services, Food and Drug Administration (USDHHS), Center for Drug Evaluation and Research (CDER), Chemistry. *Reviewer Guidance: Validation of Chromatographic Methods*. CMC-3. (November 1994, revised 5–14–2007 to include graphics). http://www.fda.gov/cder/guidance/cmc3_rev.pdf

U.S. Department of Health and Human Services, Food and Drug Administration (USDHHS), Center for Drug Evaluation and Research (CDER), ICH-Quality. *Guideline for Industry: Text on Validation of Analytical Procedures*, ICH-Q2A (March 1995). http://www.fda.gov/cder/guidance/ichq2a.pdf

U.S. Department of Health and Human Services, Food and Drug Administration (USDHHS), Center for Drug Evaluation and Research (CDER), Center for Biologics Evaluation and Research (CBER). *Guidance for Industry, Q2B Validation of Analytical Procedures: Methodology*. ICH-Q2B (November 1996). http://www.fda.gov/cder/guidance/1320fnl.pdf

U.S. Department of Health and Human Services, Food and Drug Administration (USDHHS), Center for Drug Evaluation and Research (CDER), Center for Veterinary Medicine (CVM). *Guidance for Industry: Bioanalytical Method Validation*. BP. (May 2001). http://www.fda.gov/cder/guidance/4252fnl.pdf

Van Heukelem, L. and Thomas, C.S. (2001). Computer-assisted high-performance liquid chromatography method development with applications to the isolation and analysis of phytoplankton pigments. *J. Chromatogr. A*. **910**, 31–49.

Van Heukelem, L., Thomas, C.S. and Glibert, P. (2002). *Sources of variability in chlorophyll analysis by fluorometry and high-performance liquid chromatography in a SIMBIOS inter-calibration exercise*. NASA Technical Memorandum 2002–211606, Greenbelt: NASA Goddard Space Flight Center.

Zapata, M. and Garrido, J.L. (1991). Influence of injection conditions in reversed-phase high performance liquid chromatography of chlorophyll and carotenoids. *Chromatographia* **31**, 589–94.

Appendix 5A

A symbology and vocabulary for an HPLC lexicon

STANFORD B. HOOKER AND LAURIE VAN HEUKELEM

The purpose of this appendix is to define terms, symbols and parameters that facilitate unambiguous communication and discussion of the many steps involved in quantitative analysis (by HPLC) of phytoplankton pigments in marine systems. More specifically, this appendix is necessary for the full understanding of Chapter 5 (this volume). The symbology presented here does not encompass all pigments currently known to marine systems, but it is designed with sufficient flexibility to allow all pigments – even those not yet identified – to be uniquely accommodated. At the outset of each HPLC round-robin inter-comparison sponsored by the National Aeronautics and Space Administration (NASA), participants listed pigments they quantified routinely (Hooker *et al.*, 2000, 2005, and 2009). While these lists differed, all included a core set of common pigments, which provided a foundation for evaluating accuracy among laboratories and identifying and quantifying uncertainties associated with a diversity of laboratory practices and procedures. Evaluations of the common pigment subset led to advancements in quantitative accuracy, with an important potential for improving the utility of large pigment databases, because the pigments in this subset are the most frequently analysed and typically the most abundant in natural systems. The emphasis here on core phytoplankton pigments does not negate the importance of other more unique pigments, because knowledge gained regarding methodological uncertainties in the analysis of any subset of pigments is usually pertinent to the quantitative analysis of other pigments.

The need for this appendix was established in the earliest NASA round robins, which exposed a scientific oddity: although all of the laboratories were familiar with and used Jeffrey *et al.* (1997) as a reference for HPLC terminology, none of the laboratories involved were sharing information on a recurring basis using unequivocally unique formulistic communications, that is, *a precise lexicon expressed with a unique symbology and vocabulary was not available to document all aspects of an HPLC method, particularly the formulations necessary to describe the many computational steps in a method.* The lack of nuance in the original symbology – and, thus, vocabulary – used by HPLC analysts

Phytoplankton Pigments: Characterization, Chemotaxonomy and Applications in Oceanography, eds. Suzanne Roy, Carole A. Llewellyn, Einar Skarstad Egeland and Geir Johnsen. Published by Cambridge University Press.

meant some important procedures were described using unnecessarily vague language. The inaccurate determination of extraction volume, for example, is a frequent occurrence, and in one HPLC round robin (Van Heukelem *et al.*, 2002), 25% of the laboratories calculated the extraction volume incorrectly; this mistake was also made by a laboratory participating in a subsequent round robin (Hooker *et al.*, 2005) and was of the order of a 5% overestimation of pigment concentration. In more than one of these cases, miscalculations were the result of interpretation errors by laboratories trying to implement a particular methodology.

In some parts of this appendix, abbreviations from the SCOR Working Group are used for pigment presentations, but the majority of the presentation uses a more compact lexicon (see below; note that these lexicon abbreviations differ from pigment abbreviations used in the rest of this volume). This lexicon was developed to satisfy the diversity of presentation requirements spanning text, tables and formulae inherent in documenting scientific investigations. Producing compact and easily deciphered formulae is particularly important for representing clearly the many nuances and differences between methods, and for producing a statistical description of an experiment, like a round robin. Regardless of the intended use of the vocabulary and symbology, it is important to remember that formulations require an understanding of the practices and procedures they represent. It is not enough to simply master the lexicon, particularly when dealing with literature for which the level of detail is not in keeping with the specificity presented here – some deeper investigation will probably be needed to establish whether or not the parameters involved are properly defined or represented.

As already noted, the variety of methods assembled to execute a round robin usually means that some pigments are analysed by only a few laboratories, whereas others are analysed by everyone. The latter constitute a core group of pigments that are routinely useful to many aspects of marine studies and, following the nomenclature of Claustre *et al.* (2004), are here referred to as the *primary pigments* (PPig) and are composed of the three total chlorophylls plus nine carotenoids. The *secondary pigments* are the individual pigments used to create a primary pigment composed of separate contributions (e.g. the total chlorophylls). Noting that at least three laboratories must quantity a pigment for the results to be statistically useful, a further classification is usually performed wherein the *tertiary pigments* are those pigments not included in the composition of the primary and secondary pigments for which three or more laboratories provided quantitations; and the *ancillary pigments* are those remaining pigments only analysed by one or two laboratories. Although this nomenclature implies some precedence or ranking, this is only true from the perspective of round robins based on marine phytoplankton pigments for which certain pigments are routinely used more often than others (e.g. chlorophyll *a*).

The pigments given a unique symbology here were selected from amongst many possibilities on the basis of several unique characteristics: a) they are frequently reported by laboratories and are typically the most commonly abundant in natural systems, b) they are from living (or produced by living) or senescent phytoplankton, c) they are important to an understanding of phytoplankton diversity and abundance, or d) they are important because they present a potential for chromatographic co-elution when analysed using HPLC methods. The symbology can be expanded, using the same structure, to include a myriad of other pigments. The symbols used to indicate the concentration (C) of an individual primary pigment are as follows:

C_{T_a} Total chlorophyll *a*
C_{T_b} Total chlorophyll *b*
C_{T_c} Total chlorophyll *c*
C_C Carotenes
C_A Alloxanthin
C_B 19′-Butanoyloxyfucoxanthin
C_{Dd} Diadinoxanthin
C_{Dt} Diatoxanthin
C_F Fucoxanthin
C_H 19′-Hexanoyloxyfucoxanthin
C_P Peridinin
C_Z Zeaxanthin

These are the same 12 pigments given in the topmost portion of Table 5A.1 (for the latter, note the common use of square brackets and a compact abbreviation to also indicate pigment concentrations). The first three primary pigments are the (total, T) pigment associations for the chlorophylls and the other nine are all carotenoids, of which Carotenes is a sum.

The concentrations of the secondary pigments that are needed to produce the primary pigments are as follows:

C_a Chlorophyll *a*
C_{D_a} Divinyl chlorophyll *a*
C_{C_a} Chlorophyllide *a*
C_b Chlorophyll *b*
C_{D_b} Divinyl chlorophyll *b*
C_{C_b} Chlorophyllide *b*
C_{c_1} Chlorophyll c_1
C_{c_2} Chlorophyll c_2
C_{c_3} Chlorophyll c_3
C_{Mg} Mg-2,4-divinyl pheoporphyrin a_5 monomethyl ester
$C_{\beta\beta}$ β,β-Carotene
$C_{\beta\varepsilon}$ β,ε-Carotene

These pigments are not the full subset of secondary pigments considered in NASA round robins, but are representative of those most important for marine studies. The concentrations of the tertiary and ancillary pigments are as follows:

C_{An} Antheraxanthin
C_{As} Astaxanthin
C_{Cn} Canthaxanthin

Table 5A.1. *The (individual) primary pigments plus the higher-order data products they are used to create: sums, ratios and indices. Each data type is shown with the corresponding variable forms, names and calculation formulae (if applicable). Applicable allomers and epimers are not shown explicitly, but are assumed to be a part of the appropriate calculations. The variable forms indicate pigment concentrations (square brackets) and are patterned after the nomenclature established by the SCOR Working Group 78 (Jeffrey* et al., *1997); the abbreviated forms are shown in parentheses. Note that the pigment indices (also considered as macrovariables) are a combination of a pigment sum and ratio.*

Variable	Primary pigment (PPig)	Calculation
[TChl *a*]	Total chlorophyll *a*[†]	[Chlide *a*] + [DVChl *a*] + [Chl *a*]
[TChl *b*]	Total chlorophyll *b*[†]	[Chlide *b*] + [DVChl *b*] + [Chl *b*]
[TChl *c*]	Total chlorophyll *c*[†]	[Chl c_1] + [Chl c_2][‡] + [Chl c_3]
[Caro]	Carotenes[†] (Caro)	[ββ-Car] + [βε-Car]
[Allo]	Alloxanthin (Allo)	
[But]	19′-Butanoyloxyfucoxanthin (But-fuco)	
[Diad]	Diadinoxanthin (Diadino)	
[Diato]	Diatoxanthin (Diato)	
[Fuco]	Fucoxanthin (Fuco)	
[Hex]	19′-Hexanoyloxyfucoxanthin (Hex-fuco)	
[Peri]	Peridinin (Peri)	
[Zea]	Zeaxanthin (Zea)	
Variable	*Pigment Sum*	*Calculation*
[TChl]	Total Chlorophyll (TChl)	[TChl *a*] + [TChl *b*] + [TChl *c*]
[PPC]	Photoprotective Carotenoids (PPC)	[Allo] + [Diad] + [Diato] + [Zea] + [Caro]
[PSC]	Photosynthetic Carotenoids (PSC)	[But] + [Fuco] + [Hex] + [Peri]
[PSP]	Photosynthetic Pigments (PSP)	[PSC] + [TChl]
[TAcc]	Total Accessory Pigments (TAcc)	[PPC] + [PSC] + [TChl *b*] + [TChl *c*]
[TPig]	Total Pigments (TPig)	[TAcc] + [TChl *a*]
[DP]	Total Diagnostic Pigments (DP)	[PSC] + [Allo] + [Zea] + [TChl *b*]
Variable	*Pigment Ratio*	*Calculation*
[TAcc]/[TChl *a*]	The [TAcc] to [TChl *a*] ratio	[TAcc]/[TChl *a*]
[TChl *a*]/[TPig]	The [TChl *a*] to [TPig] ratio	[TChl *a*]/[TPig]
[PPC]/[TPig]	The [PPC] to [TPig] ratio	[PPC]/[TPig]
[PSC]/[TPig]	The [PSC] to [TPig] ratio	[PSC]/[TPig]
[PSP]/[TPig]	The [PSP] to [TPig] ratio	[PSP]/[TPig]

Table 5A.1. (*cont.*)

Variable	Primary pigment (PPig)	Calculation
Variable	*Pigment Index*	*Calculation*
[mPF]	Microplankton proportion factor (MPF)	([Fuco] + [Peri])/[DP]
[nPF]	Nanoplankton proportion factor (NPF)	([Hex] + [But] + [Allo])/[DP]
[pPF]	Picoplankton proportion factor (PPF)	([Zea] + [TChl *b*])/[DP]

† Considered as individual pigments, although computed as sums by most methods.
‡ For many methods used for routine analysis, the contribution of MgDVP is included, and Chl c_1 and Chl c_2 are not chromatographically separated.

C_{Cr} Crocoxanthin
C_{Dc} Diadinochrome
C_{Dn} Dinoxanthin
C_E Echinenone
C_G Gyroxanthin diester
C_L Lutein
C_M Monadoxanthin
C_{Pb_a} Pheophorbide *a*
C_{Pb_b} Pheophorbide *b*
C_{Pt_a} Pheophytin *a*
C_{Pt_b} Pheophytin *b*
C_P Prasinoxanthin
C_{Py_a} Pyropheophytin *a*
C_S Siphonein
C_{Va} Vaucheriaxanthin (ester)
C_V Violaxanthin

The grouping of pigments to form sums permits the formulation of variables useful to different investigations. For example, the pooling of photosynthetic and photoprotective carotenoids (PSC and PPC, respectively) is useful to photophysiological studies (Bidigare *et al.*, 1987) and the total amount of accessory (non-chlorophyll *a*) pigments (TAcc) is useful in remote-sensing investigations (Trees *et al.*, 2000). The ratios derived from these pooled variables, e.g. [PSC]/[TChl *a*], are dimensionless, and have the advantage of automatically *scaling* the comparison of results from different areas and pigment concentrations.

A useful pigment sum is the total diagnostic pigments (DP), which was introduced by Claustre (1994) to estimate a pigment-derived analogue to the *f*-ratio (the ratio of new to total production) developed by Eppley and Peterson (1979). The use of DP was extended by Vidussi *et al.* (2001) and Uitz *et al.* (2006) to derive size-equivalent pigment indices that roughly correspond to the biomass proportions of pico-, nano- and micro-phytoplankton, which are denoted [pPF], [nPF], and [mPF] respectively, and are also referred to as *macrovariables*. These are composed of pigment sums and are ratios, so they should be particularly useful in reconciling inquiries applied to databases from different oceanic regimes.

Together with the individual primary pigments, the pigment sums, ratios and indices used in NASA round-robin inter-comparisons are presented in Table 5A.1. Note that [TChl *a*], [TChl *b*] and [TChl *c*] do not represent individual pigment concentrations – each represents a group of pigments roughly characterized by the same absorption spectra (including some degradation products). These chlorophyll sums allow comparison of results originating from HPLC methods that differ in the way the pigments within the same family are quantified (e.g. chlorophyll *c* types) or whose extraction procedures may or may not generate degradation forms (e.g. chlorophyllide *a*). Perhaps most importantly, these sums permit comparison of methods that differ in their capability for differentiating monovinyl from divinyl forms.

The definitions of the sums in Table 5A.1 – and, thus, the ratios that use them – are deterministic (or finite) and sometimes differ from the original definitions presented in the literature. For example, [TAcc] is composed of 11 primary pigments and not all of the possible accessory pigments a method can produce, because this differs between methods; similarly [PPC] does not include contributions from lutein and violaxanthin, because many methods do not quantify these pigments. From the point of view of a round-robin inter-comparison, exact definitions are required, because this allows unequivocal comparison of a parameter. Indeed the recommendation proposed here is that all principal parameters should have unique formulations based on commonly quantitated (primary) pigments, so that databases and methods can be inter-compared and utilized with equal efficacy.

Alternative definitions based on the individual capabilities of a method should be identified by slightly different symbology. For example, the total accessory pigments produced by an individual method could be represented by [ΣAcc], and the total of all pigments by [ΣPig]. Note that from a purist point of view, no one method can quantify all pigments needed to properly compute the pigment sums and ratios – many of which will be at inconsequential concentrations – but many methods could provide the primary pigments involved, so it makes sense to agree on definitions based on the latter. Note that in the current era of emerging large databases, investigators are going to create the sums and ratios they need, and if a particular pigment is not available from a particular method, a recurring practice is to create the sum assuming zero concentrations for the missing pigments.

The simplicity of the arguments for finite definitions makes it easy to imagine counterarguments against this rather global approach based on smaller-scale instances. For example, the formulations for the pigment indices might require alternative definitions for open ocean environments versus coastal and shelf waters, because of differences in community composition (and, thus, a need for different pigment sums). More exact formulations for the indices based on location and trophic status could – and for some inquiries, should – be devised. The point here is to establish a global set of definitions for exact inter-comparisons, which are defined within a symbology and vocabulary framework that supports any needed refinements or nuances. The competence and applicability of the global approach has already been demonstrated in the literature (Uitz *et al.*, 2006) and is not considered a limitation.

The symbology presented here is used primarily to represent the final pigment concentrations for each field sample, because this is the way most laboratories report their result, and to facilitate statistical analysis of the data. The basic relationships for calculating the concentration of an individual pigment emerge in a straightforward fashion from the calibration methodology, but as the nuances of an individual method are incorporated, the formulations become more complicated – and it is the details of the nuances that frequently lead to the aforementioned interpretation errors. Ignoring the specific details of the basic HPLC processes, because they are presented in detail by Jeffrey *et al.* (1997) and Bidigare *et al.* (2003), the formulation for quantitation begins with the terms describing the calibration of the HPLC system:

$$\tilde{C}_{P_i} = \hat{A}_{P_i} R_{P_i}, \tag{5A.1}$$

where $\tilde{C}_{P_i}$ is the amount of pigment injected (usually in units of nanograms), $\hat{A}_{P_i}$ is the area of the parent peak (and associated isomers in some instances) for pigment P_i (usually in milli-absorbance units[1] or microvolts as a function of time) and R_{P_i} is the response factor. The latter is the calibration coefficient for the HPLC system, and it takes on a separate value for each pigment being quantitated. For the general problem, the response factor is denoted R, but for the specific problem of a particular pigment, it is denoted R_{P_i}. The R values are usually expressed as the amount of pigment divided by the peak area.

The formulation given in (5A.1) is based on a single-point calibration wherein one or more injections of a calibration standard at a known concentration is injected onto the HPLC column. An alternative approach is to create a dilution series of the pigment standard, inject these one at a time, and then fit the response of the HPLC system to a linear function ($y = mx + b$) using least-squares analysis (this is also referred to as a *multi-point* calibration). In this case, pigment amount is computed as

$$\tilde{C}_{P_i} = \frac{\hat{A}_{P_i} - b_{P_i}}{m_{P_i}}, \tag{5A.2}$$

[1] A milli-absorbance unit is denoted mAU.

where m_{P_i} is the slope (equating change in peak area with change in amount) and b_{P_i} is the y-intercept.

The formulation given in (5A.2) can be expressed to follow (5A.1) as follows:

$$\tilde{C}_{P_i} = \hat{A}_{P_i}\left[\frac{1-(b_{P_i}/\hat{A}_{P_i})}{m_{P_i}}\right], \tag{5A.3}$$

where the equivalent R_{P_i} for (5A.1) is given by the terms in brackets. If the linear regression is forced through zero, $b_{P_i} = 0$, and (5A.3) becomes

$$\tilde{C}_{P_i} = \frac{\hat{A}_{P_i}}{m_{P_i}} \tag{5A.4}$$

and $R_{P_i} = 1/m_{P_i}$ (note that the inverse slope is change in amount divided by change in peak area, which matches the definition for R). In this context, it is convenient to reconsider the definition of R_{P_i}, which some authors have done (Bidigare *et al.*, 2003), as the inverse of the original definition, that is, $F_{P_i} = 1/R_{P_i}$ and (5A.1) becomes

$$\tilde{C}_{P_i} = \frac{\hat{A}_{P_i}}{F_{P_i}}. \tag{5A.5}$$

The advantage of this approach is that F_{P_i} follows directly from the slope of the linear calibration curve and, for the common case of forcing the slope through zero, $F_{P_i} = m_{P_i}$. For the purposes of this presentation, which is derived from the experiences of the NASA round-robin community, the majority of the methods used the original definition of R, so it is retained hereafter.

The governing equation for the determination of pigment concentration can be expressed as

$$\tilde{C}_{P_i} = \frac{V_x}{V_f}\frac{\tilde{C}_{P_i}}{V_c}, \tag{5A.6}$$

where V_x is the extraction volume, V_c is the volume of sample extract injected onto the HPLC column (measured in the same units as V_x) and V_f is the volume of water filtered in the field (usually through a glass-fibre filter with a 0.7 μm pore size) to create the sample (measured in litres).

There are two common procedures associated with the use of an internal standard, and they are distinguished here by the number of laboratory steps involved: a) the extraction solvent and internal standard are contained together in a mixture (prepared beforehand), which is added to the sample in one step, or b) the extraction solvent and internal standard are added separately in two steps.

In the one-step approach, a volume of solvent and internal standard is mixed together in a batch, and a small portion of the mixed volume, V_m (typically 3 mL), is added to the sample. In the two-step approach, a volume of the extraction solvent, V_e, is added to the sample (e.g. 3 mL) followed by a small volume of internal standard, V_s (e.g. 50 μL). The filter, now soaking in the solvent–standard mixture, is disrupted (most commonly with a sonic probe), clarified (to remove filter debris) and a volume of the clarified sample extract, V_c, is injected onto the HPLC column.

The advantage of using an internal standard is V_x (5A.6) can be computed more precisely by correcting for the presence of residual water retained on the filter (plus any variations in volume caused by evaporation) by using a) the peak area of the internal standard when it is injected onto the HPLC column ($\hat{A}_c$) prior to its addition to the sample, and b) the peak area of the internal standard in the sample ($\hat{A}_s$). In the one-step approach, $\hat{A}_c$ is determined by injecting the solvent–standard mixture onto the HPLC column, whereas for the two-step approach, the standard is injected directly onto the column.

For the one-step approach, the internal standard is diluted by the extraction solvent, so Equation (5A.6) is rewritten as

$$\tilde{C}_{P_i} = \frac{\hat{A}_{c1}}{\hat{A}_{s_1}} \frac{V_m}{V_f} \frac{\breve{C}_{P_i}}{V_c} \quad (5A.7)$$

that is, $V_{x_1} = V_m \hat{A}_{c_1}/\hat{A}_{s_1}$, where the '1' in the subscripts indicates the one-step methodology. For the two-step approach, Equation (5A.6) is rewritten as

$$\tilde{C}_{P_i} = \frac{\hat{A}_{c2}}{\hat{A}_{s_2}} \frac{V_s}{V_f} \frac{\breve{C}_{P_i}}{V_c} \quad (5A.8)$$

and $V_{x_2} = V_s \hat{A}_{c_2}/\hat{A}_{s_2}$ (the '2' in the subscripts indicates two-step methodology).

If an internal standard is not used, an estimate of the volume of water retained on the filter, V_w, is added to the volume of extraction solvent, V_e, so Equation (5A.6) is rewritten as

$$\tilde{C}_{P_i} = \frac{V_e + V_w}{V_f} \frac{\breve{C}_{P_i}}{V_c} \quad (5A.9)$$

and $V_{x'} = V_e + V_w$. For a 25 mm filter, water retention is usually assumed to be 0.2 mL (Bidigare *et al.*, 2003).

The formulations presented in Equations (5A.7)–(5A.9) document alternative possibilities for specifying the V_x term in Equation (5A.6). The uncertainties associated with the three representations (V_{x_1}, V_{x_2} and $V_{x'}$) will necessarily be different, so the distinctions are important, particularly when it comes to properly documenting a method. Within the context of two analysts discussing their individual methods, a lack of distinction in the symbology – and, thus, the vocabulary – can lead to significant misunderstandings about the protocols being used.

An understanding of the factors affecting peak area integration is important because peak area is a recurring variable in the calculation equations, (5A.1)–(5A.3) and (5A.7)–(5A.8). Broad, asymmetrical peaks yield poorer results than sharp narrow peaks, in part because the separation between pigments is compromised, and because differentiation of the peak beginning and ending from the baseline slope is more difficult to determine.

The separation or resolution, R_s, between peaks is commonly calculated based on peak widths at half-height or at the baseline. For this example, the calculation is based on the baseline:

$$R_s = 2\frac{t_{R_2} - t_{R_1}}{\hat{w}_{B_2} + \hat{w}_{B_1}}, \tag{5A.10}$$

where t_{R_1} and t_{R_2} are the retention times of peaks 1 and 2, respectively, and $\hat{w}_{B_1}$ and $\hat{w}_{B_2}$ are the corresponding widths at their bases (Wright, 1997).

Solvent strength is described by the so-called *polarity index* (PI) and denoted, P_I. The higher the P_I value, the weaker, or more polar, the solvent is in relation to the type of stationary phase being used. When a mobile phase is composed of more than one solvent, P_I is calculated as in the following example for a 70:30, methanol:buffer solution (Bidlingmeyer, 1992):

$$\begin{aligned} P_I &= (0.70 \times 6.6) + (0.30 \times 9.0) \\ &= 7.3, \end{aligned} \tag{5A.11}$$

where 6.6 is the PI of pure methanol, and 9.0 is the PI of water.

Concentrations of primary pigment standards are determined spectrophotometrically based on the principles of the Lambert-Beer law, which states that the fraction of the incident light at a particular wavelength λ that is absorbed by a solution depends on the thickness of the sample, the concentration of the absorbing compound in the solution and the chemical nature of the absorbing compound (Segel, 1968). This relationship can be expressed as

$$A(\lambda) = a(\lambda) l_c C, \tag{5A.12}$$

where $A(\lambda)$ is absorbance, $a(\lambda)$ is the absorption coefficient (a constant), l_c is the thickness of the sample in centimetres (the pathlength of the cuvette being used) and C is concentration. To determine concentration from a measured absorbance, Equation (5A.12) is rewritten as

$$C = \frac{A(\lambda)}{a(\lambda)\, l_c}, \tag{5A.13}$$

where the units for C depend on the expression of $a(\lambda)$. For example, for consistency with current NASA round-robin inter-comparisons regarding absorptivity, if the concentration is expressed in molarity, a becomes the molar absorption coefficient (ε) and if the concentration is expressed as grams per litre, a is the specific absorption coefficient (α, also denoted d in the Data sheets at the end of this volume, according to IUPAC nomenclature). If concentration is expressed in per cent weight per volume (usually in units of grams per 100 mL), a becomes $a_{1\%}$ (Segel, 1968). Usually, a 1 cm pathlength is used, so $l_c = 1$ cm in most cases. Absorption coefficients vary depending on wavelength and the solvent in which the compound is suspended, and they are always provided with the solvent and wavelength used.

Primary pigment standards used to calibrate HPLC systems are either a) purchased in solution (with concentrations provided by the manufacturer), b) isolated from natural sources, or c) purchased in solid form. In the latter two cases, pigments are suspended in the solvents specified for use with the absorption coefficients selected by the laboratory involved, and absorbance is measured spectrophotometrically at the wavelength specified with the selected absorption coefficient. Assuming the specific absorption coefficient is used, determination of the concentrations of pigment standards requires the wavelength of maximum absorbance for the particular pigment (which is specified for use with the absorption coefficient) and a correction measurement for the absorbance of the pigment at 750 nm:

$$C_{P_i} = \frac{A(\lambda_m) - A(750)}{\alpha(\lambda_m)\ l_c}, \tag{5A.14}$$

where λ_m is the corresponding wavelength of maximum absorbance.

In the analytical approach adopted in NASA round robins for field samples, no one laboratory (or result) is presumed more correct than another – all methods are considered properly validated by the individual analysts. Furthermore, there is no absolute truth for field samples, so an unbiased approach is needed to inter-compare the methods. The first step in developing an unbiased analysis is to calculate the average concentration for each pigment in each sample as a function of the contributing laboratories:

$$\bar{C}^{L_j}_{P_i}(S_k) = \frac{1}{N_R}\sum_{l=1}^{N_R} C^{L_j}_{P_i}(S_{k,l}), \tag{5A.15}$$

where P_i identifies the pigment or pigment association, L_j is the laboratory (or method) code, $S_{k,l}$ sets the sampling station (or site) and the replicate number, with the k index for the former and l for the latter and N_R is the total number of replicates (usually three for NASA round robins). The collection of replicates for a particular sampling station is usually referred to as a 'batch'. Only one value for each batch of replicates is reported for each sampling site, and this is generically referred to as a 'sample' (so the number of samples equals the number of batches or stations). This is done to mimic the usual practice of having only one realization of a pigment per station (understanding that in most field campaigns, the characterization of a sampling site or station usually includes data from different depths, but, again, each depth is usually characterized using a solitary value for each pigment).

Averages of a sample across the methods reporting a particular pigment in a sample are used to estimate the true value of the pigment for each sample:

$$\bar{C}^{A}_{P_i}(S_k) = \frac{1}{N_L}\sum_{j=1}^{N_L} \bar{C}^{L_j}_{P_i}(S_k), \tag{5A.16}$$

where the superscript A denotes an average across all (applicable) methods, and N_L is the number of laboratories quantitating a pigment.

The unbiased per cent difference (UPD), ψ, for each pigment of the individual laboratories with respect to the average values is then calculated for each sample as

$$\psi^{L_j}_{P_i}(S_k) = 100\frac{C^{L_j}_{P_i}(S_k) - \bar{C}^{A}_{P_i}(S_k)}{\bar{C}^{A}_{P_i}(S_k)}. \tag{5A.17}$$

Note that the formulation in Equation (5A.17) provides a relative per cent difference (RPD), because it is signed: a positive ψ value indicates that the pigment concentration for a particular laboratory is greater than the average for that pigment (a negative value indicates that the laboratory pigment concentration is less than the average). Although $\bar{C}^{A}_{P_i}$ is not considered truth, it is the reference value or proxy for truth by which the performance of methods with respect to one another is quantified.

When RPD values for methods that do not present any trend relative to the average consensus are summed, however, there is the risk of destroying some or all of the variance in the data. To preserve an appropriate measurement of the variance in the data, absolute UPD values, $|\psi|$, are averaged over the number of samples (N_S) to give the average absolute per cent difference (APD) of each laboratory for each pigment across all the samples:

$$|\bar{\psi}|^{L_j}_{P_i} = \frac{1}{N_S}\sum_{k=1}^{N_S}\left|\psi^{L_j}_{P_i}(S_k)\right|, \tag{5A.18}$$

where S_k is the kth sample number associated with pigment P_i. For NASA round robins, 12 or 24 batches of triplicates are collected, from which 12 or 24 samples are quantitated, respectively.

Absolute values are used in the overall averages, so positive and negative ψ values do not cancel out and artificially lower the average difference. The latter is particularly important for pigments with low concentrations, but also in terms of a general philosophy: the ψ values are the primary measure of dispersion between the methods, so it is important to ensure that they are not underestimated.

Another useful parameter is the average of the $|\bar{\psi}|$ values for a particular pigment across the number of laboratories (N_L) reporting the pigment:

$$|\bar{\psi}|^{A}_{P_i} = \frac{1}{N_L}\sum_{j=1}^{N_L}|\bar{\psi}|^{L_j}_{P_i}, \tag{5A.19}$$

where the A code indicates that all the laboratories are averaged (and $\bar{\psi}^{A}_{P_i}$ values are formed in a similar fashion from the UPD values). In general, Equation (5A.19) is only computed for the primary pigments, because they are the only ones routinely quantitated by all laboratories.

To examine the replicate data for each method more closely, the coefficient of variation (ξ) is used, which is expressed as the per cent ratio of the standard deviation (σ) in each batch of replicates with respect to the average concentration for the pigment in the batch ($\bar{C}_{P_i}$):

$$\xi_{P_i}^{L_j}(S_k) = 100\frac{\sigma_{P_i}^{L_j}(S_k)}{\bar{C}_{P_i}^{L_j}(S_k)}, \tag{5A.20}$$

where S_k is the kth sample number. Individual ξ values are computed for each pigment, for each sample and for each method; and then all the ξ values for a particular method are averaged to yield an average precision ($\bar{\xi}$) for the method and pigment:

$$\bar{\xi}_{P_i}^{L_j} = \frac{1}{N}\sum_{k=1}^{N_S}\xi_{P_i}^{L_j}(S_k). \tag{5A.21}$$

The formulations presented in Equations (5A.17)–(5A.21) are for field samples, but they are applicable to the use of laboratory standards or mixes as inter-comparison samples by redefining the indexing limits and setting S_k to match the laboratory standards.

References

Bidigare, R.R., Smith, R.C., Baker, K.S. and Marra, J. (1987). Oceanic primary production estimates from measurements of spectral irradiance and pigment concentrations. *Global Biogeochem. Cycles* **1**, 171–86.

Bidigare, R.R., Van Heukelem, L. and Trees, C.C. (2003). HPLC phytoplankton pigments: sampling, laboratory methods and quality assurance procedures. In *Ocean Optics Protocols for Satellite Ocean Color Sensor Validation, Revision 4, Volume V: Biogeochemical and Bio-optical Measurements and Data Analysis Protocols*. NASA Technical Memorandum 2003–211621/Rev4-Vol.V, ed. J.L. Mueller, G.S. Fargion and C.L. McClain. Greenbelt: NASA Goddard Space Flight Center, pp. 5–14.

Bidlingmeyer, B.A. (1992). *Practical HPLC Methodology and Applications*. New York: John Wiley & Sons.

Claustre, H. (1994). The trophic status of various oceanic provinces as revealed by phytoplankton pigment signatures. *Limnol. Oceanogr.* **39**, 1206–10.

Claustre, H., Hooker, S.B., Van Heukelem, L., Berthon, J.F., Barlow, R., Ras, J., Sessions, H., Targa, C., Thomas, C.S., van der Linde, D., Marty, J.-C. (2004). An intercomparison of HPLC phytoplankton pigment methods using *in situ* samples: application to remote sensing and database activities. *Mar. Chem.* **85**, 41–61.

Eppley, R.W. and Peterson, B.J. (1979). Particulate organic matter flux and planktonic new production in the deep ocean. *Nature* **282**, 677–80.

Hooker, S.B., Claustre, H, Ras, J., Van Heukelem, L., Berthon, J.-F., Targa, C., van der Linde, D., Barlow, R. and Sessions, H. (2000). *The First SeaWiFS HPLC Analysis Round-Robin Experiment (SeaHARRE-1)*, SeaWiFS Postlaunch

Technical Report Series. S. B. Hooker and E. R. Firestone (eds). NASA Technical Memorandum 2000–206892, vol. 14. Greenbelt: NASA Goddard Space Flight Center.

Hooker, S.B., Van Heukelem, L., Thomas, C.S., Claustre, H., Ras, J., Barlow, R., Sessions, H., Schlüter, L., Perl, J., Trees, C., Stuart, V., Head, E., Clementson, L., Fishwick, J., Llewellyn, C. and Aiken, J. (2005). *The Second SeaWiFS HPLC Analysis Round-Robin Experiment (SeaHARRE-2).* NASA Technical Memorandum 2005–212785, Greenbelt: NASA Goddard Space Flight Center.

Hooker, S.B., Van Heukelem, L., Thomas, C.S., Claustre, H., Ras, J., Schlüter, L., Clementson, L., Van der Linde, D., Eker-Develi, E., Berthon, J.-F., Barlow, R., Sessions, H., Ismail, H. and Perl, J. (2009). *The Third SeaWiFS HPLC Analysis Round-Robin Experiment (SeaHARRE-3).* NASA Technical Memorandum 2009–215849, Greenbelt: NASA Goddard Space Flight Center.

Jeffrey, S.W., Mantoura, R. F. C. and Wright, S.W. (eds.) (1997). *Phytoplankton Pigments in Oceanography: Guidelines to Modern Methods.* Paris: UNESCO Publishing.

Segel, I.H. (1968). *Enzymes. Biochemical Calculations.* New York: John Wiley & Sons.

Trees, C.C., Clark, D.K., Bidigare, R.R., Ondrusek, M.E. and Mueller, J.L. (2000). Accessory pigments versus chlorophyll *a* concentrations within the euphotic zone: a ubiquitous relationship. *Limnol. Oceanogr.* **45**, 1130–43.

Uitz, J., Claustre, H., Morel, A. and Hooker, S.B. (2006). Vertical distribution of phytoplankton communities in open ocean: an assessment based on surface chlorophyll. *J. Geophys. Res.* **111**, C08005, doi:10.1029/2005JC003207.

Van Heukelem, L., Thomas, C.S. and Glibert, P. (2002). *Sources of variability in chlorophyll analysis by fluorometry and high-performance liquid chromatography in a SIMBIOS inter-calibration exercise.* NASA Technical Memorandum 2002–211606, Greenbelt: NASA Goddard Space Flight Center.

Vidussi, F., Claustre, H., Manca, B.M., Luchetta, A. and Marty, J.-C. (2001). Phytoplankton pigment distribution in relation to upper thermocline circulation in the eastern Mediterranean Sea during winter. *J. Geophys. Res.* **106**, 19939–56, doi:10.1029/1999JC000308.

Wright, S.W. (1997) Appendix H: Summary of terms and equations used to evaluate HPLC chromatograms. In: *Phytoplankton Pigments in Oceanography: Guidelines to Modern Methods*, ed. S.W. Jeffrey, R. F. C. Mantoura and S.W. Wright. Paris: UNESCO Publishing, pp. 622–30.

6

Quantitative interpretation of chemotaxonomic pigment data

HARRY W. HIGGINS, SIMON W. WRIGHT[1] AND LOUISE SCHLÜTER

6.1 Introduction

The use of pigments for the quantitative chemotaxonomic analysis of phytoplankton populations began in the 1970s when thin layer chromatography revealed a large diversity of pigments in the phytoplankton (Jeffrey, 1974), many of which appeared to be restricted to certain algal taxa and could be sensitively detected even in the presence of protozoa, bacteria and detritus, in which they are absent. Since that time, developments in phytoplankton systematics, particularly through DNA analysis (e.g. Karlson *et al.*, 2010), have revealed a much greater taxonomic diversity in phytoplankton than previously imagined, while simultaneous improvements in chromatography have led to the identification of >70 pigments in 45 pigment patterns (tabulated in Jeffrey *et al.*, Chapter 1, this volume). This provides substantial additional power in pigment analysis but also hugely complicates interpretation of pigment data. Relatively few pigments are now regarded as unambiguous markers – most are distributed across several taxa.

The pigment composition of microalgae is strongly influenced by several environmental factors that complicate interpretation of field data. A full synopsis is beyond the scope of this chapter, but known influences and key references include: irradiance (Johnsen *et al.*, 1994; Goericke and Montoya, 1998; Schlüter *et al.*, 2000; Rodríguez *et al.*, 2006a), spectral distribution of light (Wood, 1985), ultraviolet (Gerber and Häder, 1994); day length (Sakshaug and Andresen, 1986), diurnal cycle (Tukaj *et al.*, 2003), nutrient status (Goericke and Montoya, 1998; Henriksen *et al.*, 2002; Stæhr *et al.*, 2004; Hou *et al.*, 2007), iron concentration (van Leeuwe and Stefels, 1998; DiTullio *et al.*, 2007; Hopkinson *et al.*, 2007), mixing regime (Brunet *et al.*, 2003; Thompson *et al.*, 2007) and growth phase (Wilhelm and Manns, 1991; Henriksen *et al.*, 2002; Redalje *et al.*, 2008). The pigment content can vary qualitatively between members of a genus, or even between strains of single species (Stolte *et al.*, 2000;

[1] Corresponding author (Simon.Wright@aad.gov.au).

Phytoplankton Pigments: Characterization, Chemotaxonomy and Applications in Oceanography, eds. Suzanne Roy, Carole A. Llewellyn, Einar Skarstad Egeland and Geir Johnsen. Published by Cambridge University Press.

Zapata *et al.*, 2004; Laza-Martinez *et al.*, 2007). Thus the pigment content of a field population cannot be accurately predicted even if one knows the species present.

Nevertheless, pigments remain as valuable chemotaxonomic markers for algal taxa, particularly for the pico- and nano-phytoplankton, which are difficult to identify microscopically. The strength of pigment analysis for surveys is in identifying and mapping populations within a data set, whether they are regions within a geographic survey or a succession of populations in a temporal study.

This chapter considers interpretation of pigment field data – the status of various pigment markers, the tools available for quantitative interpretation, what is known of the variability of pigment content in the field and how interpretations of pigment data compare with other techniques.

6.2 Qualitative assessment of data

The pigment analyst's prime tool is the tabulated chemotaxonomic distribution of marker pigments (e.g. Chapter 1, this volume). However, such tables are inherently incomplete and of inconsistent reliability, with as yet unrecognized taxa missing, and some pigment types based on very few strains of unknown general applicability. While the presence of pigments in a particular strain may be reliable, the absence of pigments is less so because minor peaks may be unidentified or unreported. Finally, the possible symbiotic acquisition of chloroplasts, particularly by dinoflagellates or ciliates, should not be ignored. Microscopy and/or other techniques should be used where possible to support interpretation of pigment data. Nevertheless, much taxonomic information can be deduced from pigment data, albeit with some ambiguities.

The complex distribution of chemotaxonomic pigments requires a structured approach to interpretation of chromatograms. The relative concentrations of chlorophyll (Chl) *b* and Chls *c* indicate the relative importance of taxa from the green and red lineages, facilitating subsequent interpretation.

6.2.1 Specific markers for algal types

Certain pigments are restricted to particular taxa and are (relatively) unambiguous markers for them, albeit with caveats as noted below (Types *sensu* Jeffrey *et al.*, Chapter 1, this volume):

- DVChl *a*, DVChl *b*: prochlorophytes = cyanobacteria type 4.
- Siphonaxanthin esters: prasinophytes type 2B.
- Prasinoxanthin (Pras): prasinophytes type 3A; with uriolide/micromonal group: prasinophytes type 3B.
- Peridinin (Peri): dinoflagellates type 1.
- Alloxanthin (Allo): cryptophytes, but commonly found in cryptophycean symbionts of the ciliate *Myrionecta rubra* (*Mesodinium rubrum*) (Hibberd, 1977) and dinoflagellates type 4, and rarely in chlorophytes type 1.

- Gyroxanthin diesters: significant peaks in dinoflagellates type 2, proposed as markers for toxic species such as *Gymnodinium breve* (= *Karenia brevis*) (Millie *et al.*, 1997) however a similar pigment is found in pelagophytes type 1 and some haptophytes (Bjørnland *et al.*, 2003; Zapata, 2005).
- Markers for minor groups: Eutreptiellanone: euglenophytes; Vaucheriaxanthin esters: eustigmatophytes and chrysophytes; Bacteriochlorophyll *a*: proteobacteria; Heteroxanthin: xanthophytes.
- Non-polar chlorophyll *c* pigments: haptophytes and dinoflagellates.
  - Chl c_2-MGDG [18:4/14:0] can be a major pigment in *Emiliania huxleyi* and is found in haptophytes types 3–8 *sensu* Zapata *et al.* (2004), but has not been found in dinoflagellates.
  - Chl c_2-MGDG [14:0/14:0] was regarded as a defining pigment of haptophytes type 7 (genus *Chrysochromulina*, Zapata *et al.*, 2004) but traces have since been found in two dinoflagellates (Laza-Martinez *et al.*, 2007). For use in CHEMTAX, see Rodríguez *et al.* (2006b).
  - Chl c_1-MGDG pigments are restricted to haptophytes types 3–5.
- Polar chlorophyll *c* pigments
  - MgDVP is a trace intermediate in virtually all taxa, but is a major marker pigment in prasinophytes type 3. Chl $c_{3(\mathrm{CS\text{-}170})}$ has been found only in some prasinophytes type 3.
  - Chls c_1, c_2 and c_3 are quite widespread in the chromophyte algae (diatoms, haptophytes and chrysophytes (including pelagophytes)). Jeffrey *et al.* (Chapter 1) list several new Chl *c* types that may be useful markers.
- Zeaxanthin (Zea) is a widespread minor pigment but is a useful marker for Cyanobacteria when they are dominant in tropical and subtropical waters. In polar waters where cyanobacteria are virtually absent, the major source of zeaxanthin may be bacteria (Wright *et al.*, 2009).
- 4-keto-myxoxanthophyll and its ester were found to be useful markers for the toxic cyanobacterium *Nodularia spumigena* in the Baltic Sea (Schlüter *et al.*, 2004, 2008).
- If Chl *b* is present, chlorophytes and prasinophytes type 1 can be distinguished by their relative ratios of lutein to Chl *b* (Lutein:Chl *b* = 0.30–1.77, 0–0.18, respectively, Schlüter and Møhlenberg, 2003). Prasinophytes generally have higher zeaxanthin: lutein ratios than chlorophytes. Micromonal, micromonol, and the 'unidentifed carotenoid from *Micromonas pusilla*' (UnkMp, Wright and Jeffrey, 1997; Zapata *et al.*, 2000) distinguish prasinophytes type 3 from other green algae. UnkMp is useful as it elutes after Chl *a* in a relatively clear part of the chromatogram.
- Fucoxanthin (Fuco) and its derivatives are very difficult to allocate, since they are major components in many taxa, including diatoms, haptophytes, dinoflagellates, chrysophytes, pelagophytes, raphidophytes and bolidophytes – many of which may be co-dominant. Thus Fuco, 19′-butanoyloxyfucoxanthin (But-fuco), and 19′-hexanoyloxyfucoxanthin (Hex-fuco) generally have multiple sources within a population. Moreover, the ratios of these pigments and of their derivatives vary widely within given

pigment types (e.g. Zapata *et al.*, 2004; Van Lenning *et al.*, 2003). Recently recognized keto-derivatives 4-ketofucoxanthin (haptophytes type 5), and 4-keto-Hex-fuco (haptophytes types 6–8) offer more specificity, but have already been discovered in dinoflagellates (De Salas *et al.*, 2004).

- Dinoflagellates are generally difficult to distinguish using pigments, apart from type 1, with the unambiguous markers peridinin (Peri) and dinoxanthin. All other pigmented types have acquired chloroplasts and their pigments from other taxa. Heterotrophic dinoflagellates may be dominant yet invisible to pigment analysis, except for what they have consumed!
- Diadinoxanthin and diatoxanthin have little chemotaxonomic value since they are so widespread taxonomically, however their relative concentrations change rapidly with irradiance (Demers *et al.*, 1991; see also Chapter 11) and hence they are valuable indicators of light history (Welschmeyer and Hoepffner, 1986; Claustre, 1994) and vertical mixing velocities (Brunet *et al.*, 2003), but only if samples can be filtered and frozen within minutes of collection, due to their rapid interconversion.
- It should be noted that while some pigments (e.g. MV-Chl c_3, Chl c_2-like *Pavlova gyrans*-type and 4-ketofucoxanthin) are extremely useful in determining the presence of particular taxa, low field concentrations (low cellular content or low taxa abundance), incomplete chromatographic resolution or limited information on the universal presence of the pigment in all members of the given taxa and the the effect of environmental factors on Pigment:Chl *a* ratios may preclude their use in quantitave analysis (Zapata *et al.*, 2004; Laza-Martinez *et al.*, 2007).

6.3 Non-taxonomic interpretation of pigment data sets

Before considering methods for taxonomic interpretation, two alternative approaches will be described.

6.3.1 Pigment based size classes

The pigment index (F_P) was introduced as an index of trophic status, based on seven diagnostic pigments (DP): Fuco, Peri, Hex-fuco, But-fuco, Zea, TChl *b* (= Chl *b* + DVChl *b*) and Allo (Claustre, 1994), where

$$F_P = (\text{Fuco} + \text{Peri})/\Sigma\,\text{DP}.$$

This ratio was based on the fact that diatoms and dinoflagellates typically flourished in regions of new production, whereas other taxa were more typical of recycling communities, and a rough taxonomic allocation of pigments: Fuco (diatoms), Peri (dinoflagellates), Hex-fuco and But-fuco (prymnesiophytes and chrysophytes), Zea (cyanobacteria), TChl *b* (chlorophytes and prochlorophytes) and Allo (cryptophytes). It was analogous to the *f*-ratio (new production/total production) of Eppley and Peterson (1979) with a similar magnitude.

The pigment index was reinterpreted as the proportion of microplankton by Vidussi *et al.* (2001), who introduced similar categories for nanoplankton and picoplankton, later modified by Uitz *et al.* (2006), who used multiple linear regression of a global database to determine weighting factors for each DP that were then used to calculate a weighted sum of diagnostic pigments (Σ DP_w) and fractions of three size classes, thus:

$$\begin{aligned}
\Sigma\, DP_w &= 1.41[\text{Fuco}] + 1{:}41[\text{Peri}] + 1.27[\text{Hex-fuco}] + 0.35[\text{But-fuco}] \\
&\quad + 0.60[\text{Allo}] + 1.01[\text{TChl}\,b] + 0.86[\text{Zea}] \\
f_{micro} &= (1.41[\text{Fuco}] + 1.41[\text{Peri}])/\Sigma\, DP_w \\
f_{nano} &= (1.27[\text{Hex-fuco}] + 0.35[\text{But-fuco}] + 0.60[\text{Allo}])/\Sigma\, DP_w \\
f_{pico} &= (1.01[\text{TChl}\,b] + 0.86[\text{Zea}])/\Sigma\, DP_w
\end{aligned}$$

The authors recognised that these are quasi-size classes, involving many assumptions of the taxonomic affiliations of pigments and the size distribution of phytoplankton taxa. In particular, they do not recognise the proliferation of picoeukaryotes, many of which contain Fuco and Hex-fuco (see Section 6.6.3). Indeed, substantial proportions of these pigments plus alloxanthin were found in the 0.4–2 μm size range in the Southern Ocean (Uitz *et al.*, 2009). Nevertheless, the equations gave good predictions of total Chl *a* (> 10 μm), but were less successful in the < 10 μm category.

This technique has not been widely tested, particularly in local applicability of the weighting factors derived from the global database. Also the weighting factors are based on pigment concentrations integrated over the euphotic zone and thus are not directly applicable to vertical profiles. However, since global measurements of phytoplankton functional types are urgently required (Le Quéré *et al.*, 2005), this technique seems worthy of further exploration and, if necessary, local calibration.

6.3.2 Ecological similarity indices

In a thorough analysis of a four-year time series from the English Channel, Sherrard *et al.* (2006) applied six similarity indices to HPLC pigment concentrations and to phytoplankton biomass data based on microscopy. Highly significant correlations were found between indices based on pigments and microscopy, and the technique was suggested as a useful tool for rapid and reproducible environmental monitoring of phytoplankton communities, particularly in relation to eutrophication. This method uses raw pigment data directly, without prior interpretation, and is thus ideally suited to detecting changes in communities, although other techniques will be required to determine what those changes are.

6.4 Mathematical tools for taxonomic interpretation of pigment data sets

Five methods have been employed to determine the quantitative composition of complex phytoplankton populations from chemotaxonomic pigment data with unpredictable marker pigment:Chl *a* ratios:

(1) *Multiple linear regression* has been employed to determine the relationship between single pigment markers and Chl *a* (Gieskes and Kraay, 1983) and remains in common use. This method is statistically sound but cannot account for the majority of markers that are shared between several taxa. However, it remains a valuable tool for initial exploration of pigment data, or for interpretation of very large data sets, particularly in relation to remote sensing (e.g. Uitz *et al.*, 2006).

(2) *Inverse simultaneous equations* (ISE) address the problem of shared markers by calculating the contribution of each taxon to total Chl *a* from the concentrations of marker pigments while subtracting contributions to those marker pigments by other groups (Everitt *et al.*, 1990; Letelier *et al.*, 1993; Bidigare and Ondrusek, 1996; Vidussi *et al.*, 2000). For instance, as one of multiple simultaneous equations, Letelier *et al.* (1993) calculated the contribution of cyanobacteria to chlorophyll *a* thus:

$$[\mathrm{Chl}\, a]_{\mathrm{cyano}} = 2.1\{[\mathrm{Zea}] - 0.07([\mathrm{Chl}\, b] - 2.5[\mathrm{Pras}])\}$$

This assumed a Chl *a*:Zea ratio of 2.1 in cyanobacteria but had to correct for the concentration of zeaxanthin for contributions of chlorophytes and prasinophytes using the second half of the equation. Similar equations were required for each of the major algal taxa that shared marker pigments. The various ratios were modified by inverse methods to optimise total Chl *a*. This approach is mathematically sound but very cumbersome since a new set of equations must be developed for each new population encountered.

(3) *Excel Solver*. The Microsoft Excel module 'Solver' provides a simple tool for optimising pigment:Chl *a* ratios for constructing simultaneous equations as above. This has been used as a preliminary to CHEMTAX calculations (Not *et al.*, 2008).

(4) *CHEMTAX software* (Mackey *et al.*, 1996) is similar in principle to the ISE method in that the contributions of algal taxa are calculated using assumed 'best guess' ratios that are subsequently modified by the software. Rather than building sets of equations, the operator constructs a matrix of pigment:Chl *a* ratios for each algal type expected in the sample set (Table 6.1) to calculate the total Chl *a* content of each sample. Pigment ratios are ideally based on those of local cultures and environmental conditions, but are normally based on literature values or previous experience in the region. A 'ratio limits' matrix is also constructed that

Table 6.1. *Example matrix of pigment:Chl* a *ratios for a range of algal types, which may be used as 'seed' values in a CHEMTAX calculation for a subantarctic data set. Ratios are based on average values from Table 6.2.*

	Pigment													
Class	*Chl* c_3	*Chl* c_2	*Chl* c_1	*Peri*	*But-fuco*	*Fuco*	*Pras*	*Viola*	*Hex-fuco*	*Allo*	*Zea*	*Lut*	*Chl* b	*Chl* a
Cyanobacteria	0	0	0	0	0	0	0	0	0	0	0.64	0	0	1
Chlorophytes	0	0	0	0	0	0	0	0.049	0	0	0.032	0.17	0.32	1
Prasinophytes	0	0	0	0	0	0	0.25	0.054	0	0	0.058	0.021	0.73	1
Cryptophytes	0	0.20	0	0	0	0	0	0	0	0.38	0	0	0	1
Diatoms-1	0	0.23	0.025	0	0	0.62	0	0	0	0	0	0	0	1
Diatoms-2	0.08	0.28	0	0	0	0.99	0	0	0	0	0	0	0	1
Dinoflagellates-1	0	0.22	0	0.56	0	0	0	0	0	0	0	0	0	1
Dinoflagellates-2	0.04	0.09	0	0	0.06	0.19	0	0	0.18	0	0	0	0	1
Haptophytes-6	0.18	0.21	0	0	0.005	0.23	0	0	0.47	0	0	0	0	1
Haptophytes-8	0.17	0.19	0	0	0.10	0.3	0	0	0.37	0	0	0	0	1

limits the extent to which each pigment can be adjusted (see Section 6.4.1, *Assumptions and constraints* below). CHEMTAX then iteratively adjusts each of the pigment:Chl *a* ratios so that the difference between the observed and the calculated Chl *a* concentrations are minimised across the data set. The algorithm is discussed in Section 6.4.2, *Reaching the optimum solution*.

The great advantage of CHEMTAX over ISE is the ease with which the taxonomic composition of the populations can be developed and altered – by simply adding or subtracting a row from the matrix – unlike ISE where every equation may require modification. Also, individual pigments can be enabled or disabled prior to each analysis without modifying the matrix itself, allowing the operator to test multiple scenarios during exploration of the data. Increases in computing power since the development of CHEMTAX now make it practical to test multiple scenarios, to undertake multiple randomised starting matrices (e.g. 50 trials, Wright *et al.*, 2009) and to derive standard deviations of the estimates.

(5) *Bayesian method.* A Bayesian compositional estimator (BCE) was recently-published based on Bayesian inference within the R software platform (Van den Meersche *et al.*, 2008). This differs from CHEMTAX in that it assumes a probability distribution for the input ratio matrix (rather than random within limits) and uses a Markov chain Monte Carlo algorithm to generate results with full posterior probability distribution, allowing means, confidence intervals etc. to be derived. In this respect it seems more statistically robust than CHEMTAX from which standard deviations are derived by analysing results across multiple runs. It also offers the opportunity of incorporating other data sets such as microscopy that cannot be dealt with in current versions of CHEMTAX.

The BCE is computationally demanding, requiring 10^5–10^6 iterations for each model run, and is more suitable for small data sets, even a single sample. The 40-sample Southern Ocean data set used to compare BCE with CHEMTAX was regarded as huge and the authors advised only using a few samples per run. Nevertheless, BCE obtained very similar results to CHEMTAX on the same data, which is encouraging for both techniques.

The philosophy of BCE is the reverse of CHEMTAX. Van den Meersche *et al.* (2008) state that 'attempts to estimate biomarker ratios from field data should be discouraged...' (whereas this is the role of CHEMTAX) '...unless sampling is performed accordingly...[with good taxonomy from microscopy etc.]'. The Bayesian estimator is driven more by knowledge of input ratios than by the field data set, and increasing sample numbers degraded output rather than improving it. Thus it appears to be more suited at this stage to small studies with good knowledge of the pigment characteristics of local populations.

6.4.1 Assumptions and constraints of inverse simultaneous equations and CHEMTAX

In order to obtain a physiological and ecological meaningful result, certain assumptions and constraints must be considered.

(1) *Positivity constraint.* The taxa abundances and pigment:Chl *a* ratios must be ≥ 0.
(2) *Normalisation constraint.* Pigment data and pigment:Chl *a* ratios in CHEMTAX are normalised so that the sum of the taxa-specific Chl *a* contributions will always equal the observed sample Chl *a*, but this means that the residual errors (calculated – observed Chl *a*) are also relative making it difficult to compare different scenarios or to compare the absolute fit between different data sets.
(3) *Solution space.* Since pigment ratios are varied automatically in methods described in Section 6.4, (2)–(5), pigment ratios, if unconstrained, may venture outside realistic physiological and ecological pigment values (see Goericke and Montoya, 1998). In CHEMTAX V1 (Mackey *et al.*, 1996), bounds of 500% from the few available literature values were recommended. With more data available, tighter bounds can often be set, as will be discussed in the following section (see also Section 6.4.2, *Reaching the optimum solution*).
(4) *Pigment ratios are constant across all samples of the data set.* This basic assumption of all of the mathematical techniques employed is rarely true given the effects of light (and hence depth), nutrients and other factors on pigment ratios (see Section 6.5). In practice, this assumption can generally be satisfied by subdividing the data set by cluster analysis and/or depth binning, provided that sample numbers are sufficient.
(5) *Realistic initial (or seed) pigment ratios.* As CHEMTAX and ISE hunt for optimum ratios within the chosen constraints, the search is faster and more reliable if the starting ratios are close to the ultimate goal. Strategies and resources are discussed in Table 6.2 and Section 6.4.2.
(6) *Variations in the abundances of different algal groups are not correlated.* In most field data sets there is usually some degree of co-variance where, for example, there are parallel increases in the abundances of several algal taxa as a sub-surface chlorophyll *a* maximum or a front is approached, or following a nutrient intrusion. Where there is correlation between pigments from unrelated taxa, other correlations must be treated with caution (Wright and Jeffrey, 2006). In these situations, estimations of the contribution of different taxa to total Chl *a* from multiple linear (or nonlinear) regressions and possibly ISE may be subject to bias (Goericke and Montoya, 1998) similar to the bias inherent in estimates of phytoplankton carbon based on regressions of Chl *a* and POC (Banse, 1977). Such co-variance can be handled by CHEMTAX, but it complicates the initial development and evaluation (Mackey *et al.*, 1996).

Table 6.2. *Pigment:Chl* a *ratios for pigment types in culture and in the field, showing mean ratios for low, medium and high light (LL, ML, HL, respectively), minimum, maximum, mean and standard deviation for all data, and the number of observations for each data set* (n). *'–e' stands for ester(s) Separate tables are given for a, b, Cyanobacteria; c, Chlorophytes; d, Prasinophytes; e, Euglenophytes; f, Cryptophytes; g, Diatoms; h, Synurophytes; i, Dictyochophytes and Pelagophytes; j, Raphidophytes and Eustigmatophytes; k–m, Haptophytes; n, Xanthophytes; p, References. Values in italics are derived from <3 observations.*

6.2a. Cyanobacteria Type 1

Taxa	*Data*	*n*	*Aph*	*Myxo*	*Kmyxo*	*Kmyxo-e*	*Calo*	*Zea*	*Oscil*	*Cantha*	*Echin*	*Cryp*	*ββ-Car*
CYANO-1													
Culture	LL	14	0.077	*0.043*	*0.131*	0.006		0.051	0.012	0.031	0.051	*0.008*	0.067
"	ML	6	0.119	*0.186*	*0.217*	0.029		0.185	0.040	0.053	0.086	*0.019*	0.066
"	HL	7	0.140	*0.262*	*0.314*	0.045		0.227	0.071	0.076	0.082	*0.026*	0.077
"	Min	44	0.006	0.002	0.035	0.006	0.001	0.004	0.009		0.003	0.007	0.006
"	Max	44	0.196	0.538	0.314	0.220	0.099	0.863	0.120	0.132	0.122	0.026	0.247
"	**Mean**	**44**	**0.118**	**0.120**	**0.133**	**0.073**	**0.050**	**0.143**	**0.036**	**0.043**	**0.066**	**0.016**	**0.065**
"	Std dev	44	0.061	0.137	0.092	0.066	0.069	0.214	0.030	0.033	0.034	0.007	0.067
Field	LL	3						0.179			0.194		
"	ML	1						*0.138*			*0.183*		
"	HL	11						0.214			0.210		
"	Min	18						0.018			0.096		
"	Max	18						0.617			0.269		
"	**Mean**	**18**						**0.215**			**0.191**		
"	Std dev	18						0.135			0.042		

Examples: Freshwater and marine colonial forms *Aphanizomenon flos-aquae*, *Oscillatoria* sp., *Trichodesmium* sp. Reference nos. 1–11 (see Table 6.2p)

6.2b. Cyanobacteria Type 2

Taxa	*Data*	*n*	*Myxo*	*Zea*	*DVChl b*	*Echin*	*Cryp*	*βε-Car*	*ββ-Car*
CYANO-2									
Culture	LL	9	*0.041*	0.247		*0.035*	0.001		0.122
”	ML	9	*0.080*	0.659		*0.052*	0.001		0.143
”	HL	8	*0.163*	0.852		*0.063*	0.002		0.233
”	Min	31	0.041	0.076		0.035	0.001		0.010
”	Max	31	0.163	1.721		0.063	0.002		0.654
”	**Mean**	**31**	**0.086**	**0.636**		**0.052**	**0.001**		**0.201**
”	Std dev	31	0.054	0.477		0.012	0.001		0.173
Field	LL	16		0.358					*0.140*
”	ML	17		0.498					0.091
”	HL	51		0.868					0.134
”	Min	101		0.210					0.060
”	Max	101		2.500					0.200
”	**Mean**	**101**		**0.656**					**0.118**
”	Std dev	101		0.555					0.043
Examples: *Synechococcus* sp., *Synechocystis* sp. Reference nos. 1, 3, 4, 10, 12–46 (see Table 6.2p)									
CYANO-4									
Culture	Mean	0	No Data						
Field	LL	9		0.334	1.808			*0.300*	
”	ML	7		0.330	1.711			*0.147*	*0.016*
”	HL	16		0.533	0.494			0.157	0.038
”	Min	44		0.050	0.070			0.144	0.007
”	Max	44		1.000	3.500			0.300	0.038

Table 6.2. (*cont.*)

Taxa	*Data*	*n*	*Myxo*	*Zea*	*DVChl b*	*Echin*	*Cryp*	*βε-Car*	*ββ-Car*
”	**Mean**	**44**		**0.389**	**0.994**			**0.178**	**0.027**
	Std dev	44		0.254	0.815			0.061	0.015

Example: *Prochlorococcus marinus*. Reference nos. 10, 13, 16, 19, 21, 23, 26, 34, 35, 37, 46, 47 (see Table 6.2p)

6.2c. Chlorophytes

Type	*Data*	*n*	*Neo*	*Viola*	*Anth*	*Asta*	*Lut*	*Zea*	*Chl b*	*βε-Car*	*ββ-Car*
CHLORO-1											
Culture	LL	14	0.078	0.050	0.006	0.006	0.131	0.006	0.348		0.024
”	ML	8	0.047	0.047	0.019	0.009	0.165	0.027	0.330		0.031
”	HL	12	0.080	0.037	0.024	0.022	0.180	0.076	0.268		0.035
”	Min	43	0.001	0.009	0.001	0.006	0.093	0.001	0.178	0	0.001
”	Max	43	0.224	0.094	0.038	0.022	0.338	0.138	0.546	0	0.079
”	**Mean**	**43**	**0.066**	**0.049**	**0.014**	**0.012**	**0.171**	**0.032**	**0.315**		**0.033**
”	Std dev	43	0.046	0.027	0.013	0.009	0.054	0.034	0.083		0.027
Field	LL	18	0.029	0.078			0.129	0.027	0.328	*0.004*	
”	ML	17	0.035	0.076			0.147	0.043	0.332	0.003	*0.068*
”	HL	58	0.036	0.069			0.187	0.047	0.339	0.005	*0.141*
”	Min	114	0.004	0.005	0	0	0.020	0.007	0.050	0.002	0.043
”	Max	114	0.144	0.470	0	0	0.710	0.140	1.280	0.028	0.157
”	**Mean**	**114**	**0.041**	**0.067**			**0.172**	**0.039**	**0.334**	**0.007**	**0.092**
”	Std dev	114	0.031	0.076			0.134	0.036	0.210	0.009	0.047

Example: *Dunaliella tertiolecta*. Reference nos. 1–3, 5–8, 10, 11, 13, 14, 16–21, 25–32, 34–46, 48–51 (see Table 6.2p)

6.2d. Prasinophytes

Type	*Data*	*n*	*MgDVP*	*Siph*	*Uri*	*Neo*	*Pras*	*Viola*	*Anth*	*Lut*	*Zea*	*Chl b*	*βε-Car*	*ββ-Car*
PRASINO-1														
Culture	LL	3				0.091		0.139		0.008	0.012	0.591		*0.073*
”	ML	9	0.008			0.063		0.111	0.023	0.079	0.020	0.699	0.027	0.097
”	HL	3				0.071		0.211		0.052	0.056	0.535		*0.143*
”	Min	17	0.001			0.021		0.024	0.002	0.005	0.002	0.480		0.025
”	Max	17	0.015			0.103		0.255	0.051	0.382	0.087	1.313	0.041	0.191
”	**Mean**	**17**	**0.008**			**0.072**		**0.138**	**0.023**	**0.057**	**0.026**	**0.631**	**0.027**	**0.102**
”	Std dev	17	0.007			0.025		0.064	0.020	0.091	0.021	0.189	0.019	0.045
Field	LL	0												
”	ML	0												
”	HL	3				*0.031*		*0.078*			0.011	0.686		
”	Min	3				*0.031*		*0.078*				0.301		
”	Max	3				*0.031*		*0.078*			0.024	0.958		
”	**Mean**	**3**				***0.031***		***0.078***			**0.011**	**0.686**		
”	Std dev	3									0.012	0.343		

Examples: *Nephroselmis olivaceae*, *Tetraselmis* sp. Reference nos. 24, 25, 30, 33, 43, 51, 52 (see Table 6.2p)

Type	*Data*	*n*	*MgDVP*	*Siph*	*Uri*	*Neo*	*Pras*	*Viola*	*Anth*	*Lut*	*Zea*	*Chl b*	*βε-Car*	*ββ-Car*
PRASINO-2														
Culture	LL	1	*0.023*	*0.170*		*0.056*		*0.055*	*0.017*	*0.038*	*0.065*	*0.786*	*0.034*	*0.271*
”	ML	1	*0.022*	*0.081*		*0.033*		*0.057*	*0.006*	*0.096*	*0.005*	*0.812*	*0.020*	*0.062*
”	HL	0												

Table 6.2. (*cont.*)

Type	*Data*	*n*	*MgDVP*	*Siph*	*Uri*	*Neo*	*Pras*	*Viola*	*Anth*	*Lut*	*Zea*	*Chl b*	*βε-Car*	*ββ-Car*
"	Min	2	*0.022*	*0.081*		*0.033*		*0.055*	*0.006*	*0.038*	*0.005*	*0.786*	*0.020*	*0.062*
"	Max	2	*0.023*	*0.170*		*0.056*		*0.057*	*0.017*	*0.096*	*0.065*	*0.812*	*0.034*	*0.271*
"	**Mean**	**2**	**0.023**	**0.126**		**0.045**		**0.056**	**0.012**	**0.067**	**0.035**	**0.799**	**0.027**	**0.167**
"	Std dev	2	0.001	0.063		0.016		0.001	0.008	0.041	0.042	0.018	0.010	0.148
Field	Mean	1						*0.069*				*0.977*		
Examples: *Pyramimonas parkeae*, *Tetraselmis* spp. Reference nos. 52, 53 (see Table 6.2p)														
PRASINO-3														
Culture	LL	11	0.066		0.083	0.071	0.246	0.041	*0.055*	0.004	0.020	0.858	0.015	0.030
"	ML	21	0.067		0.077	0.051	0.189	0.070	0.019	0.016	0.030	0.664	0.010	0.047
"	HL	6	0.041			0.067	0.262	0.020	*0.022*	0.071	0.164	0.550		0.108
"	Min	41	0.003		0.047	0.018	0.024	0.008	0.001	0.001		0.131		0.011
"	Max	41	0.136		0.102	0.142	0.881	0.149	0.055	0.180	0.353	1.032	0.032	0.108
"	**Mean**	**41**	**0.062**		**0.078**	**0.063**	**0.245**	**0.054**	**0.021**	**0.021**	**0.058**	**0.704**	**0.012**	**0.049**
"	Std dev	41	0.027		0.012	0.031	0.154	0.038	0.017	0.039	0.078	0.172	0.010	0.028
Field	LL	14			*0.210*	0.116	0.241	0.072		0.008	0.030	0.953		
"	ML	12				0.088	0.215	0.088		0.014	0.075	0.852		
"	HL	37	*0.028*		*0.206*	0.078	0.248	0.126		0.009	0.059	0.764		
"	Min	77	*0.028*		*0.206*	0.040	0.044	0.031		0.002		0.020	*0.031*	*0.004*
"	Max	77	*0.028*		*0.210*	0.151	0.710	0.360		0.050	0.320	2.111	*0.031*	*0.004*
"	**Mean**	**77**	***0.028***		***0.208***	**0.093**	**0.222**	**0.099**		**0.011**	**0.057**	**0.911**	***0.031***	***0.004***
"	Std dev	77			0.003	0.038	0.133	0.061		0.009	0.064	0.580		

Examples: *Pycnococcus provasolii*, *Micromonas pusilla*. Reference nos. 10, 13, 14, 16, 18, 20, 22, 24–26, 28–30, 32–35, 37, 42–46, 49, 51–54 (see Table 6.2p)

6.2e. Euglenophytes

Type	*Data*	*n*	*Neo*	*Viola*	*Diadino*	*Diato*	*Lut*	*Zea*	*Cantha*	*Siph-e*	*Chl b*	*Echin*	*ββ-Car*
EUGLENO-1													
Culture	LL	4	0.059	0.002	0.155	*0.009*		*0.004*	*0.009*	*0.066*	0.354	*0.015*	0.028
”	ML	5	0.080	0.002	0.211	0.037	0.039	*0.039*	*0.018*	*0.080*	0.477	*0.031*	*0.017*
”	HL	2	*0.054*	*0.012*	*0.309*	*0.017*		*0.104*	*0.001*	*0.081*	*0.208*	*0.033*	*0.019*
”	Min	16	0.032	0	0.080	0.009	0.022	0.002	0.001	0.048	0.187	0.015	0.003
”	Max	16	0.138	0.012	0.635	0.058	0.055	0.104	0.060	0.119	0.828	0.070	0.053
”	**Mean**	**16**	**0.068**	**0.006**	**0.251**	**0.026**	**0.039**	**0.042**	**0.022**	**0.080**	**0.377**	**0.037**	**0.021**
”	Std dev	16	0.026	0.005	0.151	0.019	0.023	0.038	0.026	0.023	0.253	0.023	0.016
Field	LL	7	0.013		0.193	*0.001*					0.346		
”	ML	7	0.022	*0.012*	0.173	*0.001*		*0.104*			0.292		
”	HL	14	0.028	*0.007*	0.215	*0.001*		*0.056*			0.302		
”	Min	39	0.010	0.007	0.124	0.001		0.056			0.198		
”	Max	39	0.072	0.012	0.400	0.001		0.104			0.420		
”	**Mean**	**39**	**0.021**	**0.009**	**0.213**	**0.001**		**0.072**			**0.330**		
”	Std dev	39	0.018	0.003	0.063	0		0.028			0.082		

Examples: *Euglena gracilis*, *Eutreptiella gymnastica*. Reference nos. 1–3, 6, 7, 11, 17, 20, 25, 30, 34, 35, 38, 43, 55, 56, 57 (see Table 6.2p)

6.2f. Cryptophytes

Type	*Data*	*n*	*Chl* c_2	*Allo*	*βε-Car*
CRYPTO-1					
Culture	LL	7	0.348	0.295	0.021
”	ML	8	0.219	0.387	0.021
”	HL	6	0.204	0.506	0.018
”	Min	29	0.050	0.163	0.003
”	Max	29	0.571	0.791	0.059
”	**Mean**	**29**	**0.196**	**0.379**	**0.024**
”	Std dev	29	0.194	0.147	0.023
Field	LL	18	0.060	0.211	0.014
”	ML	17	0.063	0.219	0.025
”	HL	63	0.115	0.277	0.028
”	Min	119	0.055	0.104	0.002
”	Max	119	0.212	0.790	0.060
”	**Mean**	**119**	**0.104**	**0.253**	**0.023**
”	Std dev	119	0.063	0.106	0.018

Example: *Chroomonas salina*. Reference nos. 1, 3, 6–8, 10–14, 16–21, 24–37, 40, 42–46, 48–50, 53, 55, 57 (see Table 6.2p)

6.2g. Diatoms

Type	*Data*	*n*	*Chl c_3*	*Chl c_2*	*Chl c_1*	*Chl c_1+c_2*	*Fuco*	*Viola**	*Diadino*	*Diato*	*Zea**	*ββ-Car*
DIATOM-1												
Culture	LL	19		*0.310*			0.498	*0.004*	0.108	0.008	*0.004*	0.022
"	ML	9					0.563	*0.013*	0.165	0.068	*0.013*	0.026
"	HL	17		*0.154*			0.514	*0.009*	0.201	0.157	*0.050*	0.046
"	Min	75		0.012	0.000		0.191	0.004	0.011	0.002	0.004	0.001
"	Max	75		0.310	0.057		1.710	0.013	0.784	0.296	0.050	0.111
"	**Mean**	**75**		**0.045**†	**0.023**†		**0.623**	**0.009**	**0.163**	**0.054**	**0.021**	**0.026**
"	Std dev	75		0.069	0.018		0.256	0.004	0.156	0.070	0.020	0.027
Field	LL	17		*0.165*	*0.107*		0.947		0.137	*0.016*		
"	ML	18		*0.093*	*0.025*		0.805		0.096	*0.016*		*0.009*
"	HL	62		0.205	0.107		0.776		0.188	*0.018*		*0.025*
"	Min	118		0.093	0.012		0.22		0.015	0.016		0.008
"	Max	118		0.300	0.210		3.100	0.003	0.690	0.063	0.008	0.030
"	**Mean**	**118**		**0.179**	**0.087**		**0.775**	**0.001**	**0.163**	**0.028**	**0.003**	**0.019**
"	Std dev	118		0.056	0.070		0.478	0.002	0.111	0.023	0.005	0.010

Example: *Chaetoceros didymus*. Reference nos. 1–3, 5–8, 10–14, 16–21, 24–40, 42–46, 48–51, 53, 57, 58 (see Table 6.2p). *Not normally present
† Values anomalously low (see Section 6.5.1)

Type	*Data*	*n*	*Chl c_3*	*Chl c_2*	*Chl c_1*	*Chl c_1+c_2*	*Fuco*	*Viola**	*Diadino*	*Diato*	*Zea**	*ββ-Car*
DIATOM-2												
Culture	Mean	0	No data									
Field	LL	2	*0.066*	*0.299*			*1.100*					
"	ML	0										
"	HL	5	0.089	0.276			0.958					
"	Min	7	0.016	0.130			0.830					
"	Max	7	0.267	0.375			1.267					
"	**Mean**	**7**	**0.083**	**0.284**			**0.998**					
"	Std dev	7	0.085	0.081			0.162					

Example: *Pseudonitzschia* sp., *Rhizoselenia setigera*. Reference nos. 47, 59 (see Table 6.2p)

6.2h. Synurophytes

Type	*Data*	*n*	*Chl c_3*	*Chl c_1+c_2*	*Fuco*	*Viola*	*Diadino*	*Anth*	*Diato*	*Zea*	*ββ-Car*
SYNURO-1*											
Culture	LL	3	0.042	0.032	0.274	0.076	0.005		0.005	*0.014*	0.002
”	ML	3	0.045	0.032	0.288	0.071	0.018		0.017	*0.004*	0.004
”	HL	3	0.049	0.032	0.286	0.041	0.025		0.055	*0.129*	0.004
”	Min	12	0.030	0.008	0.204	0.017	0.003	0	0.002	0.004	0.001
”	Max	12	0.062	0.054	0.338	0.166	0.037	0	0.118	0.129	0.007
”	**Mean**	**12**	**0.046**	**0.029**	**0.287**	**0.070**	**0.016**		**0.025**	**0.049**	**0.003**
”	Std dev	12	0.011	0.019	0.045	0.055	0.009		0.033	0.069	0.002
Field	LL	3			0.167	0.040		*0.011*		*0.004*	
”	ML	1			*0.121*	*0.032*				*0.006*	
”	HL	5			0.216	0.057		0.011		*0.005*	
”	Min	11	0	0.042	0.077	0.023	*0.020*	0.010	0	0.004	0
”	Max	11	0	0.042	0.375	0.116	*0.020*	0.012	0	0.021	0
”	**Mean**	**11**		**0.042**	**0.213**	**0.059**	**0.020**	**0.011**		**0.009**	
”	Std dev	11			0.082	0.028		0.001		0.008	

Example: *Mallomonas* sp. Reference nos. 1, 3, 6, 48, 60 (see Table 6.2p)

*All freshwater data

6.2i. Dictyochophytes and Pelagophytes

Type	*Data*	*n*	*Chl c_3*	*Chl c_2*	*Chl c_1+c_2*	*But-fuco*	*Fuco*	*Hex-fuco*	*Diadino*	*Diato*	*ββ-Car*
DICTYO-1											
Culture	LL	1			*0.132*		*0.344*		*0.042*	*0.001*	*0.025*
”	ML	2			*0.100*	0	*0.440*		*0.108*	*0.011*	*0.038*
”	HL	1			*0.089*		*0.285*		*0.157*	*0.014*	*0.046*
”	Min	7	0	0	0.089	0	0.285	0	0.042	0.001	0.025
”	Max	7	0	0	0.132	0	0.588	0	0.202	0.019	0.046
”	**Mean**	**7**			**0.108**	**0**	**0.348**		**0.112**	**0.009**	**0.035**
”	Std dev	7			0.015		0.107		0.059	0.007	0.007
Field	**Mean**	**1**	***0.059***	***0.129***		***0.677***	***0.097***				
Example: *Pseudochattonella farcimen*. Reference nos. 17, 41, 43, 57 (see Table 6.2p)											
PELAGO-1											
Culture	LL	1	*0.171*		*0.387*	*0.240*	*1.241*		*0.097*	*0.022*	*0.021*
”	ML	2	*0.155*		*0.390*	*0.761*	*0.632*		*0.304*	*0.047*	*0.018*
”	HL	1	*0.197*		*0.414*	*0.702*	*0.317*		*0.153*	*0.199*	*0.023*
”	Min	7	0.155	0	0.387	0.240	0.317	0	0.097	0.022	0.018
”	Max	7	0.311	0	0.546	1.246	1.373	0	0.438	0.199	0.024
”	**Mean**	**7**	**0.230**		**0.465**	**0.658**	**0.779**		**0.191**	**0.074**	**0.021**
”	Std dev	7	0.064		0.076	0.350	0.417		0.115	0.065	0.003
Field	LL	10	0.143	*0.316*	*0.316*	1.165	0.425	*0.008*	0.073		
”	ML	10	0.180		*0.370*	0.866	0.354	*0.009*	0.117		*0.009*
”	HL	33	0.154	0.288	0.285	0.831	0.337	0.025	0.298		*0.012*
”	Min	69	0.040	0.037	0.120	0.231	0.048	0	0.020	*0.057*	0.009
”	Max	69	0.550	0.470	0.470	3.099	1.241	0.527	0.690	*0.057*	0.012
”	**Mean**	**69**	**0.149**	**0.275**	**0.289**	**0.847**	**0.365**	**0.067**	**0.182**	***0.057***	**0.011**
”	Std dev	69	0.118	0.125	0.091	0.550	0.231	0.141	0.175		0.002
Example: *Pelagococcus subviridis*. Reference nos. 10, 13, 14, 16, 18, 19, 21, 26, 27, 29, 30–37, 42–44, 46, 47 (see Table 6.2p)											

6.2j. Raphidophytes and Eustigmatophytes

Type	*Data*	*n*	*MgDVP*	*Chl c_2*	*Chl c_1*	*Fuco*	*Viola*	*Diadino*	*Zea*	*Echin*	*ββ-Car*	*Vauch*
RAPHIDO-1												
Culture	LL	1	*0.001*	*0.068*	*0.015*	*0.289*	*0.023*		*0.002*			
”	ML	2				*0.702*	*0.293*	*0.046*	*0.064*			
”	HL	1	*0.002*	*0.060*	*0.011*	*0.273*	*0.016*		*0.121*			
”	Min	4	0.001	*0.060*	*0.011*	0.273	0.016	*0.046*	0.002			
”	Max	4	0.002	*0.068*	*0.015*	0.831	0.319	*0.046*	0.121			
”	**Mean**	**4**	**0.002**	**0.064**	**0.013**	**0.492**	**0.156**	**0.046**	**0.062**			
”	Std dev	4	0.001	0.006	0.003	0.265	0.159		0.060			
Field	**Mean**	**0**	**No data**									
EUSTIG-1												
Culture	Min	3					0.147			0.002	0.022	0.047
”	Max	3					0.166			0.003	0.026	0.050
”	**Mean**	**3**					**0.155**			**0.003**	**0.024**	**0.049**
”	Std dev	3					0.010			0.000	0.002	0.002
Field	**Mean**	**0**	**No data**									

Examples: *Chattonella marina*, *Nannochloropsis* spp. (respectively). Reference nos. 30, 54, 57 (see Table 6.2p)

6.2k. Haptophytes Type 1 & 2

Type	*Data*	*n*	*Chl* $c_{2\ Pg}$	*MgDVP*	*Chl* c_2	*Chl* c_1	*Fuco*	*Diadino*	*Diato*	*ββ-Car*
HAPTO-1										
Culture	LL	4		0.004	0.038	0.058	0.330	0.348	0.017	0.061
”	ML	8		0.005	0.042	0.059	0.303	*0.475*	*0.019*	*0.105*
”	HL	1		*0.022*	*0.063*	*0.097*	*0.235*	*0.158*	*0.016*	*0.033*
”	Min	13		0.000	0.010	0.020	0.200	0.158	0.010	0.010
”	Max	13		0.022	0.063	0.102	0.469	0.475	0.020	0.105
”	**Mean**	**13**		**0.006**	**0.042**	**0.061**	**0.306**	**0.337**	**0.017**	**0.064**
”	Std dev	13		0.007	0.014	0.025	0.073	0.114	0.004	0.044
Field	**Mean**	**0**	**No data**							
Example: *Pavlova lutheri*. Reference nos. 61, 62, 63 (see Table 6.2p)										
HAPTO-2										
Culture	LL	5	0.037	0.003	0.033	0.092	0.519	0.289	0.027	0.093
”	ML	9	0.034	0.005	0.023	0.066	0.426	*0.416*	*0.069*	*0.092*
”	HL	1	*0.012*	*0.015*	*0.018*	*0.054*	*0.181*	*0.098*	*0.030*	*0.025*
”	Min	15	0.012	0.000	0.016	0.045	0.181	0.098	0.020	0.010
”	Max	15	0.057	0.015	0.056	0.120	0.735	0.594	0.104	0.240
”	**Mean**	**15**	**0.034**	**0.005**	**0.026**	**0.074**	**0.440**	**0.297**	**0.038**	**0.083**
”	Std dev	15	0.011	0.006	0.011	0.022	0.141	0.143	0.028	0.078
Field	**Mean**	**0**	**No data**							
Example: *Pavlova gyrans*. Reference nos. 30, 61, 62, 63 (see Table 6.2p)										

6.2l. Haptophytes Type 3–5

Type	*Data*	*n*	*Chl c_3*	*MVChl c_3*	*MgDVP*	*Chl c_2*	*Chl c_1*	*Kfuco*	*Fuco*	*Viola*	*Diadino*	*Diato*	*Zea*	*ββ-Car*	*Chl c_2-MGDG [14/18]*
HAPTO-3															
Culture	LL	2			*0.012*	*0.355*	*0.085*		*0.503*		*0.215*	*0.009*		*0.061*	*0.066*
”	ML	10			0.005	0.120	0.058		0.378		0.307	0.020		0.049	0.028
”	HL	2			*0.028*	*0.313*	*0.111*		*0.239*		*0.139*	*0.037*		*0.044*	*0.029*
”	Min	17			0.002	0.051	0.032		0.195		0.061	0.001		0.028	0.014
”	Max	17			0.028	0.355	0.111		0.801		0.514	0.083		0.093	0.066
”	**Mean**	**17**			**0.009**	**0.163**	**0.066**		**0.347**		**0.179**	**0.025**		**0.046**	**0.032**
”	Std dev	17			0.009	0.124	0.028		0.189		0.155	0.026		0.020	0.018
Field	**Mean**	**0**	**No data**												

Example: *Dicrateria inornata*. Reference nos. 43, 61, 63 (see Table 6.2p)

Type	*Data*	*n*	*Chl c_3*	*MVChl c_3*	*MgDVP*	*Chl c_2*	*Chl c_1*	*Kfuco*	*Fuco*	*Viola*	*Diadino*	*Diato*	*Zea*	*ββ-Car*	*Chl c_2-MGDG [14/18]*
HAPTO-4															
Culture	LL	0													
”	ML	6	0.081	0.003	0.006	0.057	0.092		0.495						0.025
”	HL	0													
”	Min	6	0.065	0.003	0.003	0.027	0.071		0.411						0.021
”	Max	6	0.099	0.003	0.010	0.107	0.110		0.569						0.032

”	**Mean**	**6**	**0.081**	**0.003**	**0.006**	**0.057**	**0.092**		**0.495**					**0.025**
”	Std dev	6	0.013		0.003	0.028	0.016		0.061					0.005
Field	**Mean**	**0**	**No data**											

Example: *Prymnesium parvum*. Reference no. 63 (see Table 6.2p)

HAPTO-5														
Culture	LL	1	*0.064*		*0.007*	*0.058*	*0.072*		*0.490*	*0.002*	*0.133*	*0.030*	*0.003*	*0.020*
”	ML	3	0.049		0.006	0.105	0.164	0.057	0.394					0.018
”	HL	1	*0.018*			*0.009*	*0.016*		*0.263*	*0.011*	*0.318*	*0.665*	*0.022*	*0.004*
”	Min	5	0.018		0.004	0.009	0.016	0.033	0.263	0.002	0.133	0.030	0.003	0.004
”	Max	5	0.064		0.009	0.111	0.215	0.099	0.490	0.011	0.318	0.665	0.022	0.020
”	**Mean**	**5**	**0.046**		**0.006**	**0.077**	**0.116**	**0.057**	**0.387**	**0.007**	**0.226**	**0.348**	**0.013**	**0.016**
”	Std dev	5	0.017		0.002	0.043	0.075	0.037	0.086	0.006	0.131	0.449	0.013	0.007
Field	**Mean**	**0**	**No data**											

Example: *Ochrosphaera verrucosa*. Reference nos. 54, 63 (see Table 6.2p)

6.2m. Haptophytes Type 6–8

Type	*Data*	*n*	*Chl c_3*	*MVChl c_3*	*MgDVP*	*Chl c_2*	*Chl c_1*	*But-fuco*	*Fuco*	*Hex-kfuco*	*Hex-fuco*	*Diadino*	*Diato*	*βε-Car*	*ββ-Car*	*Chl c_2-MGDG [14/18]*	*Chl c_2-MGDG [14/14]*
HAPTO-6																	
Culture	LL	4	*0.229*		*0.018*	*0.341*			0.340	*0.125*	0.234	0.094	0.011	*0.013*	0.023	*0.099*	
”	ML	21	0.176	0.016	0.007	0.192		0.005	0.263	0.149	0.562	0.127	0.036		0.025	0.089	
”	HL	6	*0.130*	0.042	*0.022*	*0.290*			0.072	*0.029*	0.248	0.102	0.051		0.020	*0.043*	
”	Min	34	0.110	0.005	0.002	0.145	0	0	0.006	0.029	0.037	0.026	0.004	*0.013*	0.008	0.043	0
”	Max	34	0.229	0.050	0.022	0.341	0	0.008	0.722	0.313	1.507	0.316	0.169	*0.013*	0.049	0.099	0
”	**Mean**	**34**	**0.177**	**0.018**	**0.009**	**0.209**		**0.005**	**0.229**	**0.137**	**0.470**	**0.137**	**0.047**	***0.013***	**0.026**	**0.087**	
”	Std dev	34	0.031	0.015	0.006	0.066		0.004	0.186	0.082	0.344	0.094	0.054		0.015	0.014	
Field	LL	14	0.150			*0.171*		0.016	0.224		1.342	0.099				*0.097*	
”	ML	15	0.132			0.180	0	0.015	0.186		1.138	0.100		0.007	0.013		
”	HL	42	0.155	*0.119*	0	0.157	0	0.019	0.195		1.054	0.122		0.014	0.030	*0.090*	0
”	Min	86	0.033	*0.119*	0	0.080	0	0	0	0	0.228	0.015	0	0.006	0.012	0.082	0
”	Max	86	0.300	*0.119*	0	0.230	0	0.198	0.722	0	2.066	0.196	0	0.014	0.030	0.097	0
”	**Mean**	**86**	**0.146**	***0.119***	**0**	**0.168**	**0**	**0.015**	**0.195**		**1.214**	**0.109**		**0.010**	**0.021**	**0.091**	**0**
”	Std dev	86	0.087			0.048	0	0.028	0.172		0.595	0.046		0.004	0.010	0.008	

Example: *Emiliania huxleyi*. Reference nos. 13, 15, 16, 18, 19, 21–24, 26–29, 31–35, 37, 40, 42–45, 47, 61, 63–65 (see Table 6.2p)

HAPTO-7

Culture	LL	7	0.215	*0.017*	0.010	0.227	*0.009*	*0.023*	0.420	0.100	0.682	0.181	0.005	0.025	0.030	*0.091*	0.110
"	ML	15	0.210	*0.028*	0.015	0.198	*0.009*	0.009	0.436	0.215	0.543	0.267	0.015	0.017	0.037	0.097	0.081
"	HL	6	0.171	*0.030*	0.020	0.251	*0.025*	*0.013*	0.259	0.037	0.491	0.167	0.086	*0.007*	0.022	*0.085*	0.131
"	Min	29	0.060	0.008	0.003	0.094	0.009	0	0.004	0.016	0.007	0.070	0.005	0.007	0.013	0.002	0.027
"	Max	29	0.346	0.041	0.045	0.400	0.025	0.024	1.184	0.674	1.315	0.644	0.365	0.045	0.068	0.285	0.300
"	**Mean**	**29**	**0.202**	**0.026**	**0.015**	**0.216**	**0.014**	**0.011**	**0.388**	**0.151**	**0.567**	**0.201**	**0.061**	**0.020**	**0.030**	**0.094**	**0.103**
"	Std dev	29	0.052	0.011	0.011	0.071	0.009	0.008	0.365	0.160	0.309	0.136	0.115	0.011	0.015	0.084	0.078
Field	LL	2	*0.079*			*0.144*		*0.009*	*0.125*		*1.077*						*0.134*
"	ML	0															
"	HL	4	0.171			0.190	0	0.004	0.244		0.706						0.109
"	Min	8	0.058	0	0	0.110	0	0	0.009	0	0	0	0	0	0	*0.034*	0
"	Max	8	0.296	0	0	0.422	0	0.062	0.465	0	1.281	0	0	0	0	*0.034*	0.188
"	**Mean**	**8**	**0.136**			**0.198**	**0**	**0.012**	**0.199**		**0.794**					***0.034***	**0.101**
"	Std dev	8	0.088			0.101	0	0.021	0.168		0.386						0.073

Example: *Chrysochromulina hirta*. Reference nos. 25, 54, 61, 63, 64 (see Table 6.2p)

HAPTO-8

Culture	LL	4	0.222	*0.010*	0.071	0.232		0.043	0.555	0.049	0.250	0.125	0.006	*0.027*	0.028	0.069	
"	ML	22	0.176	0.018	0.007	0.179		0.129	0.274	0.062	0.422	0.216	0.039	*0.024*	0.053	0.055	
"	HL	4	0.192	*0.020*	0.068	0.270		0.082	0.229	0.049	0.288	0.183	0.043		0.041	0.066	
"	Min	33	0.021	0.010	0.002	0.074	0	0	0.011	0.005	0	0.006	0.006	0.023	0.001	0.013	0
"	Max	33	0.319	0.025	0.190	0.290	0	0.272	1.404	0.141	1.426	0.563	0.090	0.028	0.102	0.099	0
"	**Mean**	**33**	**0.171**	**0.016**	**0.029**	**0.192**		**0.103**	**0.300**	**0.058**	**0.371**	**0.141**	**0.029**	**0.025**	**0.032**	**0.058**	

Table 6.2. (*cont.*)

Type	*Data*	*n*	*Chl c_3*	*MVChl c_3*	*MgDVP*	*Chl c_2*	*Chl c_1*	*But-fuco*	*Fuco*	*Hex-kfuco*	*Hex-fuco*	*Diadino*	*Diato*	*βε-Car*	*ββ-Car*	*Chl c_2-MGDG [14/18]*	*Chl c_2-MGDG [14/14]*
"	Std dev	33	0.075	0.008	0.057	0.061		0.091	0.321	0.038	0.341	0.160	0.028	0.002	0.032	0.025	
Field	LL	5	0.423					0.688	1.032		0.900	0.113					
"	ML	5	0.215					0.655	0.577		0.668	*0.085*					
"	HL	19	0.262	*0.106*	0	0.199	0	0.263	0.374		0.644	0.150					0
"	Min	33	0.047	*0.106*	0	0.041	0	0.010	0.010	0	0.090	0.048	0.054	*0.006*	*0.002*	*0.049*	0
"	Max	33	0.550	*0.106*	0	0.413	0	0.980	2.150	0	1.706	0.250	0.054	*0.006*	*0.002*	*0.054*	0
"	**Mean**	**33**	**0.276**	***0.106***	**0**	**0.167**	**0**	**0.373**	**0.476**		**0.684**	**0.128**	**0.054**	***0.006***	***0.002***	***0.052***	**0**
"	Std dev	33	0.137			0.134	0	0.298	0.504		0.313	0.059				0.004	

Example: *Phaeocystis pouchetii*. Reference nos. 10, 16, 24, 27, 28, 30, 33, 36, 38, 40, 43, 45, 49, 54, 61, 63, 64, 66 (see Table 6.2p)

6.2n. Dinoflagellates

Type	*Data*	*n*	*Chl* c_3	*MgDVP*	*Chl* c_2	*Peri*	*But-fuco*	*Fuco*	*Hex-fuco*	*Dino*	*Diadino*	*Diato*	*Gyro-e*	*ββ-Car*
Dino-1														
Culture	LL	7		*0.004*	0.213	0.513				*0.049*	0.199	0.009		0.032
”	ML	7			0.272	0.654				*0.051*	0.274	0.029		0.040
”	HL	6		*0.008*	0.177	0.516				*0.059*	0.262	0.138		0.013
”	Min	28	0	*0.004*	0.090	0.285	0	0	0	0.029	0.123	0	0	0.002
”	Max	28	0	*0.008*	0.568	1.028	0	0	0	0.083	0.456	0.242	0	0.119
”	**Mean**	**28**		**0.006**	**0.218**	**0.558**				**0.054**	**0.253**	**0.046**		**0.026**
”	Std dev	28		0.003	0.131	0.183				0.023	0.098	0.066		0.034
Field	LL	16			*0.111*	0.838					0.164	*0.020*		
”	ML	16			*0.532*	0.851					0.165	*0.020*		
”	HL	56		0	0.218	0.730	0	0	0	*0.094*	0.170	*0.022*		
”	Min	110	0	0	0.043	0.165	0	0	0	*0.094*	0	0.020	0	*0.022*
”	Max	110	0	0	0.532	1.625	0	0	0	*0.094*	0.380	0.243	0	*0.022*
”	**Mean**	**110**		**0**	**0.245**	**0.804**	**0**	**0**	**0**	*0.094*	**0.177**	**0.076**		***0.022***
”	Std Dev	110		0	0.132	0.302	0	0	0	0	0.087	0.111		

Example: *Amphidinium carterae*. Reference nos. 1, 3, 6–8, 10–14, 16–21, 24–38, 40, 42–46, 48–50, 54, 55, 57 (see Table 6.2p)

Type	*Data*	*n*	*Chl* c_3	*MgDVP*	*Chl* c_2	*Peri*	*But-fuco*	*Fuco*	*Hex-fuco*	*Dino*	*Diadino*	*Diato*	*Gyro-e*	*ββ-Car*
Dino-2														
Culture	LL	2	*0.050*	*0.004*	*0.094*		*0.055*	*0.247*	*0.178*		*0.051*	*0.041*	*0.076*	
”	ML	0												
”	HL	1	*0.020*	*0.008*	*0.091*		*0.067*	*0.076*	*0.197*		*0.040*	*0.285*	*0.075*	
”	Min	3	0.020	0.004	0.063	0	*0.055*	0.076	*0.178*	0	0.024	0.026	*0.075*	0
”	Max	3	0.050	0.008	0.124	0	*0.067*	0.267	*0.197*	0	0.078	0.285	*0.076*	0
”	**Mean**	**3**	**0.035**	**0.006**	**0.093**		***0.061***	**0.190**	***0.188***		**0.047**	**0.122**	***0.076***	
”	Std dev	3	0.021	0.003	0.031		*0.008*	0.101	*0.013*		0.028	0.142	*0.001*	

Table 6.2. (*cont.*)

Type	*Data*	*n*	*Chl* c_3	*MgDVP*	*Chl* c_2	*Peri*	*But-fuco*	*Fuco*	*Hex-fuco*	*Dino*	*Diadino*	*Diato*	*Gyro-e*	*ββ-Car*
Field	LL	0												
”	ML	2	*0.179*		*0.126*	0	*0.081*	*0.300*	*0.194*				*0.050*	
”	HL	4	0.262		*0.144*		*0.070*	0.226	0.101		0.079		*0.039*	
”	Min	7	0.067	0	0.086	0	0.040	0.030	0.066	0	0.030	0	0.027	0
”	Max	7	0.310	0	0.179	0	0.122	0.310	0.248	0	0.171	0	0.050	0
”	**Mean**	**7**	**0.205**		**0.125**	**0**	**0.079**	**0.219**	**0.135**		**0.079**		**0.043**	
”	Std dev	7	0.108		0.040	0	0.037	0.103	0.059		0.079		0.011	

Example: *Gymnodinium breve*. Reference nos. 2, 22, 27, 31, 40, 54, 67, 68 (see Table 6.2p)

6.2o. Xanthophytes

Type	*Data*	*n*	*Chl* c_2	*Chl* c_1+c_2	*Total Chl c*	*Fuco*	*Viola*	*Diadino*	*Diato*	*Zea*	*Cantha*	*Echin*	*ββ-Car*	*Vauch*
XANTHO-1														
Culture	Mean	1					*0.048*	*0.135*	*0.019*	*0.002*	*0.001*	*0.003*	*0.016*	*0.038*
Field	Mean	1	*0.123*	*0.123*	*0.123*	*0.392*	*0*	*0.186*		*0*				

Example: *Vaucheria* sp. Reference nos. 5, 6 (see Table 6.2p)

6.2p. References for Tables 6.2a–o

Ref.	*Study*	*Habitat*	*Region*
1	Berger, 2005	Freshwater	Northern German dimictic lakes
2	Brotas and Plante-Cuny, 2003	Estuary microphytobenthos	Tagus estuary mudflats
3	Descy *et al.*, 2000	Freshwater	Wisconsin Lakes, USA
4	Dijkman *et al.*, 2010	Coastal mats	North Sea – barrier island Schiermonnikoog
5	Fietz *et al.*, 2005	Freshwater	Lake Baikal – cultures
6	Guisande *et al.*, 2008	Freshwater	Amazonian, Andean, and Caribbean lakes, lagoons, and swamps
7	Lionard *et al.*, 2008	Estuary	Schelde estuary Belgium/Netherlands
8	Marinho and Rodrigues, 2003	Freshwater	Eutrophic reservoir, southeastern Brazil
9	Schlüter *et al.*, 2004	Coastal	Baltic Sea and cultures
10	Veldhuis and Kraay, 2004	Oceanic	Sub-tropical Atlantic
11	Cartaxana *et al.*, 2009	Estuary	Brackish coastal lagoon, Algarve, Portugal
12	Adolf *et al.*, 2006	Estuary	Chesapeake Bay, USA
13	Bel Hassen *et al.*, 2009	Coastal	Gulf of Gabes, Mediterranean
14	Carreto *et al.*, 2003	Oceanic/coastal	Rio de La Plata, Argentina
15	Devilla *et al.*, 2005	Coastal	Plymouth coastal waters – algal biocides
16	DiTullio *et al.*, 2005	Oceanic	Eastern equatorial Pacific and Peru upwelling
17	Eker-Develi *et al.*, 2008	Coastal	Southern Baltic Sea
18	Fujiki *et al.*, 2009	Oceanic	Subarctic North Pacific
19	Furuya *et al.*, 2003	Oceanic/coastal	East China Sea
20	Gameiro *et al.*, 2007	Estuary	Tagus estuary, Portugal
21	Gibb *et al.*, 2001	Oceanic	NE Atlantic
22	Guisande *et al.*, 2002	Estuary	Ría de Vigo, NW Spain

Table 6.2. (*cont.*)

Ref.	*Study*	*Habitat*	*Region*
23	Hashihama *et al.*, 2008a	Coastal	Sagami Bay, Japan
24	Havskum *et al.*, 2004	Estuary (mecocosm)	Mesocosms, Denmark
25	Henriksen *et al.*, 2002	Coastal	Coastal temperate – Denmark
26	Higgins and Mackey, 2000	Oceanic	Western equatorial Pacific
27	Irigoien *et al.*, 2004	Coastal	English Channel, Station L4
28	Ishikawa *et al.*, 2002	Oceanic	Southern Ocean
29	Lee *et al.*, 2009	Coastal	Jeju Island, Korea
30	Lewitus *et al.*, 2005	Estuary	SE USA
31	Llewellyn *et al.*, 2005	Coastal	English Channel
32	Lohrenz *et al.*, 2003	Coastal	North Carolina inner shelf – Chesapeake Bay estuary outflow
33	Lovejoy *et al.*, 2007	Oceanic	Arctic seas
34	Mackey *et al.*, 1998	Oceanic	Western equatorial Pacific
35	Marty *et al.*, 2008	Oceanic	NW Mediterranean Sea
36	Meyer-Harms *et al.*, 1999	Oceanic	Norwegian Sea spring bloom
37	Miki *et al.*, 2008	Oceanic	Sulu Sea, western Pacific
38	Muylaert *et al.*, 2006	Coastal	Belgian coastal waters
39	Nair, 2007	Oceanic	Cultures
40	Örnólfsdóttir *et al.*, 2003	Estuary/coastal	Galveston Bay and west coast Florida
41	Rodríguez *et al.*, 2003a	Oceanic	Bay of Biscay
42	Sato *et al.*, 2007	Coastal	Otsuchi Bay, Japan
43	Schlüter *et al.*, 2000	Coastal	Coastal phytoplankton and cultures
44	Suzuki *et al.*, 2002	Oceanic	Subarctic Pacific and the Bering Sea

45	Wright and van den Enden, 2000	Oceanic	East Antarctic marginal ice zone – BROKE 96
46	Zhu *et al*., 2009	Estuary	Changjiang (Yangtze River) estuary
47	Rodríguez *et al*., 2006b	Coastal	NW Iberian upwelling
48	Buchaca *et al*., 2005	Freshwater	Oligotrophic Pyrenees lake
49	Hashihama *et al*., 2008b	Oceanic/coastal	E. Antarctic marginal ice zone (140° E)
50	Meyer *et al*., 2002	Oceanic	
51	Staehr *et al*., 2002	Coastal	Northern European
52	Latasa *et al*., 2004	Oceanic/coastal	Various
53	Rodríguez *et al*., 2002	Oceanic	Bransfield Straits, Antarctica
54	Rodríguez *et al*., 2006a	Oceanic	
55	Ansotegui *et al*., 2001	Estuary	Spanish, Urdaibai Est. (Bay of Biscay)
56	Bel Hassen *et al*., 2008	Coastal	Gulf of Gabes, Mediterranean
57	Seoane *et al*., 2006	Estuary	Spanish Nervion River estuary
58	Yao *et al*., 2006	Oceanic/coastal	South China Sea and Jiaozhou Bay
59	Wright *et al*., 2009	Coastal	Antarctic ice edge
60	Schlüter *et al*., 2006	Freshwater	Lake phytoplankton
61	Seoane *et al*., 2009	Estuary	Nervion River estuary, Bay of Biscay
62	Van Lenning *et al*., 2003	Coastal	Various
63	Zapata *et al*., 2004	Coastal/oceanic	Various
64	Carreto *et al*., 2008	Oceanic/coastal	Rio de La Plata, Argentina
65	Leonardos and Harris, 2006	Oceanic	Various
66	Antajan *et al*., 2004	Coastal	Belgium, coastal waters
67	De Salas *et al*., 2003	Coastal	Tasmania, South Africa
68	De Salas *et al*., 2004	Coastal	Tasmania

6.4.2 Reaching the optimum solution

Both ISE and CHEMTAX employ a 'steepest descent' algorithm to optimise the pigment:Chl *a* ratios used in calculating the taxonomic content of a population. They aim to minimise the difference between observed Chl *a* and total Chl *a* calculated from the product of marker pigment concentrations in each sample and the pigment: Chl *a* ratio for each pigment. In CHEMTAX terms, each non-zero element of the initial ratio matrix F_0 ('best guess' values) is varied in turn by a specified factor (typically 10%) and the calculated Chl *a* content and RMS residual (difference from observed Chl *a*) are calculated each time. The variation causing the biggest decrease in RMS residual is kept, giving a new ratio matrix F_1. Each element of F_1 is then varied in turn, with the variation giving the biggest decrease in RMS residual being kept, and so on. This results in a monotonically decreasing RMS residual until a minimum is reached.

The steepest descent algorithm has been likened to trying to find the lowest point in a landscape by dropping a blind parachutist from a plane with the simple instruction to 'walk downhill until you can walk downhill no more'. If the parachutist lands in a smooth basin, then the simple strategy outlined above will be successful. If, however, he lands outside the basin, then it will fail. The basin represents the solution space referred to in point *c* (above) and in CHEMTAX terms, the ratio limits matrix. If the basin has a rough surface (i.e. through noise in the data), the parachutist may well be caught in a local minimum. This scenario may be accommodated by asking the parachutist to initially move using large steps, and to take successively smaller steps only when large steps produce no further gain. Thus he may step out of local holes to find the global minimum. This strategy is employed by default in CHEMTAX where the step size is decreased by a factor of 1.3 when no improvement is found. Consider a second well-defined local minimum in the basin. If the initial step size is insufficient, the parachutist may well be trapped in the local minimum, but if it is too large he may step beyond the basin and become lost. In very rough terrain (e.g. a mountain range) it is very likely that a local, rather than a global, minimum will be found. The solution is to drop dozens of parachutists from slightly different locations within the search area, each following the modified strategy above, and to choose the location with the lowest elevation. If all of the parachutists arrive at the same location, then one would have confidence that they had found the global minimum. If, however, their endpoints were to be widely scattered, this would indicate that the result was unsound. This approach is easily implemented in the current version of CHEMTAX (v.1.95) by setting up multiple (20–60) F_0 matrices, and directing CHEMTAX to use each of them in turn (along with a 'successive runs' strategy similar to Latasa, 2007 for each matrix), as described in the following section.

6.4.3 Guide to quantitative chemotaxonomic interpretation of pigment data

Since interpreting a pigment data set can be quite challenging, the following step-by-step guide is offered as a strategy to facilitate the process.

- Examine the pigment data for specific markers for algal types (Section 6.2.1).
- Examine available complementary data:
  - *Microscopy data.* At least qualitative (down to species level if possible e.g. Laza-Martinez *et al.*, 2007); note if dinoflagellates are present (there may be non-pigmented heterotrophic and non-peridinin types in addition to peridinin containing type 1 dinoflagellates). Epifluorescence microscopy will distinguish heterotrophic organisms as well as those with phycobiliproteins.
  - *Flow cytometry and FlowCAM data.* Profiles of fluorescence per cell are useful in determining the mixed layer depth and photoacclimation characteristics (for constraining or adjusting pigment:Chl *a* ratios) while photoacclimation kinetics represent a dynamic tracer for vertical mixing (Dusenberry *et al.*, 2001). Profiles of taxa-specific fluorescence (fluorescence per cell × cell abundance; Marie *et al.*, 2005) are a useful comparison for profiles of calculated taxa-specific Chl *a* (including total picoeukaryote Chl *a* when summed over the respective taxa) or other taxa-specific pigments e.g. zeaxanthin from *Synechococcus* and *Prochlorococcus* as determined by CHEMTAX (Mackey *et al.*, 1996), or by an alternate Microsoft Excel Solver approach developed by Latasa (Not *et al.*, 2008). FlowCAM provide cell count and size data with digital images for cell identification (for larger cells), cell Chl and phycoerythrin fluorescence data (e.g. See *et al.*, 2005).
  - *In situ and in vivo fluorometry data. In situ* fluorescence profiles can be used for interpolation between discrete CTD samples. Multi-wavelength *in situ* fluorometers (e.g. BBE Fluoroprobe, Ghadouani and Smith, 2005; See *et al.*, 2005; Smith *et al.*, 2007) can provide some taxonomic data. Underway (flow-through) *in vivo* fluorescence data can help to define water masses. Delayed fluorescence excitation spectroscopy can be useful for phytoplankton abundance and composition, and the effect of environmental factors on pigments (e.g. PAR and temperature, Greisberger and Teubner, 2007).
  - *Environmental data.* Surface, *in situ* and spectral irradiance data. CTD and other hydrographic data: mixed layer depth, mixing rates, nutrients and other water mass properties.
  - *Remote-sensing data.* Large-scale water mass dynamics, surface Chl *a* and river plumes.
  - *Productivity and grazing data.* Including PhytoPAM (e.g. Thompson *et al.*, 2007) and FRRF data (e.g. Corno *et al.*, 2005; Pemberton *et al.*, 2007).

Cluster analysis can be used to sub-divide pigment data into clusters representing ecological or geographic provinces by water mass properties (e.g. Lohrenz *et al.*,

2003), inferred pigment size associations (e.g. Vidussi *et al.*, 2001; Bricaud *et al.*, 2004; Dandonneau *et al.*, 2006; Uitz *et al.*, 2006) or relative pigment concentrations (e.g. Furuya *et al.*, 2003; Vidussi *et al.*, 2004; Dandonneau and Niang, 2007; Not *et al.*, 2008) – particularly useful if biological or physical boundaries are ill-defined or do not coincide.

6.4.3.1 Pigment data exploration

- Multiple linear regression (MLR) is an aid to help determine (for each cluster) which taxa to consider in subsequent quantitative chemotaxonomic analysis (Schlüter and Møhlenberg, 2003), likely initial or 'seed' values of pigment: Chl *a* ratios and the effect of nutrients (DiTullio *et al.*, 2007). Some insights into phytoplankton composition can be obtained by plotting and performing MLR on major pigments against Chl *a*; green algal pigments (lutein, prasinoxanthin, siphonaxanthin esters and zeaxanthin) against Chl *b*; fucoxanthin and its derivatives against Chl c_2; Hex-fuco against Chl c_3 and fucoxanthin against Hex-fuco (see e.g. Schlüter and Møhlenberg, 2003; DiTullio *et al.*, 2007; Laza-Martinez *et al.*, 2007).
- Testing the correlation between two pigments that are not found in the same organism, e.g. Hex-fuco and Chl *b* will reveal their relative independence and the credibility of other correlations (see Section 6.4.1 (6)).
- If sampled quickly, depth profile plots of diatoxanthin:Chl *a*, diadinoxanthin: Chl *a*, (diatoxanthin + diadinoxanthin):Chl *a* and diatoxanthin:(diatoxanthin + diadinoxanthin) ratios provide an index of light history and allow calculation of mixing velocities (Garde and Cailliau, 2000; Stolte *et al.*, 2000; Riegman and Kraay, 2001; Thompson *et al.*, 2007).

6.4.3.2 CHEMTAX analysis

- *Sub-grouping.* If necessary, coherent sub-sets of samples should be defined, with a similar (as near as practical homogeneous) array of environmental conditions (light, nutrients, stratification or mixing regimes, water mass properties, etc.) to provide constancy of pigment ratios across all samples of a data group. Such groups may be based on cluster analysis (above), or environmental attributes such as depth (if stratified, sub-groups should represent at least the surface mixed layer (light field and nutrient status), the DCM and below the DCM, but preferably one should divide the set into depth strata if there are sufficient samples), water masses (e.g. across fronts and other current, coastal, estuarine, river, lake or biological boundaries), temporal variation (season, inter-annual) and/or size fractionation data. In CHEMTAX, the minimum number of samples for a sub-group is usually between 5 and 10 samples (Mackey *et al.*, 1997) depending on the data set. When sub-group sample numbers are small, the accuracy of taxa estimates may be improved by analysing the data in overlapping sub-groups (to increase the numbers in each) and averaging the results. Where pigment data spans more than one ecological domain but the small number of samples precludes division into data sub-sets or the boundary conditions are ill-defined, it is sometimes possible to obtain effective biological separation by including distinct taxa that are characteristic of each domain into an analysis of the combined data set (e.g. Carreto *et al.*, 2003, 2008; Veldhuis and Kraay, 2004).

- *Initial Pigment: Chl a ratio and ratio limit matrices* should be constructed for each of the sub-groups, considering all available complementary data. Ideally, starting ratios would be obtained from cultures collected at the study site, as exemplified by Laza-Martinez *et al.* (2007), although more usually pigment ratios are based on literature values (see Table 6.2). Occasionally, near unialgal blooms in the study region can be used to determine *in situ* pigment ratios for analysis of the field data and for comparison and validation of pigment ratios from laboratory cultures (e.g. *Gymnodinium* sp. (Carreto *et al.*, 2001), *Karenia brevis* (Redalje *et al.*, 2008), *Prochlorococcus* (Goericke *et al.*, 2000), *Synechococcus* and *Prochlorococcus* (DiTullio *et al.*, 2005)). Where taxonomic detail is not sufficient or where pigment:Chl *a* ratios appropriate for the local environmental conditions cannot be deduced, pigment groups have been combined (e.g. fucoxanthin containing or chemotaxonomic groups; Ansotegui *et al.*, 2001, 2003; Rodríguez *et al.*, 2002).
- *Preliminary CHEMTAX analysis.* A preliminary CHEMTAX run of each sub-group may produce ratios that change by large amounts (e.g. $> 100\%$), which may indicate outliers in the data group, inappropriate data group definition (taxa considered or sample allocation) or inappropriate initial 'seed' pigment ratios. Different scenarios may be tested, for instance, including different taxonomic groupings or pigment types. Data grouping, pigment suite taxa definitions, initial pigment: Chl *a* ratios or ratio limits may require adjustment until CHEMTAX adjustments of the pigment:Chl *a* ratios (and the taxa abundances) are not excessive.
- *Comprehensive CHEMTAX analysis.* As discussed in Section 6.4.2, it is recommended to perform multiple CHEMTAX analyses on each depth bin, from randomized starting points, to avoid being trapped in local minima and to test the confidence of the global minimum. In practice, one constructs the normal pigment: Chl *a* ratio matrix (F_0), then makes multiple (routinely 60) copies of that matrix while multiplying each cell of the initial table by a randomly determined factor F, where $F = 1 + S \times (R - 0.5)$, S is a scaling factor (e.g. 0.7) and R is a random number between 0 and 1 generated separately for each cell using the Microsoft Excel RAND function. Each of these ratio matrices is used as the starting point for a CHEMTAX optimization analysis. The solution with the smallest residual for each sub-group is used for the estimated taxonomic abundance. The best 10% of results ($n = 6$) are then chosen to calculate the average and the standard deviation of the abundance estimates. In such analyses, one may plot the average percentage of each taxon versus the RMS residual and see that the results tend to converge on a final solution. Thus one may have some confidence in the result.

When sub-groups are derived from a depth profile, it is useful to plot pigment ratios with depth, manually smooth them, and generate a new set of F_0 matrices for a final set of runs. This helps taxa with low abundance (and little statistical weight) to be adequately optimised. Additionally, Latasa (2007) recommended repeatedly using the output ratio matrix from one run as the input matrix for the next run until the calculated ratios (and taxa abundances) converge to stable final values (when plotted against the RMS residual).

Two issues concerning the randomised trial strategy should be considered. First, the scaling factor S acts to globally broaden the ratio limit boundaries set above, in this case by +/− 35%, allowing greater scope. Secondly, adding random noise to pigment ratios in taxa with similar pigment complements (e.g. haptophytes type 6 and 8) may cause role reversals, e.g. the modified type 6 may more closely resemble type 8 than the type 6 category, and is optimised accordingly. The output data should be checked carefully for such occurrences. The plot of average % taxa versus RMS (described above) may show sharp alternating shifts between taxa. If this is the case, the reversed taxa should be renamed, the particular matrix eliminated or the scaling factor reduced when adding random noise. This should be considered when setting the ratio limits.

- *Publication of CHEMTAX (or ISE) estimates.* At least the final pigment:Chl *a* ratio matrices (calculated best fit) and the calculation constraints (e.g. ratio boundaries, Goericke and Montoya, 1998) should be published, and preferably also the initial 'seed' values, as a basis for future studies. A database of the final pigment:Chl *a* ratios, species data (if available) and the associated environmental parameters will potentially be of great benefit in further improving pigment-derived estimates of algal community structure.

6.5 Variability of marker pigment:Chl *a* from cultures and field studies

The variation of pigment:Chl *a* ratios due to genetic diversity and environmental factors presents a challenge for quantitative interpretation of pigment field data. In this section we review the ranges of ratios found in cultures and field studies as a starting point for interpretation and as a basis for determining whether calculated pigment:Chl *a* ratios are realistic.

Pigment:Chl *a* ratios for freshwater, estuarine, coastal and oceanic environments from 66 culture and field studies have been collated (Table 6.2), with minima, maxima, means and standard deviations. For field studies, ratios are final best fit after optimisation by CHEMTAX or ISE. Since the ambient algal light field (incident light, depth and mixing regime) is a major factor influencing algal pigment ratios, data are also tabulated for low light (LL), medium light (ML) or high light (HL) sub-groups (approximately < 50, 50–100, > 100 μmol photons m^{-2} sec^{-2}, respectively), as defined in the original papers or deduced by the current authors from published data, where available. Other variables such as nutritional status could not be explicitly summarized here, but are discussed below.

The following caveats apply to Table 6.2. The database is affected by technical issues such as which pigments were chromatographically resolved, quantified and reported in a study. In field studies, the degree to which errors were minimised during quantitative chemotaxonomic analysis (by following procedures similar to the guidelines proposed in this chapter) was often not reported and the reliability of taxonomic identification was often unclear, clearly affecting the reliability of data. In addition, the sample size was uneven (n = 1 to 119, for each pigment type) – low numbers resulting in high variability and potentially masking possible trends with changes in the light level. Some pigments were often not reported, reducing the sample number in some cases to zero. This is particularly so in field studies where minor pigments could not be discerned or taxonomically attributed. Since space did not permit listing 'n' for each pigment in each pigment type, ratios have been listed in italics if based on less than three instances. The statistics of the pigment ratios in the database are affected by variation in response to environmental conditions (considered below) and different regional strains of a given phytoplankton taxon (Laza-Martinez *et al.*, 2007; Wright and Jeffrey, 2006). For example, most strains of *Phaeocystis* spp. contain significant quantities of Hex-fuco (e.g. Zapata *et al.*, 2004), but many northern European strains lack this pigment (e.g. Claustre *et al.*, 1990; Antajan *et al.*, 2004). It is therefore advantageous to determine pigment ratios using cultures of local isolates (Mackey *et al.*, 1996; Irigoien *et al.*, 2004). Where cultures are not available, Table 6.2 may provide a useful guide to pigment ratios in the field.

6.5.1 Pigment:Chl a *ratios in culture versus field*

Tables 6.2a–o show a good agreement between pigment:Chl *a* ratios in culture and the field, with close agreement of mean values in many cases, notably Cyanobacteria Type 2, Chlorophytes and Euglenophytes (Tables 6.2 b, c, e) respectively. In other cases, major pigments normally overlapped in observed ranges if not in mean values. Given that ratios from cultures are generally obtained from healthy log-phase cells, whereas field samples may derive from any stage of a bloom or from unfavourable environments, the similarity of calculated field ratios with those of cultures is encouraging. This supports the use of CHEMTAX (from which most of the ratios in Table 6.2 were derived) and the general approach to interpretation of field data (although it is acknowledged that there was often a considerable range in the values encountered in culture and field).

Some differences were noted: Dictyochophytes differed in the pigment composition between field and culture (Table 6.2i), but only in relation to pigments that are known to be variable in occurrence (Jeffrey *et al.*, Chapter 1), suggesting that different organisms were involved in the various studies. In haptophytes type 6

(Table 6.2m), Hex-fuco:Chl *a* ratios were significantly higher ($p < 0.001$) in field studies than in culture, although their ranges were wide in both cases and mostly overlapped. Such organisms, particularly the coccolithophorid *Emiliania huxleyi*, are often significant components of the phytoplankton, and thus the basis of such variation merits further study. Ratios for Chls c_1 and c_2 in culture seem anomalously low (Table 6.2g). There were relatively few studies that resolved this pair, and they were dominated by some cultures from the coastal South China Sea that do not appear to be typical. Most other cultures had ratios more like those of the field studies.

6.5.2 Irradiance

Phytoplankton adjust their cellular content of photosynthetic pigments to maximise the efficiency of photosynthesis under different light regimes, and synthesise photoprotective pigments to minimise damage by excess radiation (Falkowski and Raven, 1997). Significant increases in ratios of photoprotective pigments were apparent in culture data with increasing light for nearly all phytoplankton types (Tables 6.2a–n), particularly zeaxanthin, diatoxanthin and diadinoxanthin. Corresponding increases were normally seen in field data, but were generally smaller. On the other hand, photosynthetic pigment:Chl *a* ratios either remained relatively constant, or sometimes decreased with increasing irradiance (e.g. Chl *b*, Table 6.2d; Chl c_2, Tables 6.2f, i; Fuco and Hex-fuco, Tables 6.2i, k, l, m).

These patterns are in close accordance with previous observations. Algal cellular concentrations of light-harvesting pigments (LHP) and Chl *a* generally co-vary in response to changes in irradiance (Falkowski and Raven, 1997; Goericke and Montoya, 1998; Johnsen and Sakshaug, 1993; Rodríguez *et al.*, 2006a) with changes in the corresponding LHP:Chl *a* ratios less than the change in the LHP intracellular concentration (and typically less than a factor of 2), variable in direction and strain dependent (Goericke and Montoya, 1998). By contrast, with exposure to high light, the cellular concentration of photoprotective carotenoids (PPC) remains constant or increases, whereas the Chl *a* content per cell decreases; therefore, the PPC:Chl *a* ratio is expected to increase as a response to high light. Variation of PPC:Chl *a* ratios with change in light are likely to be greater than for LHP:Chl *a* ratios.

Changes in pigment ratios with irradiance are difficult to predict. Culture studies have shown that with increasing irradiance levels, the Fuco:Chl *a* ratio in diatoms may decrease (Schlüter *et al.*, 2000, 2006; Nair, 2007), stay the same (Schlüter *et al.*, 2000; Stæhr *et al.*, 2002) or increase (Brotas and Plante-Cuny, 2003; Nair, 2007). Similar variation has been seen in the field (e.g. Wright and van den Enden, 2000; Berger, 2005; Rodríguez *et al.*, 2006b). Also some pigment:Chl *a* ratios can be substantially different under a fluctuating light regime compared to growth under constant light for the same photoperiod (Nicklisch and Woitke, 1999; Fietz and Nicklisch, 2002).

While individual strains of phytoplankton probably exhibit similar pigment patterns to those observed in culture when distributed on a vertical light gradient through the water column, observed patterns in the field may differ as they represent the sum of multiple strains, ecotypes and species that are adapted to different light environments (Moore *et al.*, 1998, 2002; Moore and Chisholm, 1999; Rocap *et al.*, 2002; Katano *et al.*, 2004; Veldhuis and Kraay, 2004). Furthermore, where water column stability in stratified oceanic conditions permits photoacclimation to develop, there may be additional nutrient gradients that further confound pigment patterns (Henriksen *et al.*, 2002; Stæhr *et al.*, 2002; Suzuki *et al.*, 2002; Havskum *et al.*, 2004). In regions where there is seasonal mixing to the bottom of the water column (e.g. English Channel, Llewellyn *et al.*, 2005) and shallow areas with high concentrations of suspended sediments (e.g. the Belgian coastal zone of the North Sea, Muylaert *et al.*, 2006), pigment ratios even in near-surface populations may result from low average light levels.

Goericke and Montoya (1998) suggested that pigment:Chl *a* ratios with significant variability in response to irradiance (e.g. Chls *c* and photoprotective pigments) should be avoided in iterative methods for determining taxa-specific Chl *a* abundance. This problem can be minimised if sample numbers permit the data set to be subdivided adequately to account for such variation (e.g. by depth binning in a stratified water column, where photoacclimation may be evident).

6.5.3 Nutrients

Nutrient levels can also affect algal pigment:Chl *a* ratios and hence the accuracy of iterative quantitative chemotaxonomic methods that use them. In general, nitrogen is considered as the primary macro-nutrient limiting oceanic waters, although phosphorous limitation has been reported in regional areas including e.g. the Pacific Ocean (Karl, 1998) and the Sargasso Sea (Wu *et al.*, 2006). In high nutrient, low chlorophyll (HNLC) oceanic waters (e.g. the Southern Ocean), iron is limiting and can also affect pigment:Chl *a* ratios (DiTullio *et al.*, 2007). In freshwater environments, phosphorous is usually the primary limiting macronutrient (Descy *et al.*, 2009).

The effect of nutrients and irradiance on algal pigment ratios has been examined in several culture studies (e.g. Wilhelm and Manns, 1991; Latasa and Berdalet, 1994; Goericke and Montoya, 1998; Henriksen *et al.*, 2002; Stæhr *et al.*, 2002; Leonardos and Geider, 2004; Lewitus *et al.*, 2005; Nair, 2007). As Chl *a* contains nitrogen, nitrogen limitation (and HL) usually decreases the intracellular Chl *a* concentration (although Leonardos and Geider (2004) found no significant variation of Chl *a* in the diatom *Chaetoceros muelleri* grown under varying nitrogen and phosphorous conditions). However, the effect on pigment:Chl *a* ratios can be variable due to the corresponding effect of nutrient limitation on the other pigment. Although changes in pigment:Chl *a* ratios in nutrient-limited exponentially growing batch cultures may only be small (compared to nutrient-replete conditions), the transition to stationary conditions (usually as a result of nutrient or possibly light limitation) generally leads to substantial increases in most pigment:Chl *a* ratios, primarily due to a decrease in the intracellular Chl *a* (as carotenoids do not contain nitrogen and changes in their intracellular concentrations are smaller).

The direction and magnitude of the effect of nutrient limitation on algal pigment: Chl *a* ratios is also species dependent, highlighting the need to identify local algal species to improve the accuracy of quantitative chemotaxonomic estimates in field studies. For example, Goericke and Montoya (1998) used nitrate-limited continuous cultures of seven phytoplankton species and found that fucoxanthin:Chl *a* ratios increased with decreasing growth rate in three species, while it decreased in two other species. Henriksen *et al.* (2002) found that the magnitude of changes in a particular pigment ratio due to nutrient limitation varied considerably between algal groups (including species and strains): there were minor variations in peridinin:Chl *a* in the dinoflagellate *Scrippsiella* sp., significant changes in Fuco: Chl *a* ratios in the diatom *Ditylum brightwellii*, while alloxanthin:Chl *a* increased up to nine-fold in the cryptophyte *Rhodomonas marina*. Peridinin:Chl *a* ratios changed little in the dinoflagellate *Heterocapsa* sp. under nutrient limitation (Latasa and Berdalet, 1994) but up to 40% in the dinoflagellate *Amphidinium klebsii* (Wilhelm and Manns, 1991).

Responses of pigment:Chl *a* ratios in field populations to trophic status were examined by Descy *et al.* (2009) who used several freshwater algal pigment data sets (> 2,000 samples) and marine data from the highly stratified western equatorial Pacific to calculate the mean mixed layer pigment:Chl *a* ratios of several common algal taxa. There was a broad consistency in pigment:Chl *a* ratios over the wide range in trophic status, but as with culture studies, responses were species dependent and variable. The most dramatic variation observed across the trophic gradient was for the ratio of the photoprotective pigment zeaxanthin to Chl *a* in cyanobacteria type 2 (*Synechococcus*) which responded strongly to the turbidity and coloured dissolved organic matter (CDOM) in more eutrophic environments. This illustrates a more general interaction between nutrient status and light environment (through phytoplankton self-shading and production of CDOM) that makes prediction of pigment ratios more complex.

The varied and, in some cases, strong pigment responses to nutrient limitation suggests that it is important to accommodate potential nutrient responses in interpretation of field data, for instance by subdividing data according to nutrient status. Nutrient responses were exploited in a recent study by Wright *et al.* (2010) who created separate CHEMTAX categories for high-iron and low-iron haptophytes (based on ratios from DiTullio *et al.*, 2007) and found evidence for export-driven iron limitation of Antarctic surface waters that led to deep chlorophyll maxima.

6.6 Comparison with results from microscopy and other techniques

6.6.1 Relative strengths and weaknesses of chemotaxonomy and microscopy

Since the introduction of the pigment chemotaxonomic method, there have been many investigations on its reliability for quantitative phytoplankton measurements and monitoring (reviewed in Wright and Jeffrey, 2006). Comparison of pigment chemotaxonomy with microscopy is difficult since the methods measure different parameters and have different units, cell number or mg Carbon l^{-1} versus μg Chl *a* l^{-1}, appropriate to studies of carbon flux or primary production respectively, with natural variation in the ratio between cell carbon and Chl *a* content.

For many years, microscopy was the only method for identifying phytoplankton. For quantitative measurements, the algal cells were counted and measured under the microscope, the volumes of the cells were calculated by assigning simple geometric forms to the cells (Hillebrand *et al.*, 1999) and converted to cell carbon using published carbon:volume ratios (Montagnes *et al.*, 1994; Menden-Deuer and Lessard, 2000). Microscopy can produce reliable identification of large cells, but often pico- and nanoplankton cells are unidentified or even overlooked (e.g. Roy *et al.*, 1996; Schlüter *et al.*, 2000; Ansotegui *et al.*, 2001; Trottet *et al.*, 2007). Furthermore, it relies greatly on the skill of the taxonomist as well as the number of cells counted and measured. For 95% confidence limits, 400 cells must be counted *for each species* to obtain +/− 10% precision (Cassell, 1965). Typically, counting precision ranges from 20–50% depending on the number of cells counted (Wright and Jeffrey, 2006), and determining average cell volumes can be difficult where cells have a large size range, e.g. diatoms. Such poor precision on microscopic counts and volumes poses a problem if they are regarded as 'true' values for the phytoplankton composition and biomass.

High performance liquid chromatography pigment analyses can be carried out with ~1% precision, but the process of subsampling, filtering, extraction, pipetting, etc., causes the precision on analyses of replicate samples to be poorer. Thus, an intercomparison exercise with eight participating laboratories showed an average coefficient of variation (CV) of 7.4% for primary pigments (Hooker *et al.*, 2005; see also Chapter 5, this volume). When calculating the taxonomic proportion from the pigment concentrations, the precision will be poorer (Wright and Jeffrey, 2006). The chemotaxonomic method allows analysis of picoplankton and nanoplankton fractions that are often indistinguishable using microscopy, even with an electron microscope, through the use of pigment markers. In areas where pico- and nano-sized algae prevail, e.g. in oligotrophic oceans where 80% of the algae species are < 3 μm (Goericke, 1998), the traditional microscopy method is inadequate.

Pigment chemotaxonomy and microscopy are thus complementary, the former providing rapid analyses (>10 × microscopy), with high precision but poor taxonomic reliability, except on small cells, whereas microscopy is very slow and labour intensive, but provides good taxonomic reliability, except on small cells. The relative merits of the techniques were evident in a mesocosm experiment studying the phytoplankton response to temperature (Lassen *et al.*, 2010), monitored using inverted microscopy and HPLC. Generally, microscopy confirmed the results from HPLC and CHEMTAX, however microscopy could not identify the smallest flagellates, and these were underestimated when compared to the relevant phytoplankton groups from pigment analysis. Microscopic counts of the different groups had CVs ranging from 25–140%, with no significant difference detectable between the different temperature treatments. On the other hand, the CVs from the chemotaxonomic method increased significantly (t-test, $p < 0.05$) for all groups from 7–20% at 6 °C, to 14–50% at 9 °C, and 20–62% at 12 °C, indicating that the natural phytoplankton populations in the enclosures varied much more at higher temperatures. The differences were confirmed by fluorometry and nutrient measurements. These variations could not be detected by microscopy but were apparent with the higher precision of chemotaxonomy.

Microscopy nonetheless provides valuable taxonomic information that is critical for qualitative evaluation of chemotaxonomic results. Some knowledge of the phytoplankton communities of the area sampled is essential for trustworthy results of the calculated biomasses (Wright *et al.*, 1996; Henriksen *et al.*, 2002). Information from microscopy should be used to determine the major algal types for inclusion in CHEMTAX matrices (see e.g. Carreto *et al.*, 2003; Ansotegui, *et al.*, 2003; Irigoien *et al.*, 2004; Laza-Martinez *et al.*, 2007). Thus algal cells with different pigment signatures may be identified, e.g. dinoflagellates lacking peridinin (e.g. Meyer-Harms and Pollehne, 1998; Rodríguez *et al.*, 2003b; Irigoien *et al.*, 2004), several different subtypes of haptophytes with different pigment content (Jeffrey and Wright, 1994; Zapata *et al.*, 2004) or prasinophytes (Egeland *et al.*, 1997; Latasa *et al.*, 2004) (see Chapter 1, this volume). Such cells may otherwise introduce errors in chemotaxonomic analyses. Symbiotic associations pose an ongoing risk of complete misidentification if microscopy is neglected – examples include cyanobacteria in diatoms (Hallegraeff and Jeffrey, 1984), diatoms in dinoflagellates (Takano *et al.*, 2008) and cryptophytes in ciliates (Hibberd, 1977) or dinoflagellates (Hackett *et al.*, 2003).

It is seldom feasible to microscopically screen all samples that have been analysed by HPLC, particularly if picoplankton are abundant. Screening representative samples plus those that show a relative increase in chlorophyll *a* biomass (indicative of bloom formation) or other pigment anomalies will avoid missing taxa or using inappropriate ratios (Irigoien *et al.*, 2004; Wright and Jeffrey, 2006).

6.6.2 Verification of pigment chemotaxonomy

As a result of the lack of more precise methods for quantification of phytoplankton, microscopy has been used in many studies for verifying the results of quantitative chemotaxonomy (reviewed, Wright and Jeffrey, 2006), with generally good agreement found between the two methods in oceanic environments (e.g. Wright *et al.*, 1996;

Schlüter *et al.*, 2000; Gall *et al.*, 2001; Garibotti *et al.*, 2003) and lakes (Greisberger and Teubner, 2007; Guisande *et al.*, 2008). When large discrepancies between biomass estimations from microscopy and CHEMTAX have been found, some of the problems were ascribed to microscopy, but some of the inconsistencies were also ascribed to the phytoplankton groups chosen and their pigment ratios loaded into CHEMTAX (Llewellyn *et al.*, 2005; Lewitus *et al.*, 2005). Havskum *et al.* (2004) found good agreement between epifluorescence microscopy and flow cytometry on the biomass of cyanobacteria, but indicated that the pigment ratios used for cyanobacteria in CHEMTAX were erroneous and underestimated the contribution of this group. Picoplankton, particularly *Prochlorococcus* and *Synechococcus*, can be characterised by flow cytometry which, unlike microscopy, has excellent cell counting statistics. The results of CHEMTAX and flow cytometry are, however, not directly comparable since several different ecotypes of *Prochlorococcus* and *Synechococcus* exist which have different pigment ratios and possibly also different pigment to carbon ratios (Mackey *et al.*, 2002). Schlüter *et al.* (2000) found good agreement between microscopy and CHEMTAX in two data sets that were dominated by large cells, noting though some sensitivity to input ratios in cyanobacteria and prymnesiophytes, but in another set that was dominated by small cells, CHEMTAX disagreed with microscopic data from two laboratories, which also disagreed with each other.

The biomass of diatoms determined by microscopy has been much higher than the results of CHEMTAX in several studies (Schlüter and Møhlenberg, 2003; Havskum *et al.*, 2004; Vidussi *et al.*, 2004). This seems to be caused by a relatively higher carbon: chlorophyll *a* ratio in larger diatoms (Breton *et al.*, 2000) and in diatoms which have turned into a stationary growth phase (Llewellyn and Gibb, 2000). Henriksen *et al.* (2002) found that using pigment ratios derived from cultures in exponential growth phase at medium or low light produced better agreement with carbon biomasses derived from microscopy, than ratios from stationary phase at high light.

By comparing results of microscopy to results of CHEMTAX, it has been shown that the biomass of phytoplankton groups lacking unique diagnostic pigments can be determined with CHEMTAX using special ratios for these groups (e.g. Wright *et al.*, 1996; Schlüter *et al.*, 2000). Species with special pigment signatures can also be determined by CHEMTAX. Muylaert *et al.* (2006) found a close relationship between *Phaeocystis* cell counts and *Phaeocystis* chlorophyll *a* biomass in the North Sea estimated using chlorophyll c_3 as diagnostic pigment, which was shown to be correlated to biomass of *Phaeocystis* in the English Channel (Breton *et al.*, 2000). A new diagnostic pigment, gyroxanthin diester, was used to resolve the relative contribution of the toxic dinoflagellate *Karenia brevis* by using CHEMTAX (Örnólfsdóttir *et al.*, 2003), and the specific diagnostic pigment, 4-keto-myxoxanthophyll-like, was used to determine the biomass of the toxic cyanobacteria in the Baltic Sea (Schlüter *et al.*, 2004). However, Veldhuis and Kraay (2004) were not as successful in resolving the two cyanobacteria *Trichodesmium* and *Synechococcus* using CHEMTAX, compared with results from microscopy and flow cytometry; the pigment signatures and ratios of

Trichodesmium and *Synechococcus* were apparently too close (although resolution of *Trichodesmium* and *Synechococcus* may have been possible if the data set had been sub-divided into depth bins and analysed by CHEMTAX separately). In several studies, diagnostic pigments have revealed the presence of groups that were not detected by the microscope method, and their biomass was determined by CHEMTAX: e.g. dinoflagellates, prasinophytes, haptophytes, pelagophytes and cryptophytes (Ansotegui *et al.*, 2001; Schlüter and Møhlenberg, 2003; Rodríguez *et al.*, 2003b). Cells may have been too small to be detected by inverted microscopy or inadequately preserved (Gieskes and Kraay, 1983; Simon *et al.*, 1994).

6.6.3 Other techniques

New instruments for monitoring phytoplankton communities include the FlowCAM, a continuous imaging flow cytometer, suited to the identification of larger phytoplankton, and the Fluoroprobe, which measures fluorescence and allocates phytoplankton into major algal groups. Like microscopy, these methods are not directly comparable to quantitative chemotaxonomy, since biomass determinations are not measured in the same units. In a study on the utility of these new intruments, See *et al.* (2005) designated the chemotaxonomic method as 'a traditional method' and used it together with microscopy for comparison with results of FlowCAM and Fluoroprobe. Both the FlowCAM and the Fluoroprobe were unable to distinguish several algal classes and the chemotaxonomic method showed its superiority in assigning the smallest nanoplankton to haptophytes (See *et al.*, 2005). Nevertheless, these instruments may be useful aids for rapid mapping of populations. The recent development of molecular approaches, particularly analysis of DNA from genes encoding rRNA and RuBPC, has revolutionised analysis of phytoplankton populations (Díez *et al.*, 2001; Moon-van der Staay *et al.*, 2001; Not *et al.*, 2008), allowing characterization even of unculturable organisms to beyond the species level. This has revealed a huge and previously unrecognised diversity in the phytoplankton, particularly in the eukaryotic picoplankton (reviewed, Vaulot *et al.*, 2008; Worden and Not, 2008). For instance, Le Gall *et al.* (2008) isolated 212 strains by sorting flow cytometry, among which analysis of the 18S RNA gene found 38 strains of *Pelagomonas calceolata*. Liu *et al.* (2009) recently reported 674 new DNA sequences, all from pico-haptophytes, showing the global importance of such organisms and explaining the regular occurrence of Hex-fuco in oceanic pigment samples. Moreover, there appears to be a rapid turnover of diversity, with little overlap between occurrences of 176 picoeukaryote clones on successive sampling dates from the Mediterranean (Massana *et al.*, 2004). Such taxonomic power belittles that of chemotaxonomy or microscopy, yet it is not without problems. In a multidisciplinary study of phytoplankton of the Indian Ocean that may represent the future of such studies, Not *et al.* (2008) combined inverted and epifluorescence microscopy, flow cytometry, pigment analysis, denaturing gel gradient electrophoresis (DGGE), 18S rDNA clone libraries and fluorescent *in situ* hybridisation (FISH), and found considerable differences between techniques, attributed in part to selectivity of DNA amplification techniques. Additionally, much of the newly discovered diversity remains as gene sequences with little knowledge of the biology or ecological significance of the source organisms, or indeed their pigment content.

6.7 Conclusions

In spite of the introduction of powerful new techniques for phytoplankton analysis, pigment chemotaxonomy remains a useful and powerful technique for analysis of phytoplankton populations. Clearly it is limited in its ability to describe the complexity recently discovered with phytoplankton, however, it can reliably distinguish phytoplankton taxonomic groups with high precision, provided that microscopy or other techniques are used to avoid potential pitfalls of taxa with atypical pigments. It is thus very appropriate for ecological studies, allowing measurements of responses to short-term perturbations (e.g. hurricane Katrina, Pinckney *et al.*, 2009), seasonal changes (e.g. Llewellyn *et al.*, 2005), climate change or fine-scale mapping (e.g. Wright *et al.*, 2010). In many cases the functional characteristics of species are more important than their number (Ives and Carpenter, 2007). Indeed, Le Quéré *et al.* (2005) urge the study of phytoplankton functional types as a basis for global biogeochemical models, and list pigment HPLC as a key measurement for several of the groups.

Improvements in our knowledge of the taxonomic distribution of marker pigments through pigment analysis of new cultures, coupled with use of DNA clone libraries will surely facilitate interpretation of field data from such studies.

Acknowledgements

We thank Jean-Pierre Descy, University of Namur, Belgium, and many users of CHEMTAX for provision of data and Lee Belbin, Blatant Fabrications P/L, for the parachutist analogy. This work was supported by the Australian Government's Cooperative Research Centres Programme through the Antarctic Climate and Ecosystems Cooperative Research Centre (SWW), and CSIRO Marine and Atmospheric Research (HWH).

References

Adolf, J. E., Yeager, C. L., Miller, W. D., Mallonee, M. E. and Harding, L. W. Jr. (2006). Environmental forcing of phytoplankton floral composition, biomass, and primary productivity in Chesapeake Bay, USA. *Estuar. Coast. Shelf Sci.* **67**, 108–22.

Ansotegui, A., Trigueros, J. M. and Orive, E. (2001). The use of pigment signatures to assess phytoplankton assemblage structure in estuarine waters. *Estuar. Coast. Shelf Sci.* **52**, 689–703.

Ansotegui, A., Sarobe, A., Trigueros, J. M., Urrutxurtu, I. and Orive, E. (2003). Size distribution of algal pigments and phytoplankton assemblages in a coastal–estuarine environment: contribution of small eukaryotic algae. *J. Plankton Res.* **25**, 341–55.

Antajan, E., Chrétiennot-Dinet, M.-J., Leblanc, C., Daro, M. H. and Lancelot, C. (2004). 19′–hexanoyloxyfucoxanthin may not be the appropriate pigment to trace occurrence and fate of *Phaeocystis*: the case of *P. globosa* in Belgian coastal waters. *J. Sea Res.* **52**, 165–77.

Banse, K. (1977). Determining the carbon-to-chlorophyll ratio of natural phytoplankton. *Mar. Biol.* **41**, 199–212.

Bel Hassen, M., Drira, Z., Hamza, A., Ayadi, H., Akrout, F. and Issaoui, H. (2008). Summer phytoplankton pigments and community composition related to water mass properties in the Gulf of Gabes. *Estuar. Coast. Shelf Sci.* **77**, 645–56.

Bel Hassen, M., Drira, Z., Hamza, A., Ayadi, H., Akrout, F, Messaoudi, S., Issaoui, H., Aleya L. and Bouaïn A. (2009). Phytoplankton dynamics related to water mass properties in the Gulf of Gabes: Ecological implications. *J. Mar. Syst.* **75**, 216–26.

Berger, S. A. (2005). *Environmental variables and plankton communities in the pelagic of lakes: enclosure experiment and comparative lake survey*. Dissertation zur Erlangung des Doktorgrades, der Ludwig-Maximilians-Universität Munich, Germany.

Bidigare, R. R. and Ondrusek, M. E. (1996). Spatial and temporal variability of phytoplankton pigment distributions in the central equatorial Pacific Ocean. *Deep-Sea Res. II* **43**, 809–33.

Bjørnland, T., Haxo, F. T. and Liaaen-Jensen, S. (2003). Carotenoids of the Florida red tide dinoflagellate *Karenia brevis*. *Biochem. Syst. Ecol.* **31**, 1147–62.

Breton, E., Brunet, C., Sautour, B. and Brylinski, J-M. (2000). Annual variations of phytoplankton biomass in the Eastern English Channel: comparison by pigment signatures and microscopic counts. *J. Plankton Res.* **22**, 1423–40.

Bricaud, A., Claustre, H., Ras, J. and Oubelkheir, K. (2004). Natural variability of phytoplanktonic absorption in oceanic waters: Influence of the size structure of algal populations. *J. Geophys. Res.* **109**, C11010, doi:10.1029/2004JC002419.

Brotas, V. and Plante-Cuny, M-R. (2003). The use of HPLC pigment analysis to study microphytobenthos communities. *Acta Oecol.* **24**, S109–15.

Brunet, C., Casotti, R., Aronne, B. and Vantrepotte, V. (2003). Measured photophysiological parameters used as tools to estimate vertical water movements in the coastal Mediterranean. *J. Plankton Res.* **25**, 1413–25.

Buchaca, T., Felip, M. and Catalan, J. (2005). A comparison of HPLC pigment analyses and biovolume estimates of phytoplankton groups in an oligotrophic lake. *J. Plankton Res.* **27**, 91–101.

Carreto, J. I., Seguel, M., Montoya, N. G., Clément, A. and Carignan, M. O. (2001). Pigment profile of the ichthyotoxic dinoflagellate *Gymnodinium* sp. from a massive bloom in southern Chile. *J. Plankton Res.* **23**, 1171–75.

Carreto, J. I., Montoya, N. G., Benavides, H. R., Guerrero, R. and Carignan, M. O. (2003). Characterization of spring phytoplankton communities in the Río de La Plata maritime front using pigment signatures and cell microscopy. *Mar. Biol.* **143**, 1013–27.

Carreto, J. I., Montoya, N. G., Akselman, R., Carignan, M. O., Silva, R. I. and Colleoni, D. A. C. (2008). Algal pigment patterns and phytoplankton assemblages in different water masses of the Río de la Plata maritime front. *Cont. Shelf Res.* **28**, 1589–606.

Cartaxana, P., Mendes, C. R. and Brotas, V. (2009). Phytoplankton and ecological assessment of brackish and freshwater coastal lagoons in the Algarve, Portugal. *Lakes & Reservoirs: Research & Management* **14**, 221–30.

Cassell, E. A. (1965). Rapid graphical method for estimating the precision of direct microscopic counting data. *Appl. Microbiol.* **13**, 293–96.

Claustre, H. (1994). The trophic status of various oceanic provinces as revealed by phytoplankton pigment signatures. *Limnol. Oceanogr.* **39**, 1206–10.

Claustre, H., Poulet, S. A., Williams, R., Marty, J.-C., Coombs, S., Ben Mlih, F., Hapette, A. M. and Martin-Jezéquel, V. (1990). A biochemical investigation of a *Phaeocystis* sp. bloom in the Irish Sea. *J. Mar. Biol. Assoc. UK* **70**, 197–207.

Corno, G., Letelier, R. M., Abbott, M. R. and Karl, D. M. (2005). Assessing primary production variability in the North Pacific Subtropical Gyre: a comparison of Fast Repetition Rate Fluorometry and ^{14}C measurements. *J. Phycol.* **42**, 51–60.

Dandonneau, Y., Montel, Y., Blanchot, J., Giraudeau, J. and Neveux, J. (2006). Temporal variability in phytoplankton pigments, picoplankton and coccolithophores along a transect through the North Atlantic and tropical southwestern Pacific. *Deep-Sea Res. I* **53**, 689–712.

Dandonneau, Y. and Niang, A. (2007). Assemblages of phytoplankton pigments along a shipping line through the North Atlantic and tropical Pacific. *Prog. Oceanogr.* **73**, 127–44.

Demers, S., Roy, S., Gagnon, R. and Vignault, C. (1991). Rapid light-induced changes in cell fluorescence and in xanthophyll-cycle pigments of *Alexandrium excavatum* (Dinophyceae) and *Thalassiosira pseudonana* (Bacillariophyceae): A photo-protection mechanism. *Mar. Ecol. Prog. Ser.* **76**, 185–93.

De Salas, M. F., Bolch, C. J. S., Botes, L., Nash, G., Wright, S. W. and Hallegraeff, G. M. (2003). *Takayama* gen. nov. (Gymnodiniales, Dinophyceae), a new genus of unarmored dinoflagellates with sigmoid apical grooves, including the description of two new species. *J. Phycol.* **39**, 1233–46.

De Salas, M. F., Bolch, C. J. S. and Hallegraeff, G. M. (2004). *Karenia umbrella* sp. nov. (Gymnodiniales, Dinophyceae), a potentially ichthyotoxic dinoflagellate species from Tasmania, Australia. *Phycologia* **43**, 166–75.

Descy, J-P., Higgins, H. W., Mackey, D. J., Hurley, J. P. and Frost, T. M. (2000). Pigment ratios and phytoplankton assessment in northern Wisconsin lakes. *J. Phycol.* **36**, 274–86.

Descy, J-P., Sarmento, H. and Higgins, H. W. (2009). Variation of pigment ratios of phytoplankton across aquatic environments. *Eur. J. Phycol.* **44**, 319–30.

Devilla, R. A., Brown, M. T., Donkin, M. and Readman, J. W. (2005). The effects of a PSII inhibitor on phytoplankton community structure as assessed by HPLC pigment analyses, microscopy and flow cytometry. *Aquat. Toxicol.* **71**, 25–38.

Díez, B., Pedrós-Alió, C. and Massana R. (2001). Study of genetic diversity of eukaryotic picoplankton in different oceanic regions by small-subunit rRNA gene cloning and sequencing. *Appl. Environ. Microbiol.* **67**, 2932–41.

Dijkman, N. A., Boschker, H. T. S., Stal, L. J. and Kromkamp, J. C. (2010). Composition and heterogeneity of the microbial community in a coastal microbial mat as revealed by the analysis of pigments and phospholipid-derived fatty acids. *J. Sea Res.* **63**, 62–70.

DiTullio, G. R., Geesey, M. E., Maucher, J. M., Alm, M. B., Riseman, S. F. and Bruland, K. W. (2005). Influence of iron on algal community composition and physiological status in the Peru upwelling system. *Limnol. Oceanogr.* **50**, 1887–907.

DiTullio, G. R., Garcia, N., Riseman, S. F. and Sedwick, P. N. (2007). Effects of iron concentration on pigment composition in *Phaeocystis antarctica* grown at low irradiance. *Biogeochemistry* **83**, 71–81.

Dusenberry, J. A., Olson, R. J. and Chisholm, S. W. (2001). Photoacclimation kinetics of single-cell fluorescence in laboratory and field populations of *Prochlorococcus*. *Deep-Sea Res. I* **48**, 1443–58.

Egeland, E. S., Guillard, R. R. L. and Liaaen-Jensen, S. (1997). Additional carotenoid prototype representatives and a general chemosystematic evaluation of carotenoids in Prasinophyceae (Chlorophyta). *Phytochemistry* **44**, 1087–97.

Eker-Develi, E., Berthon, J.-F. and van der Linde, D. (2008). Phytoplankton class determination by microscopic and HPLC-CHEMTAX analyses in the southern Baltic Sea. *Mar. Ecol. Prog. Ser.* **359**, 69–87.

Eppley, R. W. and Peterson, B. J. (1979). Particulate organic flux and planktonic new production in the deep ocean. *Nature* **282**, 677–80.

Everitt, D. A., Wright, S. W., Volkman, J. K., Thomas, D. P. and Lindstrom, E. (1990). Phytoplankton community compositions in the western equatorial Pacific determined from chlorophyll and carotenoid pigment distributions. *Deep-Sea Res.* **37**, 975–97.

Falkowski, P. G. and Raven, J. (eds.) (1997). *Aquatic Photosynthesis*. Oxford: Blackwell.

Fietz, S. and Nicklisch, A. (2002). Acclimation of the diatom *Stephanodiscus neoastraea* and the cyanobacterium *Planktothrix agardhii* to simulated natural light fluctuations. *Photosynth. Res.* **72**, 95–106.

Fietz, S., Bleiß, W., Hepperle, D., Koppitz, H., Krienitz, L. and Nicklisch, A. (2005). First record of *Nannochloropsis limnetica* (Eustigmatophyceae) in the autotrophic picoplankton from Lake Baikal. *J. Phycol.* **41**, 780–90.

Fujiki, T., Matsumoto, K., Honda, M. C., Kawakami, H. and Watanabe, S. (2009). Phytoplankton composition in the subarctic North Pacific during autumn 2005. *J. Plankton Res.* **31**, 179–91.

Furuya, K., Hayashi, M, Yabushita, Y. and Ishikawa, A. (2003). Phytoplankton dynamics in the East China Sea in spring and summer as revealed by HPLC-derived pigment signatures. *Deep-Sea Res. II* **50**, 367–87.

Gall, M. P., Boyd, P. W., Hall, J., Sa, K. A. and Chang, H. (2001). Phytoplankton processes. Part 1: Community structure during the Southern Ocean Iron RElease Experiment (SOIREE). *Deep-Sea Res. II* **48**, 2551–70.

Gameiro, C., Cartaxana, P. and Brotas, V. (2007). Environmental drivers of phytoplankton distribution and composition in Tagus Estuary, Portugal. *Estuar. Coast. Shelf Sci.* **75**, 21–34.

Garde, K. and Cailliau, C. (2000). The impact of UV-B radiation and different PAR intensities on growth, uptake of ^{14}C, excretion of DOC, cell volume, and pigmentation in the marine prymnesiophyte, *Emiliania huxleyi*. *J. Exp. Mar. Biol. Ecol.* **147**, 99–112.

Garibotti, I. A., Vernet, M., Kozlowski, W. A. and Ferrario, M. E. (2003). Composition and biomass of phytoplankton assemblages in coastal Antarctic waters: a comparison of chemotaxonomic and microscopic analyses. *Mar. Ecol. Prog. Ser.* **247**, 27–42.

Gerber, S. and Häder, D.-P. (1994). Effects of enhanced UV-B irradiation on the red coloured freshwater flagellate *Euglena sanguine*. *FEMS Microbiol. Ecol.* **13**, 177–84.

Ghadouani, A. and Smith, R. E. H. (2005). Phytoplankton distribution in Lake Erie as assessed by a new *in situ* spectrofluorometric technique. *J. Great Lakes Res.* **31**(Supplement 2), 154–67.

Gibb, S. W., Cummings, D. G., Irigoien, X., Barlow, R. G. and Mantoura, R. F. C. (2001). Phytoplankton pigment chemotaxonomy of the northeastern Atlantic. *Deep-Sea Res. II* **48**, 795–823.

Gieskes, W. W. C. and Kraay, G. W. (1983). Dominance of Cryptophyceae during the phytoplankton spring bloom in the central North Sea detected by HPLC analysis of pigments. *Mar. Biol.* **75**, 179–85.

Goericke, R. (1998). Response of phytoplankton community structure and taxon-specific growth rates to seasonally varying physical forcing in the Sargasso Sea off Bermuda. *Limnol. Oceanogr.* **43**, 921–35.

Goericke, R. and Montoya, J. P. (1998). Estimating the contribution of microalgal taxa to chlorophyll *a* in the field – variations of pigments ratios under nutrient – and light-limited growth. *Mar. Ecol. Prog. Ser.* **169**, 97–112.

Goericke, R., Olson, R. J. and Shalapyonok, A. (2000). A novel niche for *Prochlorococcus* sp. in low-light suboxic environments in the Arabian Sea and the Eastern Tropical North Pacific. *Deep-Sea Res. I* **47**, 1183–205.

Greisberger, S. and Teubner, K. (2007). Does pigment composition reflect phytoplankton community structure in differing temperature and light conditions in a deep alpine lake? An approach using HPLC and delayed fluorescence techniques. *J. Phycol.* **43**, 1108–19.

Guisande, C., Maneiro, I., Riveiro, I., Barreiro, A. and Pazos, Y. (2002). Estimation of copepod trophic niche in the field using amino acids and marker pigments. *Mar. Ecol. Prog. Ser.* **239**, 147–56.

Guisande, C., Barreiro, A., Acuña, A., Marciales, L. J., Hernández, E., Torres, A. M., Aranguren, N., López, W., Duque, S. R., Gallo, L. J., Aguirre, N., Mogollón, M., Palacio, J. and Rueda-Delgado, G. (2008). Testing of the CHEMTAX program in contrasting neotropical lakes, lagoons, and swamps. *Limnol. Oceanogr. Methods* **6**, 643–52.

Hackett, J. D., Maranda, L., Su Yoon, H. and Bhattacharya, D. (2003). Phylogenetic evidence for the cryptophyte origin of the plastid of *Dinophysis* (Dinophysiales, Dinophyceae). *J. Phycol.* **39**, 440–48.

Hallegraeff, G. M. and Jeffrey, S. W. (1984). Tropical phytoplankton species and pigments of continental shelf waters of north and north-west Australia. *Mar. Ecol. Prog. Ser.* **20**, 59–74.

Hashihama, F., Hirawake, T., Kudoh, S., Kanda, J., Furuya, K., Yamaguchi, Y. and Ishimaru, T. (2008a). Size fraction and class composition of phytoplankton in the Antarctic marginal ice zone along the 140°E meridian during February–March 2003. *Polar Sci.* **2**, 109–20.

Hashihama, F., Horimoto, N., Kanda, J., Furuya, K., Ishimaru, T. and Saino, T. (2008b). Temporal variation in phytoplankton composition related to water mass properties in the central part of Sagami Bay. *J. Oceanogr.* **64**, 23–37.

Havskum, H., Schlüter, L., Scharek, R., Berdalet, E. and Jacquet, S. (2004). Routine quantification of phytoplankton groups – microscopy or pigment analyses? *Mar. Ecol. Prog. Ser.* **273**, 31–42.

Henriksen, P., Riemann, B., Kaas, H., Sørensen, H. M. and Sørensen, H. L. (2002). Effects of nutrient-limitation and irradiance on marine phytoplankton pigments. *J. Plankton Res.* **24**, 835–58.

Hibberd, D. J. (1977). Observations on the ultrastructure of the cryptomonad endosymbiont of the red-water ciliate *Mesodinium rubrum*. *J. Mar. Biol. Assoc. UK* **57**, 45–61.

Higgins, H. W. and Mackey, D. J. (2000). Algal class abundances, estimated from chlorophyll and carotenoid pigments, in the western Equatorial Pacific under El Niño and non-El Niño conditions. *Deep-Sea Res. I* **47**, 1461–83.

Hillebrand, H., Durselen, C.-D., Kirschtel, D., Pollinger, U. and Zohary, T. (1999). Biovolume calculation for pelagic and benthic microalgae. *J. Phycol.* **35**, 403–24.

Hooker, S. B., Van Heukelem, L., Thomas, C. S., Claustre, H., Ras, J., Barlow, R., Sessions, H., Schlüter, L., Perl, J., Trees, C., Stuart, V., Head, E., Clementson, L., Fishwick, J., Llewellyn, C. and Aiken, J. (2005). *The Second SeaWiFS HPLC Analysis Round-Robin Experiment (SeaHARRE-2). NASA Technical Memorandum 2005–212785*. Greenbelt: NASA Goddard Space Flight Center.

Hopkinson, B. M., Mitchell, B. G., Reynolds, R. A., Wang, H., Selph, K. E., Measures, C. I., Hewes, C. D., Holm-Hansen, O. and Barbeau, K. A. (2007). Iron limitation across chlorophyll gradients in the southern Drake Passage: phytoplankton responses to iron addition and photosynthetic indicators of iron stress. *Limnol. Oceanogr.* **52**, 2540–54.

Hou, J.-J., Huang, B.-Q., Cao, Z.-R., Chen, J.-X. and Hong, H.-S. (2007). Effects of nutrient limitation on pigments in *Thalassiosira weissflogii* and *Prorocentrum donghaiense*. *J. Integr. Plant Biol.* **49**, 686–97.

Irigoien, X., Meyer, B., Harris, R. and Harbour, D. (2004). Using HPLC pigment analysis to investigate phytoplankton taxonomy: the importance of knowing your species. *Helgoland Mar. Res.* **58**, 77–82.

Ishikawa, A., Wright, S. W., van den Enden, R. L., Davidson, A. T. and Marchant, H. J. (2002). Abundance, size structure and community composition of phytoplankton in the Southern Ocean in the austral summer 1999–2000. *Polar Biosci.* **15**, 11–26.

Ives, A. R. and Carpenter, S. R. (2007). Stability and diversity of ecosystems. *Science* **317**, 58–61.

Jeffrey, S. W. (1974). Profiles of photosynthetic pigments in the ocean using thin-layer chromatography. *Mar. Biol.* **26**, 101–10.

Jeffrey, S. W. and Wright, S. W. (1994). Photosynthetic pigments in the Haptophyta. In *The Haptophyte Algae*, ed. J. C. Green and B. S. C. Leadbeater. Oxford: Clarendon Press, pp. 111–32.

Johnsen, G. and Sakshaug, E. (1993). Bio-optical characteristics and photoadaptive responses in the toxic and bloom-forming dinoflagellates *Gyrodinium aureolum*, *Gymnodinium galatheanum*, and two strains of *Prorocentrum minimum*. *J. Phycol.* **29**, 627–42.

Johnsen, G., Samset, O., Granskog, L. and Sakshaug, E. (1994). *In vivo* absorption characteristics in 10 classes of bloom-forming phytoplankton: taxonomic characteristics and responses to photoadaptation by means of discriminant and HPLC analysis. *Mar. Ecol. Prog. Ser.* **105**, 149–57.

Karl, D. M. (1998). The changing sea: Long-term biogeochemical variability in the subtropical north pacific. *US JGOFS Newslett.* **9** (3), 7–9.

Karlson, B, Cusak, C. and Bresnan, E. (2010). *Microscopic and Molecular Methods for Quantitative Phytoplankton Analysis.* Intergovernmental Oceanographic Commission, Manuals and Guides, no. 55. Paris: Unesco Publishing.

Katano, T., Hirose, M. and Nakano, S. (2004). Discrimination of two phycoerythrin-pigment types of *Synechococcus* and their seasonal succession in the Uwa Sea. *Microbes Environ.* **19**, 7–12.

Lassen, M. K., Nielsen, K. D., Richardson, K., Garde, K. and Schlüter, L. (2010). The effects of temperature increases on a temperate phytoplankton community – A mesocosm climate change scenario. *J. Exp. Mar. Biol. Ecol.* **383**, 79–88.

Latasa, M. (2007). Improving estimations of phytoplankton class abundances using CHEMTAX. *Mar. Ecol. Prog. Ser.* **329**, 13–21.

Latasa, M. and Berdalet, E. (1994). Effect of nitrogen or phosphorus starvation on pigment composition of cultured *Heterocapsa* sp. *J. Plankton Res.* **16**, 83–94.

Latasa, M., Scharek, R., Le Gall, F. and Guillou, L. (2004). Pigment suites and taxonomic groups in Prasinophyceae. *J. Phycol.* **40**, 1149–55.

Laza-Martinez, A., Seoane, S., Zapata, M. and Orive, E. (2007). Phytoplankton pigment patterns in a temperate estuary: from unialgal cultures to natural assemblages. *J. Plankton Res.* **29**, 913–29.

Lee, Y.-W., Lee, J.-M. and Kim G. (2009). Identifying sharp hydrographical changes in phytoplankton community structure using HPLC pigment signatures in coastal waters along Jeju Island, Korea. *Ocean Sci. J.* **44**, 1–10.

Le Gall, F., Rigaut-Jalabert, F., Marie, D., Garczarek, L., Viprey, M., Gobet, A. and Vaulot, D. (2008). Picoplankton diversity in the South-East Pacific Ocean from cultures. *Biogeosciences* **5**, 203–14.

Leonardos, N. and Geider, R. J. (2004). Responses of elemental and biochemical composition of *Chaetoceros muelleri* to growth under varying light and nitrate: phosphate supply ratios and their influence on critical N:P. *Limnol. Oceanogr.* **49**, 2105–14.

Leonardos, N. and Harris, G. N. (2006). Comparative effects of light on pigments of two strains of *Emiliania huxleyi* (Haptophyta). *J. Phycol.* **42**, 1217–24.

Letelier, R. M., Bidigare, R. R., Hebel, D. V., Ondrusek, M., Winn, C. D. and Carl, D. M. (1993). Temporal variability of phytoplankton community structure based on pigment analysis. *Limnol. Oceanogr.* **38**, 1420–37.

Le Quéré, C., Harrison, S. P., Prentice, I. C., Buitenhuis, E. T., Aumont, O., Bopp, L., Claustre, H., Da Cunha, L. C., Geider, R., Giraud, X., Klaas, C., Kohfeld, K. E., Legendre, L., Manizza, M., Platt, T., Rivkin, R. B., Sathyendranath, S., Uitz, J., Watson, A. J. and Wolf-Gladrow, D. (2005). Ecosystem dynamics based on plankton functional types for global ocean biogeochemistry models. *Global Change Biol.* **11**, 2016–40.

Lewitus, A. J., White, D. L., Tymowski, R. G., Geesey, M. E., Hymel, S. N. and Noble, P. A. (2005). Adapting the CHEMTAX method for assessing phytoplankton taxonomic composition in southeastern U.S. estuaries. *Estuaries* **28**, 160–72.

Lionard, M., Muylaert, K., Tackx, M. and Vyverman, M. (2008). Evaluation of the performance of HPLC–CHEMTAX analysis for determining phytoplankton biomass and composition in a turbid estuary (Schelde, Belgium). *Estuar. Coast. Shelf Sci.* **76**, 809–17.

Liu, H., Probert, I., Uitz, J., Claustre, H., Aris-Brosou, S., Frada, M., Not, F. and de Vargas, C. (2009). Extreme diversity in noncalcifying haptophytes explains a major pigment paradox in open oceans. *Proc. Natl Acad. Sci. USA* **106**, 12803–08.

Llewellyn, C. A. and Gibb, S. W. (2000). Intra-class variability in the carbon, pigment and biomineral content of prymnesiophytes and diatoms. *Mar. Ecol. Prog. Ser.* **193**, 33–44.

Llewellyn, C. A., Fishwick, J. R. and Blackford, J. C. (2005). Phytoplankton community assemblage in the English Channel: a comparison using chlorophyll *a* derived from HPLC-CHEMTAX and carbon derived from microscopy cell counts. *J. Plankton Res.* **27**, 103–19.

Lohrenz, S. E., Carroll, C. L., Weidemann, A. D. and Tuel, M. (2003). Variations in phytoplankton pigments, size structure and community composition related to wind forcing and water mass properties on the North Carolina inner shelf. *Cont. Shelf Res.* **23**, 1447–64.

Lovejoy, C., Vincent, W. F., Bonilla, S., Roy, S., Martineau, M.-J., Terrado, R., Potvin, M., Massana, R. and Pedrós-Alió, C. (2007). Distribution, phylogeny, and growth of cold-adapted picoprasinophytes in Arctic seas. *J. Phycol.* **43**, 78–89.

Mackey, M. D., Mackey, D. J., Higgins, H. W. and Wright, S. W. (1996). CHEMTAX – a program for estimating class abundances from chemical markers: application to HPLC measurements of phytoplankton. *Mar. Ecol. Prog. Ser.* **144**, 265–83.

Mackey, D. J., Parslow, J. S., Griffiths, F. B., Higgins, H. W. and Tilbrook, B. (1997). Phytoplankton productivity and the carbon cycle in the western Equatorial Pacific under El Niño and non-El Niño conditions. *Deep-Sea Res. II* **44**, 1951–78.

Mackey, D. J., Higgins, H. W., Mackey, M. D. and Holdsworth, D. (1998). Algal class abundances in the western equatorial Pacific: estimation from HPLC measurements of chloroplast pigments using CHEMTAX. *Deep-Sea Res. I* **45**, 1441–68.

Mackey, D. J., Blanchot, J., Higgins, H. W. and Neveux, J. (2002). Phytoplankton abundances and community structure in the equatorial Pacific. *Deep-Sea Res. II* **49**, 2561–82.

Marie, D., Simon, N. and Vaulot, D. (2005). Phytoplankton cell counting by flow cytometry, Chapter 17. In *Algal Culturing Techniques*, ed. R. A. Andersen. Burlington: Elsevier Academic Press, pp. 253–67.

Marinho, M. M. and Rodrigues, S. V. (2003). Phytoplankton of an eutrophic tropical reservoir: comparison of biomass estimated from counts with chlorophyll-*a* biomass from HPLC measurements. *Hydrobiologia* **505**, 77–88.

Marty, J-C., Garcia, N. and Raimbault, P. (2008). Phytoplankton dynamics and primary production under late summer conditions in the NW Mediterranean Sea. *Deep-Sea Res. I* **55**, 1131–49.

Massana, R., Castresana, J., Balague, V., Guillou, L., Romari, K., Groisillier, A., Valentin, K. and Pedrós-Alió, C. (2004). Phylogenetic and ecological analysis of novel marine stramenopiles. *Appl. Environ. Microbiol.* **70**, 3528–34.

Menden-Deuer, S. and Lessard, E. J. (2000). Carbon to volume relationships for dinoflagellates, diatoms and other protist plankton. *Limnol. Oceanogr.* **45**, 569–79.

Meyer, B., Irigoien, X., Graeve, M., Head, R. N. and Harris, R. P. (2002). Feeding rates and selectivity among nauplii, copepodites and adult females of *Calanus finmarchicus* and *Calanus helgolandicus*. *Helgol. Mar. Res.* **56**, 169–76.

Meyer-Harms, B. and Pollehne, F. (1998). Alloxanthin in *Dinophysis norvegica* (Dinophysiales, Dinophyceae) from the Baltic Sea. *J. Phycol.* **34**, 280–85.

Meyer-Harms, B., Irigoien, X., Head, R. and Harris, R. (1999). Selective feeding on natural phytoplankton by *Calanus finmarchicus* before, during, and after the 1997 spring bloom in the Norwegian Sea. *Limnol. Oceanogr.* **44**, 154–65.

Miki, M., Ramaiah, N, Takeda, S. and Furuya, K. (2008). Phytoplankton dynamics associated with the Monsoon in the Sulu Sea as revealed by pigment signature. *J. Oceanogr.* **64**, 663–73.

Millie, D. F., Schofield, O. M. Kirkpatrick, G. J., Johnsen, G., Tester, P. A. and Vinyard, B. T. (1997). Detection of harmful algal blooms using photopigments and absorption signatures: A case study of the Florida red tide dinoflagellate, *Gymnodinium breve*. *Limnol. Oceanogr.* **42**, 1240–51.

Montagnes, D. J. S., Berges, J. A., Harrison, P. J. and Taylor, F. J. R. (1994). Estimating carbon, nitrogen, protein, and chlorophyll *a* from cell volume in marine phytoplankton. *Limnol. Oceanogr.* **39**, 1044–60.

Moon-van der Staay, S. Y., De Wachter, R. and Vaulot, D. (2001). Oceanic 18S rDNA sequences from picoplankton reveal unsuspected eukaryotic diversity. *Nature* **409**, 607–10.

Moore, L. R., Rocap, G. and Chisholm, S. W. (1998). Physiology and molecular phylogeny of coexisting *Prochlorococcus* ecotypes. *Nature* **393**, 464–67.

Moore, L. R. and Chisholm, S. W. (1999). Photophysiology of the marine cyanobacterium *Prochlorococcus*: Ecotypic differences among cultured isolates. *Limnol. Oceanogr.* **44**, 628–38.

Moore, L. R., Post, A. F., Rocap, G. and Chisholm, S. W. (2002). Utilization of different nitrogen sources by the marine cyanobacteria *Prochlorococcus* and *Synechococcus*. *Limnol. Oceanogr.* **47**, 989–96.

Muylaert, K., Gonzales, R., Franck, M., Lionard, M., van der Zee, C., Cattrijsse, A., Sabbe, K., Chou, L. and Vyverman, W. (2006). Spatial variation in phytoplankton dynamics in the Belgian coastal zone of the North Sea studied by microscopy, HPLC-CHEMTAX and underway fluorescence recordings. *J. Sea Res.* **55**, 253–65.

Nair, A. (2007). *Bio-optical properties of some phytoplankton functional types*. Ph.D. dissertation, Dalhousie University, Halifax, Nova Scotia, Canada.

Nicklisch, A. and Woitke, P. (1999). Pigment content of selected planktonic algae in response to simulated natural light fluctuations and a short photoperiod. *Int. Rev. Hydrobiol.* **84**, 479–95.

Not, F., Latasa, M., Scharek, R., Viprey, M., Karleskind, P., Balague, V., Ontoria-Oviedo, I., Cumino, A., Goetze, E., Vaulot, D. and Massana, R. (2008). Protistan assemblages across the Indian Ocean, with a specific emphasis on the picoeukaryotes. *Deep-Sea Res. I* **55**, 1456–73.

Örnólfsdóttir, E. B., Pinckney, J. L. and Tester, P. A. (2003). Quantification of the relative abundance of the toxic dinoflagellate, *Karenia brevis* (Dinophyta), using unique photopigments. *J. Phycol.* **39**, 449–57.

Pemberton, K. L., Smith, R. E. H., Silsbe, G. M., Howell, T. and Watson, S. B. (2007). Controls on phytoplankton physiology in Lake Ontario during the late summer: Evidence from new fluorescence methods. *Can. J. Fish. Aquat. Sci.* **64**, 58–73.

Pinckney, J. L., Wee, J. L, Hou, A. and Walker, N. D. (2009). Phytoplankton community structure responses to urban effluent inputs following Hurricanes Katrina and Rita. *Mar. Ecol. Prog. Ser.* **387**, 137–46.

Redalje, D. G., Lohrenz, S. E., Natter, M. J., Tuel, M. D., Kirkpatrick, G. J., Millie, D. F., Fahnenstield, G. L. and Van Dolah, F. M. (2008). The growth dynamics of *Karenia brevis* within discrete blooms on the West Florida Shelf. *Cont. Shelf Res.* **28**, 24–44.

Riegman, R. and Kraay, G. W. (2001). Phytoplankton community structure derived from HPLC analysis of pigments in the Faroe-Shetland Channel during summer 1999: the distribution of taxonomic groups in relation to physical/chemical conditions in the photic zone. *J. Plankton Res.* **23**, 191–205.

Rocap, G., Distel, D. L., Waterbury, J. B. and Chisholm, S. W. (2002). Resolution of *Prochlorococcus* and *Synechococcus* ecotypes by using 16S–23S ribosomal DNA internal transcribed spacer sequences. *Appl. Environ. Microbiol.* **68**, 1180–91.

Rodríguez, F., Varela, M. and Zapata, M. (2002). Phytoplankton assemblages in the Gerlache and Bransfield Straits (Antarctica Peninsula) determined by light microscopy and CHEMTAX analysis of HPLC pigment data. *Deep-Sea Res. II* **49**, 723–47.

Rodríguez, F., Pazos, Y., Maneiro, J. and Zapata, M. (2003a). Temporal variation in phytoplankton assemblages and pigment composition in a fixed station of the Ría of Pontevedra (NW Spain). *Estuar. Coast. Shelf Sci.* **58**, 499–515.

Rodríguez, F., Varela, M., Fernández, E. and Zapata, M. (2003b). Phytoplankton and pigment distributions in an anticyclonic Slope Water Oceanic eDDY (SWODDY) in the southern Bay of Biscay. *Mar. Biol.* **143**, 995–1011.

Rodríguez, F., Chauton, M., Johnsen, G., Andresen, K., Olsen, L. M. and Zapata, M. (2006a). Photoacclimation in phytoplankton: implications for biomass estimates, pigment functionality and chemotaxonomy. *Mar. Biol.* **148**, 963–71.

Rodríguez, F., Garrido, J. L., Crespo, B. G., Arbones, B. and Figueiras, F. G. (2006b). Size-fractionated phytoplankton pigment groups in the NW Iberian upwelling system: impact of the Iberian Poleward Current. *Mar. Ecol. Prog. Ser.* **323**, 59–73.

Roy, S., Chanut, J.-P., Gosselin, M. and Sime-Ngando, T. (1996). Characterization of phytoplankton communities in the lower St. Lawrence Estuary using HPLC-detected pigments and cell microscopy. *Mar. Ecol. Prog. Ser.* **142**, 55–73.

Sakshaug, E. and Andresen, K. (1986). Effect of light regime upon growth rate and chemical composition of a clone of *Skeletonema costatum* from the Trondheimsfjord, Norway. *J. Plankton Res.* **8**, 619–37.

Sato, M., Yoshikawa, T., Takeda, S. and Furuya, K. (2007). Application of the size-fractionation method to simultaneous estimation of clearance rates by heterotrophic flagellates and ciliates of pico- and nanophytoplankton. *J. Exp. Mar. Biol. Ecol.* **349**, 334–43.

Schlüter, L., Møhlenberg, F., Havskum, H. and Larsen, S. (2000). The use of phytoplankton pigments for identifying and quantifying phytoplankton groups in coastal areas: testing the influence of light and nutrients on pigment/chlorophyll *a* ratios. *Mar. Ecol. Prog. Ser.* **192**, 49–63.

Schlüter, L. and Møhlenberg, F. (2003). Detecting presence of phytoplankton groups with non-specific pigment signatures. *J. Appl. Phycol.* **15**, 465–76.

Schlüter, L., Garde, K. and Kaas, H. (2004). A 4-keto-myxoxanthophyll-like pigment is a diagnostic pigment for the toxic cyanobacteria *Nodularia spumigena* in the Baltic Sea. *Mar. Ecol. Prog. Ser.* **275**, 69–78.

Schlüter, L., Lauridsen, T. L., Krogh, G. and Jørgensen, T. (2006). Identification and quantification of phytoplankton groups in lakes using new pigment ratios – a comparison between pigment analysis by HPLC and microscopy. *Freshwat. Biol.* **51**, 1474–85.

Schlüter, L., Lutnæs, B. F., Liaaen-Jensen, S., Garde, K., Kaas, H., Jameson, I. and Blackburn, S. (2008). Correlation of the content of hepatotoxin nodularin and glycosidic carotenoids, 4-ketomyxol-2′-fucoside and novel 1′-*O*-methyl-4-ketomyxol-2′-fucoside, in 20 strains of the cyanobacterium *Nodularia spumigena*. *Biochem. System. Ecol.* **36**, 749–57.

See, J. H., Campbell, L., Richardson, T. L., Pinckney, J. L. and Shen, R. (2005). Combining new technologies for determination of phytoplankton community structure in the northern Gulf of Mexico. *J. Phycol.* **41**, 305–10.

Seoane, S., Laza, A. and Orive, E. (2006). Monitoring phytoplankton assemblages in estuarine waters: The application of pigment analysis and microscopy to size-fractionated samples. *Estuar. Coast. Shelf Sci.* **67**, 343–54.

Seoane, S., Zapata, M. and Orive, E. (2009). Growth rates and pigment patterns of haptophytes isolated from estuarine waters. *J. Sea Res.* **62**, 286–94.

Sherrard, N. J., Nimmo, M. and Llewellyn, C. A. (2006). Combining HPLC pigment markers and ecological similarity indices to assess phytoplankton community structure: an environmental tool for eutrophication? *Sci. Total Environ.* **361**, 97–110.

Simon, N., Barlow, R. G., Marie, D., Partensky, F. and Vaulot, D. (1994). Characterization of oceanic photosynthetic picoeucaryotes by flow cytometry. *J. Phycol.* **30**, 922–35.

Smith, R. E. H., Parrish, C. C., Depew, D. C. and Ghadouani, A. (2007). Spatial patterns of seston concentration and biochemical composition between nearshore and offshore waters of a Great Lake. *Freshwater Biol.* **52**, 2196–210.

Stæhr, P. A., Henriksen, P. and Markager, S. (2002). Photoacclimation of four marine phytoplankton species to irradiance and nutrient availability. *Mar. Ecol. Prog. Ser.* **238**, 47–59.

Stæhr, P. A., Markager, S. and Sand-Jensen, K. (2004). Pigment specific *in vivo* light absorption of phytoplankton from estuarine, coastal and oceanic waters. *Mar. Ecol. Prog. Ser.* **275**, 115–28.

Stolte, W., Kraay, G. W., Noordeloos, A. A. M. and Riegman, R. (2000). Genetic and physiological variation in pigment composition of *Emiliania huxleyi* (Prymnesiophyceae) and the potential use of its pigment ratios as a quantitative physiological marker. *J. Phycol.* **36**, 529–39.

Suzuki, K., Minami, C., Liu, H., Saino, T. (2002). Temporal and spatial patterns of chemotaxonomic algal pigments in the subarctic Pacific and the Bering Sea during the early summer of 1999. *Deep-Sea Res. II* **49**, 5685–704.

Takano, Y., Hansen, G., Fujita, D. and Horiguchi, T. (2008). Serial replacement of diatom endosymbionts in two freshwater dinoflagellates, *Peridiniopsis* spp. (Peridiniales, Dinophyceae). *Phycologia* **47**, 41–53.

Thompson, P. A., Pesant, S. and Waite, A. M. (2007). Contrasting the vertical differences in the phytoplankton biology of a dipole pair of eddies in the south-eastern Indian Ocean. *Deep-Sea Res. II* **54**, 1003–28.

Trottet, A., Roy, S., Tamigneaux, E. and Lovejoy, C. (2007). Importance of heterotrophic planktonic communities in a mussel culture environment: the Grande Entrée lagoon, Magdalen Islands (Québec, Canada). *Mar. Biol.* **151**, 377–92.

Tukaj, Z., Matusiak-Mikulin, K., Lewandowska, J. and Szurkowski, J. (2003). Changes in the pigment patterns and the photosynthetic activity during a light-induced cell cycle of the green alga *Scenedesmus armatus*. *Plant Physiol. Biochem.* **41**, 337–44.

Uitz, J., Claustre, H., Morel, A. and Hooker, S. B. (2006). Vertical distribution of phytoplankton communities in open ocean: An assessment based on surface chlorophyll. *J. Geophys. Res.* **111**, C08005, doi:10.1029/2005JC003207.

Uitz, J., Claustre, H., Griffiths, F. B., Ras, J., Garcia, N. and Sandroni, V. (2009). A phytoplankton class-specific primary production model applied to the Kerguelen Islands region (Southern Ocean). *Deep-Sea Res. I* **56**, 541–60.

Van den Meersche, K., Soetaert, K. and Middelburg, J. J. (2008). A Bayesian compositional estimator for microbial taxonomy based on biomarkers. *Limnol. Oceanogr. Methods* **6**, 190–99.

van Leeuwe, M. A. and Stefels, J. (1998). Effects of iron and light stress on the biochemical composition of Antarctic *Phaeocystis* sp. (Prymnesiophyceae). II. Pigment composition. *J. Phycol.* **34**, 496–503.

Van Lenning, K., Latasa, M., Estrada, M., Sáez, A. G., Medlin, L., Probert, I., Véron, B. and Young, B. (2003). Pigment signatures and phylogenetic relationships of the Pavlovophyceae (Haptophyta). *J. Phycol.* **39**, 379–89.

Vaulot, D., Eikrem, W., Viprey, M. and Moreau, H. (2008). The diversity of small eukaryotic phytoplankton (≤ 3 μm) in marine ecosystems. *FEMS Microbiol. Rev.* **32**, 795–820.

Veldhuis, M. J. W. and Kraay, G. W. (2004). Phytoplankton in the subtropical Atlantic Ocean: towards a better assessment of biomass and composition. *Deep-Sea Res. I* **51**, 507–30.

Vidussi, F., Marty, J.-C. and Chiavérini, J. (2000). Phytoplankton pigment variations during the transition from spring bloom to oligotrophy in the northwestern Mediterranean sea. *Deep-Sea Res. I* **47**, 423–45.

Vidussi, F., Claustre, H., Manca, B. B., Luchetta, A. and Marty, J.-C. (2001). Phytoplankton pigment distribution in relation to upper thermocline circulation in the eastern Mediterranean Sea during winter. *J. Geophys. Res.* **106**, 19939–56.

Vidussi, F., Roy, S., Lovejoy, C., Gammelgaard, M., Thomsen, H., Booth, B., Tremblay, J. E. and Mostajir, B. (2004). Spatial and temporal variability of the phytoplankton community structure in the North Water Polynya, investigated using pigment biomarkers. *Can. J. Fish. Aquat. Sci.* **61**, 2038–52.

Welschmeyer, A. and Hoepffner, N. (1986). Rapid xanthophyll cycling: An *in situ* tracer for mixing in the upper ocean. *EOS Trans. AGU* **67**, 969.

Wilhelm, C. and Manns, L. (1991). Changes in pigmentation of phytoplankton species during growth and stationary phases – consequences for the reliability of pigment based methods of biomass determination. *J. Appl. Phycol.* **3**, 305–10.

Wood, M. A. (1985). Adaptation of photosynthetic apparatus of marine ultraplankton to natural light fields. *Nature* **316**, 253–55.

Worden, A. Z. and Not, F. (2008) Ecology and diversity of picoeukaryotes. In *Microbial Ecology of the Oceans*, ed. D. Kirchman, 2nd edn. Hoboken: John Wiley and Sons.

Wright, S. W. and Jeffrey, S. W. (1997). High resolution HPLC system for chlorophylls and carotenoids of marine phytoplankton. In *Phytoplankton Pigments in Oceanography: Guidelines to Modern Methods*, ed. S. W. Jeffrey, R. F. C. Mantoura and S. W. Wright. Paris: UNESCO Publishing, pp. 327–41.

Wright, S. W. and Jeffrey, S. W. (2006). Pigment markers for phytoplankton production. In *Marine Organic Matter: Biomarkers, Isotopes and DNA*, ed. J. K. Volkman. The Handbook of Environmental Chemistry vol. 2, Part N. Berlin: Springer, pp. 71–104.

Wright, S. W. and van den Enden, R. L. (2000). Phytoplankton populations off east Antarctica in relation to stratification/mixing regimes: CHEMTAX analysis of HPLC pigment profiles (BROKE survey, Jan–Mar 1996). *Deep-Sea Res. II* **47**, 2363–400.

Wright, S. W., Thomas, D. P., Marchant, H. J., Higgins, H. W., Mackey, M. D. and Mackey, D. J. (1996). Analysis of phytoplankton in the Australian sector of the Southern Ocean: comparisons of microscopy and size frequency data with interpretations of pigment HPLC data using the 'CHEMTAX' matrix factorisation program. *Mar. Ecol. Prog. Ser.* **144**, 285–98.

Wright, S. W., Ishikawa, A., Marchant, H. J., Davidson, A. T., van den Enden, R. L. and Nash, G. (2009). Composition and significance of picophytoplankton in Antarctic waters. *Polar Biol.* **32**, 797–808.

Wright, S. W., van den Enden, R. L., Pearce, I., Davidson, A. T. Scott, F. J. and Westwood, K. J. (2010). Phytoplankton community structure and stocks in the Southern Ocean (30–80°E) determined by CHEMTAX analysis of HPLC pigment signatures. *Deep-Sea Res. II* **57**, 758–78.

Wu, J., Sunda, W., Boyle, E. A. and Karl, D. M. (2006). Phosphate depletion in the Western North Atlantic Ocean. *Science* **289**, 759–62.

Yao, P., Yu, Z-G. and Deng, C-M. (2006). Pigment signatures of some diatoms isolated from China Seas. *Acta Oceanol. Sinica* **23**, 108–18.

Zapata, M. (2005). Recent advances in pigment analysis as applied to picoplankton. *Vie et Milieu* **55**, 233–48.

Zapata, M., Rodríguez, F. and Garrido, J. L. (2000). Separation of chlorophylls and carotenoids from marine phytoplankton: a new HPLC method using a reversed phase C_8 column and pyridine-containing mobile phases. *Mar. Ecol. Prog. Ser.* **195**, 29–45.

Zapata, M., Jeffrey, S. W., Rodríguez, F., Clementson, L., Garrido, J. L. and Wright, S. W. (2004). Pigment variability in 37 species (65 strains) of Haptophyta: implications for phylogeny and oceanography. *Mar. Ecol. Prog. Ser.* **270**, 83–102.

Zhu, Z.-Y., Ng, W.-M., Liu, S.-M., Zhang, J., Chen, J.-C. and Wu, Y. (2009). Estuarine phytoplankton dynamics and shift of limiting factors: A study in the Changjiang (Yangtze River) Estuary and adjacent area. *Estuar. Coast. Shelf Sci.* **84**, 393–401.

7

Liquid chromatography-mass spectrometry for pigment analysis

RUTH L. AIRS AND JOSÉ L. GARRIDO

7.1 LC-MS analysis of chlorophylls and carotenoids: introduction

Because of improvements in HPLC methods for the analysis of phytoplankton pigments, pigments that are unidentified but spectrally related to known compounds are frequently reported in microalgal cultures as well as in natural distributions (Jeffrey and Wright, 1994; Vaulot *et al.*, 1994; Garrido and Zapata, 1998; Airs *et al.*, 2001a; Carreto *et al.*, 2001; Zapata *et al.*, 2004). Distinctions between common chlorophylls and carotenoids can be ascertained during HPLC from on-line UV/visible (UV/Vis) spectra, or co-elution with authentic standards. Several chlorophyll types however occur as suites of compounds. A diverse array of chlorophylls *c* have been observed in marine microalgal cultures as free acids or esterified by a range of non-polar groups (Garrido and Zapata, 1998; Garrido *et al.*, 2000; Zapata *et al.*, 2001, 2006). Photosynthetic bacteria belonging to the genera Chlorobiaceae produce suites of bacteriochlorophylls differing both in the degree of alkylation at positions C-8 and/or C-12 (Smith and Bobe, 1987; Airs and Keely, 2002), and/or the alcohol esterified to the propionic acid group at C-17 (Caple *et al.*, 1978; Otte *et al.*, 1993; Airs *et al.*, 2001b). Furthermore, chlorophyll *a* undergoes a range of transformation and alteration reactions when the phytoplankton cell is compromised and as detrital material is transported through the water column; these reactions are potentially indicative of specific environmental conditions or processes (Head and Horne, 1993; Head *et al.*, 1994; Veldhuis *et al.*, 2001; Walker and Keely, 2004). Several carotenoid types also exist as suites of compounds, for example fucoxanthin esters (Airs and Llewellyn, 2006). Characterisation of novel compounds involves rigorous chemical and analytical techniques following preparative isolation of individual components (Egeland *et al.*, 2000). Such an approach can be impractical when unknowns are present in low relative abundance. Liquid chromatography-mass spectrometry (LC-MS) permits acquisition of structural data during a single chromatographic run. Molecular mass information, used in conjunction with PDA UV/vis spectra is often sufficient for the assignment of components. HPLC coupled to tandem

Phytoplankton Pigments: Characterization, Chemotaxonomy and Applications in Oceanography, eds. Suzanne Roy, Carole A. Llewellyn, Einar Skarstad Egeland and Geir Johnsen. Published by Cambridge University Press.

MS (LC-tandem MS) adds a further dimension and can be used to identify structural differences that do not affect the UV/Vis absorption properties, or to distinguish ions with the same mass-to-charge ratio (isobaric ions). Furthermore, LC-tandem MS is particularly powerful if MS/MS spectra of an unknown compound are compared with a structurally related, identified compound.

7.2 Description of instrumentation

Mass spectrometers consist of three modules: an ion source or ionisation chamber, a mass analyser and a detector. This section describes ionisation techniques and mass analysers used in pigment analysis for the novice mass spectrometrist. The reader is directed to comprehensive LC-MS texts (e.g. Cole, 1997; Hoffmann and Stroobant, 2001; McMaster, 2005) for more detailed descriptions.

7.2.1 Modes of ionisation used in pigment analysis

Fast atom bombardment (FAB)

Mass spectrometry requires formation of gaseous ions, which, for thermally labile compounds such as chlorophylls and carotenoids, presented a problem because they could not be introduced to an ionisation chamber in gaseous form without first derivatising them to increase their volatility. Fast atom bombardment represented a breakthrough in mass spectrometry for the analysis of pigments because it provided a method of producing gas phase ions directly from thermally labile and non-volatile compounds, and is suited to the analysis of both carotenoids and chlorins (Keely and Maxwell, 1990; Van Breemen *et al.*, 1993; Garrido and Zapata, 1998). During FAB ionisation, a high velocity atom beam of an inert gas (e.g. xenon) is directed onto the sample within a matrix (e.g. glycerol), causing desorption of protonated or deprotonated molecules from the sample. The matrix needs to be a relatively involatile solvent that is capable of dissolving the sample. Continuous flow-FAB (cf-FAB) is a method of introducing the sample to the FAB probe in a continuous flow of solvent via a fused capillary which enables coupling to HPLC. The matrix can be added to the HPLC mobile phase or introduced post column. Continuous flow-FAB results in reduced background signal from the matrix compared to FAB as the ratio of matrix/sample is lower. Although cf-FAB produces abundant molecular ions, its disadvantages included an optimum flow rate of 5–10 $\mu L\ min^{-1}$ and fouling of the source from some matrices requiring regular cleaning (Caprioli, 1990; van Breemen, 1997). For many applications, FAB has now been superseded by the atmospheric pressure ionisation techniques of APCI and ESI which dominate in current applications of LC-MS.

Atmospheric pressure chemical ionisation (APCI)

Atmospheric pressure chemical ionisation is a soft ionisation technique used successfully to analyse chlorophylls (Airs and Keely, 2000; Jie *et al.*, 2002; Walker *et al.*,

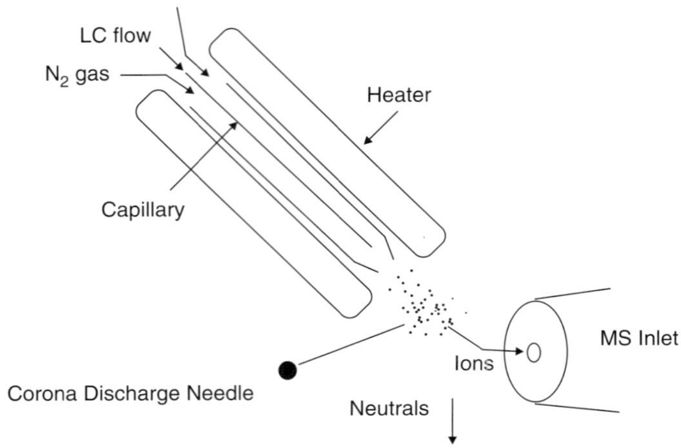

Figure 7.1. Schematic diagram of APCI source.

2003), bacteriochlorophylls (Airs and Keely, 2002; Squier *et al.*, 2004; Wilson *et al.*, 2005) and many microalgal carotenoids (van Breemen *et al.*, 1996; Airs *et al.*, 2001a; Frassanito *et al.*, 2005; Airs and Llewellyn, 2006). This technique allows for direct use of flow rates typical of standard bore HPLC. The APCI source normally comprises a heated glass capillary surrounded by a sheath flow of nitrogen gas, with an additional or auxiliary flow of nitrogen present, as well as a corona discharge needle (Figure 7.1). The HPLC eluent and sample molecules are volatilised in the heated capillary and the eluent vapours are ionised by the corona discharge at the needle. The analyte molecules are ionised in the gas phase usually by proton transfer to (negative ion mode) or from (positive ion mode) the ionised eluent vapours, forming $[M-H]^-$ or $[M+H]^+$ ions. Solvent adducts or radical cations can also be observed. For APCI, the analyte and eluent buffers must be volatile, and the mobile phase suitable for ionisation.

Electrospray ionisation (ESI)

The electrospray probe (Figure 7.2) consists of a metallic capillary surrounded by a nitrogen flow. A voltage is applied between the capillary tip and the sampling cone. Typically, electrospray works well at low flow rates (1–500 μL min^{-1}), and flow splitters have been used in combination with higher LC flow rates (1000 μL min^{-1}) to provide a low enough flow rate to form a stable spray. Some modern electrospray sources, however, are designed to form a stable spray at flow rates up to 1 mL min^{-1}. For electrospray, the analyte molecules are commonly ionised in solution. This is normally achieved by controlling the pH of the mobile phase. For neutral molecules, such as carotenoids, ionisation is achieved by charge transfer from the solvent molecules. At the tip of the capillary, the surface of the droplet containing the ionised analyte becomes charged due to the potential difference between the capillary and the

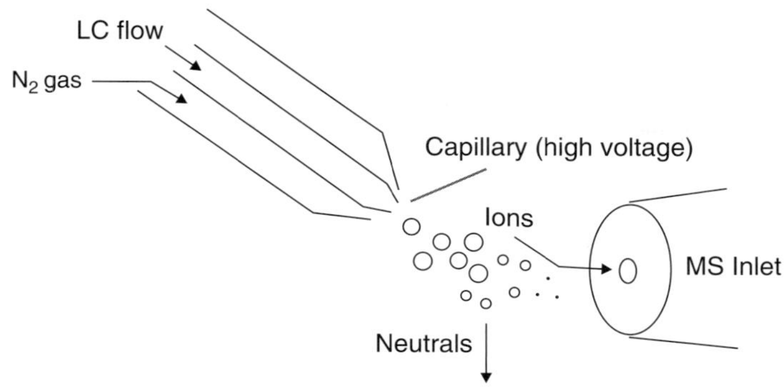

Figure 7.2. Schematic diagram of ESI source.

sampling cone. As the solvent evaporates (this is aided by some source heating) the droplet shrinks and the density of charge increases. The repulsion forces between the charges increase until the droplet becomes unstable and undergoes coulombic explosion. This process repeats until charged analyte molecules evaporate from the droplet: these can be singly or multiply charged and adducts are commonly observed during electrospray ionisation.

Matrix-assisted laser desorption ionisation (MALDI)

MALDI is a soft ionisation technique which is most commonly used for high molecular weight biomolecules (e.g. proteins and peptides). The analyte is mixed with a matrix with absorption characteristics matched to the output wavelength of the laser. Common matrices include sinapinic acid and α-cyano-4-hydroxy-cinnamic acid mixed with an organic solvent (e.g. acetonitrile) or water. The analyte–matrix solution is spotted onto a MALDI plate, the solvents vaporise, leaving the analyte molecules spread throughout the recrystallised matrix. The laser causes ionisation and desorption of molecules from the MALDI spot into the gas phase. Protonated, deprotonated and radical cations may be observed. The MALDI technique has been applied to the analysis of chlorophylls and their derivatives (Suzuki *et al.*, 2009). Terthiophene was selected as a suitable matrix, and the spectra produced gave ions corresponding to the parent molecule and corresponding chlorophyllide resulting from loss of the phytyl chain (Suzuki *et al.*, 2009). The technique has also been used to examine plant carotenoids, both in extracts and whole cells (Fraser *et al.*, 2007). It has also been applied to pigments and proteins in chlorosomes and whole cells of photosynthetic bacteria (Persson *et al.*, 2000). LC-MALDI systems are available where the matrix is added automatically post-column and spotted onto different positions on the MALDI plate or target. MALDI is well suited to time of flight mass analysers due to their large mass range and compatibility with a pulsed output.

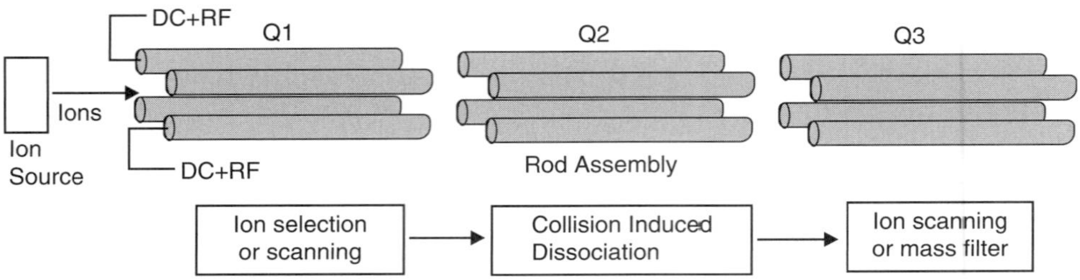

Figure 7.3. Schematic diagram of quadrupole mass analyser.

7.2.2 Mass analysers

Mass analysers commonly used in LC-MS instruments are quadrupole (Q), ion trap, time of flight (TOF) and combinations such as triple quadrupoles and QTOFs. Other mass analysers that are less common in LC-MS instruments are not covered here.

Quadrupole/triple quadrupole

A quadrupole mass analyser, sometimes referred to as a linear ion trap, consists of four parallel metal rods that have direct current (DC) and alternating radiofrequency (RF) potentials applied to them (Figure 7.3). The applied voltages affect the trajectory of ions travelling down the flight path through the centre of the four rods. For a particular combination of RF and DC voltages, only ions of a certain mass-to-charge ratio (m/z) will be stable within the interquadrupolar region and reach the detector. Other ions have unstable trajectories and will collide with the rods. This allows monitoring of a particular m/z value, or scanning by varying the voltages.

A triple quadrupole instrument (Figure 7.3) consists of a linear series of three quadrupoles. The first and third quadrupoles act as mass filters, and the middle quadrupole is used as a collision cell. In a product ion scan, the first quadrupole filters the mass-to-charge ratio of interest, and these ions undergo collision-induced dissociation (CID) in the second quadrupole. Product ions are then passed to the third quadrupole, where they are scanned. Other scanning modes include precursor ion and neutral loss scans. In a precursor ion scan, the first quadrupole scans the ions entering from the source, the ions are fragmented in the second quadrupole and the third quadrupole acts as a mass filter for a predefined mass-to-charge ratio. Using this mode, a specific product ion is detected and the corresponding precursor ion can be identified. In a neutral loss scan, the first and third quadrupoles act as scanning analysers and are linked and offset by a predefined neutral loss. Fragmentation takes place in the second quadrupole and only spectra that show the specified neutral loss are indicated. Triple quadrupole systems are effectively limited to acquiring MS/MS spectra only of specified precursor or product masses or specific neutral losses.

The sensitivity and specificity of a triple quadrupole, however, makes it particularly suited for detecting and quantifying known compounds.

Ion trap

The quadrupole ion trap, also referred to simply as an ion trap, traps ions through the action of three hyperbolic electrodes: the ring electrode and the entrance and exit endcap electrodes. The radiofrequency (RF) potential applied to the ring electrode produces a 3D quadrupolar potential field within the trap. This traps the ions in a stable oscillating trajectory, the exact motion of the ions being dependent on the voltages applied and the individual *m*/*z* ratios of the ions. For detection, the trapping potentials are altered to destabilise ion motion and result in ejection of the ions from the trap. The ions are usually ejected in order of increasing *m*/*z* by a gradual change in the trapping potentials.

Ion traps are ideally suited to performing multistage MS^n experiments for structural elucidation studies. Ions of a particular *m*/*z* can be isolated in the trap by ejecting all of the other ions. The isolated ions are then fragmented by collision-induced dissociation with a collision gas (usually helium). The product ions can then be scanned out to produce an MS^2 spectrum. An MS^3 spectrum is obtained by first isolating and dissociating the precursor ion, and then isolating and dissociating one of the product ions observed in the MS^2 spectrum, before scanning out the product ions. The sequential isolation and fragmentation steps can be repeated n times to obtain MS^n mass spectra, providing a wealth of fragmentation data or fragmentation trees. The limiting factor in the number of MS stages that can be carried out is the ion current which is reduced by each MS^n stage. The detailed fragmentation data available using an ion trap can be used to understand how a particular suite of compounds fragment with a view to assigning closely related structures.

Time of flight (TOF)

Ions in a time of flight analyser are accelerated by a potential and then travel a set distance to the detector. The velocity of an ion depends on its mass-to-charge ratio, with heavier particles reaching lower speeds. Time of flight analysers require a pulsed ion beam. Continuous ion sources, such as electrospray ionisation are generally interfaced to the TOF mass analyser by orthogonal extraction (Dawson and Guilhaus, 1989) in which the ions are introduced into the mass analyser in a direction perpendicular to the direction of flight. The advantages of a TOF analyser are good response and high resolution, permitting accurate mass measurements and therefore determination of molecular formulae. A disadvantage is a lack of MS^n capability. Hybrid instruments are available that combine the quadrupole or ion trap with a TOF analyser, providing the benefit of MS/MS selectivity and flexibility for collision experiments with the high response, resolution and accurate mass capability of a TOF analyser.

7.3 Approaches to LC-MS analysis

7.3.1 Compatibility of HPLC methods with MS

HPLC methods designed to separate phytoplankton pigments commonly use a salt or buffer for ion suppression or ion pairing, enabling retention and resolution of early eluting acidic components, namely chlorophyll *c* pigments, chlorophyllides and phaeophorbides. Ammonium acetate is often the salt of choice, typically used as a component in the primary eluent at a concentration of 0.5–1 M (Mantoura and Llewellyn, 1983; Zapata *et al.*, 1987; Wright *et al.*, 1991; van Heukelem *et al.*, 1994; Barlow *et al.*, 1997). Although ammonium acetate is volatile and therefore compatible with LC-MS, a high concentration (0.5 M) in APCI results in reduced response (Airs, 2001). Much lower concentrations (0.01 M) have been used for pigment HPLC methodology designed for application to LC-MS (Airs *et al.*, 2001a), in combination with methylation for retention of free acids. Pyridine has also been used in pigment HPLC (Garrido and Zapata, 1996, 1997; Garrido *et al.*, 2003), affording increased retention of free acid pigments, formally due to suppression of ionic interactions with silanols by the pyridinium ion. Although pyridine-containing mobile phase has been used for LC-MS (Garrido and Zapata, 1998), the ease of protonation of the pyridine molecule in the mass spectrometer source means much of the ion current is directed to forming protonated pyridine molecules, observed at m/z 80, resulting in decreased response for the target analytes.

Other solvents commonly used for pigment HPLC methods, including methanol, water, acetone, acetonitrile and ethyl acetate, are compatible with LC-MS (see Table 7.1). Normal phase solvents have also been used successfully during APCI-LC-MS for pigment analysis: the products of the allomerisation reaction of chlorophyll *a* have been studied by normal phase HPLC, utilising hexane containing 1.5–10% methanol/isopropanol as the mobile phase, in conjunction with APCI-tandem mass spectrometry (Jie *et al.*, 2002), providing a rapid and efficient method for structure assignment for these compounds which are not readily separated by RP-HPLC.

7.3.2 Ionisation methods for the analysis of chlorophylls: APCI and ESI

APCI and ESI have superseded other ionisation methods originally associated with LC-MS, e.g. the moving belt interface combined with electron ionisation or chemical ionisation, and continuous flow-FAB (cf-FAB), although much early work on LC-MS of both chlorophylls and carotenoids was achieved using cf-FAB (van Breemen *et al.*, 1991, 1993; van Breemen, 1997). APCI has been more widely cited than ESI as an ionisation method for the LC-MS analysis of chlorophylls, bacteriochlorophylls and their derivatives. Chlorophylls (with the exception of chlorophylls *c*) are likely to be more amenable to APCI than electrospray as they are not usually chromatographed as ions, they encompass non-polar structures as well as more polar forms,

Table 7.1. *Comparison of APCI and ESI ionisation modes.*

	APCI	ESI
Type of compound	Less polar; thermally stable	Ionic; very polar; thermally labile
Flow rate	HPLC flow rates	Best response at low flow rate (1–500 μL min^{-1}). Modern sources can cope with higher flow rates (up to 1 mL min^{-1}), but response may be compromised
pH	As for HPLC	Acidic for basic compounds, basic for acidic compounds
Buffers	Volatile	Volatile, low concentration (< 50 mM)
Common solvents/ modifiers	Methanol, water, acetonitrile[1], acetone, ethyl acetate, ammonium acetate	Formic acid (< 2%), acetic acid (< 2%), trifluoroacetic acid (forms adducts, lingers) methanol, water, ammonium acetate, ammonium formate, ammonium hydrogen carbonate

[1]Mobile phase compositions comprising high proportions of acetonitrile (80% or higher) cause corona voltage drop and loss of signal in APCI

and the ESI solvent modifiers that can be used are restricted due to the ease of demetalation of chlorophylls under acidic conditions. One of the earlier applications of LC-MS to pigment analysis, however, used ESI successfully for the analysis of a natural chlorin mixture (Chillier *et al.*, 1994). More recently, ESI has been applied to the analysis of chlorophylls of natural populations of *Prochlorococcus* (Goericke *et al.*, 2000), chlorophyll transformation products (Goericke *et al.*, 1999; Villanueva and Hastings, 2000) and to the analysis of chlorophylls *c* (Bertrand *et al.*, 2005). Under APCI the mechanism of ionisation involves formation of a charged adduct, usually by protonation. The nitrogen atoms of chlorophylls provide a suitable site for protonation, particularly in demetallated species, resulting in higher ionisation efficiencies of pheophytins than chlorophylls (Airs and Keely, 2000). Accordingly, enhanced sensitivity in APCI-LC-MS of chlorophylls can be achieved by converting the chlorophylls to pheophytins post-column using an addition of formic acid to the solvent flow (Airs and Keely, 2000). This acts to remove the central magnesium of chlorophylls in post chromatographic separation, but before the compounds enter the mass spectrometer. This approach can be particularly useful for the on-line LC-MS analysis of chlorophylls *c*, which exhibit very low ionisation efficiencies in the metallated form. Post-column addition of acid can also be of benefit for studying chlorophylls present in low abundance.

The full mass spectra of chlorophylls and their derivatives generally exhibit abundant protonated molecules ($[M+H]^+$) under both APCI (Harris *et al.*, 1995; Airs *et al.*, 2000; Riffé-Chalard *et al.*, 2000; Squier *et al.*, 2004) and ESI conditions

(Goericke *et al.*, 1999; Goericke *et al.*, 2000; Villanueva and Hastings, 2000; Bertrand *et al.*, 2005), although radical cations have also been observed under ESI (Frassanito *et al.*, 2005, 2006). As well as positive ion, negative ion APCI has been applied to the analysis of tetrapyrrole pigments and alteration products (Eckardt *et al.*, 1991; Verzegnassi *et al.*, 1999, 2000; Aydin *et al.*, 2003), yielding abundant molecular anions (M^-), and no significant fragmentation. Under positive ion APCI, some fragmentation of the parent ion can be observed in full MS (e.g. Talbot *et al.*, 1999a; Verzegnassi *et al.*, 2000; Airs *et al.*, 2001b; Aydin *et al.*, 2003). The chlorophylls *c*, which have porphyrin structures, have been found to render good $[M\text{-}H]^-$ ions under ESI conditions (Bertrand *et al.*, 2005), and are difficult to ionise under APCI conditions.

7.3.3 Fragmentation of chlorophylls and bacteriochlorophylls: MS/MS

The typical mass spectral fragmentations from the periphery of chlorophylls have been recognised since desorption ionisation MS, FAB and LC-MS studies in the early 1980s and 1990s (Hunt *et al.*, 1981; Hunt and Michalski, 1991; Eckardt *et al.*, 1990; Keely and Maxwell, 1990). Additional information is obtained using LC-MS/MS: firstly, sequential fragmentation permits construction of fragmentation maps, and secondly, isolation and fragmentation of a specific *m/z* value ensures that fragments only originate from the particular precursor ion. Chlorophyll *a*, its related pyro and demetaled derivatives and derivatives that have undergone oxidation on ring E give rise to a loss of 278 Daltons (Da) under MS/MS, corresponding to loss of the phytyl side chain as an alkene following proton transfer to the macrocycle (Keely and Maxwell, 1990; Figure 7.4). Typically, the ion arising from the loss of 278 Da forms the base peak in MS/MS, although a high collision energy can increase the relative abundance of other ions in the MS/MS spectrum. Other prominent ions observed in MS/MS spectra of chlorophylls with an intact carbomethoxy group arise from losses of 60 Da and 32 Da, corresponding to the loss of the entire CO_2Me group at $C13^2$ with proton transfer and the loss of methanol, respectively (Figure 7.4). These losses can also be observed directly from the protonated molecule. Pyro derivatives (e.g. pyrochlorophyll *a*), which lack the $C13^2$ CO_2Me substituent, do not exhibit losses of 32 or 60 Da under MS/MS. Notably, carbomethoxy substituents have been observed via product ions formed by losses of both 58 and 59 Da, indicating that this group can be lost via more than one mechanism (Harris *et al.*, 1995). After loss of the phytyl ester and $C13^2$ carbomethoxy group, a loss of 72 Daltons may be observed, attributed to loss of the propionic acid side chain at C-17.

The MS/MS spectra of chlorophyll *b* and its derivatives differ from that of chlorophyll *a* in the presence of ions arising from the loss of 28 Daltons (Gauthier-Jaques *et al.*, 2001), observed from both $[M+H]^+$ and $[M+H\text{-phytyl}]^+$. As this loss is not observed from chlorophyll *a*, the loss is assigned as arising from the aldehyde

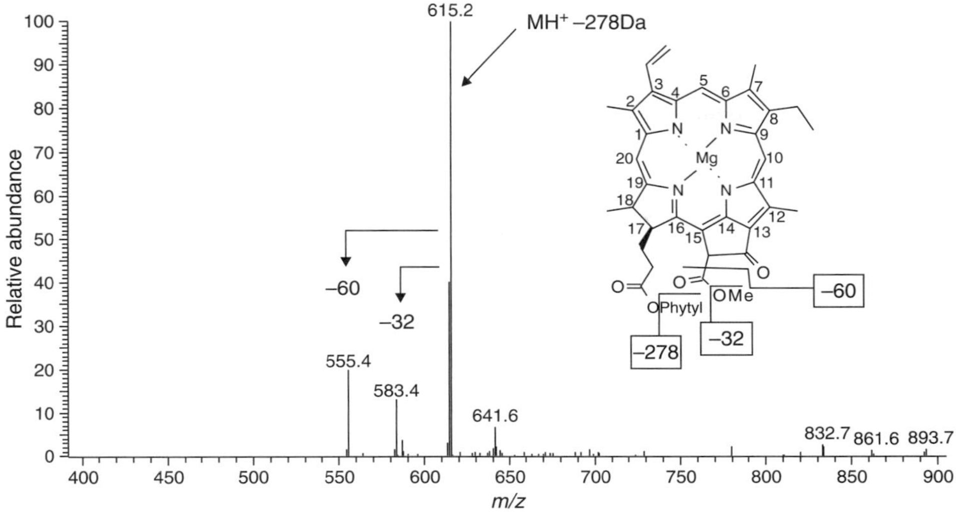

Figure 7.4. MS^2 spectrum of chlorophyll *a* (MH^+ 893), obtained during APCI-LC-MS^n analysis of a phytoplankton pigment extract.

functionality at C-7, the only structural difference between chlorophylls *a* and *b*. The loss is also observed from bacteriochlorophyll *e* which also has an aldehyde functionality at C-7. LC-MS/MS spectra of chlorophylls *c* are less common, due to a lower ionisation efficiency of porphyrins than chlorins. Porphyrins are more planar than chlorins due to their extended pi system, and the lower observed ionisation efficiency probably results from a lower availability of ring nitrogens for protonation (Airs and Keely, 2000).

Goericke *et al.* (2000) have published fragment ions of chlorophylls *c* obtained during ESI mass spectrometry of isolated pigments. The MS^2 spectra of chlorophylls c_2 and c_3, obtained during online LC-MS/MS analysis of a pigment extract are shown in Figure 7.5. Although the extract was analysed using post-column addition of formic acid, which would be expected to cause demetalation of chlorophylls prior to ionisation (Airs and Keely, 2000), the chlorophylls *c* were observed in the full mass spectrum with the central magnesium intact, at *m/z* 609 and *m/z* 653 for chlorophylls c_2 and c_3, respectively. This resistance to demetalation is attributed to the increased planarity of porphyrins compared to chlorins. In spite of not inducing demetalation of the chlorophyll *c* pigments, the addition of formic acid post-column still had a positive effect on the ionisation efficiency of the chlorophylls *c*, possibly due to the increased abundance of H_3O^+ for proton transfer reactions. The MS^2 spectrum of chlorophyll c_2 (Figure 7.5a) exhibited losses of 18 and 32 Da from the protonated molecule at *m/z* 609. Dehydration of the protonated molecule of chlorophylls without a hydroxyl group is unusual, but the loss observed for chlorophyll *c* can be rationalised by considering the structural differences between porphyrins and chlorophylls.

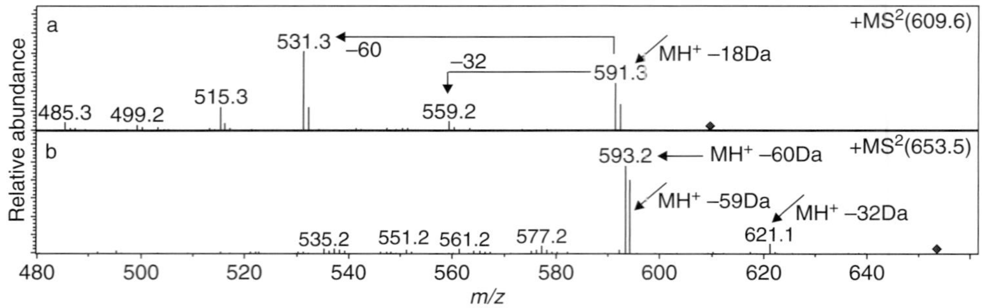

Figure 7.5. MS^2 spectra of (a) chlorophyll c_2 (MH$^+$ 609) and (b) chlorophyll c_3 (MH$^+$ 653).

The planar nature of porphyrins, such as chlorophylls *c*, would presumably make the ring nitrogens less available for protonation (compare Airs and Keely, 2000), thereby creating a shift towards oxygen as a candidate for proton acceptor during ionisation. A protonated carboxylic acid group is likely to favour fragmentation via dehydration. Interestingly, the MS^2 spectrum of chlorophyll c_3 (Figure 7.5b) gave prominent product ions arising from losses of 32 Da, 59 and 60 Da from $[M+H]^+$, but no loss of 18 Da. The absence of an ion arising from dehydration in the spectrum of chlorophyll c_3 is surprising, given its presence in the MS^2 spectrum of chlorophyll c_2. It is possible that the presence of a second carbomethoxy group in chlorophyll c_3 effectively reduced the number of molecules protonated at the carboxylic acid position, thereby reducing the likelihood of dehydration. The carbomethoxy groups of chlorophyll c_3 appear to be lost by different mechanisms, giving rise to both fragments of 59 and 60 Da (Figure 7.5b). Chlorophyll c_3 exhibits carbomethoxy substituents at positions C-7 and C-13^2 (see Figure 7.4 for structure numbering). From the MS^n spectra of chlorophyll *a*, the loss of a carbomethoxy substituent at position C13^2 is commonly observed as a loss of 60 Da with proton transfer from the macrocycle. The carbomethoxy group at position C-7 may be lost via a different mechanism without H-transfer (Harris *et al.*, 1995) due to the lack of an adjacent proton at position C-7. Under negative ion ESI, the product ion spectrum of chlorophyll c_2 was dominated by losses of 44, 32 and 60 Da (Bertrand *et al.*, 2005). A loss of 18 Da was not observed, indicating ionisation by deprotonation drives the fragmentation pathway away from dehydration.

Bacteriochlorophylls *c*, *d* and *e* are produced exclusively by obligately anaerobic phototrophs (Chlorobiaceae), and as such, they make particularly diagnostic markers for the groups of organisms that produce them, but their distribution in oceanic environments is limited. Chlorobiaceae form an important community, however, at the chemocline of the Black Sea (Manske *et al.*, 2005), and bacteriochlorophyll-derived pigment signatures from Antarctic sediments have been used to study past environmental change (Squier *et al.*, 2002, 2005). In spite of their limited distribution in the oceanic environment, an understanding of the behaviour of bacteriochlorophyll structures during LC-MS/MS and how they differ from

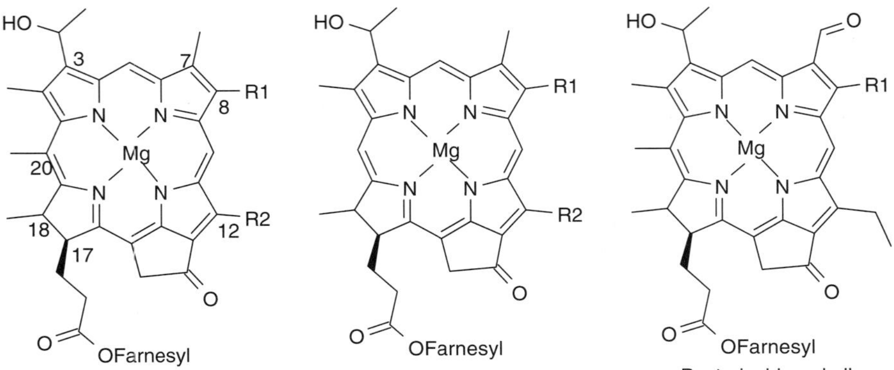

Bacteriochlorophyll *c*

R1: Et, *n*-Pr, *i*-Bu; R2: Et, Me

Bacteriochlorophyll *d*

R1: Et, *n*-Pr, *i*-Bu, *neo*-pent; R2: Et, Me

Bacteriochlorophyll *e*

R1: Et, *n*-Pr, *i*-Bu

Figure 7.6. Structures of bacteriochlorophylls *c*, *d* and *e*.

chlorophylls produced by eukaryotes contributes to our understanding of fragmentation pathways and provides additional information for the structural assignment of unknown chlorins. LC-MS/MS is particularly important for the analysis of bacteriochlorophylls *c*, *d* and *e* as they exist as suites of compounds that are indistinguishable by UV/vis detection. Unlike chlorophyll *a*, the bacteriochlorophylls of Chlorobiaceae do not exhibit a $C13^2$ carbomethoxy group, and the common homologues are esterified by farnesol (Figure 7.6). Additionally, bacteriochlorophylls *c*, *d* and *e* possess a hydroxyethyl substituent at C-3, where chlorophyll *a* possesses a vinyl group. Consequently, these structural differences create distinct MS/MS spectra.

The MS^n spectra of bacteriochorophylls have been covered in detail by Airs and Keely (2002), but the main features will be summarised here. The hydroxyethyl group at C-3 has been observed to fragment in two ways. Firstly, by dehydration, yielding an ion at $[M+H-18]^+$. This ion can be observed in the full MS spectrum (Figure 7.7a), as well as a minor ion in MS^2, occurring directly from the protonated molecule. Notably, the dehydration occurs more readily when the analysis is carried out without post-column addition of acid (Figure 7.7b), suggesting that it is largely dependent on the site of protonation (formation of a charged adduct) on the bacteriochlorophyll during ionisation. By limiting the fragmentations that occur during ionisation, post-column addition of acid focuses the ion current into the protonated molecule, providing additional benefit for MS/MS studies. The C-3 hydroxyethyl substituent has also been observed to fragment via the loss of 44 Da, assigned as loss of the entire substituent with proton transfer back to the macrocycle (Chillier *et al.*, 1994; Airs and Keely, 2002; Wilson *et al.*, 2005). The absence of a loss of 44 Da from $[M+H]^+$ of chlorophylls *a* and *b* supports the assignment.

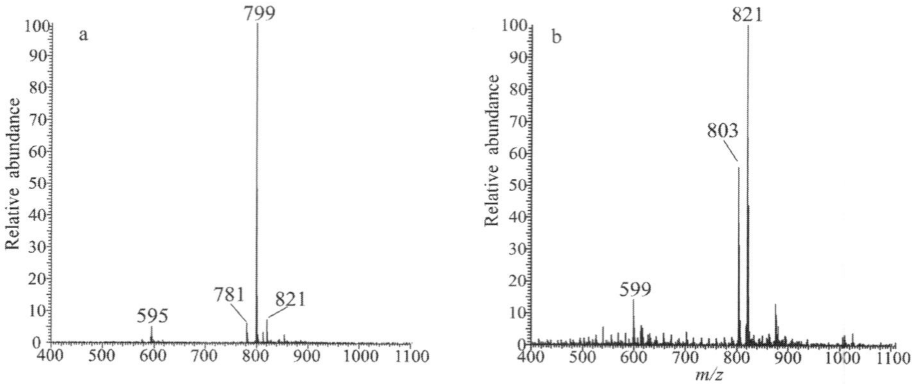

Figure 7.7. Full mass spectra of bacteriochlorophyll e_2 from *Chlorobium phaeobacteroides* CL1401 extract (a) with post-column addition of acid ($[M+3H-Mg]^+$ 799), and (b) without post-column addition of acid ($[M+H]^+$ 821).

Farnesyl-esterified bacteriochlorophyll homologues, commonly referred to as primary homologues (Borrego and García-Gil, 1994), show fragmentations in MS^2 with the most prominent ion arising from the loss of 204 Da from $[M+H]^+$ (Figure 7.8), corresponding to the loss of farnesyl as an alkene following proton transfer to the macrocycle, analogous to the loss of phytyl from chlorophyll *a*. Accordingly, information regarding the mass of both the esterifying alcohol and the macrocycle can be obtained from the MS^2 spectrum.

By contrast, LC-MS^n analysis of secondary homologues, revealed a different fragmentation pattern. The second fragmentation pattern was characterised by a high relative abundance of ions arising from expulsion of small molecules from the macrocycle, and low relative abundance of ions resulting from loss of the esterifying alcohol (Figure 7.9). The lower relative abundance of fragment ions resulting from loss of the esterifying alcohol of secondary homologues can be rationalised by considering the relative stabilities of the product ions. Farnesyl is lost by the same mechanism as the phytyl chain of chlorophyll *a* and its derivatives (Keely and Maxwell, 1990). Thus, loss of phytyl and farnesyl is facilitated by conjugation of the double bond formed on fragmentation with a double bond already present in the side chain. The relative ease of loss of the farnesyl side chain (reflected in the ion abundance in MS^2 spectra) is attributed to the stability of the product ion. The esterifying alcohols of the secondary homologues of *Chlorobium phaeobacteroides* have been identified as a range of straight chain and branched alcohols from C_{12}–C_{17} (Airs *et al.*, 2001b). These alcohols obtain no such stabilisation energy on fragmentation, making the fragmentation less facile.

Because of the highly functionalised nature of the macrocycles of bacteriochlorophylls, fragmentations occurring from the C-8 and C-12 side chains (see structure of bacteriochlorophyll *c* in Figure 7.6) are not seen during the first few sequences of

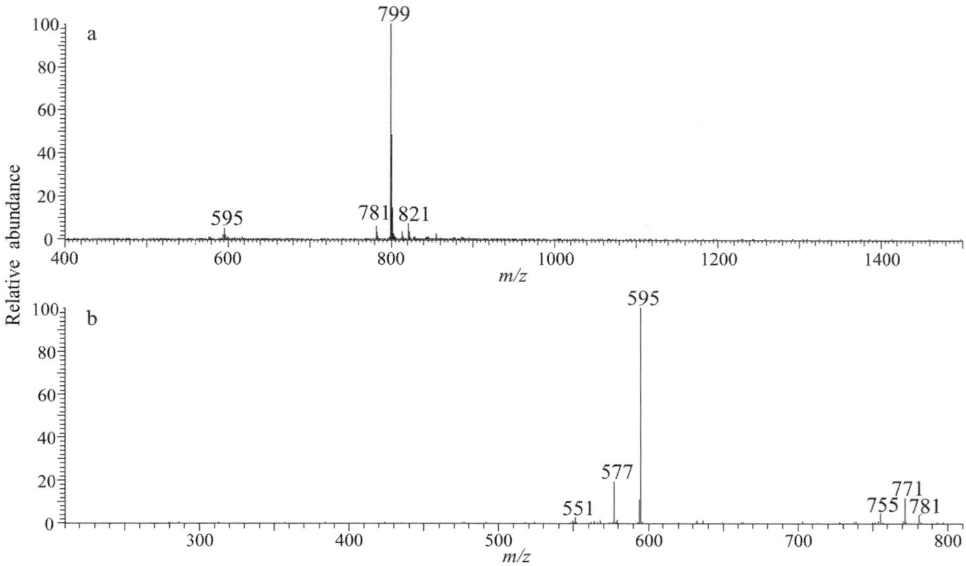

Figure 7.8. (a) Full mass spectrum and (b) MS^2 spectrum of bacteriochlorophyll homologue e_2 obtained on-line during APCI-LC-MS^n analysis of *Chlorobium phaeobacteroides* CL1401 extract.

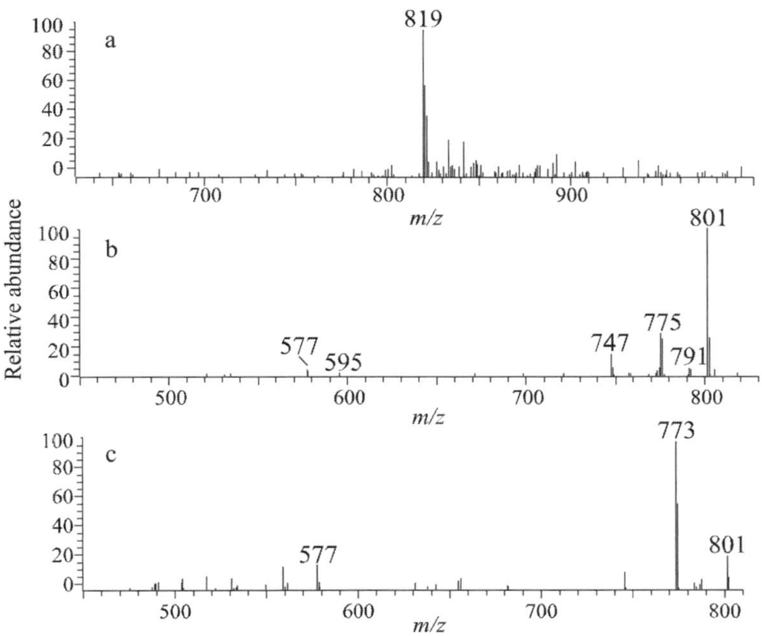

Figure 7.9. (a) Full mass spectrum, (b) MS^2 spectrum and (c) MS^3 spectrum of a bacteriochlorophyll *e* secondary homologue.

MS/MS fragmentation. Furthermore, in spectra where they start to appear (typically MS^6 or MS^7), the nature of the fragmentation is difficult to deduce when the structural isomerism of the molecule is in question. Previous studies using APCI-LC-MS/MS of porphyrins with extended alkyl substituents have shown them to fragment via a β-cleavage, and those with ethyl and methyl substituents by α-cleavage (Rosell-Melé *et al.*, 1999). APCI-LC-MS^n applied to the study of bacteriochlorophylls from cultures revealed fragmentations of the C-8 and C-12 alkyl substituents to be consistent with the observations for the alkyl porphyrins (Rosell-Melé *et al.*, 1999). The bacteriochlorophyll extended side chains have been most successfully studied, however, by first converting the bacteriochlorophylls to their bacteriopheophorbides (Wilson *et al.*, 2005), whereby losses corresponding to C-8 and C-12 were observed in MS^5 and MS^6. The alkyl substituents were lost as alkyl radicals, forming odd electron ions, contrasting with the even electron ions observed in MS^2–MS^4 (Wilson *et al.*, 2005).

7.3.4 LC-MS/MS analysis of chlorophylls from aquatic environments

Application of LC-MS/MS to the analysis of chlorophyll distributions has permitted detailed investigations based on diagnostic compounds. For example, steryl chlorin esters provide a distinct signature of zooplankton grazing on phytoplankton (Eckardt *et al.*, 1991), and LC-tandem mass spectrometry has permitted the discrimination of compounds that exhibit indistinguishable UV/Vis spectra in lacustrine, marine and Antarctic sediments (Cariou-Le Gall *et al.*, 1998; Riffé-Chalard *et al.*, 2000; Villanueva and Hastings, 2000; Airs *et al.*, 2001a; Squier *et al.*, 2002, 2005; Chikaraishi *et al.*, 2007), as well as zooplankton fecal pellets (Harris *et al.*, 1995) including studies of production and transformation studies (Talbot *et al.*, 1999a, 1999b, 2000). Similarly, carotenol chlorin esters, produced during crustacean grazing on diatoms, have been discovered and studied using LC-MS/MS (Goericke *et al.*, 1999; Chen *et al.*, 2003).

The diagnostic fragmentation pattern of chlorophyll *a* underpinned the recognition of novel alkyl sulfide derivatives of chlorophyll *a* in sediment from an Antarctic lake (Squier *et al.*, 2004). In fact, LC-MS/MS has been critical to using pigment distributions as markers for environmental change in Antarctic sediments (Squier *et al.*, 2002, 2005), for profiling sedimentary chlorophyll derivatives to relate their formation to diagenetic processes (Villanueva and Hastings, 2000; Aydin *et al.*, 2003) and to reconstruct lake phytoplankton assemblages of the past (Soma *et al.*, 2007; Romero-Viana *et al.*, 2009).

The diversity of products formed by oxidation of chlorophyll in the natural environment creates an absolute requirement for LC-MS/MS to study their distributions. LC-MS/MS has been used to identify chlorophyll and bacteriochlorophyll oxidation products (Naylor and Keely, 1998; Airs *et al.*, 2000; Walker *et al.*, 2003; Wilson *et al.*, 2004), monitor reaction pathways in laboratory model studies

(Rahmani *et al.*, 1993; Walker *et al.*, 2002) and to determine the distributions, and thus infer the significance, of oxidation products formed during the disruption, re-establishment, growth and senescence of a phytoplankton bloom (Walker and Keely, 2004).

In the analysis of bacteriochlorophylls, LC-MS/MS has permitted assignment of novel structures (Airs *et al.*, 2001b; Gich *et al.*, 2003) and assignment of the alkyl substituents at positions C-8 and C-12 of the macrocycle (Wilson *et al.*, 2005), which producer organisms vary according to environmental growth conditions. LC-MS has also been used to distinguish the esterifying alcohol of chlorophyll and bacteriochlorophyll in mutants of *Chlorobium tepidum* (Harada *et al.*, 2008).

7.3.5 Ionisation methods for the analysis of carotenoids

Although electron ionisation has been an important technique in carotenoid mass spectrometry for several decades, and is still widely used, it has limitations because the technique requires sample vaporisation prior to ionisation, which is a disadvantage in the analysis of thermally labile and non-volatile carotenoids (van Breemen, 1995). In addition, in electron ionisation spectra the molecular ions are typically minor or are absent, hence it is often necessary to use a second, softer ionisation technique, e.g. FAB, in order to provide molecular ion information (Egeland *et al.*, 2000). Fast atom bombardment ionisation has been used extensively for the analysis of carotenoids, and has been shown to produce both protonated and sodiated molecules during ionisation (Garrido and Zapata, 1998), as well as radical cations (van Breemen *et al.*, 1993; van Breemen, 1996) and generally shows little fragmentation (van Breemen, 1996; Garrido and Zapata, 1998). The limitations of FAB ionisation for carotenoid analysis during LC-MS are the low flow rates required and high maintenance level (van Breemen *et al.*, 1993; van Breemen, 1996).

APCI and ESI are now the ionisation methods of choice for analysis of carotenoids by LC-MS (Řezanka *et al.*, 2009). Semi-empirical calculations of ionisation efficiency of carotenoids demonstrate that the presence of heteroatoms increases ionisation efficiency (Guaratini *et al.*, 2005), with β, β-carotene showing a lower calculated ionisation efficiency than the xanthophylls (Guaratini *et al.*, 2005). Under ESI, xanthophylls have been shown to form both molecular ions and protonated molecules in positive ion mode (van Breemen, 1995; Guaratini *et al.*, 2005), whereas carotenes form molecular ions only (van Breemen, 1995). Similarly, in negative ion mode, deprotonated molecules of polar xanthophylls have been observed (Frassanito *et al.*, 2005, 2006) whereas carotenes do not ionise well (van Breemen, 1995). Guaratini *et al.* (2005) presented evidence that the conjugation extension can be an important effect for radical ionisation of β,β-carotene and xanthophylls during ESI, and that the presence of heteroatoms influenced the acid–base ionisation, independent of the calculated proton affinity of the molecule. With xanthophylls, the balance of molecular and protonated species could be affected by H^+ concentration.

Fucoxanthin derivatives appear to form protonated molecules and sodiated molecules under positive ion ESI, although they exhibit a high degree of fragmentation from the protonated molecule (Zapata *et al.*, 2004). Sodiated molecules have commonly been cited during positive ion ESI (Frassanito *et al.*, 2005, 2006, 2008). Under APCI, molecular ions and protonated molecules of xanthophylls and carotenes have both been observed under positive ion conditions (van Breemen *et al.*, 1996; Lacker *et al.*, 1999). Similarly, under negative ion APCI, both $M^{-\bullet}$ and $[M-H]^-$ can be observed (van Breemen *et al.*, 1996; Breithaupt, 2004). The full mass spectra of carotenoids under ESI or APCI conditions are generally dominated by abundant molecular or pseudo-molecular ions, with some exceptions. For example, lutein readily eliminates water from the allylic hydroxyl group under ionisation to give a prominent base peak at $[M+H-18]^+$ (van Breemen *et al.*, 1996; Dachtler *et al.*, 2001). Under negative ion APCI, however, a prominent ion is observed corresponding to $[M-H]^-$ for lutein, indicating that dehydration of lutein in full MS is greatly assisted by protonation. Fucoxanthin and its derivatives give rise to relatively complex spectra in full MS, exhibiting prominent ions at $[M+H]^+$, $[M+Na]^+$, $[M+H-H_2O]^+$ and $[M+H-60]^+$ (Garrido and Zapata, 1998; Zapata *et al.*, 2004; Airs and Llewellyn, 2006). Notably, when a source of sodium is applied to the system, the sodiated molecule dominates the full mass spectrum, fragmentation is minimised and the base peak signal intensity increases by approximately 35% (Airs and Llewellyn, 2006).

7.3.6 *Fragmentation of the carotenoids: MS/MS*

Much of our knowledge of the fragmentation of carotenoids has come from electron ionisation or chemical ionisation mass spectrometry. Electron ionisation generates ions by electron bombardment and typically produces spectra showing considerable fragmentation of the parent ion. Chemical ionisation generates ions by ionic reactions with an ionised reagent gas (e.g. ammonia). Fragmentation is less in chemical ionisation and spectra are therefore usually simpler. Characteristic losses from the polyene chain observed during electron ionisation, e.g. expulsion of toluene (*m/z* 92), were observed during early APCI studies of carotene (van Breemen *et al.*, 1996; Lacker *et al.*, 1999), evident in both full MS and in-source collision-induced dissociation (CID) MS/MS spectra. During a comprehensive survey of phytoplankton carotenoids using APCI-MS^n obtained using an ion trap instrument, a minor ion corresponding to the loss of toluene ($[M+H-92]^+$) was frequently evident in the full mass spectrum of carotenes and xanthophylls, but under collision-induced dissociation in the ion trap, loss of toluene was less frequently observed (Airs and Llewellyn, unpublished data). The expulsion of toluene is likely to occur therefore during heating in the APCI source, and to a lesser extent during collision-induced dissociation. In the case of xanthophylls, e.g. zeaxanthin, an ion in MS^2 at $[M+H-92]^+$

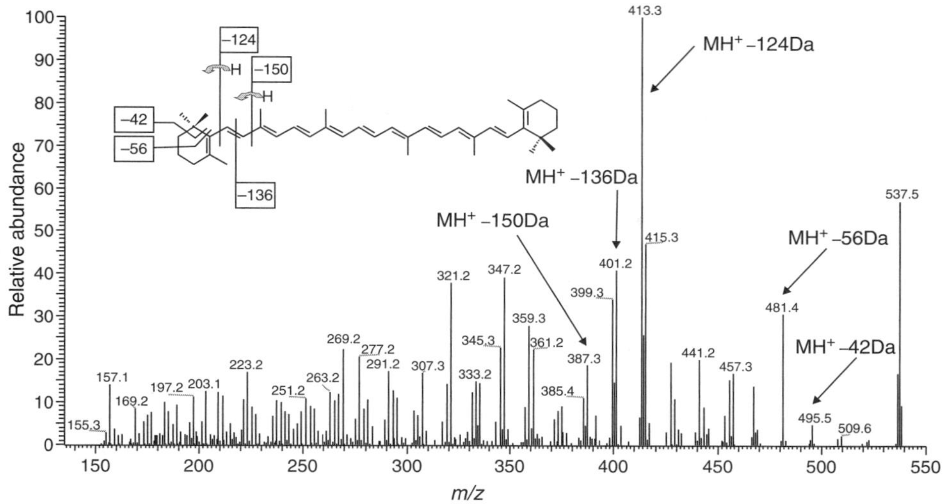

Figure 7.10. MS^2 spectrum of β,β-carotene (MH^+ 537).

may arise from two sequential losses of 18 and a loss of 56 from the cyclic end group. Using ESI-MS/MS during product ion scanning with a triple quadrupole instrument, loss of 92 Da was observed from the sodiated molecules of astaxanthin mono- and di-esters (Frassanito *et al.*, 2008). Other commonly observed losses from xanthophylls under electron ionisation, e.g. elimination of water from alcohols, acetic acid from acetates or methanol from methyl ethers, form prominent ions under MS/MS conditions (Goericke *et al.*, 1999; Chen *et al.*, 2003; Zapata *et al.*, 2004; Squier *et al.*, 2005; Airs and Llewellyn, 2006). As carotenes lack the oxygen-containing functional groups of xanthophylls, their MS^2 spectra are dominated by losses from the endocyclic rings and polyene chain, whereas the MS^2 spectra of xanthophylls are generally characterised by a prominent ion arising from the loss of an oxygen-containing functional group, and ions of lower relative abundance arising from fission of carbon–carbon bonds. Under APCI conditions, the MS^2 spectrum of β-carotene exhibited many of the same ions reported during chemical ionisation of β-carotene (Enzell and Back, 1995). The ions at *m/z* 347, 281, 255, 243, 203, 189 and 177 were evident in both spectra. Notably, ions at *m/z* 123, 137 and 149 were present in the chemical ionisation spectrum, whereas the APCI MS^2 spectrum revealed losses of 124, 136 and 150 Da (resulting from proton transfer to the neutral fragment), giving rise to ions at m/z 413, 401 and 387, respectively (Figure 7.10).

End group fragmentations commonly observed under electron ionisation conditions (Enzell and Back, 1995) can also be observed under LC-MS conditions. For example, under APCI, crocoxanthin gives rise to a loss of 56 Da from $[M+H]^+$ in MS^2, assigned to a retro Diels Alder loss from the ε-ring, mirroring the characteristic loss observed in electron ionisation (Enzell and Back, 1995).

More detailed fragmentation of simple carotenoids can be illustrated by examining the MS/MS spectra of structurally related carotenoids. Fucoxanthin and fucoxanthinol differ in the 3′ position substituent: a hydroxyl group for fucoxanthinol and an acetate group for fucoxanthin. Under APCI and ESI conditions, the fucoxanthins readily form sodiated molecules if sodium is present (Garrido and Zapata, 1998; Zapata *et al.*, 2004; Airs and Llewellyn, 2006). MS/MS studies of the sodiated molecule can be advantageous for fucoxanthin, its esters and similar compounds such as gyroxanthin and siphonaxanthin esters, as response for detection is improved (Airs and Llewellyn, 2006), and the sodiated molecules give rise to diagnostic losses/product ions in MS^2, whereas $[M+H]^+$ fragments less well, and $[M+H-H_2O]^+$ gives rise to fewer diagnostic product ions. The MS^2 spectra of the sodiated ions of fucoxanthin and fucoxanthinol differ in the relative abundance of the ion corresponding to $[M+Na-H_2O]^+$ (Figure 7.11). In the MS^2 spectrum of fucoxanthinol, the ion arising from the dehydration was base peak, whereas the ion arising from loss of the entire epoxidated ring was base peak in the MS^2 spectrum of fucoxanthin. This difference was attributed to the presence of three hydroxyl groups on fucoxanthinol, compared to two on fucoxanthin. The MS^2 spectra may also be distinguished by the presence of ions arising from loss of 60 Da for fucoxanthin, but not for fucoxanthinol, which lacks an acetate group. Notably, the MS^2 spectra for both compounds show prominent ions arising from M+Na−92. The presence of ions in the MS^3 spectra, arising from loss of 92 Da from the MS^2 base peak ion in both cases, eliminates, in this case, the possibility of the product ions arising from sequential losses from the endocyclic rings. Instead, they are attributed to the expulsion of toluene from the polyene chain.

The fucoxanthin-related suite of compounds can be extended to examine the fatty acid esters, for example 19′-hexanoyloxyfucoxanthin and its 4-keto form, 4-keto-19′-hexanoyloxyfucoxanthin. The MS^2 spectra of both compounds are dominated by ions arising from loss of 116 Da, assigned to the loss of the hexanoic acid side chain (Figure 7.12; Airs and Llewellyn, 2006). Ions arising from loss of the epoxidated endocyclic ring, dehydration and loss of acetic acid are also evident, but in much lower relative abundance compared to fucoxanthin and fucoxanthinol. The MS^3 spectra of 19′-hexanoyloxyfucoxanthin and 4-keto-19′-hexanoyloxyfucoxanthin (or 19′-hexanoyloxy-4-ketofucoxanthin, according to recent rules of chemical nomenclature) are more similar to the MS^2 spectra of fucoxanthin, exhibiting ions arising from the loss of water, acetic acid and the entire epoxidated endocyclic ring. The presence of the keto substituent is evident from the loss due to expulsion of the epoxidated endocyclic ring (168 Da), compared to loss of 154 Da for 19′-hexanoyloxyfucoxanthin (Figure 7.12). Similarly, the length of the fatty acid chain is reflected in the loss giving rise to the MS^2 base peak. For example, 19′-pentanoyloxyfucoxanthin gave rise to a loss of 102 Da on resonance excitation of $M+Na^+$, 14 Da less than hexanoic acid. In this manner, a suite of fucoxanthin esters has been assigned in *E. huxleyi* (Airs and Llewellyn, 2006).

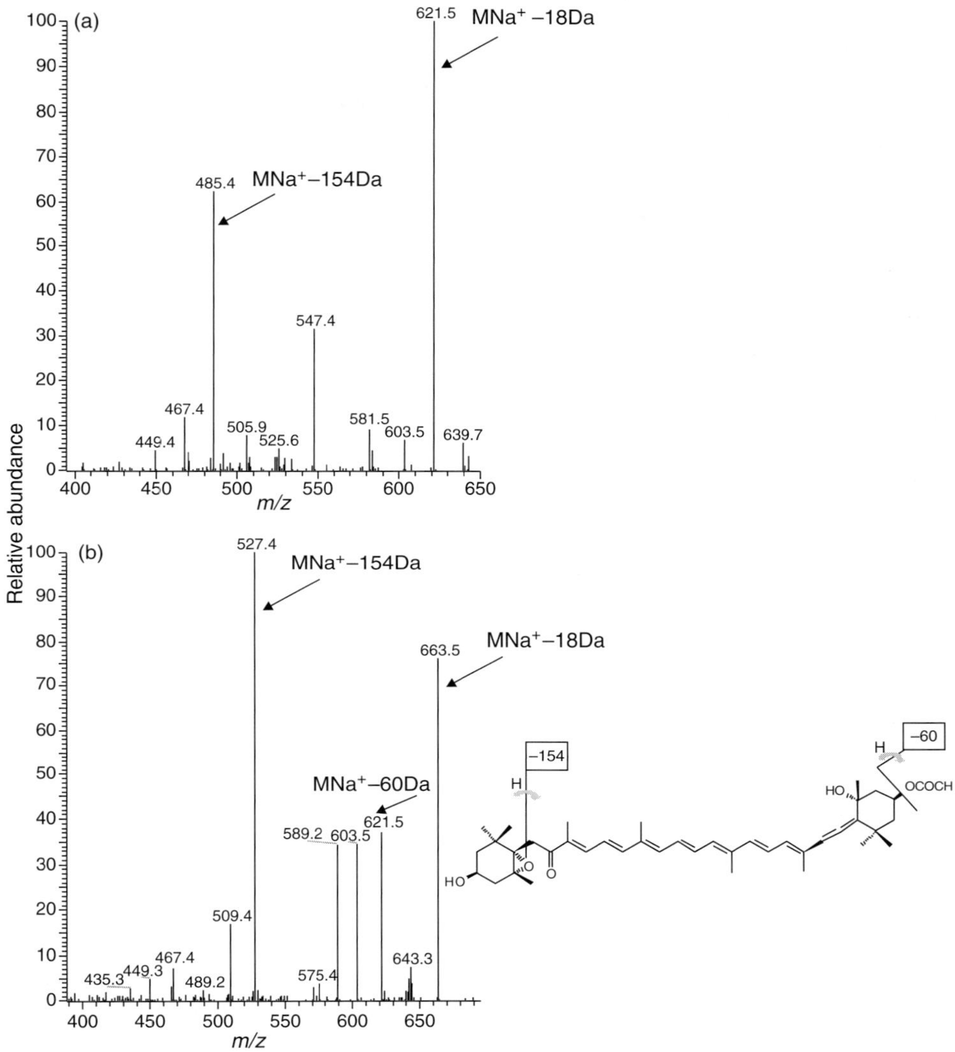

Figure 7.11. MS^2 spectra of (a) fucoxanthinol (MNa^+ 639) and (b) fucoxanthin (MNa^+ 681).

The benefit of LC-MS/MS becomes more evident when structural variations result in isobaric precursor ions. For example, 19′-heptanoyloxyfucoxanthin and 4-keto-19′-hexanoyloxyfucoxanthin are indistinguishable by their UV/Vis spectra, and both give rise to $M+Na^+$ at *m*/*z* 809. Isolation and fragmentation in the ion trap however, reveals structural variations in both the esterifying acid (116 and 130 Da; Figures 7.12 and 7.13) and the mass of the epoxidated endocyclic ring (168 and 154 Da;

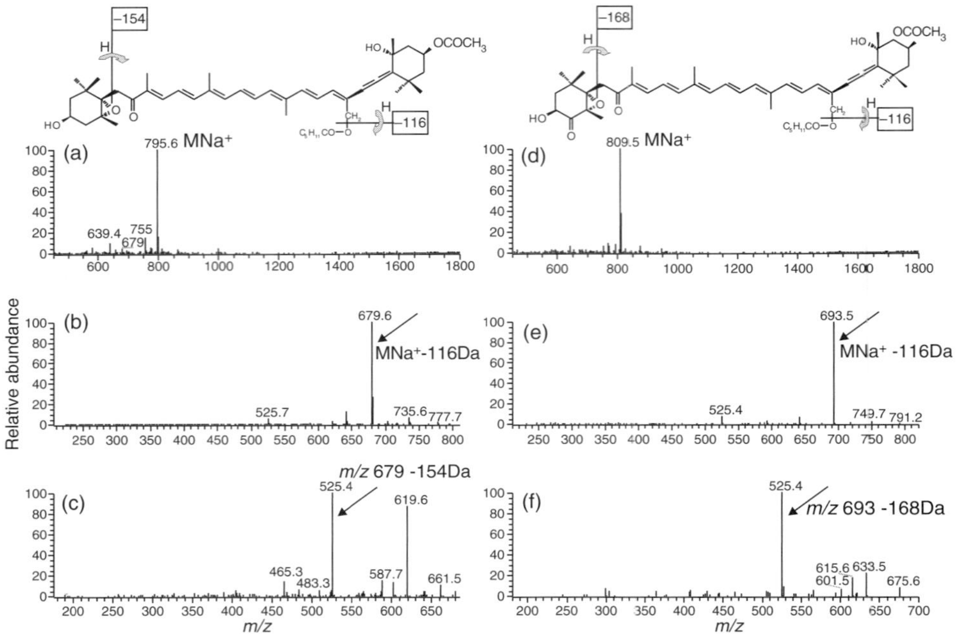

Figure 7.12. Full MS (a), MS^2 (b) and MS^3 (c) of 19′-hexanoyloxyfucoxanthin; and Full MS (d), MS^2 (e) and MS^3 (f) of 4-keto-19′-hexanoyloxyfucoxanthin.

Figures 7.12 and 7.13). Notably, the two compounds share a common product ion at *m/z* 525 (Figure 7.12) corresponding to the conserved structural moiety common to both compounds.

7.3.7 LC-MS analysis of carotenoids from the aquatic environment

LC-MS/MS has been important for assigning novel structures in microalgal cultures, particularly those present in low relative abundance (Garrido and Zapata, 1998; Zapata *et al.*, 2004; Airs and Llewellyn, 2006; Graham *et al.*, 2008). It can also play an important role in compound confirmation during routine analysis of phytoplankton samples, adding valuable molecular ion information to retention time and PDA data. LC-MS/MS has been used to analyse carotenoids as a component of biochemical signatures of microalgae (Frassanito *et al.*, 2005), and applied to the analysis of astaxanthin esters in dinoflagellates (Frassanito *et al.*, 2006) and krill (Grynbaum *et al.*, 2005). Analysis of astaxanthin esters by LC-MS is also important for understanding the biotechnological production of astaxanthin (Grewe and Griehl, 2008). LC-MS has also been applied to sediment samples to aid assignments of carotenoids and/or their diagenetic transformation products (Romero-Viana *et al.*, 2009; Tani *et al.*, 2009).

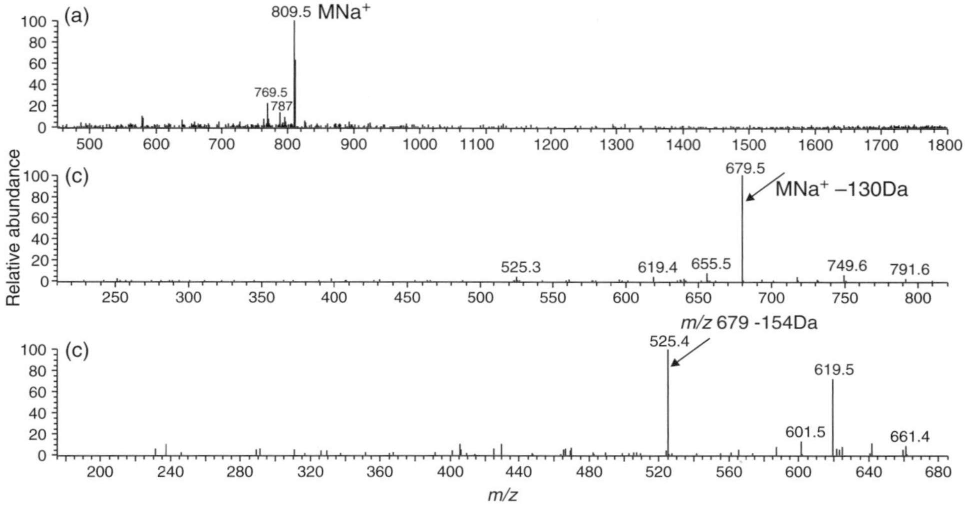

Figure 7.13. Full MS (a) MS^2 (b) and MS^3 (c) of 19′-heptanoyloxyfucoxanthin.

The benefit of the application of LC-MS/MS to pigment analysis is considerable. From characterisation of unknowns in algal cultures, component confirmation in phytoplankton samples and more elaborate alteration product analysis in sediments, LC-MS/MS is an important tool in helping us to unravel the structural complexity of photosynthetic pigments.

Abbreviations

APCI	Atmospheric pressure chemical ionisation
cf	Continuous flow
CID	Collision-induced dissociation
Da	Daltons
DC	Direct current
ESI	Electrospray ionisation
FAB	Fast atom bombardment
LC	Liquid chromatography
MALDI	Matrix-assisted laser desorption ionisation
Me	Methyl
m/z	Mass-to-charge ratio
PDA	Photodiode array detector
Q	Quadrupole
RF	Alternating radiofrequency
TOF	Time of flight

References

Airs, R. L. (2001). *Chlorophylls of phototrophic prokaryotes: analytical developments and significance of natural distributions*. Ph.D. Thesis, University of York, UK.

Airs, R. L. and Keely, B. J. (2000). A novel approach for sensitivity enhancement in atmospheric pressure chemical ionisation liquid chromatography/mass spectrometry of chlorophylls. *Rapid Commun. Mass Spectrom.* **14**, 125–28.

Airs, R. L. and Keely, B. J. (2002). Atmospheric pressure chemical ionisation liquid chromatography/mass spectrometry of bacteriochlorophylls from Chlorobiaceae: characteristic fragmentations. *Rapid Commun. Mass Spectrom.* **16**, 453–61.

Airs, R. L. and Llewellyn, C. A. (2006). Improved detection and characterization of fucoxanthin-type carotenoids: novel pigments in *Emiliania huxleyi* (Prymnesiophyceae). *J. Phycol.* **42**, 391–99.

Airs, R. L., Jie, C. and Keely, B. J. (2000). A novel sedimentary chlorin: structural evidence for a chlorophyll origin for aetioporphyrins. *Org. Geochem.* **31**, 1253–56.

Airs, R. L., Atkinson, J. E. and Keely, B. J. (2001a). Development and application of a high resolution liquid chromatography method for the analysis of complex pigment distributions. *J. Chromatogr. A* **917**, 167–77.

Airs, R. L., Borrego, C. M., Garcia-Gil, J. and Keely, B. J. (2001b). Identification of the bacteriochlorophyll homologues of *Chlorobium phaeobacteroides* strain UdG6053 grown at low light intensity. *Photosynth. Res.* **70**, 221–29.

Aydin, N., Daher, S. and Gülaçar, F. O. (2003). On the sedimentary occurrence of chlorophyllone *a*. *Chemosphere* **52**, 937–42.

Barlow, R. G., Cummings, D. G. and Gibb, S. W. (1997). Improved resolution of mono-and divinyl chlorophylls *a* and *b* and zeaxanthin and lutein in phytoplankton extracts using reverse phase C-8 HPLC. *Mar. Ecol. Prog. Ser.* **161**, 303–07.

Bertrand, M., Garrido, J. L. and Schoefs, B. (2005). Analysis of photosynthetic pigments: An update. In *Handbook of Photosynthesis*, ed. M. Pessarakli. 2nd edn, Boca Raton: CRC Press, pp. 657–68.

Borrego, C. and García-Gil, L. J. (1994). Separation of bacteriochlorophyll homologues from green photosynthetic sulfur bacteria by reversed-phase HPLC. *Photosynth. Res.* **41**, 157–63.

Breithaupt, D. E. (2004). Identification and quantification of astaxanthin esters in shrimp (*Pandalus borealis*) and in a microalga (*Haematococcus pluvialis*) by liquid chromatography mass spectrometry using negative ion atmospheric pressure chemical ionisation. *J. Agric. Food Chem.* **52**, 3870–75.

Caple, M. B., Chow, H. and Strouse, C. E. (1978). Photosynthetic pigments of green sulphur bacteria: The esterifying alcohols of bacteriochlorophylls *c* from *Chlorobium limicola*. *J. Biol. Chem.* **253**, 6730–37.

Caprioli, R. M. (1990). *Continuous-Flow Fast Atom Bombardment Mass Spectrometry*. Chichester: John Wiley and Sons.

Cariou-Le Gall, V., Rosell-Melé, A. and Maxwell, J. R. (1998). Data report: characterization of distributions of photosynthetic pigments in sapropels from holes 966D and 969C[1]. *Proc. Ocean Drilling Program, Scientific Results* **160**, 297–302.

Carreto, J. I., Seguel, M., Montoya, N. G., Clement, A. and Carignan, M. O. (2001). Pigment profile of the ichthyotoxic dinoflagellate *Gymnodinium* sp. from a massive bloom in southern Chile. *J. Plankton Res.* **23**, 1171–75.

Chen, N. H., Bianchi, T. S. and Bland, J. M. (2003). Novel decomposition products of chlorophyll-*a* in continental shelf (Louisiana shelf) sediments: Formation and transformation of carotenol chlorin esters. *Geochim. Cosmochim. Acta* **67**, 2027–42.

Chikaraishi, Y., Matsumoto, K., Kitazato, H. and Ohkouchi, N. (2007). Sources and transformation processes of pheopigments. Stable carbon and hydrogen isotopic evidence from Lake Haruna, Japan. *Org. Geochem.* **38**, 985–1001.

Chillier, X. F. D., Van Berkel, G. J., Gülaçar, F. O. and Buchs, A. (1994). Characterisation of chlorins within a natural chlorine mixture using electrospray/ion trap mass spectrometry. *Org. Mass Spectrom.* **29**, 672–78.

Cole, R. B. (1997). *Electrospray Ionisation Mass Spectrometry. Fundamentals, Instrumentation and Applications*. New York: John Wiley and Sons.

Dachtler, M., Glaser, T., Kohler, K. and Albert, K. (2001). Combined HPLC-MS and HPLC-NMR on-line coupling for the separation and determination of lutein and zeaxanthin stereoisomers in spinach and retina. *Anal. Chem.* **73**, 667–74.

Dawson, J. H. J. and Guilhaus, M. (1989). Orthogonal-acceleration time-of-flight mass spectrometer. *Rapid Commun. Mass Spectrom.* **3**, 155–59.

Eckardt, C. B., Carter, J. F. and Maxwell, J. R. (1990). Combined liquid chromatography/mass spectrometry of tetrapyrroles of sedimentary significance. *Energy and Fuels* **4**, 741–47.

Eckardt, C. B., Keely, B. J. and Maxwell, J. R. (1991). Identification of chlorophyll transformation products in a lake sediment by combined liquid chromatography-mass spectrometry. *J. Chomatogr.* **557**, 271–88.

Egeland, E. S., Garrido, J. L., Zapata, M., Maestro, M. A. and Liaaen-Jensen, S. (2000). Algal carotenoids. Part 64. Structure and chemistry of 4-keto-19′-hexanoyloxyfucoxanthin with a novel carotenoid end group. *J. Chem. Soc. Perkin Trans. 1*, 1223–30.

Enzell, C. R. and Back, S. (1995). Mass Spectrometry. In *Carotenoids, vol. 1B: Spectroscopy*, ed. G. Britton, S. Liaaen-Jensen and H. Pfander. Basel: Birkhauser, pp. 261–320.

Fraser, P. D., Enfissi, E. M. A., Goodfellow, M., Eguchi, T. and Bramley, P. M. (2007). Metabolite profiling of plant carotenoids using the matrix-assisted laser desorption ionization time-of-flight mass spectrometry. *The Plant Journal* **49**, 552–64.

Frassanito, R., Cantonati, M., Tardio, M., Mancini, I. and Guella, G. (2005). On-line identification of secondary metabolites in freshwater microalgae and cyanobacteria by combined liquid chromatography-photodiode array detection-mass spectrometric techniques. *J. Chromatogr. A* **1082**, 33–42.

Frassanito, R., Flaim, G., Mancini, I. and Guella, G. (2006). High production of unexpected carotenoids in Dinophyceae. Astaxanthin esters from the freshwater dinoflagellate *Tovellia sanguinea*. *Biochem. System. Ecol.* **34**, 843–53.

Frassanito, R., Cantonati, M., Flaim, G., Mancini, I. and Guella, G. (2008). A new method for the identification and the structural characterisation of carotenoid esters in freshwater microorganisms by liquid chromatography/electrospray ionisation tandem mass spectrometry. *Rapid Commun. Mass Spectrom.* **22**, 3531–39.

Garrido, J. L. and Zapata, M. (1996). Ion-pair reversed phase high-performance liquid chromatography of algal chlorophylls. *J. Chromatogr. A* **738**, 285–89.

Garrido, J. L. and Zapata, M. (1997). Reversed phase high performance liquid chromatographic separation of mono- and divinyl chlorophyll forms using pyridine-containing mobile phases and a polymeric octadecyl silica column. *Chromatographia* **44**, 43–49.

Garrido, J. L. and Zapata, M. (1998). Detection of new pigments from *Emiliania huxleyi* (Prymnesiophyceae) by high-performance liquid chromatography, liquid chromatography-mass spectrometry, visible spectroscopy and fast atom bombardment mass spectrometry. *J. Phycol.* **34**, 70–78.

Garrido, J. L., Otero, J., Maestro, M. A. and Zapata, M. (2000). The main nonpolar chlorophyll *c* from *Emiliania huxleyi* (Prymnesiophyceae) is a chlorophyll c_2-monogalactosyldiacylglyceride ester: a mass spectrometry study. *J. Phycol.* **36**, 497–505.

Garrido, J. L., Rodríguez, F., Campaña, E. and Zapata, M. (2003). Rapid separation of chlorophyll *a* and *b* and their demetallated and dephytylated derivatives using a monolithic silica C_{18} column and a pyridine-containing mobile phase. *J. Chromatogr. A* **994**, 85–92.

Gauthier-Jaques, A., Bortlik, K., Hau, J. and Fay, F. B. (2001). Improved method to track chlorophyll degradation *J. Agric. Food Chem.* **49**, 1117–22.

Gich, F., Airs, R. L., Danielsen, M., Keely, B. J., Abella, C. A., Garcia-Gil, J., Miller, M. and Borrego, C. M. (2003). Characterization of the chlorosome antenna of the filamentous anoxygenic phototrophic bacterium *Chloronema* sp. strain UdG9001. *Arch. Microbiol.* **180**, 417–26.

Goericke, R., Shankle, A. and Repeta, D. J. (1999). Novel carotenol chlorin esters in marine sediments and water column particulate matter. *Geochim. Cosmochim. Acta* **63**, 2825–34.

Goericke, R., Olson, R. J. and Shalapyonok, A. (2000). A novel niche for *Prochlorococcus* sp. in low-light suboxic environments in the Arabian Sea and the Eastern Tropical North Pacific. *Deep-Sea Res.* **47**, 1183–205.

Graham, J. E., Lecomte, J. T. J. and Bryant, D. A. (2008). Synechoxanthin, an aromatic C-40 xanthophyll that is a major carotenoid in the cyanobacterium *Synechococcus sp.* PCC 7002. *J. Nat. Prod.* **71**, 1647–50.

Grewe, C. and Griehl, C. (2008). Time- and media-dependent secondary carotenoid accumulation in *Haematococcus pluvalis*. *Biotech. J.* **3**, 1232–44.

Grynbaum, M. D., Hentschel, P., Putzbach, K., Rehbein, J., Krucker, M., Nicholson, G. and Albert, K. (2005). Unambiguous detection of astaxanthin and astaxanthin fatty acid esters in krill (*Euphasia superba* Dana). *J. Sep. Sci.* **28**, 1685–93.

Guaratini, T., Vessecchi, R., Pinto, E., Colepicolo, P. and Lopes, N. P. (2005). Balance of xanthophylls molecular and protonated molecular ions in electrospray ionisation. *J. Mass Spectrom.* **40**, 963–68.

Harada, J., Miyago, S., Mizoguchi, T., Azai, C., Inoue, K., Tamiaki, H. and Oh-Oka, H. (2008). Accumulation of chlorophyllous pigments esterified with the geranylgeranyl group and photosynthetic competence in the CT2256-deleted mutant of the green sulphur bacterium *Chlorobium tepidum*. *Photochem. Photobiol. Sci.* **7**, 1179–87.

Harris, P. G., Carter, J. F., Head, R. N., Harris, R. P., Eglinton, G. and Maxwell, J. R. (1995). Identification of chlorophyll transformation products in zooplankton faecal pellets and marine sediment extracts by liquid chromatography/mass spectrometry atmospheric pressure chemical ionisation. *Rapid Commun. Mass Spectrom.* **9**, 1177–83.

Head, E. J. H. and Horne, E. P. W. (1993). Pigment transformation and vertical flux in an area of convergence in the North Atlantic. *Deep-Sea Res. II* **40**, 329–46.

Head, E. J. H., Hargrave, B. T. and Subba Rao, D. V. (1994). Accumulation of phaeophorbide *a*-like pigment in sediment traps during late stages of a spring bloom: A product of dying algae? *Limnol. Oceanogr.* **39**, 176–81.

Hoffmann, E. and Stroobant, V. (2001). *Mass Spectrometry, Principles and Applications*. New York: John Wiley and Sons.

Hunt, J. E. and Michalski, T. J. (1991). Desorption-ionisation mass spectrometry of chlorophylls. In *Chlorophylls*, ed. H. Scheer. Boca Raton: CRC Press, pp. 835–53.

Hunt, J. E., Macfarlane, R. D., Katz, J. J. and Dougherty, R. C. (1981). High energy fragmentation of chlorophyll *a* and its deuterated analogue by californium-252 plasma desorption mass spectrometry. *J. Am. Chem. Soc.* **103**, 6775–78.

Jeffrey, S. W. and Wright, S. W. (1994). Photosynthetic pigments in the Haptophyta. In *The Haptophyte Algae*, ed. J. C. Green and B. S. C. Leadbeater. Oxford: Clarendon Press, pp. 111–32.

Jie, C., Walker, J. S. and Keely, B. J. (2002). Atmospheric pressure chemical ionisation-normal phase-liquid chromatography-mass spectrometry and tandem mass spectrometry of chlorophyll *a* allomers. *Rapid Commun. Mass Spectrom.* **16**, 473–79.

Keely, B. J. and Maxwell, J. R. (1990). Fast atom bombardment mass spectrometric and tandem mass spectrometric studies of some functionalised tetrapyrroles derived from chlorophylls *a* and *b*. *Energy and Fuels*, **4**, 737–41.

Lacker, T., Strohschein, S. and Albert, K. (1999). Separation and identification of various carotenoids by C30 reversed-phase high performance liquid chromatography coupled to UV and atmospheric pressure chemical ionization mass spectrometric detection. *J. Chromatogr. A* **854**, 37–44.

Manske, A. K., Glaeser, J., Kuypers, M. M. M. and Overmann, J. (2005). Physiology and phylogeny of green sulphur bacteria forming a monospecific phototrophic assemblage at a depth of 100 meters in the Black Sea. *Appl. Environ. Microbiol.* **71**, 8049–60.

Mantoura, R. F. C. and Llewellyn, C. A. (1983). The rapid determination of algal chlorophyll and carotenoid pigments and their breakdown products in natural waters by reverse-phase high performance liquid chromatography. *Anal. Chim. Acta* **151**, 297–314.

McMaster, M. C. (2005). *LC-MS A Practical Users Guide*. New York: John Wiley and Sons.

Naylor, C. C. and Keely, B. J. (1998). Sedimentary purpurins: oxidative transformation products of chlorophylls. *Org. Geochem.* **28**, 417–22.

Otte, S. C. M., Jan van de Meent, E., van Veelen, P. A., Pundsnes, A. S. and Amesz, J. (1993). Identification of the major chlorosomal bacteriochlorophylls of the green sulphur bacteria *Chlorobium vibrioforme* and *Chlorobium phaeovibrioides*: their function in lateral energy transfer. *Photosynth. Res.* **35**, 159–69.

Persson, S., Sönksen, C. P., Frigaard, N. U., Cox, R. P., Roepstorff, P. and Miller, M. (2000). Pigments and proteins in green bacterial chlorosomes studies by matrix-assisted laser desorption ionization mass spectrometry. *Eur. J. Biochem.* **267**, 450–56.

Rahmani, A., Eckardt, C. B., Brereton, R. G. and Maxwell, J. R. (1993). The use of liquid chromatography-mass-spectrometry to monitor the allomerization reactions of chlorophyll *a* and pheophytin *a* – identification of the allomers of pheophytin *a*. *Photochem. Photobiol.* **57**, 1048–52.

Riffé-Chalard, C., Verzegnassi, L. and Gülaçar, F. O. (2000). A new series of steryl chlorin esters: phaeophorbide *a* steryl esters in an oxic surface sediment. *Org. Geochem.* **31**, 1703–12.

Řezanka, T., Olšovská, J., Sobotka, M. and Sigler, K. (2009). The use of APCI-MS with HPLC and other separation techniques for identification of carotenoids and related compounds. *Current Anal. Chem.* **5**, 1–25.

Romero-Viana, L., Keely, B. J., Camacho, A., Vicente, E., Miracle, M. R. (2009). Photoautotrophic community changes in Lagunillo del Tejo (Spain) in response to lake level fluctuation: Two centuries of sedimentary pigment records. *Org. Geochem.* **40**, 376–86.

Rosell-Melé, A., Carter, J. F. and Maxwell, J. R. (1999). Liquid chromatography/tandem mass spectrometry of free base alkyl porphyrins for the characterisation of the macrocyclic substituents in components of complex mixtures. *Rapid Commun. Mass Spectrom.* **13**, 568–73.

Smith, K. M. and Bobe, F. W. (1987). Light adaptation of bacteriochlorophyll *d* producing bacteria by enzymic methylation of their antenna pigments. *J. Chem. Soc. Chem. Commun.* **4**, 276–77.

Soma, Y., Tani, Y., Soma, M., Mitake, H., Kurihana, R., Hashomoto, S., Watanabe, T. and Nakamura, T. (2007). Sedimentary steryl chlorin esters (SCEs) and other photosynthetic pigments as indicators of paleolimnological change over the last 28,000 years from the Buguldeika saddle of Lake Baikal. *J. Paleolimnol.* **37**, 163–75.

Squier, A. H., Hodgson, D. A. and Keely, B. J. (2002). Sedimentary pigments as markers for environmental change in an Antarctic lake. *Org. Geochem.* **33**, 1655–65.

Squier, A. H., Hodgson, D. A. and Keely, B. J. (2004). Structures and profiles of novel sulfur-linked chlorophyll derivatives in an Antarctic lake sediment. *Org. Geochem.* **35**, 1309–18.

Squier, A. H., Hodgson, D. A. and Keely, B. J. (2005). Evidence of late Quaternary environmental change in a continental east Antarctic lake from lacustrine sedimentary pigment distributions. *Antarct. Sci.* **17**, 361–76.

Suzuki, T., Midonoya, H. and Shioi, Y. (2009). Analysis of chlorophylls and their derivatives by matrix-assisted laser desorption/ionization-time-of-flight mass spectrometry. *Anal. Biochem.* **390**, 57–62.

Talbot, H. M., Head, R. N., Harris, R. P. and Maxwell, J. R. (1999a). Distribution and stability of steryl chlorine esters in copepod faecal pellets from diatom grazing. *Org. Geochem.* **30**, 1163–74.

Talbot, H. M., Head, R. N., Harris, R. P. and Maxwell, J. R. (1999b). Steryl esters of pyrophaeophorbide *b* a sedimentary sink for chlorophyll *b*. *Org. Geochem.* **30**, 1403–10.

Talbot, H. M., Head, R. N., Harris, R. P. and Maxwell, J. R. (2000). Discrimination against 4-methyl sterol uptake during steryl chlorine ester production by copepods. *Org. Geochem.* **31**, 871–80.

Tani, Y., Matsumoto, G. I., Soma, M., Soma, Y., Hashimoto, S. and Kawai, T. (2009). Photosynthetic pigments in sediment core HDP-04 from Lake Hovsgol

Mongolia, and their implication for changes in algal productivity and lake environment for the last 1 Ma. *Quatern. Int.* **205**, 74–83.

Vaulot, D., Birrien, J. L., Marie, D., Casotti, R., Veldhuis, M. J. W., Kraay, G. W. and Chrétiennot-Dinet, M. J. (1994). Morphology, ploidy, pigment composition, and genome size of cultured strains of *Phaeocystis* (Prymnesiophyceae). *J. Phycol.* **30**, 1022–35.

Van Breemen, R. B. (1995). Electrospray liquid chromatography mass spectrometry of carotenoids. *Anal. Chem.* **67**, 2004–09.

Van Breemen, R. B. (1996). Innovations in carotenoid analysis using LC-MS. *Anal. Chem.* **68**, 299A–304A.

Van Breemen, R. B. (1997). Liquid chromatography/mass spectrometry of carotenoids. *Pure Appl. Chem.* **69**, 2061–66.

Van Breemen, R. B., Canjura, F. L. and Schwartz, S. J. (1991). High performance liquid chromatography continuous flow fast atom bombardment mass spectrometry of chlorophyll derivatives. *J. Chromatogr.* **542**, 373–83.

Van Breemen, R. B., Schmitz, H. H. and Schwartz, S. J. (1993). Continuous-flow fast atom bombardment liquid chromatography/mass spectrometry of carotenoids. *Anal. Chem.* **65**, 965–69.

Van Breemen, R. B., Huang, R. R., Tan, Y., Sander, L. C. and Schilling, A. B. (1996). Liquid chromatography/mass spectrometry of carotenoids using atmospheric pressure chemical ionisation. *J. Mass Spectrom.* **31**, 975–81.

Van Heukelem, L., Lewitus, A. J., Kana, T. M. and Craft, N. E. (1994). Improved separations of phytoplankton pigments using temperature-controlled high-performance liquid chromatography. *Mar. Ecol. Prog. Ser.* **114**, 303–13.

Veldhuis, M. J. W., Kraay, G. W. and Timmermans, K. R. (2001). Cell death in phytoplankton: correlation between changes in membrane permeability, photosynthetic activity, pigmentation and growth. *Eur. J. Phycol.* **36**, 167–77.

Verzegnassi, L., Riffé-Chalard, C., Kloeti, W. and Gülaçar, F. O. (1999). The analysis of tetrapyrrolic pigments and derivatives in sediments by high performance liquid chromatography atmospheric pressure chemical ionisation mass spectrometry. *Fresen. J. Anal. Chem.* **364**, 249–53.

Verzegnassi, L., Riffé-Chalard, C. and Gülaçar, F. O. (2000). Rapid identification of Mg-chelated chlorins by on-line high performance liquid chromatography/atmospheric pressure chemical ionisation mass spectrometry. *Rapid Commun. Mass Spectrom.* **14**, 590–94.

Villanueva, J. and Hastings, D. W. (2000). A century-scale record of the preservation of chlorophyll and its transformation products in anoxic sediments. *Geochim. Cosmochim. Acta* **64**, 2281–94.

Walker, J. S. and Keely, B. J. (2004). Distribution and significance of chlorophyll derivatives and oxidation products during the spring phytoplankton bloom in the Celtic Sea April 2002. *Org. Geochem.* **35**, 1289–98.

Walker, J. S., Squier, A. H., Hodgson, D. A. and Keely, B. J. (2002). Origin and significance of 13^2-hydroxychlorophyll derivatives in sediments. *Org. Geochem.* **33**, 1667–74.

Walker, J. S., Jie, C. and Keely, B. J. (2003). Identification of diastereomeric chlorophyll allomers by atmospheric pressure chemical ionisation liquid chromatography/tandem mass spectrometry. *Rapid Commun. Mass Spectrom.* **17**, 1125–31.

Wilson, M. A., Airs, R. L., Atkinson, J. E. and Keely, B. J. (2004). Bacterioviridins: novel sedimentary chlorins providing evidence for oxidative processes affecting palaeobacterial communities. *Org. Geochem.* **35**, 199–202.

Wilson, M. A., Hodgson, D. A. and Keely, B. J. (2005). Atmospheric pressure chemical ionisation-liquid chromatography/multistage mass spectrometry for assignment of sedimentary bacteriochlorophyll derivatives. *Rapid Commun. Mass Spectrom.* **19**, 38–46.

Wright, S. W., Jeffrey, S. W., Mantoura, R. F. C., Llewellyn, C. A., Bjørnland, T., Repeta, D. and Welschmeyer, N. (1991). Improved HPLC method for the analysis of chlorophylls and carotenoids from marine phytoplankton. *Mar. Ecol. Prog. Ser.* **77**, 183–96.

Zapata, M., Ayala, A. M., Franco, J. M. and Garrido, J. L. (1987). Separation of chlorophylls and their degradation products by reversed-phase high performance liquid chromatography. *Chromatographia* **23**, 26–30.

Zapata, M., Edvardsen, B., Rodríguez, F., Maestro, M. A. and Garrido, J. L. (2001). Chlorophyll c_2 monogalactosyldiacylglyceride ester (chl c_2-MGDG). A novel marker pigment for *Chrysochromulina* species (Haptophyta). *Mar. Ecol. Prog. Ser.* **219**, 85–98.

Zapata, M., Jeffrey, S. W., Wright, S. W., Rodríguez, F., Garrido, J. L. and Clementson, L. (2004). Photosynthetic pigments in 37 species (65 strains) of Haptophyta: implications for oceanography and chemotaxonomy. *Mar. Ecol. Prog. Ser.* **270**, 83–102.

Zapata, M., Garrido, J. L. and Jeffrey, S. W. (2006). Chlorophyll *c* pigments: Current status. In *Advances in Photosynthesis and Respiration, vol. 25, Chlorophylls and Bacteriochlorophylls: Biochemistry, Biophysics, Functions and Applications*, ed. B. Grimm, R. J. Porra, W. Rüdiger and H. Scheer. Dordrecht: Springer, pp. 39–53.

8

Multivariate analysis of extracted pigments using spectrophotometric and spectrofluorometric methods

JACQUES NEVEUX, JUKKA SEPPÄLÄ AND YVES DANDONNEAU

8.1 Introduction

The spectral absorption and fluorescence properties of chlorophylls and pheopigments have been exploited in the past to determine the concentration of a few extracted pigments using simultaneous equations (the so-called di- or trichromatic methods). Spectrophotometric, fluorometric and spectrofluorometric techniques for pigment extracts from oceanographic samples were reviewed and compared in the 1997 volume edited by Jeffrey *et al.*, *Phytoplankton Pigments in Oceanography* (Chapters 4, 14 and Appendix F in Jeffrey *et al.*, 1997).

Recently, advanced chemometric methods developed for multi-component analysis have been applied to the analysis of pigment extracts (Neveux and Lantoine, 1993; Moberg *et al.*, 2001; Naqvi *et al.*, 2004). The use of full spectrum techniques enhances the information acquired from phytoplankton samples, and chemometric methods have been shown to be a valuable tool to extract accurate pigment concentrations from this information (Moberg *et al.*, 2001). The major advantage of these methods is that a greater number of pigments can be determined even if they have overlapping spectra, because a greater part of the information contained in the absorption or fluorescence spectra is being employed than was the case with the di- or trichromatic equations implemented in earlier work. Recent studies pointed out that simultaneous equations such as those used in trichromatic methods may yield inaccurate results because of interference from compounds other than the two or three pigments assessed with these methods (Naqvi *et al.*, 1997, 2004; Küpper *et al.*, 2007). This was acknowledged in the 1997 volume (Jeffrey and Welschmeyer, 1997; Humphrey and Jeffrey, 1997).

Multivariate techniques have also been applied in the analysis of *in vivo* phytoplankton absorption and fluorescence spectra. These have included spectral reconstruction techniques, which require a-priori knowledge of the pigments present in the samples (e.g. Bidigare *et al.*, 1987), and spectral decomposition techniques, which attempt to identify the various components of algal absorption using

Phytoplankton Pigments: Characterization, Chemotaxonomy and Applications in Oceanography, eds. Suzanne Roy, Carole A. Llewellyn, Einar Skarstad Egeland and Geir Johnsen. Published by Cambridge University Press.

mathematical techniques such as Gaussian curve fitting (e.g. Hoepffner and Sathyendranath, 1991). Several multivariate analysis methods derived from chemometrics are now used for phytoplankton species discrimination from *in vivo* absorption or fluorescence spectra (e.g. Seppälä and Olli, 2008), including the use of three-dimensional fluorescence spectroscopy (excitation–emission matrices: EEM, e.g., Zhang *et al.*, 2009). These developments are covered in greater detail in Chapters 13 and 14 of this volume as well as in Babin *et al.* (2008).

HPLC is often recommended for pigment studies since it provides, qualitatively and quantitatively, complete information on major phytoplankton pigments. However, HPLC is not available to all, and sometimes easy-to-use methods are preferred. With spectroscopy, pigment samples can be analyzed onboard with minimal technical assistance, and a lot of samples can be processed at relatively low cost. The development of multi-component data analysis has profited from improved statistical techniques, fast-scanning spectroscopy, single measurement EEM fluorometry (Jiji *et al.*, 1999) and software computing capability. The purpose of this chapter is to present basic aspects of multi-component analysis for pigment extracts, both for spectrophotometry and spectrofluorometry, and to give a few examples of these methods using oceanographic samples collected from the tropical Pacific.

8.2 Presentation of multi-component analysis methods

Absorption spectroscopic pigment quantification methods rely on Beer's law, which states that spectroscopic response at a given wavelength is linearly related to the concentration of the pigment, and that contributions from different pigments are additive. Individual contributions from each pigment to a spectrum can be obtained using their reference spectra and thereby can their concentration be determined. Typical methods to estimate concentrations include classical least squares (CLS) and non-negative least squares (NNLS) regressions. In oceanography, these methods have been applied to fluorescence excitation–emission matrices of extracted pigments (CLS: Neveux and Lantoine, 1993; NNLS: Tenório *et al.* 2005).

However, knowledge of all major pigments contributing to the absorbance or fluorescence signal may be lacking for field samples, and background noise may be large and variable, invalidating the use of CLS. In such cases, spectral data may be compressed to a few, independent variables (like eigenvectors in principal component analysis, PCA) that are subsequently related by regression to the concentration of pigment components. These multivariate factor-based regression methods include principal component regression (PCR) and partial least squares (PLS) regression. Seppälä and Olli (2008) provide a comparison between PLS, PCR and CLS regressions in a study of spectral *in vivo* fluorescence of natural phytoplankton from the Baltic Sea. Parallel factor analysis (PARAFAC), which can be thought of as a three-way version of PCA where the data are decomposed into tri-linear components, is particularly useful for excitation–emission spectrofluorometric matrices (Bro, 1997;

Table 8.1. *List of the multi-component analysis methods presented in this chapter and comparison with traditional methods.*

Method	Pigments analysed	Statistical treatment	Spectral range	Advantages	Disadvantages	Key references
SPECTROPHOTOMETRY						
Di- or trichromatic equations	Chl *a*, *b*, *c*, Chl *a*, Phe *a*	Solving a few simultaneous equations	1–3 wavelengths	Easy to use	Limited number of pigments determined; inaccurate if presence of interfering pigments	Jeffrey and Humphrey (1975); Chap. 4 and Appendix F in Jeffrey *et al.* (1997); Porra, Appendix 8A (this volume); Porra (2006); Lorenzen (1967)
Spectral reconstruction method	Chl *a*, *b*, $c_{1,2}$, total carotenoids	Fits absorption spectra of standard solutions to sample spectrum using CLS	Whole Vis absorption spectra	Improves accuracy by using a large number of wavelengths	Needs *a priori* knowledge of pigments present in sample	Naqvi *et al.* (1997, 2004)
Gauss-peak spectra method	Chl *a*, *b*, *c*, Phe *a*, *b*, *c*, and several individual carotenoids	Fits a series of Gaussian peaks, one for each pigment part of the sample mix	Whole Vis absorption spectra	Improves accuracy by using a large number of wavelengths; adjustable parameters allow for wavelength deviations, baseline changes, etc.; simultaneous quantification of a large number of pigments	Needs *a priori* knowledge of pigments present in sample	Küpper *et al.* (2000, 2007)

Table 8.1. (*cont.*)

Method	Pigments analysed	Statistical treatment	Spectral range	Advantages	Disadvantages	Key references
Multi-component analysis with factor-based regression	Chl *a*, *b*, *c*, Phe *a*, *b*, *c*	PLS (PLS-1 for a single predicted variable and PLS-2 for many)	Whole Vis absorption spectra	No need for *a priori* knowledge of pigment components present in sample	Number of eigenvectors must be optimised; good calibration data set is required	Moberg *et al.* (2000), Moberg and Karlberg (2001)
SPECTROFLUOROMETRY						
Selected emission and excitation wavelength combinations	Chl *a*, *b*, *c*, Phe *a*, *b*, *c*	Solving a few simultaneous equations	5–6 wavelength combinations	Easy to use	Limited number of pigments determined; inaccurate if other compounds interfere	Boto and Bunt (1978); Neveux and Panouse (1987)
Spectral reconstruction method, multiple linear regression	Chl *a*, *b*, $c_{1,2}$, c_3, DV-Chl *a*, *b*, Phe *a*, *b*, $c_{1,2}$, c_3, DV-Phe *a*, *b*	CLS	Fluorescence excitation–emission matrix	Improves accuracy by using a large number of wavelengths	Needs *a priori* knowledge of pigments present in sample; may yield a negative concentration for certain pigments	Neveux and Lantoine (1993); Neveux *et al.* (2003); Moberg *et al.* (2001)
	Chl *a*, *b*, $c_{1,2}$, c_3, DV-Chl *a*, *b*, Phe *a*, *b*, $c_{1,2}$, c_3, DV-Phe *a*, *b*	NNLS	Fluorescence excitation–emission matrix	Avoids overestimation of naturally present pigments by suppressing negative calculated	Needs *a priori* knowledge of pigment components present in sample	Tenório *et al.* (2005)

				pigment concentrations		
Multi-component analysis with factor-based regression	Chl *a*, *b*, *c*, Phe *a*, *b*, *c*	PARAFAC	Fluorescence excitation–emission matrix	No need for *a priori* knowledge of major pigment components present in sample	Number of eigenvectors must be optimised; good calibration data set is required; no guarantee that factors correspond to pigments of interest	Moberg *et al.* (2001)

Andersen and Bro, 2003). It has been used in the study of dissolved organic matter in aquatic environments (Stedmon and Markager, 2005) and in the spectrofluorometric determination of extracted chlorophylls and pheopigments (Moberg *et al.*, 2001).

Table 8.1 lists a number of these methods which have been used in the last decade for multi-component analysis of pigment extracts, in comparison with more traditional approaches. Detailed information on multivariate calibration techniques can be found in Martens and Næs (1989) and Duckworth (1998).

8.3 Multi-component spectrophotometric methods

Several multivariate methods for the analysis of spectrophotometric data have been developed to assess chlorophylls and/or carotenoids in pigment extracts. These include (1) spectral reconstruction methods, similar to CLS regression (Naqvi *et al.*, 1997; 2004), (2) Gauss-peak spectra methods that reconstruct the sample absorption spectrum based on a combination of Gaussian peaks corresponding to the spectrum of each individual pigment present in the sample (Küpper *et al.*, 2007) and (3) factorial analysis such as PLS regressions (Moberg *et al.*, 2000; Moberg and Karlberg, 2001). The difference between the first two methods lies in the use of directly measured spectra of pigment standards in a linear array to simulate the sample spectrum in the spectral reconstruction method, while the Gauss-peak spectra method uses a description of each pigment spectrum by a series of Gaussian peaks (see Küpper *et al.*, 2007 for a comparison of these approaches). Some of these methods will be reviewed here.

8.3.1 Spectral reconstruction method (SRC)

In SRC the whole absorption spectrum (350–800 nm) is used and the concentrations of various pigments are estimated after calibration with spectra of pure solutions from each of these pigments. Spectral reconstruction simulates the absorption spectrum of the sample mixture by a linear combination of the absorption spectra of all its constituents. Standard and sample spectra must be recorded with the same spectrophotometer using the same bandwidth and the same solvent (Naqvi *et al.*, 2004). The method has been applied to analyze pigments (chlorophyll (Chl) *a*, Chl *b* and carotenoids) in the light-harvesting complex associated with photosystem II of higher plants (Naqvi *et al.*, 1997). The approach is two-fold: first, the Chl *a* and *b* concentrations are determined using the red part of the absorption spectrum where only chlorophylls absorb. Then, the absorption of chlorophylls is deduced for the overall spectral range and it is subtracted from the sample spectrum. The difference corresponds to the absorption by carotenoids which may then be decomposed into major components (here, lutein, violaxanthin and neoxanthin) by analyzing the 350–550 nm region (see the original paper for specific problems relating to this determination, Naqvi *et al.*, 1997).

Computationally, this method uses CLS regression and can be generalised to any pigment extract provided that all major pigments present are known *a priori* and that pure standards are available for each pigment. Moreover, it can be used directly on the whole spectrum without analyzing the red and the blue bands separately.

For a mixture of *n* pigments (the unknowns), the absorbance (also called optical density) at a given wavelength $A_{(\lambda)}$ can be expressed as:

$$A_{(\lambda)} = k_1(\lambda) \cdot C1 + k_2(\lambda) \cdot C2 + \cdots\cdots + k_n(\lambda) \cdot Cn + H, \quad (8.1)$$

where $C1 \ldots Cn$ represent the concentration and $k_1(\lambda) \ldots k_n(\lambda)$ the specific (l g^{-1}) or molar absorption coefficient (l $mole^{-1}$) for each pigment in solution, and H is an intercept term including background signals and errors relating to the model.

The maximum number of wavelengths (m) is not limited, and high resolution spectra can be used. The set of m equations obtained can be formulated (in matrix form) as

$$\mathrm{A} = \mathrm{KC}, \quad (8.2)$$

where A is a vector of m measured absorbances for a sample, K is a matrix with dimensions (m, n+1) containing the specific absorption coefficient of each pigment at each wavelength and C is a vector of $n + 1$ concentrations (including H).

Classical least squares analysis determines the pigment concentrations $C1 \ldots Cn$ that minimize the sum of the squares of the errors associated with each wavelength using the model function D,

$$D = \sum_{i=1}^{i=m} (Ao - Ac)^2_{\lambda i}, \quad (8.3)$$

where Ao and Ac represent the measured and predicted absorbance at each wavelength, respectively.

Using matrix algebra, all of the pigment concentrations are approximated by resolving the general equation

$$\mathrm{C} = \left(\mathrm{K^tK}\right)^{-1}\mathrm{K^tA}, \quad (8.4)$$

where $\mathrm{K^t}$ is a transpose of matrix K.

For a given sample, the residuals D can be used to assess the quality of the fit when using different combinations of the pigments involved. The resolution of the equation set by CLS gives numerical solutions in all cases, but it may happen that solutions for some pigments are negative. This may be due to (1) modification of spectral properties of a given pigment related to the presence of other substances, (2) the presence of unknown absorbing compounds that absorb significantly and that have not been included in the calibration and (3) measurements which are done outside the range where the relationship between absorbance and concentration is linear. Negative results may also be due to noisy signals or to changes in the background signal. Negative solutions are not a serious problem if their absolute values are low with regard to the concentration of the major pigment in the extract (generally Chl *a*); but if they are relatively high, then the concentrations of the other pigments become erroneous. Non-negative least squares approximation (NNLS) can be applied to remove pigments with negative concentrations from the analysis and reduce some pitfalls in the least squares fit.

A weakness of CLS and NNLS methodology is that for optimal fit, all the major pigments present in the extract should be included and ignoring one or more important pigments will yield erroneous predictions for other pigments. This implies good prior knowledge of the phytoplankton communities where sampling is done. Additionally, these methods are sensitive to measurement noise, baseline shifts and modifications of spectral properties of components due to chemical or physical interactions, since the equations interpret these shifts in terms of pigments. To circumvent the limitations of CLS and NNLS methods, regression methods based on factorial analysis can be used (Section 8.3.3).

8.3.2 Gauss-peak spectra method (GPS)

The Gauss-peak spectra method (GPS) has been used in the past to analyze the different forms of chlorophyll in chloroplast fractions (French, 1971) and to decompose the absorption spectra measured for marine particles collected on GF/F filters (Hoepffner and Sathyendranath, 1991; Stuart *et al.*, 1998; Ficek *et al.*, 2004). It has also recently been applied to the analysis of chlorophylls and carotenoids in crude extracts of higher plants, *Euglena*, brown algae and cyanobacteria (Küpper *et al.*, 2007). In this method, the absorption spectrum of each pigment P_j is described by the sum of q Gaussian peaks. For n pigments, we use the following equation for absorbance at each wavelength λ:

$$A(\lambda) = a_o + b(\lambda) + s(\lambda) + \sum_{j=1}^{j=n} P_j(\lambda), \quad (8.5)$$

where a_o represents the offset value, and $b(\lambda)$ and $s(\lambda)$ functions simulate wavelength-dependent baseline drift and scattering by particles (for turbid samples). The contribution of each pigment to the absorption, $P_j(\lambda)$ is given by

$$P_j(\lambda) = C_j * k_j * G_j(\lambda), \quad (8.6)$$

$$\text{where } G_j(\lambda) = \sum_{i=1}^{i=q} a_i * exp\left[-0.5 * \left(\frac{\lambda - \lambda dev - \lambda_{ic}}{\sigma_i * w_i}\right)^2\right]. \quad (8.7)$$

$G_j(\lambda)$ and q represent respectively, the function and the number of Gaussian peaks describing the absorption spectrum of the pigment j; k_j is the extinction coefficient at the absorption maximum; a_i is the amplitude of peak i (in the spectrum normalized to maximum absorbance = 1); σ_i is the peak half-width (nm) and λ_{ic} is the centre of the peak q (nm).

Concentrations are determined by minimizing the squares of the difference between the sample and the fitting spectra (using the Levenberg–Marquardt method for nonlinear least squares minimization). The equations include some correction factors: λdev corrects for the wavelength inaccuracy of the spectrophotometer and w_i corrects for slight differences in temperature between standard and sample measurement, and the presence of water contamination ($< 2\%$).

In a supplement to their 2007 paper, Küpper *et al.* provided (1) GPS fits and complete GPS equations for Chl *a*, *b*, metal (Cd, Cu or Zn)-substituted Chl *a* and *b* and various carotenoids in 100% acetone, and (2) a ready-to-use fitting library for the data analysis program SigmaPlot (version 5 and above). According to Küpper *et al.* (2007), the equations are valid for any spectrophotometer that has a spectral bandwidth ≤ 1 nm and linear wavelength dispersion.

8.3.3 Multivariate calibration model with partial least squares regression (PLS)

Partial least squares regression (PLS) is a factor-based regression method that does not require knowledge of the complete sample composition. Furthermore, the analysis can be done for only a subset of components present in the samples. While CLS solves pigment concentrations by linear combination of spectra obtained from a pure solution of each pigment, the calibration for the factor-based regression methods is performed with a training set of standard solutions composed of mixtures of the different pigments at various relative concentrations. PLS is a decomposition technique that is closely related to principal component analysis (PCA) and principal component regression (PCR). In PCA, the spectral data are centred and normalized, and further reduced to (1) independent eigenvectors (spectral components that describe most of the spectral variability in all of the standard samples) and (2) a score matrix (sample-specific weighting factors for various eigenvectors such that their product describes the original spectral data).

The PCA model can be expressed as,

$$A = SF + E_A, \tag{8.8}$$

where A is a (p,m) matrix of absorbances for p samples at m wavelengths, S is a (p,f) matrix of scores values where f is the number of extracted eigenvectors, F is a (f,m) matrix of eigenvectors and E_A is an error matrix. To minimize prediction errors, the number of extracted eigenvectors must be optimized, so that the solution contains variations related to the important components, but spectral noise is largely avoided.

The matrix of concentrations does not intervene in the calculation. The score matrix of PCA reflects the concentrations of different compounds, and in PCR the scores are regressed with known concentrations of calibration samples, yielding a set of regression coefficients. These regression coefficients, eigenvectors and sample spectra are used in the analysis of the concentrations of unknown samples.

In PLS, the correlation between concentration and spectral data is used in model calculations, and there is no separate regression step. Both concentration data and spectral data are decomposed into separate eigenvectors and scores. The covariance of scores is maximized thereby optimizing the predictive power of the model. PLS can be run for one (PLS-1) or several variables or pigments (PLS-2). The key problems in PCR and PLS are (1) the requirement for a comprehensive calibration data set, (2) the abstract nature of models, (3) the need for complex calculations, and (4) the need to optimize the number of factors used in the analyses.

Partial least squares regression has been applied to the analysis of Chl *a*, *b*, *c* and their corresponding pheophytins (Phe *a*, *b* and pheoporphyrin *c*) in artificial mixtures. In Moberg *et al.* (2000) the calibration was performed with 16 mixtures of the above six pigments. The mixtures were prepared according to Brereton (1997), avoiding covariance between concentrations and maximizing the independence of regression coefficient estimates. Moreover, concentrations representative of natural water samples were used. According to Moberg *et al.* (2000), prediction of pigment concentrations is better with this method than with more classical spectrophotometric methods (Lorenzen, 1967; Jeffrey and Humphrey, 1975). The approach has been validated (Moberg and Karlberg, 2001) using six algal species representing six different phytoplankton groups by comparison of results obtained by HPLC. Predictions by the model were consistent with an *a priori* knowledge of the pigment composition and were not strongly affected when another instrument was used instead of the one employed for calibration.

8.4 Multi-component spectrofluorometric methods

Fluorometric techniques are more sensitive than spectrophotometric methods and they do not involve the time-consuming filtration of large volumes of water, which increases the risk of artificial degradation. For chlorophylls, spectrofluorometry offers more analytical capabilities than spectrophotometry, because both excitation and emission spectra can be used. However, spectrofluorometry is linear only up to a certain concentration, and standard and sample solutions should not exceed an absorbance value of 0.02 (using 1 cm pathlength cells), regardless of the excitation (absorption) and emission wavelength. Under these conditions, competition for the absorption of blue light by the different pigments can be neglected, preventing the loss of linearity between fluorescence and chlorophyll concentration, as well as re-absorption of light emitted by chlorophylls. The following section presents multivariate spectrofluorometric methods that have been applied to oceanographic or algal culture samples.

8.4.1 Spectrofluorometric analysis of chlorophylls and pheopigments using CLS and NNLS

Fluorescence emitted at the wavelength λ' when excited at the wavelength λ $[(F)_{\lambda,\lambda'}]$ can be expressed similarly to equation (8.1) as:

$$F_{(\lambda,\lambda')} = k_{1(\lambda,\lambda')} \cdot C1 + k_{2(\lambda,\lambda')} \cdot C2 + \cdots + k_{n(\lambda,\lambda')} \cdot Cn + H \qquad (8.9)$$

where $C1\ldots Cn$ represent the concentrations and $k_{1(\lambda,\lambda')}\ldots k_{n(\lambda,\lambda')}$ the specific ($l\ g^{-1}$) or molar ($l\ mole^{-1}$) fluorescence coefficients for each pigment in solution. Equations (8.2–8.4) can similarly be used if absorbance is replaced by fluorescence.

The fluorescence excitation and emission spectra are dependent on the optical and electronic settings of the spectrofluorometer (light source, gratings, mirrors and detector). Quantum corrections are required to obtain spectra independent of the instrument, for example a fluorescence excitation spectrum that matches the absorption spectrum (see Chapter 11). For determination of pigment concentrations in extracts, quantum corrections are not necessary since spectral distortions occur in the same proportion for the different pigments. Most importantly, one must (1) have an instrument with a stable behaviour in the long term which reduces the need for frequent calibration, and (2) check regularly the alignment of the monochromator and the signal reproducibility with stable standards.

In Neveux and Lantoine (1993), CLS was applied to relatively few (24) coupled wavelengths (excitation, emission) to determine ten unknowns (Chl *a*, *b*, *c*, divinyl (DV)-Chl *a*, DV-Chl *b* and their corresponding pheophytins). More recently, CLS was extended to fluorescence measured at 806 coupled wavelengths (Neveux *et al.*, 2003) by recording a series of fluorescence emission spectra between 615 and 715 nm for an excitation range varying from 390 to 480 nm (3 nm increments). Calibration of the spectrofluorometer consisted in the determination of specific fluorescence coefficients for each pigment at each combination of excitation and emission wavelengths. These coefficients were determined from solutions of six different concentrations by linear regression between fluorescence and concentration. Most of the regression coefficients were greater than 0.99. A blank correction corresponding to the signal of the pure solvent was subtracted from the fluorescence of standards and samples. Figure 8.1 illustrates the three-dimensional fluorescence spectra (not quantum-corrected) for 90% acetone solutions containing 1 $\mu g\ l^{-1}$ of the different chlorophylls which were recorded on a Hitachi F4500 spectrofluorometer. The same method was also applied to natural samples, but including a standard of Chl c_3 (from DHI Lab Products, Hørsholm, Denmark) in the analysis and using a NNLS procedure (Tenório *et al.*, 2005). The pigments used in the calibration correspond to the main chlorophylls found in aerobic marine and freshwater ecosystems. A complementary calibration may be required to obtain a reliable pigment concentration in specific cases, such as with relatively high concentrations of bacteriochlorophylls in

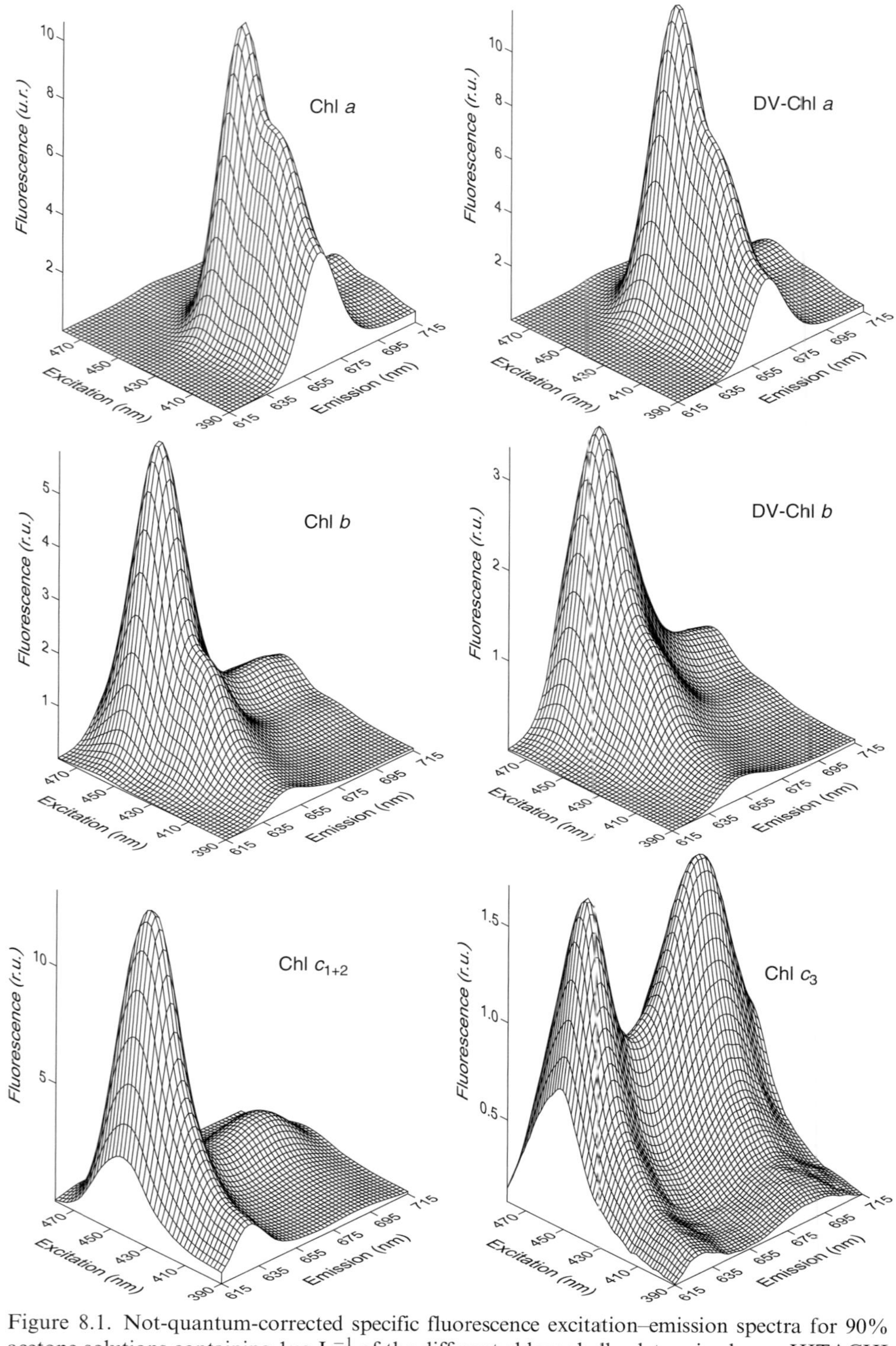

Figure 8.1. Not-quantum-corrected specific fluorescence excitation–emission spectra for 90% acetone solutions containing 1μg L^{-1} of the different chlorophylls, determined on a HITACHI F4500 spectrofluorometer. These reference spectra were used in Neveux *et al.* (2003) and Tenório *et al.* (2005).

anaerobic ecosystems (Dumestre *et al.*, 1999) or unusual red fluorescent compounds associated with zooplankton (Champalbert *et al.*, 2003).

8.4.2 PARAFAC modelling for the determination of chlorophylls and pheopigments

PARAFAC, a generalization of PCA, is a decomposition method for multi-way data (such as EEM fluorescence of pigment extracts). Instead of obtaining one score and one loading matrix as in PCA, it produces two loading matrices corresponding to the excitation and emission spectra of the different components and a score matrix for the samples (Moberg *et al.*, 2001). The PARAFAC model is based on spectral properties of standards composed of mixtures of the different compounds that have to be analyzed. After decomposition in PARAFAC components, the known concentrations of the standards are regressed onto the obtained score matrix as in principal component regression. The loading matrices are then used to calculate the score matrix and the concentrations of the compounds in the sample. A clear advantage of PARAFAC is that quantification of an analyte is possible even in the presence of unknown interfering components in the sample. A tutorial description and discussion of the PARAFAC model can be found in Bro (1997). Moberg *et al.* (2001) used the PARAFAC approach on a data set comprising 4686 fluorescence data points obtained by recording a series of fluorescence emission spectra between 600 and 730 nm for an excitation range varying from 360 to 500 nm (2 nm increments). For spectral decomposition and calibration, they used a set of nine mixtures of Chl *a*, Chl *b*, Chl *c*, Phe *a*, Phe *b* and pheoporphyrin *c*. According to Moberg *et al.* (2001), PARAFAC performs better than CLS in predicting pigment concentrations when compared to the HPLC pigment determination.

8.5 Methods comparison

Comparisons of pigment determination by some of the multivariate methods reviewed in this chapter with more traditional methods (spectrophotometry, fluorometry, HPLC) are relatively scarce. They have been undertaken on solutions of pigment standards, extracts from algal cultures (Moberg *et al.*, 2000; Moberg and Karlberg, 2001; Küpper *et al.*, 2007) or a few natural samples (Moberg and Karlberg, 2001; Moberg *et al.*, 2001). Multivariate methods improve the prediction in Chl *a* compared to the traditional simple spectrophotometric methods (Moberg *et al.*, 2000). The Gauss-peak spectra method appears better than the SRC method and was found to be a good alternative to HPLC for algal cultures (Küpper *et al.*, 2007). Unsurprisingly, errors in determination are generally larger on minor components than on major ones.

Comparison between the spectrofluorometric (SPF) methods (CLS, NNLS) and traditional HPLC methods is illustrated here using data collected during two

Table 8.2. *Methods used in the comparison of pigment determination. N is the number of samples.*

Cruise	Methods	N	Solvent for extraction	Storage	Delay before processing	References
Ewing 99/12	SPF	686	90% acetone	None	Immediate	Neveux *et al.*, 2003 Tenório *et al.*, 2005
	HPLC	27		Liquid N_2	6 months	Van Heukelem and Thomas, 2001
Bissecote 2006	SPF	161	90% acetone	Liquid N_2	1 month	Tenório *et al.*, 2005
	HPLC	49			7 months	Zapata *et al.*, 2000

oceanographic programs in tropical waters, the Maurice Ewing 99/12 (EW; North-Australian waters, November 1999) and the Bissecote (BC; New-Caledonia, February 2006) field programs. This comparison is clearly not as extensive as the methods comparison presented in Chapter 14 of Jeffrey *et al.* (1997), but it highlights the usefulness and limitations of the methods examined. For each method (Table 8.2), samples taken from the same sampling bottle were processed separately. All Chl *a* derivatives (allomers, epimers and chlorophyllide *a*) were assimilated as Chl *a* for the comparison between SPF and HPLC since SPF cannot discriminate between these pigments. Similarly, the sum of Chl *b* and DV-Chl *b* (= total, or TChl *b*) was used for comparison between SPF and HPLC because the HPLC method used did not separate these pigments.

8.5.1 Comparison of CLS and NNLS spectrofluorometric derived data

The CLS and NNLS spectrofluorometric data analysis methods were compared using results from three samples collected during the EW program (0 and 100 m depth at offshore station 26; 0 m at coastal station 28). Station 26 was characterized by a fairly high relative proportion of DV-Chl *a* (30–50% of the total Chl *a*: TChl *a* = Chl *a* + DV-Chl *a*), while Station 28 showed much lower values (less than 12%, Table 8.3). For station 26, CLS and NNLS gave similar results for the major pigments: Chl *a*, DV-Chl *a*, Chl c_{1+2}, and DV-Chl *b* (100 m sample only), but a significant difference was observed for Chl *b* from the surface sample for the two methods. Inclusion of Chl c_3 in the data treatment slightly reduced the concentrations of the other accessory chlorophylls, but this did not strongly modify the Chl *a* and DV-Chl *a* concentrations. At station 28, *Trichodesmium* spp., a filamentous cyanobacterium which contains Chl *a* and no other chlorophylls, clearly dominated the autotrophic community and low accessory chlorophylls to TChl *a* ratio were expected. Except for Chl *a* and Chl c_{1+2}, significant differences were again observed

Table 8.3. *Illustration of the difference between two multivariate analysis methods for spectrofluorometric data from pigment extracts. Values shown are pigment concentrations (ng l^{-1}) for all but the last column, and %DV-Chl* a *(= DV-Chl* a**100/(Chl* a *+ DV-Chl* a*)) for the last column, calculated by CLS or NNLS in samples taken during the Maurice Ewing 99/12 cruise (November 1999) in North-Australian waters (two depths at station St. 26 and surface sample at St. 28).*

Method	Chl *a*	Chl *b*	Chl c_{1+2}	Phe *a*	Phe *b*	DV-Chl *a*	DV-Chl *b*	DV-Phe *a*	DV-Phe *b*	Chl c_3	%DV-Chl *a*
St. 26, 0 m											
CLS	56	12	9	1	−1	53	1	1	−5	ND	48
NNLS	57	7	9	2	0	50	4	0	0	ND	46
CLS+	57	11	8	1	0	53	1	0	−4	10	47
NNLS+	58	7	8	3	0	49	4	0	0	10	45
St. 26, 100 m											
CLS	82	25	18	8	−4	56	109	5	7	ND	41
NNLS	79	25	18	8	0	58	109	7	3	ND	42
CLS+	85	21	15	11	0	52	108	5	6	28	38
NNLS+	85	21	15	11	0	52	108	4	7	30	38
St. 28, 0 m											
CLS	830	61	20	42	−28	107	−22	−4	−75	ND	11
NNLS	821	2	21	26	0	64	0	39	0	ND	7
CLS+	846	65	18	32	−41	110	−26	−13	−35	146	11
NNLS+	848		17	53	0	53	0	0	0	71	6

+ = Chl c_3 and Phe c_3 included in the analysis; ND = not determined; Phe c_{1+2} and Phe c_3 concentrations were generally insignificant and are omitted in the table

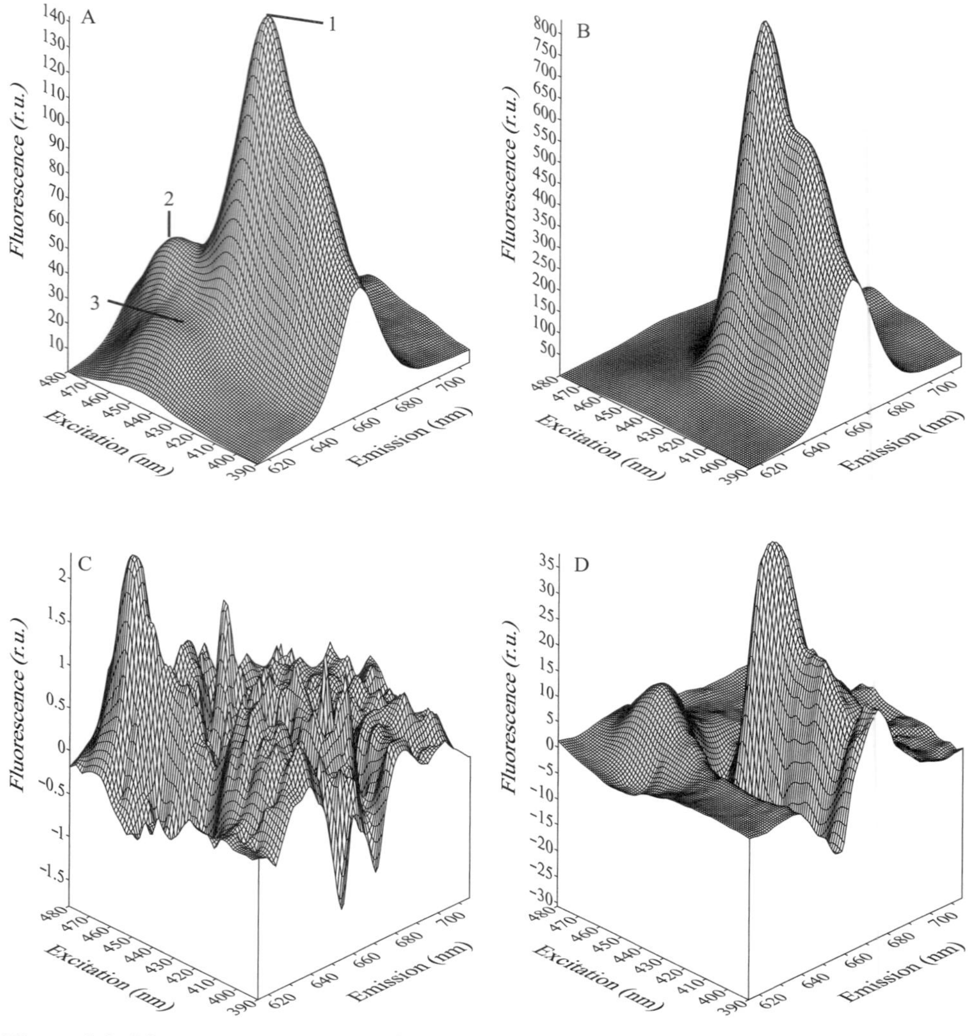

Figure 8.2. Not-quantum-corrected fluorescence excitation–emission spectra of two 90% acetone sample extracts (Ewing 99/12 cruise): (a) station 26, 100 m, (b) station 28, 0 m. Matrices of the residuals determined after NNLS calculation of pigment concentration: (c) station 26, 100 m, (d) station 28, 0 m. Peak 1: Chl *a*-type pigments; peak 2: Chl *b*-type pigments; peak 3: Chl *c*-type pigments.

between CLS and NNLS. NNLS notably reduced the Chl *b*:TChl *a* ratio from around 0.07 to near 0. Negative values were obtained with CLS for Phe *b*, DV-Chl *b*, DV-Phe *a* and DV-Phe *b*. These negative values resulted in an increase in the Chl *b* concentration and, here, NNLS seemed to give a more realistic value for Chl *b* (the reconstructed fluorescence at the excitation and emission maxima of Chl *b* by CLS clearly appeared too high with regard to the observed value, contrary to the result obtained by NNLS). Also, NNLS significantly lowered the calculated concentration of DV-Chl *a*. The three-dimensional fluorescence spectra of the residuals show that the fit between the observed and calculated spectra was not as good at station 28 compared with station 26 (Figure 8.2). Hence, both methods gave adequate results for Chl *a* and Chl c_{1+2}, but NNLS was better for the estimation of Chl *b*.

For mixtures of pigments showing large overlapping spectra, the reliability of spectrofluorometric data depends on consideration in the calibration of all spectrally representative pigments in the extracts, if using CLS or NNLS methods. When this hypothesis is fulfilled, the accuracy in determining the concentration of a specific pigment also depends on the relative contribution of this pigment to the fluorescence excitation–emission matrix; hence it depends on its relative concentration, specific absorption coefficient, and fluorescence quantum yield. For example, based on size fractionation of natural samples, the discrimination between DV-Chl *a* and Chl *a* from 24 pairs of selected excitation and emission wavelengths when using CLS (Neveux and Lantoine, 1993) is most accurate when the level of DV-Chl *a* is above 10% of TChl *a* (Lantoine, 1995). Consequently, when DV-Chl *a* is below this level, accurate discrimination between DV-Chl *a* and Chl *a* is difficult to reach by this calibration and calculation procedure.

8.5.2 Comparison of spectrofluorometric and HPLC methods

The comparison between spectrofluorometric (NNLS) and HPLC methods is examined here using the BC (n = 49) and EW (n = 27) data sets separately. Even though the water samples and pigment extracts for both types of analysis were not identical strictly speaking (same stations, but different sampling bottles), this exercise can show whether similar results would be obtained with the two types of methods for samples typically taken during an oceanographic cruise. For each pigment, the comparison was restricted to samples where the pigment to TChl *a* ratio was higher than a threshold corresponding to 0.05 for Chl *b* and Chl c_3, 0.08 for DV-Chl *a* and 0.02 for Chl c_{1+2}. Below these thresholds, the contribution of these pigments to the fluorescence spectra is typically minor, and estimates by SPF are considered unreliable, as mentioned above.

The BC samples showed a relatively close agreement between SPF and HPLC for TChl *a*, Chl *a*, DV-Chl *a*, Chl c_{1+2} and TChl *b* (Figure 8.3A–D, F). Nevertheless, HPLC indicated a significantly lower concentration of Chl c_{1+2} than SPF. For Chl c_3, a significant correlation was observed, but here also, HPLC gave values three times lower on average than SPF (Figure 8.3H). Tight relationships were also observed for TChl *a*, Chl *a*, DV-Chl *a*, Chl c_{1+2} in the EW samples (not shown), but in contrast with BC samples, here Chl c_{1+2} concentrations from HPLC were not significantly lower than from SPF. For Chl *b*, the determination coefficient R^2 was relatively low (Figure 8.3E) and HPLC did not detect Chl *b* in some samples when

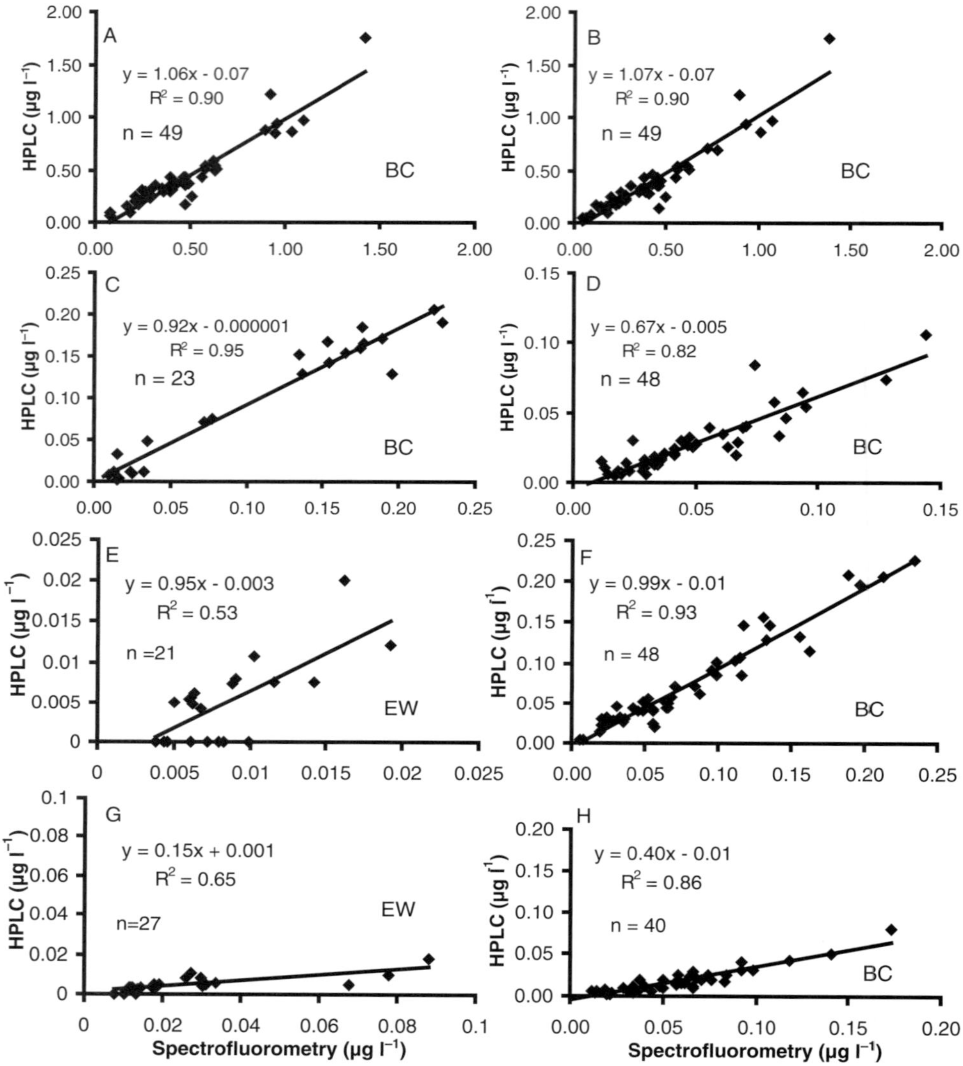

Figure 8.3. Relationships between chlorophylls measured by HPLC (HPLC Chl *a* includes chlorophyllide *a* and allomerized Chl *a*) and spectrofluorometry (NNLS) during the Bissecote (BC) and the Ewing 99/12 (EW) cruises: (A) BC TChl *a* (Chl *a* + DV-Chl *a*); (B) BC Chl *a*; (C) BC DV-Chl *a*; (D) BC Chl c_{1+2}; (E) EW Chl *b*; (F) BC Chl *b*; (G) EW Chl c_3, and (H) BC Chl c_3.

SPF indicated a TChl *b*/TChl *a* ratio > 0.05. Worse relationships than for BC samples may be due to the lower TChl *b*/TChl *a* ratios in the EW samples (0.05–0.09) compared to the BC samples (0.07–0.33), which suggest that the above-mentioned threshold may be too low. A relatively bad relationship was also found for Chl c_3 in the EW samples (Figure 8.3G), with SPF values six to eight times higher than determined by HPLC. The Chl c_3 concentration generally appeared to be overestimated by the present SPF method. This is probably due to the relatively low fluorescence quantum yield of Chl c_3 (specific fluorescence coefficient seven-fold lower than that of Chl c_{1+2} at their respective excitation–emission maxima), which gives it a relatively weak contribution to global fluorescence spectra and a large exposition to instrumental errors.

The above comparisons show that SPF with CLS or NNLS calculation can provide reliable information on biomass (TChl *a*), community structure (Chl *b*, Chl c_{1+2}, Chl c_3, DV-Chl *a* and *b*) and degradation state (pheopigments) of phytoplankton. In pelagic ecosystems, there is no reason to obtain a large divergence between SPF and HPLC for TChl *a*, except if chlorophyllide *a* and allomerized chlorophylls represent a high percentage of TChl *a*. However, chlorophyllide *a* and allomerized Chl *a* generally result from the artificial degradation of Chl *a* during sample handling (filtration, extraction, storage: Suzuki and Fujita, 1986; Jeffrey and Hallegraeff, 1987, Neveux, 1988). If this occurs, separation of these pigments to determine TChl *a* is not appropriate.

Provided that caution is exercised in the use of the different techniques, discrepancies between HPLC and non separation methods in the determination of Chl *a* are certainly lower than previously published in the 1980s. HPLC Chl *a* has often been recommended for satellite ocean colour sensor validation, but SPF can be accurate enough for this validation (Dandonneau *et al.*, 2004). Whether the accessory chlorophylls are considered or not in the SPF calculation does not strongly affect TChl *a* concentrations (mean $\pm$ 4% in oligotrophic waters: Neveux and Lantoine, 1993).

8.6 Recommendations and future considerations

Spectroscopic methods have a number of advantages over HPLC methods for the analysis of pigments in seawater or algae. Data acquisition is generally easier, the cost is lower and smaller sample volumes are required (particularly for spectrofluorometric methods). Ship-board operation, however, requires relatively stable conditions, generally only available on large-sized vessels, and the response stability of the instrument must be frequently checked. Furthermore, when using a spectrofluorometer, the ratio mode must be used to compensate for variations in the response related to fluctuations of excitation light with time.

Data analysis with the multivariate methods described in this chapter provides simultaneous quantification of a large number of pigments and is more accurate than classical methods such as trichromatic equations. Drawbacks include difficulties with the determination of minor components in sample extracts or for pigments with almost identical absorption spectra. Some of the multivariate methods require a-priori knowledge of major pigment components in sample extracts. For spectrofluorometry, calibration of the instrument is crucial, but the commercial availability of several pigment standards has made this easier. The factor-based regression methods (PLS, PARAFAC) show interesting potential, and are increasingly used both for the analysis of pigment extracts as well as for phytoplankton species discrimination from *in vivo* absorption and fluorescence data (e.g. Stæhr and Cullen, 2003; Seppälä and Olli, 2008). These methods in combination with HPLC could enhance the confidence and robustness of the assessment of pigments in field or experimental studies. Chemometrics can also help to determine the concentration of co-eluting pigments in chromatographic systems (Kavianpour and Brereton, 1998).

Acknowledgements

We thank the chief scientists who invited Jacques Neveux on the Ewing 99/12 (Douglas Capone, University of Southern California) and the Bissecote (Sylvain Ouillon, IRD-Nouméa, UR 103) cruises. We thank Karine Escoubeyrou for technical assistance in HPLC pigment analysis of the Bissecote samples, and Ajit Subramaniam (Lamont Doherty Earth Observatory) for kindly providing HPLC data for the Ewing cruise. Financial support from CNRS-INSU and ACI 'Sciences de la terre' is acknowledged. We thank the anonymous reviewers and the editors of this volume for suggestions for improvements to this chapter.

Abbreviations

BC	Bissecote cruise
CLS	Classical least squares
EEM	Excitation–emission matrix
EW	Ewing 99/12 cruise
GPS	Gauss-peak spectra
NNLS	Non-negative least squares
PARAFAC	Parallel factor analysis
PCA	Principal component analysis
PCR	Principal component regression
PLS	Partial least squares
SPF	Spectrofluorometry
SRC	Spectral reconstruction

References

Andersen, C. M. and Bro, R. (2003). Practical aspects of PARAFAC modeling of fluorescence excitation-emission data. *J. Chemom.* **17**, 200–15.

Babin, M., Roesler, C. S. and Cullen, J. J. (2008). *Real-time Coastal Observing Systems for Marine Ecosystem Dynamics and Harmful Algal Blooms*. Paris: UNESCO Publishing.

Bidigare, R. R., Smith, R. C., Baker, K. S. and Marra, J. (1987). Oceanic primary production estimates from measurements of spectral irradiance and pigment concentrations. *Global Biogeochem. Cycles* **1**, 171–86.

Boto, K. G. and Bunt, J. S. (1978). Selective excitation fluorometry for the determination of chlorophylls and pheophytins. *Analyt. Chem.* **50**, 392–95.

Brereton, R. G. (1997). Multilevel multifactor designs for multivariate calibration. *Analyst* **122**, 1521–29.

Bro, R. (1997). PARAFAC, tutorial and applications. *Chemom. Intel. Lab. Syst.* **38**, 149–71.

Champalbert, G., Neveux, J., Gaudy, R. and Le Borgne, R. (2003). Diel variations of copepod feeding and grazing impact in the high-nutrient, low-chlorophyll zone of the equatorial Pacific Ocean (0°; 3°S, 180°). *J. Geophys. Res.* **108**(C12), 8145, doi: 10.1029/2001JC000810.

Dandonneau, Y., Deschamps, P.-Y., Nicolas, J.-M., Loisel, H., Blanchot, J., Montel, Y., Thieuleux, F. and Becu, G. (2004). Seasonal and interannual variability of ocean color and composition of phytoplankton communities in the North Atlantic, equatorial Pacific and South Pacific. *Deep Sea Res. II* **51**, 303–18.

Duckworth, J. H. (1998). Spectroscopic quantitative analysis. In *Applied Spectroscopy, a Compact Reference for Practitioners*, ed. J. Workman Jr. and A. W. Springsteen. San Diego: Academic Press, pp. 93–190.

Dumestre, J. F., Vaquer, A., Gosse, P., Richard, S. and Labroue, L. (1999). Bacterial ecology of a young equatorial hydroelectric reservoir (Petit Saut, French Guiana), Evidence of reduced compound exhaustion and bacterial community adaptation. *Hydrobiologia* **400**, 75–83.

Ficek, D., Kaczmarek, S., Stoń-Egiert, J., Woźniak, B., Majchrowski, R. and Dera, J. (2004). Spectra of light absorption by phytoplankton pigments in the Baltic; conclusions to be drawn from a Gaussian analysis of empirical data. *Oceanologia* **46**, 533–55.

French, C. S. (1971). The distribution and action in photosynthesis of several forms of chlorophylls. *Proc. Nat. Acad. Sci. USA* **68**, 2893–97.

Hoepffner, N. and Sathyendranath, S. (1991). Effect of pigment composition on absorption properties of phytoplankton. *Mar. Ecol. Prog. Ser.* **73**, 11–23.

Humphrey, G. F. and Jeffrey, S. W. (1997). Tests of accuracy of spectrophotometric equations for the simultaneous determination of chlorophylls a, b, c_1 and c_2. In *Phytoplankton Pigments in Oceanography: Guidelines to Modern Methods*, ed. S. W. Jeffrey, R. F. C. Mantoura and S. W. Wright. Paris: UNESCO Publishing, pp. 616–21.

Jeffrey, S. W. and Humphrey, G. F. (1975). New spectrophotometric equations for determining chlorophylls a, b, c_1 and c_2 in higher plants, algae and natural populations. *Biochem. Physiol. Pflanzen* **167**, 191–94.

Jeffrey, S. W. and Hallegraeff, G. M. (1987). Chlorophyllase distribution in ten classes of phytoplankton: a problem for chlorophyll analysis. *Mar. Ecol. Prog. Ser.* **35**, 293–304.

Jeffrey, S. W. and Welschmeyer, N. A. (1997). Spectrophotometric and fluorometric equations in common use in oceanography. In *Phytoplankton Pigments in Oceanography: Guidelines to Modern Methods*, ed. S. W. Jeffrey, R. F. C. Mantoura and S. W. Wright. Paris: UNESCO Publishing, pp. 597–615.

Jeffrey, S. W., Mantoura, R. F. C. and Wright, S. W. (1997) (Eds.). *Phytoplankton Pigments in Oceanography: Guidelines to Modern Methods*. Paris: UNESCO Publishing.

Jiji, R. D., Cooper, G. A. and Booksh, K. S. (1999). Excitation-emission matrix fluorescence based determination of carbamate pesticides and polycyclic aromatic hydrocarbons. *Anal. Chim. Acta* **397**, 61–72.

Kavianpour, K. and Brereton, R. G. (1998). Chemometrics methods for determination of selective regions in diode array detection high performance liquid chromatography of mixtures: application to chlorophyll *a* allomers. *Analyst* **123**, 2035–42.

Küpper, H., Spiller, M. and Küpper, F. C. (2000). Photometric method for the quantification of chlorophylls and their derivatives in complex mixtures: fitting with Gauss-peak spectra. *Anal. Biochem.* **286**, 247–56.

Küpper, H., Seibert, S. and Parameswaran, A. (2007). A fast, sensitive and inexpensive alternative to analytical pigment HPLC: quantification of chlorophylls and carotenoids in crude extracts by fitting with Gauss-peak spectra. *Anal. Chem.* **79**, 7611–27.

Lantoine, F. (1995). *Caractérisation et Distribution des Différentes Populations du Picoplancton (Picoeucaryotes, Synechococcus spp., Prochlorococcus spp.) dans Diverses Situations Trophiques (Atlantique Tropical, Golfe du Lion)*. Ph.D. thesis, University of Paris-VI.

Lorenzen, C. J. (1967). Determination of chlorophyll and pheophytin: spectrophotometric equations. *Limnol. Oceanogr.* **12**, 343–46.

Martens, H. and Næs, T. (1989). *Multivariate Calibration*. Chichester: John Wiley & Sons.

Moberg, L. and Karlberg, B. (2001). Validation of multivariate calibration method for the determination of chlorophyll *a*, *b* and *c* and their corresponding pheopigments. *Anal. Chim. Acta* **450**, 143–53.

Moberg, L., Karlberg, B., Blomqvist, S. and Larsson, U. (2000). Comparison between a new application of multivariate regression and current spectroscopy methods for the determination of chlorophylls and their corresponding pheopigments. *Anal. Chim. Acta* **411**, 137–43.

Moberg, L., Robertson, G. and Karlberg, B. (2001). Spectrofluorometric determination of chlorophylls and pheopigments using parallel factor analysis. *Talanta* **54**, 161–70.

Naqvi, K. R., Melø, T. B. and Raju, B. B. (1997). Assaying the chromophore composition of photosynthetic systems by spectral reconstruction: Application to the light-harvesting complex (LHC II) and the total pigment of higher plants. *Spectrochim. Acta A* **53**, 2229–34.

Naqvi, K. R., Hassan, T. Hj. and Naqvi, Y. A. (2004). Expeditious implementation of two new methods for analysing the pigment composition of photosynthetic specimens. *Spectrochim. Acta Part A* **60**, 2783–91.

Neveux, J. (1988). Extraction of chlorophylls from marine phytoplankton. *Verh. Internat. Verein. Limnol.* **23**, 928–32.

Neveux, J. and Lantoine, F. (1993). Spectrofluorometric assay of chlorophylls and pheophytins using the least squares approximation technique. *Deep-Sea Res.* **40**, 1747–65.

Neveux, J. and Panouse, M. (1987). Spectrofluorometric determination of chlorophylls and pheophytins. *Archiv. Hydrobiol.*, **109**, 567–81.

Neveux, J., Dupouy, C., Blanchot, J., Le Bouteiller, A., Landry, M. R. and Brown, S. L. (2003). Diel dynamics of chlorophylls in high-nutrient, low-chlorophyll waters of the equatorial Pacific (180°): interactions of growth, grazing, physiological responses and mixing. *J. Geophys. Res.* **108**(C12), 8140, doi:10.1029/2000JC000747.

Porra, R. J. (2006). Spectrophotometric assays for plant, algal and bacterial chlorophylls. *Adv. Photos. Resp.* **25**, 95–107.

Seppälä, J. and Olli, K. (2008). Multivariate analysis of phytoplankton spectral *in vivo* fluorescence: estimation of phytoplankton biomass during a mesocosm study in the Baltic Sea. *Mar. Ecol. Prog. Ser.* **370**, 69–85.

Stæhr, P. A. and Cullen, J. J. (2003). Detection of *Karenia mikimotoi* by spectral absorption signatures. *J. Plankton Res.* **25**, 1237–49.

Stedmon, C. A. and Markager, S. (2005). Resolving the variability in dissolved organic matter fluorescence in a temperate estuary and its catchment using PARAFAC analysis. *Limnol. Oceanogr.* **50**, 686–97.

Stuart, V., Sathyendranath, S., Platt, T., Maas, H. and Irwin, B. D. (1998). Pigments and species composition of natural phytoplankton populations: effect on the absorption spectrum. *J. Plankton Res.* **20**, 187–217.

Suzuki, R. and Fujita, Y. (1986). Chlorophyll decomposition in *Sketetonema costatum*: a problem in chlorophyll determination of water samples. *Mar. Ecol. Prog. Ser.* **28**, 81–85.

Tenório, M. M. B., Le Borgne, R., Rodier, M. and Neveux, J. (2005). The impact of terrigeneous inputs on the Bay of Ouinné (New Caledonia) phytoplankton communities: a spectrofluorometric and microscopic approach. *Est. Coast. Shelf Sci.* **64**, 531–45.

Van Heukelem, L. and Thomas, C. S. (2001). Computer-assisted high-performance liquid chromatography method development with applications to the isolation and analysis of phytoplankton pigments. *J. Chromatogr. A* **910**, 31–49.

Zapata, M., Rodríguez, F. and Garrido, J. L. (2000). Separation of chlorophylls and carotenoids from marine phytoplankton: a new HPLC method using a reverse phase C_8 column and pyridine-containing mobile phases. *Mar. Ecol. Prog. Ser.* **195**, 29–45.

Zhang, F., Su, R., Wang, X., Wang, L., He, J., Cai, M., Luo, W. and Zheng, Z. (2009). A fluorometric method for the discrimination of harmful algal bloom species developed by wavelet analysis. *J. Exp. Mar. Biol. Ecol.* **368**, 37–43.

Appendix 8A

A proven simultaneous equation assay for chlorophylls *a* and *b* using aqueous acetone and similar assays for recalcitrant algae

ROBERT J. PORRA

8A.1 Introduction

The success of any spectrophotometric determination of the concentration of any chlorophyll (Chl) turns on the accuracy of the extinction coefficients used, since concentration is the quotient of absorbance (*A*), also known as extinction (E), at a given wavelength divided by the extinction coefficient of the Chl at the same wavelength, normally the wavelength of a major absorption peak. This also applies to the determination of a mixture of two Chls by the '*simultaneous equation*' method.

The determination of accurate extinction coefficients of Chls *a* and *b* is made difficult because pure Chls are readily oxidized and photolabile, especially in monomeric solutions in organic solvents. They are also difficult to store: even solid samples as thin smears inside sealed, aluminium foil-covered flasks can only be stored for a few weeks under O_2-free N_2 or Ar at −15 °C. Monomeric Chl solutions in organic solvents for spectrophotometric assay must be kept in the dark and measured quickly.

In this appendix a proven method for assaying Chls *a* and *b*, using buffered aqueous 80% acetone (pH 7.8) as extractrant, is described together with two methods for extracting and assaying the same Chls from some algae known as 'recalcitrant' since they are difficult to extract with most organic solvents.

8A.2 History of Arnon's simultaneous equation method

Historically, one of the most used assays for Chls *a* and *b* in higher plants and green algae has been that of Arnon (1949). The Chl *a* and *b*-containing cells to be analysed were extracted with aq. 80% acetone and the absorbances, $A^{663\ \mathrm{nm}}$ and $A^{645\ \mathrm{nm}}$, of the extract were measured in a cuvette of 1 cm light path. The two wavelengths chosen were the absorption maxima of the major red (Q_y) peaks in this solvent at 663 and 645 nm for Chls *a* and *b*, respectively, as determined by MacKinney (1941). These two absorbance measurements were then inserted into two simultaneous equations (see Arnon, 1949) and solved to give the individual concentrations of Chls *a* and *b*.

Phytoplankton Pigments: Characterization, Chemotaxonomy and Applications in Oceanography, eds. Suzanne Roy, Carole A. Llewellyn, Einar Skarstad Egeland and Geir Johnsen. Published by Cambridge University Press.

The development of these simultaneous equations by standard mathematical techniques is comprehensively described by Arnon (1949) and requires accurate knowledge of four extinction coefficients; namely, of Chls *a* and *b* at both 663 and 645 nm.

The development of the equations by Arnon (1949) was impeccable, but his equations were grossly inaccurate because the specific extinction coefficients ($d = 1\ g^{-1}\ cm^{-1}$) supplied by MacKinney (1941) at all four wavelengths were too low and involved very significant errors of −4.6 and −19.4% for Chl *a* at 663 and 645 nm, respectively, and of −4.0 and −12.0% for Chl *b* at 663 and 645 nm, respectively (see Porra, 1991, 2002). These inaccuracies were subsequently noted by many laboratories (Vernon, 1960; Ziegler and Egle, 1965; Lichtenthaler, 1987, Porra *et al.*, 1989; Wellburn, 1994) but, inexplicably, use of the Arnon's equations has continued in some laboratories for many years.

8A.3 Accurate simultaneous equations for use with aqueous 80% acetone extractant

Accurate specific (d[1]) and millimolar (ε_{mM}) extinction coefficients ($\varepsilon_{mM} = 1\ mmole^{-1}\ cm^{-1}$) of Chls *a* and *b* were determined at their Q_y absorption maxima in buffered aqueous 80% acetone, and in a number of other solvents, using freshly purified Chl *a* and *b* solutions (Porra *et al.*, 1989). The concentrations of stock solutions were verified by Mg atomic absorption spectroscopy. The aqueous 80% acetone was buffered at pH 7.8 with 2.5 mM sodium phosphate or, preferably, 50 mM Hepes-KOH to prevent removal of the central Mg atom, which occurs readily at low pH, caused by the release of metabolic acids during the extraction process. Spectrophotometric measurements were carried out in stoppered cuvettes to prevent evaporation during the recording of spectra and the spectrophotometer was zeroed at 750 nm to eliminate errors caused by non-specific absorption such as turbidity. With a modern recording spectrophotometer, the absorption maxima of Chls *a* and *b* were found to be at 663.6 and 646.6 nm (compare Porra *et al.*, 1989), though this needs to be confirmed for each spectrophotometer used. Because aqueous dilution of 10% in organic solvents can move the absorption maxima by as much as 1 nm to longer wavelengths, the Chls are extracted in buffered aqueous 80% acetone so that tissue water (i.e. cell sap) will not significantly alter the 20% aqueous component of the solvent. Experimentally, it was found that the Chl *b* peak occurred 17 nm to shorter wavelengths of the readily discernible Chl *a* peak: in mixtures the Chl *b* absorption appears only as a shoulder without an easily discernible peak. The extinction coefficients obtained in buffered aqueous 80% acetone are shown in Table 8A.1.

Table 8A.2 shows the simultaneous equations derived from the d and ε_{mM} extinction coefficients listed in Table 8A.1. The equations using d and ε_{mM} extinction coefficients give Chl concentrations in $\mu g\ mL^{-1}$ and $nmol\ mL^{-1}$, respectively.

[1] The symbol for specific absorption coefficient has changed from α to d, according to the IUPAC compendium of Analytical Nomenclature.

Table 8A.1. *Accurate specific* (d) *and millimolar* (ε_{mM}) *extinction coefficients for Chls* a *and* b *in buffered aqueous 80% acetone (pH 7.8) as determined by Porra* et al. *(1989). The spectrophotometer was zeroed at 750 nm before recording spectra between 750 and 500 nm so that all coefficients listed are difference coefficients between the Q_y maximum wavelength specified and 750 nm. Each coefficient is the mean of three measurements: for the standard deviations see Porra* et al. *(1989).*

		Difference extinction coefficients			
		Chl *a*		Chl *b*	
Solvent	Wavelength (nm)	Millimolar (ε_{mM})	Specific (*d*)	Millimolar (ε_{mM})	Specific (*d*)
Buffered aqueous 80% acetone (pH 7.8)	663.6 minus 750	76.79	85.95	9.79	10.78
	646.6 minus 750	18.58	20.79	47.04	51.84

Table 8A.2. *Simultaneous equations for the determination of Chls* a *and* b *concentrations in buffered aqueous 80% acetone (pH 7.8) as derived by Porra* et al. *(1989) using the accurate extinction coefficients presented in Table 8A.1.*

Solvent	Equations for Chl concentrations (n mol L^{-1})	Equations for Chl concentrations (μg mL^{-1})
Buffered aq. 80% acetone (pH 7.8)	$[\text{Chl}\,a] = 13.71\,A^{663.6} - 2.85\,A^{646.6}$	$[\text{Chl}\,a] = 12.25\,A^{663.6} - 2.55\,A^{646.6}$
	$[\text{Chl}\,b] = 22.39\,A^{646.6} - 5.42\,A^{663.6}$	$[\text{Chl}\,b] = 20.31\,A^{646.6} - 4.91\,A^{663.6}$
	$[\text{Chl}\,a + b] = 19.54\,A^{646.6} + 8.29\,A^{663.6}$	$[\text{Chl}\,a + b] = 17.76\,A^{646.6} + 7.34\,A^{663.6}$

Reliable extinction coefficients obtained by Porra *et al.* (1989) and in other laboratories for an extended range of solvents including *N,N′*-dimethylformamide (DMF), methanol, dimethylsulphoxide (DMSO) and chloroform, as well as the simultaneous equations derived from them have been published (Porra, 2006). To quickly correct results obtained with the Arnon equations, a simple and accurate method has been described (Porra *et al.*, 1989; Porra, 2002, 2006).

8A.4 Extraction methods

Although many shortcuts, including simple prolonged immersion in DMF or DMSO, have been tried, the only reliable method for exhaustive extraction of Chls from leaves is by repeatedly grinding leaves in the solvent with a pestle and mortar: leathery leaves may require to be finely cut with scissors before grinding. Algal cells are repeatedly and exhaustively extracted with solvent in a Potter-Elvehjem homogenizer.

8A.5 The accuracy of the simultaneous equations used with buffered aqueous 80% acetone

The accuracy of the simultaneous equations used with buffered aqueous 80% acetone (see Table 8A.2) has been elegantly confirmed in molecular modelling of plant light harvesting complex II (LHC II). In early studies, Butler and Kühlbrandt (1988), using Arnon's equations, calculated that each purified LHC II pigment polypeptide contained a total of 15 Chl molecules with a Chl *a*/*b* ratio of 1.15 indicating 8 Chl *a* and 7 Chl *b*. When these results were recalculated using the equations of Porra *et al.* (1989) a total of 14 Chl molecules per polypeptide was obtained with a Chl *a*/*b* ratio of 1.3 indicating 8 Chl *a* and 6 Chl *b*. This agrees exactly with the 8 Chls *a* and 6 Chl *b* observed in the LHC II crystal structures prepared from spinach leaves by Liu *et al.* (2004) and from pea leaves by Kühlbrandt's group (see Standfuss *et al.*, 2005) as determined by X-ray crystallography at 2.72 Å resolution (Liu *et al.*, 2004) and 2.5 Å (Standfuss *et al.*, 2005) where Chl *b* can be clearly distinguished from Chl *a*.

8A.6 Two simultaneous equation techniques specifically designed for use with recalcitrant algae

The Chls of some freshwater and marine algae, dubbed 'recalcitrant algae', are difficult to extract with solvents normally used in extraction of Chls from higher plants and two techniques have been developed to overcome this problem in *Nannochloris atomus* cells (Porra, 1990a, b).

8A.6.1 Chlorophyll extraction with aqueous 85% methanol containing 2% KOH and 1.5 mM sodium dithionite

In the first technique, the *Nannochloris atomus* cells were extracted with 2% KOH in aqueous 85% methanol containing 1.5 mM dithionite with continuous shaking for 20 min at 60 °C in the dark (Porra, 1990a). The alkaline conditions, which presumably enhance extraction by hydrolysis of structural proteins of the cell walls or membranes, also transesterifies the phytol ester of Chls *a* and *b* to a methyl ester and opens the isocyclic ring, to form Mg-rhodochlorins (Mg-RChlns) *a* and *b* with 13-methylcarboxlate and 15-methylacetate side chains (for carbon numbering see Figure 2.1B, Chapter 2, this volume) The Q_y peaks of Mg-RChlns *a* and *b* occur at 641.2 and 623.2 nm, respectively. The sodium dithionite (or sodium ascorbate, mercaptoethanol or dithiothreitol) was added to prevent further oxidation (allomerization) reactions during alkaline extraction (Porra, 1990a). In this alkaline methanolic solvent the ε_{mM} coefficients for Mg-RChln *a* were determined as 54.61 and 14.45 and for RChln *b* were 5.73 and 19.81 at 641.2 and 623.2 nm, respectively. The simultaneous equations derived from these coefficients for Chls *a* and *b* concentrations measured as their respective Mg-RChlns in nmol mL^{-1} are,

$$[\text{Chl}\,a] = 19.83 \cdot A^{641.2} - 5.74 \cdot A^{623.2} \tag{8A.1}$$

$$[\text{Chl}\,b] = 54.66 \cdot A^{623.2} - 14.6A^{641.2}. \tag{8A.2}$$

Specific (*d*) coefficients and the simultaneous equation derived from them giving Chl *a* and *b* concentrations in μg mL^{-1} are also available (see Porra, 1990a).

8A.6.2 Chlorophyll extraction with aqueous 85% methanol containing 1.5 mM sodium dithionite or sodium ascorbate

It was serendipitously discovered that if KOH was omitted from the above extractant method, Chls *a* and *b* could be extracted unchanged from *Nannochloris atomus* cells by shaking for 20 min in the dark at 60 °C in aqueous 85% methanol, but only if 1.5 mM sodium dithionite or sodium ascorbate was present (Porra, 1990b). Reductant was needed to prevent oxidation (allomerization) of the Chls during the alkaline extraction process, but Thompson and Preston (1968) reported that reductants (dithiothreitol) weakened the cell walls of the green alga, *Cladophora rupestris*, and ascribed this phenomenon to the breaking of disulfide linkages in the structural proteins of the cell wall: this phenomenon may explain the ease of Chl extraction.

In aqueous 85% methanol containing reductant, the Q_y peak maxima were at 664 and 650 nm, respectively. The ε_{mM} coefficients for Chls *a* were determined as 68.18 and 27.81 and for Chl *b* were 18.19 and 36.88 at 664 and 650 nm, respectively (Porra, 1990b). The simultaneous equations derived from these coefficients for Chls *a* and *b* in nmol mL^{-1} are,

$$[\text{Chl}\,a] = 18.36 \cdot A^{664} - 9.06 \cdot A^{650} \tag{8A.3}$$

$$[\text{Chl}\,b] = 39.44 \cdot A^{650} - 13.85 \cdot A^{664}. \tag{8A.4}$$

Specific (*d*) coefficients and the simultaneous equation derived from them giving Chl *a* and *b* concentrations in μg mL^{-1} are also available (see Porra, 1990b).

References

Arnon, D. I. (1949). Copper enzymes in isolated chloroplasts. Polyphenoloxidase in *Beta vulgaris*. *Plant Physiol.* **24**, 1–15.

Butler, P. J. G. and Kühlbrandt, W. (1988). Determination of the aggregate size in detergent solution of the light-harvesting chlorophyll *a*/*b*-protein complex from chloroplast membranes. *Proc. Natl. Acad. Sci. USA* **85**, 3797–801.

Lichtenthaler, H. K. (1987). Chlorophylls and carotenoids: pigments of photosynthetic membranes. *Methods Enzymol.* **148**, 350–82.

Liu, Z., Yan, H., Wang, K., Kuang, T., Zang, J., Gul, L., An, X. and Chang, W. (2004). Crystal structure of spinach major light harvesting complex at 2.72Å resolution. *Nature* **428**, 287–92.

MacKinney, G. (1941). Absorption of light by chlorophyll solutions. *J. Biol. Chem.* **140**, 315–22.

Porra, R. J. (1990a). The assay of chlorophylls *a* and *b* converted to their respective magnesium-rhodochlorin derivatives by extraction from recalcitrant algal cells with aqueous alkaline methanol: prevention of allomerization with reductants. *Biochim. Biophys. Acta* **1015**, 493–502.

Porra, R. J. (1990b). A simple method for extracting chlorophylls from the recalcitrant alga, *Nannochloris atomus*, without formation of spectroscopically-different magnesium-rhodochlorin derivatives. *Biochim. Biophys. Acta* **1019**, 137–41.

Porra, R. J. (1991). Recent advances and reassessment in chlorophyll extraction and assay procedures for terrestrial, aquatic and marine organisms, including recalcitrant algae. In *Chlorophylls*, ed. H. Scheer. Boca Raton: CRC Press, pp. 31–57.

Porra, R. J. (2002). The chequered history of the development and use of simultaneous equations for the accurate determination of chlorophylls *a* and *b*. *Photosynth. Res.* **73**, 149–56.

Porra, R. J. (2006). Spectrometric assays for plant, algal and bacterial chlorophylls. In *Chlorophylls and bacteriochlorophylls: biochemistry, biophysics, functions and applications*, ed. B. Grimm, R. J. Porra, W. Rüdiger and H. Scheer. Dordrecht: Springer, pp. 95–107.

Porra, R. J., Thompson, W. A. and Kriedemann, P. A. (1989). Determination of accurate extinction coefficients and simultaneous equations for assaying chlorophylls *a* and *b* extracted with four different solvents: verification of the concentration of chlorophyll standards by atomic absorption spectroscopy. *Biochim. Biophys. Acta* **975**, 384–94.

Standfuss, J., Terwisscha van Scheftinga, A. C., Lamborghini, M. and Kühlbrandt, W. (2005). Mechanisms of photoprotection and nonphotochemical quenching in pea light-harvesting complex at 2.5Å resolution. *EMBO J.* **24**, 919–28.

Thompson, E. W. and Preston, R. D. (1968). Evidence for a structural role of protein in algal cell walls. *J. Exp. Bot.* **19**, 690–97.

Vernon, L. P. (1960). Spectrophotometric determination of chlorophylls and pheophytins in plant extracts. *Anal. Chem.* **32**, 1144–50.

Wellburn, A. R. (1994). The spectral determination of chlorophylls *a* and *b*, as well as total carotenoids using various solvents with spectrophotometers of different resolution. *J. Plant Physiol.* **144**, 307–13.

Ziegler, R. and Egle, K. (1965). Zur quantitativen Analyse der Chloroplastenpigmente. 1. Kritische 1 berprüfung der spektralphotometrischen Chlorophyll Bestimmung. *Beitr. Biol. Pflanzen* **41**, 11–37.

Part III

Water-soluble 'pigments'

9

Phycobiliproteins

KAI-HONG ZHAO, ROBERT. J. PORRA AND HUGO SCHEER

9.1 Introduction

Phycobiliproteins are the major light-harvesting pigments of cyanobacteria, red algae, glaucocystophytes (cyanelles) and cryptophytes (MacColl and Guard-Friar, 1987; Sidler, 1994). They are characterized by linear tetrapyrrolic chromophores, known as bilins, that are covalently bound to cysteines of the apoproteins via thioether bonds, and they harvest light for photosynthesis efficiently in the 'green gap' where chlorophylls absorb only poorly (Sidler, 1994). Unlike isolated chlorophyll chromophores, free bilins are photophysically unsuited as photoreceptors: they absorb light only poorly and their excited states are very short lived, thereby leading to rapid conversion of excitation energy to heat (Scheer, 1982; Braslavsky *et al.*, 1983; Falk, 1989). These properties also prevail in denatured biliproteins. The photophysical properties of native biliproteins are, by contrast, much more favourable: the light absorption of the chromophores is increased by almost one order of magnitude and the excited lifetimes by four orders of magnitude, which, in combination, render them excellent photoreceptors. The absorption of individual chromophores can, moreover, be shifted by almost 100 nm, and also the circular dichroism of biliproteins is modulated drastically during folding (see Scheer, 2003 and Kupka and Scheer, 2008 for leading references). The underlying nature of these molecular adaptations, which are still only partly understood, consists mainly of extensive chromophore protein interactions by which the chromophore conformation and dynamics are modulated. Covalent binding to the apoproteins appears to be important in assisting these interactions. Although cysteine mutants indicate that covalent binding is not absolutely necessary for function (Gindt *et al.*, 1994; Jorissen *et al.*, 2002; Inomata *et al.*, 2006), it does assist functional optimization (Gindt *et al.*, 1994) because it stabilizes both the labile chromophores (Scheer, 1982) and proteins (Anderson and Toole, 1998; Shen *et al.*, 2008a, b). In cyanobacteria and red algae, up to four bilin chromophores are post-translationally attached, via thioether bonds,

Phytoplankton Pigments: Characterization, Chemotaxonomy and Applications in Oceanography, eds. Suzanne Roy, Carole A. Llewellyn, Einar Skarstad Egeland and Geir Johnsen. Published by Cambridge University Press.

to specific cysteines of up to a dozen or even more individual proteins (Sidler, 1994); further, an additional modification in β-subunits is the methylation of a conserved asparagine-72 (Swanson and Glazer, 1990; Schluchter *et al.*, 2010). Chromophore attachment also appears to be a pre-requisite for the assembly of phycobilisomes (PBS) (Anderson and Toole, 1998) which are the light-harvesting antennae of blue-green algae and of both red and cryptophyte algae.

The first part of this review (Sections 9.2 to 9.5) gives a brief overview of the structures and biosynthesis of phycobiliproteins. The combination of chromophore variations, covalent chromophore binding and extensive interactions between chromophore and protein in the native biliproteins require special analytical tools and strategies that will be outlined in the second part of this review (Section 9.6). The literature cited in this review is necessarily incomplete, but reviews or recent publications are favoured, which should allow good access to previous work.

9.2 Structures of phycobiliproteins

Phycobiliproteins from cyanobacteria, glaucocystophytes (cyanelles), and red algae are a large, monophyletic family of homologous heterodimeric proteins. Both the α- and β-subunits, which are also homologous to each other, consist of both a globin-type core that carries the chromophore(s) and an N-terminal extension that is mainly involved in subunit aggregation to αβ protomers (Figure 9.1A). These heterodimers (often referred to as 'monomers') can further aggregate to form ring-shaped or, strictly speaking, triangular-shaped 'trimers' (heterohexamers) and 'hexamers' (heterododecamers) (Sidler, 1994) (see Figure 9.1A). The 'hexamers' constitute the building blocks of a unique extra-membranous antenna complex, known as a phycobilisome (see Figure 9.1B) (Gantt and Conti, 1966; Scheer, 1982; Wehrmeyer, 1983; Glazer, 1984; Schirmer *et al.*, 1985; Gantt, 1986; Ficner and Huber, 1993; Brejc *et al.*, 1995; Ritter *et al.*, 1999; Stec *et al.*, 1999; Wang *et al.*, 2001; Adir *et al.*, 2002; Nield *et al.*, 2003; Doust *et al.*, 2004; Adir, 2005; Schmidt *et al.*, 2007). The hexamers are arranged in short stacks that form the phycobilisome core, or in longer rods that protrude from the former. This supramolecular organization is mainly due to linker proteins which are located, as a central backbone, in the inner hole of the ring-shaped biliproteins (Glazer, 1994; Sidler, 1994; Reuter *et al.*, 1999; Gantt *et al.*, 2003; Grossman *et al.*, 2003). While most of the linkers are colourless, at least two of them also carry –covalently bound– bilin chromophores.[1] One is the core-membrane linker L_{CM} (= ApcE) which attaches the phycobilisome to the photosynthetic membrane; it also serves as a terminal donor of excitation energy, captured by the hundreds of chromophores of the phycobilisome, to the reaction centers (Gindt *et al.*, 1994).

[1] There is evidence that the inner holes of biliproteins in the PBS can also harbor structurally similar, but functionally different proteins, like a photoactive or photoprotective orange carotenoid protein (OCP) which is involved in photoprotection (Wilson *et al.*, 2008), or the ferredoxin-NADP$^+$ oxidoreductase (FNR) which is involved in photosynthetic electron transport from photosystem I and, possibly, PBS binding to the photosynthetic membrane (van Thor *et al.*, 1999; Gomez-Lojero *et al.*, 2003; Morsy *et al.*, 2008, Blot *et al.*, 2009).

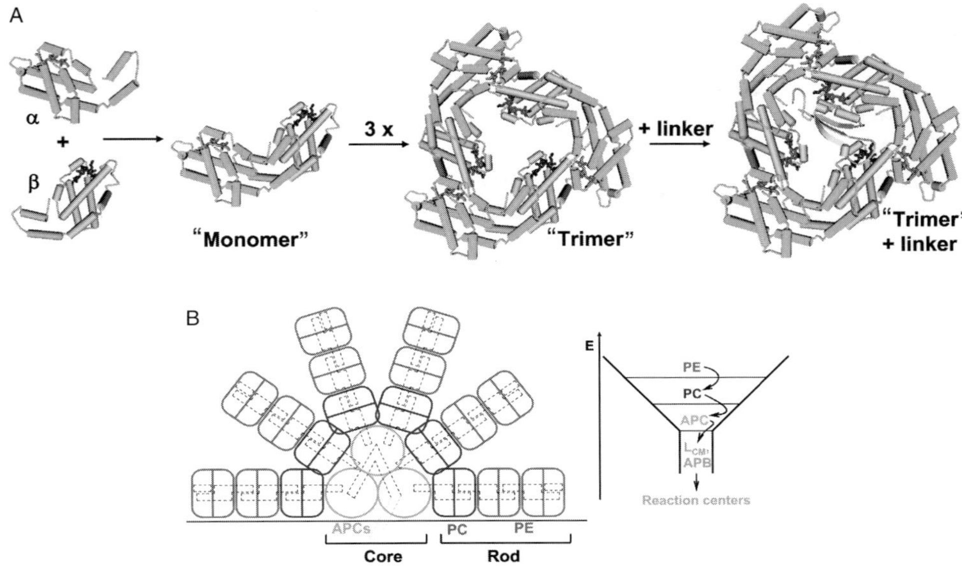

Figure 9.1. (A) Assembly of phycobiliprotein trimer. The model shown is from the only structure showing a phycobiliprotein-linker complex, i.e. APC L_C from *Mastigocladus laminosus* (Reuter *et al.*, 1999). Proteins are displayed schematically (α subunit light green, β-subunit light blue, linker L_C grey), the chromophores as stick models in darker colours. Note the proximity of L_C to two of the three β-chromophores. (B) Structural (left) and energetic funnel-models (right) of a phycobilisome. The structural model is of a hemidiscoidal PBS containing a tricylindrical core composed of APCs (greenish blue), and six rods, each containing a PC hexamer (blue) and two PE hexamers (red). The linkers in the center of the rods and in the core are indicated by the dashed shapes (black). The horizontal solid black line indicates the membrane surface. The same color coding is used in the energetic funnel model, where the vertical error indicates the ^1S excitation energy of the chromophores. Energy is transferred from the higher energy chromophores to the lower energetic ones in the PBS, and finally, via the terminal emitters, L_{CM} and allophycocyanin B (APB), to the reaction centers in the photosynthetic membrane. See colour plate section.

The other exception is the γ-subunit of class II and some class I phycoerythrins (see below), where chromophores enhance light-harvesting by the phycobilisome in the green spectral region (Nagy *et al.*, 1985; Fairchild and Glazer, 1994a; Sidler, 1994; Everroad *et al.*, 2006). *Prochlorococcus* species lack phycobilisomes, and contain only few biliproteins (Hess *et al.*, 2001; Ting *et al.*, 2001; Steglich *et al.*, 2005), and occasionally structures have been reported in cyanobacteria that are reminiscent of incomplete phycobilisomes (see for example Reuter *et al.*, 1994; Marquardt *et al.*, 1997).[2]

[2] The evolution of PBS appears more complex, as that of the contained biliproteins. Recently, an independent evolution of rods and cores has been suggested for PBS of marine *Synechococcus* (Six *et al.*, 2007).

A second type of phycobiliproteins is found in cryptophyte algae, where they are located at the inner surface of the photosynthetic membrane (Gantt *et al.*, 1971; Wehrmeyer, 1988; Wedemayer *et al.*, 1991; Sidler, 1994; Glazer and Wedemayer, 1995; Wedemayer *et al.*, 1996; Doust *et al.*, 2004). The β-subunits of cryptophyte biliproteins are phylogenetically related to red-algal phycoerythrins, the α-subunits are much shorter and probably of a different origin (Sidler and Zuber, 1988; Glazer and Wedemayer, 1995). The protomeric unit is an $\alpha\alpha'\beta_2$ hetero-tetramer, but little is known about their macromolecular organization.

Less-characterized biliproteins, which are variants of unknown function (Montgomery *et al.*, 2004), include the phycoerythrins (PE) of the phycobilisome-less branch of cyanobacteria such as the marine *Prochlorococcus* species (Hess *et al.*, 2001). Also an open question is the involvement of cyanobacterial phages in biliprotein biosynthesis and lateral gene transfer (Six *et al.*, 2007) (Dammeyer *et al.*, 2008). Last, but not least, there is a third, phylogenetically unrelated family of biliproteins, the phytochromes (Lamparter, 2004; Rockwell *et al.*, 2006) and phytochrome-like cyanobacteriochromes (Ishizuka *et al.*, 2007), which are sensory photoreceptors in plants, algae and many bacteria. Phytochromes carry only a single bilin chromophore at one of two alternative binding sites. Z-E-isomerization of the chromophore at the Δ15,16 double bond is the primary reaction of phytochromes (Rüdiger and Thümmler, 1991). Currently, there are several structures determined for the red-absorbing form that carry the Z-isomer (Wagner *et al.*, 2005; Scheerer *et al.*, 2006; Wagner *et al.*, 2007; Yang *et al.*, 2007; Essen *et al.*, 2008), and one for the stable far-red absorbing form carrying the 15E-isomer (Yang *et al.*, 2008). Both isomers have also been characterized for another photoactive biliprotein from cyanobacteria, α-phycoerythrocyanin (PEC) (Dürring *et al.*, 1990; Schmidt *et al.*, 2006, 2007). The latter shows the same photoisomerization of the Δ15,16 double bond, but spectral shifts occur in the opposite direction to that of phytochrome, and this photochemistry is lost when α-PEC is incorporated into the phycobilisome (Zhao *et al.*, 1995). The cyanobacteriochromes carry one or more photo-active chromophores. In the simple cases, they also show a 15Z/E isomerization as the primary photoreaction, but more complex cases have been described (Ikeuchi and Ishizuka, 2008).

Individual phycobiliprotein subunits carry 1–4 chromophores, with the number of binding sites increasing from allophycocyanin (APC) through phycocyanin (PC) and phycoerythrocyanin (PEC) to both phycoerythrin (PE) and cryptophyte biliproteins (see Sidler, 1994; Six *et al.*, 2007, and references therein). One conserved binding site, Cys-84, is present in all cyanobacterial and red algal phycobiliproteins, and also in the β-subunits of cryptophyte biliproteins. Additional binding sites evolved in the globin domain by insertions near the C-terminus, around position 150, and towards the N-terminus, near position 50. Most chromophores are attached by a single thioether bond at $C\text{-}3^1$, but a second linkage is present in some PEs where a phycoerythrobilin (PEB, see Abbreviations and Figure 9.2 for chromophore structures) or phycourobilin (PUB) is bound at $C\text{-}3^1$ to Cys-β50 and at $C\text{-}18^1$ to Cys-β61

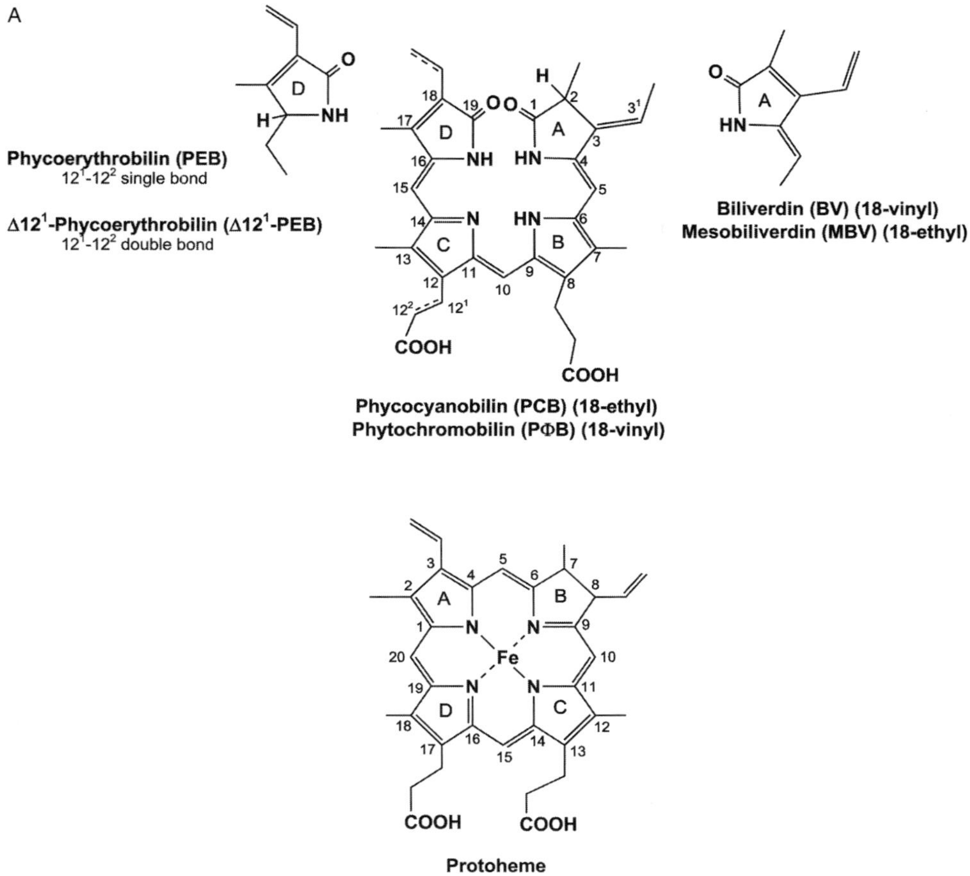

Figure 9.2. Structures of phycobiliprotein chromophores. (A) Free bilins in their thermodynamically most stable cyclic conformation; not shown is the helical character caused by steric hindrance of the terminal imine-oxygens, and protoheme. Note the different ring notation and carbon numbering for cyclic (protoheme) and open-chain tetrapyrroles (BV) as approved by the International Union of Pure and Applied Chemistry (IUPAC, (Moss, 1988) and http://www.chem.qmul.ac.uk/iupac/tetrapyrrole/). PCB and PΦB are shown in full, and only deviating partial structures for the other chromophores. Δ12^1-PEB and Δ12^1-DBV are also termed 'bilin 584' and 'bilin 618', respectively, according to the absorption maxima of the protein-bound chromophores in acidic urea (see Wedemayer *et al.*, 1996). Colours approximate those of the respective chromophores. (B) Protein-bound chromophores with their thioether bond(s) to the apoprotein, in extended conformations that are typical for native phycobiliproteins. Colours approximate those of the respective chromophores. BV, the common biosynthetic precursor, is the chromophore of bacterial (class II) phytochromes. Phycobilisome-containing organisms (cyanobacteria, glaucocystophytes (cyanelles), red algae, *Galdieria sulphuraria*) have PCB, PEB, PVB and PUB as chromophores; cryptophytes contain also the others. Δ12^1-PEB and Δ12^1-DBV are also termed 'bilin 584' and 'bilin 618', respectively, according to the absorption maxima of the protein-bound chromophores in acidic urea. A doubly linked DBV is bound to cysteines at C-3^2 and C-18^1 (see Wedemayer *et al.*, 1996). See colour plate section.

B

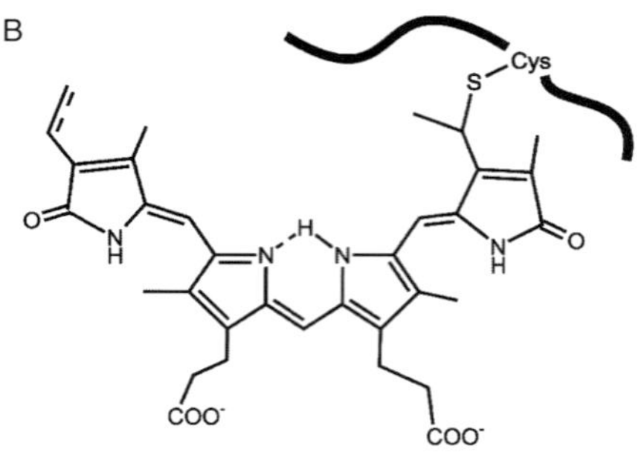

3^1-Cys - biliverdin (BV, vinyl at ring D)
3^1-Cys - mesobiliverdin (MBV, ethyl at ring D)

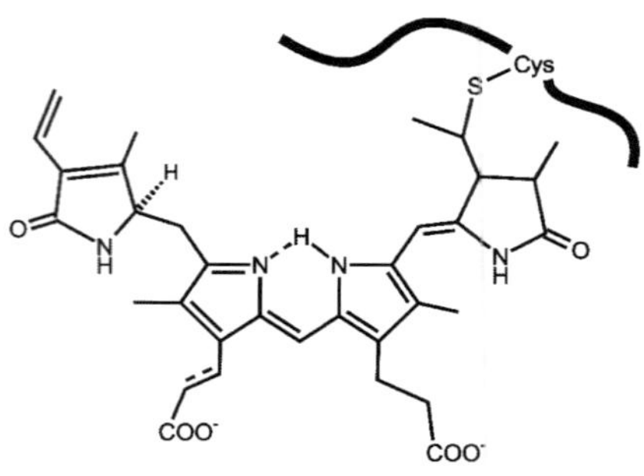

3^1-Cys - 15,16-dihydrobiliverdin (DBV)[a]
3^1-Cys - $\Delta 12^1$-15,16-dihydrobiliverdin ($\Delta 12^1$-DBV)

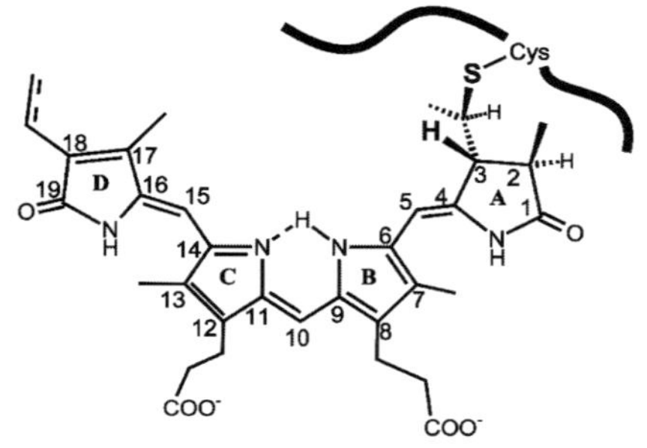

3^1-Cys-phycocyanobilin (PCB; ethyl at ring D)
3^1-Cys-phytochromobilin (PΦB; vinyl at ring D)

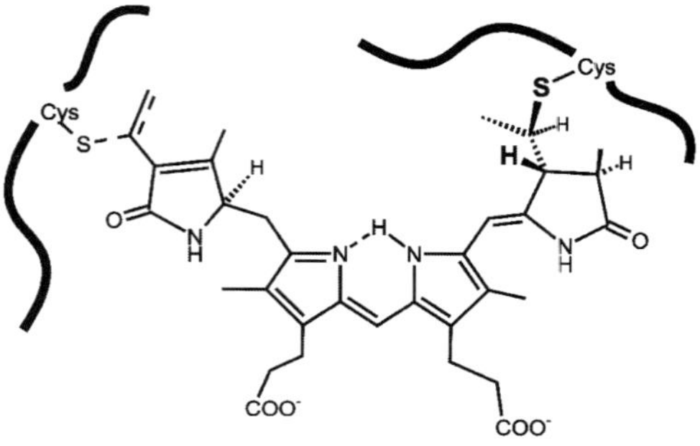

3^1(,18^1-Di)-Cys - phycoerythrobilin (PEB)
3^1(,18^1-Di)-Cys - $\Delta 12^1$-phycoerythrobilin ($\Delta 12^1$-PEB)

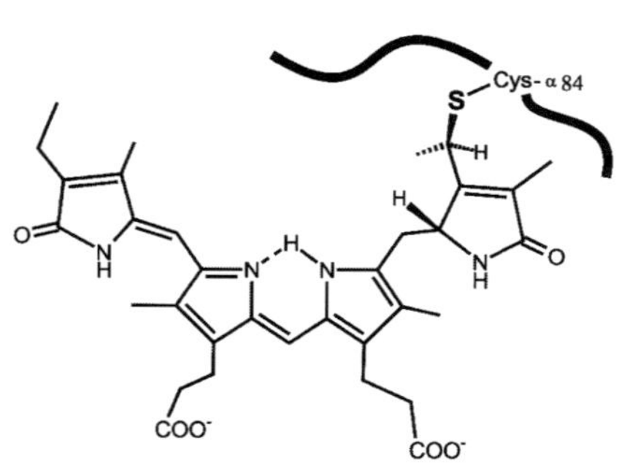

3^1-Cys - phycoviolobilin (PVB) (15-Z)

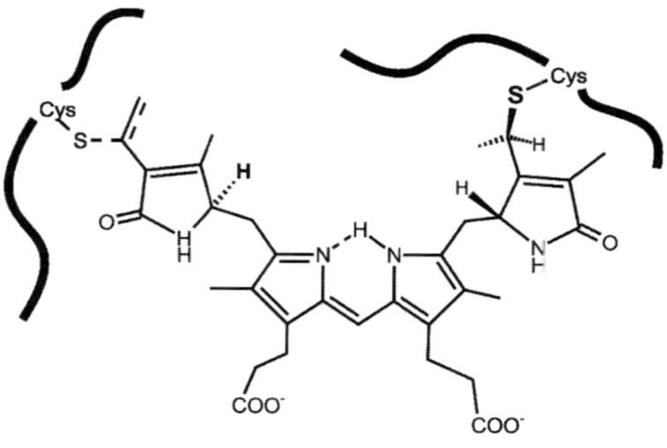

3^1(,18^1-Di)-Cys - phycourobilin (PUB)

Figure 9.2. (*cont.*)

(Ficner and Huber, 1993). New asymmetric C-atoms with well-defined configurations are generated during the addition reaction (Schirmer *et al.*, 1987; Adir *et al.*, 2002). Homologous proteins can carry different chromophores at certain sites and even chromophore exchange has been reported on a particular protein in response to

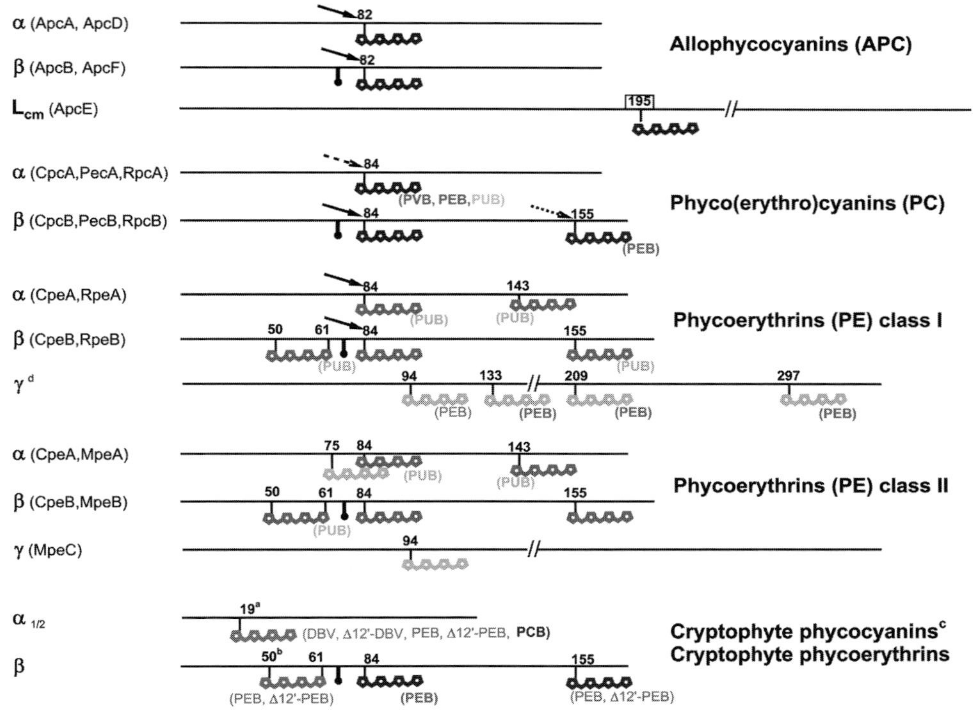

Figure 9.3. Post-translational modifications of biliproteins. Colours are indicative of the prevailing chromophores (blue = PCB, red = PEB, purple = PVB, orange = PUB, see colour plate section), alternative chromophores are indicated by their abbreviations (see Figure 9.2). Numbers give the (consensus) position of binding cysteines. Solid arrows denote chromophore attachment by CpcS-type lyases, dotted arrows CpcT-type lyases and dashed arrows CpcE/F-type lyases (see text for details). The boxed number at ApcE indicates autocatalytic attachment. The vertical bars with knobs indicate γ-methylated asparagine-β72. Not shown is a third PE, termed PE III, that has been identified in a high-light *Prochlorococcus marinus* strain; it carries only a single chromophore on the α-subunit, and none on the β-subunit (Hess *et al.*, 1996). a) MBV, b) DBV, c) main structure phycocyanin 645, d) only in red-algal b- (and possibly B-) phycoerythrin. See colour plate section.

changed light quality (Everroad *et al.*, 2006): in particular, in marine species; PUB (λ_{max}= 500 nm) can replace PEB ($\lambda_{max} \sim$ 540 nm) and PEB can replace phycocyanobilin (PCB) ($\lambda_{max} \sim$ 620 nm) thus further expanding the absorption spectrum and thereby adapting the strains to the prevailing blue and green light environment of clear waters. The largest variations are found at Cys-α84 of different cyanobacterial phyco(erythro)cyanins, to which four different chromophores, including the photoactive phycoviolobilin (PVB) can be bound, and at Cys-α19 of cryptophyte biliproteins (Figure 9.3). While cyanobacterial biliproteins generally only carry up to two different chromophores, a trichromatic R-type PC has recently been characterized from oceanic *Synechococcus* that has PUB at Cys-α84, PCB at Cys-β84 and PEB at Cys-β155, thereby covering the entire 'gap' between the Chl*a* absorption maxima (Blot *et al.*, 2009). It probably optimizes energy transfer from the PUB-rich PEs in the rods, to the PCB-carrying APCs of the core that serve as the energy donor to the reaction centers.

In addition to covalent binding, there exist extensive non-covalent interactions between the apoprotein and the chromophore that are essential for biliprotein function. Free bilins adopt a rather flexible, cyclic-helical conformation (Falk, 1989), whereas native biliprotein chromophores assume, by contrast, rigid and extended conformations (Scheer, 1982; Ficner and Huber, 1993; Ritter *et al.*, 1999; Stec *et al.*, 1999; Wang *et al.*, 2001; Adir *et al.*, 2002; Nield *et al.*, 2003; Doust *et al.*, 2004; Wagner *et al.*, 2005; Inomata *et al.*, 2006; Schmidt *et al.*, 2007). Variations of this basic conformation contribute to the fine-tuning of the spectral absorption range by which light-harvesting or photosensing is optimized. In the phycobilisome, this fine-tuning involves binding-site specific interactions and more distant interactions with linker proteins (Reuter *et al.*, 1999) and with other chromophores (see below).

9.3 Biosynthesis of phycobilin chromophores

Chromophore biosynthesis starts from protoheme (Porra *et al.*, 1997; Frankenberg and Lagarias, 2003a). The protoheme macrocycle is cleaved between rings A and B[3] by a heme oxygenase (Beale, 1993; Cornejo *et al.*, 1998) to produce equimolar amounts of biliverdin IXα (BV) (Figure 9.2), Fe^{2+} and CO. BV is then reduced at one or two out of three different positions by ferredoxin-dependent bilin reductases: these are reductions of the 18-vinyl group on ring D, of the Δ15,16-double bond between rings C and D, and reductive isomerization of ring A yielding the 3-ethylidene group characteristic of PCB, PΦB and PEB (Dammeyer and Frankenberg-Dinkel, 2008). PCB is generated by a single reductase (PcyA) that catalyzes two of these reductions, namely, at ring A and at the 18-vinyl group (Frankenberg and Lagarias, 2003b), but not the reduction of the Δ15,16-double bond. Phytochromobilin (PΦB) reductase (Hy2) catalyzes only reduction of ring A, leading to the typical chromophore of plant phytochromes (Kohchi *et al.*, 2001) while a dedicated, but as yet unknown, reductase catalyzes only the reduction of the 18-vinyl group to yield mesobiliverdin (MBV), which is one of the chromophores of cryptophyte biliprotein (Sidler, 1994; Wedemayer *et al.*, 1996). In PEB biosynthesis; two enzymes (PebA and PebB) act in sequence (Frankenberg *et al.*, 2001; Dammeyer and Frankenberg-Dinkel, 2006). The intermediate, Δ15,16-dihydrobiliverdin (DBV), is present in cryptophyte biliproteins (Wedemayer *et al.*, 1996), but the enzyme leading

[3] See Figure 9.2 for the different carbon numbering and ring notation, by which rings A, B, C, and D of cyclic tetrapyrroles become rings D, A, B and C, respectively, of linear tetrapyrroles.

to its formation is unknown in these organisms. Interestingly, certain *Prochlorococcus* phages contain a bifunctional reductase capable of reducing BV directly to PEB (Dammeyer *et al.*, 2008). PCB and PEB are the only free chromophores found in cyanobacteria and red algae, but generally in low amounts (Beale, 1994; Cornejo and Beale, 1997), while two other chromophores, namely, PVB and PUB, are only found bound in biliproteins, but not as free chromophores. A larger variety of chromophores is found in cryptophyte biliproteins (Glazer and Wedemayer, 1995; Wedemayer *et al.*, 1996), but their biosynthesis has not yet been investigated.

The bilin chromophores are subsequently attached covalently to the apoproteins and, the two missing chromophores (PVB, PUB), from PCB and PEB, respectively, generated are during attachment by a simultaneous isomerization reaction (Zhao *et al.*, 2002). Currently, three modes of chromophore attachment have been identified:

(1) Most apoproteins can bind phycobilins (PCB, PEB) spontaneously; this process, however, is generally of low fidelity: side reactions, including oxidations and, possibly, incorrect stereochemistry, lead to product mixtures (Arciero *et al.*, 1988; Fairchild and Glazer, 1994a; Schluchter and Glazer, 1999). Spontaneous chromophore binding constitutes a considerable problem in binding studies because it interferes, especially in vitro, with lyase assays, and is not easily distinguished from truly autocatalytic lyase activities (Schluchter and Bryant, 2002; Böhm *et al.*, 2007; Zhao *et al.*, 2007a). As an example, ApcA binds PCB spontaneously (Hu *et al.*, 2006; Zhao *et al.*, 2007a), but the product is spectrally distinct from the native α-subunit of APC.
(2) True autocatalytic attachment is defined here as the spontaneous attachment leading to chromoproteins that are spectroscopically, biochemically and functionally indistinguishable from the respective native forms isolated from the parent organism. Correct chromophore binding, namely, a true autocatalytic lyase activity, is currently regularly identified only among phytochromes (Wu and Lagarias, 2000; Zhao *et al.*, 2004; Inomata *et al.*, 2006). It is the exception among phycobiliproteins and has only been found for ApcE (boxed site in Figure 9.3): by addition of PCB, it could be reconstituted to the core-membrane linker, L_{CM} (Zhao *et al.*, 2005).
(3) The most ubiquitous mode of attachment in phycobiliproteins is by enzymic catalysis involving lyases, of which *three* types have so far been characterized (Scheer and Zhao, 2008). *Firstly*, there are the heterodimeric E/F-type lyases that are highly specific for a single binding site. They attach PCB to cysteine-84 of the α-subunit of CPC and PEC: in the latter case, the lyase acts simultaneously as an isomerase (Fairchild and Glazer, 1994b; Schluchter and Glazer, 1999; Zhao *et al.*, 2002). *Secondly*, there are the phylogenetically unrelated S/U-type lyases that have a much broader specificity and attach PCB to all cysteine-84 binding sites of APC, CPC and PEC, with the exception of the α-84 sites served by E/F-type lyases; they, possibly, also serve the same sites of phycoerythrin (Shen *et al.*, 2004;

Zhao *et al.*, 2007a; Saunée *et al.*, 2008; Shen *et al.*, 2008b). *Thirdly*, the T-type lyases serve the β-155 binding sites of CPC and PEC and, possibly, other sites in phycoerythrins (Shen *et al.*, 2004; Shen *et al.*, 2006; Zhao *et al.*, 2007a). Additional genes, whose products have not been thoroughly tested, have been found for all three types of lyases (Kahn *et al.*, 1997; Six *et al.*, 2007; Zhao *et al.*, 2007a; Shen *et al.*, 2008b).

Until recently, no lyases have been found for the attachment of the PUB chromophore which is widespread in marine cyanobacteria and red algae. Recently, based on a genomic approach (Everroad *et al.*, 2006), a monomeric lyase has been identified that attaches PEB to Cys-α84 of a trichromatic R-PC and concomitantly isomerizes it to PUB (Blot *et al.*, 2009). Its N- and C-termini are most homologous to the two subunits of the isomerizing PecE/F lyase that generates PVB from PCB during the attachment reaction: the two lyases, when expressed in *E. coli*, are interchangeable. R-PC, however, contributes only little to blue light absorption, and it remains to be seen if PUB at other binding sites and, in particular, in the much more abundant PEs, is formed by similar lyases. Also unknown are the lyases of cryptophyte biliproteins.

9.4 Optical spectroscopy of phycobiliproteins

Originally, phycobiliproteins were classified according to their colour: for example, the blue phycocyanins ('cyanos' = Greek 'blue') absorb around 600 nm, the blue-green allophycocyanins ('allo' = Greek 'other') absorb around 650 nm, and the red phycoerythrins ('erythros' = Greek 'red') absorb around 550 nm. Due to the diversity of chromophores and the presence of chromophorylated linkers, this clear-cut distinction is no longer valid and has been supplemented by phylogenetic information (Sidler, 1994; Six *et al.*, 2007): the prefix 'C' characterizes cyanobacterial pigments, the prefix 'M' is often used for biliproteins from marine cyanobacteria and the prefix 'R' characterizes red-algal biliproteins, while 'B' or 'b' specifies members of the *Bangiales* order of red algae. Cryptophyte biliproteins, which carry a particularly varied range of chromophores, are generally characterized by their main absorption maxima, for example, PC-645 (Glazer and Wedemayer, 1995).

Typical phycobiliprotein spectra are shown in Figure 9.4. Spectra of phycoerythrins (see Figure 9.4A) cover the spectral range from ~475–580 nm. All spectra are structured, even in the case of C-PE (spectrum 3) which contains only a single chromophore type (PEB), thus confirming that spectral variations are caused by site-specific chromophore–protein interactions. The spectra become broader and more complex if PEB is supplemented by a second chromophore type, such as PUB (see spectrum 1 in Figure 9.4A). Figure 9.4B shows spectra of phycocyanins (spectra 4, 5 and 6) and an allophycocyanin (spectrum 7): these pigments contain mainly PCB as chromophores, supplemented with PVB in the case of PEC, and with

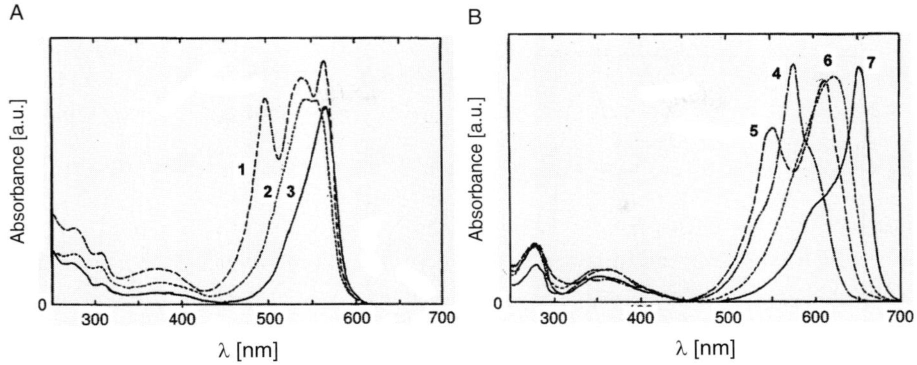

Figure 9.4. Type spectra of cyanobacterial and red algal phycoerythrins (A) (1 = R-PE, 2 = B-PE, 3 = C-PE), and of phycocyanins; (B) (4 = PEC, 5 = R-PC and 6 = C-PC) and an allophycocyanin (7). Modified from Mimuro and Kikuchi (2003).

PEB in the case of R-PC. The minimum pigmentation of phycobilisome-containing organisms is one (or more) PCs that make up the rods, and several APCs in the core. Such pigmentation is seen in some cyanobacteria, in the glaucocystophytes and in some red algae like *Galdieria sulphuraria*. In most organisms, it is supplemented by PEs or, less frequently, by PEC in the distal parts of the rods, thereby extending light-harvesting to the green spectral region. The spectral absorption of a PEC-containing cyanobacterium, *Mastigocladus laminosus*, is shown in Figure 9.5A, B. The absorption of the PVB chromophore at site α-84 of PEC is clearly seen as a shoulder around 575 nm, but its intensity is variable (see Section 9.4) and is particularly high in dense cultures grown under high CO_2 (Wiegand *et al.*, 2002). Spectra of the individual biliproteins (linker-free trimers) are shown in Figure 9.5B. The biliproteins from this freshwater cyanobacterium have only low absorption below 550 nm. By contrast, marine cyanobacteria such as *Synechococcus* WH8102, are rich in PEs with high PUB contents, which extend their phycobilisome absorption down to ~470 nm: the spectra of some of its pigments are shown in Figure 9.5C.

It has already been mentioned that the spectra of the bilin chromophores are extensively modified by non-covalent interactions with apoproteins; this is exemplified for one chromophore, PCB, in Figure 9.6A. These interactions are entirely due to non-covalent interactions since the spectra of denatured PCB-containing biliproteins are practically identical, both qualitatively and quantitatively, to those of free PCB, except for a blue-shift of ~20 nm caused by loss of the (generally 3-ethylidene) double-bond due to the cysteine addition (compare structures in Figure 9.2). The non-covalent protein chromophore interactions can be classified by at least seven mechanisms (see below), which generally occur in combination. The subject is only briefly summarized here, but for details the reader is referred to (Scheer, 2003)

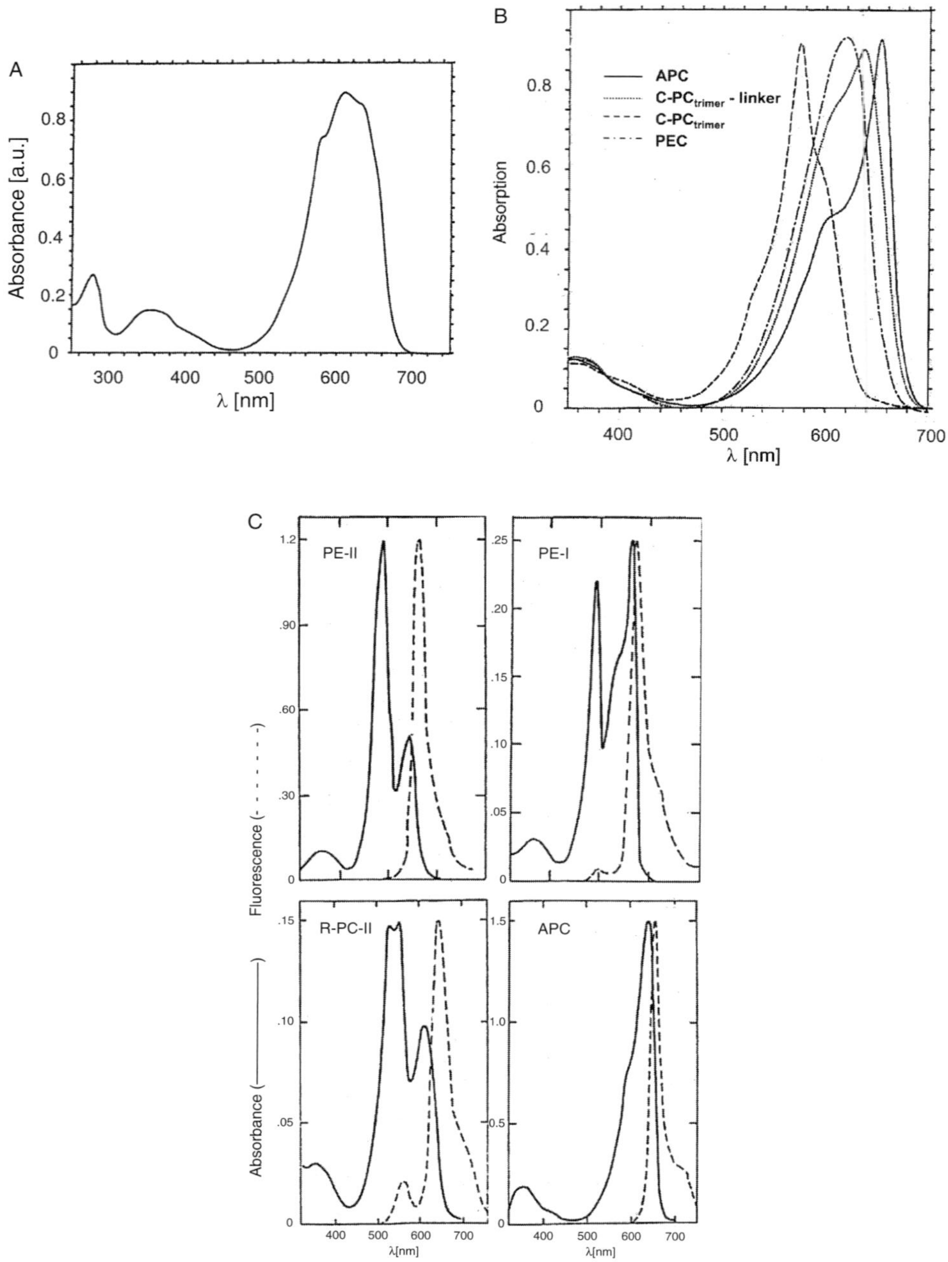

Figure 9.5. Spectrum of low-light grown phycobilisomes from the freshwater cyanobacterium, *Mastigocladus laminosus* (A), and examples of individual biliproteins contained in *Mastigocladus laminosus* (B), and in the marine cyanobacterium, *Synechococcus* WH8103 (C). Modified from Sidler (1994) and Ong and Glazer (1988). Fluorescence was excited at the short wavelength side of the lowest energy absorption band.

and references therein, and to the various crystal structures available in the protein database (http://www.rcsb.org/pdb/).

(1) Practically all bilin chromophores are protonated (Kneip *et al.*, 1998; Hahn *et al.*, 2007) by an aspartic acid that is spatially close to the nitrogen atoms of rings B and C, and conserved in all structures including even those of phytochromes (Schirmer *et al.*, 1987; Dürring *et al.*, 1990; Brejc *et al.*, 1995; Ritter *et al.*, 1999; Wang *et al.*, 2001; Adir *et al.*, 2002; Nield *et al.*, 2003). Its acidity is probably modulated by a nearby arginine residue.
(2) The chromophore conformation is extended (Figure 9.2B) (Schirmer *et al.*, 1987; Dürring *et al.*, 1990; Brejc *et al.*, 1995; Ritter *et al.*, 1999; Wang *et al.*, 2001; Adir *et al.*, 2002; Nield *et al.*, 2003); this is the major cause for the large absorption increase in native biliproteins, as compared to the denatured pigments (Figure 9.6A).
(3) The chromophores are held rather rigidly by the apoprotein, which is a major factor for extending the excited state lifetimes of the chromophores and, thereby, enhancing their function as photoreceptors (Schmidt *et al.*, 2007).
(4) During the attachment reaction, new asymmetric C-atoms are created that can differ in stereochemistry at different binding sites (Schirmer *et al.*, 1988; Adir *et al.*, 2002; Nield *et al.*, 2003).
(5) Interactions with specific amino acid residues at any particular binding site cause spectral heterogeneity which is important for light-harvesting. A special case of such interactions is the influence of post-translational methylation of asparagine-β72 (Dürring *et al.*, 1988; Swanson and Glazer, 1990; Thomas *et al.*, 1993; Adir and Lerner, 2003; Schluchter *et al.*, 2010).
(6) These interactions can be further modulated by the presence of linker proteins. This subject is only poorly understood: there is only a single high-resolution crystal structure known of a phycobilin-linker complex (Reuter *et al.*, 1999). In this case, the interaction is only indirect, by interaction of the linker with amino acids near the β-84 binding site.
(7) Last, but not least, there are also excitonic chromophore–chromophore interactions. Because of the relatively large inter-chromophore distances they are negligible in the protomers ('monomers'), but moderate couplings are obvious in the circular dichroism spectra of aggregates. An example is the C-PC 'trimer' (Debreczeny *et al.*, 1995; Kupka and Scheer, 2008), where chromophores on different 'monomers' are separated by ~2 nm (Schirmer *et al.*, 1987).

The variety of these interactions renders the biliprotein chromophores very versatile, because they allow for considerable modulation of their photophysical properties and, thereby, enhance their biological function. The extent of these modulations is exemplified for the bound PCB chromophore (Figure 9.6A) where the extinction coefficient varies by more than an order of magnitude, the absorption maxima shift over a range of ~100 nm and the excited state lifetimes (not shown) vary by more than four orders of magnitude.

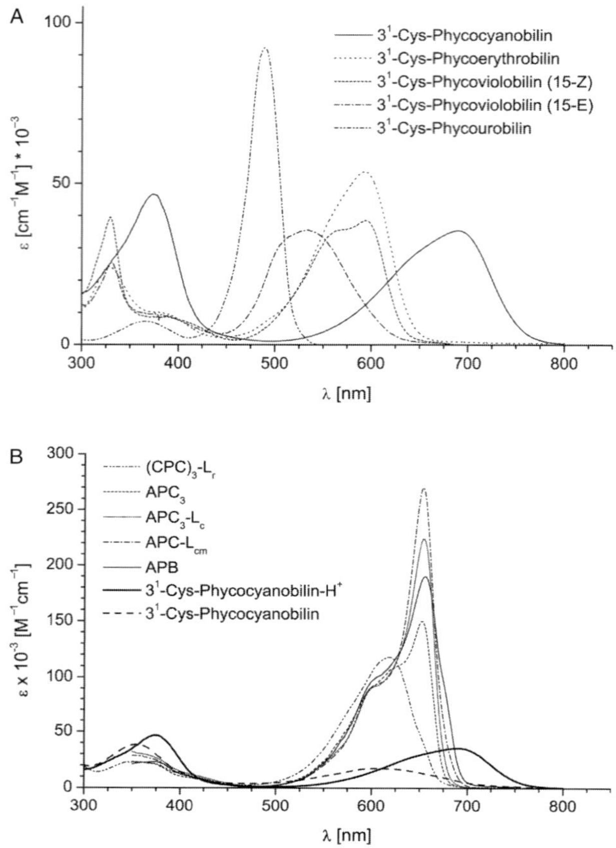

Figure 9.6. Spectral variations of the protein-bound PCB chromophore in different environments (A), and chromophore spectra of cyanobacterial phycobiliproteins in 8 M urea, pH 1.5 (B). Modified from Scheer (2003), with kind permission from Springer Science+Business Media.

A practical consequence of these variations is the difficulty in judging from the absorption spectra both the type and number of chromophores present in a phycobiliprotein. Since the variations are caused by non-covalent interactions, they are, however, abolished in the denatured biliproteins; denaturation is, therefore, the method of choice for obtaining qualitative and quantitative information of the chromophores present. However, even in the denatured state (as well as with free chromophores) the spectra can vary considerably depending on the solvent conditions. These spectra are particularly sensitive to pH (Scheer, 1976; Scheer, 1982; Braslavsky *et al.*, 1983; Falk, 1989); for example, protonation of PCB causes the long-wavelength band to be red-shifted by ~50 nm and its intensity to be doubled

Table 9.1. *Visible absorption maxima and extinction coefficients of the four cyanobacterial chromophores. Spectra refer to the singly cysteine-bound chromophores (via C-3^1) in biliproteins that are denatured in 8 M urea titrated to pH 1.9 with HCl, or to the respective bilipeptides in 10 mM trifluoroacetic acid. The values vary a little in the literature, the error limits are ± 10% or better.*

Chromophore	λ_{max} (vis) [nm]	ε [M^{-1} cm^{-1}]	Reference
PCB	663	35,500	(Glazer and Fang, 1973)
PVB	590	38,600	(Bishop *et al.*, 1987)
PEB	550	42,800	(Glazer and Hixson, 1975)
PUB	495	104,000	(Glazer and Hixson, 1977)

when compared to the neutral chromophore. Because both free phycobilins and denatured biliproteins are also very reactive, especially at high pH and/or in the presence of metal ions, the solvent used for spectral investigations of phycobiliprotein chromophores is of great practical importance: the solvent of choice is an acidic aqueous solution of 8 M urea, pH 1.5. Quantitative spectra of the four cyanobacterial chromophores in this solvent are shown in Figure 9.6B; all extinction coefficients relate to a single chromophore (see also Table 9.1).

9.5 Functions of phycobiliproteins

The major function of phycobiliproteins is light-harvesting for photosynthesis. This function is most thoroughly explored in the phycobilisome containing organisms, in particular, several cyanobacteria (see Glazer, 1985; Scheer, 2003 for leading references). The phycobilisome constitutes an energetic funnel, in which morphology and energetics are matched to allow for energy transfer from any one of the hundreds of chromophores to the reaction centres with quantum efficiencies near 100% (see Figure 9.1B). The high-energy chromophores (PUB, PEB, PVB) are located in the core-distal sections of the phycobilisome rods, the intermediate-energy chromophores (PCB) in the core-proximal sections of the rods, the low energy chromophores (PCB in APC) in the core, and the even lower energy chromophores (PCB in L_{CM} and APC-B) are nearest to the reaction centres in the membrane: the positioning of these chromophores of various energies is perfect for energy transfer to the reaction centers. The phycobilisome is, furthermore, an adaptive structure (Grossman *et al.*, 1993, 2003; Tandeau de Marsac *et al.*, 1993; Glazer, 1994;). Its composition and thereby, in general, its light-harvesting capacities can be regulated, for example, by the prevailing light-intensity and -quality, nutrient supplies and other factors. A particularly striking case is complementary chromatic adaptation[4] (Tandeau de Marsac, 1977) by which, for example, in *Tolypothrix tenuis* (= *Fremyella diplosiphon* = *Calothrix* PCC7601) the red phycoerythrins (i.e. green-light absorbing) are favoured by culturing in green light but are replaced by greenish-blue phycocyanins (i.e. red-light absorbing) in red-light cultures. Several other types of light acclimations have been characterized.

Nutrient acclimations and/or adaptations have been studied, in particular under limiting N- and S- concentrations. Because of the large number of amino acids required per chromophore (as compared to chlorophyll proteins, see Scheer, 2003), as well as biliproteins, which amount to 50% of the total proteins in cyanobacteria, phycobiliproteins are a nitrogen source that is rapidly degraded under N-starvation but readily re-synthesized if sufficient accessible N is supplied. Another case-study is the replacement, under S-limiting conditions, of the common apoproteins of biliproteins by homologues low in methionines and cysteines, other than those at binding sites (Mazel *et al.*, 1988; Mazel and Marlière, 1989).

Because of their low phototoxicity, phycobiliproteins are the only antenna pigments lacking protective carotenoids. However, a phycobilisome-associated carotenoid-protein in *Synechocystis* PCC 6803 has been characterized that may be involved in photoprotection and regulating energy transfer from phycobilisomes to the reaction centers in the photosynthetic membrane (Kirilovsky, 2007; Wilson *et al.*, 2008).

In contrast to phytochromes and related photoreceptors, phycobiliproteins are photochemically inactive, at least in their aggregated states, which comprise the phycobilisome. Residual photochemistry has been observed in many isolated biliproteins under disaggregating or partially denaturing conditions, probably due to imperfect fixation of the chromophores (Björn and Björn, 1980; Scheer, 1982). An exception is phycoerythrocyanin which, in its monomeric state, shows a pronounced photochemistry reminiscent of phytochrome (Zhao *et al.*, 1995; Schmidt *et al.*, 2007). This photochemistry is also due to a Δ15-Z/E-isomerization, but of the PVB chromophore attached to cysteine-α84, and there is presently no function associated with this transformation. The same chromophore is probably present in a phytochrome-like photoreceptor, PixJ from *Thermosynechococcus elongatus*, where it is involved in phototaxis regulation (Ishizuka *et al.*, 2007), and other sensory photoreceptors are emerging that carry bilin chromophores that are yet to be characterized (Kehoe and Gutu, 2006; Hirose *et al.*, 2008; Ikeuchi and Ishizuka, 2008). Furthermore, there is evidence of other roles for biliproteins, for example in Fe-homeostasis and regulation of tetrapyrrole biosynthesis (Franklin *et al.*, 2003; Frankenberg and Lagarias, 2003a; Dammeyer and Frankenberg-Dinkel, 2008). Additional functions are also indicated by the presence of phycobilin-reductases in phages infecting *Prochlorococcus* (Dammeyer *et al.*, 2008). This group of cyanobacteria uses chlorophylls for

[4] Strictly speaking, these responses should be termed acclimations, that is short-term physiological adjustments, while adaptations refer to long-term genetic adjustments for a given organism to survive in a given habitat. However, in particular for the acclimation to changing light conditions, the term 'complementary chromatic adaptation' (CCA) has been established, and therefore maintained in this review.

light-harvesting and lacks phycobilisomes, but often has small amounts of biliproteins. In view of the small quantities, and since they also lack APC, their contribution to light-harvesting seems minor. It may be that related biliproteins with yet unknown function(s) are also present in the 'classical' cyanobacteria, but these would be much more difficult to discern against the overwhelming amounts of light-harvesting biliproteins. If such biliproteins exist and act as photoreceptors, they would probably need to interact with signaling proteins which, in the much larger and more complex organized phytochromes and cyanobacteriochromes are already integrated in a single protein chain.

The function of cryptophyte phycobiliproteins is also light-harvesting for photosynthesis but much less is known about their organization or details of their energy transfer reactions. They lack phycobilisomes, they also lack allophycocyanins; generally, they have only a single phycobiliprotein that carries, however, an unusually large number of different chromophores. The cryptophyte biliproteins are located on the inner surface of the thylakoids. The hetero-tetrameric $\alpha\alpha'\beta_2$ protomer may already be the functional unit. Recent studies indicate that there is an efficient intra-protomeric energy transfer from high-energy to low-energy phycobilin chromophores present in the individual protomer and then on to the reaction centers, without the need for APC. This is true for both cryptophyte PEs and PCs, which both carry several different chromophores spanning a broad spectral range (van der Weij-De Wit *et al.*, 2006; van der Weij-De Wit *et al.*, 2008). Interestingly, the location of the chromophore at lowest energy in the $\alpha\alpha'\beta_2$ protomer seems to be variable (Wedemayer *et al.*, 1996).

9.6 Some useful information and procedures

The following information and procedures may be useful in characterizing phycobiliproteins present in phycobilisomes. Additional information of this type can be found in Glazer (1988), Schluchter and Bryant (2002), and Scheer (1985).

9.6.1 Spectroscopy tips

(1) Absorption spectra of native phycobiliproteins can be distorted, due to their intense fluorescence. The degree of distortion depends on the geometry of the spectrophotometer, but can be particularly high with diode array detectors and with spectrophotometers designed for measuring turbid samples. Distorted spectra are generally characterized by flattened tops to the absorption bands. When in doubt, it is recommended to run a series of dilutions and compare the normalized spectra: if dilution results in no further change of the spectral shape then the spectrum can be considered correct or undistorted. Dilution with phosphate buffer pH 7.5 ($\geq$ 50 mM) avoids dissociation of biliprotein aggregates, which would result in further spectral changes (see below).

(2) Fluorescence spectra can be distorted by self-absorption of the samples. They should be recorded at low concentrations (maximum absorption of the sample in the longest red band should be ≤ 0.05). Again, if in doubt, a dilution series is recommended, using phosphate buffer pH 7.5 (≥ 50 mM) to avoid dissociation of oligomers.
(3) Circular dichroism spectra should be registered with samples that have maximum absorptions ≤ 0.8 in the region of interest. For measurements in the far UV (e.g. when evaluating the secondary structure of the protein), it is recommended to use cuvettes with 1 mm path length. Usually, samples that are suitable for measurement in the visible spectral range ($A_{max} = \sim 0.8$ in a 1 cm cuvette), can be transferred without further dilution to a 1 mm cuvette for good results in the far-UV.
(4) Phycobilisomes are stabilized by high concentrations of phosphate buffer (pH 7.5). In most cases, 0.75 M is sufficient but even higher concentrations may be necessary in cases such as the phycobilisome of *Mastigocladus laminosus* which commence disaggregation at a concentration of 1 M.
(5) Phycobiliprotein aggregates (e.g. trimers) can dissociate at high dilution and low ionic strength, but only partly. Pure monomers can be obtained in 4 M urea.

9.6.2 Phycobilisome isolation

Cells of cyanobacteria, or phycobilisome-containing (micro)algae, are suspended in high-ionic strength potassium phosphate buffer pH 7.5 (≥ 0.75 M) and broken by multiple passages through a French press (Glazer, 1988). The high buffer concentration has to be maintained throughout the subsequent isolation process. Cell debris is removed from the resulting extract by low-speed centrifugation ($\sim$10,000 $\times$ g). Intact phycobilisomes are then isolated by sucrose-density centrifugation in an ultracentrifuge: they concentrate in a dense band at high sucrose concentration and are generally well separated from the less dense fragments. When there is no well-defined band, a second sucrose density centrifugation is recommended. Some phycobilisomes already dissociate in 0.75 M phosphate buffer (an example is *Mastigocladus laminosus)*; in this case, the isolation should be attempted at higher buffer concentrations.

9.6.3 Phycobilisome dissociation and separation of individual biliproteins

Phycobilisomes dissociate readily to 'trimers and 'hexamers' when the ionic strength of the phosphate buffer is reduced to 50 mM: the resulting mixture can be fractionated by ion-exchange chromatography. In the case of complex mixtures, preparative isoelectric focusing is the method of choice (Köst-Reyes *et al.*, 1988; Reuter *et al.*, 1999). While many linker proteins are retained in the process, larger aggregates can be isolated at intermediate ionic strength, but this requires considerable fine-tuning of the dissociation conditions and the subsequent fractionations (see Gottschalk *et al.*, 1991; Glauser *et al.*, 1993; Reuter *et al.*, 1999). Isolated fractions can be precipitated by and stored in $(NH_4)_2SO_4$ (70% w/v) at 6°C. For removal of linker proteins, individual fractions are pelleted by centrifugation (26,000 × g), resuspended in distilled water and desalted over Sephadex G25. Solid urea is then added to a final concentration of 4 M, the solution kept for 1 h at room temperature and then chromatographed over a Sepharose column (potassium phosphate buffer, 50 mM, pH 7). The desired fractions are desalted and precipitated as above.

9.6.4 Isolation of biliprotein subunits

Biliprotein subunits can only be reproducibly separated under denaturing conditions. Two alternative methods are available. The first is preparative isoelectric focusing in the presence of urea (8 M) (Köst-Reyes *et al.*, 1988; Zhao *et al.*, 1995; Parbel *et al.*, 1997). The isoelectric point of most subunits is between 5 and 6. Since the chromophores are prone to oxidative degradation under these near neutral conditions, the procedure has to be carried out under an inert gas, preferably argon, which is denser than oxygen. An alternative method is ion-exchange chromatography in dilute formic acid (Füglistaller *et al.*, 1983); however, in our hands, separations by this method are less reliable.

9.6.5 Identification of individual chromophores by denaturation in acidic urea

As explained in Section 9.4, the spectroscopic properties of bilin chromophores are extensively modulated in the native protein complexes. A reliable solvent to obtain characteristic absorption spectra of chromophores, uncoupled from the protein, is aqueous urea (8 M) after titration to pH 1.5 with concentrated HCl. Under these conditions, the proteins are fully denatured, and the chromophores protonated. Absorption maxima and extinction coefficients are shown in Figure 9.6B and are also listed in Table 9.1.

9.6.6 Extinction coefficients of native chromoproteins

Three methods are available (Storf *et al.*, 2001) to determine extinction coefficients of known chromophores in newly isolated, modified or reconstituted biliproteins. In each case, the absorption spectrum of the solution of the native (or reconstituted) biliprotein is recorded before applying one of the three treatments described below. Extinction coefficients of the native chromophores are then determined relative to the known chromophore in acid (Table 9.1), after correcting for any volume changes during the treatments. The method is applicable only for chromophores with known extinction coefficients, and may require spectral deconvolution if more than one chromophore type is present in the biliprotein. Each determination should be repeated at least three times.

(1) In method 1, the biliprotein solution is treated three times with 1 mg of TPCK (tosyl phenylalanyl chloromethyl ketone)-treated trypsin over a period of 16 h at ambient temperature and pH 6.5. The solution is subsequently acidified to pH 1.5 by addition of HCl and the spectrum is again recorded. The chromophore must be protected from oxidation during proteolysis, for example, under argon.
(2) In method 2, the chromoprotein is acidified with formic acid (1%), then treated three time with 5 mg of pepsin over a period of 16 h at ambient temperature and, finally, the spectrum is recorded again. Chromophore oxidation is less likely under the acidic conditions used.
(3) In method 3, the chromoprotein is denatured by adding urea (to 8 M) and HCl (to pH 1.5) and, subsequently, the spectrum recorded 15 min later. This is generally our method of choice, as acid conditions reduce chromophore oxidation and also because it is relatively fast, thereby reducing the time that the chromophore is exposed, unprotected by the native protein, and hence is in a vulnerable state. This method, however, may not be applicable where a chromophore is doubly linked to the protein and, therefore, potentially not fully uncoupled.

9.6.7 Identification of covalently bound chromophores in SDS-polyacrylamide gels by Zn^{++}-staining and fluorescence detection

This method (Berkelman and Lagarias, 1986; Raps, 1990) relies on the formation of fluorescent Zn^{++}-complexes with chromophores that are uncoupled from the protein, even if still covalently attached to it and, therefore, adopt an otherwise non-fluorescent cyclic-helical conformation. The present authors are not aware of any publication addressing potential problems of the method with doubly linked chromophores. Denaturing polyacrylamide gel electrophoresis is performed according to Lämmli (1970). The gel is then incubated for 30 min in an aqueous solution of Zn-acetate (1.3 M), washed in double-distilled water and placed under a UV lamp (365 nm). Bilin chromophores show orange fluorescence, which can be photographically recorded. The gel is then conventionally protein-stained, for example, with Coomassie blue, and the two records compared, or overlaid.

9.6.8 Identification of binding sites by tryptic digestion followed by mass spectrometry

Currently, this is the best method to determine binding site sequences and molecular masses of chromophores. In our hands, tryptic digest is by far superior to peptic digestion, especially if care is taken to avoid any degradation of the chromophore. It is reliable with cysteine-bound chromophores, which is the universal mode of chromophore binding in phycobiliproteins. However, more labile bonds may be encountered, for example, in enzymes that transiently bind chromophores (Tu *et al.*, 2008) and, in these cases, the acidic conditions required in mass spectroscopy (both MALDI and ESI ionization) may be too harsh.

We have obtained good results with the following procedure (Zhao *et al.*, 2007a,b): chromoproteins (10 μM) are digested with trypsin (40 μM) in potassium phosphate buffer (100 mM, pH 7.0) for several hours. After desalting with Sep-Pak cartridges, the digest is fractionated by HPLC, using a C_{18} RP column and gradient elution (with solvent ratio A:B from 80:20 to 60:40; where solvent A = 0.1% formic acid and solvent B = acetonitrile containing 0.1% formic acid). The isolated chromopeptides are analyzed by mass spectrometry in positive ion mode with a nano-ESI source. The chromophore can be seen by MS-MS spectroscopy. In the case of PCB, the molecular ion is prominent, while ring D is readily lost with PEB, resulting in a $(M\text{-}121)^+$ peak (unpublished).

9.6.9 Isolation of chromophores (O'Carra and Oh'Eocha, 1966)

All reactions should be carried out under inert gas (e.g. argon), and the isolated pigments are stored under argon at −20°C. For extended storage, ampoules sealed under vacuum are recommended.

Phycocyanobilin (PCB): Arthrospira (Spirulina) platensis (spray-dried, 40 g) is washed with 500 mL hot methanol for 15 min and then filtered: this is repeated three times to extract most of the chlorophyll and carotenoids. The extracted bacteria are then refluxed, with stirring, in methanol (400 mL) containing sodium ascorbate (1% w/v) for 16 h under argon. The resulting suspension is filtered under reduced pressure. The blue filtrate is then mixed with dilute HCl (0.1 M, 400 mL), prior to extracting three times with diethylether (100 mL) to remove residual chlorophyll *a* and carotenoids. PCB is then extracted from the residual aqueous methanol/HCl solution with $CHCl_3$ (4 times), and the blue extract dried with NaCl prior to removal of the $CHCl_3$ in a rotary evaporator. It is purified by open-column (flash) chromatography on reversed phase-bound silica C_8 (ICN) using 50 mM potassium phosphate buffer, pH 2.1 (60%) containing 2-propanol (40%). Lyophilized bacteria (40 g) yield ~60 mg of crude PCB (55% yield with respect to the phycocyanin plus allophycocyanin content), and 27 mg (25% yield) after further purification. See below for an alternative method that is particularly applicable to small-scale extraction.

Phycoerythrobilin (PEB) is similarly isolated from dried *Porphyra yezoensis* (available in Asian food shops). The algae (50 g) are cut to small pieces, suspended in water (1 l, 30 min), and then pressed in two layers of filter cloth to remove the water. The wet mass is washed three times with hot methanol (1 l), and the residue refluxed for 16 h in methanol (1.6 L) containing sodium ascorbate (1% w/v). The resulting dark-red slurry is filtered under reduced pressure, the filtrate concentrated to 400 mL in a rotary evaporator and then purified as described above for PCB, but using 50 mM potassium phosphate buffer, pH 2.1 (70%) containing 2-propanol (30%). Note that the crude extract also contains PCB, which is removed by the chromatography step.

Phytochromobilin (PΦB) is prepared from PEB by oxidation with $FeCl_3$ (M. Maisch and W. Rüdiger, private communication). PEB (15 mg) is dissolved in dimethylsulphoxide (30 mL). CH_2Cl_2 (30 mL) is added and the solution gassed with argon at 65°C for 30 min. Then 750 μL of a $FeCl_3$ solution (2.5 mg mL^{-1}) in concentrated HCl are added, and the mixture refluxed (65°C) under argon for 1.5 h. After cooling to ambient temperature, dilute HCl (200 mL, 0.1 M) is added and this solution is extracted three times with CH_2Cl_2 (100 mL). The combined CH_2Cl_2 extracts are dried with NaCl, the solvent evaporated in a rotary evaporator and the pigment purified by chromatography (see PCB preparation) using 50 mM potassium phosphate buffer, pH 2.1 containing 2-propanol (35%).

Biliverdin (BV) is prepared from commercial bilirubin (BR) by oxidation with 2,3-dichloro-5,6-dicyano-*p*-benzoquinone (Manitto and Monti, 1979). Bilirubin (Fluka) (50 mg) is dissolved in dimethylsulphoxide (30 mL) and gassed with argon for 15 min. Ten mL of a solution of 2,3-dichloro-5,6-dicyano-*p*-benzoquinone in dimethylsulphoxide (5 mg mL^{-1}) are slowly added over a period of 15 min. After a further 15 min, this mixture is partitioned in a separatory funnel between $CHCl_3$ (500 mL) and water (350 mL). The $CHCl_3$ phase is washed three times with water, dried with NaCl, and the solvent removed in a rotary evaporator. The pigment is then purified by chromatography on silica plates (silica H, 0.75 mm, 20 × 20 cm) using the upper phase of a toluene–acetic acid–water mixture (5:5:1). The dark green band containing BV is scraped off, the pigment eluted with acetic acid prior to transferring to a separating funnel and partitioning between water and $CHCl_3$. The $CHCl_3$ phase is dried with NaCl, and the solvent removed in a rotary evaporator.

PVB and PUB These free bilins are unknown in organisms containing the respective chromoproteins. Protein-bound PVB is generated from free PCB by isomerization during the lyase-catalyzed attachment reaction, and protein-bound PUB, at least in one case, has been shown to be generated by a similar lyase-catalyzed isomerization from free PEB (see Section 9.3). These free bilins can not be obtained by the methods described above for chromophore cleavage from biliproteins. Free PVB is obtained by a modification of the method used to split the heme prosthetic group from cytochrome c (Fisher *et al.*, 1973), that is, by treating α-PEC with Ag_2SO_4 (Storf, 2004).

α-PEC (5 mg) is dissolved in water (3 mL). This solution is added to a solution of Ag_2SO_4 (80 mg) in water/acetic acid (8 mL, 9:1 v/v), and the mixture incubated at 40°C under argon for 2 h. After mixing with CH_2Cl_2 (10 mL) and centrifugation (10,000 × g, 10 min), the CH_2Cl_2 phase is dried with NaCl, the solvent removed in a rotary evaporator. PVB is then purified by chromatography (see PCB). No method has been published for generating free PUB from biliproteins.

Extraction of chromophores from biliproteins (modified from Schram and Kroes, 1971). Since methanol extraction is not suitable for small-scale extractions, trifluoroacetic acid (TFA) in hydrobromic acid (HBr) is recommended. 50 mg of the biliprotein are dissolved in TFA (30 mL), and gaseous HBr introduced for 30 min through a Pasteur pipette such that individual bubbles can be well distinguished. After mixing with CH_2Cl_2 (5 mL) and centrifugation (10,000 × g, 10 min), the CH_2Cl_2 phase is dried with NaCl and the solvent removed in a rotary evaporator. The free chromophores are isolated by chromatography (see PCB). This method can be downscaled.

Dimethylesters of bilins are prepared by refluxing with boron trifluoride in methanol. The free di-acid, for example PCB (45 mg), is dissolved in warm argon-saturated methanol (30 mL) and 11 mL of BF_3 in methanol (20%) are added prior to flushing the mixture with argon and then refluxing for 20 min. After cooling, the mixture is shaken in a separating funnel with $CHCl_3$ (30 mL) and water (50 mL). The $CHCl_3$ phase is washed three times with water, dried with NaCl and the solvent removed in a rotary evaporator. The resulting dimethylester is purified on silica gel (60H, Merck) using $CHCl_3$/2-propanol (98:2) as solvent.

Hydrolysis of dimethylesters is difficult due to the lability of bilins. The following method has been reported by Lindner *et al.* (1998) and detailed by Lindner (2000). It is used for preparative purposes, and needs downscaling for smaller samples. In this case, purification would be done by chromatography while carefully protecting from oxygen and light (see PCB). Ion exchange resin (8 g, Dowex 50WX8–200, Aldrich) is added to a solution of the bilin dimethylester (12.4 mg) in TFA/water (20 mL, 1:1 v/v). This mixture is stirred in the dark at ambient temperature; the reaction is followed by thin-layer chromatography (RP-C18, acetonitrile/water 1:1 v/v). On completion of hydrolysis after approximately 48 h, the resin is removed by filtration through sintered glass (G4) and washed alternately with water and with a $CHCl_3$/methanol mixture (98:2 v/v) until the resin is colourless. The combined blue filtrate is washed with a saturated solution of NaCl in water until the pH of the aqueous phase is 6–7, dried over Na_2SO_4 and the solvent removed in a rotary evaporator. The blue residue is then crystallized from $CHCl_3$/n-hexane (98:2 v/v) at −21°C.

9.6.10 Reconstitution of phycobiliproteins from chromophores and apoproteins

Four successful methods for reconstitution of phycobiliproteins are described below:

(1) *Autocatalytic in vitro reconstitution* of full length and truncated core-membrane linker, L_{CM}, from His-tagged apoprotein, ApcE, is expressed in *E. coli* (Zhao *et al.*, 2005).

ApcE (20 μM) in potassium phosphate buffer (20 mM, pH 7.2) containing urea (4 M), sucrose (0.2 M), NaCl (0.1 M) and mercaptoethanol (0.1 mM) is mixed with PCB (20 μM) in the same buffer, and incubated in the dark for 3 h at room temperature, or for 30 min at 35°C. Urea is required in the reconstitution mixture to keep the apoproteins in solution. Reconstitution yields are ~70% (based on PCB content). The mixture is then centrifuged for 15 min at 15,000 × g to remove any particulate matter before investigating the supernatant by UV-Vis absorption, fluorescence and circular dichroism (CD) spectroscopy. For spectral measurements, the reconstituted product, L_{CM}, is purified by a second Ni-affinity chromatography. After elution from the affinity column, the samples are dialyzed twice against the respective buffer containing mercaptoethanol (1 mM) and EDTA (5 mM).

(2) *In vitro reconstitution of C-PC α-subunit catalyzed by the lyase, CpcE/F* (Zhao *et al.*, 2006a, b).

CpcE and CpcF (5–20 μM each) and His_6-tagged CpcA (5–20 μM) are dissolved in potassium phosphate buffer (15–20 mM, pH 7.2) containing NaCl (150 mM) and $MgCl_2$ (5 mM). PCB (final concentration 5–20 μM) is added as a concentrated dimethylsulphoxide solution: the final concentration of dimethylsulphoxide is 1% (v/v). The mixture is incubated at 37°C for 1 h in the dark and the product isolated and purified as described immediately above (see (1)) before quantification by fluorescence emission at 640 nm (Fairchild and Glazer, 1994b).

(3) *In vitro reconstitution of PEC α-subunit catalyzed by the lyase, PecE/F* (Zhao *et al.*, 2002).

This procedure produces photoreversibly active α-PEC that is almost free of the PCB-addition product, P641. His_6-PecA (20 μM), PecE (20 μM) and PecF (10 μM) are dissolved in a mixture of potassium phosphate buffer (20 mM) and Tris/HCl buffer (100 mM; pH 7.5) containing NaCl (150 mM), 2-mercaptoethanol (5 mM), $MnCl_2$ (3 mM) and Triton X-100 (1%). PCB (final concentration 25 μM) is added in a concentrated dimethylsulphoxide solution (final dimethylsulphoxide concentration $\leq$ 1%) and the mixture incubated at 37°C for 1 h. The product is isolated and purified as in (1) and quantified by the reversible photochemistry of α-PEC (Zhao *et al.*, 1995).

(4) *Reconstitution in E. coli of APCs and of CpcB and PecB at cysteine-84 binding sites, catalyzed by the lyase, CpcS* (Zhao *et al.*, 2007a).

This reconstitution is suitable for poorly soluble apoproteins (or lyases) and it also reduces the spontaneous, but faulty, chromophore addition. By replacing CpcS by CpcT, reconstitution of PCB can be carried out to the cysteine-155 binding sites of CpcB or PecB. For cloning strategies, see original literature, such as (Tooley and Glazer, 2002; Zhao *et al.*, 2007a). The pET-plasmid containing the acceptor protein is transformed together with one containing the genes for PCB biosynthesis (*hol* and *pcyA*) and with pCDF containing the lyase gene, cpcS, into *E. coli* BL21(DE3). Cells are grown at 20°C in LB medium containing kanamycin (20 μg mL^{-1}), chloromycetin (17 μg mL^{-1}), and streptomycin (25 μg mL^{-1}). After a 12 h induction period with isopropyl β-D-thiogalactoside (1 mM), the cells are collected by centrifugation, washed twice with double-distilled water, and stored at −20°C ready for use. Cells are lysed and the tagged proteins isolated by Ni^{++}-chromatography (Zhao *et al.*, 2006a). If necessary, the affinity enriched proteins can be further purified by FPLC (Amersham-Pharmacia) over a Superdex 75 column developed with potassium phosphate buffer (50 mM, pH 7.0) containing NaCl (150 mM), or over a DEAE-fast flow column developed with a gradient (0–1 M NaCl) in potassium phosphate buffer (20 mM, pH 7.0).

9.6.11 Analysis of phycobiliproteins

The major in vivo absorption maxima of phycobiliproteins (~500 to ~630 nm) are in a spectral region that has little overlap with other photosynthetic pigments. The only exceptions are Chl*c* and carotenoids with red-shifted spectra. A distinction from the former is straightforward because phycobiliproteins lack the intense Soret band (between 440 and 455 nm) that is characteristic of the *c*-type chlorophylls. A distinction from carotenoids can be difficult in vivo, in particular with strains that are rich in phycourobilin; a situation that is common in marine environments. In this case, fluorescence spectroscopy can be helpful, because individual biliproteins show, even in phycobilisomes, weak fluorescence signals while carotenoids are silent. Six *et al.* (2007) described three pigment types of marine *Synechococcus* species containing mainly PC (λ_{max} ~630 nm), PE I (λ_{max} ~565 nm), or PE II (λ_{max} ~545 and ~500 nm). The relative intensities of the two PE peaks can vary considerably, furthermore, acclimatization to the prevailing light climate and other environmental conditions vary the absorption spectra of cyanobacteria still further.

For a more detailed investigation it is recommended to study isolated phycobilisomes (Section 9.6.2) or the mixture of biliproteins (Section 9.6.3) by fluorescence spectroscopy. The water solubility, lack of a Soret band, and brilliant fluorescence are safe diagnostic characteristics of biliproteins. The excitation spectra for phycocyanin (λ_{em} ~650 nm; no excitation peak < 580 nm) and the phycoerythrins (PE I, λ_{em} ~575 nm; PE II, λ_{em} ~565 nm; no excitation peaks < 470 nm) already give good information without a further separation. For a more detailed analysis, the individual biliproteins have to be separated, for example by ion-exchange chromatography (Section 9.6.3) and investigated individually both in the native and the denatured state (Sections 9.6.5 and 9.6.6). Denaturation is, in particular, necessary for quantitative analysis of the chromophores present in the different biliproteins.

9.7 Concluding remarks

In the first part of this chapter, Sections 9.1 to 9.5, we have described the general properties of bilins, biliproteins and their functional photosynthetic complexes. In the second part, Sections 9.6.1 to 9.6.9, we have described methods to break down the phycobilisome complex and biliproteins into their component parts for the purpose of identification. For the sake of completeness, Section 9.6.10 describes various methods to reassemble the individual components to form the phycobilisome, as potential tools to experimentally approach the assembly of the photosynthetic phycobilisome complex.

Acknowledgements

Work of the authors was funded by the Volkswagen Stiftung (Partnership grant I/77900) (HS and KHZ), the Deutsche Forschungsgemeinschaft (SFB 533, TPA1) (HS), the National Natural Science Foundation of China for the grant (30670489) (KHZ), and the Program for New Century Excellent Talents in University for the grant (NCET-04–0717, P.R. China) (KHZ). R.J. Porra thanks CSIRO-Plant Industry (Canberra) for an Honorary Fellowship and for the use of the Black Mountain Library and CSIRO communication facilities.

Abbreviations

1 Chromophores

BV	biliverdin IXα
DBV	15,16-dihydrobiliverdin
MBV	mesobiliverdin
PCB	phycocyanobilin
PEB	phycoerythrobilin
PΦB	phytochromobilin
PUB	phycourobilin
PVB	phycoviolobilin (also known as phycobiliviolin)

2 Proteins

APC	Allophycocyanin
ApcE	apoprotein of core-membrane linker L_{CM}
CpcA, CpcB	α- and β-apoproteins of cyanobacterial PC (for PC see below)
CpcE/F	α-CPC lyase
CpeA, CpeB	α- and β-apoproteins of cyanobacterial PE (for PE see below)
CpcS/CpeS	phycobilin:Cys-84-phycobiliprotein lyase
CpcT/CpeT	phycobilin:Cys-155-phycobiliprotein lyase
HO1	heme oxygenase 1
PC	phycocyanin (prefixes stand for B: bangiales, C: cyanobacteria, M: marine (cyanobacteria), R: rhodophyte)
PcyA	(3Z)-phycocyanobilin:ferredoxin oxido-reductase
PE	phycoerythrin (see PC for prefixes)
PebA	15,16-dihydrobiliverdin:ferredoxin oxido-reductase
PebB	3Z-phycoerythrobilin:ferredoxin oxido-reductase
PEC	phycoerythrocyanin
PecE/F	phycoviolobilin:α-PEC lyase

References

Adir, N. (2005). Elucidation of the molecular structures of components of the phycobilisome: reconstructing a giant. *Photosynth. Res.* **85**, 15–32.

Adir, N. and Lerner, N. (2003). The crystal structure of a novel unmethylated form of c-phycocyanin, a possible connector between cores and rods in phycobilisomes. *J. Biol. Chem.* **278**, 25926–32.

Adir, N., Vainer, R. and Lerner, N. (2002). Refined structure of c-phycocyanin from the cyanobacterium *Synechococcus vulcanus* at 1.6 Å: insights into the role of solvent molecules in thermal stability and co-factor structure. *Biochim. Biophys. Acta* **1556**, 168–74.

Anderson, L.K. and Toole, C.M. (1998). A model for early events in the assembly pathway of cyanobacterial phycobilisomes. *Mol. Microbiol.* **30**, 467–74.

Arciero, D.M., Bryant, D.A. and Glazer, A.N. (1988). In vitro attachment of bilins to apophycocyanin. 1. Specific covalent adduct formation at cysteinyl residues involved in phycocyanobilin binding in C-phycocyanin. *J. Biol. Chem.* **263**, 18343–49.

Beale, S.I. (1993). Biosynthesis of phycobilins. *Chem. Rev.* **93**, 785–802.

Beale, S.I. (1994). Biosynthesis of cyanobacterial tetrapyrrole pigments: Hemes, chlorophylls and phycobilins. In *The Molecular Biology of Cyanobacteria*, ed. D.A. Bryant. Dordrecht: Kluwer, pp. 519–58.

Berkelman, T. and Lagarias, J.C. (1986). Visualization of bilin-linked peptides and proteins in polyacrylamide gels. *Anal. Biochem.* **156**, 194–201.

Bishop, J.E., Rapoport, H., Klotz, A.V., Chan, C.F., Glazer, A.N., Füglistaller, P., and Zuber, H. (1987). Chromopeptides from phycoerythrocyanin. Structure and linkage of the three bilin groups. *J. Am. Chem. Soc.* **109**, 875–81.

Björn, L.O. and Björn, G.S. (1980). Photochromic pigments and photoregulation in blue-green algae. *Photochem. Photobiol.* **32**, 849–52.

Blot, N., Wu, X.J., Thomas, J.C., Zhang, J., Garczarek, L., Böhm, S., Tu, J.M., Zhou, M., Ploscher, M., Eichacker, L., Partensky, F., Scheer, H. and Zhao, K. H. (2009). Phycourobilin in trichromatic phycocyanin from oceanic cyanobacteria is formed post-translationally by a phycoerythrobilin lyase-isomerase. *J. Biol. Chem.* **284**, 9290–98.

Braslavsky, S.E., Holzwarth, A.R. and Schaffner, K. (1983). Solution conformations, photophysics, and photochemistry of bili-pigments – bilirubin and biliverdin dimethylesters and related linear tetrapyrroles. *Angew. Chem. Int. Ed. Engl.* **22**, 656–74.

Brejc, K., Ficner, R., Huber, R. and Steinbacher, S. (1995). Isolation, crystallization, crystal structure analysis and refinement of allophycocyanin from the cyanobacterium *Spirulina platensis* at 2.3 Å resolution. *J. Mol. Biol.* **249**, 424–40.

Böhm, S., Endres, S., Scheer, H. and Zhao, K.-H. (2007). Biliprotein chromophore attachment: chaperone-like function of PecE-subunit of α-phycoerythrocyanin lyase. *J. Biol. Chem.* **282**, 25357–66.

Cornejo, J. and Beale, S.I. (1997). Phycobilin biosynthetic reactions in extracts of cyanobacteria. *Photosynth. Res.* **51**, 223–30.

Cornejo, J., Willows, R.D. and Beale, S.I. (1998). Phytobilin biosynthesis: cloning and expression of a gene encoding soluble ferredoxin-dependent heme oxygenase from *Synechocystis* sp. PCC 6803. *Plant J.* **15**, 99–107.

Dammeyer, T. and Frankenberg-Dinkel, N. (2006). Insights into phycoerythrobilin biosynthesis point toward metabolic channeling. *J. Biol. Chem.* **281**, 27081–89.

Dammeyer, T. and Frankenberg-Dinkel, N. (2008). Function and distribution of bilin biosynthesis enzymes in photosynthetic organisms. *Photochem. Photobiol. Sci.* **7**, 1121–30.

Dammeyer, T., Bagby, S.C., Sullivan, M.B., Chisholm, S.W. and Frankenberg-Dinkel, N. (2008). Efficient phage-mediated pigment biosynthesis in oceanic cyanobacteria. *Curr. Biol.* **18**, 442–48.

Debreczeny, M.P., Sauer, K., Zhou, J. and Bryant, D.A. (1995). Comparison of calculated and experimentally resolved rate constants for excitation energy transfer in C-phycocyanin. 2. trimers. *J. Phys. Chem.* **99**, 8420–31.

Doust, A.B., Marai, C.N.J., Harrop, S.J., Wilk, K.E., Curmi, P.M.G. and Scholes, G.D. (2004). Developing a structure–function model for the cryptophyte phycoerythrin 545 using ultrahigh resolution crystallography and ultrafast laser spectroscopy. *J. Mol. Biol.* **344**, 135–53.

Dürring, M., Huber, R. and Bode, W. (1988). The structure of gamma-N-methylasparagine in C-phycocyanin from *Mastigocladus laminosus* and *Agmenellum quadruplicatum*. *FEBS Lett.* **236**, 167–70.

Dürring, M., Huber, R., Bode, W., Rümbeli, R. and Zuber, H. (1990). Refined three-dimensional structure of phycoerythrocyanin from the cyanobacterium *Mastigocladus laminosus* at 2.7 Å. *J. Mol. Biol.* **211**, 633–44.

Essen, L.-O., Mailliet, J. and Hughes, J. (2008). The structure of a complete phytochrome sensory module in the Pr ground state. *Proc. Natl. Acad. Sci. USA* **105**: 14709–14.

Everroad, C., Six, C., Partensky, F., Thomas, J.C., Holtzendorff, J. and Wood, A. M. (2006). Biochemical bases of type IV chromatic adaptation in marine *Synechococcus* spp. *J. Bacteriol.* **188**, 3345–56.

Fairchild, C.D. and Glazer, A.N. (1994a). Nonenzymatic bilin addition to the gamma subunit of an apophycoerythrin. *J. Biol. Chem.* **269**, 28988–96.

Fairchild, C.D. and Glazer, A.N. (1994b). Oligomeric structure, enzyme kinetics, and substrate specificity of the phycocyanin alpha subunit phycocyanobilin lyase. *J. Biol. Chem.* **269**, 8686–94.

Falk, H. (ed.) (1989). *The Chemistry of Linear Oligopyrroles and Bile Pigments.* Vienna: Springer.

Ficner, R. and Huber, R. (1993). Refined crystal structure of phycoerythrin from *Porphyridium cruentum* at 2.3-nm resolution and localization of the γ subunit. *Eur. J. Biochem.* **218**, 103–06.

Fisher, W.R., Taniuchi, H. and Anfinsen, C.B. (1973). On the role of heme in the formation of the structure of cytochrome c. *J. Biol. Chem.* **248**, 3188–95.

Frankenberg, N. and Lagarias, J.C. (2003a). Biosynthesis and biological functions of bilins. In *The Porphyrin Handbook*, ed. K.M. Kadish, K.M. Smith and R. Guilard. Amsterdam: Academic Press, pp. 211–36.

Frankenberg, N. and Lagarias, J.C. (2003b). Phycocyanobilin:ferredoxin oxidoreductase of *Anabaena sp.* PCC 7120. Biochemical and spectroscopic. *J. Biol. Chem.* **278**, 9219–26.

Frankenberg, N., Mukougawa, K., Kohchi, T. and Lagarias, J.C. (2001). Functional genomic analysis of the HY2 family of ferredoxin-dependent bilin reductases from oxygenic photosynthetic organisms. *Plant Cell* **13**, 965–78.

Franklin, K.A., Linley, P.J., Montgomery, B.L., Lagarias, J.C., Thomas, B., Jackson, S.D. and Terry, M.J. (2003). Misregulation of tetrapyrrole biosynthesis in transgenic tobacco seedlings expressing mammalian biliverdin reductase. *The Plant J.* **35**, 717–28.

Füglistaller, P., Suter, F. and Zuber, H. (1983). The complete amino-acid-sequence of both subunits of phycoerythrocyanin from the thermophilic cyanobacterium *Mastigocladus laminosus. Hoppe Seyler's Z. Physiol. Chem.* **364**, 691–712.

Gantt, E. (1986). Phycobilisomes. In *Photosynthesis III: Photosynthetic Membranes and Light-Harvesting Systems*, ed. L.A. Staehelin and C.J. Arntzen. Berlin: Springer, pp. 260–68.

Gantt, E. and Conti, S.F. (1966). Granules associated with the chloroplast lamellae of *Porphyridium cruentum. J. Cell Biol.* **29**, 423–34.

Gantt, E., Edwards, M.R. and Provasoli, L. (1971). Chloroplast structure of the Cryptophyceae. Evidence for phycobiliproteins within intrathylakoidal spaces. *J. Cell Biol.* **48**, 280–90.

Gantt, E., Grabowski, B. and Cunningham, F.X. (2003). Antenna systems of red algae: phycobilisomes with photosystem II and chlorophyll complexes with Photosystem I. In *Light-Harvesting Antennas in Photosynthesis*, ed. B. Green and W. Parson. Dordrecht: Kluwer, pp. 307–22.

Gindt, Y.M., Zhou, J.H., Bryant, D.A. and Sauer, K. (1994). Spectroscopic studies of phycobilisome subcore preparations lacking key core chromophores: Assignment of excited state energies to the L_{cm}, β^{18} and α^{AP-B} chromophores. *Biochim. Biophys. Acta* **1186**, 153–62.

Glauser, M., Sidler, W. and Zuber, H. (1993). Isolation, characterization and reconstitution of phycobiliprotein rod-core linker polypeptide complexes from the phycobilisome of *Mastigocladus laminosus. Photochem. Photobiol.* **57**, 344–51.

Glazer, A.N. (1984). Phycobilisome – A macromolecular complex optimized for light energy transfer. *Biochim. Biophys. Acta* **768**, 29–51.

Glazer, A.N. (1985). Light harvesting by phycobilisomes. *Ann. Rev. Biophys. Biophys. Chem.* **14**, 47–77.

Glazer, A.N. (1988). Phycobiliproteins. *Meth. Enzymol.* **167**, 291–303.

Glazer, A.N. (1994). Adaptive variations in phycobilisome structure. *Adv. Mol. Cell Biol.* **10**, 119–49.

Glazer, A.N. and Fang, S. (1973). Chromophore content of blue-green algal phycobiliproteins. *J. Biol. Chem.* **248**, 659–62.

Glazer, A.N. and Hixson, C.S. (1975). Characterization of R-Phycocyanin. Chromophore content of R-Phycocyanin and C-Phycoerythrin. *J. Biol. Chem.* **250**, 5487–95.

Glazer, A.N. and Hixson, C.S. (1977). Subunit structure and chromophore composition of Rhodophytan phycoerythrins. *J. Biol. Chem.* **252**, 32–42.

Glazer, A.N. and Wedemayer, G.J. (1995). Cryptomonad biliproteins – an evolutionary perspective. *Photosynth. Res.* **46**, 93–105.

Gomez-Lojero, C., Perez-Gomez, B., Shen, G., Schluchter, W.M. and Bryant, D.A. (2003). Interaction of ferredoxin:NADP+ oxidoreductase with phycobilisomes and phycobilisome substructures of the cyanobacterium *Synechococcus sp.* strain PCC 7002. *Biochemistry* **42**, 13800–11.

Gottschalk, L., Fischer, R., Lottspeich, F. and Scheer, H. (1991). Origin of the red-shifted absorption in phycocyanin 632 from *Mastigocladus laminosus. Photochem. Photobiol.* **54**, 283–88.

Grossman, A.R., Schaefer, M.R., Chiang, G.G. and Collier, J.L. (1993). The phycobilisome, a light-harvesting complex responsive to environmental conditions. *Microbiol. Rev.* **57**, 725–49.

Grossman, A.R., van Waasbergen, L.G. and Kehoe, D. (2003). Environmental regulation of phycobilisome biosynthesis. In *Light-Harvesting Antennas in Photosynthesis*, ed. B. Green and W. Parson. Dordrecht: Kluwer, pp. 471–93.

Hahn, J., Kühne, R. and Schmieder, P. (2007). Solution-state ^{15}N NMR spectroscopic study of α-C-phycocyanin: implications for the structure of the chromophore-binding pocket of the cyanobacterial phytochrome Cph1. *ChemBioChem* **8**, 2249–55.

Hess, W.R., Partensky, F., van der Staay, G.W., Garcia-Fernandez, J.M., Borner, T. and Vaulot, D. (1996). Coexistence of phycoerythrin and a chlorophyll *a*/*b* antenna in a marine prokaryote. *Proc. Natl. Acad. Sci. USA* **93**, 11126–30.

Hess, W.R., Rocap, G., Ting, C.S., Larimer, F., Stilwagen, S., Lamerdin, J. and Chisholm, S.W. (2001). The photosynthetic apparatus of *Prochlorococcus*: Insights through comparative genomics. *Photosynth. Res.* **70**, 53–71.

Hirose, Y., Shimada, T., Narikawa, R., Katayama, M. and Ikeuchi, M. (2008). Cyanobacteriochrome CcaS is the green light receptor that induces the expression of phycobilisome linker protein. *Proc. Natl. Acad. Sci. USA* **105**, 9528–33.

Hu, I.C., Lee, T.R., Lin, H.F., Chiueh, C.C. and Lyu, P.C. (2006). Biosynthesis of fluorescent allophycocyanin α-subunits by autocatalytic bilin attachment. *Biochemistry* **45**, 7092–99.

Ikeuchi, M. and Ishizuka, T. (2008). Cyanobacteriochromes: a new superfamily of tetrapyrrole-binding photoreceptors in cyanobacteria. *Photochem. Photobiol. Sci.* **7**, 1159–67.

Inomata, K., Noack, S., Hammam, M.A., Khawn, H., Kinoshita, H., Murata, Y., Michael, N., Scheerer, P., Krauss, N. and Lamparter, T. (2006). Assembly of synthetic locked chromophores with *Agrobacterium* phytochromes Agp1 and Agp2. *J. Biol. Chem.* **281**, 28162–73.

Ishizuka, T., Narikawa, R., Kohchi, T., Katayama, M. and Ikeuchi, M. (2007). Cyanobacteriochrome TePixJ of *Thermosynechococcus elongatus* harbors phycoviolobilin as a chromophore. *Plant Cell Physiol.* **48**, 1385–90.

Jorissen, H.J., Quest, B., Lindner, I., Tandeau de Marsac, N. and Gärtner, W. (2002). Phytochromes with noncovalently bound chromophores: the ability of apophytochromes to direct tetrapyrrole photoisomerization, *Photochem. Photobiol.* **75**, 554–59.

Kahn, K., Mazel, D., Houmard, J., Tandeau de Marsac, N. and Schaefer, M.R. (1997). A role for cpeYZ in cyanobacterial phycoerythrin biosynthesis. *J. Bacteriol.* **179**, 998–1006.

Kehoe, D.M. and Gutu, A. (2006). Responding to color: The regulation of complementary chromatic adaptation. *Ann. Rev. Plant Biol.* **57**, 127–50.

Kirilovsky, D. (2007). Photoprotection in cyanobacteria: the orange carotenoid protein (OCP)-related non-photochemical-quenching mechanism. *Photosynth. Res.* **93**, 7–16.

Kneip, C., Parbel, A., Förstendorf, H., Scheer, H., Siebert, F. and Hildebrandt, P. (1998). FT-NIR-resonance Raman spectroscopic study of the α-subunit of phycoerythrocyanin and phycocyanin from the cyanobacterium *Mastigocladus laminosus*. *J. Raman. Spectr.* **29**, 939–44.

Kohchi, T., Mukougawa, K., Frankenberg, N., Masuda, M., Yokota, A. and Lagarias, J.C. (2001). The *Arabidopsis HY2* gene encodes phytochromobilin synthase, a ferredoxin-dependent biliverdin reductase. *Plant Cell* **13**, 425–36.

Köst-Reyes, E., Schneider, S., John, W., Fischer, R., Scheer, H. and Köst, H.-P. (1988). Fast preparative isoelectric focusing of phycocyanin subunits in layers of granulated gels. *Electrophoresis* **8**, 335–36.

Kupka, M. and Scheer, H. (2008). Unfolding of C-phycocyanin followed by loss of non-covalent chromophore-protein interactions: 1. Equilibrium experiments. *Biochim. Biophys. Acta* **1777**, 94–103.

Lämmli, U.K. (1970). Cleavage of structural proteins during the assembly of the head of bacteriophage T4. *Nature* **227**, 680–85.

Lamparter, T. (2004). Evolution of cyanobacterial and plant phytochromes. *FEBS Lett.* **573**, 1–5.

Lindner, I. (2000). *Totalsynthese neuartiger Chromophore des pflanzlichen Photorezeptors Phytochrom und Charakterisierung der biochemischen und spektralen Eigenschaften der damit generierten Chromoproteine*. Ph.D. dissertation, Universität Duisburg, Germany.

Lindner, I., Knipp, B., Braslavsky, S.E., Gärtner, W. and Schaffner, K. (1998). A novel chromophore selectively modifies the spectral properties of one of the two stable states of the plant photoreceptor phytochrome. *Angew. Chem. Int. Ed.* **37**, 1843–46.

MacColl, R. and Guard-Friar, D. (eds.) (1987). *Phycobiliproteins*, Boca Raton: CRC Press.

Manitto, P. and Monti, D. (1979). Oxidation of bilirubin with chloranil: A simple method for preparing isomerically pure biliverdin. *Experientia* **35**, 9–10.

Marquardt, J., Senger, H., Miyashita, H., Miyachi, S. and Morschel, E. (1997). Isolation and characterization of biliprotein aggregates from *Acaryochloris marina*, a *Prochloron*-like prokaryote containing mainly chlorophyll *d*. *FEBS Lett.* **410**, 428–32.

Mazel, D., Houmard, J. and De Marsac, N.T. (1988). A multigene family in *Calothrix sp.* PCC 7601 encodes Phycocyanin, the major component of the cyanobacterial light-harvesting antenna. *Mol. Gen. Genet.* **211**, 296–304.

Mazel, D. and Marlière, P. (1989). Adaptive eradication of methionine and cysteine from cyanobacterial light-harvesting proteins. *Nature* **341**, 245–48.

Mimuro, M. and Kikuchi, H. (2003). Antenna systems and energy transfer in Cyanophyta and Rhodophyta. In *Light-Harvesting Antennas in Photosynthesis*, ed. B. Green and W. Parson. Dordrecht: Kluwer, pp. 281–306.

Montgomery, B.L., Casey, E.S., Grossman, A.R. and Kehoe, D.M. (2004). AplA, a member of a new class of phycobiliproteins lacking a traditional role in photosynthetic light harvesting. *J. Bacteriol.* **186**, 7420–28.

Morsy, F.M., Nakajima, M., Yoshida, T., Fujiwara, T., Sakamoto, T. and Wada, K. (2008). Subcellular localization of ferredoxin-$NADP^+$ oxidoreductase in phycobilisome retaining oxygenic photosynthetic organisms. *Photosynth. Res.* **95**, 73–85.

Moss, G.P. (1988). IUPAC-IUB Joint commission biochemical nomenclature (JCBN). Nomenclature of tetrapyrroles. Recommendation 1986. *Eur. J. Biochem.* **178**, 277–328.

Nagy, J.O., Bishop, J.E., Klotz, A.V., Glazer, A.N. and Rapoport, H. (1985). Bilin attachment sites in the alpha-subunits, beta-subunits and gamma-subunits of R-phycoerythrin – structural studies on singly and doubly linked phycourobilins. *J. Biol. Chem.* **260**, 4864–68.

Nield, J., Rizkallah, P.J., Barber, J. and Chayen, N.E. (2003). The 1.45 Å three-dimensional structure of C-phycocyanin from the thermophilic cyanobacterium *Synechococcus elongatus*. *J. Struct. Biol.* **141**, 149–55.

O'Carra, P. and Oh'Eocha, C. (1966). Bilins released from algae and biliproteins by methanol extraction. *Phytochemistry* **5**, 993–97.

Ong, L.J. and Glazer, A.N. (1988). Sructural studies of phycobiliproteins in unicellular marine cyanobacteria. In *Light-Energy Transduction in Photosynthesis: Higher Plant and Bacterial Models*, ed. S.E. Stevens and D.A. Bryant. Rockville: American Society of Plant Physiologists, pp. 102–21.

Parbel, A., Zhao, K.H., Breton, J. and Scheer, H. (1997). Chromophore assignment in phycoerythrocyanin from *Mastigocladus laminosus*. *Photosynth. Res.* **54**, 25–34.

Porra, R.J., Pfündel, E.E. and Engel, N. (1997). Metabolism and function of photosynthetic pigments. In *Phytoplankton Pigments in Oceanography: Guidelines to Modern Methods*, ed. S.W. Jeffrey, R.F.C. Mantoura and S.W. Wright. Paris: UNESCO Publishing pp. 85–126.

Raps, S. (1990). Differentiation between phycobiliprotein and colorless linker polypeptides by fluorescence in the presence of $ZnSO_4$. *Plant Physiol.* **92**, 358–62.

Reuter, W., Westermann, M., Brass, S., Ernst, A., Böger, P. and Wehrmeyer, W. (1994). Structure, composition, and assembly of paracrystalline phycobiliproteins in *Synechocystis* sp. strain BO 8402 and of phycobilisomes in the derivative strain BO 9201. *J. Bacteriol.* **176**, 896–904.

Reuter, W., Wiegand, G., Huber, R. and Than, M.E. (1999). Structural analysis at 2.2 Å of orthorhombic crystals presents the asymmetry of the allophycocyanin-linker complex, AP $L_c^{7.8}$, from phycobilisomes of *Mastigocladus laminosus*. *Proc. Natl. Acad. Sci. USA* **96**, 1363–68.

Ritter, S., Hiller, R.G., Wrench, P.M., Welte, W. and Diederichs, K. (1999). Crystal structure of a phycourobilin-containing phycoerythrin at 1.90 Å resolution. *J. Struct. Biol.* **126**, 86–97.

Rockwell, N.C., Su, Y.-S. and Lagarias, J.C. (2006). Phytochrome structure and signaling mechanisms. *Ann. Rev. Plant Biol.* **57**, 837–58.

Rüdiger, W. and Thümmler, F. (1991). Phytochrome, the visual pigment of plants. *Angew. Chem. Int. Ed. Engl.* **30**, 1216–28.

Saunée, N.A., Williams, S.R., Bryant, D.A. and Schluchter, W.M. (2008). Biogenesis of phycobiliproteins. II. CpcS-I and CpcU comprise the heterodimeric bilin lyase that attaches phycocyanobilin to Cys-82 of β–phycocyanin and Cys-81 of allophycocyanin subunits in *Synechococcus sp.* PCC 7002. *J. Biol.Chem.* **283**, 7513–22.

Scheer, H. (1976). Studies on plant bile pigments: Characterization of a model for the phytochrome Pr chromophore. *Z. Naturforsch.* **31**c, 413–17.

Scheer, H. (1982). Phycobiliproteins: Molecular aspects of photosynthetic antenna systems. In *Light Reaction Path of Photosynthesis*, ed. F.K. Fong. Berlin: Springer, pp. 7–45.

Scheer, H. (1985). Model compounds for the phytochrome chromophore. In *Techniques in Photomorphogenesis*, ed. H. Smith and M.G. Holmes. New York: Academic Press, pp. 227–56.

Scheer, H. (2003). The pigments. In *Light-Harvesting Antennas in Photosynthesis*, ed. B. Green and W. Parson. Dordrecht: Kluwer, pp. 29–81.

Scheer, H. and Zhao, K.-H. (2008). Biliprotein maturation: the chromophore attachment. *Mol. Microbiol.* **68**, 263–76.

Scheerer, P., Michael, N., Park, J.H., Noack, S., Forster, C., Hammam, M.A., Inomata, K., Choe, H.W., Lamparter, T. and Krauss, N. (2006). Crystallization and preliminary X-ray crystallographic analysis of the N-terminal photosensory module of phytochrome Agp1, a biliverdin-binding photoreceptor from *Agrobacterium tumefaciens*. *J. Struct. Biol.* **153**, 97–102.

Schirmer, T., Bode, W. and Huber, R. (1987). Refined three-dimensional structures of two cyanobacterial C-phycocyanins at 2.1 and 2.5 Å resolution – A common principle of phycobilin-protein interaction. *J. Mol. Biol.* **196**, 677–95.

Schirmer, T., Bode, W. and Huber, R. (1988). High-resolution crystal structure of C-Phycocyanin and polarized optical spectra of single crystals. In *Photosynthetic Light-Harvesting Systems: Organization and Function*, ed. H. Scheer and S. Schneider. Berlin: De Gruyter, pp. 195–201.

Schirmer, T., Bode, W., Huber, R., Sidler, W. and Zuber, H. (1985). X-Ray crystallographic structure of the light-harvesting biliprotein C-phycocyanin from the thermophilic cyanobacterium *Mastigocladus laminosus* and its resemblance to globin structures. *J. Mol. Biol.* **184**, 257–77.

Schluchter, W.M., Shen, G., Alvey, R.M., Biswas, A., Saunée, N., Williams, S.R., Miller, C.A. and Bryant, D.A. (2010). Phycobiliprotein biosynthesis in cyanobacteria: structure and function of enzymes involved in post-translational modification. In *Recent Advances in Phototrophic Prokaryotes*, ed. P.C. Hallenbeck. New York: Springer, pp. 211–28.

Schluchter, W.M. and Bryant, D.A. (2002). Analysis and reconstitution of phycobiliproteins: methods for the characterization of bilin attachment reactions. In *Heme, Chlorophyll, and Bilins*, ed. A.G. Smith and M. Witty. Totowa: Humana Press, pp. 311–34.

Schluchter, W.M. and Glazer, A.N. (1999). Biosynthesis of phycobiliproteins in Cyanobacteria. In *The Phototrophic Prokaryotes*, ed. G.A. Peschek, W. Löffelhardt, and G. Schmetterer. New York: Kluwer Academic/Plenum Publishers, pp. 83–95.

Schmidt, M., Krasselt, A. and Reuter, W. (2006). Local protein flexibility as a prerequisite for reversible chromophore isomerization in alpha-phycoerythrocyanin. *Biochim. Biophys. Acta* **1764**, 55–62.

Schmidt, M., Patel, A., Zhao, Y. and Reuter, W. (2007). Structural basis for the photochemistry of α-Phycoerythrocyanin. *Biochemistry* **46**, 416–23.

Schram, B.L. and Kroes, H.H. (1971). Structure of phycocyanobilin. *Eur. J. Biochem.* **19**, 581–94.

Shen, G., Saunee, N.A., Gallo, E., Begovic, Z., Schluchter, W.M. and Bryant, D.A. (2004). Identification of novel phycobiliprotein lyases in cyanobacteria. In *PS 2004 Light-Harvesting Systems Workshop*, ed. R.A. Niederman, R.E. Blankenship, H. Frank, B. Robert and R. van Grondelle. Sainte-Adèle, Québec, Canada, pp. 14–15.

Shen, G., Saunee, N.A., Williams, S.R., Gallo, E.F., Schluchter, W.M. and Bryant, D.A. (2006). Identification and characterization of a new class of bilin lyase: the cpcT gene encodes a bilin lyase responsible for attachment of phycocyanobilin to Cys-153 on the β-subunit of phycocyanin in *Synechococcus sp.* PCC 7002. *J. Biol. Chem.* **281**, 17768–78.

Shen, G., Leonard, H.S., Schluchter, W.M. and Bryant, D.A. (2008a). CpcM posttranslationally methylates asparagine-71/72 of phycobiliprotein beta subunits in *Synechococcus* sp. strain PCC 7002 and *Synechocystis* sp. strain PCC 6803. *J. Bacteriol.* **190**, 4808–17.

Shen, G., Schluchter, W.M. and Bryant, D.A. (2008b). Biogenesis of phycobiliproteins. I. *cpcS* and *cpcU* mutants of the cyanobacterium, *Synechococcus sp.* PCC 7002 define a heterodimeric phycocyanobilin lyase specific for β-phycocyanin and allophycocyanin subunits. *J. Biol. Chem.* **283**, 7503–12.

Sidler, W.A. (1994). Phycobilisome and phycobiliprotein structures. In *The Molecular Biology of Cyanobacteria*, ed. D.A. Bryant. Dordrecht: Kluwer, pp. 139–216.

Sidler, W. and Zuber, H. (1988). Structural and phylogenetic relationships of Phycoerythrins from cyanobacteria, red algae and Cryptophyceae. In *Photosynthetic Light-Harvesting Systems, Organization and Function*, ed. H. Scheer and S. Schneider. Berlin: De Gruyter, pp. 49–61.

Six, C., Thomas, J.C., Garczarek, L., Ostrowski, M., Dufresne, A., Blot, N., Scanlan, D.J. and Partensky, F. (2007). Diversity and evolution of phycobilisomes in marine *Synechococcus* spp.: a comparative genomics study. *Genome Biol.* **8**, R259.1–22.

Stec, B., Troxler, R.F. and Teeter, M.M. (1999). Crystal structure of C-phycocyanin from *Cyanidium caldarium* provides a new perspective on phycobilisome assembly. *Biophys. J.* **76**, 2912–21.

Steglich, C., Frankenberg-Dinkel, N., Penno, S. and Hess, W.R. (2005). A green light-absorbing phycoerythrin is present in the high-light-adapted marine cyanobacterium *Prochlorococcus sp.* MED4. *Environ. Microbiol.* **7**, 1611–18.

Storf, M. (2004). *Überexpression und Rekonstitution der alpha-Untereinheit des Phycoerytrocyanins aus Fischerella PCC 7603*. In Chromophorbindung und Photochemie der α-Untereinheit des Phycoerythrocyanins aus *Mastigocladus laminosus*. Ph.D. thesis Ludwig-Maximilians-Universität. München, Page 27.

Storf, M., Parbel, A., Meyer, M., Strohmann, B., Scheer, H., Deng, M., Zheng, M., Zhou, M. and Zhao, K. (2001). Chromophore attachment to biliproteins: Specificity of PecE/PecF, a lyase/isomerase for the photoactive 3^1-Cys-α 84-phycoviolobilin chromophore of phycoerythrocyanin. *Biochemistry* **40**, 12444–56.

Swanson, R.V. and Glazer, A.N. (1990). Phycobiliprotein methylation: Effect of the γ-*N*-methylasparagine residue on energy transfer in phycocyanin and the phycobilisome. *J. Mol. Biol.* **214**, 787–96.

Tandeau de Marsac, N. (1977). Occurence and nature of chromatic adaptation in cyanobacteria. *J. Bacteriol.* **130**, 82–91.

Tandeau de Marsac, N., Houmard, M. and Houmard, J. (1993). Adaption of cyanobacteria to environmental stimuli: new steps towards molecular mechanism. *FEMS Microbiol. Rev.* **104**, 119–90.

Thomas, B.A., Bricker, T.M. and Klotz, A.V. (1993). Post-translational methylation of phycobilisomes and oxygen evolution efficiency in cyanobacteria. *Biochim. Biophys. Acta* **1143**, 104–08.

Ting, C.S., Rocap, G., King, J. and Chisholm, S.W. (2001). Phycobiliprotein genes of the marine photosynthetic prokaryote *Prochlorococcus*: evidence for rapid evolution of genetic heterogeneity. *Microbiology* **147**, 3171–82.

Tooley, A.J. and Glazer, A.N. (2002). Biosynthesis of the cyanobacterial light-harvesting polypeptide phycoerythrocyanin holo-α subunit in a heterologous host. *J. Bacteriol.* **184**, 4666–71.

Tu, J.-M., Kupka, M., Böhm, S., Plöscher, M., Eichacker, L., Zhao, K.-H. and Scheer, H. (2008). Intermediate binding of phycocyanobilin to the lyase, CpeS1, and transfer to apoprotein. *Photosynth. Res.* **95**, 163–68.

Van der Weij-De Wit, C.D., Doust, A.B., van Stokkum, I.H., Dekker, J.P., Wilk, K.E., Curmi, P.M., Scholes, G.D. and van Grondelle, R. (2006). How energy funnels from the phycoerythrin antenna complex to photosystem I and photosystem II in cryptophyte *Rhodomonas* CS24 cells. *J. Phys. Chem. B* **110**, 25066–73.

van der Weij-De Wit, C.D., Doust, A.B., van Stokkum, I.H., Dekker, J.P., Wilk, K.E., Curmi, P.M. and van Grondelle, R. (2008). Phycocyanin sensitizes both photosystem I and photosystem II in cryptophyte *Chroomonas* CCMP270 cells. *Biophys. J.* **94**, 2423–33.

van Thor, J.J., Gruters, O.W.M., Matthijs, H.C.P. and Hellingwerf, K.J. (1999). Localization and function of ferredoxin:$NADP^+$ reductase, bound to the phycobilisomes of *Synechocystis*. *Embo J.* **18**, 4128–36.

Wagner, J.R., Brunzelle, J.S., Forest, K.T. and Vierstra, R.D. (2005). A light-sensing knot revealed by the structure of the chromophore-binding domain of phytochrome. *Nature* **438**, 325–31.

Wagner, J.R., Zhang, J., Brunzelle, J.S., Vierstra, R.D. and Forest, K.T. (2007). High resolution structure of *Deinococcus* bacteriophytochrome yields new insights into phytochrome architecture and evolution. *J. Biol. Chem.* **282**, 12298–309.

Wang, X.Q., Li, L.N., Chang, W.R., Zhang, J.P., Gui, L.L., Guo, B.J. and Liang, D.C. (2001). Structure of C-phycocyanin from *Spirulina platensis* at 2.2

Å resolution: a novel monoclinic crystal form for phycobiliproteins in phycobilisomes. *Acta Cryst. D Biol. Cryst.* **57**, 784–92.

Wedemayer, G.J., Wemmer, D.E. and Glazer, A.N. (1991). Phycobilins of cryptophycean algae. *J. Biol. Chem.* **266**, 4731–41.

Wedemayer, G.J., Kidd, D.G. and Glazer, A.N. (1996). Cryptomonad biliproteins: Bilin types and locations. *Photosynth. Res.* **48**, 163–70.

Wehrmeyer, W. (1983). Phycobiliproteins and phycobiliprotein organization in the photosynthetic apparatus of cyanobacteria, red algae, and cryptophytes. In *Proteins and Nucleic Acids in Plant Systems*, ed. U. Jense and D.E. Fairbrother. Berlin: Springer-Verlag, pp. 143–67.

Wehrmeyer, W. (1988). Structure of Cryptophyte photosynthetic membranes. In *Photosynthetic Light-Harvesting Systems, Organization and Function*, ed. H. Scheer and S. Schneider. Berlin: De Gruyter, pp. 35–49.

Wiegand, G., Parbel, A., Seifert, M.H., Holak, T.A. and Reuter, W. (2002). Purification, crystallization, NMR spectroscopy and biochemical analyses of α-phycoerythrocyanin peptides. *Eur. J. Biochem.* **269**, 5046–55.

Wilson, A., Punginelli, C., Gall, A., Bonetti, C., Alexandre, M., Routaboul, J.-M., Kerfeld, C.A., van Grondelle, R., Robert, B., Kennis, J.T.M. and Kirilovsky, D. (2008). A photoactive carotenoid protein acting as light intensity sensor. *Proc. Natl. Acad. Sci. USA* **105**, 12075–80.

Wu, S.-H. and Lagarias, J.C. (2000). Defining the bilin lyase domain: lessons from the extended phytochrome superfamily. *Biochemistry* **39**, 13487–95.

Yang, X., Stojkovic, E.A., Kuk, J. and Moffat, K. (2007). Crystal structure of the chromophore binding domain of an unusual bacteriophytochrome, RpBphP3, reveals residues that modulate photoconversion. *Proc. Natl. Acad. Sci. USA* **104**, 12571–76.

Yang, X., Kuk, J. and Moffat, K. (2008). Crystal structure of *Pseudomonas aeruginosa* bacteriophytochrome: photoconversion and signal transduction. *Proc. Natl. Acad. Sci. USA* **105**, 14715–20.

Zhao, K.H., Haessner, R., Cmiel, E. and Scheer, H. (1995). Type I reversible photochemistry of phycoerythrocyanin involves *Z*/*E*-isomerization of α-84 phycoviolobilin chromophore. *Biochim. Biophys. Acta* **1228**, 235–43.

Zhao, K.-H., Wu, D., Wang, L., Zhou, M., Storf, M., Bubenzer, C., Strohmann, B. and Scheer, H. (2002). Characterization of phycoviolobilin phycoerythrocyanin-α84-cystein-lyase-(isomerizing) from *Mastigocladus laminosus*. *Eur. J. Biochem.* **269**, 4542–50.

Zhao, K.-H., Ran, Y., Li, M., Sun, Y.-N., Zhou, M., Storf, M., Kupka, M., Böhm, S., Bubenzer, C. and Scheer, H. (2004). Photochromic biliproteins from the cyanobacterium *Anabaena* sp. PCC 7120: lyase activities, chromophore exchange and photochromism in phytochrome and phycoerythrocyanin. *Biochemistry* **43**, 11576–88.

Zhao, K.H., Su, P., Böhm, S., Song, B., Zhou, M., Bubenzer, C. and Scheer, H. (2005). Reconstitution of phycobilisome core-membrane linker, L_{cm}, by autocatalytic chromophore binding to ApcE. *Biochim. Biophys. Acta* **1706**, 81–87.

Zhao, K.H., Su, P., Li, J.A., Tu, J.M., Zhou, M., Bubenzer, C. and Scheer, H. (2006a). Chromophore attachment to phycobiliprotein β-subunits: phycocyanobilin:cysteine-β84 phycobiliprotein lyase activity of CpeS-like protein from *Anabaena* sp. PCC7120. *J. Biol. Chem.* **281**, 8573–81.

Zhao, K.-H., Wu, D., Zhang, L., Zhou, M., Böhm, S., Bubenzer, C. and Scheer, H. (2006b). Chromophore attachment in phycocyanin: functional amino acids of phycocyanobilin – α-phycocyanin lyase and evidence for chromophore binding. *FEBS J.* **273**, 1262–74.

Zhao, K.-H., Su, P., Tu, J.M., Wang, X., Liu, H., Plöscher, M., Eichacker, L., Yang, B., Zhou, M. and Scheer, H. (2007a). Phycobilin:cystein-84 biliprotein lyase, a near-universal lyase for cysteine-84-binding sites in cyanobacterial phycobiliproteins. *Proc. Natl. Acad. Sci. USA* **104**, 14300–05.

Zhao, K.-H., Zhang, J., Tu, J.M., Böhm, S., Plöscher, M., Eichacker, L., Bubenzer, C., Scheer, H., Wang, X. and Zhou, M. (2007b). Lyase activities of CpcS and CpcT-like proteins from *Nostoc* sp. PCC7120, and sequential reconstitution of binding sites of phycoerythrocyanin and phycocyanin β-subunits. *J. Biol. Chem.* **282**, 34093–103.

10

UV-absorbing 'pigments': mycosporine-like amino acids

JOSE I. CARRETO, SUZANNE ROY, KENIA WHITEHEAD, CAROLE A. LLEWELLYN AND MARIO O. CARIGNAN

10.1 Description and role of MAAs

It is now generally believed that natural ultraviolet-A radiation (UV-A, 315–400nm) and ultraviolet-B radiation (UV-B, 280–315nm) are strong environmental factors affecting both productivity and community structure in marine and terrestrial ecosystems (de Mora *et al.*, 2000). Reduction of the stratospheric ozone layer, which has caused an increase in the UV-B flux to the Earth's surface in recent years (Farman *et al.*, 1985), could result in increased levels of UV-induced damage for most living organisms (Vincent and Neale, 2000), producing a great impact on the photosynthetic carbon fixation by plants and, consequently, on the global climate change (UNEP, 2006). At the beginning of the evolution of life on Earth, UV-B flux rates clearly exceeded the present values (Cockell and Horneck, 2001) resulting in the evolution of several protection strategies to counteract the negative effects of UV radiation (Roy, 2000). One of the adaptations whereby phytoplankton can reduce UV-induced damage is the synthesis of compounds that can absorb the damaging wavelengths and dissipate the absorbed energy without generating phototoxic reactive intermediates. A variety of such compounds have been found in aquatic and terrestrial plants (Rozema *et al.*, 2002).

As early as 1938 there were observations of UV-absorbing compounds in marine algae (Kalle, 1938; referenced in Sivalingam *et al.*, 1974). This was followed in 1969 by reports of UV-absorbing substances (named S-320) in water extracts from several species of corals and a cyanobacterium (most likely *Trichodesmium*) from the Great Barrier Reef (Shibata, 1969). Mycosporine-like amino acids (MAAs) from marine organisms were first isolated and characterized by Hirata and co-workers (Hirata *et al.*, 1979). They isolated and characterized mycosporine-glycine from the tropical zoanthid *Palythoa tuberculosa* (Ito and Hirata, 1977), a compound previously isolated from mycelia of sporulating fungi (Favre-Bonvin *et al.*, 1976), and then described several related imine derivatives of mycosporines (Hirata *et al.*, 1979). Since then, more than 20 closely related MAA compounds have been isolated and characterized from several plants and marine animals (Figure 10.1 and Table 10.1).

Phytoplankton Pigments: Characterization, Chemotaxonomy and Applications in Oceanography, eds. Suzanne Roy, Carole A. Llewellyn, Einar Skarstad Egeland and Geir Johnsen. Published by Cambridge University Press.

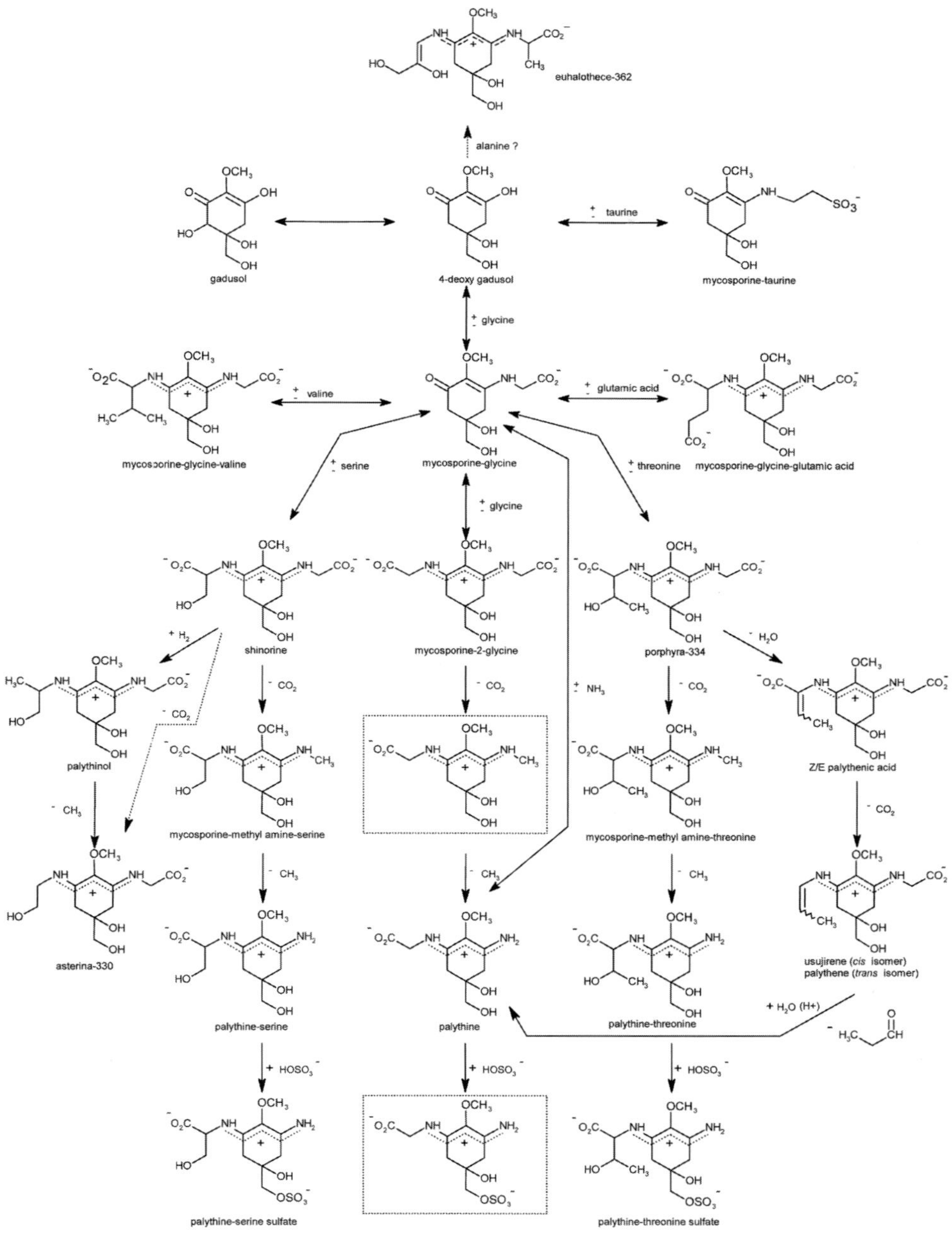

Figure 10.1. Structural relationships between the different MAAs and their feasible chemical and/or biochemical conversions.

Table 10.1. *Sources of isolation and physico-chemical properties of common MAAs.*

MAAs	λ_{max} (nm)	ε (L mol^{-1} cm^{-1})	Mol. weight	Characteristic MS fragments	Original source	References
Asterina-330	330 (H_2O)	43800	288	289 ($[M+H]^+$), 273, 243, 230, 213, 199, 197, 186, 168, 150, 137 **(3)** 289 ($[M+H]^+$), 274, 230, 212, 186 **(4)**	*Asterina pectinifera* **(1)**	**(1, 2, 3, 4)**
Dehydroxylusujirene	356		267	268 ($[M+H]^+$), 224, 209 **(5)**	*Synechocystis* sp. **(5)**	**(5)**
Euhalothece 362	362		330	331($[M+H]^+$), 316, 272, 242, 228, 213 **(6)**	*Euhalothece* sp. **(6)**	**(6)**
M-320	320 (H_2O)		559	560 ($[M+H]^+$) **(7)**	*Alexandrium tamarense* **(7)**	**(7)**
M-335/360	335 (360) (H_2O)		598	M^+ **(7)**	*Alexandrium tamarense* **(7)**	**(7)**
Mycosporine-2-glycine	332		302	303 ($[M+H]^+$), 288, 244, 200, 185, 164, 151 **(6)**	*Anthopleura elegantissima* **(8)**	**(6, 8)**
Mycosporine-glutamic acid-glycine	330 (MeOH)	43900 (MeOH)	374	M^+ **(9)**	*Dysidea herbacea* **(9)**	**(9)**
Mycosporine-glycine	310 (H_2O)	28800 (MeOH)	245		*Palythoa tuberculosa* **(10)**	**(9, 10)**
Mycosporine-methylamine-serine	325	16600	288	289 ($[M+H]^+$), 274 **(11)**	*Pocillopora eydouxi* **(11)**	**(11)**
Mycosporine-methylamine-threonine	330	33000	302	303 ($[M+H]^+$) **(12)**	*Styllopora pistillata* **(12)**	**(12)**

Mycosporine-taurine	309		295	296 ($[M+H]^+$) **(8)**	*Anthopleura elegantissima* **(8)**	**(8)**
Palythene	360	50000	284	285 ($[M+H]^+$), 241, 223, 205, 197, 195, 193, 179, 166, 149, 137 **(3)**	*Palythoa tuberculosa* **(13)**	**(3, 13)**
Palythine	320 (H_2O)	35500 **(14)** 36200 **(13)**	244	245 ($[M+H]^+$), 230, 209, 199, 186, 184, 162, 155, 150, 137 **(3)** 245 ($[M+H]^+$), 230, 227, 209, 186 **(4)**	*Chondrus yendoi* **(14)**	**(3, 4, 13, 14)**
Palythine-serine	320	10500	274	275 ($[M+H]^+$), 260 **(11)**	*Pocillopora eydouxi* **(11)**	**(11)**
Palythine-serine sulfate	321		353	353 ($[M]^-$) **(15)**	*Styllopora pistillata* **(15)**	**(15)**
Palythine-threonine sulfate	321		367	367 ($[M]^-$) **(15)**	*Styllopora pistillata* **(15)**	**(15)**
Palythine-threonine	320		288	289 ($[M+H]^+$), 245, 169	*Pocillopora capitata* **(16)**	**(16)**
Palythinol	332	43500	302	303 ($[M+H]^+$), 288, 243, 231, 199, 197, 186, 168, 150, 137 **(3)** 303 ($[M+H]^+$), 288, 244, 230, 226, 186 **(4)**	*Palythoa tuberculosa* **(13)**	**(3, 4, 13)**
Porphyra-334	334 (H_2O)	42300	346	347 ($[M+H]^+$), 303, 243, 200, 197, 186, 185, 168, 151, 137 **(3)**	*Porphyra tenera* **(17)**	**(3, 17)**
Shinorine	333 (H_2O)	44668	332	333 ($[M+H]^+$), 255, 241, 230, 211, 197, 186, 185, 168, 137 **(3)** 333 ($[M+H]^+$), 318, 274, 230, 212, 186 **(4)**	*Chondrus yendoi* **(18)**	**(3, 4, 18)**

Table 10.1. (*cont.*)

MAAs	λ_{max} (nm)	ε (L mol^{-1} cm^{-1})	Mol. weight	Characteristic MS fragments	Original source	References
Shinorine methyl ester	333 (H_2O)		346	347 ($[M+H]^+$) **(7)**	*Alexandrium tamarense* **(7)**	**(7)**
Usujirene	357 (H_2O)		284	285 ($[M+H]^+$) **(19)**	*Palmaria palmata* **(19)**	**(19)**
Z-palythenic acid	337 (H_2O)	29200	328	310 ($[M\text{-}H_2O]^+$) **(20)** 329 ($[M+H]^+$), 314, 296, 283, 268, 251, 241, 237, 225, 197, 193, 182, 175, 150, 138 **(3)**	*Halocynthia roretzi* **(20)**	**(3, 20)**

λ_{max} = maximum wavelength absorption; ε = molar extinction coefficient. **(1)** Nakamura *et al.*, 1981; **(2)** Bandaranayake, 1998; **(3)** Whitehead *et al.*, 2001; **(4)** Cardozo *et al.*, 2006; **(5)** Zhang *et al.*, 2007; **(6)** Volkmann *et al.*, 2006; **(7)** Carreto *et al.*, 2001; **(8)** Stochaj *et al.*, 1994; **(9)** Bandaranayake *et al.*, 1996; **(10)** Ito and Hirata, 1977; **(11)** Teai *et al.*, 1997; **(12)** Wu Won *et al.*, 1995; **(13)** Takano *et al.*, 1978; **(14)** Tsujino *et al.*, 1978; **(15)** Wu Won *et al.*, 1997; **(16)** Carignan *et al.*, 2009; **(17)** Takano *et al.*, 1979; **(18)** Tsujino *et al.*, 1980; **(19)** Sekikawa *et al.*, 1986; **(20)** Kobayashi *et al.*, 1981.

Wide ranging studies indicate that these compounds occur in virtually all taxa of marine and freshwater cyanobacteria and algae, in invertebrate–microbial symbioses and in metazoans (Karentz *et al.*, 1991; Gröniger *et al.*, 2000; Karentz, 2001).

The fungal mycosporines can be considered as enamino ketones which possess a common cyclohexenone ring system linked with an amino acid or an amino alcohol (Bandaranayake, 1998). Recently, several mycosporines previously described in terrestrial fungi, were found in terrestrial cyanobacteria (*Oscillatoriales*) (Volkmann and Gorbushina, 2006) and aquatic yeasts (Sommaruga *et al.*, 2004).

In contrast to fungal metabolites and with only two exceptions (mycosporine-glycine and mycosporine-taurine), UV-absorbing compounds of marine organisms are imine derivatives of mycosporines (MAAs) which contain an amino cyclohexenimine ring linked to an amino acid, amino alcohol or amino group, having absorption maxima between 320–360 nm (Bandaranayake, 1998). The molecular structure of marine mycosporines and MAAs are shown in Figure 10.1. Generally a glycine subunit is present on the C3 of the cyclohexenimine ring. However, in some corals glycine has been replaced by methyl amine (Teai *et al.*, 1997; 1998). Some MAAs isolated from corals also contains sulfate esters (Wu Won *et al.*, 1997). Recently a rare novel MAA, containing the amino acid alanine: euhalothece-362, was isolated from the unicellular cyanobacterium *Euhalothece* sp. (Volkmann *et al.*, 2006). Another novel MAA tentatively identified as dehydroxyl-usujirene was isolated from the cyanobacteria *Synechocystis* sp. (Zhang *et al.*, 2007). Several forms of atypical MAAs have been isolated from the dinoflagellate *Alexandrium* species; one of them (M-333) appears to be a mono-methyl ester of shinorine (Carreto *et al.*, 2001).

Several hypotheses about the role of MAAs in biological systems have been formulated: (a) they may protect the cells from UV photo-damage by playing a sunscreen role (Shibata, 1969); (b) they may act as antioxidants (Dunlap and Yamamoto, 1995); (c) they may contribute to osmotic regulation (Oren, 1997); (d) they may act as regulatory metabolites of sporulation and germination in fungi (Arpin and Bouillant, 1981); and (f) they may act as transducers of UV wavelengths to wavelengths utilizable for photosynthesis (Sinha *et al.*, 1998). However, experimental evidence has indicated that the major functions of MAAs in marine algae are to serve as photo-protective UV filters and/or to act as antioxidants.

Evidence for the photo-protective function of MAAs comes from a series of results showing a direct relationship between MAA concentrations and the protection of some invertebrates (Shick and Dunlap, 2002; Moeller *et al.*, 2005) and of photosynthesis, growth and motility of algae (Neale *et al.*, 1998; Karsten *et al.*, 1999; Klisch *et al.*, 2001). Direct evidence for their protective function in phytoplankton has been demonstrated where high amounts of intracellular MAAs diminish the inhibitory effect of UV radiation on photosynthesis (Neale *et al.*, 1998), or protect the motility of the dinoflagellate species *Gyrodinium dorsum* (Klisch *et al.*, 2001). The experiments performed by Neale *et al.* (1998) provided the first irrefutable evidence that MAAs act as spectrally specific UV sunscreens and are direct protectors in the red tide dinoflagellate *Akashiwo sanguinea* (= *Gymnodinium sanguineum*). Laurion *et al.* (2003, 2004) showed a high degree of MAA packaging in some dinoflagellates that may increase the protection efficiency of MAAs for specific cellular targets. Recently, Ferroni *et al.* (2010) concluded that the complement of UV-screening compounds MAAs and scytonemin is responsible for the outstanding lack of UV-sensitivity of photosynthesis in the cyanobacterium *Nostoc flagelliforme*.

The lack of fluorescence and radical production observed for porphyra-334 (Conde *et al.*, 2000) and shinorine (Conde *et al.*, 2004), together with their high degree of photostability in vitro support the photoprotective and effective sunscreen role assigned to these MAAs. In addition, oxocarbonyl-MAAs such as mycosporine-glycine (Dunlap and Yamamoto, 1995; Suhn *et al.*, 2003) and mycosporine-taurine (Zhang *et al.*, 2007) have an antioxidant activity capable of protecting against reactive oxygen species (ROS) damage. Although it is well documented that MAAs may not provide complete protection against the effects of UV radiation (Lesser, 1996a, 1996b; Carignan *et al.*, 2002), their synthesis and accumulation are frequently part of an overall strategy to diminish the damaging effects of UV radiation (Roy, 2000).

10.2 Distribution of MAAs in marine phytoplankton

The distribution of MAAs in marine and terrestrial organisms has been reviewed previously by Gröniger *et al.* (2000) and Karentz (2001). Since these reviews were published, chromatographic methods have improved and mass spectrometric (MS) methods have been developed, providing a more thorough and reliable assessment of MAA distribution in phytoplankton. Here we compile and update this distribution (Table 10.2) and present a brief overview.

One of the earliest studies where high UV absorbance in marine phytoplankton was observed, was the work of Shibata (1969) on a filamentous marine cyanobacteria (most likely *Trichodesmium*) living in the Great Barrier Reef. However, the first study on MAA composition in phytoplankton was that of Carreto *et al.* (1990a) on dinoflagellates. In this study, the authors showed that the bloom-forming dinoflagellate *Alexandrium excavatum* (= *A. tamarense*) grown at high light intensity produced large amounts of a complex mixture of MAAs. In phytoplankton extracts from surface waters in the English Channel, Llewellyn and Harbour (2003) showed that dinoflagellates provided a background of MAAs throughout the year.

The most comprehensive study of UV-absorbing compounds in temperate phytoplankton used spectrophotometry to screen 152 species and 206 strains across 12 algal classes (Jeffrey *et al.*, 1999). High levels of UV-absorbing compounds were found in cryptophytes, dinoflagellates, haptophytes and raphidophytes with highest levels in surface bloom-forming dinoflagellates. Lower levels were reported in diatoms, chlorophytes, cyanobacteria, euglenophytes, eustigmatophytes, rhodophytes, some dinoflagellates and some prymnesiophytes (Jeffrey *et al.*, 1999). These class

Table 10.2. *Distribution of MAAs in marine phytoplankton. Myc = mycosporine. Darker symbols indicate greater abundance.*

Species	Strain number or location	Shinorine	Myc-2-glycine	Palythine	Asterina-330	Porphyra-334	Myc-methyl-serine	Myc-glycine	Myc-glycine:valine (?)	M-333	Palythinol	*Z*-palythenic acid	*E*-palythenic acid (?)	Myc-taurine	M-320	Usujirene	Palythene	M-335/360	M-328/360	Unknown	Reference
Dinophyceae																					
Alexandrium catenella	Clone CC08	x		x		x	x	x		x	x	x			x	x	**X**	x		x	Carreto *et al.* (2001)
Alexandrium catenella	East Sound, Washington			x		**X**		?				x									Whitehead and Hedges (2002)
Alexandrium excavatum	Clone AM1	x		x	x	**X**		x				x				x	x				Carreto *et al.* (1990a)
Alexandrium minutum	1960 clone AL2V	x		x			x	x		x	x				x		x	x	x	x	Carreto *et al.* (2001)
Alexandrium tamarense	MDQ 1096	x		x		**X**	x	x		x	x	x			x	x	**X**	x		x	Carreto *et al.* (2001, 2005)
Alexandrium tamarense	CCMP1771	x		x		**X**		x			x	x				x?	x				Whitehead and Hedges (2002)
Alexandrium tamarense	CCMP1771	x		x		x		x		x		x				x	**X**			x	Laurion *et al.* (2003)
Alexandrium tamarense	CCMP115	**X**		x		**X**		x								x	**X**				Frame (2004)
Amphidinium carterae	University of Washington							?													Hannach and Sigleo (1998)
Amphidinium sp.	Belau			x				x													Banaszak *et al.* (2000)
Ceratocorys horrida	UW Botany UTEX LB 2499	x			x	**X**					x										Whitehead and Hedges (2002)
Gloeodinium viscum	Gulf of Aqaba					x											x				Banaszak *et al.* (2000)
Gymnodinium catenatum	CS-309/3	x				**X**		**X**													Jeffrey *et al.* (1999)

Table 10.2. (*cont.*)

Species	Strain number or location	Shinorine	Myc-2-glycine	Palythine	Asterina-330	Porphyra-334	Myc-methyl-serine	Myc-glycine	Myc-glycine:valine (?)	M-333	Palythinol	*Z*-palythenic acid	*E*-palythenic acid (?)	Myc-taurine	M-320	Usujirene	Palythene	M-335/360	M-328/360	Unknown	Reference
Gymnodinium linucheae	CC3;1740	x				x		x													Banaszak *et al.* (2000)
Gymnodinium sanguineum	CCMP1740	x				**X**		?			x										Whitehead and Hedges (2002)
Gymnodinium sanguineum				x		**X**		x									x				Neale *et al.* (1998)
Gyrodinium dorsum		x		x		**X**														x	Klisch *et al.* (2001)
Heterocapsa triquetra	CCMP449	**X**		x		**X**	x			x						x	x			x	Laurion *et al.* (2003)
Karenia brevis	Charlotte Harbour	x				x	**X**				x										Frame (2004)
Lingulodinium polyedra				x		**X**			x		x				x?		x				Vernet and Whitehead (1996)
Noctiluca scintillans	Coastal waters Mar del Plata	x		x		**X**		x			x		x							x	Carreto *et al.* (2005)
Prorocentrum micans	PRORO III	x			x	**X**		**X**													Lesser (1996b)
Prorocentrum minimum	CC6	x		x																	Sinha *et al.* (1998)
Scrippsiella sweeneyae	NIES-684	**X**		x		x		**X**									x				Taira *et al.* (2004)
Symbiodinium corculorum	350-A	**X**				x		x													Banaszak *et al.* (2000)
Symbiodinium meandrine	130-A	**X**						x													Banaszak *et al.* (2000)
Symbiodinium microadriaticum		x				**X**		x													Banaszak and Trench (1995)
Symbiodinium pilosum	185-A	x				x		x													Banaszak *et al.* (2000)
Symbiodinium sp. (*)		x		x		**X**		**X**													Banaszak *et al.* (2006)

Woloszynskia sp.	CS-341	x		x	**X**	x				Jeffrey *et al.* (1999)
Haptophytes										
Emiliania huxleyi	CCMP370	x		x	x					Carreto *et al.* (2005)
Isochrysis sp.						?				Hannach and Sigleo (1998)
Phaeocystis cf. *antarctica*							x			Bidigare *et al.* (1996)
Phaeocystis cf. *antarctica*	Alfred Wegener Institute								x	Riegger and Robinson (1997)
Phaeocystis antarctica		x				x	x			Moisan and Mitchell (2001)
Phaeocystis antarctica	Antarctica	x			x	x	**X**			Newman *et al.* (2000)
Pavlova gyrans	University of Washington	?			?	?			x	Hannach and Sigleo (1998)
Bacillariophyceae										
Chaetoceros affinis	CS-98									Jeffrey *et al.* (1999)
Chaetoceros sp.	Alfred Wegener Institute	x			**X**					Riegger and Robinson (1997)
Corethron criophilum	Palmer Station, Antarctica	x	x	x	x			x		Helbling *et al.* (1996)
Coscinodiscus centralis	Alfred Wegener Institute	x			**X**					Riegger and Robinson (1997)
Ditylum brightwellii	Puget Sound, Washington									Whitehead and Hedges (2002)
Fragilariopsis cylindricus	Alfred Wegener Institute	x			x					Riegger and Robinson (1997)
Fragilariopsis cylindricus	Palmer Station, Antarctica	x			x		x			Helbling *et al.* (1996)
Fragilariopsis linearis	Alfred Wegener Institute	x			**X**					Riegger and Robinson (1997)

Table 10.2. (*cont.*)

Species	Strain number or location	Shinorine	Myc-2-glycine	Palythine	Asterina-330	Porphyra-334	Myc-methyl-serine	Myc-glycine	Myc-glycine:valine (?)	M-333	Palythinol	*Z*-palythenic acid	*E*-palythenic acid (?)	Myc-taurine	M-320	Usujirene	Palythene	M-335/360	M-328/360	Unknown	Reference
Porosira glacialis	Alfred Wegener Institute	x				**X**															Riegger and Robinson (1997)
Porosira pseudodenticulata	Alfred Wegener Institute	x				**X**		x													Riegger and Robinson (1997)
Proboscia inermis	Alfred Wegener Institute	x				**X**															Riegger and Robinson (1997)
Pseudonitzschia sp.	Palmer Station, Antarctica					x															Helbling *et al.* (1996)
Pseudonitzschia multiseries	Coastal waters Mar del Plata	x	x	x		x								x							Carreto *et al.* (2005)
Stellarima microtrias	Alfred Wegener Institute	x				**X**															Riegger and Robinson (1997)
Thalassiosira antarctica	Alfred Wegener Institute	x				**X**															Riegger and Robinson (1997)
Thalassiosira sp.	Palmer Station, Antarctica	x				x															Helbling *et al.* (1996)
Thalassiosira sp.	Antarctica	x				x															Hernando *et al.* (2002)
Thalassiosira tumida	Alfred Wegener Institute	x				**X**															Riegger and Robinson (1997)

Thalassiosira weissflogii	University of Washington					?				Hannach and Sigleo (1998)
Raphidophyceae										
Chattonella marina	UTCMPL01	**X**				x	x			Marshall and Newman (2002)
Fibrocapsa sp.	CS-220		x	x	**X**	x				Jeffrey *et al.* (1999)
Heterosigma carterae	CS-169	x		x						Jeffrey *et al.* (1999)
Chlorophyceae										
Dunaliella tertiolecta	University of Washington					?				Hannach and Sigleo (1998)
Prasinophyceae										
Pyramimonas parkerae	University of Washington					?			x	Hannach and Sigleo (1998)
Cyanophyceae										
Cyanobacterial mat	Brattina Island, Antarctica				x	**X**				Whitehead and Hedges (2002)
Nodularia baltica	Baltic Sea	x			x					Sinha *et al.* (2003)
Nodularia harveyana	Baltic Sea	x			x					Sinha *et al.* (2003)
Nodularia spumigera	Baltic Sea	x			x					Sinha *et al.* (2003)
Prochloron sp.	Palau	x								Dionisio-Sese *et al.* (1997)
Trichodesmium sp.	Caribbean Sea	**X**		**X**	x	x		x		Subramaniam *et al.* (1999)

*This study was carried out with *Symbiodinium* sp. freshly isolated from the cuidarian *Deudrogyra Cylindricus*.

distributions are broadly consistent with studies of individually identified MAA compounds. These MAA compounds have been reported in over 50 species of marine phytoplankton, with most studies typically focussing on one or a small number of species in relation to UV impact and induction.

Table 10.2 summarizes the distribution of MAAs found in these species. However, MAA abundance and composition is dependent on the intensity, duration and spectral composition of the incident light, the nutrient status and the time of sampling in the diurnal cycle (see Section 10.3). Furthermore, recent improvements in methodology (see Section 10.5) have shown that distributions of MAAs are more complex than have hitherto been reported. Therefore prudence should be used when referencing to earlier studies, particularly where there are no chromatograms showing separation of the various compounds. Consistent with previous findings, the highest levels and diversity of MAAs are found among surface bloom-forming species (Table 10.2).

10.3 Biosynthesis, trophic transfer and extra-cellular release

The biosynthesis of MAAs in marine algae and phototropic symbioses, have been reviewed by Shick and Dunlap (2002), and readers should consult this review for detailed information. Here a summary will be presented and we will focus on studies published since this review. Inhibition of MAA synthesis in the coral *Stylophora pistillata* by the addition of glyphosate (*N*-phosphonomethyl-glycine) provided the first indirect evidence that in marine organisms, the synthesis of MAAs started from the shikimate pathway (Shick *et al.*, 1999). The central ring of the MAAs present in cyanobacteria was also shown to have its origin in the shikimate pathway, suggesting that the biosynthesis of MAAs is similar or possibly identical in eukaryotic and prokaryotic organisms (Portwich and Garcia-Pichel, 2003). The incorporation of ^{14}C-glycine in the lateral chain of mycosporine-glycine provides direct evidence that mycosporine-glycine is the direct metabolic precursor of shinorine (Portwich and Garcia-Pichel, 2003). Possibly, mycosporine-glycine condensation with an amino acid would be a common reaction in the generation of bi-substituted MAAs, since most of them contain a glycine residue (Figure 10.1). On the other hand, the high diversity of MAAs present in marine algae and phototrophic symbiotic organisms is probably derived from the biotransformation of mycosporine-glycine, shinorine, porphyra-334, and other MAAs bi-substituted with amino acids (Shick, 2004; Callone *et al.*, 2006; Carignan *et al.*, 2009). Shick (2004) proposed a differential response in terms of the types of MAAs produced, making the distinction between primary MAAs (mycosporine-glycine, porphyra-334, shinorine) and secondary MAAs, synthesized from the primary compounds. A two-stage process for the induction of MAAs in the toxic dinoflagellate *Alexandrium tamarense* proposed by Callone *et al.* (2006) was consistent with Shick's (2004) model. Callone *et al.* (2006) observed an increase in primary MAAs, notably porphyra-334, within the first 2 h of exposure to higher photosynthetically active radiation (PAR, 400–700 nm) irradiances, after which they decreased and were replaced by increasing levels of palythene, a secondary MAA.

10.3.1 Induction

Carreto *et al.* (1989) were the first to show that an increase in irradiance resulted in increased concentrations of MAAs, and that UV-A and blue light stimulated the synthesis of MAAs in free-living dinoflagellates (Carreto *et al.*, 1990a, b). Following this, similar results were found for several species of dinoflagellates (Neale *et al.*, 1998; Carreto *et al.*, 2002; Taira *et al.*, 2004). However UV-B wavelengths (specifically 310–320 nm) were more effective in inducing MAA accumulation in the dinoflagellate *Gyrodinium dorsum* (Klisch and Häder, 2002). In Antarctic diatoms, UV-A and blue light were effective in inducing MAA synthesis (Riegger and Robinson, 1997; Hannach and Sigleo, 1998; Hernando *et al.*, 2002), while UV-B + UV-A were more effective for haptophytes (Riegger and Robinson, 1997; Hannach and Sigleo, 1998). In cyanobacteria, UV-B wavelengths were most efficient in stimulating MAA synthesis (Castenholz and Garcia-Pichel, 2000; Sinha *et al.*, 2001). Corals on the other hand, needed a combination of UV-B and UV-A in addition to PAR to stimulate this synthesis (Shick *et al.*, 1999; Lesser, 2000). Hoyer *et al.* (2002) stressed the fact that no consistent induction pattern could be found for red macroalgae from Antarctica. They observed three induction patterns among 8 out of 18 species that showed an induction of MAAs: (1) responding to the full radiation spectrum, (2) responding to PAR + UV-A, with no effect of additional UV-B, and (3) similar to (2) except that additional UV-B caused a decrease in MAAs. Hence, it may be premature to extend results of the previous studies to whole group patterns. Physiological acclimation and adaptation may play a greater role than previously suspected (Zudaire and Roy, 2001; Carignan *et al.*, 2002). Indeed, strains of the dinoflagellate *A. tamarense* from different locations produced different amounts and types of MAAs even when exposed to the same light and UV radiation conditions (Laurion and Roy, 2009).

Wavelength dependence for MAA synthesis indicates that specific photoreceptors are present (Carreto *et al.*, 1990a, b; Franklin *et al.*, 2001). However, these have not yet been identified. Kräbs *et al.* (2004) determined the action spectrum for the synthesis of MAAs in the marine red macroalga *Chondrus crispus*. They suggested an unidentified UV-A-type photoreceptor with a major absorption peak at 340 nm, a second maximum at 320 nm and a smaller third peak at 400 nm. Other suggestions for UV-B photoreceptors include a reduced pterin in cyanobacteria based on action spectra for MAA induction peaking at 310 nm (Portwich and Garcia-Pichel, 2000). Shick (2004) discussed that there may be no need for a specific UV-B photoreceptor if the primary response to UV-B is via ROS, and if the induction of MAAs responds to the increases in ROS. Once this level is reached, ROS would accumulate and they would secondarily control the synthesis of MAAs, as suggested by Shick (2004).

Independently of the induction mechanisms, response kinetics seem to vary substantially among organisms, with a rapid response time for some (within the first few hours of exposure to higher irradiances in dinoflagellates: Carreto *et al.*, 1990b; Klisch and Häder, 2002; Taira *et al.*, 2004; Callone *et al.*, 2006) and a much slower response for others (several days to weeks in corals: Shick (2004) and in macroalgae: Karsten *et al.*, 1999). Stress factors other than light or UV radiation have been found to stimulate the synthesis of MAAs. Osmotic stress, alone or in combination with UV-B radiation, can induce the synthesis of MAAs in some cyanobacteria (Portwich and Garcia-Pichel, 1999). In contrast, nitrogen limitation can significantly decrease the synthesis of MAAs (Litchman *et al.*, 2002). The composition of MAAs is also highly influenced by the nitrogen status. In the dinoflagellates *Alexandrium tamarense* and *Karenia brevis*, nitrogen-starved cells have a much higher percentage of mycosporine-glycine (Frame, 2004).

10.3.2 Trophic transfer

As products of the shikimate pathway, MAAs are expected to be synthesized only by bacteria and algae; other organisms acquire and metabolize these compounds by trophic transfer, symbiotic or bacterial association. Carroll and Shick (1996) first demonstrated that non-symbiotic marine invertebrates obtain MAAs from benthic algae. However, the composition of MAAs accumulated by non-symbiotic consumers differs from that in the diets and the selective uptake of MAAs from food, suggesting specific transporters in the gut (Shick and Dunlap, 2002). Similar results have been reported for the planktonic crustacean *Euphausia superba* fed on the colonial stage of the prymnesiophyte *Phaeocystis antarctica* (Newman *et al.*, 2000). Whitehead *et al.* (2001) compared the MAA composition of an Antarctic phytoplankton assemblage, an herbivorous pteropod (*Limacina helicina*), and a predatory pteropod (*Clione antarctica*) that feeds exclusively on *L. helicina*. They observed that the number of MAAs was greater in the pteropods versus the natural phytoplankton assemblage, and that an apparent bioaccumulation was taking place during the transfer to each trophic step.

In some symbiotic associations the symbionts are responsible for the synthesis of MAAs (Banaszak *et al.*, 2000; Banaszak *et al.*, 2006; Sonntag *et al.*, 2007). However, the genes encoding enzymes for the shikimate pathway in the starlet sea anemone *Nematostella vectensis* have been analyzed and molecular evidence has indicated the horizontal transfer of ancestral genes of the shikimic acid pathway into the *N. vectensis* genome, from both bacterial and eukaryotic (dinoflagellate) donors (Starcevic *et al.*, 2008). On the other hand, in some cases the freshly isolated zooxanthellae from MAA-containing holobionts lack MAAs indicating that the MAAs may be acquired from diet (Shick *et al.*, 2002). Compared to other symbiotic marine invertebrate phyla, cnidarians possess the highest diversity of MAAs with up to 12 MAAs identified in a single coral, *Stylophora pistillata* (Shick *et al.*, 1999;

Carreto *et al.*, 2005; Carignan *et al.*, 2009). However, the *Symbiodinium* species appears to be restricted to producing five MAAs (Banaszak *et al.*, 2006, Shick, 2004). Therefore, the host tissue and conceivably their associated bacteria, can bioconvert the MAAs produced by the zooxanthellae to yield an array of secondary MAAs (Shick, 2004). Alternatively, the host may accumulate MAAs from its external diet (Shick and Dunlap, 2002; Shick, 2004) or once within the host tissue, MAAs may be metabolized by bacterial transformation or acid hydrolysis. Contrarily, when the host contains fewer MAAs than the symbiont fraction, it is possible that some MAAs are not efficiently translocated to the host (Banaszak *et al.*, 2006) or catabolized once inside the host tissue.

Recently, Sommaruga *et al.* (2006) showed that the MAA composition in the benthic ciliate *Maristentor dinoferus*, which hosts symbiotic dinoflagellates of the genus *Symbiodinium*, resembled the composition found in other dinoflagellate symbioses, with the exception of the presence of palythenic acid.

10.3.3 Extra-cellular release

Vernet and Whitehead (1996) were the first to show that MAAs were excreted from the dinoflagellate *Lingulodinium polyedra*, by exposing algal cultures to PAR and UV radiation. Whitehead and Vernet (2000) examined the seasonal abundance of MAAs and the UV absorption in particulate and dissolved material in La Jolla Bay (California). Maximum particulate UV absorption was observed in the spring, following a large red tide of *L. polyedra*, while maximum dissolved UV absorption occurred in early summer.

More recently, Steinberg *et al.* (2004) examined the production of chromophoric dissolved organic matter (CDOM) by populations of various planktonic groups from oligotrophic waters. They noted that only the cyanobacteria *Trichodesmium* produced CDOM with UV peaks at 325 nm and a shoulder at 360 nm, characteristic of MAAs (Table 10.1). Morrison and Nelson (2004) realized a multiyear seasonal study of phytoplankton absorption in the UV range at the Bermuda Atlantic Time-series Study site. Maximum values were observed during summer in surface waters, with peaks between 313 and 335 nm. Monthly averaged absorption by dissolved material (a_{CDOM}) peaked at the same time as UV phytoplankton absorption, suggesting that MAAs are at certain times and locations an important source of CDOM in oceanic waters. Interestingly, the summer maximum in UV phytoplankton absorption coincided in this work with a predominantly prokaryotic picoplankton community.

Detecting the presence of MAAs in dissolved material suggests that these compounds are not degraded too rapidly in surface waters. A high photostability of MAAs in aqueous solutions was observed in the in vitro study by Conde *et al.* (2000, 2003). Recently, Whitehead and Hedges (2005) showed that the presence of light, oxygen and a strong photosensitizing agent were necessary for MAA photodegradation.

10.4 MAAs and bioptics

Relatively few bio-optical studies have focussed on the UV waveband (see Morel *et al.*, 2007). However, interest in the UV waveband is increasing, in part because harmful algae such as the red-tide dinoflagellates produce abundant MAAs and can release them into their environment, making the detection of these compounds interesting for remote sensing of harmful blooms. Kahru and Mitchell (1998) were the first to examine this possibility in detail during a massive bloom of the red-tide dinoflagellate *L. polyedra* off southern California. Their results showed that the chlorophyll (Chl)-specific absorption by particulate matter (a_p(340 nm) Chl^{-1}) increased more than 10 times, while the absorption by dissolved organic substances (a_g(340 nm)) doubled for the red-tide samples compared to the region outside of the bloom. The authors noted that increased absorption by both dissolved and particulate material affected the reflectance ratios used in operational algorithms to estimate chlorophyll *a* from remotely sensed information. An overestimate of chlorophyll *a* values by one order of magnitude was observed for the high-bloom stations. The use of UV wavebands may be difficult for satellite observations due to exponentially increasing Rayleigh scattering at shorter wavelengths, but it may be useful for aircraft or mooring detection of harmful algal blooms. Current ocean colour sensors do not include wavebands below 400 nm, except for the Japanese Global Imager (GLI) sensor mounted on the Advanced Earth Observing Satellite II (ADEOS II) which includes a band at 375 nm. Ocean colour detection (see Chapters 13 and 14, this volume) is moving increasingly towards hyperspectral instruments that should include the UV waveband and this may provide useful information for the detection of harmful algal blooms (Stumpf, 2006).

10.5 Methodology, extraction and separation of MAAs

10.5.1 Filtration

Since the GF/F filter is considered to retain close to 100% of phytoplankton pigments in oceanographic samples of all types, no further re-evaluation of filtration methods appears to be necessary (Mantoura *et al.*, 1997). However, recent results indicated that whereas pigments are not lost during filtration, the water-soluble MAAs of some dinoflagellates seem to be released out of the cells through intact membranes, ending up in the liquid with the moist filter (Sosik, 1999; Laurion *et al.*, 2003; Frame, 2004). Cultures of the colonial cyanobacteria *Trichodesmium* subjected to variable stirring conditions also showed varying UV absorbance (Subramaniam *et al.*, 1999). Special care is thus recommended when handling samples containing MAAs to reduce losses of these compounds.

10.5.2 Storage of filtered samples

There are no systematic studies on preservation of filtered phytoplankton MAA samples for HPLC analysis. However, it has been noted that under all storage conditions including refrigeration, freezer (−20°C and −80°C) and liquid nitrogen, the UV absorption of the filters increases – due to the release of MAAs on the filters – over a period of hours to months (Sosik, 1999; Laurion *et al.*, 2003; Frame 2004). Artefacts are easily minimized by analyzing samples immediately after filtering. Nevertheless, liquid nitrogen (−196°C) or ultra-cold freezer (−90°C) stored samples show minimal changes (Frame, 2004). It should be kept in mind that when thawing the filters some of the MAAs may be lost if the liquid from the filter is not included in the extraction. Placing the frozen filter in the test tube in which extraction will be performed before it thaws can minimize this loss.

10.5.3 Extraction

Methanol or aqueous methanol are the most widely used solvents for extraction of MAAs (Carreto *et al.*, 2005 and references therein). Sonication in 100% methanol followed by filtration to remove debris appears to be a practical and efficient (> 95%) technique of extraction for algal cultures and marine phytoplankton assemblages (Carreto *et al.*, 2005; Whitehead and Hedges, 2002). It would be advantageous if the extract could be split and used for both pigment and MAA analysis. Tartarotti and Sommaruga (2002) reported that in lyophilized red macroalgae and freshwater phytoplankton assemblages, the mean total concentration of MAAs obtained in 25% aqueous methanol at 45°C during 2 h, were respectively ~13 and ~3 times higher than in extractions made with 100% methanol at 4°C. However, extraction at this condition can result in the transformation or degradation of some labile MAAs (Tartarotti and Sommaruga, 2002; Carreto *et al.*, 2005). High extraction efficiency (> 90%) was reported with 100% methanol if lyophilized red macroalgae were previously soaked with water in the dark at 4°C overnight (Carreto *et al.*, 2005). Nevertheless, extraction efficiency of the method used and stability of the more labile MAAs should be determined for the type of organisms to be analyzed.

10.5.4 Stability of the extracts

While MAAs showed a high degree of photostability in aqueous solution, there has been evidence that bacteria may be responsible for MAA transformation or degradation (Dunlap and Shick, 1998). The HPLC methods based on reversed phase usually require evaporation of the extract under reduced pressure prior to HPLC injection and re-dissolution of the residue with water or water-acidified mobile phase. In these cases, it is recommended that injection of the recomposed extracts be done as soon as possible to avoid these bacterial artefacts (Carreto *et al.*, 2005). In contrast, MAA solutions in 100% methanol were found to be very stable at ambient temperature for 24 h (Carreto *et al.*, 2005).

10.5.5 Interferences and clean-up

UV-absorbing substances unrelated to MAAs, were also present in extracts of phytoplankton assemblages and other marine organisms (Bandaranayake *et al.*, 1996; Newman *et al.*, 2000). Evaporation of the methanol extract under reduced pressure prior to HPLC injection and re-dissolution of the residue with water or water-acidified mobile phase can remove water insoluble materials (Shick *et al.*, 1999; Carreto *et al.*, 2001). Solid phase extraction on C18 cartridges gave good results (Teai *et al.*, 1997; Shick *et al.*, 1999) and specifically removed substances such as pigments and lipids that not only affect sample analysis, but also significantly shorten the life of HPLC columns. Ultrafiltration is another alternative which, in addition to the removal of water-insoluble material, removes molecules larger than the molecular weight cut-off rating of the membrane. Nevertheless, highly polar, low molecular weight UV-absorbing substances unrelated to MAAs (Bandaranayake *et al.*, 1996) are generally present in extracts of marine organisms. As these interferences are not removed by solid phase extraction or ultrafiltration, care should be taken to ensure that peak absorption is devoid of contaminants with absorption that overlaps detection wavelength.

10.5.6 HPLC separation

Techniques for the separation of MAAs include those of Nakamura *et al.* (1982); Dunlap and Chalker (1986) and their modifications (Stochaj *et al.*, 1994; Shick *et al.*, 1999; Whitehead and Hedges, 2002) and Carreto *et al.* (2001; 2005). The classical HPLC method was based on reverse-phase low silanol-free group octadecylsilica (C_{18}) columns and isocratic elution with 0.02% acetic acid as mobile phase (Nakamura *et al.*, 1982). Dunlap and Chalker (1986) first introduced the use of monomeric octylsilica (C_8) columns. As has been discussed in Carreto *et al.* (2005), these methods and their further modifications achieved good separation of several compounds (Shick *et al.*, 1999), but none of these was able to separate in a single run the strongly acidic MAAs and the more weakly acidic compounds. Recently a high-resolution HPLC method based on reverse-phase C_{18} column and trifluoroacetic acid and an ammonium- containing mobile phase was developed (Carreto *et al.*, 2005). The method is selective enough to resolve in a single run a mixture of over 20 MAAs, including the critical and highly polar compounds shinorine, mycosporine-2-glycine, and palythine-serine, the medium polarity pair palythenic acid and shinorine methyl ester (M-333), and the low polarity isomeric pair usujirene and palythene (Figure 10.2).

10.5.7 MAA detection, identification and quantification

Mycosporine-like amino acids are usually identified based on their retention time during HPLC and their characteristic UV absorption spectra obtained via diode

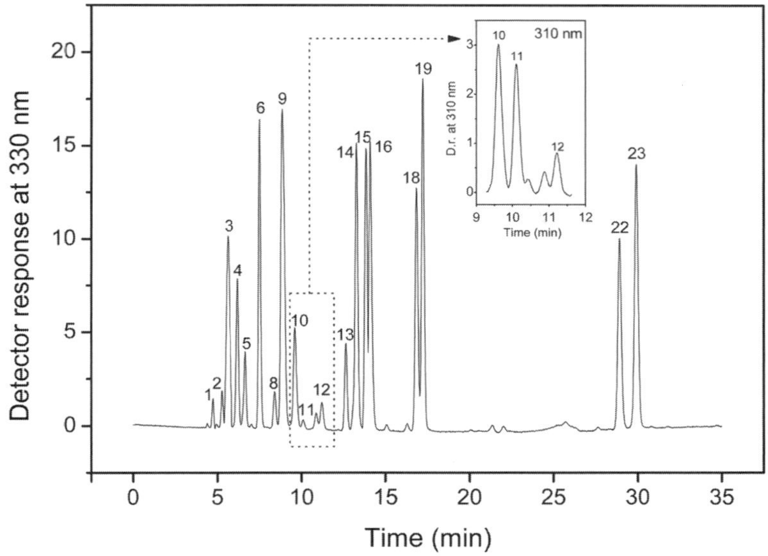

Figure 10.2. HPLC separation of MAAs covering three polarity ranges (modified from Carreto *et al.*, 2005). The insert shows the detection of mycosporine-glycine at 310 nm. Peak identification numbers: 1) palythine-serine sulfate; 2) mycosporine sulfate ester; 3) shinorine; 4) mycosporine-2-glycine; 5) palythine-serine; 6) palythine; 8) asterina-330; 9) porphyra-334; 10) mycosporine-methyl amine-serine; 11) mycosporine-glycine; 12) unknown; 13) palythine-threonine; 14) palythinol; 15) Z-palythenic acid; 16) shinorine methyl ester; 18) mycosporine-palythine-threonine; 19) mycosporine-taurine; 22) usujirene; 23) palythene.

array detection (DAD). Although DAD allows the fast acquisition of UV-Vis absorption spectra, the lack of fine spectral absorption, the influence of pH and – for some specific MAAs – wavelength absorption maxima identical or only a few nm apart (Table 10.1), makes it very difficult to distinguish MAA compounds based on absorption spectra only. In a similar way to that described for phytoplankton pigments (Mantoura and Repeta, 1997), MAAs could be quantified using their molar (ε) or specific extinction coefficients (*d*). However for some MAAs the ε is unknown. In these cases, the use of the ε of a MAA with similar chemical structure is suggested.

Mass spectrometry detection (HPLC-MS) can make an invaluable contribution towards the identification of MAAs because of its high sensitivity and the availability of powerful tandem mass spectrometric techniques (Whitehead and Hedges, 2002, 2003; Cardozo *et al.*, 2006; Volkman and Gorbushina, 2006; Carignan *et al.*, 2009). The mass spectral approach adds to LC/DAD the ability to distinguish essentially all known MAAs based on their specific retention times, wavelength maxima and molecular weights. Moreover, MS and tandem MS represent useful analytical tools for structural elucidation, especially in combination with soft ionization methods, like electrospray ionization (ESI). Whitehead *et al.* (2001) introduced LC-MS detection and quantification of MAAs using positive ionization via ESI at atmospheric pressure (API-ESI) to produce protonated molecular ions $[M+H]^+$. The use of controlled voltage (40 V) produces the highest ionization levels and minimal fragmentation of the $[M+H]^+$ ion. Concentrations were calculated on the integrated peak area of the $[M+H]^+$ chromatogram for each individual MAA extracted from the total ion current (TIC) (Whitehead and Hedges, 2002). These authors report that detection of MAAs using this technique was about 100 times more sensitive than by UV absorption. If single ion monitoring (SIM) is used, this sensitivity could be improved by a factor of 10–100. However, there are differences in ionization sensitivities between different MAAs and their quantification requires system calibration with authentic standards (Whitehead and Hedges, 2002).

The application of ESI-MS/MS methods represents a powerful tool, which is useful for structural elucidation using characteristic fragmentation patterns. Whitehead and Hedges (2002) observed that a higher cone voltage than that needed to obtain the molecular ion (> 40 V) produced an increased extent of MAA fragmentation. This fragmentation initiated small radical or neutral losses from the quasi-molecular ion $[M+H]^+$. This pattern, plus the presence of several common ions in all MAAs analyzed, should prove useful for identifying unknown MAAs (Whitehead and Hedges, 2002; 2003; Carignan *et al.*, 2009). Recently, based on a comparison of MS/MS spectra between deuterated and non-deuterated compounds, Cardozo *et al.* (2006) proposed a general fragmentation mechanism for some MAAs. The ε at their wavelength of maximum absorption and the characteristic mass fragments for MAA identification are listed in Table 10.1.

10.5.8 Recommendations for standards

The lack of commercially available MAA standards in both detection methods (DAD and MS) makes confirmation of identity and quantification difficult. Most reference MAAs must be obtained from biological sources, a process that is costly and time consuming. The fully characterized MAAs and the biological reference material and methods used for their identification are listed in Table 10.1.

In many cases the original biological reference materials are not available and therefore several alternative sources of MAAs should be used to prepare standards. When selecting the biological material it is important to recognize that MAAs found in a given cyanobacterium, alga, symbiotic consortium or metazoan depend on several factors (see Section 10.3). The original literature gives valuable information when formulating plans for preparation of reference MAAs. Chromatographic techniques involving gel permeation, ion-exchange resins, norite A, carbon and cellulose columns, preparative thin-layer chromatography on silica gel and semi-preparative HPLC have all been used in different combinations in the isolation and purification of MAAs (Bandaranayake, 1998 and references therein). Obviously, the technical details for the isolation of a specific MAA will depend on its polarity and the biological source used. The amounts and diversity of interference substances are higher in marine animals than in algae, and in many cases more sophisticated methods are needed. In our experience, purification on Dowex 50W followed by semi-preparative reverse-phase HPLC is the most convenient combination and gives the best results (Tsujino *et al.*, 1980; Bandaranayake, 1998; Carreto *et al.*, 2001).

Addendum

Recently, a MAA biosynthetic gene cluster was identified in a cyanobacterium, clarifying the origin of MAAs and revealing two unprecedented enzymatic strategies for imine formation (Balskus and Walsh (2010). The genetic and molecular basis for sunscreen biosynthesis in cyanobacteria. *Science* **329**, 1653–56).

Acknowledgements

We are grateful to Dr. G. Johnsen and two anonymous reviewers who provided helpful comments and suggestions on the manuscript.

Abbreviations and symbols

a_{CDOM}	Absorption by CDOM
ADEOS	Advanced Earth Observing Satellite
API-ESI	Atmospheric pressure – electrospray ionization
ESI	Electrospray ionization
ε	Molar extinction coefficient
GLI	Japanese Global Imager
ROS	Reactive oxygen species
SIM	Single ion monitoring
TIC	Total ion current
UV-A	Ultraviolet-A (315–400nm)
UV-B	Ultraviolet-B (280–315nm)

References

Arpin, N. and Bouillant, M.L. (1981). Light and mycosporines. In *The Fungal Spore, Morphogenetic Controls*, ed. H.R. Turiang Hohl. London: Academic Press, pp. 159–73.

Banaszak, A.T. and Trench, R.K. (1995). Effects of ultraviolet (UV) radiation on microalgal-invertebrate symbiosis. II. The synthesis of mycosporine-like amino acids in response to exposure to UV in *Anthopleura elegantissima* and *Cassiopeia xamachana*. *J. Exp. Mar. Biol. Ecol.* **194**, 233–50.

Banaszak, A.T., LaJeunesse, T.C. and Trench, R.K. (2000). The synthesis of mycosporine-like amino acids (MAAs) by cultured, symbiotic dinoflagellates. *J. Exp. Mar. Biol. Ecol.* **249**, 219–33.

Banaszak, A.T., Santos, M.G., LaJeunesse, T.C. and Lesser, M.P. (2006). The distribution of mycosporine-like amino acids (MAAs) and the phylogenetic identity of symbiotic dinoflagellates in cnidarian hosts from the Mexican Caribbean. *J. Exp. Mar. Biol. Ecol.* **337**, 131–46.

Bandaranayake, W.M. (1998). Mycosporines. Are they nature's sunscreens? *Nat. Prod. Rep.* **15**, 159–72.

Bandaranayake, W.M., Bemis, J.E. and Bourne, D.J. (1996). Ultraviolet absorbing pigments from the marine sponge *Dysidea herbacea*: isolation and structure of a new mycosporine. *Comp. Biochem. Physiol.* **115C**, 281–86.

Bidigare, R.R., Iriarte, J.L., Kang, S.-H., Karentz, D., Ondrusek, M.E. and Fryxell, G.A. (1996). Phytoplankton: quantitative and qualitative assessments. In *Foundations for Ecological Research West of the Antarctic Peninsula*, vol. 70, ed. R.M. Ross, E.E. Hofmann, L.B. Quentin. Washington: AGU, pp. 173–98.

Callone, A.I., Carignan, M.O., Montoya, N.G. and Carreto, J.I. (2006). Biotransformation of mycosporine-like amino acids (MAAs) in the toxic dinoflagellate *Alexandrium tamarense*. *J. Photochem. Photobiol. B: Biol.* **84**, 204–12.

Cardozo, K.H.M., Carvalho, V.M., Pinto, E. and Colepicolo, P. (2006). Fragmentation of mycosporine-like amino acids by hydrogen/deuterium exchange and electrospray ionisation tandem mass spectrometry. *Rap. Comm. Mass. Spec.* **20**, 253–58.

Carignan, M.O., Montoya, N.G. and Carreto, J.I. (2002). Long-term effects of ultraviolet radiation on the composition of pigment and mycosporine-like amino acids (MAAs). In *Aquaculture, Environment and Marine Phytoplankton*, ed. G. Arzul. Actes Colloque, vol. 34, Brest: IFREMER, pp. 191–207.

Carignan, M.O., Cardozo, K.H.M., Oliveira-Silva, D., Colepicolo, P. and Carreto J.I. (2009). Palythine-threonine, a major novel mycosporine-like amino acid (MAA) isolated from the hermatypic coral *Pocillopora capitata*. *J. Photochem. Photobiol B: Biol.* **94**, 191–200.

Carreto, J.I., De Marco, S. and Lutz, V.A. (1989). UV-absorbing pigments in the dinoflagellates *Alexandrium excavatum* and *Prorocentrum micans*: Effects of light intensity. In *Red Tides: Biology, Environmental Science and Toxicology*, ed. T. Okaichi, D.M. Anderson and T. Nemoto. New York: Elsevier, pp. 37–40.

Carreto, J.I., Carignan, M.O., Daleo, G. and De Marco, S.G. (1990a). Occurrence of mycosporine-like amino acids in the red-tide dinoflagellate *Alexandrium excavatum*: UV photoprotective compounds? *J. Plankton Res.* **12**, 909–21.

Carreto, J.I., Lutz, V.A., De Marco, S.G. and Carignan, M.O. (1990b). Fluence and wavelength dependence of mycosporine-like amino acid synthesis in the dinoflagellate *Alexandrium excavatum.* In *Toxic Marine Phytoplankton*, ed. E. Graneli, L. Edler and D.M. Anderson. New York: Elsevier, pp. 275–79.

Carreto, J.I., Carignan, M.O. and Montoya, N.G. (2001). Comparative studies on mycosporine-like amino acids, paralytic shellfish toxins and, pigment profiles of the toxic dinoflagellates *Alexandrium tamarense*, *A. catenella* and *A. minutum*. *Mar. Ecol. Prog. Ser.* **223**, 49–60.

Carreto, J.I., Carignan, M.O. and Montoya, N.G. (2002). Short-term effects of ultraviolet radiation on the dinoflagellate *Alexandrium catenella*. Pigment bleaching and MAAs synthesis inhibition. In *Aquaculture, Environment and Marine Phytoplankton*, ed. G. Arzul. Actes Colloque, vol. 34, Brest: IFREMER, pp. 173–90.

Carreto, J.I., Carignan, M.O. and Montoya, N.G. (2005). A high-resolution reverse-phase liquid chromatography method for the analysis of mycosporine-like amino acids (MAAs) in marine organisms. *Mar. Biol.* **146**, 237–52.

Carroll, A.K. and Shick, J.M. (1996). Dietary accumulation of mycosporine-like amino acids (MAAs) by the green sea urchin (*Strongylocentrotrus droebachiensis*). *Mar. Biol.* **124**, 561–69.

Castenholz, R.W. and Garcia-Pichel, F. (2000). Cyanobacterial responses to UV-radiation. In *The Ecology of Cyanobacteria: their Diversity in Time and Space*, ed. B.A. Whitton and M. Potts. Dordrecht: Kluwer, pp. 591–611.

Cockell, C.S. and Horneck, G. (2001). The history of the UV radiation climate on earth–Theoretical and space-based observations. *Photochem. Photobiol.* **73**, 447–51.

Conde, F.R., Churio, M.S. and Previtali, C.M. (2000). The photoprotector mechanism of mycosporine-like amino acids. Excited state properties and photostability of porphyra-334 in aqueous solution. *J. Photochem. Photobiol. B: Biol.* **56**, 139–44.

Conde, F.R., Carignan, M.O., Churio, M.S. and Carreto, J.I. (2003). *In vitro cis-trans* photoisomerization of palythene and usujirene. Implications on the in vivo transformation of mycosporine-like amino acids. *Photochem. Photobiol.* **77**, 146–50.

Conde, F.R., Churio, M.S. and Previtali, C.M. (2004). The deactivation pathways of the excited-states of the mycosporine-like amino acids shinorine and porphyra-334 in aqueous solution. *Photochem. Photobiol. Sci.* **3**, 960–67.

De Mora, S., Demers, S. and Vernet, M. (eds.) (2000). *The Effects of UV Radiation in the Marine Environment*. Cambridge: Cambridge University Press.

Dionisio-Sese, M.L., Ishikura, M., Maruyama, T. and Miyachi, S. (1997). UV-absorbing substances in the tunic of a colonial ascidian protect its symbiont, *Prochloron* sp., from damage by UV-B radiation. *Mar. Biol.* **128**, 455–61.

Dunlap, W.C. and Chalker, B.E. (1986). Identification and quantification of near-UV absorbing compounds (S-320) in a hermatypic scleractinian. *Coral Reefs* **5**, 1–5.

Dunlap, W.C. and Shick, J.M. (1998). Ultraviolet radiation-absorbing mycosporine-like aminoacids in coral reef organisms: a biochemical and environmental perspective. *J. Phycol.* **34**, 418–30.

Dunlap, W.C. and Yamamoto, Y. (1995). Small-molecule antioxidants in marine organisms: Antioxidant activity of mycosporine-glycine. *Comp. Biochem. Physiol.* **112**, 105–14.

Farman, J.C., Gardiner, B.G. and Shanklin, J.D. (1985). Large losses of total ozone in Antarctica reveals seasonal ClOx/NOx interactions. *Nature* **315**, 207–10.

Favre-Bonvin, J., Arpin, N. and Brevard, C. (1976). Structure de la mycosporine (P-310). *Can. J. Chem.* **54**, 1105–13.

Ferroni, L., Klisch, M., Pancaldi, S. and Häder, D. (2010). Complementary UV-absorption of mycosporine-like amino acids and scytonemin is responsible for UV-insensitivity of photosynthesis in *Nostoc flagelliforme*. *Mar. Drugs* **8**, 106–21.

Frame, E.R. (2004). *Mycosporine-like Amino Acids (MAAs) in Bloom Forming Phytoplankton: the Influence of Nitrogen, Ultraviolet Radiation and Species Composition*. Ph.D. thesis. University of California, San Diego, USA.

Franklin, L., Kräbs, G. and Kuhlenkamp, R. (2001). Blue light and UV-A radiation control the synthesis of mycosporine-like amino acids in *Chondrus crispus* (Florideophyceae). *J. Phycol.* **37**, 257–70.

Gröniger, A., Sinha, R.P., Klish, M. and Häder, D.-P. (2000). Photoprotective compounds in cyanobacteria, phytoplankton and macroalgae – a database. *J. Photochem. Photobiol. B: Biol.* **58**, 115–22.

Hannach, G. and Sigleo, A.C. (1998). Photoinduction of UV-absorbing compounds in six species of marine phytoplankton. *Mar. Ecol. Prog. Ser.* **174**, 207–22.

Helbling, E.W., Chalker, B.E., Dunlap, W.C., Holm-Hansen, O. and Villafañe, V.E. (1996). Photoacclimation of Antarctic marine diatoms to solar ultraviolet radiation. *J. Exp. Mar. Biol. Ecol.* **204**, 85–101.

Hernando, M., Carreto, J.I., Carignan, M.O., Ferreyra, G.A. and Gross, C. (2002). Effects of solar radiation on growth and mycosporine-like amino acids content in *Thalassiosira* sp., an Antarctic diatom. *Polar Biol.* **25**, 12–20.

Hirata, Y., Uemura, D., Ueda, K. and Takano, S. (1979). Several compounds from *Palythoa tuberculosa* (Coelenterata). *Pure Appl. Chem.* **51**, 1875–83.

Hoyer, K., Karsten, U. and Wiencke, C. (2002). Induction of sunscreen compounds in Antarctic macroalgae by different radiation conditions. *Mar. Biol.* **141**, 619–27.

Ito, S. and Hirata, Y. (1977). Isolation and structure of a mycosporine from the zoanthidian *Palythoa tuberculosa*. *Tetrahedron Lett.* **28**, 2429–30.

Jeffrey, S.W., MacTavish, H.S., Dunlap, W.C., Vesk, M. and Groenewould, K. (1999). Occurrence of UVA and UVB-absorbing compounds in 152 species (206 strains) of marine microalgae. *Mar. Ecol. Prog. Ser.* **189**, 35–51.

Kahru, M. and Mitchell, B.G. (1998). Spectral reflectance and absorption of a massive red tide off southern California. *J. Geophys. Res.* **103**, 21601–09.

Karentz, D. (2001). Chemical defences of marine organisms against solar radiation exposure: UV-absorbing mycosporine-like amino acids and scytonemin. In *Marine Chemical Ecology*, ed. J.B. Mc-Clintock and J. Baker. Boca Raton: CRC Press, pp. 481–520.

Karentz, D., Mc Euen, F.S., Land, M.V. and Dunlap, W.C. (1991). Survey of mycosporine-like amino acid compounds in Antarctic marine organisms: potential protection from ultraviolet exposure. *Mar. Biol.* **108**, 157–66.

Karsten, U., Bischof, K., Hanelt, D., Tüg, H. and Wienke, C. (1999). The effect of ultraviolet radiation on photosynthesis and ultraviolet-absorbing substances in the endemic macroalgae *Devaleraea ramentacea* (Rhodophyta). *Physiol. Plant.* **105**, 58–66.

Klisch, M. and Häder, D.-P. (2002). Wavelength dependence of mycosporine-like amino acid synthesis in *Gyrodinium dorsum*. *J. Photochem. Photobiol. B: Biol.* **66**, 60–66.

Klisch, M., Sinha, R.P., Richter, P.E. and Häder, D.-P. (2001). Mycosporine-like amino acids (MAAs) protect against UV damage in *Gyrodinium dorsum* Kofoid. *J. Plant. Physiol.* **158**, 1449–54.

Kobayashi, J., Nakamura, H. and Hirata, Y. (1981). Isolation and structure of a UV-absorbing substance from the ascidian *Halocynthia roretzi*. *Tetrahedron Lett.* **22**, 3001–02.

Kräbs, G., Watanabe, M. and Wiencke, C. (2004). A monochromatic action spectrum for the photoinduction of the UV-absorbing mycosporine-like amino acid shinorine in the red alga *Chondrus crispus*. *Photochem. Photobiol.* **79**, 515–19.

Laurion, I., Blouin, F. and Roy, S. (2003). The quantitative filter technique for measuring phytoplankton absorption: interference by MAAs in the UV waveband. *Limnol. Oceanogr. Methods* **1**, 1–9.

Laurion, I., Blouin, F. and Roy, S. (2004). Packaging of mycosporine-like amino acids in dinoflagellates. *Mar. Ecol. Prog. Ser.* **279**, 297–303.

Laurion, I. and Roy, S. (2009). Growth and photoprotection in three dinoflagellates (including two strains of *Alexandrium tamarense*) and one diatom exposed to four weeks of natural and enhanced ultraviolet-B radiation. *J. Phycol.* **45**, 16–33.

Lesser, M.P. (1996a). Elevated temperatures and ultraviolet radiation cause oxidative stress and inhibit photosynthesis in symbiotic dinoflagellates. *Limnol. Oceanogr.* **41**, 271–83.

Lesser, M.P. (1996b). Acclimation of phytoplankton to UV-B radiation: Oxidative stress and photoinhibition of photosynthesis are not prevented by UV-absorbing compounds in the dinoflagellate *Prorocentrum micans*. *Mar. Ecol. Prog. Ser.* **132**, 287–97.

Lesser, M.P. (2000). Depth-dependent photoacclimatization to solar ultraviolet radiation in the Caribbean coral *Montastraea faveolata*. *Mar. Ecol. Prog. Ser.* **192**, 137–51.

Litchman, E., Neale, P.J. and Banaszak, A.T. (2002). Increased sensitivity to ultraviolet radiation in nitrogen-limited dinoflagellates: Photoprotection and repair. *Limnol. Oceanogr.* **47**, 86–94.

Llewellyn, C.A. and Harbour, D.S. (2003). A temporal study of mycosporine-like amino acids in surface water phytoplankton from the English Channel and correlation with solar irradiation. *J. Mar. Biol. Ass. U.K.* **83**, 1–9.

Mantoura, R.F.C. and Repeta, D.J. (1997). Calibration methods for HPLC. In *Phytoplankton Pigments in Oceanography: Guidelines to Modern Methods*, ed. S.W. Jeffrey, R.F.C. Mantoura and S.W. Wright. Paris: UNESCO Publishing, pp. 407–28.

Mantoura, R.F.C., Wright, S.W., Jeffrey, S.W., Barlow, R.G. and Cummings, D.E. (1997). Filtration and storage of pigments from microalgae. In *Phytoplankton Pigments in Oceanography: Guidelines to Modern Methods*, ed. S.W. Jeffrey, R.F.C. Mantoura and S.W. Wright. Paris: UNESCO Publishing, pp. 283–306.

Marshall, J.A. and Newman, S. (2002). Differences in photoprotective pigment production between Japanese and Australian strains of *Chattonella marina* (Raphidophyceae). *J. Exp. Mar. Biol. Ecol.* **272**, 13–27.

Moeller, R.E., Gilroy, S., Williamson, C.E., Grad, G. and Sommaruga, R. (2005). Dietary acquisition of photoprotective compounds (mycosporine-like amino acids, carotenoids) and acclimation to ultraviolet radiation in a freshwater copepod. *Limnol. Oceanogr.* **50**, 427–39.

Moisan, T.A. and Mitchell, B.G. (2001). UV absorption by mycosporine-like amino acids in *Phaeocystis antarctica* Karsten induced by photosynthetically available radiation. *Mar. Biol.* **138**, 217–27.

Morel, A., Gentili, B., Claustre, H., Babin, M., Bricaud, A., Ras, J. and Tièche, F. (2007). Optical properties of the ‘clearest’ natural waters. *Limnol. Oceanogr.* **52**, 217–29.

Morrison, J.R. and Nelson, N.B. (2004). Seasonal study of phytoplankton UV absorption at the Bermuda Atlantic Time-series Study (BATS) site. *Limnol. Oceanogr.* **49**, 215–24.

Nakamura, H., Kobayashi, J. and Hirata, Y. (1981). Isolation and structure of a 330 nm UV-absorbing substance, asterina-330 from the starfish *Asterina pectinifera*. *Chem. Lett. Jpn* **10**, 1413–14.

Nakamura, H., Kobayashi, J. and Hirata, Y. (1982). Separation of mycosporine-like amino acids in marine organisms using reversed-phase high-performance liquid chromatography. *J. Chromatogr.* **250**, 113–18.

Neale, P.J., Banaszak, A.T. and Jarriel, C.R. (1998). Ultraviolet sunscreens in *Gymnodinium sanguineum* (Dinophyceae): mycosporine-like amino acids protect against inhibition of photosynthesis. *J. Phycol.* **34**, 928–38.

Newman, S.J., Dunlap, W.C., Nicol, S. and Ritz, D. (2000). Antarctic krill (*Euphausia superba*) acquire a UV-absorbing mycosporine-like amino acid from dietary algae. *J. Exp. Mar. Biol. Ecol.* **255**, 93–110.

Oren, A. (1997). Mycosporine-like amino acids as osmotic solutes in a community of halophilic cyanobacteria. *Geomicrobiol, J.* **14**, 231–40.

Portwich, A. and Garcia-Pichel, F. (1999). Ultraviolet and osmotic stress induce and regulate the synthesis of mycosporines in the cyanobacterium *Chlorogloeopsis* PCC 6912. *Arch. Microbiol.* **172**, 187–92.

Portwich, A. and Garcia-Pichel, F. (2000). A novel prokaryotic UV-B photoreceptor in the cyanobacterium *Chlorogloeopsis* PCC 6912. *Photochem. Photobiol.* **71**, 493–98.

Portwich, A. and Garcia-Pichel, F. (2003). Biosynthetic pathway of mycosporines (mycosporine-like amino acids) in the cyanobacterium *Chlorogloeopsis* sp. strain PCC 6912. *Phycologia* **42**, 384–92.

Riegger, L. and Robinson, D. (1997). Photoinduction of UV-absorbing compounds in Antarctic diatoms and *Phaeocystis antarctica*. *Mar. Ecol. Prog. Ser.* **160**, 13–25.

Roy, S. (2000). Strategies for the minimisation of UV-induced damage. In *The Effects of UV Radiation in the Marine Environment*, ed. S. de Mora, S. Demers and M. Vernet. Cambridge: Cambridge University Press, pp. 177–205.

Rozema, J., Björn, L.O., Bornman, J.F., Gaberščik, A., Häder, D.-P., Trošt, T., Germ, M., Klisch, M., Gröniger, A., Sinha, R.P., Lebert, M., He, Y.-Y., Buffoni-Hall, R., de Bakker, N.V.J., van de Staaij, J. and Meijkamp, B.B. (2002). The role of UV-B radiation in aquatic and terrestrial ecosystems-an experimental and functional analysis of the evolution of UV-absorbing compounds. *J. Photochem. Photobiol B: Biol.* **66**, 2–12.

Sekikawa, I., Kubota, C., Hiraoki, T. and Tsujino, I. (1986). Isolation and structure of a 357 nm UV absorbing substance, usujirene, from the red alga *Palmaria palmata* (L.) O. Kuntze. *Jpn. J. Phycol.* **34**, 185–88.

Shibata, K. (1969). Pigments and a UV-absorbing substance in corals and a blue-green alga living in the Great Barrier Reef. *Plant Cell Physiol.* **10**, 325–35.

Shick, J.M. (2004). The continuity and intensity of ultraviolet irradiation affect the kinetics of biosynthesis, accumulation, and conversion of mycosporine-like amino acids (MAAs) in the coral *Stylophora pistillata*. *Limnol. Oceanogr.* **49**, 442–58.

Shick, J.M. and Dunlap, W.C. (2002). Mycosporine-like amino acids and related gadusols: biosynthesis, accumulation, and UV-protective functions in aquatic organisms. *Ann. Rev. Physiol.* **64**, 223–62.

Shick, J.M., Romaine-Lioud, S., Ferrier-Pagès, C. and Gattuso, J.P. (1999). Ultraviolet-B radiation stimulates shikimate pathway-dependent accumulation of mycosporine-like amino acids in the coral *Stylophora pistillata* despite decreases in its population of symbiotic dinoflagellates. *Limnol. Oceanogr.* **44**, 1667–82.

Shick, J.M., Dunlap, W.C., Pearse, J.S. and Pearse, A.V. (2002). Mycosporine-like amino acid content in four species of sea anemones in the genus *Anthopleura* reflects phylogenetic but not environmental or symbiotic relationships. *Biol. Bull.* **203**, 315–30.

Sinha, R.P., Klisch, M., Gröninger, A. and Häder, D.-P. (1998). Ultraviolet-absorbing/screening substances in cyanobacteria, phytoplankton and macroalgae. *J. Photochem. Photobiol. B: Biol.* **47**, 83–94.

Sinha, R.P., Klisch, M., Helbling, E.W. and Häder, D.-P. (2001). Induction of mycosporine-like amino acids (MAAs) in cyanobacteria by solar ultraviolet-B radiation. *J. Photochem. Photobiol. B: Biol.* **60**, 129–35.

Sinha, R.P., Ambasht, N.K., Sinha, J.P., Klisch, M. and Häder, D.-P. (2003). UV-B-induced synthesis of mycosporine-like amino acids in three strains of *Nodularia* (cyanobacteria). *J. Photochem. Photobiol. B: Biol.* **71**, 51–58.

Sivalingam, P.M., Ikawa, T., Yokohama, Y. and Nisizawa, K. (1974). Distribution of a 334 UV-absorbing-substance in algae, with special regard of its possible physiological roles. *Bot. Mar.* **17**, 23–29.

Sommaruga, R., Libkind, D., van Broock, M. and Whitehead, K. (2004). Mycosporine-glutaminol-glucoside, a UV-absorbing compound of two *Rhodotorula* yeast species. *Yeast* **21**, 1077–81.

Sommaruga, R., Whitehead, K., Shick, J.M. and Lobban, C.S. (2006). Mycosporine-like amino acids in the zooxanthella-ciliate symbiosis *Maristentor dinoferus*. *Protist* **157**, 185–91.

Sonntag, B., Summerer, M. and Sommaruga, R. (2007). Sources of mycosporine-like amino acids in planktonic *Chlorella*-bearing ciliates (Ciliophora). *Freshw. Biol.* **52**, 1476–85.

Sosik, H.M. (1999). Storage of marine particulate samples for light-absorption measurements. *Limnol. Oceanogr.* **44**, 1139–41.

Starcevic, A., Akthar, S., Dunlap, W.C., Shick, J.M., Hranueli, D., Cullum, J. and Long, P.F. (2008). Enzymes of the shikimic acid pathway encoded in the genome of a basal metazoan, *Nematostella vectensis*, have microbial origins. *Proc. Nat. Acad. Sci. USA* **105**, 2533–37.

Steinberg, D.K., Nelson, N.B., Carlson, C.A. and Prusak, A.C. (2004). Production of chromophoric dissolved organic matter (CDOM) in the open ocean by zooplankton and the colonial cyanobacterium *Trichodesmium* spp. *Mar. Ecol. Prog. Ser.* **267**, 45–56.

Stochaj, W.R., Dunlap, W.C. and Shick, J.M. (1994). Two new UV-absorbing mycosporine-like amino acids from the sea anemone *Anthopleura elegantissima* and the effects of zooxanthellae and spectral irradiance on chemical composition and content. *Mar. Biol.* **118**, 149–56.

Stumpf, R.P. (2006). *Forecasting harmful algal blooms: the roles of optical oceanography and remote sensing*. Ocean Optics XVIII Conf., Montreal, Canada, 9–13 Oct. 2006.

Subramanian, A., Carpenter, E.J., Karentz, D. and Falkowski, P.G. (1999). Bio-optical properties of the marine diazotrophic cyanobacteria *Trichodesmium* spp. I. Absorption and photosynthetic action spectra. *Limnol. Oceanogr.* **44**, 608–17.

Suhn, H., Lee, H. and Jung, J. (2003). Mycosporine glycine protects biological systems against photodynamic damage by quenching singlet oxygen with a high efficiency. *Photochem. Photobiol.* **78**, 109–13.

Taira, H., Aoki, S., Yamanoha, Y. and Taguchi, S. (2004). Daily variation in cellular content of UV-absorbing compounds mycosporine-like amino acids in the marine dinoflagellate *Scrippsiella sweeneyae*. *J. Photochem. Photobiol. B: Biol.* **75**, 145–55.

Takano, S., Uemura, D. and Hirata, Y. (1978). Isolation and structure of two new amino acids, palythinol and palythene, from the zoanthid *Palythoa tuberculosa*. *Tetrahedron Lett.* **49**, 4909–12.

Takano, S., Uemura, D. and Hirata, Y. (1979). Isolation and structure of a 334 nm UV absorbing substance, porphyra-334 from the red algae *Porphyra tenera* Kjellman. *Chem. Lett. Jpn* **8**, 419–20.

Tartarotti, B. and Sommaruga, R. (2002). The effect of different methanol concentrations and temperatures on the extraction of mycosporine-like amino acids (MAAs) in algae and zooplankton. *Arch. Hydrobiol.* **154**, 691–703.

Teai, T., Raharivelomanana, P., Bianchini, J.P., Faura, R., Martín, P.M.V. and Cambon, A. (1997). Structure de deux nouvelles iminomycosporines isolées de *Pocillopora eydouxy*. *Tetrahedron Lett.* **38**, 5799–800.

Teai, T., Drollet, J.H., Bianchini, J.P., Cambon, A. and Martín, P.M.V. (1998). Occurrence of ultraviolet radiation-absorbing mycosporine-like amino acids in coral mucus and whole corals of French Polynesia. *Mar. Freshwat. Res.* **49**, 127–32.

Tsujino, I., Yabe, K., Sekikawa, I. and Hamanaka, N. (1978). Isolation and structure of a mycosporine from the red alga *Chondrus yendoi*. *Tetrahedron Lett.* **16**, 1401–02.

Tsujino, I., Yabe, K. and Sekikawa, I. (1980). Isolation and structure of a new amino acid, shinorine from the red alga *Chondrus yendoi* Yamada et Mikami. *Bot. Mar.* **23**, 65–68.

UNEP: United Nations Environment Programme, Environmental Effects Assessment Panel (2006). Environmental effects of ozone depletion and its interactions with climate change: Progress report, 2005. *Photochem. Photobiol. Sci.* **5**, 13–24.

Vernet, M. and Whitehead, K. (1996). Release of ultraviolet-absorbing compounds by the red-tide dinoflagellate *Lingulodinium polyedra*. *Mar. Biol.* **127**, 35–44.

Vincent, W.F. and Neale, P.J. (2000). Mechanisms of UV damage to aquatic organisms. In *The Effects of UV Radiation in the Marine Environment*, ed. S. de Mora, S. Demers and M. Vernet. Cambridge: Cambridge University Press, pp. 149–76.

Volkmann, M. and Gorbushina, A.A. (2006). A broadly applicable method for extraction and characterization of mycosporines and mycosporine-like amino acids of terrestrial, marine and freshwater origin. *Microbiol. Lett.* **255**, 286–95.

Volkmann, M., Gorbushina, A.A., Kedar, L. and Oren, A. (2006). Structure of euhalothece-362, a novel red-shifted mycosporine-like amino acid, from a halophilic cyanobacterium (*Euhalothece* sp.). *Microbiol. Lett.* **258**, 50–54.

Whitehead, K. and Hedges, J.I. (2002). Analysis of mycosporine-like amino acids in plankton by liquid chromatography electrospray ionization mass spectrometry. *Mar. Chem.* **80**, 27–39.

Whitehead, K. and Hedges, J.I. (2003). Electrospray ionization tandem mass spectrometric and electron impact mass spectrometric characterization of mycosporine-like amino acids. *Rapid Commun. Mass Spectrom.* **17**, 2133–38.

Whitehead, K. and Hedges, J.I. (2005). Photodegradation and photosensitization of mycosporine-like aminoacids. *J. Photochem. Photobiol. B: Biol.* **80**, 115–21.

Whitehead, K. and Vernet, M. (2000). Influence of mycosporine-like amino acids (MAAs) on UV absorption by particulate and dissolved organic matter in La Jolla Bay. *Limnol. Oceanogr.* **45**, 1788–96.

Whitehead, K., Karentz, D. and Hedges, J.I. (2001). Mycosporine-like amino acids (MAAs) in phytoplankton, a herbivorous pteropod (*Limacina helicina*), and its pteropod predator (*Clione antarctica*) in McMurdo Bay, Antarctica. *Mar. Biol.* **139**, 1013–19.

Wu Won, J.J., Rideout, J.A. and Chalker, B.E. (1995). Isolation and structure of a novel mycosporine-like amino acid from the reef-building corals *Pocillopora damicornis* and *Stylophora pistillata*. *Tetrahedron Lett.* **36**, 5255–56.

Wu Won, J.J., Chalker, B.E. and Rideout, J.A. (1997). Two new UV-absorbing compounds from *Stylophora pistillata*: sulfate esters of mycosporine-like amino acids. *Tetrahedron Lett.* **38**, 2525–26.

Zhang, L., Li, L. and Wu, Q. (2007). Protective effects of mycosporine-like amino acids of *Synechocystis* sp. PCC 6803 and their partial characterization. *J. Photochem. Photobiol. B: Biol.* **86**, 240–45.

Zudaire, L. and Roy, S. (2001). Photoprotection and long-term acclimation to UV radiation in the marine diatom *Thalassiosira weissflogii*. *J. Photochem. Photobiol. B: Biol.* **62**, 26–34.

Part IV

Selected pigment applications in oceanography

11

Pigments and photoacclimation processes

CHRISTOPHE BRUNET, GEIR JOHNSEN, JOHANN LAVAUD AND SUZANNE ROY

11.1 Introduction

This chapter reviews the nature of pigment variations in phytoplankton in response to changes in light regime (irradiance, spectral composition and day length). These changes belonging to processes called acclimation and/or adaptation maximize the evolutionary fitness of a species, within the constraints set by the environmental conditions (Raven and Geider, 2003). In general, adaptation indicates long-term evolutionary outcome based on the genes present in a given species (genetic adaptation) while acclimation denotes adjustments in response to variation in key environmental variables (physiological acclimation).

Photo-acclimation corresponds to a mosaic of processes involving many cellular components and occurring over a broad range of time scales, from seconds to days. These processes, covering many physiological, biochemical, biophysical and biological changes, allow the optimization of cell activities, such as photosynthesis, respiration, growth and division, when faced with changing irradiance (e.g. Herzig and Dubinsky, 1993; Anning *et al.*, 2000; Raven and Geider, 2003). This is an important issue in phytoplankton ecology because of the fluctuating light environment experienced by pelagic algae, related to daylight variations together with the exponential decrease of light and the vertical – active or passive – movements of algae along the water column.

In order to cope with these never-ending fluctuations in light-regime, marine phytoplankton can adjust their pigment pool, which mainly consists of two functional categories, namely pigments used for light harvesting and for photoprotection. Many accessory pigments constituting the light-harvesting complexes are photosynthetically active i.e. they are able to transfer the energy absorbed from sunlight to the photosynthetic reaction centers (RC) of photosystems (PS) II and I. They are called light-harvesting pigments and include the photosynthetic carotenoids. However, some carotenoids are not involved in photosynthesis and do not transfer the absorbed energy to the RC. These non-photosynthetically active carotenoids are called photoprotective carotenoids (PPC).

Phytoplankton Pigments: Characterization, Chemotaxonomy and Applications in Oceanography, eds. Suzanne Roy, Carole A. Llewellyn, Einar Skarstad Egeland and Geir Johnsen. Published by Cambridge University Press.

The function and dynamics of long-term (hours-days) and short-term (minutes-hours) photo-acclimation are described in the following two sections (11.2 and 11.3, respectively). The long-term photo-acclimation response mainly consists in changes of structure and composition of the photosystems while the short-term photo-acclimation process mainly concerns the xanthophyll cycle (XC) activation and the associated non-photochemical fluorescence quenching (NPQ). In the fourth section (11.4), the ecophysiological variability of XC and its use as a biological tracer in oceanographic studies is examined.

11.2 Long-term photoacclimative processes

In general, photoacclimation in a living cell is characterized by (i) changes in the amounts and ratios of light-harvesting pigments and photoprotective carotenoids, (ii) photosynthetic parameters, (iii) enzymatic activities involved in photosynthesis and respiration, and finally, (iv) cell volume and chemical composition (Falkowski and LaRoche, 1991). In this section, we will focus on changes in chloroplasts, light-harvesting complexes, pigment composition and function.

11.2.1 Chloroplast size, number, morphology and distribution

The different phytoplankton classes and pigment groups show huge differences in chloroplast size, numbers, morphology and distribution (Kirk and Tilney-Bassett, 1978; Larkum and Vesk, 2003). The chloroplast number is species specific and varies from one chloroplast per cell to more than 100. As an example, some species of the diatom genus *Chaetoceros* only contain one, while others have two, six or more than ten chloroplasts per cell. The species-specific differences in chloroplast size (typically 0.2–2 μm in length) and morphology (shape and structure) in a given species are also affected by the light climate (irradiance, spectral composition and day length). Light-induced chloroplast changes in a given species will especially affect light-harvesting and utilization (Raven and Geider, 2003), seen as changes in intracellular self-shading (the package effect, see Chapter 13, this volume) and the optical signature from the chloroplast (colour, optical density, and *in vivo* fluorescence emission: Falkowski and Chen, 2003; Sakshaug and Johnsen, 2005; Johnsen and Sakshaug, 2007). Typically, low light (LL)-acclimated cells have chloroplasts evenly distributed in the cells (large light-harvesting surface), while high light (HL)-acclimated cells have condensed chloroplasts (small light-absorbing surface, Blatt *et al.*, 1981).

11.2.2 Light-harvesting complexes and thylakoid membranes

The majority of pigments in phytoplankton cells are situated inside the thylakoid membranes as discrete light-harvesting complexes, made up of pigment–protein complexes. The build-up of the different light-harvesting complexes in the different

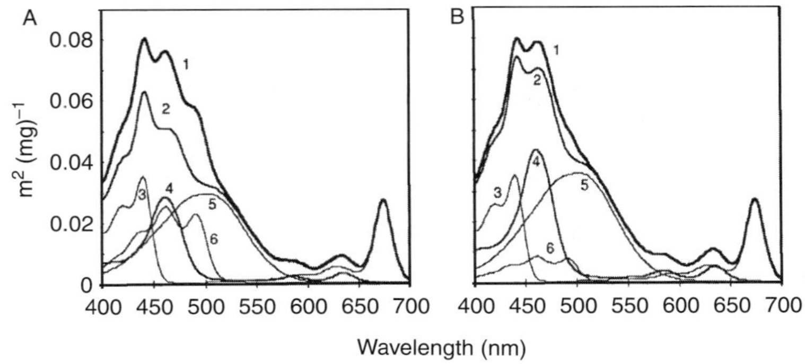

Figure 11.1. Fractional pigment-specific absorption and the effect of the light-harvesting pigments and PPC in (A) high light- and (B) low light-acclimated cells of the dinoflagellate *Prorocentrum minimum*. 1: total pigments; 2: photosynthetic pigments (total pigments minus diadinoxanthin); 3: Chl *a*; 4: Chl c_2; 5: peridinin; 6: diadinoxanthin (from Johnsen *et al.* 1994).

phytoplankton classes is discussed in Chapter 13. The light-harvesting complexes in different phytoplankton classes differ in thylakoid membrane organization and in energy regulation mechanisms (Green *et al.*, 2003). Most green algal thylakoids have both stacked and unstacked membrane regions, but often with the same PSI:PSII ratio in both regions (Berthos and Gibbs, 1998). In contrast, the chromophytes comprising Bacillariophyceae, Cryptophyceae, Dinophyceae, Pelagophyceae, Eustigmatophyceae, Chrysophyceae, Bolidophyceae, Pinguiophyceae, Raphidophyceae, Dictyochophyceae and Haptophyta (comprising the two classes Coccolithophyceae and Pavlovophyceae) do not have lateral segregation of PSI and PSII (Green *et al.*, 2003). Lateral segregation of PSI and PSII in Rhodophyta and in most Chlorophyta is related to the ability to perform state transitions (see Chapter 13).

The majority of chromophytes contain large light-harvesting complexes associated with PSII where the fraction of light-harvesting pigments and PPC is tuned as a response to light history. The fraction of pigment-specific absorption of light is highly dependent on the photo-acclimation status of algae and on water colour (Johnsen *et al.*, 1994; Schofield *et al.*, 1996; Stolte *et al.*, 2000; Johnsen and Sakshaug, 2007; Figure 11.1). Generally HL-acclimated cells are characterized by low light-harvesting pigment content and a corresponding high amount of PPC, while the inverse relationship is observed for LL-acclimated cells (Johnsen *et al.*, 1997; Stolte *et al.*, 2000; Falkowski and Chen, 2003; Rodríguez *et al.*, 2006).

Isolated and functional light-harvesting complexes (that are able to transfer light energy to acceptor chlorophyll *a*) differ in LL and HL-acclimated cells of a given species (Johnsen *et al.*, 1994; 1997). For example, HL-acclimated dinoflagellates may have a significant amount of diadinoxanthin in the antenna (denoted chlorophyll *a*-chlorophyll *c*-peridinin protein = ACP) that does not transfer light energy to chlorophyll *a*. This can be seen by using spectral absorption coefficients (indicating the total amount of light absorbed) and fluorescence excitation spectra (scaled at the red peak of chlorophyll *a*, assuming 100% light energy transfer efficiency) indicating the fraction of light that is received by acceptor chlorophyll *a*. This analysis shows that ACP antennae from HL acclimated cells with high photoprotective carotenoids (PPC, diadinoxanthin and diatoxanthin) relative to chlorophyll *a* (49% w:w) have low relative light transfer efficiency at the absorption peaks of the PPCs at 490 (15–20%) and 460 nm (20–30%, Johnsen *et al.*, 1997). In contrast, ACP antennae from LL-acclimated cells with 15% PPC relative to chlorophyll *a*, reach 70–80% relative light energy transfer efficiency at 490 and 460 nm. Johnsen *et al.* (1997) also showed that wavelengths from 550–700 nm, where PPCs do not absorb, have close to 100% relative light energy transfer efficiency.

The energy regulation mechanisms in the light-harvesting complexes, which differ between phycobiliprotein-, chlorophyll *c*- and chlorophyll *b*-containing phytoplankton, include state transitions (e.g. phosphorylation and de-phosphorylation causing movement of light-harvesting complexes in the PS II) and activation of the xanthophyll cycle caused by changes in pH (ΔpH, see Section 11.3). State transitions and the xanthophyll cycle affect the effective absorption cross-section of PSII, σ_{PSII}, defined as the efficiency of absorbed quanta to drive a photochemical reaction in angstroms or tenths of nm (nm^2 quanta^{-1}), reflecting the quantum yield of charge separation (Falkowski and Chen, 2003).

In contrast, the spectrally integrated chlorophyll *a*-specific absorption coefficient (400–700 nm), $\bar{a}^*_\varphi$ (= optical cross-section, units: m^2 (mg chlorophyll *a*)$^{-1}$) indicates the total pigment–protein absorption by the cell. Both $\bar{a}^*_\varphi$ and σ_{PSII} are spectrally dependent, inducing pigment-specific differences in photosynthetic efficiency as a function of wavelength. The effective absorption cross-section of PSII, σ_{PSII}, can be determined for a single wavelength and extrapolated to other wavelengths from a knowledge of the PSII-fluorescence excitation spectra (Johnsen *et al.*, 1997; Falkowski and Chen, 2003; Johnsen and Sakshaug, 2007). Changes in the cross-sectional area of PSII and PSI due to re-association of mobile light-harvesting complexes of PS II (i.e. state transitions) contribute to < 20% change in cross-sectional areas (Larkum, 2003). This indicates that it is the light-harvesting complexes and their pigments that contribute to the major fraction of photoacclimation in phytoplankton and that state transitions (which occur on a scale of minutes) only contribute to a small fraction in absorption cross-section changes (Larkum, 2003; Raven and Geider, 2003).

11.2.3 Pigment composition and function

The high pigment diversity in light-harvesting complexes, relative to the more conservative pigment composition in PSII and PSI, is responsible for the high pigment-group specific differences in light-harvesting and utilization (see Chapter 1). In most bloom-forming phytoplankton, some light-harvesting pigments are of high importance; these include peridinin, prasinoxanthin, violaxanthin and the fucoxanthins (including acyloxy derivatives). From the ratios between photosynthetic pigments, distinct pigment groups have been defined in the Prasinophyceae (e.g. Hooks *et al.*, 1988; Egeland *et al.*, 1995; Zingone *et al.*, 2002) or in the Haptophyta (e.g. Stolte *et al.*, 2000; Zapata *et al.*, 2004).

Generally, chlorophyll *a* is not an important pigment in light-harvesting. Its major role is to receive light energy from donor pigments in light-harvesting complexes and it is one of the key molecules in the photochemical conversion of light energy to chemically bonded energy in the RC. Usually pigment data are normalized to chlorophyll *a* as an estimation of biomass. Since cellular chlorophyll *a* content in a given species is highly light-regime dependent, the less variable particulate organic carbon (POC) is a better biomass indicator (Johnsen and Sakshaug, 1993; Brunet *et al.*, 1996; Rodríguez *et al.*, 2006). The interpretation of light-harvesting pigments and PPC in HL and LL conditions is therefore highly dependent on the type of biomass normalization (chlorophyll *a* versus POC, Rodríguez *et al.*, 2006; Johnsen and Sakshaug, 2007). The chlorophyll *a*:C ratio (w:w) is low in high light (and long day length) and high in low light. Thus, the chlorophyll *a*:C ratio indicates the photo-acclimation status and is termed the photo-acclimation index (Sakshaug *et al.*, 1997). The photo-acclimation index can be described as a function of absorbed quanta (Nielsen and Sakshaug, 1993). Averaged chlorophyll *a*:C ratios for HL- and LL-acclimated cells of 10 phytoplankton classes were 0.020 and 0.043, respectively (Johnsen and Sakshaug, 2007). Since chlorophyll *a* and light-harvesting pigments co-vary (Johnsen *et al.*, 1997; Rodríguez *et al.*, 2006; Johnsen and Sakshaug, 2007), the variation in light-harvesting pigments:C ratio will follow the corresponding variation in chlorophyll *a*:C ratio. In general, there are some common long-term photo-acclimation characteristics. The faster the growth rate, the faster the acclimation process, since cells are dependent on rapid division to adjust to the key environmental variables. Light history also has a relevant effect on the overall acclimative response of the cells (e.g. Anning *et al.*, 2000) and can be a determinant factor for the kinetics of some processes and pigment variations (Anning *et al.*, 2000; Dimier *et al.*, 2007b).

11.3 The xanthophyll cycle and short-term photoacclimation

11.3.1 Role and regulation of the xanthophyll cycle

When chlorophyll *a* molecules of the light-harvesting complexes antennae absorb light, they enter a singlet-state excitation denoted ^{1}chlorophyll *a** from which energy is deactivated following several pathways. Most of the excitation energy is used to drive photochemistry, through charge separation within the reaction center of photosystems, with some associated leaks: re-emission of the energy via chlorophyll fluorescence and heat. There is nevertheless a non-negligible part of the energy that can be dissipated through the 'triplet valve' thereby forming triplet-state excitation: chlorophyll a + light $\rightarrow$ ^{1}chlorophyll a^* $\rightarrow$ ^{3}chlorophyll a^*. This pathway depends on the lifetime of ^{1}chlorophyll a^* which itself depends on the other deactivation pathways. When the light absorbed is in excess (i.e. under a high light exposure) and the ability of the photosynthetic machinery to use the excitation energy via photochemistry is at its maximum, the yield for chlorophyll *a* fluorescence increases and the probability of ^{3}chlorophyll a^* formation increases. This situation is critical since ^{3}chlorophyll a^* can react with oxygen (O_2) within the PS II reaction center, generating reactive O_2 species such as singlet $^1O_2^*$ which are very harmful for proteins, pigments and lipids and lead to a decrease in the rate of photosynthesis. Photosynthetic organisms are able to maintain a low steady state of ^{3}chlorophyll a^* generation through several rapid 'photoprotective' mechanisms which help to minimize the production of reactive oxygen species. Non-photochemical fluorescence quenching (NPQ) is believed to be the most important of these processes and the carotenoid xanthophylls play a central role in photoprotection, especially via the xanthophyll cycle (XC).

The xanthophyll cycle involves the enzymatic de-epoxidation/epoxidation of acetylenic xanthophylls, synthesized from β,β-carotene (Lohr and Wilhelm, 2001), as a function of absorbed quanta. There are two groups of organisms which can be defined on the basis of the pigments involved in the XC (Table 11.1). A first group includes as the main XC the two-step de-epoxidation of violaxanthin (Viola) into zeaxanthin (Zea) via antheraxanthin (Anth). A second group includes the one-step de-epoxidation of diadinoxanthin (Diadino) into diatoxanthin (Diato), Diato showing the same degree of de-epoxidation as Zea. A third group includes the phyla in which there is no XC but an accumulation of Zea directly from β,β-carotene under high light exposure (a few species of red macroalgae show a XC, Raven and Geider, 2003). Within group 1, some prasinophytes have been shown to be unable to convert Viola further than Anth (Goss *et al.*, 1998) and some green macroalgae have no XC (Raven and Geider, 2003). In addition, a second XC, which is not always minor, has been reported in several plant species and in the chlorophyte *Chlamydomonas* involving the de-epoxidation of lutein-epoxide into lutein under certain circumstances like prolonged high light stress (Rascher and Nedbal, 2006).

Additionally, in the green macroalga *Caulerpa*, a secondary XC involving conversion between lutein and siphonaxanthin (biosynthetically related to siphonein) has been reported (Raniello *et al.*, 2006). Within the second group, there are also some phyla showing a second XC. These include some Heterokontophyta (Bacillariophyceae, Chrysophyceae, Xanthophyceae), Haptophyta and Dinophyta, showing the Viola cycle under prolonged high light stress (Lohr and Wilhelm, 2001). It is still unclear whether the temporary accumulation of Zea under high light conditions is only an unavoidable consequence of the properties of the XC or if it has a real physiological significance by increasing the photoprotective ability of the chloroplast. Interestingly, for an unknown reason, some very close phyla among the Heterokontophytes evolved towards the Viola cycle as the main XC (the brown algae) while others (like the diatoms) evolved towards the Diadino cycle.

Table 11.1. *Distribution of the major photosynthetic phyla according to the nature of their main xanthophyll cycle (XC). Group 3 shows no XC but accumulation of Zea under an excess of light. H stands for Heterokontophyta. Some phyla, indicated by *, have a secondary XC (see text). Viola, violaxanthin; Anth, antheraxanthin; Zea, zeaxanthin; Diadino, diadinoxanthin; Diato, diatoxanthin.*

Group 1 Viola/Anth/Zea	Group 2 Diadino/Diato	Group 3 No XC but Zea accumulation
Embryophyta *	Bacillariophyceae* (H)	Cyanophyta
Pteridophyta	Xanthophyceae* (H)	Rhodophyta (most species)
Bryophyta	Haptophyta *	Glaucocystophyta
Chlorophyta *	Dinophyta*	Cryptophyta
Phaeophyceae (H)	Raphidophyta	Chlorophyta (some species)
Eustigmatophyceae (H)	Euglenophyta	
Chrysophyceae* (H)		
Rhodophyta (some species)		

The regulation and operation of the XC (Figure 11.2) has been described in detail earlier, especially for the Viola cycle (Latowski *et al.*, 2004). The de-epoxidation/epoxidation events are ensured by two enzymes, a Viola de-epoxidase (VDE) and a Zea epoxidase (ZEP) which are part of the few (in contrast to animals) lipocalin proteins known in plants. Viola de-epoxidase is localized on the lumen side and can bind/unbind to the thlyakoid membrane as a function of the lumenal pH, its optimal pH being around 5–6. Zea epoxidase is localized on the stroma side; its pH optimum is 7.5. In addition, VDE needs the acid form of ascorbate as a co-factor and ZEP needs NADPH, H^+ and oxygen. The operation of the XC results in competition of the activity of these two enzymes as a function of the build-up of the transthylakoid proton gradient, which is driven by the irradiance-dependent photosynthetic electron transport rate and subsequent change in lumenal and stromal pH. In summary, when the irradiance is moderate to high, the lumenal pH drops down in value to 4.5 to 6.5. When the irradiance decreases to darkness, the de-epoxidation becomes weaker and finally stops, while the reverse epoxidation reaction, which is ten times slower, becomes dominant (note that ZEP is also active under high light).

Hence, accumulation of the photoprotective de-epoxidized xanthophylls Zea and Diato under an excess of light is dependent on the activity of the two enzymes which indirectly depends (via the change in pH and availability of the co-factors) on the irradiance. In the organisms displaying the Diadino cycle, a similar mechanism has been described with some special features. The Diadino de-epoxidase (DDE) has been shown to be able to de-epoxidize Diadino as well as Viola, which has been used as an explanation for the presence of the two XC in several phyla (Jakob *et al.*, 2001). Nevertheless, two genes have recently been found in the genome of the diatom

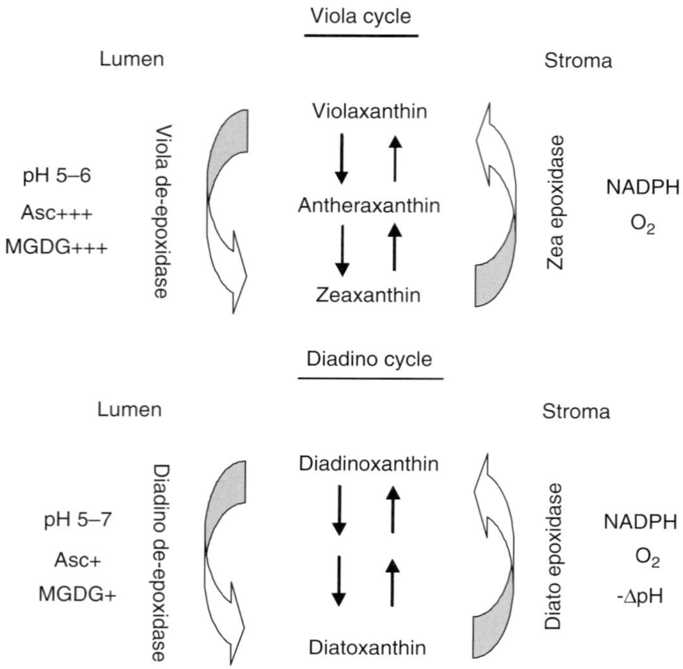

Figure 11.2. The xanthophyll cycles and their characteristics (modified from Wilhelm *et al.*, 2006). Asc, Ascorbate; MGDG, Monogalactosyldiacylglycerol lipids. Co-factor requirement for the enzymes is shown as well as the pH optimum. The '+' signs indicate the requirement for optimal de-epoxidase activity. '-ΔpH' means that the Dt epoxidase is inhibited by the high stromal pH under high light exposure.

Phaeodactylum tricornutum that encode for two homologues of the VDE ('VDE-like' genes) in addition to the gene coding for DDE (A. Gruber, personal communication, 2009). In contrast to VDE, the pH optimum of DDE is shifted towards higher pH and is active even at pH values about 7 (Jakob *et al.*, 2001). Consequently, Diadino de-epoxidation can already be triggered by a weak lumen acidification induced by, for example, chlororespiration (Jakob *et al.*, 1999). This also means that Diadino de-epoxidation already occurs for lower light intensities and shorter illumination times than for Viola de-epoxidation. Additionally, a recent study (Grouneva *et al.*, 2006) showed that DDE requires a much lower ascorbate concentration than VDE to be fully effective. Finally, DDE requires a lower concentration of lipids MGDG (monogalactosyldiacylglycerol) to drive efficient de-epoxidation, meaning that higher Dd amounts can be converted under high light (Goss *et al.*, 2005). Regarding the analogue of ZEP, the Diato epoxidase also shows an interesting characteristic: it is inactivated under excess light, which completely switches the equilibrium of the XC towards Diato accumulation (Goss *et al.*, 2006). Taken all together, these special features of the Diadino cycle explain the surprising efficiency and rapidity of accumulation of Diato in large amounts (Lavaud *et al.*, 2002b; Lavaud *et al.*, 2004). Finally, in diatoms, a species-dependent *de novo* synthesis of Diato under prolonged high light stress allows the cells to increase their capacity for photoprotection (Lavaud *et al.*, 2004).

11.3.2 The xanthophyll cycle and non-photochemical quenching

The NPQ process takes place in the light-harvesting complexes of the PS II. Its role is to dissipate as heat or reallocate part of the excitation energy before it reaches the reaction center when light has been absorbed in excess during a light exposure which exceeds the ability of the photosynthetic machinery to use all of the energy for photochemistry. Non-photochemical quenching reduces the lifetime of ^{1}chlorophyll a^* and, as a consequence, the quantum yield of chlorophyll *a* fluorescence as well as the quantum yield of photochemistry. It can be divided into three components: qE, energy-dependent quenching which is regulated by the build-up of a trans-thylakoid ΔpH and the operation of the XC; qT, state-transition quenching which allows reallocation of part of the energy absorbed from PSII to PSI; and qI, photoinhibitory quenching. Here, we will focus on the qE component. This has been investigated up to both molecular (down to a few tenths of nm) and gene levels, especially in higher plants and chlorophytes (Cogdell, 2006; Jung and Niyogi, 2006), far less in other eukaryotic algae and cyanobacteria.

The first correlation between qE and the accumulation of de-epoxidized xanthophylls under high light was observed in higher plants and chlorophytes (see Demmig-Adams, 1990). Later it was also reported in diatoms and dinoflagellates (Sakshaug *et al.*, 1987; Demers *et al.*, 1991; Olaizola and Yamamoto, 1994), chrysophytes and euglenophytes (Lichtlé *et al.*, 1995; Casper-Lindley and Bjorkman, 1998), rhodophytes (Ritz *et al.*, 1999) and more recently in cyanobacteria (Bailey *et al.*, 2005) and picoplanktonic chlorophyta (Dimier *et al.*, 2007b). A linear relationship between the operation of the XC, and the subsequent accumulation of zeaxanthin, antheraxanthin and diatoxanthin, the development of qE and the quenching of chlorophyll *a* fluorescence has been described in detail earlier (Gilmore and Yamamoto, 1991; Lavaud *et al.*, 2002b). The model for the qE mechanism is well understood in higher plants and chlorophytes (Holt *et al.*, 2004; Horton *et al.*, 2005; Cogdell, 2006). In summary, it implies a feed-back reaction from the linear electron transport via the build-up of a trans-thylakoid ΔpH and subsequent acidification of the lumen of the thylakoid: the higher the irradiance, the higher the electron transport and coupled translocation of protons, the higher the accumulation of protons into the lumen. This acidification has two consequences: the protonation of specific sites of a special light-harvesting complexes protein identified as PsbS in higher plants and the activation of VDE for synthesis of Zea. Both events enable the light-harvesting complexes' antenna to switch from a light-harvesting to a dissipative mode where excess excitation energy is converted into heat while chlorophyll *a* fluorescence is quenched.

The other group in which the qE mechanism and its relationship with the XC has been investigated in detail are the diatoms (Olaizola *et al.*, 1994; Lavaud *et al.*, 2002b). Energy-dependent quenching can be up to 4–5 times higher in diatoms than in higher plants (Lavaud *et al.*, 2002b; Ruban *et al.*, 2004), making it the most important rapid photoprotective process. It has been argued that this is due to the absence of state transitions in diatoms (Owens, 1986). Other differences have been listed in detail earlier (Wilhelm *et al.*, 2006). They include a different organization of the light-harvesting complexes (Büchel, 2003; Guglielmi *et al.*, 2005) and especially the absence of the protein PsbS, a different localization of the xanthophylls within the light-harvesting complexes (Lavaud *et al.*, 2003; Beer *et al.*, 2006), a capacity for accumulating large amounts of xanthophylls and a different composition and regulation of the XC. Additionally, qE appears to be more tightly associated with the XC and the accumulation of Diato than in higher plants with Zea (Lavaud *et al.*, 2002b), so that both the trans-thylakoid ΔpH and XC have a strong role in finely regulating qE (Lavaud *et al.*, 2002a; Ruban *et al.*, 2004; Goss *et al.*, 2006). Part of the qE process in diatoms remains to be elucidated, however recent advances (Lavaud and Kroth, 2006) contributed to the development of a first mechanistic model (Goss *et al.*, 2006).

The rhodophytes and cyanobacteria have extrinsic light-harvesting complex system (i.e. phycobilisomes) and also show a qE process even though the amplitude is weak. The same remark holds true for *Prochlorococcus* and its intrinsic phycocyanobilin antenna system (Bailey *et al.*, 2005). In these organisms, qE does not depend on a XC but on high light-dependent Zea accumulation. Actually a XC has been observed in some species of red algae, but no link with qE has been reported (Ursi *et al.*, 2003). In cyanobacteria, the process is believed to be a thermo-optic mechanism driven by blue light and taking place in the phycobilisomes where it involves a special carotenoid-binding protein and the pigments Zea and/or myxoxanthophyll (Cadoret *et al.*, 2004; Wilson *et al.*, 2006). It has been argued that qE in cyanobacteria would serve to adjust the energy transfer within the phycobilisomes of an already acclimated system to environmental stress(es) (high light, iron deficiency), but would not serve to cope with rapid fluctuations of irradiance as in higher plants and eukaryotic algae.

11.4 The xanthophyll cycle and the ecological properties of phytoplankton

11.4.1 Ecophysiology and environmental modulation of the XC

There are striking peculiarities of the algal XC activity compared to terrestrial plants, including a high degree of variation in structure and activity of the XC among phytoplankton belonging to different taxa (Casper-Lindley and Bjorkman, 1998; Lavaud *et al.*, 2004, Masojídek *et al.*, 1999, 2004; Wilhelm *et al.*, 2006). In addition, the XC activity depends on the physiological and nutritional state of the algae (Latasa and Berdalet, 1994; Staehr *et al.*, 2002). This diversity may affect the survival and the growth of species under high light and therefore their competitive ability, affecting the patterns of algal succession in phytoplankton communities (Demers *et al.*, 1991; Lavaud *et al.*, 2004). As an example of this, XC and qE were shown to be involved in seasonal succession of diatoms in estuaries (Serodio *et al.*, 2005), while a high qE appears to contribute to the domination of some diatom species in aquatic habitats where the light environment fluctuates strongly (Mitrovic *et al.*, 2003). Claustre *et al.* (1994) suggested than the higher content of Diadino per unit of chlorophyll *a* in diatoms confers an adaptive advantage in allowing a fast acclimation along sharp gradients of light. Resuspension of benthic diatoms could also be tracked from a 12 h periodicity in Diadino/chlorophyll *a* at a coastal station in the English Channel (Brunet and Lizon, 2003). Altogether, these *in situ* data support the observation of an increased Diadino content in diatoms exposed to fluctuating high light (Lavaud *et al.*, 2002b).

It has been hypothesized that the habitat characteristics account for the observed differences in photoacclimative responses between species, including the long-term variations in cellular chlorophyll *a* content (Sakshaug *et al.*, 1987), or the potential activity and efficiency of XC (Lavaud *et al.*, 2004; Lavaud *et al.*, 2007; Dimier *et al.*, 2007b, 2009a, b). As an example, strains isolated from estuaries show a higher (2.5 to 5 times) and faster qE than strains isolated from the open ocean or from coastal ecosystems (Lavaud *et al.*, 2007). The difference in photoprotection ability between open ocean and coastal *Thalassiosira* species may also be due to an adaptation to low or high iron concentrations, respectively (Strzepek and Harrison, 2004).

Few studies have investigated the XC in phytoplankton groups other than diatoms (Casper-Lindley and Bjorkman, 1998; Moisan *et al.*, 1998; Evens *et al.*, 2001; Harris *et al.*, 2005), limiting the value of comparative ecological and/or evolutionary interpretations within phytoplankton. Some recent studies have focussed on picoplanktonic species such as cyanobacteria (Cadoret *et al.*, 2004; Bailey *et al.*, 2005; Wilson *et al.*, 2006) or picoeukaryotes (Dimier *et al.*, 2007a, 2009a, b). In spite of the caution needed when generalizing results from single strains, there seems to be an effect of small size on the reactivity of the XC when compared to larger cells (Dimier *et al.*, 2009b). On the other hand, picoeukaryotes are also able to adopt an alternative photoprotective strategy, by rapidly modifying their chlorophyll *a* content (in less than one hour, Brunet *et al.*, 2006, 2007).

The 'light history' of the cell, which is the sum of past light variations experienced by the cell, is a key factor to consider when interpreting the dynamics of photoresponses (Moisan *et al.*, 1998; Anning *et al.*, 2000; Lavaud *et al.*, 2002b). Although this aspect is very difficult, if not impossible, to assess for natural populations, evidence from cultures points to a major role of past light experience on the type and kinetics of light-induced reactions. As an example, previous acclimation to high or medium irradiance strongly influences the short-term photoprotective response to a further increase in irradiance, when compared to cells acclimated to lower light (Casper-Lindley and Bjorkman, 1998; Moisan *et al.*, 1998; Dimier *et al.*, 2007b).

In general, the physiological condition of algal cells has a strong influence on the functioning of the XC. Several authors have observed an increase in the XC pigments content in stressed phytoplankton with no parallel change in light conditions (Latasa and Berdalet, 1994; Brunet *et al.*, 1996; Staehr *et al.*, 2002). An increase of Diato has also been observed in diatoms exposed to toxic polyunsaturated aldehydes (Casotti *et al.*, 2005), while inhibition of Diato epoxidation has been caused by cadmium (Bertrand *et al.*, 2001). These responses might be interpreted in relation to the potential antioxidant role of Diato, as it has been hypothesized for Zea in higher plants (Strzalka *et al.*, 2003).

11.4.2 The xanthophyll cycle and UV radiation

Depletion of the stratospheric ozone layer has resulted in increased UV-B radiation (280–315 nm) which can penetrate down to 30 meters in ocean waters (1% irradiance depth for 305 nm, Tedetti and Sempéré, 2006) and can potentially damage aquatic organisms, notably phytoplankton (e.g. Vincent and Roy, 1993). Two of the major targets of UV damage in phytoplankton are the PSII reaction center complex and the carbon-fixing enzyme Rubisco (Vincent and Neale, 2000). This damage to both the light and dark reactions of photosynthesis will reduce photochemical use of light energy and thus increase excess light energy, favoring the formation of dangerous reactive oxygen species (ROS) within the cell. Thus mechanisms that can protect against excess light energy, such as the xanthophyll cycle, should be solicited. Indeed, a number of recent studies have shown a stimulation of the XC when algae were exposed to enhanced UV-B under a field or light-simulated environment (with realistic levels of photosynthetically active radiation, PAR, and UV-A: 315–400 nm). These studies covered a range of algal groups, including diatoms (Goss *et al.*, 1999; Zudaire and Roy, 2001), dinoflagellates (with intra-specific, strain-related differences: Laurion and Roy, 2009), haptophytes (Döhler and Haas, 1995; Buma *et al.*, 2000), eustigmatophytes (Sobrino *et al.*, 2005), natural phytoplankton (Döhler and Hagmeier, 1997), and green macroalgae (Choo *et al.*, 2005). In most of these cases, an increase in the de-epoxidized pigment (Diato or Zea) was observed, while the epoxidized parent (Diadino or Viola) decreased, and this correlated with a decrease in photochemical efficiency (Fv/Fm) and an increase in non-photochemical quenching (Goss *et al.*, 1999). In one study, both Diadino and Diato increased, which may be attributed to the longer duration of the experiment (4 days, Buma *et al.*, 2000). Sobrino *et al.* (2005) determined the biological weighting function (BWF, or action spectrum) for xanthophyll de-epoxidation induced by UV radiation (in the presence of realistic PAR levels), showing that both increased irradiance and inclusion of lower wavelengths in the UV range led to more extensive de-epoxidation. The BWF was similar in shape to the

BWF for UV inhibition of photosynthesis, but with a 22-fold lower effectiveness. These results indicate that for the species used by Sobrino *et al.* (2005), stimulation of the xanthophyll cycle occurred upon UV exposure, but this was not sufficient to fully prevent UV inhibition of photosynthesis. Therefore, although xanthophyll de-epoxidation is affected by UV radiation, its main function is related to protection from excess PAR (Sobrino *et al.*, 2005).

A second group of studies reported no effect on the concentration of xanthophyll pigments upon exposure to UV radiation, although information on the level of de-epoxidation was not always available. These included studies on diatoms (Buma *et al.*, 1996), including benthic species from oyster ponds (Rech *et al.*, 2005) and the Antarctic *Chaetoceros brevis* (van de Poll *et al.*, 2005), chlorophytes (Lütz *et al.*, 1997; Roleda *et al.*, 2009), Antarctic ice phytoplankton (Schofield *et al.*, 1995), a toxic dinoflagellate (Evens *et al.*, 2001) and coastal phytoplankton communities exposed to enhanced UV-B in floating mesocosms (Mohovic *et al.*, 2006; Roy *et al.*, 2006). The lack of response was attributed to factors such as interspecific differences, the production of UV-absorbing mucilage, or greater effects of PAR and UV-A. The physiological condition of the cells influenced this response, with a stimulation of the XC when condition declined (Mohovic *et al.*, 2006). The detection of UV-B-induced effects on photoprotective pigments in other studies was caused by a non-realistic spectral balance, according to van de Poll *et al.* (2005). This, however, does not apply to all cases (e.g. Sobrino *et al.*, 2005).

A last group of studies has shown that UV can inhibit the xanthophyll cycle, generally causing a decrease in Diato concentration. Pfündel *et al.* (1992) showed that UV-B could inhibit the enzyme VDE in isolated pea chloroplasts, preventing the transformation of Anth into the de-epoxidized Zea. UV-A and UV-B damage to the xanthophyll cycle was observed in cultures of the haptophytes *Pavlova* spp. (Döhler and Lohmann, 1995), the chlorophyte *Dunaliella tertiolecta* (Döhler *et al.* 1997), and of three marine diatoms (Lohmann *et al.*, 1998), including the Antarctic diatom *Chaetoceros brevis* (Janknegt *et al.*, 2008). Mewes and Richter (2002) reported a UV-B-dependent decrease in Diato in cultured diatoms caused by an increase in the epoxidation reaction transforming Diato into Diadino. The stimulation of Diato epoxidase by UV-B may be related to the loss of the pH gradient across the thylakoid membrane which could reduce the affinity of Diato to its binding site, making it more accessible to Diato epoxidase. A UV-B-induced loss of the de-epoxidized pigment was also observed in studies by Garde and Cailliau (2000) on the haptophyte *Emiliania huxleyi*, and on green macroalgae (*Ulva lactuca*) exposed to natural sunlight with selective exclusion of UV radiation using screening foils (Bischof *et al.*, 2002, 2003). The diminished activity (or reversal) of the xanthophyll cycle has also been attributed to enhanced production of ROS (Lichtenthaler, 1998) which can reduce the relative content of Diato in cells of the marine, near-bottom diatom *Cylindrotheca closterium* subjected to enhanced UV-B during simulated emersion (Rijstenbil, 2005). This decrease in Diato was

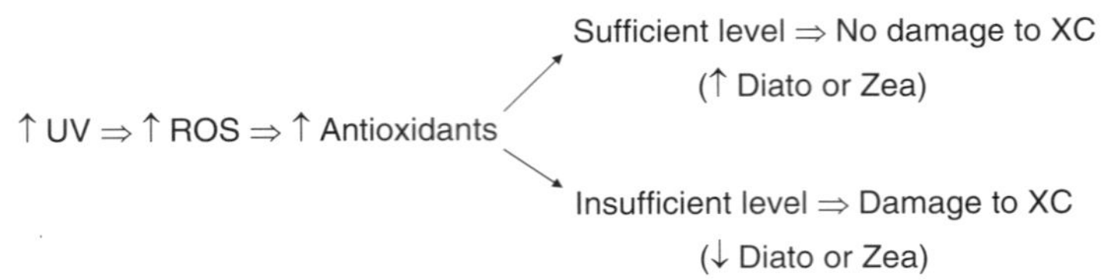

Figure 11.3. A scheme of possible UV-B effects on the XC.

related to UV-A more than to UV-B, and increased salinity exacerbated this reaction. Bischof *et al.* (2003) also related the reduction in XC activity to increased production of ROS. Hence both direct (cellular UV targets) and indirect (ROS-related) UV effects can damage the xanthophyll cycle and reduce the diatoxanthin content.

Apparently contradictory conclusions can thus be reached from these recent studies, with the enzyme-mediated xanthophyll cycle being either stimulated by UV-B or a potential UV target. However, detailed examination of the spectral and irradiance conditions reveals that most of the phytoplankton studies reporting a stimulation of the XC were done under a spectral balance (PAR:UV-A:UV-B) relatively close to that found in nature, with relatively low UV-B levels, while studies reporting inhibition of the XC were often done under more damaging spectral conditions, sometimes with low PAR levels, no UV-A and with total UV-B dose often higher than 10 kJ m^{-2}. While these studies are useful to unravel the mechanisms of UV damage to the XC, they are less useful to predict effects under natural environmental conditions. Interestingly, there are a few cases where UV-B damage to the XC has been observed under ecologically relevant conditions (natural sunlight and high PAR irradiances). These include studies on partly or fully sessile organisms such as benthic diatoms (Rijstenbil, 2005) and macroalgae (e.g. Bischof *et al.*, 2002). Epilithon and other sessile organisms may be more sensitive to UV radiation than free-floating phytoplankton perhaps because they are unable to physically avoid UV stress (Bothwell *et al.*, 1994; Vinebrooke and Leavitt, 1999). Increased oxidative damage could also affect the response of the XC to UV stress (Bischof *et al.*, 2002; Rijstenbil, 2005). Conceivably, organisms where antioxidant levels are low and oxidative stress is high could show UV damage to the XC (Figure 11.3). Acclimation is another factor which explains some of the variable responses of the XC to UV exposure, with different photoprotection mechanisms alternating through time (Zudaire and Roy, 2001). Lastly, the range of responses to environmental UV radiation is quite large in phytoplankton (Neale *et al.*, 1998), accounting for the variability in size, physiological condition and prior light history. One example of this is the different UV responses observed before, during, and after a bloom of diatoms inside a mesocosm, where the amount of Diato retained after 24 h of surface exposure was related to the fraction of inactive PSII reaction centers and was influenced by nitrate limitation (Bouchard *et al.*, 2008). Hence there is not a unique response of the xanthophyll cycle to UV radiation and understanding the overall stress condition of the cells can help elucidate these responses.

11.4.3 The XC and the dynamics of water masses

Several studies have investigated marine ecosystems at different time scales with the aim of characterizing the algal responses to the surrounding light environment through the analysis of the XC pigments (e.g. for coastal sites: Brunet *et al.*, 1993; Moline, 1998; Brunet *et al.*, 2003; Fujiki *et al.*, 2003; Muller and Wasmund, 2003; for frontal systems: Claustre *et al.*, 1994; Brunet *et al.*, 2003; and for offshore regions: Bidigare *et al.*, 1987; Olaizola *et al.*, 1992; Kashino *et al.*, 2002; Brunet *et al.*, 2006, 2007).

In situ studies have used different indicators of the photoprotective state of the cells: (Diadino + Diato)/chlorophyll *a*, Diadino/chlorophyll *a*, Diato/chlorophyll *a* or also Diato/(Diato + Diadino). For any of these, when Diadino is included in the numerator, a long-term process is implied (Bidigare *et al.*, 1987), because Diadino reacts at much longer time scales than Diato (Lavaud *et al.*, 2002b; Dimier *et al.*, 2007b). Instead, Diato is formed at very short time scales, and it represents a clear indicator of fast activation of photoprotection. The DES (de-epoxidation state) index (= Diato/(Diato + Diadino)) is often more useful than the ratio Diato/chlorophyll *a* to infer the photoprotective state of a natural phytoplankton community, since normalization by chlorophyll *a* may introduce a bias due to the natural variability of chlorophyll *a* related to the physiological state and the light history of cells. In addition, dividing by chlorophyll *a* may be misleading if chlorophyll *a* includes algal biomass with no XC (e.g. cyanobacteria) or with other XC (Viola-cycle in chlorophytes).

Time is a key factor affecting algal physiology. On a yearly scale, photoprotection in terms of any of the above mentioned indices appears to be directly correlated with the day length and the seasonal increase in total irradiance (Brunet *et al.*, 1993; Moline, 1998; Fujiki *et al.*, 2003). On a diel scale, Diato/chlorophyll *a* and Diato/(Diadino + Diato) increase during the daytime and peak around noon (Brunet *et al.*, 1993), while (Diadino + Diato)/chlorophyll *a* peaks later, due to the longer reactivity time scales of Diadino (Moline, 1998). A highly significant relationship with irradiance is generally obtained, with sinusoidal variations of Diato/chlorophyll *a* in the upper layer (Brunet *et al.*, 2008).

Phytoplankton is continuously subjected to variations in environmental parameters caused by passive displacement due to water mass dynamics, mixing or sinking (Lewis *et al.*, 1984a), or to active displacement due to cell migration. Within a certain time range, mixing triggers short-term photoacclimative responses (MacIntyre *et al.*, 2000), which can be traced and used to monitor physical processes such as upwelling, downwelling or water mass properties such as transparency.

In the case of rapid mixing, cells may experience light variations faster than their ability to photoacclimate, while in the case of slow mixing, cells may have

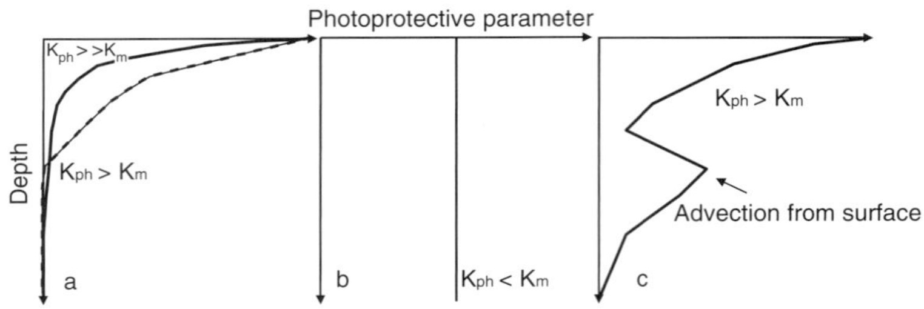

Figure 11.4. Relationship between photoprotective parameter and vertical mixing in the euphotic layer of the water column. (a) Photoprotection rate (K_{ph}) is higher than vertical mixing rate (K_m) with $K_{ph} >> K_m$ (straight line) and $K_{ph} > K_m$ (dashed line); (b) photoprotection rate (K_{ph}) is lower than the vertical mixing rate (K_m); (c) photoprotection rate (K_{ph}) is higher than vertical mixing rate (K_m) with an advection of surface phytoplankton to a deeper layer in the euphotic zone. Modified from Claustre *et al.* (1994).

the time to acclimate to the average light level (Figure 11.4). Indeed, the acclimation rates depend on which parameter is considered, since some responses may require seconds or minutes to be activated (e.g. chlorophyll *a* fluorescence or the XC), while others (e.g. chlorophyll *a* content, absorption capacity, photosynthetic parameters) require a much longer time (MacIntyre *et al.*, 2000; Brunet *et al.*, 2003). The vertical distribution of fast-reacting photo-dependent parameters (e.g. XC or fluorescence) generally presents a decrease from the very surface to the bottom of the upper mixed layer, as expected from their role in high irradiance protection (Welschmeyer and Hoepffner, 1986; Olaizola *et al.*, 1992; Claustre *et al.*, 1994; Moline, 1998; Brunet *et al.*, 2003, 2006, 2007). However, at times, in actively mixed water columns, homogeneous profiles of XC pigments have been observed, due to mixing velocities faster than photoacclimation reaction times (Brunet *et al.*, 1993).

Other examples of relationships between physical forcing and algal response come from the use of XC pigments (e.g. Claustre *et al.*, 1994; Brunet *et al.*, 2003), cell autofluorescence measured by flow cytometry (Dusenberry, 2000) or variable fluorescence (Oliver *et al.*, 2003). Many of these authors have used the kinetics of phytoplankton photoresponses to estimate mixing velocities. For example, Claustre *et al.* (1994) and Brunet *et al.* (2003, 2008) estimated, in two different regions of the Mediterranean Sea, vertical mixing velocities between 5×10^{-4} and 7×10^{-4} m s^{-1}, which are realistic values for the areas investigated (see also Falkowski, 1983; Dusenberry, 2000). Thompson *et al.* (2007) found significant differences in vertical mixing velocities (5×10^{-3} versus 8×10^{-3} m s^{-1}) using XC pigments distribution between two strongly dynamic eddies in the south-eastern Indian Ocean.

The value of these investigations lies in the fact that mixing rates are very difficult parameters to measure directly. From their results on XC pigments dynamics, Brunet *et al.* (2003, 2008) compared the percentage variations of different photodependent parameters at the surface or at the bottom of the mixed layer, and they were able to establish a decreasing hierarchy in the velocity of photodependent indicators. From the inferred kinetic coefficients, a threshold value of 4% h^{-1} was estimated as the value below which photoacclimation reactions were not significant with respect to the physical dynamics. From this, the vertical eddy diffusivity at the time of sampling could be estimated to be 1.75×10^{-2} m^2 s^{-1} (Brunet *et al.*, 2003), calculated according to Lewis *et al.* (1984b) and Cullen and Lewis (1988). This example shows how the analysis of the vertical profiles of photo-dependent parameters may provide valuable insights into the effects of hydrodynamism on algal physiology.

In order to use pigments to infer physical properties and dynamics of water masses, knowledge of the kinetics of changes in photo-physiological parameters is needed. These can be retrieved from laboratory (e.g. Falkowski, 1983) or *in situ*-simulated experiments using natural phytoplankton communities incubated under natural light (deck incubations), subjected to shifts in light intensities. The latter may provide useful terms of comparison to interpret *in situ* observations, but the appropriate sampling pace must be chosen to obtain statistically sound data (Claustre *et al.*, 1994; Brunet *et al.*, 2003). During the incubations, the photoacclimative dynamics is generally described by a first-order kinetic equation, from which the kinetic coefficient K is retrieved (Falkowski, 1983; Claustre *et al.*, 1994). Alternatively, the use of a logistic model allows the consideration of hysteresis, i.e. the influence of light experienced by the cell before sampling (light history) which may have a crucial role in the cell response, modulating its kinetics according to the sign of the light change (Cullen and Lewis, 1988). This model has been successfully applied by Cullen and Lewis (1988) and Dusenberry (2000) but remains to be tested on XC pigments dynamics.

In general, caution must be adopted when using photoprotective pigments as markers of phytoplankton dynamics in the water column. First, the approach presented is only valid for algae using Diato and Diadino in the XC and not for those using Viola and Zea. This is because Zea is also part of the constitutive antenna system of prokaryote algae. Second, it is recommended to use Diato – not Diadino – as a tracer of short-term photoacclimation, while the ratio Diadino/chlorophyll *a* is a good alternative when long-term photoacclimation processes are considered, due to the different kinetics of transformation inside the XC. Indeed, it is recommended to use Diato/(Diadino + Diato) as an indicator of the de-epoxidation state, excluding any normalization by chlorophyll *a*.

Addendum

An updated and detailed review on the XC-related photoprotection has been recently published (Goss, R. and Jakob, T. (2010). Regulation and function of xanthophyll cycle-dependent photoprotection in algae. *Photosynth. Res.* **106**, 103–22).

Acknowledgements

Christophe Brunet would like to thank the Stazione Zoologica A. Dohrn for having made possible his research and for grant support to Ph.D. students. Two anonymous referees and Dr. Carole Llewellyn are gratefully acknowledged for their helpful comments on the manuscript.

Abbreviations

a_{CDOM}	Absorption by CDOM
ACP	Chlorophyll *a*-chlorophyll *c*-peridinin protein
ADEOS	Advanced Earth Observing Satellite
a_p	Absorption by particulate matter
a_g	Absorption by dissolved substances (gelbstoff)
API	Atmospheric pressure ionization
DDE	Diadinoxanthin de-epoxidase
DEP	Diadinoxanthin epoxidase
DES	De-epoxidation state
ESI	Electrospray ionization
GLI	Global Imager sensor
NPQ	Non-photochemical fluorescence quenching
ROS	Reactive oxygen species
SIM	Single ion monitoring
TIC	Total ion current
VDE	Violaxanthin de-epoxidase
XCP	Xanthophyll-cycling pigments
ZEP	Zeaxanthin epoxidase

References

Anning, T., MacIntyre, H. L., Pratt, S. M., Sammes P. J., Gibb, S. and Geider, R. J. (2000). Photoacclimation in the marine diatom *Skeletonema costatum*. *Limnol. Oceanogr.* **45**, 1807–17.

Bailey, S., Mann, N. H., Robinson, C. and Scanlan, D. J. (2005). The occurrence of rapidly reversible non-photochemical quenching of chlorophyll *a* fluorescence in cyanobacteria. *FEBS Lett.* **579**, 275–80.

Beer, A., Gundermann, K., Beckmann, J. and Büchel, C. (2006). Subunit composition and pigmentation of fucoxanthin-chlorophyll proteins in diatoms: Evidence for a subunit involved in diadinoxanthin and diatoxanthin binding. *Biochemistry* **45**, 13046–53.

Berthos, N. R. and Gibbs, S. P. (1998). Evidence for a lack of photosystem segregation in *Chlamydomonas reinhardtii* (Chlorophyceae). *J. Phycol.* **34**, 1009–16.

Bertrand, M., Schoefs, B., Rohacek, K. and Molnar, I. (2001). Cadmium inhibits epoxidation of diatoxanthin to diadinoxanthin in the xanthophyll cycle of the marine diatom *Phaeodactylum tricornutum*. *FEBS Lett.* **508**, 153–56.

Bidigare, R. R., Smith, R. C., Baker, K. S. and Marra, J. (1987). Oceanic primary production estimates from measurements of spectral irradiance and pigment concentrations. *Global Biogeochem. Cycles* **1**, 171–86.

Bischof, K., Kräbs, G., Wiencke, C. and Hanelt, D. (2002). Solar ultraviolet radiation affects the activity of ribulose-1,5-biphosphate carboxylase-oxygenase and the composition of photosynthetic and xanthophyll cycle pigments in the intertidal green alga *Ulva lactuca* L. *Planta* **215**, 502–09.

Bischof, K., Janknegt, P.J., Buma, A.G.J., Rijstenbil, J.W., Peralta, G. and Breeman, A.M. (2003). Oxidative stress and enzymatic scavenging of superoxide radicals induced by solar UV-B radiation in *Ulva* canopies from southern Spain. *Sci. Mar.* **67**, 353–59.

Blatt, M.R., Weisenseel, M.H. and Haupt, W. (1981). A light-dependent current associated with chloroplast aggregation in the alga *Vaucheria sessilis*. *Planta* **152**, 513–26.

Bothwell, M.L., Sherbot, D.M.J. and Pollock, C.M. (1994). Ecosystem response to solar ultraviolet-B radiation: influence of trophic-level interactions. *Science* **265**, 97–100.

Bouchard, J.N., Longhi, M.L., Roy, S., Campbell, D.A. and Ferreyra, G. (2008). Interaction of nitrogen status and UVB sensitivity in a temperate phytoplankton assemblage. *J. Exp. Mar. Biol. Ecol.* **359**, 67–76.

Brunet, C. and Lizon, F. (2003). Tidal and diel periodicities of size fractionated phytoplankton pigment signatures at an offshore station in the South-Eastern English Channel. *Estuar. Coast. Shelf Sci.* **56**, 835–45.

Brunet, C., Brylinski, J.M. and Lemoine, Y. (1993). *In situ* variations of the xanthophylls diadinoxanthin and diatoxanthin: photoadaptation and relationships with an hydrodynamical system of the Eastern English Channel. *Mar. Ecol. Prog. Ser.* **102**, 69–77.

Brunet, C., Davoult, D. and Casotti, R. (1996). Physiological reactions to a change in light regime in cultured *Skeletonema costatum* (Bacillariophyta): implications to estimation of phytoplankton biomass. *Hydrobiologia* **333**, 87–94.

Brunet, C., Casotti, R., Aronne, B. and Vantrepotte, V. (2003). Measured photophysiological parameters used as tools to estimate vertical water movements in the coastal Mediterranean. *J. Plankton Res.* **25**, 1413–25.

Brunet, C., Casotti, R., Vantrepotte, V., Corato, F. and Conversano, F. (2006). Picophytoplankton diversity and photophysiology in the Strait of Sicily (Mediterranean Sea) in summer. I. Mesoscale variations. *Aquat. Microb. Ecol.* **44**, 127–41.

Brunet, C., Casotti, R., Vantrepotte, V. and Conversano, F. (2007). Vertical variability and diel dynamics of picophytoplankton in the Strait of Sicily, Mediterranean Sea, in summer. *Mar. Ecol. Prog. Ser.* **346**, 15–26.

Brunet, C., Casotti, R. and Vantrepotte, V. (2008). Phytoplankton diel and vertical variability in photobiological responses at a coastal station in the Mediterranean Sea. *J. Plankton Res.* **30**, 645–54.

Büchel, C. (2003). Fucoxanthin-chlorophyll proteins in diatoms: 18 and 19 kDa subunits assemble into different oligomeric states. *Biochemistry* **42**, 13027–34.

Buma, A.G.J., Zemmelink, H.J., Sjollema, K. and Gieskes, W.W.C. (1996). UVB radiation modifies protein and photosynthetic pigment content, volume and ultrastructure of marine diatoms. *Mar. Ecol. Prog. Ser.* **142**, 47–54.

Buma, A.G.J., van Oijen, T., van de Poll, W., Veldhuis, M.J.W. and Gieskes, W.W. C. (2000). The sensitivity of *Emiliania huxleyi* (Prymnesiophyceae) to ultraviolet-B radiation. *J. Phycol.* **36**, 296–303.

Cadoret, J.C., Démoulière, R., Lavaud, J., van Gorkom, H., Houmard, J. and Etienne, A.-L. (2004). Dissipation of excess energy triggered by blue light in cyanobacteria with CP43' (isiA). *Biochim. Biophys. Acta* **1659**, 100–04.

Casotti, R., Mazza, S., Brunet, C., Vantrepotte, V., Ianora, A. and Miralto A. (2005). Growth inhibition and toxicity of the diatom aldehyde 2-*trans*-4-*trans* decadienal on *Thalassiosira weissflogii* (Bacillariophyceae). *J. Phycol.* **41**, 7–20.

Casper-Lindley, C. and Bjorkman, O. (1998). Fluorescence quenching in 4 unicellular alage with different light-harvesting and xanthophyll-cycle pigments. *Photosynth. Res.* **56**, 277–89.

Choo, K. S., Nilsson, J., Pedersén, M. and Snoeijs, P. (2005). Photosynthesis, carbon uptake and antioxidant defence in two coexisting filamentous green algae under different stress conditions. *Mar. Ecol. Prog. Ser.* **292**, 127–38.

Claustre, H., Kerhervé, P., Marty, J. C. and Prieur, L. (1994). Phytoplankton photoadaptation related to some frontal physical processes. *J. Mar. Systems* **5**, 251–65.

Cogdell, R. J. (2006). The structural basis of non-photochemical quenching revealed? *Trends Plant Sci.* **11**, 59–60.

Cullen, J. J. and Lewis, M. R. (1988). The kinetics of algal photoadaptation in the content of vertical mixing. *J. Plankton Res.* **10**, 1039–63.

Demers, S., Roy, S., Gagnon, R. and Vignault, C. (1991). Rapid light-induced changes in cell fluorescence and in xantophyll-cycle pigments of *Alexandrium excavatum* (Dinophyceae) and *Thalassiosira pseudonana* (Bacillariophyceae) a photo-protection mechanism. *Mar. Ecol. Prog. Ser.* **76**, 185–93.

Demmig-Adams, B. (1990). Carotenoids and photoprotection in plants: a role for the xanthophyll zeaxanthin. *Biochim. Biophys. Acta* **1020**, 1–24.

Dimier, C., Corato, F., Saviello, G. and Brunet, C. (2007a). Photophysiological properties of the marine picoeukaryote *Picochlorum* RCC237 (Trebouxiophyceae, Chlorophyta). *J. Phycol.* **45**, 275–83.

Dimier, C., Corato, F., Tramontano, F. and Brunet, C. (2007b). Photoprotection and xanthophyll cycle activity in three diatoms. *J. Phycol.* **43**, 937–47.

Dimier, C., Brunet, C., Geider, R. and Raven, J. (2009a). Growth and photoregulation dynamics of the picoeukaryote *Pelagomonas calceolata* in fluctuating light. *Limnol. Oceanogr.* **54**, 823–36.

Dimier, C., Saviello, G., Tramontano, F. and Brunet, C. (2009b). Comparative ecophysiology of the xanthophyll cycle in six marine phytoplanktonic species. *Protist* **160**, 397–411.

Döhler, G. and Haas, F. T. (1995). UV effects on chlorophylls and carotenoids of the haptophycean alga *Pavlova*. *Photosynthetica* **31**, 157–60.

Döhler, G. and Hagmeier, E. (1997). UV effects on pigments and assimilation of ^{15}N-ammonium and ^{15}N-nitrate by natural marine phytoplankton of the North Sea. *Bot. Acta* **110**, 481–88.

Döhler, G. and Lohmann, M. (1995). Impact of UV radiation of different wavebands on the pigmentation of the haptophycean *Pavlova*. *J. Photochem. Photobiol. B: Biol.* **27**, 265–70.

Döhler, G., Drebes, G. and Lohmann, M. (1997). Effects of UV-A and UV-B radiation on pigments, free amino acids and adenylate content of *Dunaliella tertiolecta* Butcher (Chlorophyta). *J. Photochem. Photobiol. B: Biol.* **40**, 126–31.

Dusenberry, J. A. (2000). Steady-state single cell model simulations of photoacclimation in a vertically mixed layer: implications for biological tracer studies and primary productivity. *J. Marine Syst.* **24**, 201–20.

Egeland, E.S., Eikrem, W., Throndsen, J., Wilhelm, C., Zapata, M. and Liaaen-Jensen, S. (1995). Carotenoids from further prasinophytes. *Biochem. Syst. Ecol.* **23**, 747–55; *errata* **24**, 179–80.

Evens, T. J., Kirkpatrick, G. J., Millie, D. F., Chapman, D. J. and Schofield, O. M.E. (2001). Photophysiological responses of the toxic red-tide dinoflagellate *Gymnodinium breve* (Dinophyceae) under natural sunlight. *J. Plankton Res.* **23**, 1177–93.

Falkowski, P. G. (1983). Light-shade adaptation and vertical mixing of marine phytoplankton: a comparative field study. *J. Mar. Res.* **41**, 215–37.

Falkowski, P. G. and Chen, Y.-B. (2003). Photoacclimation of light harvesting systems in eukaryotic algae. In *Light-harvesting Antennas in Photosynthesis*, ed. B. R. Green and W. W. Parson. Dordrecht: Kluwer Academic Publishers, pp. 423–47.

Falkowski, P. G. and LaRoche, J. (1991). Acclimation to spectral irradiance in algae. *J. Phycol.* **27**, 8–14.

Fujiki, T., Toda, T., Kikuchi, T. and Taguchi, S. (2003). Photoprotective response of xanthophyll pigments during phytoplankton blooms in Sagami Bay, Japan. *J. Plankton Res.* **25**, 317–22.

Garde, K. and Cailliau, C. (2000). The impact of UV-B radiation and different PAR intensities on growth, uptake of ^{14}C, excretion of DOC, cell volume, and pigmentation in the marine prymnesiophyte, *Emiliania huxleyi*. *J. Exp. Mar. Biol. Ecol.* **247**, 99–112.

Gilmore, A. M. and Yamamoto, H. Y. (1991). Zeaxanthin formation and energy-dependent fluorescence quenching in pea chloroplasts under artificially mediated linear and cyclic electron transport. *Plant Physiol.* **96**, 635–43.

Goss, R., Bohme, K. and Wilhelm, C. (1998). The xanthophyll cycle of *Mantoniella squamata* converts violaxanthin into antheraxanthin but not to zeaxanthin-consequences for the mechanism of enhanced non-photochemical energy dissipation. *Planta* **205**, 613–21.

Goss, R., Mewes, H. and Wilhelm, C. (1999). Stimulation of the diadinoxanthin cycle by UV-B radiation in the diatom *Phaeodactylum tricornutum*. *Photosynth. Res.* **59**, 73–80.

Goss, R., Lohr, M., Latowski, D., Grzyb, J., Vieler, A., Wilhelm, C. and Strzalka, K. (2005). Role of hexagonal structure-forming lipids in diadinoxanthin and violaxanthin solubilization and de-epoxidation. *Biochemistry* **44**, 4028–36.

Goss, R., Pinto, E. A., Wilhelm, C. and Richter, M. (2006). The importance of a highly active and Δ-pH-regulated diatoxanthin epoxidase for the regulation of the PS II antenna function in diadinoxanthin cycle containing algae. *J. Plant Physiol.* **163**, 1008–21.

Green, B. R., Anderson, J. M. and Parson, W. W. (2003). Photosynthetic membranes and their light-harvesting antennas. In *Light-harvesting Antennas in Photosynthesis*, ed. B. R. Green and W. W. Parson. Dordrecht: Kluwer Academic Publishers, pp. 1–28.

Grouneva, I., Jakob, T., Wilhelm, C. and Goss, R. (2006). Influence of ascorbate and pH on the activity of the xanthophyll cycle-enzyme diadinoxanthin de-epoxidase. *Physiol. Plant.* **126**, 205–11.

Guglielmi, G., Lavaud, J., Rousseau, B., Etienne, A.-L., Houmard, J. and Ruban, A.V. (2005). The light-harvesting antenna of the diatom *Phaeodactylum tricornutum* – Evidence for a diadinoxanthin-binding subcomplex. *FEBS Lett.* **272**, 4339–48.

Harris, G. N., Scanlan, D. J. and Geider, R. J. (2005). Acclimation of *Emiliania huxleyi* (Prymnesiophyceae) to photon flux density. *J. Phycol.* **41**, 851–62.

Herzig, R. and Dubinsky, D. (1993). Effect of photoacclimation on the energy partitioning between cyclic and non-cyclic photophosphorylation. *New Phytol.* **123**, 665–72.

Holt, N. E., Fleming, G. R. and Niyogi, K. K. (2004). Toward an understanding of the mechanism of the nonphotochemical quenching in green plants. *Biochemistry* **43**, 8281–89.

Hooks, C. E., Bidigare, R. R., Keller, M. D. and Guillard, R. R.L. (1988). Coccoid eukaryotic marine ultraplankters with four different HPLC pigment signatures. *J. Phycol.* **24**, 571–80.

Horton, P., Wentworth, M. and Ruban, A. (2005). Control of the light-harvesting function of chloroplast membranes: The LHCII-aggregation model for non-photochemical quenching. *FEBS Lett.* **579**, 4201–06.

Jakob, T., Goss, R. and Wilhelm, C. (1999). Activation of diadinoxanthin de-epoxidase due to a chlororespiratory proton gradient in the dark in the diatom *Phaeodactylum tricornutum. Plant Biol.* **1**, 76–82.

Jakob, T., Goss, R. and Wilhelm, C. (2001). Unusual pH-dependence of diadinoxanthin de-epoxidase activation causes chlororespiratory induced accumulation of diatoxanthin in the diatom *Phaeodactylum tricornutum. J. Plant Physiol.* **158**, 383–90.

Janknegt, P.J, van de Poll, W. H., Visser, R. J.W., Rijstenbil, J. W. and Buma, A.G.J. (2008). Oxidative stress responses in the marine antarctic diatom *Chaetoceros brevis* (Bacillariophyceae) during photoacclimation. *J. Phycol.* **44**, 957–66.

Johnsen, G. and Sakshaug, E. (1993). Bio-optical characteristics and photoadaptive responses in the toxic and bloom-forming dinoflagellates *Gyrodinium aureolum, Gymnodinium galatheanum*, and two strains of *Prorocentrum minimum. J. Phycol.* **29**, 627–42.

Johnsen, G. and Sakshaug, E. (2007). Bio-optical characteristics of PSII and PSI in 33 species (13 pigment-groups) of marine phytoplankton, and the relevance for PAM and FRR fluorometry. *J. Phycol.* **43**, 1236–51.

Johnsen, G., Nelson, N. B., Jovine, R. V.M. and Prézelin, B. B. (1994). Chromoprotein- and pigment-dependent modeling of spectral light absorption in two dinoflagellates, *Prorocentrum minimum* and *Heterocapsa pygmaea. Mar. Ecol. Prog. Ser.* **114**, 245–58.

Johnsen, G., Prézelin, B. B. and Jovine, R. V.M. (1997). Fluorescence excitation spectra and light utilization in two red tide dinoflagellates. *Limnol. Oceanogr.* **42**, 1166–77.

Jung, H. S. and Niyogi, K. K. (2006). Molecular analysis of photoprotection of photosynthesis. In *Photoprotection, Photoinhibition, Gene Regulation And Environment*, ed. B. Demmig-Adams, W. W.I. Adams and A. K. Mattoo. Dordrecht: Kluwer Academic Publishers, pp. 127–43.

Kashino, Y., Kudoh, S., Hayashi, Y., Suzuki, Y., Odate, T., Hirawake, T., Satoh, K. and Fukuchi, M. (2002). Strategies of phytoplankton to perform effective photosynthesis in the North water. *Deep- Sea Res. II* **49**, 5049–61.

Kirk, J. T.O. and Tilney-Bassett, R. A.E. (1978). *The Plastids. Their Chemistry, Structure, Growth and Inheritance*, 2nd edn. Amsterdam: Elsevier/North-Holland Biomedical Press.

Larkum, A. W. (2003). Light-harvesting systems in algae. In *Photosynthesis in Algae*, ed. A. W. Larkum, S. E. Douglas and J. A. Raven. Dordrecht: Kluwer Academic Publishers, pp. 277–304.

Larkum, A.W. and Vesk, M. (2003). Algal plastids: Their fine structure and properties. In *Photosynthesis in Algae*, ed. A. W. Larkum, S. E. Douglas and J. A. Raven. Dordrecht: Kluwer Academic Publishers, pp. 11–28.

Latasa, M. and Berdalet, E. (1994). Effect of nitrogen or phosphorus starvation on pigment composition of cultured *Heterocapsa* sp. *J. Plankton Res.* **16**, 83–94.

Latowski, D., Grzyb, J. and Strzalka, K. (2004). The xanthophyll cycle – molecular mechanism and physiological significance. *Acta Physiol. Plant.* **26**, 197–212.

Laurion, I. and Roy, S. (2009). Growth and photoprotection in three dinoflagellates (including two strains of *Alexandrium tamarense*) and one diatom exposed to four weeks of natural and enhanced ultraviolet-b radiation. *J. Phycol.* **45**, 16–33.

Lavaud, J. and Kroth, P. (2006). In diatoms, the transthylakoid proton gradient regulates the photoprotective non-photochemical fluorescence quenching beyond its control on the xanthophyll cycle. *Plant Cell Physiol.* **47**, 1010–16.

Lavaud, J., Rousseau, B. and Etienne, A. L. (2002a). In diatoms, a transthylakoidal proton gradient alone is not sufficient for non-photochemical fluorescence quenching. *FEBS Lett.* **523**, 163–66.

Lavaud, J., Rousseau, B., van Gorkom, H. J. and Etienne, A. L. (2002b). Influence of the diadinoxanthin pool size on photoprotection in the marine planktonic diatom *Phaeodactylum tricornutum*. *Plant Physiol.* **129**, 1398–406.

Lavaud, J., Rousseau, B. and Etienne, A. L. (2003). Enrichment of the light-harvesting complex in diadinoxanthin and implications for the nonphotochemical quenching fluorescence quenching in diatoms. *Biochemistry* **42**, 5802–08.

Lavaud, J., Rousseau, B. and Etienne, A. L. (2004). General features of photoprotection by energy dissipation in planktonic diatoms (Bacillariophyceae). *J. Phycol.* **40**, 130–37.

Lavaud, J., Strzepek, R. F. and Kroth, P. G. (2007). Photoprotection capacity differs among diatoms: possible consequences on the spatial distribution of diatoms related to fluctuations in the underwater light climate. *Limnol. Oceanogr.* **52**, 1188–94.

Lewis, M. R., Cullen, J. J. and Platt, T. (1984a). Relationships between vertical mixing and photoadaptation of phytoplankton: similarity criteria. *Mar. Ecol. Prog. Ser.* **15**, 141–49.

Lewis, M. R., Horne, E. P.W., Cullen, J. J., Oakey, N. S. and Platt, T. (1984b). Turbulent motions may control phytoplankton photosynthesis in the upper ocean. *Nature* **311**, 49–50.

Lichtenthaler, H. K. (1998). The stress concept in plants: an introduction. *Ann. N.Y. Acad. Sci.* **851**, 187–98.

Lichtlé, C., Arsalane, W., Duval, J. C. and Passaquet, C. (1995). Characterization of the light-harvesting complex of *Gyraudyopsis stellifer* (Chrysophyceae) and effect of light stress. *J. Phycol.* **31**, 380–87.

Lohmann, M., Döhler, G., Huckenbeck, N. and Verdini, S. (1998). Effects of UV radiation of different wavebands on pigmentation, ^{15}N-ammonium uptake, amino acid pools and adenylate contents of marine diatoms. *Mar. Biol.* **130**, 501–07.

Lohr, M. and Wilhelm, C. (2001). Xanthophyll synthesis in diatoms: quantification of putative intermediates and comparison of pigment conversion kinetics with rate constants derived from a model. *Planta* **212**, 382–91.

Lütz, C., Seidlitz, H. K. and Meindl, U. (1997). Physiological and structural changes in the chloroplast of the green alga *Microsterias denticulata* induced by UV-B stimulation. *Plant Ecol.* **128**, 55–64.

MacIntyre, H. L., Kana, T. M. and Geider, R. J. (2000). The effect of water motion on short-term rates of photosynthesis by marine phytoplankton. *Trends Plant Sci.* **5**, 12–17.

Masojídek, J., Torzillo, G., Koblížek, M., Kopecki, J., Bernardini, P., Sacchi, A. and Komenda, J. (1999). Photoadaptation of two members of the Chlorophyta (*Scenedesmus* and *Chlorella*) in laboratory and outdoor cultures: changes in chlorophyll fluorescence quenching and the xanthophyll cycle. *Planta* **209**, 126–35.

Masojídek, J., Kopecká, J., Koblížek, M. and Torzillo, G. (2004). The xanthophyll cycle in green algae (Chlorophyta): its role in the photosynthetic apparatus. *Plant Biol.* **6**, 342–49.

Mewes, H. and Richter, M. (2002). Supplementary ultraviolet-B radiation induces a rapid reversal of the diadinoxanthin cycle in the strong-light exposed diatom *Phaeodactylum tricornutum*. *Plant Physiol.* **130**, 1527–35.

Mitrovic, S. M., Howden, C. G., Bowling, L. C. and Buckney, R. T. (2003). Unusual allometry between *in situ* growth of freshwater phytoplankton under static and fluctuating light environments: possible implications for dominance. *J. Plankton Res.* **25**, 517–26.

Moisan, T. A., Olaizola, M. and Mitchell, B. G. (1998). Xanthophyll cycling in *Phaeocystis antarctica*: changes in cellular fluorescence. *Mar. Ecol. Prog. Ser.* **169**, 113–21.

Mohovic, B., Gianesella, S. M.F., Laurion, I. and Roy, S. (2006). Ultraviolet-B photoprotection efficiency of mesocosm-enclosed natural phytoplankton communities from different latitudes: Rimouski (Canada) and Ubatuba (Brazil). *Photochem. Photobiol.* **82**, 952–61.

Moline, M. A. (1998). Photoadaptative response during the development of a coastal Antarctic diatom bloom and relationship to water column stability. *Limnol. Oceanogr.* **43**, 146–53.

Muller, A. M. and Wasmund, N. (2003). Photophysiology of surface phytoplankton communities in a transect from the mouth of the Peene-Strom to the Arkona Sea (Baltic). *Int. Rev. Hydrobiol.* **88**, 482–97.

Neale, P.J., Cullen, J. J. and Davis, R. F. (1998). Inhibition of marine photosynthesis by ultraviolet radiation: variable sensitivity of phytoplankton in the Weddell-Scotia Sea during the austral spring. *Limnol. Oceanogr.* **43**, 433–48.

Nielsen, M. and Sakshaug, E. (1993). Adaptation to spectrally different light regimes: studies of the growth rate and carbon to chlorophyll *a* ratio of the marine diatom *Skeletonema costatum*. *Limnol. Oceanogr.* **38**, 1576–81.

Olaizola, M. and Yamamoto, H. Y. (1994). Short term response of the diadinoxanthin cycle and fluorescence yield to high irradiance in *Chaetoceros muelleri* (Bacillariophyceae). *J. Phycol.* **30**, 606–12.

Olaizola, M., Bienfang, P. K. and Ziemann, D. A. (1992). Pigment analysis of phytoplankton during a subarctic bloom: xanthophyll cycling. *J. Exp. Mar. Biol. Ecol.* **158**, 59–74.

Olaizola, M., LaRoche, J., Kolber, Z. and Falkowski, P. G. (1994). Non-photochemical quenching and the diadinoxanthin cycle in a marine diatom. *Photosynth. Res.* **41**, 357–70.

Oliver, R. L., Whittington, J., Lorenz, Z. and Webster, I. T. (2003). The influence of vertical mixing on the photoinhibition of variable chlorophyll *a* fluorescence and its inclusion in a model of phytoplankton photosynthesis. *J. Plankton Res.* **25**, 1107–29.

Owens, T. G. (1986). Light-harvesting function in the diatom *Phaeodactylum tricornutum* II. Distribution of excitation energy between the photosystems. *Plant Physiol.* **80**, 739–46.

Pfündel, E. E., Pan, R. S. and Dilley, R. A. (1992). Inhibition of violaxanthin de-epoxidation by ultraviolet-B radiation in isolated chloroplasts and intact leaves. *Plant Physiol.* **98**, 1372–80.

Raniello, R., Lorenti, M., Brunet, C. and Buia, M. C. (2006). Photoacclimation of the invasive alga *Caulerpa racemosa* var. *cylindrica* to depth and daylight patterns and a putative role for siphonaxanthin. *Mar. Ecol.* **27**, 20–30.

Rascher, U. and Nedbal, L. (2006). Dynamics of photosynthesis in fluctuating light. *Curr. Opin. Plant Biol.* **9**, 671–78.

Raven, J. A. and Geider, R. J. (2003). Adaptation, acclimation and regulation in algal photosynthesis. In *Photosynthesis in Algae*, ed. A. W. Larkum, S. E. Douglas and J. A. Raven. Dordrecht: Kluwer Academic Publishers, pp. 386–412.

Rech, M., Mouget, J. L., Morant-Manceau, A., Rosa, P. and Tremblin, G. (2005). Long-term acclimation to UV radiation: effects on growth, photosynthesis and carbonic anhydrase activity in marine diatoms. *Bot. Mar.* **48**, 407–20.

Rijstenbil, J. W. (2005). UV- and salinity-induced oxidative effects in the marine diatom *Cylindrotheca closterium* during simulated emersion. *Mar. Biol.* **147**, 1063–73.

Ritz, M., Neverov, K. V. and Etienne, A. L. (1999). ΔpH-dependent fluorescence quenching and its photoprotective role in the unicellular red alga *Rhodella violacea*. *Photosynthetica* **37**, 267–80.

Rodríguez, F., Chauton, M., Johnsen, G., Andresen, K., Olsen, L. M. and Zapata, M. (2006). Photoacclimation in phytoplankton: implications for biomass estimates, pigment functionality and chemotaxonomy. *Mar. Biol.* **148**, 963–71.

Roleda, M.Y., Campana, G. L., Wiencke, W., Hanelt, D., Quartino, M. L. and Wulff, A. (2009). Sensitivity of Antarctic *Urospora penicilliformis* (Ulotrichales, Chlorophyta) to ultraviolet radiaton is life-stage dependent. *J. Phycol.* **45**, 600–09.

Roy, S., Mohovic, B., Gianesella, S. M.F., Schloss, I., Ferrario, M. and Demers, S. (2006). Effects of enhanced UV-B on pigment-based phytoplankton biomass and composition of mesocosm-enclosed natural marine communities from three latitudes. *Photochem. Photobiol.* **82**, 909–22.

Ruban, A. V., Lavaud, J., Rousseau, B., Guglielmi, G., Horton, P. and Etienne, A. L. (2004). The super-excess energy dissipation in diatom algae: comparative analysis with higher plants. *Photosynth. Res.* **82**, 165–75.

Sakshaug, E. and Johnsen, G. (2005). Absorption, fluorescence excitation and photoacclimation in phytoplankton. In *Algal Cultures, Analogues of Blooms and Applications*, ed. D. V. Subba Rao. Enfield: Science Publishers, pp. 685–714.

Sakshaug, E., Demers, S. and Yentsch, C. M. (1987). *Thalassiosira oceanica* and *T. pseudonana*: Two different photoadaptational responses. *Mar. Ecol. Prog. Ser.* **41**, 275–82.

Sakshaug, E., Bricaud, A., Dandonneau, Y., Falkowski, P. G., Kiefer, D. A., Legendre, L., Morel, A., Parslow, J. and Takahashi, M. (1997). Parameters of photosynthesis: definitions, theory and interpretation of results. *J. Plankton Res.* **19**, 1637–70.

Schofield, O., Kroon, B. M.A. and Prézelin, B. B. (1995). Impact of ultraviolet-B radiation on photosystem II activity and its relationship to the inhibition of carbon fixation rates for Antarctic ice algae communities. *J. Phycol.* **31**, 703–15.

Schofield, O., Prézelin, B. B. and Johnsen, G. (1996). Wavelength-dependency of maximum quantum yield of carbon fixation for two red tide dinoflagellates, *Heterocapsa pygmaea* and *Prorocentrum minimum*. *J. Phycol.* **32**, 574–83.

Serodio, J., Cruz, S., Vieira, S. and Brotas, V. (2005). Non-photochemical quenching of chlorophyll fluorescence and operation of the xanthophyll cycle in estuarine microphytobenthos. *J. Exp. Mar. Biol. Ecol.* **326**, 157–69.

Sobrino, C., Neale, P. J., Montero, O. and Lubián, L. M. (2005). Biological weighting function for xanthophyll de-epoxidation induced by ultraviolet radiation. *Physiol. Plant.* **125**, 41–51.

Staehr, P. A., Henriksen, P. and Markager, S. (2002). Photoacclimation of four marine phytoplankton species to irradiance and nutrient availability. *Mar. Ecol. Prog. Ser.* **238**, 47–59.

Stolte, W., Kraay, G. W., Noordeloos, A. A.M. and Riegman, R. (2000). Genetic and physiological variation in pigment composition of *Emiliania huxleyi* (Prymnesiophyceae) and the potential use of its pigment rations as a quantitative physiological marker. *J. Phycol.* **36**, 529–39.

Strzalka, K., Kostecka-Gugala, A. and Latowski, D. (2003). Carotenoids and environmental stress in plants: significance of carotenoid-mediated modulation of membrane physical properties. *Russ. J. Plant Physiol.* **50**, 168–72.

Strzepek, R. F. and Harrison, P. J. (2004). Photosynthetic architecture differs in coastal and oceanic diatoms. *Nature* **431**, 689–92.

Tedetti, M. and Sempéré, R. (2006). Penetration of ultraviolet radiation in the marine environment. A review. *Photochem. Photobiol.* **82**, 389–97.

Thompson, P. A., Pesant, S. and Waite, A. M. (2007). Contrasting the vertical differences in the phytoplankton biology of a dipole pair of eddies in the south-eastern Indian Ocean. *Deep-Sea Res. II* **54**, 1003–28.

Ursi, S., Pedersén, M., Plastino, E. and Snoeijs, P. (2003). Intraspecific variation of photosynthesis, respiration and photoprotective carotenoids in *Gracilaria birdae* Gracilariales (Rhodophyta). *Mar. Biol.* **142**, 997–1007.

van de Poll, W. H., van Leeuwe, M. A., Roggeveld, J. and Buma, A. G.J. (2005). Nutrient limitation and high irradiance acclimation reduce PAR and UV-induced viability loss in the Antarctic diatom *Chaetoceros brevis* (Bacillariophyceae). *J. Phycol.* **41**, 840–50.

Vincent, W. F. and Roy, S. (1993). Solar UV-B effects on aquatic primary production: damage, repair and recovery. *Environ. Rev.* **1**, 1–12.

Vincent, W. F. and Neale, P. J. (2000). Mechanisms of UV damage to aquatic organisms. In *The Effects of UV Radiation in the Marine Environment*, ed. S. de Mora, S. Demers and M. Vernet. Cambridge: Cambridge University Press, pp. 149–76.

Vinebrooke, R. D. and Leavitt, P. R. (1999). Differential responses of littoral communities to ultraviolet radiation in an alpine lake. *Ecology* **80**, 223–27.

Welschmeyer, N. A. and Hoepffner, N. (1986). Rapid xanthophyll cycling: an *in situ* tracer for mixing in the upper ocean. *EOS Trans. AGU*, **67**, 969.

Wilhelm, C., Büchel, C., Fisahn, J., Goss, R., Jakob, T., LaRoche, J., Lavaud, J., Lohr, M., Riebesell, U., Stehfest, K. and Kroth, P. (2006). The regulation of carbon and nutrient assimilation in diatoms is significantly different from green algae. *Protist* **157**, 91–124.

Wilson, A., Ajlani, G., Verbavatz, J. M., Vass, I., Kerfeld, C. A. and Kirilovsky, D. (2006). A soluble carotenoid protein involved in phycobilisome-related energy dissipation in cyanobacteria. *Plant Cell* **18**, 992–1007.

Zapata, M., Jeffrey, S. W., Wright, S. W., Rodríguez, F., Garrido, J. L. and Clementson, L. (2004). Photosynthetic pigments in 37 species (65 strains) of Haptophyta: implications for oceanography and chemotaxonomy. *Mar. Ecol. Prog. Ser.* **270**, 83–102.

Zingone, A., Borra, M., Brunet, C., Forlani, G., Kooistra, W. H.C.F. and Procaccini, G. (2002). Phylogenetic position of *Crustomastix stigmatica* sp. nov. and *Dolichomastix tenuilepsis* in relation to the Mamiellales (Prasinophyceae, Chlorophyta). *J. Phycol.* **38**, 1024–39.

Zudaire, L. and Roy, S. (2001). Photoprotection and long-term acclimation to UV radiation in the marine diatom *Thalassiosira weissflogii*. *J. Photochem. Photobiol. B: Biol.* **62**, 26–34.

12

Pigment-based measurements of phytoplankton rates

ANDRÉS GUTIÉRREZ-RODRÍGUEZ AND MIKEL LATASA

The most direct way to measure phytoplankton growth/production rate is to quantify changes in biomass over a period of time, something difficult to achieve methodologically in nature. A second and serious problem is the choice of the biomass unit for measurements. Many different macromolecules have been suggested (e.g. proteins, nucleic acids, ATP, lipids and pigments). However, carbon (C) is commonly the standard unit for biomass and the desired currency for productivity models (MacIntyre *et al.*, 2000). Because carbon is common to all organisms, estimation of phytoplankton rates based on carbon measurement is not straightforward. Among macromolecules, pigments are specific to phytoplankton. Their specificity and distinct optical properties make pigments a useful tool for studying phytoplankton processes.

Phytoplankton dynamics reflect the balance between production and losses, notably through population growth and grazing. It is, however, difficult to determine these rates simultaneously, and significant differences can be observed in the short term (hours, days), on the scale of phytoplankton generation times. Active growth is accompanied by a parallel grazing pressure exercised mostly by microzooplankton. This chapter reviews the pigment-based methods used to measure both production and grazing rates, focusing on incubation-dependent methods: the ^{14}C-pigment labeling technique and the serial dilution bioassay. Finally we will briefly comment on other approaches such as satellite imagery and fast repetition rate fluorometry (FRRF), which are free of incubation-derived artifacts.

12.1 Pigment labelling method

The rate of primary production is the most measured biological rate in oceanography. Primary productivity involves light absorption, uptake of CO_2 and production of O_2. In pelagic systems, it is defined as the rate of C uptake by phytoplankton (see Box 12.1). A detailed description of primary production, definitions and measurement methods can be found in Falkowski and Raven (2007).

Phytoplankton Pigments: Characterization, Chemotaxonomy and Applications in Oceanography, eds. Suzanne Roy, Carole A. Llewellyn, Einar Skarstad Egeland and Geir Johnsen. Published by Cambridge University Press.

BOX 12.1 C-based definition of primary production (Karl *et al.*, 2002)

The gross rate of primary production is the rate of organic carbon produced through photosynthesis from the reduction of carbon dioxide and comprises dissolved, particulate and respired carbon. In general, it is estimated by measuring the oxygen (O_2 or $^{18}O_2$) originating from the reduction of CO_2. An approximation can be reached with the ^{14}C method when production is much larger than respiration and the incubation period is very short (~minutes).

The net rate of primary production is gross primary production minus the respiration of the autotrophs. It can be estimated with the ^{14}C method, measuring the incorporation into the particulate and dissolved pools (~2–12 h incubations).

Finally, net community production rate refers to gross primary production rate minus respiration rate from both autotrophic and heterotrophic organisms. It is measured by means of the oxygen method.

The pigment-based method most similar to C uptake is the *pigment labelling method*. In 1981 Redalje and Laws introduced the so-called ^{14}C-pigment labelling technique for estimating phytoplankton growth rates and carbon biomass. The basic conceptual framework is as follows: phytoplankton growing in the light and exposed to ^{14}C-DIC will assimilate the dissolved inorganic carbon (DIC) and incorporate it in a balanced way to all cellular constituents, including chlorophyll *a* (Chl *a*). At the end of a few hours' incubation, the Chl *a* pool should be uniformly labeled and the specific activity of carbon in the Chl *a* molecule ($R^*_{\mathrm{Chl}\,a}$ = disintegrations per minute (dpm) • μg C-Chl a^{-1}) should equal the specific activity of the phytoplankton cellular carbon (R^*_{Cp} = dpm • μg C-cell^{-1}). The veracity of this critical assumption, $R^*_{\mathrm{Chl}\,a} = R^*_{\mathrm{Cp}}$, depends, among other factors, on the incubation time and the phytoplankton growth rate (i.e. balanced growth). Redalje and Laws (1981) concluded that incubations shorter than 12 hours could meet this requirement even in cultures with relatively slow growth rates ($\mu = 0.3$ d^{-1}). When this assumption is valid, this method allows the estimation of carbon concentration (C_p) and carbon-specific growth rate (μ) in phytoplankton. The problem with this approach is that zooplankton grazing and bacterial production co-occurring during incubations of field samples cause an overestimation of C_p and an underestimation of the growth rate. Laws (1984) developed an approach to quantify this error, which is substantial when phytoplankton carbon gains (production) and subsequent transfers (by grazing or excretion) are balanced, or when incubation time is greater than phytoplankton doubling time.

Welschmeyer and Lorenzen (1984) reformulated the simple one-pool model (P-model) of isotopic labelling based on a different conceptual model and demonstrated that there was no need to estimate carbon fixation rates to calculate specific growth rates and thus, circumvented the problem of carbon cycling during

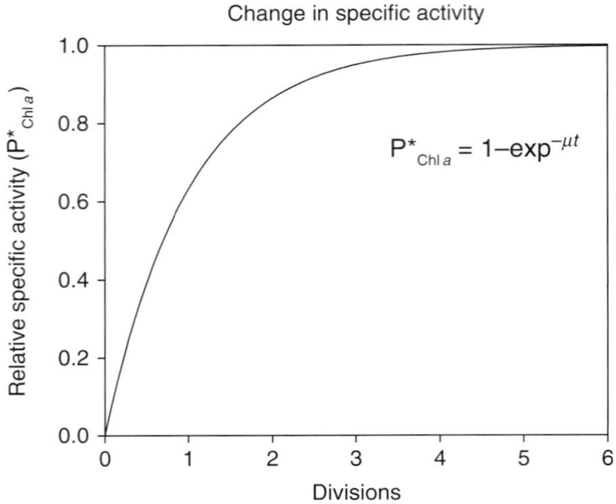

Figure 12.1. Representation of phytoplankton carbon labelling dynamics. The relative specific activity of phytoplankton carbon ($P^*_{\mathrm{Chl}\,a}$) changes with the number of phytoplankton cell divisions. Modified from Welschmeyer and Lorenzen (1984).

incubation. They postulated and checked experimentally that the newly synthesized Chl *a* specific activity ($R^*_{\mathrm{Chl}\,a}$) equals the specific activity of inorganic carbon supplied (R^*_{DIC}), i.e. that the relative specific activity of Chl *a* carbon ($P^*_{\mathrm{Chl}\,a} = R^*_{\mathrm{Chl}\,a}/R^*_{\mathrm{DIC}}$) tends asymptotically to one as a function of phytoplankton doubling time as described by the following expression (see Figure 12.1):

$$P^*_{\mathrm{Chl}\,a} = 1 - \exp^{(-\mu t)} = 1 - 2^{(-\mu t/\ln 2)}. \tag{12.1}$$

Rearrangement of Eq. (12.1) yields Eq. (12.2),

$$\mu = -\ln\left[1 - \left(1.05\,R^*_{\mathrm{Chl}\,a}/R^*_{\mathrm{DIC}}\right)\right]/t \tag{12.2}$$

which shows how growth rate (μ) can be estimated from $R^*_{\mathrm{Chl}\,a}$ and R^*_{DIC}, measurements that are independent of the ^{14}C activity of total filtered particulate matter at the end of the incubation. The term 1.05 accounts for isotope discrimination. In consequence, as long as grazers do not discriminate between labelled and unlabelled preys, this reformulation of the Chl *a*-labelling technique provides a way of measuring phytoplankton specific growth rates despite the co-occurrence of grazing during the incubation (Welschmeyer and Lorenzen, 1984). A detailed description of this method can be found in Redalje (1993).

Contrary to previous studies showing that the specific activity of Chl *a* equals that of cellular carbon after incubation periods shorter than 12 hours (Redalje and Laws, 1981; Redalje, 1983; Welschmeyer and Lorenzen, 1984), Jespersen *et al.* (1992) reported that ^{14}C was more rapidly incorporated into Chl *a* than into phytoplankton particulate carbon, leading to underestimation of C_p and overestimation of μ. However, Goericke (1992) rebutted this argument on the basis of the radiochemical interference from colourless radioactive compounds when pigments are isolated with 'simple' HPLC protocols. Acidification or saponification of isolated Chl *a* and carotenoids and subsequent re-isolation by HPLC have been shown to effectively solve this problem when present (Goericke and Welschmeyer, 1993a; 1993b). Routine collection and measurement of the radioactivity of the eluent before and after the peak of interest is recommended to test for possible contamination (Goericke, 1992). An alternative protocol was introduced by Pinckney *et al.* (1996) which substituted the collection of individual pigments and later scintillation counting by an in-line flow scintillation counting system coupled to the HPLC. The problem of possible contamination by colourless labelled compounds persists in this procedure, and pertinent control measures should be observed.

In 1989, Gieskes and Kraay presented an extension of the Chl *a*-labelling technique to carotenoids. They combined the Chl *a*-labelling technique and the chromatographic advances in algal pigment separation (Mantoura and Llewellyn, 1983; Gieskes and Kraay, 1986a, b) to measure the growth rates of various phytoplankton groups based on ^{14}C-labelling of taxon-specific pigments. This pioneer study showed the potential of the pigment labelling method to measure specific growth rates of different phytoplankton groups coexisting under the same environmental conditions. However, potential problems of the method linked to the assumption of balanced growth of the cells (Shuter, 1979; $\mu_c = \mu_{Chl\,a} = \mu_{pigment}$) also arose from this study (see Box 12.2).

The P-model of Welschmeyer and Lorenzen (1984) underestimated the true growth rate when the final degree of isotopic labelling was low due to slow growth rates and/or a short incubation time. The P-model corresponded to a simple one-pool model, and did not consider pigment precursors. Goericke and Welschmeyer (1992a, b) modified the original P-model in a new precursor-product model (F→P model), which introduced the effect of pigment precursor turnover rates on the labelling kinetics of the pigment product. In this extensive work, they studied the labelling kinetics of pigments (Chl *a* and carotenoids) and their precursor pools by measuring their turnover rate. They concluded that (1) the F→P model successfully described ^{14}C labelling of pigments, (2) the specific synthesis rate of pigments could only be reliably estimated from pigment labelling if the turnover of pigments was zero (as was the case for the different pigments analyzed) and (3) taxon-specific carbon growth rates could only be estimated from pigment growth rates if cell growth was balanced. These statements were later checked in the field (Goericke and Welschmeyer, 1993a) and extended to a number of algal species cultured in the laboratory (Goericke and Welschmeyer, 1993b). The application of the *pigment labelling method* to field studies requires fulfillment of the second and third points mentioned above and the empirical determination of the turnover rate of each carotenoid precursor. The use of an *a priori* laboratory-determined precursor

BOX 12.2 Estimates of pigment-based primary production rates in steady and non-steady state

Two types of non-steady states should be distinguished for the phytoplankton community. The first is when populations are changing because of a 'natural' effect, i.e. during the build-up of a bloom or caused by strong mixing. Under those conditions, phytoplankton can be out of balance at the population level affecting cell densities, or at the individual level affecting cell physiology.

The second type of non-steady state is that induced during the incubations required to measure most rates. We will emphasize typical artifacts and precautions needed to perform incubation experiments. Because phytoplankton growth and grazing mortality are not equilibrated over the diel cycle, the goal is to measure these rates over a period of one day (24 h). Experiments lasting less than a day measure features of phytoplankton physiology and/or zooplankton grazing behavior. It then becomes necessary to know enough about the physiology/ecology of the studied community to parameterize and extrapolate the short-term measures to a full-day cycle.

If incubations are performed over a diel cycle, artifacts derived from bottle confinement become more evident. The problems inherent in bottle incubations such as the change of hydrodynamic regime at cell-scale or the lack of short-term variability in irradiance due to vertical motion in the water column will not be addressed here. Instead we will focus on effects that translate into photoacclimation problems and how they can be avoided or taken into account to improve rate estimates. Pigment contents readily respond to changes in phytoplankton physiology, which, in turn, is driven by irradiance and nutrient availability.

The fact that pigments are most readily affected by incubation artifacts makes them a canary-in-a-coal-mine tool, i.e. other processes affected by irradiance are altered but their measurement does not record the changes affecting the phytoplankton response. McManus (1995) showed the substantial impact that photoacclimation could have on growth rates estimated from changes in pigment concentration in dilution experiments. Latasa *et al.* (1997, 2005) evaluated possible photoacclimation processes occurring during incubations based on the different synthesis rates expected for photoprotective and photosynthetic pigments under changing light conditions: higher synthesis rates of photoprotective pigments relative to photosynthetic pigments would indicate that cells were receiving higher light levels than those to which they were acclimated *in situ*, and vice versa. An alternative approach to evaluate photoacclimation during incubations takes advantage of flow cytometry to extrapolate changes in pigment/cell of the phytoplankton community from changes in fluorescence per cell of different picophytoplankton groups (Brown *et al.*, 1999; Stelfox-Widdicombe *et al.*, 2000; Quevedo and Anadón, 2001; Landry *et al.*, 2003).

turnover rate introduces an unknown systematic error to the estimation of μ. This error should be bracketed considering the possible values of pigment precursor turnover rates and the uncertainty transferred to the estimated μ (Goericke, 1998). Extension of the Chl *a*-labeling technique to carotenoid labelling has relatively seldom been applied in the field, but has proved useful to gain insights into the factors controlling phytoplankton growth in a wide range of environmental conditions (Welschmeyer *et al.*, 1991; Goericke and Welschmeyer, 1993b; Goericke, 1998, 2002; Pinckney *et al.*, 2001; Delizo *et al.*, 2007).

12.2 Serial dilution method

In 1982 Landry and Hassett presented a new experimental method to estimate microzooplankton grazing and phytoplankton growth rates. This method does not properly account for grazing exerted by larger predators because it is carried out in small bottles (< 4 l) where large zooplankton are excluded, and experimental samples are usually obtained with techniques not adequate for sampling large zooplankton. The method is based on the release of grazing pressure by dilution of the natural seawater sample with filtered (particle-free) seawater. Three assumptions are made:

(1) Growth of individual phytoplankton cells is assumed to be independent of cell density because cell physiology or growth potential is not affected by dilution.
(2) Grazing is assumed to be proportional to the encounter rate between predator and prey. Because the encounter rate is an inverse linear function of dilution, grazing should decrease in proportion to increased dilution.
(3) Growth of phytoplankton is assumed to be properly described by the equation

$$P_{(t)} = P_0 \exp^{(\mu-g)t}, \tag{12.3}$$

where μ and g are instantaneous coefficients of phytoplankton growth and grazing mortality ($\mu - g = k$, instantaneous apparent growth rate), t is incubation time, and $P_{(t)}$ and P_0 are phytoplankton densities at the end and beginning of the incubation, respectively.

Knowing the fraction of undiluted water (D) in each incubation bottle, the apparent growth rate (k) can be calculated as,

$$k = 1/t \ln\left(P_{(t)}/(P_0 D)\right). \tag{12.4}$$

Given the assumptions above, where grazing mortality is reduced with dilution and phytoplankton growth rate is unaffected, apparent growth rate should increase with dilution. An example of a standard expected plot is shown in Figure 12.2.

The first assumption can be compromised under some conditions (extreme oligotrophic or eutrophic conditions). In oligotrophic systems, where nutrient supply to phytoplankton is mainly sustained by grazing mediated regeneration processes, cell physiology may be affected in the more diluted bottles due to decreased nutrient regeneration rates associated with lower grazing. The opposite scenario may be found in systems with high phytoplankton biomass sustained by non-recycled nutrient sources where sample confinement could short-cut physically driven nutrient

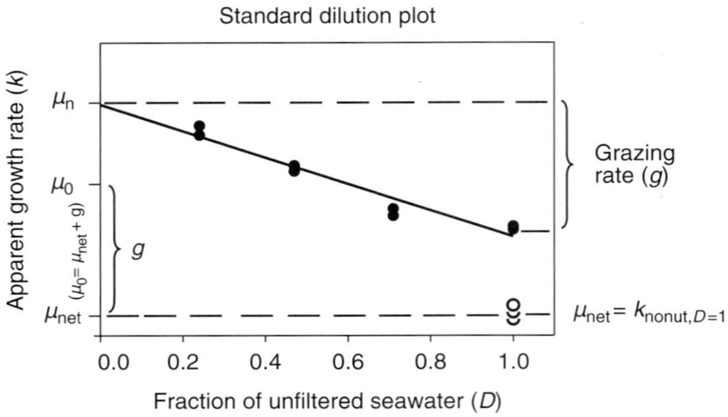

Figure 12.2. Standard dilution plot between the apparent growth rate (k) and the fraction of undiluted water (D). White and black dots are non-amended and nutrient-amended treatments, respectively. Intrinsic growth rate in nutrient-amended conditions (μ_n) is the y-intercept of the regression. Grazing is the difference between μ_n and the net growth rate in the nutrient-amended non-diluted bottles. *In situ* intrinsic growth rate (μ_o) is the sum of apparent growth rate in the non-amended non-diluted treatments ($\mu_{net} = k_{nonut,\ D=1}$) plus grazing.

supply, with the least diluted treatments being the most affected. In either case, inorganic nutrients should be added to ensure nutrient sufficiency and to fulfill the first assumption. Moreover, parallel nutrient amended and unamended incubations can provide insights on the internal and external sources of inorganic nutrients for phytoplankton growth (Andersen *et al.*, 1991). Chavez *et al.* (1991) tested the first assumption in the equatorial Pacific adding ^{14}C-DIC into the different dilution treatments. Results showed first, that ^{14}C-derived phytoplankton growth rates matched those estimated with the *dilution method*; and second, they were virtually identical in the different dilution bottles. Following the recommended nutrient amendment strategy, microzooplankton grazing rates can be estimated from the nutrient amended serial dilution and *in situ* phytoplankton growth rates can be estimated from the unamended bottles (see Figure 12.2) as,

$$\mu = k_{nonut,\ D=1} + g. \tag{12.5}$$

In this approach, it is assumed that grazing rates under ambient nutrients are not different from those under nutrient enriched conditions.

For the second assumption, linearity between grazing rate and dilution requires that the clearance rate of microzooplankton (F, volume swept clear time^{-1} predator^{-1}) be constant among different dilution treatments, independent of prey density. Gallegos (1989) modified the method for eutrophic systems, where grazing saturates at ambient prey concentrations and nonlinear relations are observed between apparent growth rate and dilution. Growth and grazing rates were estimated here using only three very dilute treatments in order to avoid prey saturation.

This method, known as the '3-points approach', allows an accurate estimation of growth and grazing rates when non-linear relations are observed. Evans and Paranjape (1992) suggested a general nonlinear model that includes the possibility of grazing saturation. Landry *et al.* (1995) proposed a modified experimental protocol to overcome problems associated with violation of the second assumption, and demonstrated that the dilution factor (D) was a good proxy for relative grazing in the equatorial Pacific. However, linearity has been shown to fail in a significant number of systems and different fitting procedures have been adopted to obtain accurate estimates of growth and grazing rates (Rivkin *et al.*, 1999; Redden *et al.*, 2002; Worden and Binder, 2003, Moigis, 2006).

Non-linear responses can also arise from differences in net growth rate of microzooplankton among dilution treatments. Dolan *et al.* (2000) and Dolan and McKeon, (2005) argued that a high mortality of ciliates in the most diluted treatments leads to systematic overestimation of microzooplankton grazing and derived phytoplankton intrinsic growth rates. However, even in the extreme case of a complete absence of microzooplankton in the most diluted treatments, phytoplankton apparent growth rate will never exceed phytoplankton intrinsic growth rate. Therefore, it is unclear why a higher ciliate mortality in these most diluted treatments should imply an overestimation of grazing rates. To the authors' knowledge, only a few studies have analyzed the growth response of microzooplankton to the dilution gradient (Agis *et al.*, 2007; First *et al.*, 2007). These studies revealed a wide variety of responses across systems with different trophic status. However, the effect of a distinct microzooplankton growth rate on estimated herbivory rates, when substantial, does not show a systematic trend. From our point of view, grazing estimates should be more sensitive to changes in average grazer concentration derived from trophic cascade processes in the less diluted treatments (e.g. truncation of the copepod-microzooplankton link). A valuable discussion regarding this and other aspects of the method can be found in Dolan and McKeon (2005) and Landry and Calbet (2005).

The resolution of the *dilution method* can be increased by using a more specific phytoplankton biomarker than Chl *a*. Burkill *et al.* (1987) combined the *dilution method* with HPLC pigment analysis to estimate growth and grazing rates of major phytoplankton groups from changes in the concentration of their pigment markers. This methodology has proved useful in estimating growth and grazing rates of major phytoplankton groups across distinct pelagic ecosystems, from the equator (Latasa *et al.*, 1997; Landry *et al.*, 2000) to the poles (Strom and Welschmeyer, 1991) and from the nutrient-rich Arabian Sea (Brown *et al.*, 2002) and temperate estuary (McManus and Ederington-Cantrell, 1992) to the oligotrophic Mediterranean Sea (Latasa *et al.*, 2005).

The use of pigment disappearance as a grazing index assumes that pigment breakdown is synonymous with cell grazing, thus pigment degradation rate inside the predator vacuoles (digestion) must be very fast and not different among pigments (Burkill *et al.*, 1987). Partial degradation of ingested pigments (Klein *et al.*, 1986) could underestimate growth and grazing rates (Barlow *et al.*, 1988). In laboratory studies, consistently high chlorophyll degradation (>94% pigment losses relative to cell losses) contrasted with the more variable carotenoid degradation, non-systematically related to its identity (Strom *et al.*, 1998). Comparing growth and grazing estimates obtained from pigments and from cell disappearance, Waterhouse and Welschmeyer (1995) concluded that pigment derived rates can underestimate true growth and grazing rates. However, their conclusions should be approached with caution since incubations were carried out under dark conditions. Because light might enhance pigment destruction (Klein *et al.*, 1986; Strom, 2001), sunset to sunset incubations are recommended to allow photooxidation of ingested material and minimize the contribution of partially digested pigments to pigment concentration estimates.

12.2.1 From pigment to C in dilution experiments

In order to convert Chl *a*-based biomass changes to carbon, the C:Chl *a* needs to be known. The dynamic response of this ratio to light, nutrient status and temperature has been captured in the regulatory model proposed by Geider *et al.* (1998). Estimation of C:Chl *a* in the field is not trivial and in spite of the variety of techniques applied (microscopy, flow cytometry, ^{14}C-labelling), it remains one of the major source of uncertainty for the estimation of primary production (Banse, 2002a). In addition to C:Chl *a* estimates obtained from pigment labelling, a new approach was introduced by Obayashi and Tanoue (2002). Chl *a* synthesis derived from serial dilution experiments and the production of particulate organic carbon in phytoplankton, estimated from ^{14}C experiments, were combined to assess the C:Chl *a* ratio of 'newly' synthesized organic matter. This approach allowed the transformation of Chl *a* synthesis and destruction rates, based on Equations 12.6 and 12.7 (modified from Landry *et al.*, 2000, based on Frost, 1972), into carbon specific rates.

$$\mathrm{Chl}\,a_{\mathrm{synthesized}} = \mu^{*}\mathrm{Chl}\,a_0{}^{*}\left[\exp^{(\mu-g)t} - 1\right]/(\mu - g) \qquad (12.6)$$

$$\mathrm{Chl}\,a_{\mathrm{destroyed}} = g^{*}\mathrm{Chl}\,a_0{}^{*}\left[\exp^{(\mu-g)t} - 1\right]/(\mu - g), \qquad (12.7)$$

where Chl a_0 is initial Chl *a* concentration.

Latasa *et al.* (2005) further extended this approach to assess carbon fluxes through the major phytoplankton groups identified by their pigment markers. The application of CHEMTAX algorithms (see Chapter 6, this volume) allowed calculation of the biomass of major phytoplankton groups in terms of Chl *a*, which in combination with pigment-based growth rates yielded Chl *a*-based fluxes for each phytoplankton group. In this case, the transformation of Chl *a* to carbon equivalents was estimated from the slope of a model II linear regression because rates of Chl *a* synthesis and carbon uptake were proportional. However, when the relation between these rates varies substantially among experiments, estimation of C:Chl *a* for each particular experiment is recommended. The transformation from Chl *a* to carbon fluxes could

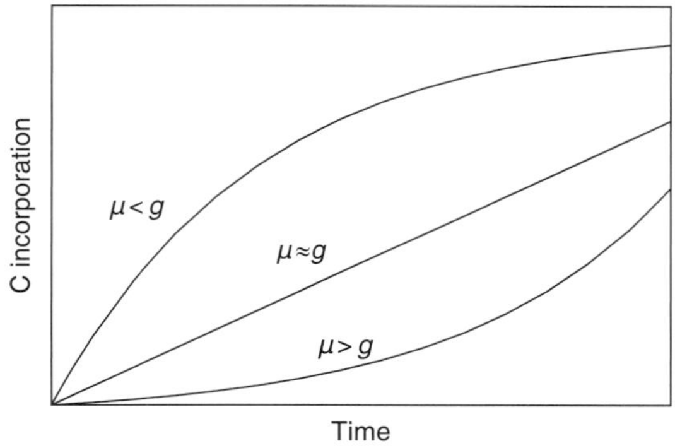

Figure 12.3. Simulated time dynamics of C incorporation according to the balance between growth (μ) and grazing (g). Scales are different for the different curves to emphasize their shape. Taken from Latasa *et al.* (2005). © 2005 by the American Society of Limnology and Oceanography, Inc.

then be made with the calculated C:Chl *a* from the whole phytoplankton community. Further considerations regarding possible differences in nutrient status among phytoplankton groups allow estimation of group-specific C:Chl *a* and associated carbon fluxes. The authors investigated the effect of growth and grazing imbalance on estimates of rates of primary production and concluded that comparison of instantaneous growth rates with primary production rates is best suited for C:Chl *a* estimations, but it does not properly reflect the dynamics of C assimilation when growth and grazing are not balanced (Figure 12.3).

12.2.2 Carbon-labelling versus dilution

The few parallel measurements of growth rates based on carbon labelling and dilution experiments have shown good agreement between both estimates. Welschmeyer *et al.* (1991) and Strom and Welschmeyer (1991) estimated taxon-specific growth rates in the high-nutrient low-chlorophyll (HNLC) waters of the Subarctic Pacific using pigment labelling and dilution experiments, respectively. In spite of different incubation times (five versus one day for labelling and dilution experiments, respectively) both approaches yielded consistent results. In the Arabian Sea, Goericke (2002) estimated growth rates which were also in good agreement with values obtained from dilution experiments (Caron and Dennett, 1999).

12.3 Emerging views from pigment-taxa approaches to estimate phytoplankton rates

12.3.1 Highly dynamic equilibrium

One of the major advantages of the *dilution method* is that it provides simultaneous estimates of growth and grazing rates, facilitating the evaluation of grazing impact upon phytoplankton stock. Thus, the percentage of the standing stock turned over daily (% d^{-1}) can be estimated from phytoplankton concentration and growth and grazing rates (Stelfox-Widdicombe *et al.*, 2000). A number of field studies comprising a wide variety of systems have reported percentage turnover rates of phytoplankton ranging from 30–120% d^{-1} (Landry *et al.*, 1984; Strom and Welschmeyer, 1991; McManus and Ederington-Cantrell, 1992; Verity *et al.*, 1993, 1996; Stelfox-Widdicombe *et al.*, 2000, 2004; Fileman and Burkill, 2001; Calbet *et al.*, 2008; Chen *et al.*, 2009). This implies that phytoplankton stock is renewed by grazing every 1–2 days. Comparison of this number with the average turnover time of 19 years estimated for terrestrial primary producers (Field *et al.*, 1998) stresses the highly dynamic character of pelagic systems.

12.3.2 Important role of microzooplankton even during blooms

Microzooplankton comprise protists and metazoans < 200 μm (Dussart, 1965). This group is generally assumed to be the main consumer of phytoplankton primary production (Calbet and Landry, 2004; Irigoien *et al.*, 2005). Their high growth rates, comparable to those exhibited by their prey, along with high feeding rates, confer on them the ability to exert an intense grazing pressure upon phytoplankton, being capable of consuming > 100% of phytoplankton daily production. Based on size-dependent trophic web structure, microzooplankton grazing is assumed to be limited to small cells. Thus, its impact is thought to be more important in relatively stable oligotrophic systems dominated by pico and nanophytoplankton compared to non-steady situations such as blooms. However, microzooplankton may be important consumers even during bloom situations dominated by large cells (Neuer and Cowles, 1994; Strom *et al.*, 2001; Banse, 2002b). In addition to these community approaches, different studies have shown that some microzooplankton groups are capable of preying upon large cells (Monger and Landry, 1991; Strom and Loukos, 1998). Dinoflagellates, by virtue of their unique and diverse feeding mechanisms, can feed on cells up to five times larger than themselves (Jacobson and Anderson, 1986; Hansen and Calado, 1999; Sherr and Sherr, 2007) and can thus impact even larger cell phytoplankton blooms (e.g. Landry *et al.*, 2008).

12.3.3 Close coupling between growth and grazing

A close coupling between growth and grazing rates has repeatedly been observed in pigment-based dilution experiments either when considering the dynamics of bulk Chl *a* or that of taxon-specific pigment markers: pigments with higher synthesis rates also exhibit higher destruction rates (Figure 12.4A, B). The ecological or methodological nature of this correlation has recently been shown to depend on the range of phytoplankton growth and grazing rate values and the methodological error of the state variable (cell counts or Chl *a* for example, Gutiérrez-Rodríguez *et al.*, 2009).

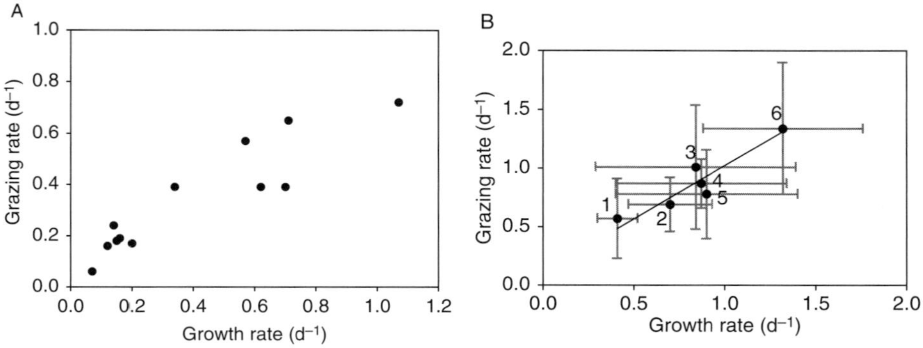

Figure 12.4. (A) Phytoplankton growth and grazing rates estimated from dilution experiments in a highly oligotrophic area of the subtropical North East Atlantic (modified from Quevedo and Anadón, 2001). (B) Average growth and grazing rates of major phytoplankton groups defined by their pigment markers assessed with HPLC analyzed dilution experiments. 1=Green algae, 2=Dinophyceae, 3=Prymnesiophyceae, 4= Phytoplankton, 5=Cyanobacteria, 6=Diatoms (built from Table 3 in Latasa *et al.*, 2005).

This study concluded that the correlation between growth and grazing observed in a coastal Mediterranean site reflected a true ecological correlation. This result has profound implications regarding the mechanisms underlying growth and grazing rates. A positive relation between prey abundance and grazing would imply that the mechanisms controlling the balance between growth and grazing operate at the population level. Usually, however, this does not apply: pigment-specific grazing rates do not show significant correlations with pigment concentrations but instead with pigment synthesis rates. Based on reports supporting a positive feedback between growth and grazing rates (Stoecker *et al.*, 1986, Cowles *et al.*, 1988; Stoecker, 1988; Verity, 1988; Butler *et al.*, 1989), a close interaction between grazer and prey at the level of individual cells may explain the growth and grazing balance observed in the field (Strom and Welschmeyer, 1991; Olson and Strom, 2002; Latasa *et al.*, 2005; Strom *et al.*, 2007; Chen *et al.*, 2009). Faster growing phytoplankton groups may have a higher quality nutrient content and could promote a healthier grazer population capable of exerting a more intense grazing pressure upon phytoplankton (Stoecker *et al.*, 1986; John and Davidson, 2001).

12.4 Other methodologies

The different methodologies to estimate the rate of primary production treated here (^{14}C-uptake, carbon labelling and serial dilution) all share a common important drawback: the need for incubation. In addition to problems associated with bottle effects, the temporal and spatial resolution of these techniques is rather limited. The occurrence of short transient physical phenomena such as fronts, eddies, mesoscale processes, stratification following turbulence are all of major importance for the functioning of pelagic systems. However, given their variable temporal and spatial occurrence, it is difficult to study them with the techniques described above and additional and complementary approaches are needed. Two such approaches are primary production estimates from satellites and from active fluorescence such as the fast repetition rate fluorometry method (FRRF). These approaches operate on spatial and temporal scales unresolved by incubation methods. They will be presented briefly below.

12.4.1. Satellite imagery

The clear advantage of using satellite imagery is the huge spatial and temporal coverage provided by this approach. A major problem, on the other hand, is the difficulty in interpreting the limited information recorded for relatively few wavelengths. Spectral shifts in the upwelling radiance detected by satellite ocean colour sensors allow the estimation of chlorophyll concentration and the modeling of primary productivity rates (Campbell *et al.*, 2002). Here, we will summarize the advantages and caveats in the estimates of satellite-based estimates of primary productivity. The reader is referred to Chapter 14 for more information on satellite applications and phytoplankton studies.

Models to estimate primary production from satellite-derived Chl *a* can be classified as empirical and semi-analytical. The first empirical models were based on the statistical relationship between satellite-derived surface chlorophyll concentration and *in situ* depth-integrated rates of primary production measurements (Smith, 1981; Platt and Herman, 1983). These models were geographically restricted to the area where the parameterization of the algorithms was performed. Later, semi-analytical algorithms incorporated information on phytoplankton physiology. The algorithms progressed in complexity from uniform biomass distribution and linear relation between photosynthesis and available light (Lewis *et al.*, 1986; Platt, 1986) to non-uniform biomass distribution and light spectral effects on photosynthesis (Platt and Sathyendranath, 1988; Morel and André, 1991; Sathyendranath *et al.*, 1995; Antoine and Morel, 1996). Given the significant progress in the characterization of chlorophyll vertical distributions and descriptions of underwater light fields, the limited temporal and spatial resolution in the determination of photosynthetic characteristics remains as the major limitation of primary productivity algorithms (Antoine and Morel, 1996; Behrenfeld and Falkowski, 1997a, b; Behrenfeld *et al.*, 2005; Carr *et al.*, 2006).

Satellite-based estimation of primary production is an active field of research and new approaches have been proposed. One is the satellite-derived estimates of C:Chl *a* which could provide information on phytoplankton physiology (Behrenfeld *et al.*, 2005; Westberry *et al.*, 2008). In addition, the launch of the Moderate Resolution Imaging Spectroradiometer instrument (MODIS) allows the remote measurement of sun-induced fluorescence (Esaias *et al.*, 1998). In theory, near-surface fluorescence and quantum yield of fluorescence can be assessed, providing global coverage for a parameter related to phytoplankton physiology (Kiefer and Reynolds, 1992; Letelier *et al.*, 1997; Huot *et al.*, 2005). However, estimation of the quantum yield of fluorescence is not straightforward and requires complex algorithms (Abbott and Letelier, 1999). In spite of the parallel progress in technology and primary production algorithms, the combination of *in situ* measurements and satellite estimates is still necessary for large-scale primary productivity surveys.

12.4.2 FRRF-based primary productivity

Fast repetition rate fluorometry (FRRF) is one of several non-invasive, highly sensitive, active fluorescence techniques that allows instantaneous *in situ* measurements of photosynthetic processes (Kolber *et al.*, 1998). Physiological characteristics can be measured with a temporal and spatial resolution unprecedented in aquatic research (Suggett *et al.*, 2009 and references therein). The method consists of a series of brief (~1–2 μs) sub-saturating excitation pulses that cumulatively saturate the reaction centers within a single turnover inducing a detectable transient fluorescence. These fluorescence dynamics can be related to the photochemical activity of PSII, from which photosynthetic electron transfer rates (ETR) and the oxygenic gross photosynthetic rate ($P^{Chl\ a}$) can be estimated (Falkowski and Kolber, 1995; Gorbunov *et al.*, 2001; Kromkamp and Forster, 2003; Suggett *et al.*, 2006a; see Box 12.3). Transformation of ETR into oxygen evolution or carbon fixation equivalents requires a series of assumptions, such as photosynthetic unit size and the electron yield of oxygen production (see Box 12.3), each of which introduces a degree of uncertainty. Several attempts have been made to clarify the relation between FRRF and ^{14}C or oxygen derived primary production estimates (Kolber and Falkowski, 1993; Suggett *et al.*, 2001, 2003, 2006a; Moore *et al.*, 2003; Raateoja *et al.*, 2004; Smyth *et al.*, 2004; Corno *et al.*, 2006; Melrose *et al.*, 2006; see also Chapters 11 and 13, this volume).

The difficulty in reconciling FRRF and ^{14}C-uptake measurements has recently been addressed by Suggett *et al.*, (2006b). First, both processes are physically and temporally separated. A considerable number of intermediate energy-consuming processes (light respiration, electron cycling around PSII, inorganic nutrient assimilation...) uncouple estimates of ETR and C-uptake. Second, ETR provided by FRRF is expressed per functional photosystem II reaction center (RCII). Thus, the size of the photosynthethic unit (n_{PSII}, mol RCII (mol Chl $a)^{-1}$) must be known in order to convert productivity rates into equivalents comparable to classical estimates, generally normalized to Chl *a* or carbon units. The high natural variability exhibited by n_{PSII} together with the difficulty of *in situ* estimates, constitute a major limitation in the interpretation of primary productivity yield based on FRRF measurements.

In addition to its potential applicability for primary production estimates, FRRF provides information on the photosynthetic activity at physiologically relevant time

BOX 12.3. **FRRF-derived fluorescence parameters, physiological properties and primary productivity rates, following Kromkamp and Forster (2003), Suggett *et al.* (2003, 2006a). A detailed description of the equations and derivations can be found in these papers. If not specified, fluorescence parameters are given in instrument units (FRRF). PSII and RCII refer to photosystem II and reaction centers of PSII, respectively. See also Chapter 11 (Section 11.1.2) and Chapter 13 (Section 13.3.1), this volume, for further discussion and other approaches.**

Symbol	Term (derivation and units)
F_0	Minimum fluorescence under dark acclimation
F_0'	Minimum fluorescence under actinic light
F_m	Maximum fluorescence after dark acclimation
F_m'	Maximum fluorescence under actinic light
F'	Steady-state fluorescence in the FRRF open chamber
F_v/F_m	Maximum PSII photochemical efficiency under dark acclimation $(F_m - F_0)/F_m$ (dimensionless)
F_v'/F_m'	Maximum PSII photochemical efficiency under actinic light $(F_m' - F_0')/F_m'$ (dimensionless)
F_q'/F_m'	PSII photochemical efficiency under actinic light $(F_m' - F')/F_m'$ (dimensionless)
σ_{PSII}	PSII effective absorption cross-section under dark acclimation (Å^2 photon^{-1})
σ_{PSII}'	PSII effective absorption cross-section under actinic light (Å^2 photon^{-1})
n_{PSII}	Photosynthetic unit size (mol RCII (mol Chl *a*)$^{-1}$)
Φ_e	Electron yield of oxygen production (mol O_2 (mol electron)$^{-1}$)
q_P	Photochemical quenching $(F_q'/F_m')/(F_v'/F_m')$ (dimensionless)
NPQ	Non-photochemical quenching $(F_m - F_m')/F_m'$ (dimensionless)
ETR	Electron transport rate (mol electron (mol Chl *a* s)$^{-1}$) ETR = E σ_{PSII}' n_{PSII} (F_q'/F_v')
$P^{\mathrm{Chl}\,a}$	Oxygenic photosynthetic rate (mol O_2 (g Chl *a* h)$^{-1}$) $P^{\mathrm{Chl}\,a} = E\ \sigma_{PSII}'\ n_{PSII}\ (F_q'/F_v')\ \Phi_e\ 0.0243$ (the constant 0.0243 accounts for the conversion of E from μmol photon m^{-2} s^{-1} to mol photons m^{-2} h^{-1}, σ_{PSII}' from Å^2 (photon)$^{-1}$ to m^2 (mol photon)$^{-1}$ and n_{PSII} from mol O_2 (mol Chl *a*)$^{-1}$ to mol O_2 (g Chl *a*)$^{-1}$).

scales such as the functional absorption cross-section, maximum quantum yield of photochemistry, and photochemical and non-photochemical quenching (see Box 12.3). In this sense, FRRF has proved useful for extending physiological characterizations of phytoplankton from local to basin scales (Raateoja *et al.*, 2004; Moore *et al.*, 2005, 2006; Behrenfeld *et al.*, 2006; Suggett *et al.*, 2006b).

In conclusion, all the methods described in this chapter have inherent strengths and weaknesses. A multiple approach combining FRRF measurements, classical incubation methods, *in situ* and satellite-derived estimates should offer insights into processes governing primary productivity over a wide range of spatial and temporal scales, from the local instantaneous response of cells to irradiance changes associated with waves or passing clouds to the integrated global response of phytoplankton to climate forcing.

Addendum

A comparison of phytoplankton carbon production and consumption by microzooplankton during spring bloom and post-bloom conditions in the NW Mediterranean revealed similar grazing impact on phytoplankton daily production (Gutiérrez-Rodríguez, A., Latasa, M., Estrada, M., Vidal, M. and Marrasé, C. (2010). *Deep-Sea Res. I* **57**, 486–500), highlighting the dominant role of microzooplankton grazing as a loss term for primary production even during spring bloom events typically dominated by large diatoms.

In an even more recent study (Gutiérrez-Rodríguez, A., Latasa, M., Agustí, S. and Duarte, C. M. *Deep-Sea Res. I*, in press), phytoplankton growth and microzooplankton grazing rates were shown to remain tightly coupled even across dramatically contrasting ecosystems such as the Canary upwelling and north Atlantic subtropical gyre. This highlights the strong regulatory control exerted by microzooplankton upon phytoplankton production across very distinctive pelagic ecosystems.

Abbreviations and symbols

(other than those defined in Box 12.3)

ATP	Adenosine triphosphate
C	Carbon
C_p	Carbon concentration of phytoplankton
D	Fraction of undiluted water in an incubation bottle
dpm	Disintegrations per minute
DIC	Dissolved inorganic carbon
F	Volume swept clear time^{-1} predator^{-1}
FRRF	Fast repetition rate fluorometry
g	Grazing mortality coefficient
HNLC	High-nutrient low-chlorophyll region
k	Instantaneous apparent growth rate
$P_{(t)}$	Phytoplankton densities at the end of the incubation
P_0	Phytoplankton densities at the beginning of the incubation
PSII	Photosystem II
RCII	Reaction centers associated with PSII
R^*	Specific activity
t	Incubation time
μ	Specific growth rate

References

Abbott, M. R. and Letelier, R. M. (1999). *Algorithm theoretical basis document – Chlorophyll fluorescence (MODIS product number 20). NASA.*

Agis, M., Granda, A. and Dolan, J. R. (2007). A cautionary note: Examples of possible microbial community dynamics in dilution grazing experiments. *J. Exp. Mar. Biol. Ecol.* **341**, 176–83.

Andersen, T., Schartau, A. K.L. and Paasche, E. (1991). Quantifying external and internal nitrogen and phosphorous pools, as well as nitrogen and phosphorous supplied through remineralization, in coastal marine plankton by means of a dilution technique. *Mar. Ecol. Prog. Ser.* **69**, 67–80.

Antoine, D. and Morel, A. (1996). Oceanic primary production.1. Adaptation of a spectral light-photosynthesis model in view of application to satellite chlorophyll observations. *Global Biogeochem. Cycles* **10**, 43–55.

Banse, K. (2002a). Should we continue to measure ^{14}C-uptake by phytoplankton for another 50 years. *Limnol. Oceanogr. Bull.* **11**, 45–46.

Banse, K. (2002b). Steemann Nielsen and the Zooplankton. *Hydrobiologia* **480**, 15–28.

Barlow, R. G., Burkill, P. H. and Mantoura, R. F.C. (1988). Grazing and degradation of algal pigments by marine protozoan *Oxyrrhis marina*. *J. Exp. Mar. Biol. Ecol.* **119**, 119–29.

Behrenfeld, M. J. and Falkowski, P. G. (1997a). A consumer's guide to phytoplankton primary productivity models. *Limnol. Oceanogr.* **42**, 1479–91.

Behrenfeld, M. J. and Falkowski, P. G. (1997b). Photosynthetic rates derived from satellite-based chlorophyll concentration. *Limnol. Oceanogr.* **42**, 1–20.

Behrenfeld, M. J., Boss, E., Siegel, D. A. and Shea, D. M. (2005). Carbon-based ocean productivity and phytoplankton physiology from space. *Global Biogeochem. Cycles* **19**, GB1006, doi:1010.1029/2004GB002299.

Behrenfeld, M. J., Worthington, K., Sherrell, R. M., Chavez, F. P., Strutton, P., McPhaden, M. and Shea, D. M. (2006). Controls on tropical Pacific Ocean productivity revealed through nutrient stress diagnostics. *Nature* **442**, 1025–28.

Brown, S. L., Landry, M. R., Barber, R. T., Campbell, L., Garrison, D. L. and Gowing, M. M. (1999). Picophytoplankton dynamics and production in the Arabian Sea during the 1995 Southwest Monsoon. *Deep-Sea Res. II* **46**, 1745–68.

Brown, S. L., Landry, M. R., Christensen, S., Garrison, D., Gowing, M. M., Bidigare, R. R. and Campbell, L. (2002). Microbial community dynamics and taxon-specific phytoplankton production in the Arabian Sea during the 1995 monsoon seasons. *Deep-Sea Res. II* **49**, 2345–76.

Burkill, P. H., Mantoura, R. F.C., Llewellyn, C. A. and Owens, N. J.D. (1987). Microzooplankton grazing and selectivity of phytoplankton in coastal waters. *Mar. Biol.* **93**, 581–90.

Butler, N. M., Suttle, C. A. and Neill, W. E. (1989). Discrimination by fresh-water zooplankton between single algal cells differing in nutritional-status. *Oecologia* **78**, 368–72.

Calbet, A. and Landry, M. R. (2004). Phytoplankton growth, microzooplankton grazing, and carbon cycling in marine systems. *Limnol. Oceanogr.* **49**, 51–57.

Calbet, A., Trepat, I., Almeda, R., Salo, V., Saiz, E., Movilla, J. I., Alcaraz, M., Yebra, L. and Simo, R. (2008). Impact of micro- and nanograzers on phytoplankton assessed by standard and size-fractionated dilution grazing experiments. *Aquat. Microb. Ecol.* **50**, 145–56.

Campbell, J., Antoine, D., Armstrong, R., Arrigo, K., Balch, W., Barber, R., Behrenfeld, M., Bidigare, R., Bishop, J., Carr, M. E., Esaias, W., Falkowski, P., Hoepffner, N., Iverson, R., Kiefer, D., Lohrenz, S., Marra, J., Morel, A., Ryan, J., Vedernikov, V., Waters, K., Yentsch, C. and Yoder, J. (2002). Comparison of algorithms for estimating ocean primary production from surface chlorophyll, temperature, and irradiance. *Global Biogeochem. Cycles* **16** (3) GB1035, doi:10.1029/2001GB001444.

Caron, D. A. and Dennett, M. R. (1999). Phytoplankton growth and mortality during the 1995 Northeast monsoon and spring intermonsoon in the Arabian Sea. *Deep-Sea Res. II* **46**, 1665–90.

Carr, M. E., Friedrichs, M. A.M., Schmeltz, M., Aita, M. N., Antoine, D., Arrigo, K. R., Asanuma, I., Aumont, O., Barber, R., Behrenfeld, M., Bidigare, R., Buitenhuis, E. T., Campbell, J., Ciotti, A., Dierssen, H., Dowell, M., Dunne, J., Esaias, W., Gentili, B., Gregg, W., Groom, S., Hoepffner, N., Ishizaka, J., Kameda, T., Le Quéré, C., Lohrenz, S., Marra, J., Melin, F., Moore, K., Morel, A., Reddy, T. E., Ryan, J., Scardi, M., Smyth, T., Turpie, K., Tilstone, G., Waters, K. and Yamanaka, Y. (2006). A comparison of global estimates of marine primary production from ocean color. *Deep-Sea Res. II* **53**, 741–70.

Corno, G., Letelier, R. M., Abbott, M. R. and Karl, D. M. (2006). Assessing primary production variability in the North Pacific Subtropical Gyre: a comparison of fast repetition rate fluorometry and C-14 measurements. *J. Phycol.* **42**, 513–13.

Cowles, T. J., Olson, R. J. and Chisholm, S. W. (1988). Food selection by copepods – discrimination on the basis of food quality. *Mar. Biol.* **100**, 41–49.

Chavez, F. P., Buck, K. R., Coale, K. H., Martin, J. H., DiTullio, G. R., Welschmeyer, N. A., Jacobson, A. C. and Barber, R. T. (1991). Growth rates, grazing, sinking, and iron limitation of equatorial Pacific phytoplankton. *Limnol. Oceanogr.* **36**, 1816–33.

Chen, B. Z., Liu, H. B., Landry, M. R., Dai, M. H., Huang, B. Q. and Jun, S. (2009). Close coupling between phytoplankton growth and microzooplankton grazing in the western South China Sea. *Limnol. Oceanogr.* **54**, 1084–97.

Delizo, L., Smith, W. O. and Hall, J. (2007). Taxonomic composition and growth rates of phytoplankton assemblages at the subtropical convergence east of New Zealand. *J. Plankton Res.* **29**, 655–70.

Dolan, J. R. and McKeon, K. (2005). The reliability of grazing rate estimates from dilution experiments: have we over-estimated rates of organic carbon consumption by microzooplankton? *Ocean Science* **1**, 1–7.

Dolan, J. R., Gallegos, C. L. and Moigis, A. G. (2000). Dilution effects on microzooplankton in dilution grazing experiments. *Mar. Ecol. Prog. Ser.* **200**, 127–39.

Dussart, B. M. (1965). Les différentes catégories de plancton. *Hydrobiologia* **26**, 72–74.

Esaias, W. E., Abbott, M. R., Barton, I., Brown, O. B., Campbell, J. W., Carder, K. L., Clark, D. K., Evans, R. H., Hoge, F. E., Gordon, H. R., Balch, W. M., Letelier, R. and Minnett, P. J. (1998). An overview of MODIS capabilities for ocean science observations. *IEEE Trans. Geosci. Rem. Sensing* **36**, 1250–65.

Evans, G. T. and Paranjape, M. A. (1992) Precision of estimates of phytoplankton growth and microzooplankton grazing when the functional response of grazers may be nonlinear. *Mar. Ecol. Prog. Ser.* **80**, 285–90.

Falkowski, P. G. and Kolber, Z. (1995). Variations in chlorophyll fluorescence yields in phytoplankton in the world oceans. *Aust. J. Plant. Physiol.* **22**, 341–55.

Falkowski, P. G. and Raven, J. A (2007). *Aquatic Photosynthesis*, 2nd edn. Princeton: Princeton University Press.

Field, C. B., Behrenfeld, M. J., Randerson, J. T., Falkowski, P. G. (1998). Primary production of the biosphere: Integrating terrestrial and oceanic components. *Science* **281**, 237–40.

Fileman, E. and Burkill, P. (2001). The herbivorous impact of microzooplankton during two short-term lagrangian experiments off the NW coast of Galicia in summer 1998. *Prog. Oceanogr.* **51**, 361–83.

First, M. R., Lavrentyev, P. J. and Jochem, F. J. (2007). Patterns of microzooplankton growth in dilution experiments across a trophic gradient: Implications for herbivory studies. *Mar. Biol.* **151**, 1929–40.

Frost, B. W. (1972). Effects of size and concentration of food particles on the feeding behaviour of the marine planktonic copepod *Calanus pacificus*. *Limnol. Oceanogr.* **17**, 805–15.

Gallegos, C. L. (1989). Microzooplankton grazing on phytoplankton in the Rhode River, Maryland: nonlinear feeding kinetics. *Mar. Ecol. Prog. Ser.* **57**, 23–33.

Geider, R. J., MacIntyre, H. L. and Kana, T. M. (1998). A dynamic regulatory model of phytoplankton acclimation to light, nutrients, and temperature. *Limnol. Oceanogr.* **43**, 679–94.

Gieskes, W. W. and Kraay, G. W. (1986a). Analysis of Phytoplankton Pigments by HPLC before, during and after mass occurrence of the microflagellate *Corymbellus aureus* during the spring bloom in the open northern North-Sea in 1983. *Mar. Biol.* **92**, 45–52.

Gieskes, W. W. and Kraay, G. W. (1986b) Floristic and physiological differences between the shallow and the deep nanophytoplankton community in the euphotic zone of the open Tropical Atlantic revealed by HPLC analysis of pigments. *Mar. Biol.* **91**, 567–76.

Gieskes, W. W.C. and Kraay, G. W. (1989). Estimating the carbon-specific growth-rate of the major algal species groups in eastern Indonesian waters by C-14 labeling of taxon-specific carotenoids. *Deep-Sea Res.* **36**, 1127–39.

Goericke, R. (1992). The chlorophyll-labeling method – the radiochemical purity of chlorophyll-a. *J. Plankton Res.* **14**, 1781–85.

Goericke, R. (1998). Response of phytoplankton community structure and taxon-specific growth rates to seasonally varying physical forcing in the Sargasso Sea off Bermuda. *Limnol. Oceanogr.* **43**, 921–35.

Goericke, R. (2002). Top-down control of phytoplankton biomass and community structure in the monsoonal Arabian Sea. *Limnol. Oceanogr.* **47**, 1307–23.

Goericke, R. and Welschmeyer, N. A. (1992a). Pigment turnover in the marine diatom *Thalassiosira weissflogii*. I. The $^{14}CO_2$-labeling kinetics of chlorophyll *a*. *J. Phycol.* **28**, 498–507.

Goericke, R. and Welschmeyer, N. A. (1992b). Pigment turnover in the marine diatom *Thalassiosira weissflogii*. II. The $^{14}CO_2$-labeling kinetics of carotenoids. *J. Phycol.* **28**, 507–17.

Goericke, R. and Welschmeyer, N. A. (1993a). The chlorophyll-labeling method: measuring specific rates of chlorophyll *a* synthesis in cultures and in open ocean. *Limnol. Oceanogr.* **38**, 80–95.

Goericke, R. and Welschmeyer, N. A. (1993b). The carotenoid-labelling method: measuring specific rates of carotenoid synthesis in natural phytoplankton communites. *Mar. Ecol. Prog. Ser.* **98**, 157–71.

Gorbunov, M. Y., Kolber, Z. S., Lesser, M. P. and Falkowski, P. G. (2001). Photosynthesis and photoprotection in symbiotic corals. *Limnol. Oceanogr.* **46**, 75–85.

Gutiérrez-Rodríguez, A., Latasa, M., Mourre, B. and Laws, E. A. (2009). Coupling between phytoplankton growth and microzooplankton grazing in dilution experiments: potential artefacts. *Mar. Ecol. Prog. Ser.* **383**, 1–9.

Hansen, P. J. and Calado, A. J. (1999). Phagotrophic mechanisms and prey selection in free-living dinoflagellates. *J. Eukaryot. Microbiol.* **46**, 382–89.

Huot, Y., Brown, C. A. and Cullen, J. J. (2005). New algorithms for MODIS sun-induced chlorophyll fluorescence and a comparison with present data products. *Limnol. Oceanogr. Methods* **3**, 108–30.

Irigoien, X., Flynn, K. J. and Harris, R. P. (2005). Phytoplankton blooms: a 'loophole' in microzooplankton grazing impact? *J. Plankton Res.* **27**, 313–21.

Jacobson, D. M. and Anderson, D. M. (1986). Thecate heterotrophic dinoflagellates – feeding-behavior and mechanisms. *J. Phycol.* **22**, 249–58.

Jespersen, A. M., Nielsen, J., Riemann, B. and Sondergaard, M. (1992). Carbon-specific phytoplankton growth-rates – a comparison of methods. *J. Plankton Res.* **14**, 637–48.

John, E. H. and Davidson, K. (2001). Prey selectivity and the influence of prey carbon: nitrogen ratio on microflagellate grazing. *J. Exp. Mar. Biol. Ecol.* **260**, 93–111.

Karl, D. M., Bidigare, R. R. and Letelier, R. M. (2002). Sustained and aperiodic variability in organic matter production and phototrophic microbial community structure in the North Pacific subtropical gyre. In *Phytoplankton Productivity*, ed. P. J. Le B. Williams, D. N. Thomas and C. S. Reynolds. Oxford: Blackwell Science, pp. 222–64.

Kiefer, D. A. and Reynolds, R. A. (1992). Advances in understanding phytoplankton fluorescence and photosynthesis. In *Primary Productivity and Biogeochenical Cycles in the Sea*, ed. P. G. Falkowski and A. D. Woodhead. New York: Plenum, pp. 155–74.

Klein, B., Gieskes, W. W.C. and Kraay, G. G. (1986). Digestion of chlorophylls and carotenoids by the marine protozoan *Oxyrrhis marina* studied by HPLC analysis of algal pigments. *J. Plankton Res.* **8**, 827–36.

Kolber, Z. S. and Falkowski, P. G. (1993). Use of active fluorescence to estimate phytoplankton photosynthesis in situ. *Limnol. Oceanogr.* **38**, 1646–65.

Kolber, Z. S., Prasil, O. and Falkowski, P. G. (1998). Measurements of variable chlorophyll fluorescence using fast repetition rate techniques: defining methodology and experimental protocols. *Biochim. Biophys. Acta-Bioenergetics* **1367**, 88–106.

Kromkamp, J. C. and Forster, R. M. (2003). The use of variable fluorescence measurements in aquatic ecosystems: differences between multiple and single turnover measuring protocols and suggested terminology. *Eur. J. Phycol.* **38**, 103–12.

Landry, M. R. and Calbet, A. (2005). Reality checks on microbial food web interactions in dilution experiments: responses to the comments of Dolan and McKeon. *Ocean Sci.* **1**, 39–44.

Landry, M. R. and Hasset, R. P. (1982). Estimating the grazing impact of marine microzooplankton. *Mar. Biol.* **67**, 283–88.

Landry, M. R., Haas, L. W. and Fagerness, V. L. (1984). Dynamics of microbial plankton communities – Experiments in Kaneohe Bay, Hawaii. *Mar. Ecol. Prog. Ser.* **16**, 127–33.

Landry, M. R., Kirshtein, J. and Constantinou, J. (1995). A refined dilution technique for measuring the community grazing impact of microzooplankton, with experimental tests in the central equatorial Pacific. *Mar. Ecol. Prog. Ser.* **120**, 53–63.

Landry, M. R., Constantinou, J., Latasa, M., Brown, S. L., Bidigare, R. R. and Ondrusek, M. E. (2000). Biological response to iron fertilization in the eastern equatorial Pacific (IronEx II). III. Dynamics of phytoplankton growth and microzooplankton grazing. *Mar. Ecol. Prog. Ser.* **201**, 57–72.

Landry, M. R., Brown, S. L., Neveux, J., Dupouy, C., Blanchot, J., Christensen, S. and Bidigare, R. R. (2003). Phytoplankton growth and microzooplankton grazing in high-nutrient, low-chlorophyll waters of the equatorial Pacific: Community and taxon-specific rate estimates from pigment and flow cytometric analyses. *J. Geophys. Res.* **108**, 8142.

Landry, M. R., Brown, S. L., Rii, Y. M., Selph, K. E., Bidigare, R. R., Yang, E. J. and Simmons, M. P. (2008). Depth-stratified phytoplankton dynamics in Cyclone Opal, a subtropical mesoscale eddy. *Deep-Sea Res. II* **55**, 1348–59.

Latasa, M., Landry, M. R., Schlüter, L. and Bidigare, R. R. (1997). Pigment-specific growth and grazing rates of phytoplankton in the central equatorial Pacific. *Limnol. Oceanogr.* **42**, 289–98.

Latasa, M., Moran, X., Scharek, R. and Estrada, M. (2005). Estimating the carbon flux through main phytoplankton groups in the northwestern Mediterranean. *Limnol. Oceanogr.* **50**, 1447–58.

Laws, E. A. (1984). Improved estimates of phytoplankton carbon based on ^{14}C incorporation into chlorophyll *a*. *J. Theor. Biol.* **110**, 425–34.

Letelier, R. M., Abbott, M. R. and Karl, D. M. (1997). Chlorophyll natural fluorescence response to upwelling events in the Southern Ocean. *Geophys. Res. Lett.* **24**, 409–12.

Lewis, M. R., Harrison, W. G., Oakey, N. S., Hebert, D. and Platt, T. (1986). Vertical nitrate fluxes in the oligotrophic ocean. *Science* **234**, 870–73.

MacIntyre, H. L., Kana, T. M. and Geider, R. J. (2000). The effect of water motion on short-term rates of photosynthesis by marine phytoplankton. *Trends Plant. Sci.* **5**, 12–17.

Mantoura, R. F.C. and Llewellyn, C. A. (1983). The rapid determination of algal chlorophyll and carotenoid pigments and their breakdown products in natural waters by reverse-phase high-performance liquid chromatography. *Anal. Chim. Acta* **151**, 297–314.

McManus, G. B. (1995). Phytoplankton abundance and pigment changes during simulated in situ dilution experiments in estuarine waters: possible artifacts caused by algal light adaptation. *J. Plankton Res.* **17**, 1705–16.

McManus, G. B. and Ederington-Cantrell, M. C. (1992). Phytoplankton pigments and growth rates, and microzooplankton grazing in a large temperate estuary. *Mar. Ecol. Prog. Ser.* **87**, 77–85.

Melrose, D. C., Oviatt, C. A., O'Reilly, J. E. and Berman, M. S. (2006). Comparisons of fast repetition rate fluorescence estimated primary production and ^{14}C uptake by phytoplankton. *Mar. Ecol. Prog. Ser.* **311**, 37–46.

Moigis, A. G. (2006). The clearance rate of microzooplankton as the key element for describing estimated non-linear dilution plots demonstrated by a model. *Mar. Biol.* **149**, 743–62.

Monger, B. C. and Landry, M. R. (1991). Prey-size dependency of grazing by free-living marine flagellates, *Mar. Ecol. Prog. Ser.* **74**, 239–48.

Moore, C. M., Suggett, D., Holligan, P. M., Sharples, J., Abraham, E. R., Lucas, M. I., Rippeth, T. P., Fisher, N. R., Simpson, J. H. and Hydes, D. J. (2003). Physical controls on phytoplankton physiology and production at a shelf sea front: a fast repetition-rate fluorometer based field study. *Mar. Ecol. Prog. Ser.* **259**, 29–45.

Moore, C. M., Lucas, M. I., Sanders, R. and Davidson, R. (2005). Basin-scale variability of phytoplankton bio-optical characteristics in relation to bloom state and community structure in the Northeast Atlantic. *Deep-Sea Res. I* **52**, 401–19.

Moore, C. M., Suggett, D. J., Hickman, A. E., Kim, Y. N., Tweddle, J. F., Sharples, J., Geider, R. J. and Holligan, P. M. (2006). Phytoplankton photoacclimation and photoadaptation in response to environmental gradients in a shelf sea. *Limnol. Oceanogr.* **51**, 936–49.

Morel, A. and André, J. M. (1991). Pigment distribution and primary production in the Western Mediterranean as derived and modeled from Coastal Zone Color Scanner observations. *J. Geophys. Res. Oceans* **96**, 12685–98.

Neuer, S. and Cowles, T. J. (1994). Protist herbivory in the Oregon upwelling system. *Mar. Ecol. Prog. Ser.* **113**, 147–62.

Obayashi, Y. and Tanoue, E. (2002). Growth and mortality rates of phytoplankton in the northwestern North Pacific estimated by the dilution method and HPLC pigment analysis. *J. Exp. Mar. Biol. Ecol.* **280**, 33–52.

Olson, M. B. and Strom, S. L. (2002). Phytoplankton growth, microzooplankton herbivory and community structure in the southeast Bering Sea: insight into the formation and temporal persistence of an *Emiliania huxleyi* bloom. *Deep-Sea Res. II* **49**, 5969–90.

Pinckney, J. L., Millie, D. F., Howe, K. E., Paerl, H. W. and Hurley, J. P. (1996). Flow scintillation counting of ^{14}C-labeled microalgal photosynthetic pigments. *J. Plankton Res.* **18**, 1867–80.

Pinckney, J. L., Richardson, T. L., Millie, D. F. and Paerl, H. W. (2001). Application of photopigment biomarkers for quantifying microalgal community composition and in situ growth rates. *Org. Geochem.* **32**, 585–95.

Platt, T. (1986). Primary production of the ocean water column as a function of surface light-intensity – algorithms for remote-sensing. *Deep-Sea Res.* **33**, 149–63.

Platt, T. and Herman, A. W. (1983). Remote-sensing of phytoplankton in the sea – Surface-layer chlorophyll as an estimate of water-column chlorophyll and primary production. *Int. J. Remote Sens.* **4**, 343–51.

Platt, T. and Sathyendranath, S. (1988). Oceanic primary production – Estimation by remote-sensing at local and regional scales. *Science* **241**, 1613–20.

Quevedo, M. and Anadón, R. (2001). Protist control of phytoplankton growth in the subtropical north-east Atlantic. *Mar. Ecol. Prog. Ser.* **221**, 29–38.

Raateoja, M., Seppala, J. and Kuosa, H. (2004). Bio-optical modelling of primary production in the SW Finnish coastal zone, Baltic Sea: fast repetition rate fluorometry in Case 2 waters. *Mar. Ecol. Prog. Ser.* **267**, 9–26.

Redalje, D. G. (1983). Phytoplankton carbon biomass and specific growth-rates determined with the labeled chlorophyll *a* technique. *Mar. Ecol. Prog. Ser.* **11**, 217–25.

Redalje, D. G. (1993). The labeled chlorophyll *a* technique for determining photoautotrophic carbon specific growth rates and biomass. In: *Handbook of Methods in Aquatic Microbial Ecology*, ed. P. F. Kemp. Boca Raton: Lewis Publishers, pp. 563–72.

Redalje, D. G. and Laws, E. A. (1981). A new method for estimating phytoplankton growth-rates and carbon biomass. *Mar. Biol.* **62**, 73–79.

Redden, A. M., Sanderson, B. G. and Rissik, D. (2002). Extending the analysis of the dilution method to obtain the phytoplankton concentration at which microzooplankton grazing becomes saturated. *Mar. Ecol. Prog. Ser.* **226**, 27–33.

Rivkin, R. B., Putland, J. N., Anderson, M. R. and Deibel, D. (1999). Microzooplankton bacterivory and herbivory in the NE subarctic Pacific. *Deep-Sea Res. II* **46**, 2579–618.

Sathyendranath, S., Longhurst, A., Caverhill, C. M. and Platt, T. (1995). Regionally and seasonally differentiated primary production in the North Atlantic. *Deep-Sea Res. I* **42**, 1773–802.

Sherr, E. B. and Sherr, B. F. (2007). Heterotrophic dinoflagellates: a significant component of microzooplankton biomass and major grazers of diatoms in the sea. *Mar. Ecol. Prog. Ser.* **352**, 187–97.

Shuter, B. (1979). A model of physiological adaptation in unicellular algae. *J. Theor. Biol.* **78**, 519–52.

Smith, R. C. (1981). Remote-sensing and depth distribution of ocean chlorophyll. *Mar. Ecol. Prog. Ser.* **5**, 359–61.

Smyth, T., Pemberton, K., Aiken, J. and Geider, R. (2004). A methodology to determine primary production and phytoplankton photosynthetic parameters from fast repetition rate fluorometry. *J. Plankton Res.* **26**, 1337–50.

Stelfox-Widdicombe, C. E., Edwards, E. S., Burkill, P. H. and Sleigh, M. A. (2000). Microzooplankton grazing activity in the temperate and sub-tropical NE Atlantic: summer 1996. *Mar. Ecol. Prog. Ser.* **208**, 1–12.

Stelfox-Widdicombe, C., Archer, S., Burkill, P. and Stefels, J. (2004). Microzooplankton grazing in *Phaeocystis* and diatom-dominated waters in the southern North Sea in spring. *J. Sea Res.* **51**, 37–51.

Stoecker, D. K. (1988). Are marine planktonic ciliates suspension-feeders. *J. Protozool.* **35**, 252–55.

Stoecker, D. K., Cucci, T. L., Hulburt, E. M. and Yentsch, C. M. (1986). Selective feeding by *Balanion* sp. (Ciliata: Balanionidae) on phytoplankton that best support its growth. *J. Exp. Mar. Biol. Ecol.* **95**, 113–30.

Strom, S. L. (2001). Light-aided digestion, grazing and growth in herbivorous protists. *Aquat. Microb. Ecol.* **23**, 253–61.

Strom, S. L. and Loukos, H. (1998). Selective feeding by protozoa: model and experimental behaviors and their consequences for population stability. *J. Plankton Res.* **20**, 831–46.

Strom, S. L. and Welschmeyer, N. A. (1991). Pigment-specific rates of phytoplankton growth and microzooplankton grazing in the open subarctic Pacific Ocean. *Limnol. Oceanogr.* **36**, 50–63.

Strom, S. L., Morello, T. A. and Bright, K. J. (1998). Protozoan size influences algal pigment degradation during grazing. *Mar. Ecol. Prog. Ser.* **164**, 189–97.

Strom, S. L., Brainard, M. A., Holmes, J. L. and Olson, M. B. (2001). Phytoplankton blooms are strongly impacted by microzooplankton grazing in coastal North Pacific waters. *Mar. Biol.* **138**, 355–68.

Strom, S. L., Macri, E. L. and Olson B. (2007). Microzooplankton grazing in the coastal Gulf of Alaska: Variations in top-down control of phytoplankton. *Limnol. Oceanogr.* **52**, 1480–94.

Suggett, D., Kraay, G., Holligan, P., Davey, M., Aiken, J. and Geider, R. (2001). Assessment of photosynthesis in a spring cyanobacterial bloom by use of a fast repetition rate fluorometer. *Limnol. Oceanogr.* **46**, 802–10.

Suggett, D., Oxborough, K., Baker, N., MacIntyre, H., Kana, T. and Geider, R. J. (2003). Fast repetition rate and pulse amplitude modulation chlorophyll *a* fluorescence measurements for assessment of photosynthetic electron transport in marine phytoplankton. *Eur. J. Phycol.* **38**, 371–84.

Suggett, D. J., Maberly, S. C. and Geider, R. J. (2006a). Gross photosynthesis and lake community metabolism during the spring phytoplankton bloom. *Limnol. Oceanogr.* **51**, 2064–76.

Suggett, D. J., Moore, C. M., Marañon, E., Omachi, C., Varela, R. A., Aiken, J. and Holligan, P. M. (2006b). Photosynthetic electron turnover in the tropical and subtropical Atlantic Ocean. *Deep-Sea Res. II* **53**, 1573–92.

Suggett D., Moore, C. M., Hickman, A. E. and Geider, R. J. (2009). Interpretation of fast repetition rate (FRR) fluorescence: signatures of phytoplankton community structure versus physiological state. *Mar. Ecol. Prog. Ser.* **376**, 1–19.

Verity, P. G. (1988). Chemosensory behaviour in marine planktonic ciliates. *Bull. Mar. Sci.* **43**, 772–82.

Verity, P. G., Stoecker, D. K., Sieracki, M. E. and Nelson, J. R. (1993). Grazing, growth and mortality of microzooplankton during the 1989 North Atlantic spring bloom at 47N, 18W. *Deep-Sea Res. I* **40**, 1793–814.

Verity, P. G., Stoecker, D. K., Sieracki, M. E. and Nelson, J. R. (1996). Microzooplankton grazing of primary production at 140 degrees W in the equatorial Pacific. *Deep-Sea Res. II* **43**, 1227–55.

Waterhouse, T. Y. and Welschmeyer, N. A. (1995). Taxon-specific analysis of microzooplankton grazing rates and phytoplankton growth-rates. *Limnol. Oceanogr.* **40**, 827–34.

Welschmeyer, N. A. and Lorenzen, C. J. (1984). ^{14}C-labeling of phytoplankton carbon and chlorophyll-*a* carbon – Determination of specific growth-rates. *Limnol. Oceanogr.* **29**, 135–45.

Welschmeyer, N., Goericke, R., Strom, S. and Peterson, W. (1991). Phytoplankton growth and herbivory in the Sub-Arctic Pacific – A chemotaxonomic analysis. *Limnol. Oceanogr.* **36**, 1631–49.

Westberry, T., Behrenfeld, M. J., Siegel, D. A. and Boss, E. (2008). Carbon-based primary productivity modeling with vertically resolved photoacclimation. *Global Biogeochem. Cycles* **22**, GB2024, doi:10.1029/2007GB003078

Worden, A. Z. and Binder, B. J. (2003). Application of dilution experiments for measuring growth and mortality rates among *Prochlorococcus* and *Synechococcus* populations in oligotrophic environments. *Aquat. Microb. Ecol.* **30**, 159–74.

13

In vivo bio-optical properties of phytoplankton pigments

GEIR JOHNSEN, ANNICK BRICAUD, NORMAN NELSON, BARBARA B. PRÉZELIN AND ROBERT R. BIDIGARE

13.1 Introduction

In this chapter we focus on spectral *in vivo* bio-optical (absorption, scattering and fluorescence) characteristics of phytoplankton. As such, the importance of measuring *in vivo* bio-optical properties of a culture, population or mixed community of phytoplankton is that the optical signatures provide us with important taxonomic, phylogenic and eco-physiological information. The optical signature in the PAR region (400–700 nm) of phytoplankton is central to understand the processes affecting the optical properties of the water column, the potential rate of primary production, phytoplankton community structure, and phytoplankton physiology and photo-ecology. This chapter gives important bio-optical information for interpretation of data from remote sensing and *in situ* monitoring of phytoplankton blooms (Chapter 14, this volume) and as an aid to study photo-acclimation processes (Chapter 11).

Light, or more precisely, a flux of photons (quanta) hitting a living phytoplankton cell can either be absorbed, scattered, transmitted, emitted as fluorescence/heat or induce photochemistry. The absorption spectrum (PAR) of phytoplankton is a composite signature of cellular photosynthetic and photoprotective pigments, with some contribution from intermediary electron transport components (e.g. flavins, quinines and cytochromes). The nature of the pigmentation varies significantly among phytoplankton groups and with environmental conditions.

The *in vivo* absorption spectrum of a phytoplankton cell (see Section 13.2) is a measure of its potential spectral absorption not its actual cellular absorption *in situ*. The actual absorbed radiation by a cell will also be a function of the environmental light field (e.g. surface white light versus dim blue-green light at depth) and the manner in which the phytoplankton cell interacts with the surrounding light field, which are both largely determined by properties of the cell surface, size and shape as well as the distribution of pigments within the cell.

Phytoplankton Pigments: Characterization, Chemotaxonomy and Applications in Oceanography, eds. Suzanne Roy, Carole A. Llewellyn, Einar Skarstad Egeland and Geir Johnsen. Published by Cambridge University Press.

In vivo Chl *a* fluorescence (Section 13.3), the re-emission of unused absorbed light energy, has been utilized in a variety of ways to estimate phytoplankton biomass, distribution and photosynthetic function. Another more indirect indicator of phytoplankton biomass and physiological state is the production and release of coloured dissolved organic matter (CDOM, Section 13.4) has many roles in phytoplankton community dynamics as well as in biogeochemical cycling in the upper ocean.

To function in photosynthesis, pigments (chlorophylls, carotenoids and phycobiliproteins) are bound in various combinations to specific proteins and localized in a highly organized fashion in the thylakoid membranes of phytoplankton. Knowledge of the structural and functional nature of these light-harvesting complexes (LHC) as well as their changing abundances is of considerable taxonomic, phylogenetic and physiological value (Section 13.5).

The eukaryotic phytoplankton can be further divided in three major pigment and bio-optical groups (Johnsen and Sakshaug, 2007). The first is the Chl *c*-containing chromophytes; Bacillariophyceae, Dinophyceae, Chrysophyceae, Coccolithophyceae, Pavlovophyceae, Raphidophyceae, Dictyochophyceae, Cryptophyceae). The second group is the Chl *b*-containing phytoplankton; Prasinophyceae (two major pigment groups), Eugleophyceae and Chlorophyceae. The third is phycobiliprotein containing cyanobacteria, cryptophytes and prochlorophytes (Table 13.1).

Phytoplankon pigments are commonly considered as reaction centre pigments of photosystems I and II (PS I, PSII), light-harvesting pigments (LHP) or photoprotective carotenoids, PPC (Chapter 11). The LHP absorb available quanta of different colours present in the water column (spectral irradiance), and transfer the light energy by inductive resonance to the photosystems that induce electron flow that leads to the production of reducing power (NADPH) and chemical energy (ATP). The LHP are capable of highly efficient energy transfer (Govindjee, 1995; Green and Parson, 2003; Larkum, 2003). The PPC are capable of light-harvesting functions, but with much lower efficiency than those of LHP (Johnsen *et al.*, 1997).

13.2 *In vivo* absorption and scattering properties

13.2.1 Theory

An extensive description of the theoretical aspects concerning absorption, scattering and attenuation properties of phytoplankton can be found e.g. in Morel and Bricaud (1986) or Morel (2008). We will report here only minimal information for the reader to understand the tight but complex relationships linking the optical properties of phytoplankton and the pigments they contain.

Optical properties of particles can be split into two categories: (i) the 'efficiency factors', which refer to individual particles, and (ii) the optical properties of the suspension, which are specified in terms of absorption (a), scattering (b) and attenuation (c) coefficients. The efficiency factors are not easily accessible to

Table 13.1. *Major categories of LHCs in phytoplankton. Primary absorption maxima are in bold type. All groups contain PSI and PSII core complexes with chromophores dominated by Chl* a. *For further details, see Section 13.5 and references in text.*

Group	Lipophilic LH Chl *a*-Chl *b* protein complexes embedded in thylakoid near PSII	Lipophilic LH Chl *a*-Chl *c* protein complexes embedded in thylakoid	Water-soluble LH peridinin-Chl *a* protein complexes localized in lumen	Water-soluble LH phycobilins organized in phycobilisomes and attached to PSII on stroma side of thylakoid	Water-soluble LH phycobilins, *not* organized in phycobilisomes, present in lumen
Chlorophytes*	Similar to those of higher plants. Can move laterally to PSI to equalize radiation utilization efficiency				
Diatoms		Chl c_1, Chl c_2 with large amounts of fucoxanthin			
Chrysophytes		Chl c_1, Chl c_2 with fucoxanthin			
Haptophytes		Chl c_1, Chl c_2 with fucoxanthin and/or its derivatives			
Dinoflagellates		Chromophore highly enriched in Chl c_2 (**450–455** nm), minor amounts of peridinin, associated with PPC	4:1 or 8:2 molar ratio of peridinin:Chl *a*, with peridinin absorption peak at **478–482** nm with broad shoulder *c.* 520–540 nm		

Rhodophytes		R-PE (496, 545, **565** nm heights vary with spp.); or B-PE (496, **545**, 565 nm), R-PC (555, **615** nm) or C-PC (615–620 nm); APC (650 nm)	
Cryptophytes	Contains alloxanthin; may be preferentially associated with PSI		PE (**566** nm)
Cyanobacteria		B-PE (496, **545**, 565 nm), Y-PE (**495**, 553 nm) and/or PEC (575 nm); C-PC (615–620 nm), APC (650 nm)	
Prochlorophytes	Apoproteins distinct from higher plant Chl *a*-Chl *b* complexes and part of PSII core		A few species have phycobilins but not in phycobilisomes

* The Prasinophytes and Euglenophytes (two classes in the Chlorophyta), are not included in this overview table and may be very different from the Chlorophytes.

measurement (except using micro-spectrophotometric methods; Iturriaga *et al.*, 1988; Neumuller *et al.*, 2002), but can be predicted theoretically from the characteristics (size, refractive index, shape) of algal cells, while the optical coefficients defined for a suspension can be obtained by experimentation. These two categories of optical properties are interlinked, and can be derived from each other, using additional information concerning the size distribution of cells.

Efficiency factors

If an algal cell (or any particle) is illuminated by electromagnetic radiation propagating as a plane wave, its efficiency factors for absorption (Q_a), attenuation (Q_c) and scattering (Q_b) are defined as ratios of radiative energy absorbed within, attenuated by and scattered by, respectively, this cell to the energy incident on its geometrical cross-section when measured perpendicular to the direction of propagation. The Q_a, Q_c and Q_b factors (with $Q_b = Q_c - Q_a$) can be determined using Mie–Lorentz theory (Mie, 1908). For a homogeneous and spherical cell, these factors are expressed as functions of Mie coefficients, which depend on the refractive index of the cell relative to that of the surrounding medium (water), m, and the relative size of the cell with respect to the wavelength in the surrounding medium ($\alpha = \pi\, d/\lambda$, where d is the cell diameter). When particles are absorbing, their refractive index is a complex number ($m = n - \mathrm{i}\, n'$, where the real and imaginary parts, n and n', are related to scattering and absorption, respectively).

As the exact Mie formulations are highly complex, the most widely used theoretical model is the 'anomalous diffraction approximation' (Van de Hulst, 1957), which provides analytical expressions for Q_a, Q_c and Q_b that are applicable to particles having a relative refractive index close to 1 (i.e. n close to 1 and n' close to 0, which is the case for algal cells). Note that this approximation cannot be used to derive the angular distribution of scattered light, or a partial integral over a given angular domain, such as backscattering (in this case full Mie computations have to be performed).

With this approximation, Q_a is a function of a unique parameter, $\rho' = 4\,\alpha\, n'$,

$$Q_a(\rho') = 1 + 2(\rho')^{-1}\exp(-\rho') + 2(\rho')^{-2}\,[\exp(-\rho') - 1]. \tag{13.1}$$

For a given wavelength λ, n' is related to the absorption coefficient of the matter forming the particle (cellular matter for phytoplankton), $a_{cm}(\lambda)$, through $n' = \lambda\, a_{cm}/4\pi$, therefore ρ' ($= a_{cm}\, d$) represents the optical thickness of the particle along its diameter. It follows that for any absorbing particle (and therefore for algal cells), Q_a only depends on (i) the particle size, (ii) and the (wavelength-dependent) a_{cm} coefficient. Both parameters play symmetrical roles, Q_a values being close to 0 for small and weakly pigmented cells, and increasing towards 1 for large and/or strongly pigmented cells (Figure 13.1A).

The efficiency factor for attenuation, Q_c, can also be expressed in the frame of the anomalous diffraction approximation, as a function of the parameter $\rho = 2\,\alpha\,(n-1)$ and of the ratio $\xi = n'/(n-1)$:

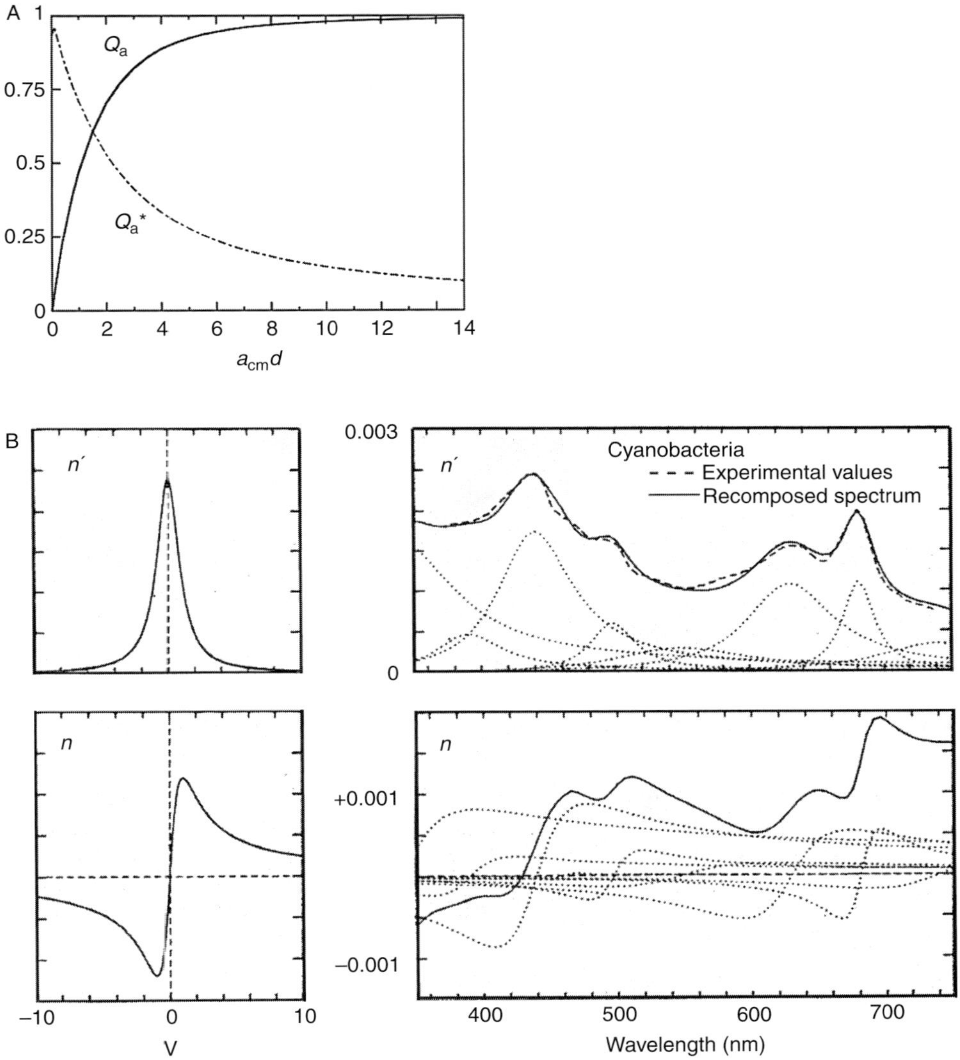

Figure 13.1. (A) Variations of the efficiency factor Q_a and of the package effect index Q_a*, versus the parameter $\rho' = a_{cm}\, d$ (where a_{cm} is the absorption coefficient of cellular matter, and d is the cell diameter) (redrawn from Morel and Bricaud, 1981). (B) Variations of the real (n) and imaginary (n') parts of the refractive index within an absorption band, according to the theory of 'anomalous dispersion'. Right: Spectral values of n' derived from absorption coefficients measured on a suspension of cyanobacteria, and associated values of n (computed by decomposing the n' spectrum into single oscillators, and summing the corresponding variations of n for each oscillator) (from Morel and Bricaud, 1986). (C) Variations of the efficiency factors for attenuation (Q_c) and scattering (Q_b) versus the parameter $\rho = 2\, \alpha\, (n-1)$, for different values of the ratio $n'/(n-1)$ (where α is the particle size relative to the wavelength in the medium, and n and n' are the real and imaginary parts of the relative refractive index). Redrawn from Morel and Bricaud (1986).

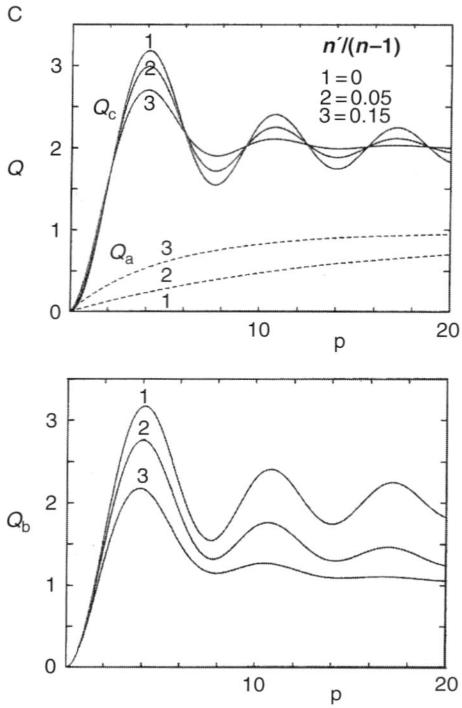

Figure 13.1. (*cont.*)

$$Q_c(\rho, \xi) = 2 - 4\exp(-\rho\tan\xi)[(\rho)^{-1}\cos\xi\sin(\rho - \xi) + (\rho)^{-2}\cos^2\xi\cos(\rho - 2\xi)] + 4(\rho)^{-2}\cos^2\xi\cos(2\xi). \tag{13.2}$$

Therefore, the efficiency factor for scattering, Q_b ($= Q_c - Q_a$) depends, in a complex way, on (i) the size of the cell (relative to the wavelength in the medium), (ii) the real part of the refractive index, n and (iii) its imaginary part, n' (Figure 13.1C). This formulation is evidence that, whereas absorption can be considered as totally independent of scattering, scattering is directly dependent on absorption (through the variations of n'). In addition, the variations in n' induce fluctuations in n (see later). Even if this simplified theory deals only with spherical and homogeneous cells, it was found to be applicable to various phytoplankton species, because of the random orientation of cells (e.g. Bricaud and Morel, 1986). Additional theories have been developed for non-spherical and heterogeneous cells (see the review by Mishchenko *et al.*, 2000).

Optical coefficients of algal suspensions

For a monodispersed population of spherical cells, the absorption, attenuation and scattering coefficients of the suspension are linked to Q_a, Q_c and Q_b, respectively, by the relationship,

$$i = (N/V)\,(\pi d^2/4)Q_i, \qquad (13.3)$$

where $i = a$, c or b, (N/V) is the cell number per unit of volume, and d is the cell diameter. For polydispersed populations, Eq. (13.3) is transformed into

$$i = (\pi/4)\int_0^\infty Q_i(d)\,\mathrm{F}(d)d^2\,d(d), \qquad (13.4)$$

where $\mathrm{F}(d)$ is the size distribution of the cells within the suspension.

The package effect

If the pigments of algal cells were uniformly distributed in the medium (as in a solution), the efficiency factor would be equal to $Q_a = (2/3)\,\rho'$, the limiting value of the Q_a function when ρ' tends towards 0. For a suspension, the relationship between Q_a and ρ' ($= a_{cm}\,d$) is nonlinear (Figure 13.1A). Because of this nonlinearity, when the cell size d (or alternatively, the absorption coefficient of cellular matter a_{cm}) increases by a given factor, Q_a increases by a smaller factor. The result is that the absorption coefficient of the suspension (a_{susp}) is always lower than the absorption coefficient of the same quantity of pigments which would be dispersed in the medium (a_{sol}). It has been shown (Morel and Bricaud, 1981) that the ratio a_{susp}/a_{sol}, noted $Q_a{}^*$, decreases according to the function,

$$Q_a{}^*(\rho') = (3/2)Q_a(\rho')/\rho'. \qquad (13.5)$$

The $Q_a{}^*$ factor (often called the 'package effect index') decreases for increasing values of a_{cm} and/or d (Figure 13.1A), which means that the package effect is the most marked for large and/or highly pigmented cells. Therefore this effect is variable from one phytoplankton population to the other, because of variations in average cell size (e.g. Nelson *et al.*, 1993; Bricaud *et al.*, 2004), and also within a given population, according to the photoacclimation state of cells, as determined by variations in the irradiance available for growth (e.g. Mitchell and Kiefer, 1988; Johnsen and Sakshaug, 1993). Note also that because $Q_a{}^*$ decreases when a_{cm} increases, the package effect induces not only a reduction in the amplitude, but also a 'flattening' of the absorption spectrum. The package effect is the main cause of the complexity in the relationship linking pigments and absorption coefficients of phytoplankton, and hinders a straightforward retrieval of pigment information from absorption spectra.

Influence of absorption on scattering

As shown by Eq. (13.2), scattering coefficients of algal suspensions are influenced by their absorption properties, through the ratio $n'/(n-1)$, since n' is directly related to a_{cm}. In addition, the electromagnetic theory predicts that the real part of the complex refractive index is modified within and in the vicinity of an absorption band ('anomalous dispersion phenomenon'). For a single absorption band, the spectral 'fluctuations' of n (with respect to its value far from the absorption band) can be derived from the n' spectral values using the (approximate) Ketteler-Helmholtz formulae (Figure 13.1B, left). As the n' spectrum of a given algal suspension can be decomposed into single oscillators, this has provided the basis to model the associated fluctuations of the n spectrum, and consequently the scattering spectrum for the same suspension (Figure 13.1B, right; see Bricaud and Morel, 1986). Such models explain why scattering spectra of algal suspensions contain 'fingerprints' of the major pigments, with usually a well-marked decrease of scattering within absorption bands. The details of pigment signatures, however, are usually less apparent, and somewhat smoothed, compared to the corresponding absorption spectra.

13.2.2 Experimental methods

Absorption properties of phytoplankton

The major problem when measuring the *in vivo* absorption properties of algal cells is to collect all photons scattered by cells, so that they are not erroneously considered as absorbed photons. Absorption coefficients of sufficiently dense phytoplankton suspensions can be determined with a classical spectrophotometer using 1-cm pathlength cuvettes, provided that most of the light scattered by the suspension can be captured, either by using an integrating sphere, or by placing the sample close to the detector (e.g. Shibata *et al.*, 1954; Bricaud *et al.*, 1983). In practice, direct measurements on suspensions are only applicable to cultured phytoplankton, as the cell density of algal populations in the natural medium is generally too low to be measured with 1-cm pathlength cuvettes. The most widely used approach therefore is to concentrate the particles on a glass-fibre filter, and measure their optical density using a blank filter as a reference, a procedure referred to as the 'quantitative filter technique' (QFT) (Mitchell, 1990). Filters constitute a highly scattering medium and as a result the absorption coefficients measured using this method need to be corrected for the pathlength amplification effect, or 'β factor', occurring within the filter (Butler, 1962). Empirical functions have been developed by various authors to correct for this effect (see e.g. Table 15.1 in Mitchell *et al.*, 2002). These corrections appear to be relatively similar for a large range of phytoplankters (Mitchell, 1990; Bricaud and Stramski, 1990; Cleveland and Weidemann, 1993; Tassan and Ferrari, 1995) but might be significantly different for picoplankton communities (Moore *et al.*, 1995; Allali *et al.*, 1997; Nelson *et al.*, 1998). Also, the use of this method in the UV domain is limited by the extracellular release of mycosporine-like amino acids during filtration (Laurion *et al.*, 2003; Chapter 10, this volume). Experimental protocols for applying the QFT (sample filter preparation, analysis, and data processing) have been described in detail by Mitchell *et al.* (2000, 2002).

Other laboratory methods have been proposed to avoid the pathlength amplification effect, e.g. concentrating suspended particles (Kirk, 1980; Weidemann and

Bannister, 1986), transferring particles from a filter to a glass slide (Allali *et al.*, 1995), or increasing the optical pathlength within samples. In this latter category, the integrating cavity absorption meter (ICAM; Pope *et al.*, 2000) and the point source integrating cavity absorption meter (PSICAM; Kirk, 1997; Röttgers *et al.*, 2005), present the advantage of allowing absorption measurements on suspensions which are free from scattering effects, because the light field within the cavity is (ideally) totally diffuse. Also recently developed is the liquid waveguide capillary cell (LWCC) coupled to a spectrometer, which allows measurements to be performed with pathlengths up to several metres. However, such measurements need to be corrected for light losses (related to scattering), using additional measurements providing the spectral dependence of scattering for the same sample (D'Sa *et al.*, 1998). In spite of these promising recent developments, the QFT remains the most routinely used method so far.

The second methodological difficulty when measuring the absorption properties of algal cells is to separate phytoplankton from other particles (e.g. inorganic and detrital particles, and heterotrophic organisms). When using the QFT, the most widely used method consists of extracting pigments by using methanol (Kishino *et al.*, 1985) or bleaching cells using sodium hypochlorite (Tassan and Ferrari, 1995; Ferrari and Tassan, 1999), and then re-measuring the absorption spectrum. It must be emphasized that such methods do not provide 'phytoplankton absorption' but rather 'pigment absorption', as the contribution of non-algal particles but also of de-pigmented algal cells has been subtracted. In addition, water-soluble pigments (e.g. phycobiliproteins, Chapter 9, this volume) are usually not or are incompletely extracted. Numerical methods are also available for partitioning particulate absorption into contributions by phytoplankton and non-algal particles (Morrow *et al.*, 1989; Roesler *et al.*, 1989; Bricaud and Stramski, 1990). As these methods include more or less constraining assumptions, it is recommended to 'calibrate' them with at least a few measurements using the above mentioned chemical methods in the investigated areas.

Scattering properties of phytoplankton

As for absorption coefficients, total scattering coefficients of cultivated phytoplankton can be measured with a spectrophotometer as the difference between attenuation and absorption measured on a suspension in a cuvette; the condition for attenuation measurements is that the acceptance angle of the detector should be extremely reduced (0.5° at the most), so that the fraction of scattered light entering the detector does not exceed a few per cent (e.g. Bricaud *et al.*, 1983; Stramski and Reynolds, 1993; Claustre *et al.*, 2002). The backscattering coefficients of various algae (Bricaud *et al.*, 1983), and more recently the angular dependence of scattering (Volten *et al.*, 1998; Voss *et al.*, 1998) have also been determined using experimental setups.

For natural populations, the pathlengths of common spectrophotometers are insufficient, but measurements can be performed e.g. by using hyperspectral absorption/attenuation meters, which have longer pathlengths (0.25 to 1 m). During the last decade, instruments have been developed to measure the volume scattering function of natural waters (Lee and Lewis, 2003). However, these highly sensitive instruments use laser sources, so measurements are limited to a few wavelengths and do not provide detailed spectral information.

As with absorption, the next step is to partition total scattering into the contributions of phytoplankton and non-algal particles. This remains problematic and is as yet unsolved. Therefore, no routine method exists for determining the scattering properties of 'algal cells only', and usually only particulate scattering coefficients are available. As non-algal particles are usually not or are weakly pigmented, spectral features observed on the scattering coefficients of total particulate matter could be interpreted in terms of algal pigment signatures. Such spectral features, however, are usually weakly marked, even in bloom situations (Stramski *et al.*, 2001, their Figure 16).

Interpretation of absorption and scattering spectra

As can be expected, most of the experimental studies aimed at deriving pigment information from optical measurements are based on absorption coefficients. The retrieval of pigment information from absorption spectra of phytoplankton is, however, not a trivial problem, because of the non-linearity of the relationship between pigment concentrations and absorption coefficients. In addition, an inherent limitation lies in the similarities existing in the spectral signatures of some pigments, which hinders their individual discrimination (Figure 13.2). As a result, absorption spectra of different populations are sometimes indistinguishable (Garver *et al.*, 1994). Finally, if the objective is not only the detection but also the quantification of pigments, other hindrances lie in the facts that (i) the weight-specific *in vivo* absorption coefficients of individual pigments are not exactly known (the assumption that they are identical to those in solvents is questionable; see e.g. Johnsen *et al.*, 1994a), and (ii) some unidentified components (pigments not detected by HPLC, or other light-absorbing compounds) are likely to modify *in vivo* absorption (Nelson *et al.*, 1993; Bricaud *et al.*, 2004; Ficek *et al.*, 2004; Section 13.3).

Notwithstanding, many efforts have been dedicated to extract pigment information from absorption spectra of phytoplankton, and various methods (or combinations of methods) have been proposed with this purpose. Hoepffner and Sathyendranath (1993) and Stuart *et al.* (1988) decomposed algal absorption spectra into Gaussian bands which were attributed to individual pigments, but the quantification of pigments was limited by the variations in the package effect for natural populations. A similar approach had been applied earlier to the spectral values of the imaginary part of the refractive index (directly related to the absorption spectrum of cellular matter, and free from the package effect) by Bricaud and Morel (1986).

Another approach, stepwise discriminant analysis, was applied to algal absorption spectra by Johnsen *et al.* (1994b) to classify absorption spectra among 31 phytoplankters from 10 classes. This method was later associated with the fourth-derivative analysis of absorption spectra, first proposed by Faust and Norris (1985)

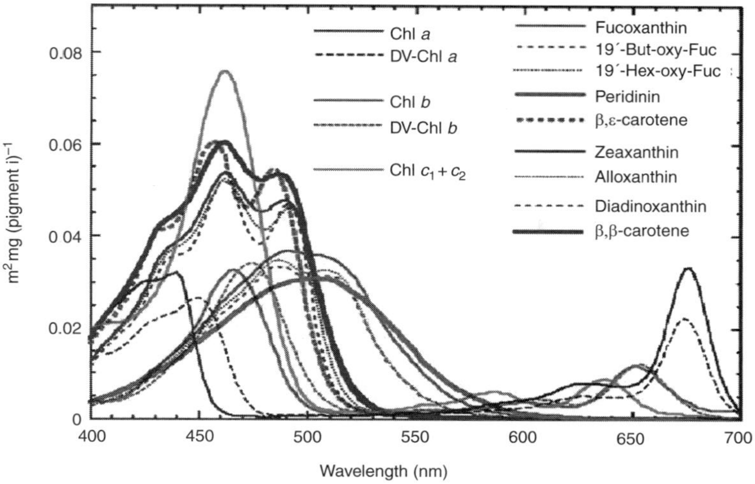

Figure 13.2. Assumed *in vivo* weight-specific absorption spectra of the main pigments, as derived from absorption spectra of individual pigments in solvent (Bricaud *et al.*, 2004). See colour plate section.

and Bidigare *et al.* (1989a), and the use of a similarity index algorithm to detect the presence of *Karenia brevis* (also known as *Gymnodinium breve* and *Ptychodiscus brevis*) in laboratory cultures (Millie *et al.*, 1997) and quantify the associated biomass in natural assemblages (Kirkpatrick *et al.*, 2000). Millie *et al.* (2002) extended this method to the discrimination of different taxonomic groups. Staehr and Cullen (2003) compared the performances of the spectral similarity index and of the multivariate partial least-squares regression technique (both applied to fourth-derivative absorption spectra), and found that the second technique was less sensitive to photoacclimation effects and to variations in the size of populations. A partial least-squares regression technique was associated with a principal component analysis by Moberg *et al.* (2002) to estimate the relative abundances of different algal classes in mixed cultures.

Derivative analysis was also used by Aguirre-Gomez *et al.* (2001), who decomposed absorption spectra into Gaussian–Lorentzian curves centred on the identified peaks, to estimate the relative contributions of pigments. Eisner *et al.* (2003) developed an index based on the slope (i.e. the first derivative) of absorption spectra in the blue-green domain to retrieve the ratio of PPC to LHP in the algal populations. Finally, neural network methods were also tested. Chazottes *et al.* (2006) used such a method, based on 'self-organizing maps' (Kohonen, 1984), to classify absorption spectra using their first derivatives, and showed that such classification can provide information about pigment concentrations. Bricaud *et al.* (2007) also showed that multi-layer perceptrons, trained on a large *in situ* database, could be used to retrieve the concentrations of the three main pigment groups (Chl *a*, *b*, *c*, PPC and LHP) from absorption spectra within reasonable accuracies (the best and poorest performances being for Chl *a* and Chl *b*, respectively).

Extracting pigment information from the scattering properties of algal cells has been seldom attempted, because, as explained above, (i) partitioning scattering coefficients into algal and non-algal components is difficult, and (ii) even if absorption bands generally induce a depression in scattering, the pigment signatures are more blurred than in absorption spectra. Some taxonomic groups can actually be identified through their scattering properties, but owing to particular physical characteristics (e.g. calcium carbonate coccoliths for coccolithophorids, or gas vacuoles for *Trichodesmium*) rather than to pigment features. More successful are the techniques based on flow cytometry to characterize different phytoplankton groups, but these are based on the combined analysis of light scattering and fluorescence measurements, the latter providing essential information on the pigments present (see review by Sosik, 2008).

13.2.3 Relationships between individual pigments and in vivo absorption properties

Historically, determination of the *in vivo* Chl *a*-specific absorption coefficient, $a^*_\varphi(\lambda)$, for natural phytoplankton populations from pigment concentrations has been difficult. Alternative approaches have been developed based on absorption reconstruction and deconvolution. The 'reconstruction' approach uses pigments to construct absorption spectra whereas the 'deconvolution' approach uses absorption spectra to estimate pigment concentrations and/or ratios. The spectral reconstruction technique (Bidigare *et al.*, 1990) was used to obtain spectral absorption coefficients, $a^*_i(\lambda)$, from knowledge of the *in vivo* absorption coefficients of the major pigment groups (Chl *a*, Chl *b*, Chl *c*), phycoerythrin (PE), light-harvesting carotenoids, PPC, and their respective volume-based concentrations as determined by HPLC (chlorophylls and carotenoids) and fluorescence spectroscopy (compare Bidigare and Ondrusek, 1996):

$$a'_\varphi(\lambda, z) = \sum_{i=1}^{n} a^*_i(\lambda) c_i(z), \qquad (13.6)$$

where $a^*_i(\lambda)$ is the weight-specific absorption coefficient for pigment group *i* and $c_i(z)$ is the concentration of pigment group *i* at depth *z* (Figure 13.2, Bricaud *et al.*, 2004). This calculation assumes that the $a'_\varphi(\lambda, z)$ spectrum is a linear combination of absorption contributions provided by 'unpackaged' pigment chromophores located within phytoplankton cells. As a consequence, the accuracy of this method is limited by (i) incomplete pigment extraction, (ii) pigment 'package' effects and (iii) differences in the intracellular absorption properties of the pigments classified within each of the pigment groups. Consequently, the uncertainties associated with this technique should be lowest when phytoplankton cell size is small and/or where pigment 'package' effects are minimal. This technique has provided reasonable approximations of $a^*_\varphi(\lambda)$ for cultures of *Prochlorococcus* (Moore *et al.*, 1995) and dinoflagellates (Nelson and Prézelin, 1990; Johnsen *et al.*, 1994a) and phytoplankton communities sampled from the Sargasso Sea (Bidigare *et al.*, 1990), and the Southern California Bight (Nelson *et al.*, 1991).

Comparisons have been made between the pigment reconstruction and filter pad methods with varying degrees of success. Bidigare *et al.* (1989b) demonstrated good agreement for a *Synechococcus* clone, if the filter pad pigment spectra were corrected for absorption by phycoerythrin and an unknown background noise likely to be associated with CDOM. Nelson and Prézelin (1990) and Johnsen *et al.* (1994a) found differences between the pigment reconstruction method for dinoflagellates and a chromoprotein spectral model, where the pigment model overestimated in the blue and underestimated in the red part of PAR compared to the chromoprotein-based model and measured *in vivo* absorption spectra which gave similar results. However, when comparisons were made with natural samples, significant differences were found (Marra *et al.*, 2000). Bricaud *et al.* (2004) speculated that these differences indicated a missing term in the reconstruction method (possibly an unidentified carotenoid). Generally, the pigment reconstruction method underestimates the filter pad/methanol wash method. It should be noted that $a^*_i(\lambda)$ for Chl *a* is the basis for scaling using the pigment reconstruction method, and there are now several indications that the red *in vivo* peak absorption coefficient with no package effect (intracellular self-shading) is close to 0.03 m^2 mg Chl a^{-1} (Johnsen *et al.*, 1994a, 1997; Bricaud *et al.*, 2004; Johnsen and Sakshaug, 2007; Figures 13.2 and 13.3A, B).

Accessory photosynthetic and photoprotective pigments

The spectral reconstruction technique has provided new insights into the role of accessory pigmentation in light absorption. For the Sargasso Sea, accessory pigments accounted for 60 and 90% of the light absorbed by resident phytoplankton at the sea surface and the base of the euphotic zone, respectively (Bidigare *et al.*, 1992). The optical depth-dependent absorption contribution provided by accessory pigmentation followed the same pattern when data collected on winter and summer cruises were compared. This result is somewhat surprising since Chl *a* distribution and water column optical properties measured during these cruises were markedly different. Absorption contributions provided by individual accessory pigment groups (Chl *b*, Chl *c*, light-harvesting carotenoids and PPC), however, showed large between-cruise differences.

The spectral reconstruction technique also allows an estimation of phytoplankton absorption that is not biased by the PPC. Several studies have documented that PPC lead to significant decreases in the maximum quantum yield of photosynthesis ($\phi_{c\ max}$) of cyanobacteria cultures (Bidigare *et al.*, 1989b), and diverse phytoplankton assemblages sampled from the tropical northeast Atlantic Ocean (Babin *et al.*, 1996) and the Arabian Sea (Bidigare *et al.*, 1997; Marra *et al.*, 2000). Highest PPC absorption contributions are typically found in the upper euphotic zone under cloud-free conditions, where PPC accumulation improves the ability to maintain efficient rates of photosynthesis that are otherwise damaging to the photochemical reaction centres. (Bidigare *et al.*, 1987; Johnsen *et al.*, 1997; Marra *et al.*, 2000). To circumvent this problem, the quantum yield can be corrected if the absorption by the 'PS-active' pigments is known or calculated. The photosynthetically active phytoplankton absorption coefficient, a_{act} (λ), can be computed via direct reconstruction, i.e. $a_{\mathrm{chl}\,a}$ (λ) + $a_{\mathrm{chl}\,b}$ (λ) + $a_{\mathrm{chl}\,c}$ (λ) + a_{pe} (λ) + a_{psc} (λ), by subtraction of the reconstructed PPC absorption coefficient (a_{ppc} (λ)) from the detrital-corrected absorption coefficient, $a^*_\varphi(\lambda)$, measured by the QFT technique (Schofield *et al.*, 1996; Bidigare *et al.*, 1997), or by multiplying $a^*_\varphi(\lambda)$ by the ratio of a_{act} (λ) to a_{ph}' (λ) (Babin *et al.*, 1996). The latter approach is advantageous since it is not biased by errors associated with pigment packaging effects. It should be noted that a_{act} (λ) for PSII absorption has also been estimated via the comparison of oxygen action, absorption and fluorescence excitation spectra obtained for phytoplankton (Haxo, 1985; Mitchell and

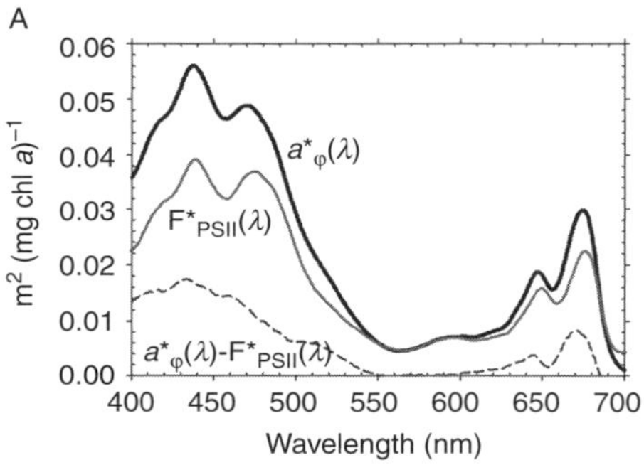

Figure 13.3. (A) Example of PSII-scaling of fluorescence excitation to absorption with average values for LL-acclimated prasinoxanthin-containing prasinophytes (*Bathycoccus prasinos, Micromonas pusilla* and *Pseudoscourfieldia marina*). *In vivo* Chl *a*-specific absorption spectra ($a^*_\varphi(\lambda)$), Chl *a*-specific PSII-scaled fl-ex spectra ($F^*_{PSII}(\lambda)$), and the corresponding difference spectra ($a^*_\varphi(\lambda)$-$F^*_{PSII}(\lambda)$) denoting the non-fluorescent fraction of photoprotective carotenoids and PSI. (B) Bio-optical characteristics of LL-acclimated chromophytes (7 PGs, upper panels), Chl *b*-containing phytoplankton (4 PGs, mid panels), and biliprotein-containing phytoplankton (2 PGs, lower panels). Left panels: $a^*_\varphi(\lambda)$; right panels: $F^*_{PSII}(\lambda)$. Pigment-groups (PG) are: PG 1 Bacillariophyceae (fucoxanthin, Chl c_{1+2}), PG 2 Dinophyceae I (peridinin, Chl c_2), PG 3 Dinophyceae II (acyl-oxy-fucoxanthins, gyroxanthin diester, Chl c_3), PG 4 Coccolithophyceae (acyl-oxy-fucoxanthins, Chl c_3), PG 5 Pavlovophyceae (fucoxanthin, Chl c_{1+2}), PG 6 Prasinophyceae I (prasinoxanthin, [3,8]-proto-chlorophyllide, Chl *b*), PG 7 Prasinophyceae II (lutein, Chl *b*), PG 8 Euglenophyceae (neoxanthin, Chl *b*), PG 9 Chlorophyceae (lutein. Chl *b*), PG 10 Chrysophyceae (fucoxanthin, Chl c_{1+2}), PG 11 Raphidophyceae (violaxanthin, Chl c_{1+2}), PG 12 Cryptophyceae (phycobiliprotein, alloxanthin, Chl c_2), and PG 13 Cyanophyceae (phycobiliproteins, zeaxanthin). (A) and (B) from Johnsen and Sakshaug (2007) with permission from Phycological Society of America. See colour plate section.

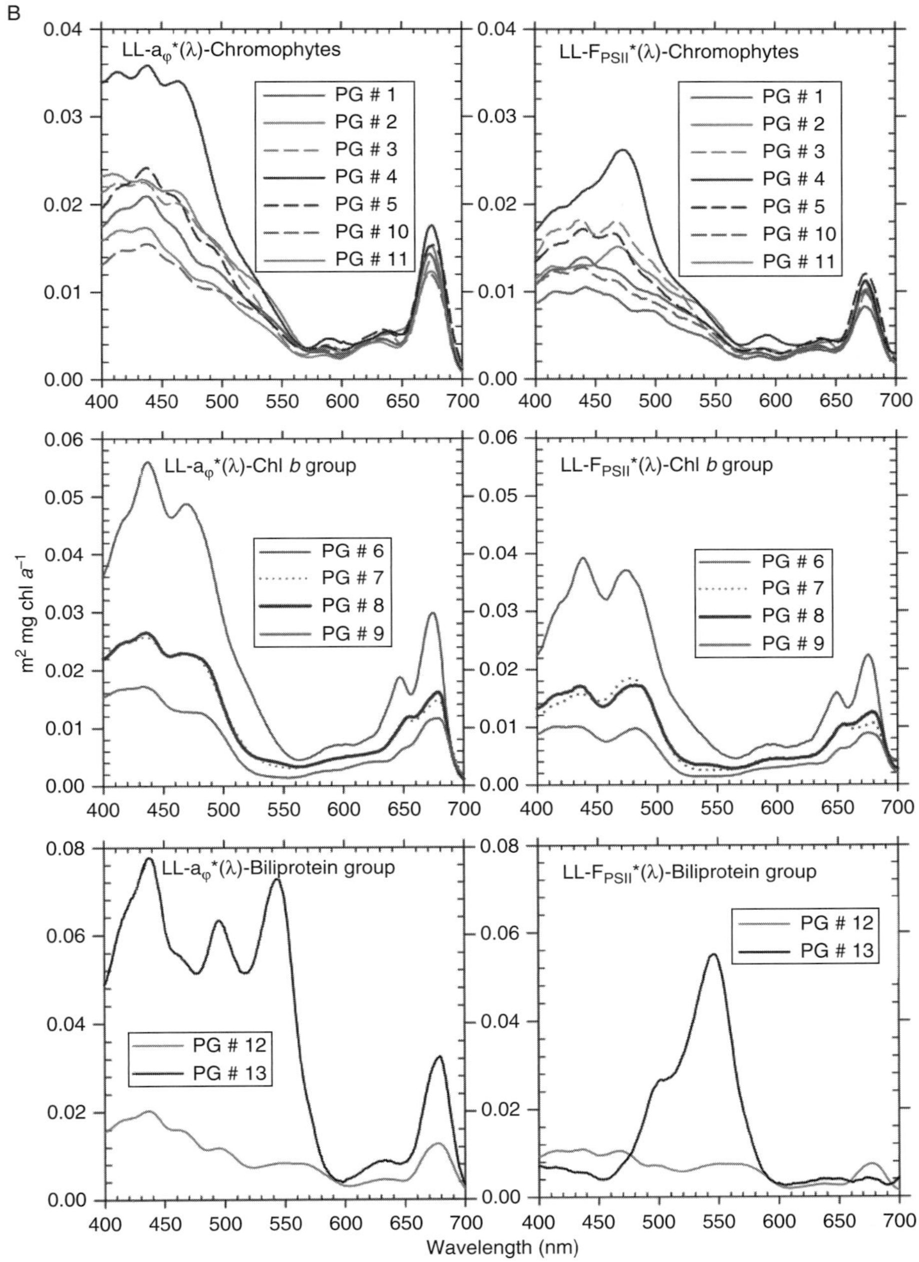

Figure 13.3. (*cont.*)

Kiefer, 1988; Neori *et al.*, 1988; Johnsen and Sakshaug, 2007) and ice algae (Bidigare *et al.*, 1993; see Section 13.3).

13.3 *In vivo* Chl *a* fluorescence excitation spectra

13.3.1 Theory

In a living phytoplankton cell, 1–5% of the light absorbed by a cell is re-emitted as Chl *a* fluorescence from the reaction centres of PSII (bonded to their associated LHC) and is responsible for ~95% of the *in vivo* fluorescence in phytoplankton (Govindjee, 1995). *In vivo*, all photosynthetic (= light-harvesting) pigments (LHC with high light-harvesting transfer efficiency) and PPC (low efficiency), transfer light to the final acceptor pigment molecule, i.e. the reaction centre Chl *a* which we measure as light emission (PSII fluorescence). This is in contrast to *in vitro* measurements, where for example, we can measure fluorescence excitation spectra characteristics (autofluorescence) from HPLC isolated Chl *a*, Chl c_1, Chl c_2 and Chl c_3 (e.g. Johnsen and Sakshaug (1993)). In living cells, Chl *a* fluorescence is an indicator of light-harvesting and utilization in phytoplankton. In view of the fact that *in vivo* Chl *a* fluorescence is extremely rich in information that is affected by several de-excitation pathways, it also becomes an ambiguous signal (Govindjee, 1995).

The *in vivo* emission peak of PSII fluorescence is centred at 685 nm (seen as deep red light by the human eye). In fact, in 1931 H. Kautsky and A. Hirsch used their eyes as instruments to detect Chl *a* fluorescence in leaves (see Govindjee, 1995). The quantum yield of Chl *a* fluorescence (ϕ_F), is the fraction of absorbed quanta re-emitted as fluorescence, or more exactly, the fraction of excited molecules that decay by fluorescence (Parson and Nagarajan, 2003; Eq. (13.7), (13.8)). ϕ_F denotes the fraction of light absorbed and utilized by PSII that is converted to Chl *a* fluorescence (emission detected at 685 nm in most fluorometers). In living and photosynthetically active cells, ϕ_F *in situ* is typically 0.5–5% re-emission per quanta absorbed, about 70% of absorbed quanta will be lost as thermal decay (heat), and the rest, ~25–30% will be used in photochemistry (Falkowski and Raven, 1997). ϕ_F *in vivo* will be at its minimum level in the dark (all functional PSII reaction centres open, i.e. oxidized). In low light conditions ϕ_F will rise (depending on the fraction of closed RC) to a maximum (3–5%) when cells are treated with saturating light intensities (maximum photosynthetic rate, all PSII reaction centres closed, i.e. reduced). The highest ϕ_F for Chl *a* fluorescence will be found *in vitro*, when extracting Chl *a* in an organic solvent (e.g. methanol). In this case Chl *a* has lost its apoproteins and is non-functional in photosynthesis (the rate constant for photochemistry, k_P, is ≈ 0, Eq. (13.7)) and a value of ϕ_F of 30% can be obtained (auto-fluorescence). Thus, *in vitro*, ϕ_F is so high that the red fluorescence can easily be seen by the human eye (eyes placed 90° to the light source illuminating pigment extract). Auto-fluorescence can also be detected in decaying cells (e.g. a post-bloom situation) with significant amounts of destroyed thylakoids. Since the fluorescence emission from phytoplankton cells competes with photochemistry for excitation energy from the sun, its measurement provides a real-time and non-invasive insight in variations in biomass, ecophysiology and photosynthetic rates (Nedbal and Koblízek, 2006).

To measure and to interpret *in vivo* and *in situ* Chl *a* fluorescence, it is important to know which key environmental variables (e.g. light climate and nutrients) induce differences in ϕ_F (Chl *a* fluorescence quenching processes). There are two major processes that modify ϕ_F, photochemical quenching (q_P) and non-photochemical quenching (q_N, see Chapter 11.2.2). Photochemical quenching lowers ϕ_F because excitation energy is used for photochemical reactions and is related to the fraction of open reaction centres in PSII, commonly measured with pulse-amplitude-modulated or fast-repetition-rate fluorometers (Falkowski and Chen, 2003; Babin, 2008). Photochemistry by open PSII reaction centres shortens the excitation lifetime to < 1000 picoseconds, lowering ϕ_F to a minimum level, F_0. A reduction of the primary quinone acceptors (Q_A^-), in closed PSII reaction centres, extends the excitation lifetime to several nanoseconds. The longer excitation lifetime in closed PSII reaction centres induces a fluorescence maximum level, F_M ($F_M \approx 5\ F_0$, Nedbal and Koblízek, 2006). Using F_0 and F_M, the maximum quantum yield of PSII-fluorescence (ϕ_{PSII}^{max}, in dark acclimated cells) can be measured (Malkin and Kok, 1966; Govindjee, 1995; Nedbal and Koblízek, 2006):

$$\phi_{PSII}^{max} = k_P^{dark}/\left(k_F + k_N^{dark} + k_P^{dark}\right) = (F_M - F_0)/F_M \equiv F_V/F_M. \tag{13.7}$$

Equation (13.7) indicates that the different rate constants (k) for primary photochemistry in PSII reaction centre (k_P), Chl *a* fluorescence emission (k_F), and non-radiative dissipation (including thermal dissipation, conversion to triplets and energy transfer in LHCII, k_N), affect ϕ_{PSII}^{max} (Govindjee, 1995; Nedbal and Koblízek, 2006; Babin, 2008). F_V denotes variable fluorescence and is related to the quantum yield of PSII charge separation. Thus ϕ_F is related to (Govindjee, 1995; Falkowski and Raven, 1997; Babin, 2008):

$$\phi_F = k_F/(k_F + k_P + k_N). \tag{13.8}$$

Non-photochemical quenching (NPQ) includes all the mechanisms that lower k_F and increase k_N, apart from q_P. Three major NPQ mechanisms are characterized by different relaxation kinetics in darkness following a period of illumination. These are the energy-dependent fluorescence quenching (qE, related to the build-up of a proton gradient causing lowering of pH in the thylakoid lumen which induces xanthophylls synthesis via a xanthophyll cycle, see PPC in Section 11.2.2), state-II-I transitions between the major LHCII and LHCI (qT, note that this mechanism occurs mainly in green algae and phycobiliprotein-containing phytoplankton, Section 11.3.2) reducing the amount of excitation energy in PSII that can de-excite to fluorescence, and finally photoinhibitory quenching (qI, mainly caused by photoinhibition and damage to the PSU, especially the D1 protein in PSII reaction centre (Falkowski *et al.*, 1995; Chapter 11, this volume).

Under actinic light (which can cause a photochemical reaction) conditions, the quantum yield of photochemistry in PSII, $\phi_{PSII}^{actinic}$, can be calculated from the steady-state fluorescence at a given irradiance (F_0') and the maximum fluorescence after a saturating pulse closing (reducing) all PSII reaction centres (F_M'). Genty *et al.* (1989) were the first to report that in steady-state photosynthesis, the product of q_P and F_V'/F_M', denotes the potential efficiency of PSII in light (Eq. (13.9)) (Govindjee, 1995; Krause and Jahns, 2003),

$$\phi_{PSII}^{actinic} = q_P \cdot F_V'/F_M' = (F_M' - F_0')/F_M' \equiv F_V'/F_M', \tag{13.9}$$

where

$$q_P = (F_m' - F')/F_v'. \tag{13.10}$$

Note that Eq. (13.9) and (13.10) are valid for steady-state fluorescence obtained in ambient light for F_M' and F' (flux of in vivo fluorescence emitted by an elementary volume) and after sample collection and isolation in the dark for F_0' (Babin, 2008).

Since in theory, variable fluorescence (F_V') may vary between 0 (F_0 state with all Q_a oxidized – all PSII reaction centres open) to 1 (F_M' state with all Q_a reduced), q_P is defined as the fraction of F_V' quenched by oxidized Q_a by utilization of excitons in photochemistry.

The relationship between q_P and q_N is crucial for the understanding of variable fluorescence (reviews by Govindjee, 1995; Falkowski and Chen, 2003; Krause and Jahns, 2003; Nedbal and Koblízek, 2006; Babin, 2008). The basics of Chl *a* fluorescence (Eqs. (13.7) to (13.10)) are not only important when using variable fluorescence instruments to estimate photosynthetic rates (e.g. using pulse amplitude modulated or fast repetition rate fluorometer techniques), but also necessary to ensure good control of method protocols and interpretation of these measurements.

13.3.2 Experimental methods of in vivo fluorescence excitation spectra

Numerous reports exist on techniques using Chl *a* fluorescence excitation spectra to measure differences between phytoplankton species, biomass, eco-physiological and photosynthetic responses (see Haxo, 1985; Neori *et al.*, 1988; Falkowski and Raven, 1997; Falkowski and Chen, 2003; Sakshaug and Johnsen, 2005; Babin, 2008). In this section we will focus on the use of *in vivo* fluorescence excitation spectra (400–700 nm) to detect pigment-specific differences in light-energy transfer to PSII in phytoplankton. An *in vivo* absorption spectrum indicates absorption of all pigments (Section 13.1). In contrast, an *in vivo* fluorescence excitation spectrum indicates the absorbed light to PSII, or more precisely, the transfer of light energy by LHPII to PSII. In the reaction centres of PSII, the water-splitting complex (evolving photosynthetic oxygen from water) is tightly connected to reaction center-Chl *a* emitting light (fluorescence) as a function of discrete wavelengths absorbed and funnelled to the reaction centres. Because of this relationship, the shape of an *in vivo* fluorescence excitation spectrum resembles that of an oxygen action spectrum (Haxo, 1985).

In vivo fluorescence excitation spectra

Pigment-group specific differences in pigment–protein complexes cause differences in in vivo fluorescence excitation spectra (Haxo, 1985; Neori *et al.*, 1988; Johnsen *et al.*, 1997; Johnsen and Sakshaug, 2007; Section 13.2). In most low light acclimated chromophytes and chlorophytes, about 70% of the light is absorbed by PSII, but when acclimated to high light conditions, this will be reduced to 55% mostly due to PPC (Johnsen and Sakshaug, 2007). In extreme cases, as in some phycobiliprotein-containing cyanobacteria, cryptophytes and rhodophytes, with $> 70\%$ of total Chl *a* associated with PSI, this imbalance can be particularly large. State II–I transitions may account for $< 20\%$ effect of in vivo PSII fluorescence signature in phycobiliprotein-containing cyanobacteria, and most chlorophytes (Larkum, 2003). In contrast, state transitions in chromophytes are absent or small (Raven and Geider, 2003; Larkum, 2003).

A major obstacle for widespread use of *in vivo* fluorescence excitation spectra has been the need for proper instrument setup, to avoid variable fluorescence (F'_V), and to ensure a proper quantum correction (using dyes or photo counters; Kopf and Heinze, 1984; Neori *et al.*, 1988; Sakshaug *et al.*, 1991; Johnsen and Sakshaug, 1993; Culver *et al.*, 1994; Lutz *et al.*, 2001). We will discuss how to scale a quantum corrected fluorescence excitation spectrum (relative scale) to a PSII-specific fluorescence excitation spectrum, obtaining the same units as the corresponding absorption spectrum in m^2 mg (Chl *a*)$^{-1}$ for different applications.

A general setup to obtain *in vivo* fluorescence excitation spectra is to put both the excitation monochromator (400 to 700 nm, in 1 nm increments) and emission monochromator (detecting emitted light at 730 nm) of the spectrofluorometer at a bandwidth of 5–10 nm (see overview of method in Johnsen and Sakshaug, 2007). A smaller bandwidth will give a more distinct pigment signature, but this lowers the signal-to-noise ratio. An infrared transmittance filter (e.g. Schott RG695) is put in front of the emission detector (photomultiplier tube) to ensure that only wavelengths > 695 nm pass through the filter. This is done to avoid stray light from the excitation monochromator (light source) and to omit reflected light from the cuvette and phytoplankton cells and debris. One must make sure that the 1-cm square fluorescence cuvettes have all sides transparent and that the cells remain at *in situ*/growth temperature by using a thermostat-controlled system (e.g. a Peltier cell).

Avoiding variable fluorescence (F'_V, inducing an unstable signal/noise ratio at a given excitation wavelength) can be obtained by ensuring that there is a high enough cell density in the cuvette (>10 μg Chl *a* l^{-1}) and that all PSII reaction centres are closed (F'_M) by using the herbicide 3-(3,4 dichlorophenyl)-1,1-dimethylurea (DCMU) at 50 μmol l^{-1} final concentration (Johnsen and Sakshaug, 1993; Johnsen *et al.*, 1997). DCMU blocks the electron transfer between Q_a and Q_b, closing all reaction centres in PSII and thus, we obtain F'_M under sub-saturating irradiances from the excitation monochromator. To ensure that the fluorescence signal is non-variable (obtaining F'_M) prior to fluorescence excitation scanning, a 1–2 min scan at 440 and 550 nm excitation light (emission detected at 730 nm) should be carried out using the same duration as the fluorescence excitation scan, after closing the PSII reaction centres with DCMU in the presence of actinic light. For field samples (often 0.1–10 μg Chl *a* l^{-1}), a 1 to 100-fold pre-concentration of cells should be carried out, using 1 G (no suction) filtration on 0.4–0.8 μm polycarbonate filter and a filtration funnel. The cells should be kept in suspension by gentle shaking and the last 3 mL of cell suspension should be poured in the fluorescence cuvette. Note that at small pore sizes, particles may clog the pores of the filter. At small pore sizes it may be difficult to get water through the filter and at large pore sizes a significant fraction of cells may be lost.

A quantum correction (usually from 400–700 nm) of a raw *in vivo* fluorescence excitation spectrum accounts for variation in spectral quality and energy output from the excitation lamp, and the spectral sensitivity of the instrument photomultiplier. For best results, use a triangular cuvette and add the quantum counter dye Basic Blue 3 from Sigma (purity > 80%) dissolved in ethylene glycol at 4.1 g l^{-1} according to Kopf and Heinze (1984, see also their purification details for different quantum counter dyes). Kopf and Heinze were among the first to carry out quantum correction across the whole visible band (PAR) for chemical applications, later the approach was adopted for phytoplankton cultures (Sakshaug *et al.*, 1991) and field experiments on sea ice-algae (Johnsen and Hegseth, 1991). Other Chl *a* fluorescent compounds can be used *in vitro* (in a buffer or in an organic solvent) to obtain a crude quantum correction. A comparison between Chl *a* extract (e.g. acetone) measured in a spectrofluorometer (raw fluorescence excitation spectrum) versus spectral absorption (spectrophotometer, optical density spectrum) can be used to quantum correct the instrument in the high absorbing blue (400–440 nm) and red parts (580–700 nm) of the PAR range (Lutz *et al.*, 1998). The weakness with this approach is that between 450 and 570 nm Chl *a* practically do not absorb light, hence there is no good quantum correction for this region. A quantum counter dye needs to have a stable fluorescence and needs to absorb efficiently in the whole spectral region of interest, in our case from 400 to 700 nm. Isolated and functional PSI have been used for the same purpose and have only low absorption at 540–565 nm range (Johnsen *et al.*, 1997).

To verify the quantum correction, HPLC-purified Chl *a*, Chl *b* or Chl *c* (highly fluorescent from 580–680 nm when excited at their respective absorption peaks from 400–500 nm), dissolved in an organic solvent (e.g. ethanol) should be measured in a

1-cm square cuvette in a spectrophotometer to obtain an *in vitro* absorbance (optical density) spectrum from 400–700 nm. The same sample should be used to acquire an *in vitro* fluorescence excitation spectrum (spectrofluorometer set up as described above). If quantum correction is done correctly, the *in vitro* absorbance and quantum corrected fluorescence excitation spectra should be similar in shape, i.e. the ratio between them as a function of wavelength should be a flat line (Johnsen and Sakshaug, 1993).

Based on bio-optical properties (400–700 nm) and pigment composition in isolated LHC, PSII and PSI, thylakoid micelles and whole cells of two dinoflagellates species with different pigment–protein build-up as model organisms, Johnsen *et al.* (1997) suggested the scaling of the fluorescence excitation spectrum to the corresponding absorption coefficients $a^*_\varphi(\lambda)$ at wavelengths where no photoprotective carotenoids absorb, i.e. between 540–700 nm. This work was the basis for a procedure to transform a relative *in vivo* and quantum corrected fluorescence excitation spectrum into a PSII-specific fluorescence excitation spectrum, $F^*_{PSII}(\lambda)$, indicating light absorbed by PSII and its corresponding LHCII (m^2 mg (Chl *a*)$^{-1}$). Later, the no-overshoot procedure, matching the spectra at 540–650 nm range, was used on 33 species of phytoplankton, comprising 10 classes (covering 13 different pigment-groups), giving information on the total amount of quanta absorbed by the cell, PSII, PSI and by photoprotective carotenoids (m^2 mg (Chl *a*)$^{-1}$ or m^2 mg C^{-1}, Johnsen and Sakshaug, 2007; Figures 13.3A, B).

13.3.3 Combined information from in vivo $a^*_\varphi(\lambda)$ and $F^*_{PSII}(\lambda)$

This section gives the overall eco-physiological information that can be obtained from $a^*_\varphi(\lambda)$ and $F^*_{PSII}(\lambda)$. When comparing shapes of *in vivo* absorption spectra with the corresponding *in vivo* fluorescence excitation spectra, cells with low content of PPC and correspondingly high content of LHP have largely the same shape. In cells acclimated to low light conditions, absorption and fluorescence excitation peaks typically resemble absorption of all pigments (absorption spectra) and the corresponding light energy transfer by LHP to PSII (fluorescence excitation spectra) in the blue (430–470 nm, chlorophylls), blue-green (460–535 nm, LHC-carotenoids), green-orange (540–590 nm, phycobiliproteins), orange (~ 585 nm, Chl *c*-group), and red (600–700 nm, all chlorophylls, except Chl c_3). In high light conditions, a significant PPC content in the LHC relative to LHP, will induce a significant reduction in light absorbed and utilized by PSII at 440, 460 and 490 nm, clearly reflected in $F^*_{PSII}(\lambda)$ (see Chapter 11; Bidigare *et al.*, 1989b; Schofield *et al.*, 1996; Johnsen *et al.*, 1997).

Points (A) to (G) below, give an overview on how to use bio-optical information, based on $a^*_\varphi(\lambda)$ and $F^*_{PSII}(\lambda)$, to obtain information on total absorption (A), light absorbed by PSII (B), absorption by PSI and PPC (C), the fraction of Chl *a* and/or quanta absorbed in different pigment proteins (D), identification of pigment groups based on fluorescence (E), the effect of PPC in light energy transfer efficiency (F), and basic information for calculation of absorbed quanta to PSII (G).

(A) $a^*_\varphi(\lambda)$, Chl *a* specific absorption coefficient, indicates total absorption by all phytoplankton pigments in a cell (Sections 13.2, 13.3; Chapter 11.1; Bricaud and Morel, 1986; Bidigare *et al.*, 1992).

(B) $F^*_{PSII}(\lambda)$, Chl *a*-specific PSII-scaled fluorescence excitation spectrum, indicates light absorbed and utilized by PSII and with the same units as $a^*_\varphi(\lambda)$ (Johnsen and Sakshaug, 2007; Figure 13.3).

(C) The difference spectrum, $a^*_\varphi(\lambda) - F^*_{PSII}(\lambda)$, indicates the non-fluorescent fraction and indicates absorption by the apparently non-fluorescent PSI and PPC (associated with LHC in PSI and PSII, Johnsen *et al.*, 1997; Figure 13.3A).

(D) The relationship between total absorption and the fraction of light absorbed and utilized by PSII and PSI and their corresponding light-harvesting complexes can be used to indicate the fraction of Chl *a* and/or quanta absorbed in PSI, PSII, LHCI, LHCII and in PPC (Johnsen *et al.*, 1997; Johnsen and Sakshaug, 2007; see Section 13.5 and Chapter 11). If the spectral irradiance $E(\lambda)$ is known, a calculation of absorbed quanta actually absorbed (from cell, pigment-protein, or PPC) can be estimated using the equation for spectrally weighted ($\overline{X}$) total absorption by all pigments ($a^*_\varphi(\lambda)$), spectrally weighted PSII absorption ($F^*_{PSII}(\lambda)$) or reconstructed *in vivo* absorption coefficient for individual pigments ($a^*_i(\lambda)$, see Eq. (13.6)). More specifically, calculating $\overline{X}$ can give us information on light-harvesting and utilization at cellular, LHC, PSII, PSI, PSC, PPC, and individual pigment level (where $\overline{X}$ can be $\overline{a}^*_\varphi$, $\overline{a}^*_{PSII}$ or $\overline{a}^*_i$, Eq. (13.11)):

$$\overline{X} = \left[\sum_{400}^{700} X(\lambda) \cdot E(\lambda) d\lambda\right] / E(\mathrm{PAR}). \tag{13.11}$$

(E) *In vivo* fluorescence excitation spectra, $F^*_{PSII}(\lambda)$, can be used to identify phytoplankton pigment groups in water masses with low phytoplankton biomass, waters with significant optical signature from CDOM and suspended matter obscuring absorption signature (see Lutz *et al.*, 1998; Chapter 14).

(F) $F^*_{PSII}(\lambda)$ is affected by PPC which perform light transfer to PSII with low efficiency, ranging from 15 to 70% of the light energy transfer efficiency in the major LHCII at the absorption peak of diadinoxanthin and diatoxanthin (PPC) – this is highly dependent on PPC content in the LHC (Johnsen *et al.*, 1997; Chapter 11).

(G) $F^*_{PSII}(\lambda)$ can be the basic input for the calculation of absorbed quanta to PSII and is species/pigment-group specific (Johnsen and Sakshaug, 2007). This information can be used to calibrate/interpret fluorescence kinetics measurements (e.g. pulse amplitude modulated (PAM) or fast repetition rate fluorometer techniques, FRRF) for estimating oxygenic primary productivity (see Chapter 12, this volume). There are two approaches to do this, the first is based on bio-optics (as mentioned above) and often used as input for PAM-based measurements (Krause and Jahns, 2003; Johnsen and Sakshaug, 2007; Hancke *et al.*, 2008a, b) and the second is based on bio-physics (and often used with FRRF measurements (Falkowski and Chen, 2003; Suggett *et al.*, 2004, 2007). The FRRF technique enables a direct biophysical estimation of both photochemical efficiency and light absorption by PSII, i.e. the functional absorption cross-section of PSII reaction centres and the photosynthetic unit size of PSII. The use of $F^*_{PSII}(\lambda)$ to estimate $\overline{a}^*_{PSII}$ is important in both approaches to fine-tune the estimation of oxygen produced per Chl *a* (or other biomass indicator) per unit time based on fluorescence kinetics.

13.4 *In vivo* absorption properties of CDOM and non-phytoplankton particles

Phytoplankton pigments are only one of several classes of light-absorbing materials found in natural waters (e.g. Kirk, 1994). Other classes include coloured (chromophoric) dissolved organic matter (CDOM, also 'gelbstoff' or 'gilvin') and non-phytoplankton particles (including non-living detritus or 'tripton', heterotrophic organisms, and inorganic sediment). Particles and dissolved material also contribute to the in-water optical properties via light scattering, in particular the highly reflective coccoliths of some species of prymnesiophytes. There is generally a correlation between the abundances of phytoplankton pigments, CDOM, and detrital particles in the global ocean, which makes possible straightforward bio-optical models of ocean colour properties where Chl *a* is the controlling variable (e.g. Morel and Maritorena, 2001). Standard algorithms for retrieving chlorophyll *a* concentration from ocean colour satellite sensors take advantage of this global correlation. But differences in the dynamics of the different classes of absorbing and scattering materials do occur, which can complicate the assessment of phytoplankton pigment absorption using spaceborne or moored ocean colour sensors.

The CDOM and particulate detritus absorption coefficients are usually determined by absorption spectroscopy on filtered seawater (for CDOM) or solvent-extracted filters loaded with marine particles, as described in Section 13.2 (for detritus). Absorption spectroscopy of CDOM usually involves measuring the optical density spectra of seawater filtrate in 10 cm cuvettes versus purified water, but in open ocean samples CDOM absorption spectra tend to be below the limit of detection in the visible wavelength range (Nelson and Siegel, 2002). To alleviate this problem researchers are experimenting with liquid waveguide spectrophotometers with optical pathlengths up to 2 m for CDOM analysis (Miller *et al.*, 2002).

Absorption spectra of CDOM and particulate detritus are typically featureless, decreasing with wavelength (Figure 13.4). CDOM spectra are often parameterized using a simple negative exponential equation, where the parameter (S, units of inverse nm) is referred to as the 'exponential slope'. This model for CDOM

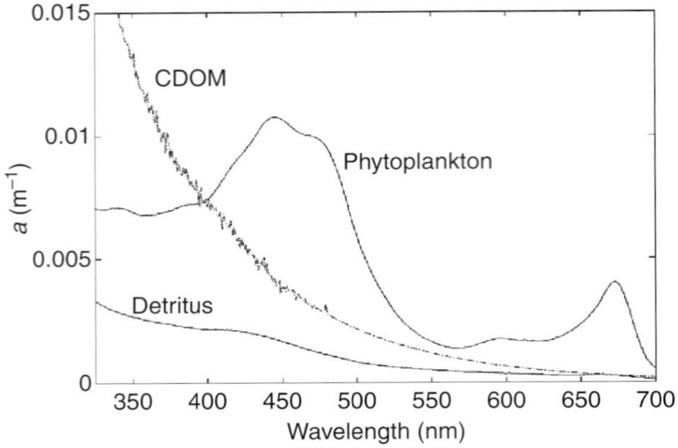

Figure 13.4. Comparison of absorption spectra (m^{-1}) of components in natural seawater, Sargasso Sea, 100 m depth, August 2000. Phytoplankton, particulate detritus and CDOM (colored dissolved organic matter). The CDOM spectrum has been reduced by a factor of 10 to show all components on the same scale. The CDOM spectrum from approximately 475 nm is an exponential fit to the shorter wavelength data.

absorption performs satisfactorily in the visible region but works well only for short wavelength intervals in the UV and over the spectrum as a whole. Comparisons between S values of CDOM spectra are only valid if the same wavelength region is considered (Twardowski *et al.*, 2004). However, similar general trends are reported in CDOM studies where S is considered. Typically, CDOM that has its origin in terrestrial waters has a smaller slope value (indicating a less-sharply curved spectrum) than open ocean CDOM (Nelson and Siegel, 2002; Nelson *et al.*, 2004). The distribution of CDOM in the photic zone of the world ocean is controlled by a balance between solar bleaching and terrestrial input or local production due to heterotrophic activity (Nelson and Siegel, 2002; Nelson *et al.*, 2004). This means the central gyres are depleted in CDOM at the surface while coastal and upwelling zones have higher levels of surface CDOM (Siegel *et al.*, 2002). Coastal zones have the highest levels of particulate load and CDOM absorption due to productivity and terrestrial input, respectively (Siegel *et al.*, 2005b).

Analysis of absorption spectra from the Sargasso Sea and the global ocean suggest that absorption by non-phytoplankton particles is small compared to phytoplankton and CDOM (Figure 13.4; Nelson *et al.*, 1998; Siegel *et al.*, 2005a), and represents less than 15% of the total particle absorption coefficient at 440 nm. Variations in the absorption spectra of non-phytoplankton particulates have not been studied systematically in the global ocean, but inspection of the available data suggests that an exponential function with a small (close to zero) value of the slope parameter may be a fair approximation for the shape of most non-phytoplankton-pigment absorption spectra. In any case, it would appear that the major contribution to ocean colour by non-phytoplankton particles comes from light scattering. In the case of coccolithophore blooms 80% of the total backscattering may be due to calcite from coccoliths (Balch *et al.*, 1996). These blooms can be visible and quantifiable from space.

In many locations (particularly in coastal or inland waters), the non-phytoplankton components can have absorption coefficients comparable to that of phytoplankton pigments in the blue wavelength region. In the global ocean, absorption by CDOM plus particulate detritus represents 30% to greater than 60% of the total non-water absorption coefficient at 440 nm (Siegel *et al.*, 2005a). Because of the nature of CDOM and detrital absorption spectra, these materials have proportionally larger impact upon the optical properties of the water column in the ultraviolet and blue-violet portions of the spectrum. These factors conspire to make retrieval of phytoplankton pigment concentration and absorption from satellite or *in situ* ocean colour sensors difficult, as phytoplankton pigments are not necessarily the dominant factor controlling ocean colour (e.g. Morel and Maritorena, 2001). One approach to solving this problem relies upon measurement of *in vivo* fluorescence of phytoplankton, using *in situ* or remote-sensing techniques (Section 13.3 and Chapter 14, this volume).

Another effective approach to retrieving phytoplankton absorption spectra from space-borne or moored ocean colour sensors takes advantage of the exact relationship between certain apparent optical properties and the absorption and backscattering coefficients of the water column derived by Gordon *et al.* (1988). In this model the ratio of upwelling radiance to downwelling irradiance (known as 'remote-sensing reflectance') is parameterized as a function of the ratio of the backscattering coefficient to the sum of the absorption coefficient and the backscattering coefficient. If the remote-sensing reflection spectrum is known (from satellite or *in situ* optical data), inversion of this function can yield the total absorption and backscattering coefficient spectra. Furthermore, the total absorption spectrum can be empirically partitioned into components representing (for example) CDOM and phytoplankton, by assuming characteristic spectral shapes for each component (Roesler and Perry, 1995; Garver and Siegel, 1997).

This general approach was used with remote-sensing reflectance data from the SeaWiFS satellite instrument by Siegel *et al.* (2002, 2005a) to map the global distribution of CDOM and detritus absorption, and particulate backscattering coefficient. Relationships between phytoplankton pigment absorption, CDOM and detritus absorption, and particulate backscatter reveal patterns that reflect phytoplankton community structure in the open ocean (Siegel *et al.*, 2005a). Similar data were used by Siegel *et al.* (2005b) to estimate the impact of CDOM upon empirical models for retrieving Chl *a* concentrations from ocean colour sensors. In high latitudes retrieval of Chl *a* concentrations via conventional empirical satellite algorithms may be high by a factor of two due to interference by CDOM and particulate detritus absorption, but the semi-analytical technique leads to significant improvement by accounting for the non-phytoplankton components.

Further application of the semi-analytical approach to ocean colour analysis yields new information in other ecologically relevant areas that complement pigment concentration data. For example, Ciotti and Bricaud (2006) used a semi-analytical approach to retrieve phytoplankton size spectra in addition to light absorption parameters from ocean colour data. These and other studies demonstrate how the optical properties of non-phytoplankton components of the water column can be employed to enhance research using phytoplankton pigment data.

13.5 Light-harvesting complexes in Chromophyta, Chlorophyta and Cyanobacteria

It has been 40 years since researchers began to accept that most if not all Chl *a* was complexed with several proteins *in vivo* (compare Thornber *et al.*, 1967a, b; Thornber, 1975; Markwell *et al.*, 1979) and not suspended as free pigments in the lipid bilayer of membranes, as had previously been assumed. This realization shifted the direction of photosynthetic research dramatically, away from a period of intense study of the organic chemistry of pigments and their solvent-dependent absorption and biophysical properties (compare Vernon and Seely, 1966; Isler *et al.*, 1971). Studies of photosynthetic membrane properties became a focus of photosynthetic research (compare Kirk and Tilney-Bassett, 1978; Larkum and Vesk, 2003) and isolation of Chl *a*-protein complexes continues to be coupled to advances in membrane fractionation techniques (compare Green and Parson, 2003). Intact and functional pigment–protein complexes that represent PSI, PSII and the diverse LHCs have been isolated and characterized with some complexes being reconstituted and even crystallized (compare Blankenship, 2002; Green and Parson, 2003; Larkum, 2003). In the 1970s and 1980s, photosynthetic research truly expanded from studies of Chl *a*-Chl *b*-containing protein complexes in higher plants and green algae to the Chl *c*-containing chromophytes (compare Green and Parson, 2003). The comparative study of Chl *a*-proteins in chromophytes has been complicated by the rich diversity of types of Chl *c* present across phytoplankton classes (Zapata *et al.*, 2006). As LHCs began to be identified, their role in determining the spectral signature of phytoplankton (Prézelin and Boczar, 1986) and underlying photoacclimation responses (Prézelin, 1987; Johnsen *et al.*, 1994b, 1997; Jovine *et al.*, 1995; Schofield *et al.*, 1996) began to be appreciated, especially as they applied to the bio-optical studies of aquatic ecosystems (Prézelin *et al.*, 1991; Bidigare *et al.*, 1992; Smith *et al.*, 1992). Knowledge of pigment-dependent photoacclimation strategies, and their direct effects on spectral photosynthesis and the photosynthesis–irradiance relationship, began to be applied to mechanistic field studies and development of spectral models of *in situ* primary productivity (Smith *et al.*, 1989, 1992; Schofield *et al.*, 1991, 1993, 1996; Lutz *et al.*, 1998; Oubelkheir *et al.*, 2005). In the last decade, molecular studies have focused on the apoprotein components of LHCs. Amino acid sequences have been used to assess genetic relationships between taxonomic groups and to develop molecular probes to identify the multigene families that encode for apoproteins and their variants (Triplett *et al.*, 1993; Green and Parson, 2003). Where algal model system studies are especially advanced, genomic studies are focusing on how changing cell conditions, endogenous regulation (e.g. biological clocks), and environmental cues regulate the nuclear transcription, cytoplasmic translation, chloroplast targeting, assembly and insertion of apoproteins into functional LHCs within thylakoid membranes (see Green and Parson, 2003; Nymark *et al.*, 2009).

Chl *a* is universally present in all oxygen-evolving photoautotrophs and the dimers of Chl *a*, P680 (PSII) and P700 (PSI), lie at the heart of the reaction centres of PSII and PSI, respectively. In their unique configuration, both P680 and P700 receive excitation energy from neighbouring Chl *a*-protein complexes that comprise the LHC associated with PSI and PII. There are no other types of accessory Chls, major LHP or phycobilins within the complex of proteins that comprise PSI and II. Research has shown that the structure and function of PSII and I are widely conserved in plants, algae and cyanobacteria. As a result, it is assumed that similar spectral signatures for isolated PSI and PSII complexes are part of the whole cell absorption properties of all phytoplankton.

In the excited state, both P680 and P700 eject an electron to electron acceptors in a process of photosynthetic charge separation (measured as F'_V, see Section 13.3.1 and Chapter 11) and are returned to their original state by receiving electrons from chemical donors, e.g. the water-splitting complex within PSII and the plastocyanin complex linked to PSI. PSII electron flow is coupled to PSI via a series of intermediary electron carriers. Plastoquinone is a mobile intermediary electron carrier and as it passes electrons out of PSII toward a Fe-S complex, it pumps additional protons into the inter-thylakoid space (lumen). Acidification of the lumen during photosynthesis is also enhanced by proton formation during the water splitting reactions of PSII. The proton motive force that results is used to drive cross-membrane production of ATP in the stroma of chloroplasts, where subsequent C-fixation chemistry occurs. Photosynthetic electron flow is complete when electrons coming out of PSI lead to the reduction of NADP to $NADPH_2$ on the stroma side of the thylakoid. Under ideal situations for linear electron flow, three ATPs are formed for every two NADPH. Minimally eight photons are required to generate one oxygen molecule from water; hence the theoretical maximum quantum yield for oxygenic photosynthesis is 0.125. Cyclic electron flow around PSI is also possible, resulting in additional ATP formation but the cessation of NADP reduction.

What varies greatly between phytoplankton pigment groups is the composition of the LHCs that pass radiative excitation energy to the photosystems (Table 13.1). In higher plants and green algae, it is the abundant presence of the light-harvesting Chl *a*-Chl *b* protein complexes that gives these plants their dominant colour. Research on the structure, function and molecular regulation of LH Chl *a*-Chl *b* protein complexes is extensive and has often led the way in advancing similar studies on the highly diverse LH Chl *a*-Chl *c* protein complexes of chromophytes (see Green and Parson, 2003). The amount of LH carotenoids present in various LH Chl *a*-Chl *b* protein complexes of green algae and LH Chl *a*-Chl *c* protein complexes of chromophytes tends to be greater than the LH Chl *a*-Chl *b* protein complexes of higher plants and perhaps related to the lower blue-green light environments in which phytoplankton often thrive. All LH protein complexes containing more than one type of chlorophyll are integrally bound within thylakoid membranes, which is why it takes the action of detergents to fractionate thylakoids and isolate them intact for further study. Many types of LH Chl *a*-protein complexes are preferentially associated with PSII under normal light conditions. Should the photon distribution between PSII and PSI become skewed, LH Chl *a*-Chl *b* proteins of higher plants and green algae are known to redistribute themselves so that a larger fraction becomes associated with PSI (see state transitions, Section 13.3.1 and Section 11.1.2). The biochemical and molecular regulation of this regulatory process is well understood for higher plants and green algae. It is not clear if similar regulatory strategies are employed universally in chromophytes where some LH Chl *a*-Chl *c* complexes have an affinity for PSI.

Another class of carotenoid complexes integrally associated with LHC are those that primarily serve photoprotective functions in high light environments (Chapter 11, this volume). Many of these PPC are capable of photosynthetic light-harvesting functions albeit with much lower efficiency than those of the major LHP associated with LHC. One way to distinguish the two is through photo-acclimation studies, as major LHP tend to increase in abundance as light levels are lowered while the opposite is true for PPC. The first category of PPCs comprise 'fast responding xanthophyll-cycle carotenoids' (response time seconds to hours; violaxanthin ↔ antheraxanthin ↔ zeaxanthin, and diadinoxanthin ↔ diatoxanthin) that undergo epoxydation and de-epoxydation in response to changes in light intensity (Demmig-Adams *et al.*, 2006; Falkowski and Chen, 2003; Section 11.2). Another category of PPCs are the 'slow responding photoprotective carotenoids' (hours to days) which do not transfer light energy to Chl *a* with high efficiency when present and whose abundances tend to increase as light levels become saturating or photoinhibitory for photosynthesis. These absorption properties may allow them to act as photoprotective sunscreens for incoming blue-green light. Examples include diadinoxanthin (Chapter 11.1, Johnsen *et al.* 1997), lutein (DallÓsto *et al.*, 2006), zeaxanthin in prokaryotes (Bidigare *et al.*, 1989b), and the 'parent carotene' β,β-carotene (Chapter 11.2.).

In contrast to the overall majority of carotenoids in phytoplankton that are situated in the thylakoids, some carotenoids are found outside the thylakoid-membranes and may affect the bio-optical properties and/or light utilization of algae significantly. These carotenoids act as photoprotective carotenoids, membrane stabilizers or they accumulate in eyespots. Examples include the accumulation of β,β-carotene and its isomers in globules between thylakoid membranes (*Dunaliella salina*, Orset and Young, 2000) or as secondary photoprotective carotenoids (astaxanthin-esters in lipid globules) outside the chloroplasts in snow algae (Bidigare *et al.*, 1993).

Some LHCs are water soluble and reside in a peripheral placement to other Chl *a*-containing LHCs associated with PS I and PSII. The best known water-soluble LHC that contains a major LH carotenoid is the brick red peridinin-Chl *a*-protein complexes (PCPs) of dinoflagellates (Prézelin and Haxo, 1976; Jovine *et al.*, 1992). PCPs account for the majority of peridinin in dinoflagellates and have characteristic absorption spectra distinct from that of pure peridinin in organic solvent. It is the dimerization of four peridinins within PCPs that extends the excited lifetime of the carotenoids and thereby allows for a 100% transfer efficiency to the monomer of Chl *a* within the same chromophore. PCP has long been known to be key to the photoadaptive strategies of dinoflagellates. PCP is coded for by a nuclear multigene family and many PCP isomers comprise the aggregate of PCP associated with distinct LH Chl *a*-Chl *c*-protein complexes in dinoflagellates. PCP is a model system for biophysical studies of carotenoid energy transfer in general and it has recently been crystallized. Note that 12 to 38% of cellular peridinin may be associated with the fat-soluble light-harvesting antenna (ACP, Johnsen *et al.*, 1997; Table 13.1) and the rest is found in PCP-complex for peridinin-containing dinoflagellates. Some peridinin-containing species, like *Prorocentrum minimum* and a clone of *Amphidinium carterae*, do not contain PCP and all the peridinin is associated with ACPs (Jovine *et al.*, 1995; Johnsen *et al.*, 1997). Such differences also reflect variations in photosynthetic responses between different pigment–protein groups of dinoflagellates (Schofield *et al.*, 1996).

The three water-soluble phycobiliproteins (see Chapter 9, this volume) that are the sole or major LHCs of cyanobacteria, cryptophytes and red algae include the red-coloured phycoerythrin (PE, absorption maximum between 490 and 570 nm), blue-coloured phycocyanin (PC, absorption maximum between 610 and 665 nm) and the violet-coloured allophycocyanin (APC, absorption maximum at 652 nm). APC is present in small quantities only but it serves as the bridge for transferring radiative excitation energy from the more abundant PE and PC to PSII.

The measurement of *in vivo* bio-optical properties in phytoplankton, as presented here, is important for several calculations of light-harvesting and utilization from the pigment level (molecule) to remote sensing of phytoplankton blooms. To interpret differences in pigment chemotaxonomy, eco-physiology and functionality of phytoplankton, *in vivo* bio-optical properties are the link between pigments, pigment-proteins, thylakoids, chloroplasts, cells, and different algal assemblages (see Chapters 11 and 14, this volume).

Currently, *in situ* and remote-sensing techniques benefit from having information on bio-optical signatures from phytoplankton, particles and CDOM from laboratory or *in situ* experiments since it helps with the interpretation of signals (e.g. are cells high-light or low-light acclimated, are cells alive or dead?). One example is given in Chapter 14, where a time-series of a remotely sensed ichthyotoxic phytoplankton bloom is presented in terms of Chl *a*. This bloom was dominated by the dictyochophyte *Verrucophora farcimen* (also known as the raphidophyte *Chattonella* cf. *verruculosa* and now known as *Pseudochattonella farcimen*, Eikrem *et al.*, 2009). The pigment signature of this bloom-forming species was similar to Chl c_3-containing coccolithophorids and dinoflagellates with tertiary chloroplasts (*Karenia brevis*, *K. mikimotoi* and *Karlodinium veneficum*, Edvardsen *et al.*, 2007; Chapter 1, this volume). The pigment signature (from HPLC) and the corresponding group-specific bio-optical signatures can not be derived from remotely sensed data and this shows that combined measurements and expertise are necessary. Future research combining measurements of pigment composition, phylogeny, bio-optical characteristics, molecular build-up of the light-harvesting complexes, electron transfer rates (using Chl *a* fluorescence kinetics such as PAM and FRR techniques), together with molecular methods (especially functional genetics) will greatly improve the understanding of *in situ* and remotely sensed phytoplankton data in the future.

Acknowledgements

We are grateful to Drs. S. Roy and C. A. Llewellyn and two anonymous reviewers who provided helpful comments and suggestions on the manuscript.

Abbreviations

Symbol	Meaning (unit)
LIGHT CLIMATE	
λ	Wavelength (nm)
E(λ)	Spectral irradiance (μmol quanta $m^{-2}s^{-1}nm^{-1}$)
E	Irradiance, PAR, (μmol quanta $m^{-2}s^{-1}$)
ABSORPTION AND SCATTERING	
a	Absorption coefficient (m^{-1})
b	Scattering coefficient (m^{-1})
c	Attenuation coefficient (m^{-1})
Q_a	Efficiency factor for absorption, Eq. (13.1) (dimensionless)
Q_b	Efficiency factor for scattering, Eq. (13.2) (dimensionless)
Q_c	Efficiency factor for attenuation, Eq. (13.2) (dimensionless)
d	Volume-equivalent diameter of a particle, Eq. (13.2) to (13.4) (m)
F(d)	Size distribution of particles, Eq. (13.4)
m	Complex refractive index of particles (relative to the medium) (dimensionless)
n	Real part of refractive index (dimensionless)
n'	Imaginary part of refractive index (dimensionless)
α	Relative size of a particle ($= \pi d/\lambda$) (dimensionless)
ρ	Phase lag suffered by the ray crossing a spherical particle along its diameter ($= 2\alpha\,(n-1)$), Eq. (13.2) (dimensionless)

Symbol	Meaning (unit)
ρ'	Optical thickness of a spherical particle along its diameter (= a_{cm} d or 4 α n'), Eq. (13.1) (dimensionless)
ξ	$n'/(n-1)$, Eq. (13.2) (dimensionless)
a_{cm}	Absorption coefficient of cellular matter (m^{-1})
a_{susp}	Absorption coefficient of a suspension (m^{-1})
a_{sol}	Absorption coefficient of a solution (m^{-1})
Q_a*	Package effect index (= a_{susp}/a_{sol}), Eq. (13.5) (dimensionless)
(N/V)	Cell number density, Eq. (13.3) (m^{-3})
β	Pathlength amplification by a glass-fibre filter (dimensionless)
$a^*_\phi(\lambda)$	*in vivo* Chl *a*-specific absorption coefficient (m^2 mg (Chl $a)^{-1}$)
$\overline{a}^*_\phi$	Light absorbed by cell (all LHC and PS), m^2 mg (Chl $a)^{-1}$, Eq. (13.11)
$a'_\phi(\lambda)$	Reconstructed $a^*_\varphi(\lambda)$ using $a^*_i(\lambda)$ in Eq. (13.6) (m^2 mg (Chl $a)^{-1}$)
$a^*_i(\lambda)$	Reconstructed in vivo absorption coefficient for individual pigment i and c_i, Eq. (13.6) (m^2 mg (Chl $a)^{-1}$)
$\overline{a}^*_i$	Light absorbed by individual pigment i in vivo, Eq. (13.11)
c_i	Concentration of pigment group i
FLUORESCENCE	
DCMU	3-(3,4 dichlorophenyl)-1,1-dimethylurea
$F^*_{PSII}(\lambda)$	Chl *a*-specific PSII scaled fl-ex spectra, m^2 mg (Chl $a)^{-1}$
$\overline{a}^*_{PSII}$	Light absorbed by PSII and LHCII, 400–700 nm, m^2 mg (Chl $a)^{-1}$, Eq. (13.11)
ϕ_F	Quantum yield of Chl *a* fluorescence, emission (685 nm) absorbed quanta^{-1} (Eq. (13.8))
RC	Reaction centre (in PSII or PSI)
LHC	Light-harvesting complex in PSII or PSI (LHCII and LHCI)
PSU	Photosynthetic unit
q_P	Photochemical quenching (PQ, Eq. (13.10))
q_N	Non-photochemical quenching (NPQ)
k	Rate constants for primary photochemistry in PSII RC (P), fluorescence emission (F), and non-radiative dissipation (N), Eq. (13.8)
ϕ^{max}_{PSII}	Maximum quantum yield of PSII-fluorescence in dark acclimated cells (Eq. (13.7))
F_0	ϕ_F at minimum level in dark acclimated cells (Eq. (13.7))
F_M	Fluorescence at maximum level in dark acclimated cells (Eq. (13.7))
F_V	Variable fluorescence in dark acclimated cells (Eq. (13.7))
$\phi^{actinic}_{PSII}$	Quantum yield of photochemistry in PSII in actinic light (F'_V/F'_M, Eq. (13.9))
F'_0	Steady-state fluorescence at a given actinic irradiance (Eq. (13.10))
F'	Flux of in vivo fluorescence emitted by an elementary volume (Eq. (13.10))
F'_M	Fluorescence at maximum level in actinic light
F'_V	Variable fluorescence in actinic light (Eq. (13.9))
Q^-_A	Primary quinone acceptor in PSII
GENERAL	
$\phi_{c\ max}$	Maximum CO_2 quantum yield, CO_2 assimilated absorbed quanta^{-1}
LHP	Light-harvesting (active) pigments

References

Aguirre-Gomez, R., Weeks, A. R. and Boxall, S. R. (2001). The identification of phytoplankton pigments from absorption spectra. *Int. J. Remote Sensing* **22**, 315–38.

Allali, K., Bricaud, A., Babin, M., Morel, A. and Chang, P. (1995). A new method for measuring spectral absorption coefficients of marine particles. *Limnol. Oceanogr.* **40**, 1526–32.

Allali, K., Bricaud, A. and Claustre, H. (1997). Spatial variations in the chlorophyll-specific absorption coefficients of phytoplankton and photosynthetically active pigments in the Equatorial Pacific. *J. Geophys. Res.* **102**, C6, 12413–23.

Babin, M. (2008). Phytoplankton fluorescence: theory, current literature and in situ measurement. In *Real-Time Coastal Observing Systems for Ecosystem Dynamics and Harmful Algal Blooms: Theory, Instrumentation and Modelling*, ed. M. Babin, C. S. Roesler and J. J. Cullen. Paris: UNESCO Publishing, pp. 237–80.

Babin, M., Morel, A., Claustre, H., Bricaud, A., Kolber, Z. and Falkowski, P. G. (1996). Nitrogen- and irradiance-dependent variations of the maximum quantum yield of carbon fixation in eutrophic, mesotrophic and oligotrophic marine systems. *Deep-Sea Res. I* **43**, 1241–72.

Balch, W. M., Kilpatrick, K. A. and Trees, C. C. (1996). The 1991 coccolithophore bloom in the central North Atlantic. 1. Optical properties and factors affecting their distribution. *Limnol. Oceanogr.* **41**, 1669–83.

Bidigare, R. R. and Ondrusek, M. E. (1996). Spatial and temporal variability of phytoplankton pigment distributions in the central equatorial Pacific Ocean. *Deep-Sea Res. II* **43**, 809–33.

Bidigare, R. R., Smith, R. C., Baker, K. S. and Marra, J. (1987). Oceanic primary production estimates from measurements of spectral irradiance and pigment concentrations. *Global Biogeochem. Cycles* **1**, 171–86.

Bidigare, R. R., Morrow, J. H. and Kiefer, D. A. (1989a). Derivative analysis of spectral absorption by photosynthetic pigments in the western Sargasso Sea. *J. Mar. Res.* **47**, 323–41.

Bidigare, R. R., Schofield, O. and Prézelin, B. B. (1989b). Influence of zeaxanthin on quantum yield of photosynthesis of *Synechococcus* clone WH7803 (DC2). *Mar. Ecol. Prog. Ser.* **56**, 177–88.

Bidigare, R. R., Ondrusek, M. E., Morrow, J. H. and Kiefer, D. A. (1990). *In vivo* absorption properties of algal pigments. Ocean Optics X, Proc. *Society of Photo-optical Instrumentation Engineers* **1302**, 290–302.

Bidigare, R. R., Prézelin, B. B. and Smith, R. C. (1992). Bio-optical models and the problems of scaling. In *Primary Productivity and Biogeochemical Cycles in the Sea*, ed. P. G. Falkowski and A. D. Woodhead. New York: Plenum Press, pp. 175–212.

Bidigare, R. R., Ondrusek, M. E., Kennicutt II, M. C., Iturriaga, R., Harvey, H. R., Macko, S. A. and Hoham, R. (1993). Evidence for a photoprotective function for secondary carotenoids of snow algae. *J. Phycol.* **29**, 427–34.

Bidigare, R. R., Latasa, M., Johnson, Z., Barber, R. T., Trees, C. C. and Balch, W. M. (1997). Observations of a *Synechococcus*-dominated cyclonic eddy in open-oceanic waters of the Arabian Sea. Ocean Optics XIII, Proc. *Society of Photo-optical Instrumentation Engineers* **2963**, 260–65.

Blankenship, R. E. (2002). *Molecular Mechanisms of Photosynthesis*. Oxford: Blackwell Science.

Bricaud, A. and Morel, A. (1986). Light attenuation and scattering by phytoplanktonic cells: a theoretical modeling. *Appl. Optics* **25**, 571–80.

Bricaud, A. and Stramski, D. (1990). Spectral absorption coefficients of living phytoplankton and non-algal biogenous matter: A comparison between the Peru upwelling area and the Sargasso Sea. *Limnol. Oceanogr.* **35**, 562–82.

Bricaud A., Morel, A. and Prieur, L. (1983). Optical efficiency factors of some phytoplankters. *Limnol. Oceanogr.* **28**, 816–32.

Bricaud, A., Claustre, H., Ras, J. and Oubelkheir, K. (2004). Natural variability of phytoplanktonic absorption in oceanic waters: influence of the size structure of algal populations. *J. Geophys. Res.* **109**, C11010, doi:10.1029/2004JC002419.

Bricaud, A., Mejia, C., Blondeau-Patissier, D., Claustre, H., Crepon, M. and Thiria, S. (2007). Retrieval of pigment concentrations and size structure of algal populations from their absorption spectra using multilayered perceptrons. *Appl. Optics* **46**, 1251–60.

Butler, W. L. (1962). Absorption of light by turbid materials. *J. Opt. Soc. Am.* **52**, 292–99.

Ciotti, A. M. and Bricaud, A. (2006). Retrievals of a size parameter for phytoplankton and spectral light absorption by colored detrital matter from water-leaving radiances at SeaWiFS channels in a continental shelf region off Brazil. *Limnol. Oceanogr. Methods* **4**, 237–53.

Claustre, H., Bricaud, A., Babin, M., Bruyant, F., Guillou, L., Le Gall, F. and Partensky, F. (2002). Diel variations in *Prochlorococcus* optical properties. *Limnol. Oceanogr.* **47**, 1637–47.

Cleveland, J. S. and Weidemann, A.-D. (1993). Quantifying absorption by marine particles: a multiple scattering correction for glass-fiber filters. *Limnol. Oceanogr.* **38**, 1321–27.

Chazottes, A., Bricaud, A., Crépon, M. and Thiria, S. (2006). Statistical analysis of a data base of absorption spectra of phytoplankton and pigment concentrations using self-organizing maps. *Appl. Optics* **45**, 8102–15.

Culver, M. E., Davis, R. F. and Perry, M. J. (1994). Instrumental considerations for deriving spectral photosynthetic absorption coefficients from total phytoplankton absorption. Ocean Optics XII. Proc. *Society of Photo-optical Instrumentation Engineers* **2258**, 123–33.

Dall'Osto, L., Lico, C., Alric, J., Giuliano, G., Havaux, M. and Bassi, R. (2006). Lutein is needed for efficient chlorophyll triplet quenching in the major LHCII antenna complex of higher plants and effective photoprotection in vivo under strong light. *BMC Plant Biol.* **6**, doi:10.1186/1471–2229–6.32.

Demmig-Adams, B., Adams III, W. W. and Mattoo, A. (eds.) (2006). *Photoprotection, Photoinhibition, Gene Regulation and Environment*. Dordrecht: Springer, 380 pp.

D'Sa, E., Lohrenz, S. E., Carroll, C. L. and Fein, H. (1998). Liquid waveguide capillary flow cell for determining absorption of scattering suspensions: comparisons with an integrating sphere. *Ocean Optics XIV*, Proc. Society of Photo-optical Instrumentation Engineers.

Edvardsen, B., Eikrem, W., Johnsen, G., Naustvoll, L., Riisberg, I., Shalchian-Tabrizi, K. and Throndsen, J. (2007). *Verrucophora farcimen* gen. et sp. nov. (Dictyochophyceae, Heterokonta) – a bloom forming ichthyotoxic flagellate from the Skagerrak, Norway. *J. Phycol.* **43**, 1054–70.

Eikrem, W., Edvardsen, B. and Throndsen, J. (2009). Renaming *Verrucophora farcimen* Eikrem, Edvardsen et Throndsen. *Phycol. Res.* **57**, 170.

Eisner, L. B., Twardowski, M. S. and Cowles, T. J. (2003). Resolving phytoplankton photoprotective: photosynthetic carotenoid ratios on fine scales using in situ spectral absorption measurements. *Limnol. Oceanogr.* **48**, 632–46.

Falkowski, P. G. and Chen, Y. B. (2003). Photoacclimation of light harvesting systems in eukaryotic algae. In *Light-Harvesting Antennas in Photosynthesis*, ed. B. R. Green and W. W. Parson. Dordrecht: Kluwer Academic Publishers, pp. 423–47.

Falkowski, P. G. and Raven, J. A. (1997). *Aquatic Photosynthesis.* Malden: Blackwell Science.

Falkowski, P. G., Green, R. and Kolber, Z. (1995). Light utilization and photoinhibition of photosynthesis in marine phytoplankton. In *Photoinhibition of Photosynthesis: from Molecular Mechanisms to the Field*, ed. N. R. Baker and J. R. Boyer. Oxford: Bios Scientific Publishers, pp. 407–32.

Faust, M. A. and Norris, K. H. (1985). In vivo spectrophotometric analysis of photosynthetic pigments in natural populations of phytoplankton. *Limnol. Oceanogr.* **30**, 1316–22.

Ferrari, G. M. and Tassan, S. (1999). A method for removal of light absorption by phytoplankton pigments using chemical oxidation. *J. Phycol.* **35**, 1090–98.

Ficek, D., Kaczmarek, S., Ston-Egiert, J., Wozniak, B., Majchrowski, R. and Dera, J. (2004). Spectra of light absorption by phytoplankton pigments in the Baltic; conclusions to be drawn from a Gaussian analysis of empirical data. *Oceanologia* **46**, 533–55.

Garver, S., Siegel, D. A. and Mitchell, B. G. (1994). Variability in near-surface particulate absorption spectra: what can a satellite ocean color imager see? *Limnol. Oceanogr.* **39**, 1349–67.

Garver, S. A. and Siegel, D. A. (1997). Inherent optical property inversion of ocean color spectra and its biogeochemical interpretation. 1. Time series from the Sargasso Sea. *J. Geophys. Res.* **102**, 18607–25.

Genty, B., Briantais, J. M. and Baker, N. R. (1989). The relationship between the quantum yield of photosynthetic electron transport and quenching of chlorophyll fluorescence. *Biochim. Biophys. Acta* **990**, 87–92.

Gordon, H. R., Brown, O. B., Evans, R., Brown, J. W., Smith, R. C., Baker, K. S. and Clark, D. K. (1988). A semi-analytical radiance model of ocean color. *J. Geophys. Res.* **93**, 10909–24.

Govindjee (1995). Sixty-three years since Kautsky: Chlorophyll *a* fluorescence. *Aust. J. Plant. Physiol.* **22**, 131–60.

Green, B. R. and Parson, W. W. (eds.) (2003). *Light-Harvesting Antennas in Photosynthesis.* Dordrecht: Kluwer Academic Publishers.

Hancke, T. B., Hancke, K., Johnsen, G. and Sakshaug, E. (2008a). Rate of O_2 production derived from PAM fluorescence: Testing three bio-optical approaches against measured O_2 production rate. *J. Phycol.* **44**, 803–13.

Hancke, K., Hancke, T., Olsen, L., Johnsen, G. and Gluud, R. (2008b). Temperature effects on microalgae photosynthesis-light responses measured by O_2-evolution, pulse amplitude modulated (PAM) fluorescence and ^{14}C-assimilation. *J. Phycol.* **44**, 501–14.

Haxo, F. T. (1985). Photosynthetic action spectrum of the coccolithophorid, *Emiliania huxleyi* (Haptophyceae): 19'-Hexanoyloxyfucoxanthin as antenna pigment. *J. Phycol.* **21**, 282–87.

Hoepffner, N. and Sathyendranath, S. (1993). Determination of the major groups of phytoplankton pigments from the absorption spectra of total particulate matter. *J. Geophys. Res.* **98**, 22789–803.

Iturriaga, R., Mitchell, B. G. and Kiefer, D. A. (1988). Microphotometric analysis of individual particles absorption spectra. *Limnol. Oceanogr.* **33**, 128–35.

Isler, O., Gutmann, H. and Solms, U. (1971). *Carotenoids.* Basel: Birkhäuser.

Johnsen, G. and Hegseth, E. N. (1991). Photoadaptation of sea-ice microalgae in the Barents Sea. *Polar Biol.* **11**, 179–84.

Johnsen, G. and Sakshaug, E. (1993). Bio-optical characteristics and photoadaptive responses in the toxic and bloom-forming dinoflagellates *Gyrodinium aureolum, Gymnodinium galatheanum*, and two strains of *Prorocentrum minimum. J. Phycol.* **29**, 627–42.

Johnsen, G. and Sakshaug, E. (2007). Bio-optical characteristics of PSII and PSI in 33 species (13 pigment groups) of marine phytoplankton, and the relevance for PAM and FRR fluorometry. *J. Phycol.* **43**, 1236–51.

Johnsen, G., Nelson, N. B., Jovine, R. V. M. and Prézelin, B. B. (1994a). Chromoprotein- and pigment-dependent modeling of spectral light absorption in two dinoflagellates, *Prorocentrum minimum* and *Heterocapsa pygmaea. Mar. Ecol. Prog. Ser.* **114**, 245–58.

Johnsen, G., Samset, O., Granskog, L. and Sakshaug, E. (1994b). *In vivo* absorption characteristics in 10 classes of bloom-forming phytoplankton: taxonomic characteristics and responses to photoadaptation by means of discriminant and HPLC analysis. *Mar. Ecol. Prog. Ser.* **105**, 149–57.

Johnsen, G., Prézelin, B. B. and Jovine, R. V. M. (1997). Fluorescence excitation spectra and light utilization in two red tide dinoflagellates. *Limnol. Oceanogr.* **42**, 1166–77.

Jovine, R. V. M., Johnsen, G. and Prézelin, B. B. (1995). Isolation of membrane bound light-harvesting-complexes from the dinoflagellates *Heterocapsa pygmaea* and *Prorocentrum minimum. Photosynth. Res.* **44**, 127–38.

Jovine, R., Triplett, E. L., Nelson, N. B. and Prézelin, B. B. (1992). Quantification of chromophore pigments, apoprotein abundance and isoelectric variants of peridinin-chlorophyll *a*-protein complexes (PCPs) in the dinoflagellate *Heterocapsa pygmaea*, grown under variable light conditions. *Plant Cell Physiol.* **33**, 733–41.

Kirk, J. T. O. (1980). Spectral absorption properties of natural waters: Contribution of the soluble and particulate fractions to light absorption in some inland waters of southeastern Australia. *Aust. J. Mar. Freshwat. Res.* **31**, 287–96.

Kirk, J. T. O. (1994). *Light and Photosynthesis in Aquatic Ecosystems.* 2nd edn. Cambridge: Cambridge University Press.

Kirk, J. T. O. (1997). Point-source integrating-cavity absorption meter: Theoretical principles and numerical modeling. *Appl. Optics* **36**, 6123–28.

Kirk, J. T. O. and Tilney-Bassett, R. A. E. (1978). *The Plastids.* Amsterdam: Elsevier/North-Holland Biomedical Press.

Kirkpatrick, G., Millie, D. F., Moline, M. A. and Schofield, O. (2000). Optical discrimination of a phytoplankton species in natural mixed populations. *Limnol. Oceanogr.* **45**, 467–71.

Kishino, M., Takahashi, M., Okami, N. and Ichimura, S. (1985). Estimation of the spectral absorption coefficients of phytoplankton in the sea. *Bull. Mar. Sci.* **37**, 634–42.

Kohonen, T. (1984). *Self-Organizing Maps*. Berlin: Springer.

Kopf, U. and Heinze, J. (1984). 2,7-Bis (diethylamino) phenazoxonium chloride as a quantum counter for emission measurements between 240 and 700 nm. *Anal. Chem.* **56**, 1931–35.

Krause, G. H. and Jahns, P. (2003). Pulse amplitude modulated chlorophyll fluorometry and its application in plant science. In *Light-Harvesting Antennas in Photosynthesis*, ed. B. R. Green and W. W. Parson. Dordrecht: Kluwer Academic Publishers, pp. 373–99.

Larkum, A. W. (2003). Light-harvesting systems in algae. In *Photosynthesis in Algae*, ed. A. W. Larkum, S. E. Douglas and J. A. Raven. Dordrecht: Kluwer Academic Publishers, pp. 277–304.

Larkum, A. W. and Vesk, M. (2003). Algal plastids. In *Photosynthesis in Algae*, ed. A. W. Larkum, S. E. Douglas and J. A. Raven. Dordrecht: Kluwer Academic Publishers, pp. 11–28.

Laurion, I., Blouin, F. and Roy, S. (2003). The quantitative filter technique for measuring phytoplankton absorption: Interference by MAAs in the UV waveband. *Limnol. Oceanogr. Methods* **1**, 1–9.

Lee, M. E. and Lewis, M. R. (2003). A new method for the measurement of the optical volume scattering function in the upper ocean. *J. Atmosph. Oceanic Technol.* **20**, 563–71.

Lutz, V. A., Sathyendranath, S., Head, E. J. H. and Li, W. K. W. (1998). Differences between *in vivo* absorption and fluorescence excitation spectra in natural samples of phytoplankton. *J. Phycol.* **34**, 214–27.

Lutz, V. A., Sathyendranath, S., Head, E. J. H. and Li, W. K. W. (2001). Changes in the *in vivo* absorption and fluorescence excitation spectra with growth irradiance in three species of phytoplankton. *J. Plankton Res.* **23**, 555–69.

Marra, J., Trees, C. C., Bidigare, R. R. and Barber, R. T. (2000). Pigment absorption and quantum yields in the Arabian Sea. *Deep-Sea Res. II* **47**, 1279–99.

Malkin, S. and Kok, B. (1966). Fluorescence induction studies in isolated chloroplasts. I. Number of components involved in the reaction and quantum yields. *Biochim. Biophys. Acta* **126**, 413–32.

Markwell, J. P., Thornber, J. P. and Boggs, R. T. (1979). Higher plant chloroplasts: evidence that all chlorophyll exists as chlorophyll-protein complexes. *Proc. Natl. Acad. Sci. USA* **76**, 1233–35.

Mie, G. (1908). Beiträge zur Optik trüber Medien, speziell kolloidalen Metallösungen. *Ann. Phys.* **25**, 377.

Miller, R. L., Belz, M., Del Castillo, C. and Trzaska, R. (2002). Determining CDOM absorption spectra in diverse coastal environments using a multiple pathlength, liquid core waveguide system. *Contin. Shelf Res.* **22**, 1301–10.

Millie, D. F., Schofield, O. M., Kirkpatrick, G. J., Johnsen, G., Tester, P. A. and Vinyard, B. T. (1997). Detection of harmful algal blooms using photopigments and absorption signatures: a case study of the Florida red-tide dinoflagellate *Gymnodinium breve*. *Limnol. Oceanogr.* **42**, 1240–51.

Millie, D. F., Schofield, O. M., Kirkpatrick, G. J., Johnsen, G. and Evens, T. J. (2002). Using absorbance and fluorescence spectra to discriminate microalgae. *J. Phycol.* **37**, 313–22.

Mishchenko, M. I., Hovenier, J. W. and Travis, L. D. (eds.) (2000). *Light Scattering by Non-Spherical Particles: Theory, Measurements, and Applications*. San Diego: Academic Press.

Mitchell, B. G. and Kiefer, D. A. (1988). Chlorophyll *a* specific absorption and fluorescence excitation spectra for light-limited phytoplankton. *Deep-Sea Res.* **35**, 639–63.

Mitchell, B. G. (1990). Algorithms for determining the absorption coefficient of aquatic particulates using the quantitative filter technique (QFT). Ocean Optics X, Proc. *Society of Photo-optical Instrumentation Engineers* **1302**, 137–48.

Mitchell, B. G., Bricaud, A., Carder, K., Cleveland, J., Ferrari, G., Gould, R., Kahru, M., Kishino, M., Maske, H., Moisan, T., Moore, L., Nelson, N., Phinney, D., Reynolds, R., Sosik, H., Stramski, D., Tassan, S., Trees, C., Weidemann, A., Wieland, J. and Vodacek, A. (2000). Determination of spectral absorption coefficients of particles, dissolved material, and phytoplankton for discrete water samples. In *Ocean Optics Protocols for Satellite Ocean Color Sensor Validation*, Rev. 2. NASA/TM-2000–209966, ed. G. S. Fargion and J. L. Mueller. Greenbelt: NASA Goddard Space Flight Center, pp. 125–53.

Mitchell, B. G., Kahru, M., Wieland, J. and Stramska, M. (2002). Determination of spectral absorption coefficients of particles, dissolved material, and phytoplankton for discrete water samples. In *Ocean Optics Protocols for Satellite Ocean Color Sensor Validation*, NASA/TM-2002–210004-/Rev.3-Vol. 2, ed. G. S. Fargion and J. L. Mueller. Greenbelt: NASA Goddard Space Flight Center, pp. 231–57.

Moberg, L., Karlberg, B., Sorensen, K. and Kallqvist, T. (2002). Assessment of phytoplankton class abundance using absorption spectra and chemometrics. *Talanta* **56**, 153–60.

Moore, L. R., Goericke, R. and Chisholm, S. W. (1995). Comparative physiology of *Synechococcus* and *Prochlorococcus*: Influence of light and temperature on growth, pigments, fluorescence and adaptive properties. *Mar. Ecol. Prog. Ser.* **116**, 259–75.

Morel, A. (2008). Introduction to optical properties in the sea: theoretical aspects. In *Real-Time Coastal Observing Systems for Ecosystem Dynamics and Harmful Algal Blooms: Theory, Instrumentation and Modelling*, ed. M. Babin, C. S. Roesler and J. J. Cullen. Paris: UNESCO Publishing, pp. 109–51.

Morel, A. and Bricaud, A. (1981). Theoretical results concerning light absorption in a discrete medium, and application to specific absorption of phytoplankton. *Deep-Sea Res.* **28**, 1375–93.

Morel, A. and Bricaud, A. (1986). Inherent properties of algal cells including picoplankton: theoretical and experimental results. In *Photosynthetic Picoplankton*, ed. T. Platt and W. K. W. Li. *Can. Bull. Fish. Aquat. Sci.* **214**, pp. 521–59.

Morrow, J. H., Chamberlain, W. S. and Kiefer, D. A. (1989). A two-component description of spectral absorption by marine particles. *Limnol. Oceanogr.* **34**, 1500–09.

Morel, A. and Maritorena, S. (2001). Bio-optical properties of oceanic waters: A reappraisal. *J. Geophys. Res.* **106**, 7163–80.

Nedbal, L. and Koblízek, M. (2006). Chlorophyll fluorescence as a reporter on *in vivo* electron transport and regulation in plants. In *Chlorophylls and Bacteriochlorophylls: Biochemistry, Biophysics, Functions and Applications*, ed. B. Grimm, R. J. Porra, W. Rüdiger and H. Scheer. Dordrecht: Springer, pp. 507–19.

Nelson, N. B. and Prézelin, B. B. (1990). Chromatic light effects and physiological modeling of absorption properties of *Heterocapsa pygmaea* (= *Glenodinium* sp.). *Mar. Ecol. Prog. Ser.* **63**, 37–46.

Nelson, N. B. and Siegel, D. A. (2002). Chromophoric DOM in the open ocean. In *Biogeochemistry of Marine Dissolved Organic Matter*, ed. D. A. Hansell and C. A. Carlson. San Diego: Academic Press, pp. 547–78.

Nelson, N. B., Prézelin, B. B. and Bidigare, R. R. (1993). Phytoplankton light absorption and the package effect in California coastal waters. *Mar. Ecol. Prog. Ser.* **94**, 217–27.

Nelson, N. B., Siegel, D. A. and Michaels, A. F. (1998). Seasonal dynamics of colored dissolved material in the Sargasso Sea. *Deep-Sea Res. I* **45**, 931–57.

Nelson, N. B., Carlson, C. A. and Steinberg, D. K. (2004). Production of chromophoric dissolved organic matter by Sargasso Sea microbes. *Mar. Chem.* **89**, 273–87.

Neori, A., Vernet, M., Holm-Hansen, O. and Haxo, F. T. (1988). Comparison of chlorophyll far-red fluorescence excitation spectra with photosynthetic oxygen action spectra for photosystem II in algae. *Mar. Ecol. Prog. Ser.* **44**, 297–302.

Neumuller, M., Cunningham, A. and McKee, D. (2002). Assessment of a microscopic photobleaching technique for measuring the spectral absorption efficiency of individual phytoplankton cells. *J. Plankton Res.* **24**, 741–46.

Nymark, M., Valle, K. C., Brembu, T., Hancke, K., Winge, P., Andresen, K., Johnsen, G. and Bones, A. M. (2009). An integrated analysis of molecular acclimation to high light in the marine diatom *Phaeodactylum tricornutum*. *PloS ONE*. **11**, e7743. doi:10.1371/journal.pone.0007743.

Orset, S. and Young, A. (2000). Exposure to low irradiances favors the synthesis of 9-cis β,β-carotene in *Dunaliella salina* (Teod.). *Plant Physiol.* **122**, 609–17.

Oubelkheir, K., Claustre, H., Babin, M. and Sciandra, A. (2005). The comparative bio-optical and biogeochemical properties of contrasted trophic regimes. *Limnol. Oceanogr.* **50**, 1795–809.

Parson, W. W. and Nagarajan, V. (2003). Optical spectroscopy in photosynthetic antennas. In *Light-Harvesting Antennas in Photosynthesis*, ed. B. R. Green and W. W. Parson. Dordrecht: Kluwer Academic Publishers, pp. 83–127.

Pope, R. M., Weidemann, A. D. and Fry, E. S. (2000). Integrating Cavity Absorption Meter measurements of dissolved substances and suspended particles in ocean water. *Dynam. Atmos. Oceans* **31**, 307–20.

Prézelin, B. B. (1987). The photosynthetic physiology of dinoflagellates. In *The Biology of Dinoflagellates* ed. F. J. R. Taylor. Oxford: Blackwell Scientific, pp. 174–223.

Prézelin, B. B. and Boczar, B. A. (1986). Molecular bases of phytoplankton spectral properties, and their potential applications to studies in optical oceanography. In *Progress in Phycological Research*, ed. F. Round and D. Chapman. Bristol: Biopress Limited, pp. 349–464.

Prézelin, B. B. and Haxo, F. T. (1976). Purification and characterization of peridinin-chlorophyll *a*-proteins from the marine dinoflagellates *Glenodinium* sp. and *Gonyaulax polyedra*. *Planta* **128**, 133–41.

Prézelin, B. B., Tilzer, M. M., Schofield, O. and Haese, C. (1991). Control of the production process of phytoplankton by the physical structure of the aquatic environment, with special reference to its optical properties. *Aquat. Sci.* **53**, 136–86.

Raven, J. A. and Geider, R. J. (2003). Adaptation, acclimation and regulation in algal photosynthesis. In *Photosynthesis in Algae*, ed. A. W. Larkum, S. E. Douglas and J. A. Raven. Dordrecht: Kluwer Academic Publishers, pp. 385–412.

Roesler, C. S. and Perry, M. J. (1995). *In-situ* phytoplankton absorption, fluorescence emission, and particulate backscattering spectra determined from reflectance. *J. Geophys. Res.* **100**, 13279–94.

Roesler, C. S., Perry, M. J. and Carder, K. L. (1989). Modeling *in situ* phytoplankton absorption from total absorption spectra in productive inland marine waters. *Limnol. Oceanogr.* **34**, 1510–23.

Röttgers, R., Schönfeld, W., Kipp, P. R. and Doerffer, R. (2005). Practical test of a point-source integrating-cavity absorption meter: the performance of different collector assemblies. *Appl. Optics* **44**, 5549–60.

Sakshaug, E. and Johnsen, G. (2005). Absorption, fluorescence excitation and photoacclimation in phytoplankton. In *Algal cultures, Analogues of Blooms and Applications*, ed. D. V. Subba Rao. Enfield: Science Publishers, pp. 687–714.

Sakshaug, E., Johnsen, G., Andresen, K. and Vernet, M. (1991). Modeling of light-dependent algal photosynthesis and growth: Experiments with the Barents Sea diatoms *Thalassiosira nordenskioeldii* and *Chaetoceros furcellatus*. *Deep-Sea Res.* **38**, 415–30.

Schofield, O., Prézelin, B. B., Bidigare, R. R. and Smith, R. C. (1993). In situ photosynthetic quantum yield: Correspondence to hydrographic and optical variability within the Southern California Bight. *Mar. Ecol. Prog. Ser.* **93**, 25–37.

Schofield, O., Prézelin, B. B. and Johnsen, G. (1996). Wavelength-dependency of maximum quantum yield of carbon fixation for two red tide dinoflagellates, *Heterocapsa pygmaea* and *Prorocentrum minimum*. *J. Phycol.* **32**, 574–83.

Schofield, O., Prézelin, B. B., Smith, R. C., Stegmann, P., Nelson, N. B., Lewis, M. R. and Baker, K. S. (1991). Variability in spectral and nonspectral measurements of photosynthetic light utilization efficiencies. *Mar. Ecol. Prog. Ser.* **78**, 253–71.

Shibata, K., Benson, A. A. and Calvin, M. (1954). The absorption spectra of suspensions of living microorganisms. *Biochim. Biophys. Acta* **15**, 461–70.

Siegel, D. A., Maritorena, S., Nelson, N. B., Hansell, D. A. and Lorenzi-Kayser, M. (2002). Global ocean distribution and dynamics of colored dissolved and detrital organic materials. *J. Geophys. Res.* **107**, 3228. doi:10.1029/2001JC000965.

Siegel, D. A., Maritorena, S., Nelson, N. B. and Behrenfeld, M. J. (2005a). Independence and interdependencies of global ocean color properties; Reassessing the bio-optical assumption. *J. Geophys. Res.* **110**, C07011, doi:10.1029/2004JC002527.

Siegel, D. A., Maritorena, S., Nelson, N. B., Behrenfeld, M. J. and McClain, C. R. (2005b). Colored dissolved organic matter and the satellite-based characterization of the ocean biosphere. *Geophys. Res. Lett.* **32**, L20605, doi:10.1029/2005GL024310.

Smith, R. C., Prézelin, B. B., Bidigare, R. R. and Baker, K. S. (1989). Bio-optical modeling of photosynthetic production in coastal waters. *Limnol. Oceanogr.* **34**, 1524–44.

Smith, R. C., Prézelin, B. B., Baker, K., Bidigare, R. R., Boucher, N. P., Coley, T., Karentz, D., MacIntyre, S., Matlick, H. A. and Menzies, D. (1992). Ozone depletion: Ultraviolet radiation and phytoplankton biology in Antarctic waters. *Science* **255**, 952–59.

Sosik, H. (2008). Characterizing seawater constituents from optical properties. In *Real-Time Coastal Observing Systems for Ecosystem Dynamics and Harmful Algal Bloom: Theory, Instrumentation and Modelling*, ed. M. Babin, C. S. Roesler and J. J. Cullen. Paris: UNESCO Publishing, pp. 281–329.

Staehr, P. A. and Cullen, J. J. (2003). Detection of *Karenia mikimotoi* by spectral absorption signatures. *J. Plankton Res.* **25**, 1237–49.

Stramski, D. and Reynolds, R. A. (1993). Diel variations in the optical properties of a marine diatom. *Limnol. Oceanogr.* **38**, 1347–64.

Stramski, D., Bricaud, A. and Morel, A. (2001). Modeling the inherent optical properties of the ocean based on the detailed composition of the planktonic community. *Appl. Optics* **40**, 2929–45.

Stuart, V., Sathyendranath, S., Platt, T., Maass, H. and Irwin, B. D. (1988). Pigment and species composition of natural phytoplankton populations: effect on the absorption spectra. *J. Plankton Res.* **20**, 187–217.

Suggett, D. J., MacIntyre, H. and Geider, R. J. (2004). Evaluation of biophysical and optical determinations of light absorption by photosystem II in phytoplankton. *Limnol. Oceanogr. Methods* **2**, 316–32.

Suggett, D. J., Le Floć H. E., Harris, G., Leonardos, N. and Geider, R. J. (2007). Different strategies of photoacclimation by two strains of *Emiliania huxleyi* (Haptophyta). *J. Phycol.* **43**, 1209–22.

Tassan, S. and Ferrari, G. M. (1995). An alternative approach to absorption measurements of aquatic particles retained on filters. *Limnol. Oceanogr.* **40**, 1358–68.

Thornber, J. P. (1975). Chlorophyll-proteins: light-harvesting and reaction center components of plants. *Ann. Rev. Plant Physiol.* **26**, 127–58.

Thornber, J. P., Gregory, R. F. P., Smith, C. A. and Bailey, J. L. (1967a). Studies on the nature of chloroplast lamella. 1. Preparation and some properties of two chlorophyll-protein complexes. *Biochemistry* **6**, 391–96.

Thornber, J. P., Stewart, J. C., Hatton, M. W. C. and Bailey J. L. (1967b). Studies on the nature of chloroplast lamella. 2. Chemical composition and further physical properties of two chlorophyll-protein complexes. *Biochemistry* **6**, 2006–14.

Triplett, E. L., Jovine, V. M., Govind, N. S., Roman, S. J., Chang, S. S. and Prézelin, B. B. (1993). Characterization of two full length cDNA sequences encoding for apoproteins of peridinin-chlorophyll *a*-protein complexes. *Molec. Mar. Biol. Biotech.* **33**, 733–41.

Twardowski, M. S., Boss, E., Sullivan, J. M. and Donaghay, P. L. (2004). Modeling the spectral shape of absorption by chromophoric dissolved organic matter. *Mar. Chem.* **89**, 69–88.

Van de Hulst, H. C. (1957). *Light scattering by small particles*. New York: John Wiley & Sons.

Vernon, L. P. and Seely, G. R. (1966). *The Chlorophylls*. New York: Academic Press.

Volten, H., de Haan, J. F., Hovenier, J. W., Schreurs, R., Vassen, W., Dekker, A. G., Hoogenboom, H. J., Charlton, F. and Wouts, R. (1998). Laboratory measurements of angular distributions of light scattered by phytoplankton and silt. *Limnol. Oceanogr.* **43**, 1180–97.

Voss, K. J., Balch, W. M. and Kilpatrick, K. A. (1998). Scattering and attenuation properties of Emiliania huxleyi cells and their detached coccoliths. *Limnol. Oceanogr.* **43**, 870–76.

Weidemann, A. D. and Bannister, T. T. (1986). Absorption and scattering coefficients in Irondequoit Bay. *Limnol. Oceanogr.* **7**, 207–17.

Zapata, M., Garrido, J. and Jeffrey, S. W. (2006). Chlorophyll c pigments: current status. In *Chlorophylls and Bacteriochlorophylls: Biochemistry, Biophysics, Functions and Applications*, ed. B. Grimm, R. J. Porra, W. Rüdiger and H. Sheer. Dordrecht: Springer, pp. 39–53.

14

Optical monitoring of phytoplankton bloom pigment signatures

GEIR JOHNSEN, MARK A. MOLINE, LASSE H. PETTERSSON, JAMES PINCKNEY, DMITRY V. POZDNYAKOV, EINAR SKARSTAD EGELAND AND OSCAR M. SCHOFIELD

14.1 Introduction

The absorption of light by algal pigments determines the cellular absorption of phytoplankton and thus contributes to the *in situ* optical signatures of coastal and offshore waters. This is the basis of a range of bio-optical approaches used for monitoring phytoplankton distribution (taxa and biomass) and is the focus of this chapter. Details regarding bio-optical signatures of phytoplankton and their spectral absorption, scattering and fluorescence characteristics are covered in Chapter 13, this volume. Details of how phytoplankton adjust their pigments in response to variation in light climate are reviewed in Chapter 11, this volume.

Phytoplankton blooms cover spatial scales that vary from patches of 1 m^2 to large blooms covering more than 1×10^6 km^2 (Franks, 1997; Smyth *et al.*, 2004; Schofield *et al.*, 2008). Related to the spatial scale is the temporal variability of these blooms (from minutes to years), depending on the physical and biological processes at a given location. The development of techniques for monitoring phytoplankton blooms at different geographical and temporal scales has evolved rapidly in recent years (Kahru and Brown, 1997; Schofield *et al.*, 1999; Babin *et al.*, 2008). Because of the wide range in scale, different methods and approaches (different sensors and corresponding platforms) are needed for monitoring water masses and associated phytoplankton blooms as a function of environmental change (i.e. temperature, salinity, circulation and light regime), biogeochemical cycling, eutrophication, ocean acidification and pollution. Globally, monitoring of phytoplankton bloom dynamics is important to estimate changes in primary productivity affecting carbon and nutrient cycling in the world oceans.

At a regional scale, fisheries, aquaculture and tourism can be affected regularly by harmful algal blooms (HABs). During HABs, coastal regions are dependent on efficient early warning and monitoring networks for rapid decision-making and response with regard to both algal toxins (e.g. moving away pens to prevent fish death or stopping the distribution of toxic mussels) and O_2 depletion (affecting all

Phytoplankton Pigments: Characterization, Chemotaxonomy and Applications in Oceanography, eds. Suzanne Roy, Carole A. Llewellyn, Einar Skarstad Egeland and Geir Johnsen. Published by Cambridge University Press.

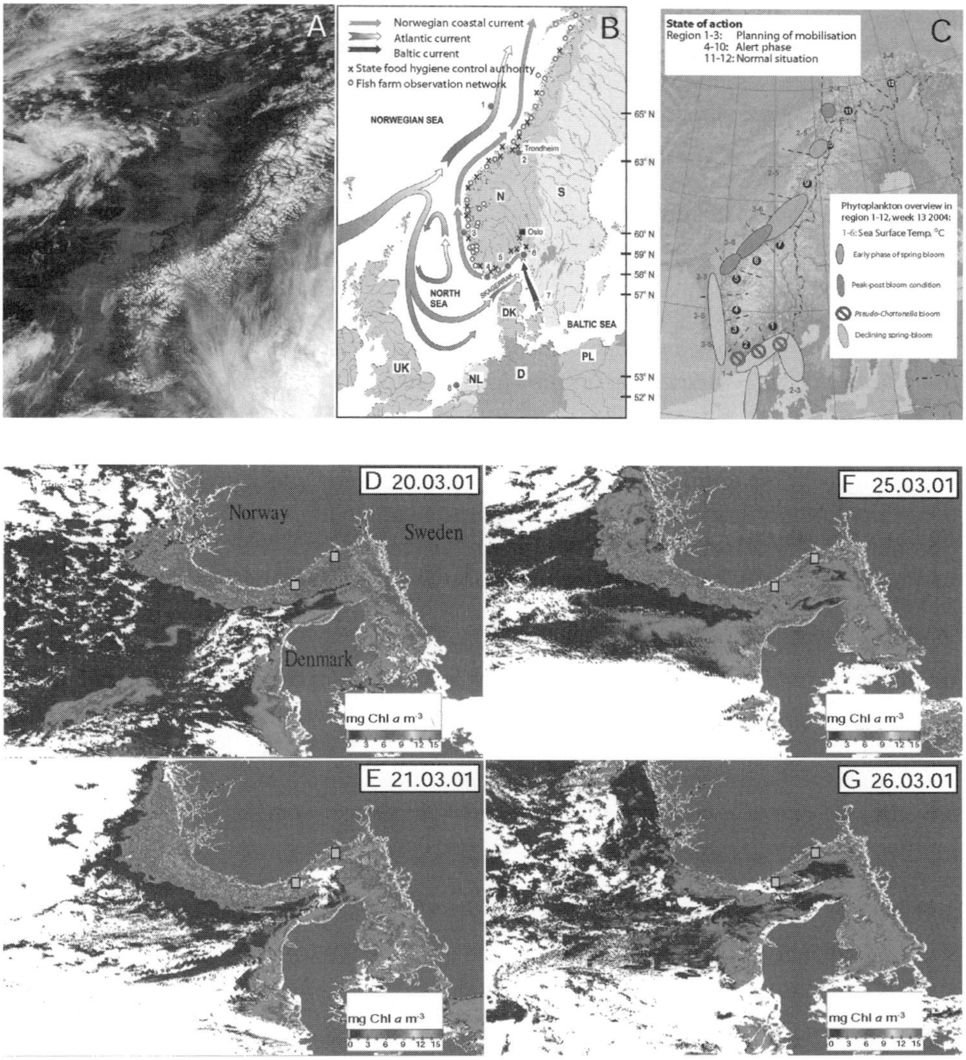

Figure 14.1. A: Phytoplankton bloom along the Norwegian coast 15 June 1998, apparently dominated by the coccolithophorid *Emiliania huxleyi*. Note that distribution of bloom follows the bottom topography and current patterns (B). SeaWiFS image (OrbView-2 satellite). NASA/Goddard Space Flight Centre. B: The Norwegian buoy (1–8) network system and sites for weekly collection of water samples for identification and enumeration of phytoplankton and mussel samples in 1993 (Johnsen *et al.*, 1997). The network is still operative, but the buoy system has been replaced by satellite images (D–G). C: Operative phytoplankton bloom situation in last week of March 2004 indicating spring-bloom in mid-Norway, see http://algeinfo.imr.no, providing the current phytoplankton (incl. HAB) situation in Norwegian waters. The coastline is divided into 12 different regions from south to north. Red numbers along the coast denote sea-surface temperature.

organisms in seawater: Figure 14.1, Tables 14.1–14.3, Appendix 14A; see Sections 14.5 and 14.6). For monitoring purposes, the inherent optical properties, which include light absorption and scattering of phytoplankton cells, are the basis for *in situ* and remote detection of biomass (measured as Chl *a*) and major taxa of phytoplankton blooms (see Section 14.2).

It is important to recognize that *in situ* optical approaches are not able to distinguish between individual phytoplankton pigments *per se*, but key marker pigments as shown in Table 14.3 can be used to distinguish between four major classes or groups (Chromophyta, Chlorophyta, prochlorophytes and phycobiliprotein-containing phytoplankton) and these can further be divided into 12 pigment-based bio-optical groupings. The 12 pigment sub-groups of phytoplankton are discussed further in Chapter 13 this volume, (see Figure 13.3), but briefly, the major groups comprise: (I) Chl *b*-containing green algae (chlorophytes, euglenophytes and prasinophytes), (II) Chl *c*-containing chromophytes (diatoms, dinoflagellates, haptophytes, chrysophytes, raphidophytes, and dictyochophytes), (III) phycobiliprotein-containing phytoplankton (cryptophytes and cyanobacteria) and (IV) divinyl Chl *a* and *b*-containing prochlorophytes (Jeffrey *et al*., 1997a, b; Johnsen *et al*., 1998; Schofield *et al*., 2004; Kirkpatrick *et al*., 2004; Johnsen and Sakshaug, 2007; Nair *et al*., 2008; Tables 14.2–14.3, Appendix 14A).

Remote sensing of phytoplankton from satellites (Sections 14.3–6), which is mainly based on optical signatures and particle size distribution data, has been used to divide phytoplankton into phytoplankton functional types (PFT) which are involved with different biogeochemical processes (compare Kahru and Brown, 1997; Falkowski *et al*., 2003; Claustre *et al*., 2004; Babin *et al*., 2005, 2008; Ciotti and Bricaud, 2006; McClain, 2009). Le Quéré *et al*. (2005) defined seven major PFT based on (a) biogeochemical role, (b) physiological and environmental requirements

Caption for Figure 14.1 (*cont.*) The warning-sign denotes an ichthyotoxic bloom of *Pseudochattonella verruculosa* (Tanabe *et al*., 2007, also known as *Verrucophora farcimen* (Edvardsen *et al*., 2007), see D–G). D–G (time-series from 20–26 March 2001): Remotely sensed time-series of Sea WiFS images of Chl *a* biomass during bloom (ichthyotoxic) dominated by the dictyochophyte *Pseudochattonella verruculosa* (previously known as the raphidophyte *Chattonella* aff. *verruculosa* as indicated by upper right panel). The massive bloom occurred while seawater temperature was at its annual low, i.e. 2–5 °C. This species bloomed regularly in spring time in Scandinavia from 1998–2007. *Pseudochattonella verruculosa* (new species and genus) has a pigment signature similar to Chl c_3-containing coccolithophorids (Edvardsen *et al*., 2007) and resembles the pigment-signature of the coccolithophorid genera *Chrysochromulina*, *Phaeocystis*, *Prymnesium* and *Emiliania*) and dinoflagellates with tertiary endosymbiont chloroplasts (*Karlodinium veneficum* and *Karenia mikimotoi*, Johnsen and Sakshaug, 1993; Rodríguez *et al*., 2006; Tables 14.2–14.4), all characteristic HAB species of this region. These blooms have caused the death of approximately 2000 metric tons of farmed Atlantic salmon in the region. Images provided by Nansen Environmental and Remote Sensing Center (NERSC) using NASA OC4.4 Chl *a* algorithm to retrieve Chl *a* concentration (available at http://HAB.nersc.no). See colour plate section.

Table 14.1. *Classifications for existing mobile platforms, their sensors and capabilities as a function of height above/under sea surface level.*

[A]Platform	Sensors	Spectral bands/Spectral range, nm	[B]Products	Range, km (Duration) Swath width
Satellite	*MERIS	15/412–1050	1,3,4,5,6	Earth
(800 km ASL)	*MODIS	36/405–14385	1,3,4,5,6	(years)
Global-local	*SeaWiFS	8/402–885	1,3,4,5,6,11	
Airplanes	[C]HI	400/400–800	3,4,5,6,9,12	1–500
(0.5–10 km ASL)	Optional		17	(1–8 h)
Local-regional	sensors			
Small AUVs	[D]WL-ECO	ex 370/em 460	4	200
(10–100 m depth)		ex 470/em 695	5	(8–20 h)
		ex 540/em 570	6	
Local-regional	[E]OCR E_d, R_u	7/412–683nm	3,4,5,6,11,13	
	[F]CTD	1200 kHz	2,14,15	
	[G]ADCP	400/400–800	7,8,16	
	[H]SSS		9	
	[C]HI		3,4,5,6,8,9,12,13	
	Optional sensors		10,17	
Gliders	Same as small AUVs	Same as small AUVs	Same as small AUVs	2000
(200–1000 m depth)				(1–6 months)
Regional				
Large AUVs	Same as small AUVs	Same as small AUVs	Same as small AUVs	100
(600–6000 m depth)				(20–70 h)
Regional				

[A] Height above sea level (ASL) or maximum depth (m) in parenthesis.

[B] Product symbols: 1 = sea surface temperature, spectroradiometer; 2 = *in situ* temperature, temperature sensor; 3 = total suspended matter, radiometer, backscatter meter; 4 = CDOM, radiometer or fluorometer; 5 = phytoplankton biomass, Chl *a*, radiometer or fluorometer;

Table 14.1. (*cont.*)

6 = phycoerythrin; 7 = current speed/direction, ADCP; 8 = zooplankton biomass, ADCP acoustic backscattered signal; 9 = structure and morphology of sea-surface/seafloor/objects of interest, side scanning sonar, HI in shallow and transparent water; 10 = water samplers for chemical and biological analyses; 11 = irradiance-PAR, irradiance sensor/modelled; 12 = full 1 nm resolution from 400–700 nm of R_u used to classify all objects of interest; 13 = spectral AOP such as attenuation coefficient and reflection ratio L_u/E_d; 14 = salinity; 15 = depth; 16 = altimeter; and 17 = platform which can easily deploy an array of *in situ* underwater optical, electro-optical, physical, and chemical sensors (see Griffiths, 2008). In air this only relates to optical sensors (e.g. camera, video and spectroradiometry).

[C] HI = Hyperspectral imagers

[D] WET Labs (Philomath, Oregon, USA) spectroradiometers ECO triplet (WL-ECO) measure Chl *a*, CDOM and backscatter

[E] Satlantic (Halifax, Nova Scotia, Canada) downwelling irradiance OCR-507I and upwelling radiance OCR-507R sensors.

[F] CTD (Conductivity-Temperature-Depth sensors from e.g. Sea-Bird Electronics, Bellevue, Washington, USA or Neil Brown Ocean Sensors, Falmouth, Maine, USA).

[G] ADCP (Acoustic Doppler Current Profilers from e.g. Teledyne RD Instruments)

[H] SSS (e.g. 300/900 kHz side scan sonar from e.g. Marine Sonic Technology, White Marsh, Virginia, USA or EdgeTech, West Wareham, Maine, USA)

Table 14.2. *Phytoplankton species reported to contain the accessory pigment gyroxanthin-diester.*

Species	References
Dinoflagellates	
Karenia brevis (=*Gymnodinium breve*) (=*Ptychodiscus brevis*)	Millie *et al.*, 1995; Bjørnland *et al.*, 2003
Karenia mikimotoi (=*Gymnodinium mikimotoi*) (=*Gyrodinium aureolum*)*	Suzuki and Ishimaru, 1992; Johnsen and Sakshaug, 1993; Hansen *et al.*, 2000
Karlodinium veneficum (=*Karlodinium micrum*) (=*Gymnodinium galatheanum*)	Bjørnland and Tangen, 1979; Johnsen and Sakshaug, 1993; Bjørnland *et al.*, 2000; Garcés *et al.*, 2006; Rodríguez *et al.*, 2006
Karlodinium armiger	Garcés *et al.*, 2006
Takayama tasmanica	de Salas *et al.*, 2003
Pelagophytes	
Pelagomonas calceolata	Bjørnland *et al.*, 2003, Zapata 2005
Coccolithophytes**	
Chrysochromulina leadbeateri	Zapata, 2005
Chrysochromulina hirta	Zapata, 2005

* Hansen *et al.* (2000) and Throndsen *et al.* (2003) report that this species-complex is now *Karenia mikimotoi*.

** Reported as gyroxanthin-diester-like

controlling biomass and productivity, (c) behaviour and (d) quantitative importance in specific regions. These seven major PFT are: (1) Pico-heterotrophs, important re-mineralizers of DOM and POM (e.g. bacteria and archea); (2) Pico-autotrophs, important in global primary production (e.g. picoeukaryotes and non-N_2-fixing photosynthetic bacteria such as *Synechococcus* and *Prochlorococcus*); (3) Phytoplankton N_2-fixers (e.g. *Trichodesmium*); (4) Phytoplankton calcifiers, responsible for $> 50\%$ of marine carbonate flux (e.g. coccolithophytes, previously known as coccolithophorids – see Chapter 1, this volume); (5) Phytoplankton DMS (dimethyl sulphide) producers, affecting the atmospheric sulfur cycle (e.g. *Phaeocystis* spp.); (6) Phytoplankton silicifiers (e.g. diatoms), contributing to most of the primary production and biomass during the spring bloom in temperate and polar waters and (7) Mixed phytoplankton (e.g. autotrophic dinoflagellates and chrysophytes), a heterogeneous size and taxonomic group.

Of these seven types, PFT 1 and 7 cannot be directly related to pigment-specific groups while PFT 2 and 3 comprise phycobiliprotein-containing cyanobacteria and divinyl Chl *a* and *b*-containing prochlorophytes, PFT 4–5 include Chl c_3-containing haptophytes (= coccolithophytes), and PFT 6 includes Chl c_{1+2} - containing

Table 14.3. *In vivo phytoplankton absorption maxima related to major key marker pigments comprising chlorophylls (Chl), photoprotective carotenoids (PPC) and light-harvesting pigments (LHP, including all Chl, light-harvesting carotenoids (LH Car) and phycobiliproteins). The indicated wavelengths are selected by discriminant analysis to optimize discrimination between classes, adapted from Johnsen* et al. *(1994), Jeffrey and Vesk (1997) (prochlorophytes), Edvardsen* et al. *(2007) (dictyochophytes).*

Phytoplankton class	481 nm	535 nm	586 nm	649 nm
	Chl, LH Car, PPC	**LHP**	**Chl *b* and *c*-group**	**Chl *b***
Chromophyta				
Diatoms	Diadino + Diato, Fuco	Fuco	Chl c_{1+2}	-
Dinoflagellates	Diadino + Diato, Peri/(Fuco + 19F)	Peri/(Fuco+19F)	Chl c_{2+3}	-
Haptophytes	Diadino + Diato, Fuco, 19F	Fuco, 19F	Chl c_{2+3}	-
Chrysophytes	Diadino + Diato, Fuco	Fuco	Chl c_2	-
Raphidophytes	Fuco, Viola, Zea	Fuco	Chl c_{1+2}	-
Dictyochophytes	Diadino + Diato, Fuco, 19F	Fuco, 19F	Chl c_{1+3}	
Chlorophyta				
Prasinophytes	Pras/Lut, Viola, Zea, Chl *b*	Pras	Chl *b*, MgDVP	Chl *b*
Euglenophytes	Diadino + Diato, Neo, Chl *b*	-	Chl *b*	Chl *b*
Chlorophytes	Viola, Zea, Lut, Chl *b*	-	Chl *b*	Chl *b*
Prochlorophytes	DV-Chl *b*, ββ-Car, Zea	-	DV-Chl *b*	Chl *b*, DV-Chl *b*
Phycobiliprotein phytoplankton				
Cryptophytes	Allo	PE	Chl c_2	-
Cyanobacteria	PUB, Zea, ββ-Car	PEB	-	-

MgDVP = Magnesium 2,4-divinyl pheoporphyrin a_5 monomethylester; DV-Chl *a* and *b* = Divinyl Chl *a* and *b*; Diadino + Diato = Diadinoxanthin + Diatoxanthin; Fuco = Fucoxanthin; Peri = Peridinin; 19F = 19′-acyloxyfucoxanthins; Pras = Prasinoxanthin; Viola = Violaxanthin; Zea = Zeaxanthin; Neo = Neoxanthin; Lut = Lutein; Allo = Alloxanthin; ββ-Car = β,β-carotene; PE = Cr-phycoerythrin 545; PUB = Phycourobilin; PEB = Phycoerythrobilin./(= or) denotes two different pigment groups.

chromophytes (see for example the lower panel of Figure 14.1 showing a toxic Chl c_3-containing chromophyte bloom, dominated by the dictyochophyte *Pseudochattonella verruculosa* (Tanabe *et al.*, 2007) aka *Verrucophora farcimen* (Edvardsen *et al.*, 2007). This alga was previously known as the raphidophyte *Chattonella* aff. *verruculosa*. Alvain *et al.* (2005) used a similar approach with *in situ* data using 22 pigments from $>$ 1100 stations. A discussion of pigment groups and their corresponding bio-optical characteristics can be found in Sections 14.5 and 14.6.

There is a range of *in situ* detection methods that are increasingly becoming standard tools for operational monitoring of phytoplankton. These methods are reviewed in Babin *et al.* (2008) and include automated submersible flow-cytometers, CCD imagers connected to flow cells (Sieracki *et al.*, 1998; Sosik, 2008), environmental sample processors for DNA probe array analyses (Scholin *et al.*, 2008), enzyme-linked immuno-sorbent assays (ELISA) for detection of group-specific toxins, molecular imprinting, and *in situ* mass spectrometry. Unlike the methods outlined in Babin *et al.* (2008), this book focuses on pigment characterization and the corresponding bio-optical properties of phytoplankton, the application of techniques to assess phytoplankton in the environment.

Bio-optical approaches currently remain the only means of collecting synoptic scale information. The focus of this chapter is on the optical methods for monitoring phytoplankton that are based primarily on discriminating phytoplankton pigments. The theoretical background to optical properties is provided in Section 14.2, and this is followed in Section 14.3 by a review of the optical sensors used in *in situ* and remote-sensing techniques. The platforms (instrument carriers) to hold *in situ* and remote-sensing instruments are detailed in Section 14.4 and some examples of monitoring phytoplankton biomass and taxa are given in Section 14.5. Increase in temporal and spatial resolution to enhance phytoplankton taxa identification, and a shift from global to local-based remote-sensing algorithms, is addressed in Section 14.6.

14.2 General optical properties of seawater and its constituents

This section firstly describes the inherent and apparent optical properties of seawater, associated optical constituents, and the corresponding optical differentiation of water masses (Case I versus Case II waters). Secondly, the principles of *in situ* classification of optically significant constituents (OSC) are described focusing on phytoplankton, CDOM and suspended matter (SM), together with remote-sensing retrieval algorithms for in-water OSC.

14.2.1 Inherent and apparent optical properties

The inherent optical properties (IOP) are those properties that depend only on the medium and thus are independent of the ambient light field. The IOP include absorption and scattering, which are additive properties of the medium. While IOP are easy to interpret, historically, they have been difficult to measure. In the last decade, however, new sensors have made these measurements significantly more straightforward (Bukata *et al.*, 1995; Bissett *et al.*, 2004; Schofield *et al.*, 2004; Dickey *et al.*, 2008). *In situ* IOP instruments consist of an artificial light source (active sensors) allowing measurements to be made 24 hours a day. For pigment-specific signals, the most relevant IOP measurements are provided by *in situ* absorption sensors. As the spectral variability of the cellular absorption is a direct reflection of the pigments present, it represents the ideal optical method for identifying specific pigment classes in the field (see Section 13.1). A key consideration for these sensors is the number of wavelengths required to discriminate between algal taxa at a given location. The absorption detectors maximize the measured absorption signal by varying the optical path length (OPL) between the light source and the detector. The more dilute the particles and dissolved material in the water, the longer the OPL required for detection, which must be configured a priori. The OPL available vary from 5 to 500 cm for a variety of instruments. There is a tradeoff, however, when considering OPL for measuring phytoplankton absorption, as other optical constituents can dominate the signal.

The apparent optical properties (AOP) are those properties that depend both on the medium and on the geometric structure of the light field, which include measurements such as the diffuse attenuation coefficient, or ocean reflectance (Kirk, 1994; Mobley, 1994; IOCCG, 2000). Historically, AOP have been the primary measurements made because they are passive (using ambient light) and relatively easy to measure (Sosik, 2008). Apparent optical property measurements, however, are difficult to interpret quantitatively with respect to bio-geo-chemo-physical properties of the water column such as Chl *a* concentration given the complex relationships between the geometric structure of the light field in three dimensions and the optical properties of the in-water constituents. Therefore, because they are easy to measure, AOP measurements are often used by empirical models to predict the parameter of interest (i.e. the depth of the euphotic zone in the water column and Chl *a* concentration).

The spectral variability of the light leaving the water is dominated by the wavelength dependency of the combined AOP spectral absorption properties of the water and its OSC, which, in turn, is partially a reflection of the pigments present in the water column. The remote-sensing reflectance is calculated from the ratio of upwelling radiance to downwelling irradiance. Since the AOP instruments (e.g. a spectroradiometer) rely on the sun as the light source, measurements can only be made during daylight hours. There are two major AOP characteristics that are commonly measured *in situ*: the spectral vertical diffuse attenuation coefficient, $K_d(\lambda)$, and the water reflectance, $R(\lambda)$. The former is calculated from the measured decrease of the downwelling light field with depth (per metre) and is often used to obtain information on the characteristics of *in situ* water components (phytoplankton, CDOM, and SM), while $R(\lambda)$ is used for remote sensing of water constituents, most often at selected bands from 400–850 nm (IOCCG, 2000). More regionally tuned bio-optical algorithms are needed to enhance the separation and identification of different in-water OSC such as particulate inorganic and organic matter (PIM and POM, respectively), CDOM, phytoplankton, and Chl *a*, in order to obtain phytoplankton pigment signature and biomass (see also Section 13.4, this volume).

14.2.2 Case I and Case II waters

For the majority of ocean waters, the optical properties are dominated by the absorption (a) and scattering (b) properties of water molecules and phytoplankton (Chapter 13). These 'clear blue' oceanic waters are designated as Case I (Preisendorfer, 1976; Jerlov, 1976; Morel and Prieur, 1977; IOCCG, 2000). Changes in the light field (intensity and spectral composition) can therefore be exploited to predict the phytoplankton biomass using a suite of empirical and/or semi-analytical models if a sensor (on *in situ* or remote-sensing platforms) can provide sufficient spectral information (Table 14.1). This is relatively straightforward in Case I waters, where water, phytoplankton and co-evolving CDOM (Section 13.1.3) are the dominant influence on the optical signatures. Many coastal waters, however, are optically complex ('green-yellow-brown coastal' Case II waters), with contributions of non-algal particles (Preisendorfer, 1976; IOCCG, 2000; Chapter 13.1.1) and predominantly terrestrially generated CDOM from fresh water run-off. In these coastal waters the complexity of the optics can be daunting to interpret, however the myriad of signals provides a great deal of information, thus more sophisticated approaches need to be applied (see Sections 14.3 and 14.6) in order to, for example, discriminate and track specific water masses and/or phytoplankton taxa in coastal waters (Moline *et al.*, 2004a, c; Oliver *et al.*, 2004; Schofield *et al.*, 2004; Babin *et al.*, 2008; Dickey *et al.*, 2008).

14.2.3 In situ *classification of optically significant constituents: phytoplankton, CDOM and suspended matter*

Within the phytoplankton, absorption by pigments is the dominant optical signal that determines the spectral signatures of the algal communities present. Correspondingly, the intracellular self-shading of pigments in phytoplankton cells (package effect) typically causes two to three times variation in the pigment-specific absorption coefficient of phytoplankton (Johnsen and Sakshaug, 2007; Chapter 13, this volume). The scattering properties of phytoplankton are species dependent and may be altered by bio-detritus, thus affecting the spectral scattering signature relative to the pigment absorption. As phytoplankton cell numbers increase, the corresponding increase in the pigment concentration skews the spectral composition of the underwater light field from blue (400–500 nm) towards green-orange (500–600 nm). *In vivo* absorption, scattering and package effect are covered in detail in Chapter 13.

Detection of optical signatures in Case I and II waters can be achieved based on the additive nature of both absorption $a(\lambda)$ (Eq. (14.1)) and scattering of light $b(\lambda)$ (Eq. (14.2)):

$$a(\lambda) = \sum_i a_i(\lambda) \tag{14.1}$$

$$b(\lambda) = \sum_i b_i(\lambda), \tag{14.2}$$

where i refers to the most commonly used optical constituents (e.g. water, Chl a, SM, CDOM). The absorption and scattering of each optical constituent can in turn be expressed as the product of its concentration, C, and absorption or scattering-specific coefficients, a^* or b^*, providing total absorption and scattering equations in terms of concentrations of individual constituents, (Eq. (14.3) and (14.4), see also Section 13.2.3, this volume):

$$a(\lambda) = a_{\mathrm{WATER}}(\lambda) + \sum_j a_j^*(\lambda) C_j \tag{14.3}$$

$$b(\lambda) = b_{\mathrm{WATER}}(\lambda) + \sum_j b_j^*(\lambda) C_j, \tag{14.4}$$

where j is typically Chl a, CDOM and SM in the total absorption relationship, and Chl a and SM in the total scattering relationship (CDOM is considered to contribute infinitesimally to scattering of light). The optical relationships detailed in Eqs. (14.1) to (14.4) are important to elucidate the separate contributions of each OSC to the total remotely sensed signal. It is this remotely sensed signal which may be 'inverted' to retrieve estimates of the concentrations of individual OSC.

14.2.4 Above water remote-sensing retrieval algorithms of optically significant constituents

The retrieval of concentrations of OSC from the water-leaving light signal can be based upon the analysis of spectral features of remote-sensing reflectance, $R_{\mathrm{rs}}(\lambda)$. R_{rs} expresses the light leaving the surface in terms of the absorption $[a(\lambda)]$ and backscattering coefficients $[b_{\mathrm{b}}(\lambda)]$, where the backscattering coefficient is a fraction of the total scattering $b(\lambda)$ in the water column (Zaneveld, 1995; Dierssen and Smith, 2000; IOCCG, 2000; Wang *et al.*, 2005; Eq. (14.5)),

$$R_{\mathrm{rs}} = \frac{L_{\mathrm{u}}(0^+,\lambda)}{E_{\mathrm{d}}(0^+,\lambda)} = \frac{f}{Q} * \frac{b_{\mathrm{b}}(\lambda)}{a(\lambda) + b_{\mathrm{b}}(\lambda)}. \tag{14.5}$$

$L_{\mathrm{u}}(0^+, \lambda)$ denotes the upwelling radiance just above the sea surface, $E_{\mathrm{d}}(0^+, \lambda)$ is the corresponding downwelling irradiance just above the sea surface, and, although somewhat variable, f/Q can be approximated using a constant value of 0.085. Note that both f and Q factors do vary (see Dierssen and Smith, 2000; Morel, 2008 and references therein) in response to the angular structure of the light field. The quantity f is a function of wavelength, water IOP (single scattering albedo and volume scattering function), solar zenith angle, aerosol optical thickness and surface roughness. Likewise, the quantity $Q(\lambda) = E_u(\lambda)/L_u(\lambda)$, where $E_u(\lambda)$ denotes upwelling irradiance, and is therefore also influenced by environmental variables and IOP (Morel and Mueller, 2002). Alternatively, a water mass may be observed by determining the irradiance reflectance $R^-(\lambda)$, which is defined as the upwelling irradiance just below the water surface, $E_u(0^-,\lambda)$, normalized to the corresponding subsurface downwelling irradiance, $E_d(0^-,\lambda)$. In practice the subsurface reflectance, $R^-(\lambda)$ is derived from measurements of the above surface reflectance $R^+(\lambda)$. R_{rs} and $R^-(\lambda)$ are interrelated and can be derived from the atmospherically corrected signal recorded by a satellite spectroradiometer.

As mentioned previously, the spectral absorption of Chl *a* can serve as a universal biomass indicator of the presence of photosynthetic phytoplankton in the water body, while the optical signatures of other pigments can be employed for identification of specific phytoplankton pigment groups (Figures 14.2–14.3). From the perspective of aircraft and satellite remote sensing, only strong optical signatures are of consequence, since any weak features are not discernible in the R_{rs} spectra (Section 14.3). Because of this, the majority of the focus in developing algorithms and products from remote sensing has been on retrieval of Chl *a* concentration. Chl *a* concentration is now a standard product available online for a number of operational satellite sensors, including SeaWiFS, MODIS and MERIS. Embedded into NASA's SeaDAS software, the SeaWiFS standard ocean Chl *a* concentration retrieval algorithm (version 4-OC4, O'Reilly *et al.*, 1998), developed and validated for open ocean waters (Case I waters), is a modified cubic polynomial function based upon the band-ratio paradigm. The OC-4 uses the ratio $R_{rs}(\lambda_i)/R_{rs}(\lambda_j)$ as the input parameter to estimate Chl *a* concentration, $C_{\mathrm{Chl}\,a}$ (Eq. (14.6)):

$$C_{\mathrm{Chl}\,a} = 10^{\left(a_0 + a_1\bar{R}^1_{rs} + a_2\bar{R}^2_{rs} + a_3\bar{R}^3_{rs}\right)} + a_4, \tag{14.6}$$

where a_k ($k = 0,\ldots,4$) are empirically derived coefficients. $\bar{R}_{RS}$ is the greatest log-ratio among $R_{rs}(443)/R_{rs}(555)$, $R_{rs}(490)/R_{rs}(555)$ and $R_{rs}(510)/R_{rs}(555)$. The MODIS standard ocean Chl *a* algorithm for open ocean waters (OC3Mo) is a modification of OC-4 with only three wavelengths (443, 488 and 550 nm), and thus two ratios of $\bar{R}_{RS}$ are used. The MERIS standard ocean Chl *a* concentration algorithm (OCMe or Algal_1 product) for Case I waters is also akin to OC-4, but $R_{rs}(\lambda)$ is replaced by $R(\lambda)$, and the R values at 443, 490 and 510 nm are normalized to $R(560)$ to compute ρ_{max}, which is the greatest log-ratio $R(\lambda_i)/R(\lambda_j)$.

The Algal_2 product, the Chl *a* concentration algorithm optimized for Case II waters, is calculated as $k_1[a_{pig}(442)]^{k_2}$, where $a_{pig}(442)$ is the pigment absorption at 442 nm, and $k_{1,2}$ are scaling factors that may be adjusted to local conditions (Morel

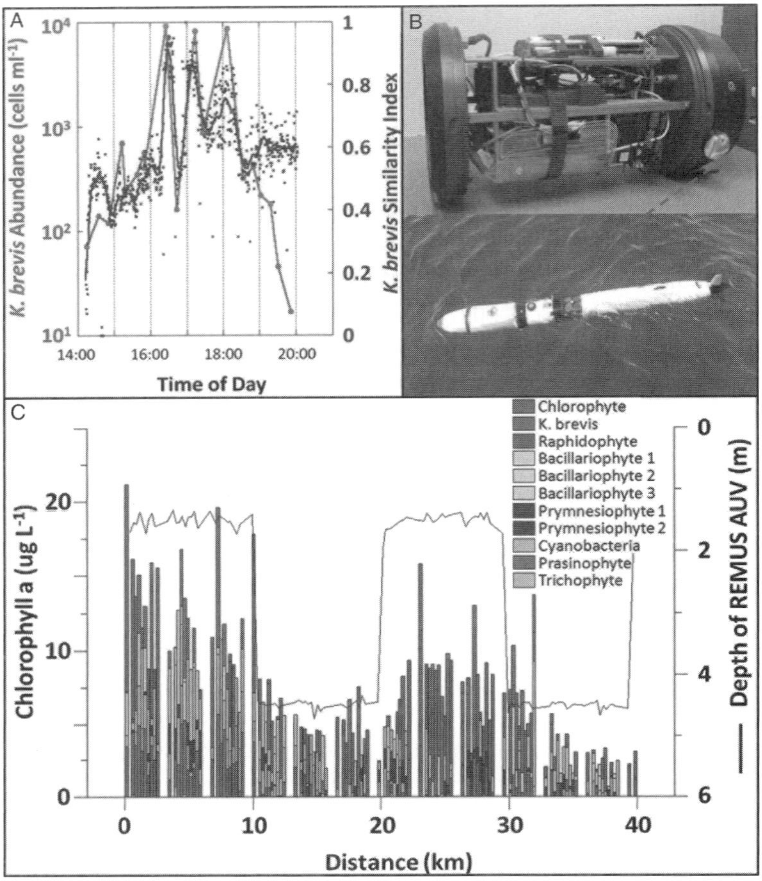

Figure 14.2. A: Surface measurements of *Karenia brevis* abundance on the West Florida Shelf taken over the day on 10 January 2003. Concurrent optically based measurements of the *K. brevis* similarity index (SI) were taken approximately every minute and corresponded well to the cell counts. The high temporal resolution of the SI measurement was able to capture rapid transitions in cell abundance. B: The 'Brevebuster' sensor for measuring the *K. brevis* SI. The sensor includes a capillary waveguide spectrometer and a series of pumps and standards for separating out the particulate fraction and CDOM. Here, the system is shown in a custom module which is integrated into a REMUS Autonomous Underwater Vehicle, AUV (below). C: Data collected on-board the REMUS AUV on 21 January, 2005. The vehicle conducted a near surface survey, first at ~ 2 m and then at 6 m depth and repeating this sequence again. The on-board Brevebuster collected the SI for *K. brevis* (red) and showed an enhanced concentration at 2 m (see Robbins *et al.*, 2006). In addition to *K. brevis*, a spectral library of multiple taxonomic groups grown at various light levels was applied to the full spectral data for each sample collected on the REMUS AUV to provide a breakdown of the major taxonomic groups of phytoplankton. This approach, based on optical differences in pigments, pigment concentrations and pigment packaging between taxonomic groups, provides information on phytoplankton community structure. The data illustrates the large differences in speciation that occurs over small spatial scales and demonstrates the benefit of optical approaches in delineating between phytoplankton taxa on relevant scales (see Kirkpatrick *et al.*, 2008). See colour plate section.

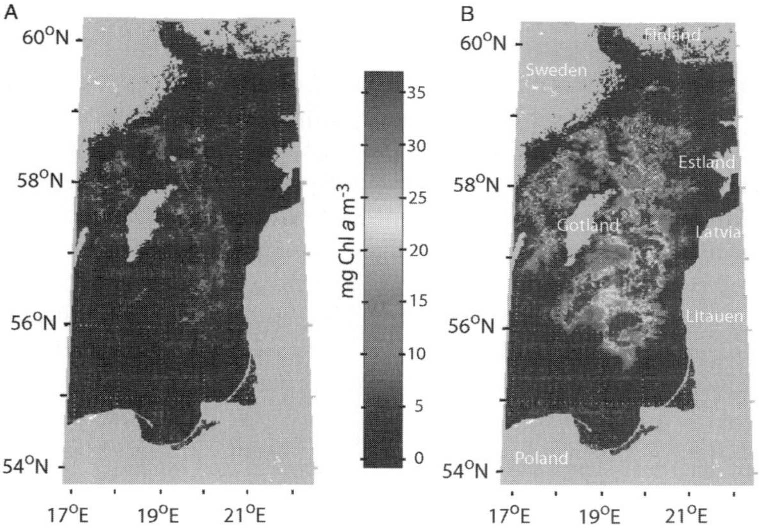

Figure 14.3. Phytoplankton biomass, seen as $C_{\text{Chl}\,a}$ (mg Chl *a* m^{-3}), distributions across the central Baltic Sea as obtained from (A) MODIS standard product, (B) MODIS data processed with a regionally tuned bio-optical algorithm. The island in the centre of the image is Gotland. Images provided by Nansen Environmental and Remote Sensing Center (NERSC). See colour plate section.

and Antoine, 2000). The major validation and tuning of these scaling factors for MERIS processing is based on numerous field investigations performed in European waters (Morel and Antoine, 2000). The principal limitation of the OC-type algorithms discussed above is that they are untenable for Case II waters (Figure 14.3). Even though continental shelf regions (often Case II waters) account for only about 10% of the world's ocean area, they represent 15–40% of the global production and export flux reaching the sea floor (McClain, 2009). Thus, Case II waters encompass many coastal and inland water bodies with economic, societal and ecological significance. In such waters, the spectral curvature of water leaving radiance $L_{\text{w}}(-0,\lambda)$, and hence $R(\lambda)$ and $R_{\text{rs}}(\lambda)$, are controlled not solely by phytoplankton, but also by coexisting SM and CDOM concentrations (IOCCG, 2000; Johnsen *et al.*, 2009). Thus, the retrieval of $C_{\text{Chl}\,a}$ can be adequately performed only in conjunction with a simultaneous determination of the two other OSC i.e. the concentrations of CDOM and SM, C_{CDOM} and C_{SM}. The OC-type algorithms are incapable of performing accurately in these optically complex waters. Moreover, since they were conceived as globally applicable, OC-type algorithms can be inaccurate in determining $C_{\text{Chl}\,a}$ when applied to specific regions of the world's oceans. Among various causes, an enhanced concentration of CDOM and SM due to shallow water mixing and sediment resuspension, or atmospheric fall-out of terrigenous particulate matter originating from dust storms may all affect phytoplankton bloom events. This is true not only for Chl *a*, but for other pigments as well (Figure 14.3).

The optical constituents (Eq. (14.1) to (14.4)) concentration vector $C = (C_{\mathrm{Chl}\ a}, C_{\mathrm{SM}}, C_{\mathrm{CDOM}})$ can be retrieved by employing one of the non-linear curve fitting multivariate procedures through iterative minimizing, for each wavelength (λ), the function $g_i(\mathrm{C})$, which is the weighted difference between remotely sensed and modelled R (either by means of Hydrolight software or a relevant parameterization) values of $R(\lambda)$ or $R_{\mathrm{rs}}(\lambda)$.

The concentration vector $\mathrm{C} = (C_{\mathrm{Chl}\ a}, C_{\mathrm{SM}}, C_{\mathrm{CDOM}})$, for OSC identified in Eq. (14.3) and (14.4), can be retrieved by employing one of the non-linear curve fitting multivariate procedures. The full spectrum residual between remotely sensed and parameterized or modelled ocean reflectance may be described as,

$$g_i(\mathrm{C}) = \sum_i \left\{ R_i^{\mathrm{spaceborne}}(\lambda)\text{-}R_i^{\mathrm{modelled}}(\lambda)/R_i^{\mathrm{spaceborne}}(\lambda) \right\}^2 \qquad (14.7)$$

and the value $g_i(\mathrm{C})$ minimized iteratively by varying C in Eq. (14.7). Among the frequently used multivariate procedures are the Levenberg-Marquardt algorithm (Bukata *et al.*, 1995), Neural Networks (Schiller and Doerffer, 1999), Simplex (Doerffer and Fischer, 1994), Statistical Regularization (Kondratyev *et al.*, 1990), and Maximum Likelihood methods (Doubovick *et al.*, 1994), or their combinations (Pozdnyakov *et al.*, 2005).

The R^{modelled} in Eq. (14.7) may be parameterized simply in the form of Eq. (14.3), (14.4) and (14.5) combined, or may be achieved by employing a 'fuller' optical model such as Hydrolight (product available at http://www.sequoiasci.com/hydrolight.html). Hydrolight can also be used to estimate the dependence of $R(\lambda)$ or $R_{\mathrm{rs}}(\lambda)$ on $a(\lambda)$ and $b_b(\lambda)$ for a wide spectrum of hydro-optical situations (Lee *et al.*, 2002; Wang *et al.*, 2005; Doerffer and Fischer, 1994; Tilstone *et al.*, 2005; Van der Woerd and Pasterkamp, 2008), which in turn may be embedded in a simpler parameterization and employed in an iterative minimization scheme.

As shown, the multivariate algorithms allow the retrieval not only of $C_{\mathrm{Chl}\,a}$, but simultaneously also the concentrations of the other optical constituents, which is important for assessing the ecological status of the targeted water body. They are mostly intended for Case II waters, but presently are also thought to be useful for Case I waters. Generally, these require more computing resources than the standard OC band ratio algorithms.

Empirical algorithms relate ratio(s) of either $R(\lambda)$, $R_{\mathrm{rs}}(\lambda)$ or $L_{\mathrm{w}}(0+, \lambda)$ at two or more wavelengths to estimate $C_{\mathrm{Chl}\,a}$, making use of statistical data from *in situ* $C_{\mathrm{Chl}\,a}$ measurements obtained simultaneously with satellite or aircraft overflight. A regression can also relate the $C_{\mathrm{Chl}\,a}$ retrieved from e.g. OC4 with the $C_{\mathrm{Chl}\,a}$ measured *in situ*. Understandably, such algorithms only hold for the conditions covered (both geographically and seasonally) and have limited applicability to ocean waters in general.

Satellite and aircraft remote sensing of water colour (i.e. the spectral distribution of the remote-sensing reflectance, R_{rs}) makes use of multi-channel sensors with a narrow spatial and spectral field of view scanning the targeted water surface. The remotely sensed signal is constituted by the sunlight that is: (1) scattered/absorbed in the intervening atmosphere, (2) specularly reflected at the water surface, (3) back-scattered in the upper part of the water column, and (4) attenuated a second time on its journey to the space or airborne sensor due to absorption and scattering by the atmosphere. Only a fraction of the light upwelling from beneath the water surface (component 3 above) carries information on the water colour and hence by inversion of the spectra, on the concentrations of the optical constituents (SM, including PIM and POM, CDOM and phytoplankton: IOCCG, 2000; Chapter 13). Thus, the above components 1, 2 and 4 constitute unwanted signals from the perspective of determining in-water concentrations of optical constituents. These unwanted signals have to be quantified and removed from the total detected signal. Some satellite sensor systems can be tilted away from the sun-glint pattern on the sea surface in order to avoid the influence of specular reflection (component 2 above). Even in this case, the atmospheric component (1 and 4) at satellite altitudes can account for more than 90% of the total signal captured by the sensor (see Gordon, 1997; Gao *et al.*, 2000). Thus, typically only 10% or less of the total remotely sensed signal is due to the water column contribution.

14.3 Current techniques for *in situ* monitoring and remote sensing of phytoplankton blooms by optical sensors

To understand, monitor and map phytoplankton blooms requires additional information on hydrodynamics, physical, biological and chemical data. This is critical to provide the scalar perspectives required to understand the transport and transformation processes occurring within natural phytoplankton populations (Schofield *et al.*, 2002; Glenn and Schofield, 2003; Schofield *et al.*, 2008). These ancillary data provide the guidance for discrete sampling required to characterize phytoplankton community composition, and help in understanding what regulates the initiation, development, maintenance and senescence of algal blooms in nature.

Pigment-based *in situ* detection of phytoplankton biomass and different bio-optical groups is most commonly done using a few wavelengths that are greatly impacted by the pigment-specific absorption maxima of the phytoplankton. Signatures from *in situ* spectral diffuse attenuation coefficients, reflectance, light beam attenuation, absorption, scattering, and Chl *a* fluorescence may be detected from remote sensing using satellites (800 km above sea surface) and aeroplane-based spectrometers (typically 1–3 km above sea surface) that detect reflected light from the sea surface. Correspondingly, *in situ* detection commonly uses fluorometers, transmissometers, radiometers, photometers, and turbidity meters (Babin *et al.*, 2008;

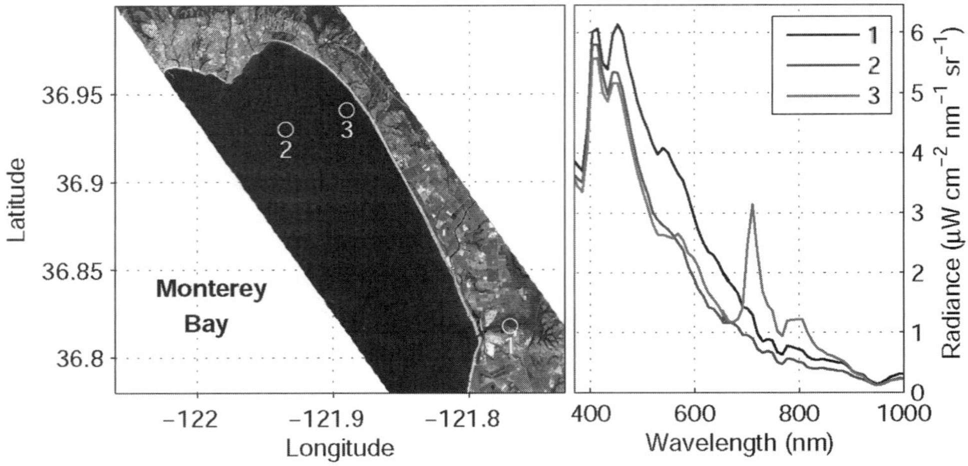

Figure 14.4. Left image: Pseudo true-colour image of an airborne remote-sensing swath across Monterey Bay, California. Data are from the Airborne Visible/Infrared Imaging Spectrophotometer (AVIRIS) on August 26, 2004. RGB bands were 711, 559 and 443 nm, respectively. Right image: The 711 nm band is effective at measuring the near-infrared signal of extreme 'red tide' blooms. Low radiance at 675 nm implies high phytoplankton biomass. Site 1 is from the river Elkhorn Slough, site 2 outside HAB and site 3 is in the middle of HAB dominated by peridinin-containing dinoflagellates such as *Akashiwo sanguiena*. Image and spectra from John Ryan, Monterey Bay Research Institute (MBARI). See colour plate section.

Dickey *et al.*, 2008; Oceanography, 2004; Moline *et al.*, 2005; Suggett *et al.*, 2010; Table 14.1, Figures 14.1–14.4).

14.3.1 In situ *sensors for optical detection of IOP and AOP*

Most optical underwater *in situ* instruments are well suited for time-series measurements of phytoplankton and particle concentrations, but are limited in the discrimination of different phytoplankton classes, pigment groups and/or species. *In situ* optical sensors can be divided into two major categories based on their characteristics and the nature of the sensors used to detect them. The two major classes of optical measurements are based on (a) IOP and (b) AOP (see also Chapter 13, Kirk, 1994; Mobley, 1994; IOCCG, 2000; Roesler and Boss, 2008; Schofield *et al.*, 2008; Sosik, 2008; MacIntyre *et al.*, 2010). For further details regarding theory, instruments and techniques, see Chapter 13, Babin *et al.* (2008) and Suggett *et al.* (2010).

Typically, *in situ* detection of OSC is done using absorption (underwater spectrophotometers), light scattering (spectrophotometers and scattering meters), light beam attenuation (transmissometers and spectrophotometers), fluorescence (fluorometers), turbidity (optical backscattering meters) and the volume scattering function, which is important for understanding R_{RS} (volume scattering and backscattering meters). These *in situ* optical instruments, when deployed on buoys (generally for time-series measurements, Fawcett *et al.*, 2006) or on autonomous underwater vehicles, need to be small sized, have low power requirements and be robust and flexible regarding deployment and data storage/transmission (Moline *et al.*, 2005; Schofield *et al.*, 2008). These instruments provide information on IOP such as absorption, light scattering, reflectance and fluorescence. Knowing the spectral characteristics of the OSC, it may be possible to delineate phytoplankton taxa using, for example, in vivo absorption fingerprints of phytoplankton, and *in situ* fluorescence detection (active sensors with their own light source) of Chl *a*, phycoerythrin or to correct for CDOM that may obscure phytoplankton pigment information (Table 14.1).

In Case II waters, the absorption by CDOM may regularly represent eight times the absorption by phytoplankton at 400–500 nm (Johnsen *et al.*, 2009). This implies that the absorption signature by phytoplankton pigments may be entirely obscured by non-algal components, even if the phytoplankton biomass is as high as 10 to 15 mg Chl *a* m^{-3}. Because of this, the use of *in situ* fluorescence excitation spectra, which are not sensitive to POM, PIM and CDOM, may be the best bio-optical approach to detect and to identify different pigment groups of phytoplankton in these systems (Chapter 13, this volume).

14.3.2 Remotely sensed fluorescence line height

The solar-stimulated Chl *a* fluorescence centred at about 685 nm can be detected in the water-leaving radiance L_w and is used for retrieving the Chl *a* concentration, $C_{Chl\ a}$, from MODIS and MERIS spectrometer data (Gower and Borstad, 2004; Huot *et al.*, 2005). This is done by employing the so-called fluorescence line height (FLH), based on a relative measure of the amount of radiance leaving the sea surface in the Chl *a* emission band (Eq. (14.8)):

$$\mathrm{FLH} = L_w(\lambda_2) - L_w(\lambda_1) - L_w(\lambda_3) - L_w(\lambda_1)\frac{(\lambda_2 - \lambda_{1)})}{(\lambda_3 - \lambda_1)} \tag{14.8}$$

where λ_2 is the sensor's band centred on the Chl *a* fluorescence, while λ_1 and λ_3 determine the baseline. Huot *et al.* (2005) point out however, a number of limitations of this standard method for MODIS Chl *a* retrieval (used here as an example), including the variation in the quantum yield of photosystem II (PSII) fluorescence (see Section 13.3.1, this volume) over the day making interpretation difficult, the use of the L_w at 412 nm in the standard retrieval algorithm, the depth of integration of Chl *a*, and the re-absorption of fluoresced light. For MERIS, the three wavelengths used are 682, 665 and 705 nm, respectively. Comparison between MERIS, SeaWiFS and MODIS sensors has been made by Zibordi *et al.* (2006) and Antoine *et al.* (2008).

FLH can be related to $C_{\mathrm{Chl}\,a}$ through a non-linear regression using regionally determined specific coefficients, as described for two regions in Huot *et al.* (2005). Huot *et al.* (2007) also showed the potential use of fluorescence in remote sensing by integrating it with reflectance and the diffuse attenuation coefficient to retrieve phytoplankton biomass (see Section 14.4). Recent innovative work on FLH as an indicator of phytoplankton physiology is promising, especially in turbid waters (McClain, 2009; Chapter 13, this volume).

14.3.3 Artificial light-stimulated spectral fluorescence

A technique that shows promise in detecting the presence of specific pigment groups and the corresponding *in situ* biomass of various algal groups is artificial light-stimulated spectral fluorescence, which examines the light energy transfer by pigments in the wavelength range from 400 to 700 nm. Light absorbed by the light-harvesting pigments (LHP) is passed on to the photosystems. A small fraction of the absorbed energy is emitted as Chl *a* fluorescence from photosystem II with a peak emission centred at 685 nm (Section 13.1.2, this volume). The rise and fall in fluorescence emission reflects the light energy transfer from the photosynthetically active pigments present within the cell. The advantage of this approach is the 100 to 200-fold higher sensitivity when compared to absorption measurements, making it useful even for dilute phytoplankton concentrations (< 0.25 mg Chl *a* m^{-3}). The technique reflects the condition of the LHP, the physiological state of the phytoplankton and is insensitive to non-algal particles (e.g. mineral particles from river run-off) and CDOM, in contrast to light absorption.

Currently, *in situ* data from multi wavelength fluorometers, such as the Algae Online Analyser, (AOA), quantify biomass, species composition and have given comparable data with HPLC derived phytoplankton pigment markers (Richardson *et al.*, 2010). A conclusion from this research was that AOA gave good first-order estimation of algal groups, but the determination of Chl *a* biomass was less robust. MacIntyre and co-workers concluded in a review of spectral fluorescence signatures (SFS) of phytoplankton that the SFS approach is robust but imprecise and that achieving $> 20\%$ algal group classification is relatively rare. They found that the highest algal group prediction power was found in waters dominated by phycobiliprotein containing taxa (MacIntyre *et al.*, 2010).

14.3.4 From global to regional algorithms – wavelength targeting

As mentioned earlier, the need for regionally tuned remote-sensing algorithms originates from local differences in CDOM and SM, primarily in Case II waters (Figure 14.3). In addition to these optical constituents, the absorption and scattering properties of the local phytoplankton also influence L_{w} and thus R_{rs}. Differences in absorption and scattering properties between algal taxa raise the possibility of providing group and possibly species-specific information. Garver *et al.* (1994) provided a systematic study of phytoplankton absorption from a variety of locations to evaluate the influence of absorption in remote-sensing applications. Using an empirical orthogonal function analysis to identify the dominant spatial and temporal patterns, only 5% of the variability in the phytoplankton absorption measured *in situ* was found to be from spectral information with 95% due to the amplitude or concentration of the sample. It was concluded that there was a low probability that remote sensing could be used as a tool to retrieve information on phytoplankton groups. This analysis, however, was conducted over the entire spectrum and was not focused on the target wavelengths of absorption by accessory pigments. An example of 'wavelength targeting' is from the work of Johnsen and colleagues, where a wavelength of 585 nm was used to target the presence of Chl c_3 in in vivo absorption and fluorescence excitation spectra, indicating potentially harmful coccolithophytes and dinoflagellates (Johnsen and Sakshaug, 1993; Johnsen *et al.*, 1994, 1998, 2007). Also, no normalization and log-transformation of *in vivo* absorption spectra, used to examine differences in spectral shape and optimizing the predictive power was conducted by Garver and colleagues (Johnsen *et al.*, 1994). Lee and Carder (2004) showed a strong correlation between *in situ* absorption measurements and radiance using hyperspectral data, demonstrating that the increase in spectral resolution and *in situ* data sets can increase the success of regional algorithms. As discussed later in this chapter, historical information of recurrent species-specific blooms can also help in tuning these algorithms. McClain (2009) and Stumpf and Tomlinson (2005) provide comprehensive overviews of ocean colour remote sensing and provide examples of the ability to reveal taxonomic information. Given that a bloom is of sufficient concentration to provide a threshold signal to noise in the spectral area of interest, there is a potential for remote-sensing retrieval of a number of phytoplankton groups based on the peak absorption of the dominant pigments (Table 14.3, Appendix 14A, Chapter 13 this volume, Sathyendranath *et al.*, 2004; Uitz *et al.*, 2008; Hirata *et al.*, 2008). As with *in situ* monitoring, most work assessing taxonomic information from blooms in remote sensing of coastal areas has focused on HAB species.

14.4 Platforms addressing the varying scales of blooms

Many of the optical instruments (e.g. spectrophotometers, turbidity meters, fluorometers and irradiance and radiance sensors) are amenable to a variety of platforms (instrument carriers). The platforms discussed below allow measurements to be made over a wide range of space and time scales; however, the combination of many platforms must be used to cover the relevant scales for phytoplankton bloom dynamics (Schofield *et al.*, 2002; Moline *et al.*, 2005; Robbins *et al.*, 2006; Dickey *et al.*, 2008; Perry *et al.*, 2008). The available platforms include satellite and airborne remote sensing, together with fixed and mobile underwater platforms. Each of these platforms allows for measurement at different scales and thus different perspectives of the environment. Remote-sensing approaches provide synoptic surface maps (Kahru and Brown, 1997; Schofield *et al.*, 2004), while fixed underwater platforms provide Eulerian sampling (Johnsen *et al.*, 1997; Berge *et al.*, 2009), and mobile assets can provide both Lagrangian and near-synoptic perspectives (Moline *et al.*, 2005; Pegau *et al.*, 2006; Dickey *et al.*, 2008; Table 14.1).

14.4.1 Moorings

Moorings, either standalone (battery powered) or powered from an electro-optical cable connected to shore, provide subsurface optical data from a fixed location (Eulerian measurements). Battery powered systems, now on many moorings allow for a significant number of sensors to be powered for sustained periods of time (see different chapters in Babin *et al.*, 2008). Data are often available near-real time through surface wireless communication networks or via cabled systems. Such moorings need ample power supply and high resolution data telemetry so that researchers are able to oversample (i.e. higher than required data sampling frequency) in time, which can be extremely valuable for documenting dynamic trends within the local ecosystems and for developing parameterizations for ecosystem models (Cullen, 2008). The ability of moored equipment to detect phytoplankton blooms by measuring the changes in the bio-optical properties in the water provides information on the development, and, also importantly, the decay of blooms (Reid *et al.*, 1987; Stramska *et al.*, 1995; Cullen and Lewis, 1995; Johnsen *et al.*, 1997; Oliver *et al.*, 2004; Schofield *et al.*, 2004, 2008). The disadvantage of these sampling strategies is that they provide no spatial information and given that these systems are expensive, only a limited number can be deployed at any given time (Cullen, 2008). The number of depths resolved is also dependent on the number of sensors one wishes to place on the mooring, which also directly influences cost and power considerations. With multiples of optical packages on a single mooring, there is also the challenge of intercalibration and differential bio-fouling, although the problems with fouling are constantly being reduced with time (Manov *et al.*, 2004).

New profiling moorings are now being developed and used that resolve the vertical structure of phytoplankton and remove the need for intercalibration (Oliver *et al.*, 2004; Devol *et al.*, 2007; Penta *et al.*, 2008), with some being able to also selectively sample the phytoplankton community for identification (Moline *et al.*, 2009). Optical towers/profiling buoys (e.g. ARGO floats) and bio-optical gliders are now equipped with radiometers, fluorometers, transmissometers and backscattering meters to identify and quantify phytoplankton groups in seawater using spectral attenuation coefficients ($K_d(\lambda)$), spectral irradiance reflectance ($R(\lambda)$), remote-sensing reflectance ($R_{rs}(\lambda)$), *in vivo* phytoplankton absorption coefficients, and CDOM absorption coefficients (Babin *et al.*, 2008; Boss *et al.*, 2008; Ras *et al.*, 2008; Johnson *et al.*, 2009).

14.4.2 Mobile underwater platforms

Another set of platforms used for the collection of subsurface optical data that are increasingly being used by the community are mobile platforms that provide Lagrangian and near-synoptic perspectives (Table 14.1, Dickey *et al.*, 2008). These platforms include profiling floats, gliders and propeller-driven autonomous underwater vehicles (AUVs), (Figure 14.2). Mobile platforms are powerful tools for assessing spatial variability of properties and processes (Brown *et al.*, 2004; Moline *et al.*, 2005), extending the reach of fixed-platform observations, and enabling both a rapid response to study intermittent events and temporal evolution of phytoplankton blooms (Robbins *et al.*, 2006; Perry *et al.*, 2008). Their proven ability to operate during severe weather expands spatial sampling regimes to times or locations where surface vessels either have difficulty operating or are not cost effective. To a great extent, the individual platform determines the sensor package. Profiling floats and gliders are low-powered buoyancy-driven platforms and as such their sensor payloads are power and size restricted. These platforms are able to operate on the order of months to years, and in the case of gliders, are able to cover 400 to 7000 km on a single battery charge. Propeller-driven vehicles on the other hand have higher power capacity for sensors and can travel at higher speeds, but are limited in their operational duration from hours to days with ranges of 50 to 600 km. The ever-increasing longevity and ever-expanding range of mobile platforms provide the infrastructural backbone that will enable them to contribute to understanding phytoplankton dynamics at every scale and to examine the connections between local, regional and global processes.

14.4.3 Above-water remote sensing

All marine applications of remotely sensed ocean-colour data are accordingly very sensitive to the quality of the method of atmospheric correction (IOCCG, 2006; Folkestad *et al.*, 2007; Stumpf and Tomlinson, 2005; Morel, 2008; Ruddick *et al.*, 2008; McClain, 2009). Even small errors in the correction procedure can cause significant inaccuracies in the estimation of in-water OSC (Figure 14.3). The spectral features of R_{rs} are determined by IOP absorption and backscattering properties, but can be affected by sun-induced emissions. For phytoplankton cells, light absorption and emission characteristics are a result of the combined composition of chlorophylls, carotenoids and phycobiliproteins, whereas backscattering is determined largely by the cell size distribution and the complex index of refraction of algal cells

Table 14.4. *Current and scheduled ocean-colour sensors and satellites. (www.ioccg.org/sensors/scheduled).*

Sensor	Agency	Satellite	Scheduled launch	Swath (km)	Resolution (m)	# of bands	Spectral coverage (nm)
GOCI	KARI/KORDI (Korea)	COMS-1 (Geostationary)	2009	2500	500	8	400–865
OCM-II	ISRO (India)	ISR-P7 (Polar)	2008	1400	1000–4000	8	400–900
OLCI	ESA (Europe)	GMES-Sentinel-3 (Polar)	2012	1120	250/1000	16	400–900
S-GLI	JAXA (Japan)	GCOM-C Japan (Polar)	2012	1150	250/1000	19	375–12,500
VIIRS	NASA/IPO	NPP (Polar)	2009	3000	370/740	22	402–11,800
VIIRS	NASA/IPO	NPOES (Polar)	2012	3000	370/740	22	402–11,800

(Chapter 13.2.1). The distribution of absorption and backscattering properties in the water column greatly influences R_{rs}, with the surface properties being exponentially weighted.

Remote sensing of ocean colour can be conducted by aircraft and satellites to provide regional maps over time scales of hours to years. The resolution varies with the platform and with the focus area (Tables 14.1, 14.4). The phytoplankton blooms have essentially patchy patterns, whose spatial frequency spectrum slope is $\sim -5/2$ in the range 10^{-2}–10^{-1} km^{-1}, and it is $\sim -5/3$ in the range 10^{-1}–10^{0} km^{-1} (Kondratyev and Pozdnyakov, 1994). Thus, the preferable at-surface pixel dimensions are 100 m $\times$ 100 m to properly resolve the meso- to small-scale ocean processes. For remote sensing of open ocean waters, the spatial resolution of about 1 km $\times$ 1 km is considered acceptable – a compromise between the area covered, repeat cycle and the manageable data volume. Current resolution from satellites ranges from 250 m to 1 km (Ladner *et al.*, 2007; Tables 14.1, 14.4), and it is down to tens of cm for aircraft ocean colour sensors (Bissett *et al.*, 2004; Klonowski *et al.*, 2007; Volent *et al.*, 2007, 2009).

Derived products are primarily based on multi-spectral information of R_{rs}, giving information of pigment absorption and scattering, where high absorption gives a corresponding low R_{rs} signal. In spite of the limited number of wavelengths on these platforms, a large number of proxies can be derived. For example, NASA's Terra and Aqua satellites both carry a MODerate resolution Imaging Spectroradiometer (MODIS), which detects water leaving reflectance at seven wavebands (at 412, 443, 488, 531, 551, 667 and 678 nm) to estimate Chl *a* concentration, Chl *a* fluorescence, total suspended matter, coccoliths (chalk platelets from coccolithophores such as *Emiliania huxleyi*) and CDOM content.

One of the greatest advantages of using an aircraft is the reduction of distance between the sensor and sea surface, thus reducing the impact of the atmospheric contamination (noise). Locally based *in situ* time-series and operational remote sensing from hyperspectral imagers (1 nm spectral resolution or better) deployed on aeroplanes or tethered balloons (1–3 km above sea surface) is one way to circumvent the problem of atmospheric correction (Lee *et al.*, 1994; Chang *et al.*, 2004; Klonowski *et al.*, 2007; Volent *et al.*, 2007; Johnsen *et al.*, 2009). In addition, local water algorithms can be applied easily and verified with corresponding *in situ* (e.g. using collected water samples) measurements of phytoplankton concentration, pigment absorption signatures, suspended matter, and CDOM. Ships of opportunity, such as the European Ferrybox sensor system, are successfully deployed for monitoring marine areas. One of the most successful airborne imagers has been the Airborne Visible/Infrared Imaging Spectrometer (AVIRIS) system. This hyperspectral sensor was one of the first to demonstrate that the pigment specific groups of phytoplankton could be delineated from ocean colour (see Richardson, 1997). AVIRIS, Hyperion and other hyperspectral imagers continue to be used to detect cyanobacteria (Andrefouet *et al.*, 2003; Brando and Phinn, 2007), mixed assemblages (Richardson and Kruse, 1999), and red tides (Ryan *et al.*, 2005, 2008; Figure 14.4). Hyperspectral radiometric sensors (including imagers) are now commercially available from a wide number of companies and have sufficient spectral resolution to identify specific phytoplankton and macro algal taxa (Craig *et al.*, 2006; Volent *et al.*, 2007, 2009).

Another airborne sensor used to evaluate phytoplankton community structure is based on active fluorescence imaging. Unlike the sun-induced fluorescence used on MODIS, the Advanced Laser Fluorometer (ALF) uses multiple lasers of different wavelengths to evaluate the emission spectra of natural populations. Chekalyuk and Hafez (2008) used the system in multiple locations between 2005 and 2007 to discriminate between three spectral types of phycoerythrin for characterization of cyanobacteria and cryptophytes in mixed phototrophic populations, and illustrating the promise of this technology for local and regional applications. These remote-sensing approaches are invaluable in providing a spatial perspective. However, the satellite signal (denoted 'first optical depth') is exponentially weighted to surface waters; therefore, *in situ* subsurface information by underwater robots equipped with sensors is also required to assess the full extent of blooms and potential biogeochemical impacts.

14.5 Case studies of optical phytoplankton monitoring

In addition to highlighting various sensors and platforms that provide optical measurements for the detection and identification of algal blooms, it is important to provide examples of how these systems are integrated towards achieving this common goal. If the full impact of blooms, with regard to their biogeochemical influence, their toxic effects, and their ecological consequences are to be realized, monitoring on the basis of Chl *a* alone will not suffice. There has been a concerted effort in the past two decades to advance beyond phytoplankton biomass measurements and to understand what regulates algal assemblages and selects for species-specific blooms. Phytoplankton community structure within a watermass has been shown to change rapidly in response to water column stabilization (Moline, 1998), sheer (Durham *et al.*, 2009), differential nutrient availability (Pinckney *et al.*, 1998, 1999), light quality/quantity (Prézelin and Boczar, 1986; Schofield *et al.*, 1991; Kroon *et al.*, 1993; Johnsen *et al.*, 1997) and selective grazing (Moline *et al.*, 2008). Variations in phytoplankton communities can differentially influence primary productivity, nutrient utilization (Moline *et al.*, 2004a), and trophic ecology of a region (Moline *et al.*, 2004c; Johnsen *et al.*, 2009; Suggett *et al.*, 2009; MacIntyre *et al.*, 2010).

14.5.1 In situ *monitoring of gyroxanthin diester in* Karenia brevis

Most autotrophic dinoflagellates contain the carotenoid peridinin as a primary accessory pigment (Jeffrey *et al.*, 1975; Prézelin, 1987; Millie *et al.*, 1993). However, some dinoflagellate species have replaced peridinin with fucoxanthin and

19′-acylofucoxanthins as primary accessory pigments (Jeffrey *et al.*, 1975; Bjørnland *et al.*, 1988; Johnsen and Sakshaug, 1993, and see Chapter 1, this volume). Six species of the fucoxanthin-containing dinoflagellates have a relatively unique carotenoid called gyroxanthin-diester (here called gyroxanthin for simplicity, Liaaen-Jensen, 1989; Bjørnland *et al.*, 2000; Table 14.2). Gyroxanthin or gyroxanthin-like carotenoids have also been reported for one pelagophyte and two haptophyte species (Bjørnland *et al.*, 2003; Zapata, 2005; Jeffrey and Wright, 2006).

A survey of 14 published studies on blooms of *Karenia brevis* revealed a mean gyroxanthin cellular concentration of 0.364 ± 0.226 pg cell^{-1} with a 95% confidence interval of 0.234–0.494 pg gyroxanthin cell^{-1}. This cellular concentration can be used as a conservative estimator for calculating the cellular abundance of *K. brevis*.

HPLC-based measurements of gyroxanthin provide a useful method for quantifying the abundance of *K. brevis* in field samples (Richardson and Pinckney, 2004). Gyroxanthin has been included in CHEMTAX analyses (see Chapter 6, this volume) of phytoplankton samples known to contain species with gyroxanthin. The ratio of the weights of gyroxanthin to Chl *a* determined using CHEMTAX (i.e. from the final ratio matrix) ranged from 0.027 (Rodríguez *et al.*, 2003) to 0.043 (Örnólfsdóttir *et al.*, 2003).

The unique *in vivo* absorption characteristics of gyroxanthin-containing *K. brevis* have also been used to quantify the *in situ* relative abundance of this species (Millie *et al.*, 1995, 1997; Stæhr and Cullen, 2003; Kirkpatrick *et al.*, 2008). Optical plankton discriminators, which measure in vivo spectral absorption, were first used in flow-through systems on ships, providing a similarity index for *K. brevis* on a scale of 0 to 1 based on the gyroxanthin contribution to the absorption spectra (Kirkpatrick *et al.*, 2000). These systems were then deployed successfully in autonomous mode on moorings, reporting data hourly via a cell phone communication link. In the most recent iteration, these sensors have been integrated into gliders and propeller-driven AUVs to map the three-dimensional distribution of *K. brevis* blooms off the west coast of Florida (Robbins *et al.*, 2006; Kirkpatrick *et al.*, 2008; Figure 14.2, Table 14.1). This approach, when combined with other on-board sensors, provides a powerful aid for interpretation of R_{RS} and as a tool for tracking and understanding the development of benign and harmful algal blooms.

Future efforts to detect and track phytoplankton groups or species, in the case of phytoplankton blooms, need a combination of both 'low tech' approaches, such as anecdotal information and local historical knowledge based on experience, and monitoring approaches, such as the use of instrumentation, sample analyses, and platforms described in this chapter. Specific knowledge of recurring phytoplankton blooms in a given region may provide reasonable assurance that the appearance of a particular pigment is directly related to a given species. Without the historical and pigment information for a given area, detection of, in this case, gyroxanthin would be difficult to interpret since it is not species-specific or even class specific, as it is found in dinoflagellates with tertiary endosymbiont chloroplasts (Bjørnland and Tangen, 1979; Johnsen and Sakshaug, 1993; Bjørnland *et al.*, 2003) and possibly in some haptophytes and pelagophyceans (Zapata, 2005; Garcés *et al.*, 2006; Table 14.2).

To use phytoplankton pigments as bio-optical markers for *in situ* operational monitoring, detection and identification, the pigments need to be present in concentrations high enough to be detected *in situ* and by remote sensing (Table 14.3, Figure 14.4). This also indicates that only the major pigment signatures may be detected (Figures 14.2–14.3, Table 14.3).

14.5.2 Satellite monitoring of phycobiliproteins in **Trichodesmium**

Using satellite imagery based on observations of absorption spectra and marine reflectance modelling, Subramaniam *et al.* (1999) provided a path for identifying surface nitrogen-fixing *Trichodesmium* blooms. A follow-on field study combining field samples and satellite observations confirmed that the cyanobacterium *Trichodesmium* could be detected using the unique reflectance due to phycoerythrin at approximately 570 nm (Subramaniam *et al.*, 2002). Critical to the detection of *Trichodesmium* is the need for a critical threshold for detection and a unique signal strong enough to overcome the other in-water optical constituents and the corresponding atmospheric contamination. With an improved algorithm, Westberry *et al.* (2005) found this threshold and sequentially applied this algorithm to the SeaWiFS data set to quantify the global occurrences of *Trichodesmium* blooms (Westberry and Siegel, 2006).

14.5.3 Satellite monitoring of Chl c_3 *and coccoliths in* **Emiliania huxleyi**

The coccolithophyte *Emiliania huxleyi* is easily detected by satellite spectroradiometers as cells covered with coccoliths and/or free coccoliths (highly reflective chalk platelets surrounding the cells) in seawater (Balch *et al.*, 1991; Smyth *et al.*, 2004). This information can also be used as tracers for different water masses, which is important for monitoring purposes (Figure 14.1). Often, remotely sensed images of *E. huxleyi* blooms lack synchronized *in situ* measurements and verification of water samples by microscopy, pigment signatures by HPLC from water samples or using *in situ* imagery or molecular tools. There have been several examples where data from remotely sensed images were incorrectly interpreted as major 'mono-specific blooms' of *E. huxleyi* (Z. Volent and G. Johnsen, unpublished). An actual phytoplankton bloom-scenario may be interpreted differently if corresponding water samples have been examined from the same remotely sensed bloom. An extensive bloom of *E. huxleyi* cells in a Norwegian fjord system in August 2004, after analysing water samples and *in situ* bio-optical sensor data, was found to be in a post-bloom phase dominated with a large amount of free coccoliths and few viable cells of *E. huxleyi*.

The water samples which by eye were dominated by light scattered by coccoliths, revealed high concentrations of large dinoflagellates dominated by *Ceratium* spp. with correspondingly high concentrations of peridinin (up to 1.5 mg m^{-3}). This emphasizes the importance of pigments as a means of identifying phytoplankton groups (Figures 14.1 and 14.3, Appendix 14A). Although satellite remote sensing has proven to be a useful tool, measurements and knowledge of pigment signature *in situ* are necessary as aids to confirm the identity and physiological state of bloom species or pigment groups present.

14.6 Future perspectives

14.6.1 Phytoplankton pigments and toxins

With the current need for coastal waters monitoring caused by a growing human population and corresponding growth in the aquaculture, fish, transport and tourist industries, the relationship between algal taxa and corresponding metabolite production, such as toxins, needs to be elucidated. The literature provides surprisingly little information regarding pigment signatures of potentially harmful phytoplankton classes and species, with most of the effort being focused on defining toxins and their respective effects (see Appendix 14A). Since both pigments and toxins can be used as chemotaxonomical markers more work needs to be done regarding pigment composition in these algae. A 'good' HAB-pigment marker should not differ significantly as a function of changes in environmental variables, such as light climate, nutrients, temperature, salinity or physiological status (Rodríguez *et al.*, 2006; Johnsen and Sakshaug, 2007). As an example, only a few species of the PSP (paralytic shellfish poisoning)-producing dinoflagellate genus *Alexandrium* have been examined with respect to pigment signature. The typical dinoflagellate marker peridinin was found in the three species already examined, while the toxin profiles were highly variable (Carreto *et al.*, 2001). Chl c_3 is found in the four HAB phytoplankton classes identified as being potentially involved in fish mortality (Appendix 14A, see details regarding Chl c_3 in other phytoplankton classes in Chapter 1, this volume). HAB species of diatoms typically belonging to the genus *Pseudo-nitzschia* may contain domoic acid causing amnesic shellfish poisoning (ASP), but no pigment information for species of this genus has been published to date.

A large fraction of harmful genera of dinoflagellates are responsible for PSP (e.g. *Alexandrium*), DSP (diarrhetic shellfish poisoning, e.g. *Dinophysis* and *Prorocentrum*), NSP (neurotoxic shellfish poisoning, e.g. *Karenia*), or produce ichthyotoxins (e.g. *Karenia* and *Karlodinium*). There is an equally high pigment diversity in sub-groups of toxic dinoflagellates, such as peridinin-containing species that are often responsible for PSP and DSP outbreaks, Chl c_3-containing species that are often ichthyotoxic (*Karlodinium* spp., *Karenia* spp., *Takayama tasmanica*, Table 14.2), and alloxanthin-containing species responsible for DSP (*Karenia cristata*, *Dinophysis norvegica*).

In addition to dinoflagellates, typical ichthyotoxic Chl c_3-HAB classes include coccolithophytes, pelagophytes, and dictyochophytes (Figure 14.1). The most globally important ichthyotoxic HABs and Chl c_3 containing classes are the coccolithophytes (previously called prymnesiophytes: see Chapter 1, this volume), dinoflagellates, dictyochophytes, and pelagophytes (Appendix 14A). The use of remote-sensing data for time-series (phytoplankton dynamics) during an ichthyotoxic bloom is shown in Figure 14.1D–G.

Chlorophytes are typically non-HAB species, with the exception of the botryoxanthin and braunixanthin-containing *Botryococcus braunii* that cause fish mortality in freshwater systems (Appendix 14A). Cyanobacteria show high toxin diversity and harmful effects on several groups of organisms. The presence of phycobiliproteins in cyanobacteria (but only in a few species of prochlorophytes) affects their bio-optical properties and can be used to differentiate cyanobacteria from other phytoplankton groups (see Claustre *et al.*, 1997; Kahru and Brown, 1997; Moline *et al.*, 2004c; Chapters 9 and 13, this volume).

For future *in situ* detection and monitoring of group-specific and species-specific blooms, characterization of their pigment signatures for potential optical detection is a logical next step and reasonable goal. This will not only advance *in situ* detection and monitoring of phytoplankton blooms, but also improve detection potential above water using remote sensing (Claustre *et al.*, 2004). For future interpretation of Chl *a* fluorescence kinetics for photosynthetic measurements (e.g. fast repetition rate fluorometry or pulse amplitude modulated fluorometry) it is important to stress that the data are sensitive to community structure and physiology (Chapter 13, this volume; Johnsen and Sakshaug, 2007; Babin *et al.*, 2008; Suggett *et al.*, 2009; MacIntyre *et al.*, 2010).

14.6.2 In situ *detection*

Because *in situ* and *in vivo* bio-optical properties of phytoplankton are pigment dependent, the discrimination success *in situ* is dependent upon the number of wavelengths, the choice of wavelengths, and the degree of interference from non-phytoplankton optical signature (Kahru and Brown, 1997; Bricaud *et al.*, 2004; Claustre *et al.*, 2004; Alvain *et al.*, 2005). The development of sensors that are able to process samples and remove non-phytoplankton signatures *in situ* to detect taxon specific absorption pigment peaks (e.g. 'Brevebuster' deployed on AUV, Robbins *et al.*, 2006; Kirkpatrick *et al.*, 2008) will greatly enhance our ability to monitor phytoplankton community structure. As mentioned above, the dependence of spectral fluorometry only on phytoplankton provides a promising technique to discriminate between pigment groups. *In situ* fluorescence excitation spectra (400–700 nm) provide distinct signatures from different pigment groups when examining the emission of Chl *a* at 730 nm (Johnsen and Sakshaug, 2007). This clearly resolves different phytoplankton pigment groups in the laboratory and should be applicable *in situ*. Solid state sensors using small bandwidth (10 nm) LEDs across the visible part of the spectrum (400–650 nm) using the major emission band of Chl *a* at 685 nm will reveal pigment differences *in situ*. Note that in vivo Chl *a* emission is significantly higher at 685 nm (gives higher sensitivity, but can not measure excitation wavelengths longer than approximately 660 nm) than at 730 nm (lower sensitivity than 685 nm, but can measure excitation wavelengths up to 700 nm) and that different instruments/techniques basically use one of the two emission settings (for details, see Chapter 13, this volume, and Suggett *et al.*, 2010).

Regardless of any new detection method or sensor developed, it will be equally important to select a suite of platforms that can deploy these sensors over relevant scales of time and space to address a given process (i.e. bloom dynamics, HABs etc). Recent improvements in field observation platforms (e.g. AUVs, gliders) expand the capabilities to sample the ocean at unprecedented scales. Development of these sensing networks will project a 3D view of phytoplankton structure from submesoscale to a regional perspective. The combination of these *in situ* approaches, in addition to the remote-sensing technologies, will also help to monitor and map the world oceans in the next decades. The future for *in situ* and remotely sensed data networks is promising, especially since the ocean colour community is highly interdisciplinary and well integrated. The use of new hyperspectral optical sensors and imagers, employing wavelength targeting on pigment-group specific absorption peaks and valleys using state-of-the-art statistical software, will enhance the predictive power for *in situ* discrimation between different pigment-groups of phytoplankton and macroalgae (Chang *et al.*, 2004; Claustre *et al.*, 2004; Babin *et al.*, 2005; Craig *et al.*, 2006; Brando and Phinn, 2007; Volent *et al.*, 2007). For excellent general reviews on in vivo and *in situ* bio-optical detection, including methods, new instruments/sensors and applications, see Babin *et al.* (2008) and Suggett *et al.* (2010).

14.6.3 Remote sensing

The use of standard remote-sensing products across all ocean types can lead to misleading results due to the optical complexity of Case II waters. This is illustrated explicitly in Figure 14.3 (left image) with no indication of the massive biomass of algae and cyanobacteria. However, application of the regionally specific NERSC/NIERSC bio-optical algorithm (applicable to local Case II waters, see above) to the same MODIS image reveals a highly pronounced bloom exactly in the expected area (Figure 14.3, right image). Sequential MODIS images processed by this algorithm in combination with the MODIS SST imagery provide a comprehensive data set to characterize this bloom (see also Figure 14.1).

In addition to algorithms tuned to regions and/or specific blooming species, it is also important to understand broad group-level composition of a phytoplankton community. The composition is important to improve primary production estimates (Claustre *et al.*, 1997; Moline *et al.*, 2004a; Tilstone *et al.*, 2005), examine trophic interactions (Moline *et al.*, 2008), and predict effects of climate change (Moline *et al.*, 2004c).

The spectral absorption and fluorescence bandwidths for algal pigments in living cells typically range between 5 to 50 nm, depending on the pigment(s) present (for details see Chapter 13, this volume). The more pigments absorb in a given spectral region (e.g. 440–480 nm), the less distinct the spectral signature of a given pigment. These features are superimposed upon the water-induced spectral curvature of R_{rs}, and in remote sensing the spectral data appear as wide-band dips accompanied by inflexions. To detect such minor, but expectedly very informative spectral curvature details, one would expect, as the number of wavelengths increases, a more detailed differentiation between different pigment groups of phytoplankton. Using nearly 400 measures of hyperspectral R_{rs} taken from numerous locations, Lee *et al.* (2007) showed that 24 wavelengths were able to capture the vast majority of the spectral information content (including accessory pigments) in the reflectance signal, based on a derivative analysis. These results suggest that perhaps full hyperspectral information is not necessarily required for future sensors, but that the wavelengths need to be selected carefully. These conclusions are similar to an *in situ* canonical correlation analysis between phytoplankton pigmentation and hyperspectral reflectance, which showed that only 25 wavelengths were found to be significant (Moline *et al.*, 2004b). Additional bands in the near infrared will be necessary to obtain the information needed for the atmospheric correction procedure (Gordon, 1997; Gao *et al.*, 2000). For practical constraints (data rates versus spatial sensor coverage), the current operational satellites operate in 8 to 15 visible spectral bands, with typical bandwidths of 5 to 15 nm. Table 14.4 lists the current and future scheduled ocean-colour sensors for the period 2009 to 2012. Starting from 2008 to 2012, these sensors meet, but only partially, the requirements in terms of spatial and spectral resolution – however most sensors provide a continuity of proven technologies already in orbit. This will provide useful long-term time series of observations.

The instruments for future *in situ* and remote sensing will be highly advanced and multifunctional. For instance, the GMES Sentinel-3 mission encompasses, *inter alia*, open ocean and coastal zones monitoring, sea surface level, wind speed and water surface roughness, sea ice, but also, land cover, land-surface temperature, and atmospheric properties. The ocean-colour and SST data products will be provided by the Sentinel-3 Ocean and Land Colour imager (OLC) fitted out with 16 channels centred at 1:413, 2:443, 3:490, 4:510, 5:560, 6:620, 7: 665, 8:681, 9:709, 10:754, 11:761, 12:779, 13: 865, 14:885, 15:900, and 16:1020 nm with a 10 nm bandwidth in the visible, but 7.5 nm for channel 8. The data will be obtained at 250 m spatial resolution in coastal waters and 1 km for the global ocean coverage. The following marine applications of channels 1–9 are assumed:1 – CDOM and detrital pigments, 2 – Chl *a* absorption max., 3 – Chl *a*, other pigments, 4 – Chl *a*, red tide, 5 – Chl *a* reference, 6 – sediment loading, 7 – Chl *a*, sediment, CDOM, 8 – Chl *a* fluorescence peak, 9 – Chl *a* fluorescence baseline. In addition to the augmentation of the global constellation of ocean-colour satellites, regional and local efforts will need to continue the use of *in situ* and remote-sensing techniques supplemented with in-water measurements to better refine our ability to detect and monitor the complexity of algal communities.

For future remote and *in situ* sensing of phytoplankton bio-optical taxonomy (Table 14.1), the combined information for pigments (Tables 14.2–4) and their corresponding bio-optical properties (Figures 14.1–4, Figures 13.2–3, this volume) will enhance the information level in the near future. Of the seven PFT, three discrete groups can be optically discriminated, namely phycobiliprotein-, divinyl Chl *a*+*b*- and Chl *c*-containing phytoplankton, comprising cyanobacteria, prochlorophytes and chromophytes, respectively (Nair *et al.*, 2008). The chromophytes can then be further divided into highly scattering groups such as the coccolithophore *Emiliania huxleyi* or low scattering cells such as several diatoms (Balch *et al.*, 1991; Smyth *et al.*, 2004; Roesler and Boss, 2008; Sosik, 2008).

With better sensors and software (Chapters 6 and 13, this volume), a general knowledge of pigments and their corresponding bio-optical properties, the combined information from remote and *in situ* approaches will probably make possible the discrimination of between 12 to 16 different bio-optical groups.

Acknowledgements

The authors thank the external reviewers for constructive help with this chapter. Comments and suggestions from our editors Drs. Suzanne Roy and Carole A. Llewellyn have been most valuable. Insightful comments and discussions with Drs. Peter Fearns and Mervyn Lynch at Curtin University, Perth, are also greatly acknowledged.

Abbreviations and symbols

$a(\lambda)$	Absorption coefficient at a specified wavelength
ALF	Advanced laser fluorometer
AOA	Algae online analyser
AOP	Apparent optical properties
ASP	Amnesic shellfish poisoning
AUV	Autonomous underwater vehicle
AVIRIS	Airborne visible/infrared imaging spectrometer
AZP	Azaspiracid shellfish poisoning
$b(\lambda)$	Scattering coefficient at a specified wavelength
$b_b(\lambda)$	Backscattering coefficient at a specified wavelength

CCD	Couple charge device
$C_{\mathrm{Chl\,a}}$	Chl *a* concentration
CFP	Ciguatera fish poisoning
DOM	Dissolved organic matter
DMS	Dimethyl sulfide
DSP	Diarrhetic shellfish poisoning
E_{d}	Downwelling irradiance
ELISA	Enzyme-linked immuno-sorbent assay
E_{u}	Upwelling irradiance
FLH	Fluorescence line height
HAB	Harmful algal blooms
IOP	Inherent optical properties
$K_{\mathrm{d}}(\lambda)$	Vertical diffuse attenuation coefficient at a specified wavelength
LHP	Light-harvesting pigments
L_{u}	Upwelling radiance
MODIS	MODerate resolution Imaging Spectrometer
NSP	Neurotoxic shellfish poisoning
OC	Ocean colour
OLC	Ocean and land colour imager
OPL	Optical path length
OSC	Optically significant constituents
PFT	Phytoplankton functional types
PIM	Particulate inorganic matter
POM	Particulate organic matter
PSP	Paralytic shellfish poisoning
RS	Remote sensing
PSII	Photosystem II
$R(\lambda)$	Water reflectance at a specified wavelength
SFS	Spectral fluorescence signatures
SM	Suspended matter

References

Alvain, S., Moulin, C., Dandonneau, Y. and Bréon, F. M. (2005). Remote sensing of phytoplankton groups in case 1 waters from global SeaWiFS imagery. *Deep-Sea Res. I* **52**, 1989–2004.

Andrefouet, S., Payri, C., Hochberg, E. J., Che, L. M., Atkinson, M. J. (2003). Airborne hyperspectral detection of microbial mat pigmentation in Rangiroa Atoll (French Polynesia). *Limnol. Oceanogr.* **48**, 426–30.

Antoine, D., d'Ortenzio, F., Hooker, S. B., Bécu, G., Gentili, B., Tailiez, D. and Scott, A. J. (2008). Assessment of uncertainty in the ocean reflectance determined by three satellite ocean color sensors (MERIS, SeaWiFS, and MODIS-A) at an offshore site in the Mediterranean Sea (BOUSSOLE project). *J. Geophys. Res.* **113**, C07013, doi:10.1029/2007JC004472.

Babin, M., Cullen, J. J., Roesler, C. S., Donaghay, P. L., Douchette, G. J., Kahru, M., Lewis, M. R., Scholin, C. A., Sieracki, M. E., Sosik, H. M. (2005). New approaches and technologies for observing harmful algal blooms. *Oceanography* **18**, 210–27.

Babin, M, Roesler, C. S., Cullen, J. J. (eds.) (2008). *Real-Time Coastal Observing Systems for Ecosystem Dynamics and Harmful Algal Blooms: Theory, Instrumentation and Modelling*. Paris: UNESCO Publishing.

Balch, W. M., Holligan, P. M., Ackleson, S. G., Voss, K. J. (1991). Biological and optical properties of mesoscale coccolithophoride blooms in the Gulf of Maine. *Limnol. Oceanogr.* **36**, 629–43.

Berge, J., Cottier, F., Last, K., Varpe, Ø., Leu, E., Søreide, J., Eiane, K., Falk-Petersen, S., Willis, K., Nygård, H., Voegedes, D., Griffiths, C., Johnsen, G., Lorenzen, D., Brierley, A. S. (2009). Diel vertical migration of Arctic zooplankton during the polar night. *Biol. Lett.* **5**, 69–72.

Bissett, W. P., Arnone, R. A., Davis, C. O., Dickey, T. D., Dye, D., Kohler, D. D. R. and Gould, R. W.Jr. (2004). From meters to kilometers: A look at ocean-color scales of variability, spatial coherence, and the need for fine-scale remote sensing in coastal ocean optics. *Oceanography* **15**, 33–43.

Bjørnland, T. and Tangen, K. (1979). Pigmentation and morphology of a marine *Gyrodinium* (Dinophyceae) with a major carotenoid different from peridinin and fucoxanthin. *J. Phycol.* **15**, 457–63.

Bjørnland, T., Guillard, R. R. L. and Liaaen-Jensen, S. (1988). *Phaeocystis* sp. clone 677-3 –a tropical marine planktonic prymnesiophyte with fucoxanthin and 19′-acyloxyfucoxanthins as chemosystematic carotenoid markers. *Biochem. Syst. Ecol.* **16**, 445–52.

Bjørnland, T., Fiksdahl, A., Skjetne, T., Krane, J. and Liaaen-Jensen, S. (2000). Gyroxanthin – the first allenic acetylenic carotenoid. *Tetrahedron* **56**, 9047–56.

Bjørnland, T., Haxo, F. T. and Liaaen-Jensen, S. (2003). Carotenoids of the Florida red tide dinoflagellate *Karenia brevis*. *Biochem. Syst. Ecol.* **31**, 1147–62.

Boss, E., Perry, M. J., Swift, D., Taylor, L., Brickley, P., Zaneveld, J. R. V. and Riser, S. (2008). Three years of ocean data from a bio-optical profiling float. *EOS, Trans. AGU* **89**, 209.

Bricaud, A., Claustre, H., Ras, J. and Oubelkheir, K. (2004). Natural variability of phytoplankton absorption in oceanic waters: Influence of size structure of algal populations. *J. Geophys. Res.* **109**, C11010, doi:10.1029/2004JC002419.

Brando, V. E. and Phinn, S. T. (2007). Coastal aquatic remote sensing applications for environmental monitoring and management. Special issues editorial. *J. Appl. Remote Sens.* **1**, 1–6.

Brown, C. A., Huot, Y., Purcell, M. J., Cullen, J. J. and Lewis, M. R. (2004). Mapping coastal optical and biogeochemical variability using an Autonomous Underwater Vehicle (AUV) and a new bio-optical inversion algorithm. *Limnol. Oceanogr. Methods* **2**, 262–81.

Bukata, R. P., Jerome, J. H., Kondratyev, K. Y. and Pozdnyakov, D. V. (1995). *Optical Properties and Remote Sensing of Inland and Coastal Waters*. Boca Raton: CRC Press.

Carreto, J. I., Carignan, M. O. and Montoya, N. G. (2001). Comparative studies on mycosporine-like amino acids, paralytic shellfish toxins and pigment profiles of the toxic dinoflagellates *Alexandrium tamarense*, *A. catenella* and *A. minutum*. *Mar. Ecol. Prog. Ser.* **223**, 49–60.

Chang, G., Mahoney, K., Briggs-Whitmire, A., Kohler, D. D. R., Mobley, C. D., Lewis, M., Moline, M. A., Boss, E., Kim, M., Philpot, W. and Dickey, T. D. (2004). The new age of hyperspectral oceanography. *Oceanography* **17**, 16–23.

Chekalyuk, A. and Hafez, M. (2008). Advanced laser fluorometry of natural aquatic environments. *Limnol. Oceanogr. Methods* **6**, 591–609.

Ciotti, A. M. and Bricaud, A. (2006). Retrievals of a size parameter for phytoplankton and spectral light absorption by colored detrital matter from water-leaving radiances at SeaWiFS channels in a continental shelf region off Brazil. *Limnol. Oceanogr. Methods* **4**, 237–53.

Claustre, H., Moline, M. A. and Prézelin, B. B. (1997). Sources of variability in the column photosynthetic cross section for Antarctic coastal waters. *J. Geophys. Res.* **102**, 25047–60.

Claustre, H., Hooker, S. B., Van Heukelem, L., Berthon, J.-F., Barlow, R., Ras, J., Sessions, H., Targa, C., Thomas, C. S., van der Linde, D., Marty, J-C. (2004). An intercomparison of HPLC phytoplankton pigment methods using in situ samples: application to remote sensing and database activities. *Mar. Chem.* **85**, 41–61.

Craig, S. E., Lohrenz, S. E., Lee, Z., Kirkpatrick, G. J., Schofield, O. and Steward, R. G. (2006). Use of hyperspectral remote sensing reflectance for detection and assessment of the harmful alga, *Karenia brevis*. *Appl. Optics* **45**, 5415–25.

Cullen, J. J. (2008). Observation and prediction of harmful algal blooms. In *Real-Time Coastal Observing Systems for Ecosystem Dynamics and Harmful Algal Blooms: Theory, Instrumentation and Modelling*, ed. M. Babin, C. S. Roesler and J. J. Cullen. Paris: UNESCO Publishing, pp. 1–41.

Cullen, J. J. and Lewis, M. R. (1995). Biological processes and optical measurements near the sea surface: Some issues relevant to remote sensing. *J. Geophys. Res.* **100** (C7), 13255–66.

de Salas, M. F., Bolch, C. J. S., Botes, L., Nash, G., Wright, S. W. and Hallegraeff, G. M. (2003). *Takayama* gen. nov. (Gymnodiniales, Dinophyceae), a new genus of unarmored dinoflagellates with sigmoid apical grooves, including the description of two new species. *J. Phycol.* **39**, 1233–46.

Devol, A. H., Ruef, W., Emerson, S. and Newton, J. (2007). In situ and remote monitoring of water quality in Puget Sound: the ORCA time-series. *Environ. Res. Eng. Manag.* **1**(39), 19–23.

Dickey, T. D., Itsweire, E. C., Moline, M. A. and Perry, M. J. (eds.) (2008). Limnology and Oceanography special issue on Autonomous and Lagrangian Platforms and Sensors (ALPS). *Limnol. Oceanogr.* **53** (part 2), 2057–360.

Dierssen, H. M. and Smith, R. C. (2000). Bio-optical properties and remote sensing ocean color algorithms for Antarctic Peninsula waters. *J. Geophys. Res.* **105**, 26301–12.

Doerffer, R. and Fischer, J. (1994). Concentrations of chlorophyll, suspended matter and gelbstoff in case II waters derived from satellite coastal zone color scanner data with inverse modelling methods. *J. Geophys. Res.* **99**, 7457–66.

Doubovick, O. V., Ostchepkov, S. L. and Lapenock, T. V. (1994). Iteration/regularization method of solution to nonlinear inverse problems and its application to interpreting water leaving radiance spectra. *Proc. Russian Academy of Sciences: Ocean and Atmospheric Physics* **30**, 106–13. (in Russian)

Durham, W. M., Kessler, J. O. and Stocker, R. (2009). Formation of thin phytoplankton layers disruption of vertical motility by shear triggers. *Science* **323**, 1067–70.

Edvardsen, B., Eikrem, W., Shalchian-Tabrizi, K., Riisberg, I., Johnsen, G., Naustvoll, L. and Throndsen, J. (2007). *Verrucophora farcimen* gen. et sp. nov. (Dictyochophyceae, Heterokonta) – a bloom forming ichthyotoxic flagellate from the Skagerrak, Norway. *J. Phycol.* **43**, 1054–70.

Falkowski, P. G., Laws, E. A., Barber, R. T. and Murray, J. W. (2003). Phytoplankton and their role in primary, new, and export production. In *Ocean Biogeochemistry: The Role of the Ocean Carbon Cycle in Global Change*, ed. M. J. R. Fasham. Berlin: Springer, pp. 99–121.

Fawcett, A., Bernard, S., Pitcher, G. C., Probyn, T. A., du Randt, A. (2006). Real-time monitoring of harmful algal blooms in the southern Benguela. *Afr. J. Mar. Sci.* **28**, 257–60.

Folkestad, A., Pettersson, L. H. and Durand, D. (2007) Inter-comparison of ocean colour data products during algal blooms in the Skagerrak. *Int. J. Rem. Sens.* **28**, 569–92.

Franks, J. S. (1997). Spatial patterns in dense algal blooms. *Limnol. Oceanogr.* **42**, 1297–305.

Gao, B.-C., Montes, M. J., Ahmad, Z. and Davis, C. O. (2000). Atmospheric correction algorithm for hyperspectral remote sensing of ocean color from space. *Appl. Optics* **39**, 887–96.

Garcés, E., Fernandez, M., Penna, A., Van Lenning, K., Gutiérrez, A. and Camp, J. (2006). Characterization of NW Mediterranean *Karlodinium* spp. (Dinophyceae) strains using morphological, molecular, chemical and physiological methodologies. *J. Phycol.* **42**, 1096–112.

Garver, S. A., Siegel, D. A. and Mitchell, B. G. (1994). Variability in near-surface particulate absorption spectra: What can a satellite ocean color imager see? *Limnol. Oceanogr.* **39**, 1349–67.

Glenn, S. M. and Schofield, O. (2003). Observing the oceans from the COOL room: our history, experience, and opinions. *Oceanography* **16**, 37–52.

Gordon, H. R. (1997). Atmospheric correction of ocean color imagery in the Earth Observing System era. *J. Geophys. Res.* **102**, 17081–106.

Gower, J. F. R. and Borstad, G. A. (2004). On the potential of MODIS and MERIS for imaging chlorophyll fluorescence from space. *Int. J. Rem. Sens.* **25**, 1459–64.

Griffiths, G. (2008). Glider and autonomous underwater vehicle observing systems. In *Real-Time Coastal Observing Systems for Ecosystem Dynamics and Harmful Algal Blooms: Theory, Instrumentation and Modelling*, ed. M. Babin, C. S. Roesler and J. J. Cullen. Paris: UNESCO Publishing, pp. 495–525.

Hansen, G., Daugbjerg, N. and Henriksen, P. (2000). Comparative study of *Gymnodinium mikimotoi* and *Gymnodinium aureolum*, comb. nov. (= *Gyrodinium aureolum*) based on morphology, pigment composition, and molecular data. *J. Phycol.* **36**, 394–410.

Hirata, T., Aiken, J., Hardman-Mountford, N. J., Smyth, T. J. and Barlow, R. G. (2008). An absorption model to determine phytoplankton size classes from satellite ocean colour. *Remote Sens. Environ.* **112**, 3153–59.

Huot, Y., Brown, C. A. and Cullen, J. J. (2005). New algorithms for MODIS sun-induced chlorophyll fluorescence and a comparison with present data products. *Limnol. Oceanogr. Methods* **3**, 108–30.

Huot, Y., Brown, C. A. and Cullen, J. J. (2007). Retrieval of phytoplankton biomass from simultaneous inversion of reflectance, the diffuse attenuation coefficient, and sun-induced fluorescence in coastal waters. *J. Geophys. Res.* **112**, C06013, doi:10.1029/2006JC003794.

IOCCG (2000). *Remote sensing of ocean colour in coastal, and other optically complex waters*, ed. S. Sathyendranath. Reports of the International Ocean-Colour Coordinating Group, no. 3. Dartmouth: IOCCG.

IOCCG (2006). *Remote Sensing of Inherent Optical Properties: Fundamentals, Tests of Algorithms, and Applications*, ed. Z.-P. Lee. Reports of the International Ocean-Colour Coordinating Group, no. 5. Dartmouth: IOCCG.

Jeffrey, S. W. and Vesk, M. (1997). Introduction to marine phytoplankton and their pigment signatures. In *Phytoplankton Pigments in Oceanography: Guidelines to Modern Methods*, ed. S. W. Jeffrey, R. F. C. Mantoura and S. W. Wright. Paris: UNESCO Publishing, pp. 37–84.

Jeffrey, S. W. and Wright, S. W. (2006). Photosynthetic pigments in marine microalgae: Insights from cultures and the sea. In *Algal Cultures, Analogues of Blooms and Applications*, ed. D. V. Subba Rao. Enfield: Science Publishers, pp. 33–90.

Jeffrey, S. W., Sielicki, M. and Haxo, F. T. (1975). Chloroplast pigment patterns in dinoflagellates. *J. Phycol.* **11**, 374–85.

Jeffrey, S. W., Mantoura, R. F. C. and Wright, S. W. (eds.) (1997). *Phytoplankton Pigments in Oceanography: Guidelines to Modern Methods*. Paris: UNESCO Publishing.

Jerlov, N. G. (1976). *Marine Optics*. New York: Elsevier.

Johnsen, G. and Sakshaug, E. (1993). Bio-optical characteristics and photoadaptive responses in the toxic and bloom-forming dinoflagellates *Gyrodinium aureolum*, *Gymnodinium galatheanum*, and two strains of *Prorocentrum minimum*. *J. Phycol.* **29**, 627–42.

Johnsen, G. and Sakshaug, E. (2007). Bio-optical characteristics of PSII and PSI in 33 species (13 pigment groups) of marine phytoplankton, and the relevance for pulse-amplitude-modulated and fast-repetition-rate fluorometry. *J. Phycol.* **43**, 1236–51.

Johnsen, G., Samset, O., Granskog, L. and Sakshaug, E. (1994). *In vivo* absorption characteristics in 10 classes of phytoplankton: Taxonomic characteristics and responses to photoadaptation by means of discriminant and HPLC analysis. *Mar. Ecol. Progr. Ser.* **105**, 149–57.

Johnsen, G., Volent, Z., Tangen, K. and Sakshaug, E. (1997). Time series of harmful and benign phytoplankton blooms in northwest European waters using the Seawatch buoy system. In *Monitoring Algal Blooms: New Techniques for Detecting Large-Scale Environmental Change*, ed. M. Kahru and C. W. Brown. New York: Springer, pp. 115–43.

Johnsen, G., Samset, O. and Sakshaug, E. (1998). Bio-optical classification of potential harmful chlorophyll c_3-containing prymnesiophyte blooms in Scandinavian waters. In *Harmful Microalgae*, ed. B. Reguera, J. Blanco, M. L. Fernández and T. Wyatt. Santiago de Compostela: Xunta de Galicia and Intergovernmental Oceanographic Commission of UNESCO, pp. 300–04.

Johnsen, G., Volent, Z., Sakshaug, E., Sigernes, F. and Pettersson, L. (2009). Remote sensing in the Barents Sea. In *Ecosystem Barents Sea*, ed. E. Sakshaug, G. Johnsen and K. Kovacs. Trondheim: Tapir Academic Press, pp. 151–68.

Johnson, K. S., Berelson, W. M., Boss, E. S., Chase, Z., Claustre, H., Emerson, S. R., Gruber, N., Körtzinger, A., Perry, M. J. and Riser, S. C. (2009). Observing biogeochemical cycles at global scales with profiling floats and gliders: prospects for a global array. *Oceanography* **22**, 216–25.

Kahru, M. and Brown, C. (eds.) (1997). *Monitoring Algal Blooms: New Techniques for Detecting Large-Scale Environmental Change*. New York: Springer.

Kirk, J. T. O. (1994). *Light and Photosynthesis in Aquatic Ecosystems*. Cambridge: Cambridge University Press.

Kirkpatrick, G. J., Millie, D. F., Moline, M. A. and Schofield, O. (2000). Optical discrimination of a phytoplankton species in natural mixed populations. *Limnol. Oceanogr*. **45**, 467–71.

Kirkpatrick, B., Fleming, L. E., Squicciarini, D., Backer, L. C., Clark, R., Abraham,W., Benson, J., Cheng, Y. S., Johnson, D., Pierce, R., Zaias, J., Bossart, G. D. and Baden, D. G. (2004). Literature review of Florida red tide: implications for human health effects. *Harmful Algae* **3**, 99–115.

Kirkpatrick, G. J., Millie, D. F., Moline, M. A., Lohrenz, S. E. and Schofield, O. M. (2008). Phytoplankton community composition observed by autonomous underwater vehicle. In *Proceedings of the XIIth International Conference on Harmful Algae*, ed. Moestrup, P. A. Tester and H. Enevoldsen. Paris: ISSHA and IOC of UNESCO, pp. 83–85.

Klonowski, W. M., Fearns, P. R. C. S. and Lynch, M. J. (2007). Retrieving key benthic cover types and bathymetry from hyperspectral imagery. *J. Appl. Remote Sens*. **1**, 011505, doi:10.1117/1.2816113.

Kondratyev, K. Ya. and Pozdnyakov, D. V. (1994). Application of remote sensing in the visible for hydrodynamic studies in lakes. *Water Pollut. Res. J. Canada* **29**, 385–402.

Kondratyev, K. Ya., Pozdnyakov, D. V. and Isakov, V. Yu. (1990). *Radiation and Hydrooptical Experiments on Lakes*. Leningrad: Nauka Press. (in Russian).

Kroon, B., Prézelin, B. B. and Schofield, O. (1993). Chromatic regulation of quantum yields values for photosystem II charge separation, oxygen evolution and carbon fixation in *Heterocapsa pygmaea* (Pyrrophyta). *J. Phycol*. **29**, 453–69.

Ladner, S. D., Sandidge, J. C., Lyon, P. E., Arnone, R. A., Gould, R. W., Lee, Z. P. and Martinolich P. M. (2007). Development of finer spatial resolution optical properties from MODIS. *Proc. SPIE* **6680**, 668013. doi:10.1117/12.736845.

Lee, Z. P. and Carder, K. L. (2004). Absorption spectrum of phytoplankton pigments derived from hyperspectral remote-sensing reflectance. *Remote Sens. Environ*. **89**, 361–68.

Lee, Z. P., Carder, K. L., Hawes, S. K., Steward, R. G., Peacock, T. G. and Davis, C. O. (1994). Model for the interpretation of hyperspectral remote-sensing reflectance. *Appl. Optics* **33**, 5721–32.

Lee, Z. P., Carder, K. L. and Arnone, R. A. (2002). Deriving inherent optical properties from water color: a multiband quasi-analytical algorithm for optically deep waters. *Appl. Optics* **41**, 5755–72.

Lee, Z. P., Carder, K., Arnone, R. and He, M. X. (2007). Determination of primary spectral bands for remote sensing of aquatic environments. *Sensors* **7**, 3428–41.

Le Quéré, C., Harrison, S. P., Prentice, C. I., Buitenhuis, E. T., Aumonts, O., Bopps, L., Claustre, H., Da Cunha, L. C., Geider, R., Giraud, X., Klaas, C., Kohfeld, K. E., Legendre, L., Manizza, M., Platt, T., Rivkin, R. B., Sathyendranath, S., Uitz, J., Watson, A. J. and Wolf-Gladrow, D. (2005). Ecosystem dynamics based on plankton functional types for global ocean biogeochemistry models. *Global Change Biol*. **11**, 2016–40.

Liaaen-Jensen, S. (1989). Studies on algal carotenoids. *Pure Appl. Chem*. **61**, 369–72.

MacIntyre, H. L., Lawrenz, E., Richardson, T. L. (2010). Taxonomic discrimination of phytoplankton by spectral fluorescence. In *Chlorophyll a Fluorescence in Aquatic Sciences: Methods and Applications*, ed. D. Suggett, M. A. Borowitzka and O. Prášil. Dordrecht: Springer, pp. 129–69.

Manov, D. V., Chang, G. C. and Dickey, T. D. (2004). Methods for reducing biofouling of moored optical sensors. *J. Atmos. Ocean. Tech.* **21**, 958–68.

McClain, C. R. (2009). A decade of satellite ocean color observations. *Annu. Rev. Mar. Sci.* **1**, 19–42.

Millie, D. F., Paerl, H. W. and Hurley, J. P. (1993). Microalgal pigment assessments of high-performance liquid chromatography: a synopsis of organismal and ecological applications. *Can. J. Fish. Aquat. Sci.* **50**, 2513–27.

Millie, D. F., Kirkpatrick, G. J. and Vinyard, B. T. (1995). Relating photosynthetic pigments and *in vivo* optical density spectra to irradiance for the Florida red-tide dinoflagellate *Gymnodinium breve*. *Mar. Ecol. Prog. Ser.* **120**, 65–75.

Millie, D. F., Schofield, O. M., Kirkpatrick, G. J., Johnsen, G., Tester, P. A. and Vinyard, B. T. (1997). Detection of harmful algal blooms using photopigments and absorption signatures: A case study of the Florida red tide dinoflagellate, *Gymnodinium breve*. *Limnol. Oceanogr.* **42**, 1240–51.

Mobley, C. D. (1994). *Light and Water: Radiative Transfer in Natural Waters*. San Diego: Academic Press.

Moline, M. A. (1998). Photoadaptive response during the development of a coastal Antarctic diatom bloom and relationship to water column stability. *Limnol. Oceanogr.* **43**, 146–53.

Moline, M. A., Arnone, R., Bergmann, T., Glenn, S., Oliver, M. J., Orrico, C., Schofield, O. and Tozzi, S. (2004a). Variability in spectral backscatter estimated from satellites and its relation to in situ measurements in optically complex coastal waters. *Int. J. Remote Sens.* **25**, 1465–68.

Moline, M. A., Blackwell, S.M., Chant, R., Oliver, M. J., Bergmann, T., Glenn, S. and Schofield, O. M. E. (2004b). Episodic physical forcing and the structure of phytoplankton communities in the coastal waters of New Jersey. *J. Geophys. Res.* **109**, C12S05 doi:10.1029/2003JC001985.

Moline, M. A., Claustre, H., Frazer, T. K., Schofield, O. and Vernet, M. (2004c). Alteration of the food web along the Antarctic Peninsula in response to a regional warming trend. *Global Change Biol.* **10**, 1973–80.

Moline, M. A., Blackwell, S. M., Allen, B., Austin, T., Forrester, N., Goldsborogh, R., Purcell, M., Stokay, R. and von Alt, C. (2005). Remote environmental monitoring units: an autonomous vehicle for characterizing coastal environments. *J. Atmos. Ocean. Technol.* **22**, 1798–809.

Moline, M. A., Frazer, T. K., Chant, R., Glenn, S., Jacoby, C. A., Reinfelder, J. R., Yost, J., Zhou, M. and Schofield, O. (2008). Biological responses in a dynamic buoyant river plume. *Oceanography* **21**, 70–89.

Moline, M. A., Blackwell, S. M., Case, J. F., Haddock, S. H. D., Herren, C. M., Orrico, C. M. and Terrill, E. (2009). Bioluminescence to reveal structure and interaction of coastal planktonic communities. *Deep-Sea Res. II* **56**, 232–45.

Morel, A. (2008). Introduction to optical properties in the sea: theoretical aspects. In *Real-Time Coastal Observing Systems for Ecosystem Dynamics and Harmful Algal Blooms: Theory, Instrumentation and Modelling*, ed. M. Babin, C. S. Roesler and J. J. Cullen. Paris: UNESCO Publishing, pp. 109–51.

Morel, A. and Antoine, D. (2000). Pigment index retrieval in Case I waters. *MERIS ATBD 2.9*, France: Laboratoire de Physique et Chimie Marines.

Morel, A. and Mueller, J. L. (2002). Normalized water-leaving radiance and remote sensing reflectance: bidirectional reflectance and other factors. In *Ocean Optics Protocols for Satellite Ocean Color Sensor Validation* vol. **2**, ed. J. L. Mueller and G. S. Fargion. NASA/Technical Memorandum 2002–210004, pp. 183–210.

Morel, A. and Prieur, L. (1977). Analysis of variations in ocean color. *Limnol. Oceanogr.* **22**, 709–22.

Nair, A., Sathyendranath, S., Platt, T., Morales, J., Stuart, V., Forget, M.-H., Devred, E. and Bouman, H. (2008). Remote sensing of phytoplankton functional types. *Remote Sens. Environ.* **112**, 3366–75.

Oceanography (2004). Special issue: Coastal ocean optics and dynamics. *Oceanography* **17**(2), 1–125.

Oliver, M. J., Schofield, O., Bergmann, T., Glenn, S., Orrico, C. and Moline, M. (2004). Deriving *in situ* phytoplankton absorption for bio-optical productivity models in turbid waters. *J. Geophys. Res.* **109**, C07S11, doi:10.1029/2002JC001627.

O'Reilly, J. E., Maritorena, S., Mitchell, B. G., Siegel, D. A., Carder, K. L., Garver, S. A., Kahru, M. and McClain, C. (1998). Ocean color chlorophyll algorithms for SeaWiFS. *J. Geophys. Res.* **103**, 24937–53.

Örnólfsdóttir, E. B., Pinckney, J. L. and Tester, P. A. (2003). Quantification of the relative abundance of the toxic dinoflagellate, *Karenia brevis* (Dinophyta), using unique photopigments. *J. Phycol.* **39**, 449–57.

Pegau, W. S., Wijesekera H. and Boyd, T. (2006). AUV based optical measurements. *EOS Trans. AGU* **87**(36), Ocean Sci. Meet. Suppl., Abstract OS24E-02.

Penta, B., Goode, W., Blum, M., Lamb, E., Shulman, I., Jolliff, J., Anderson, S. and deRada, S. (2008). Assessment of an underwater light attenuation model using bio-optical data collected in Monterey, CA. *EOS Trans. AGU* **89**(53), *Fall Meet. Suppl.*, Abstract OS43E-1301.

Perry, M. J., Sackmann, B. S., Eriksen, C. C., Lee, C. M. (2008). Seaglider observations of blooms and subsurface chlorophyll maxima off the Washington coast. *Limnol. Oceanogr.* **53**, 2169–79.

Pinckney, J. L., Paerl, H., Harrington, M. and Howe, K. (1998). Annual cycles of phytoplankton community-structure and bloom dynamics in the Neuse River Estuary, North Carolina. *Mar. Biol.* **131**, 371–82.

Pinckney, J. L., Paerl, H. and Harrington, M. (1999). Responses of the phytoplankton community growth rate to nutrient pulses in variable estuarine environments. *J. Phycol.* **35**, 1455–63.

Pozdnyakov D. V., Korosov, A. A., Grassl, H. and Pettersson, L. H. (2005). An advanced algorithm for operational retrieval of water quality parameters from satellite data in the visible. *Int. J. Remote Sens.* **26**, 2669–87.

Preisendorfer, R. W. (1976). *Hydrologic Optics. Vol. I. Introduction*. Honolulu: US Department of Commerce National Oceanic and Atmospheric Administration, Environment Research Laboratory.

Prézelin, B. B. (1987). The photosynthetic physiology of dinoflagellates. In *The Biology of Dinoflagellates. I. Dinoflagellata*. ed. F. J. R. Taylor. Oxford: Blackwell Scientific, pp. 174–223.

Prézelin, B. B. and Boczar, B. A. (1986). Molecular bases of cell absorption and fluorescence in phytoplankton: potential applications to studies in optical oceanography. In *Progress in Phycological Research*, ed. F. Round and D. Chapman. Bristol: Biopress Limited, pp. 349–464.

Ras, J., Claustre, H. and Uitz, J. (2008). Spatial variability of phytoplankton pigment distributions in the Subtropical South Pacific Ocean: comparison between *in situ* and predicted data. *Biogeosciences* **5**, 353–69.

Reid, P. C., Robinson, G. A. and Hunt, H. G. (1987). Spatial and temporal patterns of marine blooms in the Northeastern Atlantic and North Sea from the Continuous Plankton Recorder Survey. *Rapp. Pv. Reun. Cons. Int. Explor. Mer* **187**, 27–37.

Richardson, L. L. (1997). Remote sensing of algal bloom dynamics: New research fuses remote sensing of aquatic ecosystems with algal assessory pigment analysis. *BioScience* **46**, 492–501.

Richardson, L. L. and Kruse, F. A. (1999). Identification and classification of mixed phytoplankton assemblages using AVIRIS image-derived spectra. *Summaries of the Eighth JPL Airborne Earth Science Workshop*, 09–11 February 1999, Pasadena, California (JPL Publication 99–17), pp. 339–47.

Richardson, T. L. and Pinckney, J. L. (2004). Monitoring of the toxic dinoflagellate *Karenia brevis* using gyroxanthin-based detection methods. *J. Appl. Phycol.* **16**, 315–28.

Richardson, T. L., Lawrenz, E., Pinckney, J. L., Guajardo, R. C., Walker, E. A., Paerl, H. W. and MacIntyre, H. L. (2010). Spectral fluorometric characterization of phytoplankton community composition using the Algae Online Analyser. *Water Res.* **44**, 2461–72.

Robbins, I. C., Kirkpatrick, G. J., Blackwell, S. M., Hillier, J., Knight, C. A., Moline, M. A. (2006). Improved monitoring of HABs using autonomous underwater vehicles (AUV). *Harmful Algae* **5**, 749–61.

Rodríguez, F., Varela, M., Fernández, E. and Zapata, M. (2003). Phytoplankton and pigment distributions in an anticyclonic slope water oceanic eddy (SWODDY) in the southern Bay of Biscay. *Mar. Biol.* **143**, 995–1011.

Rodríguez, F., Chauton, M., Johnsen, G., Andresen, K., Olsen, L. M. and Zapata, M. (2006). Photoacclimation in phytoplankton: implications for biomass estimates, pigment functionality and chemotaxonomy. *Mar. Biol.* **148**, 963–71.

Roesler, C. S. and Boss, E. (2008). In situ measurement of inherent optical properties and potential for harmful algal detection and coastal ecosystem observations. In *Real-Time Coastal Observing Systems for Ecosystem Dynamics and Harmful Algal Blooms: Theory, Instrumentation and Modelling*, ed. M. Babin, C. S. Roesler and J. J. Cullen. Paris: UNESCO Publishing, pp. 153–206.

Ruddick, K., Lacroix, G., Park, Y., Rousseau, V., De Cauwer, V. and Sterckx, S. (2008). Overview of ocean colour: theoretical background, sensors and applicability to detection and monitoring of harmful algal blooms (capabilities and limitations). In *Real-Time Coastal Observing Systems for Ecosystem Dynamics and Harmful Algal Blooms: Theory, Instrumentation and Modelling*, ed. M. Babin, C.S. Roesler and J. J. Cullen. Paris: UNESCO Publishing, pp. 331–83.

Ryan, J., Dierssen, H. M., Kudela, R. M., Scholin, C. A., Johnson, K. S., Sullivan, J. M., Fischer, A. M., Rienecker, E. V., McEnaney, P. R. and Chavez, F. P. (2005). Coastal ocean physics and red tides. *Oceanography* **18**, 246–55.

Ryan, J., Gower, J. F. R., King, S. A., Bissett, W. P., Fisher, A. M., Kudela, R. M., Kolber, Z., Mazzillo, F., Rienecker, E. V. and Chavez, F. P. (2008). A coastal ocean extreme bloom incubator. *Geophys. Res. Lett.* **35**, L12602, doi:10.1029/2008GL034081.

Sathyendranath, S., Watts, L., Devred, E., Platt, T., Caverhill, C. and Maass, H. (2004). Discrimination of diatoms from other phytoplankton using ocean-colour data. *Mar. Ecol. Prog. Ser.* **272**, 59–68.

Schiller, H. and Doerffer, R. (1999). Neural network for emulation of an inverse model – operational derivation of Case II water properties from MERIS data. *Int. J. Remote Sens.* **20**, 1735–46.

Schofield, O., Prézelin, B. B., Stegmann, P., Nelson, N. B., Lewis, M., Smith, R. C. and Baker, K. (1991). Variability in spectral and nonspectral measurements of photosynthetic light utilization efficiencies. *Mar. Ecol. Prog. Ser.* **78**, 253–71.

Schofield, O., Grzymski, J., Bissett, P. W., Kirkpatrick, G. J., Millie, D. F., Moline, M. and Roesler, C. S. (1999). Optical monitoring and forecasting systems for harmful algal blooms: Possibility or pipe dream? *J. Phycol.* **35**, 1477–96.

Schofield, O., Bergmann, T., Bissett, W. P., Grassle, F., Haidvogel, D., Kohut, J., Moline, M. A. and Glenn, S. (2002). The long-term ecosystem observatory: an integrated coastal observatory. *IEEE J. Oceanic Eng.* **27**, 146–54.

Schofield, O., Bergmann, T., Oliver, M. J., Irwin, A., Kirkpatrick, G., Bissett, W. P., Moline, M. A. and Orrico, C. (2004). Inversion of spectral absorption in the optically complex coastal waters of the Mid-Atlantic Bight. *J. Geophys. Res.* **109**, C12S04, doi:10.1029/2003JC002071.

Schofield, O., Bosch, J., Glenn, S., Kirkpatrick, G., Kerfoot, J., Lohrenz, S., Moline, M., Oliver, M. and Bissett, P. (2008). Bio-optics in integrated ocean observing networks: potential for studying harmful algal blooms. In *Real-Time Coastal Observing Systems for Ecosystem Dynamics and Harmful Algal Blooms: Theory, Instrumentation and Modelling*, ed. M. Babin, C. S. Roesler and J. J. Cullen. Paris: UNESCO Publishing, pp. 85–107.

Scholin, C. A., Doucette, G. J. and Cembella, A. D. (2008). Prospects for developing automated systems for *in situ* detection of harmful algae and their toxins. In *Real-Time Coastal Observing Systems for Ecosystem Dynamics and Harmful Algal Blooms: Theory, Instrumentation and Modelling*, ed. M. Babin, C. S. Roesler and J. J. Cullen. Paris: UNESCO Publishing, pp. 413–62.

Sieracki, C. K., Sieracki, M. E. and Yentsch, C. S. (1998). An imaging-in-flow system for automated analysis of marine microplankton. *Mar. Ecol. Prog. Ser.* **168**, 285–96.

Smyth, T. J., Tyrrell, T. and Tarrant, B. (2004). Time series of coccolithophore activity in the Barents Sea, from twenty years of satellite imagery. *Geophys. Res. Lett.* **31**, L11302, doi:10.1029/2004GL019735.

Sosik, H. M. (2008). Characterizing seawater constituents from optical properties. In *Real-Time Coastal Observing Systems for Ecosystem Dynamics and Harmful Algal Blooms: Theory, Instrumentation and Modelling*, ed. M. Babin, C. S. Roesler and J. J. Cullen. Paris: UNESCO Publishing, pp. 281–330.

Stæhr, P. A. and Cullen, J. J. (2003). Detection of *Karenia mikimotoi* by spectral absorption signatures. *J. Plankton Res.* **25**, 1237–49.

Stramska, M., Dickey, T. D., Plueddemann, A., Weller, R., Langdon, C. and Marra, J. (1995). Bio-optical variability associated with phytoplankton dynamics in the North Atlantic Ocean during spring and summer of 1991. *J. Geophys. Res.* **100** (C4), 6621–32.

Stumpf, R. P., Tomlinson, M. C. (2005). Remote sensing of harmful algal blooms. In *Remote Sensing of Coastal Aquatic Environments: Technologies, Techniques and Applications*, ed. R. L. Miller, C. E. Del Castillo and B. A. McKee. Dordrecht: Springer, pp. 277–96.

Subramaniam, A., Carpenter, E. J. and Falkowski, P. G. (1999). Bio-optical properties of the marine diazotrophic cyanobacteria *Trichodesmium* spp. II. A reflectance model for remote sensing. *Limnol. Oceanogr.* **44**, 618–27.

Subramaniam, A., Brown, C. W., Hood, R. R., Carpenter, E. J. and Capone, D. G. (2002). Detecting *Trichodesmium* blooms in SeaWiFS imagery. *Deep-Sea Res. II* **49**, 107–21.

Suggett, D., Moore, M. C., Hickman, A. E. and Geider, R. (2009). Interpretation of fast repetition rate (FRR) fluorescence: signatures of phytoplankton community structure versus physiological state. *Mar. Ecol. Prog. Ser.* **376**, 1–19.

Suggett, D., Borowitzka, M. A. and Prášil, O. (eds.) (2010). *Chlorophyll a Fluorescence in Aquatic Sciences: Methods and Applications*. Dordrecht: Springer.

Suzuki, R. and Ishimaru, T. (1992). Characteristics of photosynthetic pigment composition of *Gymnodinium mikimotoi* Miyake et Kominami ex Oda. *J. Oceanogr.* **48**, 367–75.

Tanabe, S. H., Honda, D., Fukaya, S., Otake, I., Inagaki, Y. and Sako, Y. (2007). Proposal of *Pseudochattonella verruculosa* gen. nov., comb. nov. (Dictyochophyceae) for a former raphidophycean alga *Chattonella verruculosa*, based on 18S rDNA phylogeny and ultrastructural characteristics. *Phycol. Res.* **55**, 185–92.

Tilstone, G. H., Smyth, T. J., Gowen, R. J., Martinez-Vicente, V. and Groom, S. B. (2005). Inherent optical properties of the Irish Sea and their effect on satellite primary production algorithms. *J. Plankton Res.* **27**, 1127–48.

Throndsen, J., Hasle, G. R. and Tangen, K. (2003). *Norsk Kystplanktonflora*. Oslo: Almater Forlag AS. (in Norwegian).

Uitz, J., Huot, Y., Bruyant, F., Babin, M. and Claustre, H. (2008). Relating phytoplankton photophysiological properties to community structure on large scales. *Limnol. Oceanogr.* **53**, 614–30.

Van der Woerd, H. J. and Pasterkamp, R. (2008). HYDROPT: A fast and flexible method to retrieve chlorophyll-a from multispectral satellite observations of optically complex coastal waters. *Remote Sens. Environ.* **112**, 1795–807.

Volent, Z., Johnsen, G., and Sigernes, F. (2007). Kelp forest mapping by use of airborne hyperspectral imager. *J. Appl. Remote Sens.* **1**, 011503. doi:10.1117/1.2822611.

Volent, Z., Johnsen, G. and Sigernes, F. (2009). Microscopic hyperspectral imaging used as a bio-optical taxonomic tool for micro- and macroalgae. *Appl. Optics* **48**, 4170–76.

Wang, P., Boss, E. and Roesler, C. (2005). Uncertainties of inherent optical properties obtained from semi-analytical inversions of ocean color. *Appl. Optics* **44**, 4074–85.

Westberry, T. K. and Siegel, D. A. (2006). Spatial and temporal distribution of *Trichodesmium* blooms in the world's oceans. *Global Biogeochem. Cycles* **20**, GB4016, doi:10.1029/2005GB002673.

Westberry, T. K., Siegel, D. A. and Subramaniam, A. (2005). An improved bio-optical model for the remote sensing of *Trichodesmium* spp. blooms. *J. Geophys. Res.* **110**, C06012, doi:10.1029/2004JC002517.

Zaneveld, J. R. V. (1995). A theoretical derivation of the dependence of the remotely sensed reflectance on the inherent optical properties. *J. Geophys Res.* **100**, 13135–42.

Zapata, M. (2005). Recent advances in pigment analysis as applied to picophytoplankton. *Vie et Milieu* **55**, 233–48.

Zibordi, G., Mélin, F. and Berthon, J.-F. (2006). Comparison of SeaWiFS, MODIS and MERIS radiometric products at a coastal site. *Geophys. Res. Lett.* **33**, L06617, doi:10.1029/2006GL025778.

Appendix 14A

Pigments and toxins of harmful algae

EINAR SKARSTAD EGELAND

As there is an increasing requirement to monitor harmful algal blooms (HABs), it is useful to have knowledge on both the pigments and toxins associated with these blooms. Here, in Table 14A.1, a list is compiled of photosynthetic or mixotrophic phytoplankton species known to cause HABs in marine, brackish or freshwater with detailed information of their specific toxins and pigments. Since heterotrophic dinoflagellates, some of which are known to produce harmful blooms, do not contain photosynthetic pigments they are excluded from the list. Also excluded are species of algae found to be harmful or toxic only under laboratory conditions, these species are either not bloom forming or do not cause harm as a bloom. Many HAB species have not been examined for photosynthetic pigments and these are excluded from the HAB table below. A complete HAB species list with toxin information (with or without pigment information) is available online at www.cambridge.org/phytoplankton.

Phytoplankton Pigments: Characterization, Chemotaxonomy and Applications in Oceanography, eds. Suzanne Roy, Carole A. Llewellyn, Einar Skarstad Egeland and Geir Johnsen. Published by Cambridge University Press.

Table 14A.1. *List of phytoplankton species known to produce harmful algal blooms, their major toxins and harmful effects, and their chlorophyll and carotenoid pigments as reported in the selected references shown in the last column. Also shown are other compounds potentially useful for detection purposes, such as mycosporine-like amino acids (MAAs – for details on these, see Table 10.2 in Chapter 10, this volume). For pigment abbreviations, see the list of abbreviations at the beginning of this volume. Pigments and other compounds are listed as found in the references; others may be present, but only those reported in the references are listed (e.g. Chl* a *should be present in all of the species listed below). Toxic syndromes abbreviations are as found in Chapter 14. In the last column, IOC refers to the IOC list of Harmful Algae, see Moestrup* et al. *(2009).*

Algal species	Synonym(s)	Algal class	Toxin	Harmful effect	Chlorophyll	Carotenoid	Other pigment	Selected references
Akashiwo sanguinea	*Gymnodinium nelsonii, G. sanguineum, G. splendens*	Dinophyceae		oxygen depletion, abalone and oyster larvae mortality		ββ-Car, Diadino, Diato, Dino, Peri, peridinol	MAAs	Botes *et al.* (2003a), Johansen *et al.* (1974), Matsubara *et al.* (2007)
Alexandrium catenella	*Gessnerium catenellum, Gonyaulax catenella, G. tamarensis* var. *excavata, Protogonyaulax catenella*	Dinophyceae	decarbamoylgonyautoxins 2–3, decarbamoylsaxitoxin, gonyautoxins 1–6, neosaxitoxin, *N*-sulfocarbamoylgonyautoxins 1–4 (C1–C4)	fish mortality, PSP	Chl *a*, Chl c_2, MgDVP	ββ-Car, Diadino, Diato, Dino, Peri, peridininol	MAAs	IOC, Boczar *et al.* (1988), Carreto *et al.* (2001), Hallegraeff *et al.* (1991a), Kaga *et al.* (2006), Kim *et al.* (2005), Matsuda *et al.* (2006), Navarro *et al.* (2006)
Alexandrium minutum	*Alexandrium angustitabulatum, A. ibericum, A. lusitanicum, Pyrodinium minutum*	Dinophyceae	decarbamoylsaxitoxin, deoxygonyautoxin 4-12-ol, gonyau-toxins 1–4, neosaxi-toxin, saxitoxin	PSP	Chl *a*, Chl c_2, Chl c_3, MgDVP	ββ-Car, Diadino, Diato, Dino, Peri, peridininol	MAAs	IOC, Carreto *et al.* (2001), Chen and Chou (2001), Lim *et al.* (2007), Nascimento *et al.* (2005)

Table 14A.1. (*cont.*)

Algal species	Synonym(s)	Algal class	Toxin	Harmful effect	Chlorophyll	Carotenoid	Other pigment	Selected references
Alexandrium tamarense	*Alexandrium excavatum*, *Gessnerium tamarensis*, *Gonyaulax excavata*, *G. tamarensis*, *G. tamarensis* var. *excavata*, *Protogonyaulax tamarensis*	Dinophyceae	decarbamoylgonyautoxins 2–3, decarbamoylsaxitoxin, gonyaulaxtoxins 1–6, neosaxitoxin, saxitoxin, *N*-sulfo-carbamoylgonyautoxins 1–4 (C1–C4)	non-toxic or toxic, PSP	Chl *a*, Chl c_2, MgDVP	ββ-Car, Diadino, Diato, Dino, Peri, peridininol	MAAs	IOC, Carreto *et al.* (2001), Collins *et al.* (2009), Hallegraeff *et al.* (1991a), Kaga *et al.* (2006), Kim *et al.* (2005), Laurion and Roy (2009)
Alexandrium tamiyavanichii	*Alexandrium cohorticula*, *Gessnerium cohorticula*, *Gonyaulax cohorticula*, *Protogonyaulax cohorticula*	Dinophyceae	decarbamoylsaxitoxin, saxitoxin, *N*-sulfo-carbamoylgonyautoxins 1–4 (C1–C4)	PSP	Chl *a*	Peri		IOC, Kodama *et al.* (1988), Lim *et al.* (2006), Ogata *et al.* (1990), Ogata *et al.* (1994)
Amphidinium carterae		Dinophyceae	amphidinols 2, 4, 9–13, haemolysins 1–5	fish toxicity, haemolysis	Chl *a*, Chl c_2	ββ-Car, Diadino, Diato, Dino, P457, Peri, peridininol	MAAs	IOC, Echigoya *et al.* (2005), Hannach and Sigleo (1998), Jeffrey and Wright (1997), Johansen *et al.* (1974), Nayak *et al.* (1997)
Amphidinium operculatum	*Amphidinium klebsii*	Dinophyceae	amphidinols 1–8, 14–15	antifungal activity, haemolysis, fish mortality?	Chl *a*, Chl *c*	carotene(s), Diadino, Dino, Peri		IOC, Echigoya *et al.* (2005), Mandelli (1969), Morsy *et al.* (2005), Morsy *et al.* (2006), Wilhelm and Manns (1991)
Amphora coffeaeformis	*Amphora coffaeiformis*, *Frustulia coffeaeformis*	Bacillariophyceae	domoic acid (identity of toxic strain not proved)	non-toxic or toxic, DA	Chl *a*, Chl *c*	Diadino, Diato, Fuco		IOC, Rech *et al.* (2005), Sala *et al.* (1998), Shimizu *et al.* (1989)

Anabaena circinalis	*Anabaena hassalii*	Nostocophyceae	decarbamoylgonyautoxins 2–3, decarbamoylsaxitoxin, gonyautoxins 2–3, 5, saxitoxin, *N*-sulfocarbamoylgonyautoxins 2–3 (C1–C2)	non-toxic or toxic, PSP	Chl *a*		c-phycocyanin	Millie *et al.* (1992), Negri *et al.* (1997)
Anabaena flos-aquae		Nostocophyceae	anabaenopeptins A–B, anatoxins a–d, anatoxin-a(s), microcystin-LR, [D-Asp^3] microcystin-LR, [D-Asp^3,D-Glu $(OCH_3)_6$] microcystin-LR, [D-Glu$(OCH_3)_6$] microcystin-LR, microcystin-RR, [D-Asp^3] microcystin-RR	hepatotoxic	Chl *a*	Cantha, ββ-Car, Echin, Kmyxo, Myxo		Carmichael and Gorham (1978), Harada *et al.* (1995), Hertzberg *et al.* (1971), Leavitt and Brown (1988), Onodera *et al.* (1997), Sivonen *et al.* (1992)
Anabaena lemmermannii		Nostocophyceae	anatoxin-a(s)	neurotoxic		aphanizophyll, Cantha, ββ-Car, Cryp, Echin, Myxo, Oscil, Zea		Onodera *et al.* (1997), Schlüter *et al.* (2004)
Anabaena variabilis		Nostocophyceae	anatoxin			Cantha, ββ-Car, Echin, Myxo, Zea		Hertzberg *et al.* (1971)
Aphanizomenon flos-aquae	*Aphanizomenon* DC-1	Nostocophyceae	anatoxin-a, cylindrospermopsin, decarbamoylgonyautoxin 3, decarbamoylsaxitoxin, gonyautoxins 1, 3–5, neosaxitoxin, saxitoxin	non-toxic or toxic, hepatotoxic, neurotoxic, PSP, tadpole mass mortality	Chl *a*	aphanizophyll, Cantha, ββ-Car, β,β-carotene epoxide, Echin, Myxo		Ferreira *et al.* (2001), Hertzberg and Liaaen-Jensen (1966b), Liu *et al.* (2006a), Liu *et al.* (2006b), Preußel *et al.* (2006), Rapala *et al.* (1993)
Arthrospira fusiformis	*Spirulina fusiformis* (or *fussiformis*), *S. platensis*	Nostocophyceae	anatoxin-a, microcystin YR	non-toxic or toxic		ββ-Car, β,β-carotene-5,6-epoxide, Cryp, Echin, 3′-hydroxy-echinenone, Myxo, Oscil, Zea	c-phycocyanin, phycoerythrin	Aakermann *et al.* (1992), Ballot *et al.* (2004), Madhyastha and Vatsala (2007)

Table 14A.1. (*cont.*)

Algal species	Synonym(s)	Algal class	Toxin	Harmful effect	Chlorophyll	Carotenoid	Other pigment	Selected references
Aureococcus anophagefferens		Pelagophyceae		shellfish mass mortality	Chl *a*, Chl *c* Chl c_3	But-fuco, ββ-Car, Diadino, Diato, Fuco		Bidigare (1989), DeYoe *et al.* (1997), Hargraves and Maranda (2002)
Botryococcus braunii		Chlorophyceae	linoleic acid, α-linolenic acid, oleic acid	fish mortality	Chl *a*, Chl *b*	adonixanthin, Asta, botryoxanthins A–B, α-botryoxanthin A, braunixanthins 1–2, Cantha, ββ-Car, βε-Car, Echin, 3-hydroxy-echinenone, 3′-hydroxy-echinenone, Loro, Lut, Neo, Viola, Zea		Chiang *et al.* (2004), Grung *et al.* (1994a, 1994b), Okada (1999)
Chattonella antiqua	*Hemieutreptia antiqua*	Raphidophyceae	brevetoxins PbTx-1–3, oxidized PbTx-2	fish mortality, neurotoxic	Chl *a*	Fuco, Viola		IOC, Haque and Onoue (2002), Kohata *et al.* (1997)
Chrysochromulina leadbeateri		Prymnesiophyceae		fish mortality, haemolysis, osmotic stress in fish	Chl *a*, Chl c_2, Chl c_3, MGDG, MgDVP, MV-Chl c_3	ββ-Car, Diadino, Diato, Fuco, Hex-fuco		IOC, Johnsen *et al.* (1999), Rodriguez *et al.* (2006), Simonsen and Moestrup (1997), Zapata *et al.* (2001)
Chrysochromulina polylepis		Prymnesiophyceae	haemolysins	animal, bacterial and fish mortality,	Chl *a*, Chl c_2, Chl c_3, MGDG	ββ-Car, Diadino, Diato, Fuco, Hex-fuco, Hex-kfuco		IOC, Johnsen *et al.* (1992), Yasumoto *et al.* (1990), Zapata *et al.* (2004)

				haemolysis, osmotic stress in fish				
Cylindrospermopsis raciborskii		Nostocophyceae	cylindrospermopsin, decarbamoy lneosaxitoxin, decarbamoylsaxitoxin, deoxy-cylindrospermopsin, gonyautoxins 2–3, 6, neosaxitoxin, saxitoxin, *N*-sulfocarba-moylgonyautoxins 2–3 (C1–C2)	non-toxic or toxic, cattle mortality, hepatotoxic, PSP	Chl *a*	ββ-Car, Echin, Myxo, Zea		Bernard *et al.* (2003), Falconer and Humpage (2006), Lagos *et al.* (1999), Molica *et al.* (2002), Norris *et al.* (1999), Pomati *et al.* (2004), Thomas *et al.* (1998), Várkonyi *et al.* (2002)
Dictyocha speculum	*Distephanus speculum*	Dictyochophyceae	oxygen depletion	fish mortality	Chl *a*, Chl *c* (c_1 or c_2), Chl c_3	But-fuco, ββ-Car, Diadino, Diato, Fuco, Hex-fuco, Viola, Zea		Daugbjerg and Henriksen (2001), Henriksen *et al.* (1993)
Dinophysis acuminata	*Dinophysis boehmii, D. borealis, D. lachmanni*	Dinophyceae	dinophysistoxin 1, okadaic acid, pectenotoxin 2, 2-SA, 11, 11-SA?, 12	DSP			phycoerythrin	IOC, Blanco *et al.* (2007), Hewes *et al.* (1998), MacKenzie *et al.* (2005), Miles *et al.* (2004)
Dinophysis caudata	*Dinophysis diegensis*	Dinophyceae	dinophysistoxin 1, okadaic acid, pectenotoxin 2	DSP			phycoerythrin	IOC, Fernández *et al.* (2006), Hewes *et al.* (1998), Marasigan *et al.* (2001)
Dinophysis fortii	*Dinophysis intermedia, Dinophysis laevis, D. ovum*	Dinophyceae	dinophysistoxin 1, okadaic acid, pectenotoxin 2	DSP			phycoerythrin	IOC, Draisci *et al.* (1996), Hewes *et al.* (1998)
Dinophysis norvegica		Dinophyceae	dinophysistoxin 1, okadaic acid, pectenotoxin 2, 12	DSP	Chl *a*, Chl c_2	Allo	phycoerythrin	IOC, Geider and Gunter (1989), Meyer-Harms and Pollehne (1998), Miles *et al.* (2004)

Table 14A.1. (*cont.*)

Algal species	Synonym(s)	Algal class	Toxin	Harmful effect	Chlorophyll	Carotenoid	Other pigment	Selected references
Eucampia zodiacus		Bacillariophyceae		discoloration of *Porphyra* thalli				Nishikawa and Hori (2004)
Gambierdiscus toxicus		Dinophyceae	ciguatoxins 3C, 4A, 4B, gambieric acids A–D, gambierol, maitotoxin 1–3	antifungal activity, fish mortality, neurotoxic (ciguatera)	Chl *a*, Chl *c*	ββ-Car, Diadino, Dino, Peri		IOC, Diogène *et al.* (1994), Indelicato and Watson (1986), Morohashi *et al.* (1998), Morohashi *et al.* (2000)
Gymnodinium catenatum		Dinophyceae	decarbamoylgonyautoxin 2–3, decarbamoylsaxitoxin, deoxysaxitoxin, GC 1–3, gonyautoxins 2–6 (GTX 5–6 = B1–2), neosaxitoxin, saxitoxin, *N*-sulfo-carbamoylgonyautoxin 1–4 (C1–C4)	PSP	Chl *a*, Chl c_2	ββ-Car, Diadino, Diato, Dino, Peri	MAAs	IOC, Band-Schmidt *et al.* (2006), Hallegraeff *et al.* (1991b), Jeffrey *et al.* (1999), Montoya *et al.* (2006), Negri *et al.* (2003), Ordás *et al.* (2004), Oshima *et al.* (1993)
Gyrodinium impudicum	*Gymnodinium impudicum*	Dinophyceae	sulphated exopolysaccharide	antiviral and immuno-stimulant activity	Chl *a*, Chl c_2	ββ-Car, Diadino, Diato, Dino, Peri, peridininol		Fraga *et al.* (1995), Yim *et al.* (2003), Yim *et al.* (2005)
Heterosigma akashiwo	*Entomosigma akashiwo*, *Heterosigma carterae*, *Olisthodiscus carterae*, *Olisthodiscus luteus*	Raphidophyceae	brevetoxins PbTx-2–3, 9, oxidized PbTx-2	fish mortality	Chl *a*, Chl c_1, Chl c_2	Anth, ββ-Car, Fuco, Hex-fuco, Viola, Zea		IOC, Li *et al.* (2003), Rodríguez *et al.* (2006)

Karenia brevis	*Gymnodinium breve*, *Gymnodinium brevis*, *Ptychodiscus brevis*	Dinophyceae	brevetoxins PbTx-1–3, 5–7, 9–12, 927, tbm, hemibrevetoxin-B, hydrolytic PbTx-1–3, 7, hydrolytic oxidized PbTx-1–2, oxidized PbTx-1–2	fish and invertebrate mortality, NSP	Chl *a*, Chl *c*, Chl c_3	But-fuco, ββ-Car, βε-Car, Diadino, Diato, Fuco, Gyro-e, Hex-fuco, Hex-kfuco	MAAs	IOC, Abraham *et al.* (2006), Baden *et al.* (2005), Bidigare *et al.* 1990, Bjørnland *et al.* (2003), Frame (2004), Hernández-Becerril *et al.* (2007), Prasad and Shimizu (1989), Twiner *et al.* (2007)
Karenia cristata		Dinophyceae		human respiratory and skin distress, invertebrate mortality	Chl *a*, Chl c_2, Chl c_3	But-fuco, ββ-Car, Diadino, Fuco, Gyro-e, Hex-fuco		IOC, Botes *et al.* (2003b)
Karenia mikimotoi	*Gymnodinium mikimotoi*, *Gymnodinium nagasakiense* (often confused with *Gymnodinium aureolum* = *Gyrodinium aureolum*)	Dinophyceae	gymnocin A, B, reactive oxygen	fish and invertebrate mortality	Chl *a*, Chl $c_{1/2}$, Chl c_3	But-fuco, ββ-Car, βε-Car, Diadino, Fuco, Gyro-e, Hex-fuco		IOC, Hansen *et al.* (2000), Johnsen & Sakshaug (1993), Moestrup *et al.* 2009, Satake *et al.* (2005), Tsukano and Sasaki (2006), Yamasaki *et al.* (2004)
Karenia umbella		Dinophyceae		fish mortality	Chl *a*, Chl c_2, Chl c_3	Diadino, Diato, Fuco, Gyro-e, Hex-fuco, Hex-kfuco, Zea		IOC, de Salas *et al.* (2004)
Karlodinium armiger		Dinophyceae		fish and invertebrate mortality	Chl *a*, Chl c_2, Chl c_3, MGDG	But-fuco, ββ-Car, βε-Car, Diadino, Diato, Fuco, Gyro-e, Hex-fuco		IOC, Bergholtz *et al.* (2005), Garcés *et al.* (2006)

Table 14A.1. (*cont.*)

Algal species	Synonym(s)	Algal class	Toxin	Harmful effect	Chlorophyll	Carotenoid	Other pigment	Selected references
Karlodinium veneficum	*Gymnodinium galatheanum, Gymnodinium veneficum, Gyrodinium galatheanum, Karlodinium micrum, Woloszynskia micra*	Dinophyceae	karlotoxins 1–2,	fish and invertebrate mortality	Chl *a*, Chl c_2, Chl c_3, MgDVP	But-fuco, ββ-Car, Diadino, Diato, Fuco, Gyro-e, Hex-fuco, Viola, Zea		IOC, Adolf *et al.* (2006), Bachvaroff *et al.* (2009), Bergholtz *et al.* (2005), Bjørnland *et al.* (2003), Deeds *et al.* (2006), Johnsen & Sakshaug (1993), Kempton *et al.* (2002), Mooney *et al.* (2009)
Lingulodinium polyedrum	*Gonyaulax polyedra*	Dinophyceae	various yessotoxins and derivatives	toxic to shellfish	Chl *a*, Chl *c*	Peri	MAAs	Bowden (2006), Paz *et al.* (2004), Prézelin and Haxo (1976), Whitehead and Vernet (2000)
Limnothrix redekei	*Oscillatoria redekei*	Nostocophyceae	α-dimorphecolic acid, coriolic acid, microcystins (?)	antibacterial activity, immunosuppressant	Chl *a*	ββ-Car, Echin, Oscil, Zea		Effmert *et al.* (1991), Mundt *et al.* (2003), Nicklisch and Woitke (1999)
Microcystis aeruginosa		Nostocophyceae	aeruginosin 298-A–B, microcystilide A, microcystin-LR, YR, RR etc. microginin, micropeptins A–B, 90, microviridin B–C	embryonic damage, toxic drinking water	Chl *a*	aphanizophyll, Calo, Cantha, ββ-Car, Cryp, Echin, isozeaxanthin, Myxo, Nosto, Oscil, Zea		Bagchi (1996), Carmichael (1997), Falconer (2007), Hertzberg *et al.* (1971), Oberholster *et al.* 2004, Paerl *et al.* (1983), Park *et al.* (1993), Pichardo *et al.* (2007), Walsh *et al.* (1997), Woitke *et al.* (1997)

Microcystis wesenbergii		Nostocophyceae	microcystin-LR etc.	non-toxic or toxic	Chl *a*		allophycocyanin, phycocyanin	Braun and Bachofen (2004), Watanabe *et al.* (1988), Xing *et al.* (2007)
Nodularia spumigena	*Nodularia baltica, Nodularia litorea*	Nostocophyceae	nodularin, [Asp^1]-nodularin, [dhb^5]-nodularin, [$CMAdda^3$]-nodularin	animal mortality	Chl *a*	Cantha, ββ-Car, Cryp, Echin, Myxo	MAAs	Laamanen *et al.* (2001), Mazur-Marzek *et al.* (2006), Moffitt *et al.* (2001), Sinha *et al.* (2003)
Phaeocystis globosa		Prymnesiophyceae		haemolysis	Chl *a*, Chl c_1, Chl c_2, Chl c_3,	But-fuco, ββ-Car, βε-Car, Diadino, Diato, Fuco, Hex-fuco, Hex-kfuco		IOC, Peng *et al.* (2005), Rodríguez *et al.* (2006), Vaulot *et al.* (1994)
Phaeocystis pouchetii	*Tetraspora poucheti*	Prymnesiophyceae	*trans,trans*-deca-2,4-dienal	embryonic damage, fish mortality	Chl	But-fuco, ββ-Car, Diadino, Diato, Fuco, Hex-fuco, Zea		IOC, Hansen *et al.* (2004), Nichols *et al.* (1991)
Phormidium tenue		Nostocophyceae		toxic to mice	Chl *a*		allophycocyanin, C-phycocyanin	Mohamed *et al.* (2006), Tang and Vincent (1999)
Planktothrix agardhii	*Oscillatoria agardhii, Planktothrix mougeotii*	Nostocophyceae	microcystins dMeLR, dMeRR, RR, YR, [$ADMAdda^5$]-microcystins 1–4	non-toxic or toxic	Chl *a*	ββ-Car, β,β-caroten-5,6-epoxide, Cryp, Echin, 3'-hydroxy echinenone, Myxo, Oscil, Zea		Briand *et al.* (2002), Hertzberg *et al.* (1971), Laub *et al.* (2002), Suda *et al.* (2002)
Planktothrix rubescens	*Oscillatoria rubescens*	Nostocophyceae	anatoxin-a, [D-Asp^3]-microcystin-RR, [D-Asp^3,(*E*)-Dhb^7]-microcystin-HilR, [D-Asp^3,(*E*)-Dhb^7]-microcystin-RR, oscillapeptin J	fish and zooplankton mortality		ββ-Car, Cryp, Echin, 3'-hydroxy echinenone, Myxo, Oscil, Zea	phycocyanin, phycoerythrin	Blom *et al.* (2006), Briand *et al.* (2005), Ernst *et al.* (2007), Hertzberg and Liaaen-Jensen (1966a), Sano *et al.* (2004), Suda *et al.* (2002), Viaggiu *et al.* (2004), Welker and Erhard (2007)

Table 14A.1. (*cont.*)

Algal species	Synonym(s)	Algal class	Toxin	Harmful effect	Chlorophyll	Carotenoid	Other pigment	Selected references
Prorocentrum donghaiense	*Prorocentrum dentatum* (misidentified?), *Prorocentrum shikokuensis*	Dinophyceae	oxygen depletion	fish mortality	Chl *a*	ββ-Car, Diadino, Peri		Hou *et al*. (2007)
Prorocentrum micans		Dinophyceae		non-toxic or toxic, DSP	Chl *a*	ββ-Car, Dino, Peri, Zea	MAAs	Cassie (1981), Grabowski *et al*. (2001), Johansen *et al*. (1974), Lesser (1996), Pavela-Vrančič *et al*. (2006), Tilstone *et al*. (2010)
Prorocentrum minimum	*Exuviaella apora, Exuviaella mariae-lebouriae, Exuviaella minima, Prorocentrum cordatum, Prorocentrum cordiformis, Prorocentrum mariae-lebouriae, Prorocentrum triangulatum*	Dinophyceae	oxygen depletion	non-toxic or toxic, fish and human mortality, neurotoxic, PSP	Chl *a*, Chl c_2, MgDVP	ββ-Car, Diadino, Diato, Dino, Peri, peridininol	MAAs	IOC, Heil *et al*. (2005), Johnsen and Sakshaug (1993), Rodríguez *et al*. (2006), Sinha *et al*. (1998)
Protoceratium reticulatum	*Gonyaulax grindleyi, Peridinium reticulatum*	Dinophyceae	various yessotoxin and derivatives	non-toxic or toxic, toxic to shellfish	Chl *a*	Peri		IOC, Bowden (2006), Miles *et al*. (2005), Paz *et al*. (2007), Rein and Snyder (2006), Rodríguez *et al*. (2007), Suzuki *et al*. (2007)
Prymnesium parvum	*Prymnesium patelliferum (-ra)*	Prymnesiophyceae	prymnesin 1–2	algal and fish mortality	Chl *a*, Chl c_1, Chl c_2, Chl c_3, MGDG	ββ-Car, Diadino, Diato, Fuco, Viola		IOC, Larsen (1999), Rodríguez *et al*. (2006), Sasaki *et al*. (2006)

Pseudochattonella verruculosa	*Verrucophora farcimen, Chattonella* aff./cf. *verruculosa*	Dictyochophyceae	brevetoxins PbTx 2–3, 9	fish mortality	Chl *a*, Chl c_1, Chl c_2, Chl c_3	But-fuco, Diadino, Diato, Fuco		IOC, Bourdelais *et al.* (2002), Edvardsen *et al.* (2007)
Pseudo-nitzschia multiseries	*Nitzschia pungens* f. *multiseries, Pseudo-nitzschia pungens* f. *multiseries*	Bacillariophyceae	domoic acid	ASP			MAAs	IOC, Bates *et al.* (2004), Carreto *et al.* (2005), Hagström *et al.* (2007)
Skeletonema spp.	*Skeletonema costatum* (recently divided into several species)	Bacillariophyceae	oxygen depletion	fish gill lesions and mortality	Chl *a*, Chl c_1, Chl c_2	ββ-Car, Diadino, Diato, Fuco		Brunet *et al.* (1996), Kent *et al.* (1995)
Takayama tasmanica	*Gymnodinium pulchellum, Gyrodinium acrotrochum, Gyrodinium cladochroma*	Dinophyceae		fish mortality	Chl *a*, Chl c_2, Chl c_3	ββ-Car, Diadino, Diato, Fuco		IOC, de Salas *et al.* (2003)
Thalassiosira spp.		Bacillariophyceae		fish gill lesions and mortality	Chl *a*, Chl c_1, Chl c_2	ββ-Car, Diadino, Diato, Fuco		Kent *et al.* (1995), Yao *et al.* (2006)
Trichodesmium erythraeum		Nostocophyceae	ammonia (from decaying cells), microcystins, oxygen depletion	copepod, fish, oyster and shrimp mortality	Chl *a*	ββ-Car, Echin, Myxo, Zea	phycoerythrin	Carpenter *et al.* (1993), Negri *et al.* (2004), Ramos *et al.* (2005)
Trichodesmium thiebautii		Nostocophyceae		copepod mortality	Chl *a*	ββ-Car, Echin, Myxo, Zea	phycoerythrin	Carpenter *et al.* (1993), Hawser *et al.* (1992)
Tychonema bourrellyi	*Oscillatoria bornetii* f. *tenuis, Oscillatoria bourrellyi, Tychonema tenue*	Nostocophyceae	oxygen depletion				phycocyanin, phycoerythrin	Ganf *et al.* (1991), Heaney *et al.* (1996)

References

Aakermann, T., Skulberg, O. M. and Liaaen-Jensen, S. (1992). A comparison of the carotenoids of strains of *Oscillatoria* and *Spirulina* (Cyanobacteria). *Biochem. Syst. Ecol.* **20**, 761–69.

Abraham, A., Plakas, S. M., Wang, Z., Jester, E. L. E., El Said, K. R., Granade, H. R., Henry, M. S., Blum, P. C., Pierce, R. H. and Dickey, R. W. (2006). Characterization of polar brevetoxin derivatives isolated from *Karenia brevis* cultures and natural blooms. *Toxicon* **48**, 104–15.

Adolf, J. E., Bachvaroff, T. R., Krupatkina, D. N., Nonogaki, H., Brown, P. J. P., Lewitus, A. J., Harvey, H. R. and Place, A. R. (2006). Species specificity and potential roles of *Karlodinium micrum* toxin. *Afr. J. Mar. Sci.* **28**, 415–19.

Bachvaroff, T. R., Adolf, J. E. and Place, A. R. (2009). Strain variation in *Karlodinium veneficum* (Dinophyceae), toxin profiles, pigments, and growth characteristics. *J. Phycol.* **45**, 137–53.

Baden, D. G., Bourdelais, A. J., Jacocks, H., Michelliza, S. and Naar, J. (2005). Natural and derivative brevetoxins, historical background, multiplicity, and effects. *Environ. Health Perspect.* **113**, 621–25.

Bagchi, S. N. (1996). Cyanobacterial toxins. *J. Sci. Ind. Res.* **55**, 715–27.

Ballot, A., Krienitz, L., Kotut, K., Wiegand, C., Metcalf, J. S., Codd, G. A. and Pflugmacher, S. (2004). Cyanobacteria and cyanobacterial toxins in three alkaline Rift Valley lakes of Kenya–Lakes Bogoria, Nakuru and Elmenteita. *J. Plankton Res.* **26**, 925–35.

Band-Schmidt, C., Bustillos-Guzmán, J., Morquecho, L., Gárate-Lizárraga, I., Alonso-Rodríguez, R., Reyes-Salinas, A., Erler, K. and Luckas, B. (2006). Variations of PSP toxin profiles during different growth phases in *Gymnodinium catenatum* (Dinophyceae) strains isolated from three locations in the Gulf of California, Mexico. *J. Phycol.* **42**, 757–68.

Bates, S. S., Gaudet, J., Kaczmarska, I. and Ehrmann, J. M. (2004). Interaction between bacteria and the domoic-acid-producing diatom *Pseudo-nitzschia multiseries* (Hasle) Hasle; can bacteria produce domoic acid autonomously? *Harmful Algae* **3**, 11–20.

Bergholtz, T., Daugbjerg, N., Moestrup, Ø. and Fernández-Tejedor, M. (2005). On the identity of *Karlodinium veneficum* and description of *Karlodinium armiger* sp. nov. (Dinophyceae), based on light and electron microscopy, nuclear-encoded LSU rDNA, and pigment composition. *J. Phycol.* **42**, 170–93.

Bernard, C., Harvey, M., Briand, J. F., Biré, R., Krys, S. and Fontaine, J. J. (2003). Toxicological comparison of diverse *Cylindrospermopsis raciborskii* strains, evidence of liver damage caused by a French *C. raciborskii* strain. *Environ. Toxicol.* **18**, 176–86.

Bidigare, R. R. (1989). Photosynthetic pigment composition of the brown tide alga: unique chlorophyll and carotenoid derivatives. In *Novel Phytoplankton Blooms*, ed. E. M. Cosper, V. M. Bricelj and E. J. Carpenter. Berlin: Springer, pp. 57–75.

Bidigare, R. R., Kennicutt II, M. C, Ondrusek, M. E., Keller, M. D. and Guillard, R. R. L. (1990). Novel chlorophyll-related compounds in marine phytoplankton: distributions and geochemical implications. *Energy Fuels* **4**, 653–57.

Bjørnland, T., Haxo, F. T. and Liaaen-Jensen, S. (2003). Carotenoids of the Florida red tide dinoflagellate *Karenia brevis*. *Biochem. Syst. Ecol.* **31**, 1147–62.

Blanco, J., Álvarez, G. and Uribe, E. (2007). Identification of pectenotoxins in phytoplankton, filter feeders, and isolated cells of a *Dinophysis acuminata* with an atypical toxin profile, from Chile. *Toxicon* **49**, 710–16.

Blom, J. F., Baumann, H. I., Codd, G. A. and Jüttner, F. (2006). Sensitivity and adaption of aquatic organisms to oscillapeptin J and [D-Asp^3,(*E*)-Dhb^7] microcystin-RR. *Arch. Hydrobiol.* **167**, 547–59.

Boczar, B. A., Beitler, M. K., Liston, J., Sullivan, J. J. and Cattolico, R. A. (1988). Paralytic shellfish toxins in *Protogonyaulax tamarensis* and *Protogonyaulax catenella* in axenic culture. *Plant Physiol.* **88**, 1285–90.

Botes, L., Smit, A. J. and Cook, P. A. (2003a). The potential threat of algal blooms to the abalone (*Haliotis midae*) mariculture industry situated around the South African coast. *Harmful Algae* **2**, 247–59.

Botes, L., Sym, S. D. and Pitcher, G. C. (2003b). *Karenia cristata* sp. nov. and *Karenia bicuneiformis* sp. nov. (Gymnodiniales, Dinophyceae), two new *Karenia* species from the South African coast. *Phycologia* **42**, 563–71.

Bourdelais, A. J., Tomas, C. R., Naar, J., Kubanek, J. and Baden, D. G. (2002). New fish-killing alga in coastal Delaware produces neurotoxins. *Environ. Health Perspect.* **110**, 465–70.

Bowden, B. F. (2006). Yessotoxins–polycyclic ethers from dinoflagellates, relationships to diarrhetic shellfish toxins. *Toxin Rev.* **25**, 137–57.

Braun, E. and Bachofen, R. (2004). Homoserine-lactones and microcystin in cyanobacterial assemblages in Swiss lakes. *Hydrobiologia* **522**, 271–80.

Briand, J. F., Robillot, C., Quiblier-Llobéras, C. and Bernard, C. (2002). A perennial bloom of *Planktothrix agardhii* (Cyanobacteria) in a shallow eutrophic French lake, limnological and microcystin production studies. *Arch. Hydrobiol.* **153**, 605–22.

Briand, J.-F., Jacquet, S., Flinois, C., Avois-Jacquet, C., Maisonnette, C, Leberre, B. and Humbert, J.-F. (2005). Variation in the microcystin production of *Planktothrix rubescens* (Cyanobacteria) assessed from a four-year survey of Lac du Bourget (France) and from laboratory experiments. *Microb. Ecol.* **50**, 418–28.

Brunet, C., Davoult, D. and Casotti, R. (1996). Physiological reactions to a change in light regime in cultured *Skeletonema costatum* (Bacillariophyta), implications for estimation of phytoplankton biomass. *Hydrobiologia* **333**, 87–94.

Carmichael, W. W. (1997). The cyanotoxins. *Adv. Bot. Res.* **27**, 211–56.

Carmichael, W. W. and Gorham, P. R. (1978). Anatoxins from clones of *Anabaena flos-aquae* isolated from lakes of western Canada. *Mitt. Internat. Verein. Limnol.* **21**, 285–95.

Carpenter, E. J., O'Neil, J. M., Dawson, R., Capone, D. G., Siddiqui, P. J. A., Roenneberg, T. and Bergman, B. (1993). The tropical diazotrophic phytoplankter *Trichodesmium*, biological characteristics of two common species. *Mar. Ecol. Prog. Ser.* **95**, 295–304.

Carreto, J. I., Carignan, M. O. and Montoya, N. G. (2001). Comparative studies on mycosporine-like amino acids, paralytic shellfish toxins and pigment profiles of the toxic dinoflagellates *Alexandrium tamarense, A. catenella* and *A. minutum*. *Mar. Ecol. Prog. Ser.* **223**, 49–60.

Carreto, J. I., Carignan, M. O. and Montoya, N. G. (2005). A high-resolution reverse-phase liquid chromatography method for the analysis of mycosporine-like amino acids (MAAs) in marine organisms. *Mar. Biol.* **146**, 237–52.

Cassie, V. (1981). Non-toxic blooms of *Prorocentrum micans* (Dinophyceae) in the Karamea Bight. *N. Z. J. Mar. Freshwat. Res.* **15**, 181–84.

Chen, C.-Y. and Chou, H.-N. (2001). Ichthyotoxicity studies of milkfish *Chanos chanos* fingerlings exposed to a harmful dinoflagellate *Alexandrium minutum*. *J. Exp. Mar. Biol. Ecol.* **262**, 211–19.

Chiang, I.-Z., Huang, W.-Y. and Wu, J.-T. (2004). Allelochemicals of *Botryococcus braunii* (Chlorophyceae). *J. Phycol.* **40**, 474–80.

Collins, C., Graham, J., Brown, L., Bresnan, E., Lacaze, J.-P. and Turrell, E. A. (2009). Identification and toxicity of *Alexandrium tamarense* (Dinophyceae) in Scottish waters. *J. Phycol.* **45**, 692–703.

Daugbjerg, N. and Henriksen, P. (2001). Pigment composition and *rbc*L sequence data from the silicoflagellate *Dictyocha speculum*, a heterokont alga with pigments similar to some haptophytes. *J. Phycol.* **37**, 1110–20.

Deeds, J. R., Reimschuessel, R. and Place, A. R. (2006). Histopathological effects in fish exposed to the toxins from *Karlodinium micrum*. *J. Aquat. Animal Health* **18**, 136–48.

de Salas, M. F., Bolch, C. J. S., Botes, L., Nash, G., Wright, S. W. and Hallegraeff, G. M. (2003). *Takayama* gen. nov. (Gymnodiniales, Dinophyceae), a new genus of unarmored dinoflagellates with sigmoid apical grooves, including the description of two new species. *J. Phycol.* **39**, 1233–46.

de Salas, M. F., Bolch, C. J. S. and Hallegraeff, G. M. (2004). *Karenia umbella* sp. nov. (Gymnodiniales, Dinophyceae), a new potentially ichthyotoxic dinoflagellate species from Tasmania, Australia. *Phycologia* **43**, 166–75.

DeYoe, H. R., Stockwell, D. A., Bidigare, R. R., Latasa, M., Johnson, P. W., Hargraves, P. E. and Suttle, C. A. (1997). Description and characterization of the algal species *Aureoumbra lagunensis* gen. et sp. nov. and referral of *Aureoumbria* and *Aureococcus* to the Pelagophyceae. *J. Phycol.* **33**, 1042–48.

Diogène, G., Dubreuil, A., Breittmayer, J. P. and Puiseux-Dao, S. (1994). Cytotoxic quantification of maitotoxin-like activity from the dinoflagellate *Gambierdiscus toxicus*. *Toxic. in Vitro* **8**, 37–45.

Draisci, R., Lucentini, L., Giannetti, L., Boria, P. and Poletti, R. (1996). First report of pectenotoxin-2 (PTX-2) in algae (*Dinophysis fortii*) related to seafood poisoning in Europe. *Toxicon* **34**, 923–35.

Echigoya, R., Rhodes, L., Oshima, Y. and Satake, M. (2005). The structures of five new antifungal and hemolytic amphidinol analogs from *Amphidinium carterae* collected in New Zealand. *Harmful Algae* **4**, 383–89.

Edvardsen, B., Eikrem, W., Shalchian-Tabrizi, K., Riisberg, I., Johnsen, G., Naustvoll, L. and Throndsen, J. (2007). *Verrucophora farcimen* gen. et sp. nov. (Dictyochophyceae, Heterokonta) – a bloom-forming ichthyotoxic flagellate from the Skagerrak, Norway. *J. Phycol.* **43**, 1054–70.

Effmert, U., Mundt, S., Krause, E. and Teuscher, E. (1991). Zur immunsuppressiven Wirkung von Inhaltsstoffen aus Oscillatoria redekei van Goor (Cyanophyceae). *Pharmazie* **46**, 59.

Ernst, B., Hoeger, S. J., O'Brien, E. and Dietrich, D. R. (2007). Physiological stress and pathology in European whitefish (*Coregonus lavaretus*) induced by subchronic exposure to environmentally relevant densities of *Planktothrix rubescens*. *Aquat. Toxicol.* **82**, 15–26.

Falconer, I. R. (2007). Cyanobacterial toxins present in *Microcystis aeruginosa* extract – More than microcystins! *Toxicon* **50**, 585–88.

Falconer, I. R. and Humpage, A. R. (2006). Cyanobacterial (blue-green algal) toxins in water supplies, cylindrospermopsins. *Environ. Toxicol.* **21**, 229–304.

Fernández, M. L., Reguera, B., González-Gil, S. and Míguez, A. (2006). Pectenotoxin-2 in single-cell isolates of *Dinophysis caudata* and *Dinophysis acuta* from the Galician Rías (NW Spain). *Toxicon* **48**, 477–90.

Ferreira, F. M. B., Soler, J. M. F., Fidalgo, M. L. and Fernández-Vila, P. (2001). PSP toxins from *Aphanizomenon flos-aquae* (cyanobacteria) collected in the Crestuma-Lever reservoir (Douro River, northern Portugal). *Toxicon* **39**, 757–61.

Fraga, S., Bravo, I., Delgado, M., Franco, J. M. and Zapata, M. (1995). *Gyrodinium impudicum* sp. nov. (Dinophyceae), a non-toxic, chain-forming, red tide dinoflagellate. *Phycologia* **34**, 514–21.

Frame, E. R. (2004). *Mycosporine-like Amino Acids (MAAs) in Bloom Forming Phytoplankton: the Influence of Nitrogen, Ultraviolet Radiation and Species Composition*. Ph.D. Thesis. University of California, San Diego, USA.

Ganf, G. G., Heaney, S. I. and Corry, J. (1991). Light absorption and pigment content in natural populations and cultures of a non-gas vacuolate cyanobacterium *Oscillatoria bourrellyi (= Tychomena bourrellyi)*. *J. Plankton Res.* **13**, 1101–21.

Garcés, E., Fernandez, M., Penna, A., Lenning, K. V., Gutierrez, A., Camp, J. and Zapata, M. (2006). Characterization of NW Mediterranean *Karlodinium* spp. (Dinophyceae) strains using morphological, molecular, chemical, and physiological methodologies. *J. Phycol.* **42**, 1096–112.

Geider, R. J. and Gunter, P. A. (1989). Evidence for the presence of phycoerythrin in *Dinophysis norvegica*, a pink dinoflagellate. *Br. Phycol. J.* **24**, 195–98.

Grabowski, B., Cunningham, F. X. and Gantt, E. (2001). Chlorophyll and carotenoid binding in a simple red algal light-harvesting complex crosses phylogenetic lines. *Proc. Natl. Acad. Sci. USA* **98**, 2911–16.

Grung, M., Metzger, P., Berkaloff, C. and Liaaen-Jensen, S. (1994a). Studies on the formation and localization of primary and secondary carotenoids in the green alga *Botryococcus braunii*, including the regreening process. *Comp. Biochem. Physiol.* **107B**, 265–72.

Grung, M., Metzger, P. and Liaaen-Jensen, S. (1994b). Algal carotenoids 53; secondary carotenoids of algae 4; secondary carotenoids in the green alga *Botrycoccus braunii*, race L, new strain. *Biochem. Syst. Ecol.* **22**, 25–29.

Hagström, J. A., Granéli, E., Maneiro, I., Barreiro, A., Petermann, A. and Svensen, C. (2007). Release and degradation of amnesic shellfish poison from decaying *Pseudo-nitzschia multiseries* in presence of bacteria and organic matter. *Harmful Algae* **6**, 175–88.

Hallegraeff, G. M., Bolch, C. J., Blackburn, S. I. and Oshima, Y. (1991a). Species of the toxigenic dinoflagellate genus *Alexandrium* in southeastern Australian waters. *Bot. Mar.* **34**, 575–87.

Hallegraeff, G. M., Nichols, P. D., Volkman, J. K., Blackburn, S. I. and Everitt, D. A. (1991b). Pigments, fatty acids, and sterols of the toxic dinoflagellate *Gymnodinium catenatum*. *J. Phycol.* **27**, 591–99.

Hannach, G. and Sigleo, A. C. (1998). Photoinduction of UV-absorbing compounds in six species of marine phytoplankton. *Mar. Ecol. Prog. Ser.* **174**, 207–22.

Hansen, E., Ernstsen, A. and Eilertsen, H. C. (2004). Isolation and characterisation of a cytotoxic polyunsaturated aldehyde from the marine phytoplankter *Phaeocystis pouchetii* (Hariot) Lagerheim. *Toxicology* **199**, 207–17.

Hansen, G., Daugbjerg, N. and Henriksen, P. (2000). Comparative study of *Gymnodinium mikimotoi* and *Gymnodinium aureolum*, comb. nov. (= *Gyrodinium aureolum*) based on morphology, pigment composition, and molecular data. *J. Phycol*. **36**, 394–410.

Haque, S. M. and Onoue, Y. (2002). Variation in toxin compositions of two harmful raphidophytes, *Chattonella antiqua* and *Chattonella marina*, at different salinities. *Environ. Toxicol*. **17**, 113–18.

Harada, K.-I., Fujii, K., Shimada, T., Suzuki, M., Sano, H., Adachi, K. and Carmichael, W. W. (1995). Two cyclic peptides, anabaenopeptides, a third group of bioactive compounds from the cyanobacterium *Anabaena flos-aquae* NRC 525–17. *Tetrahedron Lett*. **36**, 1511–14.

Hargraves, P. E. and Maranda, L. (2002). Potentially toxic or harmful microalgae from the northeast coast. *Northeast. Natural*. **9**, 81–120.

Hawser, S. P., O'Neil, J. M., Roman, M. R. and Codd, G. A. (1992). Toxicity of blooms of the cyanobacterium *Trichodesmium* to zooplankton. *J. Appl. Phycol*. **4**, 79–86.

Heaney, S. I., Parker, J. E., Butterwick, C. and Clarke, K. J. (1996). Interannual variability of algal populations and their influence on lake metabolism. *Freshw. Biol*. **35**, 561–77.

Heil, C. A., Glibert, P. M. and Fan, C. (2005). *Prorocentrum minimum* (Pavillard) Schiller – A review of a harmful algal bloom species of growing worldwide importance. *Harmful Algae* **4**, 449–70.

Henriksen, P., Knipschildt, F., Moestrup, Ø. and Thomsen, H. A. (1993). Autecology, life history and toxicology of the silicoflagellate *Dictyocha speculum* (Silicoflagellata, Dictyochophyceae). *Phycologia* **32**, 29–39.

Hernández-Becerril, D. U., Alonso-Rodríguez, R., Álvarez-Góngora, C., Barón-Campis, S. A., Ceballos-Corona, G., Herrera-Silveira, J., Castillo, M. E. M. D., Juárez-Ruíz, N., Merino-Virgilio, F., Morales-Blake, A., Ochoa, J. L., Orellana-Cepeda, E., Ramírez-Camarena, C. and Rodríguez-Salvador, R. (2007). Toxic and harmful marine phytoplankton and microalgae (HABs) in Mexican coasts. *J. Environ. Sci. Health A* **42**, 1349–63.

Hertzberg, S. and Liaaen-Jensen, S. (1966a). The carotenoids of blue-green algae–I. The carotenoids of *Oscillatoria rubescensi* and an *Athrospira* sp. *Phytochemistry* **5**, 557–63.

Hertzberg, S. and Liaaen-Jensen, S. (1966b). The carotenoids of blue-green algae–II. The carotenoids of *Aphanizomenon flos-aquae*. *Phytochemistry* **5**, 565–70.

Hertzberg, S., Liaaen-Jensen, S. and Siegelman, H. W. (1971). The carotenoids of blue-green algae. *Phytochemistry* **10**, 3121–27.

Hewes, C. D., Mitchell, B. G., Moisan, T. A., Vernet, M. and Reid, F. M. H. (1998). The phycobilin signatures of chloroplasts from three dinoflagellate species, a microanalytical study of *Dinophysis caudata, D. fortii*, and *D. acuminata* (Dinophysiales, Dinophyceae). *J. Phycol*. **34**, 945–51.

Hou, J. -J., Huang, B. -Q., Cao, Z. -R., Chen, J. -X. and Hong, H. -S. (2007). Effects of nutrient limitation of pigments in *Thalassiosira weissflogii* and *Prorocentrum donghaiense*. *J. Interact. Plant Biol*. **49**, 686–97.

Indelicato, S. R. and Watson, D. A. (1986). Identification of the photosynthetic pigments of the tropical benthic dinoflagellate *Gambierdiscus toxicus*. *Mar. Fish. Rev*. **48**, 44–47.

Jeffrey, S. W. and Wright, S. W. (1997). Qualitative and quantitative HPLC analysis of SCOR reference algal cultures. In *Phytoplankton Pigments in Oceanography: Guidelines to Modern Methods*, ed. S. W. Jeffrey, R. F. C. Mantoura and S. W. Wright. Paris: UNESCO Publishing, pp. 343–60.

Jeffrey, S. W., MacTavish, H. S., Dunlap, W. C., Vesk, M. and Groenewould, K. (1999). Occurrence of UVA and UVB-absorbing compounds in 152 species (206 strains) of marine microalgae. *Mar. Ecol. Prog. Ser.* **189**, 35–51.

Johansen, J. E., Svec, W. A., Liaaen-Jensen, S. and Haxo, F. T. (1974). Carotenoids of the Dinophyceae. *Phytochemistry* **13**, 2261–71.

Johnsen, G. and Sakshaug, E. (1993). Bio-optical characteristics and photoadaptive responses in the toxic and bloom-forming dinoflagellates *Gyrodinium aureolum, Gymnodinium galatheanum*, and two strains of *Prorocentrum minimum*. *J. Phycol.* **29**, 627–42.

Johnsen, G., Sakshaug, E., Vernet, M. (1992). Pigment composition, spectral characterization and photosynthetic parameters in *Chrysochromulina polylepis*. *Mar. Ecol. Progr. Ser.* **83**, 241–49.

Johnsen, G., Dalløkken, R., Eikrem, W., Legrand, C., Aure, J. and Skjoldal, H. R. (1999). Eco-physiology, bio-optics and toxicity of the ichthyotoxic *Chrysochromulina leadbeateri* (Prymnesiophyceae). *J. Phycol.* **35**, 1465–76.

Kaga, S., Sekiguchi, K., Yoshida, M. and Ogata, T. (2006). Occurrence and toxin production of *Alexandrium* spp. (Dinophyceae) in coastal waters of Iwate Prefecture, Japan. *Nippon Suisan Gakkaishi* **76**, 1068–76.

Kempton, J. W., Lewitus, A. J., Deeds, J. R., Law, J. M. and Place, A. R. (2002). Toxicity of *Karlodinium micrum* (Dinophyceae) associated with a fish kill in a South Carolina brackish retention pond. *Harmful Algae* **1**, 233–41.

Kent, M. L., Whyte, J. N. C. and LaTrace, C. (1995). Gill lesions and mortality in seawater pen-reared Atlantic salmon *Salmo salar* associated with a dense bloom of *Skeletonema costatum* and *Thalassiosira* species. *Dis. Aquat. Org.* **22**, 77–81.

Kim, C.-J., Kim, C.-H. and Sako, Y. (2005). Paralytic shellfish poisoning toxin analysis of the genus *Alexandrium* (Dinophyceae) occurring in Korean coastal waters. *Fish. Sci.* **71**, 1–11.

Kodama, M., Ogata, T., Fukuyo, Y., Ishimaru, T., Wisessang, S., Saitanu, K., Panichyakarn, V. and Piyakarnchana, T. (1988). *Protogonyaulax cohorticula*, a toxic dinoflagellate found in the Gulf of Thailand. *Toxicon* **26**, 707–12.

Kohata, K., Watanabe, M., Yamanaka, K., Ioriya, T. and Kimura, T. (1997). A red-tide bloom of *Chattonella antiqua* (Raphidophyceae) in the Seto Inland Sea, Japan, in 1987 analysed as changes in photosynthetic pigments determined by high-performance liquid chromatography. *Wat. Res.* **31**, 2269–77.

Laamanen, M. J., Gugger, M. F., Lehtimäki, J. M., Haukka, K. and Sivonen, K. (2001). Diversity of toxic and nontoxic *Nodularia* isolates (Cyanobacteria) and filaments from the Baltic Sea. *Appl. Environ. Microbiol.* **67**, 4638–47.

Lagos, N., Onodera, H., Zagatto, P. A., Andrinolo, D., Azevedo, S. M. F. Q. and Oshima, Y. (1999). The first evidence of paralytic shellfish toxins in the freshwater cyanobacterium *Cylindrospermopsis raciborskii*, isolated from Brazil. *Toxicon* **37**, 1359–73.

Larsen, A. (1999). *Prymnesium parvum* and *P. patelliferum* (Haptophyta) – one species. *Phycologia* **38**, 541–43.

Laub, J., Henriksen, P., Brittain, S. M., Wang, J., Carmichael, W. W., Rinehart, K. L. and Moestrup, Ø. (2002). [ADMAdda^5]-Microcystins in *Planktothrix agardhii* strain PH-123 (Cyanobacteria) – importance for monitoring of microcystins in the environment. *Environ. Toxicol.* **17**, 351–57.

Laurion, I. and Roy, S. (2009). Growth and photoprotection in three dinoflagellates (including two strains of *Alexandrium tamarense*) and one diatom exposed to four weeks of natural and enhanced ultraviolet-B radiation. *J. Phycol.* **45**, 16–33.

Leavitt, P. R. and Brown, S. R. (1988). Effects of grazing by *Daphnia* on algal carotenoids, implications for paleolimnology. *J. Paleolimnol.* **1**, 201–13.

Lesser, M. P. (1996). Elevated temperatures and ultraviolet radiation cause oxidative stress and inhibit photosynthesis in symbiotic dinoflagellates. *Limnol. Oceanogr.* **41**, 271–83.

Li, D., Cong, W., Cai, Z., Shi, D. and Ouyang, F. (2003). Some psychological and biochemical changes in marine eukaryotic red tide algae *Heterosigma akashiwo* during the alleviation from iron limitation. *Plant Physiol. Biochem.* **41**, 295–301.

Lim, P.-T., Leaw, C.-P., Usup, G., Kobiyama, A., Koike, K. and Ogata, T. (2006). Effects of light and temperature on growth, nitrate uptake, and toxin production of two tropical dinoflagellates, *Alexandrium tamiyavanichii* and *Alexandrium minutum* (Dinophyceae). *J. Phycol.* **42**, 786–99.

Lim, P.-T., Sato, S., Thuoc, C. V., Tu, P. T., Huyen, N. T. M., Takata, Y., Yoshida, M., Kobiyama, A., Koike, K. and Ogata, T. (2007). Toxic *Alexandrium minutum* (Dinophyceae) from Vietnam with new gonyautoxin analogue. *Harmful Algae* **6**, 321–31.

Liu, Y., Chen, W., Li, D., Shen, Y., Li, G. and Liu, Y. (2006a). First report of aphantoxins in China–waterblooms of toxigenic *Aphanizomenon flos-aquae* in Lake Dianchi. *Ecotoxicol. Environ. Safety* **65**, 84–92.

Liu, Y., Chen, W., Li, D., Shen, Y., Liu, Y. and Song, L. (2006b). Analysis of paralytic shellfish toxins in *Aphanizomenon* DC-1 from Lake Dianchi, China. *Environ. Toxicol.* **21**, 289–95.

MacKenzie, L., Beuzenberg, V., Holland, P., McNabb, P., Suzuki, T. and Selwood, A. (2005). Pectenotoxin and okadaic acid-based toxin profiles in *Dinophysis acuta* and *Dinophysis acuminata* from New Zealand. *Harmful Algae* **4**, 75–85.

Madhyastha, H. K. and Vatsala, T. M. (2007). Pigment production in *Spirulina fussiformis* in different photophysical conditions. *Biomolec. Eng.* **24**, 301–05.

Mandelli, E. F. (1969). Carotenoid interconversion in light-dark cultures of the dinoflagellate *Amphidinium klebsii*. *J. Phycol.* **5**, 382–84.

Marasigan, A.N., Sato, S., Fukuyo, Y. and Kodoma, M. (2001). Accumulation of a high level of diarrhetic shellfish toxins in the green mussell *Perna viridis* during a bloom of *Dinophysis caudata* and *Dinophysis miles* in Sapian Bay, Panay Island, the Phillipines. *Fish. Sci.* **67**, 994–96.

Matsubara, T., Nagasoe, S., Yamasaki, Y., Shikata, T., Shimasaki, Y., Oshima, Y. and Honjo, T. (2007). Effects of temperature, salinity, and irradiance on the growth of the dinoflagellate *Akashiwo sanguinea*. *J. Exp. Mar. Biol. Ecol.* **342**, 226–30.

Matsuda, A., Nishijima, T., Fukami, K. and Adachi, M. (2006). Growth kinetics and paralytic shellfish poisoning toxin production in phosphorus-limited cultures of *Alexandrium catenella*. *Nippon Suisan Gakkaishi* **72**, 193–200.

Mazur-Marzek, H., Meriluoto, J., Pliński, M. and Szafranek, J. (2006). Characterization of nodularin variants in *Nodularia spumigena* from the Baltic

Sea using liquid chromatography/mass spectrometry/mass spectrometry. *Rapid Commun. Mass Spectrom.* **20**, 2023–32.

Meyer-Harms, B. and Pollehne, F. (1998). Alloxanthin in *Dinophysis norvegica* (Dinophysiales, Dinophyceae) from the Baltic Sea. *J. Phycol.* **34**, 280–85.

Miles, C. O., Wilkins, A. L., Samdal, I. A., Sandvik, M., Petersen, D., Quilliam, M. A., Naustvoll, L. J., Rundberget, T., Torgersen, T., Hovgaard, P., Jensen, D. J. and Cooney, J. M. (2004). A novel pectenotoxin, PTX-12, in *Dinophysis* spp. and shellfish from Norway. *Chem. Res. Toxicol.* **17**, 1423–33.

Miles, C. O., Samdal, I. A., Aasen, J. A. G., Jensen, D. J., Quilliam, M. A., Petersen, D., Briggs, L. M., Wilkins, A. L., Rise, F., Cooney, J. M. and MacKenzie, A. L. (2005). Evidence for numerous analogs of yessotoxin in *Protoceratium reticulatum*. *Harmful Algae* **4**, 1075–91.

Millie, D. F., Hersh, C. M. and Dionigi, C. P. (1992). Simazine-induced inhibition in photoacclimated populations of *Anabaena circinalis* (Cyanophyta). *J. Phycol.* **28**, 19–26.

Moestrup, Ø., Akselman, R., Cronberg, G., Elbraechter, M., Fraga, S., Halim, Y., Hansen, G., Hoppenrath, M., Larsen, J., Lundholm, N., Nguyen, L. N. and Zingone, A. (eds) (2009). In *IOC-UNESCO Taxonomic Reference List of Harmful Micro Algae*. Available online at http://www.marinespecies.org/hab/Accessed on 2010-08-31.

Moffitt, M. C., Blackburn, S. I. and Neilan, B. A. (2001). rRNA sequences reflect the ecophysiology and define the toxic cyanobacteria of the genus *Nodularia*. *Int. J. Syst. Evol. Microbiol.* **51**, 505–12.

Mohamed, Z. A., El-Sharouny, H. M. and Ali, W. S. M. (2006). Microcystin production in benthic mats of cyanobacteria in the Nile River and irrigation canals, Egypt. *Toxicon* **47**, 584–90.

Molica, R., Onodera, H., García, C., Rivas, M., Andrinolo, D., Nascimento, S., Meguro, H., Oshima, Y., Azevedo, S. and Lagos, N. (2002). Toxins in the freshwater cyanobacterium *Cylindrospermopsis raciborskii* (Cyanophyceae) isolated from Tabocas reservoir in Caruaru, Brazil, including demonstration of a new saxitoxin analogue. *Phycologia* **41**, 606–11.

Montoya, N. G., Akselman, R., Carignan, M. O. and Carreto, J. I. (2006). Pigment profile and toxin composition during a red tide of *Gymnodinium catenatum* Graham and *Myrionecta rubra* (Lohman) Jankowski in coastal waters off Mar del Plata, Argentina. *Afr. J. Mar. Sci.* **28**, 199–202.

Mooney, B. D., de Salas, M., Hallegraeff, G. M. and Place, A. R. (2009). Survey for karlotoxin production in 15 species of gymnodinioid dinoflagellates (Kareniaceae, Dinophyta). *J. Phycol.* **45**, 164–75.

Morohashi, A., Satake, M. and Yasumoto, T. (1998). The absolute configuration of gambierol, a toxic marine polyether from the dinoflagellate, *Gambierdiscus toxicus*. *Tetrahedron Lett.* **39**, 97–100.

Morohashi, A., Satake, M., Nagai, H., Oshima, Y. and Yasumoto, T. (2000). The absolute configuration of gambieric acids A–D, potent antifungal polyethers, isolated from the marine dinoflagellate *Gambierdiscus toxicus*. *Tetrahedron* **56**, 8995–9001.

Morsy, N., Matsuoka, S., Houdai, T., Matsumori, N., Adachi, S., Murata, M., Iwashita, T. and Fujita, T. (2005). Isolation and structure elucidation of a new amphidinol with truncated polyhydroxyl chain from *Amphidinium klebsii*. *Tetrahedron* **61**, 8606–10.

Morsy, N., Houdai, T., Matsuoka, S., Matsumori, N., Adachi, S., Oishi, T., Murata, M., Iwashita, T. and Fujita, T. (2006). Structures of new amphidinols with truncated polyhydroxyl chain and their membrane-permeabilizing activities. *Bioorg. Med. Chem.* **14**, 6548–54.

Mundt, S., Kreitlow, S. and Jansen, R. (2003). Fatty acids with antibacterial activity from the cyanobacterium *Oscillatoria redekei* HUB 051. *J. Appl. Phycol.* **15**, 263–67.

Nascimento, S. M., Purdie, D. A., Lilly, E. L., Larsen, J. and Morris, S. (2005). Toxin profile, pigment composition, and large subunit rDNA phylogenetic analysis of an *Alexandrium minutum* (Dinophyceae) strain isolated from the Fleet Lagoon, United Kingdom. *J. Phycol.* **41**, 343–53.

Navarro, J. M., Muños, M. G. and Contreras, A. M. (2006). Temperature as a factor regulating growth and toxin content in the dinoflagellate *Alexandrium catenella*. *Harmful Algae* **5**, 762–69.

Nayak, B. B., Karunasagar, I. and Karunasagar, I. (1997). Influence of bacteria on growth and hemolysin production by the marine dinoflagellate *Amphidinium carterae*. *Mar. Biol.* **130**, 35–39.

Negri, A. P., Jones, G. J., Blackburn, S. I., Oshima, Y. and Onodera, H. (1997). Effect of culture and bloom development and of sample storage on paralytic shellfish poisons in the cyanobacterium *Anabaena circinalis*. *J. Phycol.* **33**, 26–35.

Negri, A., Stirling, D., Quilliam, M., Blackburn, S., Bolch, C., Burton, I., Eaglesham, G., Thomas, K., Walter, J. and Willis, R. (2003). Three novel hydroxybenzoate saxitoxin analogues isolated from the dinoflagellate *Gymnodinium catenatum*. *Chem. Res. Toxicol.* **16**, 1029–33.

Negri, A. P., Bunter, O., Jones, B. and Llewellyn, L. (2004). Effects of the bloom-forming alga *Trichodesmium erythraeum* on the pearl oyster *Pinctada maxima*. *Aquaculture* **232**, 91–102.

Nichols, P. D., Skerratt, J. H., Davidson, A., Burton, H. and McMeekin, T. A. (1991). Lipids of cultured *Phaeocystis pouchetii*, signatures for food-web, biogeochemical and environmental studies in Antarctica and the Southern Ocean. *Phytochemistry* **30**, 3209–14.

Nicklisch, A. and Woitke, P. (1999). Pigment content of selected planktonic algae in response to simulated natural light fluctuations and a short photoperiod. *Internat. Rev. Hydrobiol.* **84**, 479–95.

Nishikawa, T. and Hori, Y. (2004). Effects of nitrogen, phosphorous and silicon on the growth of the diatom *Eucampia zodiacus* caused bleaching of seaweed *Porphyra* isolated from Harima-Nada, Seto Innland Sea, Japan. *Nippon Suisan Gakkaishi* **70**, 31–38.

Norris, R. L., Eaglesham, G. K., Pierens, G., Shaw, G. R., Smith, M. J., Chiswell, R. K., Seawright, A. A. and Moore, M. R. (1999). Deoxycylindrospermopsin, an analog of cylindrospermopsin from *Cylindrospermopsis raciborskii*. *Environ. Toxicol.* **14**, 163–65.

Oberholster, P. J., Botha, A-M. and Grobbelaar, J. U. (2004). *Microcystis aeruginosa*: source of toxic microcystins in drinking water. *Review. Afr. J. Biotechnol.* **3**, 159–68.

Ogata, T., Pholpunthin, P., Fukuyo, Y. and Kodama, M. (1990). Occurrence of *Alexandrium cohorticula* in Japanese coastal water. *J. Appl. Phycol.* **2**, 351–56.

Ogata, T., Kodama, M., Nomura, S., Kobayashi, M., Nozawa, T., Katoh, T. and Mimuro, M. (1994). A novel peridinin–chlorophyll *a* protein (PCP) from the marine dinoflagellate *Alexandrium cohorticula*, a high pigment content and plural spectral forms of peridinin and chlorophyll *a*. *FEBS Lett.* **356**, 367–71.

Okada, S. (1999). Production of useful hydrocarbons from microalgae. *Nippon Suisan Gakkaishi* **65**, 621–25.

Onodera, H., Oshima, Y., Henriksen, P. and Yasumoto, T. (1997). Confirmation of anatoxin-a(s), in the cyanobacterium *Anabaena lemermannii*, as the cause of bird kills in Danish lakes. *Toxicon* **35**, 1645–48.

Ordás, M.C., Fraga, S., Franco, J.M., Ordás, A. and Figueras, A. (2004). Toxin and molecular analysis of *Gymnodinium catenatum* (Dinophyceae) strains from Galicia (NW Spain) and Andalucía (S Spain). *J. Plankton Res.* **26**, 341–49.

Oshima, Y., Blackburn, S.I. and Hallegraeff, G.M. (1993). Comparative study on paralytic shellfish toxin profiles of the dinoflagellate *Gymnodinium catenatum* from three different countries. *Mar. Biol.* **116**, 471–76.

Paerl, H.W., Tucker, J. and Bland, P.T. (1983). Carotenoid enhancement and its role in maintaining blue-green algal *(Microcystis aeruginosa)* surface blooms. *Limnol. Oceanogr.* **28**, 847–57.

Park, H.-D., Watanabe, M.F., Harada, K.-I., Nagai, H., Suzuki, M., Watanabe, M. and Hayashi, H. (1993). Hepatotoxin (microcystin) and neurotoxin (anatoxin-a) contained in natural blooms and strains of cyanobacteria from Japanese freshwaters. *Nat. Toxins* **1**, 353–60.

Pavela-Vrančič, M., Ujević, I., Gladan, Ž.N. and Furey, A. (2006). Accumulation of phycotoxins in the mussel *Mytilus galloprovincialis* from the Central Adriatic Sea. *Croat. Chem. Acta* **79**, 291–97.

Paz, B., Riobó, P., Fernández, M.L., Fraga, S. and Franco, J.M. (2004). Production and release of yessotoxins by the dinoflagellates *Protoceratium reticulatum* and *Lingulodinium polyedrum* in culture. *Toxicon* **44**, 251–58.

Paz, B., Riobó, P., Ramilo, I. and Franco, J.M. (2007). Yessotoxins profile in strains of *Protoceratium reticulatum* from Spain and USA. *Toxicon* **50**, 1–17.

Peng, X.-C., Yang, W.-D., Liu, J.-S., Peng, Z.-Y., Lü, S.-H. and Ding, W.-Z. (2005). Characterization of the hemolytic properties of an extract from *Phaeocystis globosa* Scherffel. *J. Integr. Plant Biol.* **47**, 165–71.

Pichardo, S., Jos, A., Zurita, J.L., Salguero, M., Cameán, A.M. and Repetto, G. (2007). Acute and subacute toxic effects produced by microcystin-YR on the fish cell lines RTG-2 and PLHC-1. *Toxicol. in Vitro* **21**, 1460–67.

Pomati, F., Moffit, M.C., Cavaliere, R. and Neilan, B.A. (2004). Evidence for differences in the metabolism of saxitoxin and C1+2 toxins in the freshwater cyanobacterium *Cylindrospermopsis raciborskii* T3. *Biochim. Biophys. Acta* **1674**, 60–67.

Prasad, A.V.K. and Shimizu, Y. (1989). The structure of hemibrevetoxin-B, a new type of toxin in the Gulf of Mexico red tide organism. *J. Am. Chem. Soc.* **111**, 6476–77.

Preußel, K., Stüken, A., Wiedner, C., Chorus, I. and Fastner, J. (2006). First report on cylindrospermopsin producing *Aphanizomenon flos-aquae* (Cyanobacteria) isolated from two German lakes. *Toxicon* **47**, 156–62.

Prézelin, B.B., and Haxo, F.T. (1976). Purification and characterization of peridinin-chlorophyll *a*-proteins from the marine dinoflagellates *Glenodinium sp.* and *Gonyaulax polyedra*. *Planta.* **128**, 133–41.

Ramos, A. G., Martel, A., Codd, G. A., Soler, E., Coca, J., Redondo, A., Morrison, L. F., Metcalf, J. S., Ojeda, A., Suárez, S. and Petit, M. (2005). Bloom of the marine diazotrophic cyanobacterium *Trichodesmium erythraeum* in the Northwest African Upwelling. *Mar. Ecol. Prog. Ser.* **301**, 303–05.

Rapala, J., Sivonen, K., Luukkainen, R. and Niemelä, S. I. (1993). Anatoxin-a concentration in *Anabaena* and *Aphanizomenon* under different environmental conditions and comparison of growth by toxic and non-toxin *Anabaena*-strains – a laboratory study. *J. Appl. Phycol.* **5**, 581–91.

Rech, M., Mouget, J.-L., Morant-Manceau, A., Rosa, P. and Tremblin, G. (2005). Long-term acclimation to UV radiation, effects on growth, photosynthesis and carbonic anhydrase activity in marine diatoms. *Bot. Mar.* **48**, 407–20.

Rein, K. S. and Snyder, R. V. (2006). The biosynthesis of polyketide metabolites by dinoflagellates. *Adv. Appl. Microbiol.* **59**, 93–125.

Rodríguez, F., Chauton, M., Johnsen, G., Andresen, K., Olsen, L. M. and Zapata, M. (2006). Photoacclimation in phytoplankton, implications for biomass estimates, pigment functionality and chemotaxonomy. *Mar. Biol.* **148**, 963–71.

Rodríguez, J. J. G., García, M. d. C. C., Camacho, F. G., Mirón, A. S., Belarbi, E. H. and Grima, E. M. (2007). New culture approaches for yessotoxin production from the dinoflagellate *Protoceratium reticulatum*. *Biotechnol. Prog.* **23**, 339–50.

Sala, S. E., Sar, E. A. and Ferrario, M. E. (1998). Review of materials reported as containing *Amphora coffeaeformis* (Agardh) Kützing in Argentina. *Diatom Res.* **13**, 323–36.

Sano, T., Takagi, H. and Kaya, K. (2004). A Dhb-microcystin from the filamentous cyanobacterium *Planktothrix rubescens*. *Phytochemistry* **65**, 2159–62.

Sasaki, M., Takeda, N., Fuwa, H., Watanabe, R., Satake, M. and Oshima, Y. (2006). Synthesis of the JK/LM-ring model of prymnesins, potent hemolytic and ichthyotoxic polycyclic ethers isolated from the red tide alga *Prymnesium parvum*, confirmation of the relative configuration of the K/L-ring juncture. *Tetrahedron Lett.* **47**, 5687–91.

Satake, M., Tanaka, Y., Ishikura, Y., Oshima, Y., Naoki, H. and Yasumoto, T. (2005). Gymnocin-B with the largest contiguous polyether rings from the red tide dinoflagellate, *Karenia* (formerly *Gymnodinium*) *mikimotoi*. *Tetrahedron Lett.* **46**, 3537–40.

Schlüter, L., Garde, K. and Kaas, H. (2004). Detection of the toxic cyanobacteria *nodularia spumigena* by means of a 4-keto-myxoxanthophyll-like pigment in the Baltic Sea. *Mar. Ecol. Prog. Ser.* **275**, 69–78.

Sinha, R. P., Klisch, M., Gröninger, A. and Häder, D.-P. (1998). Ultraviolet-absorbing/screening substances in cyanobacteria, phytoplankton and macroalgae. *J. Photochem. Photobiol. B: Biol.* **47**, 83–94.

Shimizu, Y., Gupta, S., Masuda, K., Maranda, L., Walker, C. K. and Wang, R. (1989). Dinoflagellate and other microalgal toxins, chemistry and biochemistry. *Pure Appl. Chem.* **61**, 513–16.

Simonsen, S. and Moestrup, Ø. (1997). Toxicity tests in eight species of *Chrysochromulina* (Haptophyta). *Can. J. Bot.* **75**, 129–36.

Sinha, R. P., Ambasht, N. K., Sinha, J. P., Klisch, M. and Häder, D.-P. (2003). UV-B-induced synthesis of mycosporine-like amino acids in three strains of *Nodularia* (cyanobacteria). *J. Photochem. Photobiol. B* **71**, 51–58.

Sivonen, K., Skulberg, O. M., Namikoshi, M., Evans, W. R., Carmichael, W. W. and Rinehart, K. L. (1992). Two methyl ester derivatives of microcystins, cyclic heptapeptide hepatotoxins, isolated from *Anabaena flos-aquae* strain CYA 83/1. *Toxicon* **30**, 1465–71.

Suda, S., Watanabe, M. M., Otsuka, S., Mahakahant, A., Yongmanitchai, W., Nopartnaraporn, N., Liu, Y. and Day, J. G. (2002). Taxonomic revision of water-bloom-forming species of oscillatorioid cyanobacteria. *Int. J. Syst. Evol. Microbiol.* **52**, 1577–95.

Suzuki, T., Horie, Y., Koike, K., Satake, M., Oshima, Y., Iwataki, M. and Yoshimatsu, S. (2007). Yessotoxin analogues in several strains of *Protoceratium reticulatum* in Japan determined by liquid chromatography–hybrid triple quadrupole/linear ion trap mass spectrometry. *J. Chromatogr. A* **1142**, 172–77.

Tang, E. P. Y. and Vincent, W. F. (1999). Strategies of thermal adaption by high-latitude cyanobacteria. *New Phytol.* **142**, 315–23.

Tilstone, G, Airs, R., Martinez-Vicente, V., Widdicombe, C. and Llewellyn C. (2010). High concentrations of mycosporine like amino acids and coloured dissolved organic matter in the sea surface microlayer off the Iberian Peninsula. *Limnol. Oceanogr.* **55**, 1835–50.

Thomas, A. D., Saker, M. L., Norton, J. H. and Olsen, R. D. (1998). Cyanobacterium *Cylindrospermopsis raciborskii* as a probable cause of death in cattle in northern Queensland. *Aust. Vet. J.* **76**, 592–94.

Tsukano, C. and Sasaki, M. (2006). Structure-activity relationship studies of gymnocin-A. *Tetrahedron Lett.* **47**, 6803–07.

Twiner. M. J., Dechraoui, M. -Y. B., Wang, Z., Mikulski, C. M., Henry, M. S., Pierce, R. H. and Doucette, G. J. (2007). Extraction and analysis of lipophilic brevetoxins from the red tide dinoflagellate *Karenia brevis*. *Anal. Biochem.* **369**, 128–35.

Várkonyi, Z., Masamoto, K., Debreczeny, M., Zsiros, O., Ughy, B., Gombos, Z., Domonkos, I., Farkas, T., Wada, H. and Szalontai, B. (2002). Low-temperature-induced accumulation of xanthophylls and its structural consequences in the photosynthetic membranes of the cyanobacterium *Cylindrospermopsis raciborskii*, An FTIR spectroscopic study. *Proc. Natl. Acad. Sci. USA* **99**, 2410–15.

Vaulot, D., Birrien, J. -L., Marie, D., Casotti, R., Veldhuis, M. J. W., Kraay, G. W. and Chrétiennot-Dinet, M. -J. (1994). Morphology, ploidy, pigment composition, and genome size of cultured strains of *Phaeocystis* (Prymnesiophyceae). *J. Phycol.* **30**, 1022–35.

Viaggiu, E., Melchiorre, S., Volpi, F., Corcia, A. D., Mancini, R., Garibaldi, L., Crichigno, G. and Bruno, M. (2004). Anatoxin-a toxin in the cyanobacterium *Planktothrix rubescens* from a fishing pond in northern Italy. *Environ. Toxicol.* **19**, 191–97.

Walsh, K., Jones, G. J. and Dunstan, R. H. (1997). Effect of irradiance on fatty acid, carotenoid, total protein composition and growth of *Microcystis aeruginosa*. *Phytochemistry* **44**, 817–24.

Watanabe, M. F., Oishi, S., Harada, K.-I., Matsuura, K., Kawai, H. and Suzuki, M. (1988). Toxins contained in *Microcystis* species of cyanobacteria (blue-green algae). *Toxicon* **26**, 1017–25.

Welker, M. and Erhard, M. (2007). Consistency between chemotyping of single filaments of *Planktothrix rubescens* (cyanobacteria) by MALDI-TOF and the peptide patterns of strains determined by HPLC-MS. *J. Mass Spectrom.* **42**, 1062–68.

Whitehead, K. and Vernet, M. (2000). Influence of mycosporine-like amino acids (MAAs) on UV absorption by particulate and dissolved organic matter in La Jolla Bay. *Limnol. Oceanogr.* **45**, 1788–96.

Wilhelm, C. and Manns, L. (1991). Changes in pigmentation of phytoplankton species during growth and stationary phase – consequences for reliability of pigment-based methods of biomass determination. *J. Appl. Chem.* **3**, 305–10.

Woitke, P., Hesse, K. and Kohl, J.-G. (1997). Changes in the lipophilic photosynthetic pigment content of different *Microcystis aeruginosa* strains in response to growth irradiance. *Photosynthetica* **33**, 443–53.

Xing, W., Huang, W.-M., Li, D.-H. and Liu, Y.-D. (2007). Effects of iron on growth, pigment content, photosystem II efficiency, and siderophores production of *Microcystis aeruginosa* and *Microcystis wesenbergii*. *Curr. Microbiol.* **55**, 94–98.

Yamasaki, Y., Kim, D.-I., Matsuyama, Y., Oda, T. and Honjo, T. (2004). Production of superoxide anion and hydrogen peroxide by the red tide dinoflagellate *Karenia mikimotoi*. *J. Biosci. Bioeng.* **97**, 212–15.

Yao, P., Yu, Z. and Deng, C. (2006). Pigment signatures of some diatoms isolated from China seas. *Acta Oceanol. Sinica* **25**, 108–18.

Yasumoto, T., Underdal, B., Aune, T., Hormazabal, V., Skulberg, O. M. and Oshima, Y. (1990). Screening for hemolytic and ichthyotoxic components of *Chrysochromulina polylepis* and *Gyrodinium aureolum* from Norwegian coastal waters. In *Toxic Marine Phytoplankton*, ed. E. Granéli, B. Sundström, L. Edler and D. M. Anderson. New York: Elsevier, pp. 436–40.

Yim, J. H., Kim, S. J., Ahn, S. H. and Lee, H. K. (2003). Optimal conditions for the production of sulphated polysaccharide by marine microalga *Gyrodinium impudicum* strain KG03. *Biomolec. Eng.* **20**, 273–80.

Yim, J. H., Son, E., Pyo, S. and Lee, H. K. (2005). Novel sulphated polysaccharide derived from red-tide microalga *Gyrodinium impudicum* strain KG03 with immunostimulating activity in vivo. *Mar. Biotechnol.* **7**, 331–38.

Zapata, M., Edvardsen, B., Rodríguez, F., Maestro, M. A. and Garrido, J. L. (2001). Chlorophyll c_2 monogalactosyldiacylglyceride ester (chl c_2-MGDG). A novel marker pigment for *Chrysochromulina* species (Haptophyta). *Mar. Ecol. Prog. Ser.* **219**, 85–98.

Zapata, M., Jeffrey, S. W., Wright, S. W., Rodríguez, F., Garrido, J. L. and Clementson, L. (2004). Photosynthetic pigments in 37 species (65 strains) of Haptophyta, implications for oceanography and chemotaxonomy. *Mar. Ecol. Prog. Ser.* **270**, 83–102.

Part V

Future perspectives

15

Perspectives on future directions

CAROLE A. LLEWELLYN, SUZANNE ROY, GEIR JOHNSEN, EINAR SKARSTAD EGELAND, MATILDE CHAUTON, GUSTAFF HALLEGRAEFF, MARTIN LOHR, ULRIKE OSTER, ROBERT J. PORRA, HUGO SCHEER AND KAI-HONG ZHAO

15.1 Introduction

'We are on the verge of a golden age.'

(Quote by Martin Lohr on xanthophyll research)

This chapter presents a diverse collection of perspectives covering recent discoveries and *'crystal ball gazing'* on future directions. Detection and characterisation from a molecular level is covered through to monitoring phytoplankton dynamics and climate change at a regional and global Earth observation level. At a molecular level, perspectives are provided on our basic understanding of the role of pigments in photosynthesis and photoprotection incorporating the development of new analytical and 'omics' techniques. Applied perspectives are included on HAB detection, aquaculture and algal biotechnology. Phytoplankton pigment research continues to develop opening up many fascinating and exciting possibilities. These perspectives highlight how research on pigments acts as a linchpin across a diverse range of disciplines including microbial ecology, oceanography, limnology, remote sensing and applied phycology.

15.2 Pigments in marine bacteria and cyanobacteria – recent discoveries

New discoveries on pigments continue to highlight an increasingly complex and biodiverse aquatic environment. Since the volume by Jeffrey *et al.* (1997), the complexity and diversity of pigments across and within algal classes is becoming more apparent with the discovery of new pigments and erection of new algal classes (Chapter 1, this volume). Such discoveries are contributing to our understanding of the biodiversity and illustrate the unique ability of phytoplankton to adapt to a wide variety of niches. The discovery of new suites of pigments particularly in the marine environment continues to challenge our views on marine microbial ecology. Notable are the discoveries of pigments involved in photosynthesis in marine bacteria and cyanobacteria, including bacteriochlorophyll, proteorhodopsin and chlorophyll (Chl) *d* which are described

Phytoplankton Pigments: Characterization, Chemotaxonomy and Applications in Oceanography, eds. Suzanne Roy, Carole A. Llewellyn, Einar Skarstad Egeland and Geir Johnsen. Published by Cambridge University Press.

in the subsequent paragraphs. These discoveries are changing our perspective on the role that phytoplankton play in driving ocean productivity, biogeochemical cycles and our views on carbon budgets and energy budgets in the ocean.

Our views on carbon budgets and energy budgets were challenged in the early 2000s, when bacteriochlorophyll *a* was detected in vertical profiles throughout the surface waters of oligotrophic oceans. These and further measurements led to the conclusion that a unique type of bacterial phototrophic metabolism was occurring in photosynthetically competent aerobic anoxygenic phototrophic bacteria (AAP). AAP are facultative photoheterotrophs, metabolising organic carbon when available but capable of photosynthetic light utilisation when the organic carbon is scarce. AAP were found to be abundant in the upper open ocean (Kolber *et al.*, 2000, 2001; Lami *et al.*, 2007), contributing to 5% of the total energy cycle and up to 10% of the microbial community (Cottrell *et al.*, 2006; Sieracki *et al.*, 2006).

Our views on light-harvesting were also challenged in early 2000 with the discovery of a new type of phototrophy occurring in surface ocean bacterioplankton (Béjà *et al.*, 2000). Proteorhodopsins are protein pigments of the rhodopsin family which function as light-dependent proton pumps. Proteorhodopsins consist of an extensive family of compounds spectrally tuned to different habitats by absorbing light at different wavelengths in accordance with light available in the environment (Béjà *et al.*, 2000, 2001). Data suggests that proteorhodopsin-based phototrophy is a globally significant oceanic microbial process.

In contrast to pigments adapted to capture visible light in the surface of the oceans, Chl *d* containing cyanobacteria thriving in near-infrared light environments deeper in the water column have recently been described (Kashiyama *et al.*, 2008). Chl *d* replaces Chl *a* as the main pigment and is believed to be globally distributed in oceanic and lacustrine low light environments with high near-infrared intensity (Kashiyama *et al.*, 2008). Indeed Chl *d*-based photosynthesis needs to be properly evaluated in estimating global primary production (Kashiyama *et al.*, 2008). Even more recently another chlorophyll has been discovered, designated as Chl *f* (Chen *et al.*, 2010). Chl *f* found in stromatolites absorbs even further into the red than Chl *d* (absorbance maximum at 706 nm in methanol), suggesting that oxygenic photosynthesis can be extended further into the infrared region (Chen *et al.*, 2010). The missing Chl *e* was isolated in trace amount from feral algal populations of two members of the Xanthophyceae, namely. *Tribonema bombycinum* and *Vaucheria hamata* but has not been detected in these species when cultured: in methanol, Chl *e* absorbs at 415 and 654 nm. The chemical structure has not been determined and its status as a natural Chl or artefact is uncertain. Research on this Chl was done in the 1940 and 50s by the pigment chemist, H. H. Strain, and reviewed by Allen, (1966).

15.3 Carotenoid biosynthesis – a perspective

During the last two decades, most of the genes involved in the biosynthetic pathways of chlorophylls and carotenoids in land plants and green algae have been identified. This was greatly facilitated by the availability of powerful model organisms like tomato, *Arabidopsis* or *Chlamydomonas* with straightforward genetics and well-established molecular tools.

However, our knowledge on the biosynthesis of the plethora of pigments in aquatic environments is limited. In particular, for chromalveolate algae, comprising the major eukaryotic primary producers in the world's oceans, the biosynthesis of carotenoids is largely unexplored (Chapter 3, this volume). Recently, the genomes of the diatoms *Phaeodactylum tricornutum* and *Thalassiosira pseudonana* have been described (Bowler *et al.*, 2008; Armbrust *et al.*, 2004). Diatoms and haptoflagellates contain fucoxanthin as major light-harvesting pigment, making this carotenoid one of the most abundant xanthophylls on Earth. This alone justifies giving top priority to the elucidation of fucoxanthin biosynthesis, all the more as this will also pave the way for exploring the formation of many other related xanthophylls in chromalveolates. Moreover, understanding the genetic basis of xanthophyll biosynthesis in chromalveolates is likely to teach us a lot about the evolution of the complex carotenoid pattern in marine algae.

A detailed understanding of the biosynthesis of diagnostic xanthophylls and its regulation will be fundamental for further improvement of pigment applications in oceanography, such as the monitoring of phytoplankton dynamics, quantitative chemotaxonomy, and the *in situ* detection of photoacclimation or nutrient limitation. Aside from the larger picture of photosynthetic carotenoids in phytoplankton, we should not forget the more cryptic carotenoids in algal resting spores or in the eyespot apparatus of flagellated algae, which have been mostly neglected in spite of their vital importance for the ecological success of many marine phytoplankton species. With powerful new analytical and molecular methods at hand, we are on the verge of a golden age of research on algal xanthophylls.

15.4 Chlorophyll and bacteriochlorophyll biosynthesis – recent advances

Since the volume by Jeffrey *et al.* (1997), there have been major advances in the biosynthesis of chlorophyll and bacteriochlorophyll as described in Chapter 2 (this volume). Since the drafting of Chapter 2 (this volume) further developments have transpired on the later stages of biosynthesis of chlorophyll.

Improved access to the nitrogenase-related enzymes, DPOR (light-independent protochlorophyllide reductase) (Fujita and Bauer, 2002) and COR (chlorophyllide reductase) (Nomata *et al.*, 2006) has opened the way to detailed biochemical and biophysical studies. Based on subunit-hybridization experiments, the two enzymes have been shown to be closely related (Wätzlich *et al.*, 2009). Aside from their regioselectivity for ring D and ring B reduction, respectively, they have similar substrate specificities; especially noteworthy, is their insensitivity to modifications of ring E (Bröcker *et al.*, 2008b; Wätzlich, 2009). Electron transfer reactions from non-haem Fe-proteins to the pigments involve transient interactions among the subunits (Bröcker *et al.*, 2008a) which appear to be triggered by ATP (Bröcker *et al.*, pers. commun.).

The biosynthetic pathways of the bacteriochlorophylls are described in Chapter 2 (this volume) but challenges remain. The biosynthetic pathway to the chlorosomal bacteriochlorophylls, BChls *c*, *d* and *e*, has been rapidly elucidated using the sequenced *Chlorobaculum* (formerly *Chlorobium*) *tepidum* (see Chapter 2, this volume), but understanding regulation and final assembly mechanisms into chlorosomes remains a challenge. One point of regulation for BChl *c* biosynthesis appears to be provided by multiple Mg-chelatase subunits H, which may be involved in substrate channeling to subsequent biosynthetic enzymes (Chew *et al.*, 2009). By manipulating the biosynthetic pathway, the pigment heterogeneity of chlorosomes could be reduced. Using such 'streamlined' chlorosomes, a novel alternating syn–anti interaction model of chlorosomal bacteriochlorophylls has been advanced (Ganapathy *et al.*, 2009). An open question is still the mechanism of the de-methoxycarbonylation at ring E.

15.5 Chlorophyll degradation – a perspective

In contrast to the biosynthetic pathways of the chlorophylls (and carotenoids) which are relatively well established, the degradation pathways and fate of chlorophyll are still poorly understood. Whilst the role of phytoplankton mortality is a crucial component of microbial cycling in the upper water column and cell lysis techniques are providing new information on phytoplankton mortality, the role of chlorophyll degradation in this turnover of organic matter is still little known. Major advances in the terrestrial environment are likely to filter through into the aquatic environment to further our understanding on phytoplankton cycling and stress response. Recent studies indicate that the most rapid destruction of chlorophyll and its derivatives occurs in toxic conditions with added microorganisms, and that cellular senescence, grazing-induced cell disruption, physicochemical environment (O_2 and so on), and microbial processing all need to be considered in order to fully address chlorophyll diagenesis (Szymczak-Żyła *et al.*, 2008).

Close cooperation between organic chemists and plant physiologists continues to increase our understanding of chlorophyll degradation. While current work is focused on terrestrial plants and green algae (see Chapter 2, this volume), the rapid progress is expected to fuel interest in marine organisms. Natural breakdown products and their stereoisomers have become accessible by a biomimetic, semi-synthetic approach (Oberhuber *et al.*, 2008). Breakdown products have been introduced as diagnostic tools in phytopathology and plant physiology (Moser *et al.*, 2009a), their spontaneous oxidation products contribute to autumnal leaf colours (Moser *et al.*, 2008) and antioxidant functions have also been proposed (Moser *et al.*, 2009b). Enzymes involved in chlorophyll *a* and *b* turnover have also been implicated in the regulation of other processes: a recent example is the function of the N-terminal domain of the Chl *a* oxidase as a protease regulator (Sakuraba *et al.*, 2009).

Aside from understanding the catabolites produced from chlorophyll breakdown there have been recent discoveries on the formation of heavy metal chlorophylls (transmetalation of chlorophyll). This has potential importance in terms of pollution studies and the production of reactive oxygen species. The excellent photophysical properties of chlorophylls derive, in part, from using Mg as the central metal; heavier metals lead to more rapid excitation loss, or triplet formation. Mg-tetrapyrroles, however, are thermodynamically unstable, as witnessed by the facile demetalation and transmetalation of chlorophylls. While such reactions are largely inhibited in the intact photosynthetic apparatus by means that are only incompletely understood, this protection is not perfect. Formation of transmetalated chlorophylls (Küpper *et al.*, 2006), especially in PSII, appears to be a key step in metal toxicity, but further clarification is required. Single-cell analysis may provide a novel approach; recently, it was used to study chromium and copper-induced inhibition of photosynthesis in *Euglena gracilis* by fluorescence kinetic microscopy (Rocchetta and Küpper, 2009).

15.6 Phycobiliproteins – a perspective

The structure, function, regulation and adaptation of the phycobilisome remains one of the most challenging questions in the basic biology of cyanobacteria and red algae, and biliproteins are also of considerable practical interest. Understanding chromophore biosynthesis and the extensive post-translational modifications of biliproteins is likely to excite both future interest and progress in phycobilisome research and biliprotein applications.

After the discovery of two new lyase families (see Chapter 9, this volume), understanding the function of the numerous family members present in a single organism, as well as their interactions, timing and regulations is currently an active field of research (Blot *et al.*, 2009; Kupka *et al.*, 2009; Schluchter *et al.*, 2010). With the recent identification of the enzymes responsible for asparagine methylation (Miller *et al.*, 2008; Shen *et al.*, 2008), most, if not all, of the functional components required for producing the phycobilisome components are known, and the assembly process appears within reach. Another important line of progress is the identification of a growing number of cyanobacteriochromes. These relatives of the phytochromes are photochromic biliproteins that cover the entire visible spectral range (Narikawa *et al.*, 2009; Cornilescu *et al.*, 2008; Ikeuchi and Ishizuka, 2008). They not only appear to be the long sought-after 'phycochromes' involved in a diverse range of photoregulatory responses in cyanobacteria but, as multifunctional proteins, also appear to be involved in other regulatory processes (Ulijasz *et al.*, 2009).

Fluorescent or photochromic chromoproteins are invaluable labelling tools in biology, and are also considered for data storage and processing. The use of biliproteins has been limited because, unlike the family of fluorescent proteins such as the green fluorescent protein, GFP (Tsien, 2009), they were not accessible *in vivo* from just a single gene without the requirement to add external chromophores. A novel, modular access to biliproteins is based on expressing the structural gene fused to genes coding for chromophore biosynthesis from endogenous haem, thereby requiring the introduction of only a single gene for labelling; this has been exemplified with a red-green photoreversible biliprotein that is fluorescent in the red state, and with persistently red fluorescent proteins (Zhang *et al.*, 2010).

15.7 Adaptation and acclimation of phytoplankton to stressful environments – recent advances

The complementary and growing area of 'omics' embracing genomics, transcriptomics and metabolomics presents exciting opportunities to give better insight into phytoplankton community characterisation and function. Notably, combining molecular, metabolic and physiological measurements is proving powerful in understanding the adaptation and acclimation of phytoplankton to their environment. One striking consideration is the greater diversity of pigments in the oceans compared with on land, reflecting the selective pressure for efficient light capture in an environment where light quality and quantity is highly variable. The bewildering diversity of carotenoids is particularly evident when looking at the various xanthophylls bound to the peripheral light-harvesting proteins of algal photosystems, reflecting the selective pressure on single-celled photosynthesizers to efficiently capture light, in particular the green light prevailing in most aquatic environments (Chapter 13, this volume).

Recent advances in molecular technologies together with an understanding of the role pigments play in photosynthesis and photoprotection are providing new insights into the ability of phytoplankton to cope with highly unpredictable and rapid changes in irradiance and spectral quality (Peers *et al.*, 2009; Nymark *et al.*, 2009). A combination of non-targeted transcriptomics and metabolomic approaches has been used to study the ability of diatoms to cope with iron-limited environments (Allen *et al.*, 2008). It is apparent from these studies that chlorophyll biosynthesis and pigment metabolism play key roles in the adaptation (long-term) and acclimation (short-term) of phytoplankton to their environment and in carbon metabolism.

15.8 Underpinning technical advances

The advances in analytical chemistry techniques combined with mushrooming of molecular techniques underpin major enhancements in our knowledge and understanding of pigments across many research areas. Analytical chemistry techniques for the identification of pigments have improved significantly. Advanced chromatographic methods like Ultra-HPLC, novel stationary phases including C_{30}-columns and monolithic columns (Chapter 4, this volume) and advances in LC-MS (Chapter 7, this volume) and the emerging LC-NMR technology allow rapid separation and structure elucidation of yet unknown minor carotenoids. These minor carotenoids could be candidate biosynthetic intermediates and/or specific to growth and environmental conditions (such as in fucoxanthin and siphonaxanthin esters: Yoshii *et al.*, 2005; Airs and Llewellyn, 2006). Higher resolution MS instruments such as Matrix-Assisted Laser Desorption Ionisation – Time of Flight (MALDI-TOF: Chapter 4, this volume) and high-resolution magic angle spinning nuclear magnetic resonance spectroscopy (HR-MAS-NMR: this chapter) bring the capability of measuring whole cell metabolite profiles.

Advances in analytical biochemistry including metabolomics come along with an even larger progress in molecular biology, including genomics and transcriptomics (Dupont *et al.*, 2007). Next-generation sequencing technologies facilitate genome sequencing of algae on a larger scale, creating a wealth of genetic information that will enable *in silico* comparative genomics search for candidate genes of, for example, pigment biosynthesis. Depending on the species-specific repertoire of available molecular tools this has the potential to be complemented by reverse genetics or screening for tagged pigment mutants.

15.9 Characterising algae using HR-MAS-NMR – recent advances

As with LC-MS described in Chapter 7 (this volume), improvements in sensitivity are ensuring that nuclear magnetic resonance (NMR) and LC-NMR are becoming increasingly assessable as tools in the accurate chemical characterisation of pigments. Matrix-Assisted Laser Desorption Ionisation – Time of Flight (MALDI-TOF: Chapter 7, this volume) and high-resolution magic angle spinning nuclear magnetic resonance spectroscopy (HR-MAS-NMR) are techniques that enable characterisation of whole cells. Here we describe the potential of HR-MAS-NMR to characterise metabolites including pigments within whole algal cells.

NMR spectroscopy has become a useful analytical tool in fields outside physics and chemistry, from where it originally emerged. In short, NMR analysis is applied to nuclei that possess 'spin': ^{1}H, ^{13}C, ^{15}N, ^{31}P and ^{29}Si are examples of isotopes relevant for algae studies. In a NMR spectrum a peak is the cumulative signal from nuclei that are experiencing the same chemical environment in the sample, e.g. ^{1}H in methyl groups differ from ^{1}H along CH-chains. Conventional NMR techniques are applied to samples in either the 'solid' or 'liquid' state, e.g. to study pigment molecular structures (Egeland *et al.*, 1997). Development of high resolution (HR) techniques further extended the use to also include samples of a 'semi-solid' state (e.g. complex samples of tissue or biofluids), and with 'magic angle spinning' (MAS) it is possible to study metabolites within whole cells such as microalgae: Spinning the sample at 54.7° relative to the magnetic field removes unwanted effects related to the physics of molecules in heterogeneous samples. This opens up the way for analyses of molecules embedded in membranes or organelles, which are otherwise poorly resolved in the NMR spectrum. In other words, HR-MAS-NMR is a very potent tool in metabolomics and studies of metabolites such as pigments, carbohydrates, or fatty acids – even directly on whole cells. Alia *et al.* (2009) wrote a review of MAS-NMR as a tool to study the structures of photosynthetic complexes, and Miglietta and Lamanna (2006) discussed some of the challenges related to studies of pigments embedded in membranes. HR-MAS-NMR has been used to study the C:N ratio of the red alga *Solieria chordalis* (Bondu *et al.*, 2008) and to analyse the metabolic composition of *Thalassiosira pseudonana* (Chauton *et al.*, 2003b). HR-MAS-NMR has also been used to study Si-uptake and metabolism in *T. pseudonana* cells (Brunner *et al.*, 2009) and glucans in *Chaetoceros muelleri* (Størseth *et al.*, 2004). Data output from NMR analyses is huge and it is necessary to apply multivariate statistical analyses to be able to sort and interpret the large information output. Principal component analysis (PCA) is often used to study clustering or grouping, and this was used to sort different species of microalgae based on their ^{1}H HR-MAS-NMR spectra (Chauton *et al.*, 2003a). Gao *et al.* (2008) used time-domain (TD) NMR to study lipids in *Chlorella protothecoides*. This technique makes use of the different relaxation times of hydrogen nuclei situated in different environments: ^{1}H nuclei in carbohydrates and proteins have shorter relaxation times compared to ^{1}H nuclei in smaller, freely rotating molecules such as water and lipids. In the future, development of stronger magnets and specially adapted pulse sequence programs will increase the usefulness of NMR in metabolomics and chemotaxonomy. Smaller and transportable magnets are already developed and used to study e.g. fat in salmon, and the use of such instruments may be extended to include macroalgae as well. NMR spectroscopy is also a great tool in combination with genomics, since NMR analysis can be used to study cell regulation on the gene level which leads to differences in metabolic composition. Future work on chemotaxonomy or metabolomics may include MAS-NMR as the tool to bridge our knowledge of metabolites as it is today (when metabolites are extracted from their natural environment and studied by more conventional methods) to study both structure and functionality of metabolites in their natural surroundings in membranes, vacuoles, plastids or other cell organelles.

15.10 Recent improvements in remote sensing

During the last decade, several advances have seen the light of day for *in situ* monitoring and remote sensing of phytoplankton dynamics. Optical instruments used for *in situ* and remotely-based measurements of spectral absorption, scattering, transmission or fluorescence now have better spectral resolution. Some of these optical sensors are now 'hyperspectral' with a spectral resolution of 1 nm or better (in contrast to multispectral sensors, typically with 5–15 nm bandwidth), enhancing the ability to detect differences in phytoplankton pigment signatures (Chapters 13 and 14, this volume). Likewise, some of these sensors are hyperspectral imagers producing images in which the user can obtain reflectance spectra at 1 nm spectral resolution in the visible range *for every pixel* in a large area of interest. These imagers can be used for monitoring and mapping purposes from different platforms (instrument carriers) such as satellites, aeroplanes, ships, underwater robots or using microscopes, elucidating spectral characteristics of phytoplankton at different levels – from organelles to the mapping of large blooms (Klonowski *et al.*, 2007; Volent *et al.*, 2007, 2009). Most optical sensors manufactured later than 2000 are small, low weight, some of them are solid-state sensors (no moving parts) with low power requirements and can therefore be used in underwater robots (such as remotely operated vehicles (ROVs), autonomous underwater vehicles (AUVs) and gliders) to determine phytoplankton biomass distribution and speciation in time and space (3D distribution). In the next decades, we will get access to smaller, cheaper, more user-friendly sensors and software that will enhance monitoring and mapping of phytoplankton dynamics. These will also create a need for closer integration between biologists, chemists, physicists and engineers for more efficient monitoring and mapping systems. Theme maps (e.g. bloom of a given algal species in a given area), with daily updates, will be used the way we are using meteorological information and maps today.

15.11 The increased use of pigments with a cautionary note – a perspective

Routine monitoring of ocean health and phytoplankton dynamics is critical if we seek to forecast ocean-related risks to human health and safety, define fisheries management options, and shed light on the impact of climate variability on marine life and humans. Phytoplankton species are the 'miner's canary' of the oceans and hence are increasingly targeted by global ocean observation systems.

There has been a growing interest in pigments other than Chl *a* from the remote-sensing community. Recognition of the important biogeochemical roles played by different phytoplankton groups has stimulated research into ways to identify these groups using remote sensing. These so-called 'phytoplankton functional types' (PFTs) include nitrogen fixers (cyanobacteria), silicifiers (mostly diatoms), calcifiers (coccolithophores) and the climate-related dimethyl sulfide (DMS) gas producers (mostly dinoflagellates and haptophytes) (Le Quéré *et al.*, 2005; Nair *et al.*, 2008; Aiken *et al.*, 2009). Two major approaches have been used to derive information on PFTs from ocean-colour data: one is based on Chl *a* related abundance classes followed by back-scattering characteristics to subdivide size classes into functional classes (Aiken *et al.*, 2007). The other is based on spectral characteristics that may differ among PFTs (Sathyendranath *et al.*, 2004). Both approaches, but even more so the second, require detailed pigment information. Other approaches used in the detection and monitoring of blooms (mostly of harmful or toxic phytoplankton species) are described in Chapter 14 (this volume). The basic and current knowledge of bio-optical properties of phytoplankton pigments (absorption, scattering and fluorescence properties of living cells), important for interpretation of *in situ* and remotely sensed data, are described in Chapter 13 (this volume). Future phytoplankton pigment research and monitoring will most probably integrate pigment chemistry, their optical properties and corresponding eco-physiology using new optical sensors on underwater robots in combination with remote sensing.

Interest in pigments other than Chl *a* has also grown significantly with the development of hyperspectral imagers soon to be incorporated in satellite platforms (but already available for use in aeroplanes – see Chapter 14, this volume). In recognition of the importance of high quality pigment data for the calibration of remotely sensed ocean-colour models, NASA space agency has undertaken several intercalibration exercises (e.g. Hooker *et al.*, 2009), which resulted in a much better appreciation for quality of data and stressed the need for more careful analytical methods. This has resulted in the development of a quality assurance plan, described in Chapter 5 (this volume).

As a cautionary note though, we should be aware that the impressive developments in sophisticated analytical pigment methodology may have raised unrealistic expectations that pigments can be readily used as functional group, genus or even species markers. This, together with developments in molecular technology, could inadvertently be leading to a global loss of skills in identifying phytoplankton using microscopy. We still have a lot to learn and it is essential therefore that the hard work of acquiring high quality HPLC pigment data (notably when computed with CHEMTAX) should always be accompanied by microscopy to guarantee correct interpretation. For example, it was recently discovered that 19′-hexanoyloxy-4-keto-fucoxanthin 4-keto-19′-hexanoyloxyfucoxanthin (previously named), a putative marker for *Emiliania huxleyi*, was lacking in the unique Southern Ocean strains of *E. huxleyi*. Similarly, an abundance of previously overlooked fucoxanthin-type pigment-containing *Karlodinium* dinoflagellates was recently discovered in the Southern Ocean. Attempts to discriminate these latter dinoflagellate species using gyroxanthin diester marker pigments failed, when it became clear that several haptophyte species also produce this pigment (see Chapters 1 and 14, this volume). Finally, from a human health point of view, it is critical to determine whether an *Alexandrium tamarense* dinoflagellate bloom is composed of a non-toxic European genotype or a toxic North American genotype, something that can not be assessed via pigments alone.

15.12 Applied phycology

Algae are becoming increasingly important in our changing world where there is a requirement to seek out sustainable sources of chemicals and replacements for current petrochemical based products. Whilst not central to the theme of this book, the application of algae for practical use, applied phycology, is a rapidly growing area for which there is an increasing amount of literature including introductory reviews by Cardozo *et al.* (2007) and Raja *et al.* (2008). With, in particular, high growth rates, algae offer many advantages over conventional crops. In addition to having potential to produce bulk industry chemicals, algae contain various unique compounds of interest to the medical, pharmaceutical or cosmetic sectors. Algae also have applications as biomaterials for ceramics and adhesives and more broadly in bioremediation. There is an increasing interest in microalgae as biofuel, in eventual replacement for corn and other land plants which use up agricultural land. Here, aside from lipids, isoprenoid-based compounds are being assessed as potential biofuel targets. In terms of pigments, carotenoids and phycobilins are excellent candidates as natural colourants, and carotenoids in particular, are a good source of antioxidants. Astaxanthin is widely used as a colouring agent in commercial salmonid farming and as an antioxidant. Lutein and zeaxanthin are promoted widely for their role in preventing aged-related macular degeneration of the eyes. Porphyrins have been extensively studied as molecules in photodynamic therapy for cancer tumour treatment and several chlorophyll alteration products have been identified as having potential in this area (Dandler *et al.*, 2010).

15.13 The crystal ball

The perspectives given in this chapter highlight the many and diverse applications of pigments. What is clear is that the analysis of pigments is central to understanding the productivity of our oceans. In particular, the role that pigments play is fundamental in the process of photosynthesis and photoprotection thereby determining phytoplankton community biodiversity under a changing environmental and climatic scenario. Combining analytical techniques with advances in molecular techniques (including metabolomics and transcriptomics) will enhance further understanding on pigments in photosynthesis and photoprotection, in chemotaxonomy and in metabolic processes. A future synergy of approaches combining analytical and molecular approaches with those of bio-optics and remote-sensing techniques, where pigments are a central linchpin, paves the way forward to understanding how our planet works and how it will be affected by climate and environmental change (see for example Boyce *et al.*, 2010).

References

Aiken, J., Fishwick, J. R., Lavender, S., Barlow, R., Moore, G., Sessions, H., Bernard, S., Ras, J. and Hardman-Mountford, N. J. (2007). Validation of MERIS reflectance and chlorophyll during the BENCAL cruise October, 2002: Preliminary validation of new products for phytoplankton functional types and photosynthetic parameters. *Int. J. Rem. Sens.* **28**, 497–516.

Aiken, J., Pradhan, Y., Barlow, R., Lavender, S., Poulton, A., Holligan, P. and Hardman-Mountford, N. (2009). Phytoplankton pigments and functional types in the Atlantic Ocean: a decadal assessment, 1995–2005. *Deep-Sea Res. II* **56**, 899–917.

Airs, R. L. and Llewellyn, C. A. (2006). Improved detection and characterisation of fucoxanthin-type carotenoids: Novel pigments in *Emiliania huxleyi* (Prymnesiophyceae). *J. Phycol.* **42**, 391–99.

Alia, A., Ganapathy, S. and de Groot, H. J. M. (2009). Magic angle spinning (MAS) NMR: a new tool to study the spatial and electronic structure of photosynthetic complexes. *Photosynth. Res.* **102**, 415–25.

Allen, A. E., LaRoche J., Maheswari, U., Lommer, M., Schaner, N., Lopez, P. J., Firazzi, G., Fernie, A. R. and Bowler, C. (2008). Whole-cell response of the pennate diatom *Phaeodactylum tricorutum* to iron starvation. *PNAS* **105**, 10438–43.

Allen, M. B. (1966). Distribution of the chlorophylls. In *The Chlorophylls*, ed. L. P. Vernon and G. B. Seely. New York: Acadamic Press, pp. 511–19.

Armbrust, V., Berges, J. A., Bowler, C., Green, B., Martinez, D., et al. (2004). The genome of the diatom *Thalassiosira pseudonana*: ecology, evolution, and metabolism. *Science* **306**, 79–85.

Béjà, O., Aravind, L., Koonin, E. V., Suzuki, M. T., Hadd, A., Nguyen, L. P., Jovanovich, S., Gates, C. M., Feldman, R. A., Spudich, J. L., Spudich, E. N. and DeLong, E. F. (2000). Bacterial rhodopsin: Evidence for a new type of phototrophy in the sea. *Science* **289**, 1902–06.

Béjà, O., Spudich, E. N., Spudich, J. L., Leclerc, M. and DeLong, E. F. (2001). Proteorhodopsin phototrophy in the ocean. *Nature* **411**, 786–89.

Blot, N., Wu, X. J., Thomas, J. C., Zhang, J., Garczarek, L., Böhm, S., Tu, J. M., Zhou, M., Ploscher, M., Eichacker, L., Partensky, F., Scheer, H. and Zhao, K. H. (2009). Phycourobilin in trichromatic phycocyanin from oceanic cyanobacteria is formed post-translationally by a phycoerythrobilin lyase-isomerase. *J. Biol. Chem.* **284**, 9290–98.

Bondu, S., Kervarec, N., Deslandes, E. and Pichon, R. (2008). The use of HRMAS NMR spectroscopy to study intra-cellular carbon/nitrogen ratio of *Solieria chordalis* (Rhodophyta). *J. Appl. Phycol.* **20**, 673–79.

Bowler, C., Allen, A. E., Badger, J. H., Grimwood, J., Jabbari, K., et al. (2008). The *Phaeodactylum* genome reveals the evolutionary history of diatom genomes. *Nature* **456**, 239–44.

Boyce, D. G., Lewis, M. R. and Worm, B. (2010). Global phytoplankton decline over the past century. *Nature* **466**, 591–96.

Bröcker, M. J., Virus, S., Ganskow, S., Heathcote, P., Heinz, D. W., Schubert, W.-D., Jahn, D. and Moser, J. (2008a). ATP-driven reduction by dark-operative protochlorophyllide oxidoreductase from *Chlorobium tepidum* mechanistically resembles nitrogenase catalysis. *J. Biol. Chem.* **283**, 10559–67.

Bröcker, M. J., Wätzlich, D., Uliczka, F., Virus, S., Saggu, M., Lendzian, F., Scheer, H., Rüdiger, W., Moser, J. and Jahn, D. (2008b). Substrate recognition of nitrogenase-like dark operative protochlorophyllide oxidoreductase from *Prochlorococcus marinus*. *J. Biol. Chem.* **283**, 29873–81.

Brunner, E., Gröger, C., Lutz, K., Richthammer, P., Spinde, K. and Sumper, M. (2009). Analytical studies of silica biomineralization: towards an understanding of silica processing by diatoms. *Appl. Microbiol. Biotechnol.* **84**, 607–16.

Cardozo, K. H. M., Guaratini, T., Barros, M. P., Falcão, V. R., Tonon, A. P., Lopes, N. P., Campos, S., Torres, M. A., Souza, A. O., Colepicolo, P. and Pinto, E. (2007). Metabolites from algae with economic impact. *Comp. Biochem. Physiol. Part C* **146**, 60–78.

Chauton, M. S., Optun, O. I., Bathen, T. F., Volent, Z., Gribbestad, I. S. and Johnsen, G. (2003a). HR MAS ^{1}H NMR spectroscopy analysis of marine microalgal whole cells. *Mar. Ecol. Progr. Ser.* **256**, 57–62.

Chauton, M. S., Størseth, T. R. and Johnsen, G. (2003b). High-resolution magic angle spinning ^{1}H NMR analysis of whole cells of *Thalassiosira pseudonana* (Bacillariophyceae): Broad range analysis of metabolic composition and nutritional value. *J. Appl. Phycol.* **15**, 533–42.

Chen, M., Schliep, M., Willows, R. D., Cai, Z.-L., Neilan, B. A. and Scheer, H. (2010). A red-shifted chlorophyll. *Science* **329**, 1318–19.

Chew, A. G. M., Frigaard, N.-U. and Bryant, D. A. (2009). Mutational analysis of three bchH paralogs in (bacterio-)chlorophyll biosynthesis in *Chlorobaculum tepidum*. *Photosynth. Res.* **101**, 21–34.

Cornilescu, G., Ulijasz, A. T., Cornilescu, C. C., Markley, J. L. and Vierstra, R. D. (2008). Solution structure of a cyanobacterial phytochrome GAF domain in the red-light-absorbing ground state. *J. Mol. Biol.* **383**, 403–13.

Cottrell, M. T., Mannino, A. and Kirchman, D. L. (2006). Aerobic anoxygenic phototrophic bacteria in the Mid-Atlantic Bight and the North Pacific Gyre. *Appl. Environ. Microbiol.* **72**, 557–64.

Dandler, J., Wilhelm, B. and Scheer, H. (2010). Distribution of chlorophyll- and bacteriochlorophyll-derived photosensitizers in human blood plasma. *Photochem. Photobiol.* **86**, 182–93.

Dupont, S., Wilson, K., Obst, O., Sköld, H., Nakano, H. and Thorndyke, M. C. (2007). Marine ecological genomics: when genomics meets marine ecology. *Mar. Ecol. Prog. Ser.* **332**, 257–73.

Egeland, E. S., Guillard, R. R. L. and Liaaen Jensen, S. (1997). Additional carotenoid prototype representatives and a general chemosystematic evaluation of carotenoids in Prasinophyceae (Chlorophyta). *Phytochemistry* **44** (6), 1087–97.

Fujita, Y. and Bauer, C. E. (2002). The light-independent protochlorophyllide reductase: a nitrogenase-like enzyme catalyzing a key reaction for greening in the dark. In *The Porphyrin Handbook*, ed. K. M. Kadish, K. M. Smith and R. Guilard. Amsterdam: Academic Press, pp. 109–56.

Ganapathy, S., Oostergetel, G. T., Wawrzyniak, P. K., Reus, M., Gomez Maqueo Chew, A., Buda, F., Boekema, E. J., Bryant, D. A., Holzwarth, A. R. and de Groot, H. J. (2009). Alternating syn-anti bacteriochlorophylls form concentric helical nanotubes in chlorosomes. *Proc. Natl. Acad. Sci. USA* **106**, 8525–30.

Gao, C., Xiong, W., Zhang, Y., Yuan, W. and Wu, Q. (2008). Rapid quantitation of lipid in microalgae by time-domain nuclear magnetic resonance. *J. Microbiol. Methods* **75**, 437–40.

Hooker, S. B., Van Heukelem, L., Thomas, C. S., Claustre, H., Ras, J., Schlüter, L., Clementson, L., Van der Linde, D., Eker-Develi, E., Berthon, J.-F., Barlow, R., Sessions, H., Perl, J. and Trees, C. (2009). *The Third SeaWiFS HPLC Analysis Round-Robin Experiment (SeaHARRE-3)*. NASA Tech. Memo. 2009–215849, Greenbelt: NASA Goddard Space Flight Center.

Ikeuchi, M. and Ishizuka, T. (2008). Cyanobacteriochromes: a new superfamily of tetrapyrrole-binding photoreceptors in cyanobacteria. *Photochem. Photobiol. Sci.* **7**, 1159–67.

Kashiyama, Y., Miyashita, H., Ohkubo, N., Ogawa, O., Chikaraishi, Y., Takano, Y., Suga, H., Toyofuku, T., Nomaki, H., Kitazato, H., Nagata, T. and Ohkouchi, N. (2008). Evidence of global chlorophyll *d*. *Science* **321**, 658.

Klonowski, W. M., Fearns, P. R. C. S. and Lynch, M. J. (2007). Retrieving key benthic cover types and bathymetry from hyperspectral imagery. *J. Appl. Remote Sens.* **1**, 011505 doi:10.1117/1.2816113.

Kolber, Z. S., Van Dover, C. L., Niederman, R. A. and Falkowski, P. G. (2000). Bacterial photosynthesis in surface waters of the open ocean. *Nature* **407**, 177–79.

Kolber, Z. S., Plumley, F. G., Lang, A. S., Beatty, J. T., Blankenship, R. E., VanDover, C. L., Vetriani, C., Koblizek, M., Rathgeber, C. and Falkowski,

P. G. (2001). Contribution of aerobic photoheterotrophic bacteria to the carbon cycle in the ocean. *Science* **292**, 2492–95.

Kupka, M., Zhang, J., Fu, W.-L., Tu, J.-M., Böhm, S., Su, P., Chen, Y., Zhou, M., Scheer, H., and Zhao, K.-H. (2009). Catalytic mechanism of S-type phycobiliprotein lyase: Chaperone-like action and functional amino acid residues. *J. Biol. Chem.* **284**, 36405–14.

Küpper, H., Küpper, F. C., and Spiller, M. (2006). [Heavy metal]-chlorophylls formed *in vivo* during heavy metal stress and degradation products formed during digestion, extraction and storage of plant material. In *Chlorophylls and Bacteriochlorophylls: Biochemistry, Biophysics, Functions and Applications*, ed. B. Grimm, R. J. Porra, W. Rüdiger and H. Scheer. Dordrecht: Springer, pp. 67–77.

Lami, R., Cottrell, M. T., Ras, J., Ulloa, O., Obernosterer, I., Claustre, H., Kirchman, D. L. and Lebaron, P. (2007). High abundances of aerobic anoxygenic photosynthetic bacteria in the South Pacific Ocean. *Appl. Environ. Microbiol.* **73**, 4198–205.

Le Quéré, C., Harrison, S. P., Prentice, C. I., Buitenhuis, E. T., Aumonts, O., Bopps, L., Claustre, H., Da Cunha, L. C., Geider, R., Giraud, X., Klaas, C., Kohfeld, K. E., Legendre, L., Manizza, M., Platt, T., Rivkin, R. B., Sathyendranath, S., Uitz, J., Watson, A. J. and Wolf-Gladrow, D. (2005). Ecosystem dynamics based on plankton functional types for global ocean biogeochemistry models. *Global Change Biol.* **11**, 2016–40.

Miglietta, M. L. and Lamanna, R. (2006). ^{1}H HR-MAS NMR of carotenoids in aqueous samples and raw vegetables. *Magn. Res. Chem.* **44**, 675–85.

Miller, C. A., Leonard, H. S., Pinsky, I. G., Turner, B. M., Williams, S. R., Harrison, L. Jr., Fletcher, A. F., Shen, G., Bryant, D. A. and Schluchter, W. M. (2008). Biogenesis of Phycobiliproteins: III. CpcM is the asparagine methyltransferase for phycobiliprotein beta-subunits in cyanobacteria. *J. Biol. Chem.* **283**, 19293–300.

Moser, S., Ulrich, M., Müller, T., and Kräutler, B. (2008). A yellow chlorophyll catabolite is a pigment of the fall colours. *Photochem. Photobiol. Sci.* **7**, 1577–81.

Moser, S., Müller, T., Holzinger, A., Lütz, C., Jockusch, S., Turro, N. J. and Kräutler, B. (2009a). Fluorescent chlorophyll catabolites in bananas light up blue halos of cell death. *Proc. Natl. Acad. Sci. USA* **106**, 15538–43.

Moser, S., Müller, T., Oberhuber, M., and Kräutler, B. (2009b). Chlorophyll catabolites – Chemical and structural footprints of a fascinating biological phenomenon. *Eur. J. Org. Chem.* **2009**, 21–31.

Nair, A., Sathyendranath, S., Platt, T., Morales, J., Stuart, V., Forget, M.-H., Devred, E. and Bouman, H. (2008). Remote sensing of phytoplankton functional types. *Remote Sens. Environ.* **112**, 3366–75.

Narikawa, R., Muraki, N., Shiba, T., Ikeuchi, M., and Kurisu, G. (2009). Crystallization and preliminary X-ray studies of the chromophore-binding domain of cyanobacteriochrome AnPixJ from *Anabaena* sp. PCC 7120. *Acta Crystallogr. Sect. F Struct. Biol. Cryst. Commun.* **65**, 159–62.

Nomata, J., Mizoguchi, T., Tamiaki, H. and Fujita, Y. (2006). A second nitrogenase-like enzyme for bacteriochlorophyll biosynthesis: reconstitution of chlorophyllide *a* reductase with purified X-protein (BchX) and YZ-protein (BchY-BchZ) from *Rhodobacter capsulatus*. *J. Biol. Chem.* **281**, 15021–28.

Nymark, M., Valle, K. C., Brembu, T., Hancke, K., Winge, P., Andresen, K., Johnsen, G. and Bones, A. M. (2009). An integrated analysis of molecular acclimation to high light in the marine diatom *Phaeodactylum tricornutum*. *PloS ONE* **11**, e7743. doi:10.1371/journal.pone.0007743.

Oberhuber, M., Berghold, J. and Kräutler, B. (2008). Chlorophyll breakdown by a biomimetic route. *Angew. Chem. Int. Ed.* **47**, 3057–61.

Peers, G., Truong, T. B., Ostendorf, E., Busch, A., Elrad, D., Grossman, A. R., Hippler, M. and Niyogi, K. K. (2009). An ancient light-harvesting protein is critical for the regulation of algal photosynthesis. *Nature* **462**, 518–21.

Raja, R., Hemaiswarya, S., Kumar, N. A., Sridhar, S. and Rengasamy, R. (2008). A perspective on the biotechnological potential of microalgae. *Crit. Rev. Microbiol.* **34**, 77–88.

Rocchetta, I. and Küpper, H. (2009). Chromium- and copper-induced inhibition of photosynthesis in *Euglena gracilis* analysed on the single-cell level by fluorescence kinetic microscopy. *New Phytol.* **182**, 405–20.

Sakuraba, Y., Tanaka, R., Yamasato, A. and Tanaka, A. (2009). Determination of a chloroplast degron in the regulatory domain of chlorophyllide *a* oxygenase. *J. Biol. Chem.* **284**, 36689–99.

Sathyendranath, S., Watts, L., Devred, E., Platt, T., Caverhill, C. and Maas, H. (2004). Discrimination of diatoms from other phytoplankton using ocean-colour data. *Mar. Ecol. Prog. Ser.* **272**, 59–68.

Schluchter, W. M., Shen, G., Alvey, R. M., Biswas, A., Saunée, N., Williams, S. R., Miller, C. A. and Bryant, D. A. (2010). Phycobiliprotein biosynthesis in cyanobacteria: structure and function of enzymes involved in post-translational modification. In *Recent Advances in Phototrophic Prokaryotes*, ed. P. C. Hallenbeck. New York: Springer, pp. 211–28.

Shen, G., Leonard, H. S., Schluchter, W. M., and Bryant, D. A. (2008). CpcM posttranslationally methylates asparagine-71/72 of phycobiliprotein beta subunits in *Synechococcus* sp. strain PCC 7002 and *Synechocystis* sp. strain PCC 6803. *J. Bacteriol.* **190**, 4808–17.

Sieracki, M. E., Gilg, I. C., Their, E. C., Poulton, N. J. and Goericke, R. (2006). Distribution of planktonic aerobic anoxygenic photoheterotrophic bacteria in the northwest Atlantic. *Limnol. Oceanogr.* **51**, 38–46.

Størseth, T. R., Hansen, K., Skjermo, J. and Krane, J. (2004). Characterization of a β-D-(1→3)-glucan from the marine diatom *Chaetoceros muelleri* by high-resolution magic-angle spinning NMR spectroscopy on whole algal cells. *Carbohydrate Res.* **339**, 421–24.

Szymczak-Żyła, M., Kowalewska, G. and Louda, J. W. (2008). The influence of microorganisms on chlorophyll *a* degradation in the marine environment. *Limnol. Oceanogr.* **53**, 851–62.

Tsien, R. Y. (2009). Constructing and exploiting the fluorescent protein paintbox (Nobel Lecture). *Angew. Chem., Int. Ed.* **48**, 5612–26.

Ulijasz, A. T., Cornilescu, G., von Stetten, D., Cornilescu, C., Velazquez Escobar, F., Zhang, J., Stankey, R. J., Rivera, M., Hildebrandt, P. and Vierstra, R. D. (2009). The cyanochromes: Blue-green photoreversible photoreceptors defined by a stable double cysteine linkage to a phycoviolobilin-type chromophore. *J. Biol. Chem.* **284**, 29757–72.

Volent, Z., Johnsen, G., and Sigernes, F. (2007). Kelp forest mapping by use of airborne hyperspectral imager. *J. Appl. Remote Sens.* **1**, 011503 doi:10.1117/1.2822611.

Volent, Z., Johnsen, G. and Sigernes, F. (2009). Microscopic hyperspectral imaging used as bio-optical taxonomic tool for micro- and macroalgae. *Appl. Optics* **48**, 4170–76.

Wätzlich, D. (ed.) (2009). *Nitrogenase-Like Enzymes of Chlorophyll and Bacteriochlorophyll Biosynthesis*. Ph.D. Dissertation, Technische Universität Braunschweig.

Wätzlich, D., Brocker, M. J., Uliczka, F., Ribbe, M., Virus, S., Jahn, D. and Moser, J. (2009). Chimeric nitrogenase-like enzymes of (bacterio)chlorophyll biosynthesis. *J. Biol. Chem.* **284**, 15530–40.

Yoshii, Y., Takaichi, S., Maoka, T., Suda, S., Sekiguchi, H., Nakayama, T. and Inouye, I. (2005). Variation of siphonaxanthin series among the genus *Nephroselmis* (Prasinophyceae, Chlorophyta), including a novel primary methoxy carotenoid. *J. Phycol.* **41**, 827–34.

Zhang, J., Wu, X. J., Wang, Z. B., Chen, Y., Wang, X., Zhou, M., Scheer, H. and Zhao, K. H. (2010). Fused-gene approach to photoswitchable and fluorescent biliproteins. *Angew. Chem. Int. Ed.* **49**, 5456–58.

Part VI

Aids for practical laboratory work

Appendix A

Update on filtration, storage and extraction solvents

JAMES L. PINCKNEY, DAVID F. MILLIE AND LAURIE VAN HEUKELEM

A.1 Filtration

In Chapter 10 of Jeffrey *et al.* (1997), Whatman GF/F (or equivalent) filters (0.7 μm nominal pore size) were recommended for sample filtration. With the exception of targeted studies, GF/F filters remain the most commonly used media for routine filtration and accompanying in vitro analyses by HPLC, fluorometry and spectrophotometry. As indicated in the above-mentioned Chapter 10 (p. 284–287), many studies have compared the effectiveness of different filter types and highlighted their advantages/disadvantages. Of particular note are three relatively recent papers highlighting the limitations of GF/F filters. Knefelkamp *et al.* (2007) compared six different filter types and concluded that Whatman nylon membranes (0.2 μm pore size, 47 mm diameter) provided the most consistent results with respect to chlorophyll *a* analyses. Nucleopore filters (0.2 μm) have been reported to retain as much as four times the amount of chlorophyll *a* as GF/F filters in open ocean samples (Dickson and Wheeler, 1993). Furthermore, Lee *et al.* (1995) reported that GF/F filters retained only 13–51% of small bacterioplankton (< 0.8 μm diameter) in natural samples. In contrast, recent comparisons of filter types have reported no differences in pigment concentrations obtained using GF/F and membrane filters in a variety of aquatic habitats (Chavez *et al.*, 1995; Morán *et al.*, 1999). The choice of filter type, whether glass-fibre or membrane, should be determined by the individual investigator for their particular application. However, once a filter type is selected, it should be used uniformly for sample filtrations to insure consistency between samples.

In estuarine and coastal waters, particulate matter (seston) may result in rapid saturation and 'clogging' of filters. Continued vacuum filtration of 'clogged' filters may promote mechanical stress and induce cell lysis, potentially resulting in underestimation of the actual pigment concentrations in the sample (Goldman and Dennett, 1985; Taguchi and Laws, 1988; Richardson and Pinckney, 2004). The total time for sample filtration should not exceed 5–10 min to minimize filter saturation (Wasmund *et al.*, 2006). Filters should be removed as soon as the passage of water through the

Phytoplankton Pigments: Characterization, Chemotaxonomy and Applications in Oceanography, eds. Suzanne Roy, Carole A. Llewellyn, Einar Skarstad Egeland and Geir Johnsen. Published by Cambridge University Press.

filter is undetectable and the vacuum should never exceed 50 kPa. Although not commonly used, positive pressure filtration (7–14 kPa) reportedly allows the filtration of larger volumes of water with reduced filtration times (Gibb *et al.*, 2001; Bidigare *et al.*, 2002). Regardless of the filtration method used, multiple filters can be pooled to achieve the biomass necessary for HPLC analyses. After filtration, filters should be folded in half, blotted on absorbent paper to remove excess water, and immediately flash frozen and stored in liquid nitrogen or at −80 °C (Wright and Jeffrey, 2006).

Different filter combinations have been used to separate the phytoplankton community into discrete size categories. Most methods consist of some pairing of glass-fibre and membrane (polycarbonate) filters to achieve fractionation of two or more size categories ($\leq$ 2, 2 to 20, 20 to 200 and $>$ 200 μm) (Iriarte and Purdie, 1994; Carrick and Schelske, 1997; Riegman *et al.*, 1998; Joint *et al.*, 2002; Ansotegui *et al.*, 2003; Zapata, 2005; Seoane *et al.*, 2006; Lance *et al.*, 2007; Sun *et al.*, 2007). In many cases, pigment concentrations for some size fractions are based on difference calculations which can lead to erroneous results. Filtration protocols should include an independent measure of total pigment concentrations (e.g. whole water filtered through a GF/F) and the sum total of each of the fractions should equal that of the whole water sample. If the sum of the fractions does not equal the total biomass, then the fraction filtration protocol should be re-evaluated and corrected.

A.1.2 Sample storage

Liquid nitrogen (−196 °C) and ultra-cold freezers (−80 °C) are commonly used for long-term (weeks to months) storage of samples before extraction procedures. Dry ice (*c.* −78 °C) is an acceptable alternative for flash-freezing and short-term storage of samples when access to liquid nitrogen or an ultra-cold freezer is limited. However, CO_2 gas, which is produced as dry ice evaporates, may create acidic conditions that can rapidly transform and/or degrade pigments. Samples stored on dry ice should be kept in air-tight containers (microcentrifuge tubes, plastic bags, etc.) to minimize exposure to high concentrations of CO_2 gas. Filter samples stored in liquid nitrogen or at −80 °C remain stable for periods of up to one year (Sosik, 1999; Van Heukelem *et al.*, 2002). For long-term storage many investigators place frozen samples in air-tight plastic bags and evacuate the air in the bags to minimize pigment oxidation. Both liquid nitrogen and ultra-cold freezers appear equally suitable for flash-freezing samples immediately following filtration to prevent alteration of pigments due to rapid xanthophyll cycling (Southerland and Lewitus, 2004).

A.1.3 Extraction

The issue of which solvent system and sample preparation procedure is best for each pigment extraction remains unresolved. There is no single extraction protocol suitable for the diversity of microalgal species encountered in aquatic environments. Rather, numerous studies over the past four decades have proposed many methods that appear reliable for specific applications or algal assemblages, but have limitations when used universally. Although the sonication/DMF (dimethyl formamide) method recommended in Chapter 9 of Jeffrey *et al.* (1997) (Section 9.6, p. 279–280) is fairly robust, the toxicity of DMF has limited the widespread acceptance of this method. Researchers in Japan are a notable exception, with many scientists (e.g. Kuwahara *et al.*, 2000; Hayashi *et al.*, 2001; Aranami *et al.*, 2001; Fujiki *et al.*, 2003; Tada *et al.*, 2004; Miyaguchi *et al.*, 2006; Baek *et al.*, 2007) employing DMF as the extraction solvent (following the method described by Suzuki and Ishimaru, 1990). The use of DMF by other researchers is limited, but several investigators recently have used this solvent (e.g. Bahnwart *et al.*, 1999; Mouget *et al.*, 1999; Pilkaitytë *et al.*, 2004; Kruskopf and Flynn, 2006; Schagerl and Künzl, 2007). DMF has also been used in combination with other solvents. For example, Hagerthey *et al.* (2006) reported that a mixed solvent system of methanol/acetone/DMF/water (30:30:30:10) was more reliable for extracting pigments within periphyton (filamentous cyanobacteria, green algae and diatoms) than 90% acetone, 90% methanol and acetone/methanol/water (45:45:10) solvent combinations. The previous edition recommended methanol as an alternative solvent for pigment extraction; however, a major caveat is that prolonged storage (> 1 h) in methanol promotes the allomerization of chlorophylls (Bowles *et al.*, 1984; van Leeuwe *et al.*, 2006). Accordingly, the potential degradation effects of methanol should be considered, particularly if the use of a refrigerated autosampler or prolonged storage result in long processing times for samples.

Acetone:water (90:10) remains the most commonly used solvent mixture for extracting pigments from phytoplankton. Although other solvents may provide higher extraction efficiencies, typically acetone is the solvent of choice due to its relatively low toxicity, low cost, general applicability, availability of absorption coefficients for chlorophylls/carotenoids and comparability with common analyses for estimating pigment abundance (i.e. fluorometry and spectrophotometry; Latasa and Bidigare, 1998; Wasmund *et al.*, 2006). The dilution of extracts with water (or buffer) prior to sample injection into the HPLC is common practice for enhancing the separation and peak shapes of early eluting pigments (Vidussi *et al.*, 1996; Barlow *et al.*, 1997; Latasa and Bidigare, 1998). However, Latasa *et al.* (2001) noted that the dilution of an extract to ≤ 60:10 acetone:water or ≤ 80:20 methanol:water resulted in the immediate (< 5 min) precipitation of pigments. Injection protocols using solvent:water dilutions should be performed immediately prior to sample injection, preferably using the mixing feature of the autosampler (if available and reliable). Autosamplers that mix the sample and buffer in the injection loop immediately prior to loading onto the column minimize pigment precipitation and are necessary for protocols that require a high buffer to sample ratio for peak separation. The choice of a buffer (water, ammonium acetate, tetrabutylammonium acetate, etc.) is entirely method dependent and should be evaluated using appropriate performance metrics in the laboratory implementing the method.

The optimum ratio of phytoplankton biomass on the filter to the volume of solvent needed for extraction has not been determined. However, Grinham *et al.*

(2007) found that a sediment to solvent volume ratio of 1:2 produced the highest extraction efficiencies for benthic microalgae from a variety of sediment types. The solubility of most carotenes in acetone exceeds 200 mg l^{-1} (Craft and Soares, 1992) while chlorophylls are highly soluble in 90% acetone. Most natural samples will not approach the solubility limitations for pigment extractions. However, filtration of very dense cultures or extreme blooms may provide sufficient biomass to limit extractions based on pigment solubilities.

In Chapter 10 of Jeffrey *et al.* (1997) (p. 302), Mantoura *et al.* (1997) reported that lyophilization (freeze-drying) of samples resulted in rapid degradation of pigments and concluded that the method should not be used for sample storage. However, their conclusions were based on storage of the samples for up to one year at *c*. 22 °C following lyophilization. Recent studies have demonstrated that lyophilization, immediately followed by extraction, significantly improves extraction efficiencies for phytoplankton (van Leeuwe *et al.*, 2006), sedimentary pigments (Buffan-Dubau and Carman, 2000; Leavitt and Hodgson, 2001; Chen *et al.*, 2003; Reuss and Conley, 2005; Brotas *et al.*, 2007) and periphyton (Hagerthey *et al.*, 2006). Samples should be protected from light and dried under a hard vacuum (< 0.1 Pa, −50 °C), extracted, and analysed immediately thereafter (Reuss and Conley, 2005; van Leeuwe *et al.*, 2006). Furthermore, lyophilization minimizes potential errors associated with filter retention of water for 'wet' samples. Sediment samples as well as glass-fibre filters retain water (*c*. 145–200 μL for 25 mm Whatman GF/F) which may affect volumetric calculations and extraction solvent concentrations, particularly when small solvent volumes are used (Latasa and Bidigare, 1998; Claustre *et al.*, 2004). Variation in the water content of filters also may contribute to methodological differences between replicate samples and inflate reproducibility errors.

Many investigators employ mechanical disruption (sonication, grinding) and/or soaking (for periods up to 48 h at −20 °C) to facilitate the extraction process and enhance recovery. The effectiveness of these procedures seems to vary with types of phytoplankton, solvents used, duration of the treatment, etc. Before adopting a particular protocol, preliminary experiments should be undertaken to determine the most effective procedure for the range of samples under investigation. The three most important criteria for evaluating an extraction method are (1) the ability to completely extract all pigments from field samples irrespective of the phytoplankton species composition, (2) compatibility with the chromatographic technique (the ability to produce sharp peaks) and (3) the stability of the pigments in the extraction solvent (Wright and Jeffrey, 2006).

A.1.4 Stability of extracted samples

Acetone-extracted pigment samples experience minimal degradation when stored for up to 21 days at −15 °C (Hooker *et al.*, 2005). Total chlorophyll *a* (chlorophyllide *a*, divinyl-chlorophyll *a*, chlorophyll *a*) and carotenoids are very stable with degradation rates in the order of −0.2% d^{-1} (Hooker *et al.*, 2005). Furthermore, pigment extracts in acetone are significantly more stable than extracts in methanol (Bowles *et al.*, 1984; van Leeuwe *et al.*, 2006). Acetone extracts frozen at −15 °C for 155 days showed an average decline in major pigments of *c.* 12%, 24% for total chlorophyll (chlorophylls *a*, *b*, *c*), 8% for total carotenoids and 9% for total chlorophyll *a* (chlorophyllide *a*, divinyl-chlorophyll *a*, chlorophyll *a*) (Hooker *et al.*, 2005).

Pigment extracts are usually stored in a temperature-controlled autosampler while queued for analysis by HPLC. When acetone is used as the extraction solvent, the degradation rates are less than 2% d^{-1} for most pigments except chlorophyll *c.* (3% d^{-1}) over a time period of 50 h (Hooker *et al.*, 2005). Degradation rates are significantly higher for methanol extracts and range from 4 to 60% d^{-1}, depending on pigment type (Hooker *et al.*, 2005; van Leeuwe *et al.*, 2006). Flushing and sealing vials with argon reportedly helps stabilize methanol extracts while in the autosampler (S. Roy, pers. comm.). However, each investigator should evaluate pigment stability with their own method and instrumentation, as measurements of degradation rates within autosamplers are highly influenced by analysis precision and may be system specific.

A.1.5 Benthic microalgae (microphytobenthos)

The collection and extraction of photosynthetic pigments from microalgae and bacteria associated with sediments requires modification of the general protocols for phytoplankton. Samples are usually collected using core tubes of varying diameters (1 to 10 cm^2) and lengths. Sediment in the core tube is extruded and sectioned at precise depth intervals, placed in vials, and immediately frozen. A variation of this technique uses a 'cryolander', which freezes the sample *in situ* using liquid N_2 for subsequent high resolution sectioning with a freezing microtome (Wiltshire *et al.*, 1997). Ideally, pigment concentrations should be expressed in terms of both of weight per unit area (mg m^{-2}) and weight per dry weight of sediment (µg g dry $sediment^{-1}$). The sediment photic zone is limited to the upper few millimetres of sediment (Jørgensen and Des Marais, 1986; Kühl *et al.*, 1994) and contains the photosynthetically active biomass while deeper samples represent buried cells and pigment diagenesis products. A wide variety of solvent systems are used for pigment extraction although acetone (90–100%) seems to be the most common. Sediments typically contain enough porewater to significantly dilute the solvent and affect extraction efficiency. Lyophilization is an effective technique for removing porewater and enhancing extractions (Buffan-Dubau and Carman, 2000). Sediment samples may also be sonicated to facilitate extraction and extracts can easily be cleared using a centrifuge or microcentrifuge. Readers are referred to several recent papers for specific details and protocols for the collection and analysis of photopigments in sediment samples (Buffan-Dubau and Carman, 2000; Reuss and Conley, 2005; Grinham *et al.*, 2007; Brotas *et al.*, 2007) and periphyton (Hagerthey *et al.*, 2006).

References

Ansotegui, A., Sarobe, A., Trigueros, J. M., Urrutxurtu, I. and Orive, E. (2003). Size distribution of algal pigments and phytoplankton assemblages in a coastal-estuarine environment: contribution of small eukaryotic algae. *J. Plankton Res.* **24**, 341–55.

Aranami, K., Watanabe, S., Tsunogai, S., Hayashi, M., Furuya, K. and Nagata, T. (2001). Biochemical variation in dimethylsulfide, phytoplankton pigments and heterotrophic bacterial production in the Subarctic North Pacific during summer. *J. Oceanogr.* **57**, 315–22.

Baek, S. H., Shimode, S. and Kikuchi, T. (2007). Reproductive ecology of the dominant dinoflagellate, *Ceratium fusus*, in coastal area of Sagami Bay, Japan. *J. Oceanogr.* **63**, 35–45.

Bahnwart, M., Hübener, T. and Schubert, H. (1999). Downstream changes in phytoplankton composition and biomass in a lowland river-lake system (Warnow River, Germany). *Hydrobiologia* **391**, 99–111.

Barlow, R. G., Cummings, D. G. and Gibb, S. W. (1997). Improved resolution of mono- and divinyl chlorophylls *a* and *b* and zeaxanthin and lutein in phytoplankton extracts using reverse phase C-8 HPLC. *Mar. Ecol. Prog. Ser.* **161**, 303–07.

Bidigare, R. R., Van Heukelem, L. and Trees, C. C. (2002). HPLC phytoplankton pigments: Sampling, laboratory methods, and quality assurance procedures. In *Ocean Optics Protocols for Satellite Ocean Color Sensor, Revision 3, Volume 2,* ed. J. Mueller and G. Fargion, NASA Technical Memorandum 2002–210004. Greenbelt: NASA Goddard Space Flight Center, pp. 258–68.

Bowles, N. D., Paerl, H. W. and Tucker, J. (1984). Effective solvents and extraction periods employed in phytoplankton carotenoid and chlorophyll determinations. *Can. J. Fish. Aquat. Sci.* **42**, 1127–31.

Brotas, V., Mendes, C. R. and Cartaxana, P. (2007). Microphytobenthic biomass assessment by pigment analysis: comparison of spectrophotometry and high performance liquid chromatography methods. *Hydrobiologia* **587**, 19–24.

Buffan-Dubau, E. and Carman, K. R. (2000). Extraction of benthic microalgal pigments for HPLC analyses. *Mar. Ecol. Prog. Ser.* **204**, 293–97.

Carrick, H. J. and Schelske, C. L. (1997). Have we overlooked the importance of small phytoplankton in productive waters? *Limnol. Oceanogr.* **42**, 1613–21.

Chavez, F. P., Buck, K. R., Bidigare, R. R., Karl, D. M., Hebel, D., Latasa, M. and Campbell, L. (1995). On the chlorophyll *a* retention properties of glass-fiber GF/F filters. *Limnol. Oceanogr.* **40**, 428–33.

Chen, N., Bianchi, T. S. and Bland, J. M. (2003). Novel decomposition products of chlorophyll-*a* in continental shelf (Louisiana shelf) sediments: Formation and transformation of carotenol chlorin esters. *Geochim. Cosmochim. Acta* **67**, 2027–42.

Claustre, H., Hooker, S. B., Van Heukelem, L., Berthon, J.-F., Barlow, R., Ras, J., Sessions, H., Targa, C., Thomas, C. S., van der Linde, D. and Marty, J.-C. (2004). An intercomparison of HPLC phytoplankton pigment methods using in situ samples: application to remote sensing and database activities. *Mar. Chem.* **85**, 41–61.

Craft, N. E. and Soares, J. H. (1992). Relative solubility, stability, and absorptivity lutein and β-carotene in organic solvents. *J. Agric. Food Chem.* **40**, 431–34.

Dickson, M.-L. and Wheeler, P. A. (1993). Chlorophyll *a* concentrations in the North Pacific: Does a latitudinal gradient exist? *Limnol. Oceanogr.* **38**, 1813–18.

Fujiki, T., Toda, T., Kikuchi, T. and Taguchi, S. (2003). Photoprotective response of xanthophyll pigments during phytoplankton blooms in Sagami Bay, Japan. *J. Plankton Res.* **25**, 317–22.

Gibb, S. W., Cummings, D. G., Irigoien, X., Barlow, R. G., Fauzi, R. and Mantoura, C. (2001). Phytoplankton pigment chemotaxonomy of the northeastern Atlantic. *Deep-Sea Res. II* **48**, 795–823.

Goldman, J. C. and Dennett, M. R. (1985). Susceptibility of some marine phytoplankton species to cell breakage during filtration and post-filtration rinsing. *J. Exp. Mar. Biol. Ecol.* **86**, 47–58.

Grinham, A. R., Carruthers, T. J. B., Fisher, P. L., Udy, J. W. and Dennison, W. C. (2007). Accurately measuring the abundance of benthic microalgae in spatially variable habitats. *Limnol. Oceanogr. Methods* **5**, 119–25.

Hagerthey, S. E., Louda, J. W. and Mongkronsri, P. (2006). Evaluation of pigment extraction methods and a recommended protocol for periphyton chlorophyll *a* determination and chemotaxonomic assessment. *J. Phycol.* **42**, 1125–36.

Hayashi, M., Furuya, K. and Hattori, H. (2001). Spatial heterogeneity in distributions of chlorophyll *a* derivatives in the Subarctic North Pacific during summer. *J. Oceanogr.* **57**, 323–31.

Hooker, S. B., Van Heukelem, L., Thomas, C. S., Claustre, H., Ras, J., Barlow, R., Sessions, H., Schlüter, L., Perl, J., Trees, C., Stuart, V., Head, E., Clementson, L., Fishwick, J., Llewellyn, C. and Aiken, J. (2005). *The Second SeaWiFS HPLC Analysis Round-Robin Experiment (SeaHARRE-2).* NASA TM/2005–212785, Greenbelt: NASA Goddard Space Flight Center.

Iriarte, A. and Purdie, D. A. (1994). Size distribution of chlorophyll *a* biomass and primary production in a temperate estuary (Southampton Water): the contribution of photosynthetic picoplankton. *Mar. Ecol. Prog. Ser.* **115**, 283–97.

Jeffrey, S. W., Mantoura, R. F. C. and Wright, S. W. (eds.) (1997). *Phytoplankton Pigments in Oceanography: Guidelines to Modern Methods.* Paris: UNESCO Publishing.

Joint, I., Henriksen, P., Fonnes, G. A., Bourne, D., Thingstad, T. F. and Riemann, B. (2002). Competition for inorganic nutrients between phytoplankton and bacterioplankton in nutrient manipulated mesocosms. *Aquat. Microb. Ecol.* **29**, 145–59.

Jørgensen, B. B. and Des Marais, D. (1986). A simple fiber-optic microprobe for high resolution light measurements: Application in marine sediment. *Limnol. Oceanogr.* **31**, 1376–83.

Knefelkamp, B., Carstens, K. and Wiltshire, K. H. (2007). Comparison of different filter types on *chlorophyll-a* retention and nutrient measurements. *J. Exp. Mar. Biol. Ecol.* **345**, 61–70.

Kruskopf, M. and Flynn, K. J. (2006). Chlorophyll content and fluorescence responses cannot be used to gauge reliably phytoplankton biomass, nutrient status or growth rate. *New Phytol.* **169**, 525–36.

Kühl, M., Lassen, C. and Jørgensen, B. (1994). Light penetration and light intensity in sandy marine sediments measured with irradiance and scalar irradiance fiber-optic microprobes. *Mar. Ecol. Prog. Ser.* **105**, 139–48.

Kuwahara, V. S., Toda, T., Hamasaki, K., Kikuchi, T. and Taguchi, S. (2000). Variability in the relative penetration of ultraviolet radiation to photosynthetically available radiation in temperate coastal waters, Japan. *J. Oceanogr.* **56**, 399–408.

Lance, V. P., Hiscock, M. R., Hilting, A. K., Stuebe, D. A., Bidigare, R. R., Smith, W. O. and Barber, R. T. (2007). Primary productivity, differential size fraction and pigment composition responses in two Southern Ocean in situ iron enrichments. *Deep-Sea Res. I* **54**, 747–73.

Latasa, M. and Bidigare, R. R. (1998). A comparison of phytoplankton populations of the Arabian Sea during the Spring Intermonsoon and Southwest Monsoon of 1995 as described by HPLC-analyzed pigments. *Deep-Sea Res. II* **45**, 2133–70.

Latasa, M., van Lenning, K., Garrido, J. L., Scharek, R., Estrada, M., Rodríguez, F. and Zapata, M. (2001). Losses of chlorophylls and carotenoids in aqueous acetone and methanol extracts prepared for RPHPLC analysis of pigments. *Chromatographia* **53**, 385–91.

Leavitt, P. R. and Hodgson, D. A. (2001). 15. Sedimentary pigments. In *Tracking Environmental Change Using Lake Sediments. Volume 3: Terrestrial, Algal, and Siliceous Indicators*, ed. J. P. Smol, H. J. B. Birks and W. M. Last. Dordrecht: Kluwer Academic Publishers, pp. 292–325.

Lee, S., Kang, Y.-C. and Fuhrman, J. A. (1995). Imperfect retention of natural bacterioplankton cells by glass fiber filters. *Mar. Ecol. Prog. Ser.* **119**, 285–90.

Mantoura, R. F. C., Wright, S. W., Jeffrey, S. W., Barlow, R. G. and Cummings, D. E. (1997). Filtration and storage of pigments from microalgae. In *Phytoplankton Pigments in Oceanography: Guidelines to Modern Methods*, ed. S. W. Jeffrey, R. F. C. Mantoura and S. W. Wright. Paris: UNESCO Publishing, pp. 283–305.

Miyaguchi, H., Fujiki, T., Kikuchi, T., Kuwahara, V. S. and Toda, T. (2006). Relationship between the bloom of *Noctiluca scintillans* and environmental factors in the coastal waters of Sagami Bay, Japan. *J. Plankton Res.* **28**, 313–24.

Morán, X. A. G., Gasol, J. P., Arin, L. and Estrada, M. (1999). A comparison between glass fiber and membrane filters for the estimation of phytoplankton POC and DOC production. *Mar. Ecol. Prog. Ser.* **187**, 31–41.

Mouget, J.-L., Tremblin, G., Morant-Manceau, A., Morançais, M. and Robert, J.-M. (1999). Long-term photoacclimation of *Haslea ostrearia* (Bacillariophyta): effect of irradiance on growth rates, pigment content and photosynthesis. *Eur. J. Phycol.* **34**, 109–15.

Pilkaitytë, R., Schoor, A. and Schubert, H. (2004). Response of phytoplankton communities to salinity changes – a mesocosm approach. *Hydrobiologia* **513**, 27–38.

Reuss, N. and Conley, D. J. (2005). Effects of sediment storage conditions on pigment analyses. *Limnol. Oceanogr. Methods* **3**, 477–87.

Richardson, T. L. and Pinckney, J. L. (2004). Monitoring of the toxic dinoflagellate *Karenia brevis* using gyroxanthin-based detection methods. *Appl. Phycol.* **16**, 315–28.

Riegman, R., Flameling, I. A. and Noordeloos, A. A. M. (1998). Size-fractionated uptake of ammonium, nitrate and urea and phytoplankton growth in the North Sea during spring 1994. *Mar. Ecol. Prog. Ser.* **173**, 85–94.

Schagerl, M. and Künzl, G. (2007). Chlorophyll *a* extraction from freshwater algae – a reevaluation. *Biol. Bratislava* **62**, 270–75.

Seoane, S., Laza, A. and Orive, E. (2006). Monitoring phytoplankton assemblages in estuarine waters: The application of pigment analysis and microscopy to size-fractionated samples. *Est. Coast. Shelf Sci.* **67**, 343–54.

Sosik, H. M. (1999). Storage of marine particulate samples for light-absorption measurements. *Limnol. Oceanogr.* **44**, 1139–41.

Southerland, H. A. and Lewitus, A. J. (2004). Physiological responses of estuarine phytoplankton to ultraviolet light-induced fluoranthene toxicity. *J. Exp. Mar. Biol. Ecol.* **298**, 303–22.

Sun, J., Feng, Y., Zhang, Y. and Hutchins, D. A. (2007). Fast microzooplankton grazing on fast-growing, low-biomass phytoplankton: a case study in spring in Chesapeake Bay, Delaware Inland Bays, and Delaware Bay. *Hydrobiologia* **589**, 127–39.

Suzuki, R. and Ishimaru, T. (1990). An improved method for the determination of phytoplankton chlorophyll using *N,N*-dimethylformamide. *J. Ocean. Soc. Jpn* **46**, 190–94.

Tada, K., Yamaguchi, H. and Montani, S. (2004). Comparison of chlorophyll *a* concentrations obtained with 90% acetone and *N,N*-dimethylformamide extraction in coastal water. *J. Oceanogr.* **60**, 259–61.

Taguchi, S. and Laws, E. A. (1988). On the microparticles which pass through glass fiber filter type GF/F in coastal and open waters. *J. Plankton Res.* **10**, 999–1088.

Van Heukelem, L., Thomas, C. and Glibert, P. (2002). Sources of variability in chlorophyll analysis by fluorometry and high-performance liquid chromatography in a SIMBIOS inter-calibration exercise. NASA/TM-2002–211606. Greenbelt: NASA Goddard Space Flight Center.

van Leeuwe, M. A., Villerius, L. A., Roggeveld, J., Visser, R. J. W. and Stefels, J. (2006). An optimized method for automated analysis of algal pigments by HPLC. *Mar. Chem.* **102**, 267–75.

Vidussi, F., Claustre, H., Bustillos-Guzmàn, J., Cailliau, C. and Marty, J.-C. (1996). Determination of chlorophylls and carotenoids of marine phytoplankton: separation of chlorophyll *a* from divinyl-chlorophyll *a* and zeaxanthin from lutein. *J. Plankton Res.* **18**, 2377–82.

Wasmund, N., Topp, I. and Schories, D. (2006). Optimising the storage and extraction of chlorophyll samples. *Oceanologia* **48**, 125–44.

Wiltshire, K. H., Blackburn, J. and Paterson, D. M. (1997). The cryolander: a new method for fine-scale in situ sampling of intertidal surface sediments. *J. Sed. Res.* **97**, 977–81.

Wright, S. W. and Jeffrey, S. W. (2006). Pigment markers for phytoplankton production. In *Marine Organic Matter: Biomarkers, Isotopes and DNA*, ed. J. K. Volkman. The Handbook of Environmental Chemistry, vol. 2, part N. Berlin: Springer, pp. 71–104.

Zapata, M. (2005). Recent advances in pigment analysis as applied to picophytoplankton. *Vie et Milieu* **55**, 233–48.

Appendix B

HPLC instrument performance metrics and validation

AIMEE R. NEELEY, CRYSTAL S. THOMAS, STANFORD B. HOOKER AND LAURIE VAN HEUKELEM

Currently, there are over 90 companies that offer HPLC hardware and accessories, and more than 30 that offer complete systems. Given the myriad choices available in the marketplace, the discerning chromatographer needs to approach equipment purchases with a critical mindset and a clear understanding of what they require from an HPLC system or component. This appendix covers some of the features available in HPLC autosamplers, pumps, detectors and ovens. It is not meant to be a definitive catalog of available HPLC hardware components and design elements. Instead, it is designed to call attention to some of the features available in specific HPLC hardware that the authors of this appendix have researched in the context of how these decisions can affect one's ability to produce consistent, high quality pigment results. A thorough review of the basics and advancements in HPLC hardware is covered in the third edition of *Introduction to Modern Liquid Chromatography* (Snyder *et al.*, 2010).

To make informed decisions regarding one's needs in HPLC hardware, one must understand the component design (and software control thereof) from the perspective of its contribution to combined uncertainty. Uncertainties in pigment results related to hardware characteristics are most often associated with injectors and detectors, and to a lesser extent, column oven design and pump capabilities. The uncertainties of the latter are often related to implementation of a method (e.g. baseline disturbance).

Performance metrics should be applied to validate any chromatographic method and are a necessary component of a quality assurance plan (see Chapter 5, this volume). They can also be used to help assess and validate the robustness, accuracy and precision of HPLC system hardware. Having previously-defined performance metrics allows one to more critically assess potential new hardware purchases. When choosing hardware components, it is important to consider advantages and disadvantages of design differences available from different vendors in the context of method requirements and attainable performance metrics. Design differences can limit one's ability to change settings that can improve linearity, precision, accuracy,

Phytoplankton Pigments: Characterization, Chemotaxonomy and Applications in Oceanography, eds. Suzanne Roy, Carole A. Llewellyn, Einar Skarstad Egeland and Geir Johnsen. Published by Cambridge University Press.

and sample throughput. Manufacturer specifications are not published with pigment chromatography in mind and can vary among HPLC manufacturers, as well as within models offered from the same vendor. Manufacturer specifications should be used, therefore, only as a guideline, but are a starting point for comparison. Once purchased, a rigorous assessment of performance metrics should be conducted for each newly configured system. In Table B.1, the specifications of each hardware component are listed along with the implications of choosing a particular specification. Options that are commonly available from most HPLC manufacturers are also described in the following pages for major HPLC components.

In addition to performance metrics, the amount of attention required to maintain a working HPLC system and the availability of technical service for repairs should also be considered. It can be advantageous to choose a manufacturer that has local support available to service the instrument in an acceptable amount of time. In addition, many manufacturers offer different tiers of phone and online support. Research should be performed into the required system maintenance and downtime experienced by other chromatographers with the same hardware. A well-performing but high maintenance system may not be time or cost effective to maintain, depending on product and analysis demands.

The following sections provide an overview of hardware components needed for HPLC pigment methods, examples of variations in design features, and the implications of such features on method performance. HPLC hardware component design is ever changing, making it impossible for a document such as this to remain forever relevant. It is in this context that performance metrics are especially useful when making hardware selection choices.

B.1 Hardware components overview

No one HPLC configuration is best for every chromatographer, or even every pigment chromatographer. The specific needs of the laboratory must always be considered when purchasing hardware. The following are some factors to consider:

- How many samples need to be processed per day?
- What method(s) is likely to be implemented?
- What will be the expected range of pigment concentrations in the samples?
- How many HPLC operators will there be?
- What pigments will be analyzed?
- Will the HPLC system be used for diverse applications (i.e. other than pigment analysis)?
- What are the accuracy requirements?
- How much money is available for the initial purchase and future maintenance?

Answers to questions such as these will affect decisions on design features of the major HPLC components that are optimal for any laboratory.

Table B.1. *Hardware specifications and implications.*

Considerations for specifications	Implications	Hardware performance specifications
INJECTOR		
1. Injector capabilities[1,3,8,10]	1. Accuracy and precision of sample draw[11]	1. $\leq \pm$ 1% injector accuracy; 0.3–0.5% injector precision
2. Interchangeable sample loop configuration[5,8]	2. Allows flexibility of sample volume capacity	2. Dependent on configuration: 0.1–500 μL, up to 1500 μL
3. Interchangeable syringe size[3,4,6]	3. Size can affect accuracy and precision of sample draw	3. Hardware dependent
AUTOSAMPLER[2]		
1. Mixing functions[10]	1. Inadequate mixing can affect separation of peaks; over-mixing can cause degradation of pigments	1. Hardware dependent
2. Temperature control	2. Temperature-controlled sample trays can minimize sample evaporation and degradation	2. Temperature range 4–40°C
3. Type of tubing[9,21,22]	3. Certain types of tubing can leach out contaminant (especially with an acidic solvent) or be incompatible with other applications such as low-pressure or high-pressure systems	3. High-pressure tubing: PEEK or stainless steel Low-pressure tubing: Teflon, polyethylene, polypropylene
4. Sample tray capacity	4. Interchangeable sample trays allow flexibility in sample vial size and number	4. 35–175 vials (depending on vial size and sample volume); 96-well plates
5. Length and diameter of tubing[21,22]	5. Can affect dwell volume; longer tubing increases volume	5. Dependent on hardware and application
6. Needle rinse[5,7,2]	6. Can alleviate intersample contamination from carryover[7]	6. 0.05% carryover with needle rinse; 1% carry-over without rinse
7. Needle height[5,6]	7. Adjustment of needle height can improve precision and prevent needle damage	7. Software and hardware dependent
8. Metering head capacity	8. Determines settable range of injection volumes	8. $\leq$ 2000 μL

DETECTOR		
1. Type of detector	1. Determines available wavelength range	1. UV-vis and fluorescence detector
2. Flow cell[23]	2. Choices vary to include volume, path length and pressure maximum	2. 2–10 μL
3. Bandwidth and slit width	3. Selection of bandwidth and slit width can affect signal-to-noise ratio and absorbance	3. Number of programmable slit widths and bandwidths vary considerably between manufacturers
4. Linearity	4. See Chapter 5	4. > 2 AU
5. Accuracy[8,12,11]	5. See Chapter 5	5. ± 1 nm
6. Noise[12,13]	6. Baseline noise can affect signal detection, accuracy and precision	6. $0.5–0.8 \times 10^{-5}$ AU
7. Wavelength range selection[13]	7. A-priori determination of wavelengths necessary for detection of desired compounds	7. 190–950 nm
PUMP[15,20]		
1. Flow rates[7,17,18]	1. Flow rate influences the accuracy of gradient formation – especially at low flow rates	1. Precision: 0.05–0.2%; RSD Accuracy: ± 1.0% (10 μL min^{-1})
2. Dwell volume[16,21,14]	2. Dwell volume can delay the gradient and negatively affect early eluting peaks	2. From < 200 μL up to 13 mL
3. Tubing composition and size/fittings[19,21,22]	3. Tubing type, size, length and fitting type determined by type of pump and dwell volume requirements, as well as pressure requirements.	3. PEEK, stainless steel (see #3 of Autosampler category in this table)
4. Gradient composition accuracy and precision[17]	4. Gradient time may need modification depending on chromatography; pump must proportion the correct gradient consistently	4. Accuracy: 0.35–1.0%; Precision: 0.15–0.5%
5. Flow rate range[17]	5. Depends on type of pump (i.e. quaternary or binary); Flow rate also depends on application	5. 0–10 mL min^{-1}; 0–5 mL min^{-1}

[1] Dolan (2001a); [2] Dolan (2001b); [3] Dolan (1998); [4] Dolan (1987a); [5] Dolan (1987b); [6] Dolan (1997a); [7] Dolan (2006a); [8] Dolan (2002); [9] Dolan (1989); [10] Dolan (1997b); [11] Dolan (2004a); [12] Dolan (1996); [13] Dolan (2005); [14] Dolan (1991); [15] Dolan (1985a); [16] Dolan (2006b); [17] Dolan (2004b); [18] Dolan (1990); [19] Dolan (1988); [20] Dolan (1983); [21] Dolan (1986); [22] Dolan (1985b); [23] Dolan (1984).

B.2 Injectors

The injector is responsible for the delivery of a sample to the column. Although manual injections used to be the predominant mode, automated injections are now preferred. Most injectors consist of a six-position valve, a plunger or syringe, and a sample loop. In the LOAD position, the sample is shuttled to the sample loop. At this point, the mobile phase is bypassed away from the sample loop, and the loop is left depressurized. When the required sample volume has been drawn, the valve is switched to INJECT, and the mobile phase flushes the sample from the loop to the column. The two types of injector modes are discussed in the following sections.

B.2.1 Filled-loop mode

In filled-loop injection mode, the volume of sample injected is determined by the volume capacity of the loop; therefore, the accuracy of the sample volume is dependent on the volume accuracy of the loop. A high degree of accuracy and precision of this injection mode is only possible if the loop is completely filled (Snyder *et al.*, 2010). During sample load, the sample loop is completely filled with sample and subsequently injected onto the column. A disadvantage of this mode is that a large amount of sample is required to make certain that the loop is uniformly filled. In addition, a change of injection volume requires a loop change. The loop must be overfilled twice with sample to ensure no carryover from a previous sample and uniform distribution of the sample in the loop (Dolan, 2001a).

B.2.2 Partial-loop mode

In partial-loop injection mode, sample volume is measured by a syringe and transferred to the loop; therefore, the accuracy and precision of the syringe is critical for accurate and precise sample draw. To ensure accurate sample injection, the loop size should be large enough that the total injection volume fills less than half of the loop (Snyder *et al.*, 2010). Sample loops are easily changed, so a sample loop should be chosen based on the intended injection volume. Partial-loop injection mode is advantageous when sample volume is limiting and flexibility in injection volume is necessary.

B.3 Autosamplers

Autosamplers come with a number of options, including injection mode and autosampler design. Both components are integral in determining the accuracy and precision of the sample analysis. It is important to carefully research these options and to choose based on one's needs. Advantages and disadvantages of each autosampler design are discussed below.

B.3.1 Autosampler design

HPLC autosamplers consist of five common components: injection valve, sample loop, needle, sample vial and tray. Individual manufacturers will have their own variations on these components, and possibly additional elements as well. Autosampler features that can affect precision and accuracy include: draw speed, syringe volume, needle style, vial cap septum material and needle height. Not all hardware/software combinations allow the user to manipulate some of these parameters, so a potential buyer should take note when researching autosamplers.

There are three common types of autosampler design that determine the mode in which the sample is delivered to the column: push-to-fill, pull-to-fill and needle-in-loop. When choosing an autosampler, it is important to consider the injector capabilities, i.e. what options the operator can control. Each design brings different capabilities and problem sets, which will be discussed below. For a detailed description of autosampler design, see Snyder *et al.* (2010).

Push-to-fill design

In a push-to-fill design, the syringe draws sample from the vial and delivers sample to the loop. This design has a low volume requirement for each sample. In addition, the filling of the syringe is controlled by a stepping motor, which provides accurate and precise sample delivery. The valve in this type of design can accommodate various sample loop sizes. This design also provides the capability for random vial access, as well as thorough flushing (Dolan, 2001a).

Pull-to-fill design

The pull-to-fill design has several advantages compared to the push-to-fill: it is simple, syringe precision is not required, and dwell volume need not be considered (Dolan, 2001a). The main drawback to the pull-to-fill design is that a larger volume of sample is needed to fill the needle and associated tubing before it reaches the sample loop. Up to 1 mL of sample may be lost in the process, therefore, this design would not be advantageous to those who are limited by sample volume (Dolan, 2001a).

Needle-in-loop design

In this design, the needle is not a separate component, rather it is an integral part of the sample loop. The needle draws the sample from the vial and then directly moves back inline, injecting the entire sample into the sample loop. Because the needle is inline, no sample is wasted and carryover is minimal (Snyder *et al.*, 2010). The needle-in-loop design is particularly useful when sample volume is limiting (Dolan, 2001a).

Autosampler precision and accuracy depend greatly on the syringe and needle components. Some autosamplers have the option to install different syringe sizes for various volume capacities: e.g., 0–100 μL, 100 μL–1 mL, etc. The syringe size or volume is dependent upon the application. For instance, injection volumes of less than 100 μL would require a syringe size of 1–100 μL instead of a syringe with the volume capacity of 100–1000 μL. This phenomenon exists because every syringe has error associated with the positioning of the syringe during sample draw. The positioning error translates to volumetric error in the measurement of the sample draw. If a 1000 μL syringe is used to draw a minute volume (such as 10 μL), the associated movement and positioning of the plunger is rather small, thus contributing a larger positioning error and, therefore, a larger volumetric error. The lesson learned is to determine a priori what injection volume will be necessary for the application and adjust the syringe size accordingly (Dolan, 1997a).

Other sources of inaccuracy and imprecision include needle blockage and inadequate sample draw and mixing. Needle blockage can occur when a piece of septum from the sample vial breaks off, blocking the needle aperture, or when the sample is not filtered properly. This type of occurrence may be prevented by installing either a beveled point needle or a side port needle (Dolan and Snyder, 1989). Another way to prevent needle blockage is by the use of septa coated with PTFE on one or both sides, which lessens the occurrences of septum coring (Dolan, 1997a).

Needle height adjustments may be used to obtain optimum performance of the autosampler. The ability to adjust the needle height relative to the bottom of the vial can also be important when sample volume is limited or when different types or sizes of vials are used. If the needle height is too high, some air (instead of the entire sample volume) will be drawn. If the needle height is too low relative to the sample vial, the needle will hit the bottom of the vial, resulting in a system error and likely damage to the sample needle. Some systems incorporate a spring-loaded needle to prevent this. The ability to adjust needle height is especially advantageous when alternating between different types of vials (Dolan, 1997a).

Carryover is a common problem in chromatography. It is often caused either by residual sample remaining in the plumbing or is associated with the needle (externally and internally; Dolan, 2006a). The two most common autosampler designs, push-to-fill and needle-in-loop, each have their own unique problem set for carryover. To eliminate residual sample left in tubing (a problem in push-to-fill autosamplers), a wash solvent may be used to flush tubing between samples. This is not as much of a problem for needle-in-loop autosamplers, as the needle and loop are flushed with mobile phase during sample injection. However, another potential source of carryover is residual sample left on the outside of the needle, as well as residual sample remaining on the high pressure seal found in the needle-in-loop design during injection. The first way this problem may be mitigated is through the use of proper vial caps. Vial septa are the 'first line of defense' against carryover because, as the needle withdraws from the vial, the outside of the needle must pass back through the hole, thus wiping away some of the residual sample. Another option to combat carryover is to use needles coated with material, such as PEEK or PTFE, which can reduce carryover up to 20% (Dolan, 2006a). Platinum coated needles can also reduce carryover. The use of Delrin, Tefzel and PEEK injector rotor seals may also reduce the occurrence of residual sample on internal autosampler surfaces. Some autosamplers come with the option of a needle rinse, which can be implemented between samples to reduce carryover as a result of residual sample on the outside of the needle.

Other important considerations for autosampler design include a temperature controlled sample tray, mixing capabilities and a light-protective door. A temperature controlled sample tray can help minimize sample evaporation and degradation before analysis. The ability to mix sample with buffer in the loop instead of premixing on the benchtop also avoids a potential source of inaccuracy. For larger sample volumes, a multidraw kit that consists of a sample loop with a large diameter for sample volumes up to 5 mL may be acquired separately. Because pigment samples in the sample tray should have minimal light exposure to prevent light-induced degradation, considerations should also be made for operator-controlled autosampler light and a darkened door.

B.4 Pumps

Almost all HPLC pumps use some type of reciprocating piston (Snyder *et al.*, 2010). All but the simplest of pigment analysis methods require gradient elution; therefore, an online mixer is needed in the pump to proportion and combine the different solvents. The accuracy and precision of solvent composition is important and is dependent on an accurate solvent mixer. A high-pressure mixer has two separate, independent pumps that deliver solvents 'A' and 'B' in the proper proportions to a high-pressure mixing chamber, so solvents are combined in the mixing chamber. Greater accuracy of mixing can be achieved through a high-pressure mixer because it has a lower dwell volume due to its micromixer, which can result in increased accuracy and stability of retention times. Because mixing occurs at a high pressure, there are few problems with outgassing. Typically, because of the expense of separate pumps, high-pressure mixers are only found in binary pumps, although some manufacturers include selector valves that allow for the non-simultaneous use of more than two solvents. Binary pumps, because of their high-pressure capabilities, are necessary for high-pressure applications (Wright and Mantoura, 1997).

A low-pressure mixer has a single pump that first draws the solvents through proportioning valves to control the composition upstream of the mixing chamber, then the solvents are mixed in a low-pressure mixing chamber. Inaccuracies occur because of proportioning errors during solvent mixing as a result of outgassing (Wright and Mantoura, 1997), so a post-mixing degasser is required. Pumps that have low-pressure mixers are usually equipped with the capability to simultaneously draw from four different solvents (quaternary pump).

The inclusion of a pulse damper can minimize pump noise. Other parameters, such as dwell volume, tubing type (e.g., PEEK or stainless steel) and size and flow rate range are dependent on the intended application and should be considered. The flexibility of compressibility compensation settings (or the availability of user definition) is also an important function. When solvents are compressed to a higher pressure the volume decreases. The compressibility value is unique to every solvent. If the incorrect compressibility setting is used, the selected flow rate and the actual flow rate will differ, thereby decreasing the accuracy of the flow rate and in turn, negatively affecting retention times and gradient accuracy. Most HPLC systems either allow the user to manually select compressibility compensation values, or they have tools for automatic compensation.

The dwell volume must be considered when configuring the pump hardware in an HPLC system. Dwell volume (V_d) is the volume from tubing, connectors, guard columns, filters, solvent mixers, pulse dampers and the pump heads that exists between where the solvents of the mobile phase are mixed and the head of the column. Dwell volume can be as little as <0.2 mL and as great as 13 mL, depending on the configuration of the system. The dwell volume effectively forces a gradient delay that can negatively affect the resolution of early eluting peaks and shift the retention times of all peaks within the chromatogram. Once dwell volume has been determined, dwell time may be calculated. Dwell time (T_d) is the time necessary for the solvent gradient to reach the column and is calculated as V_d/F, where F is the flow rate. It is critically important to calculate dwell volume for every system and every method. This is particularly important when transferring methods between systems, as dwell volume can vary depending on the configuration (Dolan, 2006b) and is a common source of method transfer problems.

Obvious considerations when choosing a pump include method-specific parameters such as number of solvents used simultaneously and backpressure specifications. If a chromatographer plans to implement a method that uses three solvents concurrently, then a binary pump would be useless, as it can only pump two solvents simultaneously. If a method is implemented that uses a narrow-bore column, then a system should be chosen that can accommodate higher pressures. The newer ultra-performance HPLC (UPLC) technologies are specially designed for such applications.

B.5 Detectors

The most popular types of detectors in use in pigment chromatography today are UV-visible and fluorescence detectors. Fluorescence detectors offer the advantage of being highly sensitive to analytes, yet comparatively less sensitive to changes in temperature and flow. Because not all pigments fluoresce, UV-visible detectors are a popular option for pigment analysis; these cover a wavelength range capable of detecting known marine pigments and are available as fixed-wavelength, variable wavelength or photodiode array (PDA). When choosing a detector, there must be an understanding of the wavelengths and wavelength ranges necessary for the detection of compounds of interest (and if the instrument will be used for diverse applications). Fixed-wavelength detectors have the advantage of being relatively inexpensive and having a simple design, but if more than one wavelength is to be monitored to detect the pigments of interest, a multiple wavelength or PDA (photodiode array) detector would be necessary. PDAs and variable-wavelength detectors from different manufacturers will differ in the number of simultaneous wavelengths that can be monitored; PDAs also vary in the number of diodes present. Most chromatographers find the collection of spectral data essential. A PDA is superior to a variable-wavelength detector for this task, because a PDA scans all wavelengths simultaneously, while a variable-wavelength detector has to change wavelengths. More information on the characteristics and optimization of different detector types can be found in Snyder *et al.* (2010).

The application of performance metrics for a detector requires that the response of the detector system (including the flow cell) keeps within the assumptions of the method. For most HPLC methods used to quantitate phytoplankton samples, a linear response is assumed, because this greatly simplifies the calibration process. An example of how significant this work can be is shown in Figure B.1, which summarizes the results of calibrations with a detector that was unknowingly incompatible with the linear methodological assumptions being made. The calibrations were conducted on two different columns and the number of dilutions exceeded the normal practice. The working range established for the method was 2–200 ng inj^{-1}, and the calibration process assumed a linear response of the detector system. The main panel of Figure B.1 shows the residuals to the C_8 method linear calibration, which exhibit a significant nonlinearity throughout the working range. A properly validated method will have calibration residuals in the order of $\pm$ 2% (the yellow band in the main panel), but the C_8 method results exhibit residuals beyond minus 40% (i.e. –40%). The points greater than 200 ng inj^{-1} are not part of the calibration parameterization, but are included because the relevant patents for the flow cell being used (investigated as a result of the poor linearity seen in the data) predicted a nonlinear response beyond the working range, which is seen in the data.

The degree of nonlinearity is further explored in the inset panel in Figure B.1, which shows the red-to-blue ratio for the wavelengths used to quantitate pigments for the two methods. Under normal circumstances for the methods used, the red-to-blue ratio is supposed to be constant as a function of the injection amount, so if all the ratios are normalized by the largest amount injected during the calibration process (in this case 200 ng), the expected results should be approximately 1 (and again fall within the yellow band shown in the inset panel), except for the data beyond the working range. The ratio data show a significant departure from constancy within the working range – as much as 30% – which actually exceeds the anticipated nonlinearity beyond the working range (which is less than 20%).

The coefficients of determination for the 436, 450 and 664 nm linear regressions in Figure B.1 were 0.9900, 0.9894 and 0.9864, respectively, for the C_8 method, and

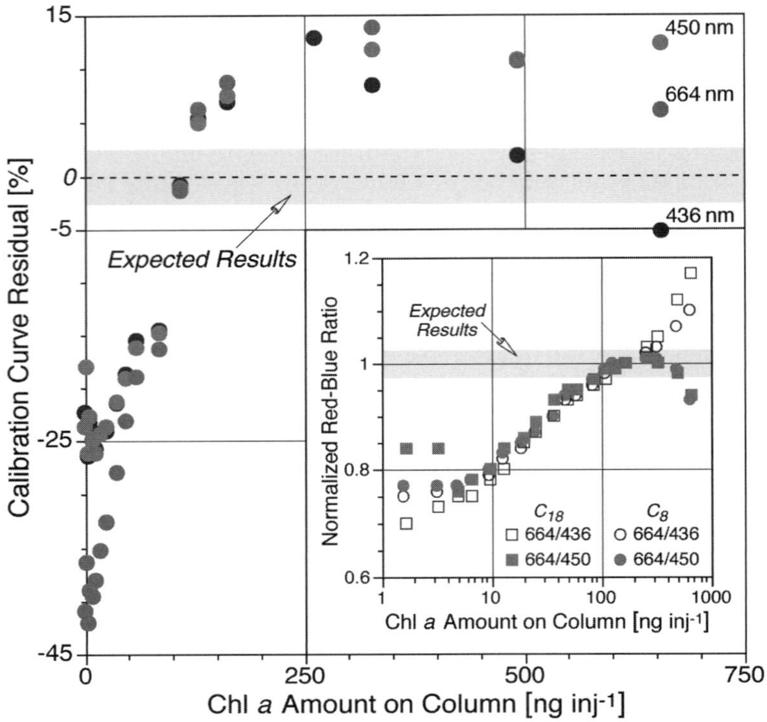

Figure B.1. Example of a detector with a nonlinear response. The main panel shows the calibration curve residuals as a function of the amount of Chl*a* on the column for three wavelengths used to quantitate pigments: 436, 450 and 664 nm. The dotted line shows expected results. The insert panel shows the degree of nonlinearity in a plot of normalized blue to red ratios (664/436 nm and 664/450 nm) for a C_8 (circles) and a C_{18} (squares) column. See colour plate section.

0.9988, 0.9981 and 0.9974, respectively, for the C_{18} method. Given the large departures from linearity shown in the residuals and the red-to-blue ratios, these results show the importance of not relying on simple regression statistics to establish detector linearity – particularly the concept of establishing a threshold and then assuming adequate linearity if the coefficient of determination exceeds the threshold (Mantoura and Repeta, 1997). A much more effective and sensitive procedure is to investigate the residuals of the fitted calibration curve to the actual data (for the analysis presented here, the residuals were computed as the actual value minus the fitted value and then dividing the difference by the fitted value) as suggested by Hooker *et al.* (2005).

In addition to choosing the most appropriate detector type, one must also investigate the availability of customization and the capabilities of the detector. As discussed in Chapter 5 of this volume and in the preceding paragraphs, linear range, working range, detection limits, and the signal-to-noise ratio (SNR) should be defined when validating an HPLC system. Information on sensitivity, noise, linearity, and response are often published by manufacturers, however, there is not necessarily a consistent way in which they are calculated or presented. For accurate comparison it is important to compare sensitivities that correspond to equivalent noise levels (Snyder and Kirkland, 1979). Detectors can have the ability to customize bandwidth, slit width, data sampling rate, flow cell size and temperature, and lamp type (depending on the detector). Every setting that is changed affects the performance of the detector. While it can be advantageous to have many options for customization of settings, parameters should be changed with a thorough understanding and investigation of what is necessary for the specific application and how said changes influence the performance of the detector. Some detectors have the option for temperature control of the flow cell, which is important because temperature changes can effect changes in the refractive index of the mobile phase.

B.6 A note on considerations for applications

Some methods require that the column be heated for optimal pigment separation and results and this would necessarily require a thermostatted column compartment. Even if a method describes the column temperature as 'ambient temperature', ambient room temperature may fluctuate during analysis, therefore a thermostatted column compartment may reduce the occurrence of aberrant retention time shifts by maintaining the column at constant temperature. When choosing a thermostatted column compartment or column oven, it is important that it provides the capability to evenly distribute temperature along the entire length of the column and preheat solvent so as to not create a temperature gradient within the column. Manufacturer specifications of the accuracy and precision of the oven temperature should also be examined.

Considerations should be made for intended future applications of the instrumentation. For research and method development applications, innovative technologies may be tested. These types of applications require longer time frames for setup and operation, in addition to increased operator attention to hardware details and maintenance. If the system will be used primarily in production mode, i.e. large batch sample analysis using an established methodology only to produce quantitative pigment data, a robust and simpler system may be ideal. These decisions should be made prior to system purchase.

In Table B.1 (Hardware specifications and implications), some of the information provided within this appendix is simplified and organized into three distinct categories for each hardware component: Injector, Autosampler, Detector and Pump. In the 'Considerations for specifications' column, available design features for each component of hardware are listed. Within the second column, 'Implications,' the implications of each design feature are listed. An improper choice for the intended application may result in poor instrument and method performance. Lastly, the third column – entitled 'Hardware performance specifications' – lists the typical performance specifications for each component of hardware found for a wide variety of HPLC system manufacturers. In addition, the footnotes accompanying many of the components may be used to retrieve more detailed information on the various topics covered within the table.

Even with the newest technologies and most advanced HPLC equipment available, the quality of an HPLC system's performance and data produced is ultimately only as good as the system operator. At the very least, even if the HPLC hardware is functioning properly, the operator must have the training and expertise to collect the correct type and amount of quality assurance and quality control data, in order for users to have confidence in the results. Comprehensive training and education in good laboratory practices and HPLC concepts, applications, instrument validation and maintenance are necessary. A multitude of resources for this type of training and education is available:

- In-house training
- Open enrollment courses
- Web-based seminars
- Workshops at conferences and tradeshows
- CD-based training
- Websites and other electronic resources
- HPLC application books
- Peer-reviewed journals and trade magazines

Some HPLC system and hardware manufacturers offer in-house training (training provided in one's home facility) that can be customized to current application needs. In-house training is beneficial because training is conducted using one's own system configuration. Open enrollment courses (classes offered at an offsite location) are conducted all over the world by a variety of companies and provide comparable training to in-house training. Alternatively, web seminars that cover a number of HPLC topics are either conducted 'live' over the internet by an instructor, or are provided as an on-demand presentation; these provide a cheaper, or sometimes free, alternative to in-house training and open enrollment classes. HPLC conferences and tradeshows are also a good source for basic concept and advance technique classes. Other comprehensive resources, such as websites, web forums, and HPLC hardware and application books are available and cover all aspects of chromatography. Web forums are particularly useful as they provide an environment for chromatographers to discuss HPLC concepts and troubleshooting among peers. All of the aforementioned resources are useful tools for understanding everything from basic HPLC concepts to advanced, innovative techniques. In addition, see Snyder *et al.* (2010) for further details about HPLC resources and training.

References

Dolan, J.W. (1983). An HPLC troubleshooting guide. *LCGC* **1**, 10–16.
Dolan, J.W. (1984). Optical detectors part I: general principles. *LCGC* **2**, 290–92.
Dolan, J.W. (1985a). Connecting tubing. *LCGC* **3**, 92–97.
Dolan, J.W. (1985b). Equipment usage. *LCGC* **3**, 956–60.
Dolan, J.W. (1986). Extracolumn effects: two case studies. *LCGC* **4**, 1086–90.
Dolan, J.W. (1987a). Troubleshooting: troubleshooting autosamplers II. *LCGC* **5**, 224–26.
Dolan, J.W. (1987b). Troubleshooting autosamplers I. *LCGC* **5**, 92–98.
Dolan, J.W. (1988). Troubleshooting LC fittings Part 1. *LCGC* **6**, 788–92.
Dolan, J.W. (1989). Reader's questions. *LCGC* **7**, 102–106.
Dolan, J.W. (1990). Pump problems. *LCGC* **8**, 916–18.
Dolan, J.W. (1991). Method reproducibility, leaks and check-valve problems. *LCGC* **9**, 88–92.
Dolan, J.W. (1996). Noise problems: a case study. *LCGC* **14**, 378–82.
Dolan, J.W. (1997a). Maintaining autosampler performance. *LCGC* **15**, 516–21.
Dolan, J.W. (1997b). Reproducibility problems. *LCGC* **15**, 1118–20.
Dolan, J.W. (1998). Autosampler precision. *LCGC* **16**, 910–14.
Dolan, J.W. (2001a). Autosamplers Part 1: Design features. *LCGC* **19**, 386–91.
Dolan, J.W. (2001b). Autosamplers Part 2: Problems and solutions. *LCGC* **19**, 478–82.
Dolan, J.W. (2002). Performance qualification of LC systems. *LCGC* **20**, 842–48.
Dolan, J.W. (2004a). System suitability. *LCGC* **22**, 430–35.
Dolan, J.W. (2004b). Gradient performance checks. *LCGC* **22**, 982–88.
Dolan, J.W. (2005). The role of the signal to noise ratio in precision and accuracy. *LCGC* **23**, 1256–60.
Dolan, J.W. (2006a). Autosampler carryover. *LCGC* **24**, 754–60.
Dolan, J.W. (2006b). Dwell volume revisited. *LCGC* **24**, 458–66.
Dolan, J.W. and Snyder, L.R. (1989). *Troubleshooting LC Systems: A Comprehensive Approach to Troubleshooting LC Equipment and Separations.* Totowa: Humana Press.
Hooker, S.B., Van Heukelem, L., Thomas, C.S., Claustre, H., Ras, J., Barlow, R., Sessions, H., Schlüter, L., Perl, J., Trees, C., Stuart, V., Head, E., Clementson, L., Fishwick, J., Llewellyn, C. and Aiken, J. (2005). *The Second SeaWiFS HPLC Analysis Round-Robin Experiment (SeaHARRE-2).* NASA TM/2005–212785, Greenbelt: NASA Goddard Space Flight Center.
Mantoura, R.F.C. and Repeta, D.J. (1997). Calibration methods for HPLC. In *Phytoplankton Pigments in Oceanography: Guidelines to Modern Methods*, ed. S.W. Jeffrey, R.F.C. Mantoura and S.W. Wright. Paris: UNESCO Publishing, pp. 407–27.
Snyder, L.R. and Kirkland, J.J. (1979). *Introduction to Modern Liquid Chromatography,* 2nd edn., Hoboken: John Wiley and Sons.
Snyder, L.R., Kirkland, J.J. and Dolan, J.W. (2010). *Introduction to Modern Liquid Chromatography,* 3rd edn., Hoboken: John Wiley and Sons.
Wright, S.W. and Mantoura, R.F.C. (1997). Guidelines for selecting and setting up an HPLC system and laboratory. In *Phytoplankton Pigments in Oceanography: Guidelines to Modern Methods*, ed. S.W. Jeffrey, R.F.C. Mantoura and S.W. Wright. Paris: UNESCO Publishing, pp. 383–406.

Appendix C

Minimum identification criteria for phytoplankton pigments

EINAR SKARSTAD EGELAND

This appendix updates Appendix 1 by Jeffrey and Mantoura (1997) in the 1997 SCOR Phytoplankton Pigment book (Jeffrey et al., 1997). The recommendations given herein recognize that identification criteria range from the ideal to the practical, depending on the nature of the sample.

When analysing algal pigments, the various pigments detected should always be compared with authentic standards. Some standards are commercially available (see Appendix E), others must be isolated from an appropriate source (see Data sheets and Appendix D).

Schiedt and Liaaen-Jensen (1995) have given minimum criteria for the identification of carotenoids:

- Obtain identical UV-visible absorption spectra as for the standard in two different solvents
- Obtain identical chromatographic properties as the standard in two different systems, also demonstrated by co-chromatography
- Obtain identical molecular mass as for the standard

These criteria were given with isolated carotenoids in mind, but should also be used for the identification of all pigments separated in an HPLC chromatogram, both for carotenoids and chlorophylls.

C.1 Identical UV-visible absorption spectra

For any 'unknown' pigment, the UV-visible absorption spectra for two different solvents or solvent mixtures (e.g. HPLC eluents) should be identical with those of an unambiguous pigment standard. Evaluated literature data may be used for comparison (see e.g. the Data sheets in this volume, or Britton *et al.*, 2004). When comparing the measured spectra of the 'unknown' compound with literature data, the following should be verified:

Phytoplankton Pigments: Characterization, Chemotaxonomy and Applications in Oceanography, eds. Suzanne Roy, Carole A. Llewellyn, Einar Skarstad Egeland and Geir Johnsen. Published by Cambridge University Press.

- The peak maxima given for a specified solvent must match the measured peak maxima for the same solvent, within a range of ± 1–2 nm.
- The shape of the pigment's UV-visible absorption spectrum should match a published spectrum (if available), and the difference between the published and the measured spectrum band ratios (chlorophylls) or fine structure (carotenoids) must not exceed ± 10% using the same solvent.

If the reference spectrum of a good pigment standard (a standard with unambiguous identity and a purity of at least 90%) has been recorded using the same spectrophotometer or using the same HPLC system on the same instrument as the 'unknown' pigment, no shift of peak wavelength nor alteration of the spectrum shape should be seen.

C.2 Identical chromatographic properties

The isolated 'unknown' pigment or an 'unknown' HPLC peak to be identified must have the same retention time as the reference standard in two different chromatographic systems. The two different systems can be using different HPLC eluent systems (using different solvents, not only different concentrations) on the same column, two different HPLC columns (e.g. one C_8 and one C_{18} column) using the same solvents, or by using one HPLC system and one TLC system. A mixture of equal amounts of the 'unknown' pigment (isolated or in extract) and pure reference pigment should be made. The mixture, when chromatographed (so-called 'co-chromatography'), must show a single peak for the unknown and the reference pigment and must not give any peak separation, peak broadening or any distortion of the peak shape.

C.3 Identical molecular mass

The 'unknown' pigment should have the correct molecular mass according to the reference pigment standard, using either an HPLC-linked mass detector (see Chapter 7, this volume) or a traditional self-standing mass spectrometer. Any fragment ions should be consistent with those of the reference pigment standard, noting though that such patterns vary greatly between the different ionization methods.

C.4 Algal cultures

When analysing algal cultures, no pigment should be regarded as *identified* unless all the above given recommendations are fulfilled. If a pigment completely matches some of the given criteria for identification, but some criteria have not been fulfilled (e.g. no mass spectrum and/or no reference standard for co-chromatography), this should be clearly stated and the pigments should be given as 'zeaxanthin-like', 'chlorophyll *d*-like' etc. Whenever some information does not match with the reference standard or the literature data used for comparison, the pigment should be given as 'unknown' with additional information.

C.5 Oceanographic samples

When analysing a large number of oceanographic samples containing numerous pigments, it is impractical to determine the identity for each pigment with the same confidence as given above. For such samples the identity of each pigment will be determined by chromatographic and absorption (and preferably also fluorescent) spectroscopic properties only, following the guidelines given above.

C.6 Pigments not separated

Be aware of any pigments overlapping in the HPLC chromatogram. In most cases (not all!) this can be detected by checking the UV-visible absorption spectra in detail. The spectrum obtained from the start, the maximum and the end of the peak should be identical (with the exception of overall intensity) regarding both the shape and wavelength. If not, the peak represents (at least) two different pigments eluting jointly in the same peak.

If, by chance, two (or more) pigments have identical retention times, they are likely to be mixed homogenously in the same peak. The best way to check for such co-eluting pigments is by analysing the sample using another HPLC system where separation can be achieved. A mass detector *may* give two (or more) molecular ions, but as several algal pigments have the same molecular mass, this may not always be the case.

In the case of overlapping peaks or impure peaks, no quantification should be done, unless the uncertainty in quantification is very clearly stated. However, if the pigments can be clearly discriminated by using a mass detector (giving two different molecular ions) or having non-overlapping absorption spectra, quantification may be achievable.

References

Britton, G., Liaaen-Jensen, S. and Pfander, H. (eds.), Mercadante, A.Z. and Egeland, E.S. (compilers) (2004). *Carotenoids. Handbook.* Basel: Birkhäuser.

Jeffrey, S.W. and Mantoura, R.F.C. (1997). Minimum criteria for identifying phytoplankton pigments. In *Phytoplankton Pigments in Oceanography: Guidelines to Modern Methods*, ed. S.W. Jeffrey, R.F.C. Mantoura and S.W. Wright. Paris: UNESCO Publishing, pp. 631–32.

Jeffrey, S.W., Mantoura, R.F.C. and Wright, S.W. (eds.) (1997). *Phytoplankton Pigments in Oceanography: Guidelines to Modern Methods.* Paris: UNESCO Publishing.

Schiedt, K. and Liaaen-Jensen, S. (1995). Isolation and analysis. In Carotenoids, volume 1A: *Isolation and Analysis*, ed. G. Britton, S. Liaaen-Jensen and H. Pfander. Basel: Birkhäuser, pp. 81–108.

Appendix D

Phytoplankton cultures for standard pigments and their suppliers

SUZANNE ROY, SIMON W. WRIGHT AND S.W. JEFFREY

D.1. SCOR reference cultures

The 1997 *Phytoplankton Pigments in Oceanography* volume by Jeffrey *et al.* (1997) identified two algal culture laboratories that could provide identical strains of reference algal cultures for pigments recommended by SCOR Working Group 78. It was suggested that the source of pigment standards should be strains of phytoplankton whose pigments had been identified chemically, since few pure pigment standards were commercially available at that time. This has since changed, with the development of purified pigment standards from DHI (Hørsholm, Denmark, http://c14.dhigroup.com/index.html, formerly VKI).

Since some pigments are still not currently available from DHI and some researchers prefer to isolate pigments from algal cultures, several additional cultures and their suppliers are recommended in the present volume.

The species recommended as authentic sources of pigments in the 1997 volume were listed in Table 6.1 (page 182) of that book. this Table is updated below (Table D.1), as are the two suppliers identified in that volume:

1 – **Australian National Algae Culture Collection (ANACC, formerly CSIRO Algal Culture Collection)**

CSIRO Marine and Atmospheric Research, GPO Box 1538, Hobart, Tasmania, 7001, Australia.

Website: http://www.cmar.csiro.au/places/Australian-National-Algae-Culture-Collection.html

Phone: (03) 6232 5316; Fax: (03) 6232 5471; email: cathy.johnston@csiro.au

Contact Person: Cathy Johnston, microalgae supply service technical officer; cost as of August 2010: $120 AUD for each 20 mL test tube; $195 for each 250 mL tissue-culture flask.

2 – **Provasoli-Guillard National Center for Culture of Marine Phytoplankton (CCMP)**

Bigelow Laboratory for Ocean Sciences, P.O. Box 475, 180 McKown Point Road, West Boothbay Harbor, Maine 04575 U.S.A. Website: http://ccmp.bigelow.org/

Phytoplankton Pigments: Characterization, Chemotaxonomy and Applications in Oceanography, eds. Suzanne Roy, Carole A. Llewellyn, Einar Skarstad Egeland and Geir Johnsen. Published by Cambridge University Press.

Phone: (001) 207 633 9630; Fax: (001) 207 633 9715; email: ccmp@bigelow.org

Cost as of August 2010: $50 USD per 15 mL starter culture for nonprofit organizations (10% discount if 6–10 cultures; 20% discount if more than 10 cultures).

Algal culture collections exist in several countries. Information can be found at the World Data Centre for Microorganisms (http://wdcm.nig.ac.jp/hpcc.html), noting that this also includes non-algal microbes. Information on particular strains and culture locations is available at http://www.straininfo.net, while more general information on species, including their taxonomy, may be found at http://www.algaebase.org.

Table D.1. *SCOR-recommended reference algal cultures as authentic sources of pigments (updated from Jeffrey* et al., *1997). Jeffrey and Wright (1997) published quantitative analyses of the listed CSIRO strains of each recommended species, grown and harvested under standard conditions, using the HPLC method of Wright* et al. *(1991). CCMP strains of many standard cultures are now available, but have not been analysed under standard conditions, see below.*

Algal species	***Algal class***	***CSIRO culture code***	***CCMP culture code***
Amphidinium cruentum	Dinophyceae	CS-212	CCMP 1314*
Dunaliella tertiolecta	Chlorophyceae	CS-175	CCMP 1320*
Emiliania huxleyi	Prymnesiophyceae	CS-57	CCMP 373
Euglena gracilis	Euglenophyceae	CS-66	– (only *Euglena* sp.)
Micromonas pusilla	Prasinophyceae	CS-86	CCMP 491
Pavlova lutheri	Prymnesiophyceae	CS-182	CCMP 1325*
Pelagococcus subviridis	Chrysophyceae	CS-99	CCMP 1429*
Phaeodactylum tricornutum	Bacillariophyceae	CS-29	CCMP 2559
Porphyridium purpureum	Rhodophyceae	CS-25	CCMP 1947*
Pycnococcus provasoli	Prasinophyceae	CS-185	CCMP 1203*
Rhodomonas (= *Chroomonas*) *salina*	Cryptophyceae	CS-174	CCMP 1319
Synechococcus sp.	Cyanobacteria	CS-197	CCMP 1334*

* not the same strain in the two culture collections

D.2. Reference microalgal cultures proposed for new algal classes and prochlorophytes

Several new algal classes and pigments have been found in the last decade. These are described in Chapter 1. Table D.2 provides culture sources of the new algal groups listed in Chapter 1 and in Tables 1.28 to 1.30. Unlike the species listed in Jeffrey and Wright (1997), the pigment compositions of most of these species have not been

Table D.2. *Reference microalgal cultures recommended for new algal classes, following the order listed in Chapter 1, this volume. Chapter 1 lists the pigments present in each algal species. Full names of the culture collections can be found in Table D.3. Only one source is indicated, even though some of the cultures can be found in several culture collections. N/A = not available.*

Algal species	***Algal class***	***Culture collection and code***
Aphanizomenon flos-aquae	Cyanobacteria (Cyano-1)	UTEX LB 2384
Synechococcus sp.	Cyanobacteria (Cyano-2)	RCC-29
Prochlorothrix hollandica	Cyanobacteria (Cyano-3)	CCAP 1490/1
Prochlorococcus marinus	Cyanobacteria (Cyano-4)	RCC 156
Acaryochloris marina	Cyanobacteria (Cyano-5)	NBRC 102967
Cyanophora paradoxa	Glaucocystophyceae (Glauco-1)	CCMP 329
Chaetoceros didymus	Bacillariophyceae (Diatom-1)	CS-2
Rhizoselenia setigera	Bacillariophyceae (Diatom-2)	CS-62
Nitzschia bilobata	Bacillariophyceae (Diatom-3)	CS-47
Bolidomonas pacifica	Bolidophyceae	RCC 205
Pseudochattonella verruculosa	Dictyochophyceae (Dictyo-1)	RCC 1082
Nannochloropsis oculata	Eustigmatophyceae	AC 849/1 or 849/7
Pelagococcus subviridis	Pelagophyceae (Pelago-1)	CS-99
Pinguiochrysis pyriformis	Pinguiophyceae	NBRC 102948
Chatonella marina	Raphidophyceae (Raphido-1)	RCC 1501
Mallomonas papillosa	Synurophyceae (Synuro-1)	CCMP 476 & 2540
Vaucheria terrestris	Xanthophyceae	CCCM 7067
Pavlova lutheri	Pavlovophyceae (Hapto-1)	CS-23
Pavlova gyrans	Pavlovophyceae (Hapto-2)	CCMP 608
Dicrateria inornata	Prymnesiophyceae (Hapto-3)	CS-254
Prymnesium parvum	Prymnesiophyceae (Hapto-4)	CS-345
Ochrosphaera verrucosa	Prymnesiophyceae (Hapto-5)	CCMP 594
Emiliania huxleyi	Prymnesiophyceae (Hapto-6)	CS-57
Chrysochromulina hirta	Prymnesiophyceae (Hapto-7)	CS-228

Table D.2. (*cont.*)

Algal species	***Algal class***	***Culture collection and code***
Phaeocystis pouchetii	Prymnesiophyceae (Hapto-8)	CS-165
Amphidinium carterae	Dinophyceae (Dino-1)	CS-212
Karlodinium veneficum (toxic)	Dinophyceae (Dino-2)	CCMP 2936
Kryptoperidinium foliaceum	Dinophyceae (Dino-3)	UTEX LB 1688
Dinophysis norvegica	Dinophyceae (Dino-4)	N/A[1]
Gymnodinium chlorophorum (= *Lepidodinium viride*)	Dinophyceae (Dino-5)	RCC 1488
Eutreptiella gymnastica	Euglenophyceae	CCMP 1594
Dunaliella tertiolecta	Chlorophyceae (Chloro-1)	RCC 6
Chlamydomonas parkeae	Chlorophyceae (Chloro-2)	NIES 440
Nephroselmis olivacea	Prasinophyceae (Prasino-1)	AC 1960/4B
Pyramimonas parkeae	Prasinophyceae (Prasino-2A)	CCMP 724
Nephroselmis pyriformis	Prasinophyceae (Prasino-2B)	CCMP 717 and 549
Pycnococcus provasolii	Prasinophyceae (Prasino-3A)	CCMP 1203
Micromonas pusilla (as source for Chl $c_{3CS\text{-}170}$)	Prasinophyceae (Prasino-3B)	CCMP 487 and 1195 CS-170
Picochlorum oklahomensis	Trebouxiophyceae	CCMP 2329
Mesostigma viride	Mesostigmatophyceae	CCMP 2046

[1] A few species of *Dinophysis* have now been successfully cultured using a ciliate prey (Park *et al.*, 2006)

Table D.3. *List of culture collections mentioned in Table D.2.*

Acronym	***Culture Collection***	***Web address***
AC	ALGOBANK- Caen, France	http://www.unicaen.fr/ufr/ibfa/algobank/home.php
CCAP	Culture Collection of Algae and Protozoa, SAMS Research Services Ltd, Scotland, UK	http://www.ccap.ac.uk/cultures/cultures.htm
CCMP	Provasoli-Guillard National Center for Culture of Marine Phyoplankton, USA	https://ccmp.bigelow.org/

Table D.3. (*cont.*)

Acronym	*Culture Collection*	*Web address*
CS	Australian National Algae Culture Collection (ANACC, formerly CSIRO Algal Culture Collection) Australia	http://www.cmar.csiro.au/places/Australian-National-Algae-Culture-Collection.html
NBRC	National Institute of Technology and Evaluation (NITE), Japan	http://www.nbrc.nite.go.jp/e/
NIES	National Institute for Environmental Studies, Microbial Culture Collection, Japan	http://mcc.nies.go.jp/
RCC	Roscoff Culture Collection, France	http://www.sb-roscoff.fr/Phyto/RCC/
UTEX	The University of Texas at Austin, The Culture Collection of Algae (UTEX) USA	http://web.biosci.utexas.edu/utex/default.aspx

validated by the present authors. Also, cultures were not grown and analysed under standard conditions. In addition, some species that were the original sources for pigments are no longer available in culture, and similar species have been recommended instead. Table D.3 gives website addresses of the culture collections mentioned in Table D.2.

References

Jeffrey, S.W. and Wright, S.W. (1997). Qualitative and quantitative HPLC analysis of SCOR reference algal cultures. In *Phytoplankton Pigments in Oceanography: Guidelines to Modern Methods*, ed. S.W. Jeffrey, R.F.C. Mantoura and S.W. Wright. UNESCO monographs on oceanographic methodology, **vol 10**. Paris, France: UNESCO Publishing, pp. 343–60.

Jeffrey, S.W., Mantoura, R.F.C. and Wright, S.W. (eds.) (1997a). *Phytoplankton Pigments in Oceanography: Guidelines to Modern Methods*. UNESCO Monographs on oceanographic methodology, **vol. 10**. Paris, France: UNESCO Publishing, 667 pp.

Park, M.G., Kim, S., Kim, H.S., Myung, G., Kang, Y.G. and Yih, W. (2006). First successful culture of the marine dinoflagellate *Dinophysis acuminata*. *Aquat. Microb. Ecol*. **45** 101–06.

Wright, S.W., Jeffrey, S.W., Mantoura, R.F.C., Llewellyn, C.A., Bjørnland, T., Repeta, D. and Welschmeyer, N. (1991). Improved HPLC method for the analysis of chlorophylls and carotenoids from marine phytoplankton. *Mar. Ecol. Prog. Ser*. **77**: 183–96.

Appendix E

Commercial suppliers of phytoplankton pigments

EINAR SKARSTAD EGELAND AND LOUISE SCHLÜTER

This appendix gives an overview of pigments commercially available at the time of writing (June 2009).

Pigments may be isolated from a well-documented natural source (see Appendix D) by means of preparative HPLC or TLC, or for the more common pigments, bought from a commercial supplier. Whatever the source of the standard, for quantitative purposes it is essential that the pigment be of the very highest purity, i.e. close to 100%. The pigment standard must not contain any *cis*-isomers of carotenoids or allomers of chlorophylls, since such impurities have their own, in most cases unknown, absorption coefficient (for absorption coefficients, see Part VII, Data sheets).

DHI (see [1] at the bottom of Table E.1) is offering solutions (2.5 mL, approx. 1 μg mL^{-1}, the specific concentration is given). These solutions may be used directly for HPLC calibration. All other companies [2–6] are selling solid pigments, mostly crystalline. For solid samples, a small amount must be completely dissolved in an appropriate solvent (one with a known extinction coefficient) and the pigment concentration is calculated from the absorbance (by use of a spectrophotometer) before HPLC calibration.

Crystalline samples may be kept for several years if stored under nitrogen or argon in airtight vials and kept frozen below –20 °C. Pigment solutions should also be stored in sealed vials in a nitrogen atmosphere below –20 °C and are stable for at least three years after purchase. The solutions should, however, be used immediately after opening the vials.

Prices are given for 1 mg crystals or 2.5 mL of about 1 μg mL^{-1} solution (DHI only), unless otherwise stated. Purity is only given in Table E.1 when it is lower than 95% of the solid material. Please be aware that prices may change and that tax and freight have not been added. Updated information should always be checked with the suppliers' web page [1–6]. Some pigments are available in different qualities or from different sources by the same company. In such cases, only the best quality is given in the table below. In some cases, more than one company are selling the same product,

Phytoplankton Pigments: Characterization, Chemotaxonomy and Applications in Oceanography, eds. Suzanne Roy, Carole A. Llewellyn, Einar Skarstad Egeland and Geir Johnsen. Published by Cambridge University Press.

e.g. pigments offered by Fluka [2] will in most cases also be available at Sigma [2]. The standards produced by DHI are also offered from Chromadex [3]. Whenever the source is known, only one supplier is given. Only pigments encountered in phytoplankton are included in the table. Some of the suppliers (e.g. CaroteNature [4]) are offering a large range of carotenoids prepared by total synthesis or by isolation. If requested, CaroteNature [4] may be willing to produce a carotenoid not given in their price list.

References

[1] www.labproducts.dhigroup.com
[2] www.sigmaaldrich.com
[3] www.chromadex.com
[4] www.carotenature.com
[5] www.vwr.com
[6] www.frontiersci.com

Table E.1. *Phytoplankton pigments available commercially.*

Pigment	Supplier	Isolated from or prepared by	Price (June 2009)	Remarks
Adonirubin	CaroteNature [4]	Synthesis	€ 370	Also available as mixture of optical isomers
Alloxanthin	DHI [1]	Cryptophyceae	€ 126	Solvent: 100% ethanol
Antheraxanthin	CaroteNature	Natural, source not given	€ 370	
	DHI	Chlorophyceae	€ 359	Solvent: 100% ethanol
Aphanizophyll	DHI	Cyanobacteria	€ 142	Solvent: 100% acetone
Astaxanthin	CaroteNature	Synthesis	€ 245	Also available as mixture of optical isomers; optical isomers and *cis*-isomers available on request
	Sigma [2]	Not given ($\geq$ 92% pure)	€ 94.60 (250 mg)	Unknown stereochemistry
	DHI	Synthetic	€ 126	Solvent: 100% acetone
Bacteriochlorophyll *a*	Frontier Scientific [6]	*Rhodopseudomonas sphaeroides*	US$ 53	Purity not given
19′-Butanoyloxyfucoxanthin	DHI	Prymnesiophyceae	€ 359	Solvent: 100% ethanol
Canthaxanthin	CaroteNature	Synthesis	€ 35	Also *cis*-isomers available
	Sigma	Not given (94% pure)	€ 52.40 (10 mg)	
	DHI	Cyanobacteria	€ 142	Solvent: 100% ethanol
β,β-Carotene (β-carotene)	CaroteNature	Synthesis	€ 25	Also *cis*-isomers available
	DHI	Cyanobacteria	€ 142	Solvent: 100% acetone
	Fluka [2]	Not given	€ 15.90 (1 g)	
	VWR [5]	Not given	€ 70 (1 g)	

β,β-Caroten-5,6-epoxide	CaroteNature	Synthesis (94% pure)	€ 315	Only as mixture of optical isomers
β,β-Caroten-4-ol (isocryptoxanthin)	CaroteNature	-	-	Only on request; only as mixture of optical isomers
β,ε-Carotene (α-carotene)	CaroteNature	Synthesis	€ 315	Also available as mixture of optical isomers; optical isomer available on request
	DHI	Cryptophyceae	€ 142	Solvent: 100% acetone
β,ε-Caroten-3-ol (zeinoxanthin)	CaroteNature	Synthesis	€ 370	'α-cryptoxanthin' is used both for β,ε-carotene-3-ol and β,ε-carotene-3′-ol
β,ψ-Carotene (γ-carotene)	CaroteNature	Synthesis	€ 245	
ε,ε-Carotene (ε-carotene)	CaroteNature	Synthesis	€ 370	Optical isomers and mixture of optical isomers available on request
ε,ψ-Carotene (δ-carotene)	CaroteNature	Synthesis	€ 315	Only as mixture of optical isomers
ψ,ψ-Carotene (lycopene)	CaroteNature	Synthesis	€ 120	*cis*-isomers available on request
	Fluka	Not given (≥ 90% pure)	€ 123	
	DHI	Synthetic	€ 126	Solvent: 100% acetone
Chlorophyll *a*	Sigma	Spinach (purity not given)	€ 97.10	Several samples received with poor quality
	DHI	Cyanobacteria	€ 126	Solvent: 90% acetone
Chlorophyll *b*	Sigma	Spinach (≥ 90% pure)	€ 109.50	
	DHI	Chlorophyceae	€ 126	Solvent: 90% acetone
Chlorophyll c_2	DHI	Cryptophyceae	€ 142	Solvent: 90% acetone
Chlorophyll c_3	DHI	Prymnesiophyceae	€ 142	Solvent: 90% acetone

Table E.1. *(cont.)*

Pigment	Supplier	Isolated from or prepared by	Price (June 2009)	Remarks
Chlorophyllide *a*	DHI	Bacillariophyceae	€ 142	Solvent: 90% acetone
Crocoxanthin	DHI	Cryptophyceae	€ 359	Solvent: 100% ethanol
Cryptoxanthin (β-cryptoxanthin) (β,β-caroten-3-ol)	CaroteNature	Synthesis	€ 245	
	DHI	Fruits	€ 126	Solvent: 100% acetone
Diadinoxanthin	DHI	Bacillariophyceae	€ 142	Solvent: 100% ethanol
Diatoxanthin	DHI	Bacillariophyceae	€ 359	Solvent: 100% ethanol
7′,8′-Dihydro-β,ψ-carotene (β-zeacarotene)	CaroteNature	Synthesis	€ 315	
Divinyl chlorophyll *a*	DHI	*Prochlorococcus* sp.	€ 142	Solvent: 90% acetone
Divinyl protochlorophyllide (MgDVP)	DHI	Prasinophyceae	€ 142	Solvent: 90% acetone
Echinenone	CaroteNature	Synthesis	€ 245	
	DHI	Cyanobacteria	€ 142	Solvent: 100% ethanol
Fucoxanthin	CaroteNature	*Fucus* sp. (94% pure)	€ 315	
	DHI	Bacillariophyceae	€ 126	Solvent: 100% ethanol
Gyroxanthin diester	DHI	Dinophyceae	€ 359	Solvent: 100% ethanol
19′-Hexanoyloxyfucoxanthin	DHI	Prymnesiophyceae	€ 142	Solvent: 100% ethanol
3-Hydroxyechinenone	CaroteNature	Synthesis	€ 370	Only as mixture of optical isomers
4′-Hydroxyechinenone	CaroteNature	Synthesis	€ 370	Only as mixture of optical isomers
Lutein (xanthophyll)	CaroteNature	Natural, source not given (94% pure)	€ 120	3′-epimer also available
	DHI	Chlorophyceae	€ 126	Solvent: 100% ethanol
	Fluka	Not given (~90% pure)	€ 171	

Mutatoxanthin	CaroteNature	Synthesis	€ 370	Common artifact in extracts, from antheraxanthin
Myxoxanthophyll	DHI	Cyanobacteria	€ 142	Solvent: 100% acetone
9′-*cis*-Neoxanthin ((9′Z)-neoxanthin)	CaroteNature	Natural, source not given	€ 370	9′-*cis* is the main isomer in plants and green algae; neoxanthin from other algae is/may be *trans*-neoxanthin
	DHI	Chlorophyceae	€ 142	Solvent: 100% ethanol
Peridinin	DHI	Dinophyceae	€ 142	Solvent: 100% ethanol
Pheophorbide *a*	DHI	Bacillariophyceae	€ 126	Solvent: 90% acetone
	Frontier Scientific	Not given	US$ 58 (50 mg)	Purity not given
Pheophytin *a*	DHI	Bacillariophyceae	€ 142	Solvent: 90% acetone
Prasinoxanthin	DHI	Prasinophyceae	€ 126	Solvent: 100% ethanol
Pyropheophorbide *a*	Frontier Scientific	Not given	US$ 47 (10 mg)	
Taraxanthin (lutein epoxide)	CaroteNature	Natural, source not given	€ 370	
7,8,7′,8′-Tetrahydro-ψ,ψ-carotene (ζ-carotene)	CaroteNature	Synthesis	€ 315	
Violaxanthin	CaroteNature	Natural, source not given	€ 370	In extracts extremely labile to acids
	DHI	Chlorophyceae	€ 142	Solvent: 100% ethanol
Zeaxanthin	CaroteNature	Synthesis	€ 245	Pure and mixture of optical isomers available on request
	DHI	Cyanobacteria	€ 142	Solvent: 100% ethanol
	Fluka	Not given	€ 362	

Part VII

Data sheets aiding identification of phytoplankton carotenoids and chlorophylls

EINAR SKARSTAD EGELAND
IN COLLABORATION WITH
JOSÉ LUIS GARRIDO, LESLEY CLEMENTSON, KJERSTI ANDRESEN, CRYSTAL S. THOMAS, MANUEL ZAPATA,
RUTH AIRS, CAROLE A. LLEWELLYN, GREGORY L. NEWMAN, FRANCISCO RODRÍGUEZ AND SUZANNE ROY

Introduction

Since the publication of 47 key phytoplankton pigment data sheets in the volume by Jeffrey *et al.* (1997b), several new algal groups and pigments have been reported. To reflect this and the increased use of mass spectrometry for phytoplankton pigment characterisation we have compiled revised and expanded data sheets documenting 47 carotenoids and 21 chlorophylls. These new data sheets complement the ones produced for the 1997 volume. They are also available online, at www.cambridge.org/phytoplankton, for ease of consultation. We do not include data sheets for the many chlorophyll transformation products found particularly in sediments; for information on these pigments readers should refer to the comprehensive review by Keely (2006). Similarly, our coverage of pigments contained in phototrophic bacteria is limited mostly to cyanobacteria found in the water column of freshwater and marine environments (see Chapter 1, this volume), hence we exclude the newly discovered chlorophyll *f* in stromatolites (Chen *et al.*, 2010). Readers interested in anoxygenic phototrophic bacteria should consult the reviews by Takaichi (1999) and Scheer (2006).

Throughout the data sheets, the information presented is as accurate as possible, with original publications being referenced where there is evidence of a high degree of purity for the pigments. In some cases, as for chlorophyll *a* or fucoxanthin, many publications from various sources with differing quality were available. In other cases, especially for recently reported structures, only one or very few articles were available. Data given for more recently discovered pigments cannot be crosschecked against other reports, and therefore the data given may be less accurate than for the more common pigments studied since the 1930s. This is of most relevance to the reported specific absorption coefficients. Accurate coefficients need to be determined from mg quantities of highly purified and dry pigments; however this has not been a priority for researchers and hence, reliable values do not exist for many pigments. Until such time as accurate specific absorption coefficients become available, we strongly recommend that the specific absorption coefficients used to derive reported pigment concentrations should be given in all publications.

In addition to all the major carotenoids and chlorophylls found in phytoplankton we include minor pigments when they have an important marker role: this is generally where their presence in not limited to one or a few phytoplankton species. The final selection for data sheets to be included in this volume was based on the availability of information provided by the contributing authors. A few pigments presented in the volume by Jeffrey *et al.* (1997) have not been included because there was no new information; readers interested in those pigments should refer to that volume. A far greater number of pigments than those presented have been reported from cultivated phytoplankton. Pigments not presented here are listed at the end of this data sheet section; for further details, see e.g. Chapter 1, this volume, Liaaen-Jensen (1998) or Britton *et al.* (2004).

Where there are various forms of a pigment (esters or glycoside derivatives), we present examples that we feel most useful for the analysis of aquatic samples. Further esters or glycosides are commented on in the 'Remarks' section of each data sheet. A few examples of furanoxide derivatives of carotenoid epoxides have also been included; these can be readily found due to traces of acids in solvents or phytoplankton material. For example, furanoxides are regularly seen in prasinoxanthin-containing phytoplankton when extracts have been concentrated or left for some time before HPLC analysis.

Names and molecular structures

The heading of each data sheet gives the common name of the pigment. In a few cases, this has been a matter of preference (e.g. β,ψ-carotene or γ-carotene), but in most cases, a pigment is generally known by only one name.

Two abbreviations are recommended for each pigment. A general abbreviation is given for common use in e.g. tables and figures. For consistency, this abbreviation generally follows that used in Jeffrey *et al.* (1997b). The other abbreviation, given in parentheses, is recommended for use in mathematical formulae, or wherever a very short abbreviation is required.

A formal, systematic name is given for each chlorophyll-type pigment, based on the most recent IUPAC recommendations (International Union of Pure and Applied Chemistry and International Union of Biochemistry, 1987, 1999, 2004; International Union of Pure and Applied Chemistry, 1993). For carotenoids, systematic names are taken from the Carotenoids Handbook (Britton *et al.*, 2004).

The molecular formula and weight of each pigment is included with the molecular structure. The structure has been confirmed by NMR in most cases. Those that have not been confirmed using NMR are commented on in the 'Remarks' section. The stereochemistry of each molecule is shown, even when it has not been unequivocally determined by means of X-ray crystallography or comparison with synthetically made carotenoids or chlorophylls.

Occurrence and accompanying pigments

For each data sheet, information is provided on the algal groups where the pigment is commonly encountered. In addition, a suggested source culture is given that might be used for cultivation of biomass for subsequent isolation of the pigment (for isolation procedures, see the first SCOR phytoplankton pigment volume (Jeffrey, 1997; Repeta and Bjørnland, 1997). Some of the pigments presented are commercially available, see Appendix E of this volume.

Common derivatives are given in addition to other pigments regularly encountered with the pigment presented. Biosynthetically related pigments, which may or may not be seen together with the pigment of interest, are also given. When the presented pigment is a derivative of a biosynthetic pigment, the parent pigment is given too. Here, the recommended biological source will give the parent pigment, which may be transformed into the presented pigment by e.g. acidification. For details see the individual sheets.

HPLC chromatograms

Two example chromatograms are presented of cultivated unialgal material, which contains the pigment presented on the data sheet. These have been graciously provided by the various contributors and are not necessarily prepared from the same material, so there will be variations in the pigments present in each of the chromatograms. The main peaks in the chromatograms are identified, using recommended abbreviations ('Unk' = unknown). The upper HPLC chromatogram, presented as system 1, uses the method published by Van Heukelem and Thomas (2001); while the lower chromatogram, given as system 2, uses the method published by Zapata *et al.* (2000). In a few cases, a third system has been referred to which uses the method published by Airs *et al.* (2001). The *x*-axis denotes elapsed time in minutes; and the *y*-axis is relative intensity of the absorbance signal at a given wavelength, usually 440 nm (this can vary according to the needs of the analyst and the absorbance maxima of the pigments of interest). For readers interested in a comparison of retention times between methods, see Wright and Jeffrey (2006).

UV-Vis spectral information

At the top of the right-hand side of each data sheet, UV-Vis (UV and visible range, covering generally from 350 to 700nm) spectral information for the 'pure' pigment is presented in frequently used solvents. The wavelength of maximum light absorption is in italics and wavelengths given in parentheses denote 'shoulders' of the spectrum. For all pigments which have more than one absorption peak, the ratio between the maxima is given (blue:red ratio for chlorophylls, % III:II values for carotenoids). Description and calculation of these parameters can be found in the Data sheets section of Jeffrey *et al.* (1997b).

To calculate pigment concentration from spectrophotometric information, it is essential to use an accurate specific absorption coefficient for the right pigment in the right solvent; otherwise the amount of pigment calculated could easily be ± 20% of the true value. Regrettably, due to the meticulous care required to prepare a pure, solid pigment in the mg range, specific absorption coefficients are seldom determined. Since the degree of purity of the compounds (and the solvents into which they are dissolved!) used when determining specific absorption coefficients is generally unknown, the coefficients available from the scientific literature can be quite variable for a given pigment/solvent pair and thus, have an undetermined validity in many cases. In general, when more than one value is available for a pigment in a certain solvent, the highest value should be used. The first author's best recommendation is given on the first line of the 'Recommended specific absorption coefficient' on the right hand page of the data sheets. When this recommendation is for solvents not compatible with common HPLC methods for pigment analysis, a second recommendation is given, which generally follows the suggestions given in the volume by Jeffrey *et al.* (1997b). Note that it should be possible to determine spectrophotometrically the absorbance of a pigment standard in the first recommended solvent and then add a small volume of this solution to most solvents used in common HPLC systems. However, if diluted pigments are needed and/or if there are solvent compatibility issues, the first solution must be concentrated to about 5% of its initial volume with a gentle stream of nitrogen, before making the volume up to the original volume with the new preferred solvent. For users who prefer $E^{1\%}{}_{1\ cm}$ or molar ε values, calculations to obtain these from *d* values can be found in Jeffrey *et al.* (1997b). A list of all specific absorption coefficients found in the literature (irrespective of their quality) can be found in Appendix F, available only online, at www.cambridge.org/phytoplankton.

For a number of pigments, specific absorption coefficients have never been determined (denoted n.d. = not determined). In such cases, it is generally recommended to use a specific absorption coefficient for a similar chlorophyll or carotenoid. If a coefficient for another pigment with an identical conjugated double bond system exists, the absorption coefficient has been assumed to be identical and was adjusted for difference in molecular weight. If no pigment with a similar conjugated double bond system exists, or where none has a reported specific absorption coefficient, a value was estimated based on compounds with similar conjugated double bond systems. These calculated or estimated values are presented in the 'Remarks' section. Note that this differs from the recommendation given in the 1997 volume, where the default value for carotenoids was 250 L $g^{-1}cm^{-1}$ (at peak wavelength), based on the specific absorption coefficient of β,β-carotene in acetone. This highlights the importance of reporting specific absorption coefficients in publications, as stated above.

Absorption spectra are illustrated for two different solvents and for the two HPLC system eluents, whenever available (again graciously offered by the named contributors at the front of this section). Other pigment absorption spectra can be found elsewhere (Jeffrey *et al.*, 1997a and Mercadante and Egeland, 2004). The *x*-axis denotes wavelength in nanometres; and the *y*-axis is relative absorbance (= optical density, dimensionless). In some spectra, the absorption maxima wavelengths differ from those given in the UV-Vis spectra table. This probably reflects differences in the calibration of the spectrophotometer used or insufficient purity.

Mass spectrometry information

Mass spectra are presented at the bottom of page two of each data sheet. A molecular ion can help confirm the identity of a pigment (see Appendix C, this volume), and fragment ions may be highly valuable in the structural characterisation of an unknown pigment. For most pigments, the traditional electron impact data are available, but other types of mass spectrometers have provided data for some of the pigments. Users should be aware that differences appear in the fragmentation pattern of a mass spectrum. This is dependent on the type of instrument used (see Chapter 7, this volume). In the data sheets where ion trap data is provided then the fragmentation data, generated using MS/MS, is given for ions derived from the precursor ion. Derived ions are denoted by an arrow from the precursor ion. Where ion trap data is given no relative intensity data can be sensibly provided.

Remarks

The final 'Remarks' section gives any other essential information, e.g. fluorescence data for chlorophylls or suggested specific absorption coefficients, where these are calculated or estimated on the basis of other pigments.

Pigments presented in data sheets

The data sheets are presented in alphabetical order in three groups: first, chlorophyll-related pigments; second, carotenes, and third, xanthophylls.
Chlorophyll-related pigments:

Bacteriochlorophyll *a*, Chlorophyll *a*, Chlorophyll *a* allomers, Chlorophyll *a* epimer, Chlorophyll *b*, Chlorophyll *b* epimer, Chlorophyll c_1, Chlorophyll c_2, Chlorophyll c_2-MGDG, Chlorophyll c_3, Chlorophyll *d*, Chlorophyllide *a*, Chlorophyllide *b*, Divinyl chlorophyll *a*, Divinyl chlorophyll *b*, Magnesium 2,4-divinylpheoporphyrin a_5 monomethyl ester (MgDVP), Monovinyl chlorophyll c_3, Pheophorbide *a*, Pheophytin *a*, Pheophytin *b*, Pyropheophorbide *a*, Pyropheophytin *a*.
Carotenes:

β,β-carotene, β,ε-carotene, β,ψ-carotene, ε,ε-carotene, ψ,ψ-carotene (lycopene).
Xanthophylls:

Alloxanthin, Antheraxanthin, Astaxanthin, Auroxanthin, 19′-Butanoyloxyfucoxanthin, Caloxanthin, Canthaxanthin, Crocoxanthin, Cryptoxanthin, Diadinoxanthin,

Diadinochrome, Diatoxanthin, Dihydrolutein, Dinoxanthin, Echinenone, Eutreptiellanone, Fucoxanthin, Gyroxanthin dodecanoate ethanoate, 19′-Hexanoyloxyfucoxanthin, 19′-Hexanoyloxy-4-ketofucoxanthin, Loroxanthin, Loroxanthin dodecenoate, Lutein, Micromonal, Micromonol, Monadoxanthin, Mutatoxanthin, Myxol quinovoside, 9′-*cis*-Neochrome, 9′-*cis*-Neoxanthin, all-*trans*-Neoxanthin, Nostoxanthin, Oscillol diquinovoside, Peridinin, Prasinoxanthin, Siphonaxanthin, Siphonaxanthin dodecenoate, Uriolide, Vaucheriaxanthin, Vaucheriaxanthin ethanoate octanoate, Violaxanthin, Zeaxanthin.

Other phytoplankton pigments

Other carotenoids and chlorophylls have also been reported from phytoplankton species, in addition to the ones given in the data sheets. These include:

adonirubin
adonirubin esters
adonixanthin esters
alloxanthin diester
anhydrodiatoxanthin
anhydromicromonal
anhydromicromonol
anhydroprasinoxanthin
anhydrouriolide
anhydrouriolide furanoxides (diastereomers)
aphanizophylls (different sugars)
apofucoxanthins (several)
astaxanthin esters (mono- and diesters)
aurochromes (diastereomers)
α-botryoxanthin
botryoxanthin A
botryoxanthin B
braunixanthin 1
braunixanthin 2
19′-butanoyloxyfucoxanthinol
19′-butanoyloxyhalocyntiaxanthin 3′-acetate
19′-butanoyloxy-4-ketofucoxanthin
ε,ε-carotene (other enantiomer)
ε,ψ-carotene
ς-carotene
β,β-carotene-2,2′-diol
β,β-carotene epoxide
β,β-carotene furanoxides (mutatochrome; diastereomers)
β,ε-carotene epoxide
β,β-caroten-2-ol
β,ε-caroten-2-ol
β,β-caroten-2-ol epoxide
β,β-caroten-2-ol furanoxides (diastereomers)
β,ε-caroten-2-ol epoxide
β,ε-caroten-2-ol furanoxides (diastereomers)
chlorophyll c_1-like *Kryptoperidinium foliaceum*-type
chlorophyll c_2-like *Pavlova gyrans*
chlorophyll c_2-MGDG (14:0/14:0) and others
chlorophyll $c_{cs\text{-}170}$
chlorophyll *f*
α-cryptoeutreptiellanone
β-cryptoeutreptiellanone
β,ε-cryptoxanthin (3′-OH)
cryptoxanthin diepoxide
cryptoxanthin difuranoxides (diastereomers)
cryptoxanthin 5,6-epoxide

cryptoxanthin epoxide furanoxides (diastereomers)
cryptoxanthin 5,8-furanoxides (diastereomers)
cryptoxanthin 5′,6′-epoxide
cryptoxanthin 5′,8′-furanoxides (diastereomers)
deepoxydinoxanthin
deepoxyneoxanthin
deepoxyuriolide
dehydrouriolide
dehydrouriolide furanoxides (diastereomers)
dihydroprasinoxanthin epoxide
dinochrome (diastereomers)
fucoxanthin acetate
fucoxanthinol
halocyntiaxanthin 3′-acetate
heteroxanthin
hexadehydro-β,β-caroten-3-ol ester
19′-hexanoyloxyfucoxanthinol
19′-hexanoyloxyhalocyntiaxanthin 3′-acetate
19′-hexanoyloxy-4-ketofucoxanthinol
3-hydroxyechinenone
3′-hydroxyechinenone
4′-hydroxyechinenone
19′-hydroxyfucoxanthin
6′-hydroxysiphonaxanthin esters
3-hydroxy-β-zeacarotene
isocryptoxanthin
4-ketofucoxanthin
4-ketomyxolxanthophyll (different sugars)
lutein epoxide (taraxanthin)
lutein furanoxides (diastereomers)
luteoxanthin
manixanthin (di-*cis*-alloxanthin)
neochromes (*trans* forms; diastereomers)
non-polar chlorophyll c_1-like *Prymnesium parvum*-like
octadehydro-β,β-carotene
19′-octanoyloxyfucoxanthin
P457
P468
pectenolone
19′-pentanoyloxyfucoxanthin
19′-pentanoyloxyfucoxanthinol
peridinin furanoxide
peridininol
peridininol furanoxides (diastereomers)
pheophorbide *b*
phytoene
phytofluene
preprasinoxanthin
prolycopene (tetra-*cis*-ψ,ψ-carotene)
pyrrhoxanthin
pyrrhoxanthinol
siphonaxanthin methyl ether
unidentified ketocarotenoid from *Eugomontia sacculata*
unidentified carotenoid ester(s) from Mamiellales
uriolide furanoxide
vaucheriaxanthin ethanoate octanoate furanoxide
vaucheriaxanthin furanoxide
β-zeacarotene
zeinoxanthin (β,ε-cryptoxanthin (3-OH))

Acknowledgements

We acknowledge the help of the Australian National Algae Culture Collection within CSIRO Marine and Atmospheric Research, with special thanks to Dr Susan Blackburn, Mr Ian Jameson and Ms Cathy Johnston. We are also grateful to Dr S.W. Jeffrey and Dr S.W. Wright for fruitful discussion and suggestions during the preparation of these data sheets. We also thank Dr Geir Johnsen and students Cesilie Røtnes Amundsen, Halvor Prytz and Solveig Lysfjord Sørensen for help with chromatograms, spectra, figures and editing.

References

Airs, R.L., Atkinson, J.E. and Keely, B.J. (2001). Development and application of a high resolution liquid chromatographic method for the analysis of complex pigment distributions. *J. Chromatogr. A* **917**, 167–77.

Britton, G., Liaaen-Jensen, S. and Pfander, H. (eds.), Mercadente, A.Z. and Egeland, E.S. (compilers) (2004). *Carotenoids. Handbook.* Basel: Birkhäuser.

Chen, M., Schliep, M., Willows, R.D., Cai, Z.-L., Neilan, B.A. and Scheer, H. (2010). A red-shifted chlorophyll. *Science* **329**, 1318–19.

International Union of Pure and Applied Chemistry (1993). *A Guide to IUPAC Nomenclature of Organic Compounds*, ed. R. Panico, W.H. Powell and J.-C. Richer. Oxford: Blackwell Science.

International Union of Pure and Applied Chemistry (1999). Revised section F: Natural products and related compounds. *Pure Appl. Chem.* **71**, 587–643.

International Union of Pure and Applied Chemistry (2004). Preferred names in the nomenclature of organic compounds, provisional recommendations. http://old.iupac.org/reports/provisional/abstract04/BB-prs310305/CompleteDraft.pdf

International Union of Pure and Applied Chemistry and International Union of Biochemistry (1987). Nomenclature of tetrapyrroles. *Pure Appl. Chem.* **59**, 779–832.

Jeffrey, S.W. (1997). Preparation of chlorophyll standards. In *Phytoplankton Pigments in Oceanography: Guidelines to Modern Methods*, ed. S.W. Jeffrey, R.F.C. Mantoura and S.W. Wright. Paris: UNESCO Publishing, pp. 207–38.

Jeffrey, S.W., Mantoura, R.F.C. and Bjørnland, T. (1997a). Data for the identification of 47 key phytoplankton pigments. In *Phytoplankton Pigments in Oceanography: Guidelines to Modern Methods*, ed. S.W. Jeffrey, R.F.C. Mantoura and S.W. Wright. Paris: UNESCO Publishing, pp. 449–559.

Jeffrey, S.W., Mantoura, R.F.C. and Wright, S.W. (1997b). *Phytoplankton Pigments in Oceanography: Guidelines to Modern Methods.* Paris: UNESCO Publishing, 661 pp.

Keely, B.J. (2006). Geochemistry of chlorophylls. In *Chlorophylls and Bacteriochlorophylls: Biochemistry, Biophysics, Functions and Applications*, ed. B. Grimm, R.J. Porra, W. Rüdiger and H. Scheer. Dordrecht: Springer, pp. 535–61.

Liaaen-Jensen, S. (1998). Carotenoids in chemosystematics. In *Carotenoids Vol. 3: Biosynthesis and Metabolism*, ed. G. Britton, S. Liaaen-Jensen and H. Pfander. Basel: Birkhäuser Verlag, pp. 217–47.

Repeta, D.J. and Bjørnland, T. (1997). Preparation of carotenoid standards. In *Phytoplankton Pigments in Oceanography: Guidelines to Modern Methods*, ed. S.W. Jeffrey, R.F.C. Mantoura and S.W. Wright. Paris: UNESCO Publishing, pp. 239–60.

Scheer, H. (2006). An overview of chlorophylls and bacteriochlorophylls: biochemistry, biophysics, functions and applications. In *Chlorophylls and*

Bacteriochlorophylls: Biochemistry, Biophysics, Functions and Applications, ed. B. Grimm, R.J. Porra, W. Rüdiger and H. Scheer. Dordrecht: Springer, pp. 1–26.

Takaichi, S. (1999). Carotenoids and carotenogenesis in anoxygenic photosynthetic bacteria. In *The Photochemistry of Carotenoids*, ed. H.A. Frank, A.J. Young, G. Britton and R.J. Cogdell. Dordrecht: Kluwer Academic Publishers, pp. 39–69.

Van Heukelem, L. and Thomas, C.S. (2001). Computer-assisted high-performance liquid chromatography method development with applications to the isolation and analysis of phytoplankton pigments. *J. Chromatogr.* **910**, 31–49.

Wright, S.W. and Jeffrey, S.W. (2006). Pigment markers for phytoplankton production. In *Marine Organic Matter: Biomarkers, Isotopes and DNA*, ed. J. Volkman. The Handbook of Environmental Chemistry, vol. 2, Part N. Berlin: Springer, pp. 71–104.

Zapata, M., Rodríguez, F. and Garrido, J.L. (2000). Separation of chlorophylls and carotenoids from marine phytoplankton: a new HPLC method using a reversed phase C_8 column and pyridine-containing mobile phases. *Mar. Ecol. Prog. Ser.* **195**, 29–45.

1 Chlorophylls

Bacteriochlorophyll *a*

Recommended abbreviation: BChl *a* (BCa)

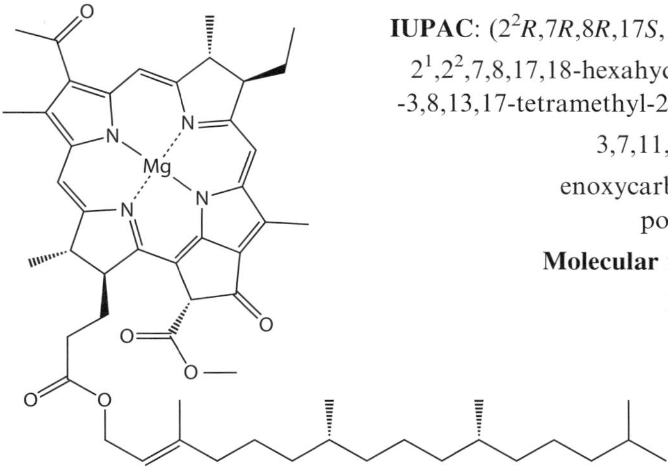

IUPAC: (2^2R,$7R$,$8R$,$17S$,$18S$)-12-Ethanoyl-7-ethyl-2^1,2^2,7,8,17,18-hexahydro-2^2-(methoxycarbonyl)-3,8,13,17-tetramethyl-2^1-oxo-18-{2-[(2*E*,7*R*,11*R*)-3,7,11,15-tetramethylhexadec-2-enoxycarbonyl]ethyl}cyclopenta[*at*]porphyrinatomagnesium(II)

Molecular formula: $C_{55}H_{74}N_4O_6Mg$

Molecular weight: 911.50

Biological occurrence	Chloroflexaceae, Chlorobiaceae and purple bacteria [141]
Source culture	*Rhodobacter sulfidophilus* (photosynthetic bacteria)
Alteration products	Various derivatives as for Chl *a* and Chl *b*
Biosynthetically related to	Chlide *a*, MgDVP
Occurs together with	See Table 3 in [120] for an overview

UV-Vis spectra

Solvent	λ_{max} (nm)	Band ratio (blue:red ratio)	Ref.
Acetone	358, 577, *770*	0.94	[140]
Diethyl ether	357, 392, 573, *770*	0.76	[140]
Ethanol	366, 607, *773*	0.94	[140]
Methanol	364, 530, 609, 695, *771*	0.79	[170]
Recommended specific absorption coefficient *d* (L g^{-1} cm^{-1})		75.9 (at 770 nm, acetone) [140]	

Mass spectra

Ionization technique	Mass analyser type	Diagnostic ions (m/z, rel. intensity)	Ref.
MALDI	TOF	910 $[M]^+ \rightarrow$ 632 $[M\text{-phytyl}]^+$	[156]

Remarks	Fluorescence: excitation 360 nm, emission 795 nm (ethanol) [40] Several other bacteriochlorophylls present in bacteria: see [141]

Chlorophyll *a*

Recommended abbreviation: Chl *a* (C*a*)

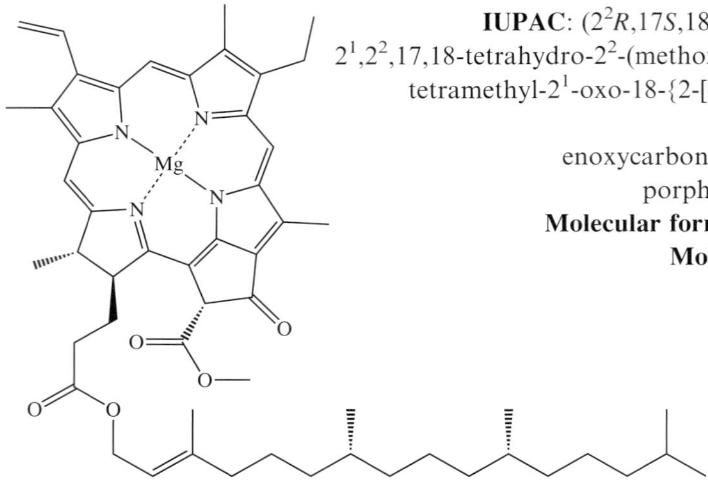

IUPAC: (2^2R,17*S*,18*S*)-12-Ethenyl-7-ethyl-2^1,2^2,17,18-tetrahydro-2^2-(methoxycarbonyl)-3,8,13,17-tetramethyl-2^1-oxo-18-{2-[(2*E*,7*R*,11*R*)-3,7,11,15-tetramethylhexadec-2-enoxycarbonyl]ethyl}cyclopenta[*at*]porphyrinatomagnesium(II)
Molecular formula: $C_{55}H_{72}N_4O_5Mg$
Molecular weight: 893.49

Biological occurrence	Characteristic pigment in all photosynthetic algae and plants
Source culture	*Chroomonas salina* (cryptophyte)
Alteration products	Chlide *a*, Pheide *a*, Phe *a*, Chl *a*′, Chl *a* allo, Pyro derivatives
Biosynthetically related to	Chl *b*, MgDVP
Occurs together with	

HPLC chromatogram of *Synechococcus* sp. (system 1)

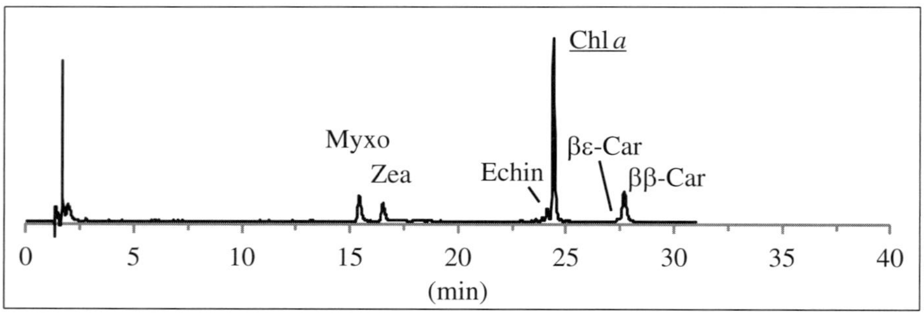

HPLC chromatogram of *Rhodomonas baltica* (system 2)

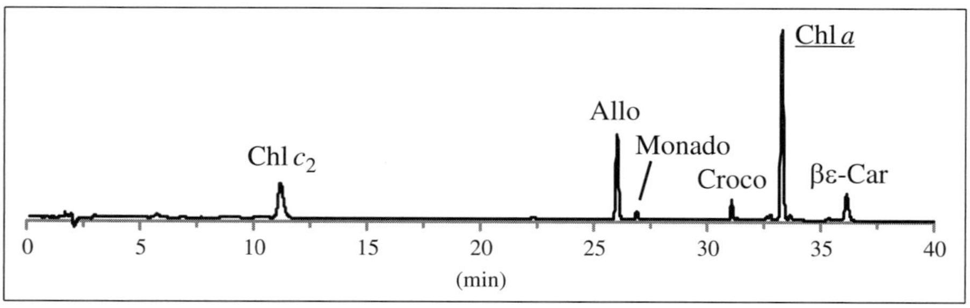

UV-Vis spectra (see also reference spectra below)

Solvent	λ_{max} (nm)	Band ratio (blue:red ratio)	Ref.
Acetone	383, 411, *430*, 534, 580, 617, 662	1.2	[145]
Diethyl ether	409, *428*, 495, 530, 575, 614, 660	1.3	[98]
95% Ethanol	414, 432, 618, 649, *664*	0.99	[121]
Methanol	432, 618, 652, *665*	0.97	[121]
Recommended specific absorption coefficient *d* (L g^{-1} cm^{-1})		129 (at 428 nm, diethyl ether) [98] 87.7 (at 664 nm, 90% acetone) [107]	

Reference spectra

In acetone

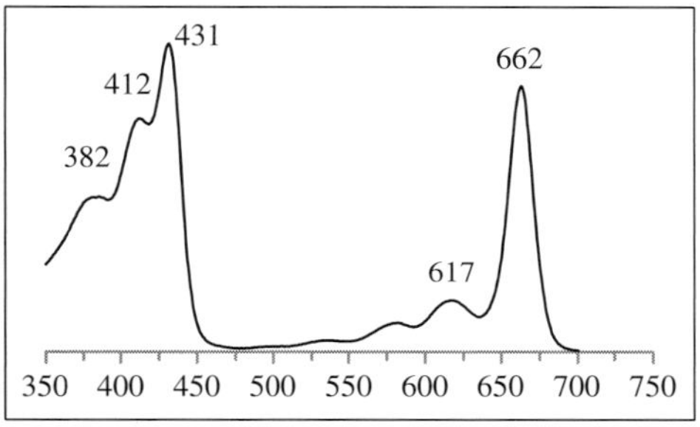

In HPLC solvent system 1

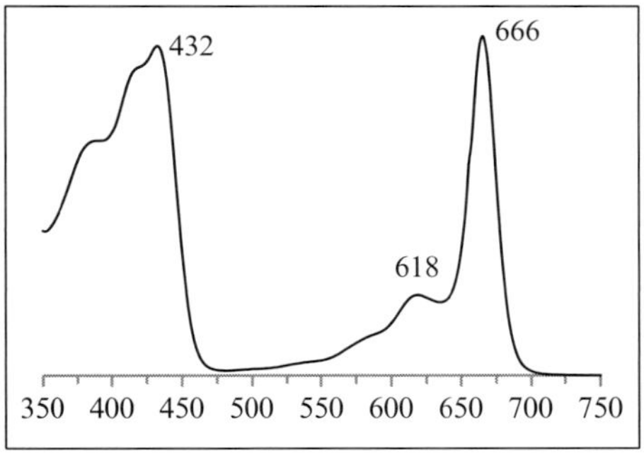

In HPLC solvent system 2

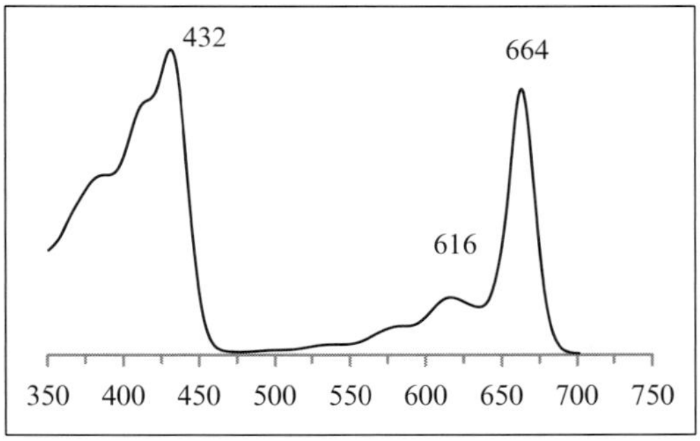

Mass spectra

Ionization technique	Mass analyser type	Diagnostic ions (m/z, rel. intensity)	Ref.
FAB	Magnetic sector	892 [M]$^+$ (25), 614 [M-phytyl]$^+$ (100), 555 [M-337]$^+$ (39), 481 [M-60–351]$^+$ (29)	[27]

Remarks Fluorescence: excitation 428 nm, emission 666 nm (diethyl ether) [26]

Chlorophyll *a* allomers

Recommended abbreviation: Chl *a* allo (C*a*al)

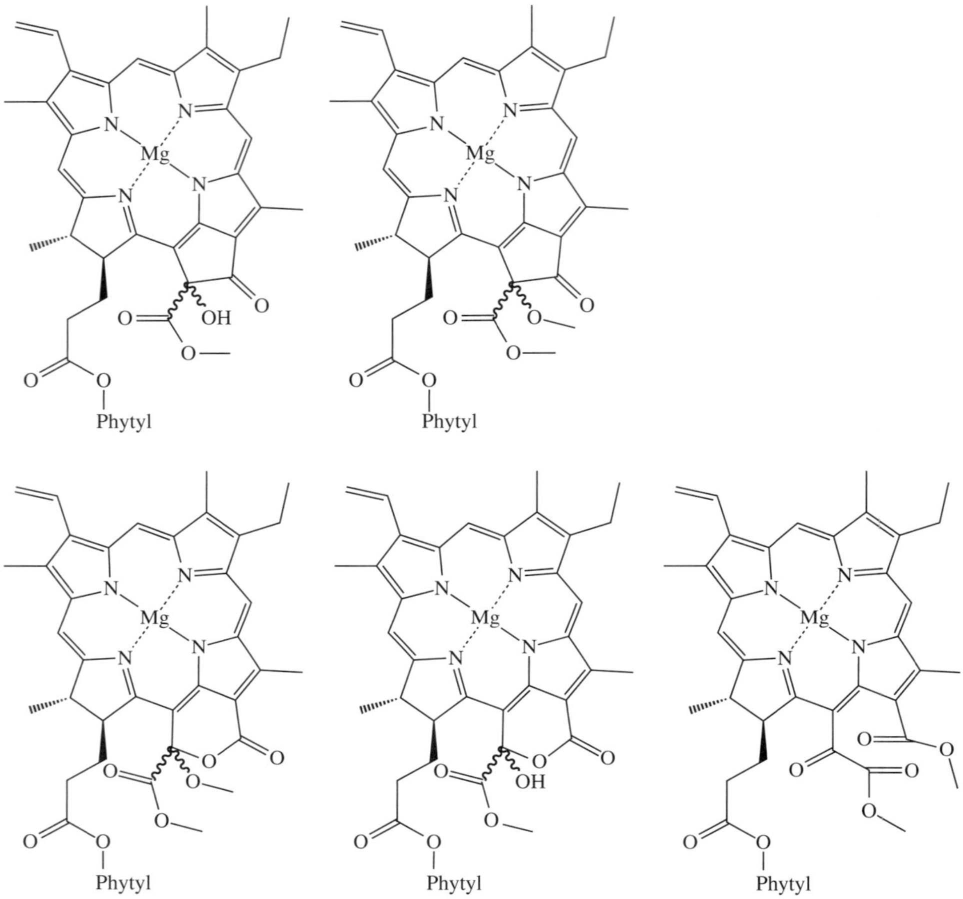

Alteration products of	Various oxidation products of Chl *a* [100, 172]
Source culture	*Chroomonas salina* (cryptomonad)
Alteration products	
Synthetically related to	Chl *a*
Occurs together with	

HPLC chromatogram of *Pavlova gyrans* (system 1)

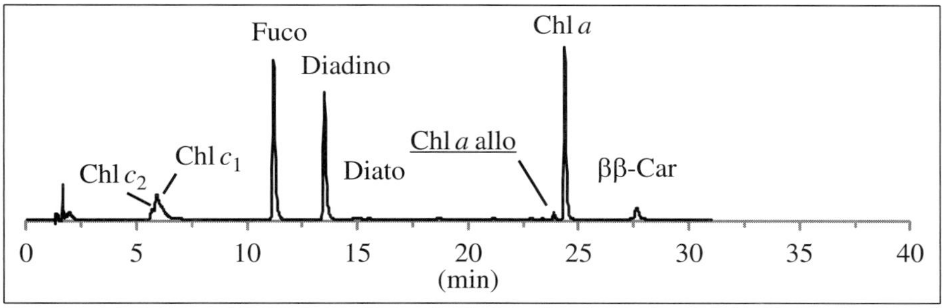

HPLC chromatogram of a coastal marine sample (system 2)

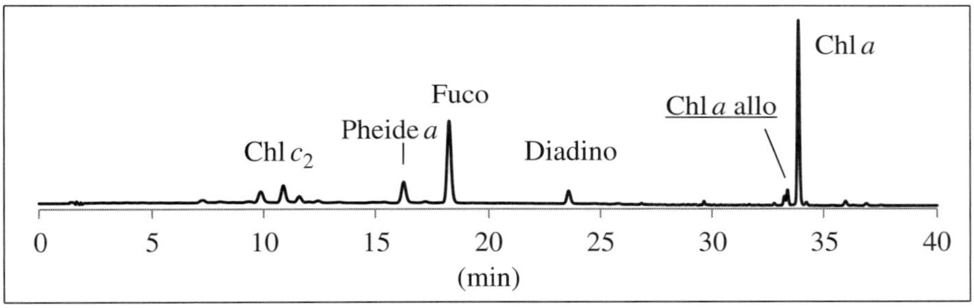

Remarks	Similar allomerization products also from other chlorophylls [100, 102, 172] *d*: assume identical to Chl *a*

Chlorophyll *a* epimer

Recommended abbreviation: Chl *a*′ (Ca′)

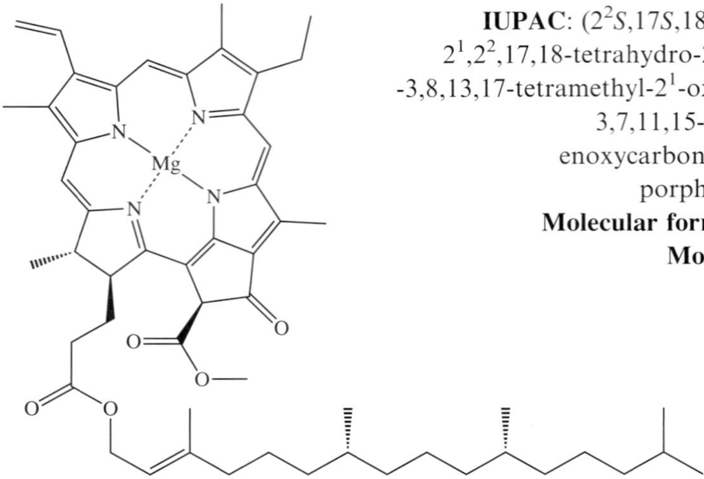

IUPAC: (2^2S,17*S*,18*S*)-12-Ethenyl-7-ethyl-2^1,2^2,17,18-tetrahydro-2^2-(methoxycarbonyl)-3,8,13,17-tetramethyl-2^1-oxo-18-{2-[(2*E*,7*R*,11*R*)-3,7,11,15-tetramethylhexadec-2-enoxycarbonyl]ethyl}cyclopenta[*at*]porphyrinatomagnesium(II)
Molecular formula: $C_{55}H_{72}N_4O_5Mg$
Molecular weight: 893.49

Alteration product of	Chlorophyll *a*; occurs in slightly acidic or basic extracts using polar solvents
Source culture	*Chroomonas salina* (cryptomonad)
Alteration products	Chl *a*, Phe *a*′, Chl *a*′ allo, Pphe *a*
(Bio)synthetically related to	Chl *a*
Occurs together with	

HPLC chromatogram of *Pavlova lutheri* (system 1)

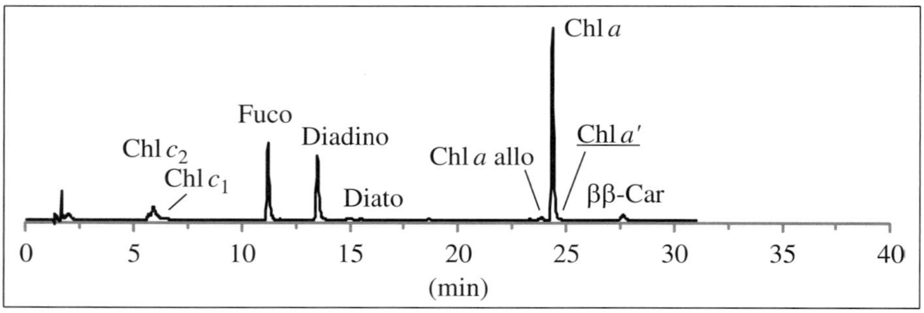

UV-Vis spectra and fluorescence spectra

For all practical purposes, identical with the parent chlorophyll

Recommended specific absorption coefficient *d* ($L\ g^{-1}\ cm^{-1}$)	Assume identical to Chl *a*

Chlorophyll *b*

Recommended abbreviation: Chl *b* (C*b*)

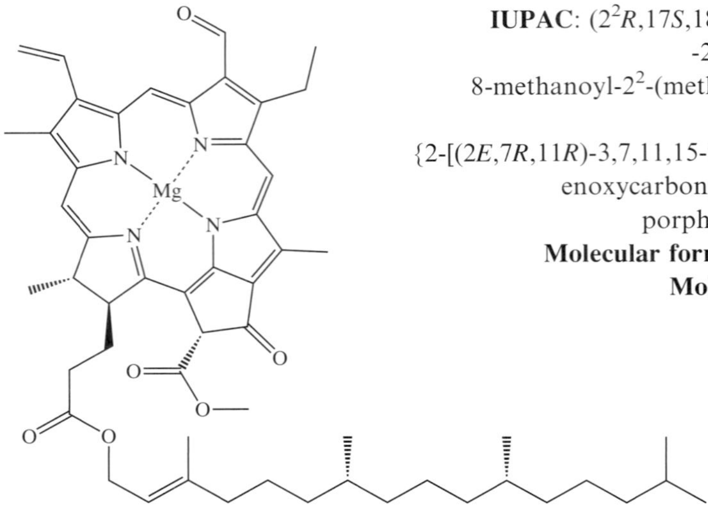

IUPAC: (2^2R,17*S*,18*S*)-12-Ethenyl-7-ethyl-2^1,2^2,17,18-tetrahydro-8-methanoyl-2^2-(methoxycarbonyl)-3,13,17-trimethyl-2^1-oxo-18-{2-[(2*E*,7*R*,11*R*)-3,7,11,15-tetramethylhexadec-2-enoxycarbonyl]ethyl}cyclopenta[*at*]porphyrinatomagnesium(II)

Molecular formula: $C_{55}H_{70}N_4O_6Mg$

Molecular weight: 907.47

Biological occurrence	Dominant pigment in all classes of green algae, in higher plants and some prochlorophytes (Cyano-3, Chapter 1)
Source culture	*Dunaliella tertiolecta* (chlorophyte), *Pycnococcus provasolii* (prasinophyte)
Alteration products	Chlide *b*, Pheide *b*, Phe *b*, Chl *b'*, Chl *b* allo and pyro deriv.
Biosynthetically related to	Chl *a*, MgDVP
Occurs together with	Lut, Zea, Viola, *c*-Neo

HPLC chromatogram of *Dunaliella tertiolecta* (system 1)

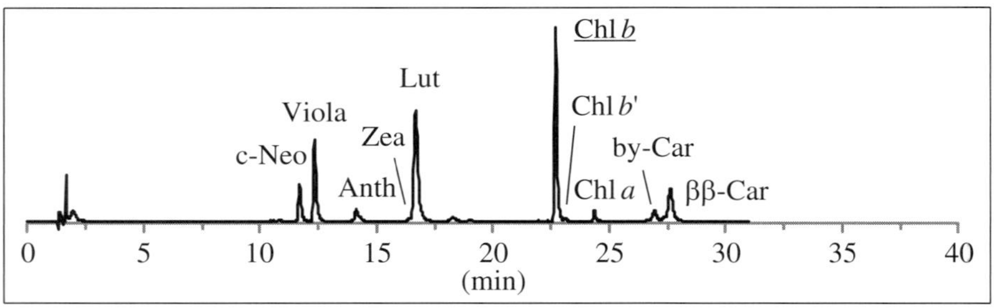

HPLC chromatogram of *Codium tomentosum* (system 2)

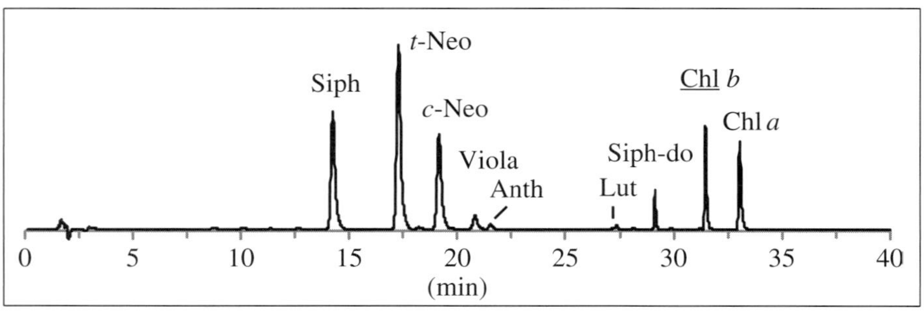

UV-Vis spectra (see also reference spectra below)

Solvent	λ_{max} (nm)	Band ratio (blue:red ratio)	Ref.
Acetone	*457*, 597, 646	2.8	[145]
Diethyl ether	428, *453*, 593, 642	2.8	[98]
95% Ethanol	*464*, 601, 649	2.6	[121]
Methanol	*469*, 603, 652	2.7	[121]
Recommended specific absorption coefficient *d* ($L\ g^{-1}\ cm^{-1}$)		176 (at 453 nm, diethyl ether) [98] 51.4 (at 647 nm. 90% acetone) [107]	

Reference spectra

For spectrum in acetone, see [109]

In 90% acetone

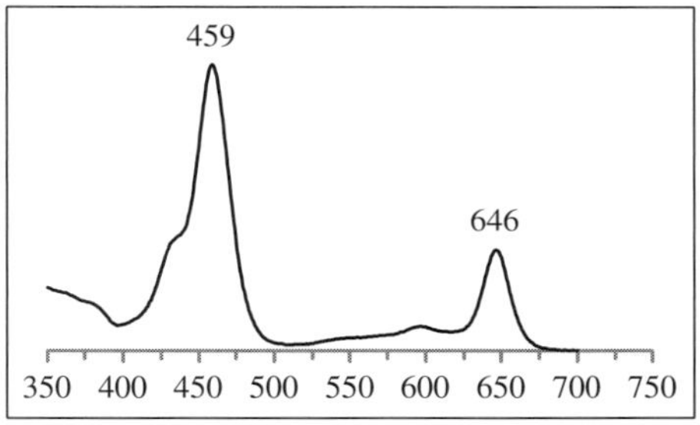

In HPLC solvent system 1

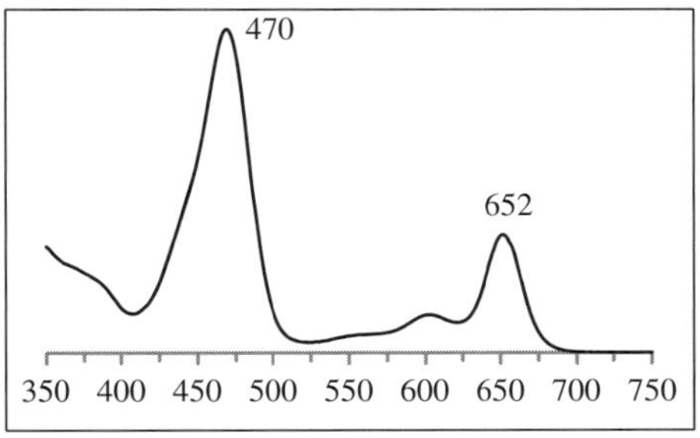

In HPLC solvent system 2

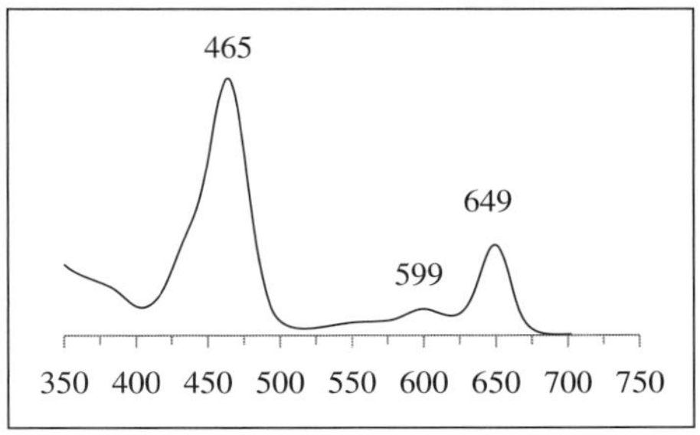

Mass spectra

Ionization technique	Mass analyser type	Diagnostic ions (m/z, rel. intensity)	Ref.
FAB	Magnetic sector	906 $[M]^+$ (67), 628 $[M\text{-phytyl}]^+$ (100), 569 $[M\text{-}337]^+$ (30), 495 $[M\text{-}60\text{–}351]^+$ (39)	[27]

Remarks Fluorescence: excitation 453 nm, emission 646 nm (diethyl ether) [26]

Chlorophyll *b* epimer

Recommended abbreviation: Chl *b*′ (C*b*′)

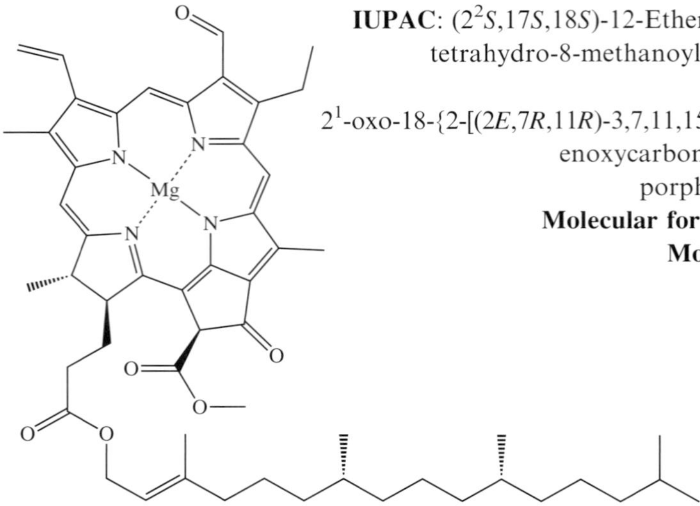

IUPAC: (2^2S,17*S*,18*S*)-12-Ethenyl-7-ethyl-2^1,2^2,17,18-tetrahydro-8-methanoyl-2^2-(methoxycarbonyl)-3,13,17-trimethyl-2^1-oxo-18-{2-[(2*E*,7*R*,11*R*)-3,7,11,15-tetramethylhexadec-2-enoxycarbonyl]ethyl}cyclopenta[*at*]porphyrinatomagnesium(II)

Molecular formula: $C_{55}H_{70}N_4O_6Mg$

Molecular weight: 907.47

Alteration product of	Chlorophyll *b*; occurs in slightly acidic or basic extracts using polar solvents
Source culture	*Dunaliella tertiolecta* (chlorophyte), *Pycnococcus provasolii* (prasinophyte)
Alteration products	Chl *b*, Phe *b*′, Chl *b*′ allo, Pphe *b*
(Bio)synthetically related to	Chl *b*
Occurs together with	

UV-Vis spectra and fluorescence spectra

For all practical purposes, identical with the parent chlorophyll

Recommended specific absorption coefficient *d* (L g^{-1} cm^{-1})	Assume identical to Chl *b*

Chlorophyll c_1

Recommended abbreviation: Chl c_1 (Cc_1)

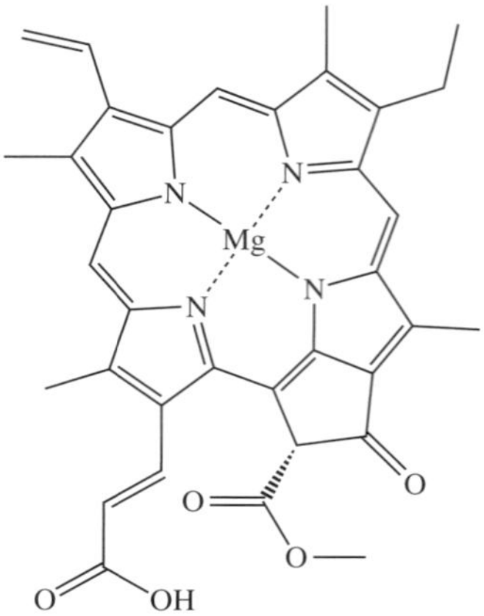

IUPAC: (2^2R,18^1E)-18-(3-Carboxyethenyl)-12-ethenyl-7-ethyl-2^1,2^2-dihydro-2^2-(methoxycarbonyl)-3,8,13,17-tetramethyl-2^1-oxocyclopenta[*at*]porphyrinatomagnesium(II)

Molecular formula: $C_{35}H_{30}N_4O_5Mg$
Molecular weight: 610.94

Biological occurrence	Minor pigment in most types of algae within the 'red algal lineage' (see Chapter 1, this volume)
Source culture	*Mallomonas papillosa* (synurophyte)
Alteration products	Chl c_1', corresponding pheophorbide and pyro-derivatives
Biosynthetically related to	Chl c_2, MVChl c_3
Occurs together with	Chl c_2, Fuco

HPLC chromatogram of *Pavlova gyrans* (system 1)

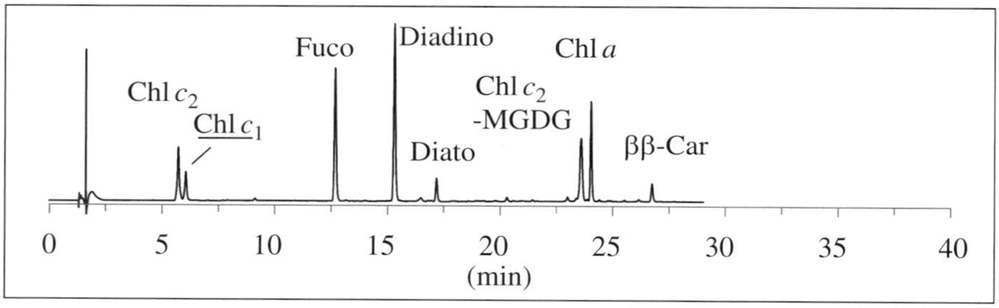

HPLC chromatogram of *Isochrysis galbana* (system 2)

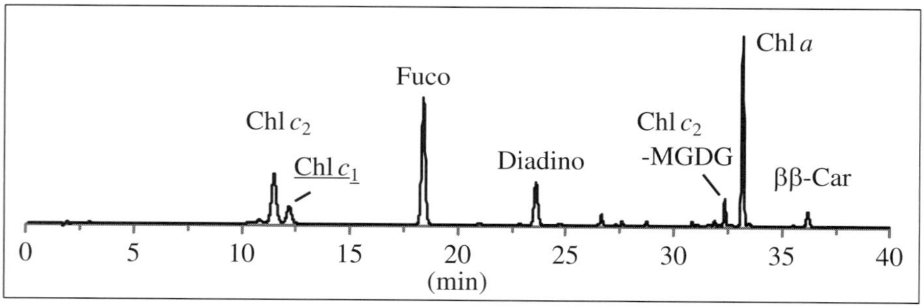

UV-Vis spectra (see also reference spectra below)

Solvent	λ_{max} **(nm)**	**Band ratio (blue:red ratio)**	**Ref.**
Acetone	*446*, 578, 629	9	[105]
Diethyl ether	*444*, 577, 626	9	[58]
Methanol	*445*, 584, 634	7	[104]
Recommended specific absorption coefficient *d* (L g^{-1} cm^{-1})		346 (at 462 nm in pyridine) [105] 318 (at 443 nm, 90% acetone + 1% pyridine) [105]	

Reference spectra
In acetone

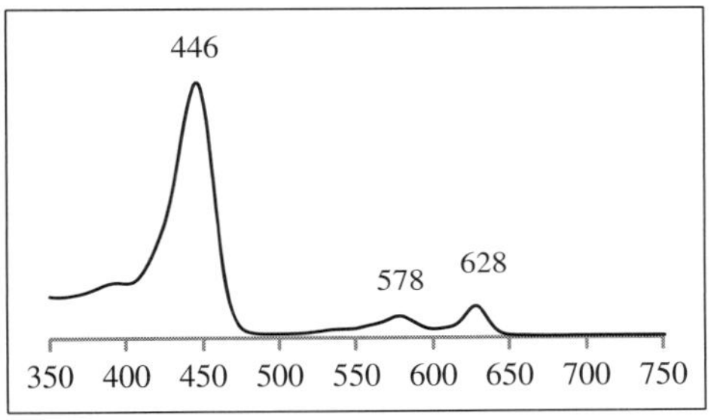

For spectrum in diethyl ether, see [109]

In HPLC solvent system 1

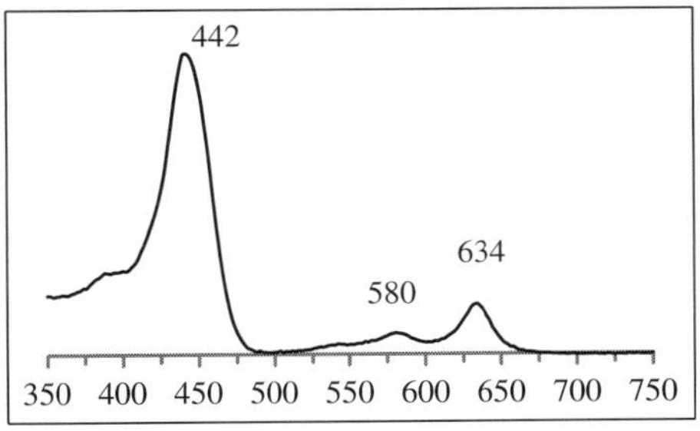

In HPLC solvent system 2

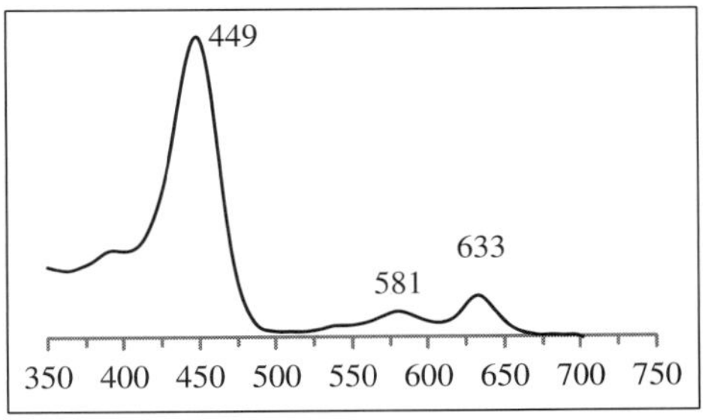

Mass spectra

Ionization technique	**Mass analyser type**	**Diagnostic ions (m/z, rel. intensity)**	**Ref.**
ESI$^+$	Ion trap	611 [M+H]$^+$; 611 → 593 [M+H-18]$^+$ (100), 567 [M+H-44]$^+$ (20), 551 [M+H-60]$^+$ (63), 534 [M+H-77]$^+$ (15)	[76]

Remarks Fluorescence: excitation 450 nm, emission peaks at 633, 694 nm (acetone) [105]

Chlorophyll c_2

Recommended abbreviation: Chl c_2 (Cc_2)

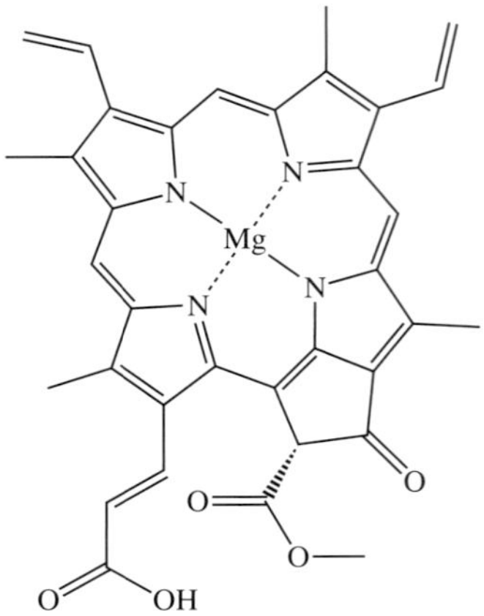

IUPAC: (2^2R,18^1E)-18-(3-Carboxyethenyl)-7,12-diethenyl-2^1,2^2-dihydro-2^2-(methoxycarbonyl)-3,8,13,17-tetramethyl-2^1-oxocyclopenta[*at*]porphyrinatomagnesium(II)

Molecular formula: $C_{35}H_{28}N_4O_5Mg$
Molecular weight: 608.93

Biological occurrence	Minor pigment in most types of algae within the 'red algal lineage' (see Chapter 1, this volume)
Source culture	*Amphidinium carterae* (dinoflagellate)
Alteration products	Chl c_2', corresponding pheophorbide and pyro-derivatives
Biosynthetically related to	Chl c_1, Chl c_3
Occurs together with	Fuco and with Chl c_1 in many cases (see Chapter 1)

HPLC chromatogram of *Gymnodinium catenatum* (system 1)

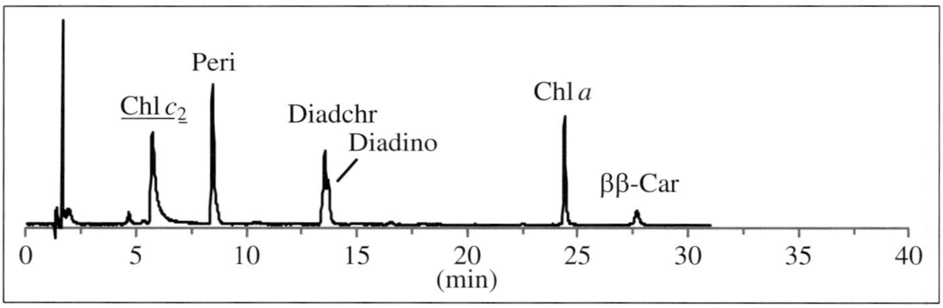

HPLC chromatogram of *Rhodomonas baltica* (system 2)

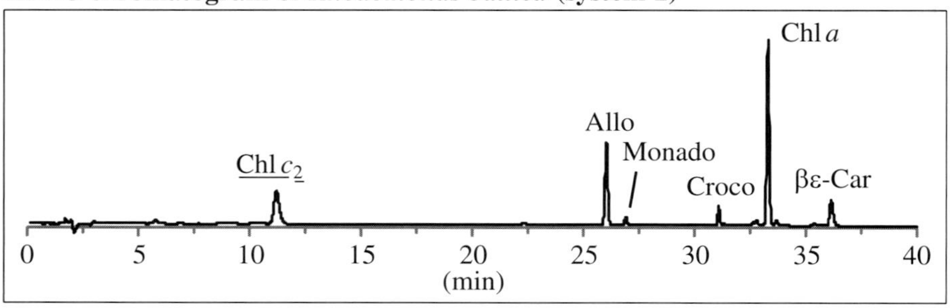

UV-Vis spectra (see also reference spectra below)

Solvent	λ_{max} (nm)	Band ratio (blue:red ratio)	Ref.
Acetone	*450*, 581, 629	11	[57]
Diethyl ether	*449*, 582, 629	14	[104]
Methanol	*452*, 587, 635	10	[104]
Recommended specific absorption coefficient *d* ($L\ g^{-1}\ cm^{-1}$)		459 (at 466 nm in pyridine) [105] 374 (at 444 nm in 90% acetone + 1% pyr.) [105]	

Reference spectra

In acetone

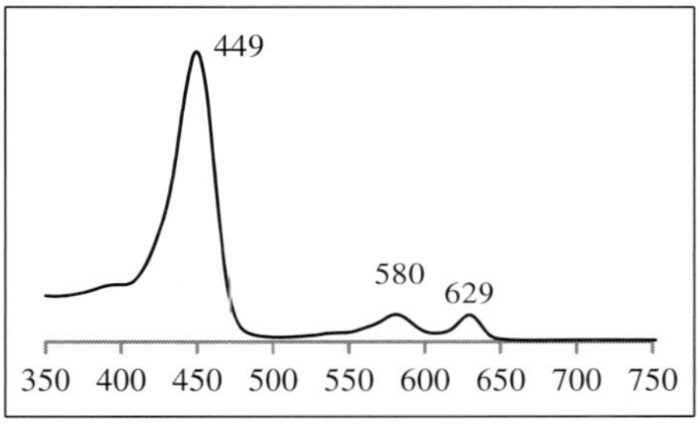

In 90% acetone + 1% pyridine

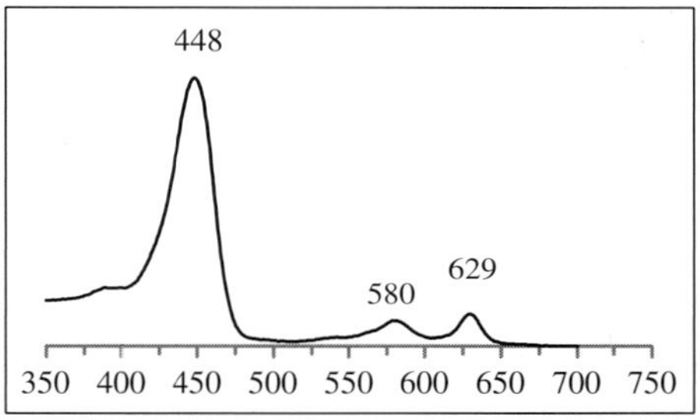

In HPLC solvent system 1

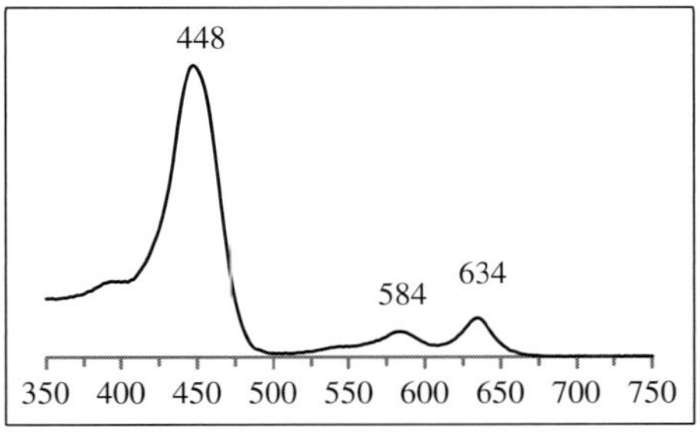

In HPLC solvent system 2

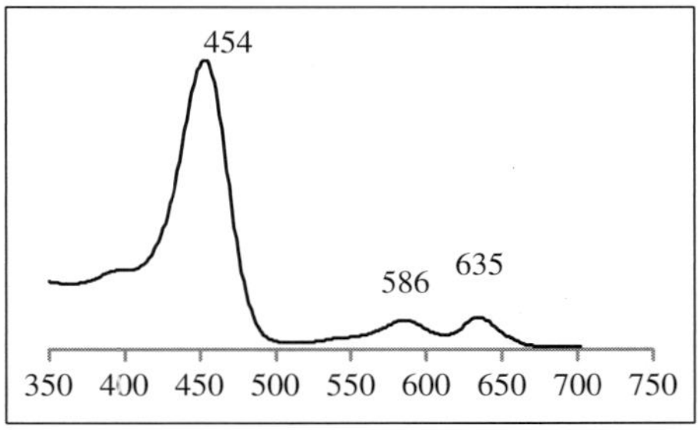

Mass spectra

Ionization technique	Mass analyser type	Diagnostic ions (m/z, rel. intensity)	Ref.
ESI^+	Ion trap	609 $[M+H]^+$; 609 → 591 $[M+H\text{-}18]^+$ (33), 565 $[M+H\text{-}44]^+$ (69), 564 $[M+H\text{-}45]^+$ (100), 549 $[M+H\text{-}60]^+$ (53), 532 $[M+H\text{-}77]^+$ (26)	[76]

Remarks Fluorescence: excitation 453 nm, emission 635, 696 nm (acetone) [105]
Several other Chl *c* pigments of unknown structure: see [179]

Chlorophyll c_2-monogalactosyldiacylglyceride ester [18:4/14:0]

Recommended abbreviation: **Chl c_2-MGDG (Cc_2M)**

IUPAC: unknown
Molecular formula: $C_{76}H_{96}N_4O_{14}Mg$
Molecular weight: 1313.90

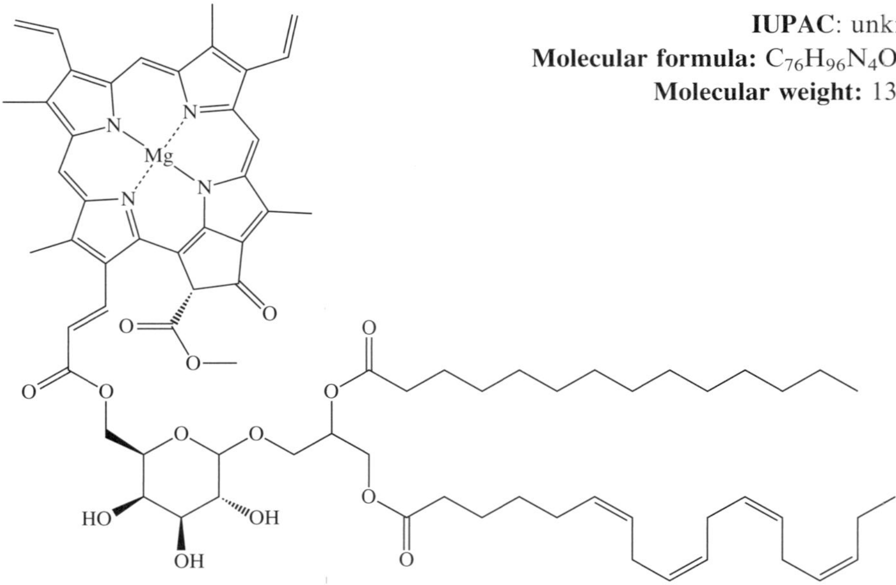

Biological occurrence	Dominant pigment in some haptophytes (see Chapter 1)
Source culture	*Emiliania huxleyi* (coccolithophyte)
Alteration products	Not known, corresponding phytoporphyrins expected
Biosynthetically related to	Chl c_2
Occurs together with	Chl c_2, Fuco, Hex-fuco, Hex-kfuco

HPLC chromatogram of *Chrysochromulina camella* (system 1)

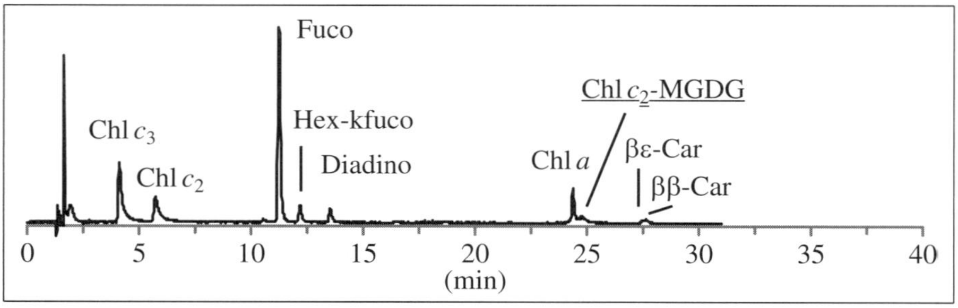

HPLC chromatogram of *Emiliania huxleyi* (system 2)

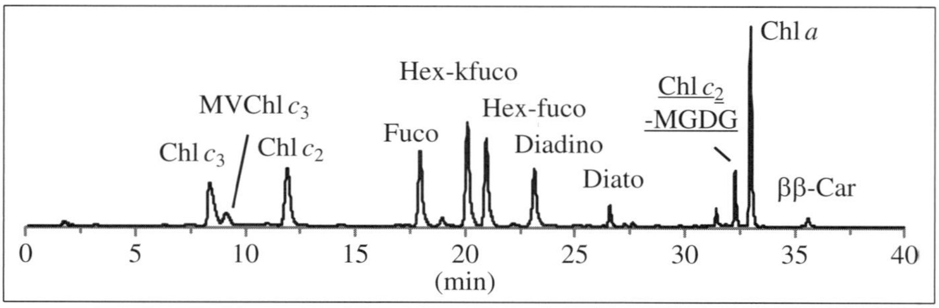

UV-Vis spectra (see also reference spectra below)

Solvent	λ_{max} (nm)	Band ratio (blue:red ratio)	Ref.
Acetone	*453*, 582, 631	10	[70]
Diethyl ether	*452*, 581, 629	13	[73]
Recommended specific absorption coefficient d (L g^{-1} cm^{-1})		n.d, see Remarks	

Reference Spectra

In HPLC solvent system 2

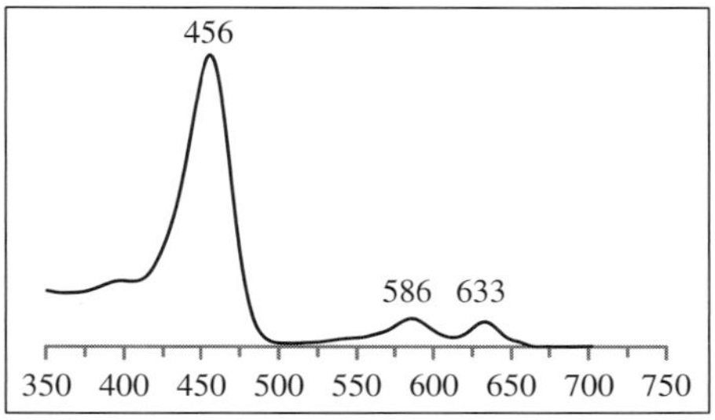

Mass spectra

Ionization technique	Mass analyser type	Diagnostic ions (m/z, rel. intensity)	Ref.
FAB	Double quadrupole	1312, 1313 [M, M+1]$^+$ → 1085, 1038, 809, 752, 693, 609, 591, 563, 549, 533, 519, 504, 477	[70]

Remarks Fluorescence: excitation 452 nm, emission 634, 695 nm (diethyl ether) [73]
$d = 213$ L g^{-1} cm^{-1} (at 453 nm in acetone; calc. from Chl c_2) is recommended, as no value has been determined for Chl c_2-MGDG [105]. Other fatty acid esters of Chl c_2-MGDG have been reported [179]. Structure not proven by NMR

Chlorophyll c_3

Recommended abbreviation: Chl c_3 (Cc_3)

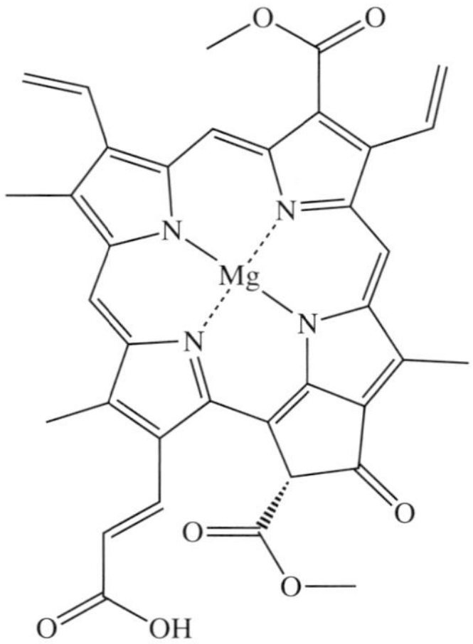

IUPAC: (2^2R,18^1E)-18-(3-Carboxyethenyl)-7,12-diethenyl-2^1,2^2-dihydro-2^2,8-di(methoxycarbonyl)-3,13,17-trimethyl-2^1-oxocyclopenta[*at*]porphyrinatomagnesium(II)

Molecular formula: $C_{36}H_{28}N_4O_7Mg$
Molecular weight: 652.94

Biological occurrence	Dominant pigment in bolidophytes, many diatoms and haptophytes and some dinoflagellates. Occasional in dictyochophytes and pelagophytes
Source culture	*Ochrosphaera neopolitana* [180], strain CS-285 [8]
Alteration products	Chl c_3', corresponding pheophorbide and pyro-derivatives
Biosynthetically related to	Chl c_2, MVChl c_3
Occurs together with	Chl c_2, Diadino, Fuco

HPLC chromatogram of *Karlodinium micrum* (system 1)

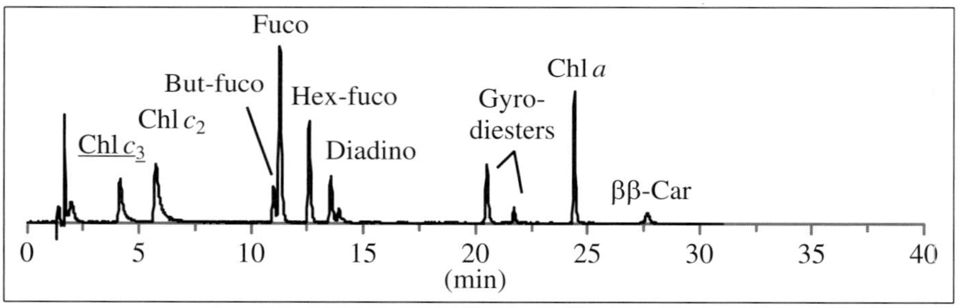

HPLC chromatogram of *Emiliania huxleyi* (system 2)

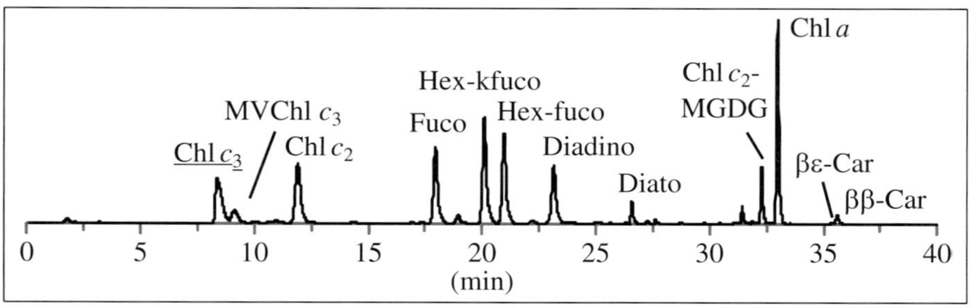

UV-Vis spectra (see also reference spectra below)

Solvent	λ_{max} (nm)	Band ratio (blue:red ratio)	Ref.
Acetone	*452*, 586, 626	28	[145]
Diethyl ether	*451*, 585, 626	32	[110]
Recommended specific absorption coefficient *d* (L g^{-1} cm^{-1})		n.d., see Remarks	

Reference spectra

In acetone

For spectrum in acetone, see [109]

In HPLC solvent system 1

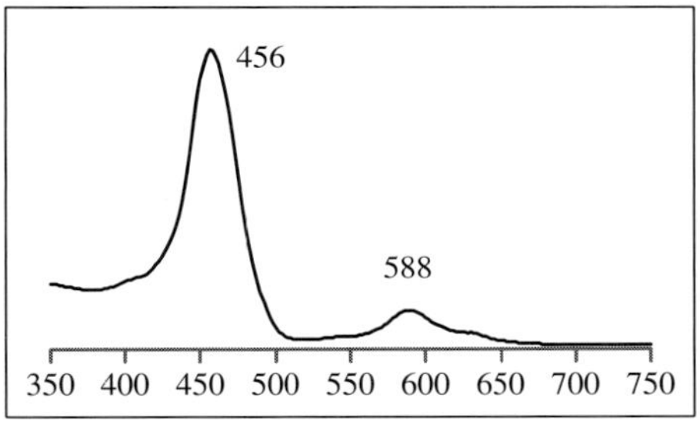

In HPLC solvent system 2

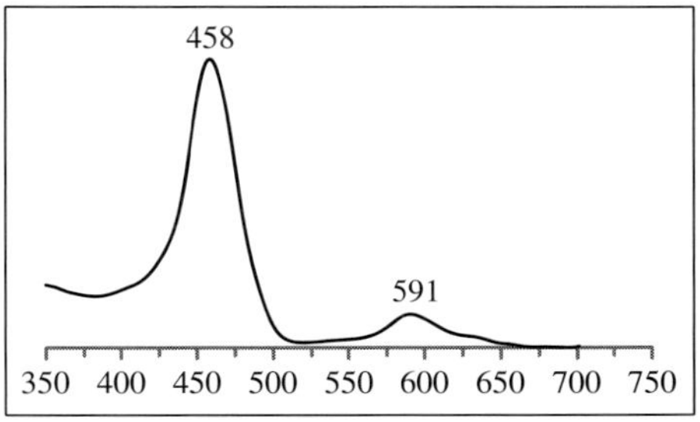

Mass spectra

Ionization technique	Mass analyser type	Diagnostic ions (m/z, rel. intensity)	Ref.
ESI$^+$	Ion trap	653 [M+H]$^+$; 653 → 635 (100), 621 (92), 609 (58), 593 (55)	[76]

Remarks Fluorescence: excitation 452 nm, emission 635, 690 nm (acetone) [145]

$d = 590$ L g^{-1}cm^{-1} (at λ_{max} in pyridine) is recommended, as no *d* has been determined for Chl c_3. In 90% acetone + 1% pyridine, use $d =$ 346 L g^{-1}cm^{-1} (at 453 nm) (mean of the values for Chl c_1 and c_2 at Soret max.) [109]

Chlorophyll *d*

Recommended abbreviation: Chl *d* (C*d*)

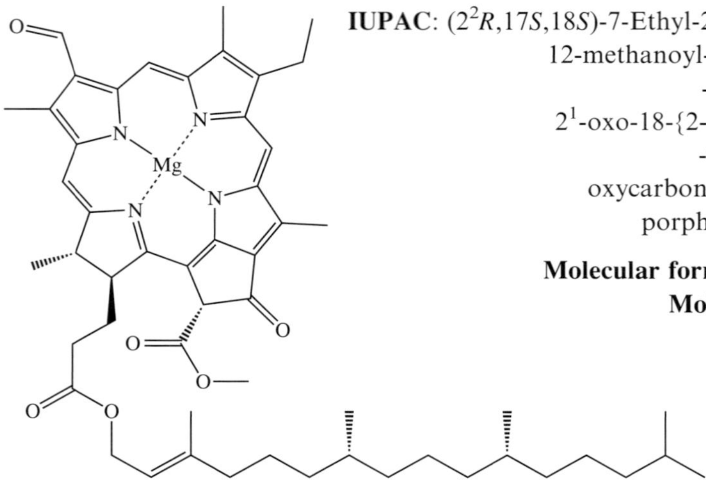

IUPAC: (2^2R,17*S*,18*S*)-7-Ethyl-2^1,2^2,17,18-tetrahydro-12-methanoyl-2^2-(methoxycarbonyl)-3,8,13,17-tetramethyl-2^1-oxo-18-{2-[(2*E*,7*R*,11*R*)-3,7,11,15-tetramethylhexadec-2-oxycarbonyl]ethyl}cyclopenta[*at*]porphyrinatomagnesium(II)

Molecular formula: $C_{54}H_{70}O_6N_4Mg$
Molecular weight: 895.46

Biological occurrence	Major chlorophyll in a few prochlorophytes (Cyano-5, Chapter 1)
Source culture	*Acaryochloris marina* (prochlorophyte/cyanobacteria)
Alteration products	Chlide *d*, Pheide *d*, Phe *d*, Chl *d*′, Chl *d* allo, Pphe *d*
Biosynthetically related to	Chl *a*, MgDVP
Occurs together with	MgDVP, βε-Car, Zea

UV-Vis spectra

Solvent	λ_{max} (nm)	Band ratio (blue:red ratio)	Ref.
Acetone	394, 445, 600, 660, *693*	0.99	[111]
Diethyl ether	392, 447, 595, 643, *688*	0.89	[147]
Methanol	400, 455, 697	n.d.	[129]
Recommended specific absorption coefficient *d* (L g^{-1} cm^{-1})		118 (at 687 nm in diethyl ether)	[97]

Mass spectra

Ionization technique	Mass analyser type	Diagnostic ions (m/z, rel. intensity)	Ref.
MALDI	TOF	894 $[M]^+$, 616; 616 → 557, 543, 530, 515, 494, 483	[161]

Remarks Fluorescence: excitation 696, emission ~752 nm (diethyl ether) [147]

Chlorophyllide *a*

Recommended abbreviation: Chlide *a* (Cd*a*)

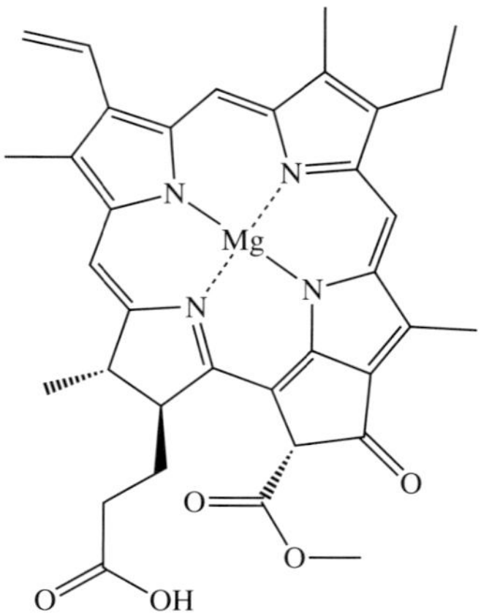

IUPAC: (2^2R,17*S*,18*S*)-18-(2-Carboxyethyl)-12-ethenyl-7-ethyl-2^1,2^2,17,18-tetrahydro-2^2-(methoxycarbonyl)-3,8,13,17-tetramethyl-2^1-oxocyclopenta[*at*]porphyrinatomagnesium(II)

Molecular formula: $C_{35}H_{34}N_4O_5Mg$
Molecular weight: 614.97

Alteration product of	Chlorophyll *a*; occurs in senescent algae, damaged diatoms, zooplankton faecal pellets. Extraction artefact for algae with highly active chlorophyllase enzyme
Source culture	*Chroomonas salina* (cryptophyte)
Alteration products	Pheide *a*
(Bio)synthetically related to	Chl *a*, Pheide *a*
Occurs together with	

HPLC chromatogram of *Guillardia theta* (system 1)

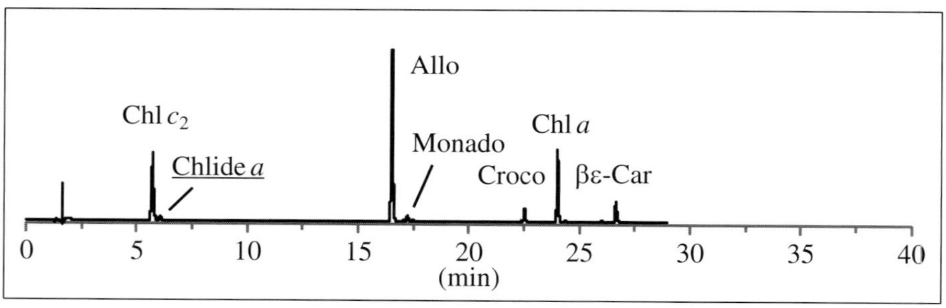

HPLC chromatogram of *Micromonas pusilla* (system 2)

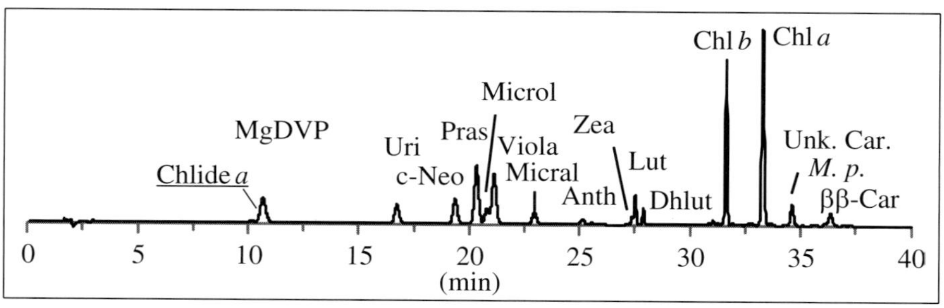

UV-Vis spectra (see also reference spectra below)

Solvent	λ_{max} (nm)	Band ratio (blue:red ratio)	Ref.
Acetone	412, *431*, 580, 617, 664	1.1	[108]
Diethyl ether	*429*, 661	n.d.	[149]
Recommended specific absorption coefficient *d* (L g^{-1} cm^{-1})		n.d., see Remarks	

Reference spectra

For spectrum in acetone, see [109]

In HPLC solvent system 1

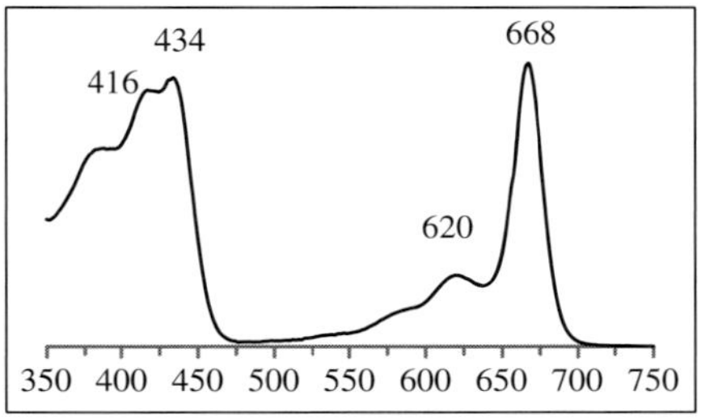

In HPLC solvent system 2

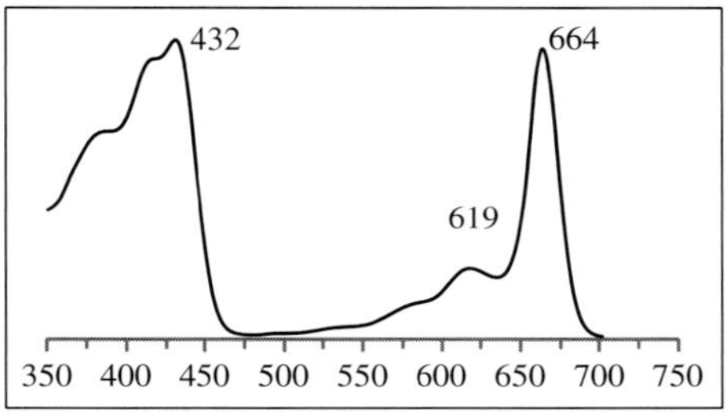

Mass spectra

Ionization technique	Mass analyser type	Diagnostic ions (m/z, rel. intensity)	Ref.
FAB	Magnetic sector	614 $[M]^+$	[27]

Remarks

Fluorescence data: excitation 426 nm, emission 667 nm (acetone) [145]

$d = 187\,L\,g^{-1}cm^{-1}$ (at λ_{max} in diethyl ether; calc. from Chl *a*) is recommended, as no *d* has been determined for Chlide *a*. In 90% acetone, use $d = 127\,L\,g^{-1}cm^{-1}$ (at 664 nm) [124, 109]

Chlorophyllide *b*

Recommended abbreviation: Chlide *b* (Cd*b*)

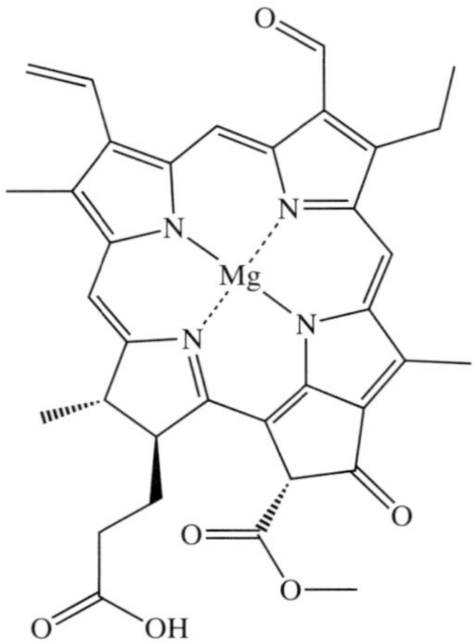

IUPAC: (2^2R,17*S*,18*S*)-18-(2-Carboxyethyl)-12-ethenyl-7-ethyl-2^1,2^2,17,18-tetrahydro-8-methanoyl-2^2-(methoxycarbonyl)-3,13,17-trimethyl-2^1-oxocyclopenta[*at*]porphyrinatomagnesium(II)

Molecular formula: $C_{35}H_{32}N_4O_6Mg$
Molecular weight: 628.96

Alteration product of	Chlorophyll *b*; occurs in senescent algae, zooplankton fecal pellets
Source culture	*Dunaliella tertiolecta* (chlorophyte), *Pycnococcus provasolii* (prasinophyte)
Alteration products	Pheide *b*
(Bio)synthetically related to	Chl *b*, Pheide *b*
Occurs together with	Chlide *a*

HPLC chromatogram (system 1)

NO DATA AVAILABLE

HPLC chromatogram of *Dunaliella tertiolecta* (system 2)

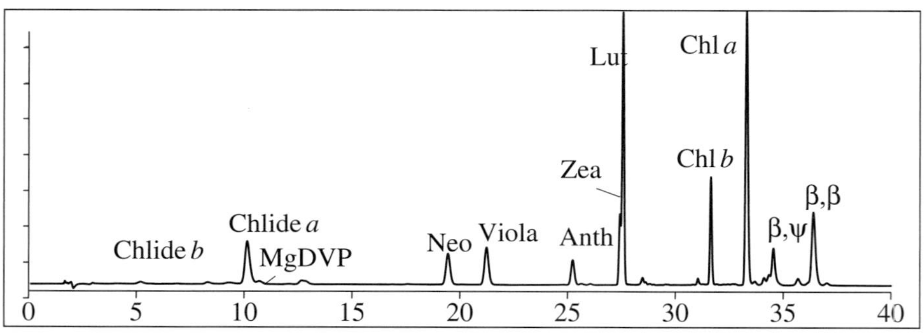

UV-Vis spectra (see also reference spectra below)

Solvent	**λ_{max} (nm)**	**Band ratio (blue:red ratio)**	**Ref.**
Acetone	*458*, 596, 646	2.9	[145]
Diethyl ether	*451*, 642	n.d.	[149]
Recommended specific absorption coefficient *d* (L g^{-1} cm^{-1})		n.d., see Remarks	

Reference spectra

For spectrum in acetone, see [109]

In HPLC solvent system 1

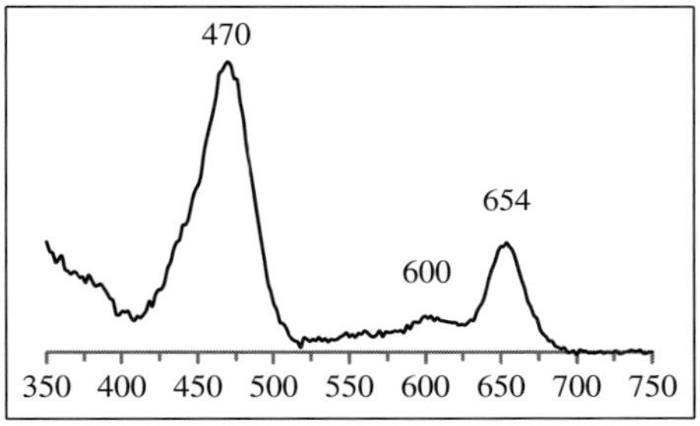

In HPLC solvent system 2

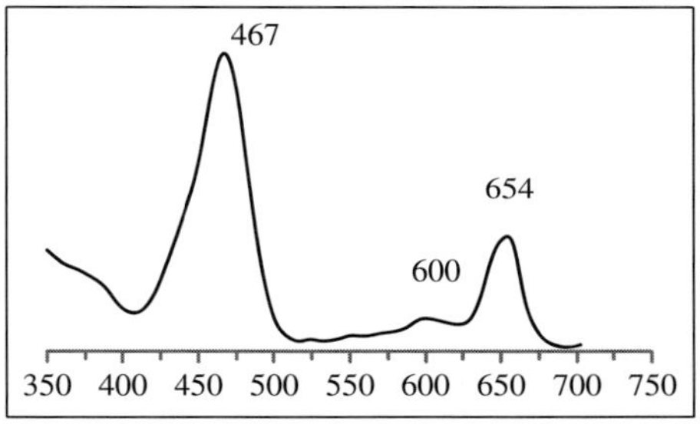

Mass spectra

Ionization technique	**Mass analyser type**	**Diagnostic ions (m/z, rel. intensity)**	**Ref.**
FAB	Magnetic sector	628 [M]$^+$	[27]

Remarks Fluorescence data: excitation 454 nm, emission 652 nm (acetone) [145]
$d = 254$ L g^{-1} cm^{-1} (at λ_{max} in diethyl ether; calc. from Chl *b*) is recommended, as no value has been determined for Chlide *b*. For 90% acetone, use $d = 74.1$ L g^{-1} cm^{-1} (at 645 nm) [124, 109]

Divinyl chlorophyll *a*

Recommended abbreviation: DVChl *a* (DC*a*)

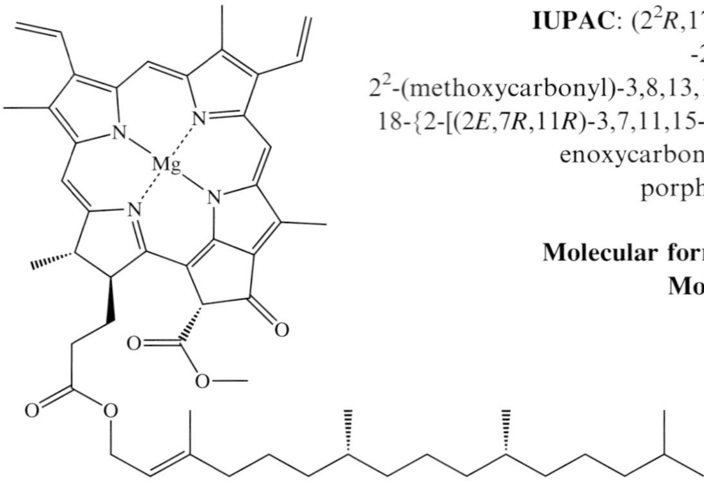

IUPAC: (2^2R,17*S*,18*S*)-7,12-Diethenyl-2^1,2^2,17,18-tetrahydro-2^2-(methoxycarbonyl)-3,8,13,17-tetramethyl-2^1-oxo-18-{2-[(2*E*,7*R*,11*R*)-3,7,11,15-tetramethylhexadec-2-enoxycarbonyl]ethyl}cyclopenta[*at*]porphyrinatomagnesium(II)

Molecular formula: $C_{55}H_{70}N_4O_5Mg$
Molecular weight: 891.47

Biological occurrence	Dominant chlorophyll in *Prochlorococcus* (prochlorophyte/cyanobacteria) (Cyano-4, Chapter 1)
Source culture	*Prochlorococcus marinus* (prochlorophyte/cyanobacteria)
Alteration products	DVChlide *a*, DVPheide *a*, DVPhe *a*, DVChl *a*′, DVChl *a* allo, DVPphe *a* probable
Biosynthetically related to	DVChl *b*, MgDVP
Occurs together with	DVChl *b*, MgDVP, βε-Car, Zea

HPLC chromatogram of *Prochlorococcus* sp. (system 1)

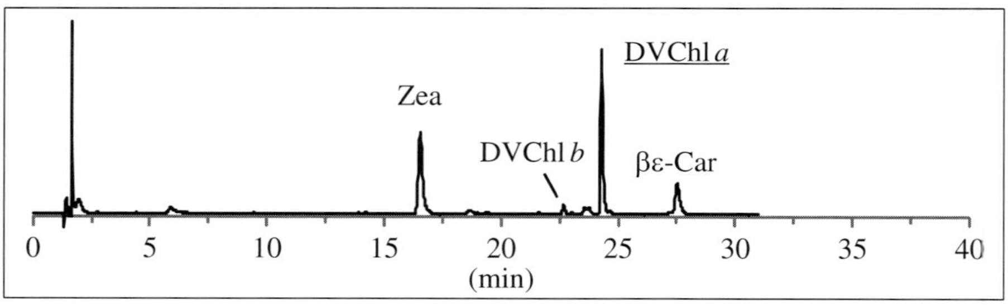

HPLC chromatogram of *Prochlorococcus* sp. (system 2)

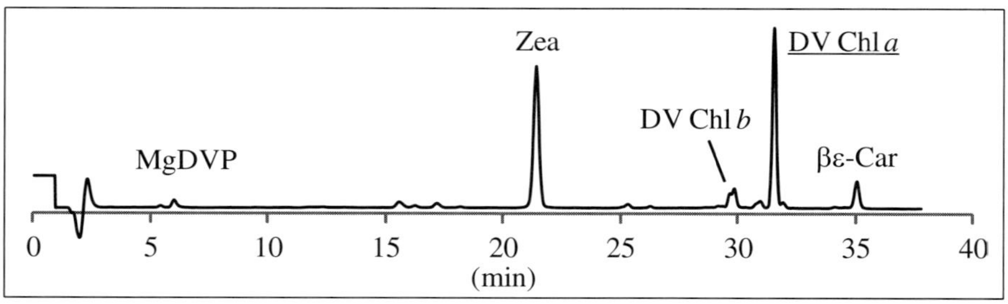

UV-Vis spectra (see also reference spectra below)

Solvent	λ_{max} (nm)	Band ratio (blue:red ratio)	Ref.
Diethyl ether	*436*, ≈615, 661	1.3	[74]
Recommended specific absorption coefficient *d* ($L\ g^{-1}\ cm^{-1}$)		n.d., see Remarks	

Reference spectra

For spectrum in acetone, see [109]

In HPLC solvent system 1

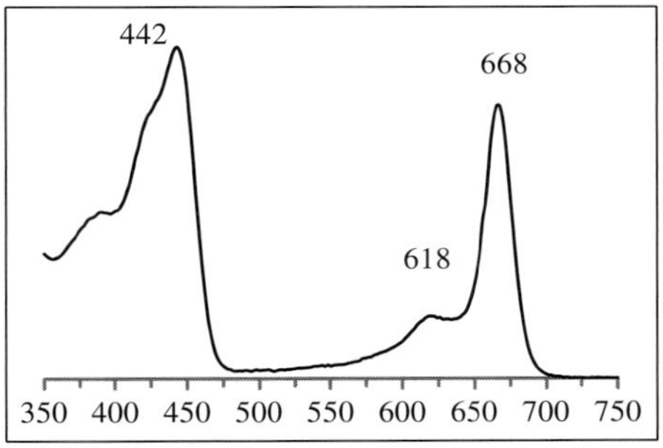

In HPLC solvent system 2

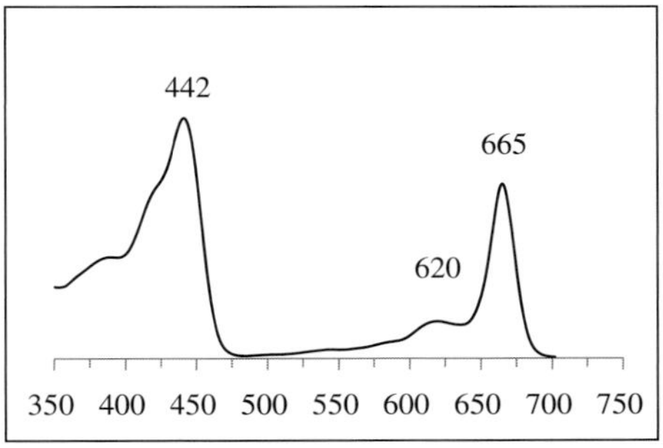

Mass spectra

Ionization technique	Mass analyser type	Diagnostic ions (m/z, rel. intensity)	Ref.
FAB	magnetic sector	891 $[M+H]^+$ → 613 $[M+H–phytyl]^+$	[13]

Remarks

Fluorescence: excitation 439 nm, emission 663, 680 nm (acetone) [11]
$d = 170\,L\,g^{-1}cm^{-1}$ (at λ_{max} in diethyl ether) is recommended, as no value has been determined for DVChl *a*. For 90% acetone, use $d = 88.3$ (at 663 nm) (calculated from Chl *a* [109])

Divinyl chlorophyll *b*

Recommended abbreviation: DVChl *b* (DC*b*)

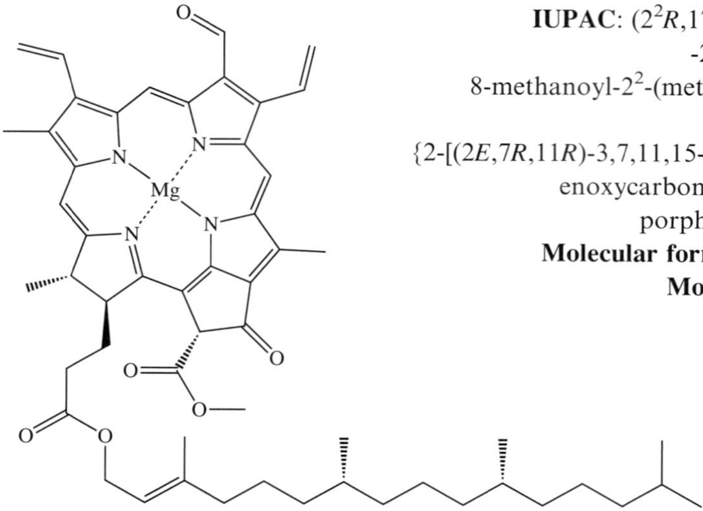

IUPAC: (2^2R,17*S*,18*S*)-7,12-Diethenyl-2^1,2^2,17,18-tetrahydro-8-methanoyl-2^2-(methoxycarbonyl)-3,13,17-trimethyl-2^1-oxo-18-{2-[(2*E*,7*R*,11*R*)-3,7,11,15-tetramethylhexadec-2-enoxycarbonyl]ethyl}cyclopenta[*at*]porphyrinatomagnesium(II)

Molecular formula: $C_{55}H_{68}N_4O_6Mg$

Molecular weight: 905.46

Biological occurrence	Dominant chlorophyll in *Prochlorococcus* (prochlorophyte/cyanobacteria) (Cyano-4, Chapter 1)
Source culture	*Prochlorococcus marinus* (prochlorophyte/cyanobacteria)
Alteration products	DVChlide *b*, DVPheide *b*, DVPhe *b*, DVChl *b′*, DVChl *b* allo, DVPphe *b* probable
Biosynthetically related to	DVChl *a*, MgDVP
Occurs together with	DVChl *a*, MgDVP, βε-Car, Zea

HPLC chromatogram of *Prochlorococcus* sp. (system 1)

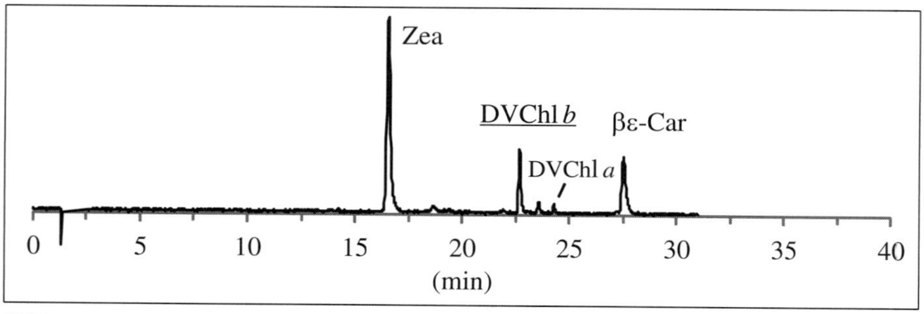

HPLC chromatogram of *Prochlorococcus* sp. (system 2)

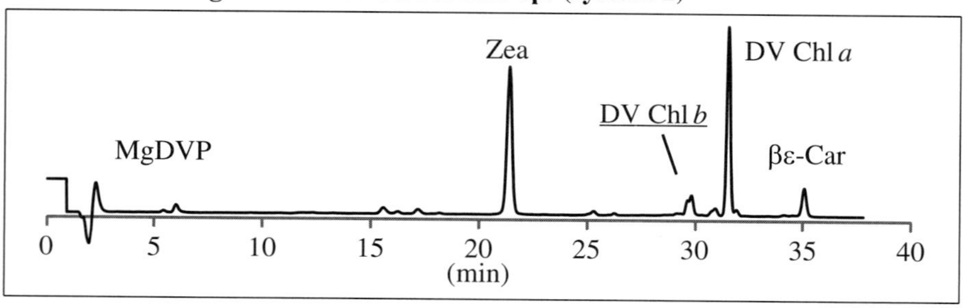

UV-Vis spectra (see also reference spectra below)

Solvent	λ_{max} (nm)	Band ratio (blue:red ratio)	Ref.
Diethyl ether	*460*, 595, 644	n.d.	[75]
Recommended specific absorption coefficient *d* (L g^{-1} cm^{-1})		n.d., see Remarks	

Reference spectra

For spectrum in acetone, see [109]

In HPLC solvent system 1

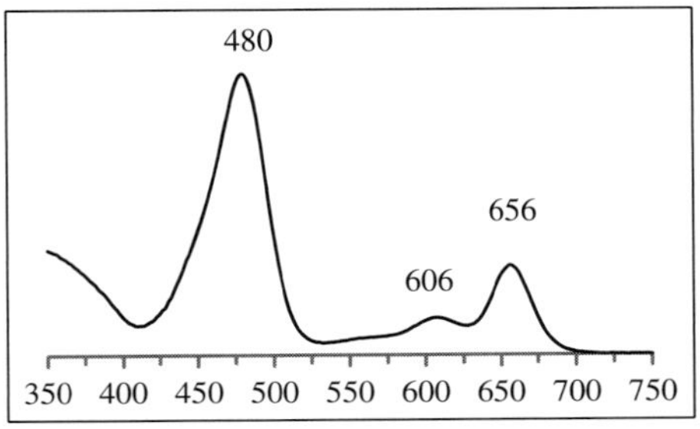

In HPLC solvent system 2

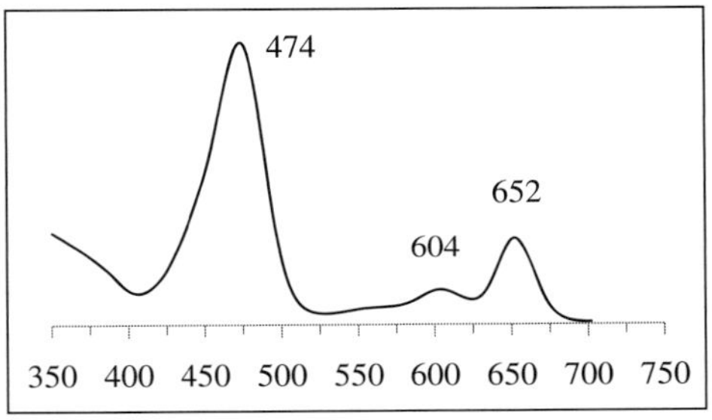

Mass spectra

Ionization technique	Mass analyser type	Diagnostic ions (m/z, rel. intensity)	Ref.
FAB	magnetic sector	905 $[M+1]^+$	[29]

Remarks	Fluorescence: excitation 462 nm, emission 652 nm (acetone) [11] $d = 230\,L\,g^{-1}\,cm^{-1}$ (at λ_{max} in diethyl ether) is recommended, as no value has been determined for DVChl *b*. In 90% acetone, use $d = 51.4$ (at 647 nm) (calculated from Chl *b* [109])

Magnesium 2,4-divinylpheoporphyrin a_5 monomethyl ester

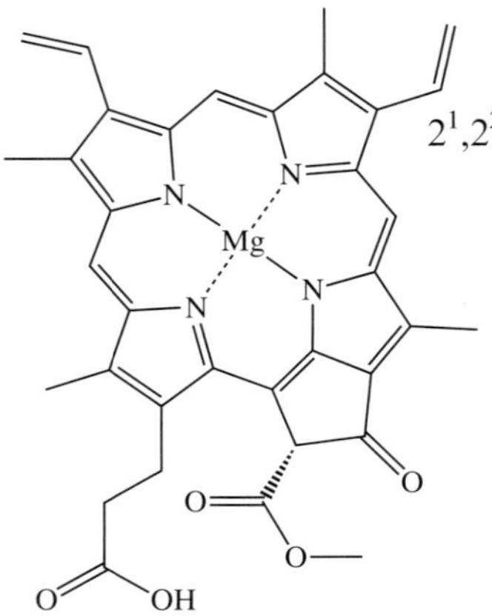

Recommended abbreviation: MgDVP (MD)
IUPAC: (2^2R)-18-(3-Carboxyethenyl)-7,12-diethenyl-2^1,2^2-dihydro-2^2-(methoxycarbonyl)-3,8,13,17-tetramethyl-2^1-oxocyclopenta[*at*]porphyrinatomagnesium(II)

Molecular formula: $C_{35}H_{30}N_4O_5Mg$
Molecular weight: 610.94

Biological occurrence	Dominant pigment in prasinoxanthin-containing algae, trace amounts in most other algal groups (Chapter 1, this volume)
Source culture	*Micromonas pusilla* (prasinophyte)
Alteration products	Corresponding epimer, corresponding pheophorbide and pyro-derivatives
Biosynthetically related to	Precursor of other chlorophylls
Occurs together with	Chl *b*, Micral, *c*-Neo, Pras, Uri, Viola

HPLC chromatogram of *Pycnococcus provasolii* (system 1)

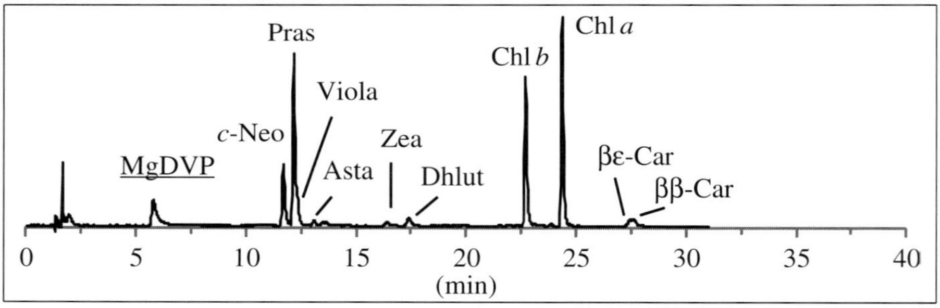

HPLC chromatogram of *Micromonas pusilla* (system 2)

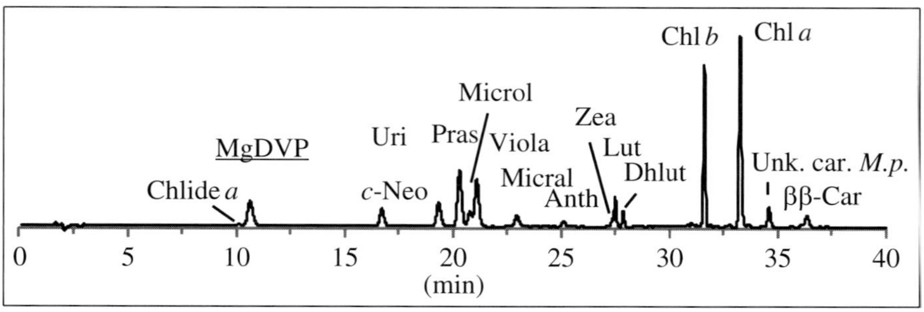

UV-Vis spectra (see also reference spectra below)

Solvent	λ_{max} (nm)	Band ratio (blue:red ratio)	Ref.
Acetone	*438*, 575, 625	9	[145]
Diethyl ether	*437*, 574, 624	10	[110]
Recommended specific absorption coefficient *d* ($L\ g^{-1}\ cm^{-1}$)		398 (at 456 nm in pyridine) [93] 394 (at 438 nm in 99% acetone + 1% pyridine [93]	

Reference spectra

For spectrum in acetone, see [109]

In HPLC solvent system 1

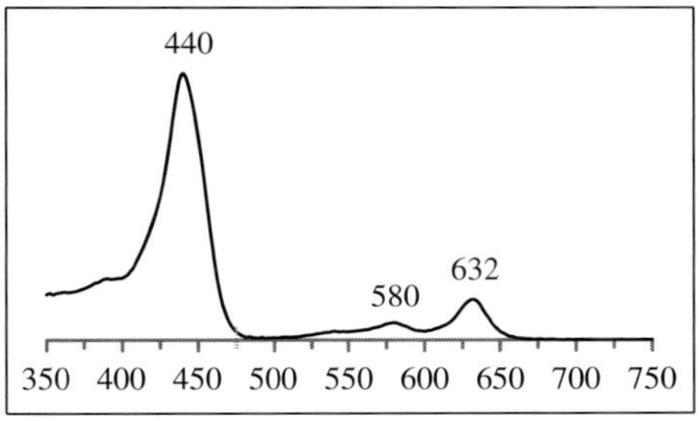

In HPLC solvent system 2

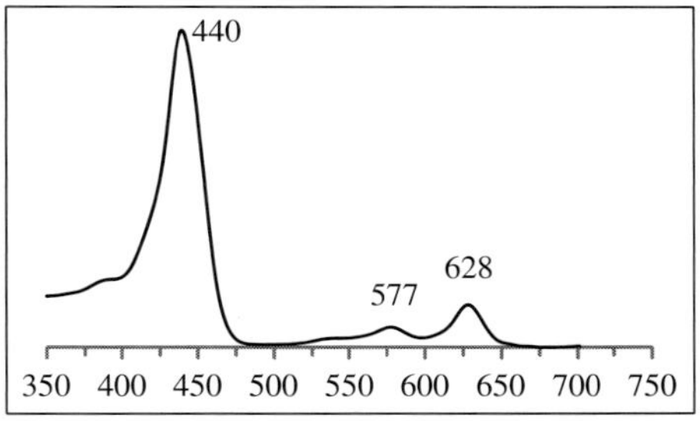

Mass spectra

Ionization technique	Mass analyser type	Diagnostic ions (m/z, rel. intensity)	Ref.
ESI^+	Ion trap	611 $[M+H]^+$; 611 → 551 $[M+H-60]^+$ → 523 $[M+H-60–28]^+$ (29), 522 (36), 509 (44), 505 (81), 495 (100), 494 (98), 492 (30), 453 (22)	[76]

Remarks Fluorescence: excitation 439 nm, emission 625 nm (diethyl ether + 1% pyridine) [93]. Various semisystematic names are given in the scientific literature for this pigment, e.g. magnesium 3,8-divinylpheoporphyrin a_5 monomethyl ester, [3,8-divinyl]protochlorophyllide *a* and [8-vinyl]-protochlorophyllide *a* [93]

Monovinyl chlorophyll c_3

Recommended abbreviation: MVChl c_3 (MC c_3)

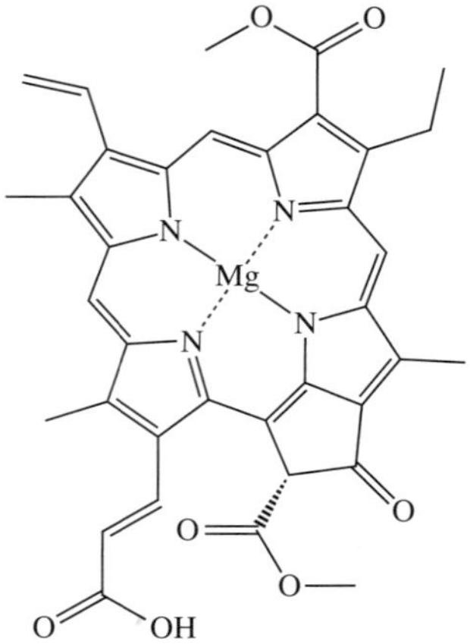

IUPAC: ($2^2R,18^1E$)-18-(3-Carboxyethenyl)-12-ethenyl-7-ethyl-$2^1,2^2$-dihydro-2^2,8-di(methoxycarbonyl)-3,13,17-trimethyl-2^1-oxocyclopenta[*at*]porphyrinatomagnesium(II)

Molecular formula: $C_{35}H_{30}N_4O_7Mg$
Molecular weight: 654.95

Biological occurrence	Minor pigment in haptophytes Pigment Type 6 (see Chapter 1)
Source culture	*Emiliania huxleyi* (coccolithophyte), strain CS-57
Alteration products	MVChl c_3', corresponding pheophorbide and pyro-derivatives
Biosynthetically related to	Chl c_1, Chl c_3
Occurs together with	Chl c_2, Chl c_3, Diadino, Fuco, Hex-fuco, Hex-kfuco

HPLC chromatogram of *Emiliania huxleyi* (system 1)

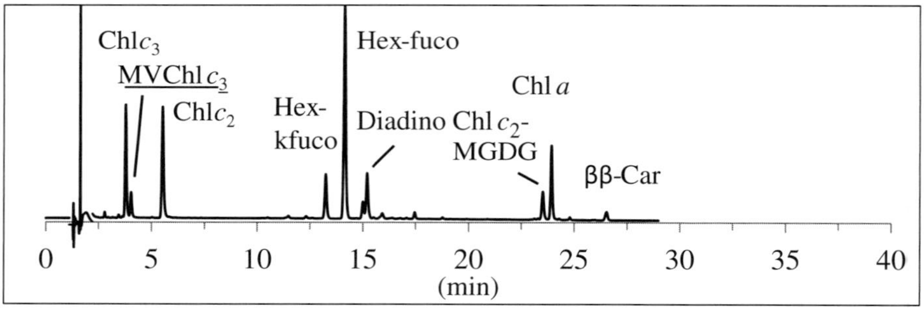

HPLC chromatogram of *Emiliania huxleyi* (system 2)

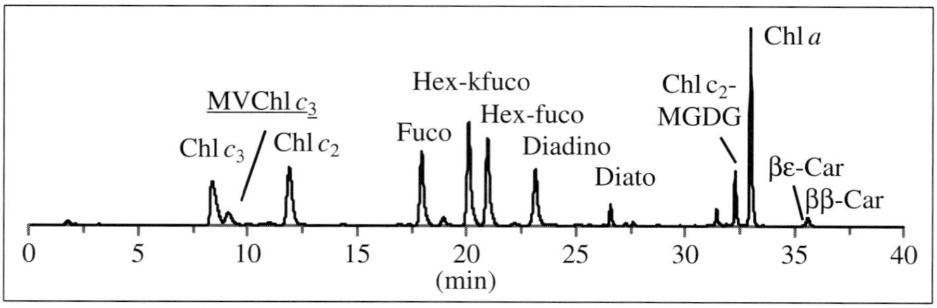

UV-Vis spectra (see also reference spectra below)

Solvent	λ_{max} (nm)	Band ratio (blue:red ratio)	Ref.
Acetone	*446*, 580, 624	25	[72]
Diethyl ether	*447*, 582, 626	24	[73]
Recommended specific absorption coefficient d **(L g^{-1} cm^{-1})**			n.d., see Remarks

Reference spectra

In acetone

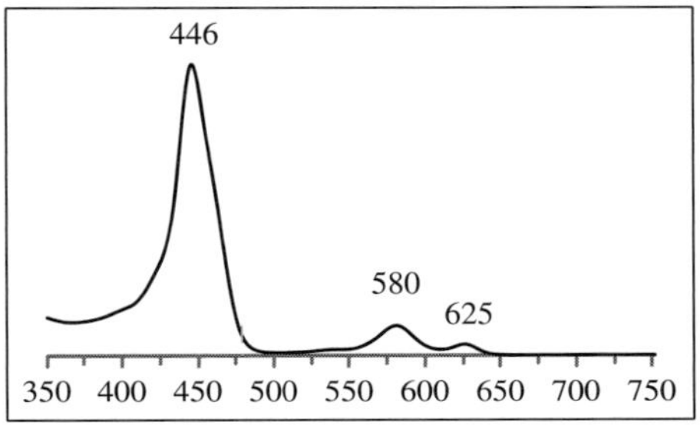

In HPLC solvent system 1

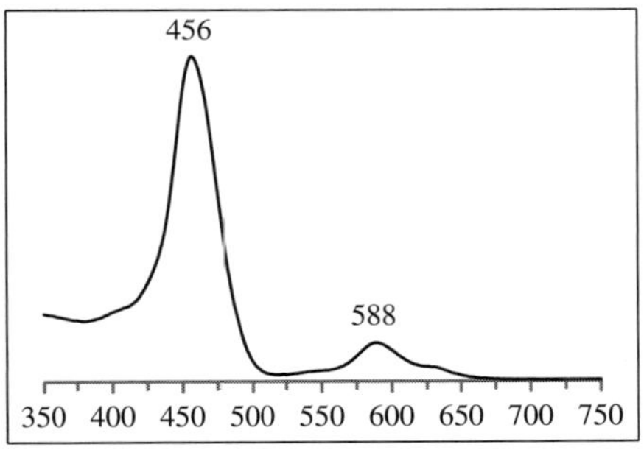

In HPLC solvent system 2

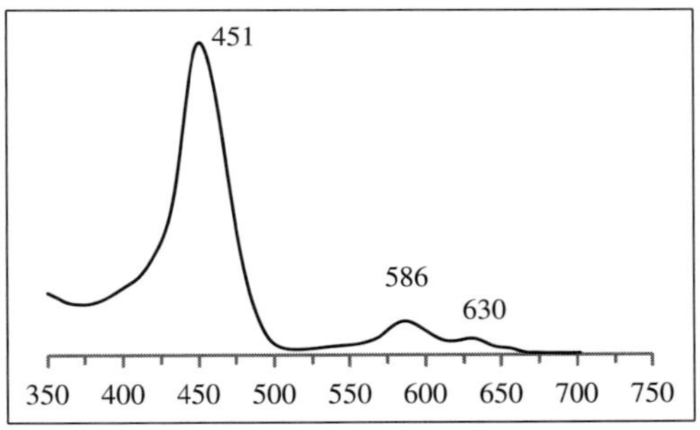

Mass spectra

Ionization technique	Mass analyser type	Diagnostic ions (m/z, rel. intensity)	Ref.
ESI$^+$	Ion trap	655 $[M+H]^+$; 655 → 637 (100), 623 (98), 611 (46), 595 (74)	[76]

Remarks	Fluorescence: excitation 445 nm, emission 629, 690 nm (acetone) [73] $d = 440$ L g^{-1} cm^{-1} (at λ_{max} in pyridine) is recommended, as no value has been determined for MVChl c_3. In 90% acetone + 1% pyridine, use $d = 346$ L g^{-1} cm^{-1} (at 453 nm) [179]. Structure not proven by NMR

Pheophorbide *a*

Recommended abbreviation: Pheide *a* (Pd*a*)

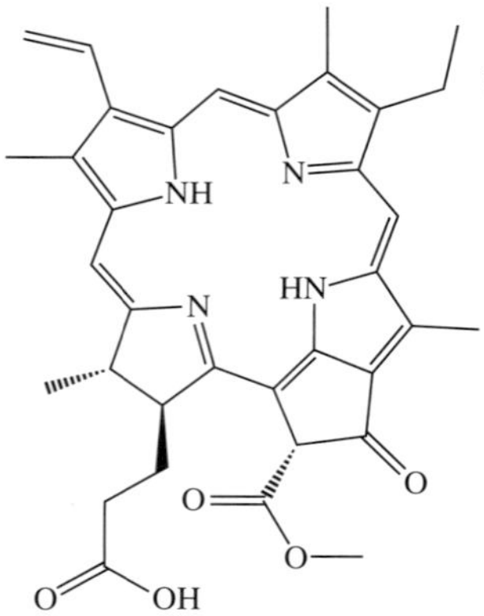

IUPAC: [(2^2R,17*S*,18*S*)-12-Ethenyl-7-ethyl-2^1,2^2,17,18-tetrahydro-2^2-(methoxycarbonyl)-3,8,13,17-tetramethyl-2^1-oxocyclopenta[*at*]porphyrin-18-yl]propanoic acid

Molecular formula: $C_{35}H_{36}N_4O_5$
Molecular weight: 592.68

Alteration product of	Chlorophyll *a*; occurs in senescent algae, zooplankton fecal pellets and sediments
Source culture	*Chroomonas salina* (cryptophyte)
Alteration products	
Synthetically related to	Chl *a*, Phe *a*
Occurs together with	

HPLC chromatogram of a coastal water sample (system 1)

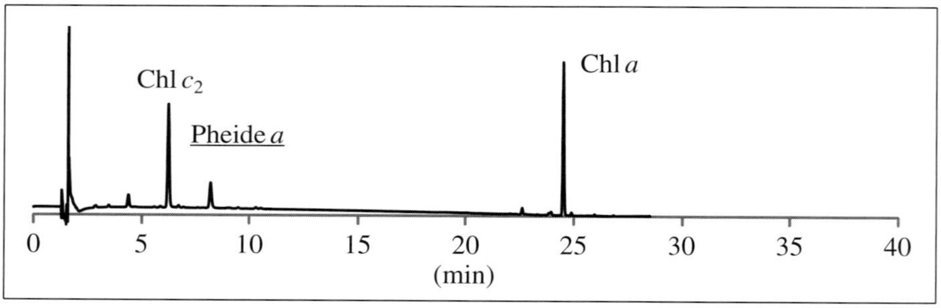

HPLC chromatogram of a coastal water sample (system 2)

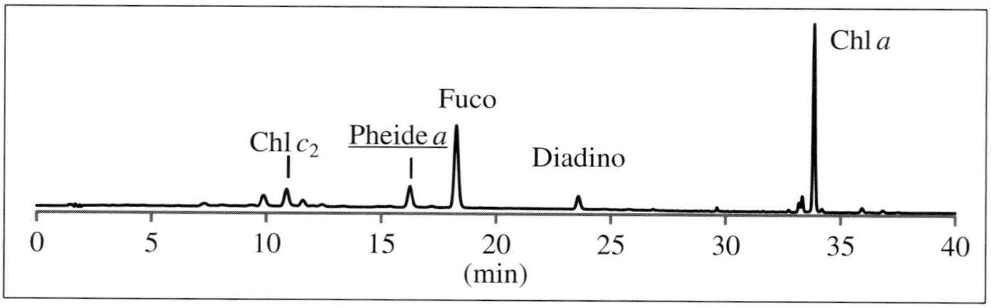

UV-Vis spectra (see also reference spectra below)

Solvent	λ_{max} (nm)	Band ratio (blue:red ratio)	Ref.
Acetone	*410*, 505, 535, 559, 608, 666	2.2	[145]
Diethyl ether	*408*, 467, 504, 533, 560, 610, 667	2.1	[99]
Recommended specific absorption coefficient *d* (L g^{-1} cm^{-1})		177 (at 411 nm in tetrahydrofuran) [101] 74.2 (at 667 nm in 90% acetone) [124, 109]	

Reference spectra

For spectrum in acetone, see [109]

In HPLC solvent system 1

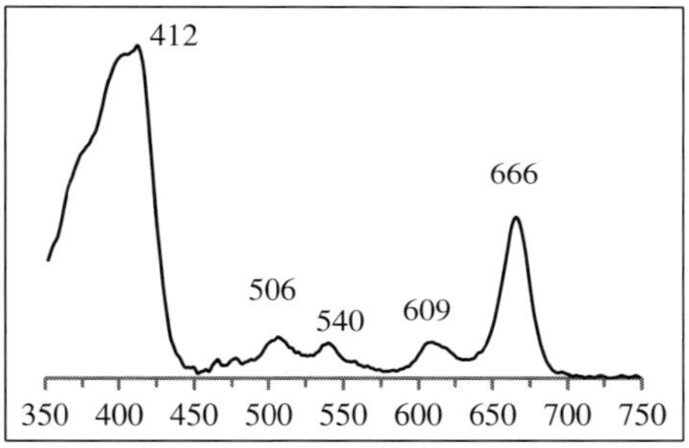

In HPLC solvent system 2

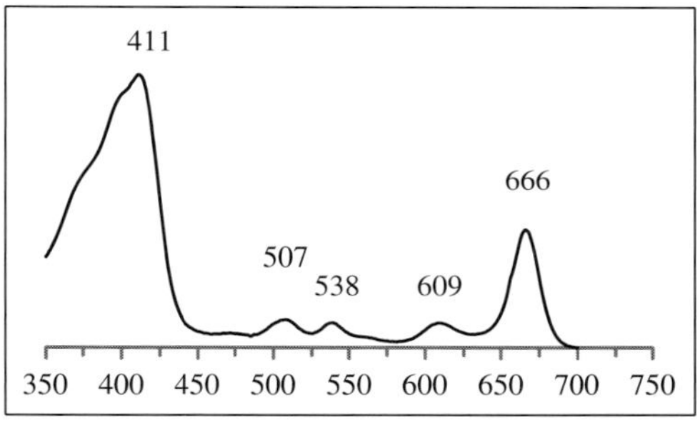

Mass spectra

Ionization technique	Mass analyser type	Diagnostic ions (m/z, rel. intensity)	Ref.
FAB	magnetic sector	593 $[M+H]^+ \rightarrow$ 575 $[M+H\text{-}18]^+$	[131]

Remarks

Fluorescence: excitation 406 nm, emission 672 nm [145]

Pheopigments are generally easier to detect using fluorescence. Several Phide *a* derivatives often present (e.g. [154]), including cyclic pheophorbide derivatives in fecal pellets and sediment [77, 168]

Pheophytin *a*

Recommended abbreviation: Phe *a* (Pha)

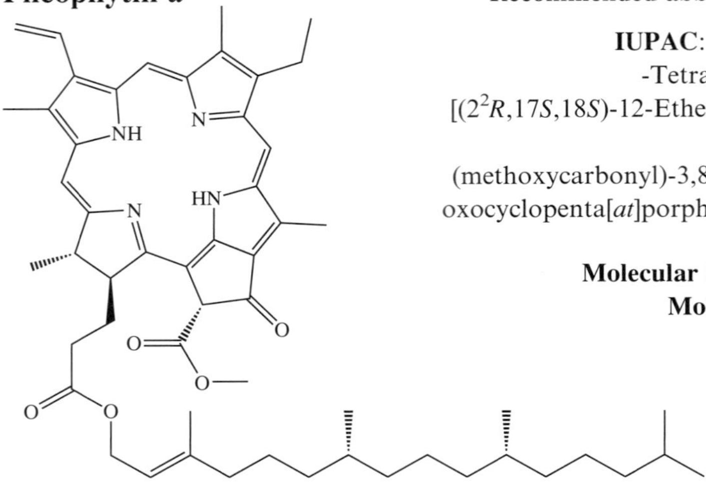

IUPAC: (2*E*,7*R*,11*R*)-3,7,11,15-Tetramethylhexadec-2-enyl [(2^2*R*,17*S*,18*S*)-12-Ethenyl-7-ethyl-2^1,2^2,17,18-tetrahydro-2^2-(methoxycarbonyl)-3,8,13,17-tetramethyl-2^1-oxocyclopenta[*at*]porphyrin-18-yl]propanoate

Molecular formula: $C_{55}H_{74}N_4O_5$
Molecular weight: 871.20

Alteration product of	Chlorophyll *a*. Found in zooplankton fecal pellets, senescent algae, sediments. The acid-catalysed demetallation also occurs in slightly acidic extracts, especially in prasinophyte extracts
Source culture	*Chroomonas salina* (cryptophyte)
Alteration products	Epimer, Allomer, Pheide *a*, Pphe *a*
(Bio)synthetically related to	Chl *a*, Pheide *a*, Pphe *a*
Occurs together with	Carotenoid furanoxides

HPLC chromatogram of *Amphidinium carterae* (system 1)

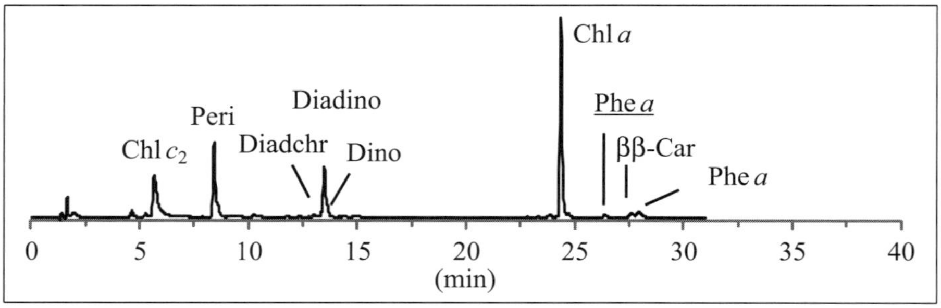

HPLC chromatogram of a coastal marine sample (system 2)

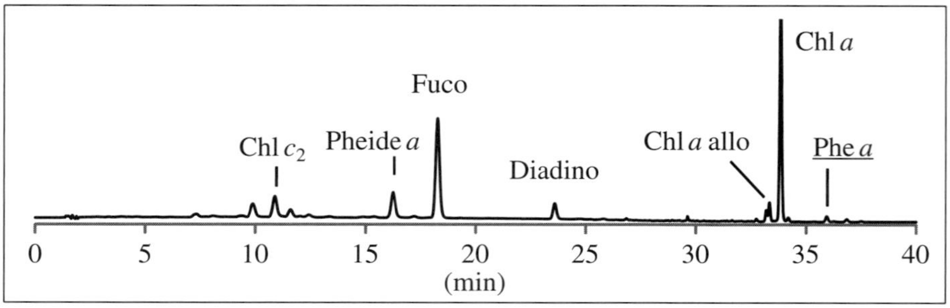

UV-Vis spectra (see also reference spectra below)

Solvent	λ_{max} (nm)	Band ratio (blue:red ratio)	Ref.
Acetone	*410*, 505, 535, 560, 610, 666	2.3	[109]
Diethyl ether	*408*, 505, 534, 610, 667	2.1	[12]
95% Ethanol	*417*, 507, 536, 564, 662	3.3	[121]
Methanol	*417*, 533, 567, 602, 654	4.1	[121]
Recommended specific absorption coefficient *d* ($L\ g^{-1}\ cm^{-1}$)		139 (at 411 nm in tetrahydrofuran) [125] 51.2 (at 667 nm in 90% acetone) [124, 109]	

Reference spectra

For spectrum in acetone, see [109]

In HPLC solvent system 1

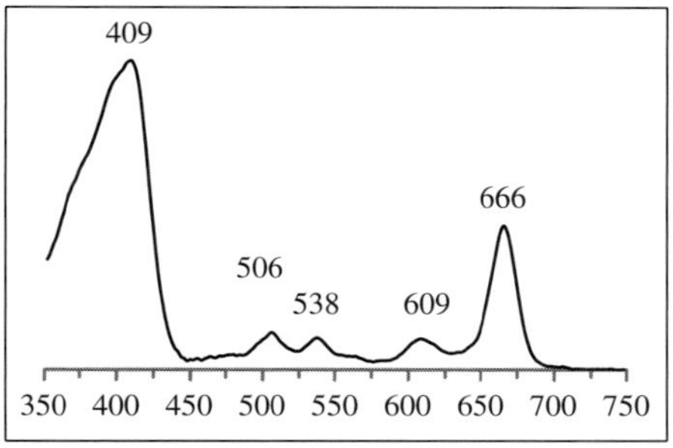

In HPLC solvent system 2

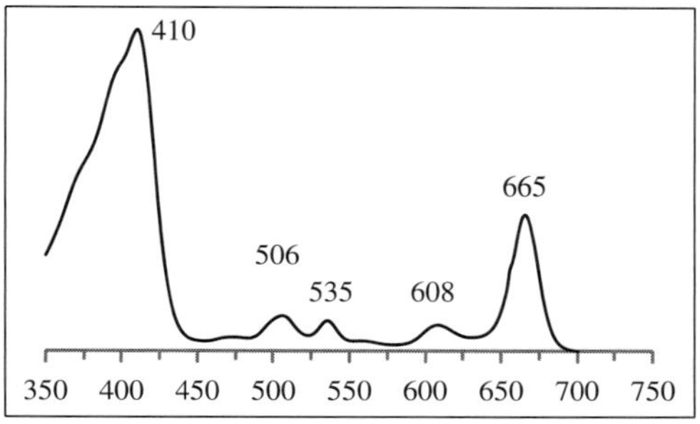

Mass spectra

Ionization technique	Mass analyser type	Diagnostic ions (m/z, rel. intensity)	Ref.
FAB	magnetic sector	870 $[M]^+$ (100), 592 $[M\text{-phytyl}]^+$ (53), 533 (41), 520 (36), 459 (83)	[27]

Remarks Fluorescence: excitation 408 nm, emission 672 nm (diethyl ether) [26]
Pheopigments are generally easier to detect using fluorescence

Pheophytin *b*

Recommended abbreviation: Phe *b* (Ph*b*)

IUPAC: (2*E*,7*R*,11*R*)-3,7,11,15-Tetramethylhexadec-2-enyl [(2^2R,17*S*,18*S*)-12-Ethenyl-7-ethyl-2^1,2^2,17,18-tetrahydro-8-methanoyl-2^2-(methoxycarbonyl)-3,13,17-trimethyl-2^1-oxocyclopenta[*at*]porphyrin-18-yl]propanoate

Molecular formula: $C_{55}H_{72}N_4O_6$
Molecular weight: 885.18

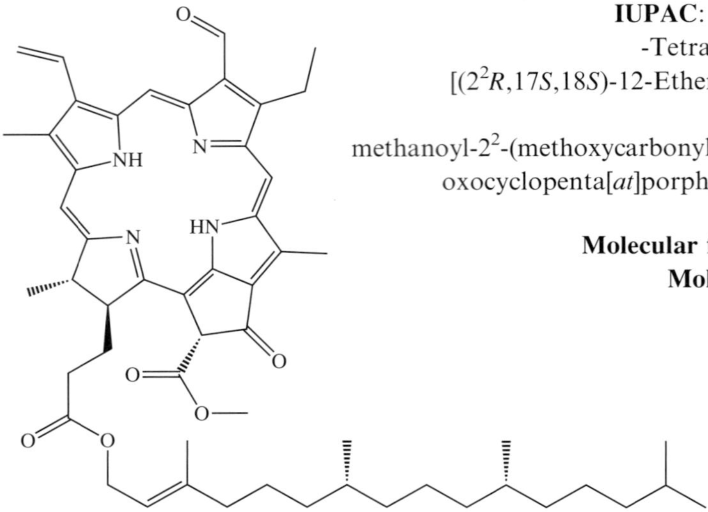

Alteration product of	Chlorophyll *b*. The acid-catalysed demetallation occurs in slightly acidic extracts, especially in prasinophyte extracts. Found in protozoan fecal pellets and terrestrial plant detritus
Source culture	*Dunaliella tertiolecta* (chlorophyte), *Pycnococcus provasolii* (prasinophyte)
Alteration products	Epimer, Allomer, Pheide *b*, Pphe *b*
Synthetically related to	Chl *b*, Pheide *b*, Pphe *b*
Occurs together with	Phe *a*, carotenoid furanoxides

HPLC chromatogram of acidified *Dunaliella tertiolecta* (system 1)

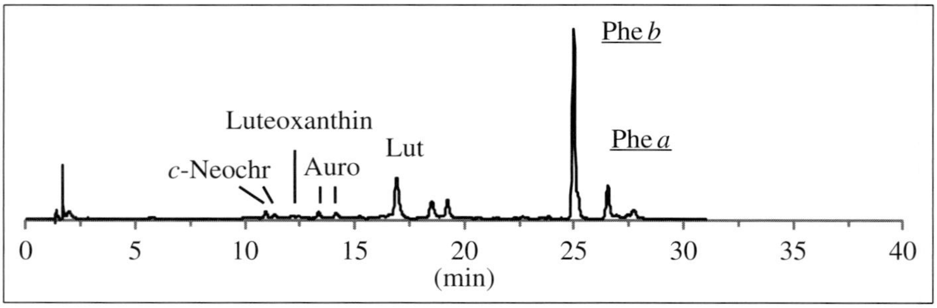

HPLC chromatogram (system 2)

NO DATA AVAILABLE

UV-Vis spectra (see also reference spectra below)

Solvent	λ_{max} **(nm)**	**Band ratio (blue:red ratio)**	**Ref.**
Acetone	*433*, 527, 597, 653	5.1	[121]
Diethyl ether	*433*, 520, 555, 599, 655	5.2	[12]
95% Ethanol	*437*, 528, 599, 654	4.3	[121]
Methanol	*420*, 535, 597, 648	4.9	[121]
Recommended specific absorption coefficient *d* (L g^{-1} cm^{-1})		244 (at 435 nm in tetrahydrofuran) [101] 31.8 (at 657 nm in 90% acetone) [109]	

Reference spectra

For spectrum in acetone, see [109]

In HPLC solvent system 1

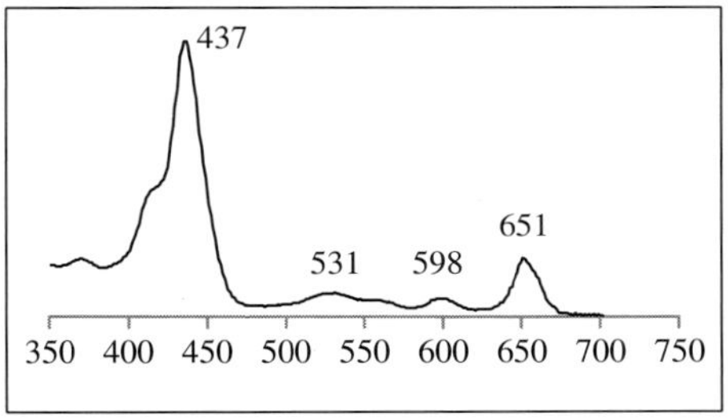

In HPLC solvent system 2

NO DATA AVAILABLE

Mass spectra

Ionization technique	**Mass analyser type**	**Diagnostic ions (m/z, rel. intensity)**	**Ref.**
FAB	magnetic sector	884 $[M]^+$ (100), 606 $[M\text{-phytyl}]^+$ (37), 547 (31), 533 (15), 473 (23)	[27]

Remarks	Fluorescence: excitation 434 nm, emission 658 nm (diethyl ether) [26] Pheopigments are generally easier to detect using fluorescence

Pyropheophorbide *a*

Recommended abbreviation: Ppheide *a* (Pd*a*)

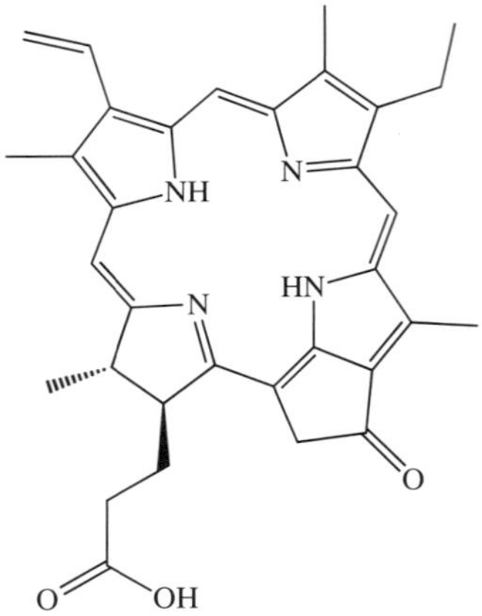

IUPAC: [(17*S*,18*S*)-12-Ethenyl-7-ethyl-2[1],2[2],17,18-tetrahydro-3,8,13,17-tetramethyl-2[1]-oxocyclopenta[*at*]porphyrin-18-yl]propanoic acid

Molecular formula: $C_{33}H_{34}N_4O_3$
Molecular weight: 534.65

Alteration product of	Chlorophyll *a*; occurs in senescent algae and zooplankton fecal pellets
Source culture	*Chroomonas salina* (cryptophyte)
Alteration products	
Synthetically related to	Chl *a*, Pheide *a*, Phe *a*, Pphe *a*
Occurs together with	Pheide *a*, Pphe *a*

HPLC chromatogram (system 1)

NO DATA AVAILABLE

HPLC chromatogram of a coastal marine sample (system 2)

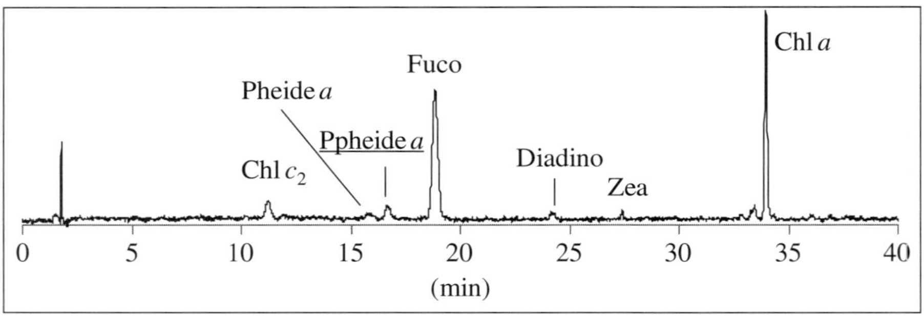

UV-Vis spectra (see also reference spectra below)

Solvent	λ_{max} (nm)	Band ratio (blue:red ratio)	Ref.
For all practical purposes, identical with pyropheophytin *a*			
Recommended specific absorption coefficient *d* (L g^{-1} cm^{-1})	93.5 (at red maximum in diethyl ether) [137]		

Reference spectra

In acetone

For spectrum in acetone, see [109]

In HPLC solvent system 1

NO DATA AVAILABLE

In HPLC solvent system 2

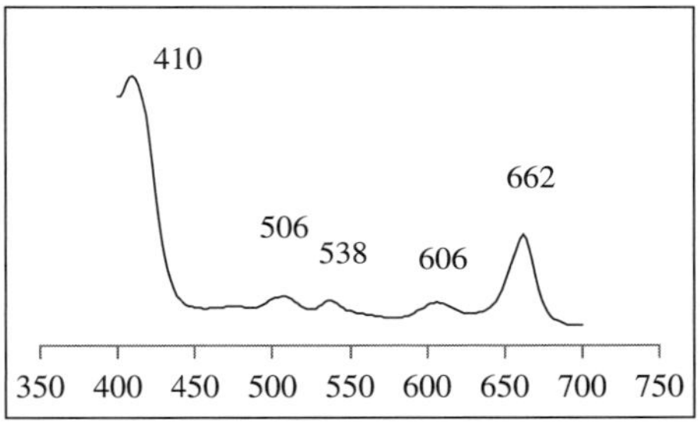

Mass spectra

Ionization technique	Mass analyser type	Diagnostic ions (m/z, rel. intensity)	Ref.
FAB	magnetic sector	534 $[M]^+$; fragments in figure, see ref.	[131]

Remarks Fluorescence: for all practical purposes, identical with pyropheophytin *a*.
Pheopigments are generally easier to detect using fluorescence

Pyropheophytin *a*

Recommended abbreviation: Pphe *a* (Py*a*)

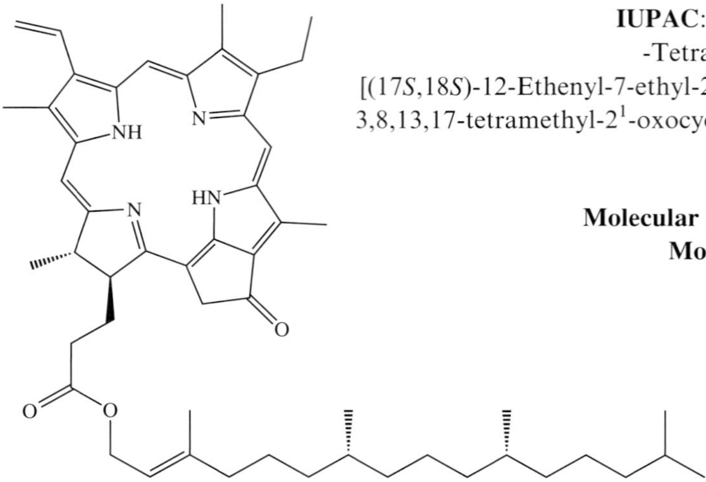

IUPAC: (2*E*,7*R*,11*R*)-3,7,11,15-Tetramethylhexadec-2-enyl [(17*S*,18*S*)-12-Ethenyl-7-ethyl-2[1],2[2],17,18-tetrahydro-3,8,13,17-tetramethyl-2[1]-oxocyclopenta[*at*]porphyrin-18-yl]propanoate

Molecular formula: $C_{53}H_{72}N_4O_3$
Molecular weight: 813.16

Alteration product of	Chlorophyll *a*; occurs in senescent algae and zooplankton fecal pellets
Source culture	*Chroomonas salina* (cryptophyte)
Alteration products	
Synthetically related to	Chl *a*, Phe *a*
Occurs together with	

HPLC chromatogram of *Amphidinium carterae* (system 1)

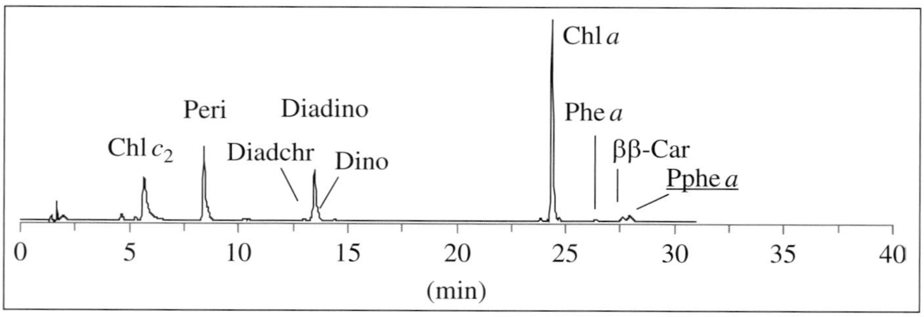

HPLC chromatogram of a sediment trap sample (system 3)

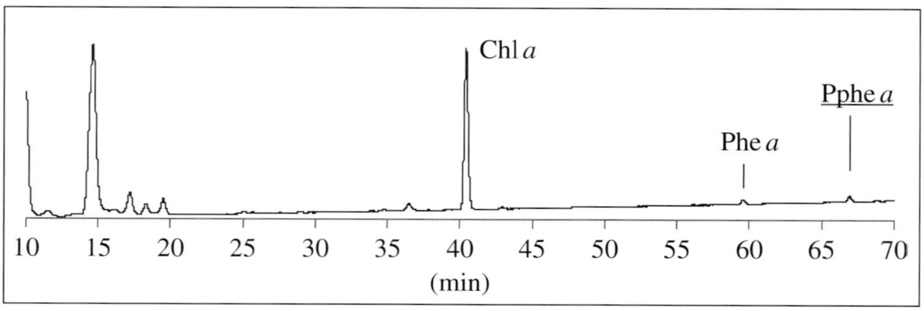

UV-Vis spectra (see also reference spectra below)

Solvent	λ_{max} (nm)	Band ratio (blue:red ratio)	Ref.
Acetone	*410*, 507, 536, 609, 667	2.3	[145]
Diethyl ether	*409*, 667	2.1	[134]
Recommended specific absorption coefficient *d* ($L\ g^{-1}\ cm^{-1}$)		60.3 (at 667 nm in diethyl ether)	[134]

Reference spectra

In acetone

For spectrum in acetone, see [109]

In HPLC solvent system 1

NO DATA AVAILABLE

In HPLC solvent system 2

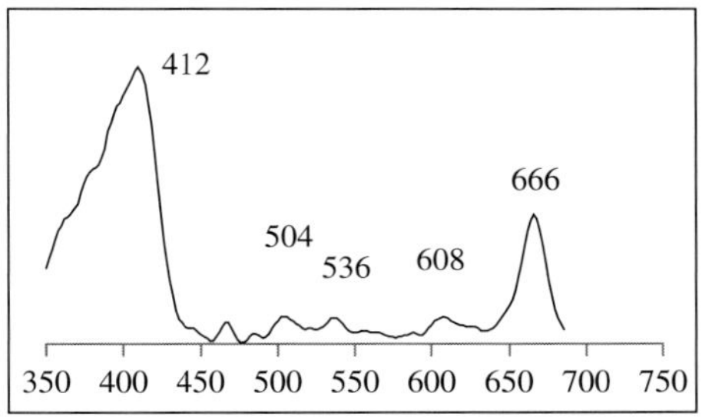

Mass spectra

Ionization technique	Mass analyser type	Diagnostic ions (m/z, rel. intensity)	Ref.
APCI	Ion trap	813 $[M+H]^+$ MS2: 353, 507	[6]

Remarks	Fluorescence: excitation 407 nm, emission 672 nm (acetone) [145] Pheopigments are generally easier to detect using fluorescence

2 Carotenes

β,β-Carotene

Recommended abbreviation: ββ-Car (ββ)

IUPAC: β,β-Carotene (trivial name: β-Carotene)

Molecular formula: $C_{40}H_{56}$

Molecular weight: 536.87

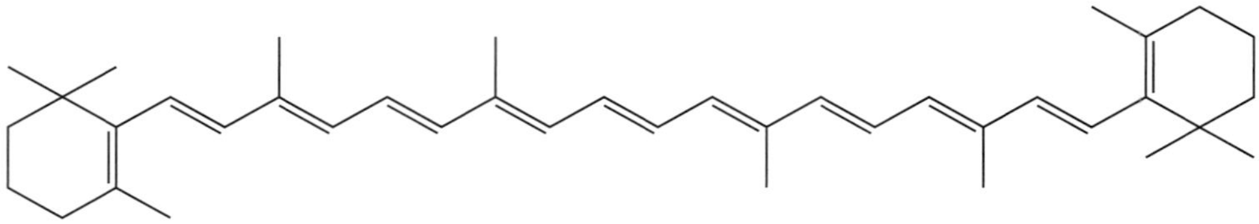

Biological occurrence	Dominant pigment in chlorophytes, prasinophytes, mesostigmatophytes, rhodophytes and one group of dinoflagellates. Minor in all other algal groups. Also present in plants (notably carrots) [78]
Source culture	*Pavlova lutheri* (Pavlovophyceae)
Alteration products	*Cis*-isomers
Biosynthetically related to	Cantha, Asta, Zea, Viola, Neo, Diato, Diadino, Allo etc.
Occurs together with	

HPLC chromatogram of *Synechococcus* sp. (system 1)

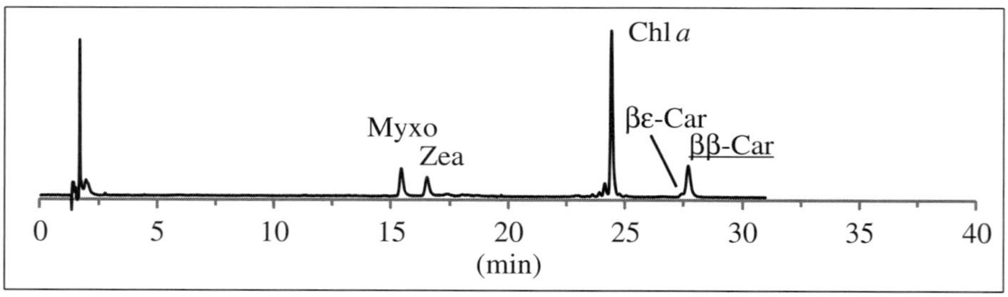

HPLC chromatogram of *Dunaliella salina* (system 2)

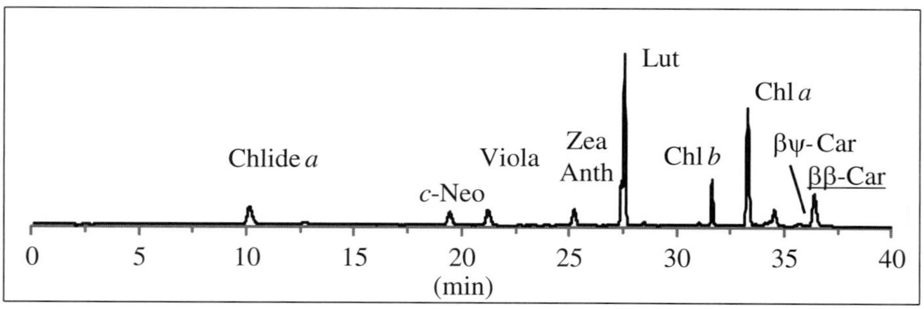

UV-Vis spectra (see also reference spectra below)

Solvent	λ_{max} **(nm)**	**Band ratio (% III:II)**	**Ref.**
Acetone	(427), *454*, 480	21	[96]
Diethyl ether	(430), *447*, 476	5	[133]
Ethanol (see Remarks)	(428), *451*, 480	27	[86]
Hexane	(425), *451*, 477	29	[96]
Methanol (see Remarks)	(429), *449*, 475	25	[160]
Recommended specific absorption coefficient d **(L g^{-1} cm^{-1})**		259 (at 453 nm, hexane) [103] 250 (at 454 nm, acetone) [96]	

Reference spectra

In acetone

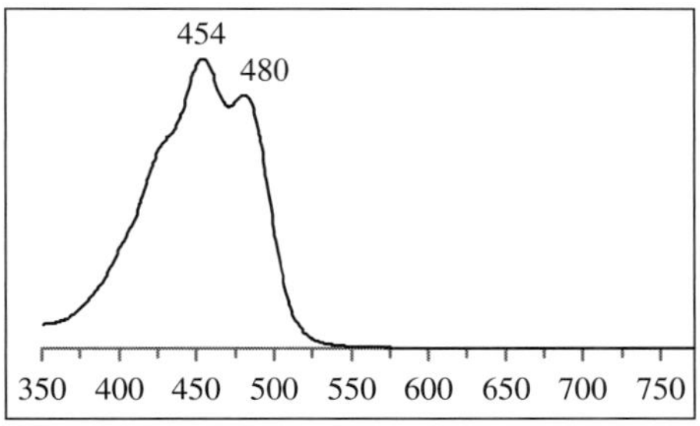

In hexane

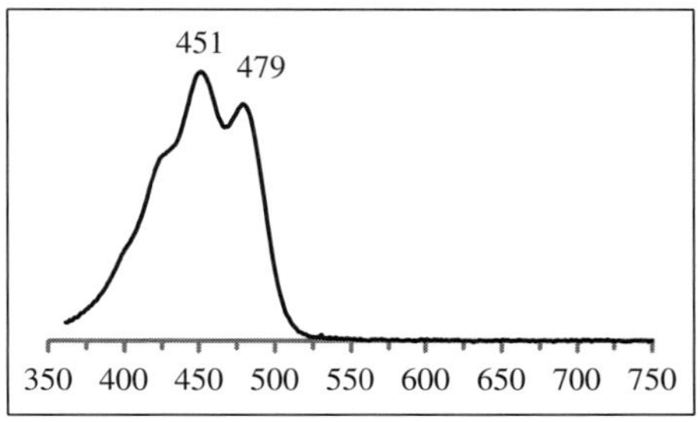

In HPLC solvent system 1

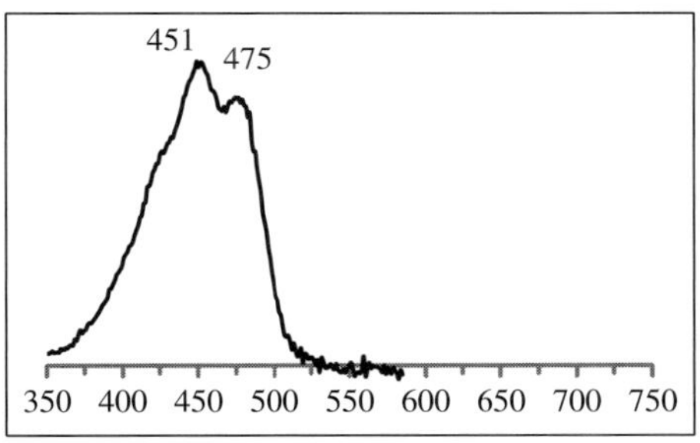

In HPLC solvent system 2

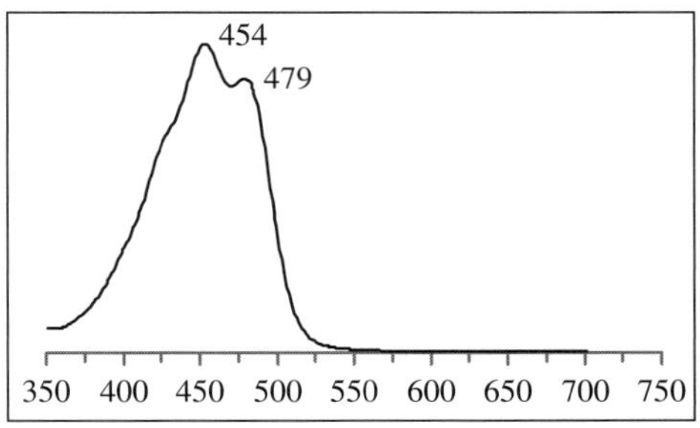

Mass spectra

Ionization technique	**Mass analyser type**	**Diagnostic ions (m/z, rel. intensity)**	**Ref.**
EI	Magnetic sector	536 $[M]^+$ (44), 444 $[M\text{-}92]^+$ (15), 430 $[M\text{-}106]^+$ (4), 119 (72), 109 (22), 91 (45), 83 (23), 69 (100)	[144]

Remarks To aid dissolving in alcohol, first dissolve in a drop or two of hexane

β,ε-Carotene **Recommended abbreviation: βε-Car (βε)**

IUPAC: (6′*R*)-β,ε-Carotene (trivial name: α-carotene)

Molecular formula: $C_{40}H_{56}$ **Molecular weight:** 536.87

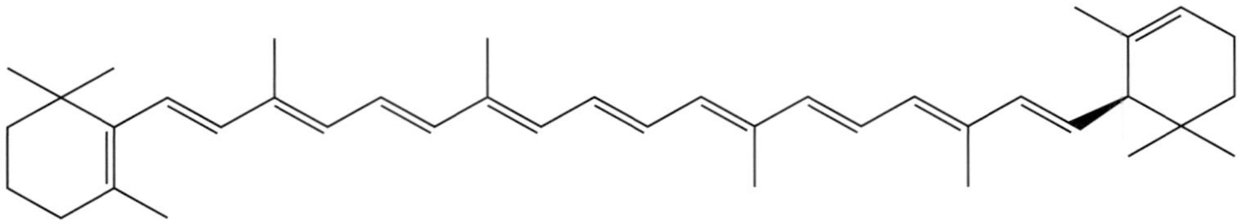

Biological occurrence	Minor or trace pigment in chlorophytes, prasinophytes, cryptophytes, some dinoflagellates, cyanobacteria
Source culture	*Chroomonas salina* (cryptomonad)
Alteration products	*Cis*-isomers
Biosynthetically related to	Lut, Loro, Siph
Occurs together with	Lut, Zea, Viola, Anth, Neo, Allo, Croco, Monado

HPLC chromatogram of *Chroomonas salina* (system 1)

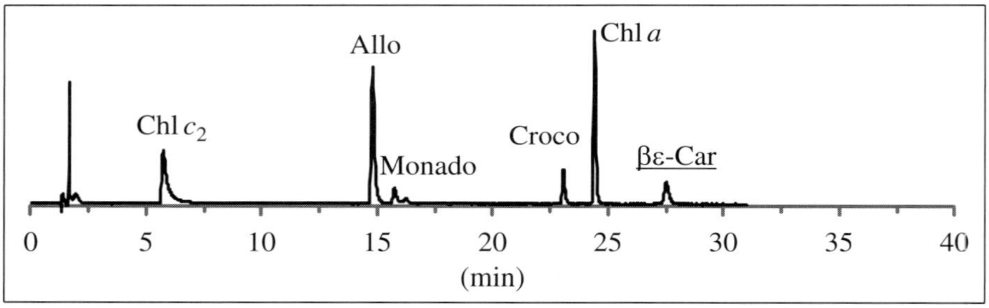

HPLC chromatogram of *Codium fragile* (system 2)

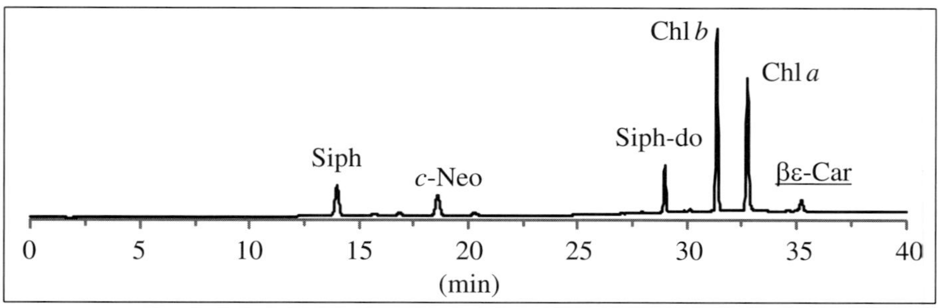

UV-Vis spectra (see also reference spectra below)

Solvent	λ_{max} (nm)	Band ratio (% III:II)	Ref.
Acetone	(424), *448*, 476	56	[96]
Ethanol (see remarks)	423, *444*, 473	61	[86]
Hexane	423, *446*, 474	71	[57]
Methanol (see remarks)	(422), *442*, 471	61	[160]
Recommended specific absorption coefficient d (L g^{-1} cm^{-1})		270 (at 446 nm, hexane) [57] 270 (at 448 nm, acetone) [96]	

Reference spectra
In hexane

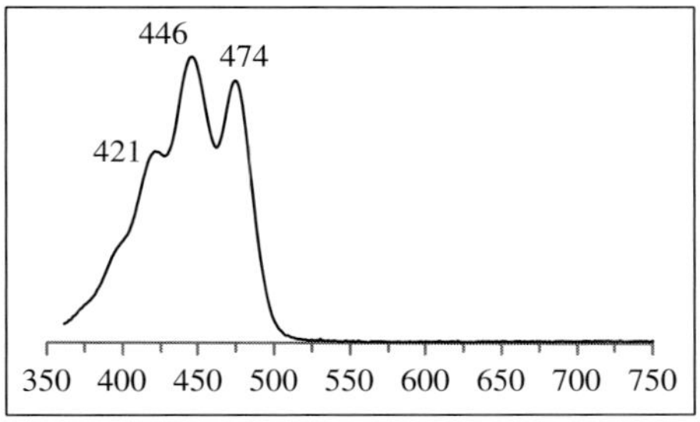

For spectrum in acetone, see [109]

In HPLC solvent system 1

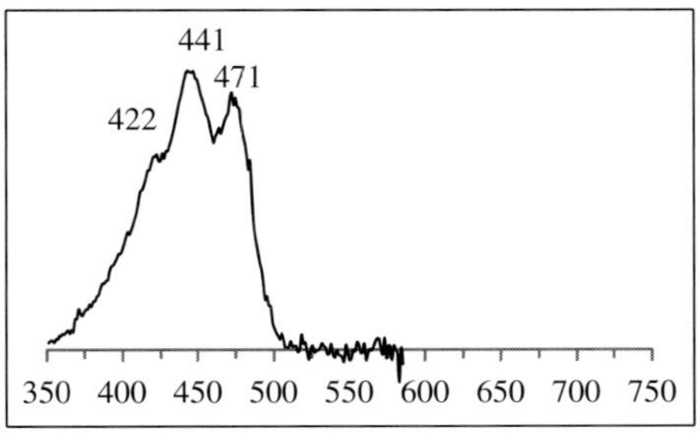

In HPLC solvent system 2

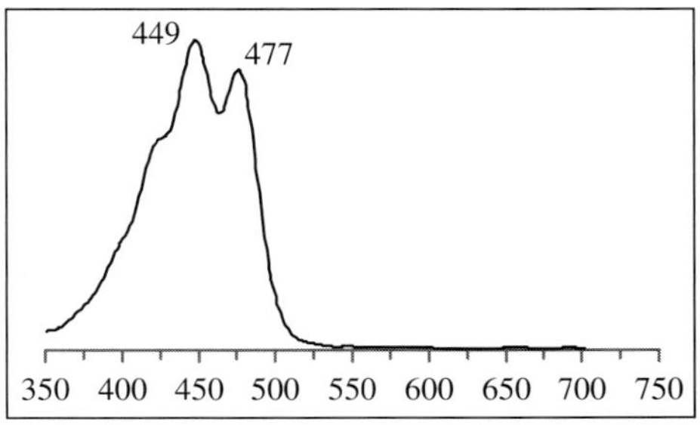

Mass spectra

Ionization technique	Mass analyser type	Diagnostic ions (m/z, rel. intensity)	Ref.
EI	Magnetic sector	536 [M]$^+$ (100), 480 [M-56]$^+$ (2), 444 [M-92]$^+$ (25), 430 [M-106]$^+$ (18), 388, 378, 374	[144]

Remarks To aid dissolving, add a drop or two of hexane before adding the alcohol

β,ψ-Carotene

Recommended abbreviation: βψ-Car (βψ)

IUPAC: β,ψ-Carotene

(trivial name: γ-carotene)

Molecular formula: $C_{40}H_{56}$

Molecular weight: 536.87

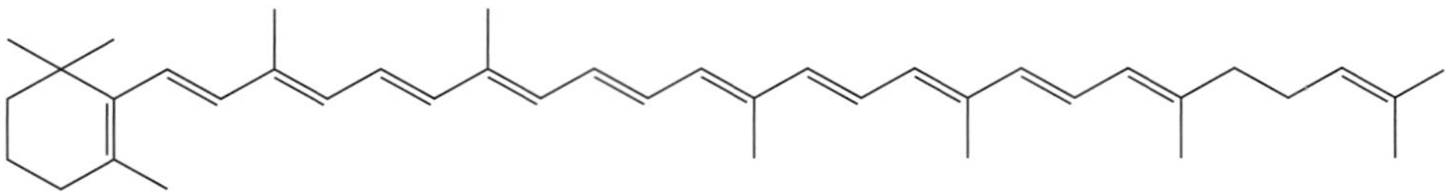

Biological occurrence	Occasionally found in some chlorophytes and prasinophytes, dinoflagellates Pigment Type 3 and mesostigmatophytes
Source culture	*Dunaliella tertiolecta* (chlorophyte)
Alteration products	*Cis*-isomers
Biosynthetically related to	Lyco, ββ-Car, βε-Car (biosynthetic precursor of ββ-Car and βε-Car)
Occurs together with	

HPLC chromatogram of *Dunaliella tertiolecta* (system 1)

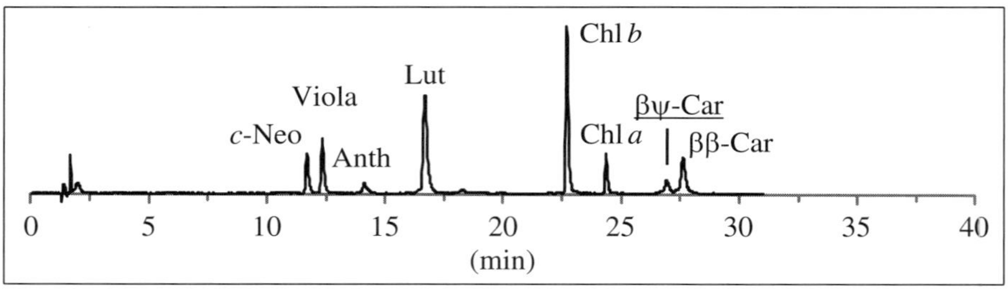

HPLC chromatogram of *Dunaliella salina* (system 2)

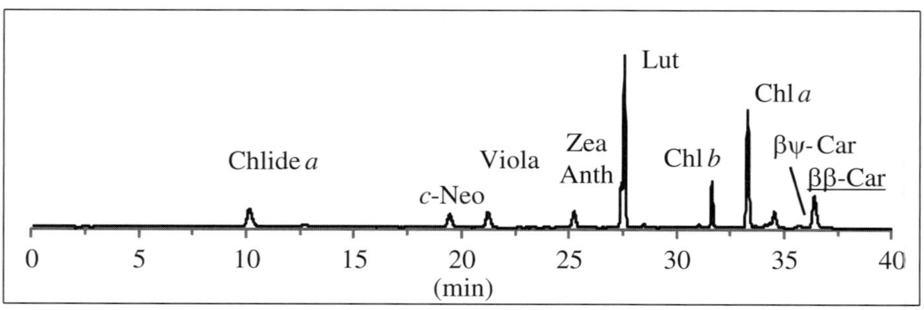

UV-Vis spectra (see also reference spectra below)

Solvent	λ_{max} (nm)	Band ratio (% III:II)	Ref.
Acetone	(439), *461*, 491	n.d.	[66]
Cyclohexane	439, *464*, 496	59	[69]
Ethanol (see remarks)	(440), *460*, 489	23	[83]
Recommended specific absorption coefficient d **(L g^{-1} cm^{-1})**		276 (at 462 nm, hexane) [181] 319 (at 459 nm, petroleum ether) [126]	

Reference spectra

For spectrum in acetone, see [109]

For spectrum in hexane, see [109]

In HPLC solvent system 1

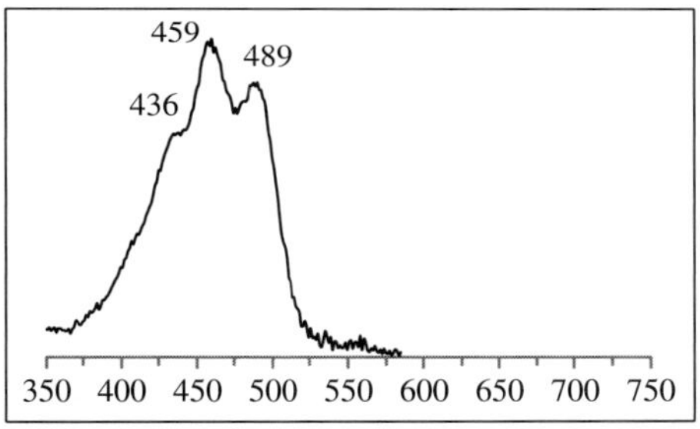

In HPLC solvent system 2

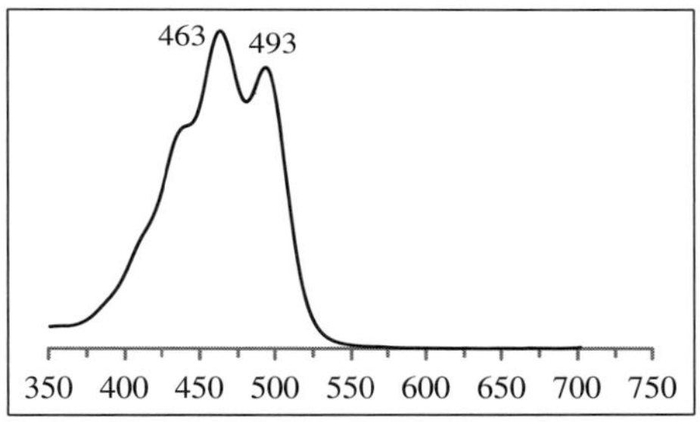

Mass spectra

Ionization technique	Mass analyser type	Diagnostic ions (m/z, rel. intensity)	Ref.
EI	Magnetic sector	536 $[M]^+$ (100), 467 $[M\text{-}69]^+$ (8), 444 $[M\text{-}92]^+$ (20), 430 $[M\text{-}106]^+$ (25), 407 (m*, M → M-69)	[30]

Remarks To aid dissolving, add a drop or two of hexane before adding the alcohol

ε,ε-Carotene

Recommended abbreviation: εε-Car (ε)

IUPAC: (6*S*,6′*S*)-ε,ε-Carotene

(trivial name: ε-carotene)

Molecular formula: $C_{40}H_{56}$

Molecular weight: 536.87

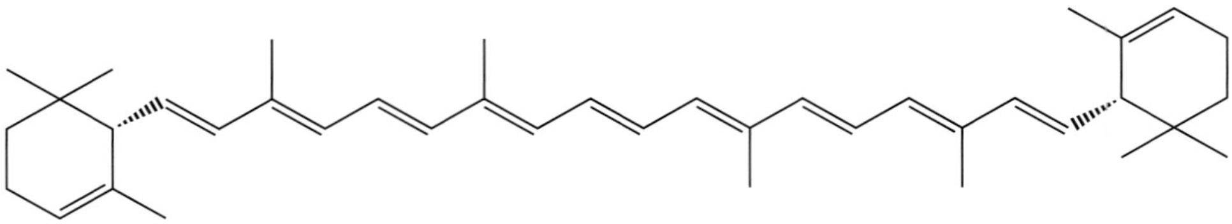

Biological occurrence	Significant pigment in pelagophytes but not always present (see Chapter 1). Occasional traces in cultures of chlorophytes, diatoms, cryptomonads and prasinophytes Pigment Type 3
Source culture	*Pelagococcus subviridis* (chrysophyte)
Alteration products	*Cis*-isomers
Biosynthetically related to	Lyco
Occurs together with	

HPLC chromatogram (system 1)

NO DATA AVAILABLE

HPLC chromatogram of *Pelagomonas calceolata* (system 2)

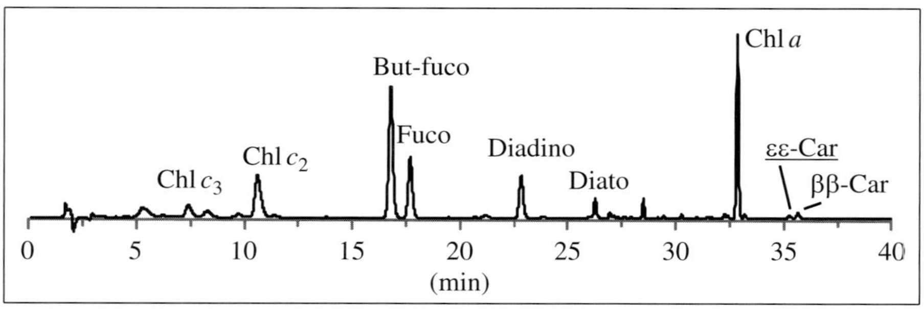

UV-Vis spectra (see also reference spectra below)

Solvent	λ_{max} (nm)	Band ratio (% III:II)	Ref.
Cyclohexane	416, 440, 470	n.d.	[157]
95% Ethanol	417, *440*, 470	96	[36]
Hexane	415, 439, *469*	101	[23]
Recommended specific absorption coefficient ***d*** **(L g^{-1} cm^{-1})**		312 (at 440 nm, petroleum ether) [143]	

Reference spectra
In hexane

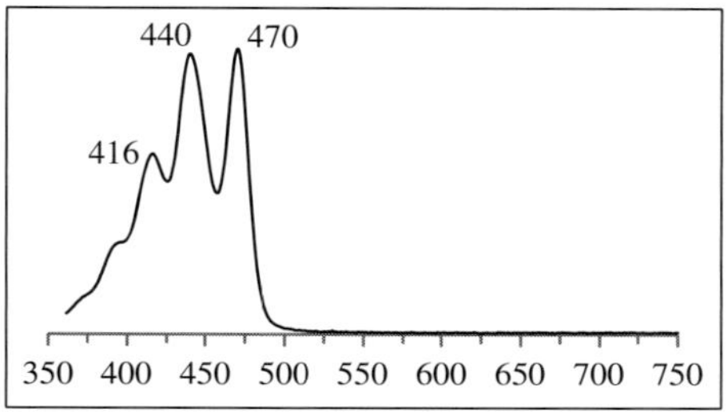

In HPLC solvent system 1

NO DATA AVAILABLE

In HPLC solvent system 2

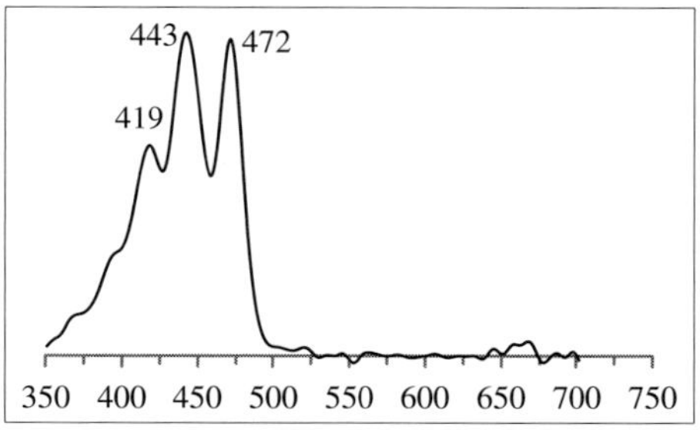

Mass spectra

Ionization technique	Mass analyser type	Diagnostic ions (m/z, rel. intensity)	Ref.
EI	Magnetic sector	536 [M]$^+$ (100), 480 [M-56]$^+$ (5), 444 [M-92]$^+$ (28), 430 [M-106]$^+$ (8), 388 [M-92–56]$^+$ (30)	[143]
Remarks	Opposite stereochemistry than βε-Car and Lut [23, 46]		

ψ,ψ-Carotene (Lycopene)

Recommended abbreviation: Lyco (Ly)

IUPAC: ψ,ψ-Carotene

Molecular formula: $C_{40}H_{56}$

Molecular weight: 536.87

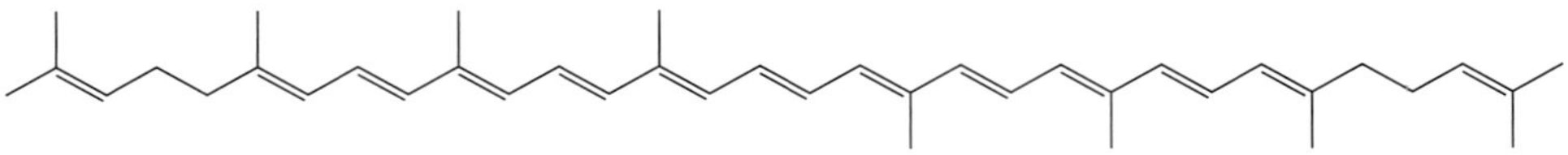

Biological occurrence	Minor or trace pigment in mesostigmatophytes (also characteristic of tomatoes)
Source culture	*Mesostigma viride* (mesostigmatophyte)
Alteration products	*Cis*-isomers
Biosynthetically related to	The biosynthetic precursor to all carotenoids
Occurs together with	

HPLC chromatogram of mixed standards (system 1)

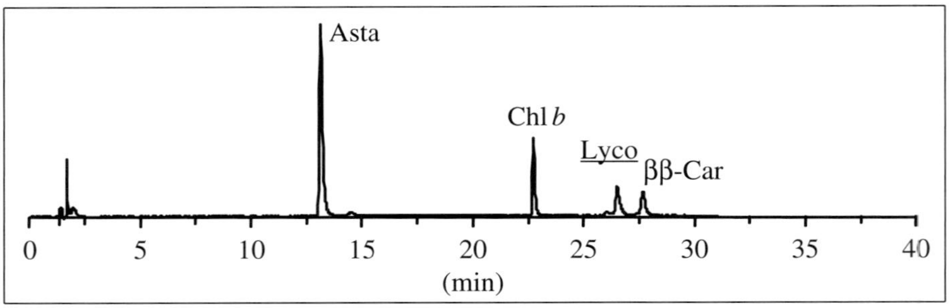

HPLC chromatogram (system 2)

NO DATA AVAILABLE

UV-Vis spectra (see also reference spectra below)

Solvent	λ_{max} (nm)	Band ratio (% III:II)	Ref.
Acetone	448, *474*, 506	84	[2]
Ethanol	443, *472*, 502	n.d.	[113]
Hexane	448, *473*, 504	n.d.	[14]
Recommended specific absorption coefficient *d* (L g^{-1} cm^{-1})		347 (at 473 nm, hexane) [181] 345 (at 474 nm, acetone) [2]	

Reference spectra

In acetone

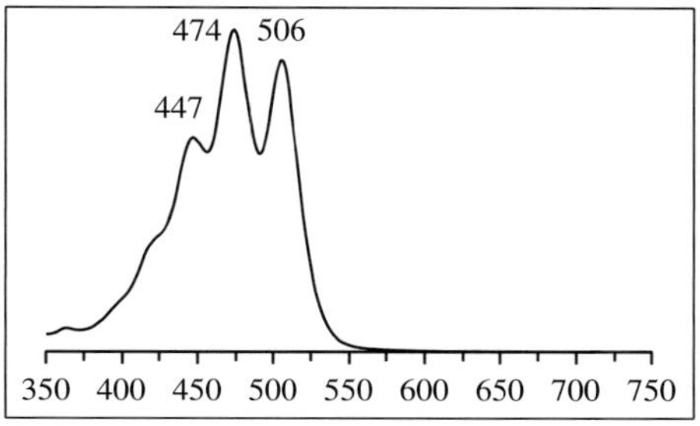

In hexane

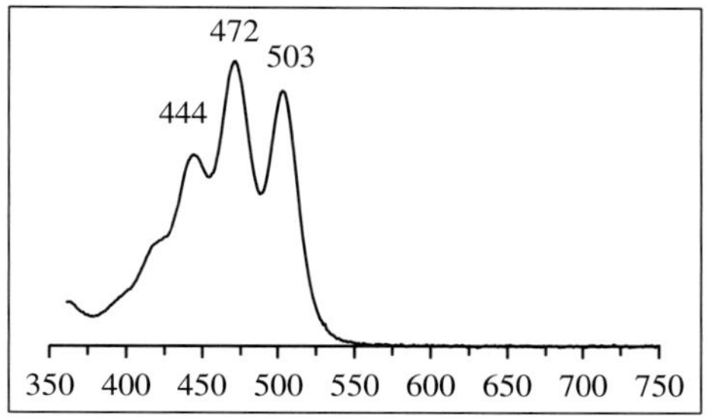

In HPLC solvent system 1

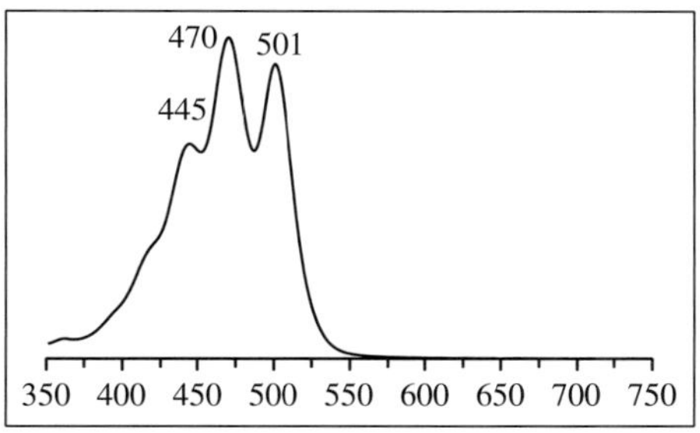

In HPLC solvent system 2

NO DATA AVAILABLE

Mass spectra

Ionization technique	Mass analyser type	Diagnostic ions (m/z, rel. intensity)	Ref.
EI	Magnetic sector	536 $[M]^+$ (22), 467 $[M-69]^+$ (4), 444 $[M-92]^+$ (2), 430 $[M-106]^+$ (4), 109 (19), 91 (47), 69 (100)	[55]

Remarks Mesostigmatophytes are found in freshwater environments

3 Xanthophylls

Alloxanthin

Recommended abbreviation: Allo (Al)

IUPAC: (3*R*,3′*R*)-7,8,7′,8′-Tetradehydro-β,β-carotene-3,3′-diol

Molecular formula: $C_{40}H_{52}O_2$

Molecular weight: 564.84

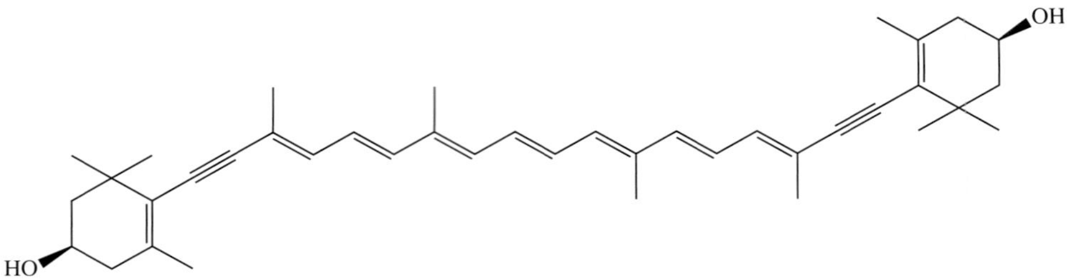

Biological occurrence	Major in cryptophytes, found in two chlorophytes (see Chapter 1, this volume)
Source culture	*Chroomonas salina* (cryptophyte)
Alteration products	Easily isomerizes to the more stable 9,9′-di*cis* isomer [43]
Biosynthetically related to	Diadino
Occurs together with	Chl c_2, βε-Car, Croco, Monado

HPLC chromatogram of *Chroomonas salina* (system 1)

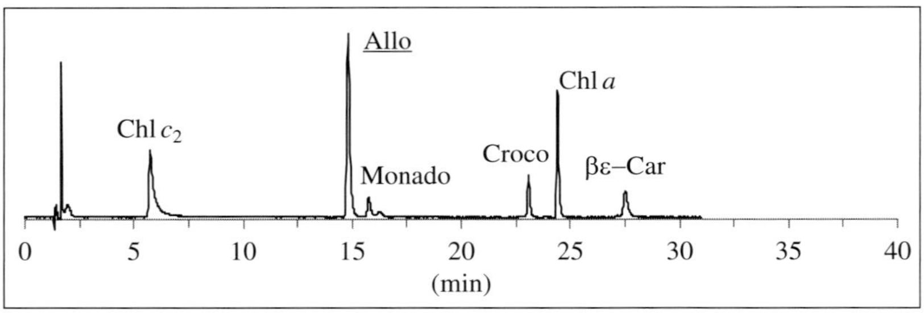

HPLC chromatogram of *Rhodomonas baltica* (system 2)

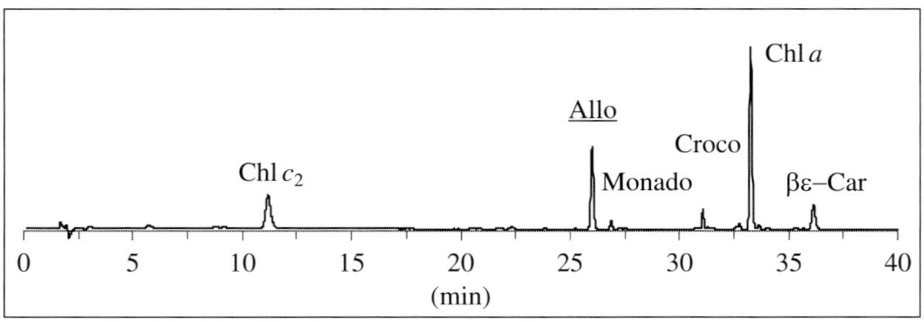

UV-Vis spectra (see also reference spectra below)

Solvent	λ_{max} (nm)	Band ratio (% III:II)	Ref.
Acetone	(428), *454*, 484	50	[145]
Diethyl ether	(430), *451*, 481	44	[133]
Ethanol	(427), *450*, 478	29	[87]
Hexane	(427), *451*, 480	n.d.	[37]
Recommended specific absorption coefficient *d* ($L\ g^{-1}\ cm^{-1}$)		216 (at 464 nm, in benzene) [43] 250 (at 454 nm, in acetone) [109]	

Reference spectra

In acetone

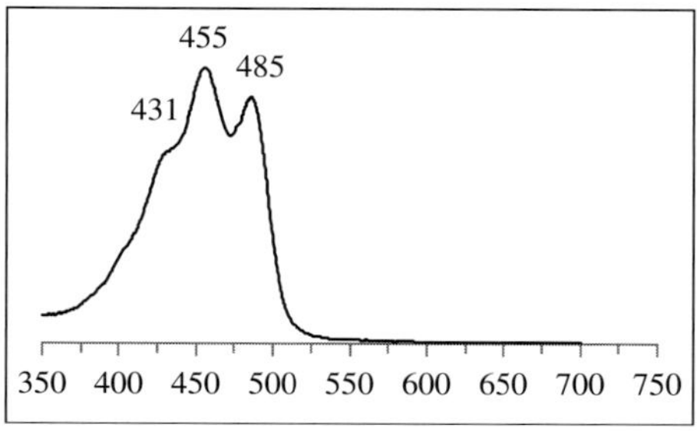

In HPLC solvent system 1

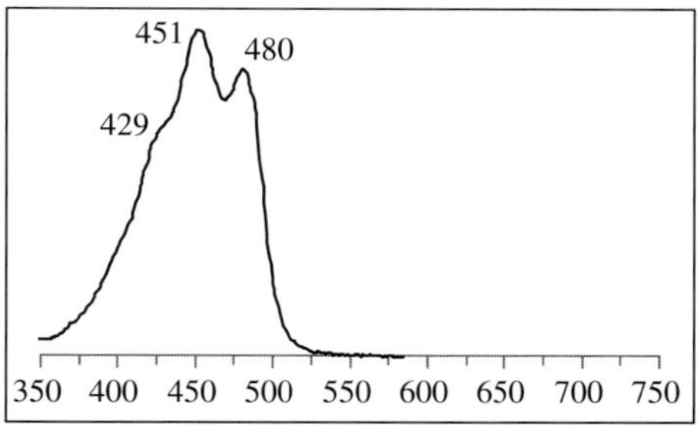

In HPLC solvent system 2

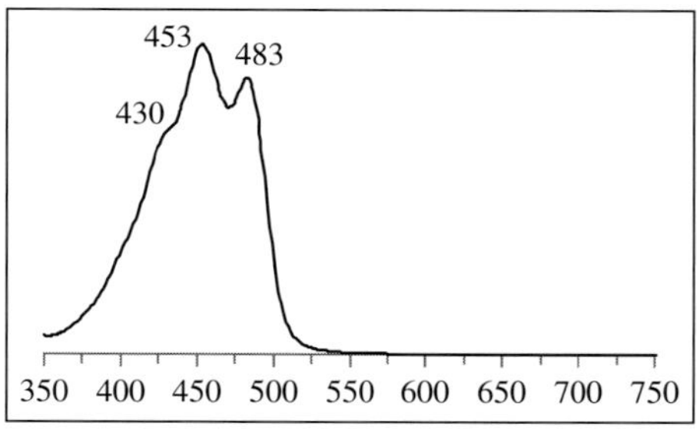

Mass spectra

Ionization technique	Mass analyser type	Diagnostic ions (m/z, rel. intensity)	Ref.
EI	Magnetic sector	564.398 $[M]^+$ (50), 549 $[M\text{-}15]^+$ (1), 546 $[M\text{-}18]^+$ (1), 119 (14), 105 (18), 91 (25), 41 (100)	[43]

Remarks

Antheraxanthin

Recommended abbreviation: Anth (An)

IUPAC: (3*S*,5*R*,6*S*,3′*R*)-5,6-Epoxy-5,6-dihydro-β,β-carotene-3,3′-diol

Molecular formula: $C_{40}H_{56}O_3$ **Molecular weight:** 584.87

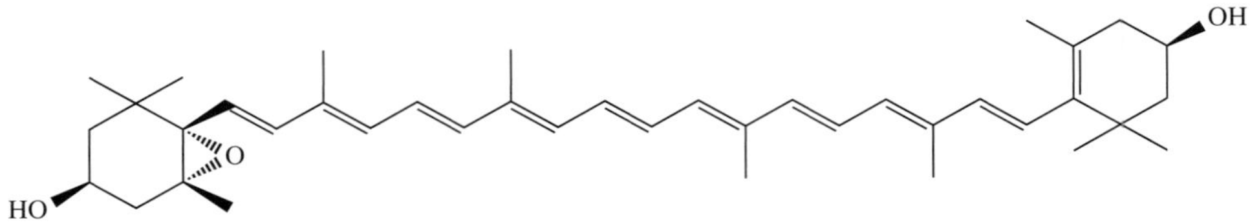

Biological occurrence	Minor pigment in chlorophytes, prasinophytes, trebouxiophytes, mesostigmatophytes, chlorarachniophytes, and some chrysophytes and eustigmatophytes. Also found in seaweeds and plants. Major in anthers of some flowers [78]
Source culture	*Dunaliella tertiolecta* (chlorophyte)
Alteration products	Undergoes rearrangement to Mutato in weakly acidic solutions. *Cis*-isomers
Biosynthetically related to	Biosynthetic intermediate between Zea and Viola
Occurs together with	Zea, Viola

HPLC chromatogram of *Dunaliella tertiolecta* (system 1)

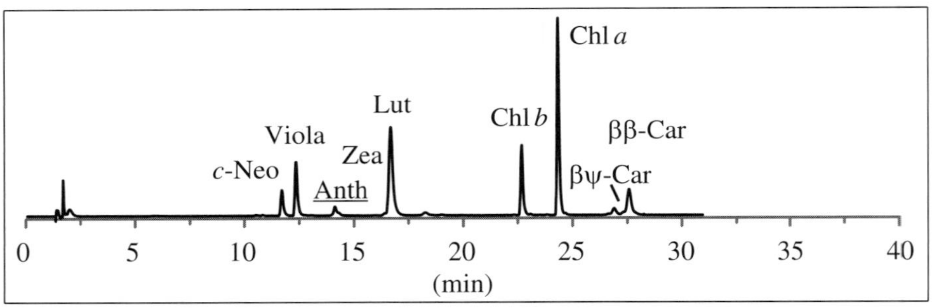

HPLC chromatogram of *Micromonas pusilla* (system 2)

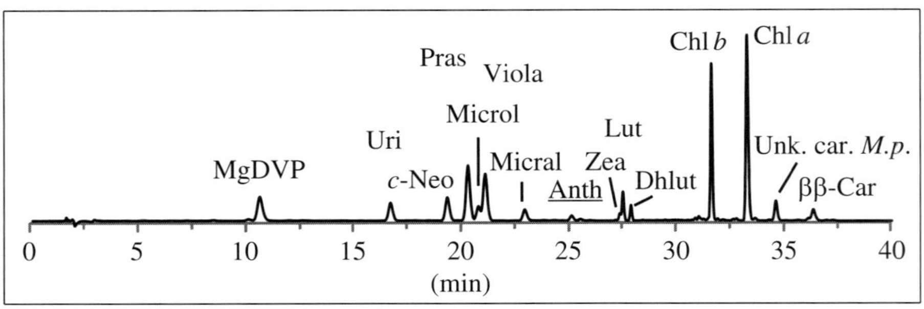

UV-Vis spectra (see also reference spectra below)

Solvent	λ_{max} (nm)	Band ratio (% III:II)	Ref.
Acetone	(422), *448*, 475	29	[72]
Ethanol	422, *444*, 472	54	[152]
Hexane	420, *444*, 472	n.d.	[85]
Recommended specific absorption coefficient *d* (L g^{-1} cm^{-1})		235 (at 446 nm, ethanol)	[84]

Reference spectra

In acetone

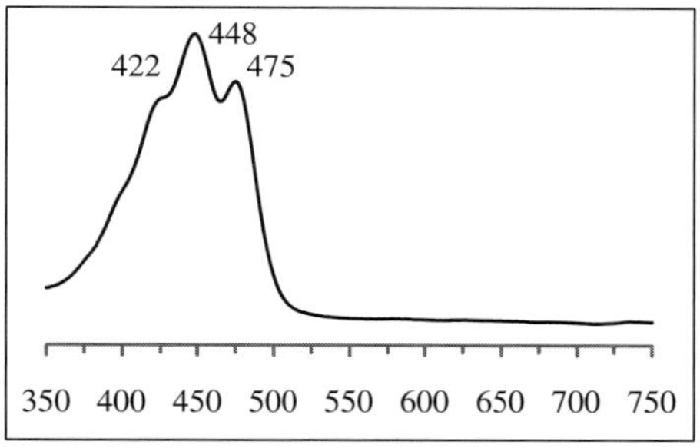

In ethanol

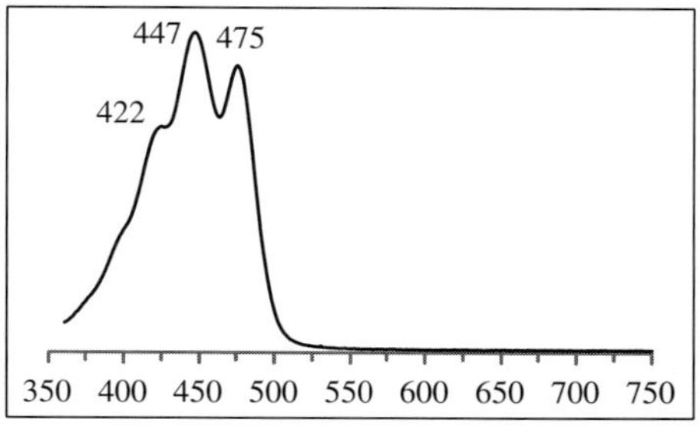

In HPLC solvent system 1

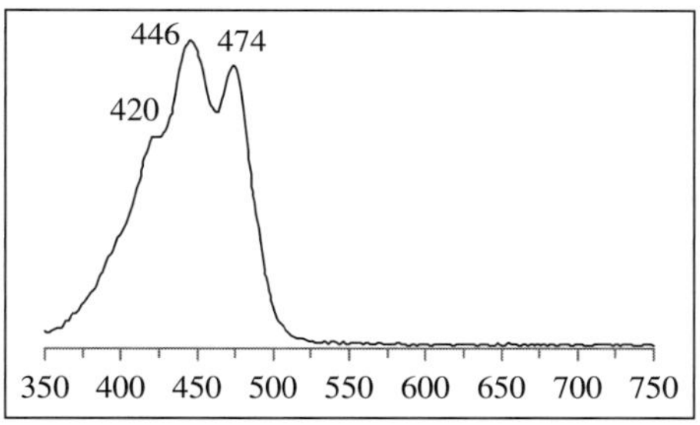

In HPLC solvent system 2

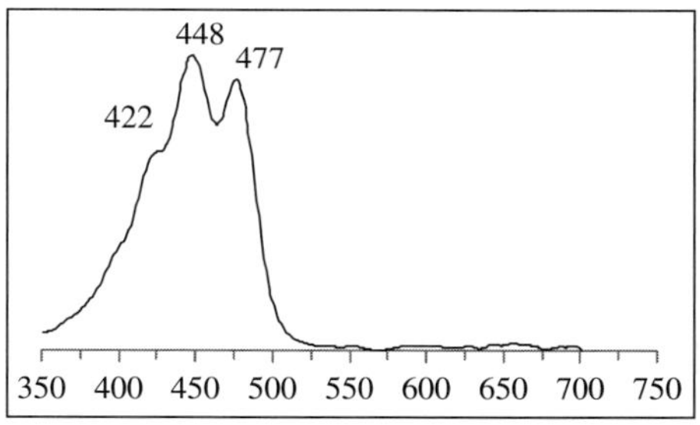

Mass spectra

Ionization technique	Mass analyser type	Diagnostic ions (m/z, rel. intensity)	Ref.
EI	Magnetic sector	584 $[M]^+$ (60), 504 $[M\text{-}80]^+$ (30), 492 $[M\text{-}92]^+$ (6), 221 (30), 181 (10), 43 (100)	[127]

Remarks Part of a 'xanthophyll cycle' (see Chapter 11, this volume). May be present in diatoms under prolonged high light stress [123]

Astaxanthin

Recommended abbreviation: Asta (As)

IUPAC: (3*S*,3′*S*)-3,3′-Dihydroxy-β,β-carotene-4,4′-dione

Molecular formula: $C_{40}H_{52}O_4$ **Molecular weight:** 596.84

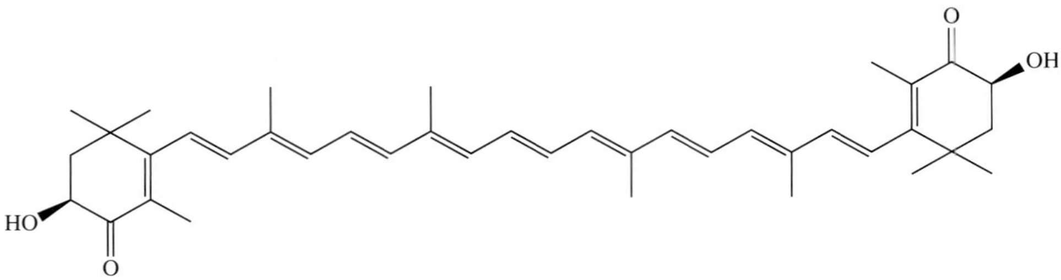

Biological occurrence	Present in some chlorophytes, but major (mainly as mono- and diesters of fatty acids) in some of these upon nitrogen starvation. Major in salmonids and crustaceans [78, 79]
Source culture	Synthetic (see Appendix E, this volume)
Alteration products	*Cis*-isomers; astacene (oxidation product)
Biosynthetically related to	Zea, Cantha
Occurs together with	

HPLC chromatogram of *Pycnococcus provasolii* (system 1)

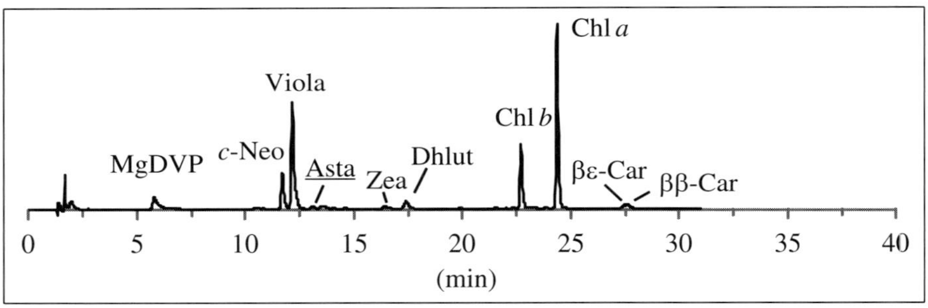

HPLC chromatogram of synthetic astaxanthin (system 2)

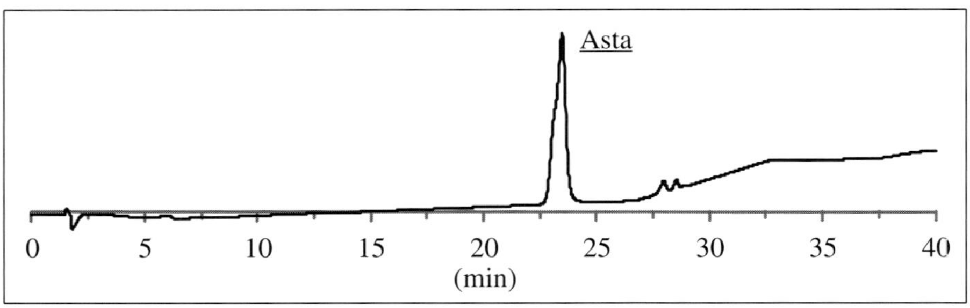

UV-Vis spectra (see also reference spectra below)

Solvent	λ_{max} (nm)	Band ratio (% III:II)	Ref.
Acetone	475	-	[82]
Ethanol	476	-	[33]
Hexane	466–467	-	[41]
Methanol	470–472	-	[41]
Recommended specific absorption coefficient *d* ($L\ g^{-1}\ cm^{-1}$)		206 (at 473 nm, methanol)	[79]

Reference spectra

In acetone

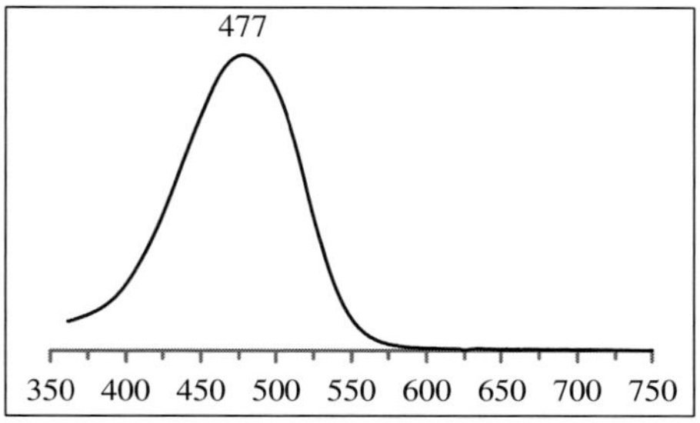

In methanol

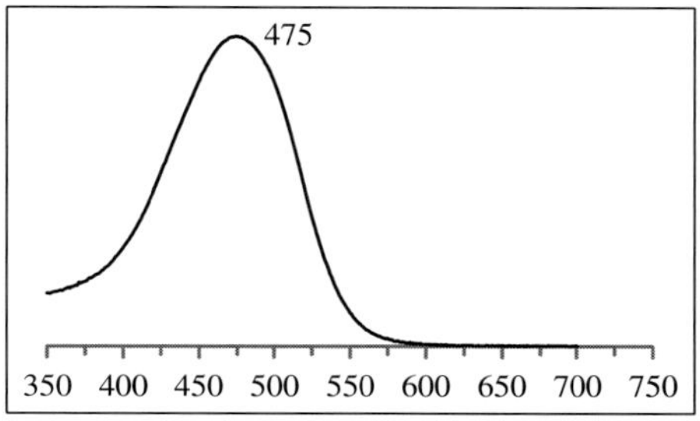

In HPLC solvent system 1

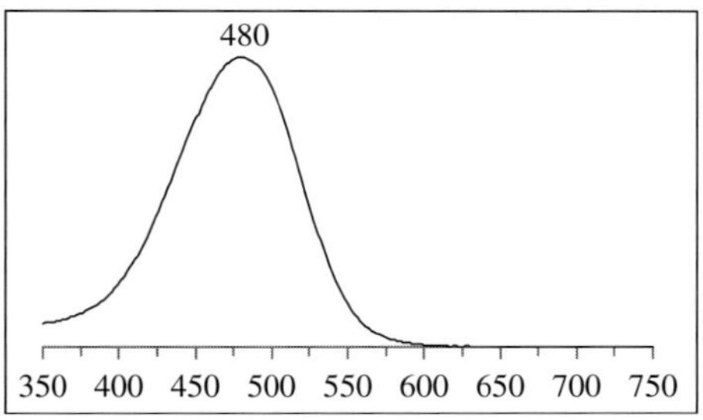

In HPLC solvent system 2

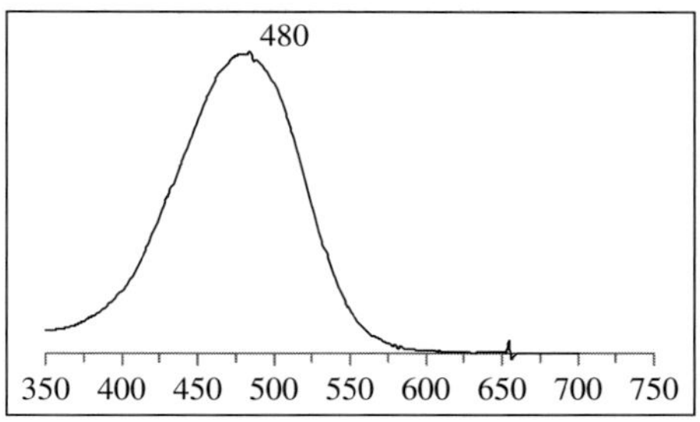

Mass spectra

Ionization technique	Mass analyser type	Diagnostic ions (m/z, rel. intensity)	Ref.
EI	Magnetic sector	596 $[M]^+$ (7), 580 $[M\text{-}16]^+$ (9), 564 $[M\text{-}16\text{–}16]^+$ (6), 133 (56), 109 (24), 91 (100)	[55]

Remarks Mixtures of (3*S*,3′*S*), (3*R*,3′*R*) and (3*R*,3′*S*)-Asta are common in aquatic animals [15]. Esterification will change polarity and hence retention time

Auroxanthin

Recommended abbreviation: Auro (Au)

IUPAC: (3*S*,5*R*,8*RS*,3′*S*,5′*R*,8′*RS*)-5,8:5′,8′-Diepoxy-5,8,5′,8′-tetrahydro-β,β-carotene-3,3′-diol

Molecular formula: $C_{40}H_{56}O_4$

Molecular weight: 600.87

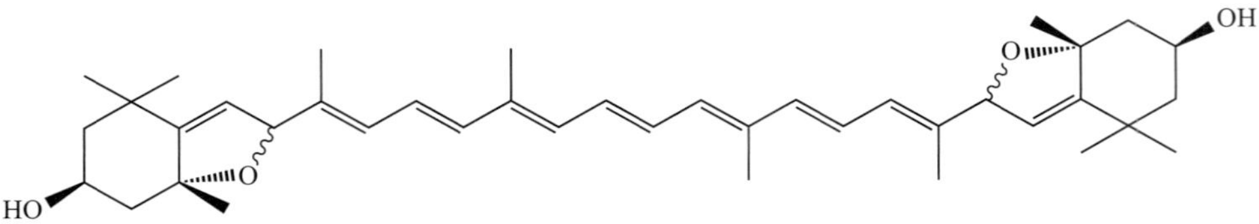

(Always occurs as a mixture of the three (3*S*,5*R*,8*R*,3′*S*,5′*R*,8′*R*), (3*S*,5*R*,8*R*,3′*S*,5′*R*,8′*S*) and (3*S*,5*R*,8*S*,3′*S*,5′*R*,8′*S*) optical isomers)

Alteration product of	Viola. The acid-catalysed rearrangement occurs in slightly acidic extracts, especially in prasinophyte extracts [91]
Source culture	*Dunaliella tertiolecta* (chlorophyte)
Alteration products	*Cis*-isomers
Synthetically related to	Viola
Occurs together with	Luteoxanthin, Neochr

HPLC chromatogram of acidified *Dunaliella tertiolecta* (system 1)

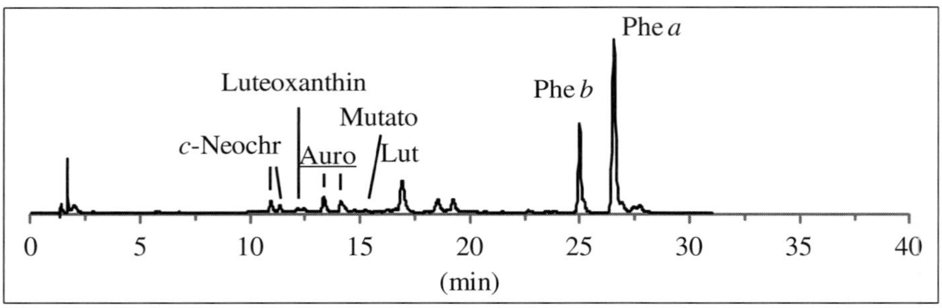

HPLC chromatogram (system 2)

NO DATA AVAILABLE

UV-Vis spectra (see also reference spectra below)

Solvent	**λ_{max} (nm)**	**Band ratio (% III:II)**	**Ref.**
Acetone	380, 401, *425*	125	[51]
Ethanol	379, *400*, 425	92	[136]
Hexane	380, 400, 425	n.d.	[167]
Recommended specific absorption coefficient *d* ($L\ g^{-1}\ cm^{-1}$)		181 (at 403 nm, ethanol) [114]	

Reference Spectra

In HPLC solvent system 1

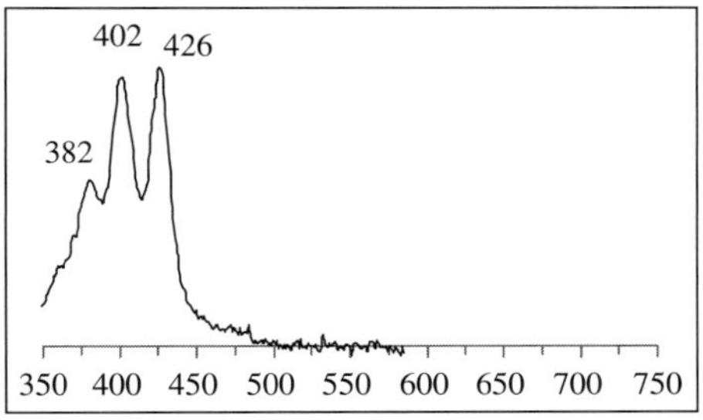

In HPLC solvent system 2

NO DATA AVAILABLE

Mass spectra

Ionization technique	**Mass analyser type**	**Diagnostic ions (m/z, rel. intensity)**	**Ref.**
EI	Magnetic sector	600 $[M]^+$ (44), 584 $[M\text{-}16]^+$ (22), 582 $[M\text{-}18]^+$ (11), 568 $[M\text{-}16\text{–}16]^+$ (12), 221 (100), 181 (68)	[51]

Remarks Turns green on silica TLC (partly ionised to a blue oxonium ion)

19′-Butanoyloxyfucoxanthin — Recommended abbreviation: But-fuco (BF)

IUPAC: (3*S*,5*R*,6*S*,3′*S*,5′*R*,6′*R*)-19′-Butanoyloxy-5,6-epoxy-3′-ethanoyloxy-3,5′-dihydroxy-6′,7′-didehydro-5,6,7,8,5′,6′-hexahydro-β,β-caroten-8-one

Molecular formula: $C_{46}H_{64}O_8$ **Molecular weight:** 745.00

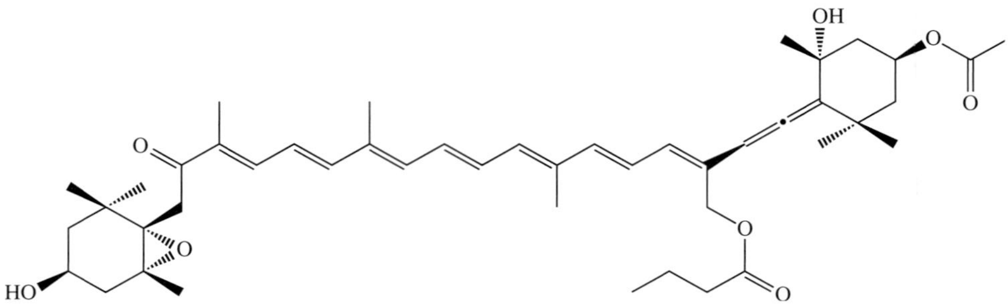

Biological occurrence	Dominant pigment in pelagophytes and dictyochophytes. Also present in some haptophytes; trace amounts in dinoflagellates Pigment Type 2 (see Chapter 1)
Source culture	*Pelagococcus subviridis* (pelagophyte)
Alteration products	*Cis*-isomers
Biosynthetically related to	Fuco, Hex-fuco, and other alkanoyloxy-derivatives of Fuco
Occurs together with	Chl c_2, Fuco, Diato, Diadino

HPLC chromatogram of *Karlodinium micrum* (system 1)

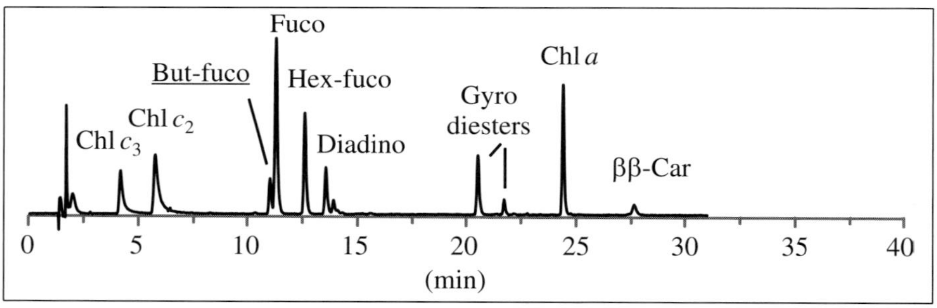

HPLC chromatogram of *Karlodinium micrum* (system 2)

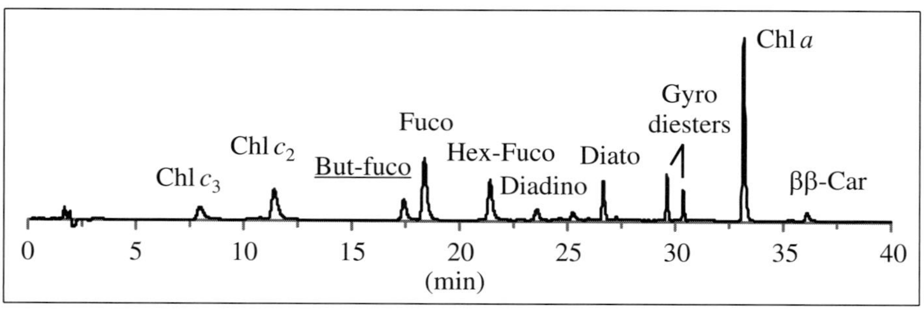

UV-Vis spectra (see also reference spectra below)

Solvent	λ_{max} (nm)	Band ratio (% III:II)	Ref.
Acetone	(423), *446*, 472	37	[23]
Ethanol	*446*, 470	18	[173]
Hexane	(426), *446*, 473	57	[23]
Recommended specific absorption coefficient *d* ($L\ g^{-1}\ cm^{-1}$)		n.d., see Remarks	

Reference spectra
In acetone

For spectrum in acetone, see [109]

In HPLC solvent system 1

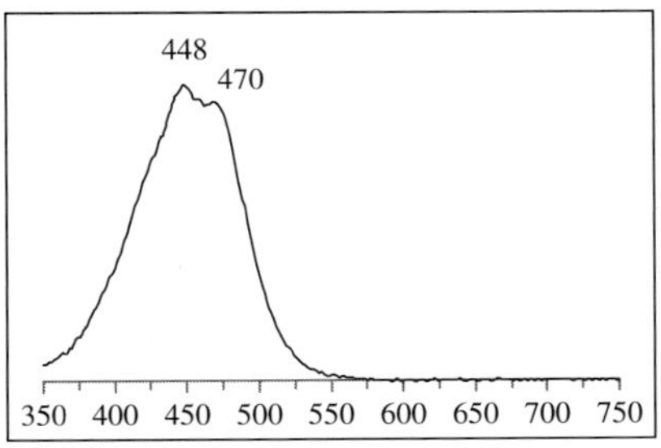

In HPLC solvent system 2

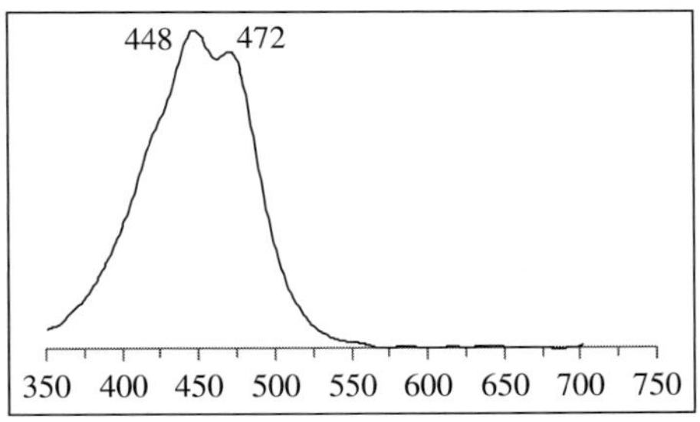

Mass spectra

Ionization technique	Mass analyser type	Diagnostic ions (m/z, rel. intensity)	Ref.
APCI	Ion trap	767 $[M+Na]^+ \rightarrow$ 679 $[M+Na\text{-}88]^+ \rightarrow$ 661 $[M+Na\text{-}88\text{–}18]^+$, 619 $[M+Na\text{-}88\text{–}60]^+$, 525	[5]

Remarks $d = 147\ L\ g^{-1}\ cm^{-1}$ (at 445 nm in acetone; calc. from Fuco) is recommended, as no value has been determined for But-fuco [109]. Several other, minor alkanoyloxyfucoxanthins have been reported [5]

Caloxanthin

Recommended abbreviation: Calo (Cal)

IUPAC: (2*R*,3*R*,3′*R*)-β,β -Carotene-2,3,3′-triol

Molecular formula: $C_{40}H_{56}O_3$ **Molecular weight:** 584.87

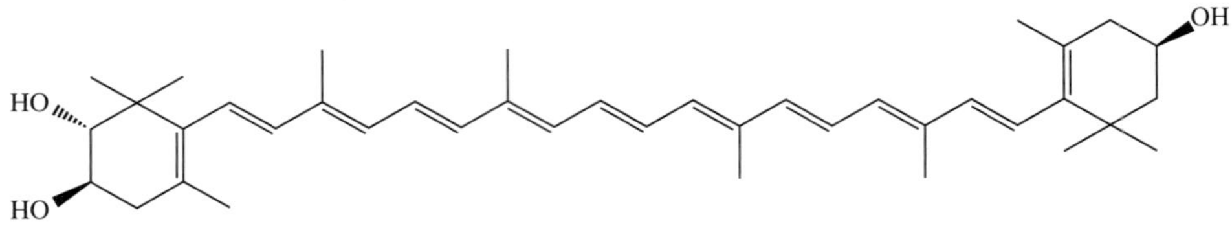

Biological occurrence	Major characteristic carotenoid from some cyanobacteria (Cyano-1, Chapter 1), e.g. *Anacystis nidulans* and *Chlorogloeopsis fritschii*
Source culture	*Chlorogloeopsis fritschii* (cyanobacteria)
Alteration products	*Cis*-isomers
Biosynthetically related to	ββ-Car, Zea, Nosto
Occurs together with	Nosto, Zea

HPLC chromatogram of a *Chlorogloeopsis fritschii* (system 3)

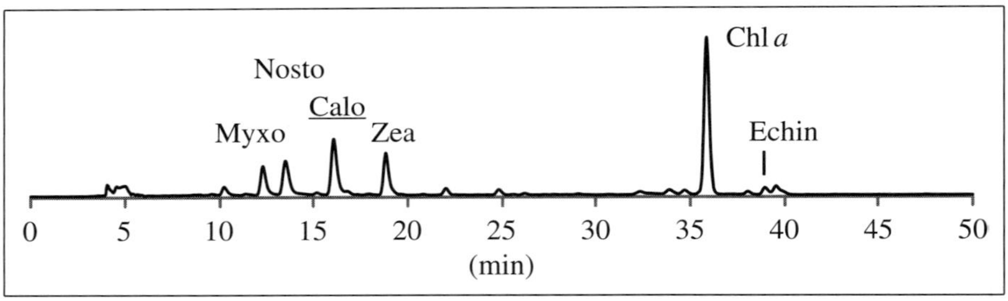

HPLC chromatogram (system 2)

NO DATA AVAILABLE

UV-Vis spectra (see also reference spectra below)

Solvent	λ_{max} **(nm)**	**Band ratio (% III:II)**	**Ref.**
Acetone	(430), *454*, 481	n.d.	[33]
Ethanol	(425), *450*, 478	n.d.	[146]
Recommended specific absorption coefficient *d* ($L\ g^{-1}\ cm^{-1}$)		n.d., see Remarks	

Reference spectra
In acetone

NO DATA AVAILABLE

In HPLC solvent system 3

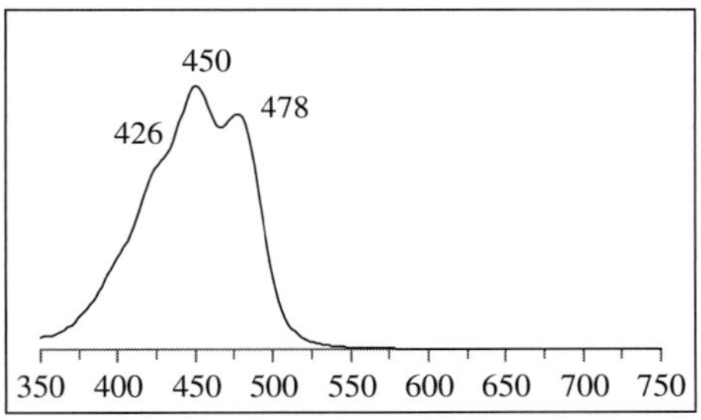

Mass spectra

Ionization technique	**Mass analyser type**	**Diagnostic ions (m/z, rel. intensity)**	**Ref.**
EI	Magnetic sector	584 $[M]^+$ (100), 566 $[M\text{-}18]^+$ (10), 492 $[M\text{-}92]^+$ (13), 478 $(M\text{-}106)^+$ (3), 133 (24), 91 (43), 83 (22), 62 (42)	[33]

Remarks $d = 238\ L\ g^{-1}\ cm^{-1}$ (at λ_{max} in ethanol; calc. from Zea) is recommended, as no value has been determined for Calo

Canthaxanthin

Recommended abbreviation: Cantha (Ct)

IUPAC: β,β-Carotene-4,4′-dione

Molecular formula: $C_{40}H_{52}O_2$

Molecular weight: 564.84

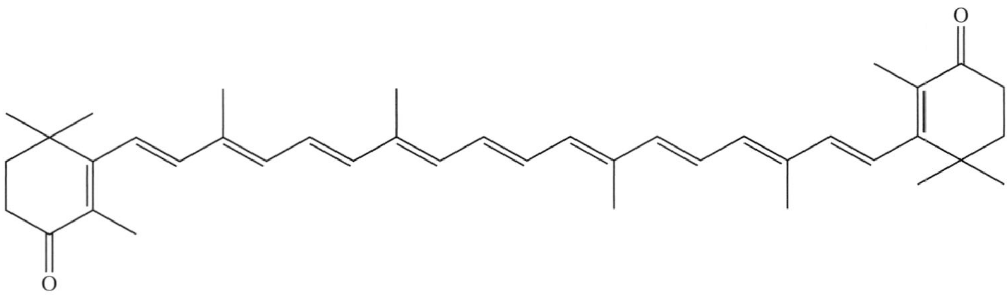

Biological occurrence	Minor or trace pigment in eustigmatophytes, cyanobacteria (Cyano-1) and some dinoflagellates. Detected in some cultures of chlorophytes, diatoms and prymnesiophytes. Major in some chlorophytes upon nitrogen starvation [78]
Source culture	Synthetic (see Appendix E, this volume)
Alteration products	*Cis*-isomers
Biosynthetically related to	ββ-Car, Echin, Asta
Occurs together with	

HPLC chromatogram of *Guillardia theta* – canthaxanthin: internal standard (system 1)

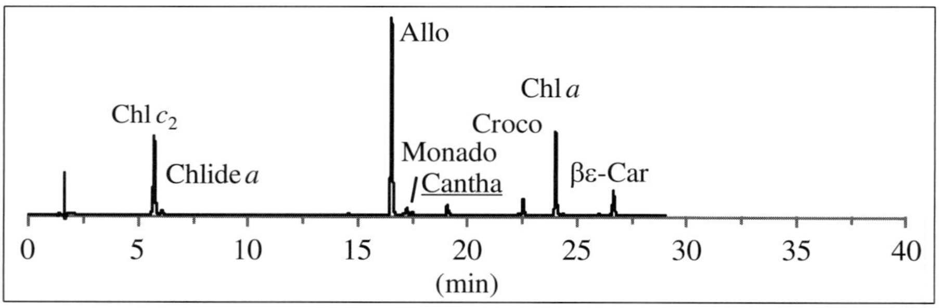

HPLC chromatogram of synthetic canthaxanthin (system 2)

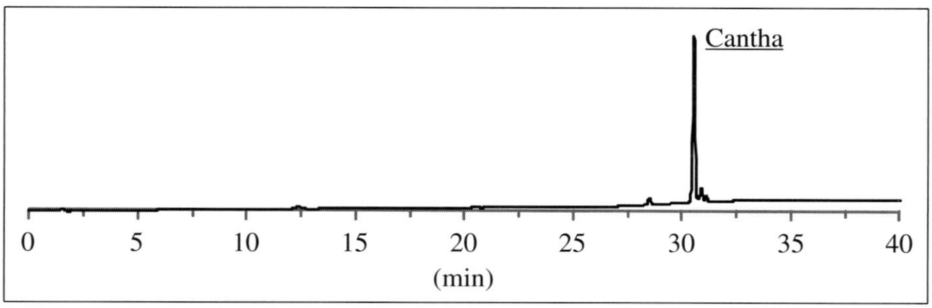

UV-Vis spectra (see also reference spectra below)

Solvent	λ_{max} (nm)	Band ratio (% III:II)	Ref.
Acetone	468	-	[82]
Ethanol	478	-	[135]
Hexane	468	-	[54]
Recommended specific absorption coefficient *d* **($L\ g^{-1}\ cm^{-1}$)**		220 (at 469 nm, cyclohexane)	[155]

Reference spectra

In acetone

For spectrum in acetone, see [109]

In methanol

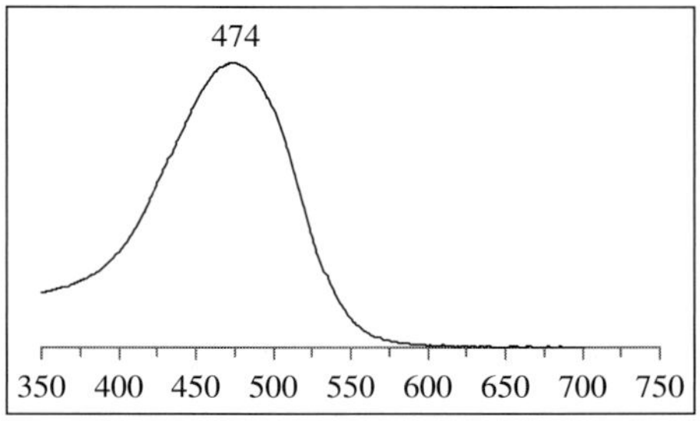

In HPLC solvent system 1

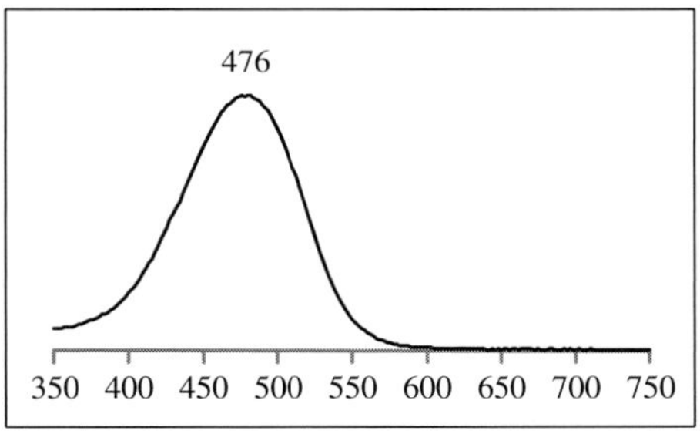

In HPLC solvent system 2

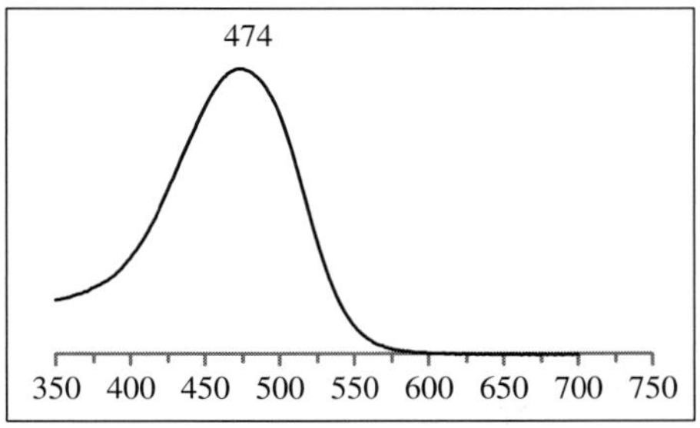

Mass spectra

Ionization technique	Mass analyser type	Diagnostic ions (m/z, rel. intensity)	Ref.
EI	Magnetic sector	564 $[M]^+$ (41), 562 $[M\text{-}2]^+$ (2), 472 $[M\text{-}92]^+$ (7), 458 $[M\text{-}106]^+$ (2), 91 (19), 83 (100)	[55]

Remarks

Crocoxanthin

Recommended abbreviation: Croco (Co)

IUPAC: (3*R*,6′*R*)-7,8-Didehydro-β,ε-caroten-3-ol

Molecular formula: $C_{40}H_{54}O$ **Molecular weight:** 550.86

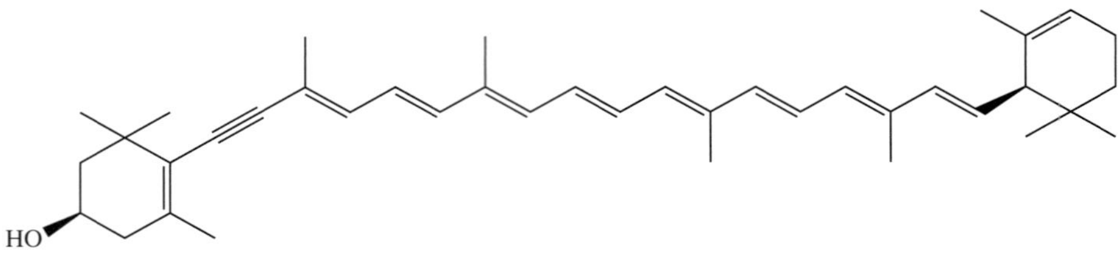

Biological occurrence	Cryptophytes only (minor carotenoid)
Source culture	*Chroomonas salina* (cryptophyte)
Alteration products	Easily isomerises to the more stable 9-*cis* isomer [43]
Biosynthetically related to	Monado
Occurs together with	Allo, Monado

HPLC chromatogram of *Chroomonas salina* (system 1)

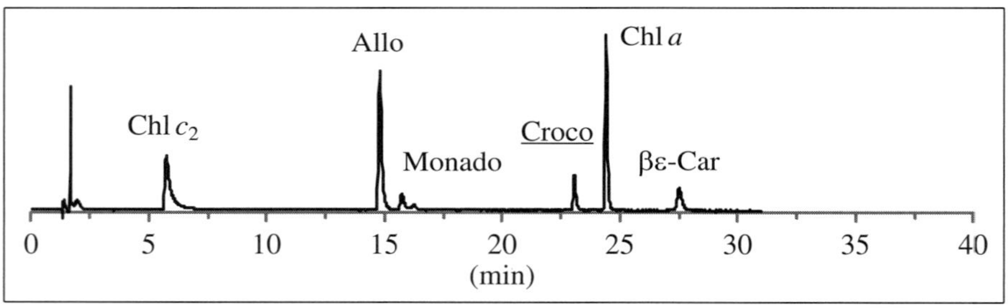

HPLC chromatogram of *Rhodomonas baltica* (system 2)

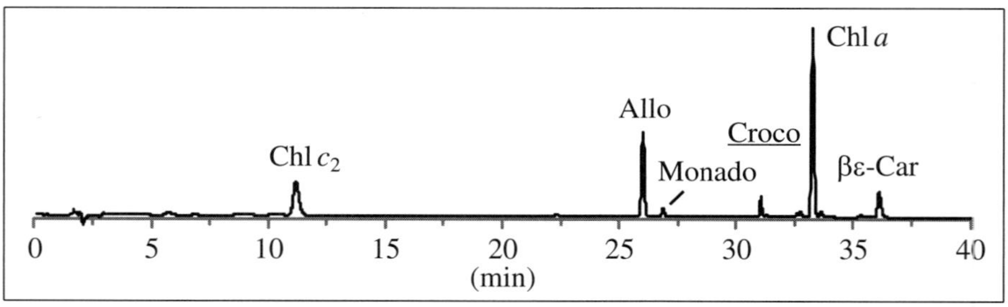

UV-Vis spectra (see also reference spectra below)

Solvent	λ_{max} (nm)	Band ratio (% III:II)	Ref.
Diethyl ether	428, *445*, 475	58	[133]
Ethanol	(421), *443*, 472	62	[87]
Hexane	422, *445*, 475	69	[35]
Recommended specific absorption coefficient *d* (L g^{-1} cm^{-1})		n.d., see Remarks	

Reference spectra

In acetone

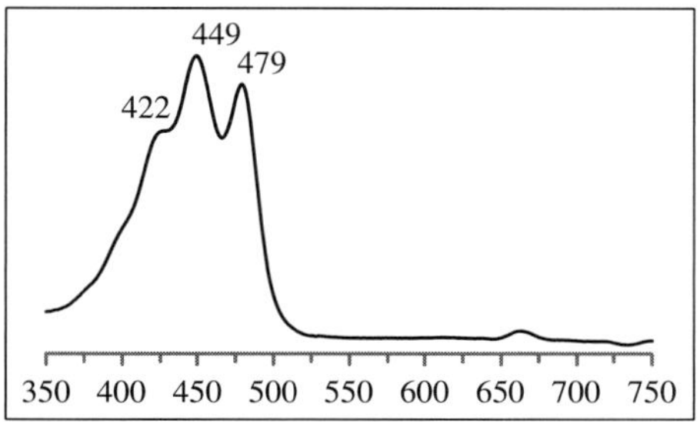

In hexane

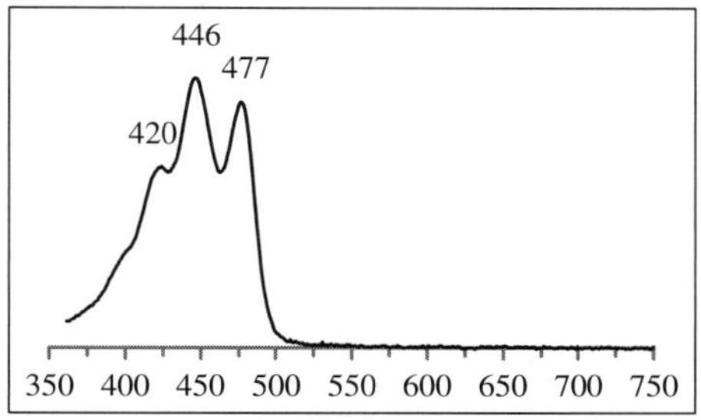

In HPLC solvent system 1

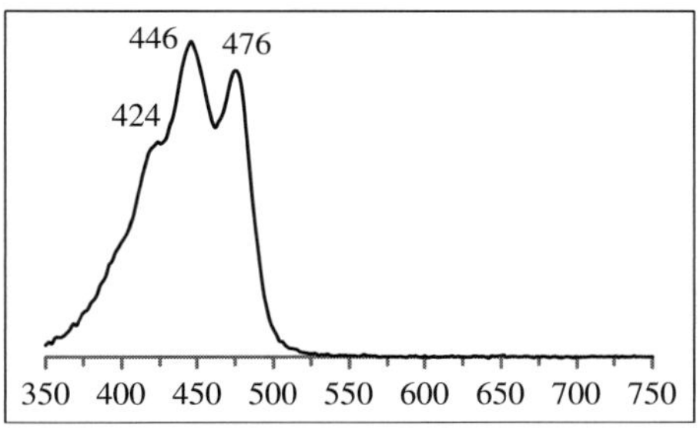

In HPLC solvent system 2

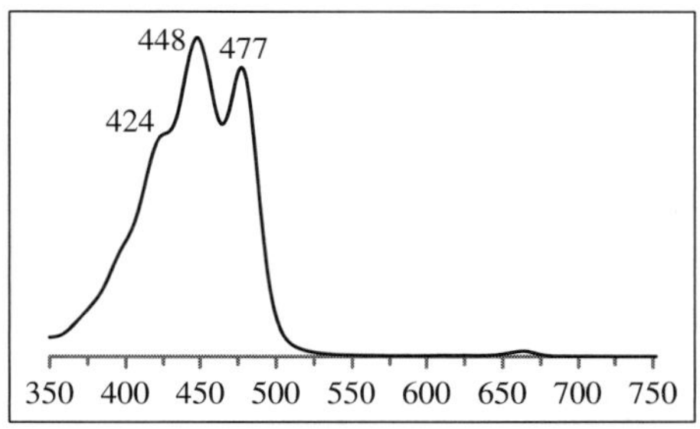

Mass spectra

Ionization technique	Mass analyser type	Diagnostic ions (m/z, rel. intensity)	Ref.
EI	Magnetic sector	550.421 $[M]^+$ (100), 535 $[M\text{-}15]^+$ (2), 532 $[M\text{-}18]^+$ (1), 494 $[M\text{-}56]^+$ (<1), 458 [M-92, 2], 444 [m*, 550 → 494], 119 (17), 105 (21), 92 (26)	[43]

Remarks $d = 237\ \mathrm{L\,g^{-1}cm^{-1}}$ (at 445 nm in hexane; calc. from Diadino) is recommended, as no value has been determined for Croco. In ethanol, use the value for β,β–carotene: $d = 250\ \mathrm{L\,g^{-1}cm^{-1}}$ (at 443 nm) [45, 109]

Cryptoxanthin

Recommended abbreviation: Cryp (Cy)

IUPAC: (3*R*)-β,β-Caroten-3-ol

Molecular formula: $C_{40}H_{56}O$

Molecular weight: 552.87

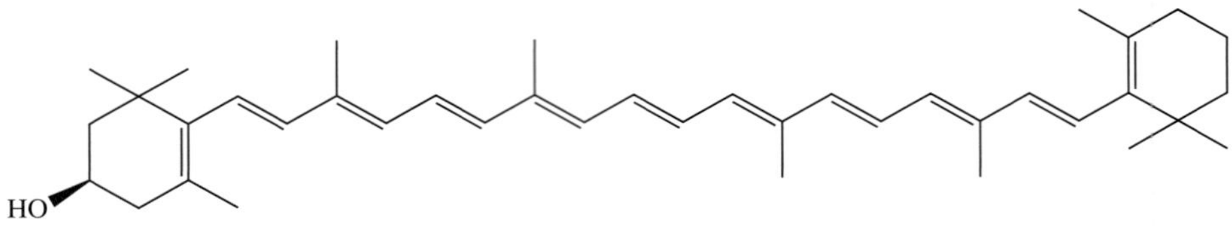

Biological occurrence	Dominant pigment in glaucocystophytes, minor pigment in some cyanobacteria (Cyano-1) and prochlorophytes (*Prochloron* spp., Cyano-3) (see Chapter 1, this volume)
Source culture	*Prochloron* spp.
Alteration products	*Cis*-isomers
Biosynthetically related to	The biosynthetic intermediate between ββ-Car and Zea
Occurs together with	ββ-Car, Zea

HPLC chromatogram (system 1)

NO DATA AVAILABLE

HPLC chromatogram of *Synechococcus* sp. (system 2)

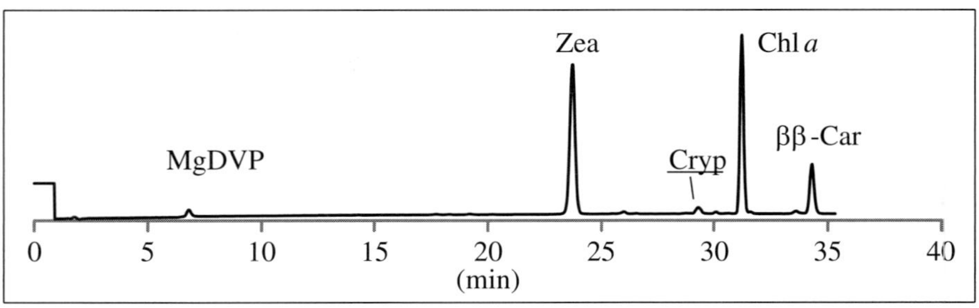

UV-Vis spectra (see also reference spectra below)

Solvent	λ_{max} (nm)	Band ratio (% III:II)	Ref.
Acetone	(427), *450*, 475	n.d.	[64]
Ethanol	(428), *449*, 473	30	[86]
Hexane	(427), *452*, 478	43	[180]
Recommended specific absorption coefficient *d* ($L\ g^{-1}\ cm^{-1}$)		246 (at 452 nm, hexane) [180] 250 (at 452 nm, ethanol) [45]	

Reference spectra
In acetone

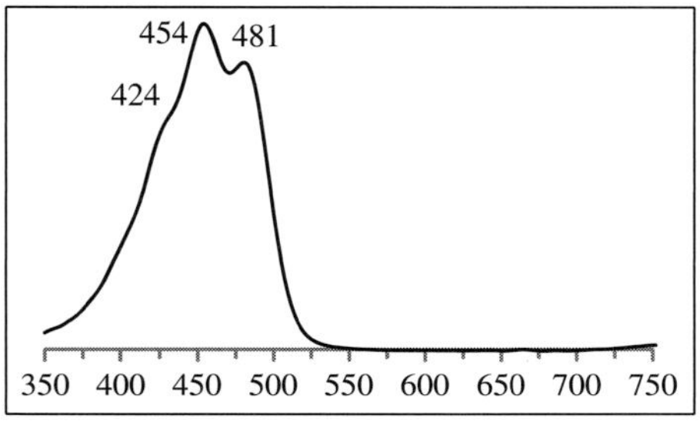

In HPLC solvent system 1

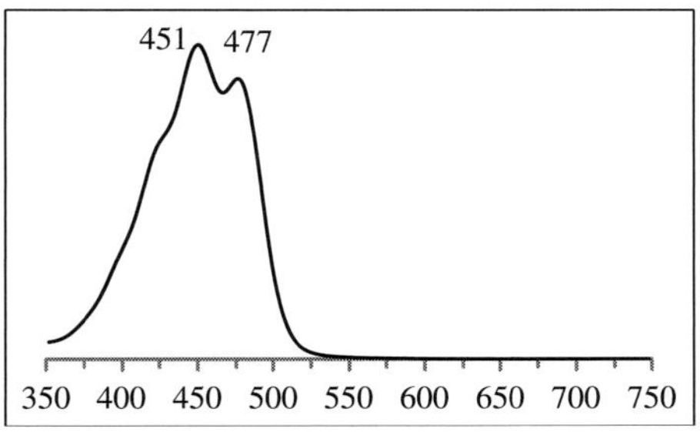

In HPLC solvent system 2

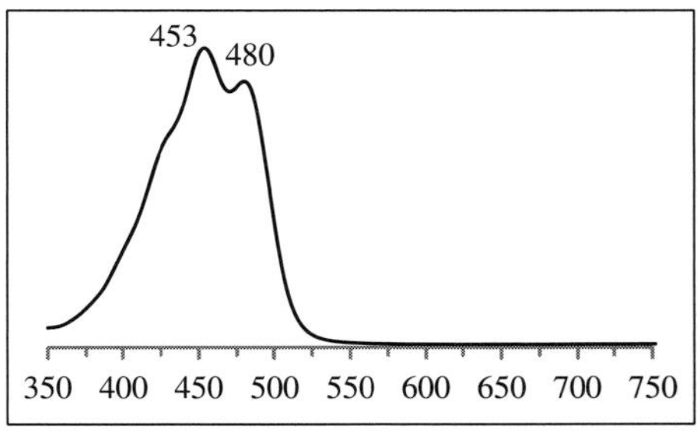

Mass spectra

Ionization technique	Mass analyser type	Diagnostic ions (m/z, rel. intensity)	Ref.
EI	Magnetic sector	552 $[M]^+$ (85), 550 $[M\text{-}2]^+$ (11), 534 $[M\text{-}18]^+$ (100), 460 $[M\text{-}92]^+$ (2), 442 $[M\text{-}18\text{–}92]^+$ (3), 267 $[M\text{-}18]^+$ (8), 105 (55)	[20]

Remarks

Diadinochrome

Recommended abbreviation: Diadchr (Ddc)

IUPAC: (3*S*,5*R*,8*RS*,3′*R*)-5,8-Epoxy-7′,8′-didehydro-5,8-dihydro-β,β-carotene-3,3′-diol

Molecular formula: $C_{40}H_{54}O_3$ **Molecular weight:** 582.85

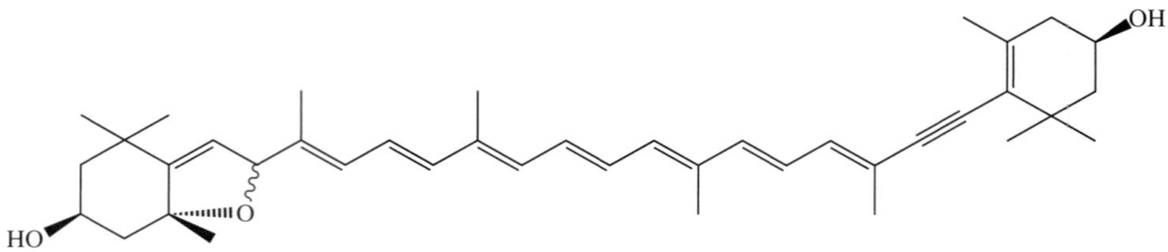

(Always occurs as a mixture of the two (3*S*,5*R*,8*R*,3′*R*) and (3*S*,5*R*,8*S*,3′*R*) optical isomers)

Alteration product of	Diadino. The acid-catalysed reaction occurs in slightly acidic extracts [91]
Source culture	*Amphidinium carterae* (dinoflagellate)
Alteration products	*Cis*-isomers
Synthetically related to	Diadino
Occurs together with	Diadino

HPLC chromatogram of *Amphidinium carterae* (system 1)

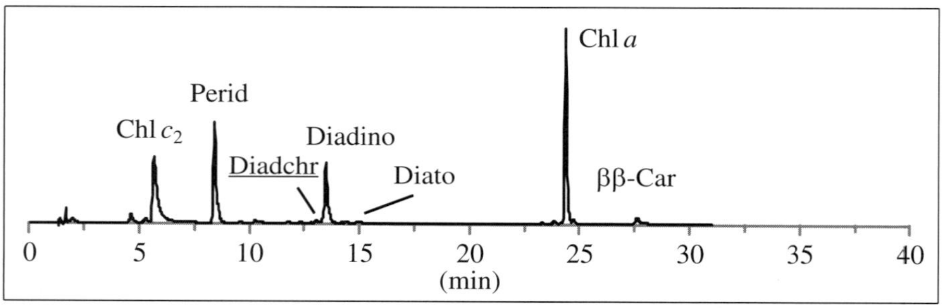

HPLC chromatogram of *Scrippsiella trochoidea* (system 2)

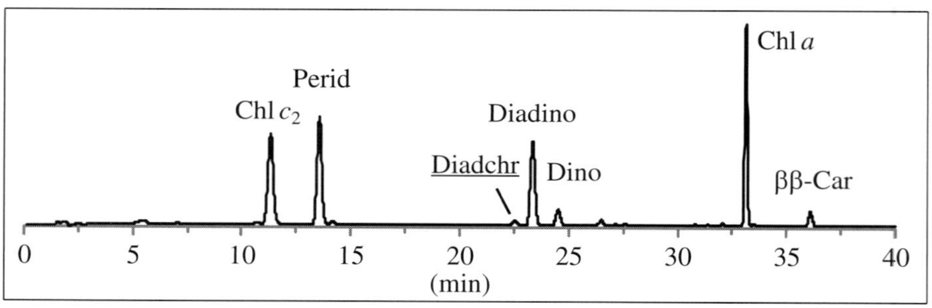

UV-Vis spectra (see also reference spectra below)

Solvent	λ_{max} (nm)	Band ratio (% III:II)	Ref.
Acetone	405, *427*, 454	48	[91]
Diethyl ether	(402), *425*, 452	n.d.	[60]
Ethanol	409, *430*, 457	n.d.	[53]
Recommended specific absorption coefficient *d* (L g^{-1} cm^{-1})		n.d., see Remarks	

Reference spectra

For spectrum in acetone, see [109]

In HPLC solvent system 1

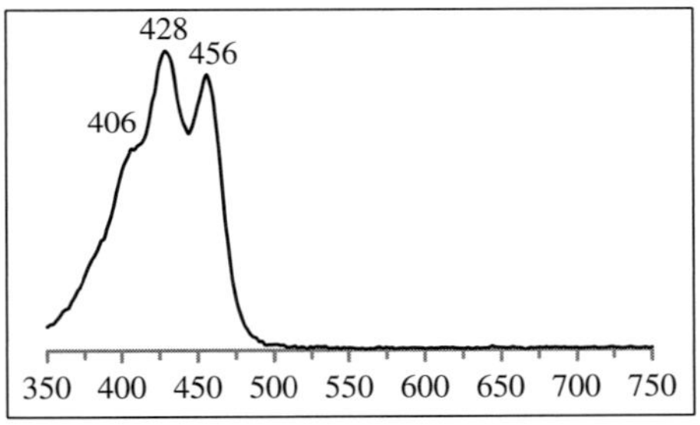

In HPLC solvent system 2

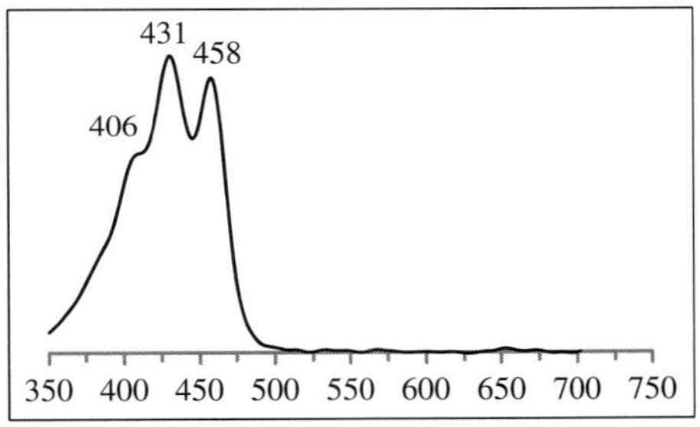

Mass spectra

Ionization technique	Mass analyser type	Diagnostic ions (m/z, rel. intensity)	Ref.
EI	Magnetic sector	582 $[M]^+$ (68), 580 $[M\text{-}2]^+$ (27), 566 $[M\text{-}16]^+$ (4), 564 $[M\text{-}18]^+$ (7), 502 $[M\text{-}80]^+$ (18), 243 (46), 221 (100), 181 (50), 165 (62)	[91]

Remarks $d = 210\,\mathrm{L\,g^{-1}cm^{-1}}$ (at 427 nm in acetone) is recommended, as no value has been determined for Diadchr

Diadinoxanthin

Recommended abbreviation: Diadino (Dd)

IUPAC: (3*S*,5*R*,6*S*,3′*R*)-5,6-Epoxy-7′,8′-didehydro-5,6-dihydro-β,β-carotene-3,3′-diol

Molecular formula: $C_{40}H_{54}O_3$ **Molecular weight:** 582.85

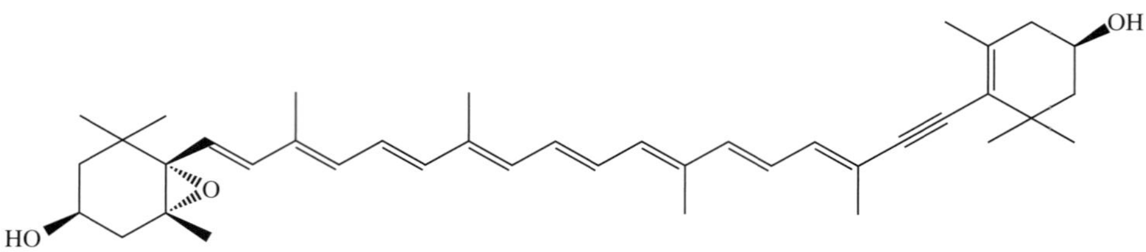

Biological occurrence	Dominant pigment in bolidophytes, diatoms, euglenophytes, haptophytes, pelagophytes, phaeothamniophytes, dictyochophytes (silicoflagellates) and some dinoflagellates. Minor in xanthophytes (see Chapter 1, this volume)
Source culture	*Amphidinium carterae* (dinoflagellate), *Phaeodactylum tricornutum* (diatom)
Alteration products	Undergoes rearrangement to Diadchr in weakly acidic solutions. *Cis*-isomers
Biosynthetically related to	Diato
Occurs together with	Diato, Fuco and derivatives, Peri

HPLC chromatogram of *Gymnodinium catenatum* (system 1)

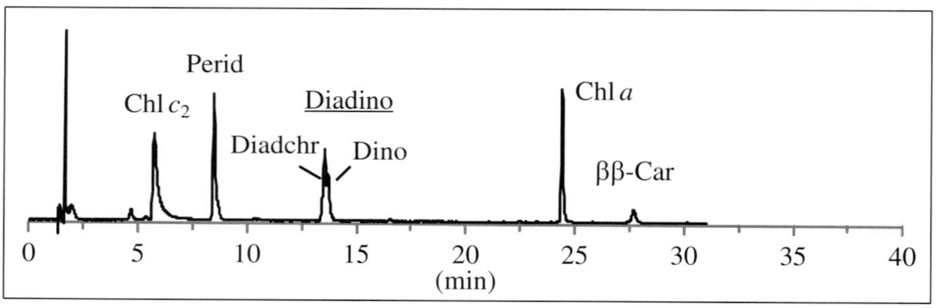

HPLC chromatogram of *Scrippsiella trochoidea* (system 2)

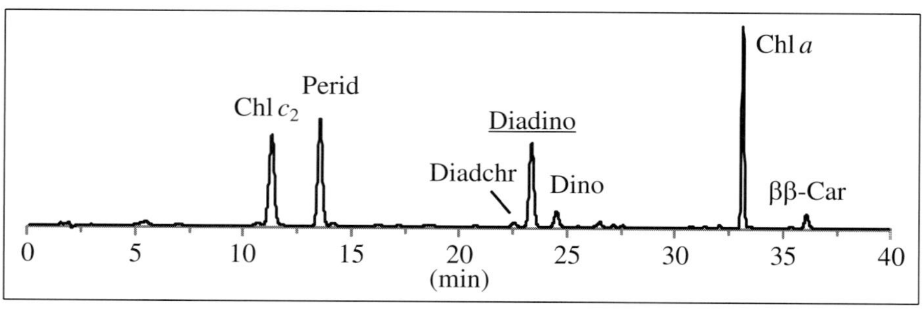

UV-Vis spectra (see also reference spectra below)

Solvent	λ_{max} (nm)	Band ratio (% III:II)	Ref.
Acetone	(428), *449*, 479	75	[25]
Diethyl ether	424, *446*, 477	n.d.	[61]
Ethanol	(424), *445*, 476	64	[122]
Hexane	(421), *445*, 475	63	[122]
Recommended specific absorption coefficient ***d*** **(L g^{-1} cm^{-1})**		224 (at 448 nm, acetone) [112]	

Reference spectra
In acetone

For spectrum in acetone, see [109]

In HPLC solvent system 1

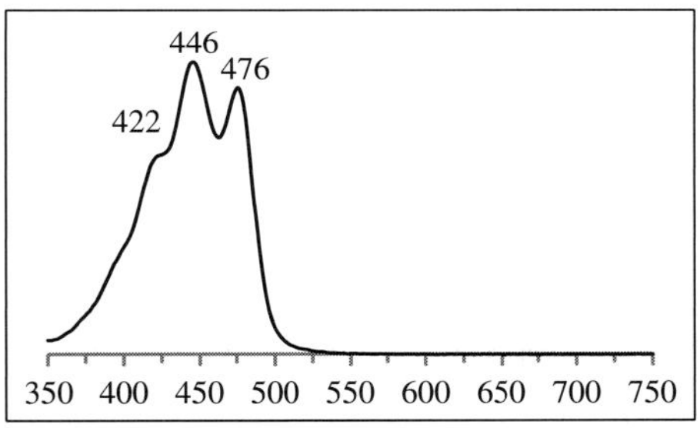

In HPLC solvent system 2

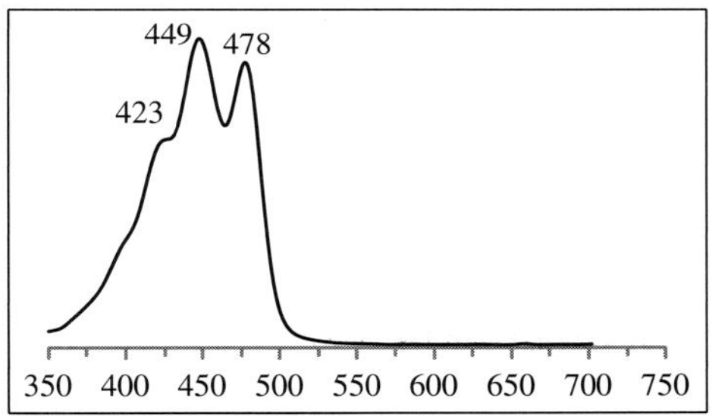

Mass spectra

Ionization technique	Mass analyser type	Diagnostic ions (m/z, rel. intensity)	Ref.
EI	Magnetic sector	582 [M]$^+$ (100), 567 [M-15]$^+$ (6), 564 [M-18]$^+$ (3), 502 [M-80]$^+$ (20), 490 [M-92]$^+$ (14), 487 [M-15–80]$^+$ (10), 475 [M-15–92]$^+$ (4), 221 (50), 181 (26)	[25]

Remarks The recommended *d* value is the average of the two values given in [112]. Part of a 'xanthophyll cycle' (see Chapter 11 this volume)

Diatoxanthin

Recommended abbreviation: Diato (Dt)

IUPAC: (3*R*,3′*R*)-7,8-Didehydro-β,β-carotene-3,3′-diol

Molecular formula: $C_{40}H_{54}O_2$ **Molecular weight:** 566.86

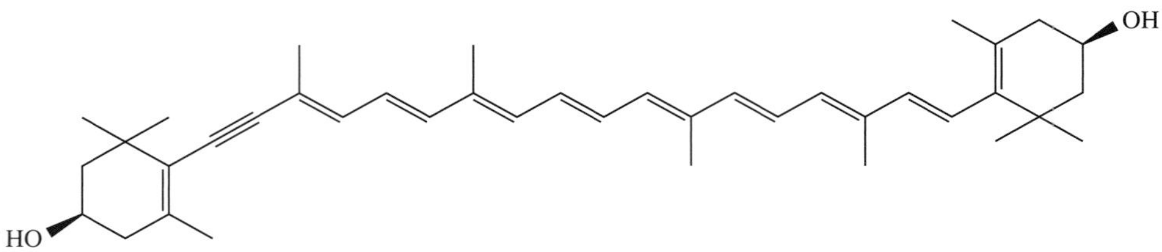

Biological occurrence	Minor pigment in euglenophytes, diatoms, bolidophytes, haptophytes, dictyochophytes, pelagophytes and some dinoflagellates
Source culture	*Amphidinium carterae* (dinoflagellate)
Alteration products	Easily isomerises to the more stable 9-*cis* isomer [43]
Biosynthetically related to	Diadino
Occurs together with	Diadino, Fuco and derivatives, Peri

HPLC chromatogram of *Pavlova gyrans* (system 1)

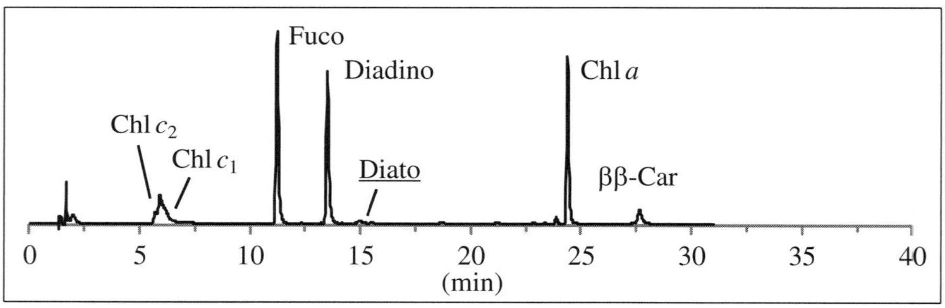

HPLC chromatogram of *Emiliania huxleyi* (system 2)

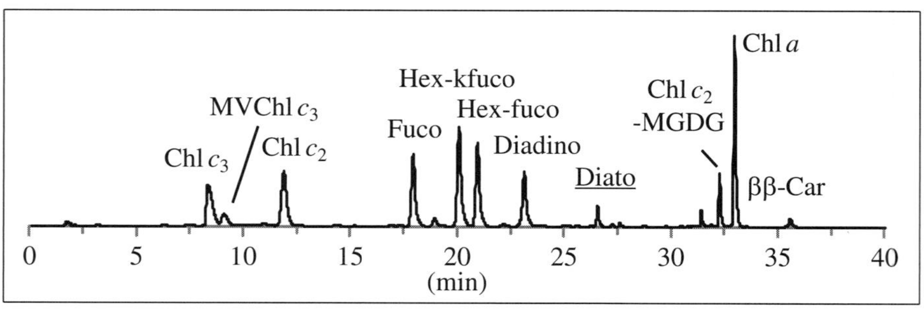

UV-Vis spectra (see also reference spectra below)

Solvent	λ_{max} (nm)	Band ratio (% III:II)	Ref.
Acetone	430, *453*, 480	42	[92]
Ethanol	(428), *452*, 478	n.d.	[130]
Hexane	(426), *451*, 480	31	[145]
Recommended specific absorption coefficient d **(L g^{-1} cm^{-1})**		272 (at 453 nm, acetone) [92]	

Reference spectra

For spectrum in acetone, see [109]

For spectrum in hexane, see [109]

In HPLC solvent system 1

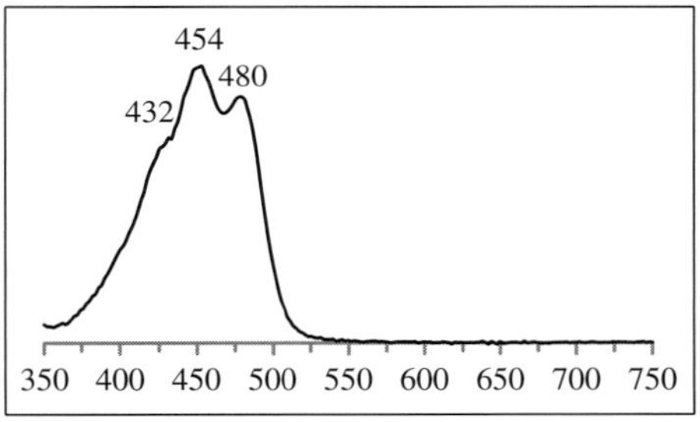

In HPLC solvent system 2

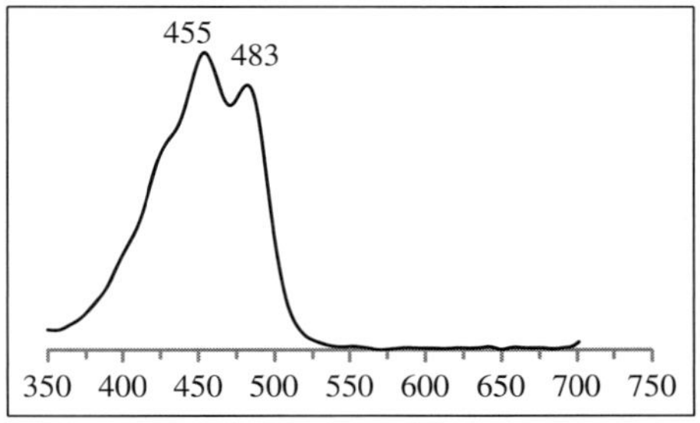

Mass spectra

Ionization technique	Mass analyser type	Diagnostic ions (m/z, rel. intensity)	Ref.
EI	Magnetic sector	566 [M]$^+$ (100), 548 [M-18]$^+$ (11), 474 [M-92]$^+$ (6), 408 [M-158]$^+$ (5), 283 (15), 119 (31), 43 (34)	[92]
Remarks	Part of a 'xanthophyll cycle' (see Chapter 11 this volume)		

Dihydrolutein

Recommended abbreviation: Dhlut (Dl)

IUPAC: (3*R*,3′*R*,6′*R*)-7′,8′-Dihydro-β,ε-carotene-3,3′-diol

Molecular formula: $C_{40}H_{58}O_2$ **Molecular weight:** 570.89

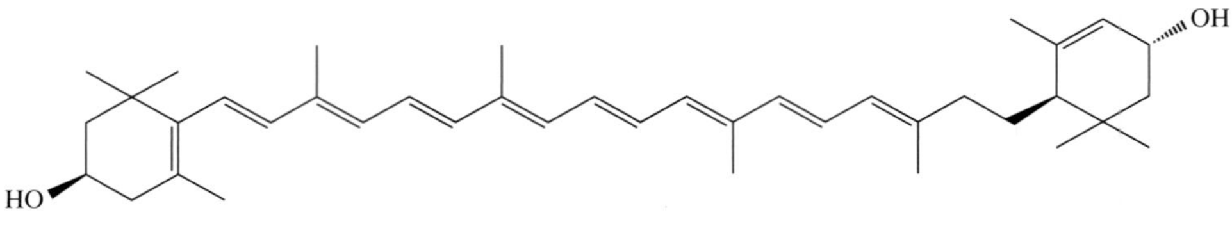

Biological occurrence	Only encountered in prasinophytes Pigment Type 3 [50]
Source culture	*Pseudoscourfieldia marina* (prasinophyte)
Alteration products	*Cis*-isomers
Biosynthetically related to	Lut (?), Microl, Micral, Uri
Occurs together with	Pras, Microl, Micral, Uri, Chl *b*, MgDVP

HPLC chromatogram of *Pycnococcus provasolii* (system 1)

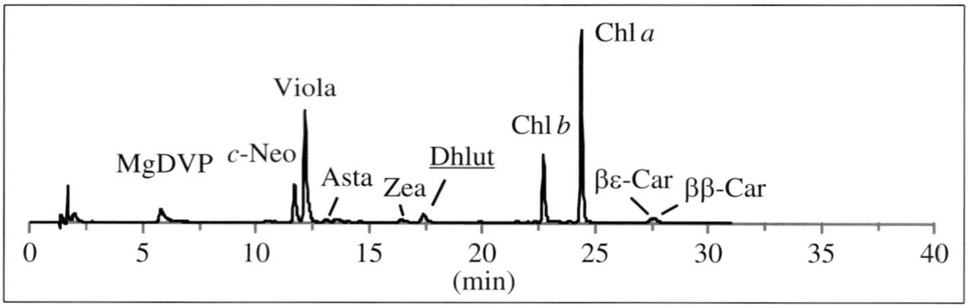

HPLC chromatogram of *Micromonas pusilla* (system 2)

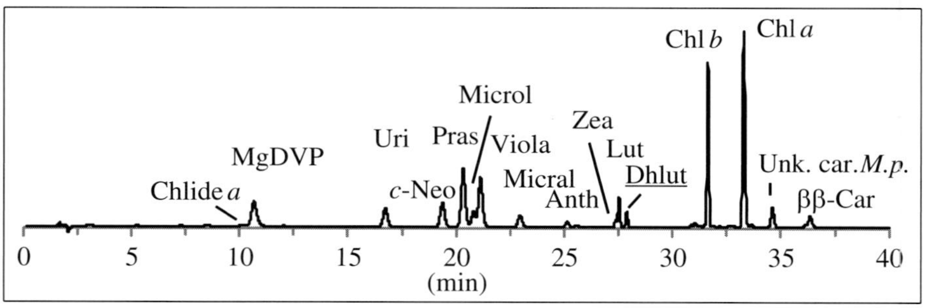

UV-Vis spectra (see also reference spectra below)

Solvent	λ_{max} (nm)	Band ratio (% III/II)	Ref.
Acetone	(402), *426*, 452	47	[50]
Recommended specific absorption coefficient *d* **(L g^{-1} cm^{-1})**		n.d., see Remarks	

Reference spectra
In acetone

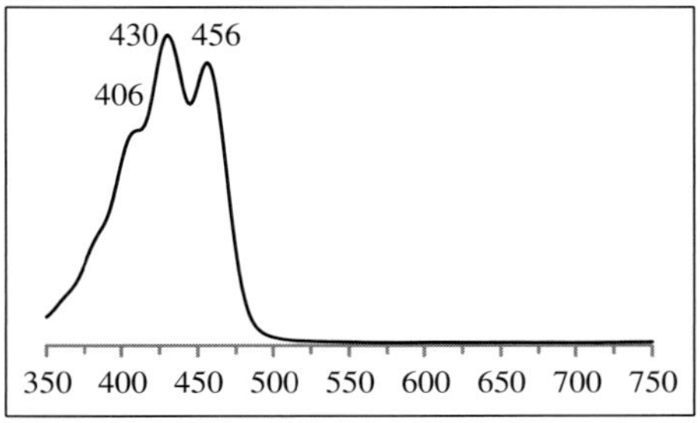

In HPLC solvent system 1

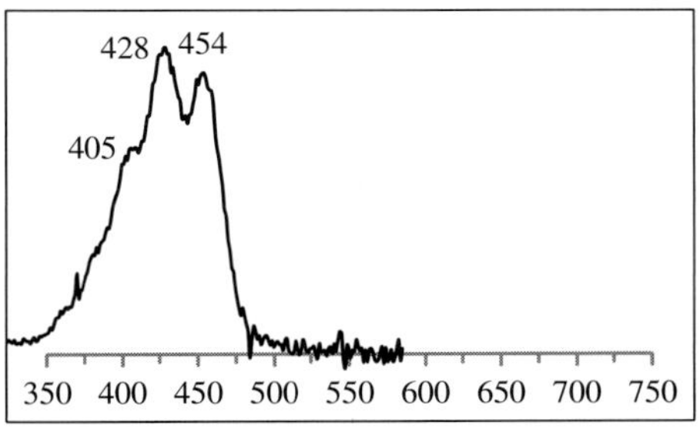

In HPLC solvent system 2

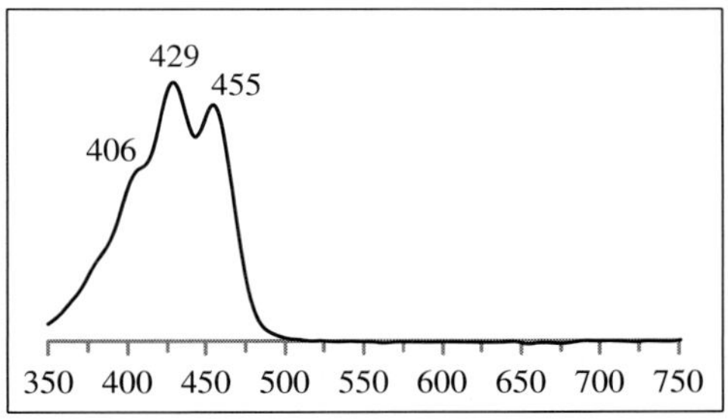

Mass spectra

Ionization technique	Mass analyser type	Diagnostic ions (m/z, rel. intensity)	Ref.
EI	Magnetic sector	570 $[M]^+$ (100), 552 $[M\text{-}18]^+$ (35), 478 $[M\text{-}92]^+$ (4)	[52]

Remarks Structure not proven by NMR
$d = 229\ L\,g^{-1}cm^{-1}$ (at λ_{max} in ethanol, calc. from Mut) is recommended, as no *d* has been determined for Dhlut

Dinoxanthin

Recommended abbreviation: Dino (Dn)

IUPAC: (3*S*,5*R*,6*S*,3′*S*,5′*R*,6′*R*)-5,6-Epoxy-3′-ethanoyloxy-6′,7′-didehydro-5,6,5′,6′-tetrahydro-β,β-carotene-3,5′-diol

Molecular formula: $C_{42}H_{58}O_5$ **Molecular weight:** 642.91

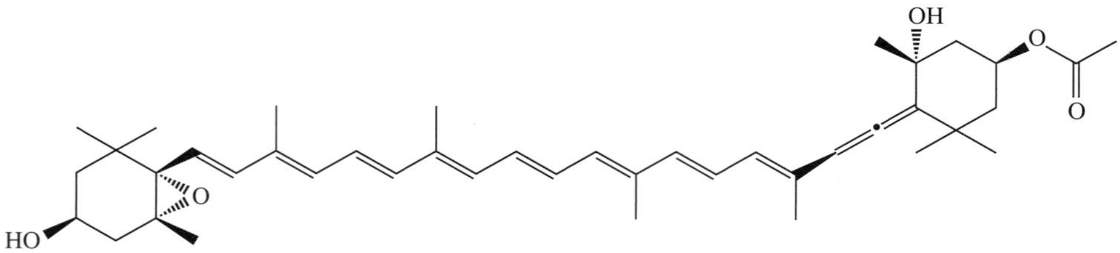

Biological occurrence	Minor pigment in Peri-containing dinoflagellates. Occasionally found in trace in various chromophyte cultures
Source culture	*Amphidinium carterae* (dinoflagellate)
Alteration products	Undergoes rearrangement to Dinochr in weakly acidic solutions. *Cis*-isomers
Biosynthetically related to	Neo, Viola
Occurs together with	Peri (if dinoflagellates)

HPLC chromatogram of *Gymnodinium catenatum* (system 1)

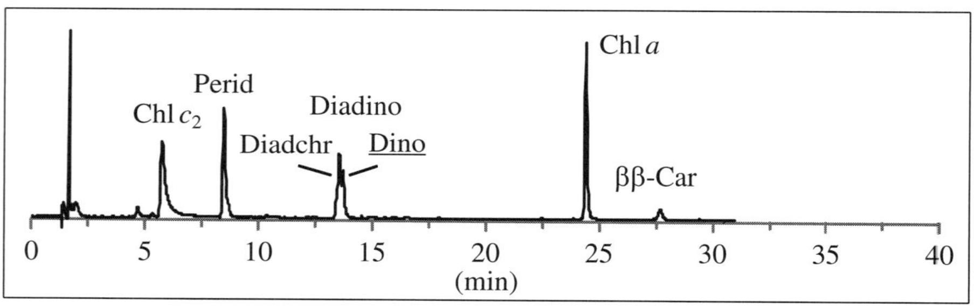

HPLC chromatogram of *Scrippsiella trochoidea* (system 2)

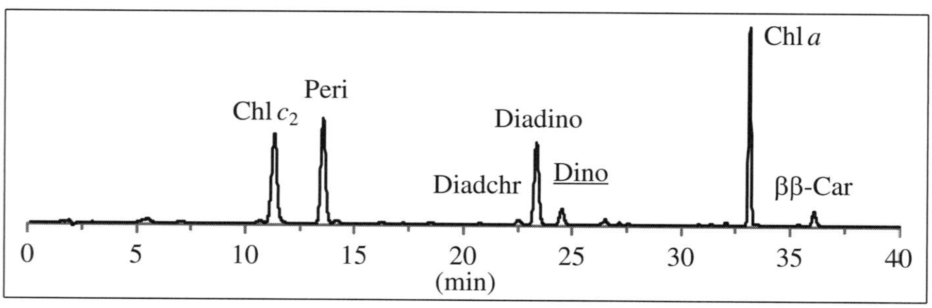

UV-Vis spectra (see also reference spectra below)

Solvent	λ_{max} (nm)	Band ratio (% III:II)	Ref.
Acetone	418, *442*, 471	86	[145]
Ethanol	417, *441*, 470	85	[145]
Hexane	416, *439*, 469	83	[122]
Methanol	416, *438*, 467	76	[122]
Recommended specific absorption coefficient *d* (L g^{-1} cm^{-1})		n.d., see Remarks	

Reference spectra

For spectrum in acetone, see [109] For spectrum in ethanol, see [109]

In HPLC solvent system 1

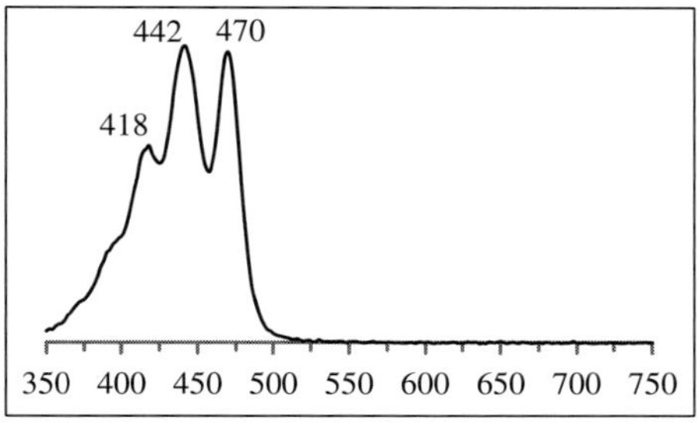

In HPLC solvent system 2

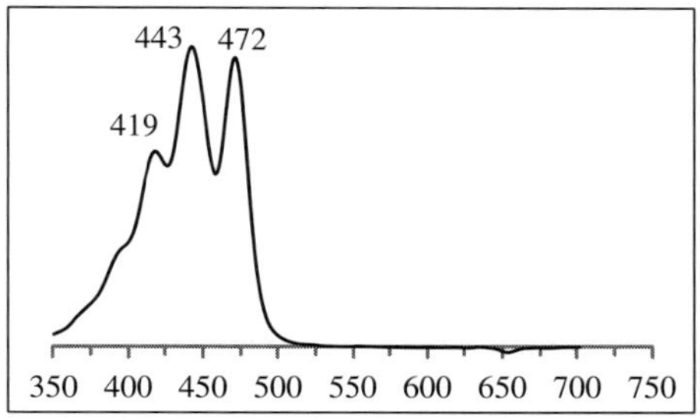

Mass spectra

Ionization technique	Mass analyser type	Diagnostic ions (m/z, rel. intensity)	Ref.
EI	Magnetic sector	642 $[M]^+$ (3), 624 $[M\text{-}18]^+$ (11), 606 $[M\text{-}18\text{–}18]^+$ (7), 221 (100), 181 (64)	[18]

Remarks As *d* has not been determined for Dino, a value based on *t*-Neo was calculated: $d = 222\,\mathrm{L\,g^{-1}cm^{-1}}$ (at 441 nm, ethanol). Note spectral similarity with Viola (but Dino is slightly red-shifted). Dinochrome is formed from Dino in slightly acidic extracts: [18, 22, 25]

Echinenone

Recommended abbreviation: Echin (Ec)

IUPAC: β,β-Caroten-4-one

Molecular formula: $C_{40}H_{54}O$

Molecular weight: 550.86

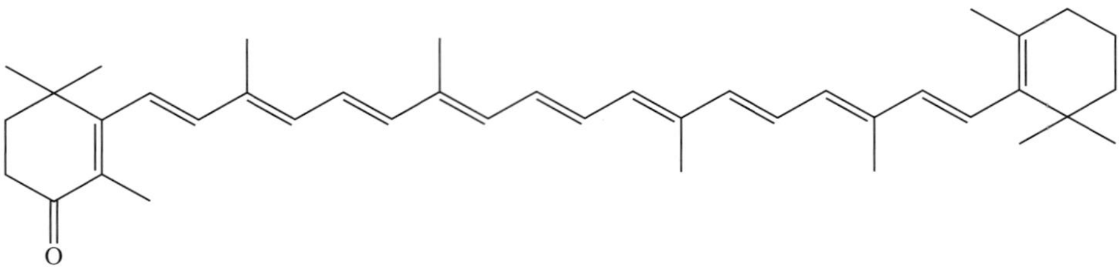

Biological occurrence	Minor pigment in some prochlorophytes (Cyano-3) and cyanobacteria (Cyano-1, Chapter 1). Occasional traces in algal cultures from several other classes
Source culture	*Oscillatoria agardhii (rubescens)* (cyanobacteria)
Alteration products	*Cis*-isomers
Biosynthetically related to	ββ-Car, Cantha, Asta
Occurs together with	

HPLC chromatogram *Nodularia spumigena* (system 1)

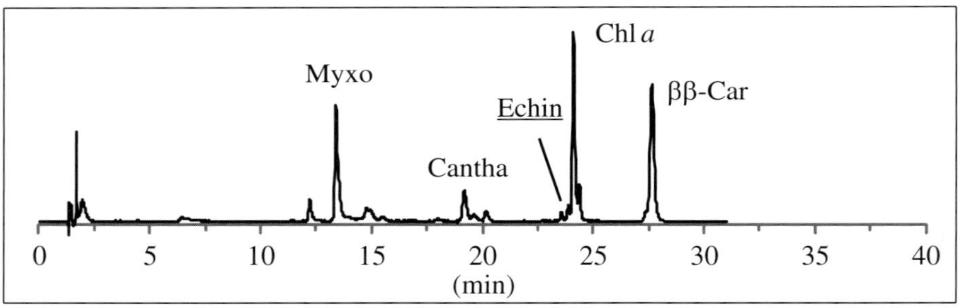

HPLC chromatogram *Chlorogloeopsis fritschii* (system 3)

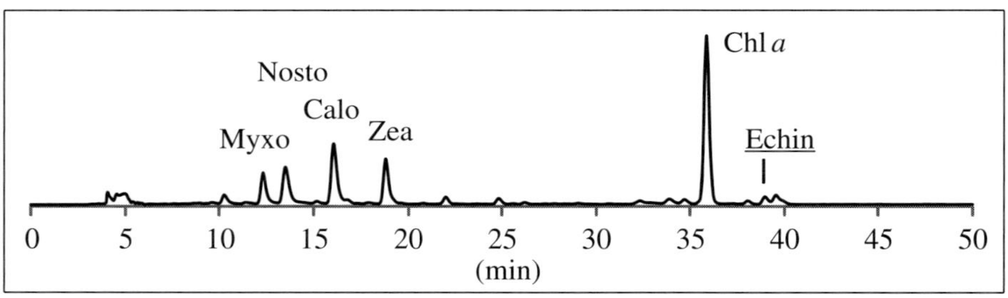

UV-Vis spectra (see also reference spectra below)

Solvent	λ_{max} (nm)	Band ratio (% III:II)	Ref.
Acetone	*459*, (475)	-	[88]
Diethyl ether	(430), *452*, (472)	-	[59]
Ethanol	461	-	[86]
Hexane	458	-	[135]
Methanol	460	-	[159]
Recommended specific absorption coefficient *d* (L g^{-1} cm^{-1})		216 (at 458 nm, hexane) [68]	

Reference spectra
In acetone

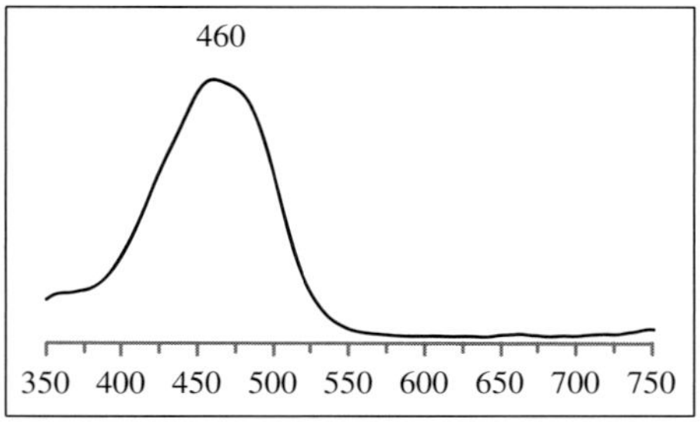

In HPLC solvent system 2

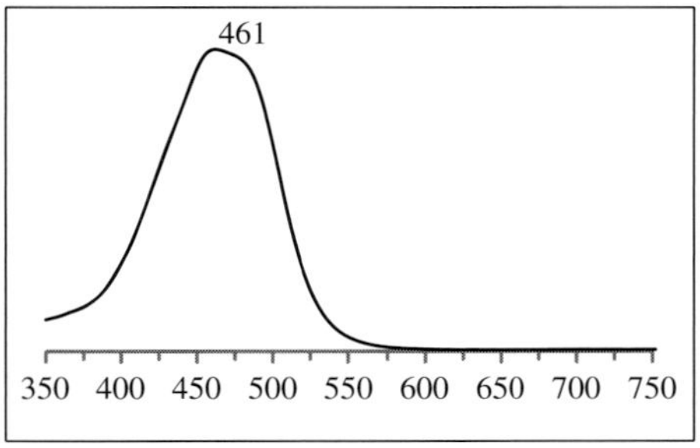

In HPLC solvent system 3

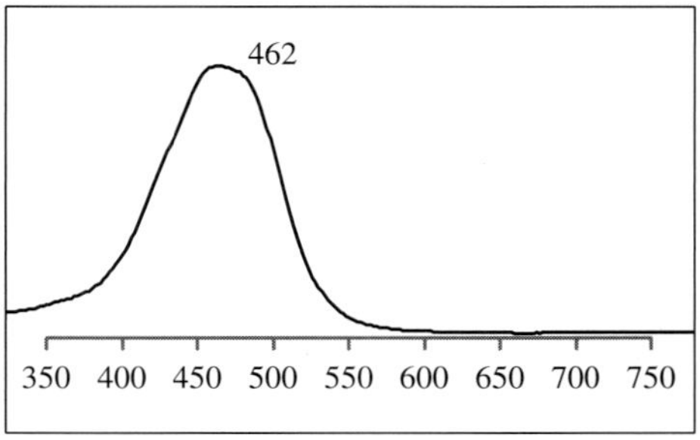

Mass spectra

Ionization technique	Mass analyser type	Diagnostic ions (m/z, rel. intensity)	Ref.
EI	Magnetic sector	550 $[M]^+$ (100), 534 $[M-16]^+$ (10), 458 $[M-92]^+$ (6), 444 $[M-106]^+$ (2), 189 (16)	[59]

Remarks

Eutreptiellanone

Recommended abbreviation: Eutr (Eu)

IUPAC: (3*S*,5*R*,6*S*)-3,6-Epoxy-3′,4′,7′,8′-tetradehydro-5,6-dihydro-β,β-caroten-4-one

Molecular formula: $C_{40}H_{50}O_2$ **Molecular weight:** 562.82

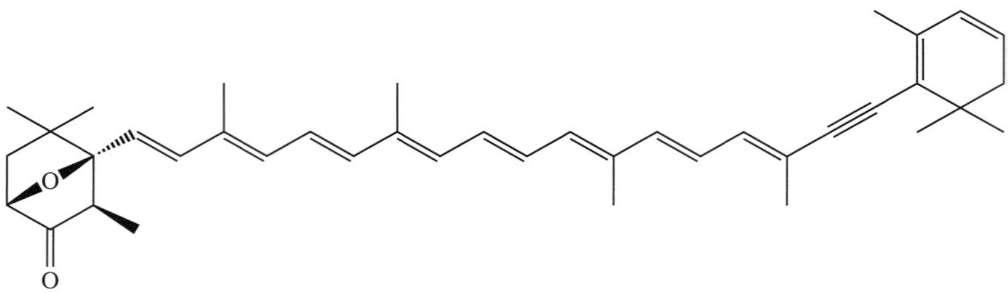

Biological occurrence	Only as major carotenoid in *Eutreptiella gymnastica*
Source culture	*Eutreptiella gymnastica* (euglenophyte)
Alteration products	*Cis*-isomers
Biosynthetically related to	Other carotenoids from *Eutreptiella gymnastica*
Occurs together with	Chl *b*, Diato, Diadino

HPLC chromatogram of *Eutreptiella* cf. *gymnastica* (system 1)

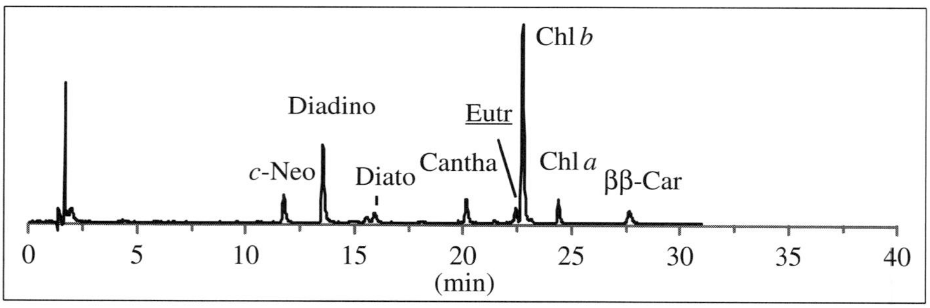

HPLC chromatogram of *Eutreptiella gymnastica* (system 2)

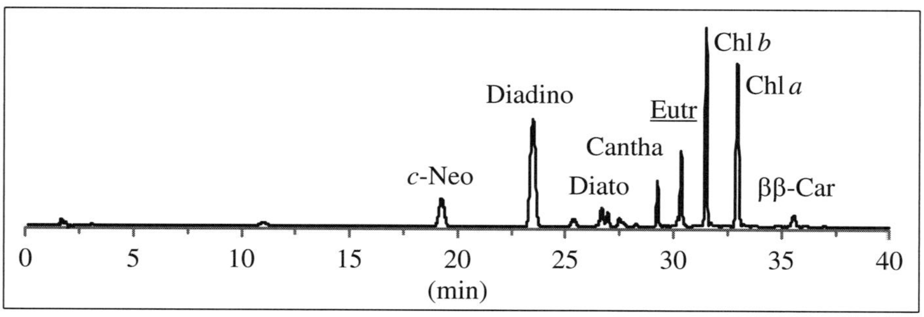

UV-Vis spectra (see also reference spectra below)

Solvent	λ_{max} (nm)	Band ratio (% III:II)	Ref.
Acetone	(441), *461*, 491	47	[17]
Diethyl ether	(434), *457*, 483	48	[58]
Petroleum ether	(438), *458*, 487	50	[17]
Recommended specific absorption coefficient *d* (L g^{-1} cm^{-1})		n.d.	

Reference spectra

In HPLC solvent system 1

NO DATA AVAILABLE

In HPLC solvent system 2

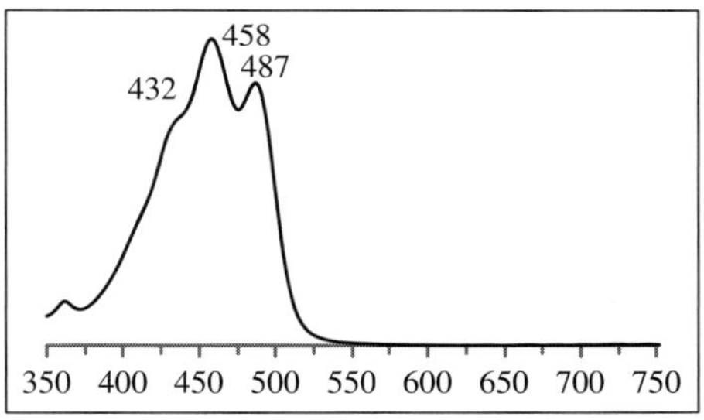

Mass spectra

Ionization technique	Mass analyser type	Diagnostic ions (m/z, rel. intensity)	Ref.
EI	Magnetic sector	562.3808 $[M]^+$ (100), 547 $[M\text{-}15]^+$ (6), 506.3188 $[M\text{-}56]^+$ (2), 491.2956 $[M\text{-}56\text{–}18]^+$ (2), 470 $[M\text{-}92]^+$ (2), 455 $[M\text{-}92\text{–}15]^+$ (2), 261 (12)	[58]

Remarks 'Unknown 1′ in ref. [17]
$d = 220\,L\,g^{-1}cm^{-1}$ (at 461 nm in acetone) is recommended, as no value has been determined for Eutr

Fucoxanthin

Recommended abbreviation: Fuco (F)

IUPAC: (3*S*,5*R*,6*S*,3′*S*,5′*R*,6′*R*)-5,6-Epoxy-3′-ethanoyloxy-3,5′-dihydroxy-6′,7′-didehydro-5,6,7,8,5′,6′-hexahydro-β,β-caroten-8-one

Molecular formula: $C_{42}H_{58}O_6$ **Molecular weight:** 658.91

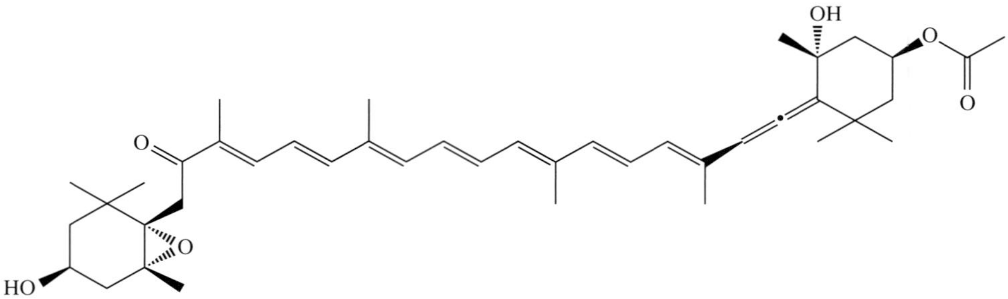

Biological occurrence	Dominant carotenoid in most algal classes of the Red Algal lineage (see Chapter 1, this volume) and brown seaweeds
Source culture	*Phaeodactylum tricornutum* (diatom)
Alteration products	*Cis*-isomers
Biosynthetically related to	Neo, Hex-fuco, But-fuco, Hex-kfuco, Kfuco
Occurs together with	Hex-fuco, But-fuco, Diato, Diadino, Viola, Zea, Chl *c*s

HPLC chromatogram of *Karlodinium micrum* (system 1)

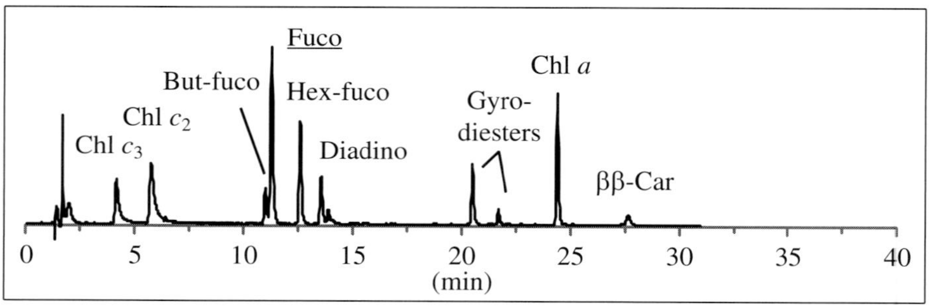

HPLC chromatogram of *Emiliania huxleyi* (system 2)

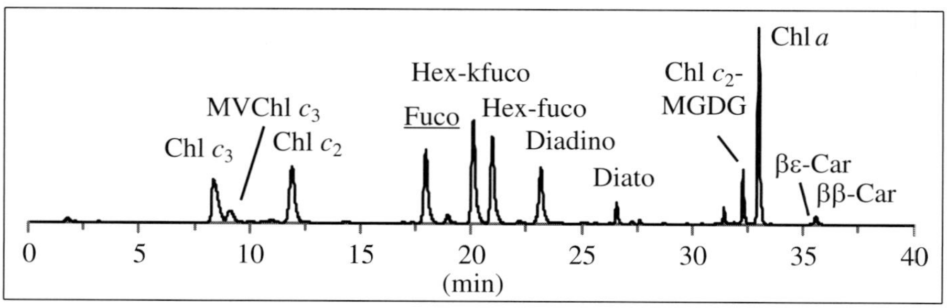

UV-Vis spectra (see also reference spectra below)

Solvent	λ_{max} (nm)	Band ratio (% III:II)	Ref.
Acetone	(420), *444*, 467	5	[89]
Diethyl ether	*446*, 470	3	[173]
Ethanol	*449*, (467)		[174]
Hexane	(428), *446*, 475	40	[89]
Recommended specific absorption coefficient ***d*** **(L g^{-1} cm^{-1})**		166 (at 443 nm, acetone) [90]	

Reference spectra

In acetone

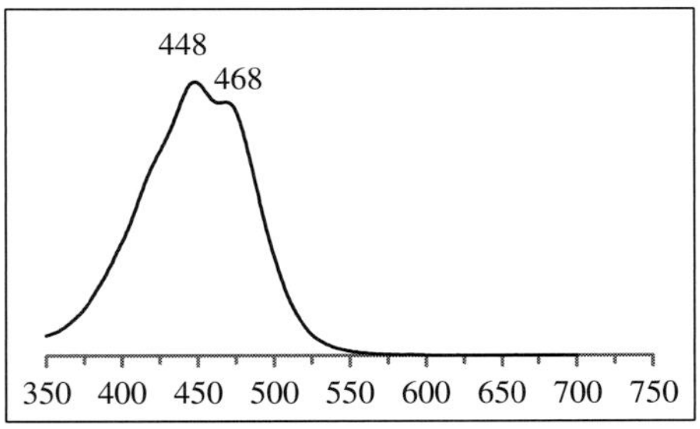

In ethanol

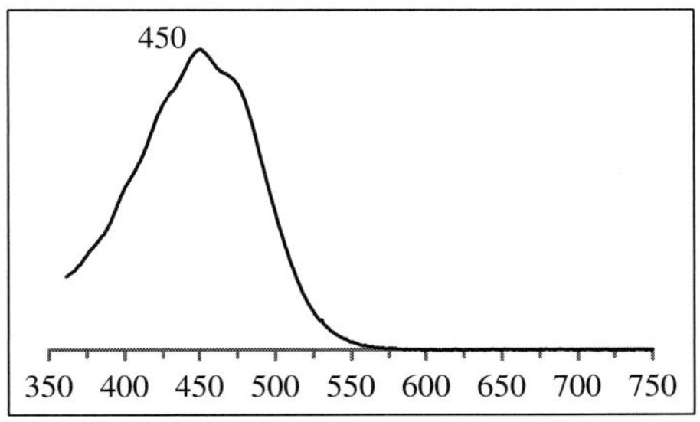

In HPLC solvent system 1

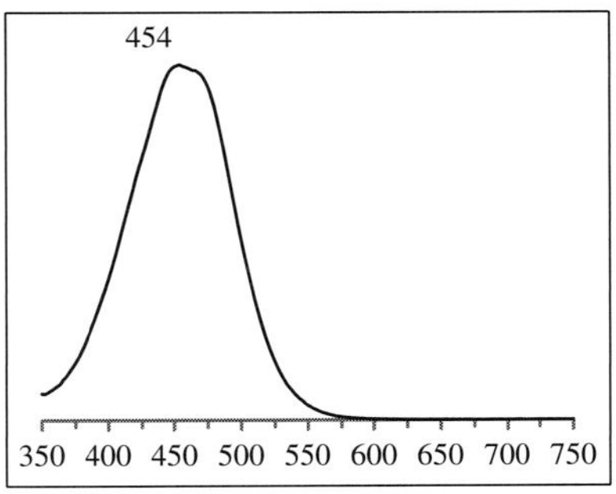

In HPLC solvent system 2

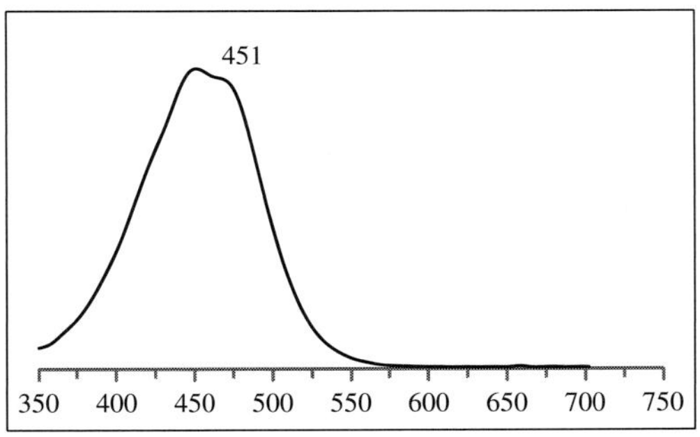

Mass spectra

Ionization technique	Mass analyser type	Diagnostic ions (m/z, rel. intensity)	Ref.
APCI	Ion trap	681 $[M+Na]^+ \rightarrow$ 527 $[M+Na\text{-}154]^+ \rightarrow$ 467 $[M+Na\text{-}154\text{–}60]^+$	[5]
Remarks	4-ketofucoxanthin has also been encountered as a minor pigment in a few haptophytes [49]		

Gyroxanthin dodecanoate ethanoate (one of several gyroxanthin diesters)

Recommended abbreviation: Gyro-de (Gd)

IUPAC: (3*S*,5*R*,6*S*,3′*S*,5′*R*,6′*R*)-19-Dodecanoyloxy-5,6-epoxy-3′-ethanoyloxy-7,8,6′,7′-tetradehydro-5,6,5′,6′-tetrahydro-β,β-carotene-3,5′-diol

Molecular formula: $C_{54}H_{78}O_7$ **Molecular weight:** 839.19

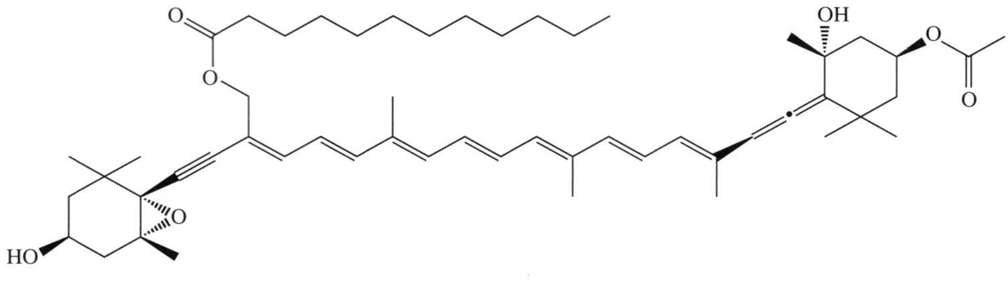

Biological occurrence	A major carotenoid in dinoflagellates Pigment Type 2, minor in pelagophytes and coccolithophytes (see Chapter 1)
Source culture	*Karenia brevis* (dinoflagellate)
Alteration products	Easily isomerises to the more stable 9-*cis* isomer [21]
Biosynthetically related to	Dino?
Occurs together with	Fuco, Diato, Diadino, various Fuco derivatives, Chl *c*s

HPLC chromatogram of *Karlodinium micrum* (system 1)

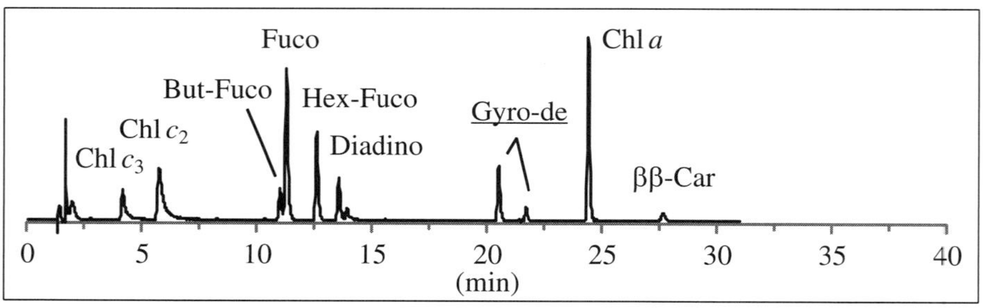

HPLC chromatogram of *Karlodinium micrum* (system 2)

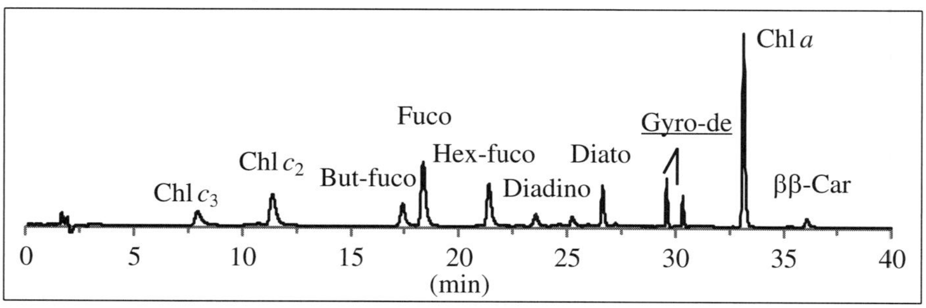

UV-Vis spectra (see also reference spectra below)

Solvent	λ_{max} (nm)	Band ratio (% III:II)	Ref.
Acetone	(423), *444*, 472	55	[21]
Hexane	418, *442*, 470	76	[21]
Recommended specific absorption coefficient *d* (L g^{-1} cm^{-1})		n.d., see Remarks	

Reference spectra
In acetone

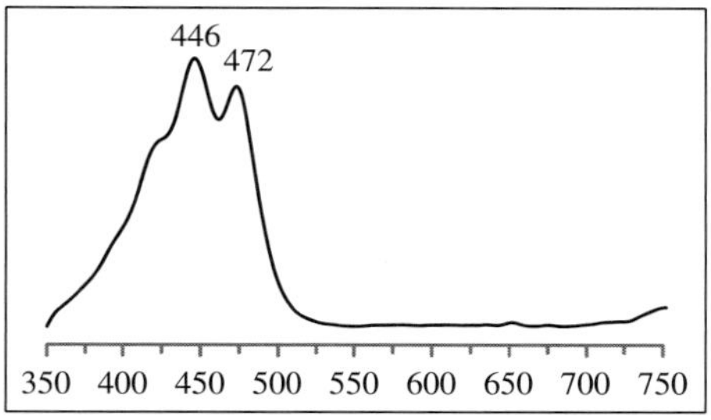

In HPLC solvent system 1

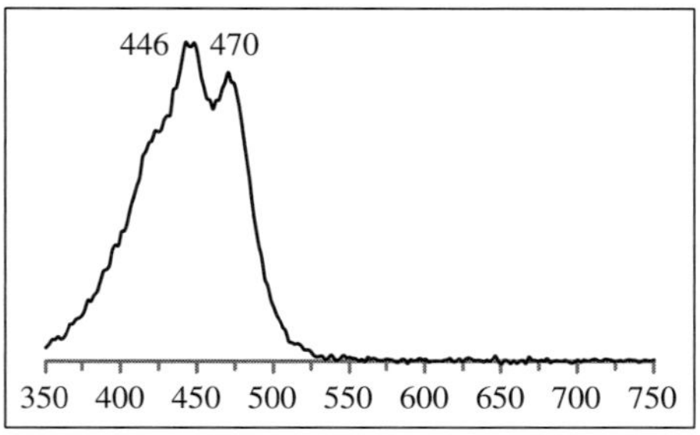

In HPLC solvent system 2

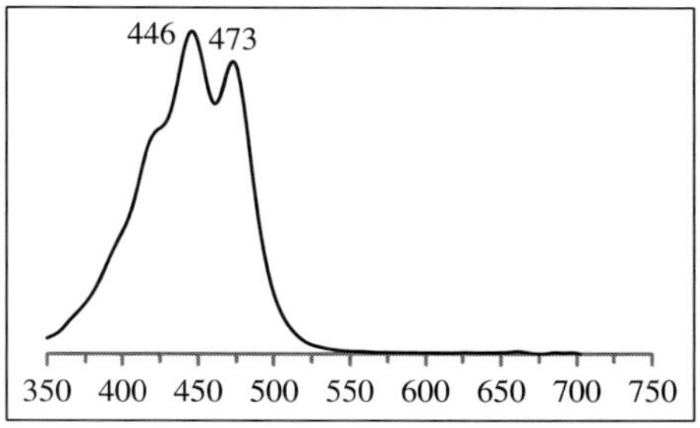

Mass spectra

Ionization technique	Mass analyser type	Diagnostic ions (m/z, rel. intensity)	Ref.
ESI$^+$	Ion trap	861 $[M+Na]^+$ → 661 $[M+Na-200]^+$ → 555 $[M+Na-200–106]^+$, 643 $[M+Na-200–18]^+$, 601 $[M+Na-200–60]^+$, 569 $[M+Na-200–92]^+$	[178]

Remarks *d* = 183 L g^{-1} cm^{-1} (at 444 nm in acetone; calc. from an average carotenoid value of 250) is recommended, as no value has been determined for Gyro-de [21] Other gyroxanthin diesters also found (see the chromatograms shown) [21, 24]

19′-Hexanoyloxyfucoxanthin **Recommended abbreviation: Hex-fuco (HF)**

IUPAC:(3*S*,5*R*,6′*S*,3′*S*,5′*R*,6′*R*)-5,6-Epoxy-3′-ethanoyloxy-19′-hexanoyloxy-3, 5′-dihydroxy -6′,7′-didehydro-5,6,7,8,5′,6′-hexahydro-β,β-caroten-8-one

Molecular formula: $C_{48}H_{68}O_8$ **Molecular weight:** 773.05

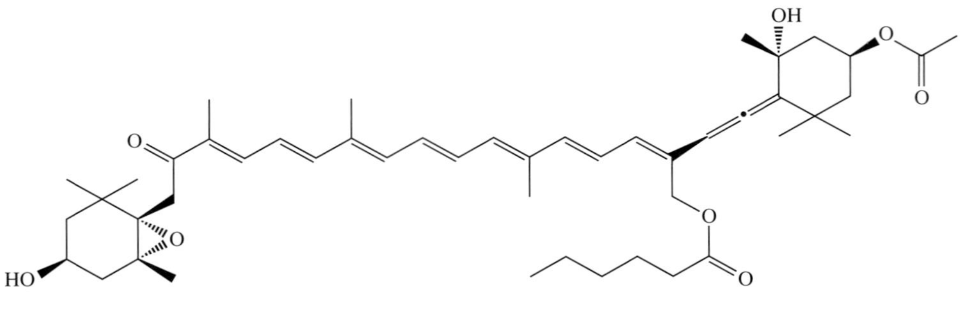

Biological occurrence	Major in many haptophytes and dinoflagellates Pigment Type 2 (see Chapter 1, this volume)
Source culture	*Emiliania huxleyi* (coccolithophyte), strain CS-57
Alteration products	*Cis*-isomers
Biosynthetically related to	Fuco, But-fuco, Hex-kfuco
Occurs together with	Fuco, other Fuco-derivatives, Diato, Diadino, Chl *cs*

HPLC chromatogram of *Emiliania huxleyi* (system 1)

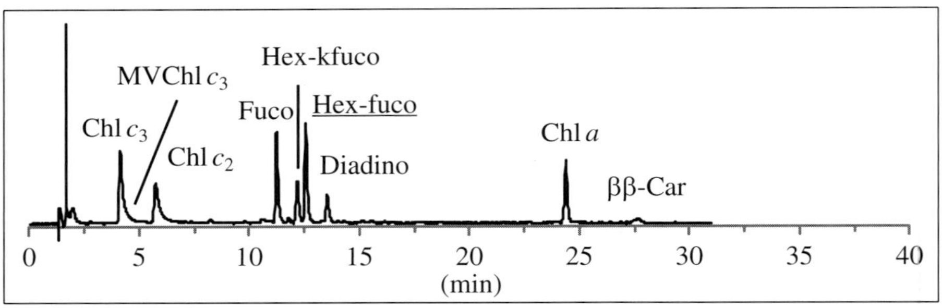

HPLC chromatogram of *Emiliania huxleyi* (system 2)

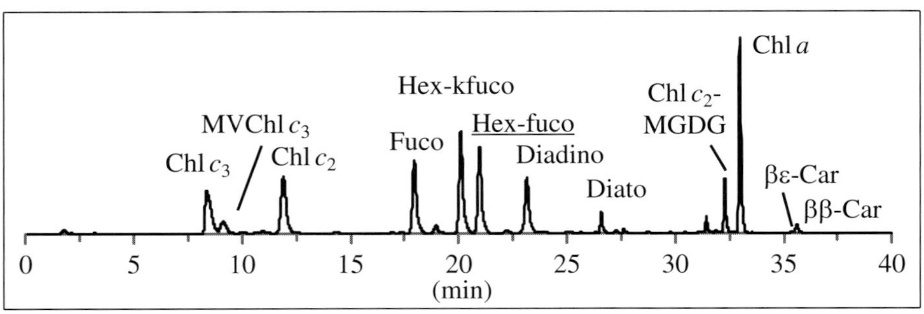

UV-Vis spectra (see also reference spectra below)

Solvent	λ_{max} (nm)	Band ratio (% III:II)	Ref.
Acetone	420, *443*, 463	40	[16]
Hexane	(423), *445*, 474	64	[22]
Methanol	443		[7]
Recommended specific absorption coefficient *d* **(L g^{-1}cm^{-1})**		n.d., see Remarks	

Reference spectra

For spectrum in acetone, see [109]

In HPLC solvent system 1

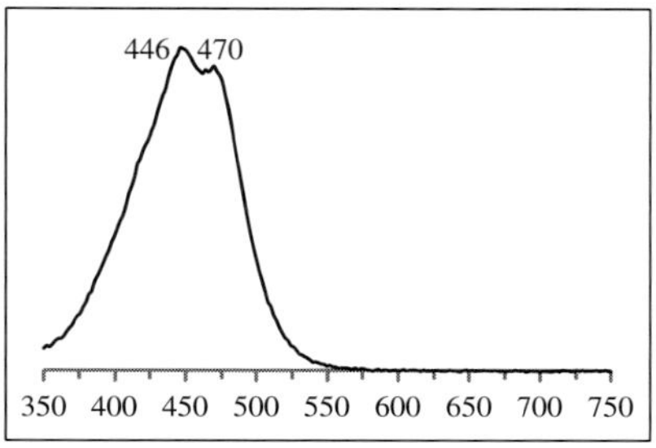

In HPLC solvent system 2

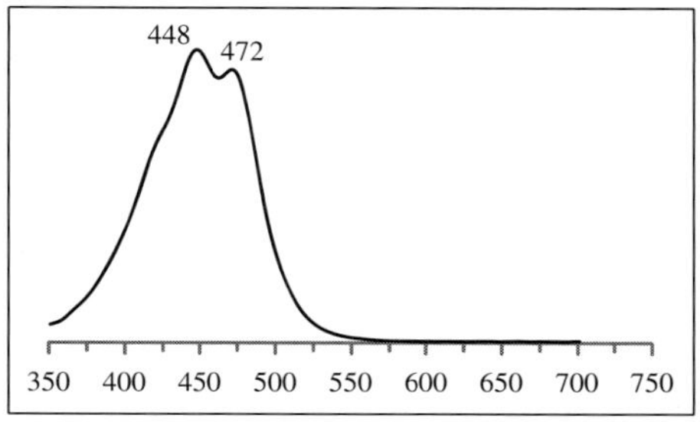

Mass spectra

Ionization technique	Mass analyser type	Diagnostic ions (m/z, rel. intensity)	Ref.
APCI	Ion trap	795 $[M+Na]^+ \rightarrow$ 679 $[M+Na\text{-}116]^+ \rightarrow$ 661 $[M+Na\text{-}116\text{–}18]^+$, 619 $[M+Na\text{-}116\text{–}60]^+$, 525	[5]

Remarks $d = 142$ L g^{-1}cm^{-1} (at 445 nm in acetone; calc. from Fuco) is recommended, as no value has been determined for Hex-fuco [109]. Several other, minor alkanoyloxyfucoxanthins have been reported [5]

19′-Hexanoyloxy-4-ketofucoxanthin **Recom. abbrev.: Hex-kfuco (HKf)**

IUPAC: (3*S*,5*S*,6*S*,3′*S*,5′*R*,6′*R*)-5,6-Epoxy-3′-ethanoyloxy-19′-hexanoyloxy-3,5′-dihydroxy-6′,7′-didehydro-5,6,7,8,5′,6′-hexahydro-β,β-carotene-4,8-dione

Molecular formula: $C_{48}H_{66}O_9$ **Molecular weight:** 787.03

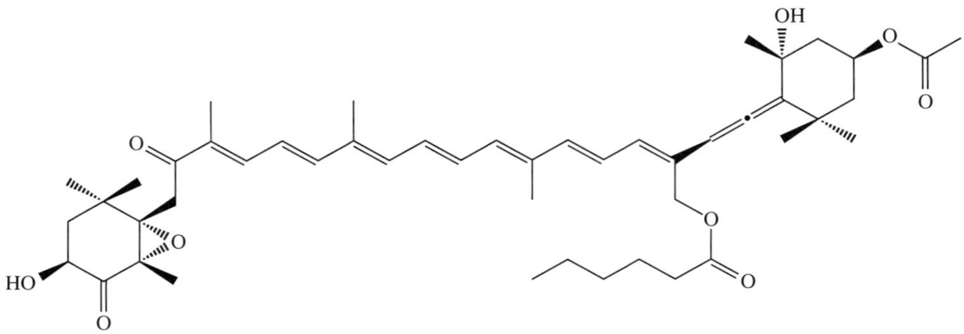

Biological occurrence	Major in some haptophytes (see Chapter 1, this volume)
Source culture	*Ochrosphaera neopolitana* (coccolithophyte) [180], strain CS-285 [8]
Alteration products	*Cis*-isomers. Degrades upon storage in acetone [49]
Biosynthetically related to	Fuco, Kfuco
Occurs together with	Fuco, Diato, Diadino, Chl *c*s

HPLC chromatogram of *Emiliania huxleyi* (system 1)

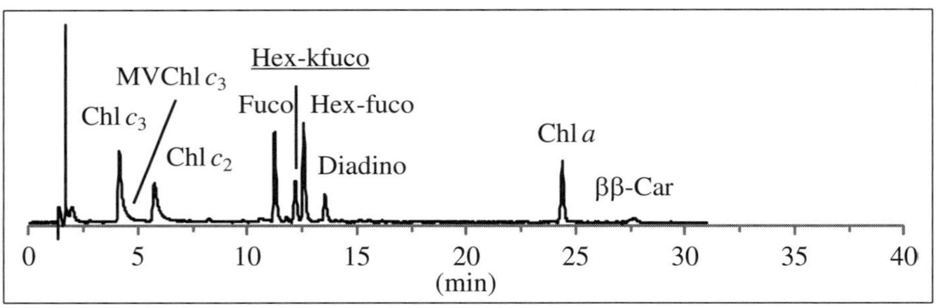

HPLC chromatogram of *Emiliania huxleyi* (system 2)

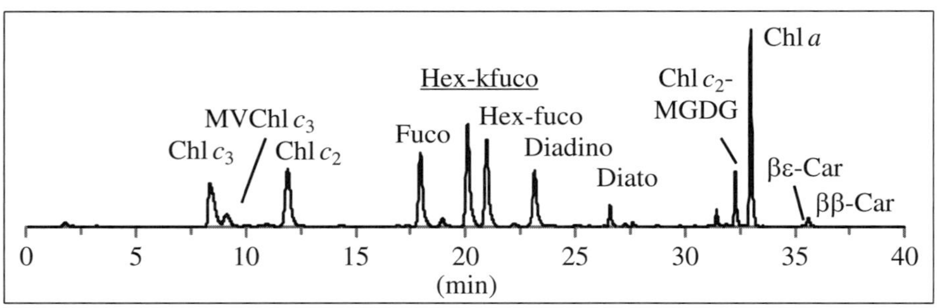

UV-Vis spectra (see also reference spectra below)

Solvent	λ_{max} (nm)	Band ratio (% III:II)	Ref.
Acetone	(420), *443*, 467	38	[49]
Methanol	*443*, 465	28	[49]
Recommended specific absorption coefficient *d* (L g^{-1} cm^{-1})		n.d., see Remarks	

Reference spectra

In acetone

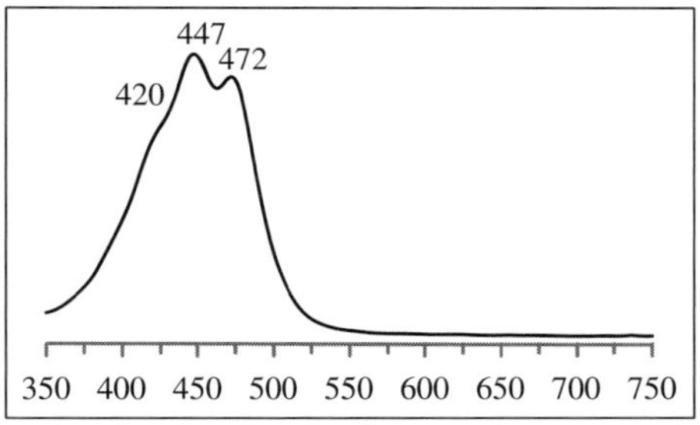

In HPLC solvent system 1

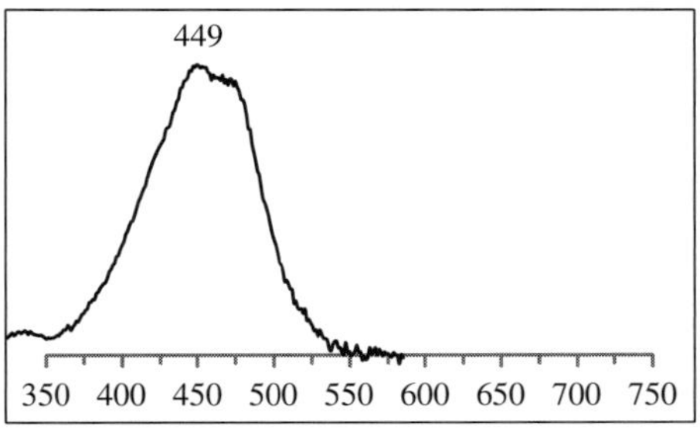

In HPLC solvent system 2

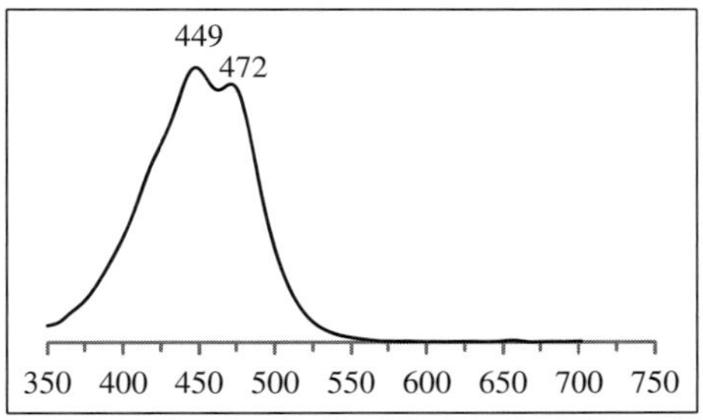

Mass spectra

Ionization technique	Mass analyser type	Diagnostic ions (m/z, rel. intensity)	Ref.
APCI	Ion trap	809 $[M+Na]^+ \rightarrow$ 693 $[M+Na\text{-}116]^+$, 675 $[M+Na\text{-}116\text{–}18]^+$, 633, 525	[5]

Remarks Former name = 4-keto-19′-hexanoyloxyfucoxanthin [49]
d = 139 L g^{-1}cm^{-1} (at 443 nm in acetone; calc. from Fuco) is recommended, as no value has been determined for Hex-kfuco.
Earlier misidentified as 19′-hexanoyloxyparacentrone 3-acetate

Loroxanthin

Recommended abbreviation: Loro (Lo)

IUPAC: (3*R*,3′*R*,6′*R*)-β,ε-Carotene-3,19,3′-triol

Molecular formula: $C_{40}H_{56}O_3$ **Molecular weight:** 584.87

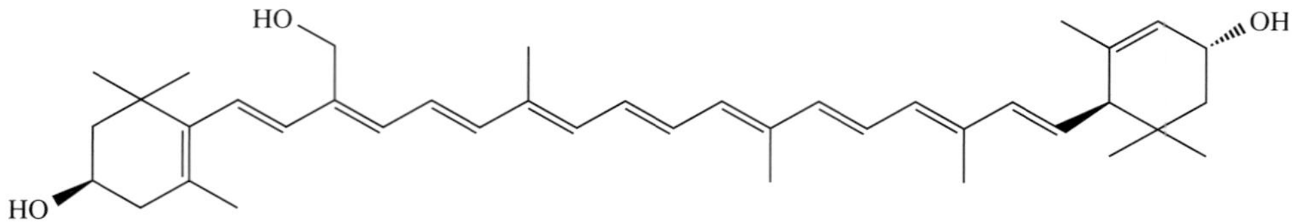

Biological occurrence	Minor pigment in some chlorophytes and prasinophytes
Source culture	*Pyramimonas parkeae* (prasinopyte) [71], CCMP 724 [164]
Alteration products	*Cis*-isomers
Biosynthetically related to	βε-Car, Lut, Loro-d, Siph
Occurs together with	βε-Car, Lut, Loro-d, Neo, Chl *b*

HPLC chromatogram of an *unidentified chlorophyte* (system 1)

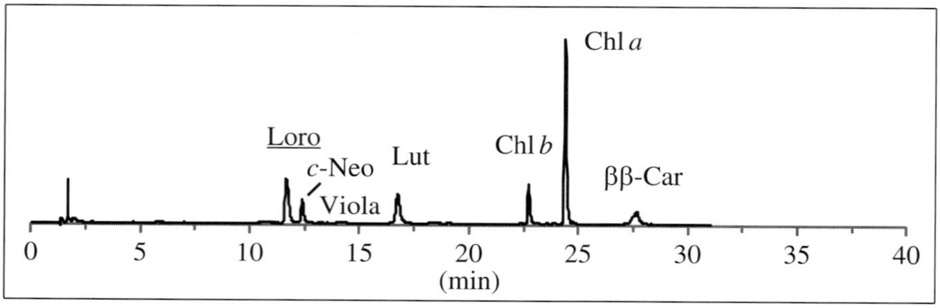

HPLC chromatogram of *Tetraselmis suecica* (system 2)

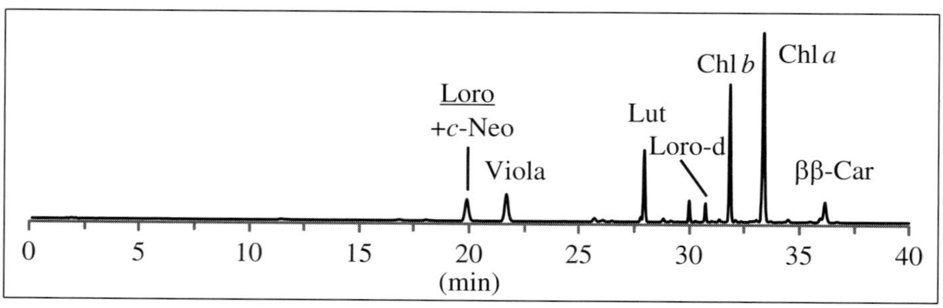

UV-Vis spectra (see also reference spectra below)

Solvent	λ_{max} (nm)	Band ratio (% III:II)	Ref.
Acetone	(424), *447*, 472	44	[132]
Diethyl ether	(422), *444*, 472	48	[71]
Ethanol	(425), *446*, 473	n.d.	[116]
Hexane	423, *445*, 473	58	[71]
Recommended specific absorption coefficient *d* (L g^{-1} cm^{-1})		n.d., see Remarks	

Reference spectra
In acetone

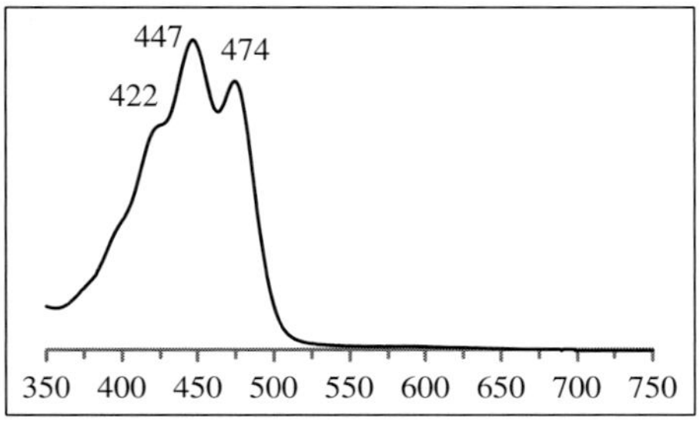

In HPLC solvent system 1

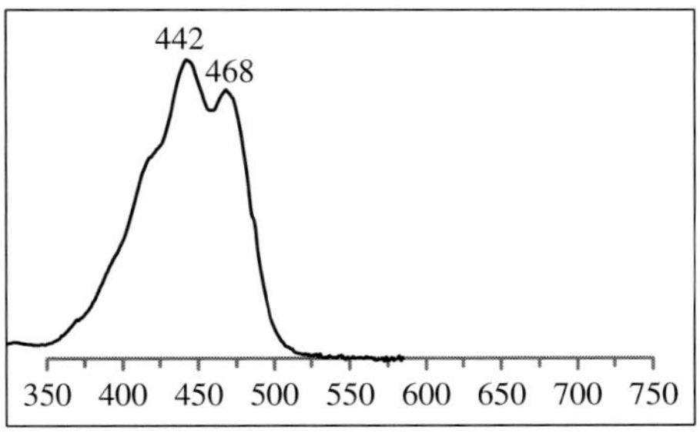

In HPLC solvent system 2

NO DATA AVAILABLE

Mass spectra

Ionization technique	Mass analyser type	Diagnostic ions (m/z, rel. intensity)	Ref.
ESI$^+$	Ion trap	584 $[M]^+ \rightarrow$ 566 $[M\text{-}18]^+$, 492 $[M\text{-}92]^+$, 474 $[M\text{-}92\text{–}18]^+$, 446 $[M\text{-}138]^+$, 428 $[M\text{-}138\text{–}18]^+$	[71]

Remarks $d = 245\ \mathrm{L\,g^{-1}\,cm^{-1}}$ (at λ_{max} in dioxane; calc. from Lut) is recommended, as no value has been determined for Loro. In ethanol, use the value for lutein corrected for the difference in molecular weight: $d = 248\ \mathrm{Lg^{-1}cm^{-1}}$ (at 445 nm)

Loroxanthin dodecenoate

Recommended abbreviation: Loro-d (Lod)

IUPAC: (3*R*,3′*R*,6′*R*)-19-(Dodec-2-enoyloxy)-β,ε-carotene-3,3′-diol

Molecular formula: $C_{52}H_{76}O_4$ **Molecular weight:** 765.16

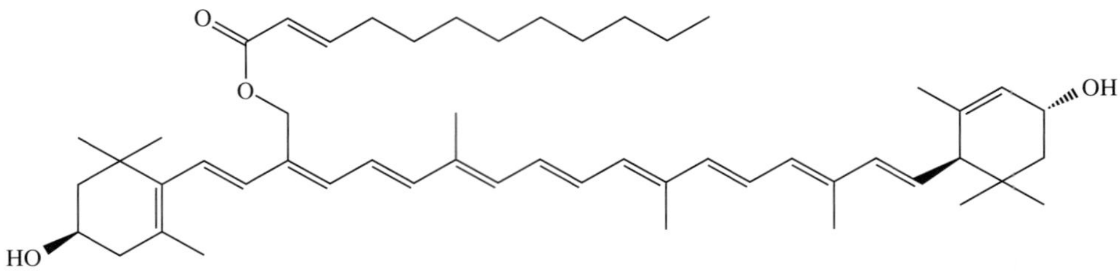

Biological occurrence	Minor in some prasinophytes, also encountered in an euglenophyte and a chlorarachniophyte
Source culture	*Pyramimonas parkeae* (prasinophyte) [139], CCMP 724 [164]
Alteration products	*Cis*-isomers
Biosynthetically related to	βε-Car, Loro, Lut, Siph esters
Occurs together with	βε-Car, Lut, Loro, Neo, Chl *b*

HPLC chromatogram (system 1)

NO DATA AVAILABLE

HPLC chromatogram of *Tetraselmis suecica* (system 2)

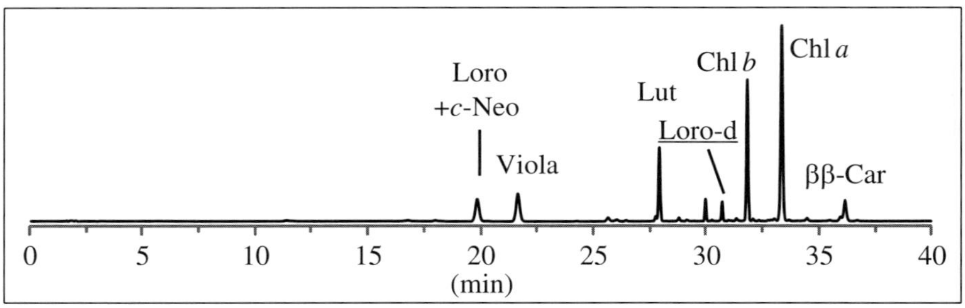

UV-Vis spectra (see also reference spectra below)

Solvent	λ_{max} (nm)	Band ratio (% III:II)	Ref.
Acetone	(423), *449*, 477	42	[71]
Hexane	422, *446*, 474	64	[71]
Methanol	*445*, 471	42	[139]
Recommended specific absorption coefficient *d* (L g^{-1} cm^{-1})		n.d., see Remarks	

Reference spectra

In acetone

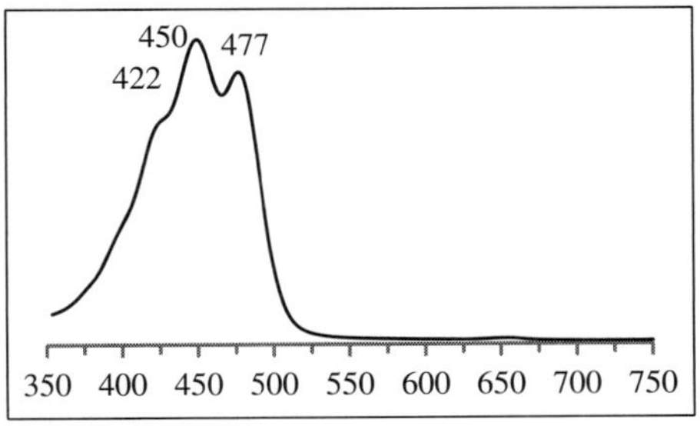

In HPLC solvent system 1

NO DATA AVAILABLE

In HPLC solvent system 2

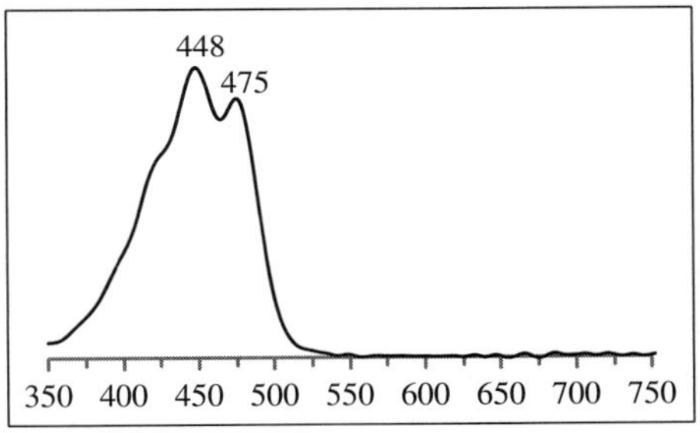

Mass spectra

Ionization technique	Mass analyser type	Diagnostic ions (m/z, rel. intensity)	Ref.
ESI$^+$	Ion trap	764 [M]$^+$, 747 [M+H-18]$^+$ → 729 [M+H-18–18]$^+$, 549 [M+H-18–198]$^+$, 531 [M+H-18–18–198]$^+$	[71]

Remarks Other alkenoylloroxanthins have also been encountered [65]

$d = 187\,\mathrm{L\,g^{-1}cm^{-1}}$ (at λ_{max} in dioxane; calc. from Lut) is recommended, as no value has been determined for Loro-d. In ethanol, use the value for lutein, corrected for the difference in molecular weight: $d = 190\,\mathrm{L\,g^{-1}cm^{-1}}$ (at λ_{max})

Lutein

Recommended abbreviation: Lut (L)

IUPAC: (3*R*,3′*R*,6′*R*)-β,ε-Carotene-3,3′-diol

Molecular formula: $C_{40}H_{56}O_2$

Molecular weight: 568.87

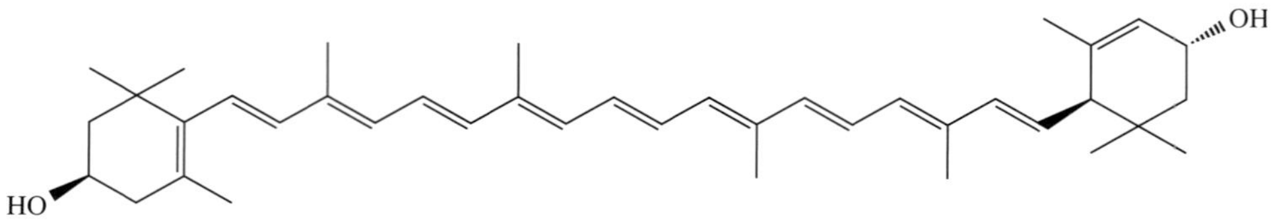

Biological occurrence	Dominant pigment in chlorophytes, chlorarachniophytes and prasinophytes. Minor in mesostigmatophytes. Also in red seaweeds, green leaves and some flowers
Source culture	*Dunaliella tertiolecta* (chlorophyte); *Tetraselmis suecica* (prasinophyte)
Alteration products	*Cis*-isomers
Biosynthetically related to	βε-Car, Loro, Siph, Loro and Siph esters, Dhlut?
Occurs together with	βε-Car, Zea, Viola, Neo, Chl *b*

HPLC chromatogram of *Dunaliella tertiolecta* (system 1)

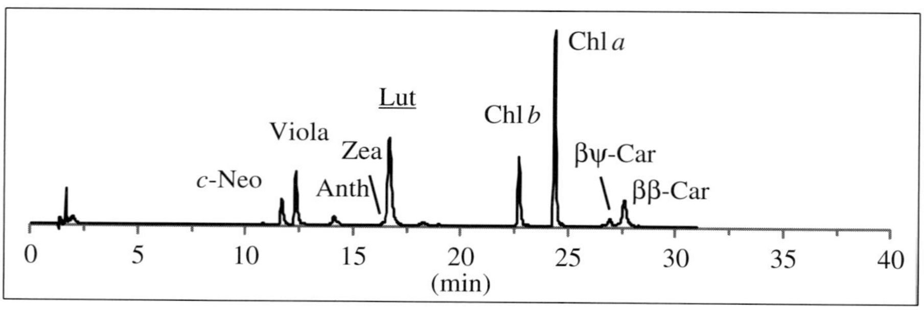

HPLC chromatogram of *Micromonas pusilla* (system 2)

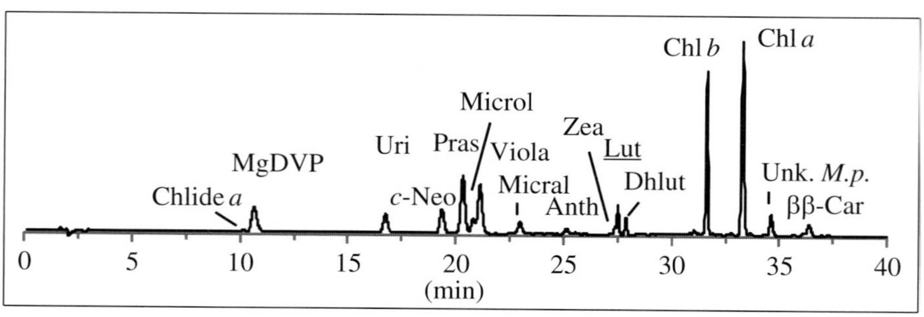

UV-Vis spectra (see also reference spectra below)

Solvent	λ_{max} **(nm)**	**Band ratio (% III:II)**	**Ref.**
Acetone	425, *448*, 476	67	[109]
Ethanol	422, *445*, 474	62	[86]
Hexane	421, *445*, 474	76	[109]
Methanol	(422), *443*, 470	62	[160]
Recommended specific absorption coefficient ***d*** **(L g^{-1} cm^{-1})**		252 (at 453 nm, dioxane) [128] 255 (at 445 nm, ethanol) [148]	

Reference spectra

For spectrum in acetone, see [109]

In ethanol

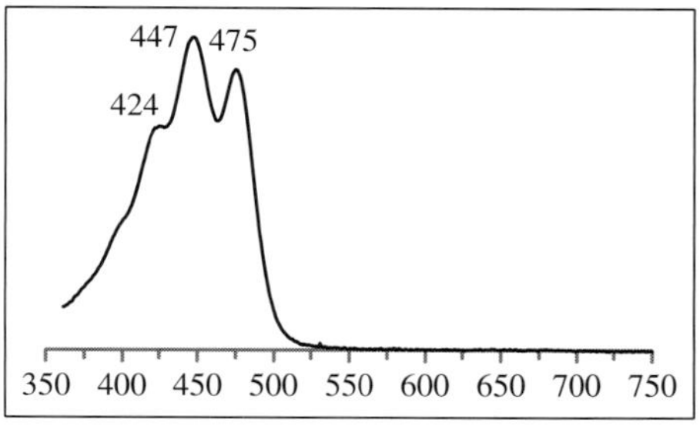

In HPLC solvent system 1

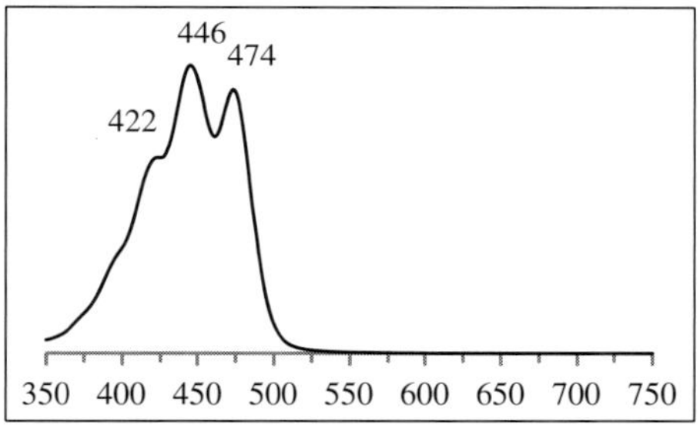

In HPLC solvent system 2

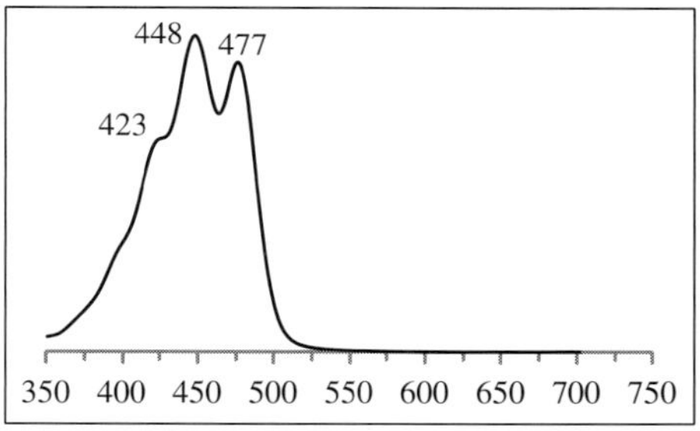

Mass spectra

Ionization technique	**Mass analyser type**	**Diagnostic ions (m/z, rel. intensity)**	**Ref.**
EI	Magnetic sector	568 $[M]^+$ (23), 550 $[M\text{-}18]^+$ (31), 512 $[M\text{-}56]^+$ (1), 476 $[M\text{-}92]^+$ (8), 462 $[M\text{-}106]^+$ (5), 458 $[M\text{-}92\text{–}18]^+$ (12), 415 $[M\text{-}153]^+$ (2), 392 (29), 324 (41), 91 (100)	[55]

Remarks The name 'Xanthophyll' is found in older literature. Lutein and lutein epoxide are part of a 'xanthophyll cycle': see Chapter 11 this volume

Micromonal

Recommended abbreviation: Micral (Mia)

IUPAC: (3*R*,3′*R*,6′*R*)-3,3′-Dihydroxy-7′,8′-dihydro-β,ε-caroten-19′-al

Molecular formula: $C_{40}H_{56}O_3$ **Molecular weight:** 584.87

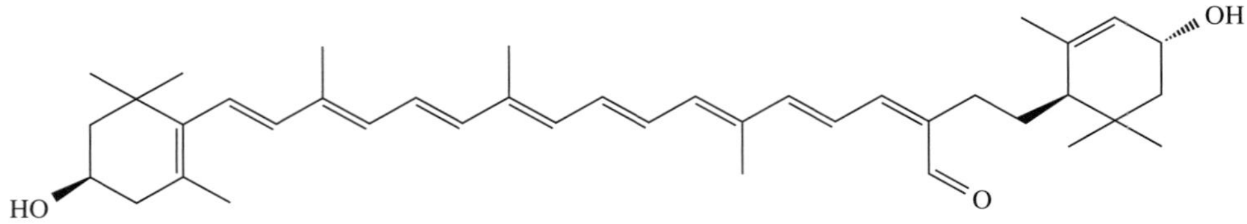

Biological occurrence	Dominant pigment in prasinophytes Pigment Type 3B (Chapter 1, this volume)
Source culture	*Micromonas pusilla* (prasinophyte)
Alteration products	*Cis*-isomers
Biosynthetically related to	Lut (?), Dhlut, Microl, Uri
Occurs together with	Pras, Microl, Dhlut, Uri, Chl *b*, MgDVP

HPLC chromatogram of *Micromonas pusilla* (system 1)

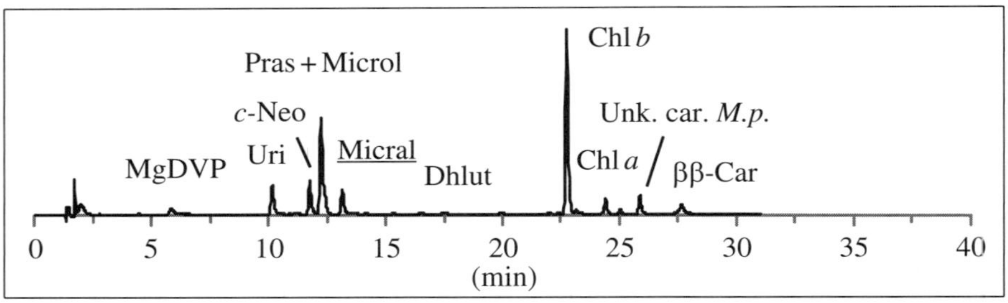

HPLC chromatogram of *Micromonas pusilla* (system 2)

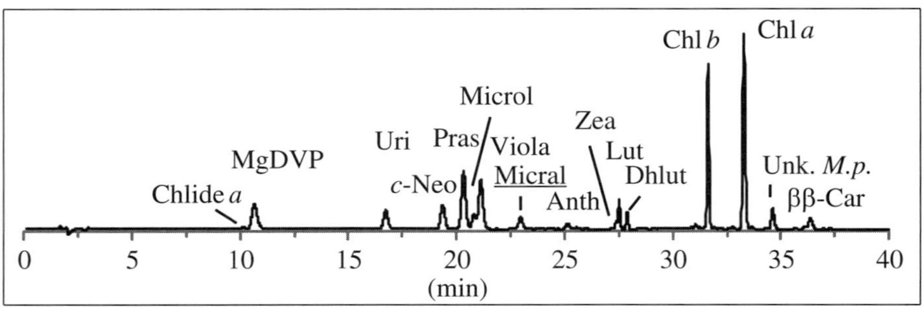

UV-Vis spectra (see also reference spectra below)

Solvent	**λ_{max} (nm)**	**Band ratio (% III:II)**	**Ref.**
Acetone	(423), *449*, (472)	-	[52]
Diethyl ether	(422), *447*, (470)	-	[52]
Hexane	(425), *449*, (474)	-	[52]
Recommended specific absorption coefficient *d* ($L\ g^{-1}\ cm^{-1}$)		n.d., see Remarks	

Reference spectra

In acetone

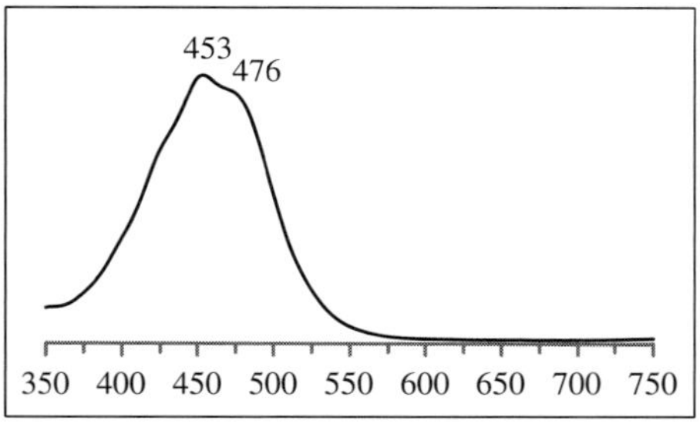

In HPLC solvent system 1

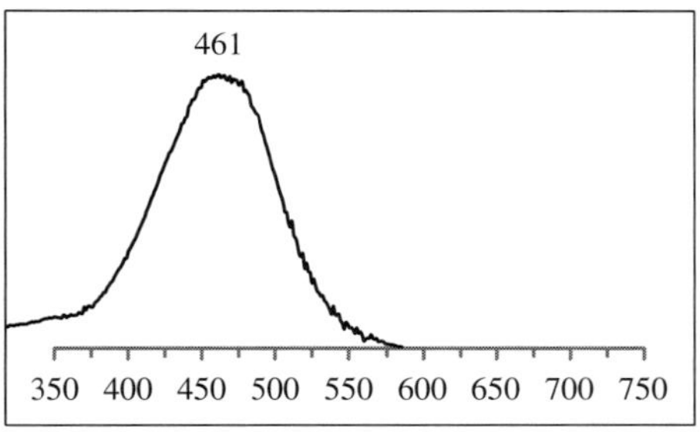

In HPLC solvent system 2

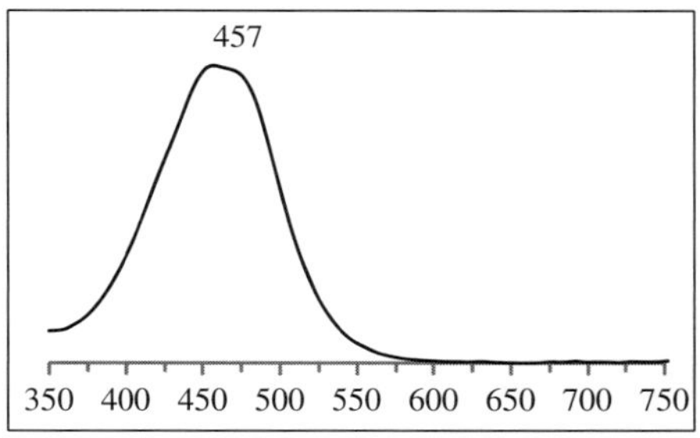

Mass spectra

Ionization technique	**Mass analyser type**	**Diagnostic ions (m/z, rel. intensity)**	**Ref.**
EI	Magnetic sector	584 $[M]^+$ (74), 566 $[M\text{-}18]^+$ (100), 548 $[M\text{-}18\text{–}18]^+$ (12), 538 $[M\text{-}46]^+$ (10), 492 $[M\text{-}92]^+$ (6), 478 $[M\text{-}106]^+$ (6), 444 $[M\text{-}138]^+$ (6), 431 $[M\text{-}153]^+$ (8)	[52]

Remarks $d = 195\ L\ g^{-1} cm^{-1}$ (at λ_{max} in ethanol) is recommended, as no value has been determined for Micral

Micromonol

Recommended abbreviation: Microl (Mio)

IUPAC: (3*R*,3′*R*,6′*R*)-7′,8′-Dihydro-β,ε-carotene-3,3′,19′-triol

Molecular formula: $C_{40}H_{58}O_3$ **Molecular weight:** 586.89

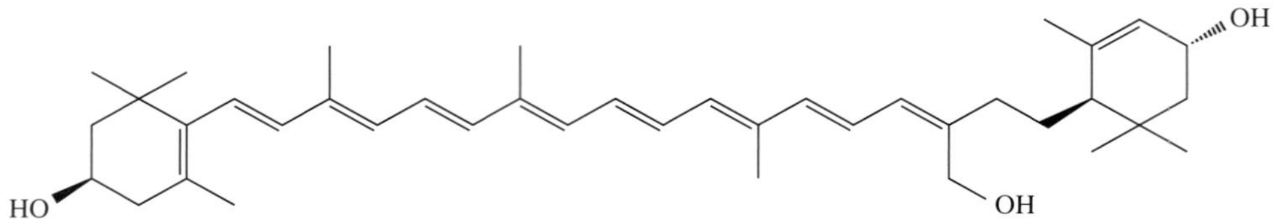

Biological occurrence	Minor pigment in prasinophytes Pigment Type 3B (Chapter 1, this volume)
Source culture	*Micromonas pusilla* (prasinophyte)
Alteration products	*Cis*-isomers
Biosynthetically related to	Lut (?), Dhlut, Micral, Uri
Occurs together with	Pras, Micral, Dhlut, Uri, Chl *b*, MgDVP

HPLC chromatogram of *Micromonas pusilla* (system 1)

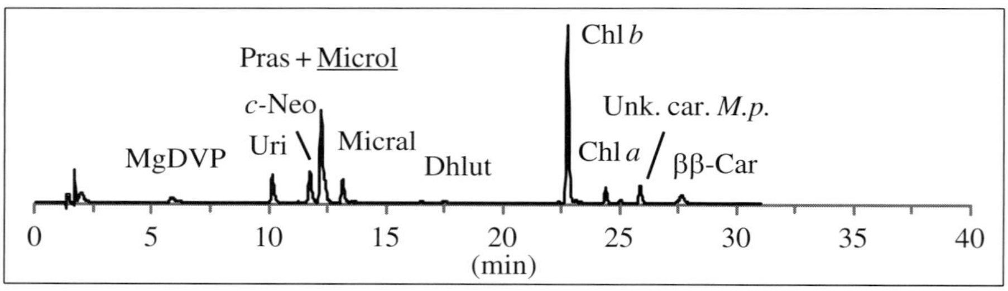

HPLC chromatogram of *Micromonas pusilla* (system 2)

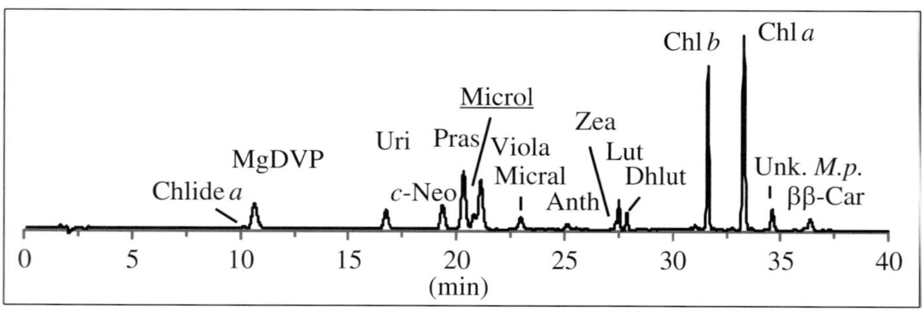

UV-Vis spectra (see also reference spectra below)

Solvent	λ_{max} (nm)	Band ratio (% III:II)	Ref.
Acetone	(406), *427*, 452	54	[52]
Ethanol	(403), *424*, 450	49	[52]
Recommended specific absorption coefficient *d* ($L\ g^{-1}\ cm^{-1}$)		n.d., see Remarks	

Reference spectra

In HPLC solvent system 1

NO DATA AVAILABLE

In HPLC solvent system 2

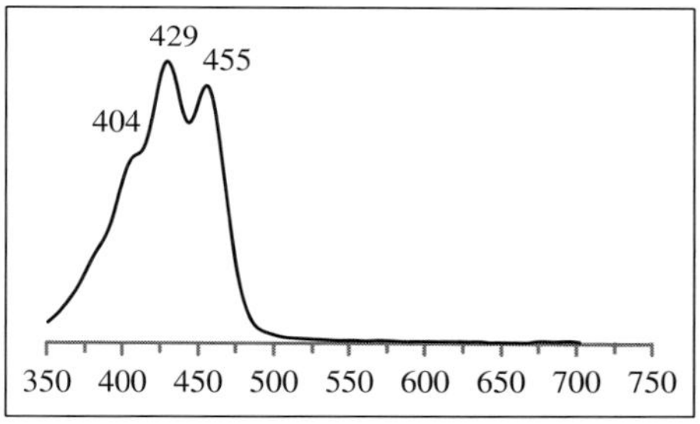

Mass spectra

Ionization technique	Mass analyser type	Diagnostic ions (m/z, rel. intensity)	Ref.
EI	Magnetic sector	586 $[M]^+$ (100), 568 $[M\text{-}18]^+$ (26), 550 $[M\text{-}18\text{–}18]^+$ (14), 532 $[M\text{-}18\text{–}18\text{–}18]^+$ (5), 494 $[M\text{-}92]^+$ (5), 480 $[M\text{-}106]^+$ (4), 446 $[M\text{-}140]^+$ (19)	[52]

Remarks $d = 223\ L\ g^{-1} cm^{-1}$ (at 424 nm in ethanol; calc. from Mutato) is recommended, as no value has been determined for Microl

Monadoxanthin

Recommended abbreviation: Monado (Mo)

IUPAC: (3*R*,3′*R*,6′*R*)-7,8-Didehydro-β,ε-carotene-3,3′-diol

Molecular formula: $C_{40}H_{54}O_2$ **Molecular weight:** 566.86

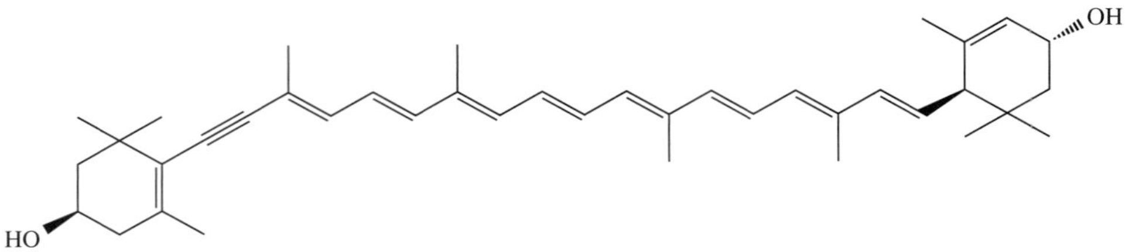

Biological occurrence	Minor pigment in cryptophytes; also present in two chlorophytes (see Chapter 1, this volume)
Source culture	*Chroomonas salina* (cryptophyte)
Alteration products	Easily isomerizes to the more stable 9-*cis* isomer [43]
Biosynthetically related to	Croco
Occurs together with	Allo, Croco

HPLC chromatogram of *Chroomonas salina* (system 1)

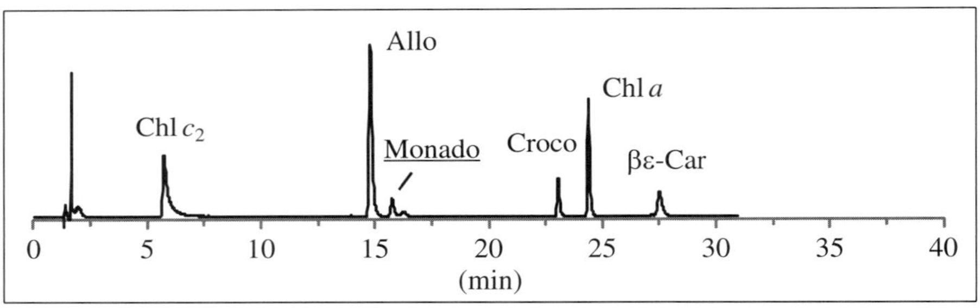

HPLC chromatogram of *Rhodomonas baltica* (system 2)

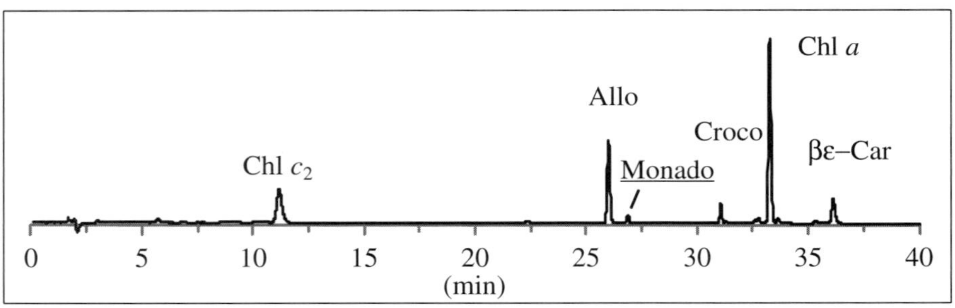

UV-Vis spectra (see also reference spectra below)

Solvent	λ_{max} (nm)	Band ratio (% III:II)	Ref.
Diethyl ether	428, *446*, 475	60	[133]
Ethanol	424, *447*, 477	n.d.	[37]
Hexane	422, *445*, 475	72	[35]
Recommended specific absorption coefficient *d* (L g^{-1} cm^{-1})		n.d., see Remarks	

Reference spectra

In acetone

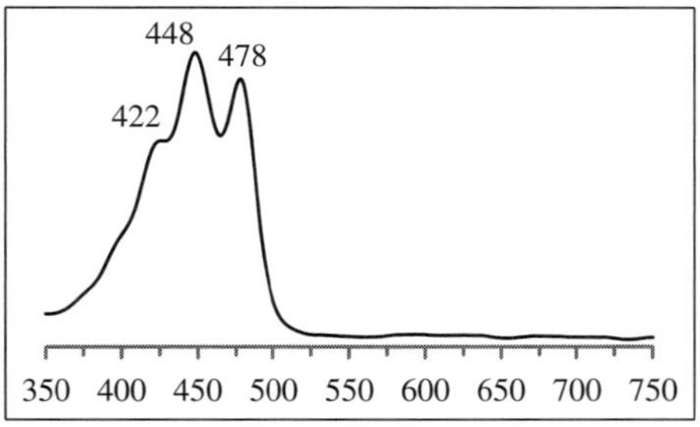

In HPLC solvent system 1

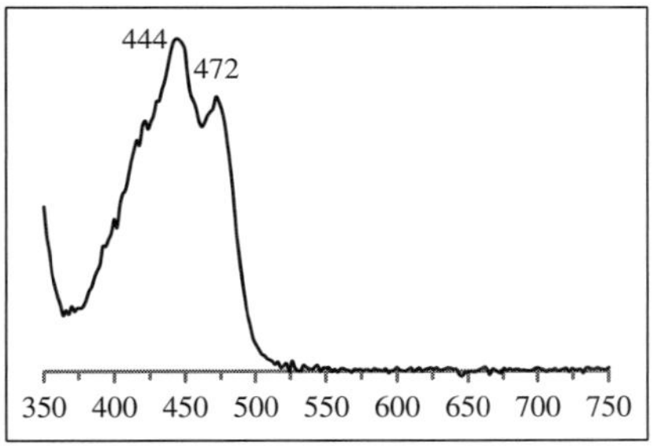

In HPLC solvent system 2

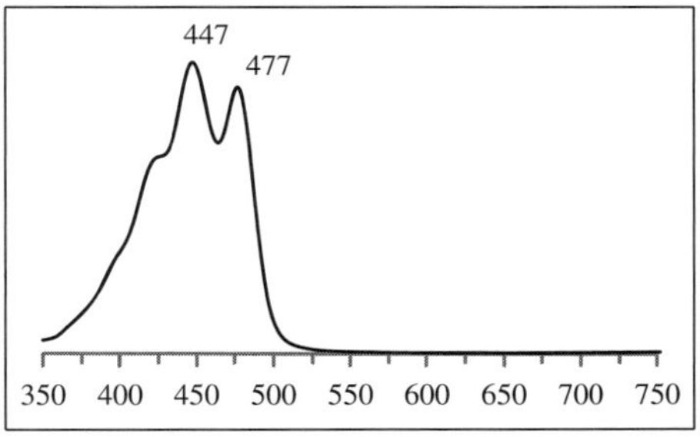

Mass spectra

Ionization technique	Mass analyser type	Diagnostic ions (m/z, rel. intensity)	Ref.
EI	Magnetic sector	566.411 $[M]^+$ (100), 548 $[M\text{-}18]^+$ (17), 533 $[M\text{-}18\text{–}15]^+$ (2), 530 $[M\text{-}18\text{–}18]^+$ (<1), 474 $[M\text{-}92]^+$ (4), 460 $[M\text{-}106]^+$ (2), 119 (34), 105 (42), 91 (56)	[43]

Remarks $d = 230\ \mathrm{L\,g^{-1}\,cm^{-1}}$ (at 445 nm in hexane; calc. from Diadino) is recommended, as no value has been determined for Monado

Mutatoxanthin

Recommended abbreviation: Mutato (Mu)

IUPAC: (3*S*,5*R*,8*RS*,3′*R*)-5,8-Epoxy-5,8-dihydro-β,β-carotene-3,3′-diol

Molecular formula: $C_{40}H_{56}O_3$ **Molecular weight:** 584.87

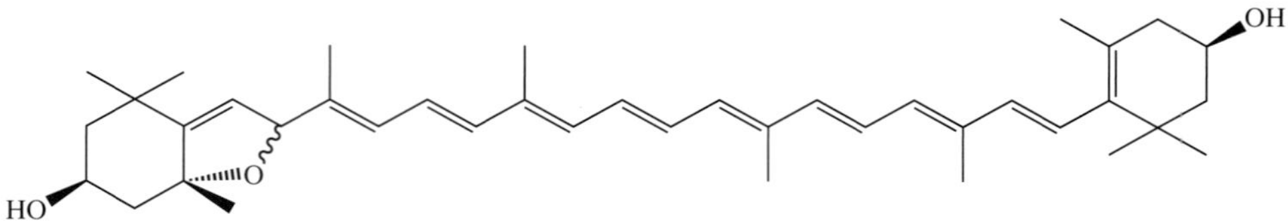

Always occurs as a mixture of the two (3*S*,5*R*,8*R*,3′*R*) and (3*S*,5*R*,8*S*,3′*R*) optical isomers)

Alteration product of	Anth. The acid-catalysed rearrangement occurs in slightly acidic extracts, especially in prasinophyte extracts [91]
Source culture	*Dunaliella tertiolecta* (chlorophyte)
Alteration products	*Cis*-isomers
Synthetically related to	Anth
Occurs together with	Zea, Auro, Neochr

HPLC chromatogram of acidified *Dunaliella tertiolecta* (system 1)

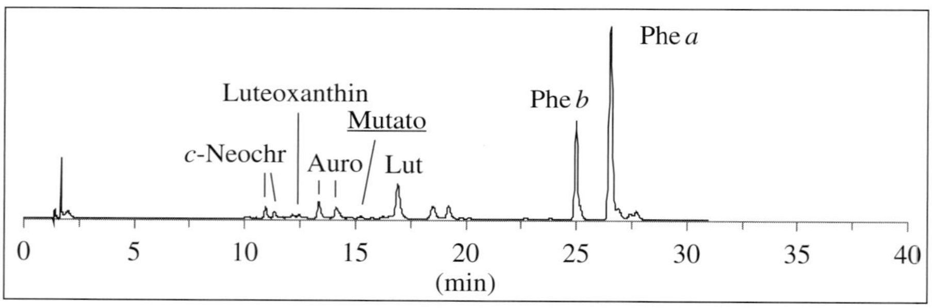

HPLC chromatogram (system 2)

NO DATA AVAILABLE

UV-Vis spectra (see also reference spectra below)

Solvent	λ_{max} (nm)	Band ratio (% III:II)	Ref.
Ethanol	404, *427*, 453	n.d.	[81]
Recommended specific absorption coefficient *d* (L g^{-1} cm^{-1})		224 (at 430 nm, ethanol) [119]	

Reference spectra

In HPLC solvent system 1

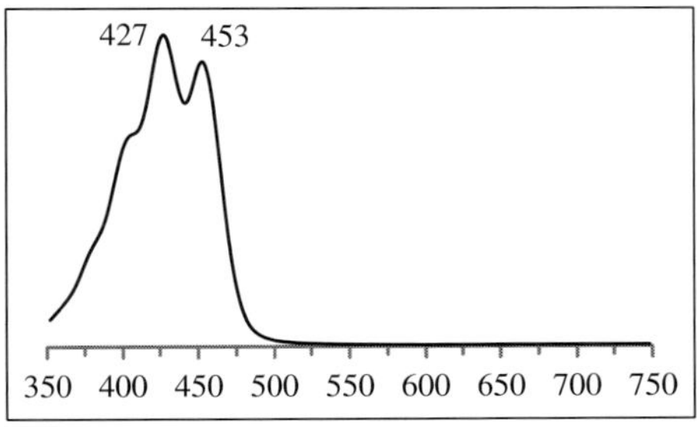

In HPLC solvent system 2

NO DATA AVAILABLE

Mass spectra

Ionization technique	Mass analyser type	Diagnostic ions (m/z, rel. intensity)	Ref.
EI	Magnetic sector	584 $[M]^+$ (6), 566 $[M-18]^+$ (4), 550 $[M-34]^+$ (2), 504 $[M-80]^+$ (7), 492 $[M-92]^+$ (2), 221 (40), 181 (26), 43 (100)	[20]

Remarks

Myxol quinovoside

Recommended abbreviation: Myxo (My)

IUPAC: (3*R*,2′*S*)-2′-(6-Deoxy-α-L-glucopyranosyloxy)-3′,4′-didehydro-1′,2′-dihydro-β,ψ-carotene-3,1′-diol (trivial name: Myxoxanthophyll – see Remarks)

Molecular formula: $C_{46}H_{66}O_7$ **Molecular weight:** 731.01

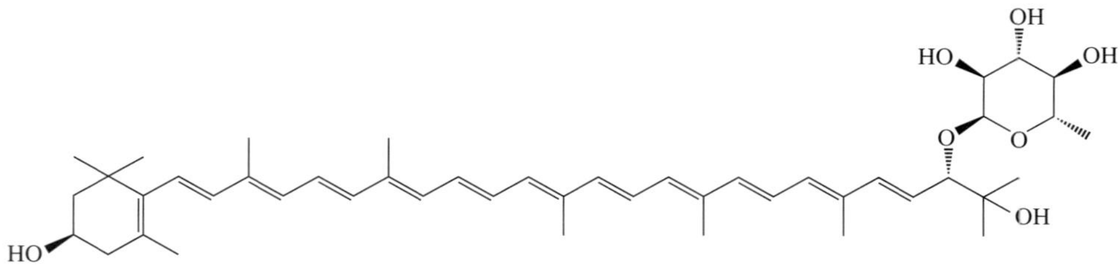

Biological occurrence	A dominant carotenoid in certain cyanobacteria (Cyano-1, Chapter 1)
Source culture	*Oscillatoria agardhii (rubescens)* (cyanobacteria)
Alteration products	*Cis*-isomers
Biosynthetically related to	βψ-Car, Kmyxo
Occurs together with	ββ-Car, Echin, Kmyxo, Oscil

HPLC chromatogram of *Synechococcus* sp. (system 1)

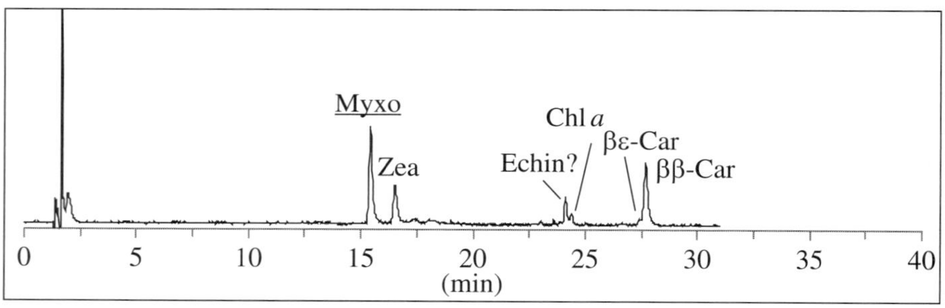

HPLC chromatogram of *Chlorogloeopsis fritschii* (system 3)

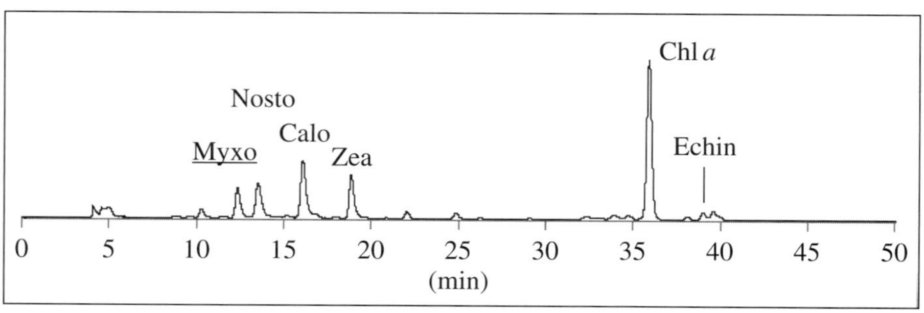

UV-Vis spectra (see also reference spectra below)

Solvent	λ_{max} (nm)	Band ratio (% III:II)	Ref.
Acetone	(450), *478*, 510	57	[94]
Methanol	446, *472*, 502	55	[1]
Recommended specific absorption coefficient *d* (L g^{-1} cm^{-1})		216 (at 478 nm, acetone) [94]	

Reference spectra
In acetone

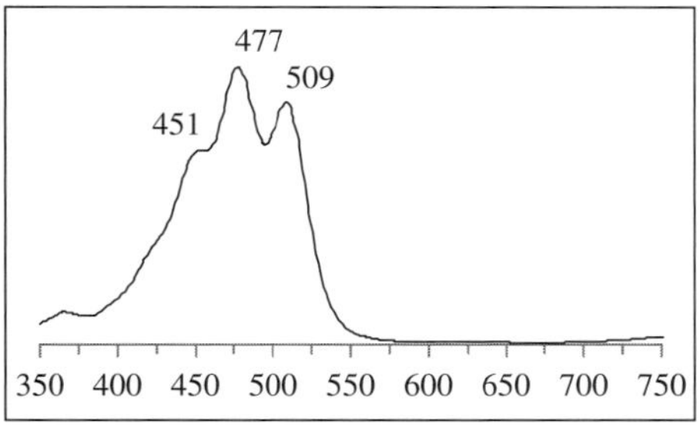

In HPLC solvent system 1

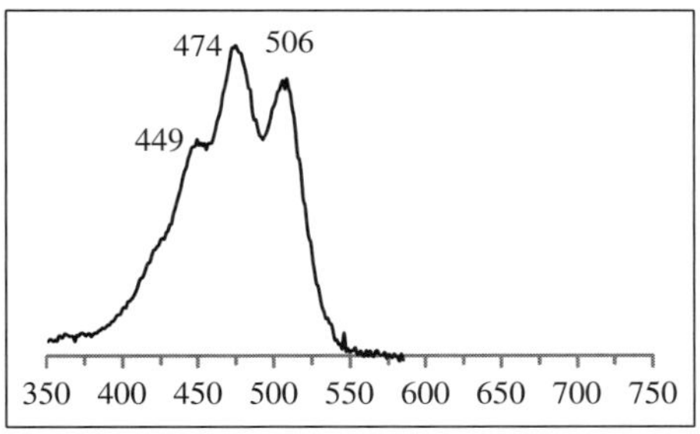

In HPLC solvent system 2

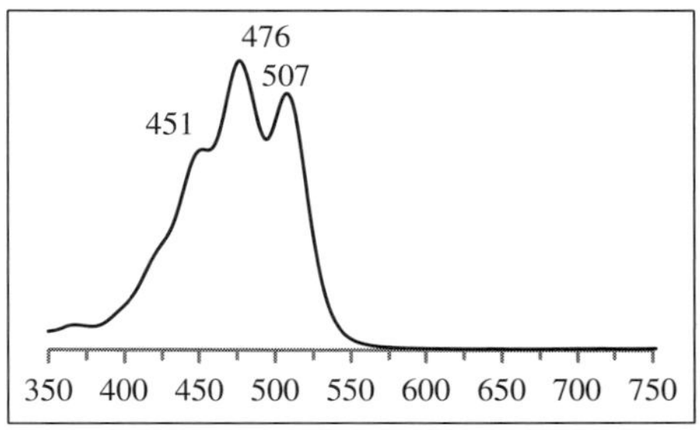

Mass spectra

Ionization technique	Mass analyser type	Diagnostic ions (m/z, rel. intensity)	Ref.
EI	Magnetic sector	730.4852 $[M]^+$ (2), 584 $[M-146]^+$ (2), 566 (7), 550 (8), 474 $[M-256]^+$ (3), 462 $[M-268]^+$ (8), 444 $[M-286]^+$ (15), 419 $[M-311]^+$ (7), 327 $[M-403]^+$ (15)	[94]

Remarks The name 'myxoxanthophyll' is recommended when the sugar moiety is unknown. Other myxol glycosides and 4-keto forms have been reported [1, 65, 142, 159]. Cyanobacterial myxoxanthophylls are unusual because they are glycosylated on the 2′-OH position of the ψ end of the molecule [80]

9′-*cis*-Neochrome **Recommended abbreviation: *c*-Neochr (*c*Nc)**

IUPAC: 9′-*cis*-(3*S*,5*R*,6*R*,3′*S*,5′*R*,8′*RS*)-5′,8′-Epoxy-6,7-didehydro-5,6,5′,8′-tetrahydro-β,β-carotene-3,5,3′-triol

Molecular formula: $C_{40}H_{56}O_4$ **Molecular weight:** 600.87

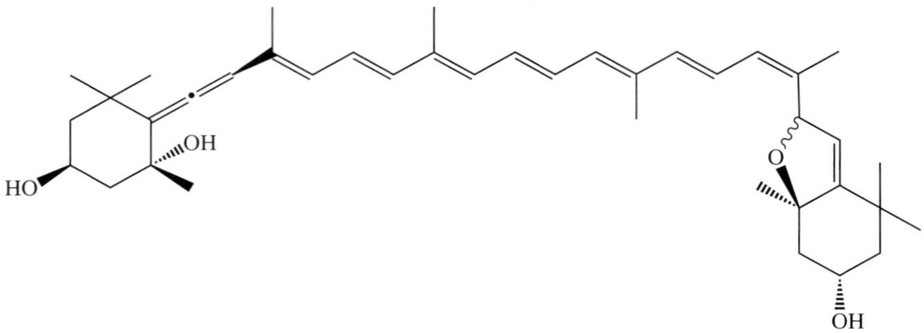

(Always occurs as a mixture of the two (3*S*,5*R*,6*R*,3′*S*,5′*R*,8′*R*) and (3*S*,5*R*,6*R*,3′*S*,5′*R*,8′*S*) optical isomers)

Alteration product of	*c*-Neo. The acid-catalysed rearrangement occurs in slightly acidic extracts, especially in prasinophyte extracts [91]
Source culture	*Dunaliella tertiolecta* (chlorophyte)
Alteration products	*Trans*-isomer, other *cis*-isomers
Synthetically related to	*c*-Neo
Occurs together with	Auro, Mutato

HPLC chromatogram of acidified *Dunaliella tertiolecta* (system 1)

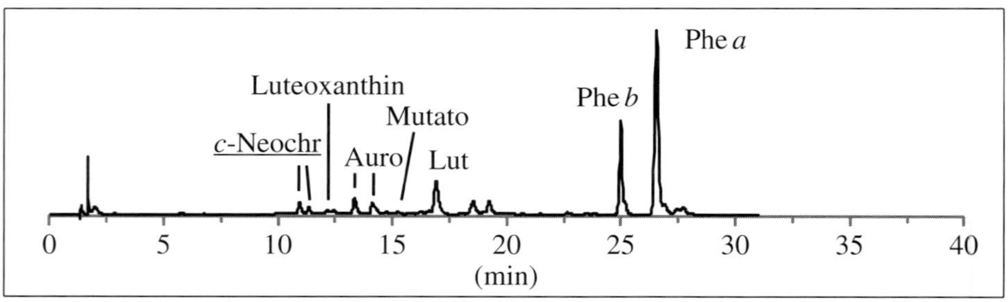

HPLC chromatogram (system 2)

NO DATA AVAILABLE

UV-Vis spectra (see also reference spectra below)

Solvent	**λ_{max} (nm)**	**Band ratio (% III:II)**	**Ref.**
Acetone	398, *422*, 450	88	[91]
Diethyl ether	399, 422, *449*	n.d.	[32]
Ethanol	401, *424*, 451	n.d.	[39]
Recommended specific absorption coefficient *d* ($L\ g^{-1}\ cm^{-1}$)		227 (at 424 nm, ethanol) [39]	

Reference spectra

In HPLC solvent system 1

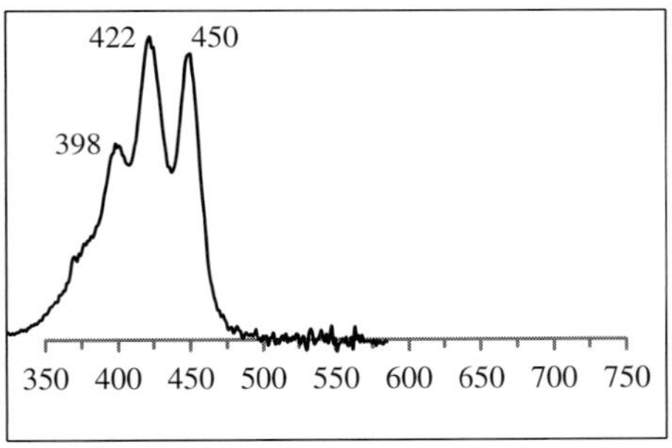

In HPLC solvent system 2

NO DATA AVAILABLE

Mass spectra

Ionization technique	**Mass analyser type**	**Diagnostic ions (m/z, rel. intensity)**	**Ref.**
	see Remarks		

Remarks For MS, see *c*-Neo

9′-*cis*-Neoxanthin **Recommended abbreviation: *c*-Neo (*c*N)**

IUPAC: 9′-*cis*-(3*S*,5*R*,6*R*,3′*S*,5′*R*,6′*S*)-5′,6′-Epoxy-6,7-didehydro-5,6,5′,6′-tetrahydro-β,β-carotene-3,5,3′-triol

Molecular formula: $C_{40}H_{56}O_4$ **Molecular weight:** 600.87

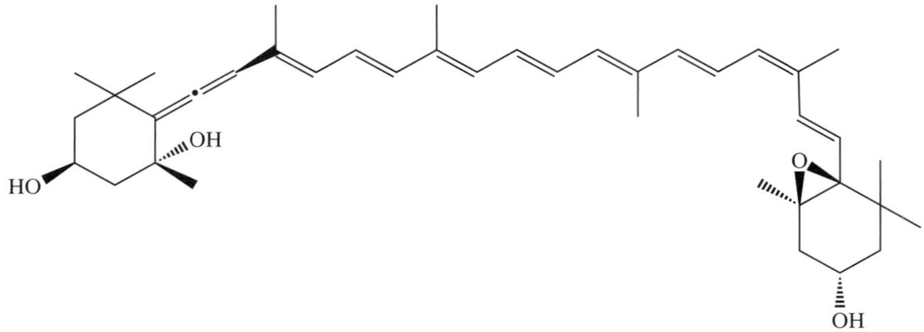

Biological occurrence	Dominant pigment in chlorophytes, prasinophytes and in dinoflagellates Pigment Type 5. Minor in euglenophytes, chlorarachniophytes and trebouxiophytes. Also in green leaves
Source culture	*Dunaliella tertiolecta* (chlorophyte)
Alteration products	Undergoes rearrangement to *c*-Neochr in weakly acidic solutions. *Trans*-isomer, *cis*-isomers
Biosynthetically related to	Zea, Anth, Viola, *t*-Neo
Occurs together with	Chl *b*, Lut, Viola, Zea

HPLC chromatogram of *Dunaliella tertiolecta* (system 1)

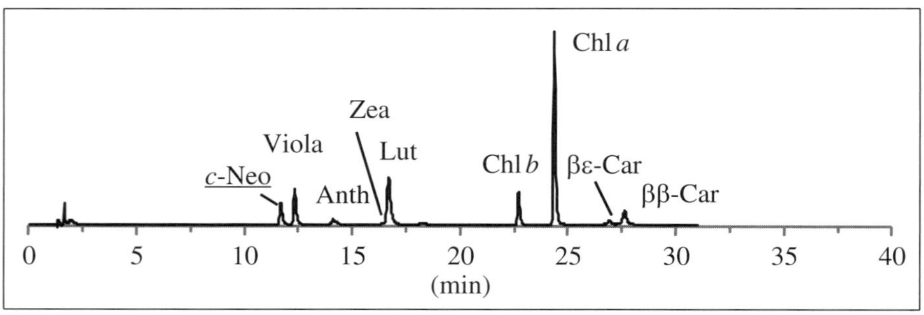

HPLC chromatogram of *Micromonas pusilla* (system 2)

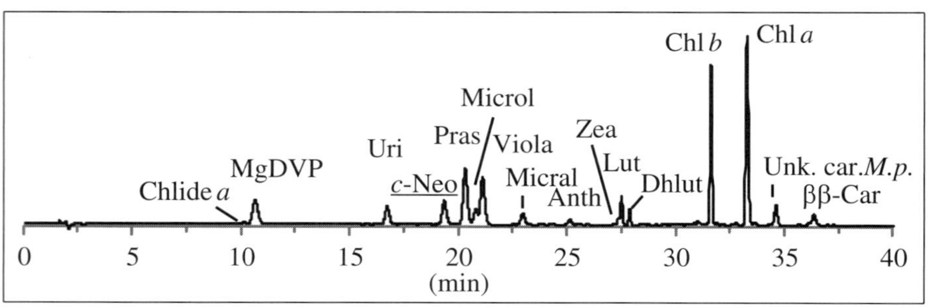

UV-Vis spectra (see also reference spectra below)

Solvent	λ_{max} (nm)	Band ratio (% III:II)	Ref.
Acetone	413, *438*, 466	87	[17]
Diethyl ether	414, *437*, 465	82	[91]
Ethanol	413, *437*, 466	87	[10]
Hexane	411, *435*, 463	97	[109]
Recommended specific absorption coefficient *d* ($L\ g^{-1}\ cm^{-1}$)		233 (at 437 nm, ethanol) [10]	

Reference spectra

For spectrum in acetone, see [109]

In ethanol

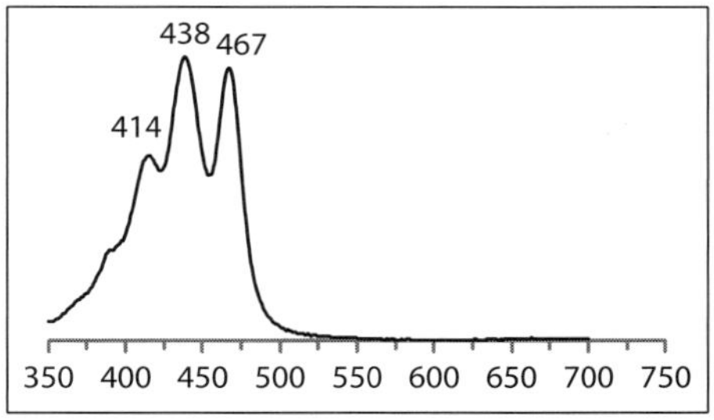

In HPLC solvent system 1

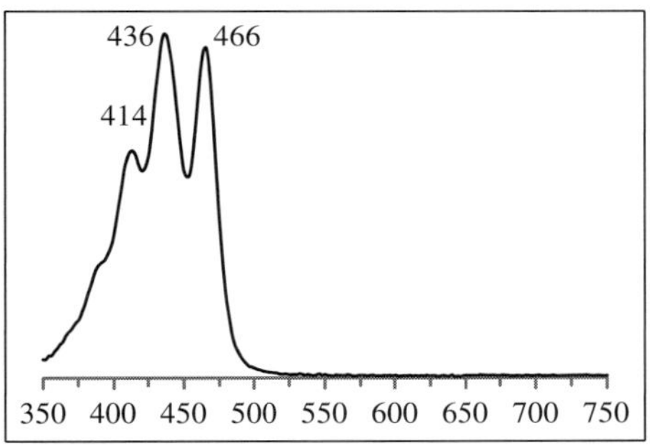

In HPLC solvent system 2

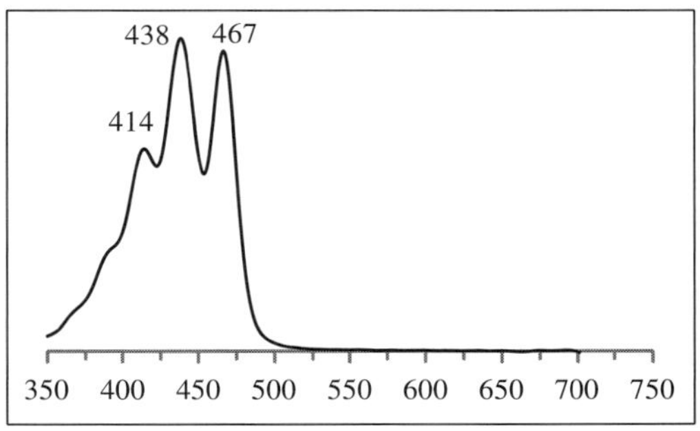

Mass spectra

Ionization technique	Mass analyser type	Diagnostic ions (m/z, rel. intensity)	Ref.
FAB	Magnetic sector	600 $[M]^+$ (100), 449 $[M\text{-}151]^+$ (6), 393 $[M\text{-}207]^+$ (20), 391 $[M\text{-}209]^+$ (12), 339 (10), 313 (9), 171 (12)	[28]

Remarks Do not confuse with all-*trans*-Neoxanthin (see its data sheet)

all-*trans*-Neoxanthin **Recommended abbreviation: *t*-Neo (*t*N)**

IUPAC: (3*S*,5*R*,6*R*,3′*S*,5′*R*,6′*S*)-5′,6′-Epoxy-6,7-didehydro-5,6,5′,6′-tetrahydro-β,β-carotene-3,5,3′-triol

Molecular formula: $C_{40}H_{56}O_4$ **Molecular weight:** 600.87

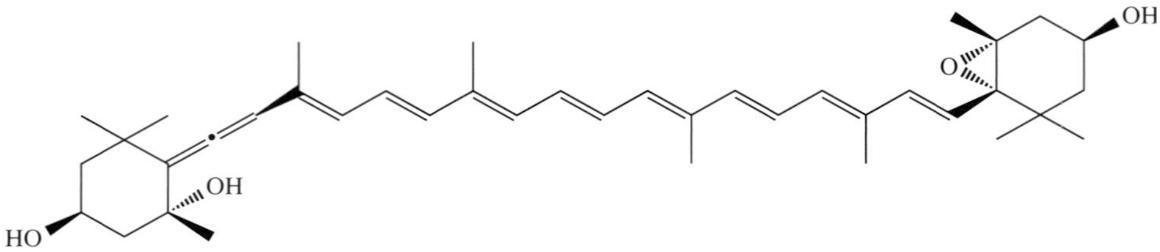

Biological occurrence	Dominant pigment in mesostigmatophytes. Also in various petals and fruits [158]
Source culture	*Mesostigma viride* (mesostigmatophyte) [177], NIES-296 [162]
Alteration products	Undergoes rearrangement to *trans*-Neochr in weakly acidic solutions. *Cis*-isomers
Biosynthetically related to	Zea, Anth, Viola, *c*-Neo
Occurs together with	Chl *b*, Siph esters, Viola

HPLC chromatogram (system 1)

NO DATA AVAILABLE

HPLC chromatogram of *Codium tomentosum* (system 2)

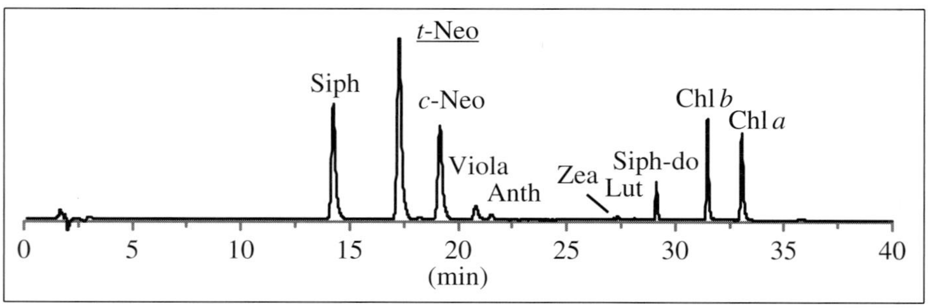

UV-Vis spectra (see also reference spectra below)

Solvent	λ_{max} **(nm)**	**Band ratio (% III:II)**	**Ref.**
Acetone	416, *440*, 469	84	[151]
Ethanol	418, *442*, 471	91	[9]
Hexane	416, *440*, 469	94	[109]
Recommended specific absorption coefficient *d* (L g^{-1}cm^{-1})		238 (at 442 nm, ethanol) [9]	

Reference spectra

In HPLC solvent system 1

NO DATA AVAILABLE

In HPLC solvent system 2

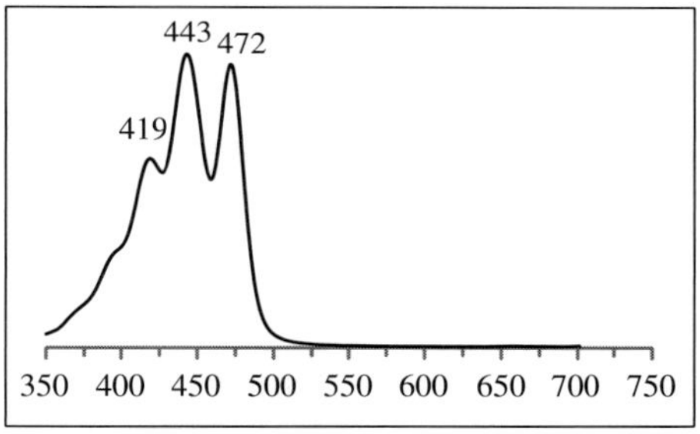

Mass spectra

Ionization technique	**Mass analyser type**	**Diagnostic ions (m/z, rel. intensity)**	**Ref.**
	see Remarks		

Remarks For MS, see *c*-Neo.
Occurs together with *c*-Neo in some prasinophytes [47]

Nostoxanthin

Recommended abbreviation: Nosto (Nos)

IUPAC: (2*R*,3*R*,2′*R*,3′*R*)-β,β-Carotene-2,3,2′,3′-tetrol

Molecular formula: $C_{40}H_{56}O_4$ **Molecular weight:** 600.87

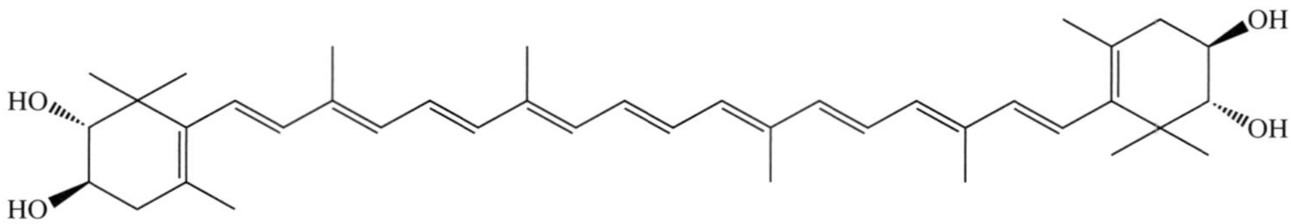

Biological occurrence	Characteristic carotenoid from some cyanobacteria (Cyano-1, Chapter 1), e.g. *Anacystis nidulans* and *Chlorogloeopsis fritschii*
Source culture	*Chlorogloeopsis fritschii* (cyanobacteria)
Alteration products	*Cis*-isomers
Biosynthetically related to	ββ-Car, Zea, Calo
Occurs together with	Calo, Zea

HPLC chromatogram of *Chlorogloeopsis fritschii* (system 3)

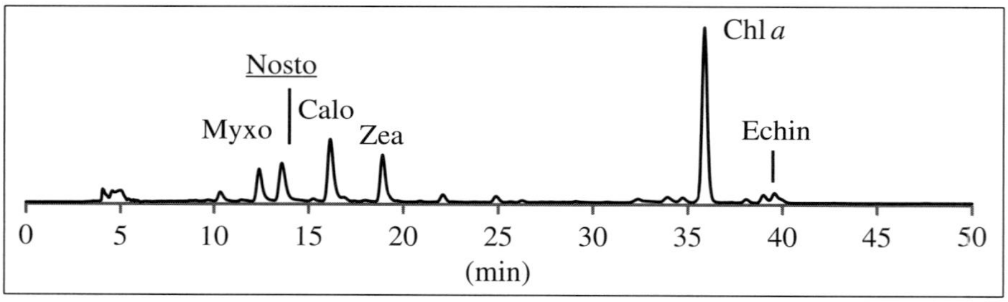

HPLC chromatogram (system 2)

NO DATA AVAILABLE

UV-Vis spectra (see also reference spectra below)

Solvent	λ_{max} (nm)	Band ratio (% III:II)	Ref.
Acetone	(431), *453*, 480	n.d.	[33]
Ethanol	429, *449*, 477	n.d.	[117]
Recommended specific absorption coefficient *d* ($L\ g^{-1}\ cm^{-1}$)		n.d., see Remarks	

Reference spectra

In acetone

NO DATA AVAILABLE

In HPLC solvent system 3

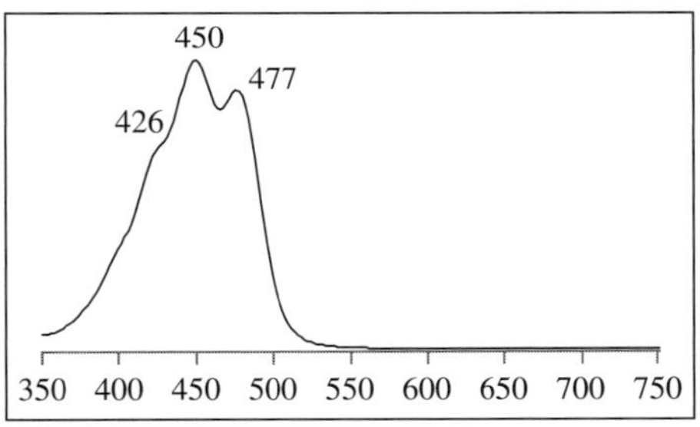

Mass spectra

Ionization technique	Mass analyser type	Diagnostic ions (m/z, rel. intensity)	Ref.
EI	Magnetic sector	600 $[M]^+$ (100), 584 $[M\text{-}16]^+$ (31), 568 $[M\text{-}16\text{–}16]^+$ (2), 508 $[M\text{-}92]^+$ (11), 494 $[M\text{-}106]^+$ (2), 492 $[M\text{-}16\text{–}92]^+$ (2), 133 (22), 91 (33), 69 (25)	[33]

Remarks: $d = 232\ L\ g^{-1}\ cm^{-1}$ (at λ_{max} in ethanol; calc. from Zea) is recommended, as no value has been determined for Nosto

Oscillol diquinovoside

Recommended abbreviation: Oscil (O)

IUPAC: 2,2′-Di-(6-deoxy-α-L-glucopyranosyloxy)-3,4,3′,4′-tetradehydro-1,2,1′,2′-tetrahydro-ψ,ψ-carotene-1,1′-diol (trivial name: Oscillaxanthin – see Remarks)

Molecular formula: $C_{52}H_{76}O_{12}$ **Molecular weight:** 893.15

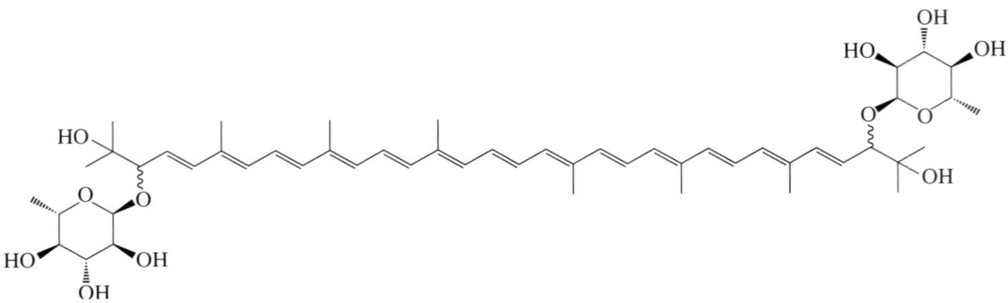

Biological occurrence	Characteristic carotenoid in some cyanobacteria (Cyano-1, Chapter 1)
Source culture	*Oscillatoria agardhii (rubescens)* (cyanobacteria)
Alteration products	*Cis*-isomers
Biosynthetically related to	Lyco
Occurs together with	ββ-Car, Echin, Myxo

HPLC chromatogram (system 1)

NO DATA AVAILABLE

HPLC chromatogram of *Oscillatoria agardhii* var. Kolbotnvatn (system 2)

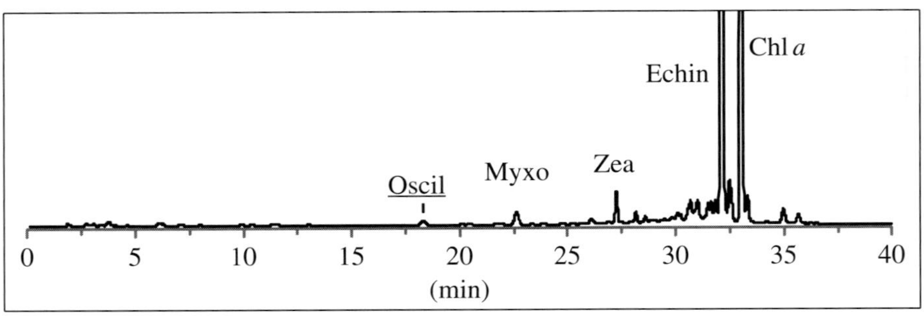

UV-Vis spectra (see also reference spectra below)

Solvent	λ_{max} (nm)	Band ratio (% III:II)	Ref.
Acetone	470, *499*, 534	n.d.	[67]
Ethanol	468, *492*, 526	59	[153]
Methanol	464, *491*, 525	42	[1]
Recommended specific absorption coefficient *d* ($L\ g^{-1}\ cm^{-1}$)		75 (at 490 nm, in 90% methanol: 10% pyridine) [95]	

Reference spectra

In HPLC solvent system 1

NO DATA AVAILABLE

In HPLC solvent system 2

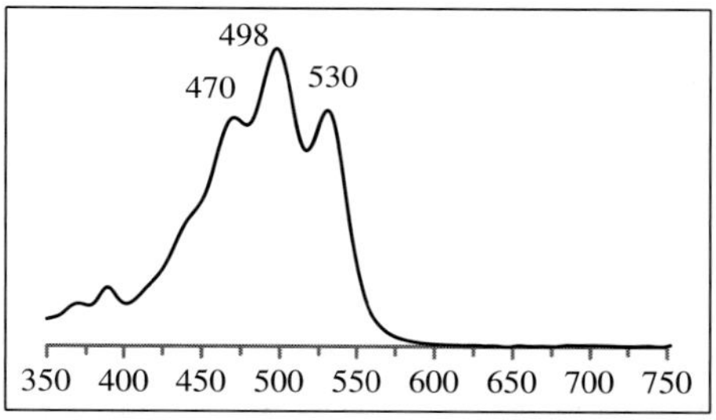

Mass spectra

Ionization technique	Mass analyser type	Diagnostic ions (m/z, rel. intensity)	Ref.
FD	Double focusing	892 $[M]^+$	[166]

Remarks The name 'oscillaxanthin' is recommended when the sugar moiety is unknown. Other oscillol diglycosides have been reported [e.g. 1, 166]. Chirality at C-2,2′ is disputed [135, 165, 166]

Peridinin

Recommended abbreviation: Peri (P)

IUPAC: (3*S*,5*R*,6*S*,3′*S*,5′*R*,6′*R*)-5,6-Epoxy-3′-ethanoyloxy-3,5′-dihydroxy-6′,7′-didehydro-5,6,5′,6′-tetrahydro-12′,13′,20′-trinor-β,β-caroten-19,11-olide

Molecular formula: $C_{39}H_{50}O_7$

Molecular weight: 630.81

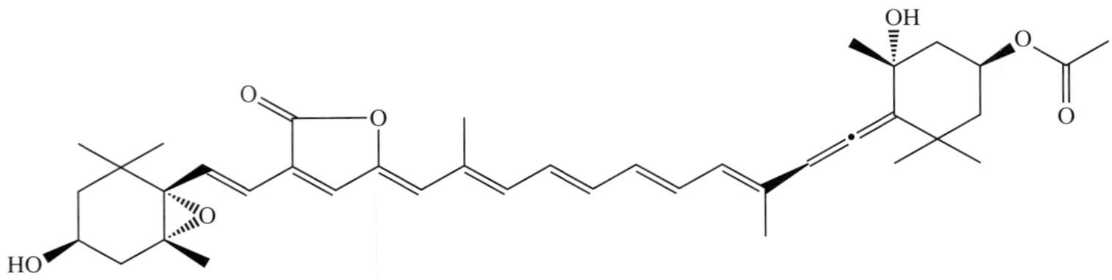

Biological occurrence	Major carotenoid in dinoflagellates Pigment Type 1 (Chapter 1)
Source culture	*Amphidinium carterae* (dinoflagellate)
Alteration products	Undergoes rearrangement to the corresponding furanoxide in acidic solutions, but Peri is more stable than other epoxides [115]. *Cis*-isomers
Biosynthetically related to	*trans*-Neo, Dino
Occurs together with	Chl c_2, Diadino, Diato, Dino, P457

HPLC chromatogram of *Gymnodinium catenatum* (system 1)

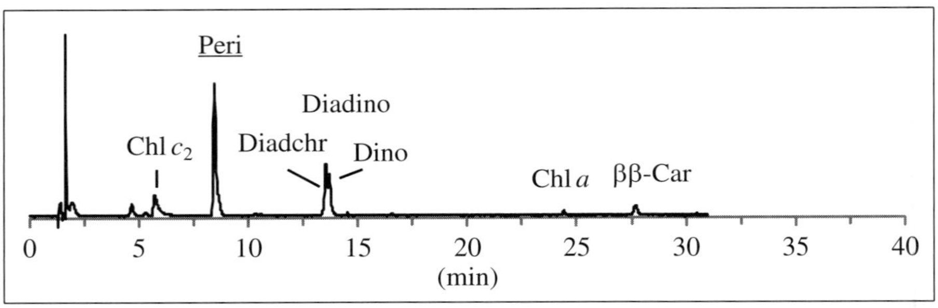

HPLC chromatogram of *Scrippsiella trochoidea* (system 2)

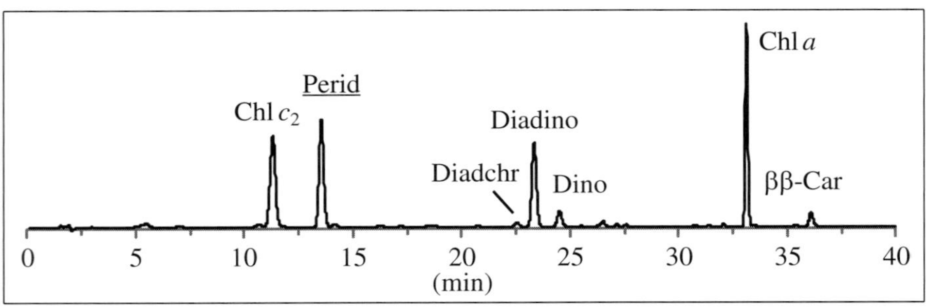

UV-Vis spectra (see also reference spectra below)

Solvent	λ_{max} (nm)	Band ratio (% III:II)	Ref.
Acetone	465	-	[118]
Ethanol	475	-	[150]
Hexane	(430), *454*, 483	74	[118]
Recommended specific absorption coefficient *d* (L g^{-1} cm^{-1})		135 (at 475 nm, ethanol) [150] 134 (at 466 nm, acetone) [106]	

Reference spectra
In acetone

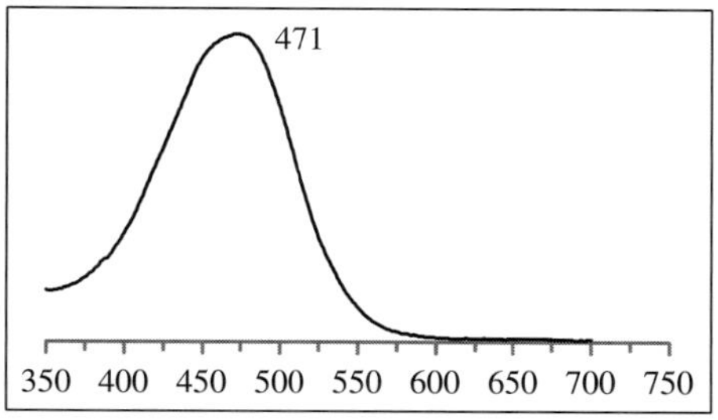

For spectrum in hexane and ethanol, see [109]

In HPLC solvent system 1

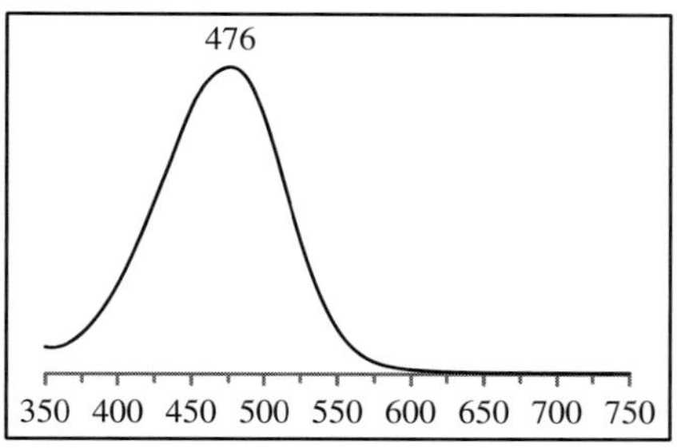

In HPLC solvent system 2

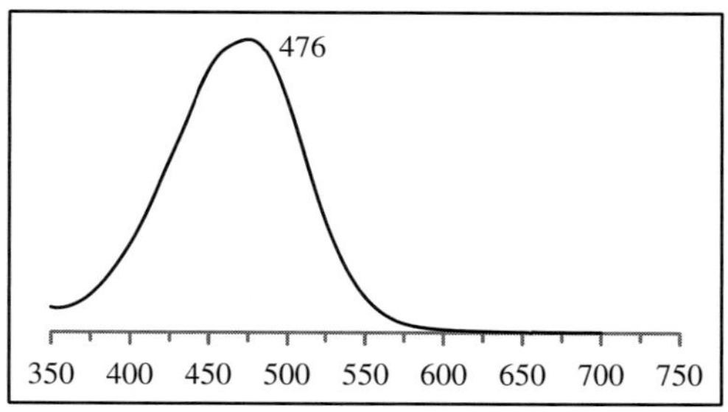

Mass spectra

Ionization technique	Mass analyser type	Diagnostic ions (m/z, rel. intensity)	Ref.
EI	Magnetic sector	630 [M]$^+$ (36), 612 [M-18]$^+$ (30), 594 [M-18–18]$^+$ (2), 570 [M-60]$^+$ (9), 552 [M-18–60]$^+$ (8), 538 [M-92]$^+$ (6), 234 (20), 221 (10), 181 (33), 169 (100)	[118]

Remarks Peri furanoxide elutes right after Peri in HPLC solvent system 1

Prasinoxanthin

Recommended abbreviation: Pras (Pr)

IUPAC: (3*S*,6*R*,3′*R*,6′*R*)-3,6,3′-Trihydroxy-7,8-dihydro-γ,ε-caroten-8-one

Molecular formula: $C_{40}H_{56}O_4$ **Molecular weight:** 600.87

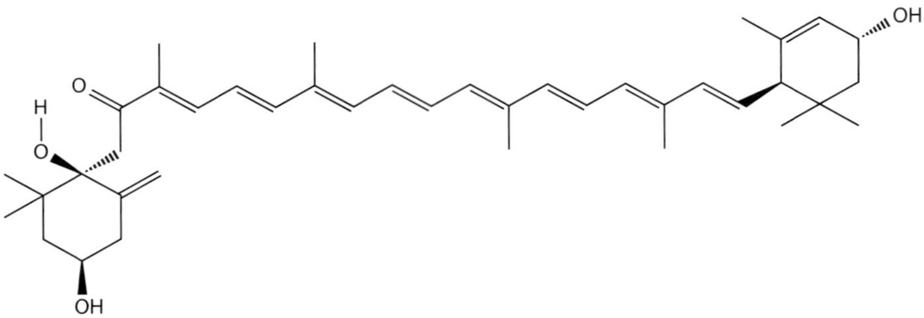

Biological occurrence	Dominant pigment in prasinophytes Pigment Types 3A and 3B. Occasionally found in dinoflagellates Pigment Type 5 (see Chapter 1, this volume)
Source culture	*Pycnococcus provasolii* (prasinophyte)
Alteration products	*Cis*-isomers
Biosynthetically related to	Lut?
Occurs together with	Chl *b*, ββ-Car, βε-Car, Dhlut, Micral, *cis*-Neo, *trans*-Neo, Uri

HPLC chromatogram of *Pycnococcus provasolii* (system 1)

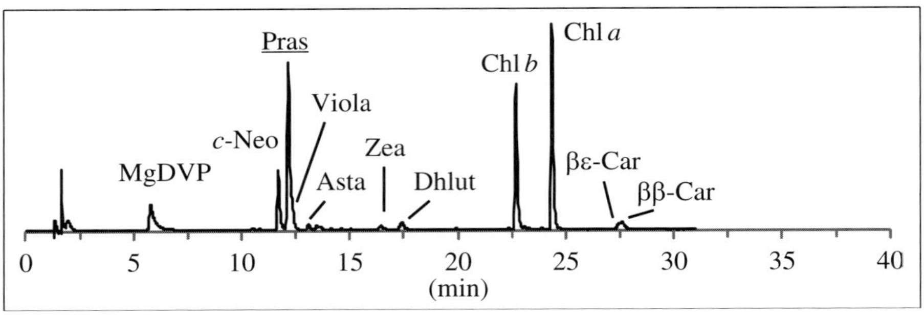

HPLC chromatogram of *Micromonas pusilla* (system 2)

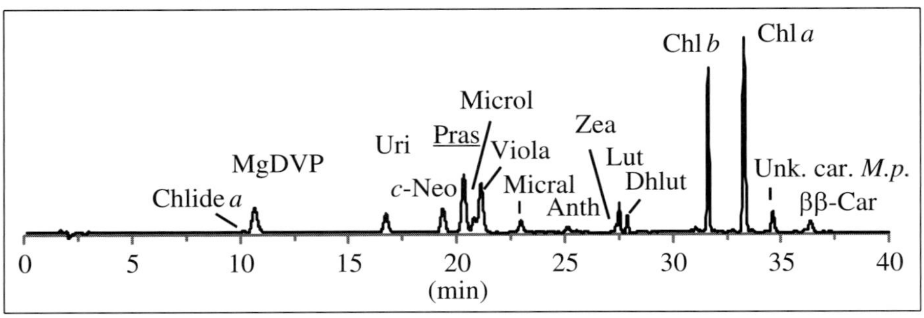

UV-Vis spectra (see also reference spectra below)

Solvent	λ_{max} (nm)	Band ratio (% III:II)	Ref.
Acetone	(427), *448*, (468)	-	[51]
Diethyl ether	*446*, (466)	-	[62]
Ethanol	*451*, 470	1	[48]
Hexane	(429), *452*, (480)	-	[52]
Recommended specific absorption coefficient *d* ($L\ g^{-1}\ cm^{-1}$)		n.d., see Remarks	

Reference spectra

For spectrum in acetone, see [109]

For spectrum in hexane, see [109]

In HPLC solvent system 1

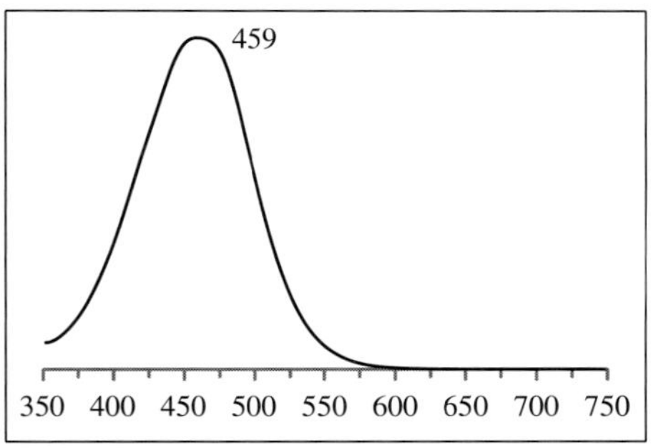

In HPLC solvent system 2

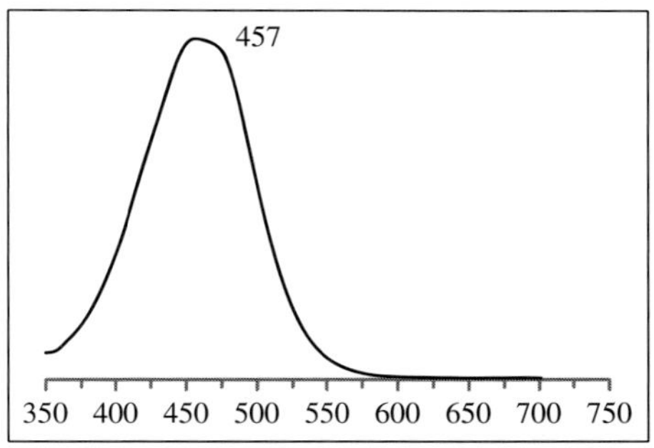

Mass spectra

Ionization technique	Mass analyser type	Diagnostic ions (m/z, rel. intensity)	Ref.
EI	Magnetic sector	600 $[M]^+$ (2), 582 $[M\text{-}18]^+$ (2), 446.3186 $[M\text{-}154]^+$ (100), 428 $[M\text{-}154\text{–}18]^+$ (21), 308 $[M\text{-}154\text{–}138]^+$ (6)	[51]

Remarks $d = 182\ L\ g^{-1} cm^{-1}$ (at λ_{max} in acetone, calc. from Fuco) is recommended, as no value has been determined for Pras

Siphonaxanthin

Recommended abbreviation: Siph (S)

IUPAC: (3*R*,3′*R*,6′*R*)-3,19,3′-Trihydroxy-7,8-dihydro-β,ε-caroten-8-one

Molecular formula: $C_{40}H_{56}O_4$ **Molecular weight:** 600.87

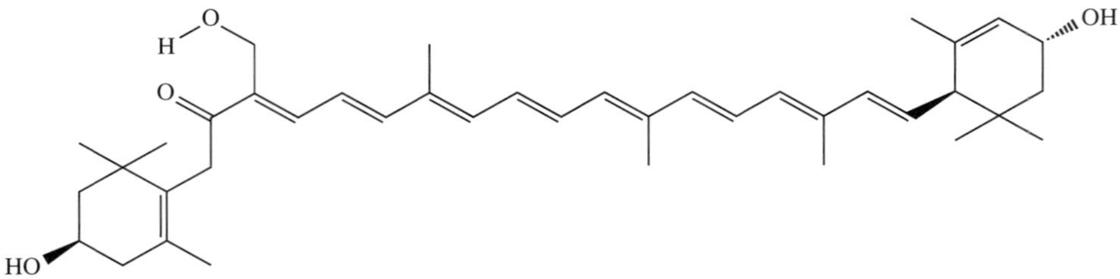

Biological occurrence	Dominant pigment in prasinophytes Pigment Type 2B and mesostigmatophytes. Minor in chlorophytes Pigment Type 2 (see Chapter 1, this volume). Also in siphonous green seaweeds
Source culture	*Mesostigma viride* (mesostigmatophyte) [177], NIES-296 [162]
Alteration products	*Cis*-isomers
Biosynthetically related to	Lut, Loro, Siph esters
Occurs together with	Chl *b*, Neo, Siph esters

HPLC chromatogram (system 1)

NO DATA AVAILABLE

HPLC chromatogram of *Codium fragile* (system 2)

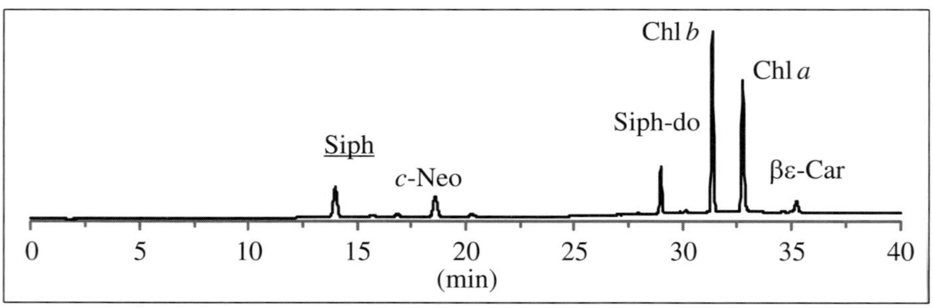

UV-Vis spectra (see also reference spectra below)

Solvent	λ_{max} (nm)	Band ratio (% III:II)	Ref.
Acetone	*441*, (461)	–	[50]
Diethyl ether	*441*, (464)	–	[58]
Ethanol	~448	–	[169]
Recommended specific absorption coefficient *d* ($L\ g^{-1}\ cm^{-1}$)		n.d., see Remarks	

Reference spectra

For spectrum in acetone, see [109]

In HPLC solvent system 1

NO DATA AVAILABLE

For spectrum in hexane, see [109]

In HPLC solvent system 2

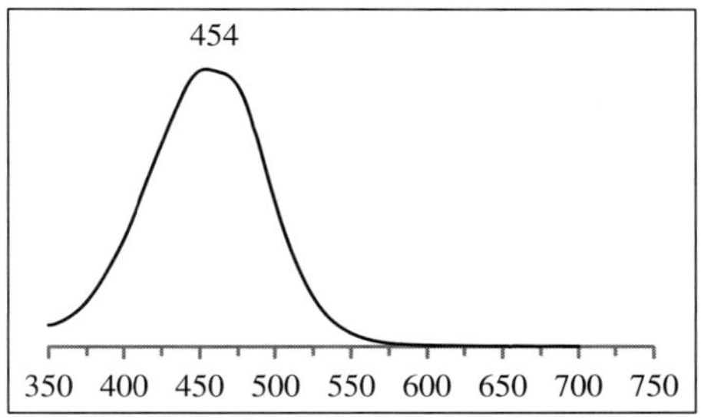

Mass spectra

Ionization technique	Mass analyser type	Diagnostic ions (m/z, rel. intensity)	Ref.
EI	Magnetic sector	600 $[M]^+$ (100), 584 $[M\text{-}16]^+$ (9), 582 $[M\text{-}18]^+$ (19), 572 $[M\text{-}28]^+$ (8), 564 $[M\text{-}18\text{–}18]^+$ (4), 494 $[M\text{-}106]^+$ (1)	[58]

Remarks $d = 182\ L\ g^{-1}\ cm^{-1}$ (at λ_{max} in acetone, calc. from Fuco) is recommended, as no value has been determined for Siph

Siphonaxanthin dodecenoate **Recommended abbreviation: Siph-do (Sdo)**

IUPAC: (3*R*,3′*R*,6′*R*)-19-(*trans*-Dodec-2-enoyloxy)-3,3′-dihydroxy-7,8-dihydro-β, ε-caroten-8-one

Molecular formula: $C_{52}H_{76}O_5$ **Molecular weight:** 781.16

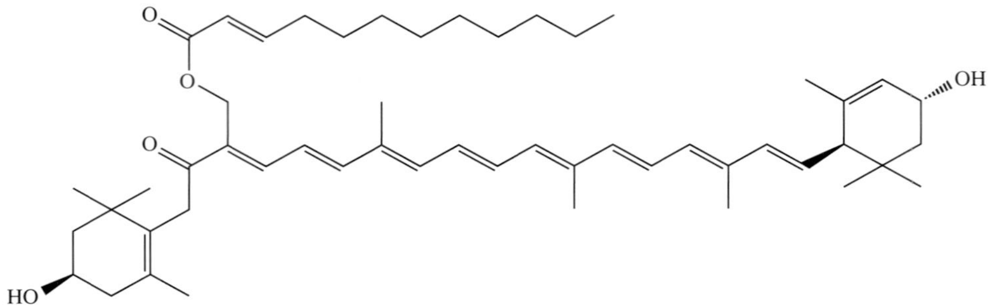

Biological occurrence	Dominant pigment (esters) in mesostigmatophytes, occasional in prasinophytes Pigment Type 2B and minor in chlorophytes Pigment Type 2. Also in siphonous green seaweeds
Source culture	*Pyramimonas amylifera* (prasinophyte) [50], PLY 246 [163]
Alteration products	*Cis*-isomers
Biosynthetically related to	Lut, Loro, Loro-d, Siph
Occurs together with	Chl *b*, Hsiph esters, Neo, Siph, Viola

HPLC chromatogram *Codium fragile* (system 2)

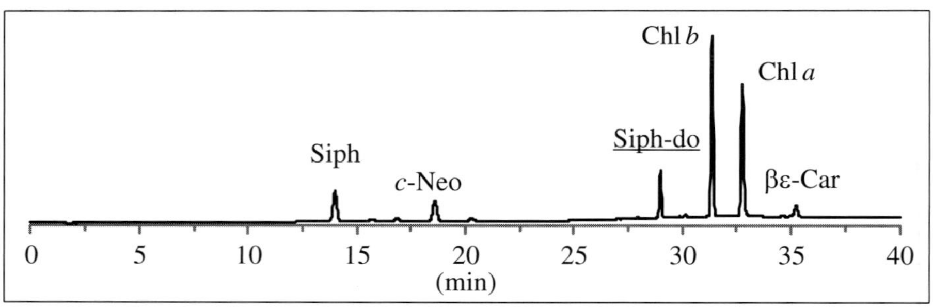

HPLC chromatogram of *Eutreptiella gymnastica* (system 3)

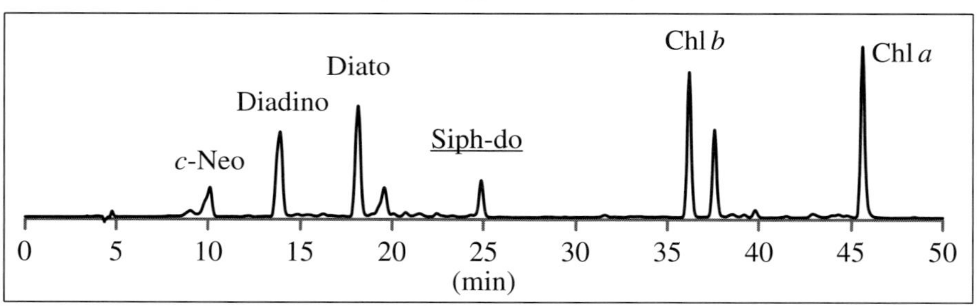

UV-Vis spectra (see also reference spectra below)

Solvent	λ_{max} **(nm)**	**Band ratio (% III:II)**	**Ref.**
Acetone	*448*, (463)	-	[50]
Diethyl ether	448	-	[58]
Ethanol	456	-	[169]
Recommended specific absorption coefficient ***d*** **(L $g^{-1}cm^{-1}$)**		n.d., see Remarks	

Reference spectra

In HPLC solvent system 2

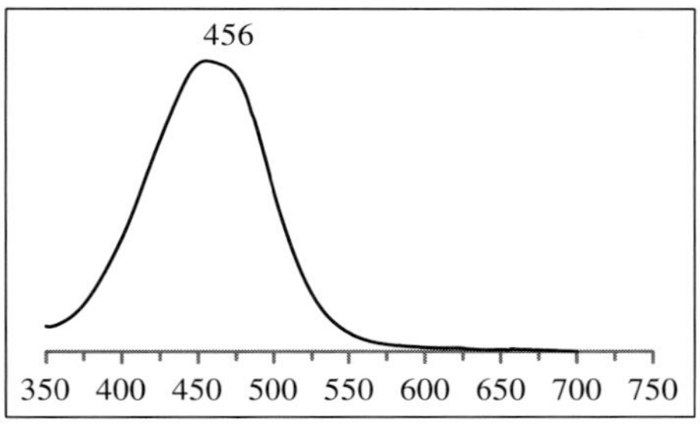

In HPLC solvent system 3

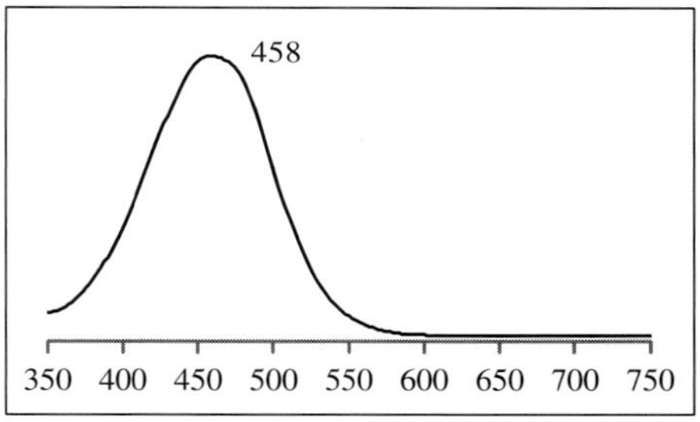

Mass spectra

Ionization technique	**Mass analyser type**	**Diagnostic ions (m/z, rel. intensity)**	**Ref.**
EI	Magnetic sector	780 $[M]^+$ (59), 762 $[M\text{-}18]^+$ (29), 688 $[M\text{-}92]^+$ (5), 582 $[M\text{-}180\text{–}18]^+$ (42), 568 $[M\text{-}180\text{–}18\text{–}18]^+$ (56), 180 (100)	[50]

Remarks The name 'Siphonein' was erroneously given to this compound in earlier studies [31]. Other esters of siphonaxanthin and 6′-hydroxysiphonaxanthin have also been encountered [50, 175, 176], including siphonein (= Siph 19-dodecanoate). $d = 140\,L\,g^{-1}cm^{-1}$ (at λ_{max} in acetone, calc. from Fuco) is recommended, as no value has been determined for Siph-do

Uriolide

Recommended abbreviation: Uri (U)

IUPAC: (3*S*,5*R*,6*S*,3′*R*,6′*R*)-5,6-Epoxy-3,3′-dihydroxy-5,6,7′,8′-tetrahydro-β,ε-caroten-19′,11′-olide

Molecular formula: $C_{40}H_{54}O_5$

Molecular weight: 614.85

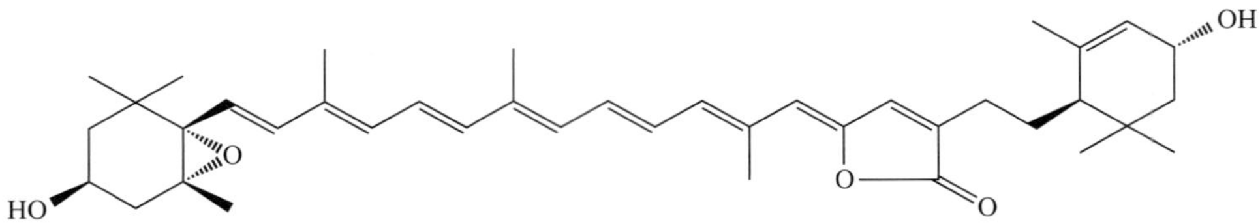

Biological occurrence	Dominant pigment in prasinophytes Pigment Type 3B (see Chapter 1, this volume)
Source culture	*Mantoniella squamata* (prasinophyte) [48], CCAP1965/1 [42]
Alteration products	Undergoes rearrangement to Uri-fur in weakly acidic solutions. *Cis*-isomers
Biosynthetically related to	Dhlut, Microl, Micral
Occurs together with	Chl *b*, MgDVP, Dhlut, Micral, Microl, Neo, Pras

HPLC chromatogram of *Micromonas pusilla* (system 1)

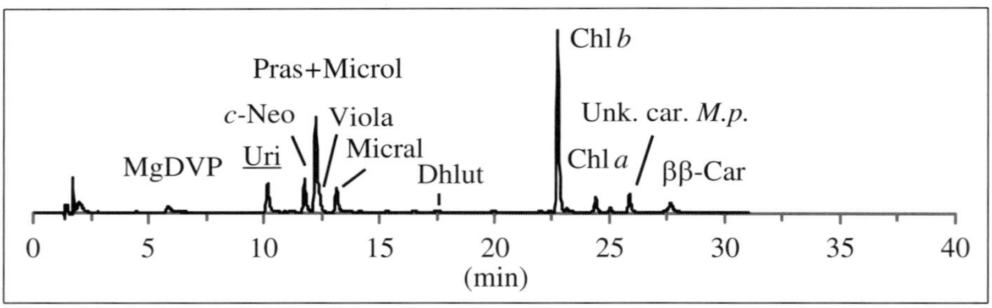

HPLC chromatogram of *Micromonas pusilla* (system 2)

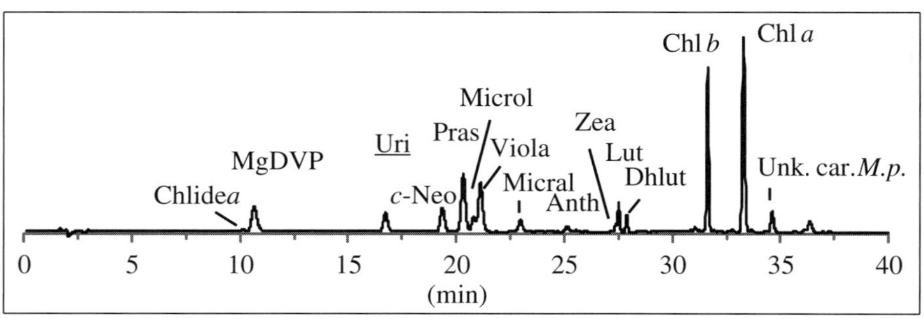

UV-Vis spectra (see also reference spectra below)

Solvent	λ_{max} (nm)	Band ratio (% III:II)	Ref.
Acetone	*448*, 472	11	[63]
Hexane	(427), *448*, 478	38	[63]
Methanol	*448*, 470	2	[63]
Recommended specific absorption coefficient *d* ($L\ g^{-1}\ cm^{-1}$)		n.d., see Remarks	

Reference spectra
In acetone

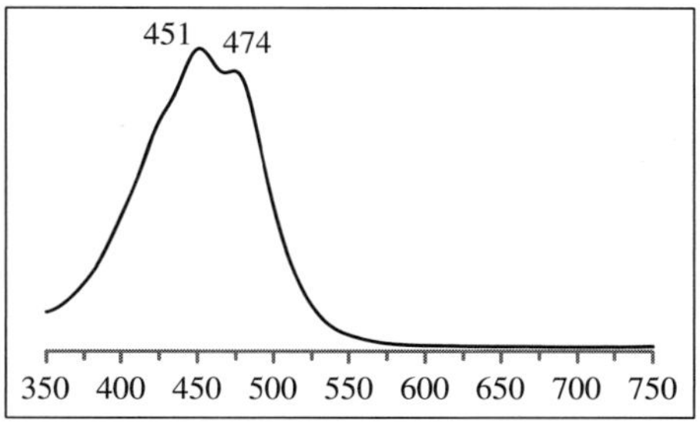

NO DATA AVAILABLE

In HPLC solvent system 1

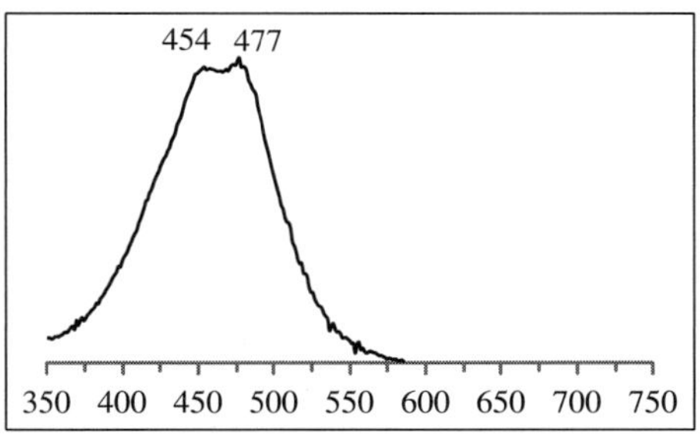

In HPLC solvent system 2

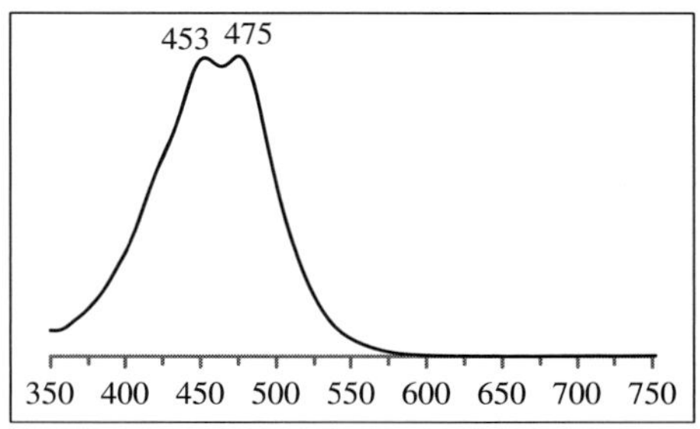

Mass spectra

Ionization technique	Mass analyser type	Diagnostic ions (m/z, rel. intensity)	Ref.
EI	Magnetic sector	614 $[M]^+$ (30), 596 $[M\text{-}18]^+$ (50), 534 $[M\text{-}80]^+$ (25), 522 $[M\text{-}92]^+$ (35), 516 $[M\text{-}80\text{–}18]^+$ (25), 504 $[M\text{-}92\text{–}18]^+$ (15), 221 (100)	[63]

Remarks $d = 140\ L\ g^{-1}\ cm^{-1}$ (at λ_{max} in ethanol) is recommended, as no value has been determined for Uri. Uriolide furanoxide forms in slightly acidic conditions, which usually occur with prasinophyte extracts [63]

Vaucheriaxanthin

Recommended abbreviation: Vauch (Va)

IUPAC: (3*S*,5*R*,6*R*,3′*S*,5′*R*,6′*S*)-5′,6′-Epoxy-6,7-didehydro-5,6,5′,6′-tetrahydro-β,β-carotene-3,5,3′,19′-tetrol

Molecular formula: $C_{40}H_{56}O_5$ **Molecular weight:** 616.87

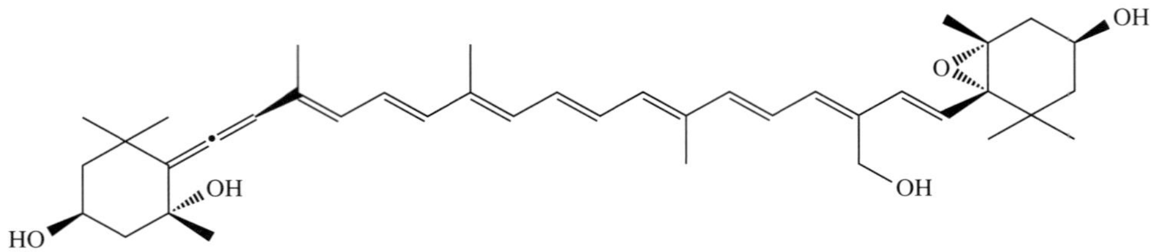

Biological occurrence	Dominant pigment in eustigmatophytes (see Chapter 1)
Source culture	*Nannochloropsis oculata* (eustigmatophyte)
Alteration products	Undergoes rearrangement to Vauchfur in weakly acidic solutions [91]. *Cis*-isomers
Biosynthetically related to	Zea, Anth, Viola, Neo, Vauch-eo, Vauch esters
Occurs together with	Vauch esters

HPLC chromatogram of *Nannochloropsis oculata* (system 1)

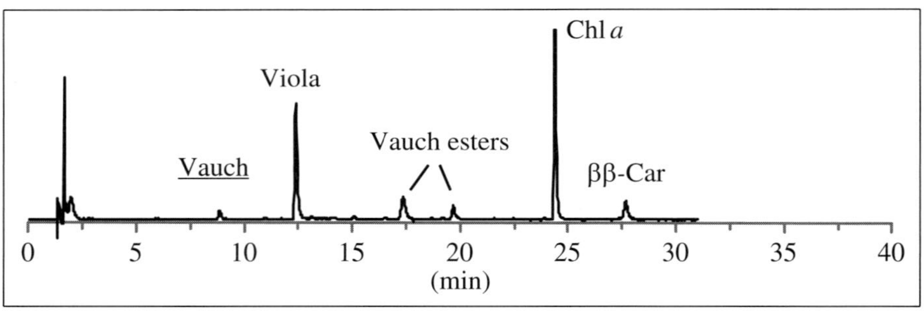

HPLC chromatogram of *Nannochloropsis gaditana* (system 2)

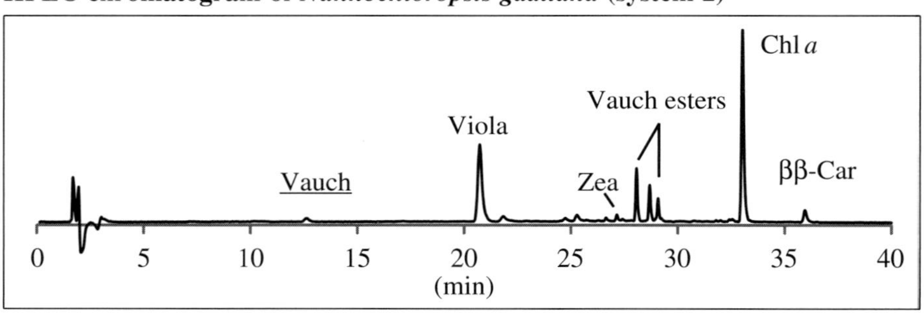

UV-Vis spectra (see also reference spectra below)

Solvent	λ_{max} (nm)	Band ratio (% III:II)	Ref.
Acetone	416, *438*, 466	54	[47]
Ethanol	419, *442*, 471	78	[152]
Recommended specific absorption coefficient *d* ($L\ g^{-1}\ cm^{-1}$)		n.d., see Remarks.	

Reference spectra

For spectrum in acetone, see [109]

For spectrum in hexane, see [109]

In HPLC solvent system 1

NO DATA AVAILABLE

In HPLC solvent system 2

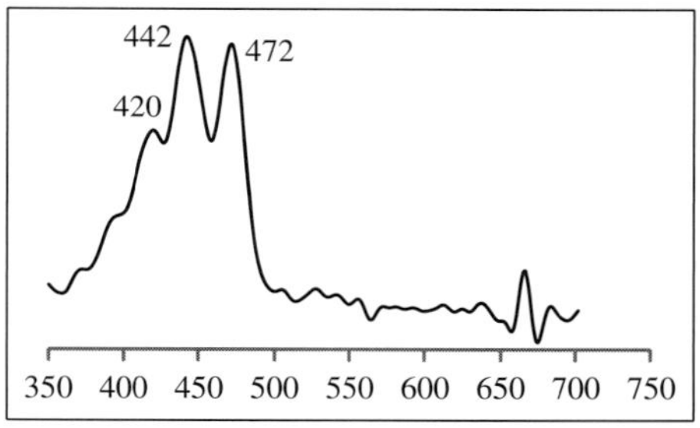

Mass spectra

Ionization technique	Mass analyser type	Diagnostic ions (m/z, rel. intensity)	Ref.
EI	Magnetic sector	616 $[M]^+$ (11), 598 $[M\text{-}18]^+$ (11), 580 $[M\text{-}18\text{–}18]^+$ (19), 562 $[M\text{-}18\text{–}18\text{–}18]^+$ (30), 544 $[M\text{-}18\text{–}18\text{–}18\text{–}18]^+$ (21), 197 (73), 181 (100)	[47]

Remarks $d = 232\ L\ g^{-1} cm^{-1}$ (at 442 nm, ethanol) is recommended (calc. from *trans*-Neo), as no value has been determined for Vauch. Vaucheriaxanthin furanoxide is formed in slightly acidic extracts [91]

Vaucheriaxanthin ethanoate octanoate **Recom. abbreviation: Vauch-eo**

IUPAC: (3*S*,5*R*,6*S*,3′*S*,5′*R*,6′*R*)-5,6-Epoxy-3′-ethanoyloxy-19-octanoyloxy-6′,7′-didehydro-5,6,5′,6′-tetrahydro-β,β-carotene-3,5′-diol

Molecular formula: $C_{50}H_{72}O_7$ **Molecular weight:** 785.10

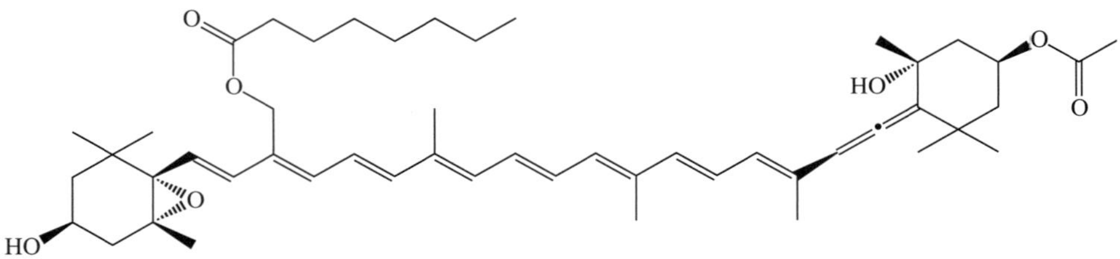

Biological occurrence	Dominant pigment in eustigmatophytes, minor in xanthophytes (see Chapter 1, this volume)
Source culture	*Nannochloropsis oculata* (eustigmatophyte)
Alteration products	Undergoes rearrangement to Vauch-eo-fur in weakly acidic solutions [91]. *Cis*-isomers
Biosynthetically related to	Zea, Anth, Viola, Neo, Vauch
Occurs together with	Vauch

HPLC chromatogram of *Nannochloropsis oculata* (system 1)

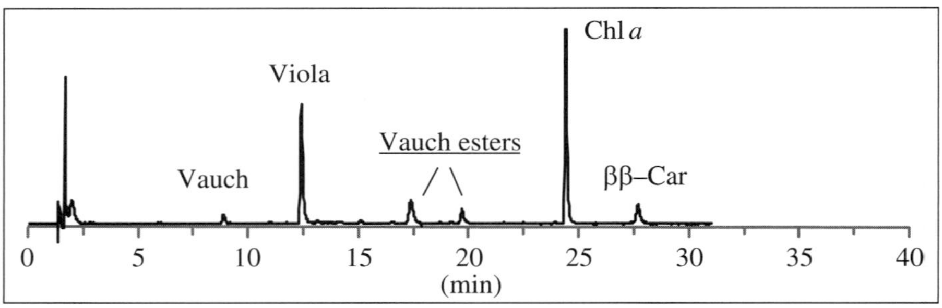

HPLC chromatogram of *Nannochloropsis oculata* (system 3)

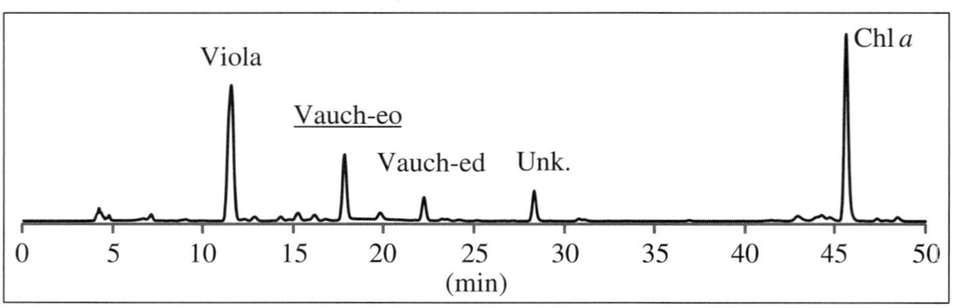

UV-Vis spectra (see also reference spectra below)

Solvent	λ_{max} (nm)	Band ratio (% III:II)	Ref.
Acetone	422, *445*, 471	33	[47]
Ethanol	419, *442*, 471	80	[152]
Hexane	419, *442*, 471	64	[109]
Recommended specific absorption coefficient ***d*** **(L g^{-1} cm^{-1})**		n.d., see Remarks.	

Reference spectra

For spectrum in acetone, see [109] For spectrum in hexane, see [109]

In HPLC solvent system 1

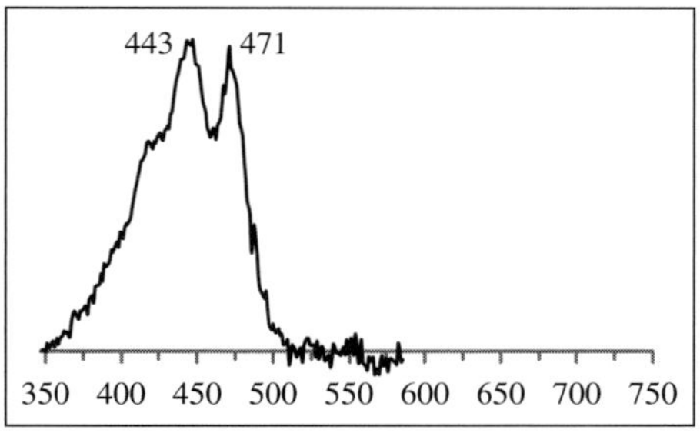

In HPLC solvent system 3

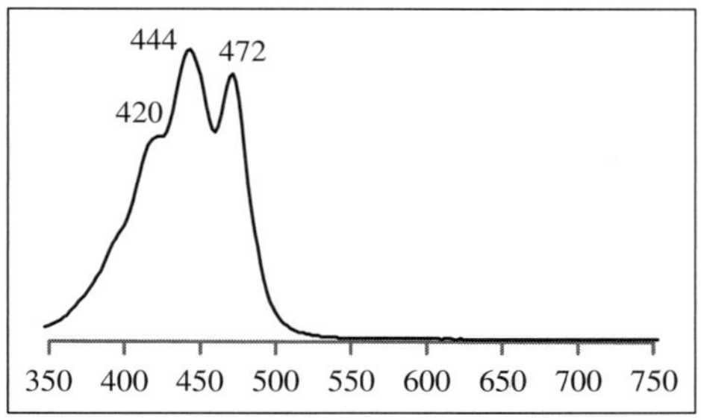

Mass spectra

Ionization technique	Mass analyser type	Diagnostic ions (m/z, rel. intensity)	Ref.
EI	Magnetic sector	784 $[M]^+$ (6), 766 $[M\text{-}18]^+$ (3), 706 $[M\text{-}18\text{–}60]^+$ (2), 640 $[M\text{-}18\text{–}126]^+$ (15), 197 (58), 181 (100)	[47]

Remarks Former name: vaucheriaxanthin 3-acetate 19′-octanoate. Found also in mixture with vaucheriaxanthin ethanoate decanoate (Vauch-ed) [47]. Other esters and their acid-catalysed forms exist: [91]. $d = 182\,\text{L g}^{-1}\text{cm}^{-1}$ (at 442 nm, ethanol) is recommended (calc. from *trans*-Neo), as no value has been determined for Vauch-eo

Violaxanthin

Recommended abbreviation: Viola (V)

IUPAC: (3*S*,5*R*,6*S*,3′*S*,5′*R*,6′*S*)-5,6:5′,6′-Diepoxy-5,6,5′,6′-tetrahydro-β,β-carotene-3,3′-diol

Molecular formula: $C_{40}H_{56}O_4$ **Molecular weight:** 600.87

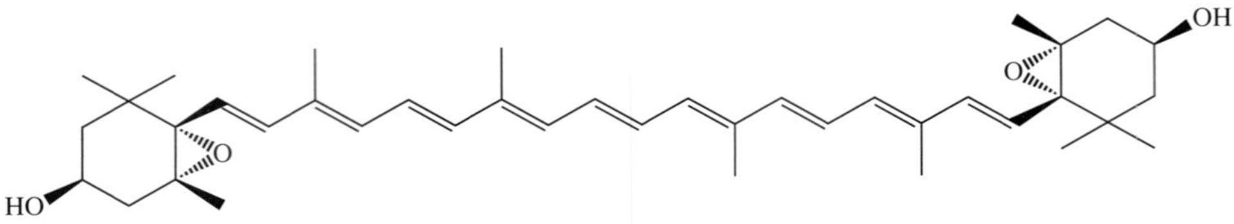

Biological occurrence	Dominant pigment in chrysophytes, eustigmatophytes, synurophytes, mesostigmatophytes, chlorophytes, prasinophytes, and dinoflagellates Pigment Type 5. Minor in chlorarachniophytes, pinguiophytes and raphidophytes (see Chapter 1, this volume). Also present in higher plants and brown seaweeds
Source culture	*Dunaliella tertiolecta* (chlorophyte)
Alteration products	Undergoes rearrangement to Luteoxanthin and Auro in weakly acidic solutions. *Cis*-isomers
Biosynthetically related to	ββ-Car, Cryp, Zea, Anth, Neo
Occurs together with	Anth, Neo

HPLC chromatogram of *Dunaliella tertiolecta* (system 1)

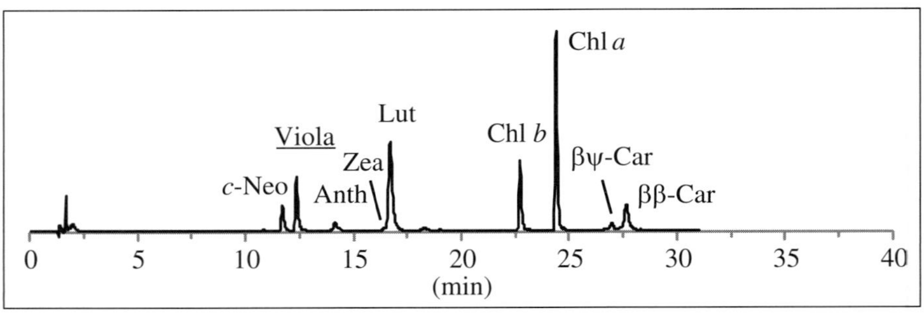

HPLC chromatogram of *Micromonas pusilla* (system 2)

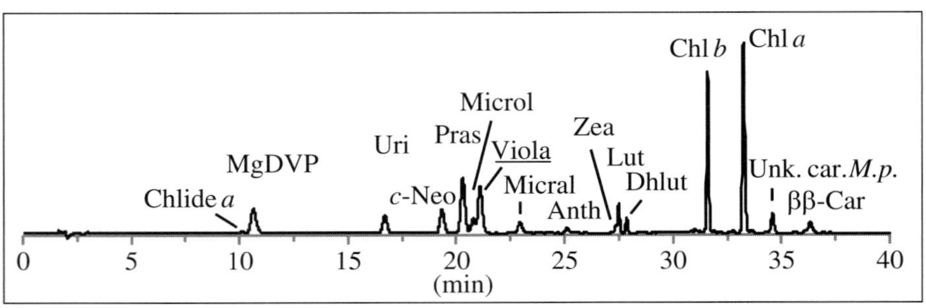

UV-Vis spectra (see also reference spectra below)

Solvent	λ_{max} (nm)	Band ratio (% III:II)	Ref.
Acetone	415, *438*, 467	79	[91]
Ethanol	417, *440*, 469	93	[152]
Hexane	417, *440*, *470*	100	[145]
Methanol	415, *436*, 466	90	[160]
Recommended specific absorption coef. *d* (L $g^{-1}cm^{-1}$)		254 (at 437 nm in diethyl ether:methylbutane:ethanol 5:5:2) [4] 255 (at 443 nm in ethanol) [44]	

Reference spectra

In acetone

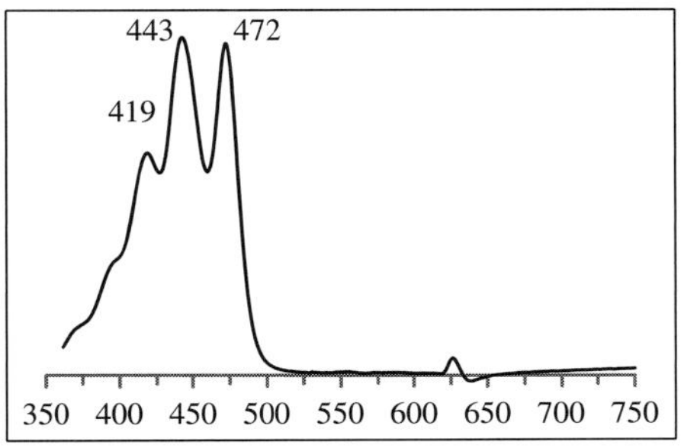

In ethanol

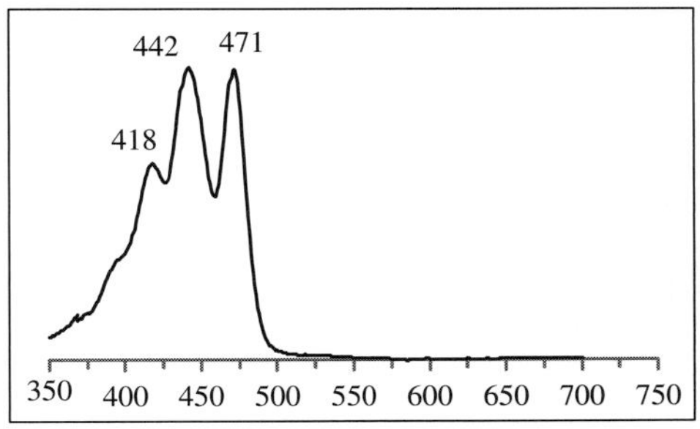

In HPLC solvent system 1

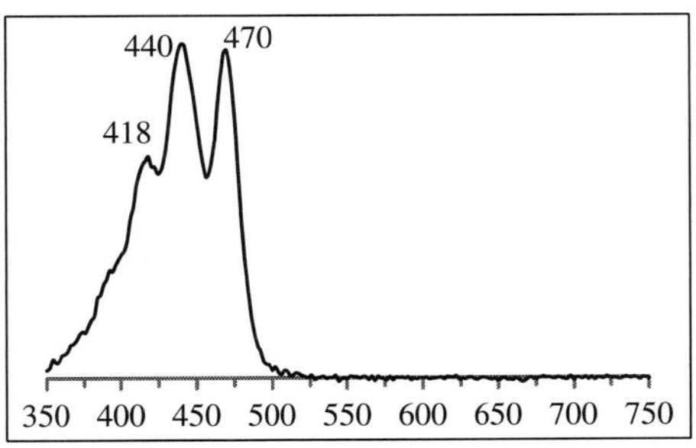

In HPLC solvent system 2

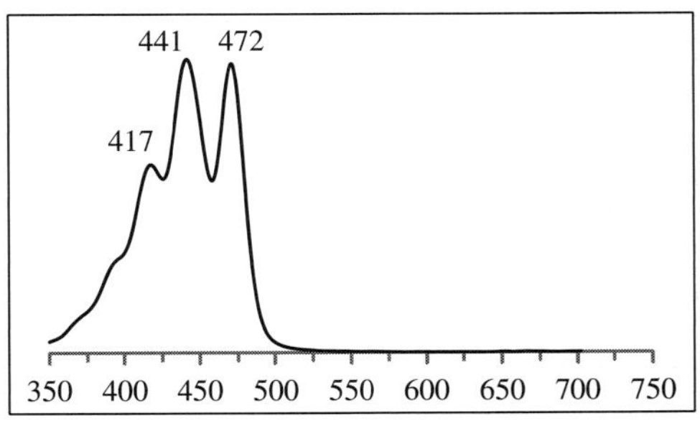

Mass spectra

Ionization technique	Mass analyser type	Diagnostic ions (m/z, rel. intensity)	Ref.
CI	Magnetic sector	601 $[M+1]^+$ (47), 583 $[M+1-18]^+$ (36), 565 $[M+1-18-18]^+$ (22), 510 (35), 509 $[M+1-92]^+$, 221 (27), 181 (100)	[4]

Remarks Part of a 'xanthophyll cycle': see Chapter 11. May be present in diatoms under prolonged high light stress [123]. Transforms into luteoxanthin and auroxanthin in slightly acidic extracts, particularly in prasinophytes [91]

Zeaxanthin

Recommended abbreviation: Zea (Z)

IUPAC: (3*R*,3′*R*)-β,β-Carotene-3,3′-diol

Molecular formula: $C_{40}H_{56}O_2$

Molecular weight: 568.87

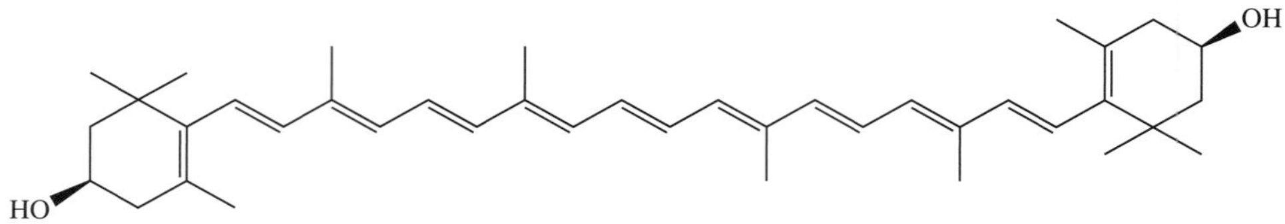

Biological occurrence	Dominant pigment in cyanobacteria, chrysophytes, eustigmatophytes, pelagophytes, rhodophytes and dinoflagellates Pigment Type 5. Minor in pinguiophytes, raphidophytes, chlorarachniophytes, chlorophytes, prasinophytes and trebouxiophytes. Occasional in dictyochophytes and dinoflagellates Pigment Type 3 (see Chapter 1, this volume)
Source culture	*Synechococcus* sp. (DC-2) (cyanobacteria)
Alteration products	*Cis*-isomers
Biosynthetically related to	ββ-Car, Cryp, Anth, Viola, Neo
Occurs together with	

HPLC chromatogram of *Prochlorococcus* sp. (system 1)

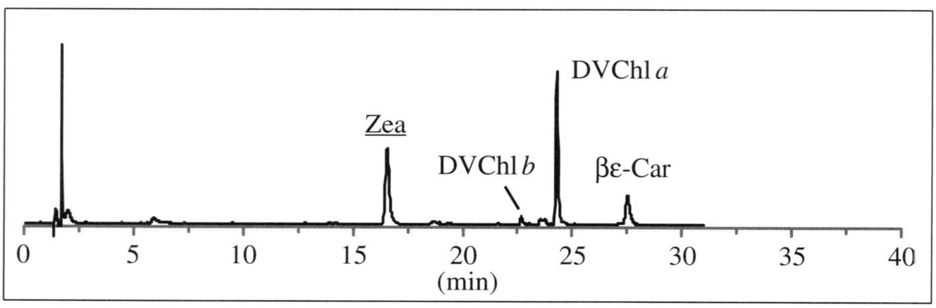

HPLC chromatogram of *Prochlorococcus* sp. (system 2)

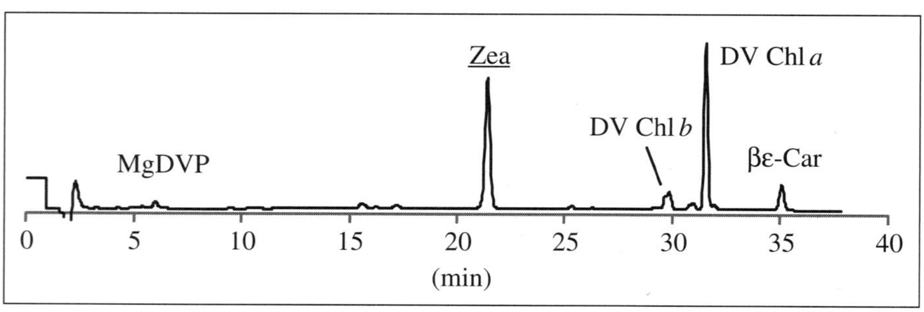

UV-Vis spectra (see also reference spectra below)

Solvent	λ_{max} (nm)	Band ratio (% III:II)	Ref.
Acetone	(428), *454*, 481	33	[145]
Ethanol	(428), *450*, 478	26	[152]
Hexane	(424), *450*, 478	46	[145]
Methanol	(429), *449*, 475	25	[160]
Recommended specific absorption coefficient *d* (L g^{-1} cm^{-1})		245 (at 453 nm, ethanol) [148]	

Reference spectra

For spectrum in acetone, see [109]

For spectrum in hexane, see [109]

In HPLC solvent system 1

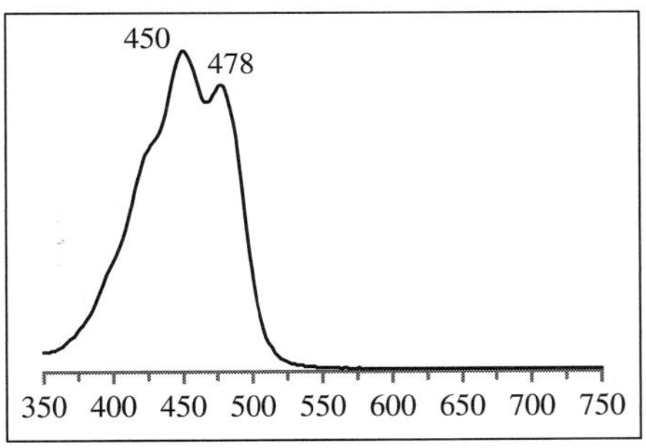

In HPLC solvent system 2

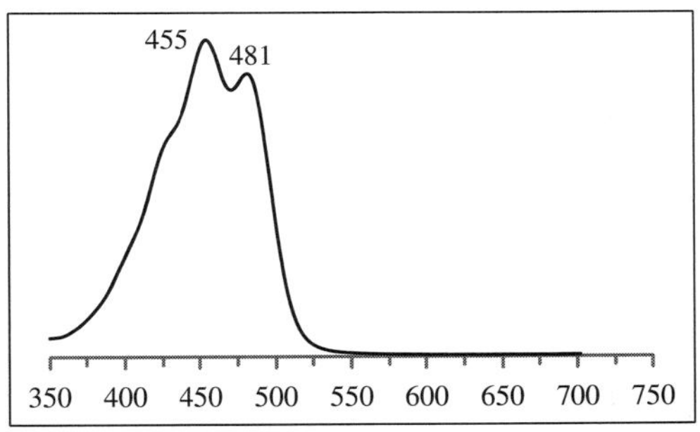

Mass spectra

Ionization technique	Mass analyser type	Diagnostic ions (m/z, rel. intensity)	Ref.
EI	Magnetic sector	568 [M]$^+$ (100), 550 [M-18]$^+$ (84), 532 [M-18–18]$^+$ (5), 489 [M-79]$^+$ (1), 476 [M-92]$^+$ (13), 462 [M-106]$^+$ (1), 458 [M-18–92]$^+$ (11), 410 [M-158]$^+$ (5)	[19]

Remarks Part of a 'xanthophyll cycle' – see Chapter 11, this volume. May be present in diatoms under prolonged high light stress [123]

References

[1] Aakermann, T., Skulberg, O. M. and Liaaen-Jensen, S. (1992). A comparison of the carotenoids of strains of *Oscillatoria* and *Spirulina* (Cyanobacteria). *Biochem. Syst. Ecol.* **20**, 761–69.

[2] Aasen, A. J. and Liaaen-Jensen, S. (1966a). The carotenoids of flexibacteria II. A new xanthophyll from *Saprospira grandis*. *Acta Chem. Scand.* **20**, 811–19.

[3] Aasen, A. J. and Liaaen-Jensen, S. (1966b). Carotenoids of flexibacteria. IV. The carotenoids of two further pigment types. *Acta Chem. Scand.* **20**, 2322–34.

[4] Acemoglu, M., Uebelhart, P., Rey, M. and Eugster, C. H. (1988). Synthese von enantiomerenreinen Violaxanthinen und verwandten Verbindungen. *Helv. Chim. Acta* **71**, 931–56.

[5] Airs, R. L. and Llewellyn, C. A. (2006). Improved detection and characterization of fucoxanthin-type carotenoids: novel pigments in *Emiliania huxleyi* (Prymnesiophyceae). *J. Phycol.* **42**, 391–99.

[6] Airs, R. L., Atkinson, J. E. and Keely, B. J. (2001). Development and application of a high resolution liquid chromatographic method for the analysis of complex pigment distributions. *J. Chromatogr. A* **917**, 167–77.

[7] Arpin, N., Svec, W. A. and Liaaen-Jensen, S. (1976). New fucoxanthin-related carotenoids from *Coccolithus huxleyi*. *Phytochemistry* **15**, 529–32.

[8] Australian National Algae Culture Collection, see http://www.cmar.csiro.au/microalgae/

[9] Baumeler, A. and Eugster, C. H. (1992). Synthese von (6*R*,all-*E*)-Neoxanthin und verwandten Allen-Carotinoiden. *Helv. Chim. Acta* **75**, 773–90.

[10] Baumeler, A., Zerbe, O., Kunz, R. and Eugster, C. H. (1994). (6*R*, 9′*Z*)-Neoxanthin: Synthese, Schmelzverhalten, Spektren und Konformationsberechnungen. *Helv. Chim. Acta* **77**, 909–930.

[11] Bazzaz, M. B. (1981). New chlorophyll *a* and *b* chromophores isolated from a mutant of *Zea mays* L. *Naturwissenschaften* **68**, 94–95.

[12] Bazzaz, M. B. (1981). New chlorophyll chromophores isolated from a chlorophyll-deficient mutant of maize. *Photobiochem. Photobiophys.* **2**, 199–207.

[13] Bazzaz, M. B. (1982). 4-Vinyl-4-desethyl chlorophyll *a*: Characterisation of a new naturally occurring chlorophyll using fast atom bombardment, field desorption and ‘in beam’ electron impact mass spectrometry. *Tetrahedron Lett.* **23**, 1211–14.

[14] Bindl, E., Lang, W. and Rau, W. (1970). Untersuchungen über die lichtabhängige Carotinoidsynthese VI. Zeitlicher Verlauf der Synthese der einzelnen Carotinoide bei Fusarium aquaeductuum unter verschiedenen Induktionsbedingungen. *Planta (Berl.)* **94**, 156–74.

[15] Bjerkeng, B. (1990). *Havbruksrelatert carotenoidkjemi*. Ph.D. thesis. Trondheim: Norges Tekniske Høgskole.

[16] Bjerkeng, B., Haugan, J. A. and Liaaen-Jensen, S. (1991) Microalgal carotenoids and chemosystematics: updating the Prymnesiophyceae. *Biol. Oceanogr.* **6**, 271–78.

[17] Bjørnland, T. (1982). Chlorophylls and carotenoids of the marine alga *Eutreptiella gymnastica*. *Phytochemistry* **21**, 1715–19.

[18] Bjørnland, T. (1990). Chromatographic separation and spectrometric characterization of native carotenoids from the marine dinoflagellate *Thoracosphaera heimii*. *Biochem. Syst. Ecol.* **18**, 307–16.

[19] Bjørnland, T. and Aguilar-Martinez, M. (1976). Carotenoids in red algae. *Phytochemistry* **15**, 291–96.

[20] Bjørnland, T., Borch, G. and Liaaen-Jensen, S. (1984). Configurational studies on red algae carotenoids. *Phytochemistry* **23**, 1711–15.

[21] Bjørnland, T., Fiksdahl, A., Skjetne, T., Krane, J. and Liaaen-Jensen, S. (2000). Gyroxanthin – the first allenic acetylenic carotenoid. *Tetrahedron* **56**, 9047–56.

[22] Bjørnland, T., Guillard, R. R. L. and Liaaen-Jensen, S. (1988). *Phaeocystis* sp. clone 677–3 – a tropical marine planktonic prymnesiophyte with fucoxanthin and 19'-acyloxyfucoxanthins as chemosystematic carotenoid markers. *Biochem. Syst. Ecol.* **16**, 445–52.

[23] Bjørnland, T., Liaaen-Jensen, S. and Throndsen, J. (1989). Carotenoids of the marine chrysophyte *Pelagococcus subviridis*. *Phytochemistry* **28**, 3347–53.

[24] Bjørnland, T., Pereira, C., Liaaen-Jensen, S., Guillard, R. R. L. and Bidigare, R. R. (1987). Pigment distribution patterns in oceanography: carotenoids of the microflagellate MC-1. Abstracts of presentations, 8th International symposium on carotenoids, Boston, p. P39.

[25] Bjørnland, T. and Tangen, K. (1979). Pigmentation and morphology of a marine *Gyrodinium* (Dinophyceae) with a major carotenoid different from peridinin and fucoxanthin. *J. Phycol.* **15**, 457–65.

[26] Boardman, N. K. and Thorne, S. W. (1971). Sensitive fluorescence method for the determination of chlorophyll *a*/chlorophyll *b* ratios. *Biochim. Biophys. Acta* **253**, 222–31.

[27] Breemen, R. B. v., Canjura, F. L. and Schwartz, S. J. (1991). Identification of chlorophyll derivatives by mass spectroscopy. *J. Agric. Food Chem.* **39**, 1452–56.

[28] Breemen, R. B. v., Schmitz, H. H. and Schwartz, S. J. (1995). Fast atom bombardment tandem mass spectrometry of carotenoids. *J. Agric. Food Chem.* **43**, 384–89.

[29] Brereton, R. G., Bazzaz, M. B., Santikarn, S. and Williams, D. H. (1983). Positive and negative ion fast atom bombardment mass spectrometric studies on chlorophylls: structure of 4-vinyl-4-desethyl chlorophyll *b*. *Tetrahedron Lett.* **24**, 5775–78.

[30] Britton, G., Goodwin, T. W., Harriman, G. E. and Lockley, W. J. S. (1977). Carotenoids of the ladybird beetle, *Coccinella septempunctata*. *Insect Biochem.* **7**, 337–45.

[31] Britton, G., Liaaen-Jensen, S. and Pfander, H. (eds). Mercadante, A. Z. and Egeland, E. S. (compilers) (2004). *Carotenoids. Handbook.* Basel: Birkhäuser.

[32] Buchecker, R., Liaaen-Jensen, S. and Eugster, C. H. (1976a). Reinvestigation of original taraxanthin samples. *Helv. Chim. Acta* **59**, 1360–64.

[33] Buchecker, R., Liaaen-Jensen, S., Borch, G. and Siegelman, H. W. (1976b). Carotenoids of *Anacystis nidulans*, structures of caloxanthin and nostoxanthin. *Phytochemistry* **15**, 1015–18.

[34] Buchwald, M. and Jencks, W. P. (1968). Optical properties of astaxanthin solutions and aggregates. *Biochemistry* **7**, 834–43.

[35] Chapman, D. J. (1966). Three new carotenoids isolated from algae. *Phytochemistry* **5**, 1331–33.

[36] Chapman, D. J. and Haxo, F. T. (1963). Identity of ε-carotene and ε_1-carotene. *Plant Cell Physiol.* **4**, 57–63.

[37] Cheng, J. Y., Don-Paul, M. and Antia, N. J. (1974). Isolation of an unusually stable *cis*-isomer of alloxanthin from a bleached autolysed culture of *Chroomonas salina* grown photoheterotrophically on glycerol. Observations on *cis-trans* isomerization of alloxanthin. *J. Protozool.* **21**, 761–68.

[38] Chereskin, B. M., Wong, Y. -S. and Castelfranco, P. A. (1982). *In vitro* synthesis of the chlorophyll isocyclic ring. *Plant Physiol.* **70**, 987–93.
[39] Cholnoky, L., Györgyfy, K., Szabolcs, J., Weedon, B. C. L. and Waigth, E. S. (1966). Foliaxanthin. *Chem. Commun.* **13**, 404–05.
[40] Connolly, J. S., Samuel, E. B. and Janzen, A. F. (1982). Effects of solvent on the fluorescence properties of bacteriochlorophyll *a*. *Photochem. Photobiol.* **36**, 565–74.
[41] Cooper, R. D. G., Davis, J. B., Leftwick, A. P., Price, C. and Weedon, B. C. L. (1975). Carotenoids and related compounds. Part XXXII. Synthesis of astaxanthin, phoenicoxanthin, hydroxyechinenone, and the corresponding diosphenols. *J. Chem. Soc., Perkin Trans. 1*, **1975**, 2195–204.
[42] Culture Collection of Algae and Protozoa, Cambridge, see http://www.ccap.ac.uk/
[43] Davies, A. J., Khare, A., Mallams, A. K., Massy-Westropp, R. A., Moss, G. P. and Weedon, B. C. L. (1984). Carotenoids and related compounds. Part 38. Synthesis of (3*RS*, 3′*RS*)-alloxanthin and other acetylenes. *J. Chem. Soc., Perkin Trans. 1* **1984**, 2147–57.
[44] Davies, B. H. (1965). Analysis of carotenoids pigments. In *Chemistry and Biochemistry of Plant Pigments*, ed. T. W. Goodwin. London: Academic Press, pp. 489–532.
[45] Davies, B. H. (1976). Carotenoids. In *Chemistry and Biochemistry of Plant Pigments*, 2nd Ed., Vol. 2, ed. T. W. Goodwin. London: Academic Press, pp. 38–165.
[46] Davies, B. H. (1979). Solved and unsolved problems of carotenoid formation. *Pure Appl. Chem.* **51**, 623–30.
[47] Egeland, E. S. (1996). *Algekarotenoider og kjemosystematikk*. Ph.D. thesis, Norwegian University of Science and Technology, Trondheim.
[48] Egeland, E. S., Eikrem, W., Throndsen, J., Wilhelm, C., Zapata, M. and Liaaen-Jensen, S. (1995). Carotenoids from further prasinophytes. *Biochem. Syst. Ecol.* **23**, 747–55. *Errata* **24**, 179–80 (1996).
[49] Egeland, E. S., Garrido, J. L., Zapata, M., Maestro, M. A. and Liaaen-Jensen, S. (2000). Algal carotenoids. Part 64. Structure and chemistry of 4-keto-19′-hexanoyloxyfucoxanthin with a novel carotenoid end group. *J. Chem. Soc., Perkin Trans. 1* **2000**, 1223–30.
[50] Egeland, E. S., Guillard, R. R. L. and Liaaen-Jensen, S. (1997). Additional carotenoid prototype representatives and a general chemosystematic evaluation of carotenoids in Prasinophyceae (Chlorophyta). *Phytochemistry* **44**, 1087–97.
[51] Egeland, E. S., Johnsen, G., Eikrem, W., Throndsen, J. and Liaaen-Jensen, S. (1995). Pigments of *Bathycoccus prasinos* (Prasinophyceae): methodological and chemosystematic implications. *J. Phycol.* **31**, 554–61.
[52] Egeland, E. S. and Liaaen-Jensen, S. (1995). Ten minor carotenoids from Prasinophyceae (Chlorophyta). *Phytochemistry* **40**, 515–20.
[53] Egger, K., Nitsche, H. and Kleinig, H. (1969). Diatoxanthin und Diadinoxanthin – Bestandteile des Xanthophyllgemisches von *Vaucheria* und *Botrydium*. *Phytochemistry* **8**, 1583–85.
[54] Entschel, R. and Karrer, P. (1958). Carotinoidsynthesen XXII. Umsetzungsprodukte des Carotins mit Bromsuccinimid. (Einführung von Äther- und Hydroxyl-Gruppen in den Kohlenwasserstoff.) *Helv. Chim. Acta* **41**, 402–13.
[55] Enzell, C. R., Francis, G. W. and Liaaen-Jensen, S. (1969). Mass spectrometric studies of carotenoids 2. A survey of fragmentation reactions. *Acta Chem. Scand.* **23**, 727–50.
[56] Eugster, C. H. and Karrer, P. (1955). Carotinoidsynthesen XV. Kristallisiertes *d,l*-α-Carotin. *Helv. Chim. Acta* **38**, 610–12.

[57] Fawley, M. W. (1989). A new form of chlorophyll *c* involved in light-harvesting. *Plant Physiol.* **91**, 727–32.

[58] Fiksdahl, A., Bjørnland, T. and Liaaen-Jensen, S. (1984). Algal carotenoids with novel end groups. *Phytochemistry* **23**, 649–55.

[59] Fiksdahl, A., Foss, P., Liaaen-Jensen, S. and Siegelman, H. W. (1983). Carotenoids of blue-green algae – 11. Carotenoids of chromatically-adapted cyanobacteria. *Comp. Biochem. Physiol.* **76 B**, 599–601.

[60] Fiksdahl, A. and Liaaen-Jensen, S. (1988). Diacetylenic carotenoids from *Euglena viridis*. *Phytochemistry* **27**, 1447–50.

[61] Fiksdahl, A., Withers, N., Guillard, R. R. L. and Liaaen-Jensen, S. (1984). Carotenoids of the Raphidophyceae – a chemosystematic contribution. *Comp. Biochem. Physiol.* **78 B**, 265–71.

[62] Foss, P., Guillard, R. R. L. and Liaaen-Jensen, S. (1984). Prasinoxanthin – a chemosystematic marker for algae. *Phytochemistry* **23**, 1629–33.

[63] Foss, P., Guillard, R. R. L. and Liaaen-Jensen, S. (1986). Carotenoids from eucaryotic ultraplankton clones (Prasinophyceae). *Phytochemistry* **25**, 119–24.

[64] Foss, P., Lewin, R. A. and Liaaen-Jensen, S. (1987). The carotenoids of *Prochloron* sp. (Prochlorophyta). *Phycologia* **26**, 142–44.

[65] Foss, P., Skulberg, O. M., Kilaas, L. and Liaaen-Jensen, S. (1986). The carbohydrate moieties bound to the carotenoids myxol and oscillol and their chemosystematic applications. *Phytochemistry* **25**, 1127–32.

[66] Francis, G. W. and Halfen, L. N. (1972). Cyanophyta. Oscillatoriaceae. γ-Carotene and lycopene in *Oscillatoria princeps*. *Phytochemistry* **11**, 2347–48.

[67] Francis, G. W., Hertzberg, S., Andersen, K. and Liaaen-Jensen, S. (1970). New carotenoid glycosides from *Oscillatoria limosa*. *Phytochemistry* **9**, 629–35.

[68] Ganguly, J., Krinsky, N. I. and Pinckard, J. H. (1956). Isolation and nature of echinenone, a provitamin A. *Arch. Biochem. Biophys.* **60**, 345–51.

[69] Garbers, C. F., Eugster, C. H. and Karrer, P. (1953). Carotinoidsynthese XIV. Synthese des γ-Carotins. *Helv. Chim. Acta* **36**, 1783–87.

[70] Garrido, J. L., Otero, J., Maestro, M. A. and Zapata, M. (2000). The main nonpolar chlorophyll *c* from *Emiliania huxleyi* (Prymnesiophyceae) is a chlorophyll c_2-monogalactosyldiacylglyceride ester: a mass spectrometry study. *J. Phycol.* **36**, 497–505.

[71] Garrido, J. L., Rodríguez, F., Zapata, M. (2009). Occurrence of loroxanthin, loroxanthin decenoate, and loroxanthin dodecenoate in *Tetraselmis* species (Prasinophyceae, Chlorophyta). *J. Phycol.* **45**, 366–74.

[72] Garrido, J. L. and Zapata, M. unpublished data

[73] Garrido, J. L., Zapata, M. and Muñiz, S. (1995). Spectral characterization of new chlorophyll *c* pigments isolated from *Emiliania huxleyi* (Prymnesiophyceae) by high-performance liquid chromatography. *J. Phycol.* **31**, 761–68.

[74] Gieskes, W. W., Kraay, G. W. (1983). Unknown chlorophyll *a* derivatives in the North Sea and the tropical Atlantic Ocean revealed by HPLC analysis. *Limnol. Oceanogr.* **28**, 757–66.

[75] Goericke, R. and Repeta, D. J. (1992). The pigments of *Prochlorococcus marinus*: The presence of divinyl chlorophyll *a* and *b* in a marine procaryote. *Limnol. Oceanogr.* **37**, 425–33.

[76] Goericke, R., Olson, R. J. and Shalapyonok, A. (2000). A novel niche for *Prochlorococcus* sp. in low-light suboxic environments in the Arabian Sea and the Eastern Tropical North Pacific. *Deep-Sea Res. I* **47**, 1183–205.

[77] Goericke, R., Strom, S. L. and Bell, M. A. (2000). Distribution and sources of cyclic pheophorbides in the marine environment. *Limnol. Oceanogr.* **45**, 200–11.
[78] Goodwin, T. W. (1980). *The Biochemistry of the Carotenoids, Vol. 1, Plants*, 2nd edn. London: Chapman and Hall.
[79] Goodwin, T. W. (1984). *The Biochemistry of the Carotenoids, Vol. 2, Animals*, 2nd edn. London: Chapman and Hall.
[80] Graham, J. E. and Bryant, D. A. (2009). The biosynthetic pathway for myxol-2′ fucoside (myxoxanthophyll) in the cyanobacterium *Synechococcus* sp. strain PCC 7002. *J. Bacteriol.* **191**, 3292–300.
[81] Gross, J., Ikan, R. and Eckhardt, G. (1983). Carotenoids of the fruit of *Averrhoa carambola*. *Phytochemistry* **22**, 1479–81.
[82] Grung, M., Metzger, P. and Liaaen-Jensen, S. (1994). Algal carotenoids 53; secondary carotenoids of algae 4; secondary carotenoids in the green alga *Botryococcus braunii*, race L, new strain. *Biochem. Syst. Ecol.* **22**, 25–29.
[83] Hager, A. and Stransky, H. (1970). Das Carotinoidmuster und die Verbretung des lichtinduzierten Xanthophyllcyclus in verschiedenen Algenklassen. *Arch. Mikrobiol.* **72**, 68–83.
[84] Hager, A. and Meyer-Bertenrath, T. (1966). Die Isolierung und quantitative Bestimmung der Carotinoide und Chlorophylle von Blättern, Algen und isolierten Chloroplasten mit Hilfe dünnschichtchromatographisher Methoden. *Planta (Berl.)* **69**, 198–217.
[85] Hager, A. and Meyer-Bertenrath, T. (1967). Die Identifizierung der an Dünnschichten getrennten Carotinoide grüner Blatter und Algen. *Planta (Berl.)* **76**, 149–68.
[86] Hager, A. and Stransky, H. (1970). Das Carotinoidmuster und die Verbreitung des lichtinduzierten Xanthophyllcyclus in verschiedenen Algenklassen III. Grünalgen. *Arch. Mikrobiol.* **72**, 68–83.
[87] Hager, A. and Stransky, H. (1970). Das Carotinoidmuster und die Verbreitung des lichtinduzierten Xanthophyllcyclus in verschiedenen Algenklassen. V. Einzelne Vertreter der Cryptophyceae, Euglenophyceae, Bacillariophyceae, Chrysophyceae und Phaeophyceae. *Arch. Mikrobiol.* **73**, 77–89.
[88] Halfen, L. N. and Francis, G. W. (1972). The influence of culture temperature on the carotenoid composition of the blue-green alga, *Anacystis nidulans*. *Arch. Mikrobiol.* **81**, 25–35.
[89] Haugan, J. A., Englert, G., Glinz, E. and Liaaen-Jensen, S. (1992). Algal carotenoids. 48. Structural assignments of geometrical isomers of fucoxanthin. *Acta Chem. Scand.* **46**, 389–95.
[90] Haugan, J. A. and Liaaen-Jensen, S. (1989). Improved isolation procedure for fucoxanthin. *Phytochemistry* **28**, 2797–98.
[91] Haugan, J. A. and Liaaen-Jensen, S. (1994). Blue carotenoids. Part 2. The chemistry of the classical colour reaction of common carotenoid 5,6-epoxides with acid. *Acta Chem. Scand.* **48**, 152–59.
[92] Haugan, J. A. and Liaaen-Jensen, S. (1994). Total synthesis of acetylenic carotenoids. 2. Synthesis of (all-*E*)-(3*R*,3′*R*)-diatoxanthin and (all-*E*)-(3*R*)-7,8-didehydrocryptoxanthin. *Acta Chem. Scand.* **48**, 899–904.
[93] Helfrich, M., Ross, A., King, Garry C., Turner, A. G. and Larkum, A. W. D. (1999). Identification of [8-vinyl]-protochlorophyllide *a* in phototrophic procaryotes and algae: chemical and spectroscopic properties. *Biochim. Biophys. Acta* **1410**, 262–72.

[94] Hertzberg, S. and Liaaen-Jensen, S. (1969). The structure of myxoxanthophyll. *Phytochemistry* **8**, 1259–80.
[95] Hertzberg, S. and Liaaen-Jensen, S. (1969). The structure of oscillaxanthin. *Phytochemistry* **8**, 1281–92.
[96] Hiyama, T., Nishimura, M. and Chance, B. (1969). Determination of carotenes by thin-layer chromatography. *Anal. Biochem.* **29**, 339–50.
[97] Holt, A. S. and Morley, H. V. (1959). A proposed structure for chlorophyll *d*. *Can. J. Chem.* **37**, 507–14.
[98] Hynninen P. H. and Lötjönen, S. (1983). Large-scale preparation of crystalline (10*S*)-chlorophylls *a* and *b*. *Synthesis* **9**, 705–08.
[99] Hynninen, P. H. (1973). Chlorophylls IV. Preparation and purification of some derivatives of chlorophylls *a* and *b*. *Acta Chem. Scand.* **27**, 1771–80.
[100] Hynninen, P. H. (1991). Chemistry of chlorophylls: modifications. In *Chlorophylls*, ed. H. Scheer. Boca Raton: CRC Press, pp. 145–209.
[101] Hynninen, P. H. and Lötjönen, S. (1980). Preparation of phorbin derivatives from chlorophyll mixture utilizing the principle of selective hydrolysis. *Synthesis* **1980(7)**, 539–41.
[102] Hyvärinen, K. and Hynninen, P. H. (1999). Liquid chromatographic separation and mass spectrometric identification of chlorophyll *b* allomers. *J. Chromatogr. A.* **837**, 107–16.
[103] Isler, O., Lindlar, H., Montavon, M., Rüegg, R. and Zeller, P. (1956). Synthesen in der Carotinoid-Reihe. 1. Mitteilung. Die technische Synthese von β-Carotin. *Helv. Chim. Acta* **39**, 249–59.
[104] Jeffrey, S. W. (1969). Properties of two spectrally different components in chlorophyll *c* preparations. *Biochim. Biophys. Acta* **177**, 456–67.
[105] Jeffrey, S. W. (1972). Preparation and some properties of crystalline chlorophyll c_1 and c_2 from marine algae. *Biochim. Biophys. Acta* **279**, 15–33.
[106] Jeffrey, S. W. and Haxo, F. T. (1968). Photosynthetic pigments of dinoflagellates (zooxanthellae) from corals and clams. *Biol. Bull.* **135**, 149–65.
[107] Jeffrey, S. W. and Humphrey, G. F. (1975). New spectrophotometric equations for determining chlorophylls *a*, *b*, c_1 and c_2 in higher plants, algae and natural phytoplankton. *Biochem. Physiol. Pflanzen* **167**, 191–94.
[108] Jeffrey, S. W. and Lorenzen, C. J., unpublished. In Jeffrey, S. W., Mantoura, R. F. C. and Bjørnland, T. (1997). Data for the identification of 47 key phytoplankton pigments. In *Phytoplankton Pigments in Oceanography: Guidelines to Modern Methods*, ed. S. W. Jeffrey, R. F. C. Mantoura and S. W. Wright. Paris: UNESCO Publishing, pp. 449–559.
[109] Jeffrey, S. W., Mantoura, R. F. C. and Bjørnland, T. (1997). Data for the identification of 47 key phytoplankton pigments. In *Phytoplankton Pigments in Oceanography: Guidelines to Modern Methods*, ed. S. W. Jeffrey, R. F. C. Mantoura and S. W. Wright. Paris: UNESCO Publishing, pp. 449–559.
[110] Jeffrey, S. W. and Wright, S. W. (1987). A new spectrally distinct component in preparations of chlorophyll *c* from the micro-alga *Emiliania huxleyi* (Prymnesiophyceae). *Biochim. Biophys. Acta* **894**, 180–88.
[111] Jensen, A. and Aasmundrud, O. (1963). Paper chromatographic characterization of chlorophylls. *Acta Chem. Scand.* **17**, 907–12.
[112] Johansen, J. E., Svec, W. A., Liaaen-Jensen, S. and Haxo, F. T. (1974). Carotenoids of the Dinophyceae. *Phytochemistry* **13**, 2261–71.

[113] Karrer, P., Eugster, C. H. and Tobler, E. (1950). Synthesen von Carotinoidfarbstoffen III. Totalsynthese des Lycopins. *Helv. Chim. Acta* **33**, 1349–52.

[114] Karrer, P. and Rutschmann, J. (1942). Auroxanthin, ein kurzwellig absorbierender Carotinfarbstoff. *Helv. Chim. Acta* **25**, 1624–27.

[115] Kjøsen, H., Norgård, S., Liaaen-Jensen, S., Svec, W. A., Strain, H. H., Wegfahrt, P., Rapoport, H. and Haxo, F. T. (1976). Algal carotenoids. XV. Structural studies on peridinin. Part 2. Supporting evidence. *Acta Chem. Scand.* **30**, 157–64.

[116] Kleinig, H., Nitsche, H. and Egger, K. (1969). The structure of siphonaxanthin. *Tetrahedron Lett.* **59**, 5139–42.

[117] Kleinig, H., Heumann, W., Meister, W. and Englert, G. (1977). Carotenoids of *Rhizobia*. I. New Carotenoids from *Rhizobium lupini*. *Helv. Chim. Acta* **60**, 254–58.

[118] Krane, J., Aakermann, T. and Liaaen-Jensen, S. (1992). Algal carotenoids 47 – NMR study of all-*trans*-peridinin, including complete ^{1}H and ^{13}C NMR assignments. *Magn. Res. Chem.* **30**, 1169–77.

[119] Krinsky, N. I. and Goldsmith, T. H. (1960). The carotenoids of the flagellated alga, *Euglena gracilis*. *Arch. Biochem. Biophys.* **91**, 271–79.

[120] Liaaen-Jensen, S. (1998). Carotenoids in biosynthesis. In *Carotenoids, Vol. 3: Biosynthesis*, ed. G. Britton, S. Liaaen-Jensen and H. Pfander. Basel: Birkhäuser, pp. 217–47.

[121] Lichtenthaler, H. K. (1987). Chlorophylls and carotenoids: pigments of photosynthetic biomembranes. *Meth. Enzymol.* **148**, 350–82.

[122] Loeblich, A. R. and Smith, V. E. (1968). Chloroplast pigments of the marine dinoflagellate *Gyrodinium resplendens*. *Lipids* **3**, 5–13.

[123] Lohr, M. and Wilhelm, C. (1999). Algae displaying the diadinoxanthin cycle also possess the violaxanthin cycle. *Proc. Natl. Acad. Sci. USA* **96**, 8784–89.

[124] Lorenzen, C. J. and Jeffrey, S. W. (1980). Determination of chlorophyll in seawater. UNESCO Technical Papers in Marine Science, No. **35**.

[125] Lötjönen, S. and Hynninen, P. H. (1983). An improved method for the preparation of (10*R*)- and (10*S*)-pheophytins *a* and *b*. *Synthesis* **1983(9)**, 708–10.

[126] Marchand, P. S., Rüegg, R., Schwieter, U., Siddons, P. T. and Weedon, B. C. L. (1965). Cartenoids and related compounds. Part XI. Synthesis of δ-carotene and ε-carotene. *J. Chem. Soc.* **46**, 2019–26.

[127] Märki-Fischer, F., Buchecker, R., Eugster, C. H., Englert, G., Noack, K. and Vecchi, M. (1982). Absolute Konfiguration von Antheraxanthin, '*cis*-Antheraxanthin' und der diastereomeren Mutatoxanthine. *Helv. Chim. Acta* **65**, 2198–11.

[128] Mayer, H. and Rüttimann, A. (1980). Synthese von optisch aktiven, natürlichen Carotinoiden und strukturell verwandten Naturprodukten IV. Synthese von (3*R*,3′*R*,6′*R*)-Lutein. *Helv. Chim. Acta* **63**, 1451–55.

[129] Miyashita, H., Ikemoto, H., Kurano, N., Adachi, K., Chihara, M. and Miyachi, S. (1996). Chlorophyll *d* as a major pigment. *Nature* **383**, 402.

[130] Nitsche, H. (1970). Reaktionen an fucoxanthin I. Überführung in diadinochrom und diatoxanthin. *Tetrahedron Lett.* **11**, 4913–16.

[131] Otsuki, A., Kaneda, Y. and Hashimoto, S. (1993). Identification and significance of pyrochlorins in fecal pellets of the marine malacostracan crustaceans *Heptacarpus rectirostris* and *Palaemon pacificus*. *Mar. Biol.* **115**, 463–67.

[132] Partali, V., Olsen, Y., Foss, P. and Liaaen-Jensen, S. (1985). Carotenoids in food chain studies – I. Zooplankton (*Daphnia magna*) response to a unialgal (*Scenedesmus acutus*) carotenoid diet, to spinach, and to yeast diets supplemented with individual carotenoids. *Comp. Biochem. Physiol.* **82 B**, 767–72.

[133] Pennington, F. C., Haxo, F. T., Borch, G. and Liaaen-Jensen, S. (1985). Carotenoids of Cryptophyceae. *Biochem. Syst. Ecol.* **13**, 215–19.

[134] Pennington, F. C., Strain, H. H., Svec, W. A. and Katz, J. J. (1964). Preparation and properties of pyrochlorophyll *a*, methyl pyrochlorophyllide *a*, pyropheophytin *a*, and methyl pyropheophorbide *a* derived from chlorophyll by decarbomethoxylation. *J. Am. Chem. Soc.* **86**, 1418–26.

[135] Petracek, F. J. and Zechmeister, L. (1956). Reaction of β-carotene with N-bromosuccinimide: The formation and conversions of some polyene ketones. *J. Am. Chem. Soc.* **78**, 1427–34.

[136] Pfeifhofer, H. W. and Grill, D. (1984). Carotinoide in Fichtennadeln 1. Qualitative Untersuchungen. *Phyton* **24**, 283–93.

[137] Roberts, D. W. A. (1969). The quantitation of pheophorbide *a*, pyropheophorbide *a*, and their esters. *Can. J. Biochem.* **47**, 721–24.

[138] Rønneberg, H., Foss, P., Ramdahl, T., Borch, G., Skulberg, O. M. and Liaaen-Jensen, S. (1980). Occurrence and chirality of oscillaxanthin. *Phytochemistry* **19**, 2167–70.

[139] Sasa, T., Takaichi, S., Hatakeyama, N. and Watanabe, M. M. (1992). A novel carotenoid ester, loroxanthin dodecenoate, from *Pyramimonas parkeae* (Prasinophyceae) and a chlorarachniophycean alga. *Plant Cell Physiol.* **33**, 921–25.

[140] Sauer, K., Smith, J. R. L. and Schultz, A. J. (1966). The dimerization of chlorophyll *a*, chlorophyll *b* and bacteriochlorophyll in solution. *J. Am. Chem. Soc.* **88**, 2681–88.

[141] Scheer, H. (2006). An overview of chlorophylls and bacteriochlorophylls: biochemistry, biophysics, functions and applications. In *Chlorophylls and bacteriochlorophylls: Biochemistry, biophysics and applications*, ed. B. Grimm, R. J. Porra, W. Rüdiger and H. Scheer. Dordrecht: Springer, pp. 1–26.

[142] Schlüter, L., Lutnaes, B. F., Liaaen-Jensen, S., Garde, K., Kaas, H., Jameson, I. and Blackburn, S. (2008). Correlation of the content of hepatotoxin nodularin and glycosidic carotenoids, 4-ketomyxol-2′-fucoside and novel 1′-*O*-methyl-4-ketomyxol-2′-fucoside, in 20 strains of the cyanobacterium *Nodularia spumigena*. *Biochem. Syst. Ecol.* **36**, 749–57.

[143] Schwieter, U., Bolliger, H. R., Chopard-dit-Jean, I. H., Englert, G., Kofler, M., König, A., Planta, C. v., Rüegg, R., Vetter, W. and Isler, O. (1965). Synthesen in der Carotinoid-Reihe. 19. Mitteilung: Physikalische Eigenschaften der Carotinoide. *Chimia* **19**, 294–302.

[144] Schwieter, U., Englert, G., Rigassi, N. and Vetter, W. (1969). Physical organic methods in carotenoid research. *Pure Appl. Chem.* **20**, 365–420.

[145] SCOR WG 78 data in Jeffrey, S. W., Mantoura, R. F. C. and Bjørnland, T. (1997). Data for the idenfication of 47 key phytoplankton pigments. In *Phytoplankton Pigments in Oceanography: Guidelines to Modern Methods*, ed. S. W. Jeffrey, R. F. C. Mantoura and S. W. Wright. Paris: UNESCO Publishing, pp. 449–559.

[146] Smallidge, R. L. and Quackenbush, F. W. (1973). β,β–carotene-2,3,3′-triol: a new carotenoids in *Anacystis nidulans*. *Phytochemistry* **12**, 2481–82.

[147] Smith, J. H. C. and Benitez, A. (1955). Chlorophylls: Analysis in plant materials. In *Modern Methods of Plant Analysis*, ed. K. Paech and M. V. Tracey. Berlin: Springer, pp. 142–96.

[148] Strain, H. H. (1938). *Leaf xanthophylls*. Washington Carnegie Institution of Washington, Publication No 490.

[149] Strain, H. H., Cope, B. T. and Svec, W. A. (1971). Analytical procedures for the isolation, identification, estimation, and investigation of the chlorophylls. *Meth. Enzymol.* **23**, 452–76.

[150] Strain, H. H., Svec, W. A., Wegfahrt, P., Rapoport, H., Haxo, F. T., Norgård, S., Kjøsen, H. and Liaaen-Jensen, S. (1976). Algal carotenoids. XIV. Structural studies on peridinin. Part 1. Structure elucidation. *Acta Chem. Scand.* **30 B**, 109–20.

[151] Strand, A., Kvernberg, K., Karlsen, A. M. and Liaaen-Jensen, S. (2000). Geometrical *E*/*Z* isomers of (6*R*)- and (6*S*)-neoxanthin and biological implications. *Biochem. Syst. Ecol.* **28**, 443–55.

[152] Stransky, H. and Hager, A. (1970). Das Carotinoidmuster und die Verbreitung des lichtinduzierten Xanthophyllcyclus in verschiedenen Algenklassen. II. Xanthophyceae. *Arch. Mikrobiol.* **71**, 164–90.

[153] Stransky, H. and Hager, A. (1970). Das Carotinoidmuster und die Verbreitung des lichtinduzierten Xanthophyllcyclus in verschiedenen Algenklassen IV. Cyanophyceae und Rhodophyceae. *Arch. Mikrobiol.* **72**, 84–96.

[154] Strom, S. L. (1993). Production of pheopigments by marine protozoa: results of laboratory experiments analysed by HPLC. *Deep-Sea Res. I* **40**, 57–80.

[155] Surmatis, J. D., Walser, A., Gibas, J., Schwieter, U. and Thommen, R. (1970). New methods for the synthesis of oxygenated carotenoids. *Helv. Chim. Acta.* **53**, 974–90.

[156] Suzuki, T., Midonoya, H. and Shioi, Y. (2009). Analysis of chlorophylls and their derivatives by master-assisted laser desorption/ionization-time-of-flight mass spectrometry. *Anal. Biochem.* **390**, 57–62.

[157] Tămaş, V. and Bodea, C. (1970). $TiCl_4$-Carotenoid complexes. VIII Cyclization of lycopene and δ-carotene. *Revue Roumaine Chimie* **15**, 819–22.

[158] Takaichi, S. and Mimuro, M. (1998). Distribution and geometric isomerism of neoxanthin in oxygenic phototrophs: 9′-*cis*, a sole molecular form. *Plant Cell Physiol.* **39**, 968–77.

[159] Takaichi, S., Mochimaru, M., Maoka, T. and Katoh, H. (2005). Myxol and 4-ketomyxol 2′-fucosides, not rhamnosides, from *Anabaena* sp. PCC 7120 and *Nostoc punctiforme* PCC 73102, and proposal for the biosynthetic pathway of carotenoids. *Plant Cell Physiol.* **46**, 497–504.

[160] Takaichi, S. and Shimada, K. (1992). Characterization of carotenoids in photosynthetic bacteria. *Meth. Enzymol.* **213**, 374–85.

[161] Tomo, T., Suzuki, T., Hirano, E., Tsuchiya, T., Miyashita, H., Dohmae, N. and Mimuro, M. (2006). Reversible absorption change of chlorophyll *d* in solutions. *Chem. Phys. Lett.* **423**, 282–87.

[162] The microbial culture collection cataloge for National Institute for Environmental Studies, see http://mcc.nies.go.jp/localeAction.do?lang=en

[163] The Plymouth Culture Collection of Marine Algae, see http://www.mba.ac.uk/culturecollection.php

[164] The Provasoli-Guillard National Center for Culture of Marine Phytoplankton catalogue, see http://ccmp.bigelow.org/

[165] Traber, B. and Pfander, H. (1996). Synthesis of (all-*E*,2*R*,2′*R*)-oscillol. *Helv. Chim. Acta* **79**, 499–506.

[166] Tsuchiya, T., Takaichi, S., Misawa, N., Maoka, T., Miyashita, H. and Mimuro, M. (2005). The cyanobacterium *Gloeobacter violaceus* PCC 7421 uses bacterial-type phytoene desaturase in carotenoid biosynthesis. *FEBS Lett.* **579**, 2125–29.

[167] Valadon, L. R. G. and Mummery, R. S. (1967). Carotenoids of certain Compositae flowers. *Phytochemistry* **6**, 983–88.

[168] Walker, J. S. and Keely, B. J.. (2004). Distribution and significance of chlorophyll derivatives and oxidation products during the spring phytoplankton bloom in the Celtic Sea April 2002. *Org. Geochem.* **35**, 1289–98.

[169] Walton, T. J., Britton, G., Goodwin, T. W., Diner, B. and Moshier, S. (1970). The structure of siphonaxanthin. *Phytochemistry* **9**, 2545–52.

[170] Weigl, J. W. (1952). Concerning the absorption spectrum of bacteriochlorophyll. *J. Am. Chem. Soc.* **75**, 999–1000.

[171] Withers, N. W., Cox, E. R., Tomas, R. and Haxo, F. T. (1977). Pigments of the dinoflagellate *Peridinium balticum* and its photosynthetic endosymbiont. *J. Phycol.* **13**, 354–58.

[172] Woolley, P. S., Moir, A. J., Hester, R. E. and Keely, B. J. (1998). A comparative study of the allomerization reaction of chlorophyll *a* and bacteriochlorophyll *a*. *J. Chem. Soc., Perkin Trans. 2*, **1998**, 1833–39.

[173] Wright, S. W. and Jeffrey, S. W. (1987). Fucoxanthin pigment markers of marine phytoplankton analysed by HPLC and HPTLC. *Mar. Ecol. Prog. Ser.* **38**, 259–66.

[174] Yamano, Y., Tode, C. and Ito, M. (1995). Carotenoids and related polyenes. Part 3. First total synthesis of fucoxanthin and halocyntiaxanthin using oxo-metallic catalyst. *J. Chem. Soc., Perkin Trans. 1*, **1995**, 1895–904.

[175] Yoshii, Y. (2006). Diversity and evolution of photosynthetic antenna systems in green plants. *Phycol. Res.* **54**, 220–29.

[176] Yoshii, Y., Takaichi, S. Maoka, T. Hanada, S. and Inouye, I. (2002). Characterization of two unique carotenoids fatty acid esters from *Pterospera cristatum* (Prasinophyceae, Chlorophyta). *J. Phycol.* **38**, 297–303.

[177] Yoshii, Y., Takaichi, S., Maoka, T. and Inouye, I. (2003). Photosynthetic pigment composition in the primitive green alga *Mesostigma viride* (Prasinophyceae): phylogenetic and evolutionary implications. *J. Phycol.* **39**, 570–76.

[178] Zapata, M. *et al.* unpublished data.

[179] Zapata, M., Garrido, J. L. and Jeffrey, S. W. (2006). Chlorophyll *c* pigments: Current status. In *Chlorophylls and Bacteriochlorophylls. Biochemistry, Biophysics, Functions and Applications.* ed. B. Grimm, R. J. Porra, W. Rüdiger and H. Scheer. Dordrecht: Springer, pp. 39–53.

[180] Zapata, M., Jeffrey, S. W., Wright, S. W., Rodríguez, F., Garrido, J. L. and Clementson, L. (2004). Photosynthetic pigments in 37 species (65 strains) of Haptophyta: implications for oceanography and chemotaxonomy. *Mar. Ecol. Prog. Ser.* **270**, 83–102.

[181] Zechmeister, L., LeRosen, A. L., Schroeder, W. A., Polgár, A. and Pauling, L. (1943). Spectral characteristics and configuration of some stereoisomeric carotenoids including prolycopene and pro-γ-carotene. *J. Am. Chem. Soc.* **65**, 1940–51.

[182] Zscheile, F. P., White, J. W., Beadle, B. W. and Roach, J. R. (1942). The preparation and absorption spectra of five pure carotenoid pigments. *Plant Physiol.* **17**, 331–46.

Index

bold = tables; italics = figures; underline = data sheet